重回海河边

天津城市设计的理论与实践探索

BACK TO THE HAIHE WATERFRONT

PRACTICE AND EXPLORATION OF URBAN DESIGN IN TIANJIN

朱雪梅　主编

江苏人民出版社

图书在版编目（CIP）数据

重回海河边 ：天津城市设计的理论与实践探索 / 朱雪梅著 . -- 南京 ：江苏人民出版社，2019.10
ISBN 978-7-214-24002-6

Ⅰ．①重… Ⅱ．①朱… Ⅲ．①城市规划－建筑设计－研究－天津 Ⅳ．① TU984.221

中国版本图书馆 CIP 数据核字 (2019) 第 215787 号

书　　名	重回海河边　天津城市设计的理论与实践探索
主　　编	朱雪梅
项目策划	陈　景
责任编辑	刘　焱
特约编辑	陈　景
美术编辑	张仅宜
出版发行	江苏人民出版社
出版社地址	南京市湖南路A楼，邮编：210009
出版社网址	http://www.jspph.com
印　　刷	天津久佳雅创印刷有限公司
开　　本	889 mm×1194 mm　1/16
印　　张	20
版　　次	2019年10月第1版　2019年10月第1次印刷
标准书号	ISBN 978-7-214-24002-6
定　　价	298.00元

（江苏人民出版社图书凡印装错误可向承印厂调换）

前言
Preface

《重回海河边》系列丛书包括《重回海河边　天津城市设计的故事》和《重回海河边　天津城市设计的理论与实践探索》两书。

《重回海河边　天津城市设计的故事》讲述了天津城市设计的故事，围绕天津海河两岸公共空间展开，集中探讨人性化的城市设计。该书通过不同人物的视角和口吻，讲述着天津城市与城市设计工作的发展变迁。这是本次城市设计试点工作的一大创新，从设计师视角转化到生活在城市的各类人的视角，将城市设计工作与人的真实生活相融合，表现出不同人物的真实生活与情感。书中人物是虚拟的，但故事是真实的。不同的人物分别代表了天津城的广大市民、参与天津城市规划建设的境内外规划设计师，通过他们的故事记述天津城市设计取得的成绩。

《重回海河边　天津城市设计的理论与实践探索》是天津城市设计的实例作品集，系统阐述了天津建卫600余年以来城市与城市规划发展演变，梳理中外城市规划思想在天津成长与实践历程，深入剖析了城市设计工作在城市各阶段发展中所起的作用。2008年，以迎接北京奥运会为契机，天津进入了城市规划与建设快速发展期。从中心城区到滨海新区，通过运用城市设计开启了城市空间形态的整体塑造和城市形象系统提升。2017年，为落实中央城市工作会议精神，天津被住房和城乡建设部列为第二批试点城市。在本轮试点工作中，天津不断转变发展观念，聚焦“以人为本”的城市设计理念和方法，在试点项目中重点关注城市居民最为关心的生态、老城、住区等问题：通过完善城市中心步行系统与公交系统，提升公共开放空间的品质，以增加天津城市中心的活力，活化城市历史街区；通过大规模增加城市公园、郊野公园和乡村郊区等绿色生活空间，恢复津沽城市生态系统，实现人与自然的和谐共生；通过天津2035年、2050年住宅理想类型和新型社区的探索，设计中国人向往的美好生活场景；通过理想城市的探察，明确城市明天发展的方向，为实现中华民族伟大复兴的中国梦描绘美好愿景。本书既是天津城市设计实践案例的汇编，更是一部新时代中国城市设计理论重点内容的展示。

Back to the Haihe Waterfront series is divided into two parts: *Back to the Haihe Waterfront—the Story of Urban Design in Tianjin and Back to the Haihe Waterfront—Practice and Exploration of Urban Design in Tianjin*.

Back to the Haihe Waterfront—The Tale of Urban Design in Tianjin tells the story of Tianjin urban design from the perspective of human beings, centering on the public space on both sides of the Haihe river in Tianjin and focusing on humanized urban design.

The book tells the development and changes of Tianjin city and urban design work through the perspectives and tones of different characters.

This is a major innovation of the urban design pilot projects. It transforms the perspective of designers to that of people living in cities, integrates urban design work with people's real life, and displays the real life and emotions of different characters.

The characters are virtual, but the story is real.

Different characters represent the general citizens of Tianjin, the overseas planning designers who participate in Tianjin urban planning and construction, and the collective portraits of several generations of urban planning designers, describing the achievements of Tianjin urban design through their stories.

Back to the Haihe Waterfront—Practice and Exploration of Urban Design in Tianjin is the portfolio of Tianjin urban design cases, elaborating the development and planning history of more than 600 years since it was built, combing Chinese and western urban planning thought grew up in Tianjin and the practice process, in-depth analysis of the urban design in urban development in different periods.

In 2008, Tianjin entered a period of rapid urban planning and construction to welcome the Beijing Olympic Games.

From the central urban area to the Binhai New District, urban design starts the overall shaping of urban spatial form and the improvement of urban image system.

In 2017, Tianjin was listed as the second batch of pilot cities by The Ministry of Housing And Urban-rural Development to implement the spirit of the Central Urban Work Conference.

In the pilot projects, Tianjin changed development concept, focusing on "people-oriented" idea and method of urban design, such as the ecology, old town, residential problem which urban residents are most concerned about: through improving the walking and bus system, to impove the quality of public open space, in order to increase the vitality of Tianjin city center and the activation of city historical and cultural blocks.

Through the large-scale increasing green spaces such in urban area, countries and rural areas, to restore the urban ecosystem in Tianjin, to achieve harmony between man and nature.

To explore the ideal housing types and new communities in 2035 and 2050 in Tianjin, and to design the beautiful life scenes that Chinese people yearn for.

Through the exploration of the ideal city, to point out the future development direction of the city, and to draw a beautiful vision for the realization of the Chinese dream of the great rejuvenation of the Chinese nation.

This book is not only a collection of practical cases of urban design in Tianjin, but also a display of the key content of urban design theory in China in the new era.

目录
Contents

海河与天津卫
THE HAIHE RIVER AND TIANJIN WEI

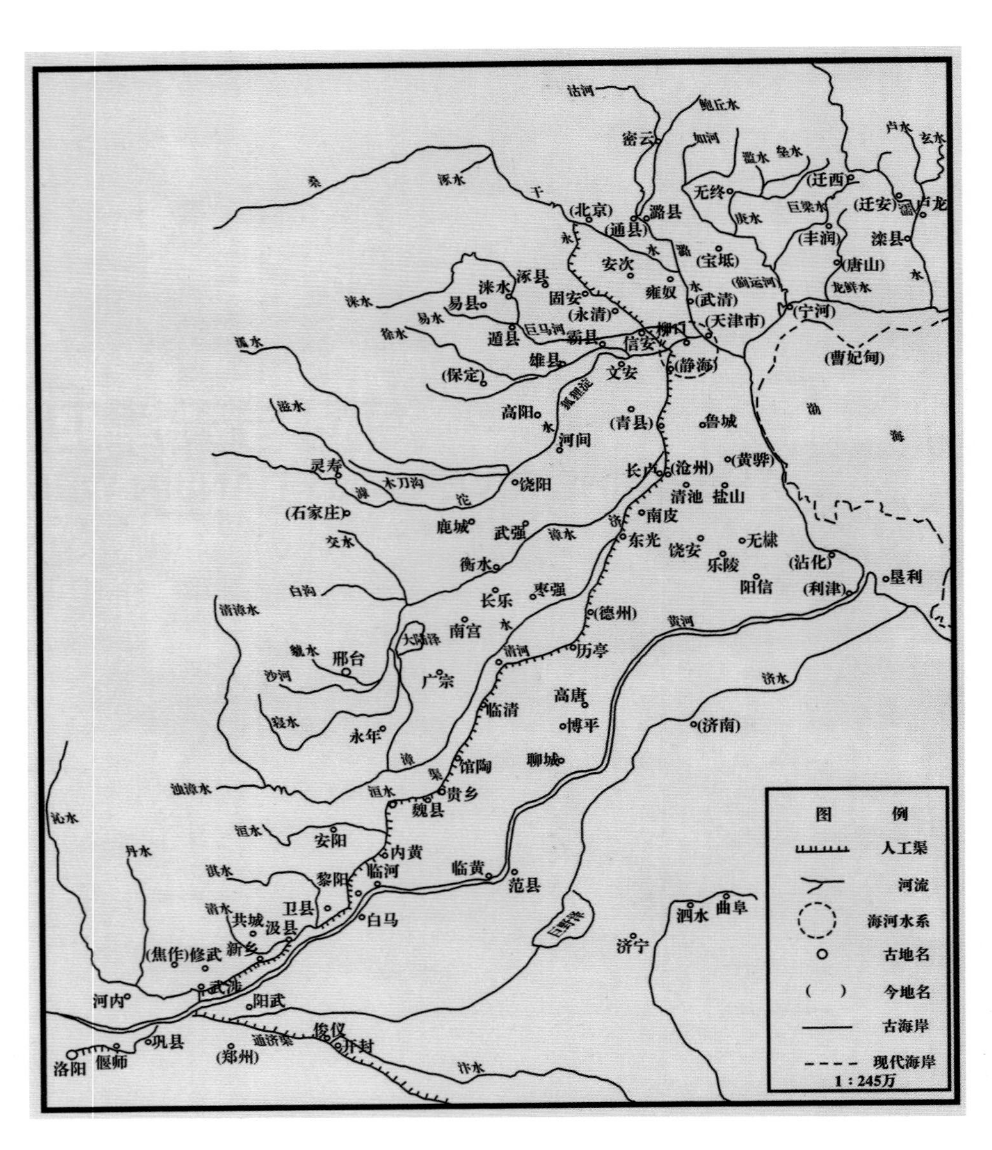

三岔河口——天津城市的发祥地

隋朝修建京杭运河后，隋炀帝为了支援北部边防和向塞外开拓，开凿了南接沁水、北达涿郡的永济渠，成为天津地区南北水路交通大动脉，海河的交通枢纽格局由此形成。南运河和北运河的交汇处，也就是今天的三岔河口，是天津城市的发祥地。

The Sancha Estuary—The Earliest Birthplace of Tianjin

After the construction of the Beijing–Hangzhou Grand Canal in the Sui Dynasty, Emperor Yangdi of the Sui Dynasty, in order to support the northern frontier defense and open up to the outside of the fortress, excavated the Yongji Canal connecting Qinshui in the south and Zhuojun in the north, which became the main artery of the north-south waterway traffic in Tianjin and formed the traffic hub pattern of the Haihe River. At the intersection of the south canal and the north canal, the Sancha Estuary, is the earliest birthplace of Tianjin.

隋代永济渠的开凿及海河水系形成示意图
The Digging of Yongji Canal in Sui Dynasty and The Formation of Haihe River System

明朝设立天津卫

明建文二年（1400），燕王朱棣在三岔河口渡过大运河，南下争夺皇位。朱棣成为皇帝后，为纪念由此起兵的“靖难之役”，在永乐二年（1404）将此地改名为天津，即“天子经过的渡口”之意。作为军事要地，在三岔河口西南的小直沽一带，筑城设卫，称天津卫。

Tianjin Wei was Established in the Ming Dynasty

In 1400, the second year of Ming Jianwen, Zhu Di, the King of Yan, crossed the Grand Canal at Sancha Estuary and travelled south to compete the throne. When Zhu Di became emperor, he changed the location's name to Tianjin in 1404, which meant the ferry crossing through which the son of heaven passed, to commemorate the Jingnan Campaign. As a military base, Tianjin began to build a city guard in the Xiaozhigu area southwest of Sancha Estuary, which is called Tianjin Wei (the Guard of Tianjin).

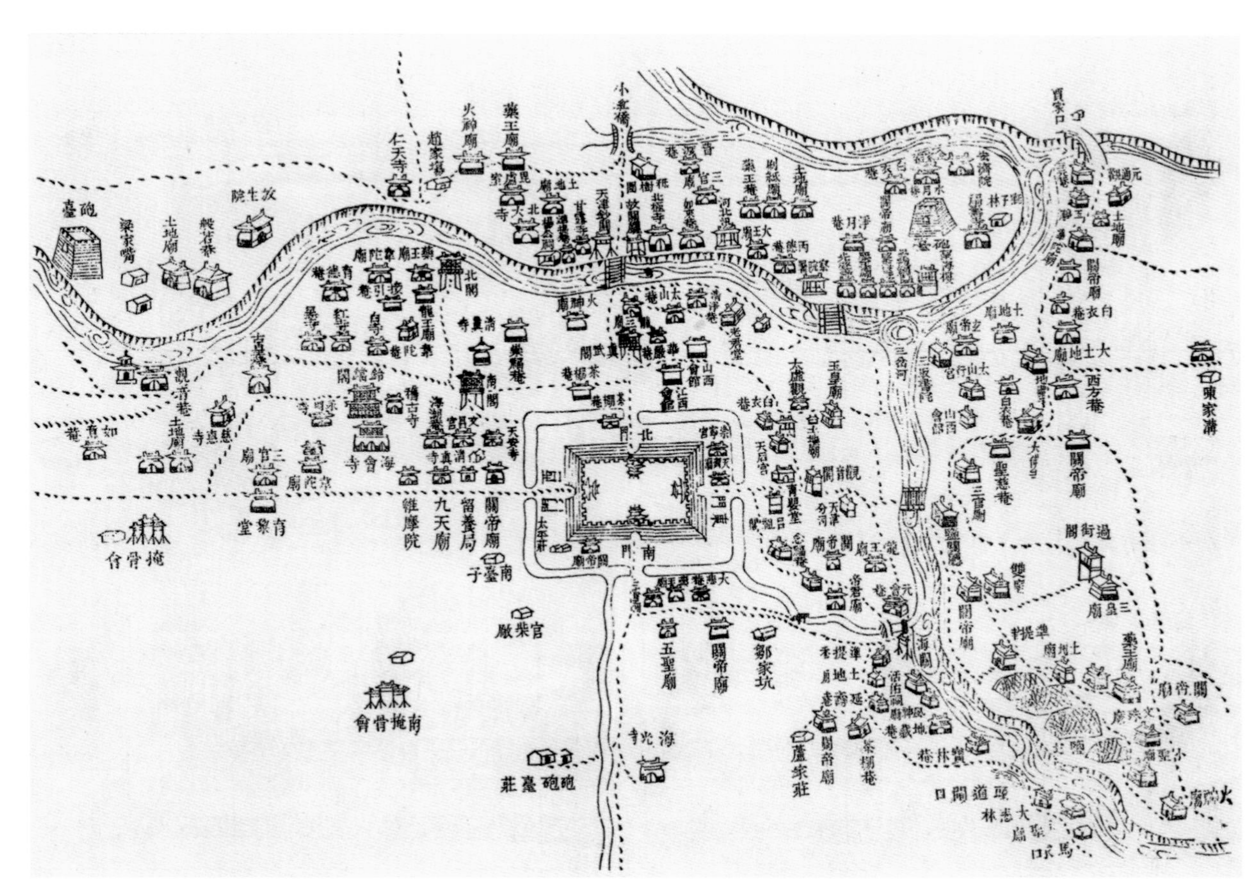

《津门保甲图——天津城厢图》（1846 年，清道光二十六年）
Baojia Atlas of Tianjin - Tianjin Wei Map (1846, The 26t Year of Qing Daoguang)

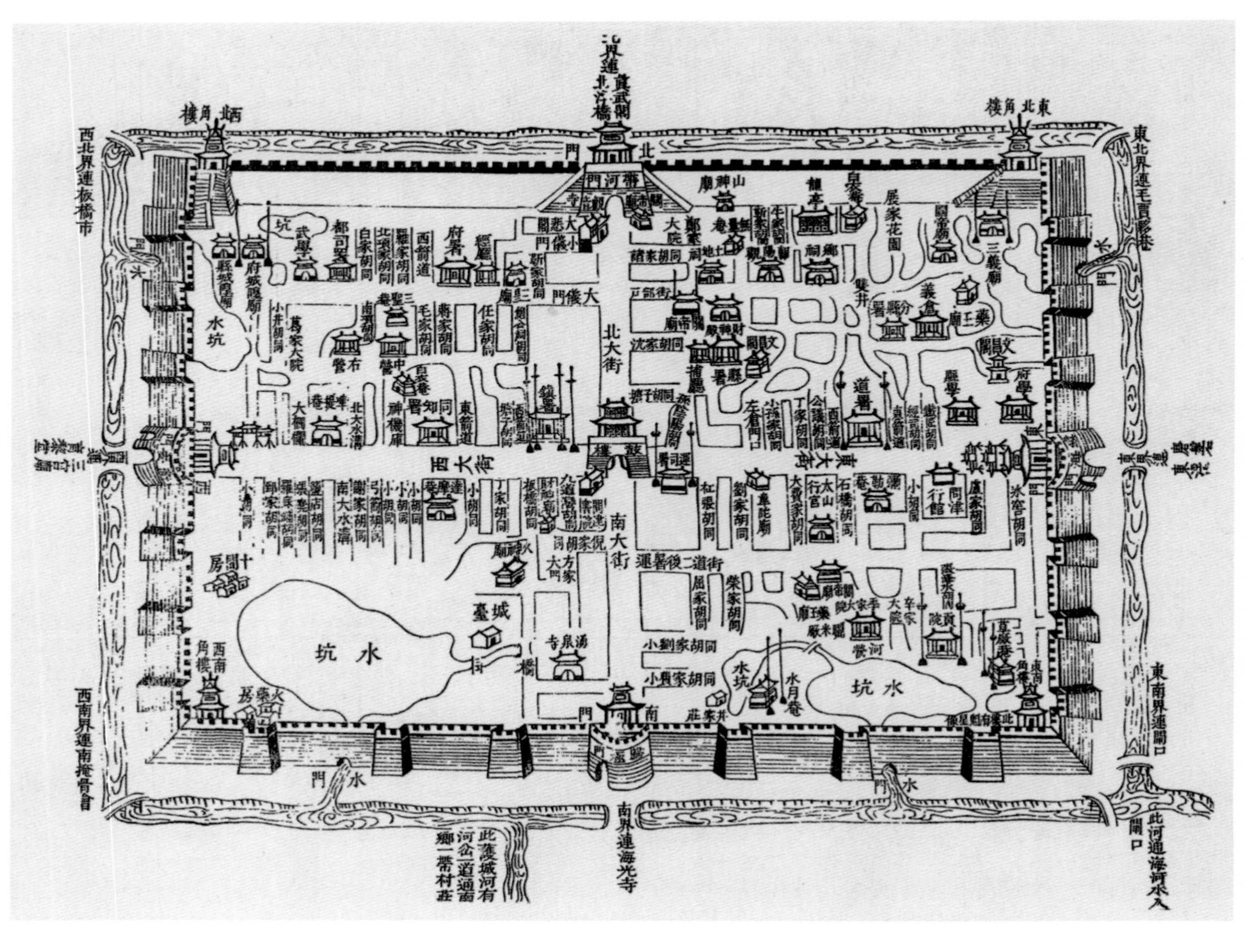

《津门保甲图——县城内图》
Baojia Altas of Tianjin - Illustration of Inner Tianjin City

天津卫城

城垣是长 1.5 千米、宽 1.0 千米的矩形，平面以鼓楼为中心形成“经纬纵横、十字方城”的空间格局。城内自然地形东北高，西南低，因此，卫城里按照“文东武西”分布，西北片是官署最为集中的区域，众多武职衙门汇聚在这里。东南片区住着很多名门望族。西南片区有老城里最低洼的区域，后来随着人口增多，成为了百姓的汇聚区。

Tianjin Wei

The town wall is a rectangle of 1.5 km by 1.0 km, and the layout is centered on the Drum Tower to form a spatial pattern of "With a network in vertical and horizontal, a square city with a cross main road". The natural terrain in the city is high in the northeast and low in the southwest. Therefore, according to the distribution of “Civial East, Military West", the northwest part is the most concentrated area of the government agencies. Many famous families live in the southeast area. The southwest has the lowest area in height in the old town. Later, with the increase of the population, it became the convergence area of the people.

三岔河口——商贾云集

明朝期间积极疏通了大运河，使浙江杭州到天津的南北水路全线畅通。大规模的漕粮运输使三岔河口成为了海河两路进入北京的咽喉。作为航运枢纽、经济活动的中心，1725 年（清雍正三年）天津卫被改为天津州，1731 年升州为府，正式设置地方政权建制。城东城北沿河弧形狭长地带，成为商贾云集、群众活动的集中之地，而军政官府衙署多设在城垣内。这种城垣内为军事政治中心、城垣外为经济活动中心的格局始终未变。

The Sancha Estuary—Commerce Gathering

During the Ming Dynasty, The Grand Canal was actively dredged, which made the north-south waterway from Hangzhou to Tianjin smooth. Large-scale transportation of water and grain has made the Sancha Estuary the throat for the Haihe River to enter Beijing. As a shipping hub and the center of economic activities, Tianjin was transformed from Tianjin Wei to Tianjin Zhou in 1725 (The 3rd Year of Qing Yongzheng) and to Tianjin Fu in 1731, which means a local government was established. The narrow arc along the river in the north and east of city has become a place where businessmen gathered and mass activities concentrated. The military and government offices are mostly located inside the city walls, where became military and political center, while the pattern of economic activity centers outside the city walls remains unchanged.

《潞河督运图》
Luhe River Supervisory Transport Map

九国租界

从 1840 年鸦片战争至 1900 年八国联军入侵的 60 年间，英、法、美等 9 个国家，先后在天津、上海、汉口等 12 个城市共设立 30 个租界。至 1902 年天津九国租界基本划定，用地近 16 平方千米，是老城厢面积的 10 倍。

Nine-nation Concessions

During the 60 years from the Opium War in 1840 to the invasion of the Eight-nation Alliance in 1900, nine countries, including Britain, France and the United States, etc. established 30 concessions in 12 cities, including Tianjin, Shanghai and Hankou. By 1902, the nine-nation concessions in Tianjin had been basically demarcated, covering nearly 16 square kilometers, 10 times the area of the old town.

天津地图（1912 年）
Map of Tianjin (1912)

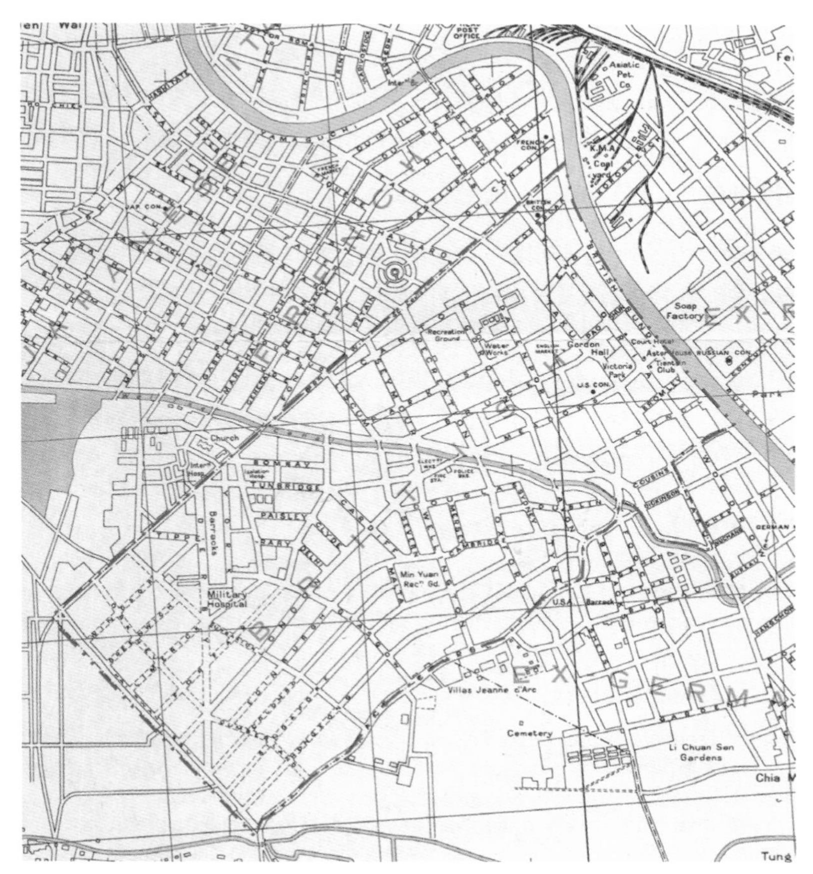

英租界地图
Map of British Concession

英租界

英国将天津城东南海河岸边紫竹林一带开辟为英国租界，经过三次扩张，东临海河，南沿马场道到佟楼，西到旧海光寺大道（今天的西康路），北沿营口道，在各国租界中规模最大。

五大道

五大道地区拥有 20 世纪二三十年代建成的具有不同国家建筑风格的花园式房屋 2 000 多所，建筑面积达 100 多万平方米，被称为万国建筑博览苑。

British Concession

The United Kingdom appropriated the Zizhulin area along the Haihe River in the southeast of Tianjin City as a British concession. After three times of expansion, it ranged from Haihe River in the east to Machang Road and Tonglou in the south, from old Haiguang Temple Avenue in the west (today's Xikang Road) to Yingkou Road in the north. Among all the concessionary territories in Tianjin, it was the largest in scale.

Wudadao

Wudadao area has more than 2000 garden-style houses with different countries architectural styles built in the 1920s and 1930s, with a total construction area of more than 1 million square meters. It is called "The World Architecture Expo".

维多利亚路（今解放北路）
Victoria Road (North Jiefang Road)

维多利亚花园
Victoria Park

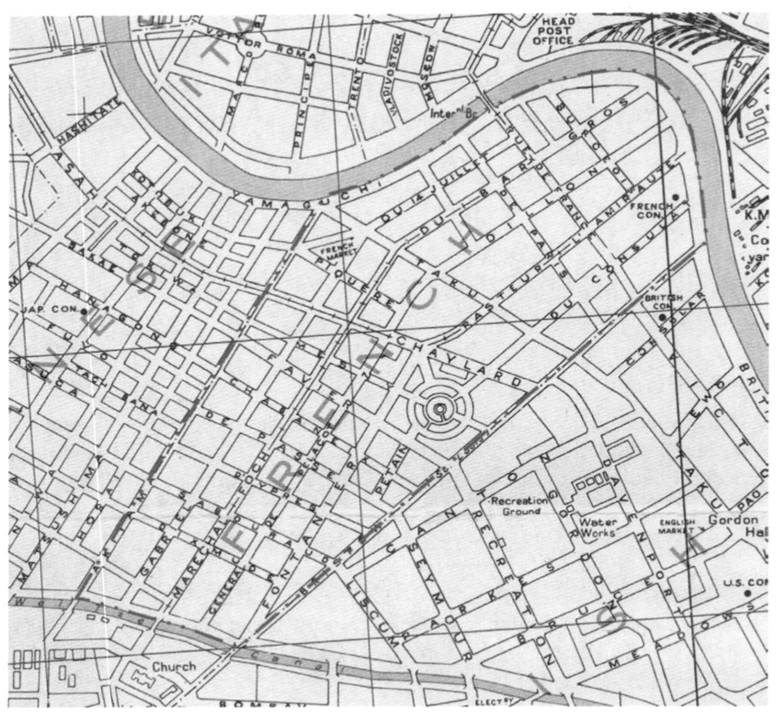

法租界地图
Map of French Concession

法租界照片
French Concession Pictures

法租界

法租界毗邻英租界，位于锦州道至营口道一带，采用欧洲传统古典主义规划手法。在空间格局上，以轴线和中心花园来控制整个区域，笔直的街道尽端多以高大的建筑做底景。

劝业场 – 梨栈地区

法租界里当时最为繁荣的是梨栈，即现在的劝业场一带。这里曾经是一大片芦苇地。今天的和平路当年就叫“梨栈大街”。“南有上海大世界，北有天津八大天。”劝业场是当时华北地区规模最大的百货商场，最多时楼内有大小商户和摊位三百多家。场内设有天华景戏院、天宫影院、天乐评戏院等八个娱乐场所，合称“八大天”。随后，位于其附近的交通饭店、惠中饭店和浙江兴业银行以及服装店、剧院、影院等相继开业，组成当时津城最大、最繁华的商业娱乐中心，素有“东方小巴黎”之称。

French Concession

French concession, adjacent to the British concession, is located between Jinzhou Avenue and Yingkou Avenue. It adopts the traditional European classical planning method. In terms of spatial pattern, the whole area is anchored by the axis and the central garden. At the end of the straight streets, the background is mostly tall buildings.

Quanyechang—Lizhan District

At that time, the most prosperous part of the French concession was Lizhan, which is now in the area of Quanyechang. Once here was a large reed field, and today's Heping Road was once called Lizhan Boulevard. "There is Shanghai's Dashijie in the south and Tianjin's Eight Big Tian's in the north." At that time, Quanyechang was the largest department store in northern China. At most, there were more than 300 businesses and stands in the building. There were eight entertainment venues including Tianhua Scenic Theatre, Tiangong Cinema and Tianle Pingju Opera Theatre, which are called the “Eight Big Tian's”. Subsequently, the Jiaotong Hotel, Huizhong Hotel, Zhejiang Industrial Bank and the clothing stores, theatres and cinemas that opened in succession formed the largest and most prosperous commercial entertainment center in Tianjin at that time, known as the "little oriental paris".

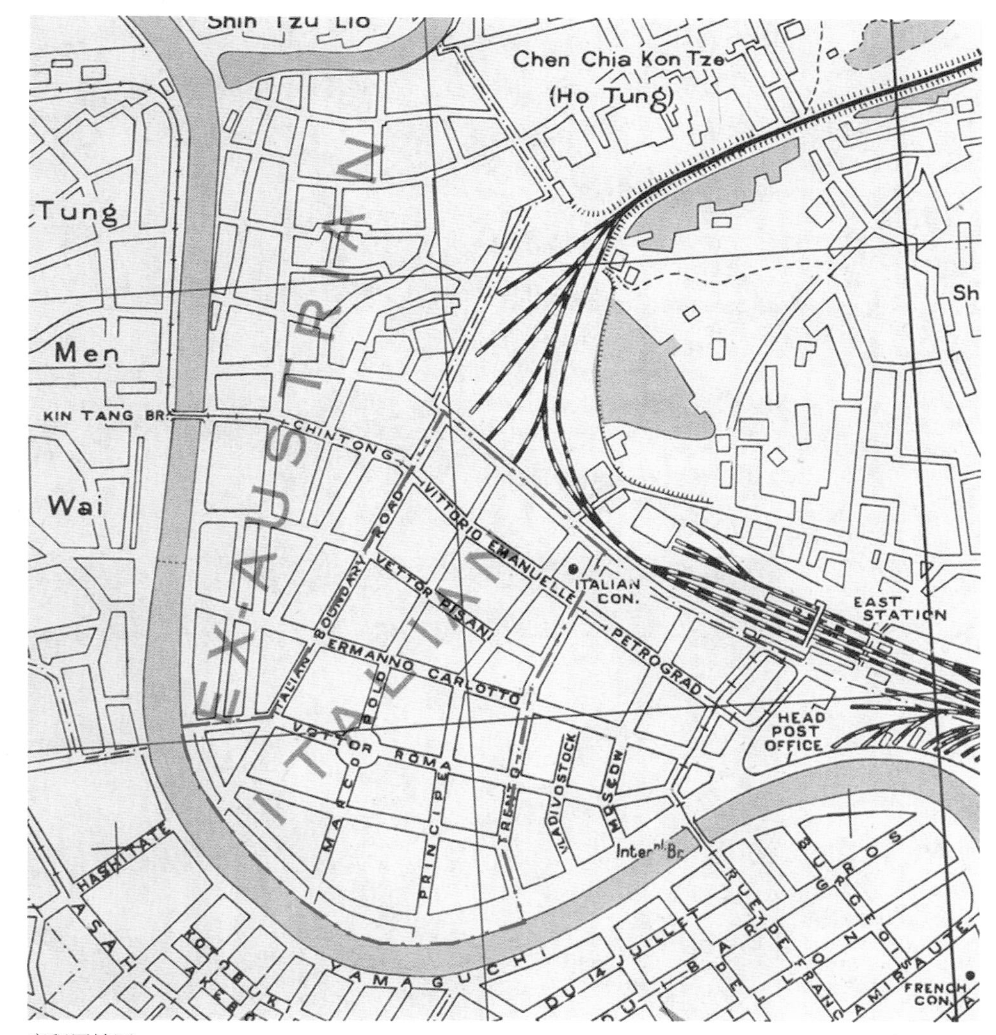

意租界地图
Map of Italian Concession

意租界

意租界位于海河北岸，它在建设之前几乎就是一个村落，规划后的网状街道，体现了现代城市规划风格，并与周边的俄租界、奥租界很好地衔接起来。

马可 · 波罗广场

中央立有高 13.6 米的柯林斯石柱，顶立和平女神像，基座四周有雕塑和喷水装置，圆形广场与四周建筑和谐统一，体现意大利风格。

Italian Concession

The Italian concession is located on the coast of Haihe River. Before construction, it was simply a village. The planned networked streets reflect the modern urban planning style and are well-connected with the Russian concession and the Austrian concession.

Marco Polo Square

There stands a Corinthian stone column of 13.6 meters high in the center, on top of it the Statue of Peace. Sculptures and water sprinklers surround the base. The circular square is in harmony with the surrounding buildings, reflecting the Italian style.

马可 · 波罗广场
Marco Polo Square

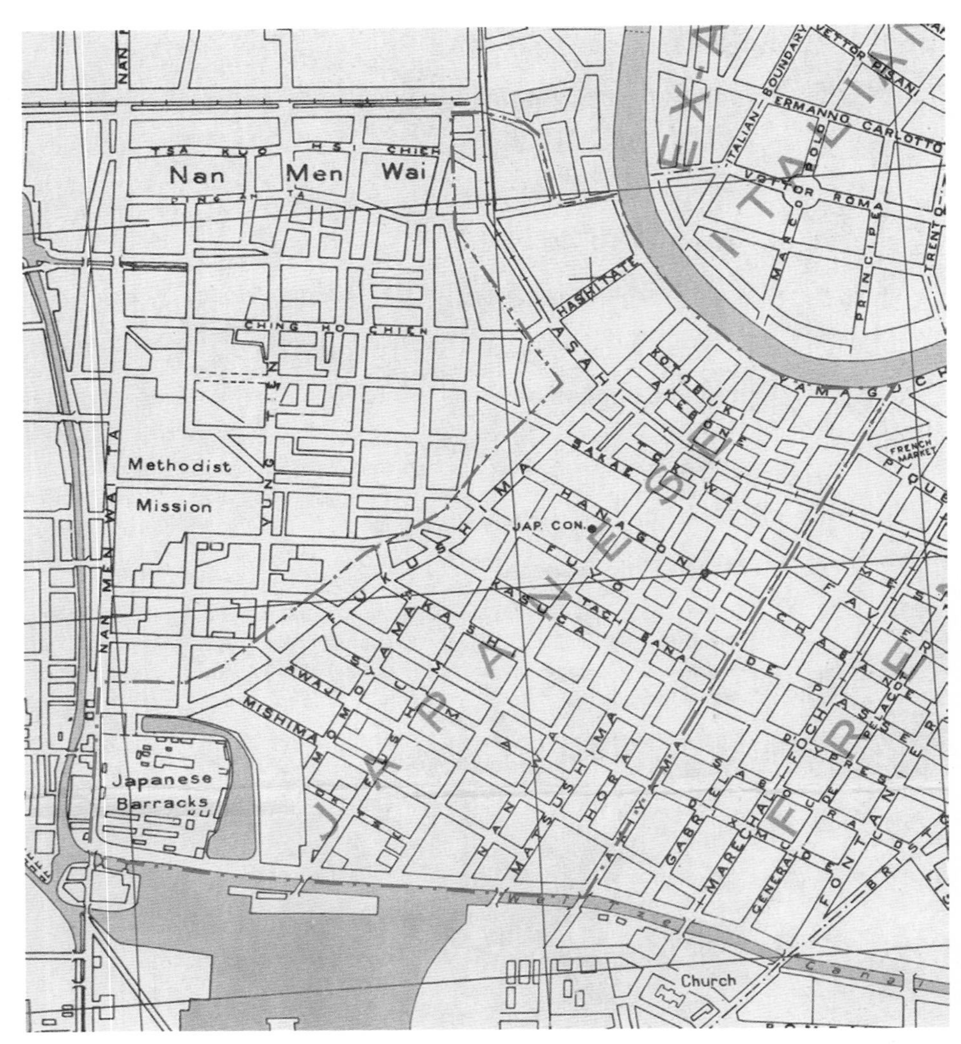

日租界地图
Map of Japanese Concession

日租界

日租界内采用日式井字街区布局，方格形街坊，形制规整，尺度较小。租界内建筑多为 2 ~ 3 层，室内“和风”，室外“洋风”。

旭街（今和平路）

中国最早的近现代商业街之一。

Japanese Concession

In the Japanese concession, the Japanese-style intersecting parallel block layout was adopted, with square blocks, in regular shape and small scale. Most of the buildings in the concession area were 2-3 stories, with the interior decorated in Japanese style and the exterior in Western style.

Xu Street (Heping Road)

One of the earliest modern shopping streets in China.

旭街
Xu Street

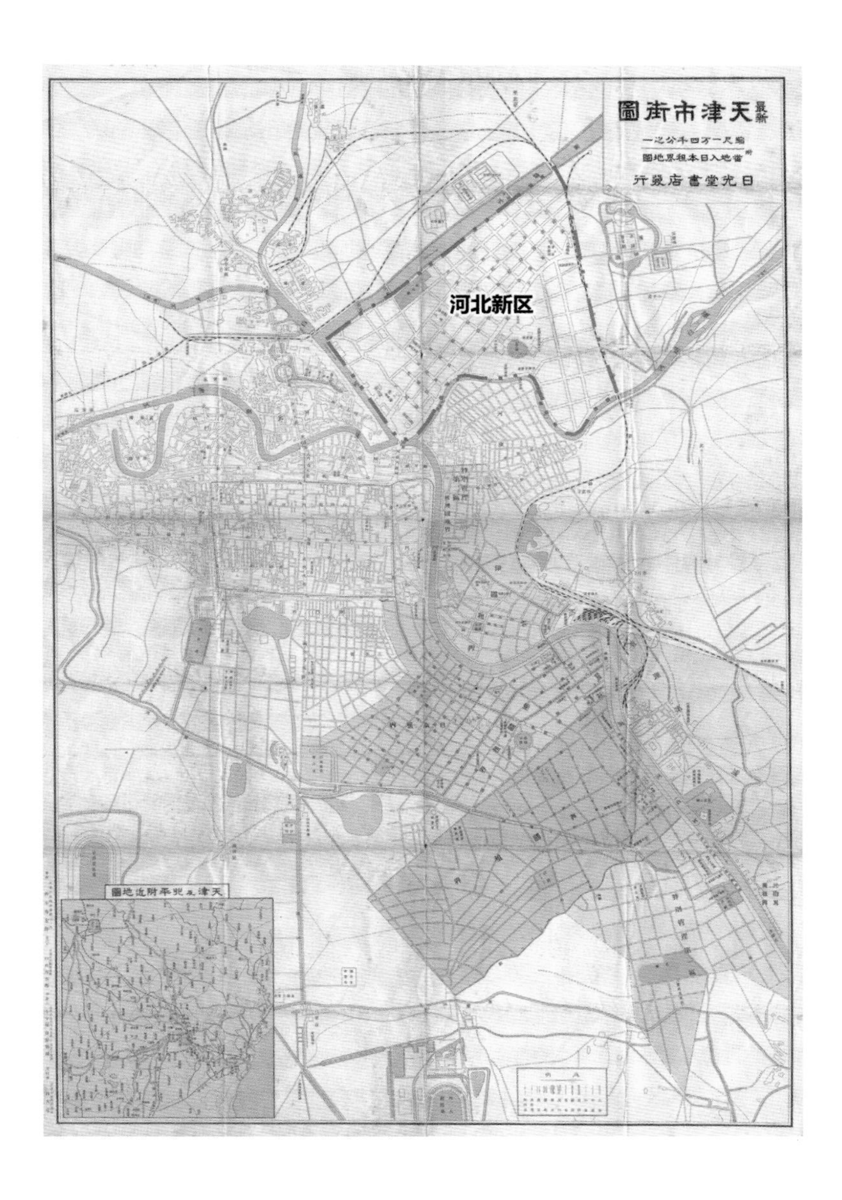

河北新区

1901 年，袁世凯接替李鸿章担任直隶总督兼北洋大臣。1902 年，他从八国联军“都统衙门”手中接管天津。因城厢地区遭到八国联军的严重破坏，“房屋尽成瓦砾”且已无发展空间，而城东南的海河两岸由于便于发展的地势，已被各国租界侵占；所以袁世凯决定推行“新政”，规划开发海河上游以北地区——河北新区，形成与各国租界相抗衡的格局。

Hebei New Area

In 1901, Yuan Shikai succeeded Li Hongzhang as the Viceroy of Zhili and Minister of Beiyang. In 1902, he took over Tianjin from the "Dutong Yamen" of the Eight-Nation Alliance. Because of the serious destruction of Tianjin City and the surrounding areas by the Eight-Nation Alliance, the houses were in ruins and there was no room for urban development. The land on both sides of the Haihe River in the southeast of the city that was easy to develop was occupied by the concessions of various countries. So Yuan Shikai decided to carry out a new plan to develop the Hebei New Area north of the upper reaches of the Haihe River to form a pattern of competing with the concessions of other countries.

河北新区位置图
Location of Hebei New Area

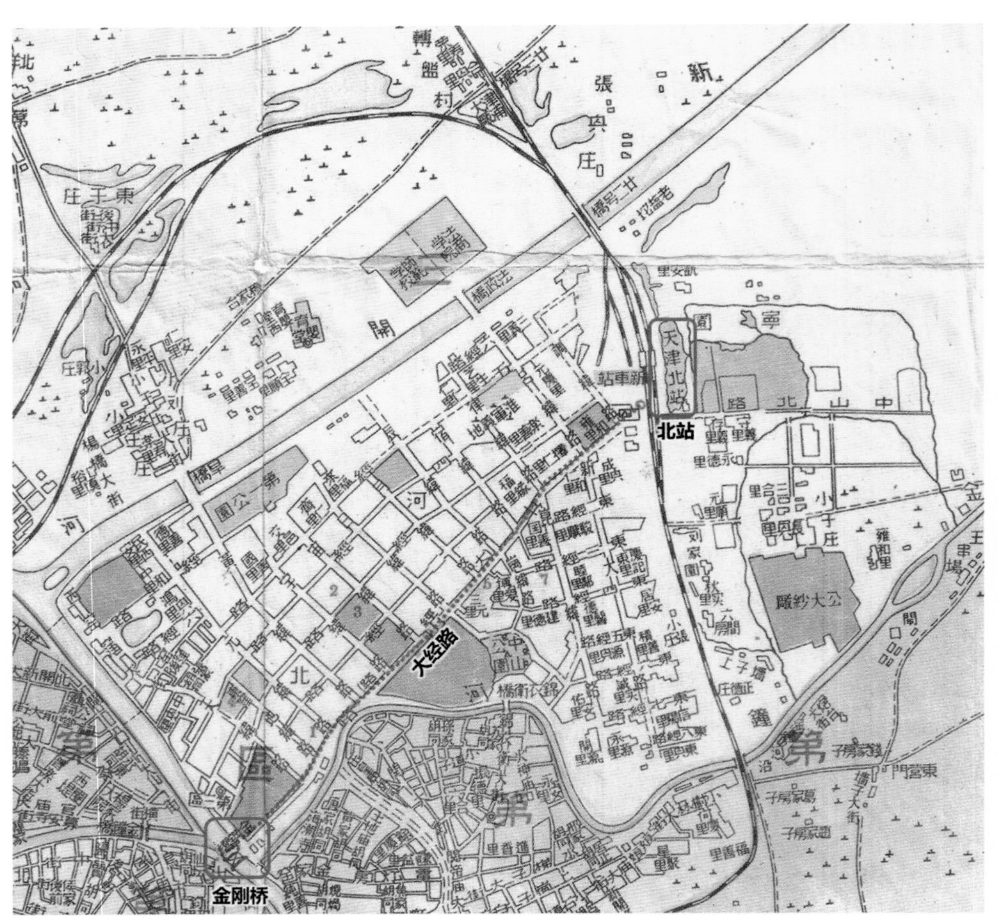

河北新区 1945 年地图
Map of Hebei New Area, 1945

大经路——以交通为导向的发展模式

河北新区建设了一条从总督衙门到新车站的大经路。1903 年，建设了开启式的金钢铁桥。大经路成了一条重要的道路，沿路也繁华起来。规划以大经路为主干线，贯穿南北，北部修建火车站——天津总站连通外省，南部修建金刚桥，与城市内部连通，充分体现了现代城市规划中以交通为导向带动城市建设的基本思路。

窄路密网

大经路以北并行的道路是四条经路，与其相交的，东西向从南至北依次是从《千字文》中取字，以天、地、元、黄、宇、宙、日、月、辰、宿、律、吕、调命名的 13 条纬路，形成了窄路密网的道路格局。

Dajing Road—A Transportation-Oriented Development Model

Hebei New Area has built a main road from Yamen (the governor's office) to the new train station. In 1903, the drawbridge Jingangtie Bridge was built. Dajing Road became an important road and businesses flourished along it. The planning takes Dajing Road as the main road that runs through the north and south, and constructs the Tianjin Railway Station in the north to connect with other provinces and Jingang Bridge in the south to connect with the inner city, which fully reflects the fundamental idea of transportation-oriented urban construction in modern urban planning.

"Narrow Road, Dense Network"

There are four parallel roads to the north of Dajing Road, which intersect with each other. From the south to the north, the east-west facing intersecting roads are named after characters from the *Thousand Character Classic*: heaven, earth, yuan, huang, yu, zeus, sun, moon, chen, su, law, lv, tune, forming a layout of "narrow road, dense network".

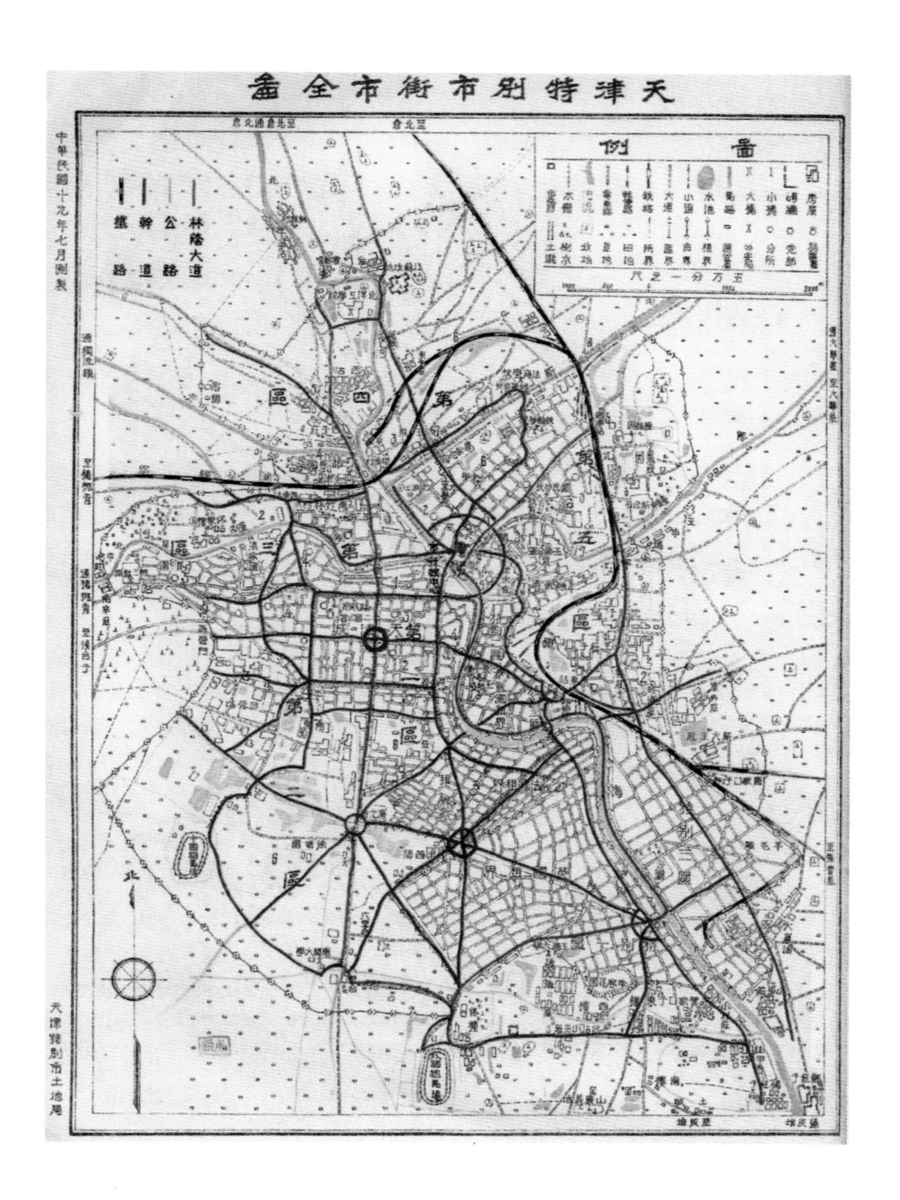

天津特别市物质建设方案：人性视角的城市设计实践

1930 年，天津特别市政府进行《天津特别市物质建设方案》的公开征选。这是天津近代城市规划史上第一部详细、全面的规划方案，也是中国首次通过竞赛征集并由中国建筑师、规划师设计的城市规划方案。

梁思成和张锐共同拟定的方案成为首选，初步明确了近代天津城市应发挥港口职能和经济中心城市的作用，体现了近代城市规划要为促进城市经济发展服务的思想。

"Material Construction Project for Tianjin Special Municipality City": Urban Design Practice from the Perspective of Human Nature

In 1930, the Tianjin special municipal government conducted a public selection for proposal of the "Material Construction Project for Tianjin Special Municipality City". This is the first detailed and comprehensive planning plan in the history of modern urban planning in Tianjin. It is also the first urban planning plan compiled by Chinese architects and urban planners through a competition in modern China.

Liang Sicheng and Zhang Rui jointly worked out the plan and was placed as the first choice. They preliminarily clarified that the modern Tianjin City should play the role of a port- and economic-centric city, which embodied the idea that modern urban planning should serve urban economic development.

1930 年《天津特别市物质建设方案》
1930, Material Construction Project for Tianjin Special Municipal City

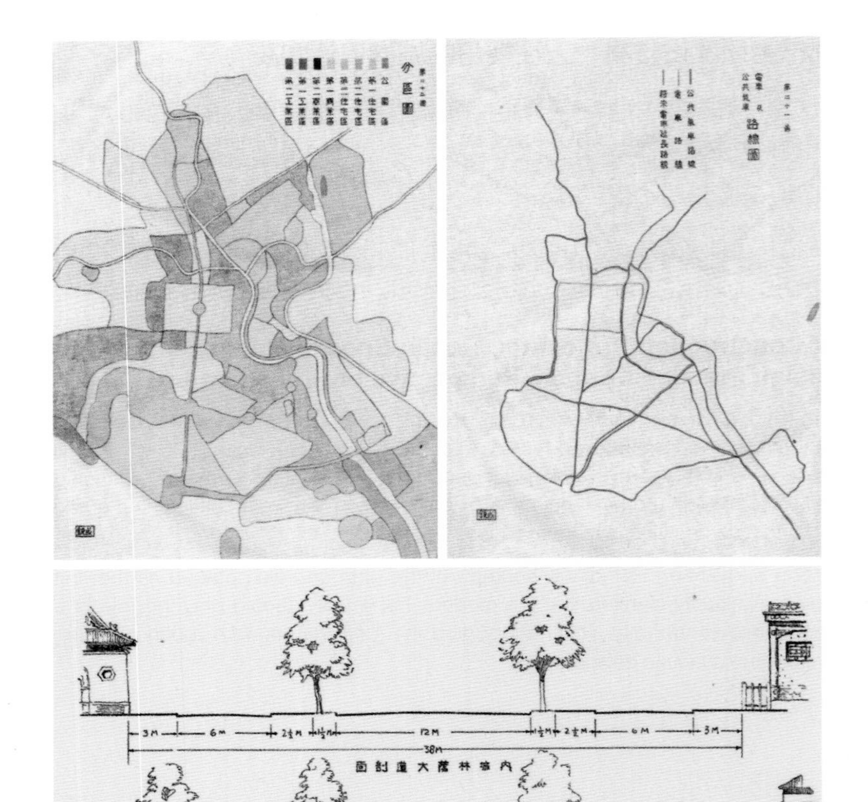

1930 年《天津特别市物质建设方案》
1930, Material Construction Project for Tianjin Special Municipal City

从城市设计出发的规划方案

该规划不仅从城市总体布局上提出方案，而且从城市设计的角度，在许多分项规划方面，提供可以指导施工设计的具体原则和若干实施办法、必要的管理规定及资金筹集办法等，增强方案实施的可能性和规划深度。

重视基本功能分区：将城市划分为公园区、住宅区、商业区、工业区等，保证良好的生活环境品质，并对各区内的建筑概念和要求分别进行说明。

重视街道空间塑造：以方便市民使用为原则，在利用原有道路的基础上，形成主干道、次干道、林荫大道、内街和公道五种城市道路形式。

重视海河两岸环境品质：对海河两岸功能、环境规划特别指出，要设置码头，配套基础设施，并强调可学习欧美城市，在河岸两侧开辟公园，美化都市环境，对公共建筑、飞机场等位置和形式都提出了具体要求。

Planning Scheme Starting From Urban Design

The plan not only suggests solutions for the overall layout of the city, but also provides specific principles and some implementation methods that can guide the construction design as well as necessary management regulations and fund-raising methods, etc. to enhance the possibility and depth of the implementation of the plan from the perspective of urban design.

Attention to basic functional zoning: dividing the city into parks, residential areas, commercial areas, industrial areas, etc. to guarantee the quality of living environment, and specifying the architectural concepts of each area separately.

Attention to the shaping of street space: guided by the principle of facilitating citizens' usage, on the basis of utilizing existing roads, five urban road forms are formed: main road, submain road, tree-lined avenue, inner street and highway.

Attention to the quality of environment on both sides of the Haihe River: the planning specifies the functions and environment of both sides of the Haihe River, suggests setting up wharfs and the supporting infrastructure, and emphasizes that we can learn from European and American cities and place parks on both sides of the river bank to beautify the urban environment, and also put forward specific requirements for the location and form of public buildings and airports.

“大天津都市计划大纲”

1937 年至 1945 年，是日本侵占天津时期，在这段时期内开始对塘沽地区进行建设。当时，华北日伪当局为了适应侵华战略需要，制定了所谓“大天津都市计划大纲”，针对天津特别市、塘沽和海河沿岸地区的港口建设、城市发展进行了规划。后来为配合港口建设编制了所谓“塘沽都市计划大纲”，明确：“塘沽市区是天津都市区的一部分，与新港建设相关联，为水陆交通中心枢纽与工业地带。”

"Outline of the Greater Tianjin Metropolitan Plan"

From 1937 to 1945, during the period of Japanese colonial rule in Tianjin, the development and construction of Tanggu area began. At that time, in order to meet the strategic needs of invading China, the Japanese puppet state in northern China formulated the "Outline of the Greater Tianjin Metropolitan Plan", which aimed at the port construction and urban development of Tianjin Special Municipality City, Tanggu and the coastal areas of Haihe River. Later, in order to coordinate with port construction, the "Tanggu Metropolitan Plan Outline" was compiled, which made it clear: "Tanggu is a part of Tianjin's metropolitan area, which is related to the construction of new port and is a hub of land and water transportation and industrial zone."

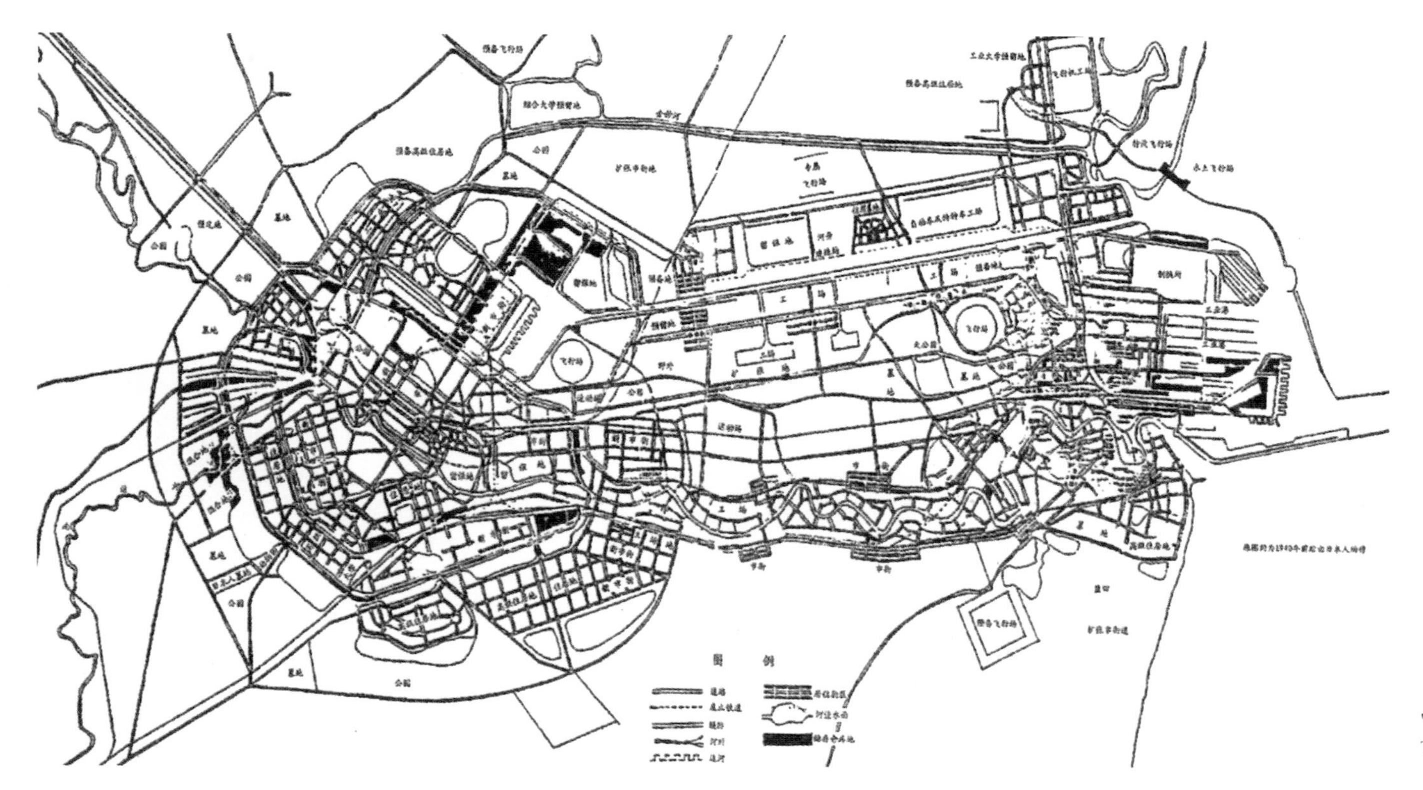

“大天津都市计划大纲”
"The Outline of the Greater Tianjin Metropolitan Plan"

天津新港

所谓“大天津都市计划大纲”出台后，计划在天津与北塘之间开凿运河入海的建港方案因为费用大，决定放弃，经研究决定，在塘沽修建新港。塘沽新港选定在海河口以北，塘沽以东，当时的筑港工程原则是：海河航道水浅，新建港区航道，使新港不受大沽沙航道水深限制；筑防沙设施，避免淤泥影响港区水域；兼顾海河航道的利用，继续发挥天津老港的作用。

中华人民共和国建立后，国家非常重视天津港口的发展。1951 年，政务院决定修建塘沽新港，当家做主的港口工人仅用了一年多时间，就圆满完成了第一期建港工程，使几乎淤死的港口重新焕发生机，于 1952 年正式开港。经过三期扩建，至 1988 年底，新港区域总面积已发展到约 200 平方千米。

Xingang, Tianjin

After the promulgation of the "Outline of the Greater Tianjin Metropolitan Plan", the plan to build a new port by digging a canal to reach the sea between Tianjin and Beitang was abandoned because of the high cost, and it was decided that a new port was to be built in Tanggu. The location for Tanggu new port was selected to be at the north of Haihe River's Estuary and east of Tanggu. At that time, the principles of port construction were as follow: the Haihe River channel was shallow, so a new port area channel would be built so that the new port was not restricted by the depth of Dagu Sand Channel; sand prevention facilities were to be built to avoid silt deposition; taking into account the utilization of the sea channel, allowing the old port in Tianjin to continue to play its role.

After the founding of the People's Republic of China, the state placed great importance in the development of Tianjin's ports. In 1951, the State Council of the Central Government decided to build Tanggu new port—Xingang. In just over a year, the port workers, who were the masters of the project, successfully completed the first phase of the port construction project, revitalized the almost silted port and officially opened the port in 1952. After three phases of expansion, by the end of 1988, the total area of the Xingang area had grown to approximately 200 square kilometers.

集装箱码头
Container Terminal

客运码头
Passenger Wharf

粮食码头
Grain Wharf

唐山大地震

1976 年 7 月 28 日 3 时 42 分 53.8 秒，河北省唐山、丰南一带发生 7.8 级强烈地震。当时，天津是仅次于唐山的重灾大城市，全市死亡人数达到 2.4 万，受伤人数达 1.6 万，直接和间接经济损失达到 75 亿元，使天津成为全国唯一遭受 8 度以上地震破坏的特大城市。

全市各类建筑遭到破坏的共计 7 052 万平方米，占原有面积的 68%。道路、桥梁等基础设施大面积受损。

全市 70 余万居民失去住房，无家可归，所有的开阔地都搭设了“临建棚”，不仅体育场、街心公园、学校操场、小广场全都成了“临建区”，就连比较宽阔的大马路两旁也见缝插针盖满了小房子，一夜之间立体的城市变成了平面的城市，先后搭盖了临建棚 23 万间，有上百万群众住在各式各样的“临建棚”之中。

Tangshan Earthquake

On July 28, 1976, at 3:42:53.8 am, a strong earthquake of magnitude 7.8 occurred in Tangshan and Fengnan of Hebei province. At that time, Tianjin was the second largest damaged city besides Tangshan, with 24,000 deaths and 16,000 injuries. Its direct and indirect economic losses amounted to 7.5 billion yuan, making Tianjin the only mega-city in the country to be damaged by earthquakes with magnitude above 8.

A total of 7.52 million square meters of all kinds of buildings in the city were damaged, accounting for 68% of the original area. Infrastructure such as roads and bridges has been damaged in a large area.

More than 700,000 residents in the city lost their houses and became homeless. All the open areas were built with "Temporary Shelters". Not only stadiums, street parks, school playgrounds and small squares became "Temporary Construction Areas". On both sides of the wider roads, small houses were covered with stitches. Overnight, the three-dimensional city became a flat city, with 230,000 temporary shelters built successively, with millions of people live in various "Temporary Shelters".

震后照片
post-earthquake photos

震后重建

面对巨大的灾难，天津市委、市政府和天津人民在党中央领导下，在各省、市、自治区和人民解放军的大力支援下，立即投入抗震救灾斗争。1977 年至 1978 年，全市以恢复生产为重点，完成工厂企业恢复重建投资 3.72 亿元，修复新建工业项目 164 个，竣工面积 73 万平方米。

在震灾恢复重建取得阶段性成果的基础上，加快修复震损住宅，改善和提高人民群众的居住条件，成为全市最重要、最紧迫的任务。从 1981 年开始，天津市进入震灾恢复重建的新阶段。

经过 3 年奋战，到 1983 年，全市共修复加固住宅 676.5 万平方米，先后重建了贵阳路、大营门、东南角、大胡同、黄纬路、大直沽后台（即“老六片”）和崇仁里、小稻地、求是里、南头窑、小西关（即“新五片”）等居民住宅区，安置震灾户和临建户 11.2 万户，安置各类住房困难户 4.13 万户。

Post-earthquake Reconstruction

Facing the huge disaster, the Tianjin Municipal Committee, the municipal government and the people of Tianjin, under the leadership of the CPC central committee and with the strong support of provinces, municipalities, autonomous regions and the People's Liberation Army, immediately devoted themselves to the earthquake relief struggle. From 1977 to 1978, the city focused on restoring production. It completed the reconstruction of factories and enterprises with an investment of 372 million yuan, 164 new industrial projects and a completed area of 730,000 square meters.

On the basis of the phased results of earthquake recovery and reconstruction, it was the most important and urgent task for the whole city to speed up the repair of earthquake-damaged houses and improve the living conditions of the people. Since 1981, Tianjin has entered a new stage of earthquake recovery and reconstruction.

After three years of hard work, by 1983, 6,765,000 square meters of residential buildings had been repaired and strengthened. Guiyang Road, Dayingmen, Dongnanjiao, Dahutong, Huangwei Road, Dazhigu Houtai (namely "Old Six Areas") and residential areas such as Chongrenli, Xiaodaodi, Qiushili, Nantouyao and Xiaoxiguan (namely "New Five Areas") had been rebuilt successively, and 112,000 earthquake-stricken households and temporary built households and 41,300 households in difficulty had been resettled.

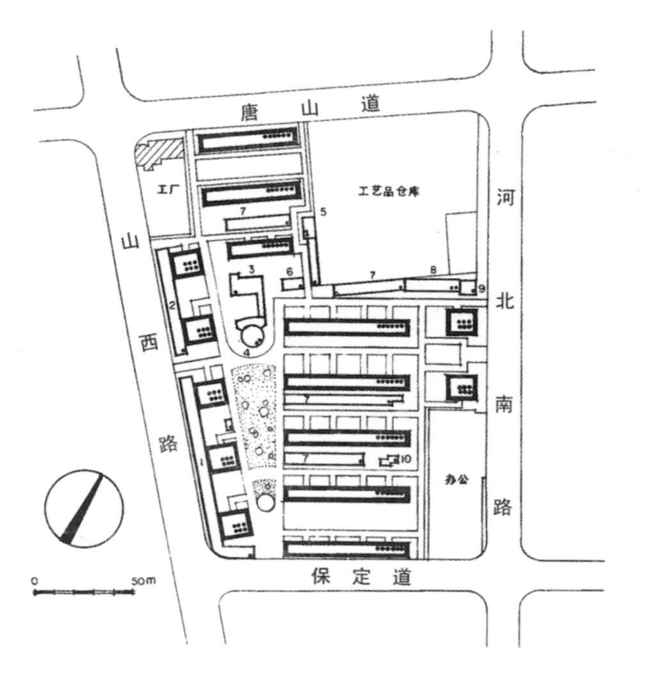

1978 年崇仁里片规划
Chongrenli District Planning in 1978

1949 年后天津工业区与居住区建设

1949 年后，天津城市经济开始逐渐恢复，大量国有企业快速发展建设，像天津棉纺厂、天津运输场等。这种大规模的国有企业，厂区大，职工多，厂子里也是一应俱全，拥有食堂、澡堂等各种设施。

中华人民共和国成立之初，天津市劳动人民的住房极其困难。1952 年，为解决老百姓的住房问题，天津市政府下决心建设一批住宅区，也就是工人新村。市政建设委员会制定《扩大城区建设计划草案》，完成了 7 个工人新村的建设选点，总建筑面积达 105 万平方米。7 个工人新村具体包括：中山门、王串场、西南楼、佟楼、吴家窑、丁字沽、唐家口。

Construction of Industrial and Residential Areas in Tianjin After 1949

After 1949, the urban economy began to recover gradually, and a large number of state-owned enterprises developed rapidly, such as Tianjin Cotton Textile Mill, Tianjin Transportation Factory and so on. This kind of large-scale state-owned enterprise has a large factory area, a large number of employees, and all kinds of facilities such as canteens and bathhouses are served in the factory.

At the beginning of liberation, the housing of the working people in Tianjin was extremely difficult. In 1952, in order to solve the housing problem of the common people, the Tianjin municipal government decided to build a number of residential areas, that is, Workers' New Villages. The municipal construction committee formulated The Draft Plan For Expanding Urban Development, completing the construction of seven new workers' villages with a total floor area of 1,050,000 square meters. Seven new workers' villages include Zhongshanmen, Wangchuanchang, Xinanlou, Tonglou, Wujiayao, Dingzigu and Tangjiakou.

丁字沽工人新村建成时的照片
Dingzigu Workers' New Village

1952 年“完成丁字沽工人新村光荣任务纪念章”
The Glory Mission Medal of Dingzigu Workers' New Village

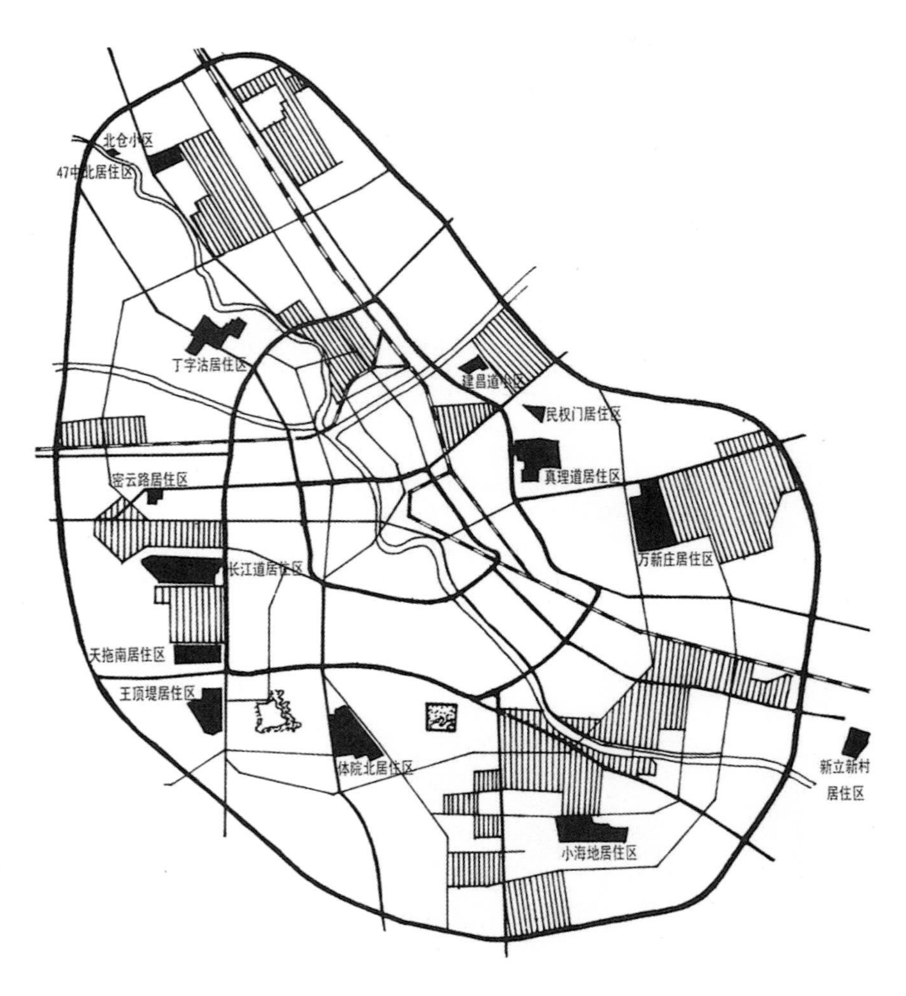

天津 14 片新居住区
14 New Residential Areas In Tianjin

14 片新居住区

震后恢复重建的同时，为了进一步体现城市总体规划要求，适当降低市中心区人口密度，使城市人口分布趋于合理，一批配套设施齐全、布局合理、建筑新颖、环境优美的新住宅区，如体院北、真理道、小海地、丁字沽、密云路、长江道、天拖南、建昌道、北仓、民权门、新立村、王顶堤、万新村等 14 个新住宅区，陆续按规划建成，总共占地 1 062.7 公顷，总建筑面积 777 万平方米，十几万户居民迁入新居。这些小区的规划建设均达到中华人民共和国成立以来天津市住宅建设的最好水平。

14 New Residential Areas

At the same time, in order to further reflect the requirements of the urban master plan, appropriately reducing the population density in the downtown area and make the urban population distribution more reasonable, a number of new residential areas with complete supporting facilities, reasonable layout, novel buildings and beautiful environment, were planned, Fourteen new residential areas, including Tuyuanbei, Zhenli Avenue, Xiaohaidi, Dingzigu, Miyun Road, Changjiang Road, Tiantuonan, Jianchang Avenue, Beicang, Minquanmen, Xinli Village, Wangdingdi and Wanxin Village, etc. had been built successively according to planning, covering a total area of 1062 hectares, with a total construction area of 7.77million square meters, and hundreds of thousands of households have moved into new homes. The planning and construction of these residential areas have reached the best level of residential construction in Tianjin since the founding of the People's Republic of China.

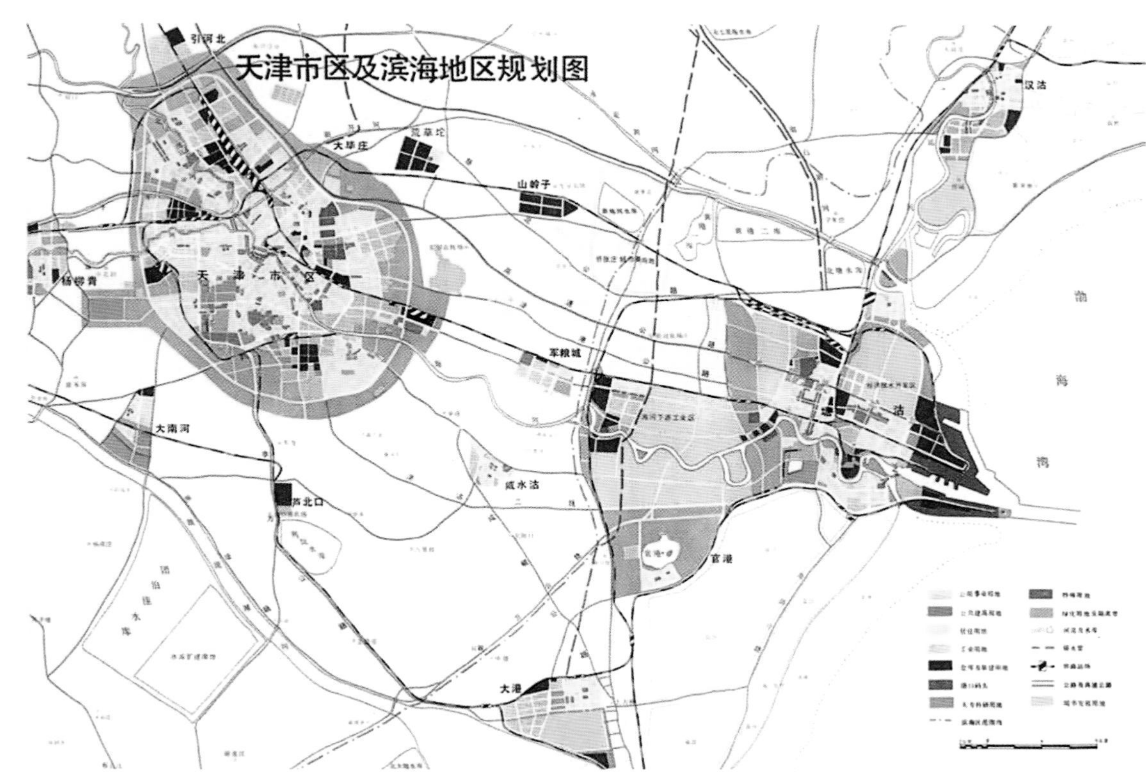

《天津市城市总体规划（1986—2000）》
Urban Master Plan of Tianjin (1986-2000)

一条扁担挑两头

1984 年，天津被列为 14 个开放的沿海港口城市之一，天津经济技术开发区正式成立，成为中国首批国家级开发区之一。

1986 年 8 月，《天津市城市总体规划》通过国家批复，这是天津历史上首个具有法律效力的城市总体规划，也是改革开放后国家首批批复的城市规划，期限为 1986 年到 2000 年，规划提出“一条扁担挑两头”的城市布局，确定了“工业东移”战略。同时，首次明确了“综合性工业基地，开放型、多功能的经济中心和现代化的港口城市”的城市定位，对此后天津若干年的发展起到了至关重要的作用。

One Pole Carries Two Ends

In 1984, Tianjin was listed as one of the 14 coastal port cities. Tianjin Economic-Technological Development Area (TEDA) was formally established and became one of the first national development areas in China.

In August 1986, "Tianjin Urban Master Plan" was approved by the State Council. It is the first city master plan with legal effect in Tianjin history. It is also the first city plan approved by the State Council after the Reform and Opening. The plan covers the period is from 1986 to 2000. The plan proposes the urban layout of "one pole carries two ends", and solidifies the strategy of "industry moving eastward". At the same time, the city's role of "comprehensive industrial base, open, multi-functional economic center and modern port city" was defined for the first time, which played a crucial role in the development of Tianjin in the following years.

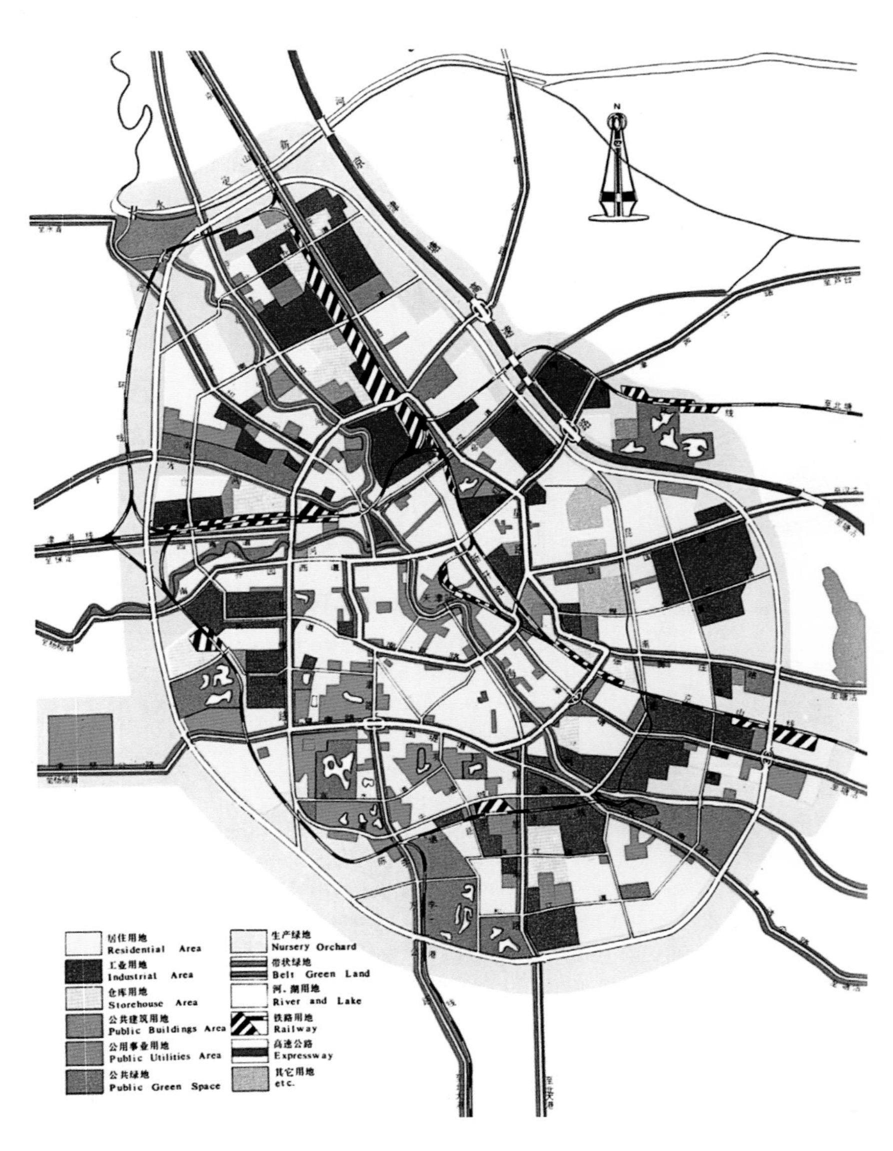

三环十四射

1986 年规划中的市区道路网，由主干道的 3 个环线和东南、西北 2 个半环线以及 14 条放射线构成环形放射路网系统的骨架，联系市区内各分区和功能分区，承担市区的主要交通量。

Three Rings and Fourteen Radiations

In 1986, the urban road network was planned. The three ring lines of the main road, two half-loop lines of the east-south and west-north, and 14 radiation lines formed the framework of the ring-radiation road network system, which connected the various zones and functional zones in the urban area, assuming the main traffic volume in the urban area.

《天津市城市总体规划（1986—2000）》
Urban Master Plan of Tianjin (1986-2000)

内环线：内环线长 15.2 千米，由 9 条路、街、道构成，线内面积 13.3 平方千米，为天津市中心区。

中环线：中环线连接市内 6 个行政区和若干个大型居住区、工业区及仓储区。在设计上西半环为 3 块板，东半环为 4 块板。与环线相交的主干道路口，采用立体交叉；与次干道相交的路口，分别采用扩大路口和渠化设计。

外环线：似彩带环绕天津四郊的外环线，全长 71.44 千米，路宽 50 米，其间有跨河大桥 4 座，立交桥 6 座。

Inner Ring Line: It is 15.2 kilometers long and consists of 9 roads, streets and avenues. Its area is 13.3 square kilometers in the central urban area of Tianjin.

Central Ring Line: It connects six administrative districts and several large residential, industrial and storage areas in the city. In the design, the west half ring has 3 plates, the east half ring has 4 plates, and the main road intersections with the ring road are three-dimensional intersections, and the intersection with the secondary main roads are enlarged intersections or channelized intersections respectively.

Outer Ring Line: Like a ribbon around the outskirts of Tianjin, it is 71.44 kilometers long and 50 meters wide. There are 4 river-crossing bridges and 6 overpasses.

内环线（南开三马路）
Inner Ring Line (Nankai Sanma Road)

中环线上的中山门蝶式立交桥（1986 年）
Zhongshanmen Butterfly Overpass on Central Ring Line (1986)

中环线于 1986 年 7 月 1 日全线竣工通车
Central Ring Line Was Completed and Opened in July 1, 1986

外环线（1987 年）
Outer Ring Line (1987)

改造“三级跳坑”住宅

“三级跳坑”住宅，是指屋内地面比院子低，院子比胡同低，胡同比路面低的住房。据 1984 年调查，天津的“三级跳坑”住宅共计 18 821 间，252 153 平方米，涉及居民 13 946 户。成片的集聚地区包括：河北区望海楼、谦益里，河西区下瓦房、北洋工房，南开区长安里、荣德里等。

1984 年底，天津在抗震救灾恢复重建任务完成之后，即开展改造“三级跳坑”住宅工作。在改造过程中，采取因地制宜的方法：对有条件实施规划的，就成片进行重建改造; 暂时无法按规划要求实施改造的，则就地翻建以改善居住条件。

1985 年底，市政府在河西区“北洋工房”改造的北洋新里召开现场会，并为纪念碑揭幕，标志“三级跳坑”改造任务胜利完成。该项目所取得的社会效益、经济效益和环境效益是不可低估的，为今后的城市住宅建设和城市旧区改造积累了可供借鉴的经验。

Renovation of "Triple Jump Pit" Dwellings

"Triple jump pit" dwelling refers to the dwelling with lower floor than courtyard,and the courtyard is lower than Hutong,the Hutoug is lower than road surface. According to the survey in 1984, there were 18,821 triple jump pit dwellings in Tianjin, 252,153 square meters, involving 13,946 households. The agglomeration areas include Wanghailou , Qianyili in Hebei District, Xiawafang and Beiyang Gongfang in Hexi District, Chang'anli and Rongdeli in Nankai District, etc.

At the end of 1984, after the completion of the task of earthquake relief and reconstruction, Tianjin began to rebuild the "triple jump pit" housing. In the process of transformation, measures should be taken according to local conditions: If possible, the reconstruction would be carried out in the whole area; if impossible to implementing the plan, the renovation of current houses would be carried out to improve the living conditions.

At the end of 1985, the Municipal Government held an on-site meeting in Beiyang Xinli, which was renovated by Beiyang Gongfang in Hexi District, and unveiled the monument, marking the successful completion of the task of "triple jump pit" renovation. The social, economic and environmental benefits achieved by this project can not be underestimated, which has accumulated useful experience for future urban residential construction and urban old area reconstruction.

未改造的“三级跳坑”住宅
Untransformed "Triple Jump Pit" Dwellings

翻修住宅

天津是一个老城市，有相当数量的陈旧住宅是危房，十分简陋，严重损坏或漏雨。1987 年，市政府将翻修住宅列为改善城市人民生活的一项重点工作，集中力量修缮严重危漏的民用房屋，改善群众居住条件。

两年时间，全市翻修住宅 30 449 间，建筑面积 543 201 平方米，取得了显著成果。首先提高了旧住宅的房质等级，恢复了房屋的原使用功能，确保安全，改善条件。其次，经过翻修的住宅，扩大了部分使用面积，通过改变布局，增加了厨房、厕所等功能，构成新型单元住宅。

Renovation of Residential Buildings

Tianjin is an old city, there are a considerable number of old houses which are dangerous, seriously damaged or leaky. In 1987, the municipal government listed the renovation of residential buildings as a key work to improve people's lives, focusing on solving serious leakage of civil housing and improving people's living conditions.

In two years, the city renovated 30,449 residential buildings, building area 543,201 square meters, and achieved remarkable results. First of all, it improved the quality grade of the old house, restored the original function of the house, ensured the safety and improves the conditions. Secondly, the renovated residence enlarged part of the use area. By changing the layout, the functions of kitchen and toilet were added to form a new type of unit residence.

西南楼 8 段改造后的住宅楼
Upgraded Residential Buildings of Xinanlou Section 8

平房改造

1987 年 5 月，天津开始对简陋平房进行较大规模的改造。

天津市平房改造采取了以就地改造为主，以规划用地挖潜建房为辅的方式；平房改造后的建筑形式以多层住宅为主，以低层住宅为辅。

就地改造：在拆除平房的原地，重新进行规划建设，因地制宜地改造市政配套设施，增建公共建筑和园林绿化，提高居住条件，改善居住环境。

规划用地挖潜建房：新征用土地或在建成的居住区挖掘部分闲置土地，建设一些低层廉价商品房、低层丛林住宅组团，以优惠价格向原居民出售，经营者将收入补贴平房改造。

Bungalow Renovation

In May 1987, Tianjin began to carry out large-scale renovation of simple bungalows.

The transformation of bungalow in Tianjin took the local transformation as the main method, and taping potential for building houses on the planned land as the supplementary way. After the transformation of bungalow, the building forms are mainly multi-storey residential building, supplemented by low-rise residential building.

In-site Renovation: To demolish the original site of bungalows, re-plan and construct, to adapt to local conditions, to renovate municipal supporting facilities, to build additional public buildings and landscape greening, and to improve living conditions and environment.

Taping Potential For Building Houses on The Planned Land: To newly requisition land or to excavate part of idle land in the built residential areas, building some low-rise and low-cost commercial housing and low-rise jungle housing groups, selling them to local residents in preferential prices, and the operators will transform the income-subsidized bungalows.

平房改造成的小二楼
A Small 2-Floor House Converted From A Bungalow

王串场 14 段住宅楼
Residential Building of Wangchuanchang Section 14

梅江居住区规划鸟瞰图
Meijiang Residential Area Planning Aerial View

华苑居住区建成照片
Photo of Huayuan Residential Area

“世纪危改”

作为中国的一个老工业基地，天津城市改造的任务十分繁重，1994 年，全市急需成片改造的危陋平房就有 738 万平方米。

经过 6 年的努力，西广开、谦德庄、中山门、南市、万德庄……这些天津普通百姓生于斯、长于斯的著名“窝棚区”全部得到改造。2000 年，天津市居民人均使用面积由 1993 年的 9.6 平方米提高到 13 平方米，住房成套率由 54%提高到 70%以上，住宅热化率从 22%提高到 48%。

这次天津历史上规模最大的平房改造工程被誉为“世纪危改”。

大型居住区建设

20 世纪 90 年代，天津一边进行大规模危陋平房与旧区改造，一边在中心城区建设了很多新居住区，东有万松居住区，南有梅江、双林居住区，西有华苑居住区，北有西横堤居住区。

"Century Shabby Dwelling Transformation"

As an old industrial base in China, the task of urban renovation in Tianjin is very arduous. Before 1994, there were 7.38 million square meters of dilapidated bungalows in urgent need of renovation as an area in this city.

After six years of efforts, these famous "shanty areas" like Xiguangkai, Qiandezhuang, Zhongshanmen, Nanshi, Wandezhuang, etc. where ordinary people in Tianjin were born and grew up have all been transformed. In 2000, the per capita living area of Tianjin residents increased from 9.6 square meters in 1993 to 13 square meters, the complete housing rate increased from 54% to more than 70%, and the residential heating rate increased from 22% to 48%.

The largest dwelling renovation project in Tianjin's history is known as "Century Shabby Dwelling Transformation"

Construction of Big Residential Districts

In the 1990s, Tianjin built many new residential districts in the central urban area, including Wansong residential district in the east, Meijiang residential district and Shuanglin residential district in the south, Huayuan residential district in the west and Xihengdi residential district in the north.

金龙起舞

THE GOLDEN DRAGON DANCE

1350 年
晓日三岔口，连樯集万艘
In the morning, thousands of boats gathered at the Sancha Estuary of the Haihe River

1404 年
明成祖朱棣将“天子渡口”定名为“天津”，并筑城天津卫
Ming Emperor Zhu Di named "Ferry of Emperor" as "Tianjin" and built the city

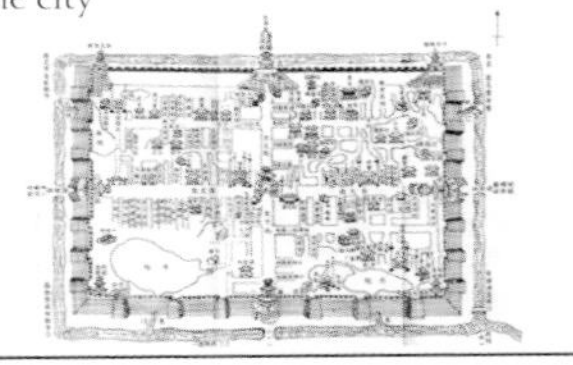

1793 年
英国特使马戛尔尼在三岔河口换乘小船，到避暑山庄觐见皇帝
The special envoy Macartney of UK changed the small boat to pay a formal visit to Emperor of Qing Dynasty

1873 年
海河边的英国领事馆
The consulate of UK was near the Haihe River

1909 年
西方轮船行驶在海河上
Western ships went on the Haihe River

1927 年
解放桥原名“万国桥”，即国际桥之意
The original name of Jiefang Bridge is World Bridge

1939 年
特大洪水淹没天津
A big flood inundated Tianjin

1956 年
海河边货运码头
A cargo terminal on the Haihe River

1959 年
中心广场及检阅台竣工
The Center Square and Platform was completed

1979 年
大规模根治海河工程基本完成
The huge project for operating the Haihe River was basically completed

1981 年
海河游泳场开放
Swimming area in the Haihe River was open

1982 年
海河边中心广场举办消夏晚会
An open evening was held on Center Square close to the Haihe River

1983 年
鸟瞰海河
Birds-eye view of the Haihe River

1985 年
“鲸鱼”游艇下水
The yacht like a whale was launched

1988 年
第三届海河杯冰上运动会
The Third Winter Games was held on the Haihe River

海河历史
History of the Haihe River

项目概况

海河是天津的母亲河，海河承载着天津的光荣与梦想。

海河以三岔河口为起点，全长 72 千米，其中城区段长 19.2 千米，为海河开发改造的重点区域。

1997 年澳大利亚政府资助南站[1]（六纬路）地区发展研究，首次提出海河沿线新的发展模式。

1998 年联合国教科文组织致函天津市政府，希望启动海河沿线研究。

2000 年委托法国康赛普特公司完成海河总体规划和环境设计。

2002 年 7 美国 EDAW 公司完成海河总体规划咨询成果。

在上述成果的基础上，天津市于 2002 年 11 月完成了《海河两岸综合开发改造规划》，又称“金龙起舞”规划，以“把海河建成国际一流的服务型经济带、文化带和景观带。弘扬海河文化，创建世界名河”为战略目标；重点改造上游 42 平方千米的沿河区域；以十项系统工程为实施重点，建设古文化街等六大节点，以市场化开发的方式经营城市。海河的规划与建设为天津开启了辉煌的新时代。

2017-2018 年天津市规划局组织开展了《海河规划评估》工作，全面评估和总结了过去十几年海河地区的建设与经验，以期海河之经验在今后的城市建设中发挥更大的作用。

1 南站：指天津河东区六纬路的老南站。——编注

Project Overview

The Haihe River is the mother river of Tianjin, bearing whose glory and dream.

Haihe River starts from the Sancha Estuary, with a total length of 72 kilometers. The urban section of that is 19.2 kilometers long, which is the most important section.

In 1997, the Australian Government funded the development research on the South Station Area and for the first time proposed a new development model along the Haihe River.

In 1998, UNESCO sent a letter to the Tianjin Municipal Government and hoped to initiate research on the area along the Haihe River.

In 2000, the French company Concept was commissioned to complete the overall planning and environmental design of the Haihe River.

In July 2002, EDAW Company completed the Haihe River overall planning consultation.

Based on the above achievements, Tianjin completed the "Comprehensive Development and Reconstruction Plan for Both Sides of the Haihe River" in November 2002, and named it "The Golden Dragon Dance", with the goal of building the Haihe River into "the world-class service-oriented economic belt, cultural belt, and landscape belt". Bringing forward the Haihe River culture and creating a world-famous river as a strategic goal. It will focus on reconstructing an area of 42 square kilometers along the upstream of the Haihe River. The ten key projects were taken as the key points of implementation. The planning and construction of the Haihe River opened the brilliant new era for Tianjin.

In 2017-2018, the Tianjin Municipal Planning Bureau organized the "Haihe River Planning Assessment" work, comprehensivly assessing and summarizing the construction and experience of the Haihe River area in the past ten years, with a view to the experience of Haihe River in the future urban construction.

改造前的海河及两岸情况

生态：海河沿岸的工业企业大多为污染严重的化工、冶金、纺织、印染等传统工业企业，由于天津市传统的雨污混流排水体系，汛期工业企业废水、城市污水随雨水大量排入海河，造成海河水质的污染。

工业：传统的钢铁、冶金、纺织、机械制造和造纸等 60 余家大中型企业分布于海河沿岸，这些企业大多是国有老企业，厂房破旧、设备陈旧、工艺落后、经济效益低下。

历史：海河两岸有一大批具有较高历史价值、建筑价值和人文价值的文物古迹，包括中国传统建筑和异国建筑，由于历史原因，历史建筑承载居民过重，私搭乱建情况严重，风貌建筑出现较严重损毁，亟待进行保护整修。

民生：海河沿岸大量危旧平房，约 400 万平方米，不仅居住空间狭小，而且居住条件较差，迫切需要提升改造。

The Situation of the Haihe River and Its Banks Before Reconstruction

Ecology: Most of the industrial enterprises along the Haihe River are the traditional industrial enterprises such as chemical industry, metallurgy, textile, printing and dyeing, which are seriously pollution. Due to the traditional mixed flow drainage system of rain and sewage in Tianjin, industrial wastewater and urban sewage discharge into The Haihe River in flood season, resulting in water pollution of the Haihe River.

Industry: Traditional iron and steel, metallurgy, textile, machinery manufacturing and papermaking and more than 60 large and medium-sized enterprises are distributed along the Haihe River. Most of these enterprises are old state-owned enterprises with dilapidated plants, obsolete equipment, backward technology and low economic benefits.

History: On both sides of the Haihe River, there are a large number of cultural relics and historic sites with high historical value, architectural value and cultural value, including Chinese traditional buildings and foreign buildings. Due to historical reasons, historical buildings are overloaded with residents, and the construction of private buildings is in serious disorder.

People's livelihood: A large number of dilapidated old bungalows along the Haihe River, about 4 million square meters, not only have narrow living space, but also have poor living conditions, which urgently need upgrading and transformation.

2000年海河两岸现状
The Haihe River in 2000

规划目标 | PLANNING OBJECTIVES

规划目标：把海河建设成为独具特色、国际一流的服务型经济带、文化带、景观带，弘扬海河文化，创建世界名河。

规划六大主题目标：

规划的第一个目标：展现悠久历史文化。保护和恢复沿岸的文物和风貌建筑，新建一批文化设施，以海河为主线，把保留的历史文化街区与新的文化设施串成一条丰富有序的历史文化资源带。

规划的第二个目标：发展滨河服务产业。调整优化海河两岸的用地结构，创建以商业、贸易、服务、文化、娱乐、金融等公共设施为主的滨河经济开发带，带动第三产业发展。

规划的第三个目标：突出亲水城市形象。以建筑、绿化、小品、灯光等其他设施来综合构建一条具有国际化大都市水准的全天候的景观带，建设亲水休闲空间，还海河于市民。

规划的第四个目标：建设城市生态依托。实施雨污分流和清淤，改善海河水质。在两岸建设绿化带，形成网络状绿化系统，形成“水系相连、水绕城转、水清船通”的非常美好的景象。

规划的第五个目标：改善道路交通系统。规划建设海河周边的路网系统，修建沿河快速路，新建跨河桥梁，形成“七横十二纵”的道路交通体系。

规划的第六个目标：改善海河旅游环境。整合开发旅游资源，打造旅游线路，形成海河旅游品牌。

（摘自《海河神韵集萃》）

Planning Objective: To build the Haihe River into a unique and world-class service economic belt, cultural belt and landscape belt, carry forward Haihe River culture, and create a world-famous river.

Six Plan Goals:

The First Goal: To show the long history and culture. We will protect and restore cultural relics and buildings along the coast, and build a number of new cultural facilities. Taking the Haihe River as the main line, we will link the preserved historical and cultural districts with the new cultural facilities to form a rich and orderly historical and cultural resource belt.

The Second Goal: To develop the riverside service industry. We will adjust and optimize the land use structure on both sides of the Haihe River, and create a riverfront economic development zone dominated by commercial, trade, service, cultural, entertainment, financial and other public facilities to drive the development of the tertiary industry.

The Third Goal: To highlight the image ofa water-friendly city. Building, greening, sketch, lighting and other facilities to form a comprehensive international metropolitan level of all-weather landscape belt, the construction of water-friendly leisure space, hand Haihe River back to the residents.

The Fourth Goal: To build an urban ecological support. The implementation of rain and sewage diversion and dredging, improve the water quality of Haihe River .Green belts will be built on both sides of the straits to form a network-like green system and create a beautiful scene of "connected water systems, water circling the city and water clearing boats".

The Fifth Goal: To improve the road traffic system. Planning and construction of road network system around Haihe River, construction of expressway along the river, new bridge across the river, the formation of "7 horizontal and 12 vertical" road traffic system.

The Sixth Goal :To improve tourism environment of the Haihe River, integrate and develop tourism resources, build tourism routes and form Haihe River tourism brand.

(From *The Spirit and Charm of the Haihe River*)

海河两岸从源头到入海口全长 72 千米，规划分为三大段落。

由三岔河口到外环线为海河的上游段

这一段的海河长度约 19.2 千米，位于中心城区，规划为公共设施、绿地、居住等功能，规划核心区用地面积 42 平方千米。规划逐步把海河沿岸现有的工业、仓储用地调整为商贸、金融、旅游和绿化用地，大力发展商贸旅游等第三产业，加强海河两岸的绿化和景观建设，切实保护海河两岸的文物古迹和风貌建筑，使海河两岸的功能更加完善，充分体现天津国际大都市的风采。

由外环线到二道闸为中游段

这一段海河长度约 18 千米，腹地面积 120 平方千米，规划为生态的海河风景旅游区和高新经济发展区。其中依托两岸良好的自然条件，建设以生态旅游观光为主的独特的滨河风景旅游区，作为天津旅游的重要拳头产品。旅游区的周边结合未来高新技术的发展，创建以生态能源、信息技术等新型产业经济为主的田园式的研究发展区域。这一段海河沿线要突出田园风光，保护好生态环境。

由二道闸至入海口为下游段

这一段海河长度约 34.8 千米，腹地面积 150 平方千米，为滨海新区范围。海河塘沽城区段规划为商业金融、居住等生活用地，海河下游要加强港口和工业岸线的开发利用，形成国际现代物流中心和世界加工制造中心，体现多功能内河港口经济区的独特景观。

The Haihe River is 72 kilometers long from the source to the estuary, and the plan is divided into three sections.

The Section From the Sancha Estuary to the Outer Ring Road is the Upper Section of the Haihe River

The length of Haihe River in this section is about 19.2 kilometers, which is located in the central urban area. It is planned as public facilities, green space, residence and other functions. The planned core area covers an area of 42 square kilometers. Planning gradually along the Haihe River existing industrial adjustment for trade, finance, tourism, warehousing land and green land, develop business tourism and other tertiary industry, to strengthen the construction of Haihe River on both sides of the greening and landscape, the effective protection of Haihe River on both sides of the cultural relics and landscape architecture, the Haihe River on both sides of the function to be more perfect, fully reflects the international metropolis charm of Tianjin.

The Section From the Outer Ring Road to the Erdao Sluice is the Middle Section of the Haihe River

The length of this section of Haihe River is about 18 kilometers, with a hinterland area of 120 square kilometers. It is planned to be an ecological Haihe River scenic tourism area and high-tech economic development area. Relying on the good natural conditions on both sides of the river, the construction of a unique riverside scenic tourism area mainly based on ecological tourism, as an important product of Tianjin tourism. The surrounding areas of the tourism area will be combined with the future development of high and new technologies to create an idyllic research and development area dominated by new industrial economy such as ecological energy and information technology. This section of Haihe River should highlight the pastoral scenery along the protection of natural ecological environment.

From the Erdao Sluice to the Estuary is the Downstream Section of the Haihe River

The length of this section of Haihe River is about 34.8 kilometers, with a hinterland area of 150 square kilometers, which is the Binhai New Area. Tanggu District of Haihe River is planned as commercial, financial, residential and other living land. The development and utilization of ports and industrial shoreline should be strengthened in the lower reaches of Haihe River to form an international modern logistics center and a world processing and manufacturing center, reflecting the unique landscape of multifunctional inland river port economic zone.

上游段规划四个功能分区

传统文化商贸区。北安桥上游海河沿岸区域，是天津市的发祥地，有着丰富的历史文化资源和富有天津特色的商贸业，重点依托大悲院、三条石、天后宫等历史文化遗产，延续天津城市的历史脉络，发展天津特色的传统商贸业和文化旅游业。

都市消费娱乐区。北安桥到赤峰桥海河沿岸的区域，包括和平路和滨江道商业区，中心广场和天津站，是天津的商业中心和交通中心枢纽，依托商业资源，用现代商业新理念，建设新型商业设施、标志性写字楼，形成天津市的中心商业区域。

中央商务金融区。赤峰桥到奉化桥海河沿岸区域，包括解放北路金融街、小白楼地区等，是天津市金融商务集中发展的区域，建设解放北路金融城、南站（六纬路）商务区、小白楼商务区。

智慧城。奉化桥到外环线海河沿岸区域。结合城市结构调整，企业东迁可腾出大量的土地，也有可开发的农用地和宝贵的自然景观，是中心城区未来建设的宝贵资源。规划重点是构造新型城市形态，建设以先进的网络技术、智能化技术为核心的高产出、高附加值的新产业区。

Functional Zoning of the Upper Section

Traditional Culture Heritage District: The upper reaches of Beian Bridge, along the Haihe River, is the birthplace of Tianjin, with the historical and cultural resources of lamps and the commercial and trade industry with Tianjin characteristics. Relying on the historical and cultural heritages such as Dabei Temple, Santiaoshi and Tianhou Temple, etc., the traditional commercial and trade industry with Tianjin characteristics and cultural tourism industry will be developed.

Urban Entertainment District: The area from Bei'an bridge to Chifeng bridge along the coast of the Haihe River, including Heping Road and Binjiang Street Business District, the Central Plaza, and Tianjin Railway Station, is the business center and transportation hub of Tianjin. Relying on modern business and new commercial concepts, new type of commercial facilities and landmark high rise offices will be built to form the center commercial area of Tianjin.

Central Business District: From Chifeng Bridge to Fenghua Bridge along the coast of Haihe River, including North Jiefang Road Financial Street, Xiaobailou, Tianjin South Station area, is an area where Tianjin financial business development is concentrated.

Smart Town District: From Fenghua Bridge to the outer ring road along the coast of Haihe River. Combined with the structural adjustment, enterprises can move east to free up a large amount of land, there are also agricultural land that can be developed and valuable natural landscape, which is a valuable resource for the future construction of the central city. The plan is focus to construct a new type of urban form and build a new industrial area with high output and high added value, with advanced network technology and intelligent technology as the core and loose urban form as the feature.

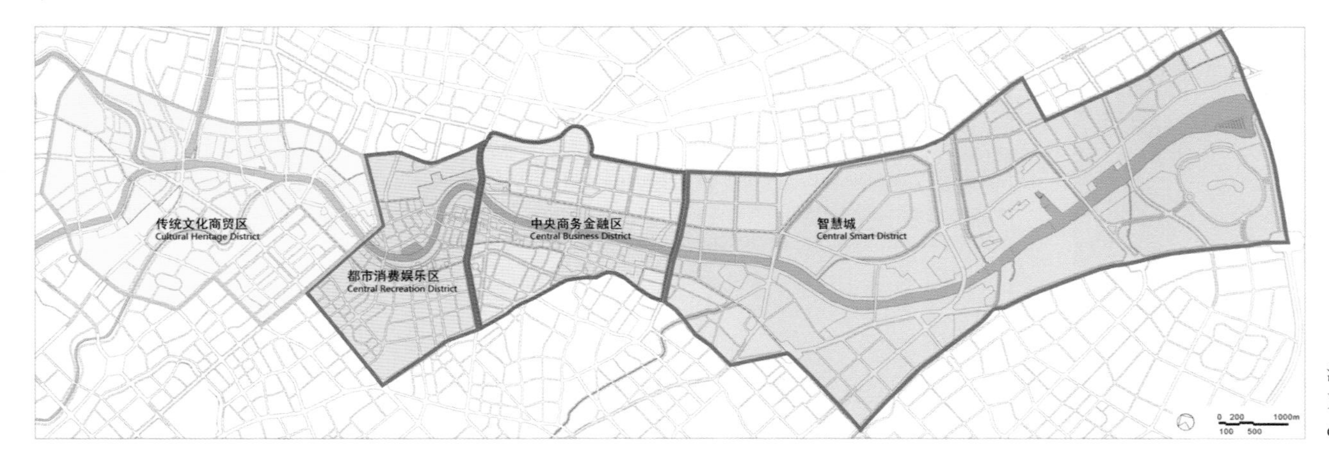

海河上游段功能分区图
Functional Zoning of the Upper Section of Haihe River

基础设施建设成果

改造海河堤岸 20 千米，改造沿线面积达到 243 万平方米。新建、改造桥梁 20 座。建设沿河公园 10 座，面积 40 万平方米。新建、改造道路 53 条，79 千米。修建水、电、气、热各类管线 632 千米，提升改造两岸灯光 8 千米。

经济开发成果（截止到 2011 年）

完成土地整理 85 块，形成净地 768 公顷，正在整理土地 759 公顷。引进项目 74 个，建筑面积 1 239 万平方米，投资额 971 亿元。建设大型商业设施 11 个，建筑面积 74.6 万平方米。建设高星级酒店 6 个，建筑面积 44.9 万平方米。建设写字楼 12 个，建筑面积 77.76 万平方米。建设展览馆等公益设施 5 个，建筑面积 4.44 万平方米。整修历史建筑 72 栋，恢复名人故居 3 处。

Achievements of Infrastructure Construction

20 kilometers of embankments along the Haihe River were renovated, covering an area of 2.43 million square meters. 20 bridges were built or renovated. 10 parks along the river with an area of 400,000 square meters were built. 53 roads were built or renovated, covering 79 kilometers. 632 kilometers pipelines of tap water, electric power, gas and heating were built. 8 kilometers the lighting along the banks were upgraded and transformed.

Achievements of Economic Development (up To 2011)

85 plots of land have been completed, forming 768 hectares of net land, and 759 hectares are under consolidation. 74 projects were introduced, with a construction area of 12.39 million square meters and an investment of 97.1 billion yuan. 11 large-scale commercial facilities were built with a floor area of 746,000 square meters. 6 high-star hotels with a construction area of 449,000 square meters were built. 12 office buildings were constructed and with construction area of 777,600 square meters. 5 public welfare facilities, including exhibition halls, were built with a floor area of 44,400 square meters. 72 historic buildings were renovated and 3 former residences of celebrities were recovered.

海河上游段建成效果
The Effect of the Upper Section of the Haihe River

土地开发

截止到 2011 年，完成土地出让 496 公顷，已落实开发项目建筑面积 1 250 万平方米，总投资 1 056 亿元。

海河上游区域棉二、棉四、棉六、天钢等 42 家大中型企业完成搬迁，特别是纺织、冶金两大行业的东移，促进了天津市产业布局的调整。

累计拆除危旧房屋 300 万平方米，安置居民 11 万户，极大地改善了拆迁居民的住房条件。

Land Development

By 2011, 496 hectares of land had been transferred, and 12.5 million square meters of construction area had been developed, with a total investment of 105.6 billion yuan.

The relocation of 42 large and medium-sized enterprises such as the Second Cotton Mill, the Fouth Cotton Mill, the Sixth Cotton Mill and the Tianjin Steel Plant in the upper section area of Haihe River, especially the eastward migration of the two major industries of textile and metallurgy, promoted the adjustment of industrial layout in Tianjin.

A total of 3 million square meters of dilapidated houses were demolished, and 110,000 households were resettled, greatly improving the housing conditions of the residents.

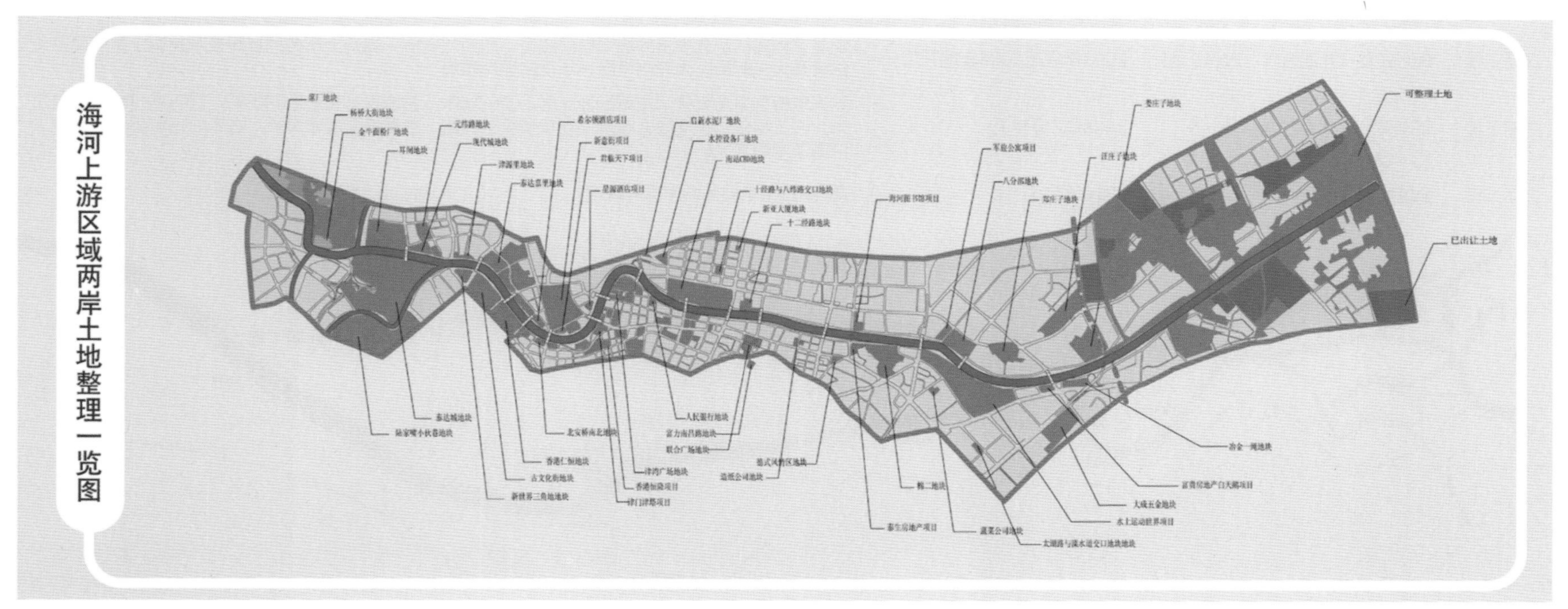

海河上游区域两岸土地整理一览图
Overview of Land Consolidation on Both Sides of the Upper Section of the Haihe River

十大工程——堤岸改造工程

海河上游 20 千米的堤岸改造全部完成。形成 15.72 千米的亲水岸线，改造沿线面积达到 243 万平方米。

天津站前广场和津湾广场是堤岸改造的重要节点。

Ten Major Projects: Embankment Reconstruction Projects

All the 20 kilometers of the upper reaches of the Haihe River have been renovated. An area of 2.43 million square meters was transformed along the 15.72 km waterfront.

The Front Square of Tianjin Railway Station and Jinwan Square were the important nodes of embankment reconstruction.

阶段	项目名称	时间节点	工程内容
一期工程	海河开发" 四二一六”改造工程	2002年底至2003年底开工	三岔河口至解放桥沿线堤岸改造，完成了起步段89千米的堤岸结构改造
二期工程	海河开发“九四五八”改造工程	2003年底至2004年底开工	完成了慈海桥至北安桥的堤岸景观建设，启动北安桥—海河大桥段的堤岸景观建设，重点建设大沽桥至大光明桥、大光明桥至光华桥的堤岸景观
三期工程	海河开发堤岸第三期工程	2006年8月开工	从光华桥到海津大桥，左岸设计长度2 629.16米，右岸设计长度2 080米，工程总长度4 709.16米。该工程分四个标段，左岸：光华桥至兰秀路、兰秀路至海津大桥；右岸：光华桥至洪泽路、洪泽路至海津大桥
近期工程	全运会堤岸改造	2016年完工	海河后5千米地区两岸堤岸改造

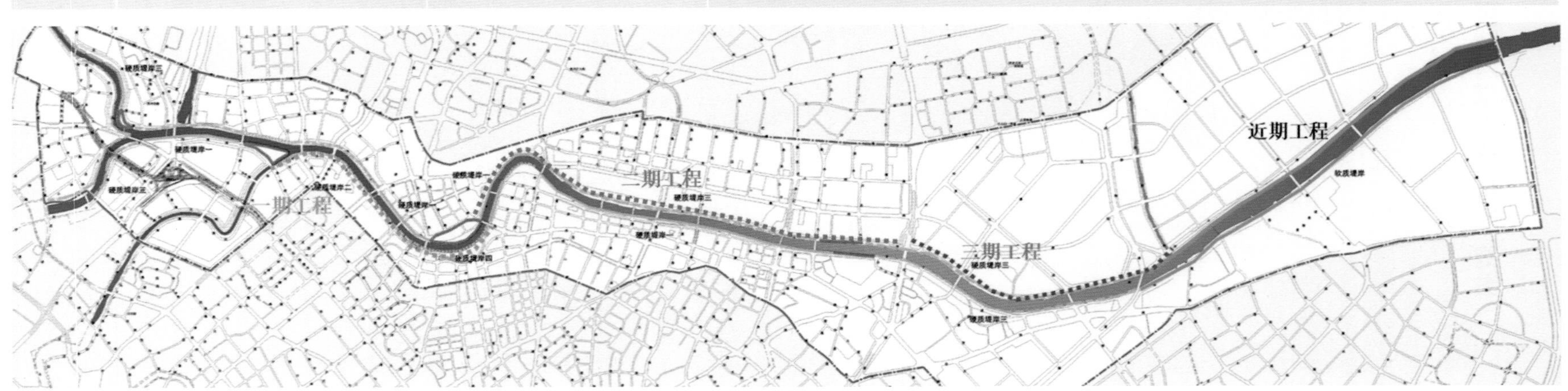

海河堤岸改造建设历程
The Phase of Haihe River Embankment Reconstruction and Construction

十大工程——道路工程

新建、改造道路 53 条，共 79 千米。在 2000—2015 年间，交通量增长 175%，但由于道路与工程的实施，跨河通道饱和度普遍下降，海河上建设的多座跨河桥梁起到了缓解交通拥堵，增强两岸联系的作用。

Ten Major Projects: Road Projects

53 roads were built or renovated, totaling 79 kilometers. Between 2000 and 2015, the traffic volume increased by 175%. However, due to the implementation of roads and engineering, the saturation of cross-river passageways generally decreased, and a number of cross-river bridges built on the Haihe River played a role in easing traffic congestion and strengthening the cross-strait links.

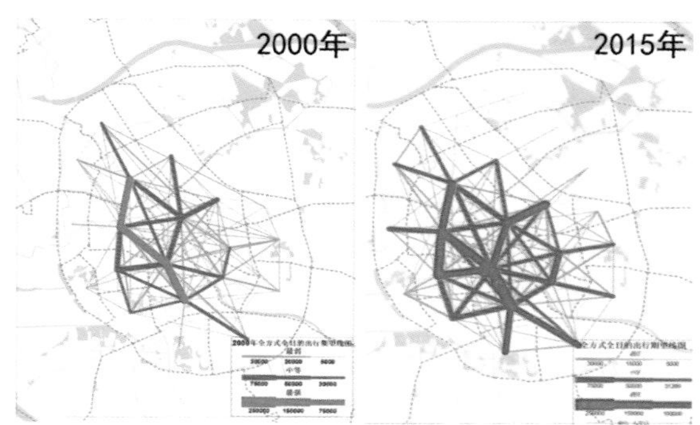

交通出行情况对比图
The Comparison Chart of Traffic Trip

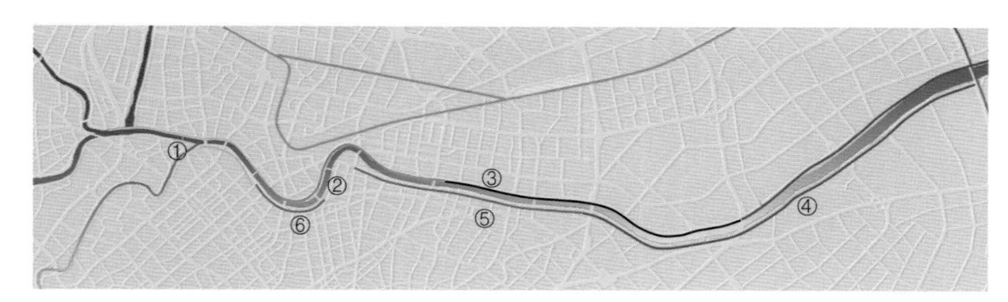

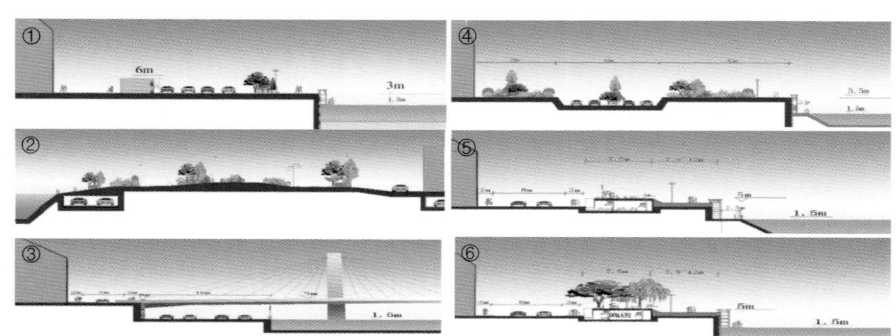

滨河道路断面分区图
Zoning Map of Riverside Road Section

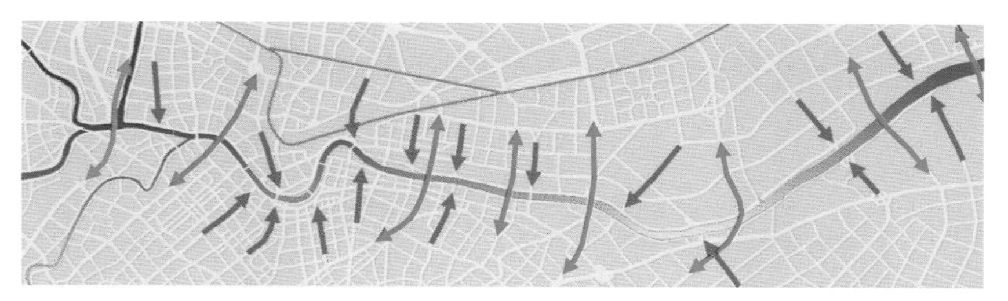

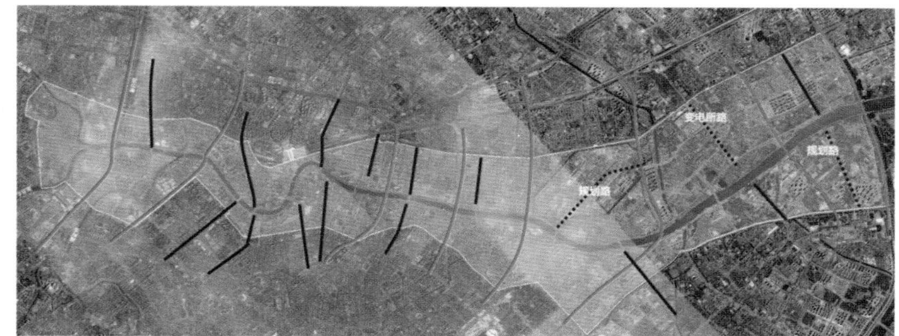

跨河交通规划图
River-Crossing Traffic Planning

序号	名称	形式	建设状态											
			2003	2004	2005	2006	2007	2008	2009	2010	2011	2012	2013	2014
1	新红桥	原有保留												
2	志成桥	新建												
3	永乐桥（慈海桥）	新建	——	开工										
4	金钢桥	抬升	——	开工										
5	狮子林桥	抬升、拓宽	完成											
6	金汤桥	重建			完成									
7	进步桥（湖南桥）	新建	——	——	——	开工	通车							
8	北安桥	抬升、拓宽	——	开工	通车									
9	大沽桥	新建	开工	建设	通车									
10	解放桥	修复、保护			开工	完成								
11	赤峰桥	老桥北侧新建	——	——	开工	建设	建设	通车						
12	保定桥（金汇桥）	新建	——	开工	建设	通车								
13	大光明桥	抬升、景观改造	——	——	——	——	开工	通车						
14	金阜桥（蚌埠桥）	新建	——	开工	建设	建设	通车							
15	奉化桥（直沽桥）	新建	——	开工	建设	建设	通车							
16	刘庄桥	改造	——	完成										
17	光华桥	拓宽	——	——	——	开工	通车							
18	国泰桥	新建	——	——	——	——	开工	建设	建设	建设	建设	通车		
19	富民桥	新建	——	——	开工	建设	通车							
20	海津大桥	原有保留												
21	春意桥	新建												通车
22	吉兆大桥	新建											通车	
23	海河桥	原有保留												

海河桥梁建设历程
The Process of Haihe Bridge Construction History

十大工程——桥梁工程

拆除桥梁 1 座，原址改造 7 座，原址新建 2 座，新增建设 11 座，原有保留 3 座。

Ten Major Projects: Bridge Projects

1 bridge was removed; 7 bridges were reconstructed on the original sites; 2 new bridges were built on the original sites; 11 new buildings were built. The original 3 bridges were reserved.

海河桥梁建设历程示意图
The Process Map of Haihe Bridge Construction History

狮子林桥共有 184 只狮子。

大沽桥获得国际桥梁最高奖——尤金 · 菲戈奖。

架设于永乐桥上的“天津之眼”成为天津最重要的城市地标之一。

百年老桥解放桥得到保留和修缮，恢复开启功能。

There are 184 stone lions on the Shizilin Bridge.

Dagu Bridge won the Eugene Figo Award, the highest international bridge award.

The ferris wheel named "Eye of Tianjin" on the Yongle Bridge has become the most important city landmark in Tianjin.

The century-old Jiefang Bridge has been preserved and repaired to restore its opening function.

狮子林桥
The Picture of Shizilin Bridge

解放桥
The Picture of Jiefang Bridge

十大工程——河道通航工程

抬升改造了狮子林桥、北安桥等 5 座桥梁，恢复了解放桥和金汤桥的开启功能，提高桥下高度，满足航道标准。

建设 8 处码头，开启 4 条游船线路。海河夜景游成为天津最有特色的旅游亮点之一。

Ten Major Projects: River Navigation Projects

Five bridges, including Shizilin Bridge and Bei'an Bridge, were renovated to meet the waterway standard, and the opening of Jiefang Bridge and Jintang Bridge were restored.

Through building eight docks and opening four cruise lines, Haihe River night tour has become one of the most distinctive tourist attractions in Tianjin.

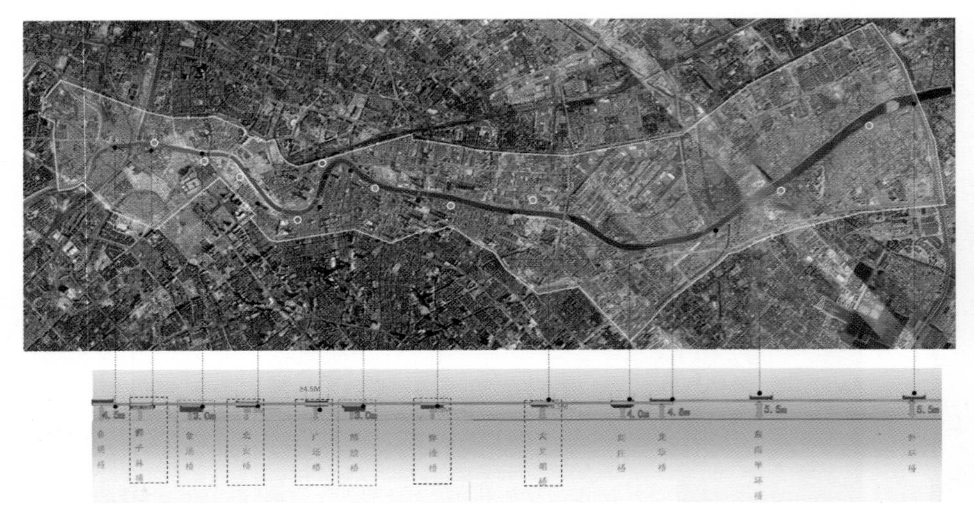

桥梁净高示意图
The Sketch Map of Bridge Net Height

海河游船
The Picture of Haihe River Cruises

十大工程——广场绿化工程

海河两岸新建绿化广场、公园面积为 40 多万平方米。

Ten Major Projects: Square Greening Project

The newly built green squares and parks on both sides of the Haihe River cover an area of more than 400,000 square meters.

广场绿化建设情况示意图
The Sketch Map of Square Greening Construction

十大工程——环境景观工程

策划中的 20 组雕塑，保留延续了 1 处，完成建设 10 处，增加 2 处。

Ten Major Projects: Environmental Landscape Projects

In the planning of 20 groups of sculptures,1 sculpture was retained, 10 sculptures were completed, and 2 sculptures were added.

天子之渡：天石舫
Tianzi Ferry: Tianshi Fang

会师金汤
Joint Forces at the Jintang Bridge

浪漫心港
Romantic Heart Port

邮路漫漫
Long Postal Route

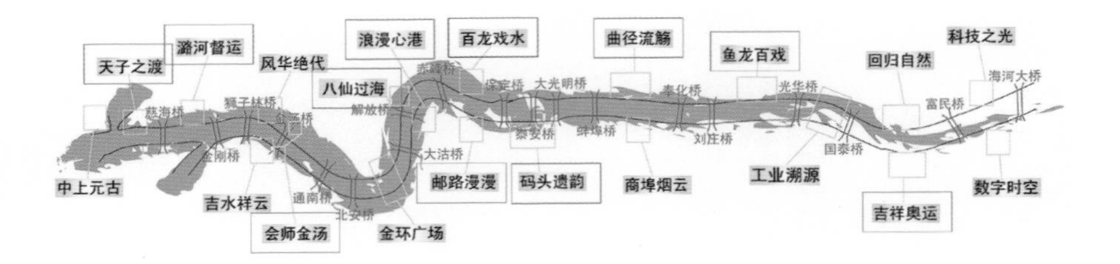

景观雕塑位置图
The Location Map of Landscape Sculptures

十大工程——灯光夜景工程

海河综合开发灯光夜景工程是海河基础设施建设中的重要部分，由多家国际著名景观设计单位规划设计，灯光照明规划结合海河两岸的重要节点和标志性建筑，确定重点照明区、照明段和层次，确定照明的总体色彩、风格和形式。

该项工程成为延续至今的长期工程，城市持续不断地营造更美的夜景。

Ten Major Projects : Lighting and Night Scene Projects

Haihe River Comprehensive Development Lighting Project is an important part in the construction of infrastructure by many international landscape design companies. Lighting planning combined with the important node and landmark buildings on both sides of the Haihe River, determine the accent lighting, lighting and level, determine the overall color, style and lighting form. The project continues to this day as a long-term project, the city continues to create a more beautiful night scene.

海河综合开发

- 堤岸灯具2万盏
- 22个项目
- 工程量120万平方米
- 76栋高层灯光点缀
- 16个夜景节点

奋战900天

- 年投资额由不足400万元提高到2 500万元
- 改造44幢、新增34幢建筑灯光设施
- 新建7座码头灯光系统
- 提升11座桥梁的灯光效果

历年夜景灯光修复工程

- 海河沿线灯光带总长已经突破25千米

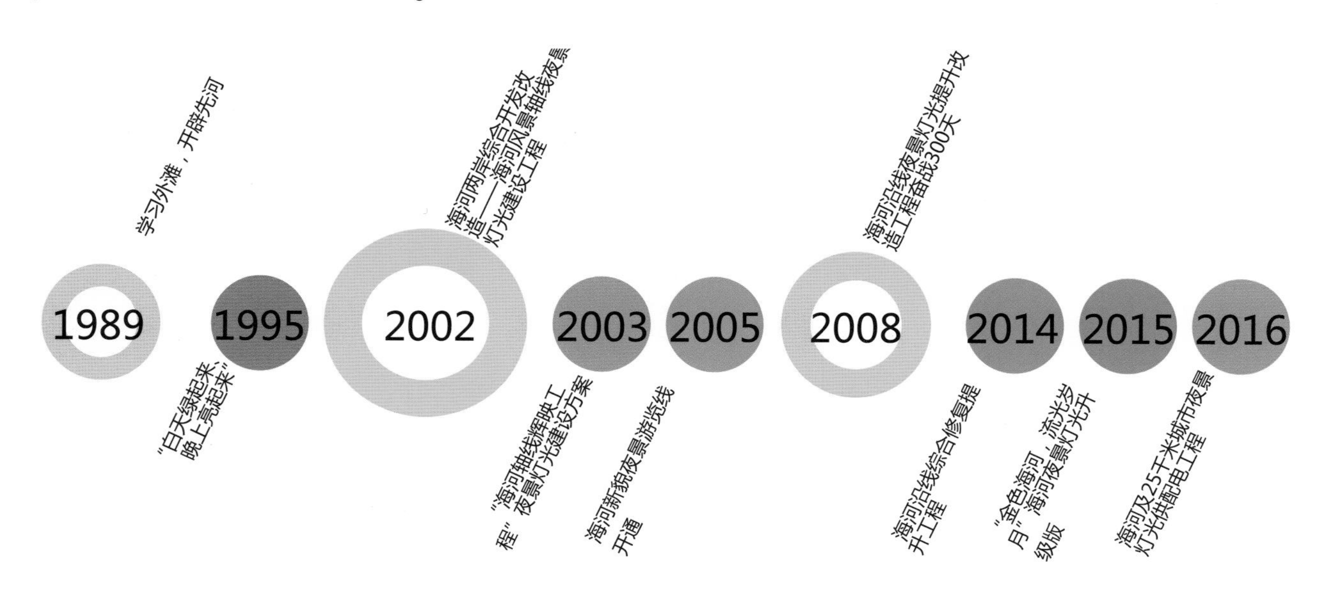

十大工程——公共建筑工程

海河两岸规划建设 10 大标志性建筑，提高海河的文化内涵。目前已建成近代工业及城市历史博物馆、天津市规划展览馆、天津民俗馆、美术学院美术馆。

Ten Major Projects : Public Construction Projects

The two sides of Haihe River are planning to build 10 landmark buildings to improve the cultural connotation of Haihe River. At present, the Modern Industrial and Urban History Museum, Tianjin Planning Exhibition Hall, Tianjin Folk Museum, Art Academy Art Museum were built.

天津民俗馆
Tianjin Folk Museum

天津市规划展览馆
Tianjin Planning Exhibition Hall

十大工程——整修置换工程

在海河两岸建筑整修置换中，特别强调以“修旧如旧”的原则整修历史建筑。

Ten Major Projects : Renovation and Replacement Projects

In the renovation and replacement of buildings on both sides of Haihe River, the principle of " restoration as the past " is particularly emphasized.

整修前后对比图
Comparison Chart Before and After Renovation

2002 年第一次国际招标——六点一桥

1. 运河经济文化商贸区
2. 海河水上运动世界
3. 大悲院经济文化商贸区
4. 古文化街海河楼商贸区
5. 海河广场及和平路地区
6. 南站（六纬路）中心商务区
7. 永乐桥

2004 年第二次国际招标——三片区

8. 意式风情区
9. 天津智慧城
10. 柳林地区

2007 年第三次国际招标——六片区

11. 西沽公园地区
12. 子牙渔湾地区
13. 小白楼商务区
14. 德式风貌区
15. 英式风貌区
16. 法式风貌区

First Round International Competition in 2002—Six Point One Bridge

1.Canal Economic and Cultural Trade Area
2.Haihe Water Sports World Area
3.Dabei Temple Economic and Cultural Trade Area
4.Ancient Culture Street and Haihelou Business Area
5.Haihe Square and Heping Road Area
6.South Station (Liuwei Road) Central Business District
7. Yongle Bridge

Second Round International Competition in 2004—Three Areas

8.Italian Style Area
9.Tianjin Smart Town
10. Liulin Area

Third Round International Competition in 2007—Six Areas

11.Xigu Park Area
12.Ziyayu Bay Area
13.Xiaobailou Business District
14.German Style Area
15.British Style Area
16.French Style Area

古文化街海河楼商贸区 Ancient Culture Street and Haihelou Business Area

基本情况

位于子牙河、南运河与海河三条河流的交汇处——三岔河口，是海河的起点。北至规划西青道延长线，东至子牙河，南至南运河，西至河北大街，规划用地面积 63.67 公顷。新建建筑面积 160 万平方米。该区域是天津的重要发祥地，是地方特色最强烈，传统文化最浓厚的地区之一。

竞标方案

共有 6 个竞标方案，中标方案旨在创造一系列的中心和广场，来展现城市每个历史阶段的特征，通过这种方式讲述一个关于城市成长的故事。这些中心都有属于自己的命名，例如：现代工业博物馆中心广场、现代商贸广场、文化中心、海运广场以及新世纪广场，每个名字都会唤起人们对一段历史时期的回忆。这些广场和中心的商业设施同时也为周边地区提供服务，加强了城市其他部分与该地段的联系。

Basic Situation

Located at the intersection of Ziya River, South Canal and Haihe River - Sancha Estuary, is the starting point of Haihe River. It is planned to extend Xiqing Road in the north, Ziya River in the east, South Canal in the south and Hebei Avenue in the west. The planned land area is 63.67 hectares. A new construction area of 1.6 million square meters. This area is the important birthplace of Tianjin, is the local characteristics of the strongest, one of the strongest traditional culture.

Competition Scheme

There are six competition plans, and the winning plan aims to create a series of centers and squares to show the characteristics of each historical stage of the city, and to tell a story about the growth of the city in this way. These centers have their own names, such as the Center Square of the Museum of Modern Industry, the Modern Square of Commerce and Trade, the Cultural Center, the Ocean Shipping Square and the New Century Square. Each name evokes memories of a historical period. These squares and the commercial facilities in the center also serve the surrounding area and strengthen the connection with the rest of the city.

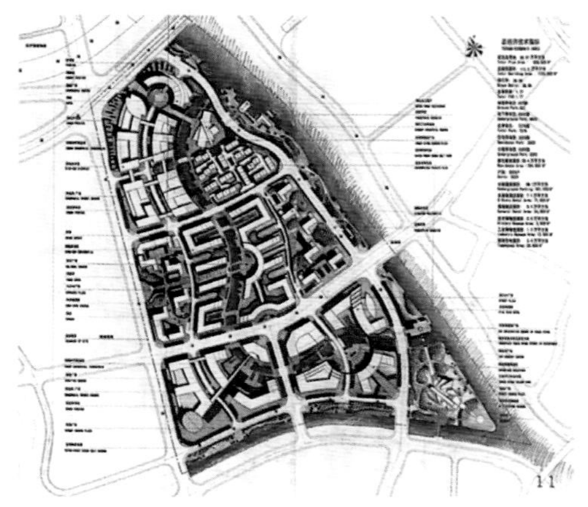

中标深化方案
Winning Plan

一号方案
Plan 1

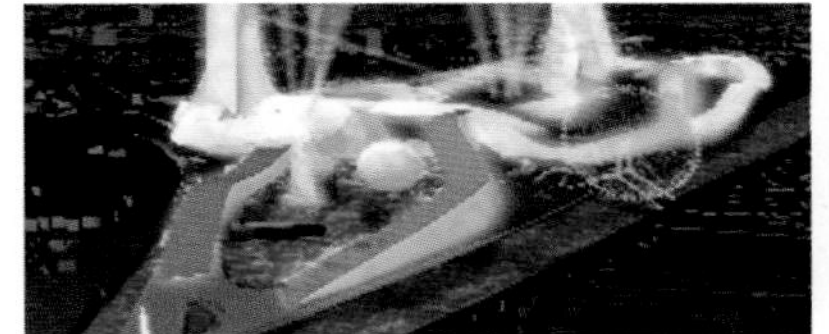

二号方案
Plan 2

三号方案
Plan 3

四号方案
Plan 4

五号方案
Plan 5

六号方案
Plan 6

基本情况

位于河西区挂甲寺地区，北至新围堤道（光华桥），东至海河，南至春阳大街，西至南北大街。规划用地面积 41.5 公顷，其中可开发用地 19.59 公顷。

竞标方案

共有 5 个竞标方案，中标方案在充分研究地形特点和规律，考虑到空间的性质和建筑物的朝向之后，以面向海河的辐射结构解决了几乎所有现状地块和建筑的矛盾，使整个区域浑然一体。设计中在大的轴线关系布局上充分考虑了周边建筑的衔接，使得城市的肌理在本区得以延续并形成亲近的秩序。为解决公共空间与居住空间性质上的冲突，设计中采用道路与水系隔离再以人行天桥连接的方式，促进了不同性质空间的共生和相辅相成。既安静、优美又亲水、方便。

Basic Situation

Located in the Guajia Temple area of Hexi District, the four ranges are: north to New Embankment Road (Guanghua Bridge), east to Haihe River, south to Chunyang Street, west to Nanbei Street. The planned land area is 41.5 hectares, of which 19.59 hectares can be developed.

Competition Scheme

There are 5 competition schemes in total. After fully studying the topographic features and rules, taking into account the nature of space and the orientation of buildings, the winning scheme adopts the radiation structure facing Haihe River to meet the contradiction between almost all existing plots and buildings. Make the whole area one. In the design, the connection of the surrounding buildings is fully considered in the layout of the large axis relationship, so that the urban texture can be continued and form a close order in this area. In order to solve the conflict between the nature of public space and residential space, the design adopts the way of separating road and water system and connecting by footbridge to solve the spatial symbiosis of different nature.

四号方案（第一名）Plan 4 （First Prize）

一号方案（第二名）Plan 1 （Second Prize）

三号方案 Plan 3

四号方案（第一名）Plan 4（First Prize）

二号方案（第三名）Plan 2（Third Prize）

五号方案 Plan 5

大悲院经济文化商贸区 | DABEI TEMPLE ECONOMIC AND CULTURAL TRADE AREA

基本情况

大悲院经济文化商贸区位于子牙河北岸。西至李公祠大街，北至七马路，南至中山路，东至元纬路。规划总用地为 41 公顷，新建建筑面积 83 万平方米。区域内的一些旧有建筑和街道布局具有近代中国开埠城市的半殖民地半封建社会的典型特征。

竞标方案

共有 4 个竞标方案，中标方案依托传统文化资源，大力开展旅游和商贸活动，赋予这一地区活跃的生命力。标志性建筑“凤凰广场”形成低层扇形的内街式商业区，更多的是依赖大悲院的庙会氛围而产生的格局，完整的大空间充分考虑引入海河的景观，产生与城市肌理“对立”且“统一”的体量。

Basic Situation

Dabei Temple Economic and Cultural Trade Area is located on the north bank of Ziya River. Ligongci Street in the west, Qima Road in the north, Zhongshan Road in the south and Yuanwei Road in the east. The total planning land is 41 hectares, with a new construction area of 830,000 square meters. Some of the old buildings and street layout in the area have the typical characteristics of the semi-colonial and semi-feudal society in modern China.

Competition Scheme

There are 4 competition schemes in total, and the winning scheme relies on traditional cultural resources to vigorously carry out tourism and business activities, endowing this region with active vitality. The low-rise fan-shaped inner street business district formed by the landmark building "Phoenix Square" is more a pattern generated by relying on the temple fair atmosphere of the Dabei Temple. The complete large space fully considers the introduction of the landscape of Haihe River and produces a volume that is "opposite" and "unified" with the urban fabric.

一号方案 Plan 1

三号方案（中标方案）Plan 3（Winning Plan）

二号方案 Plan 2

四号方案 Plan 4

基本情况

古文化街海河楼商贸区位于天津市南开区，是海河传统文化商贸区的重要节点。北至通北路，东至海河，南至水阁大街，西至东马路，规划用地面积 13.91 公顷。该地区是天津最早的漕运码头区，元代至清代曾是当时百船聚会、车水马龙、商贾云集、寸土寸金的地方。

竞标方案

共有 4 个竞标方案，中标方案充分利用古文化街现有旅游资源和传统文化、民族文化的商业资源，既继承和发扬古文化街历史上“华洋杂处、南北交融”的传统文化，又体现先进的时代文化。区内保留天后宫，整修玉皇阁、通庆里，结合新建海河楼、亲水平台等新的旅游景点，以“吉水祥云，潜龙出世”为主题，集旅游、购物、餐饮、休闲、娱乐、康体、住宿为一体，使之成为最富津味文化内涵、集聚人气的商旅长廊，形成天津人体验老风情、外埠游客了解津门、探寻津貌不可遗漏的商旅休闲区。

Basic Situation

Ancient Culture Street and Haihelou Business Area is located in Nankai District of Tianjin. It is an important node of Haihe River Traditional Culture and Trade Area. It extends north to Tongbei Road, east to Haihe River, south to Shuige Street, west to East Road, with planning land area of 13.91 hectares. This area is the earliest canal transport dock area in Tianjin. From Yuan Dynasty to Qing Dynasty, it was a place where hundreds of ships gathered, traffic was heavy, merchants gathered and land was precious.

Competition Scheme

There are 4 competition schemes in total, and the winning plan makes full use of the existing tourism resources of the Ancient Culture Street and the commercial resources of traditional culture and national culture, inherits and develops the traditional culture of "Chinese and Foreign Mix, North and South Blend" in the history of the Ancient Culture Street, and reflects the advanced culture of the times. This area keep the Tianhou Temple, the Yuhuangge, restored in Tongqingli, combined with new Haihelou, new tourist attractions, such as the level set to "auspicious water cloud, hidden dragon born" as the theme, tourism, shopping, dining, leisure, entertainment, recreation, accommodation, making it the most taste and cultural connotation.

三号方案（第一名）Plan 3(First Prize)

一号方案（第二名）Plan 2(Second Prize)

二号方案（第三名）Plan 1(Third Prize)

四号方案 Plan 4

基本情况

该区域位于天津市中心城区的核心地段，横跨海河，连接和平区和河北区，是天津市的商业中心。北至博爱道，东至规划大沽路，南至新华路，西至福安大街。规划用地面积 72.45 公顷，新建建筑面积 100 万平方米。和平广场中心商业区包括中心广场、和平广场、和平路步行商业街三部分，海河以 S 形的曲线穿越商业区。

竞标方案

共有 6 个竞标方案，中标方案将海河南岸区域重新整合为三个区域，它们各有特点，分别产生于不同层次的市场概念以及商业运作，既能提高整个区域的档次，又可形成良好的互动。以和平广场中心商业区为载体，将形成天津的五大标志：金环（连通海河两岸的环形游廊）、津门（商业办公综合体）、津塔（天津地标）、金街（和平路商业街）、金色音乐厅（提升商业中心文化内涵的音乐厅）。

Basic Situation

This area is located in the core of the downtown area of Tianjin, across the Haihe River, connecting Heping District and Hebei District, and is the commercial center of Tianjin. It extends north to Boai Road, east to planning Dagu Road, south to Xinhua Road, west to Fuan Street. The planned land area is 72.45 hectares, and the new construction area is 1 million square meters. The Peace Square Central Business District includes three parts:Central Square, Peace Square and Heping Road pedestrian commercial street. Haihe River crosses the business district in an s-shaped curve.

Competition Scheme

There are 6 competition schemes in total, and the winning scheme reintegrates the south bank area of Haihe River into three regions, each of which has its own characteristics and comes from different levels of market concepts and commercial operations, which not only improves the level of the whole region, but also forms a good interaction between them. In the Peace Square Central Business District as the carrier, will form five landmarks: Gold Rings, Tianjin Gate, Tianjin Tower, Gold Street , Golden Hall .

六号方案（第一名）Plan 6(First Prize）

三号方案（第二名）Plan 3(Second Prize）

一号方案 Plan 1

二号方案 Plan 2

四号方案 Plan 4

五号方案 Plan 5

基本情况

南站（六纬路）中心商务区位于河东区，是天津市中心商务区的重要组成部分。北至六经路，东至六纬路，南至十一经路，西至海河，规划用地面积 29.68 公顷。

竞标方案

共有 5 个竞标方案，中标方案以“综合软硬件优质服务的一体化数字中心商务区，现代化国际港口大都市的重要标志区，城市商业新形象和绿色都市亲水新概念的代表”为定位。集中在国际化、现代化、数字化和亲水化四个方面进行设计。

Basic Situation

Located in Hedong District, it is an important part of Tianjin CBD. It extends north to Liujing Road, east to Liuwei Road, south to Shiyijing road and west to Haihe River. The planned land area is 29.68 hectares.

Competition Schemethe

There are 5 competition schemes in total, the winning scheme is positioned as "An integrated digital central business district with high-quality services of software and hardware, a new image of urban commerce and a representative of the new concept of green city with water". The design focuses on four aspects: internationalization, modernization, digitalization and hydrophilic.

二号方案（第一名）Plan 2 (First Prize)

一号方案（第二名）Plan 1 (Second Prize)

三号方案 Plan 3

二号方案（第一名）Plan 2 (First Prize)

五号方案（第三名）Plan 5 (Third Prize)

四号方案 Plan 4

基本情况

天津海河综合开发建设上游段第一座新建桥梁——永乐桥是集交通、游览为一体的桥梁，直径 140 米的巨型摩天轮纵向安装于桥中间。它将超过英国泰晤士河畔号称世界上最大的“伦敦之眼”，跃居世界第一。该桥是由日本川口卫设计事务所设计的。

Basic Situation

As the first new bridge on the upstream section of Haihe River Comprehensive Development and Construction, Yongle Bridge integrates traffic, sightseeing as one. A giant ferris wheel, known as the "Eye of Tianjin", with the diameter of 140 meters, is installed in the middle of the bridge. It surpasses the "London Eye" on Thames River and becomes the world's largest ferris wheel on the river. The bridge was designed by Kawaguchi Design Office in Japan.

设计方案
Design Scheme

建设情况
Construction Situation

基本情况

天津意式风情区位于天津市河北区，是由五经路、博爱道、胜利路、建国道合围而成的四方形地区，区内拥有保存完整的具有百年历史的欧式建筑近 200 栋。

竞标方案

共有 4 个竞标方案。一号方案来自法国的 AREP 公司，二号方案来自意大利的 PDS 公司与天津规划院的联合体，三号方案来自元宏建筑设计顾问（天津）有限公司及北京都林国际工程设计咨询有限公司的联合体，四号方案来自德国的 AS&P 公司。

Basic Situation

Italian Style Area is located in Hebei District of Tianjin. It is a square area surrounded by Wujing Road, Boai Road, Shengli Road and Jianguo Road. There are nearly 200 well-preserved century-old European style buildings in the area.

Competition Scheme

There are four competition schemes. No. 1 scheme is from AREP Company in France, No. 2 scheme is from PDS Company in Italy and joint venture of Tianjin Academy of Urban Planning and Design, No. 3 scheme is from joint venture of Yuanhong Architectural Design Consulting (Tianjin) Co., Ltd and Beijing Dulin International Engineering Design Consulting Co., Ltd, and No. 4 scheme is from AS&P Company in Germany.

二号方案 Plan 2

三号方案 Plan 3

一号方案 Plan 1

四号方案 Plan 4

基本情况

该片区是《海河综合开发改造规划》中“智慧城”的北岸部分，规划以可持续发展和生态建设为主题，在这一带构造新型的城市形态，以先进的网络技术与智能化技术为核心，以松散的城市形态创造具有高产出和高附加值的新产业区。

Basic Situation

This area is the north part of " Smart Town " in the Haihe River Comprehensive Development and Reconstruction Plan, planning for sustainable development and ecological construction as the theme, construct the new urban form in this area, the construction of advanced network technology and intelligent technology as the core, a loosely urban form to create new industrial district with high output and high added value.

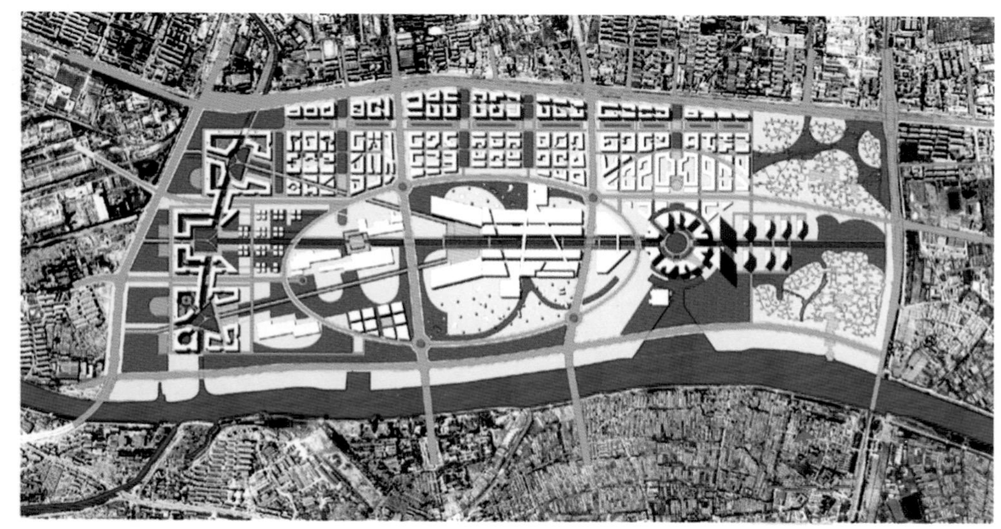

智慧城北岸城市设计
Urban Design of the North Bank of Smart Town

基本情况

该片区是《海河综合开发改造规划》中“智慧城”的南岸部分，规划方案由SBA事务所编制完成。

Basic Situation

This area is the south bank area of the "Smart Town" in the Haihe River Comprehensive Development and Reconstruction Plan, compiled by SBA.

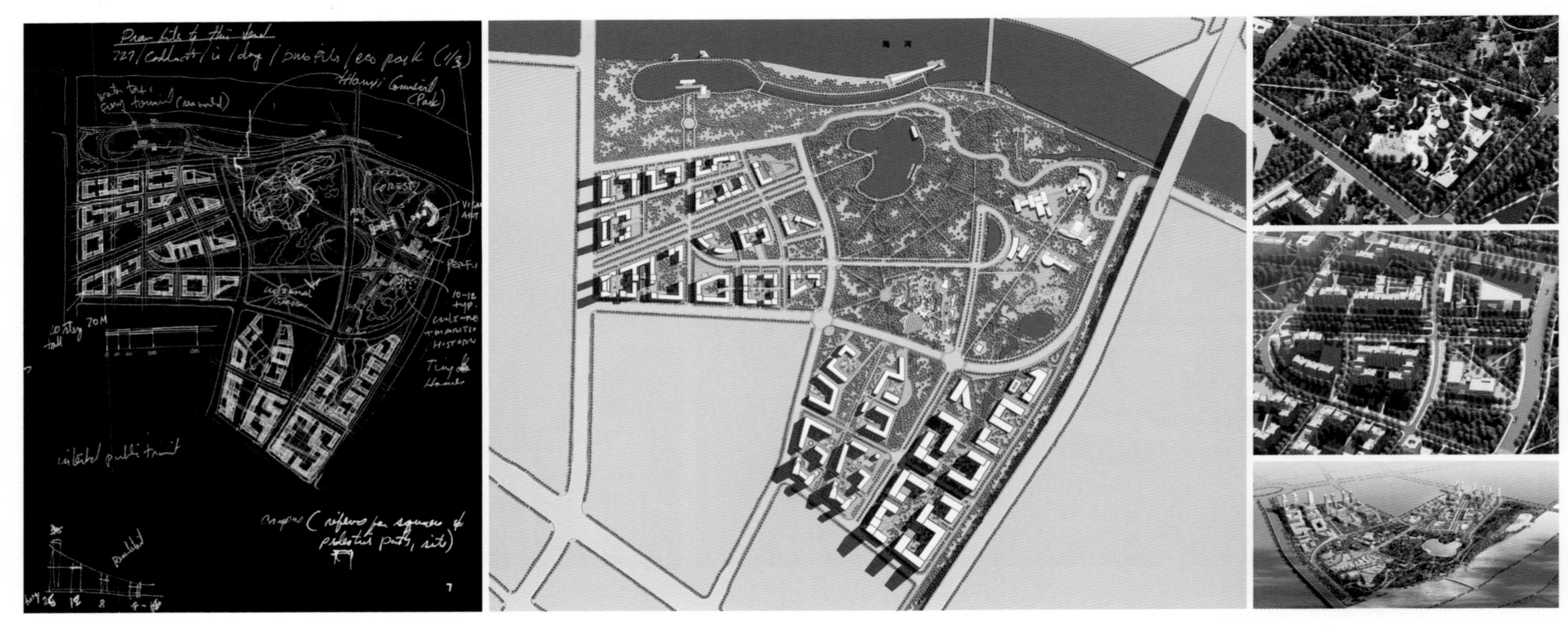

智慧城南岸城市设计
Urban Design of the South Bank of Smart Town

基本情况

西沽公园地区东至天泰路，南至新开河，西至红桥北大街，北至光荣道。用地面积约 180.75 公顷。

竞标方案

共有 4 个竞标方案。一号方案主题为创造一个社区中心与配套公共建筑群、一个风格各异的居住区、一个商业区、一个复合的传统风貌区。二号方案着重强调西沽公园地区地处三河交汇的位置，通过精心剪辑城市历史记忆，以运河及活跃的亲水开放空间缝合两岸，形成统一和谐的整体。三号方案对城市功能和用地布局进行了合理调整，将大型商业结合地铁站点综合设置，梳理了西沽公园的用地范围和边界道路，达到良好的整体功能和景观效果。四号方案中心概念为共享阳光、绿地、水岸的城市住区交往空间。

Basic Situation

Xigu Park extends east to Tiantai Road, south to Xinkai River, west to North Hongqiao Street, north to Guangrong Street. The land area is about 180.75 hectares.

Competition Scheme

There are four competition schemes. The theme of Plan 1 is to create a community center and supporting public buildings, a residential area with different styles, a business district and a compound area with traditional features. Plan 2 focuses on strengthening the position of Xigu Park area at the intersection of three rivers. By carefully editing the historical memory of the city, the two banks are stitched together by canals and active water-loving open space to form a unified and harmonious whole. Plan 3 makes reasonable adjustment to the urban function and land layout, combines the comprehensive setting of large commercial stations with subway stations, sorts out the land scope and boundary road of Xigu Park, and achieves good overall function and landscape effect. The central concept of Plan 4 is to share sunshine, green space and waterfront urban residential communication space.

一号方案
Plan 1

三号方案
Plan 3

四号方案
Plan 4

二号方案（第一名）
Plan 2 (First Prize)

基本情况

子牙渔湾项目在天津市中心城区西北部，位于海河上游段规划的传统文化商贸区内，隶属红桥区，规划区北边紧邻西沽公园。北起红桥北大街，南至子牙河，西至欢庆路，用地面积约 77.36 公顷。

竞标方案

共有 4 个竞标方案。一号方案由子牙河与人工河所形成的 T 形结构，提供多样可能的生活方式。二号方案从整体城市角度考虑城市设计问题，但对原地形、地貌破坏较大。三号规划整体结构基于地区整体的可持续发展，综合考虑功能配置、景观组织、生态环境和开发时序等方面的要求。四号方案总体布局从南到北大致分三个段落。南部沿子牙河为滨水商业带。中部为中高档居住区和商住混合社区。北部为高档居住区。

一号方案 Plan 1

二号方案 Plan 2

三号方案 Plan 3

四号方案 Plan 4

Basic Situation

Ziyayu Bay project is located in the northwest of Tianjin central city , in the Traditional Culture and Business Area planned for the upper reaches of Haihe River. It belongs to Hongqiao District and is adjacent to Xigu Park in the north of the planning area. with Ziya River in the south, Huanqing Road in the west, North Hongqiao Street in the north, covering an area of about 77.36 hectares.

Competition Scheme

There are four competition schemes. Plan 1 is a t-shaped structure formed by Ziya River and artificial river, offering a variety of possible lifestyles. Plan 2 considers the urban design from the perspective of the whole city, but it has great damage to the original topography and landform. The overall structure of Plan 3 is based on the overall sustainable development of the region, taking into account the requirements of functional configuration, landscape organization, ecological environment and development timing. The overall layout of Plan 4 is divided into three sections from south to north. The south area along the Ziya River for the waterfront commercial zone. The central area is a mixed residential and commercial area. The north area is an upscale residential area.

基本情况

小白楼地区历史上是天津市传统的商业中心，隶属和平、河西两区，是总体规划确定的中心商务区（CBD）的一部分。该地区北邻解放北路商务区，东与南站（六纬路）商务区隔海河相对，南邻德式风貌区，西接五大道风貌保护区，规划总用地面积 50.29 公顷。

竞标方案

共有 4 个竞标方案。一号方案将小白楼地区规划为富于活力、人性化的新城市空间，天津购物休闲、旅游休闲活动的标志节点，拥有历史与现代特征的“天津金谷”。二号方案设计一个立体绿色公众平台，以一种全新的方式统一组织了旧有、混杂的城市公共空间，增强了原有建筑与新建建筑之间的和谐性。三号方案在兼顾现状建筑群和实际的条件下，注入创新意念，把原区提升至一个国际级的商务中心区。在营造规划新环境时，本方案极重视历史的延续。四号方案竭力开发以地铁和公共汽车为主的公共客运，尽可能地开辟提供给自行车和行人的通道。

Basic Situation

Xiaobailou District, historically the traditional commercial center of the city, is a part of the Central Business District (CBD) defined in the master plan, which is subordinate to Heping Districts and Hexi Districts. This area is adjacent to the business district of North Jiefang Road in the north, South Station Business Distric across the Haihe River in the east, Germany Style Area in the south, and Wudadao historical area in the west. The total planned land area is 50.29 hectares.

Competition Scheme

There are four competition schemes. Plan 1 is to plan the Xiaobailou Area as a dynamic and humanized new urban space, a symbolic node of Tianjin shopping leisure and tourism leisure activities, and a "Tianjin Golden Valley" with historical and modern characteristics. Plan 2 is to design a three-dimensional green public platform, which unifies the old and mixed urban public space in a new way and enhances the harmony between the original building and the new one. Under the condition of giving consideration to both the existing buildings and the actual conditions, Plan 3 injected innovative ideas and promoted the original district to an international business center. In creating a new planning environment, the program attaches great importance to the continuation of history. Plan 4 strives to develop public transport, mainly by subway and bus, and to open up the channel for bicycle and pedestrian as much as possible.

一号方案
Plan 1

二号方案
Plan 2

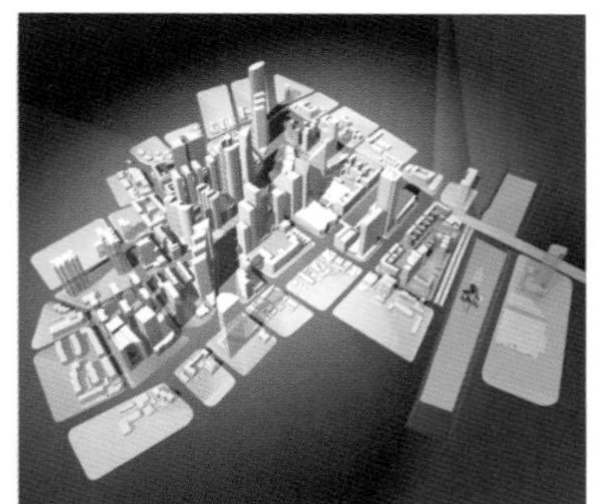
三号方案
Plan 3

四号方案（第一名）
Plan 4 (First Prize)

四号方案（第一名）
Plan 4 (First Prize)

基本情况

德式风貌区位于天津中心城区，隶属天津市河西区，紧邻中心商务区，总用地面积约 84.17 公顷。

竞标方案

共有 3 个竞标方案。一号方案在原有的历史背景下进一步延续德式风格，以反映出该地区的个性，并赋予该地区及整个城市特有的历史风貌。二号方案采用德国特色的围合式布局，沿海河建造小运河，形成沿岸绿色岛链。三号方案在保持整个区域尺度亲切，富于文化气质的前提下，沿大沽路建造两点一线的高强度开发地段，提高土地使用率，并与小白楼 CBD 相呼应。

Basic Situation

The German Style Area is located in the central city of Tianjin, affiliated to Hexi District, close to CBD, with a total land area of about 84.17 hectares.

Competition Scheme

There are three competition schemes. Plan 1 further extends the German style in the original historical background to reflect the personality of the region and endows the region and the whole city with unique historical features. Plan 2 adopts the enclosed layout with German characteristics, and builds small canals along the coastal river to form the green island chain along the coast. On the premise of keeping the whole regional scale cordial and rich in cultural temperament, Plan 3 builds a high-intensity development area along Dagu Road with two points and one line to increase the land utilization rate, and echoes the Xiaobailou CBD.

一号方案
Plan 1

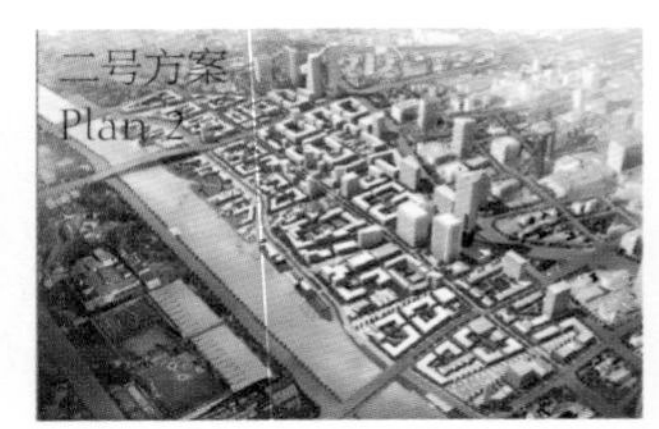
二号方案
Plan 2

三号方案
Plan 3

建设情况
Building Condition

基本情况

位于天津市中心城区的核心地段，隶属和平区，是天津中心商务区的重要组成部分，用地面积约 127.23 公顷。

竞标方案

共有 4 个竞标方案。一号方案以保护老建筑和老街区为重点，改造定位为天津工商服务产业起飞的孵化器、城市旅游观光产业永久的基石、大众运输导向的活力城区、公共投资的理想空间。二号方案本着尊重历史和自然的原则，以资源整合为出发点，重点解决交通配套问题，创造出英式风貌区独特城市景观。三号方案注重亲水区，注重对解放北路历史金融区的保护，并在此基础上建造新的金融构架，且建造“英国公园”使之成为该区新的“绿心”。四号方案以中小规模的土地开发模式为主，用渐进式的方式逐步实现。

Basic Situation

British Style Area is located in the core of the downtown area of Tianjin, affiliated to Heping District , is an important part of CBD, covering an area of 127.23 hectares.

Competition Scheme

There are four competition schemes. Plan 1 focuses on the protection of old buildings and old blocks, and is positioned as an incubator for the takeoff of Tianjin's industrial and commercial service industry, a permanent cornerstone for the city's tourism industry, a vibrant urban area oriented by mass transportation, and an ideal space for public investment. Plan 2 is based on the principle of respecting history and nature, taking resource integration as the starting point, focusing on solving the problem of supporting transportation, and creating a unique urban landscape with British style. Plan 3 focuses on the hydrophilic area, focuses on the protection of the historical financial area of North Jiefang Road, and builds a new financial framework on this basis, and builds "British Park" to become the new "green heart" of the area. The fourth plan is mainly based on the small and medium-size land development mode, which is gradually realized in an incremental way.

二号方案
Plan 2

基本情况

法式风貌区位于天津市中心城区的核心地段，隶属和平区。该区为海河两岸综合开发改造规划确定的中央金融商务区的重要组成部分，规划总用地面积 117.40 公顷。

竞标方案

共有 3 个竞标方案。一号方案优化法式风貌区的现状风貌并重组该区域城市功能，创造最具天津城市特色的都市中心与新城市空间。二号方案重点考虑城市肌理和建筑的完整性并将其保存下来。三号方案中“北五大道”方案充分体现了法式风貌区的区位特色，兼顾海河总体规划、城市历史风貌特色保护和国际化发展的要求。

Basic Situation

The French Style Area is located in the core of the downtown area of Tianjin, which belongs to Heping District. This area is an important part of CBD in the Haihe River Comprehensive Development and Reconstruction Plan, with a total planned land area of 117.40 hectares.

Competition Scheme

There are three competition schemes. Plan 1 optimizes the current situation of the French Style Area and reorganizes the regional urban functions to create the most distinctive urban center and new urban space of Tianjin. Plan 2 focuses on the preservation of urban fabric and architectural integrity. The plan of "North Wudadao" in Plan 3 fully reflects the location characteristics of the French Style Area, and gives consideration to the overall planning of Haihe River, the protection of the city's historical features and characteristics, and the requirements of international development.

一号方案（第一名）Plan 1（First Prize）

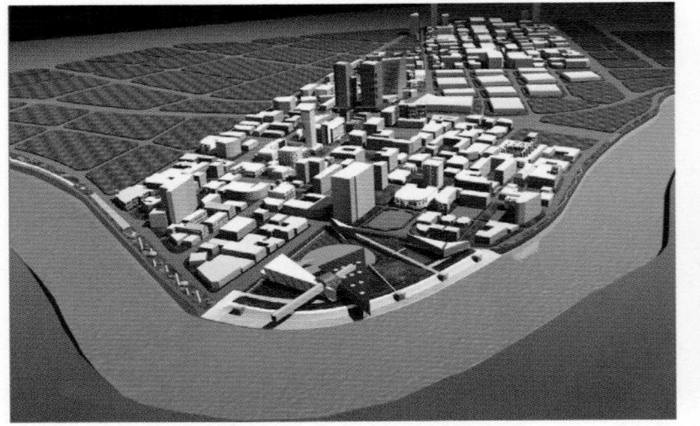

二号方案 Plan 2

三号方案 Plan 2

津湾广场

津湾广场是天津市标志性建筑群，其建筑风格多元，突出地反映了中西文化交汇的时代特征，展现了天津近代城市的功能、文化和风尚，出色的灯光设计使其成为海河夜景的重要组成部分。总占地面积 12.5 公顷，建筑面积约 76 万平方米。

Jinwan Square

Jinwan Square is the landmark building complex in Tianjin. Its diversified architectural style highlights the characteristics of the times when Chinese and western cultures meet, and reflects the functions, culture and fashion of modern Tianjin. The excellent lighting design makes it an important part of the night view of Haihe River. It covers a total area of 12.5 hectares with a construction area of about 760,000 square meters.

津湾广场 Jinwan Square

奥式商务公园

位于原奥匈帝国租界中，保留修复了三栋风貌建筑：袁氏官邸、冯国璋故居、奥匈领事馆。这三幢百年建筑以其特有的风格屹立在海河东岸，成为海河岸边一道亮丽的风景，奥式商务公园的三座建筑一律以中低层建筑为主，不但继承了奥式文化区内的原味风格，更开创了一种全新的现代化生态商务办公模式。

Austrian Style Business Park

Austrian Style Business Park is located in the former Austrian Concession, it has retained and restored three style buildings: Yuan Shikai residence, Feng Guozhang former residence and Austro-Hungarian Empire Consulate. Three century-old buildings stand on the east bank of the Haihe River with their unique styles and become a beautiful scene on the bank. The three buildings are all low-rise buildings, which not only inherit the original style of the Austrian Style Area, but also create a new modern ecological business office mode.

沙盘模型
Model

方案平面图
General Plan

建设情况
Building Condition

形象巨变

2002 年与 2017 年的天津湾。

Image Change

Jinwan Square in 2002 and 2017.

2002 年
In 2002

2017 年
In 2017

海河上游两岸鸟瞰
Overall View of the Upstream Section of Haihe River

大光明桥两岸
Both Sides Area of Daguangming Bridge

泰安道五大院地区鸟瞰
Overall View of Five Courtyards in Taian Road

发挥“杠杆作用”

海河的开发改造成功之处绝不仅限于激活海河沿线地区，而是像一个“杠杆”一样，撬动了整个天津市的发展活力。海河的道路桥梁建设使天津市的整体交通效率提升了 30%，带动城市综合竞争力走出颓势，快速提升，带动城市经济总量快速跃升，是同时期北京增速的 1.3 倍、上海增速的 1.8 倍。更重要的是，海河建设汇集了民心，极大地增加了城市自豪感与城市自信心。古语云“不谋城不足以谋一域”，海河规划正是建立在“谋城”的高度思考“一域”之事，所以它的实施才会展现出惊人的“杠杆作用”。

尊重历史文化

历史文化是城市的灵魂，要像爱惜自己的生命一样保护好城市历史文化遗产。海河规划建设中非常可贵之处就是在“大拆大建”的时期中，坚持保护历史文化。当时被普遍认为影响经济发展的历史遗存，现在全都成为重要的经济增长点，意式风情区、五大道、解放北路不仅带动了高质量的经济发展，而且也成为天津的城市名片，留住了城市记忆，留住了天津的城市灵魂。

创新运营方式

海河开发改造启动的十大工程全部是公益类建设，需要巨大的投资，如果采用常规政府投资的运营方式是完全不可能实现的。因此海河开发改造探索了全国最早的经营城市方法，政府担保大额贷款—投资建设—周边土地升值—土地出让金升值部分还贷款，应用这套方法大获成功。尽管这套运营方法可能不再适合于现在的经济社会环境，但在十五年前的确是最合适的，而且海河开发改造的这种跳出常规的思维方法，也正是我们探索适应新时代的经营方式的精髓。

Play the "Leveraging Effect"

The success of Haihe River development and transformation is not limited to activating the area along the Haihe River, but it is like a “lever”, which instigates the development of the whole city. The construction of roads and bridges in Haihe River has improved the overall traffic efficiency of Tianjin by 30%, and has driven the overall competitiveness of the city out of its rapid growth. The rapid growth of the total economic output of the city was 1.3 times the growth rate of Beijing in the same period and 1.8 times the growth rate of Shanghai. More importantly, the construction of the Haihe River has brought people's hearts together, which has greatly increased the city's pride and urban self-confidence. The old saying goes "Who doesn't study to manage a city is hardly to manage for one domain". The Haihe River Plan is based on the highly thought "one domain" of the "manage a city", so its implementation will show an amazing leveraging effect .

Respect Historical Culture

Historical culture is the soul of the city. It is necessary to protect the city‘s historical and cultural heritage as well as to cherish its own life. The very valuable aspect of the planning and construction of the Haihe River is to adhere to the protection of historical culture during the period of “demolition and construction”. At that time, it was widely believed that historical relics affecting economic development are now all important economic growth points. The Italian Style Area, Wudadao District, and North Jiefang Road not only promoted high-quality economic development, but also became Tianjin's city business cards, retaining city memories and retaining Tianjin's urban soul.

Innovative Operating Methods

The ten major projects launched from Haihe River development and transformation are all public welfare constructions and require huge investment. It is completely impossible to use a conventional government investment operation. Therefore, Haihe River development and transformation explored the earliest method of operating cities in the country. The government guaranteed large loans—construction investment—appreciation of surrounding land on surrounding—land appreciation part of the appreciation of land grants also repaid loans. The application of this method was great success. Although it may no longer be suitable for the current economic and social environment, it was indeed the most suitable 15 years ago. Moreover, the Haihe River development and transformation way of thinking that jumps out of the normal way is the essence of our approach to adapting to the new era of business.

城市设计在天津
URBAN DESIGN IN TIANJIN

天津"135"工程
Tianjin 135 Project

"135" 工程

"135" 迎奥运市容环境综合整治工程（以下简称"135"工程），以天津市成为北京奥运会协办城市为契机，对天津市容市貌进行全面提升，展现天津深厚的历史文化底蕴、独特的自然风貌、现代化大都市气息。

整个规划主要包含"一带三区五线"等系列项目设计："一带"——海河，"三区"——奥体中心、天津站、小白楼，"五线"——复康路、十一经路、卫津路、南京路、解放路，以及增加的友谊路及东南半环快速路。"一带三区五线"串接城市交通枢纽、历史风貌区、商贸中心区、科教文化区、奥运中心区等重要功能区，构成横跨东西、纵贯南北的两大轴线，形成天津城市的景观骨架。

The 135 Project

The 135 Comprehensive Renovation Project (hereinafter referred to as the 135 Project) takes Tianjin as the co-host city of the Beijing Olympic Games as an opportunity to comprehensively improve the appearance of Tianjin, showing its profound historic and cultural background, unique natural features and modern metropolitan atmosphere.

The whole plan mainly includes a series of project designs, such as "One Belt"—the Haihe River, "Three Districts"—Olympic Sports Center, Tianjin Railway Station, Xiaobailou Area, and "Five Lines"—Fukang Road, Shiyijing Road, Weijin Road, Nanjing Road, Jiefang Road, and the added Youyi Road and Southeast Half Ring Expressway. "One Belt, Three Districts and Five Lines" connect the important functional areas such as the urban transportation hub, historic scenery area, commercial and trade center area, science and education cultural area and the Olympic Center area, forming two axes and the landscape skeleton of urban Tianjin.

天津奥体中心
Tianjin Olympic Sports Center

奥体中心

奥体中心的总体定位是成为市区的一大景观亮点，不仅在比赛期间正常有序地运转，而且在赛后持续地释放其魅力，并成为天津市群众性体育文化生活的重要基地。

整治与新建相结合。必须在设计之初先做选择：哪些可以保留，哪些要加以改造，哪些要彻底重新设计。

兼顾长期与短期效应。由于奥体中心在奥运期间的窗口效应，因此必须在短时间内完成总体景观格局设计，达到较好的景观效果。但从长远而言，作为一处承载群众体育文化生活的平台，在设计中也必须考虑其景观的可持续性。

大格局，小细部。在格局上必须做到整体性强、空间清晰明确。但在竖向设计、界面处理、小空间营造乃至小品设计等方面必须做到准确、合理、独具匠心，才能使细部服务于整体，实现既定的设计目标。

Olympic Sports Center

The overall role of the Olympic sports center is to become a highlight of the urban landscape, not only to ensure its normal and orderly operation during the Olympics, but also to maintain its charm continuously, and become an important location of mass sports and cultural life in Tianjin.

Combining renovation with construction. At the beginning of the design, choices must be made: what can be retained, what should be reformed, and what should be completely redesigned.

Consider both short-term and long-term effects. Because of the attention placed on the Olympic Sports Center during the Olympic Games, the overall landscape pattern must be completed in a short time to achieve better landscape effect. But in the long run, as a platform for mass sports and cultural life, the sustainability of its landscape must also be considered in the design.

Big pattern, small details. In terms of the landscape pattern, we must provide a strong integrity and clear definitions of space. But in the aspects of vertical design, interface treatment, small space construction, and even design of items in the space, it is necessary to be accurate, reasonable and ingenious in order to make details serve the whole and achieve the established design goals.

天津站地区

天津站前广场建筑及景观改造工程是奥运环境整治工程的重要组成部分，作为奥运火炬接力的最后一站，对北京奥运会的成功举办起着关键性的作用。整体目标是将其打造成为新世纪的津门标志并为奥运火炬接力准备必要条件。

以简约大气为原则，突出地域特色，把握时代脉搏，利用先进的建造工艺与材料，创造新的津门标志。

Tianjin Station Area

Tianjin Railway Station Front Square Architecture and Landscape Renovation Project is an important part of the Olympic Environmental Renovation Project. As the last stop of the Olympic torch relay, it plays a key role in the successful hosting of the Beijing Olympic Games. The overall goal is to create a symbol of Tianjin in the new century and provide the necessary conditions for the Olympic torch relay.

With advanced design concepts and the guiding principle of simple and bold, highlighting regional characteristics, grasping the pulse of the times, using advanced construction technology and materials to create a new symbol of Tianjin.

天津站
Tianjin Railway Station Area

十一经路（卫国道至曲阜道）

从曲阜道、十一经路全线至卫国道外环线，两侧建筑与环境街道整治，全长约10.8千米。

卫国道是天津滨海国际机场进入天津市区的重要迎宾干线，是进出天津的第一门户，是展示天津的第一窗口。

通过对旧建筑尺度与色彩的斟酌与调整，使建筑与周围环境相融合和协调。在使建筑更美观的同时，为这座城市注入了新的生机与活力，给市民的生活增添了新的亮点。

Shiyijing Road (Weiguo Road to Qufu Road)

From Qufu Road and Shiyijing Road to the outer ring of Weiguo Road, buildings and environmental streets on both sides are renovated, covering a total length of about 10.8 kilometers.

Weiguo Road is an important main line for traffic from Tianjin Binhai International Airport to enter downtown Tianjin. It is the first and foremost gateway into Tianjin and the most important window to showcase Tianjin.

Through the consideration and adjustment of the scale and color of the old buildings, the buildings and the surrounding environment are integrated and coordinated. While improving the buildings' appearances, it also injects vitality and liveliness into the city and adds new highlights to the lives of the people.

十一经路整治前后立面对比
Comparison Before and After Effect of Shiyijng Road

南京路

南京路全长 3.4 千米，目标是建设成为天津市商业繁荣、环境大气、最具现代气息的城市走廊，使南京路变成一条精致而大气的交通大道、一条繁华而有序的商业大街、一条干净整洁的清新大街和一条高雅靓丽的灯光走廊。

Nanjing Road

Nanjing Road is 3.4 kilometers long. The goal is to build a city corridor with a modern atmosphere and prosperous commerce in our city, turning Nanjing Road into a delicate and atmospheric traffic road, a busy yet orderly commercial street, a clean and fresh street, and an elegant and beautiful lighting corridor.

整治前吉利大厦

整治前南京路鸟瞰

整治后道路界面

整治后吉利大厦

整治后的街头绿地

整治后的居住环境

南京路
Nanjing Road

解放北路

解放北路全长 1.8 千米，将被建设成为天津市最具欧洲古典气质的精致街道和金融机构集聚、风貌建筑荟萃的历史街区。

通过历史风貌建筑修旧如旧，现代建筑改造外观，与整条街道风格协调，保证街道界面连续性和整体建筑风格统一，改善首层建筑立面，提升绿地品质。

North Jiefang Road

North Jiefang Road is 1.8 kilometers long. It will be built into the street with the highest classical European qualities in Tianjin. It will also be a historic district where financial institutions and buildings gather together.

Through the renovation of historic buildings and reconstructing the appearance of modern buildings, the style of the whole street is coordinated, the continuity of the street interface and the unity of the overall architectural style are guaranteed, the facade of the first floor is improved, and the quality of green space is improved.

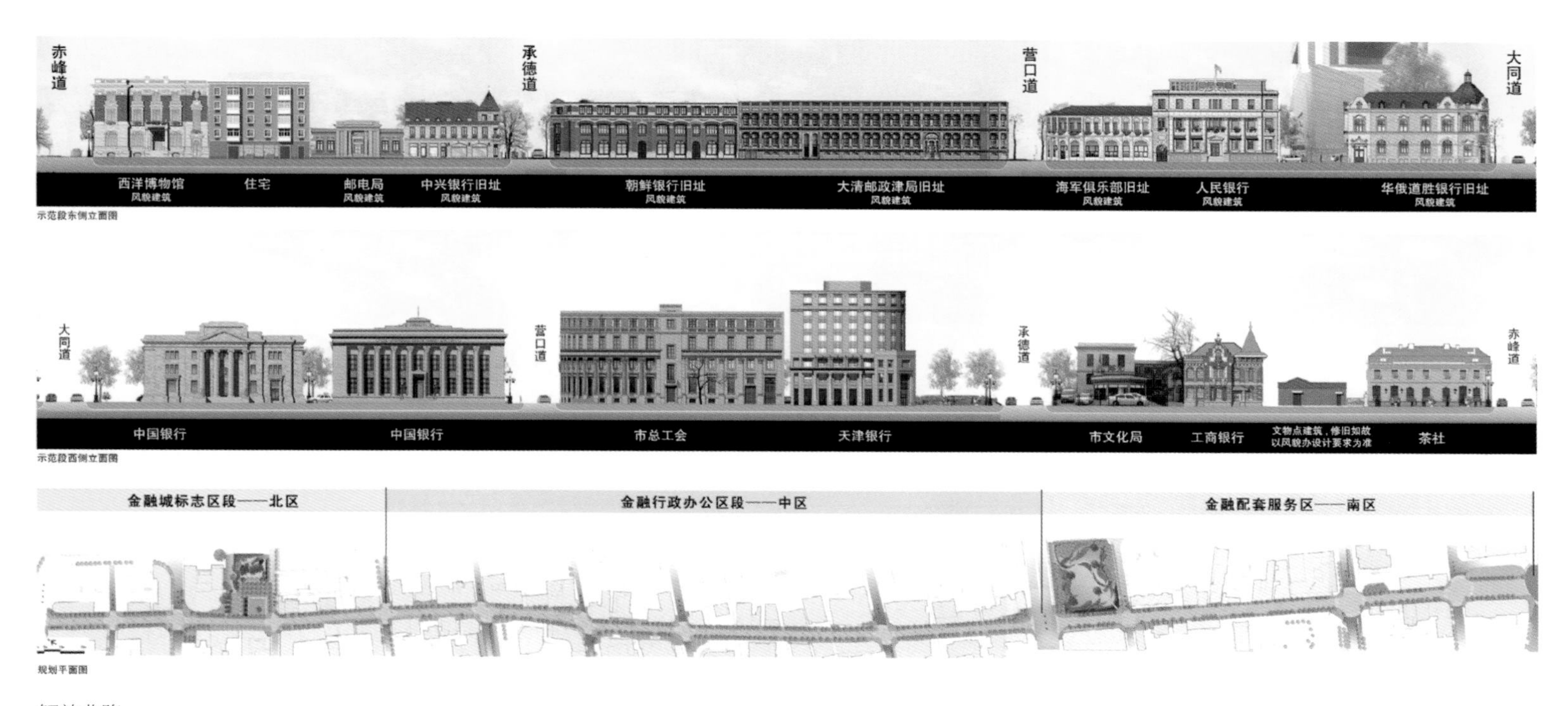

解放北路
North Jiefang Road

解放南路（曲阜道—黑牛城道）

全长 4.2 千米，区域内部分路段位于历史风貌保护区控制区与协调区范围内。规划以居住功能为主兼顾生活与交通的景观道路，以及亲切、宜人兼顾交通的现代城市生活展示区。

规划地段与历史风貌保护区相邻，规划路段景观从欧式古典风貌过渡到现代城市风格。设计上无论从建筑整修、道路断面或是街具均做到风格统一，和谐过渡。

South Jiefang Road (Qufu Road to Heiniucheng Road)

The total length is 4.2 kilometers, and several sections within the region are within the control and coordination areas of the historic reservation area. Landscape roads are planned with residential functions as the main consideration, while also being a friendly and pleasant modern city life exhibition area with both traffic and transportation.

The planned section is adjacent to the historic reservation area, and the planned section's landscape transits from European Classical style to modern urban style. In terms of design, no matter regarding building renovation, road section or street furniture, the style should be unified and harmonious transition should be achieved.

整治后政协俱乐部

改造后多层住宅效果图

改造后刀具厂效果图

电力公司立面改造设计

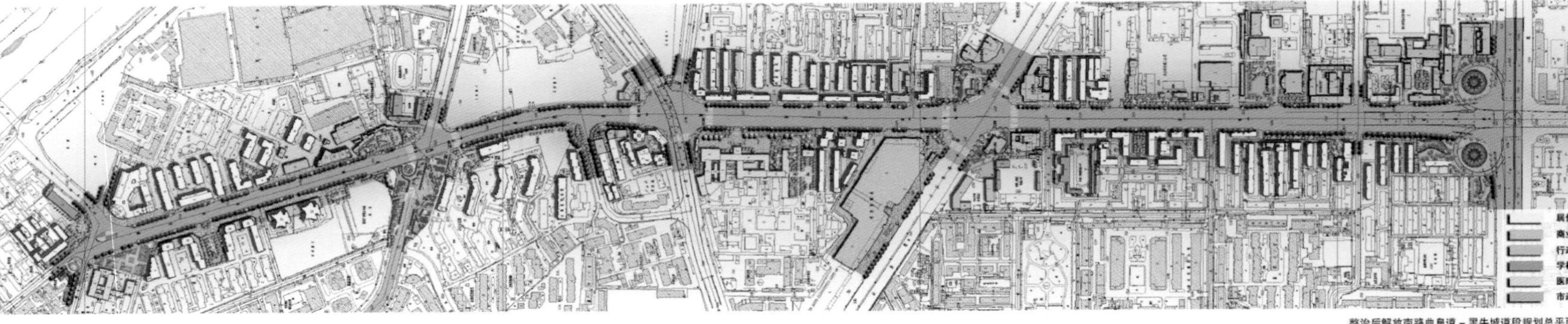

整治后解放南路曲阜道－黑牛城道段规划总平面图

解放南路
South Jiefang Road

友谊路

南起黑牛城道，北至围堤道，全长1.86千米，规划成为集商务、会展、文博、高端百货、市民广场等大型重要公共建筑汇聚的国际品质精品街区，塑造精锐开放、雅致清新、最具国际大都市品质的城市客厅。

Youyi Road

Ranging from Heiniucheng Road in the south to Weidi Road in the north, Youyi Road is 1.86 kilometers in length. It is planned to become a high-quality international residential area with large and important public buildings such as commerce, exhibitions, cultural exhibitions, high-end department stores and citizen's squares, creating a “city living room” with international metropolitan quality, which is exquisite, open, elegant and fresh.

整治前友谊路东天津珠宝街段

黑牛城道
友谊大厦
友谊花园
住宅
天津珠宝街
如家快捷酒店
西园西道
海天大厦
宾馆南道

整治后友谊路东天津珠宝街段

友谊路
Youyi Road

东南半环快速路

全长 13.8 千米，是天津快速路系统的主要组成部分，将规划成展示天津市美好市容环境的窗口，打造成规整、清洁、统一的城市快速路风景线。

以简约大气为原则，利用先进的建造工艺与材料，对沿街建筑形态进行分类处理，对现状建筑色彩进行收集，统一划分出主次色调，有针对性地对建筑进行粉刷整修。

Southeast Half Ring Expressway

It is 13.8 kilometers long and is the main component of Tianjin's expressway system. It will be planned as a window to showcase Tianjin's beautiful city environment and create a regular, clean and unified urban expressway scenic line.

Based on the principle of simplicity and boldness, advanced construction technology and materials are used to classify the architectural forms along the street, collect the current architectural colors, unify the primary and secondary tones, and carry out targeted painting and renovation of buildings.

整治前昆仑路现状规划对比（阳新里段）

整治后昆仑路现状规划对比（阳新里段）

娄庄子地区整修设计

东南半环快速路
Southeast Half Ring Expressway

项目概况

作为一条有着百年历史的街道，解放北路见证了天津这座城市的沧桑变迁，凝聚了厚重的历史文化底蕴。在城市快速发展的今天，作为天津中心城区CBD的重要组成部分，解放北路地区又扮演着不可替代的重要角色。近些年，随着城市功能的不断完善，解放北路地区的整体环境品质得到了巨大改善。

改造工作分为两部分展开：首先通过整治解放北路沿线的整体景观环境，营造富有生气的百年金融老街，带动提升周边地区的整体文化氛围；其次对解放北路周边的滨江道、承德道、大同道、长春道、哈尔滨道、赤峰道、营口道、大连道八条道路开展规划提升工作，力争将解放北路地区、五大道地区和劝业场地区打造成为中心城区现代服务业与都市旅游业的“金三角”。

Project Overview

As a street with a history of one hundred years, North Jiefang Road has witnessed the vicissitudes of Tianjin, and has condensed its rich historical and cultural heritage. Today, with the rapid development of the city, as an important part of CBD in Tianjin's central urban area, North Jiefang Road plays an irreplaceable and important role. In recent years, with the continuous improvement of urban functions, the overall environmental quality of North Jiefang Road area has been greatly improved.

The renovation work is divided into two parts: first, by renovating the overall landscape environment along the North liberation road, we can create a vibrant century-old financial street, and promote the overall cultural atmosphere of the surrounding areas; second, we can improve the overall cultural atmosphere of the riverside road, Chengde Road, Datong Road, Changchun Road, Harbin Road, Chifeng Road, Yingkou Road and Dalian Road around the North Jiefang Road are planned and upgraded to make North Jiefang Road area, Wudadao area and Quanyechang area into the "Golden Triangle" of modern service industry and urban tourism in the central city.

解放北路街道竣工实景
Completion of North Jiefang Road Street

保护性建筑改造前后对比（上：改造前；下：改造后）
Comparison of the Effect of Protective Buildings Before and After Transformation (Top Picture: Before Transformation; Bottom Picture: After Transformation)

破损建筑修补前后对比
Before and After Repairing Damaged Buildings

不协调建筑改造前后对比
Before and After Reconstruction of Uncoordinated Buildings

拆除围墙前后对比
Before and After Wall Demolition

改造建筑外檐，营造商业氛围

解放北路周边地区建筑用途以金融及商务办公为主，并有少量住宅及商业设施，规划针对建筑风貌特征不明显或建筑质量较差部分进行适当改造，更新配套设施，完善使用功能，提升环境品质。根据现状建筑保护类别、使用功能及外立面情况，规划对建筑外檐采取 4 种改造措施。保护修缮：针对保护性建筑，进行外檐清洗与修缮，修旧如旧。整修提升：对沿街破损建筑进行修补，改善形象。改造融合：对沿街与街区风貌不协调建筑重新设计，改造立面。拆除整治：对违章建筑与围墙进行拆除，恢复原貌。

调整道路断面，加宽人行便道，规划对区域内道路的断面进行调整，对部分道路取消路面停车、压缩路面宽度、加宽人行道宽度，营造尺度适宜的、连续的步行空间，以提升沿街商业氛围。

Rebuilding the Eaves of Buildings to Create A Commercial Atmosphere

In the surrounding areas of North Jiefang Road, the construction uses are mainly financial and commercial office, with a small number of residential and commercial facilities. The plan is to make appropriate renovation for the parts with unclear architectural features or poor building quality, update supporting facilities, improve the use function and improve the environmental quality. According to the types of protection, functions and facades of buildings, four transformation measures are planned for the outer eave of buildings. Protective renovation: For protective buildings, cleaning and repairing the outer eave are carried out, and the old ones are as old as the old ones. Renovation and upgrading: repair the damaged buildings along the street, improve the image. Renovation and integration: redesign the buildings that do not coordinate with the style of the street and the block, and transform the elevation. Demolition and renovation: demolition of illegal buildings and walls to restore the original appearance.

Adjusting road sections and widening sidewalks. The plan adjusts the section of roads in the region, cancels road parking, compresses the width of roadway, widens the width of pavement, and creates suitable and continuous walking space to enhance the commercial atmosphere along the street.

滨江道断面改造后效果
Effect After cross-section Reconstruction of Binjiang Road

合江路口增加街头绿地后效果
Effect After Adding Street Green Space at Hejiang Intersection

滨江道增加街头小品后效果
Effect After Adding Street Furnitures in Binjiang Road

增加绿地广场，改善街道环境

将影响街道环境的违章建筑及部分围墙进行拆除，增加街头绿地与广场；改造部分路段的铺装形式，实行交通管制，使部分城市道路可以作为城市广场等开放空间使用；提升改造街具，增设街头休息座椅，适当增加城市雕塑、街头小品等公共艺术设施。

一直以来，天津对老城区的环境品质提升工作非常重视，尤其是对解放北路这样具有深厚历史文化底蕴的地区，更是做了大量的研究与实践，并取得了显著效果。随着城市功能的逐步完善和业态品质的不断提升，解放北路及周边地区从内在功能到外在品质都将取得质的飞跃。

Increasing Green Space and Improving Street Environment

Demolition of illegal buildings and some fences affecting the street environment will increase street green space and squares; transformation of pavement forms of some sections and implementation of traffic control, so that some urban roads can be used as open space such as urban squares; upgrading and transformation of street furniture, adding street rest seats, and increasing urban carvings appropriately such as street sketches and other public art facilities.

For a long time, Tianjin has attached great importance to the improvement of environmental quality in old urban areas, especially in areas with deep historical and cultural heritage such as North Jiefang Road, which has done a lot of research and practice, and achieved remarkable results. With the gradual improvement of urban functions and the continuous improvement of business quality, North Jiefang Road and its surrounding areas will make a qualitative leap from internal function to external quality.

五大道局部鸟瞰
Partial Aerial View of Wudadao Area

项目概况

五大道是目前天津市规模最大、保存最完好的历史文化街区，它始建于1901年，历史上是英租界的高级住宅区，也是20世纪初英国花园城市理论在中国的实践。

随着城市快速发展，五大道历史文化街区面临许多亟待解决的问题：一是，对街区的历史文化特征缺乏多角度的深入研究；二是，部分更新建筑和环境品质亟待提高；三是，规划管理手段落后，缺乏精细化管理的有效措施。近年来，在五大道历史文化街区城市设计的指导下，全面深入地研究历史空间特色，并将研究成果用以规范和引导街区内部的更新建设，延续历史上既有秩序又丰富多样的空间环境特色。同时，探索和建立一套精细化的管理方法，将城市设计成果转化为可辨识、可度量、有效果的管理工具，有效促进历史街区更新改造和环境品质的提升。

Project Overview

Wudadao are the largest and most well-preserved historic and cultural blocks in Tianjin at present. It was built in 1901. Historically, it was a high-end residential area in the British concession. It was also a result of the practice of the British garden city movement in China at the beginning of the 20th century. With the rapid development of the city, Wudadao Historic and Cultural Area is facing many urgent problems to be solved.

The problems are as follow: firstly, there is a lack of in-depth studies of the historic and cultural characteristics of the blocks; secondly, there is an urgent need to improve the quality of some renewed buildings and the environment; thirdly, there is a lack of effective measures for delicacy management. In recent years, under the guidance of the urban design of Wudadao Historic and Cultural Block, the characteristics of historic space have been studied comprehensively, and the research results have been used to standardize and guide the renewal and construction of the block, thus maintaining the orderly and diverse characteristics of its spatial environment in history. At the same time, a set of refined management methods is explored and established to transform urban design results into identifiable, measurable and effective management tools, effectively promoting the renovation of historic blocks and the improvement of environmental quality.

五大道空间肌理
Space Fabric of Wudadao Area

创新方法，深入挖掘历史空间特色

城市设计在大量深入翔实的现状调研基础上，针对五大道历史文化街区内的建筑类型、街廓肌理、街道与街巷格局等方面，采用城市形态学和建筑类型学方法对历史空间特色进行全面深入的研究，探讨造就五大道独特生活品质的空间格局和特点。

尊重历史，精细编制城市设计导则

针对对历史建筑不恰当地翻新和装饰、拆除围墙、更新建筑背离五大道设计传统、沿街设施不规范等设计和建设中的具体问题，通过城市设计，严格保护五大道历史文化街区的整体环境，并进一步对每一座院落、建筑进行仔细甄别、分类，分析其构成要素，有针对性地编制城市设计导则。

Innovative Methods, In-Depth Excavation of the Characteristics of Historic Space

On the basis of a large number of in-depth and detailed investigation on the current situation of urban design, aiming at the aspects of building type, street texture, street and lane pattern in historic and cultural districts, urban morphology and architectural typology in the Wudadao Historic and Cultural Area, a comprehensive and in-depth study on the characteristics of historic space is conducted to explore the spatial pattern and characteristics that create the unique quality of life of Wudadao.

Respect History and Compile Urban Design Guidelines

Regarding specific problems in planning and construction such as the inappropriate renovation and decoration of historic buildings, the removal of fences, the renewal of building going against design traditions of Wudadao, and non-standard design of curbside facilities, it is emphasized that the overall environment of Wudadao Historic and Cultural Block should be strictly protected, and each courtyard and building should be carefully screened and classified, its constituent elements analyzed, and urban design guidelines should be formulated accordingly.

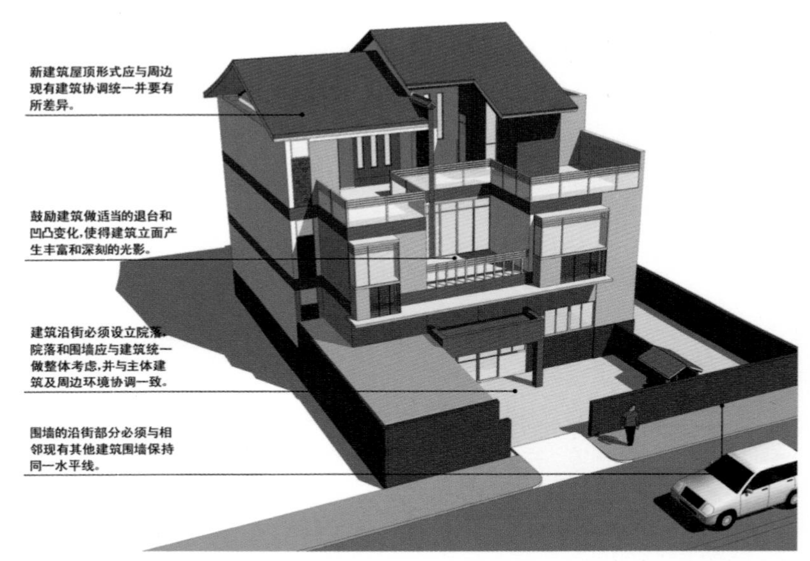

创新手段，建立三维立体化管理系统

城市设计方法具有立体化和直观性的特点，利用城市设计进行管理就是要将整体思维和立体思维引入规划管理中去。通过城市设计为五大道 2 514 幢建筑建立三维数字模型，对建设项目进行三维空间审核并进行动态监控，为全方位、立体化、精细化的规划管理提供强有力的技术支持。

指导更新，取得良好的实施效果

五大道历史文化街区城市设计的主要成果已纳入《五大道历史文化街区保护规划》，2012 年 4 月获得天津市政府批复，现已成为街区内进行各项建设活动、编制修建性详细规划、建筑设计以及各专项规划的管理依据，推动了历史街区保护水平的大幅提升。

Innovative Means to Establish Three-Dimensional Management System

Urban design methods are three-dimensional and intuitive in nature, and carrying out management through urban design is to introduce holistic thinking and three-dimensional thinking into planning management. Through the urban design, a three-dimensional digital model of 2514 buildings in Wudadao is established, and 3D space audit and dynamic monitoring of construction projects are carried out, which provides strong technical support for the comprehensive, three-dimensional and refined planning and management.

Guidance and Renewal to Achieve Good Implementation Results

The main achievements of urban design of Wudadao Historic and Cultural Area have been incorporated into the "Wudadao Historic and Cultural Area Protection Plan", which was approved by Tianjin municipal government in April 2012. It has become the management basis for various construction activities, detailed planning, architectural design and special planning in the blocks, and has promoted the level of protection and renewal of historic blocks to a great extent.

五大道建筑三维数字模型
Three-Dimensional Digital Model of Wudadao Area

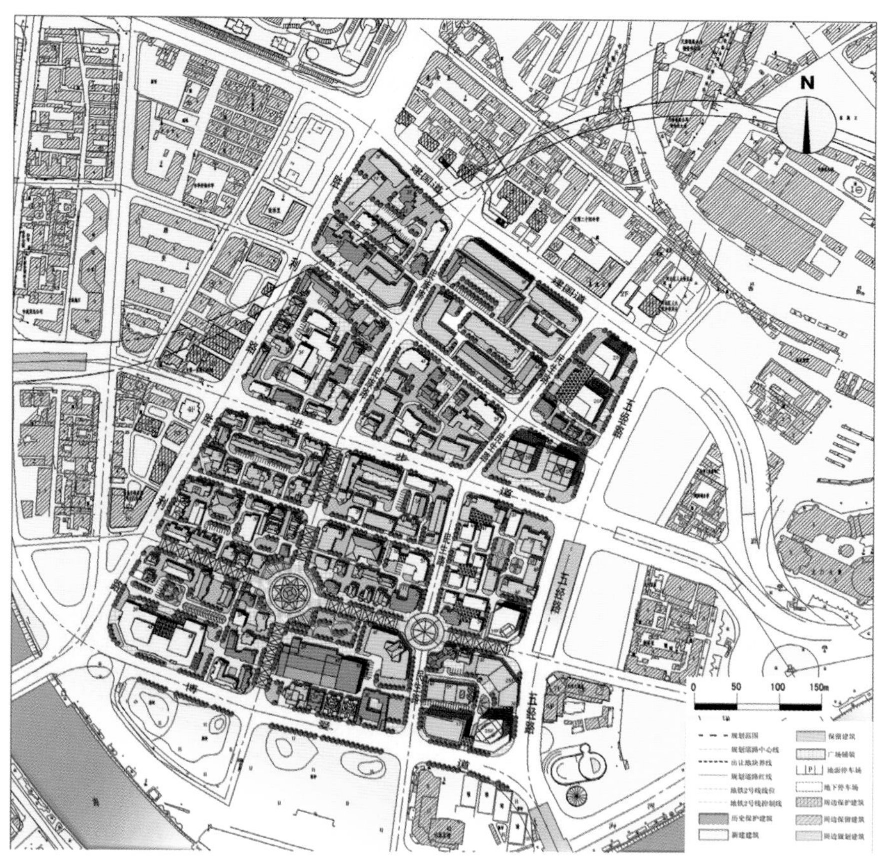

规划总平面图
Master Plan

项目概况

天津意式风情区（简称“意风区”）位于天津市中心城区的几何中心，河北区的海河东岸，由建国道、胜利路、博爱道、五经路围合而成，总占地面积28.91公顷，是《天津市城市总体规划》确定的一宫花园历史文化保护区的重要组成部分。

百年沧桑，随着时间的推移，由于战乱、地震等灾害的影响以及城市的变迁，意风区失去了往昔的神韵。2002年天津市启动海河两岸综合开发改造工程，意风区作为一个重要节点，规划编制工作适时启动。为还原意风区的本来面貌，城市设计以保护历史风貌特色和街区空间格局、保护历史环境、改善人居环境、完善市政基础设施、促进和谐发展为原则，采取保留旧有城市的结构和肌理，保留原有建筑的尺度与风格，分类保护、修旧如旧的方法进行。

Tianjin Italian Style District

Tianjin Italian Style Area (hereinafter referred to as Yifeng Area) is located in the geometric center of downtown Tianjin and the east bank of Haihe River in Hebei Area. It is enclosed by Jianguo Road, Shengli Road, Bo'ai Road and Wujing Road. It covers an area of 28.91 hactares. It is an important part of the historic and cultural reservation area of Yigong Garden determined by the "Tianjin Urban Master Plan".

With the passage of time, due to the impact of wars, earthquakes and other disasters as well as changes in the city, Yifeng Area lost its charm. In 2002, Tianjin started the comprehensive development and transformation project on both sides of the Haihe River. As an important node, the planning work of Yifeng Area started in time. In order to restore the original appearance of Yifeng Area, urban design is based on the principles of protecting historical features and spatial pattern of blocks, protecting historical environment, improving living environment, improving municipal infrastructure and promoting harmonious development. It adopts the methods of retaining the structure and texture of the old city, retaining the scale and style of the original buildings, classifying protection and repairing the old as before.

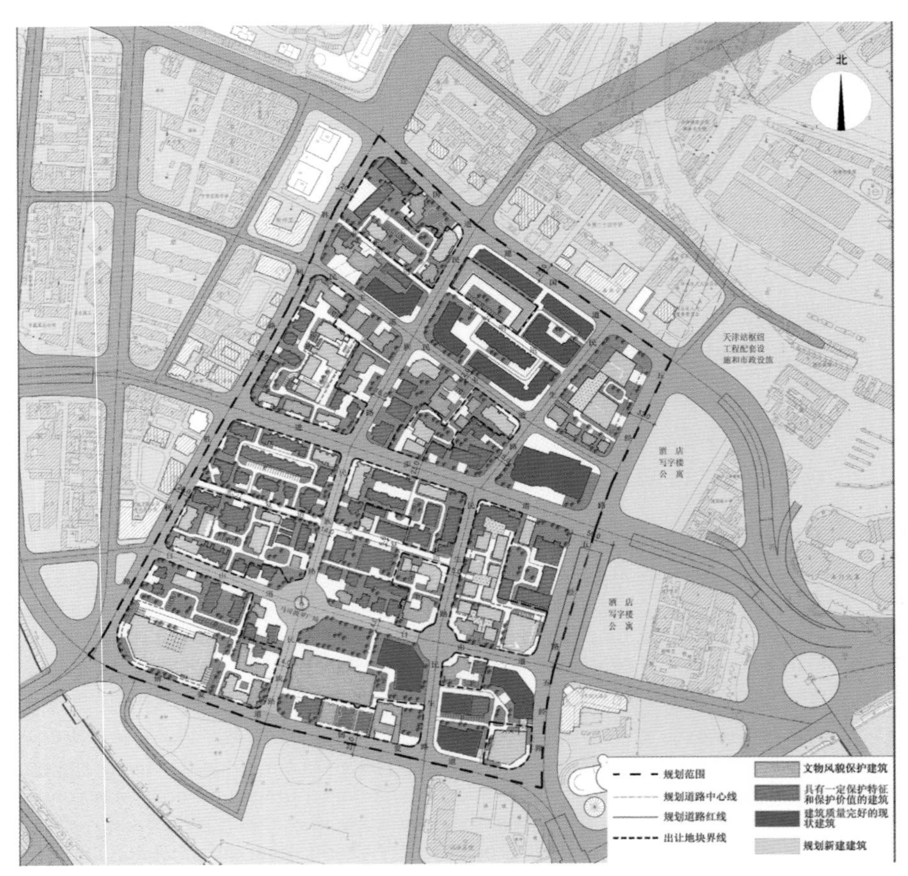

建筑保护与更新方式规划图
Building Protection and Update Method

传承历史文脉，保留街区空间格局

意风区由十四个街坊组成。通过划定不同层次的控制区域来保护历史风貌特色和街区空间格局，标志性的马可·波罗广场与十字街步行区构成控制的核心区，核心区遗存有大量的意大利风格建筑，整个街区以低层建筑为主，严格控制新建筑的风格和体量，使之与保护建筑协调统一，区内以小块石材铺就步行小径，创造了亲切、宜人的环境 。

保护环境特色，整修历史风貌建筑

意风区内遗存的意式建筑约 140 栋。首先对现状建筑进行甄别，拆除违章建筑和无保留价值的老旧建筑，保留建筑按照文保单位、历史风貌建筑、有一定风貌特征和保留价值的建筑以及质量完好建筑四个类别进行控制保护。

Inheritance of Historical Context and Preservation of Block Spatial Pattern

Yifeng Area is composed of fourteen neighborhoods. The historic features and the spatial pattern of the block are protected by delimiting the control areas at different levels. The iconic Marco Polo Square and the cross-street pedestrian area constitute the core control area. The core area contains a large number of Italian-style buildings. The whole block is dominated by low-rise buildings, and the style and volume of the new buildings are strictly controlled to make it coordinated with the protected buildings. Small stones pave footpaths in the area, creating a friendly and pleasant environment.

Protecting Environmental Characteristics and Renovating Historic Architectures

There are about 140 Italian-style buildings left in Yifeng Area. Firstly, the existing buildings are screened, the illegal buildings and the old buildings with no reservation value are demolished, and the reserved buildings are controlled and protected according to four categories: cultural protection units, historic buildings, buildings with certain features and preservation value, and buildings with good quality.

实施效果
Implementation Effect

营造城市活力，丰富功能，美化环境

意风区内用地多以商业金融业为主，辅以办公、旅游、娱乐、居住等功能，为使保护与发展具有可操作性，将现代生活融入其中，体现功能多样性，在保留意式居住社区风貌特色的前提下，赋予部分建筑以休闲购物、旅游服务、办公等功能，创造区域发展的活力。

改善人居环境，完善市政基础设施

意风区城市设计在实施过程中充分考虑了改善人居环境的城市需求，交通组织体现以人为本的原则，由围合区域街坊的城市道路承担区域内外的交通联系，马可·波罗广场和十字街区则作为步行区为行人提供舒适的漫步环境，保留原有步行尺度，使建筑与街巷相得益彰。

Creating the Vitality of the City, Enriching Regional Functions and Beautifying the Environment

The land in Yifeng Area is mainly used for commercial and financial purposes, supplemented by functions of offices, tourism, entertainment and residence in order to make protection and development operable. To integrate modern life into it and embody functional diversity, some buildings are designed to serve the functions of leisure shopping, tourism service and offices, while retaining the features of an Italian-style residential community, so as to create the vitality of regional development.

Enhancing the Habitat Environment and Perfecting Municipal Infrastructure

In the implementation of urban design in Yifeng Area, the urban demand for improving living environment is fully considered. Traffic organization fully embodies human-centered principles. Urban roads enclosing regional neighborhoods link traffic inside and outside the region. Marco Polo Square and the cross-shaped block provide a comfortable walking environment for pedestrians as pedestrian zones, retaining the original walking scale and make buildings and streets complement each other.

项目概况

天津中心城区规划范围为外环线所围合区域，具有清晰的城市边界。近年来，中心城区以城市设计为引领，全面提升了城市建设水平。城市空间不断拓展，北部新区规划将中心城区范围扩展至 433 平方千米，轨道交通网络建设将带动城市空间结构进一步优化。与此同时，中心城区面临着从“增量式”到“存量式”转变的发展新常态，需要厘清思路，进一步引导和统筹城市形态与秩序，提升城市空间品质。

本次总体城市设计，有效地衔接了中心城区总规修编和控规深化，系统性地整合了城市存量土地和现有开发建设，提升了规划管理的水平。同时，明确了城市立体化的三维空间控制引导思路，进一步优化了城市格局，完善了城市形态，保护了生态环境，引导了城市风貌特色的进一步强化。

Project Overview

The planning scope of Tianjin central urban area is the area surrounded by the outer ring road, and it has a clear urban boundary. In recent years, urban design has taken the lead in central urban areas, which has comprehensively improved the level of urban construction. Urban space continues to expand, the planning of the North Area will expand the central urban area to 433 square kilometers, and the construction of rail transit network will further optimize the urban spatial structure. At the same time, the central city is facing a new normal development from "incremental planning " to " inventory planning " transformation. It is necessary to clarify ideas, further guide and coordinate the urban form and order, and improve the quality of urban space.

The overall urban design effectively links up the revision and deepening of the central urban master plan, systematically integrates the urban inventory land and the existing development and construction, and improves the level of planning management. At the same time, it clarifies the guiding thought of three-dimensional urban spatial control, further optimizes the urban pattern, improves the urban form, protects the ecological environment, and guides the further strengthening of urban features.

海河发展轴
Haihe River Development Axis

整体鸟瞰图
Overall Bird's-Eye View

存量视角下以人为本的总体城市设计

2008 年以来，随着国家战略与天津市空间发展战略全面推进，中心城区城市格局不断优化，城市建设水平不断提升。但在从“增量式”转化为“存量式”的新常态下，有的发展模式已经不再能适应现状。例如，城市存量土地的建设与空间结构的关系不够紧密，导致城市结构松散，整体秩序不明确；城市保护与发展的矛盾持续存在；城市快速发展中新的建设项目控制引导不足，给整体的城市活力和空间质量带来负面影响；社区生活和邻里环境的营造缺失，街道活力不足等。因此，中心城区总体城市设计工作重点是针对近年来新的发展趋势，对城市形态和空间系统进行进一步优化和梳理；结合存量土地开发，进一步落位重点地区和重要节点；以人为本，针对人的需求进一步提出城市更新与活力提升策略。针对现存问题，重点从强化城市格局、重视城市保护、增添城市亮点、提升城市活力四个方面提出城市总体设计框架。

The People-oriented Overall Urban Design From the Perspective of Inventory Planning

Since 2008, with the national strategy and Tianjin's space development strategy comprehensively promoted, the urban structure of the central city has been continuously optimized, and the level of urban construction has been continuously improved. However, under the new normal from "incremental planning " to " inventory planning ", some development models are no longer able to adapt to the present situation. For example, the relationship between the construction of urban inventory land and the spatial structure is not close enough, resulting in loose urban structure and unclear overall order; the contradiction between urban protection and development persists; the rapid control of new urban construction projects is insufficiently controlled and given to the overall city. The negative impact of vitality and space quality is the lack of community life and neighborhood environment, and lack of street vitality. Therefore, the overall urban design work in the central urban area is focused on new development trends in recent years, further optimization and combing of urban forms and spatial systems; combined with the development of inventory land, further focus on key areas and important nodes. The people-oriented design is to meet people's needs and further propose urban renewal and vitality improvement strategies. In view of the existing problems, the focus is on the overall design framework of the city from four aspects: strengthening the urban pattern, attaching importance to urban protection, adding urban highlights, and improving urban vitality.

天津中心城区建设强度分布图
Construction Strength Distribution Map

城市保护——增加城市的自然延续感和时代传承感

增加城市未被开发的自然感，可以为市民增添宁静、休闲的感受。河流水系与绿化公园是城市的自然禀赋，也是最具价值的核心资源，规划在梳理十三条主要河道的基础上，构建城市绿道系统，为市民提供亲近自然的休闲场所。从人的视角强化滨水空间的形态，以滨水梯度原则控制两岸建筑，形成连续的建筑界面，在河口交汇处、河道转弯处设置地标与节点。同时，加强生态修复，建立从自然郊野向中心延伸的生态体系。构建由外环路环外六个郊野公园、“一环十一园”的大型城市公园、中环线和快速路上 15 个城市公园组成的城市公园系统，满足市民亲近自然的需求，同时开放空间也将成为市民眺望城市公共中心天际线的最佳场所。

Urban Protection: Increasing the Sense of Natural Continuity and Times Inheritance of Cities

Increasing the natural feeling of the city that has not been developed can add a sense of tranquility and leisure to the citizens. River water system and green park are the city's natural endowment and the most valuable core resource. The plan is to build the urban greenway system on the basis of combing the 13 main rivers to provide the public with a close to nature. From the perspective of people, the shape of the waterfront space is strengthened, and the construction of the two sides of the strait is controlled by the principle of waterfront gradient, forming a continuous building interface, and landmarks and nodes are set at the intersection of the estuary and the turn of the river. At the same time, strengthen ecological restoration and establish an ecosystem that extends from the natural countryside to the center. To construct a city park system consisting of six country parks outside the outer ring, a large urban park in the “One Ring and Eleven Parks ”, the Central Line and 15 urban parks on the express road to meet the needs of the citizens in close proximity to nature, and the open space will also become the best place for the people to look at the city public center's skyline.

大型开放空间分布图
Large Open Space System Diagram

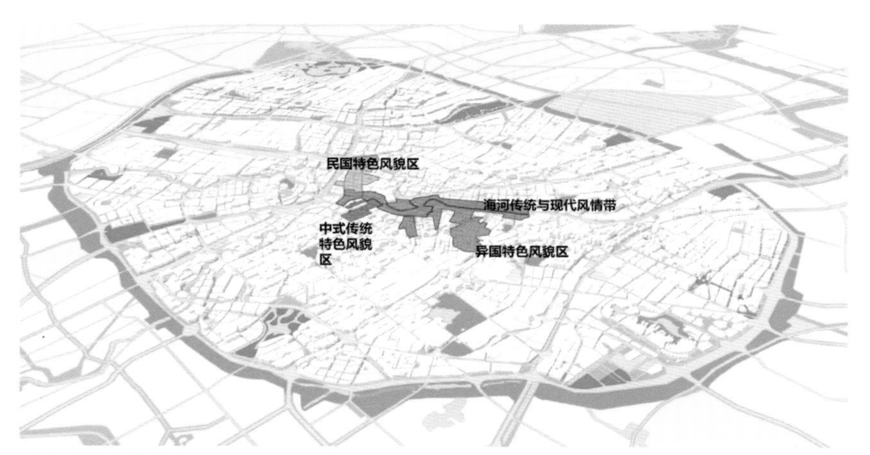

特色历史风貌分区图
Distribution Map of Characteristic Historic Style

城市亮点——在和谐统一的城市环境中增添新的时尚趣味

不断更新的重点地区与地标节点，可以为市民生活增添新的时尚趣味与视觉感受，增强城市的生命力。通过塑造引领时代潮流的城市地标，进一步强化城市结构。城市新的重点建设地区以紧凑的布局和功能复合的开发模式，塑造出宜人的街区尺度，营造出充满活力的生活街区，带动人流聚集，成为城市新的地标节点。

规划整体上结合重点地区梳理地标建筑的分布，增加建筑高度整体分布的逻辑性，沿公共中心、轨道枢纽、视觉焦点设置地标建筑，严格控制历史街区建筑高度。同时，根据 14 个开放空间的眺望观景点来安排高层建筑组群。按照视线仰角控制建筑组群的空间层次，同时考虑建筑顶部的变化，塑造优美的天际轮廓线。规划重点营造 17 条视线廊道线型空间，连通重要的开放空间与对景建筑，建立明确的指向性。

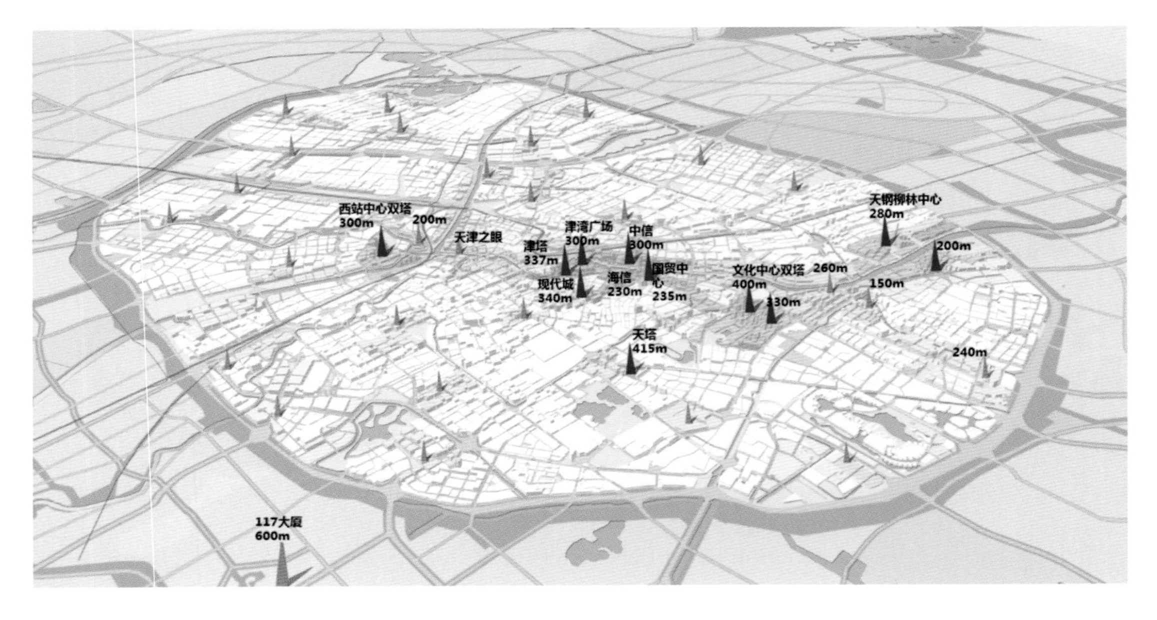

Urban Bright Spot: Adding New Fashion Interest to Harmonious and Unified Urban Environment

Renewal of key areas and landmark nodes can add new fashion interest and visual experience to the life of citizens and enhance the vitality of the city. By defining the key areas, the city landmarks leading the trend of the times will be shaped, and the urban structure will be further strengthened. With compact layout and complex development mode, the new key construction areas of the city create pleasant neighborhood scale, create vibrant living blocks, drive the gathering of people, and become new landmark nodes of the city.

The overall planning combs the distribution of landmark buildings in key areas, increases the logic of the overall distribution of building height, sets landmark buildings along public centers, subway hubs and visual focus, and strictly controls the building height of historical areas. At the same time, high-rise building groups are arranged according to 14 open space sightseeing spots. According to the elevation of line of sight, the space level of the building group is controlled, and the change of the top of the building is considered to create a beautiful skyline. The plan focuses on building 17 sightline gallery space, connecting important open space and landscape architecture, and establishing a clear orientation.

地标建筑分布图
Landmark Buildings Map

城市活力——营造具有归属感的社区街道与邻里生活

经济新常态下社区活力问题逐渐凸显，原有大街廓相互隔离的封闭小区，难以形成有活力的街道生活。尤其是新建小区都以点式高层为主，住宅空间类型单一，缺乏完善的社区、邻里环境和生活交往空间。现有的城市设计往往不重视对城市背景地区的考虑，忽视社区生活和邻里环境的营造。居民生活在各个不同的邻里中，满足人的心理归属需求是增加城市活力的基础。因此，首先应按照不同年代、不同类型划分社区，并与控规单元划分相结合，明确社区边界，增强居民归属感。同时结合不同类型的社区空间与居民生活的特点，提出不同的更新与发展策略，如封闭的商品小区应增加便民措施、沿街商业设施等，希望为市民提供长期可持续改善的社区生活环境。

此外，规划将引导舒适便捷的街道生活，从城市系统层面提高支路网密度，避免大街廓封闭小区的无序蔓延，构建窄路密网的生活空间。规划具体到每个社区单元，沿次干道与支路划定社区生活主街，成为汇集居民出行人流的主要街道，同时与轨道站等公交枢纽快速接驳，方便居民日常出行。

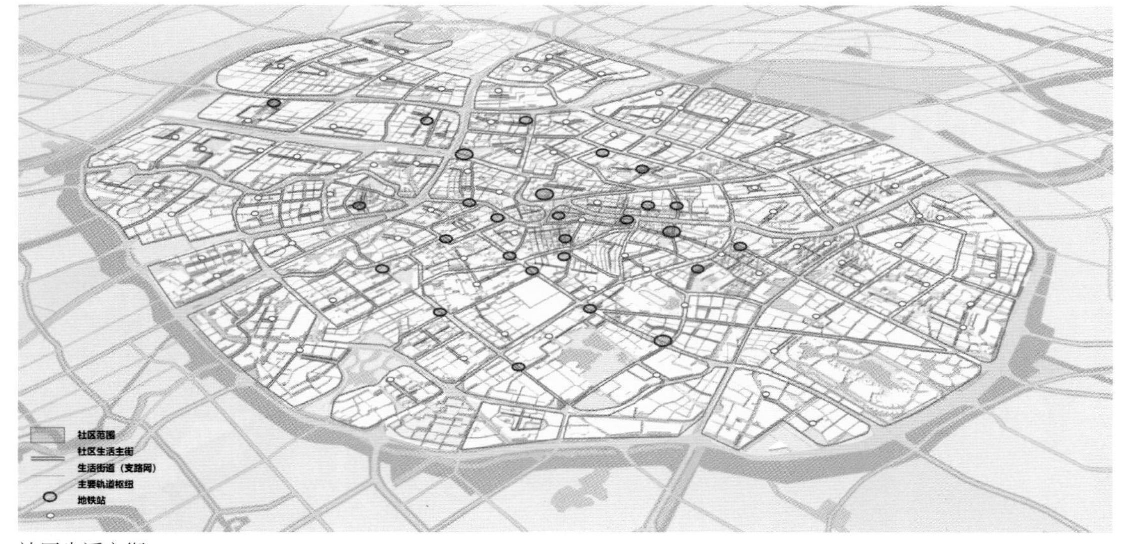

社区生活主街
Main Street of Community Life

Urban Vitality: Building Community Street and Neighborhood Life with a Sense of Belonging

Under the new normal economic situation, the problem of community vitality is becoming more and more prominent. It is difficult to form a vigorous street life in the closed community with isolated streets. Especially, new residential districts are dominated by high-rise buildings with single residential space type and lack of perfect community, neighborhood and living space. Existing urban design often neglects the consideration of urban background areas and neglects the construction of community life and neighborhood environment. Residents live in different neighborhoods. Satisfying people's psychological belonging needs is the basis of increasing urban vitality. Therefore, first of all, we should divide communities according to different ages and types, and combine them with the division of regulatory units to clarify the boundaries of communities and increase the sense of belonging. At the same time, according to the characteristics of different types of community space and resident lives, different renewal and development strategies are put forward, such as increasing convenience facilities and commercial facilities along the street in closed commercial districts, in order to provide long-term sustainable improvement of community living environment for citizens.

In addition, the planning will guide comfortable and convenient street life, increase the density of branch network from the level of urban system, avoid the disorderly spread of closed streets, and construct narrow road dense network living space. The plan extends to each community unit, along the sub-main road and branch road, delimits the main street of community life, and becomes the main street for gathering people's pedestrian flow. At the same time, it connects with the track station and other public transport hub quickly, in order to be convenient for residents to travel daily.

项目概况

2011 年底，滨海新区政府指导开展“滨海核心区城市设计全覆盖”编制工作。规划从总体、片区、单元和重点地区四个层面同步推进。

滨海新区核心区总体城市设计以双城模式审视滨海核心区的发展趋势，针对滨海新区在城市快速增长和转型阶段存在的问题，提出智慧成长、高效有序、绿色健康、文化传承、品质生活五个方面的规划目标，核心区将实现从近代工业港口重镇到国际现代都市的转化。

Project Overview

Since the end of 2011, under the command of the Binhai New District government, the compilation project of "full coverage of urban design in Binhai coastal core areas" has been carried out. Planning is carried out simultaneously on four levels: overall, regional, unit and key areas.

The overall urban design of the core area of Binhai New District is based on the two-city model. It examines the development trend of the core area of Binhai New District. In view of the problems existing in the rapid urban growth and transformation stage of Binhai New District, it puts forward five planning objectives of smart growth, high efficiency and order, eco-friendly, cultural inheritance and quality of life, and realizes the reborn of the core area from a modern industrial port town to an international modern city.

滨海新区整体鸟瞰图
Overall Aerial View of Binhai New District

“一心集聚、双轴延伸”的城市结构

强化“一心集聚”的公共中心。建立中央大道城市发展轴和海河生活服务轴，形成“黄金十字”的空间发展轴线。

以商务金融、航运服务、文化科研等现代服务业为核心，建设功能强大、运营高效、管理创新、生态环保的国际一流城市核心区。

Urban Structure of "Centralized Conglomeration and Dual-Axis Extension"

Strengthen the public center of "centralized conglomeration" and establish the Central Avenue urban development axis and the domestic service axis of Haihe River, and form the spatial development axis of "golden cross".

With modern service industries such as commerce and finance, shipping services, cultural and scientific research as the core, we will build a powerful, efficient, innovatively managed and eco-environmental friendly international first-class urban core area.

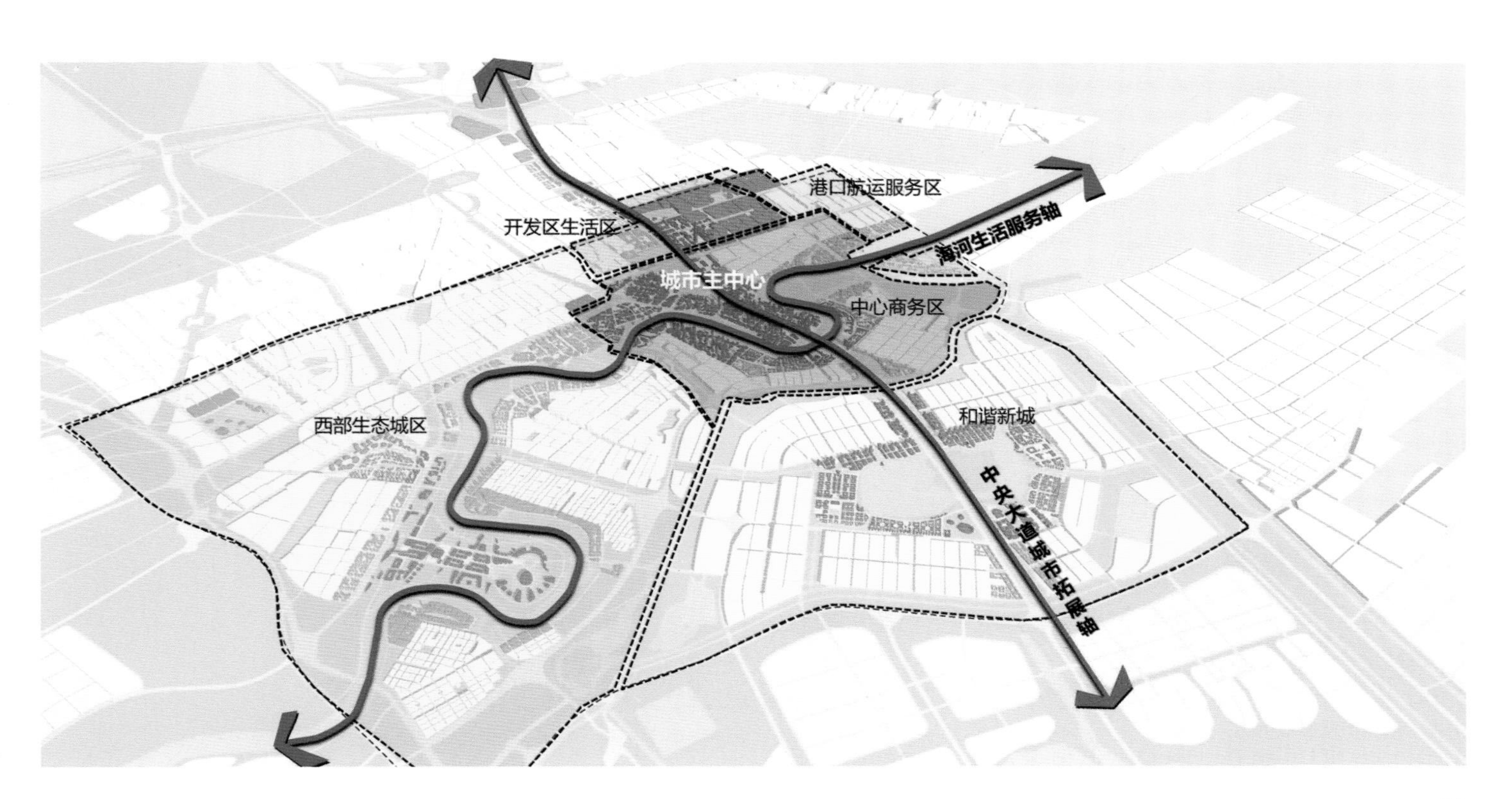

规划结构图
Planning Structural Map

海河生活服务轴鸟瞰
Aerial View of Haihe Life Service Axis

中央大道城市发展轴鸟瞰
Aerial View of Urban Development Axis of Central Avenue

“滨河面海、疏密有致”的城市形态

强化滨水地区的空间层次，形成海滨地区、入海口和河道两岸的滨水空间。优化中心区，形成点状高强度开发单元。重点突出城市公共中心和中心制高点。

Urban Form of "Reaching Riverside and Facing Sea, Dense and Compact"

Strengthen the spatial layers of waterfront areas and form waterfront space in coastal areas, estuaries and both sides of the river. Optimize the central area to form point-like high-strength development units, focusing on highlighting the urban public center and the central high points.

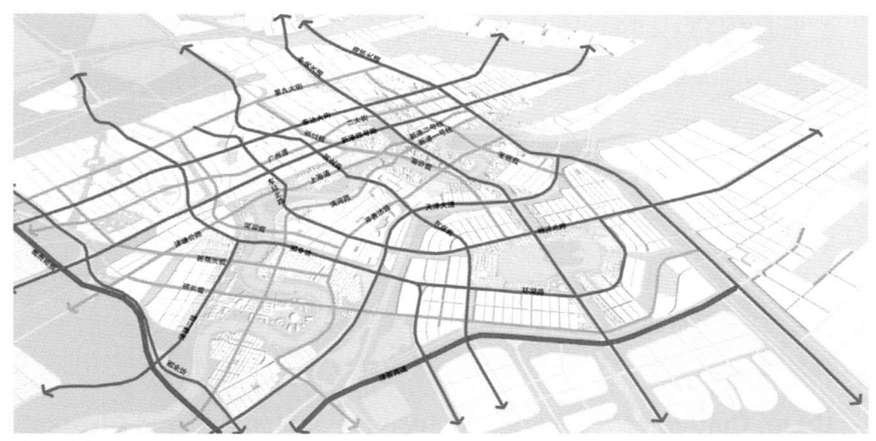

道路系统
Road System

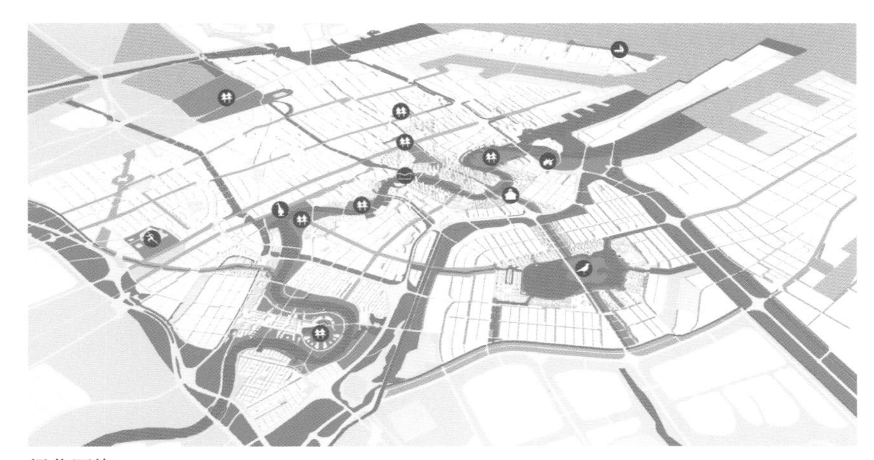

绿化网络
Greening Network

“高速环城、五横五纵”的综合交通

建立客运骨架路网，疏解集输港货运交通，实现港城分离，客货分流。优化轨道线网布局，以 TOD（以公共交通为导向的开发模式）拓展核心区城市建设，建立高效的大众运输和有序的道路网体系。

“蓝脉绿网、城景相融”的城市环境

通过疏通现有河道，加快南部生态湖及周边水网建设，形成“三横四纵、七河一湖”的生态水网系统，构建“两横五纵”的绿化网络，建设十四个主题鲜明的城市级绿地公园和若干社区公园，创造绿色健康的高品质生活。

Comprehensive Transportation of "High-Speed around the City, Five Horizons and Five Verticals"

Establish a passenger transport skeleton road network, unblock the freight traffic of ports, realize the separation of port and city, and divert passengers and cargo. Optimize the layout of track network, expand the urban construction of core area with the Transit-Oriented Development (TOD) mode, and establish efficient mass transportation and orderly road network system.

Urban Environment of "Blue Vein, Green Net and City Scenery Harmony”

Through dredging the existing rivers and speeding up the construction of ecological lakes and surrounding water networks in the south, form an ecological water network system of "three horizons and four verticals, seven rivers and one lake", construct a green coverage of "two horizons and five verticals", build fourteen urban green parks with distinctive themes and several community parks, and cultivate green and healthy high-quality lives.

历史文化景点及文物分布
Distribution of Historical and Cultural Scenic Spots and Cultural Relics

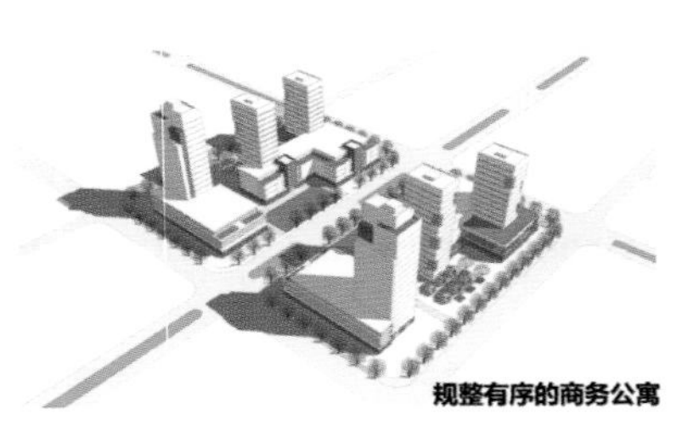

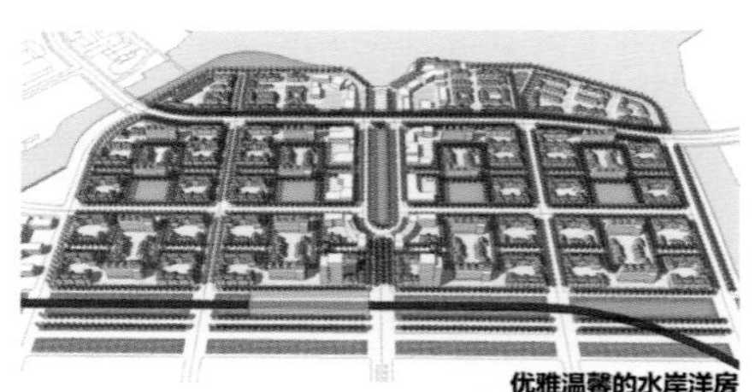

多元舒适的居住模式
Multiple Comfortable Living Mode

“开放包容、时尚多元”的城市文化

尊重有限的历史资源，划定并保护海河两岸的 30 处文物古迹和近代的工业历史遗存。形成以渔盐文化、工业文化等为主题，与滨水景观休闲设施相结合的城市滨水旅游线路。

“尺度宜人、多姿多彩”的城市生活

结合不同区位环境与街道朝向，探讨高中低开发强度下的多尺度社区建筑模式，营造多元舒适的居住模式。包括规整有序的商务公寓，亲切多元的都市合院，幽静别致的近郊联排别墅和优雅温馨的水岸洋房。

Urban Culture of "Open and Inclusive, Fashionable and Diverse"

Respect the limited historical resources, delimit and protect 30 cultural and historic relics and modern industrial historic relics on both sides of the Haihe River. Form urban waterfront tourism routes with the theme of fishery and salt culture, industrial culture and other leisure facilities combined with waterfront landscape.

Urban Life of "Pleasant Scale and Various Colors"

Combining different locational environments and street orientations, explore multi-scale community building models under various development intensities, and create a multi-dimensional and comfortable living model, including neat and orderly business apartments, cordial and diverse urban courtyards, quiet and chic suburban townhouses and elegant and warm waterfront houses.

项目概况

天津的城市空间形态从 600 年前设卫开始的老城厢，到逐水而兴的海河沿岸租界地，再到集中、延展、扩张的中心城区，城市一直围绕着功能性需求的商业商贸中心在发展演变，但中心城区的城市空间一直缺乏具有标志性和凝聚力的城市中心。

天津文化中心于2008年启动建设，并于2012年竣工投入使用。规划充分强调人民性、公共性，以大剧院、博物馆、美术馆、图书馆、科技馆、阳光乐园（青少年活动中心）等文化设施为主，辅以购物中心、银河广场、四线地铁交会枢纽等商业、休闲、交通功能，成为具有标志性和凝聚力的功能多元混合、繁华开放、舒适的城市文化活动中心。天津文化中心，以其新的城市中心形态，在空间上承接老中心的公共资源，同时新老中心在文化层面相互渗透影响。新的文化中心与现有的行政中心、接待中心，形成三角之势，共同构建中心城区的城市中心，承担综合职能，完善城市文化服务功能，打破了千百年来以行政中心为核心的常规布局方法，形成以文化为主导的中心城区，这不仅仅是对城市文脉的延续，更是对其进行了发展和新时代的解读。

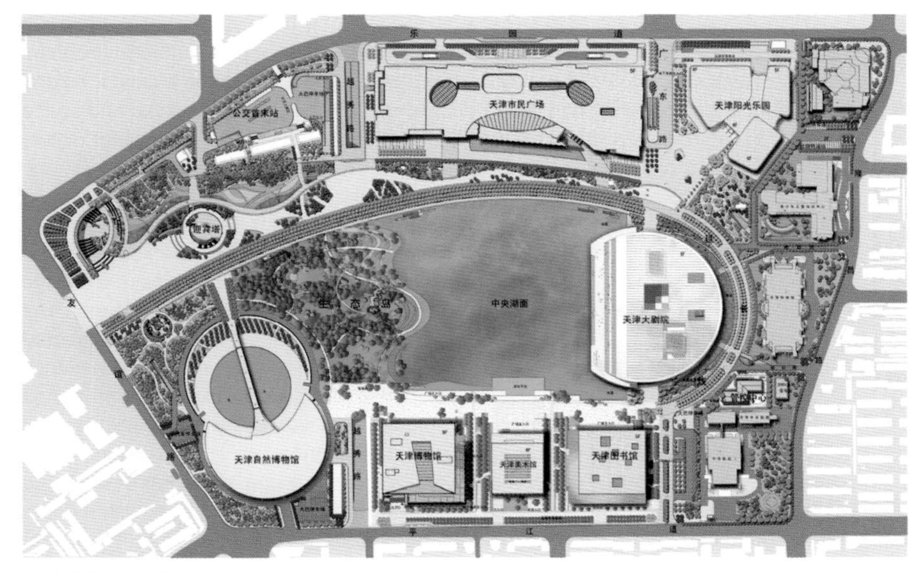

天津文化中心总平面图
General Plan of Tianjin Cultural Center

Project Overview

The urban spatial form of Tianjin has evolved from the old city seat set up 600 years ago to the waterfront concession of the Haihe River, to the central city with concentrated expansion. The city has been developing around the commercial and commercial center with functional needs. However, the urban space in the downtown area has always lacked a landmark and cohesive city center.

The Tianjin Cultural Center started construction in 2008 and was completed and put into use in 2012. It is mainly composed of cultural facilities such as the grand theatre, museum, art gallery, library, science and technology museum, and the Sunshine Paradise (Youth Activity Center), supplemented by shopping centers, Galaxy Plaza, and four subway lines interchange hub. The commercial and leisure transportation function has become a symbolic and cohesive function of multi-mix, prosperous and open and comfortable urban cultural activity center. The Tianjin Cultural Center, in the form of its new urban center, undertakes the public resources of the old center in space, while the new and old centers are infiltrated at the cultural level. The new cultural center is connected with the existing administrative center and reception center, forming a triangular trend, jointly constructing the urban center of the central city, undertaking comprehensive functions, improving the urban cultural service function, and breaking the administrative center as the core for thousands of years. The conventional layout method forms a culturally-oriented central city. This is not only a continuation of the urban context, but also an interpretation of its development and new era.

和谐有序的空间序列

为了形成和谐有序的空间体验，切实有效地把控整体空间形态和品质，城市设计确定了建筑组群中每个建筑的位置、高度、界面、主次关系、新旧关系以及空间处理要求。自然博物馆为碟状造型，从地面缓缓升起，而同样为圆形的天津大剧院则“悬浮”在大地景观之上。大剧院作为主体建筑，设于中心湖东岸，作为自然景观轴线底景，其漂浮于空中的半圆形与自然博物馆融入大地的半月形，形成了“天”与“地”的新旧对话关系：自然博物馆植根于大地，而大剧院则与空灵的天际相连。中心湖南岸文化建筑与北岸购物中心、阳光乐园的形体，限定于 100 米进深、30 米限高的基地之内，对外侧沿街界面提出严格的贴线要求，对内侧沿湖界面则要求相对灵活，并且要求将主入口设于沿湖界面，围绕中心湖营造积极的开放空间，突出大剧院的主体地位。

天津文化中心鸟瞰
Bird's-Eye View of Tianjin Cultural Center

Harmonious and Orderly Spatial Sequence

In order to form a harmonious and orderly space experience and effectively control the overall space form and quality, urban design determines the location, height, interface, primary and secondary relationship, old and new relationship and space processing requirements of each building in the building group: the Natural History Museum is a dish shape, rising slowly from the ground, and the same. For the round Tianjin Grand Theater, it is "suspended" above the landscape of the earth. As the main building, the Grand Theatre is located on the east bank of the central lake and the background of the natural landscape axis. The semi-circle floating in the sky and the half-moon of the present state of the natural museum penetrate into the earth, forming a new and old dialogue relationship between "sky" and "earth". The natural history museum is rooted in the earth, while the Grand Theatre is connected with the empty sky. In order to form a harmonious and orderly space experience and effectively control the overall space form and quality, the overall urban design determines the location, shape, height, primary and secondary relationship of each building in the cultural building group and the space processing requirements among the buildings. The form of cultural buildings along south bank, shopping malls on the north shore and Sunshine Paradise is limited to the base with a depth of 100 meters and a height of 30 meters. Strict requirements are put forward for the outer street interface, while relatively flexible for the inner lake interface, and the main entrance is required to be located at the lake interface, so as to create a positive environment around the central lake. Open space, highlighting the main position of the Grand Theatre.

和而不同的建筑风貌

天津文化中心建筑群本身既包含新建的大剧院、博物馆、美术馆、图书馆、银河购物中心（现万象城）、阳光乐园六座新建筑，同时又有区域内已有的中华剧院、天津自然博物馆。在注重整体空间组织的同时，天津文化中心的总体统筹加强了对使用者观感体验影响力最强的建筑色彩、风格、形式的统筹，从而在统一协调的体量风格要求下，形成了特色突出、个性十足的文化建筑外观，探索了强调时代感和地域特性的建筑特色——简洁洗练、沉稳庄重、新而不怪的建筑形式，完整、明晰的空间构成，适宜的结构体系，精心雕琢的细部，实现了整体与个性的平衡。城市设计要求各单体建筑外墙材料以石材为主，通过不同肌理、不同材质的搭配产生细微对比，形成文化底蕴十足的外观感受。

天津文化中心主要建筑
Main Architectures of Tianjin Cultural Center

Harmonious and Different Architectural Styles

The Tianjin Cultural Center complex itself includes six new buildings, including the new Grand Theatre, Museum, Art Gallery, Library, Galaxy Shopping Center (now The Mixc) and Sunshine Paradise. At the same time, the Chinese Theatre and Tianjin Museum of Natural History have been built in the region. While paying attention to the overall spatial organization, the overall planning of Tianjin Cultural Center strengthens the overall planning of the most prominent architectural colors, styles and forms for users' perception experience, thus forming the external feelings of the cultural buildings with outstanding characteristics and full personality under the requirements of the unified and coordinated volume style, and exploring the emphasizing time. Architectural features of contemporary and regional characteristics - concise, sober, solemn, new and not strange architectural forms, complete and clear space composition, appropriate structural system, meticulously carved details, to achieve the balance between the whole and personality. The urban design requires that the external wall materials of individual buildings should be mainly made of stone materials, which can produce slight contrast through the matching of different textures and materials to form a full cultural sense of appearance.

永续发展的可持续设计、统建共管的能源系统

因地制宜运用节能生态技术，实践可持续发展理想。在能源利用方面，为了降低投资、节能减排，依据不同业态特征、管理权属，文化中心区域集中设置了三处能源站。通过集中建设、集中管理，降低了能源系统的初期投入与运行成本，采用可再生能源技术提供冷热源，实现了节能降耗的目标。同时通过减少非空气途径排热，最大限度减少冷却塔或风冷室外机的数量，降低了对环境的压力，美化了地面景观。

生态调蓄的雨水系统

为了节约利用水资源，规划建设了生态水系统。通过雨水收集调蓄系统，每年可利用 9 万立方米雨水补给中心湖。中心湖水容量 16 万立方米，在生态岛南北两侧设有约 2 600 立方米的生态净化群落，通过物理过滤、滤料基质吸附、水生植物根区分解吸收等手段净化湖水，将水质控制在三类以上标准，实现了水资源循环利用。

Sustainable Design for Sustainable Development Building a Comanaged Energy System

Suitable for local conditions, energy-saving ecological technology, practice the ideal of sustainable development. In terms of energy utilization, in order to reduce investment, energy saving and emission reduction, three energy stations have been centralized in the cultural center area according to different business characteristics and management ownership. Through centralized construction and management, the initial investment and operation cost of energy system are reduced, and renewable energy technology is used to provide cold and heat sources, thus achieving the goal of energy saving and consumption reduction. At the same time, the number of cooling towers or air-cooled outdoor units is minimized by reducing non-air heat exhaust, which reduces the pressure on the environment and beautifies the ground landscape.

Rainwater System for Ecological Regulation and Storage

In order to save and utilize water resources, an ecological water system has been planned and constructed. Through the rainwater collection and storage system, 90,000 cubic meters of rainwater can be used to recharge the central lake every year. The water capacity of the central lake is 160,000 cubic meters. There are about 260 cubic meters ecological purification communities on the north and south sides of the ecological island. The lake water is purified by means of physical filtration, adsorption of filter media and decomposition and absorption of aquatic plant roots. The water quality is controlled by three or more standards, and the water resources are recycled.

西区能源站
West District Energy Station

地下车库采用太阳能
Solar Energy Used in Underground Garage

高度复合的交通系统

交通、市政、地下空间、防灾减灾等专项规划合并开展了综合研究。为缓解地面交通压力，规划形成了便捷通畅的立体交通网络，创造安全愉悦的地下空间，营造良好的地面景观，为充分感受自然、最大限度地节约能源，通过设计将自然光线引入文化建筑与地下的公共空间，从而实现公共交通、市政管网、地下空间、地面景观、室内外空间五位一体高度整合的系统工程。

Highly Composite Traffic System

Special plans for transportation, municipal administration, underground space and disaster prevention and mitigation have been combined to carry out comprehensive research. In order to alleviate the pressure of ground traffic, a convenient and smooth three-dimensional traffic network has been planned, safe and pleasant underground space has been created, good ground landscape has been created, and natural light has been introduced into cultural buildings and underground public space through design to fully feel nature and save energy to the maximum extent, thus realizing public traffic. Municipal pipe network, underground space, ground landscape and indoor and outdoor space are highly integrated systems engineering.

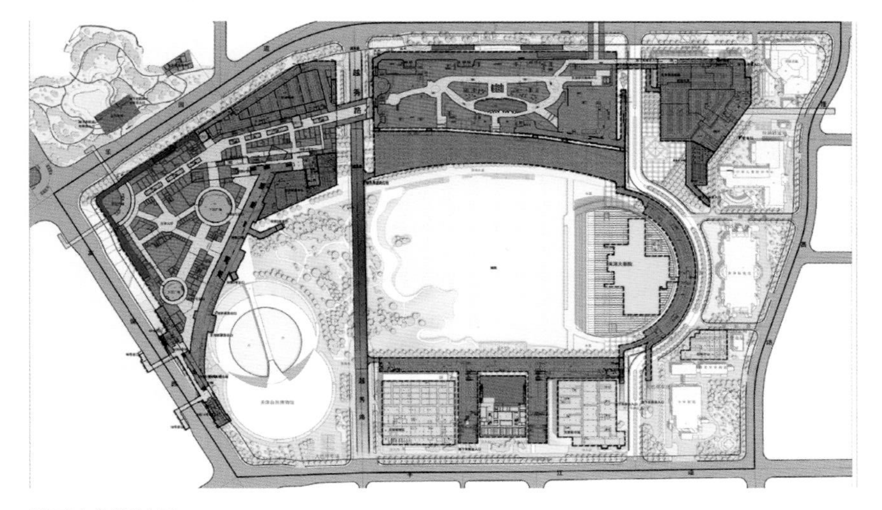

地下空间设计图
Underground Space Design Drawings

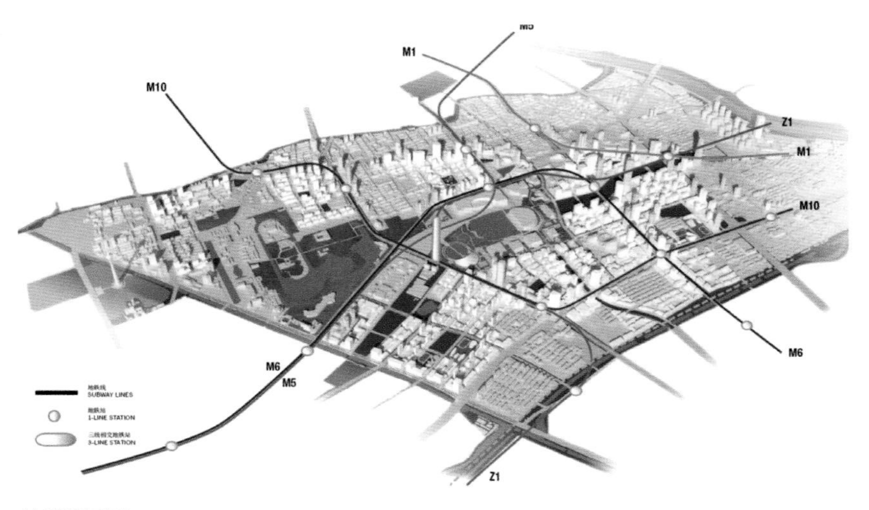

轨道线网图
Track Network Chart

五位一体的建设实施统筹

文化中心设计工作繁多复杂，40 余家不同国度与专业背景的团队参与其中。面对各种声音，规划设计组在“设计—建造”全程中，发挥城市设计的统筹与指引作用，正视争论、尊重个性、主动协调、寻求共识，汇聚众多智慧，确保一张蓝图落实到底。在实施阶段，通过规划师审查巡查、建筑师责任制、设计联席会、设计例会、论证会、施工现场协调会，构建全专业协同工作平台，基于城市设计导则，结合各专业的日常探讨与决策，在四十余家设计单位之间，实现规划师、建筑师、景观师、工程师、艺术家协同工作，在更广、更深层面，强化完善城市设计总体目标。

Five In One - The Overall Planning of Construction and Implementation

The design of cultural centers is complex, involving more than 40 teams from different countries and professional backgrounds. Facing all kinds of voices, the planning and design team plays an overall and guiding role in the whole process of "design construction", facing up to disputes, respecting individuality, actively coordinating, seeking consensus, gathering a lot of wisdom, and ensuring the implementation of a blueprint to the end. In the implementation stage, through planners' inspection and inspection, architects' responsibility system, joint design meetings, design meetings, demonstration meetings and construction site coordination meetings, a professional collaborative work platform is constructed. Based on urban design guidelines and combined with daily discussion and decision-making of various majors, planners can be realized among more than 40 design units. Architects, landscape architects, engineers and artists work together to strengthen and improve the overall goal of urban design in a broader and deeper level.

大剧院

博物馆

美术馆

图书馆

阳光乐园

银河购物中心（现为万象城购物中心）

大剧院音乐厅

博物馆内景

美术馆内景

图书馆内景

阳光乐园内景

银河购物中心（现为万象城购物中心）内景

天津文化中心建筑
Architectures of Tianjin Cultural Center

项目概况

2008 年，为适应天津经济高速增长的态势，促进城市文化繁荣发展，天津市委市政府决定在中心城区开发建设天津文化中心。

文化中心将与小白楼商业商务中心区合二为一，共同组成一个更加强大的世界级城市中心，成为完善城市功能、提升城市面貌、弘扬城市文化、促进城市发展的强有力城市“心脏”。

Project Overview

In 2008, in order to adapt to the rapid economic growth in Tianjin and promote the prosperous development of urban culture, Tianjin Municipal Committee and Municipal Government made the decision to develop and construct Tianjin Cultural Center in the central urban area.

The cultural center will be combined with the Xiaobailou commercial and business center to form a more powerful world-class city center, which will become a powerful "heart" of the city to improve its functions, enhance its appearance, promote its culture and encourage its development.

规划天际线
Skyline Planning

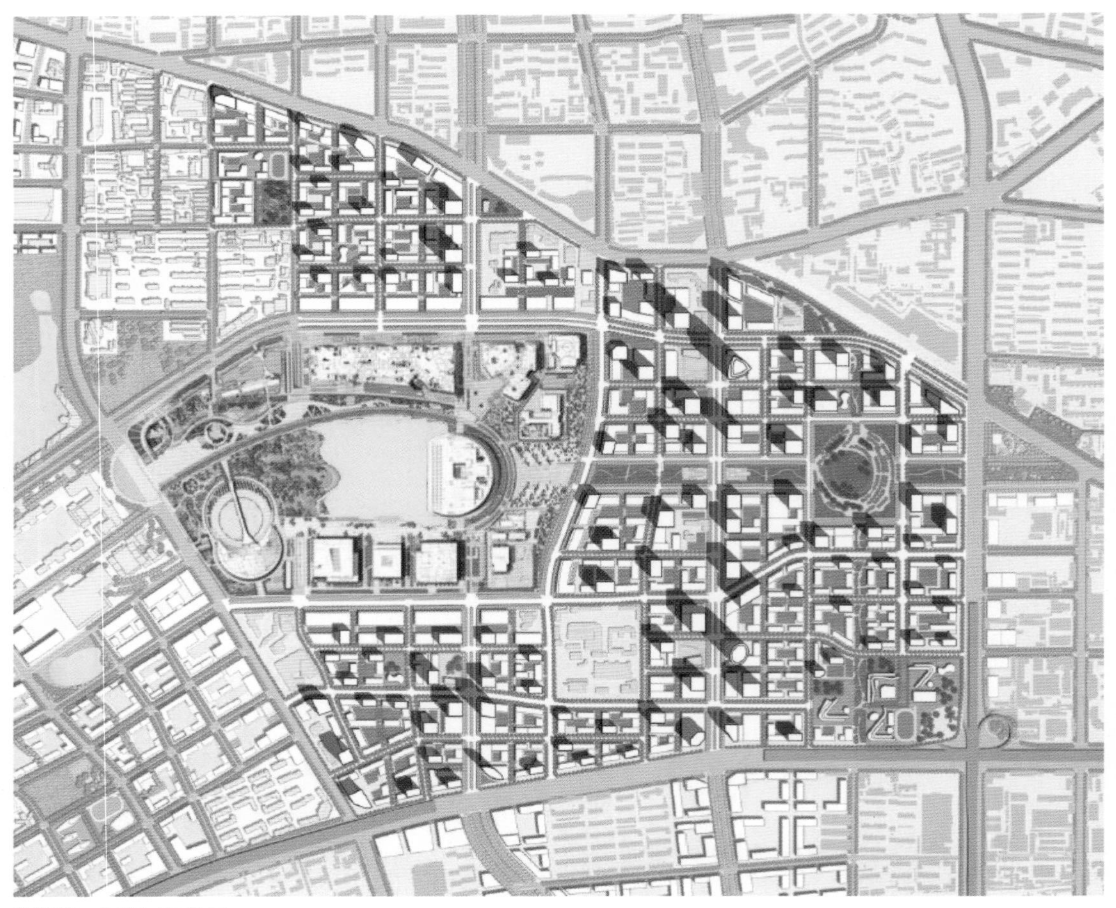

窄街廓、密路网的格局
Narrow Road and Dense Road Network Pattern

亲切舒适的城市肌理，创造理想有活力的“天津尺度”

天津的城市肌理保留了舒适的、人性的街道尺度与建筑景观，如狭长紧凑的林荫道以及亲切的中低层建筑等。在对八大里文化商务核心区的规划中，以长 130 米 × 宽 100 米的尺度为基本的街区单元，以规划的 4 ~ 8 层的建筑群围合地块，为较高的塔楼和标志性建筑预留空间。通过绿轴、南北广场建筑层次的组织，塑造出理想的“天津尺度”。小街区、密路网，增加商业地块的面街机会，更加适于步行，并为机动车交通提供更多的可选择路径，提升公共服务设施的易达性，创造出积极、健康、充满活力的城市生活。

Friendly and Comfortable Urban Texture, Creating an Ideal and Active "Tianjin Scale"

Urban fabric in Tianjin retains comfortable, humane street scales and architectural landscapes, such as narrow and compact avenues and intimate middle and low-rise buildings. In the planning of the cultural and business core area of Badali, the scale of the basic block unit is 100X130 meters. The planned 4-8 storey buildings surrounded the land reserve space for tall towers and landmark buildings. Through the organization of the green axis and the architecture of the North and South Squares, the ideal "Tianjin scale" is created. Small blocks and dense road networks, which increase the chance of commercial street plots, are more suitable for walking, and provide more alternative routes for motor vehicle traffic, improve the accessibility of public service facilities, and create a positive, healthy and vibrant urban life.

整体鸟瞰图
Overall Aerial View

高强度和多功能的规划布局，为经济和社会活动提供最根本的支撑

有活力的城市意味着有多样化的活动、人群在公共区域交汇，并且有机会对话、分享与聚会。八大里文化商务核心区由开放空间和城市道路自然划分为四个功能复合而联系紧密的区域。主要的商业开发在沿路、地铁站周围及主要开放空间发展，商业建筑围绕着绿轴和公园布置，独特的邻里居住区域分布在西北、西南及东南。每一个区域均具有混合开发的功能，在紧凑的空间内各类活动同时或交替发生着，营造出不同的活力氛围。

High-intensity and Versatile Planning Layout That Provides the Most Fundamental Support for Economic and Social Activities

A vibrant city means a diverse range of activities, where people meet in public areas, and have the opportunity to talk, share and gather. The core area of the cultural and business center of Badali is divided into four functionally combined and closely related areas by open space and urban roads. The main commercial developments are along the road, around the subway station and in the main open space. The active commercial buildings are arranged around the green axis and park, and the unique residential neighborhoods are located in the northwest, southwest and southeast of the region. Each area has a hybrid development function, and various activities occur simultaneously or alternately in a compact space to create a different vibrant atmosphere.

连续、开放、有吸引力的街道界面

长久以来，街道作为一种重要途径，对组织社会生活发挥着不可替代的作用。街道活力因交通功能而产生，并以这持续不断的交通流而保持长久，但最终为交通所累：现代主义城市规划片面强化街道的交通功能，却将其界面功能弃之不顾，使社会活动丧失了发生的载体，城市街道因此变得单调乏味。在文化中心周边地区城市设计中，我们为每一条街道赋予准确的定位——兼顾交通与界面属性，制定明确的设计控制条件，以创造出连续、开放有吸引力的街道界面。

Continuous, Open and Attractive Street Interface

For a long time, the streets have played an irreplaceable role in organizing social. The vitality of the street is caused by the traffic function, and lasts for a long time with continuous traffic flow. However, it is finally plagued by traffic: Modernist urban planning unilaterally strengthens the traffic function of the street, but disregards its interface function and loses social activities. As a result, the city streets have become tedious. In the urban design of the surrounding areas of the Cultural Center, we gave each street an accurate positioning, taking into account the traffic and interface attributes, and establishing clear design control conditions to create a continuous, open and attractive street interface.

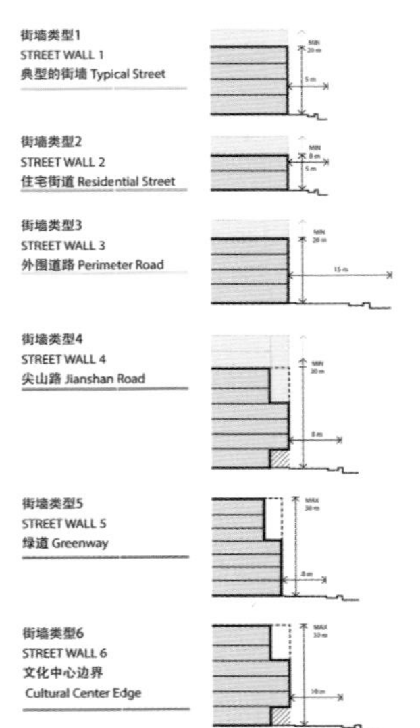

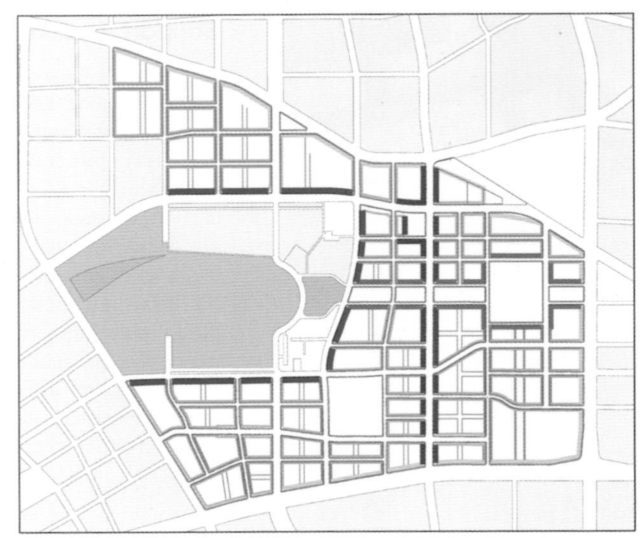

街墙类型
Types of Street Walls

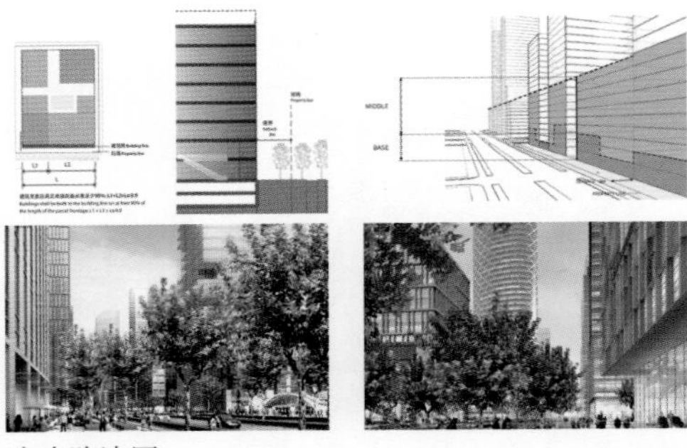

尖山路边界
Boundary of Jianshan Road

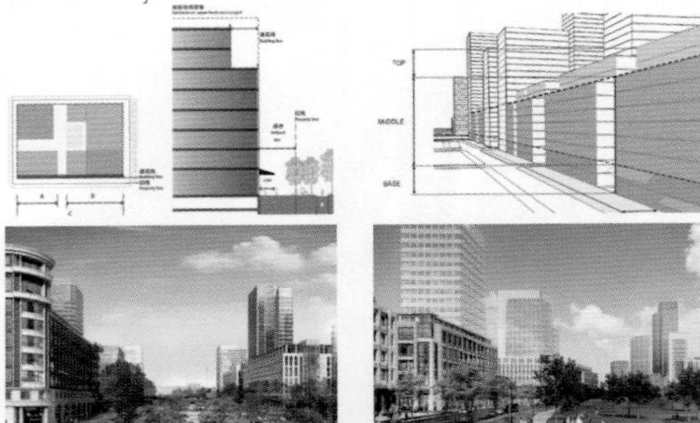

绿轴两侧边界
Boundary of Green Axis

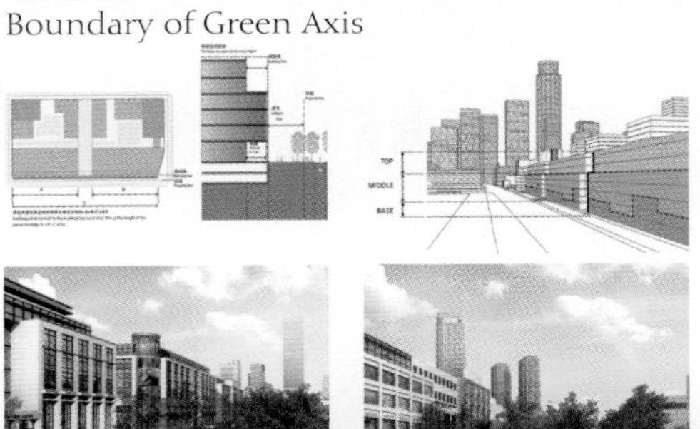

文化中心边界
Boundary of Cultural Center

通过恰当的环境设计，营造活力四射的中央绿轴

城市绿轴在当代社会由权力空间向市民空间的转化，使其成为一个指向明确的公共活动场所。将其由传统的“看”的空间发展为可进入的“用”的空间是设计主要努力的方向。文化中心周边地区的中央绿轴西端起始于天津大剧院，穿越尖山路，串联中央公园，止于东端的科技馆（规划），全长1.4千米。绿轴两侧为较窄的单行道，路外为商务办公建筑。它可以有效地沟通重要公共建筑、场所、轨道站点与周边地块，在未来将会有极大的使用率。我们欲将其打造为一条可进入的、舒适的艺术科学走廊，并结合周边功能进行分段设计，使其成为文化中心周边地区的活力中心。

文化公园 Cultural Park

下沉广场 Sunken Plaza

Create a Vibrant Central Green Axis With the Right Environment Design

The transformation of the urban green axis from the power space to the civic space in contemporary society has made it a clear public activity venue. The development of the traditional "look" space into the accessible "use" space is the main direction of design. The western end of the central green axis in the surrounding area of the Cultural Center starts at the Tianjin Grand Theatre, crosses Jianshan Road, connects to Central Park, and stops at the eastern end of the Science and Technology Museum (planning), with a total length of 1.4 kilometers. The two sides of the green axis are narrow one-way streets, and the roads are commercial office buildings. It can effectively communicate important public buildings, places, orbital sites and surrounding plots, and will have a great use in the future. We have built it into an accessible, comfortable art science corridor that is segmented with peripheral features to make it a vibrant center.

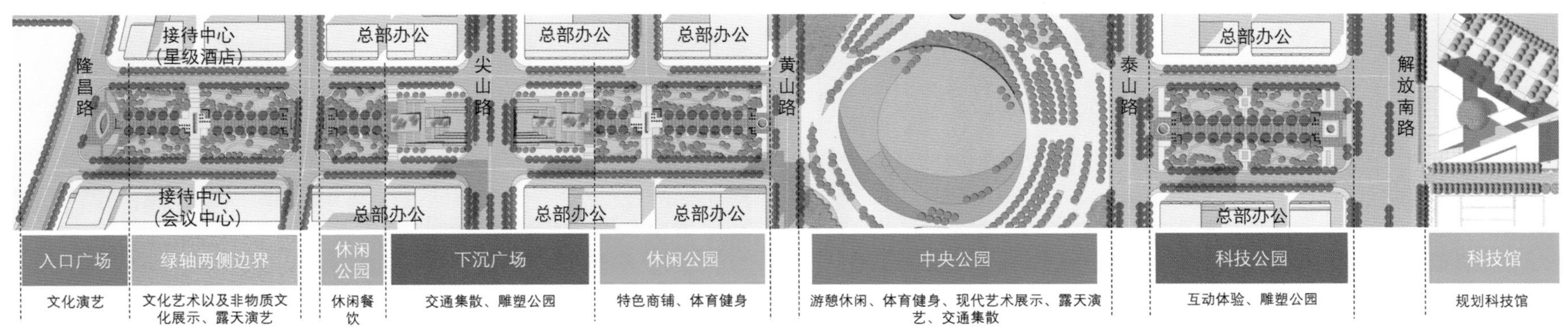

中央绿轴总平面 General Plan of Central Green Axis

建立系统性、全天候的步行空间网络

本地区的步行区以尖山路和中央绿轴为中心，形成一个相互连通、完善的系统。同时，通过以地铁公共交通节点、地下商业街区为基点的步行者空间和网络建设，打造“能够步行”的地下空间。地上、地下空间互补，并与骑楼系统共同打造全天候的步行空间网络。

Establish a Systematic, All-day Pedestrian Network

The pedestrian zone in the area is centered on Jianshan Road and the central green axis, forming an interconnected and perfect system. At the same time, through the subway public transportation node, underground commercial block as the base point of the pedestrian space and network construction, to create a "walkable" underground space. The ground and underground spaces are complementary, and together with the arcade system, an all-day pedestrian space network is built.

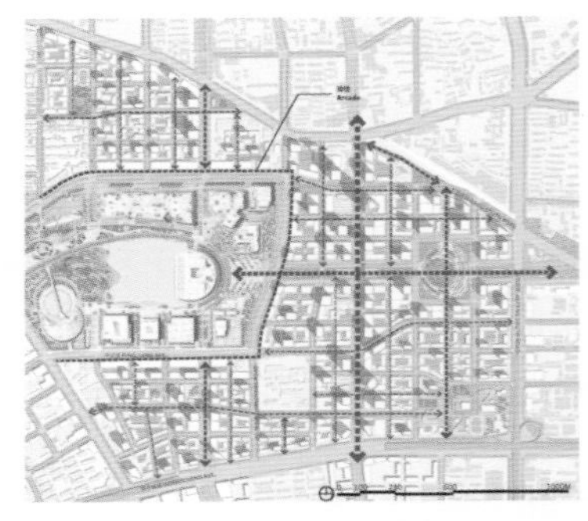

地面步行空间系统
Walking System on the Ground

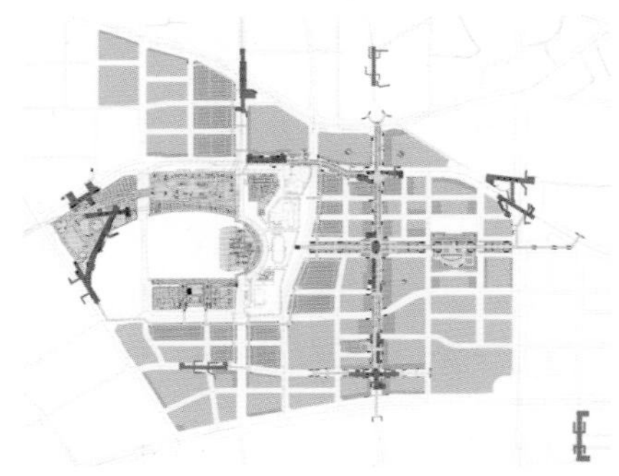

地下步行空间及商业
Walking and Commercial Space Underground

舒适的地面人行道
Comfortable Pedestrian

便捷的步行转换空间
Convenient Walking Conversion Space

通过恰当的、高品质的设计实现城市活力向地区纵深渗透

尖山路南北两端为城市广场，地下为地铁站，周边为本地区内开发强度最高的地块，因此具有极强的交通集散作用。进行恰当的平面功能布局可以对人流进行合理引导，使其成为激发周边地区城市活力的重要元素。同时地区内的普通街道同样强调步行环境的塑造，包括宽阔舒适的步行道，紧贴街道、拥有较高通透度的建筑首层界面，其将成为承载频繁的日常活动的平台。此外在由建筑紧密围合的地块内部强调首层建筑的公共性与开放性，让步行通道与城市街道连通，使城市活力向地块内部渗透。

Deep Penetration of Urban Vitality to the Region Through Appropriate and High-quality Design

The city square at the north and south ends of Jianshan Road is a subway station. The surrounding area is the highest strength development site in the area. Therefore, it has a strong traffic distribution function. The proper plane function layout can reasonably guide the flow of people and make it become an important element that inspires the vitality of the surrounding cities. At the same time, the ordinary streets in the area also emphasize the shaping of the walking environment, including wide and comfortable walking paths, close to the street, and the first floor of the building with high transparency, making it a platform for carrying frequent daily activities. In addition, the publicity and openness of the first-floor building will be emphasized within the tightly enclosed plots, and the pedestrian passages will be connected with the urban streets to make the city's vitality penetrate into the interior of the plot.

地铁出入口广场
Subway Entrance Plaza

步行环境
Walking Environment

规划背景

在公共文化服务的“1.0 时代”，文化“事业”型体制，形成各自为政的宏大空间和封闭单一的供给模式。功能单一、尺度惊人、运营低效，难以满足人民日益增长的精神文化需求。

近年来，滨海新区凭借先进制造业的优势，开发开放取得了显著的成绩。但公共文化资源仍严重不足，市民文化生活匮乏，极大地削弱了滨海新区的城市活力和吸引力。在反思传统模式的基础上，滨海文化中心以为市民营造美好文化生活为立足点，以人性尺度的文化长廊串联文化建筑形成文化综合体，开创了公共文化服务“2.0 时代”的规划设计、建设运营的全新模式。

Planning Background

In the 1.0 era of public cultural service, the cultural "enterprise" system formed a huge space with independent administration and a closed and single supply mode. It is difficult to meet the growing spiritual and cultural needs of the people due to its single function, super scale and inefficient operation.

In recent years, Binhai New Area has made remarkable achievements in its development and opening up with the advantage of advanced manufacturing industry. However, the public cultural resources are still seriously insufficient and the cultural life of the citizens is scarce, which greatly weakens the vitality and attraction of Binhai New Area. On the basis of reflecting on the traditional mode, Binhai Cultural Center aims to create a better cultural life for the citizens, and combines cultural buildings with cultural corridors of human scale to form a cultural complex, creating a new mode of planning, design, construction and operation in the era of public cultural service 2.0.

滨海新区文化中心效果图
Overall Aerial View of Binhai Cultural Center

基地现状
Current Situation of the Site

设计方案总平面图
Masterplan of Urban Design

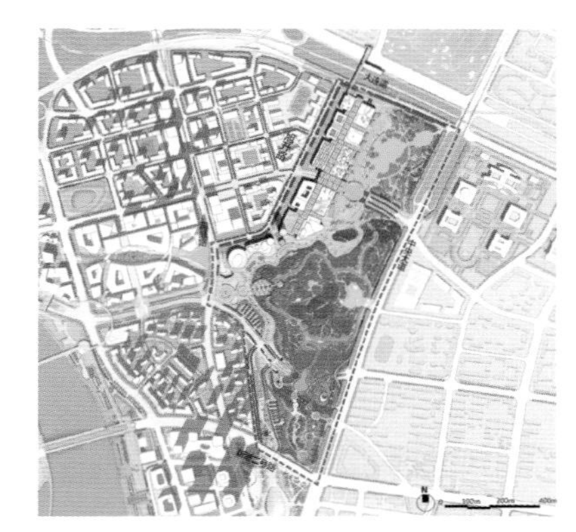

实施方案总平面图
Masterplan of Implementation

项目概况

基地为原永利碱厂旧址，该厂创办于 1914 年，是中国乃至世界制碱工业的先导，奠定了中国民族化学工业的基础。从“老碱厂”到“文化中心”，它代表了滨海新区百年工业记忆的传承延续。

滨海文化中心位于滨海新区核心区，在汇聚城市核心功能的中央大道发展轴上，是连接于家堡金融区、响螺湾商务区、天碱商业区、开发区生活区和塘沽中心区等最重要地区的枢纽。规划总用地面积 90 公顷，总建筑规模 59 万平方米，其中地上建筑面积 36.3 万平方米，地下建筑面积 22.7 万平方米。建设内容由文化长廊串联现代城市与工业探索馆、图书馆、美术馆、市民活动中心、演艺中心、艺术创意创业等的文化综合体和中央公园共同组成，将逐步实现滨海新区从核心文化功能的完善到国际化文化视野的拓展升级。

Project overview

Founded in 1914, the base is the former site of Yongli Alkali Factory, which is the forerunner of alkali industry in China and even the world, and lays the foundation of Chinese national chemical industry. From the "old alkali factory" to the "cultural center", it represents the continuation of Binhai New Area's century-old industrial memory.

Binhai cultural center is located in the core area of Binhai New Area, on the development axis of Central Avenue, which gathers the core functions of the city. It is a hub connecting the most important urban areas, such as Yujiapu Financial District, Xiangluowan Business District, Tianjian Business District, development zone living area and Tanggu central district. The total planned land area is 90 hectares, and the total construction scale is 590,000 square meters, including 363,000 square meters of above-ground construction area and 227,000 square meters of underground construction area. The construction content consists of the cultural complex and the central park that connects the cultural corridor with the modern city and industry exploration hall, library, art gallery, civic activity center, performing arts center, artistic creativity and entrepreneurship, etc., and will gradually realize the improvement of Binhai New District from the core cultural function to the expansion and upgrading of international cultural vision.

第一次国际方案征集
The First International Proposal Solicitation

第二次国际方案征集
The Second International Proposal Solicitation

以个体项目演变，推动城市整体公共效益最大化

自 2009 年确定选址以来，滨海文化中心经历了 7 年规划设计历程，40 多轮方案比选，进行了 2 次建筑国际方案征集。这不仅是对个体项目的研究探讨，更是对滨海新区核心区整体城市空间成长变化中循序渐进的思考。最终为核心区这块“寸金”之地，寻找到了最有价值的归宿，它对中央大道城市发展主轴线的形成发挥了关键性作用。在这场城市的博弈中，它协调了市民、政府、开发商等多方立场，在保证经济效益的同时，发挥最大的公共效益，还城市于市民。

The Evolution of Individual Projects Promotes the Maximization of the Overall Public Benefits of the City

Since the site selection was determined in 2009, Binhai Cultural Center has gone through seven years of planning and design process, more than 40 rounds of scheme comparison and selection, and guided two international architectural scheme solicitation. This is not only the research and discussion of individual projects, but also the gradual thinking of the overall urban space growth and change in the core area of Binhai New Area. Finally, it has found the most valuable destination for the core area, which plays a key role in the formation of the urban development main axis of Central Avenue. In this urban game, it balances the interests of the citizens, the government, developers and many other needs, while ensuring economic benefits, give full play to the public benefits, and return the city to the people.

单侧无围合

规整围合　中央围合　流线围合

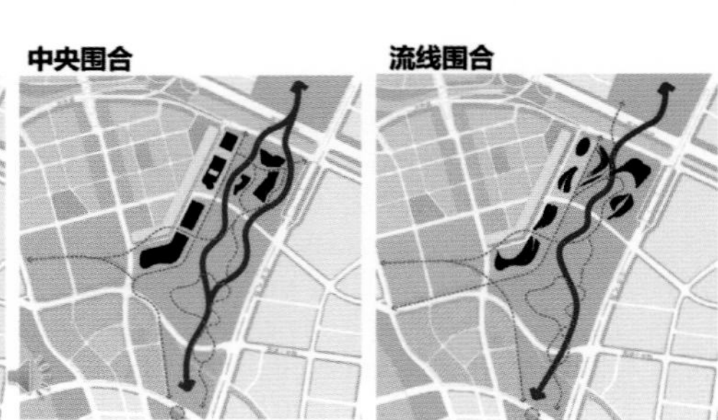

中央围合

街道

长廊

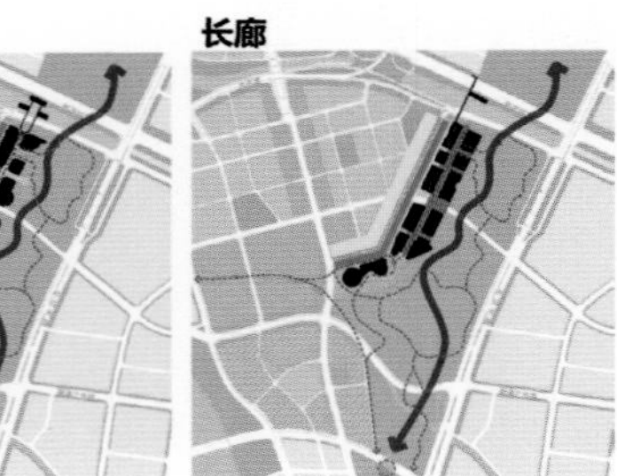

布局演变示意图
Evolution of Space

中央公园效果图
Perspectives of Central Park

尊重自然，塑造绿色城市

规划尊重自然本底，整合零散绿地，构建连通于家堡交通枢纽和海河绿色景观网络，形成长 2 700 米，宽 300 米，贯穿南北的中央公园，拥有 800 米景观展示带。

通过重构建筑群、绿地、海河的空间格局，为城市在最核心区保留了一块珍贵永恒的绿色资产，促进滨海新区核心区“三生空间”的融合。

通过交通干道的局部下穿，最大限度保持公园景观的完整连续，并打通与周边城市地标景观的视线通廊，提供最佳的城市观景平台。

Respect Nature and Create Green Urban Places

The plan respects the natural background, integrates scattered green space, and builds a green landscape network connecting Yujiapu transportation hub and Haihe Riverr, forming a central park with a length of 2700 meters and a width of 300 meters, running through the north and south, with a landscape display surface of 800 meters.

By reconstructing the spatial pattern of buildings, green space and Haihe River, the city retains a precious and eternal green asset in the most core area, and promotes the integration of three living Spaces in the core area of Binhai New Area.

Through the partial undercutting of the traffic artery, the park landscape can be kept intact and continuous to the maximum extent, and the landscape sight corridor with the surrounding city landmarks can be opened up, providing the best urban viewing platform.

以人为本，营造连续友好的步行环境

规划以文化长廊衔接区域，营造从都市到自然的 24 小时无阻断的步行连接，极大地改善了核心区步行网络和步行环境。

外部，文化综合体完全融入自然，与公园可直接通过步行相连，充分互动。内部，形成人性化尺度的文化长廊，以“伞”为主题形成文化综合体的灵魂。长廊的半室外空间更适应滨海地区冬寒、大风的气候特点，通过可持续设计，实现自然采光最大化，被动式低能耗和宜人的风环境，为市民提供四季皆宜的舒适体验。长廊与两侧建筑充分互动，如无数著名的街道那样，去承载市民最丰富的生活。

同时，依托“六线七站”的轨道网络，与轨道站点无缝衔接，创造了高效便捷的地下步行网络。

People-oriented and Create a Continuous and Friendly Walking Environment

Planning to connect the area with the cultural corridor, create a 24-hour non-blocking pedestrian connection from the city to the nature, and greatly improve the walking network and environment of the core area.

Externally, the cultural complex is fully integrated with nature, connected with the park on foot and fully interactive. Inside, it forms a cultural corridor with human scale, and takes "umbrella" as the motif to form the soul of the cultural complex. The semi outdoor space of the promenade is more suitable for the cold and windy climate of the coastal area. Through sustainable design, it maximizes the natural lighting, passive low energy consumption and pleasant wind environment, providing the residents with a comfortable experience suitable for all seasons. The promenade fully interacts with the buildings on both sides, like countless great streets, to carry the richest life of citizens.

Meanwhile, relying on the "six lines and seven stations" rail system, it seamlessly connects with rail stations and realizes an efficient and convenient underground walking network.

文化长廊分段效果图
Section of Cultural Corridor Renderings

以市场化为导向，创新公共文化服务模式

以文化长廊为核心的文化综合体模式，突破了既有体制壁垒和技术难关，实现了三个层面的创新。

以机制改革为先导，在规划之初即与市场对接，一方面整合协调近 10 个政府部门，同时引入市场力量，充分匹配需求，界定权责，形成文化事业与文化产业联动的产业网络，以公共文化服务带动文化消费。

以制度设计为支撑，在国家提倡 PPP 模式（政府和社会资本合作）之前，超前地将“政府购买公共服务”模式首次应用于公共文化设施的投资建设上，政府以土地入股与社会资本共同形成混合投资主体，进行建设、运营、管理。同时设计相应的土地供应、规划审批、管理边界划分等配套制度。

以空间模式创新为载体，在长廊的串联下，多元业态在三维空间深度融合，塑造出一个开放互动、亲民友善，互融共生的社交场所，既和谐统一，又富多样性。

Market-Oriented Innovation of Public Cultural Service Model

Urban design puts forward a cultural complex model with cultural corridor as the core. From concept to deepening implementable plan, it constantly breaks through institutional barriers and technical difficulties, and achieves three levels of innovation.

Taking the mechanism reform as the guide, we connected with the market at the beginning of the planning. On the one hand, we integrated and coordinated nearly 10 government departments, at the same time, we introduced market forces to fully match demand, define rights and responsibilities, and form an industrial network linked by cultural undertakings and cultural industries, so as to promote cultural consumption through public cultural services.

Supported by system design and before the Public-Private Partnership (PPP) model was advocated by the state, the advanced model of "government purchasing public services" was applied to the investment and construction of public cultural facilities for the first time. The government jointly formed a mixed investment subject with social capital through holding land shares to carry out construction, operation and management. At the same time, the corresponding supporting systems of land supply, planning approval and boundary division management are designed.

Based on the innovation of space mode, connected by the cultural corridor, diversified business types will be deeply integrated in three-dimensional space to create a harmonious social place which is open, interactive, friendly to the people, and mutually integrated.

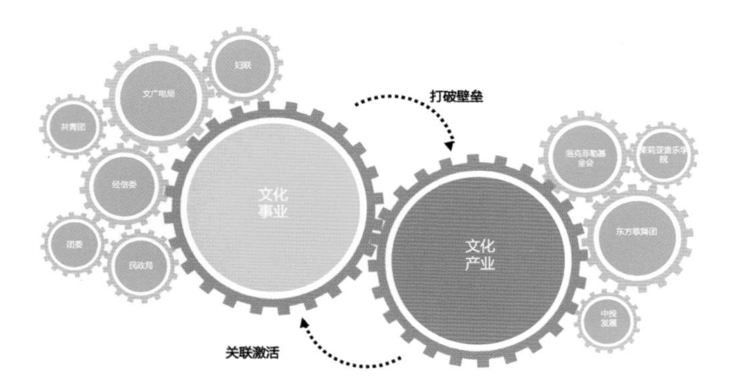

文化综合体模式 Cultural Complex Model

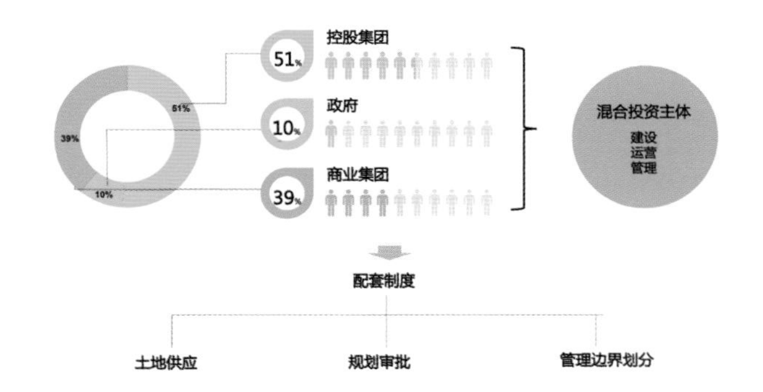

创新公共文化服务模式 Innovation of Public Cultural Service Model

高水准的集成化设计，确保精准落实

滨海文化中心是涉及建筑、景观、交通、市政、地下空间、策展等的高度集成化项目，城市设计为集成设计提供了统筹多专业的平台，建立了多维度、立体化的思维模式。通过国际工作营、设计例会、专题探讨等机制，协同五个国家、数十家设计单位。在良性互动中，城市设计在建筑群体形态、界面控制、竖向与水平联系、交通衔接等诸多方面建立明确规则，确保精准落实，使滨海新区项目组织管理水平迈向国际化。

High Level of Integrated Design to Ensure Accurate Implementation

Binhai Cultural Center is a highly integrated project involving architecture, landscape, transportation, municipal administration, underground space, curation and so on. Urban design provides a multi-disciplinary platform for integrated design, and establishes a multi-dimensional mode of thinking.

In the positive cross-disciplinary interactions, urban design establishes clear rules on the aspects of form of building groups, display view control, vertical links to the horizontal, traffic connections and many other to ensure accurate implementation.

设计方案效果图
Project Design Rendering

实施方案效果图
Project Implementation Rendering

实施效果

滨海文化中心城市设计，是一次机制创新和技术突破，以拥抱市场的姿态，激发全社会文化创新活力；是一次规划价值观的转变，从对仪式感的偏好回归到对日常生活、人文社会的关注；更是一次复杂的改革实践，以建设高品质城市空间作为公共政策，对人们的生活发挥着积极而深远的影响，对滨海新区迈向国际化现代都市具有里程碑的意义。

Implementation Effect

Binhai Cultural Center is a mechanism innovation and technological breakthrough, which stimulates the whole society's cultural innovation vitality by embracing the market. It is a transformation of planning values, from a preference for ritual to a focus on secular life and humanistic society. It is also a complex reform practice. Taking high quality urban space as the public policy, it exerts a positive and far-reaching influence on people's life and has a milestone significance for Binhai New Area to become an international modern city.

文化长廊
Cultural Corridor

滨海图书馆
Binhai Library

滨海城市与工业探索馆
Binhai City and Industrial Exploration Museum

滨海演艺中心
Binhai Performing Arts Center

滨海美术馆
Binhai Art Museum

实施效果

滨海文化中心于 2017 年正式向公众开放。其中滨海图书馆一经亮相，便成为“网红”，被誉为“中国最美图书馆”，更惊艳了国际规划建筑领域，成为现象级大事件，日接待量超 2 万人次。以文化长廊为核心的城市与工业探索馆、美术馆、演艺中心等全面投入使用。各类艺术展览、文化活动、文创产业精彩纷呈，为市民创造了丰富的文化生活，为滨海乃至整个天津带来了空前的文化影响力和社会效益。

Implementation Effect

The Binhai Cultural Center officially opened to the public in 2017. Binhai Library became "web celebrity" once it was unveiled, and was praised as "the most beautiful library in China". It also amazed the international planning and architecture field, and became a phenomenon event with daily reception of more than 20,000 people. The City and Industry Exploration Museum, Art Gallery and Performing Arts Center with the cultural corridor as the core were fully put into operation. All kinds of art exhibitions, cultural activities and cultural and creative industries are colorful, creating rich cultural life for the citizens and bringing unprecedented cultural influence and social benefits to Binhai New Area and even Tianjin.

于家堡金融区城市设计效果图
Yujiapu Financial District Rendering

规划背景

在《天津市滨海新区中心商务商业区总体规划（2005—2020）》的基础上，滨海新区不断对规划进行深化完善。2005 年确定滨海新区金融创新区选址在于家堡地区。随后进行的国际城市设计工作营和专家咨询研讨会，对于于家堡金融区等滨海新区中心商务区的范围，从总体架构到局部节点，从发展思路到分期实施，从京津城际车站选址到海河下游两岸防洪，都进行了广泛研究。最后，由美国 SOM 建筑设计公司的芝加哥分公司牵头开展了于家堡金融区的城市设计工作。

Planning Background

On the basis of the "General Plan of Tianjin Binhai New Area Central Business District (2005-2020)", Binhai New Area has been continuously deepened and improved. In 2005, the location of Financial Innovation Zone in Binhai New Area was determined to be located in Yujiapu area. Subsequently, the International Urban Design Camp and Expert Consultation Seminar conducted extensive research on the scope of the central business district of Binhai New Area, such as Yujiapu Financial District, from the overall structure to local nodes, from development ideas to phased implementation, from the location of Beijing-Tianjin Intercity Station to flood control on both sides of the lower reaches of the Haihe River. Finally, the Chicago Branch of SOM Architectural Design Company led the urban design work in Yujiapu Financial District.

规划总平面图
Master Plan

项目概况

于家堡金融区总规划地块 120 个，占地 386 万平方米，总建筑面积 950 万平方米。于家堡金融区将建设成为充满活力的多功能金融区，吸引世界级的金融和商务机构进驻。在这个崭新的中央商务区里，围绕着中心交通枢纽，将会是高密度的混合功能的建筑群。四通八达的交通网络和完善的市政设施将会从半岛延伸至周边区域，形成一座 21 世纪的可持续发展的新城，为新一代的职员及居民提供高品质的工作和生活空间。

Project Overview

The total planning sites of Yujiapu Financial Zone is 120, covering 3.86 million square meters, with a total construction area of 9.5 million square meters. Yujiapu Financial Zone will be built into a vibrant multi-functional financial zone, attracting world-class financial and business institutions. In this brand-new CBD, surrounded by the central transportation hub, will be a high density mixed-function building complex. The extensive transportation network and perfect municipal facilities will extend from the peninsula to the surrounding areas. A new sustainable development city in the 21st century will be formed to provide high quality working and living space for the new generation of employees and residents.

核心要点

层次性的街道体系将整个金融区内的通行路线连接起来，其中包括具有特色的中央大道沿南北方向贯穿整个半岛，作为机动车和步行循环的主要通廊，将半岛北面的火车站与最南端直接连通。

特色步行街位于半岛的中心，形成东西走向的主要通廊，作为金融区的主要购物街。该步行街上将云集各种商业零售店，分别位于低层或多层办公楼内。

一个位于中央大道内的新的直线形公园，与沿半岛南北轴布置的场地为市民提供了一个安全、愉快的步行连接。

Core points

The hierarchical street system links the traffic of the whole financial district, including the distinctive Central Avenue running through the whole peninsula along the north-south direction. It serves as the main corridor for motor vehicles and pedestrian cycles and directly connects the railway station in the north of the peninsula with the southernmost section.

Characteristic walking street is located in the center of the peninsula, forming the main corridor of east-west direction, as the main shopping street of the financial district. The pedestrian street will be crowded with a variety of commercial retail stores located in low-Rise or multi-storey office buildings.

A new straight-line park on Central Avenue provides a safe and pleasant walking link to the site along the north-south axis of the peninsula.

城市道路系统规划图
Urban Road System Planning

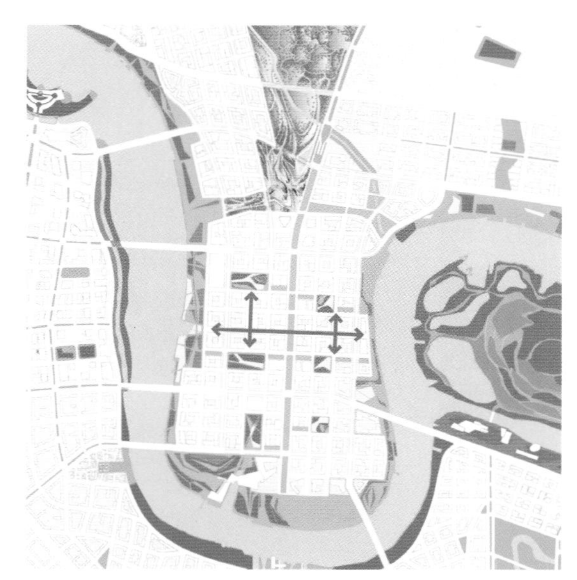

地面步行系统规划图
Ground Walking System

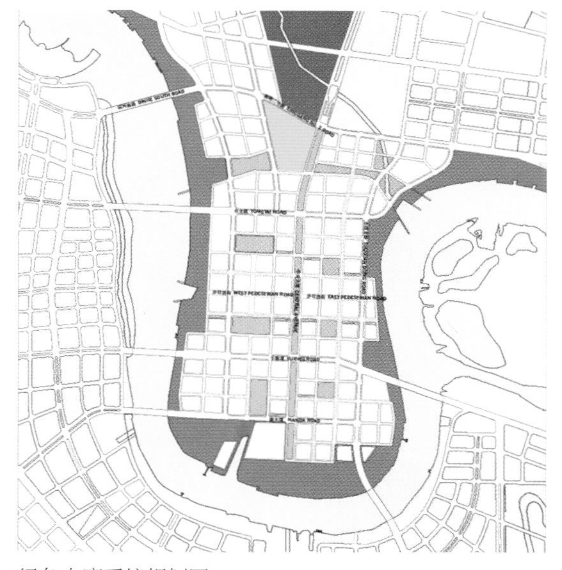

绿色走廊系统规划图
Green Corridor System

中央大道

由林荫人行道和自然的公园所组成的中央大道将成为一条贯穿于家堡南北向轴线的绿色通道。结合周边发展区域，中央大道与天碱解放路商业街连通，形成纵贯南北的商业走廊和金融机构以及服务性住宅的首选位置。

通过相关案例研究及模拟分析，确定中央大道两侧建筑间距，以保证形成最优化的街道景观空间。一条 80 米 宽的中央大道可以容纳宽为 30 ~ 40 米的线形公园，并将穿越中央大道的时间从 2 分钟缩短到 1 分钟。一条 80 米宽的中央大道可以起到连接城市的两个部分的作用，然而一条 120 米宽的街道则会把城市一分为二。

Central Avenue

The central avenue, which consists of a shady sidewalk and a natural park, will be a green passage through the north-south axis of the fort. Combining with the surrounding development areas, Central Avenue and Tianjian Jiefang Road are connected to form the preferred location of commercial corridors and financial institutions as well as service housing throughout the north and south.

Through relevant case studies and simulation analysis, the spacing width of buildings on both sides of Central Avenue is determined to ensure the formation of optimal street landscape space. An 80-metre wide Central Avenue can accommodate a 30 to 40 metres wide linear park and shorten the time of crossing Central Avenue from 2 minutes to 1 minute. An 80-metre wide Central Avenue serves as a link between the two parts of the city, while a 120-metre wide street divides the city into two parts.

中央大道及周边绿廊
Central Avenue and Surrounding Green Corridor

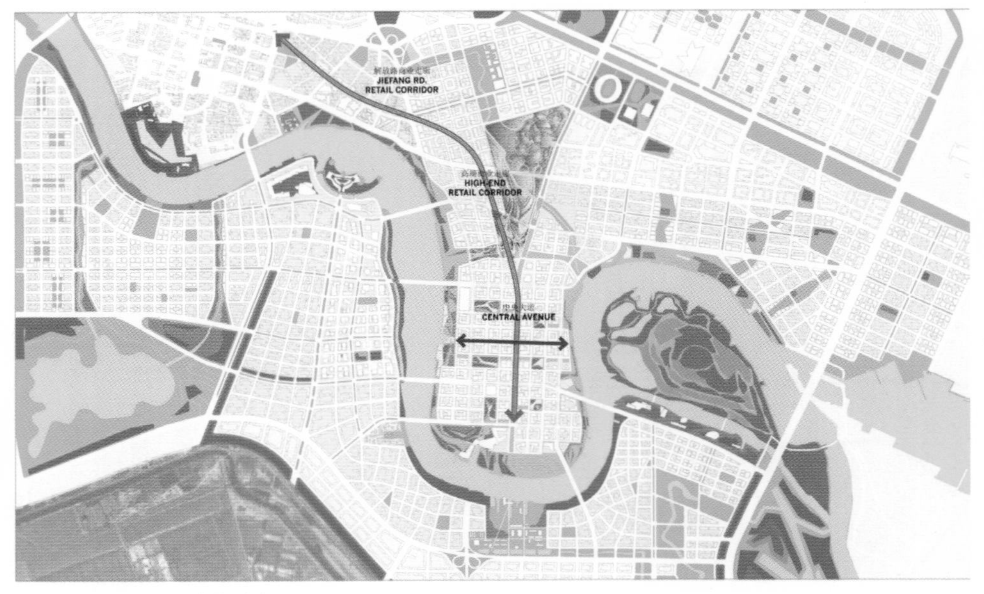

中央大道与商业走廊的融合
Integration of Central Avenue and Commercial Corridor

高铁站和交通枢纽

高铁站和交通枢纽将成为通往于家堡以及大滨海地区的大门。作为京津高铁的终点站，这座交通枢纽将成为不同运输方式的交汇连接中心。

这些运输载体包括高速铁路，三条地铁线，本地以及区域的客车，出租车以及私家车。优越的地理位置以及通达性将会使这座交通枢纽成为塘沽地区的功能核心。

High-speed Railway Station and Transportation Hub

High-speed railways and transport hubs will be the gateway to the home Fort and coastal areas. As the terminal of Beijing-Tianjin high-speed railway, this transportation hub will become the intersection and connection center of different modes of transportation.

These carriers include high-speed railway, three Metro lines, local and regional buses, taxis and private cars. Superior geographical location and accessibility will make this transportation hub the functional core of Tanggu area.

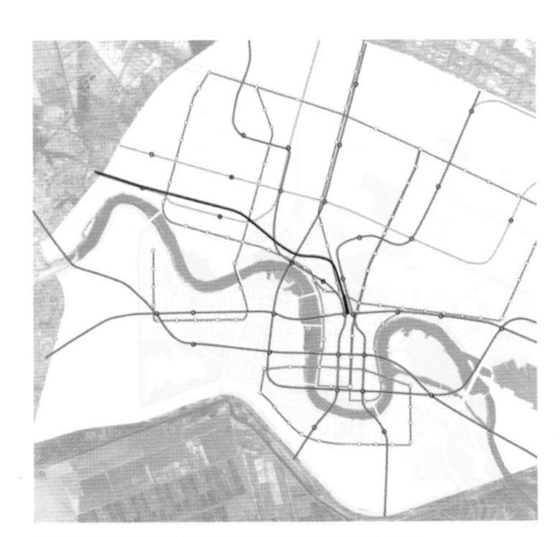

于家堡高铁站与城市轨道的换乘关系
The Transfer Relation Between Yujiapu High-Speed Rail Station and Urban Rail

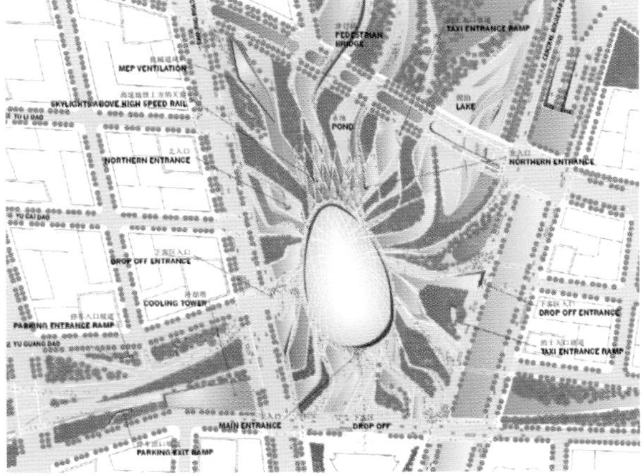

于家堡高铁站规划平面图
Yujiapu High-Speed Railway Station Planning Master Plan

于家堡高铁站
Yujiapu High-Speed Rail Station

用地格局

用地性质：不同性质的用地在层次性的街道体系之内围绕着高档次的公园和滨水开放区域而布置。金融区的土地使用规划包括办公、商业、娱乐、文化、酒店和住宅等。

容积率：以于家堡高铁站中心交通枢纽和中央大道为核心，便利的交通条件及区位优势可支撑更高的容积率。整体容积率分布呈现由中心向沿河区域逐渐降低的趋势。

Land Use Pattern

Land Use: Different types of land are arranged around high-grade parks and waterfront open areas within the hierarchical street system. Land Use Planning in Financial District public, commercial, entertainment, culture, hotels and residences.

FAR: With Yujiapu High-speed Railway Station's central transportation hub and Central Avenue as the core, convenient traffic conditions and location advantages can support higher volume ratio. The overall volume ratio distribution shows a decreasing trend from the center to the river area.

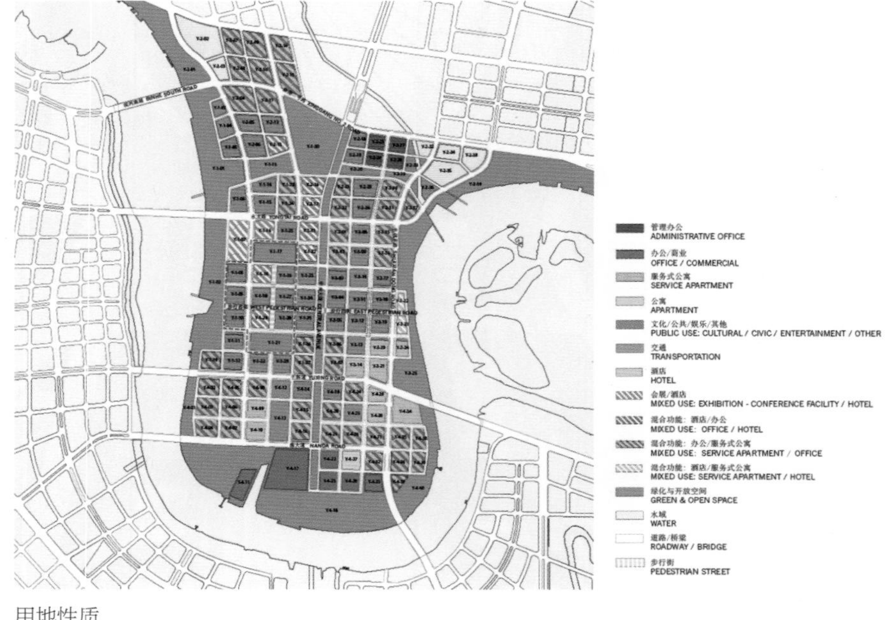

用地性质
Land Use

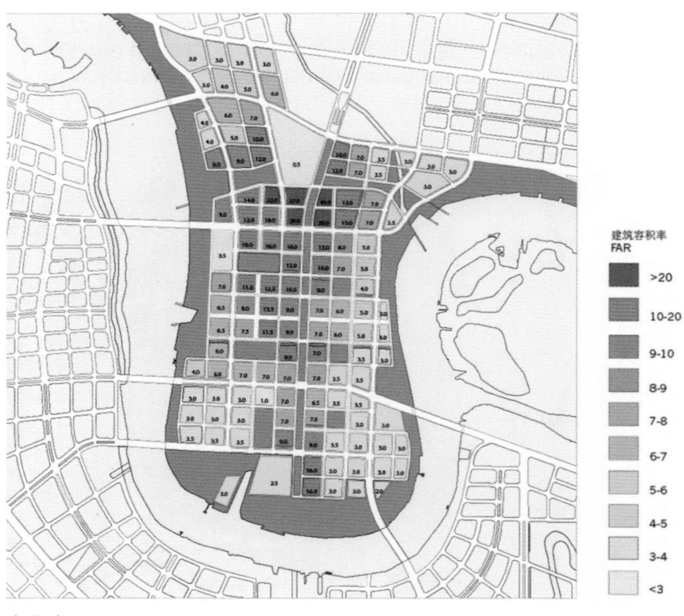

容积率
FAR

（单位：米）

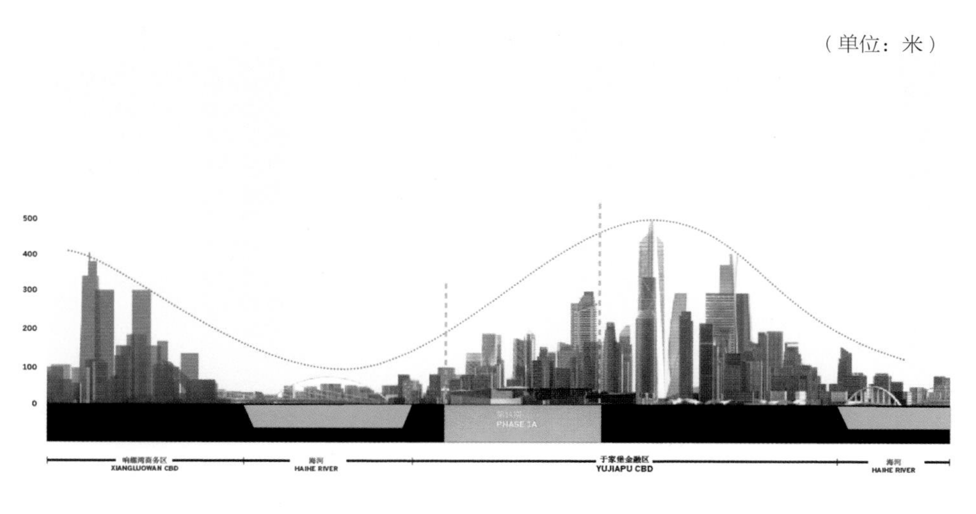

天际线

对光照和视野敏感的城市肌理和建筑形体是构造具有凝聚力的城市景观的重要因素。建筑高度将会由河滨向中心交通枢纽和中央大道逐渐升高。在这样的一个建筑形体的设计指导下，位于不同位置的建筑都将享有各自的景观视野，同时在中央商务区创造了别具一格的城市天际线。

Skyline

Urban texture and architectural form sensitive to light and vision are important factors in constructing a cohesive urban landscape. The height of the building will gradually increase from the relative low of the riverfront to the central transportation hub and Central Avenue. Under the guidance of the design of such a building form, buildings located in different locations will enjoy their own landscape vision, while creating a unique urban skyline in the Central Business District.

天际线分析图
Skyline Analysis Chart

地下空间

于家堡半岛标高将会提升至洪水位之上，以解决海河的洪水问题。提升后的地面将更有利于地下空间的发展以及市政设施的敷设。车辆隧道、地铁、地下步行空间、地下车库以及地下管线都将组合成为地下设施的一部分，并支持可持续发展的策略。

Underground Space

The Yujiapu peninsula will be raised above the flood level to solve the flood problem of the Haihe River. Upgraded ground will be more conducive to the development of underground space and the laying of municipal facilities. Vehicle tunnels, subway, pedestrian underground space, underground garage and underground pipelines will be combined as part of underground facilities and support sustainable development strategies.

地铁线路图
Metro System Map

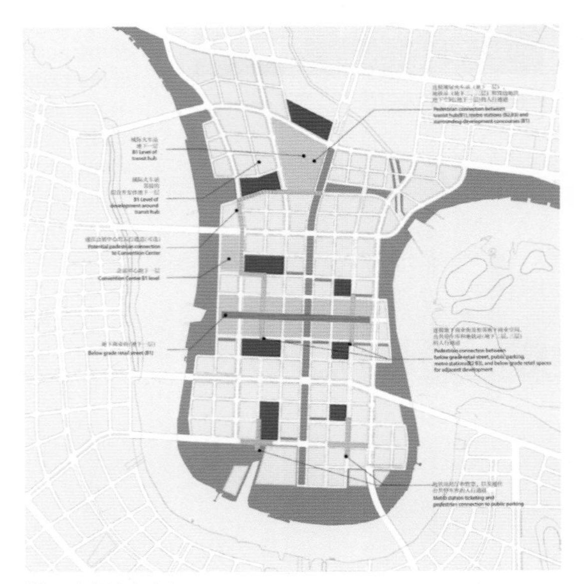

地下步行空间图
Underground Walking Space Map

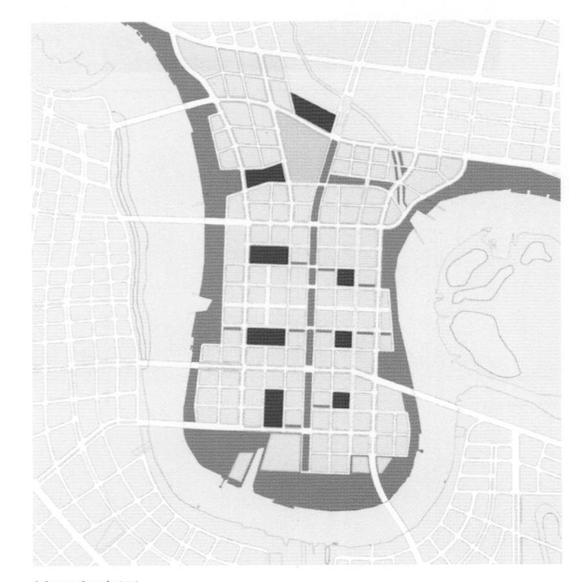

地下车库图
Underground Garage Map

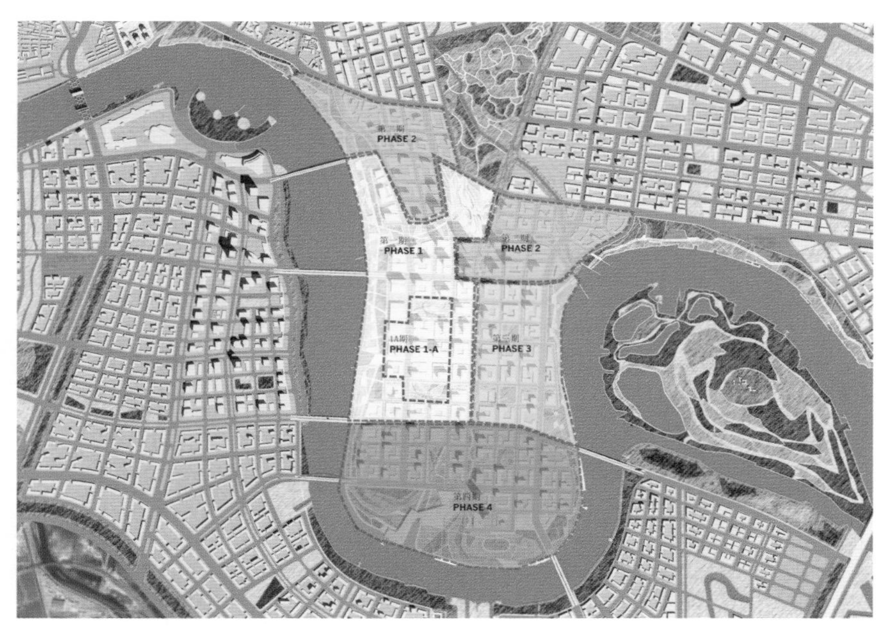

分期计划图
Stage Plan

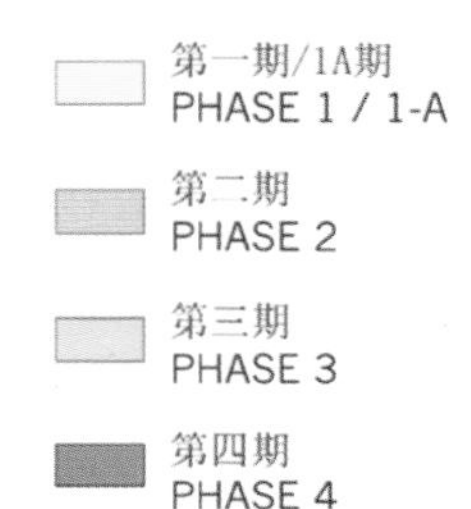

分期发展

于家堡目前将分四期规划发展。

第一期坐落在响螺湾的对岸，于家堡半岛的西部河滨。1A 期部分包括了商业、金融服务大厦以及会议中心。而整个一期的发展将包括交通枢纽以及其他办公和服务性住宅塔楼。

第二期将发展半岛北部，联系着塘沽地区与于家堡新中央商业区。第二期的地块功能包括管理办公、住宅和一般办公。

第三期包含了在半岛东岸的混合功能建筑，将延续一期的地块发展模式。

第四期将发展半岛的南部，其中服务性住宅和办公功能混合将会是这一区域的主要功能，同时，第四期将结合具有独特性文化的河滨公园和中央大道的终点来发展。

Development by stages

Yujiapu will be developed in four phases.

The first phase is located on the opposite bank of Xiangluowan, on the western bank of the Yujiapu peninsula. Phase 1A covers business, financial services building and conference center. The whole phase of development will include transportation hubs and other office and service residential towers.

The second phase will develop the northern part of the peninsula, linking Tanggu district with Yujiapu New Central Business District. The second phase of land functions includes management office, residential and general office.

The third phase includes mixed-function buildings on the eastern coast of the peninsula, which will continue the first phase of the land development model.

Phase IV will develop the southern part of the peninsula, in which service residential and office functions will be the main functions of the area. Phase IV will combine the end of riverside parks and central avenues with unique cultural characteristics.

于家堡 1A 期实施照片
Implementation Photo of Yujiapu Phase 1A

于家堡 1A 期实施照片
Implementation Photo of Yujiapu Phase 1A

项目概况

天碱老厂区位于原塘沽区与开发区之间，解放路是原塘沽区的传统商业街区。天碱解放路商业区将规划成为滨海新区最重要的商业文化中心区。

Project Overview

The old Tianjian factory is located between the old Tanggu District and TEDA. Jiefang Road is the traditional commercial block of the old Tanggu District. Tianjian Jiefang Road is planned to become the most important commercial and cultural center of Binhai New Area.

规划鸟瞰图
Planning Aerial View

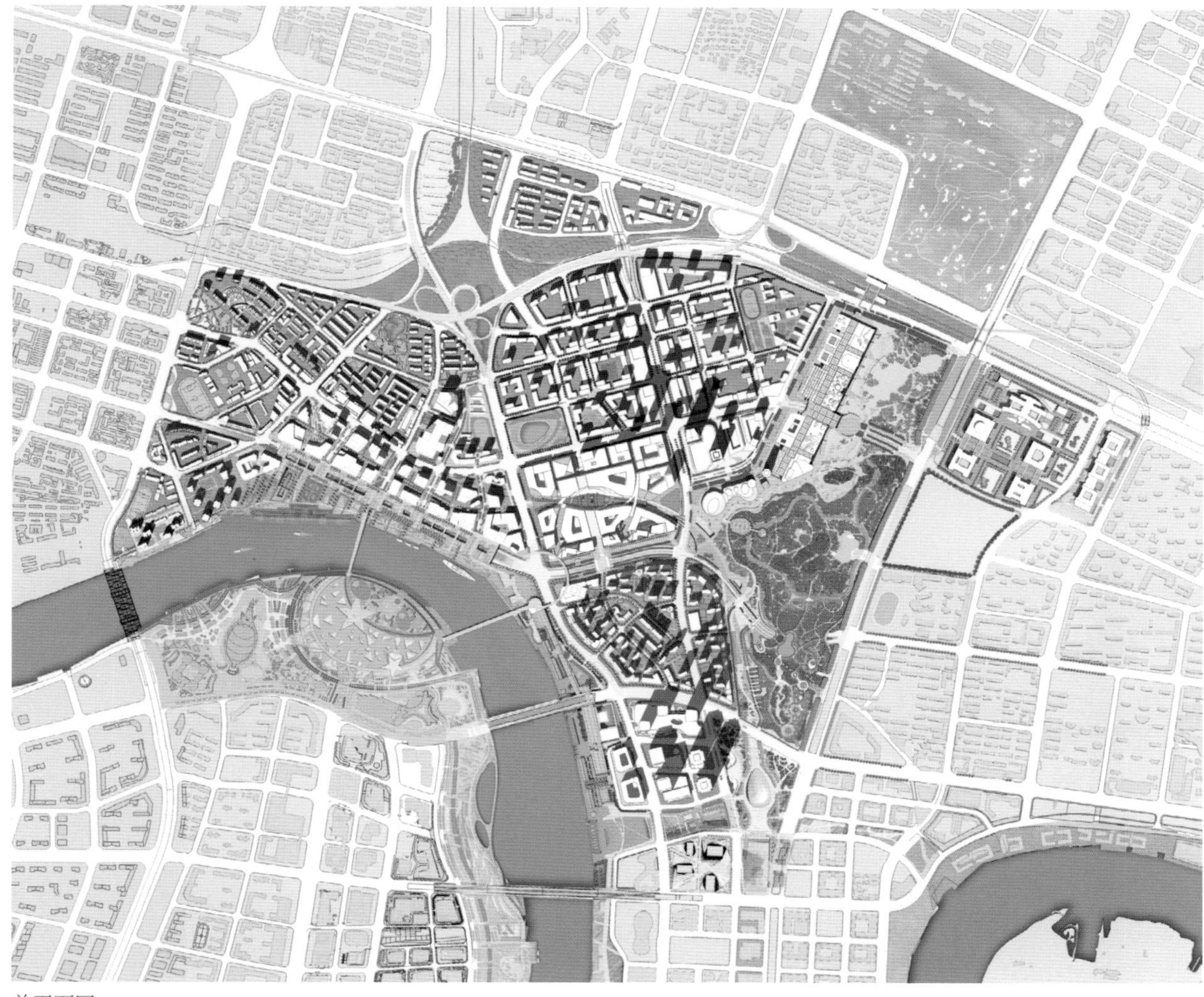

总平面图
Master Plan

功能定位

天碱解放路商业区的发展目标是成为滨海新区国际一流的商业和文化中心。

通过分析与周边区域商业业态间的关系，实现天碱解放路商业业态整体的同业差异化发展和互补性融合。

Functional Orientation

Tianjian Jiefang Road Business District is planned to be developed into an international first-class business and cultural center in Binhai New Area.

Through the analysis of the district's relationship with the commercial models in the surrounding areas, we can realize differentiation within businesses and complementary integration of the commercial models of Tianjian Jiefang Road as a whole.

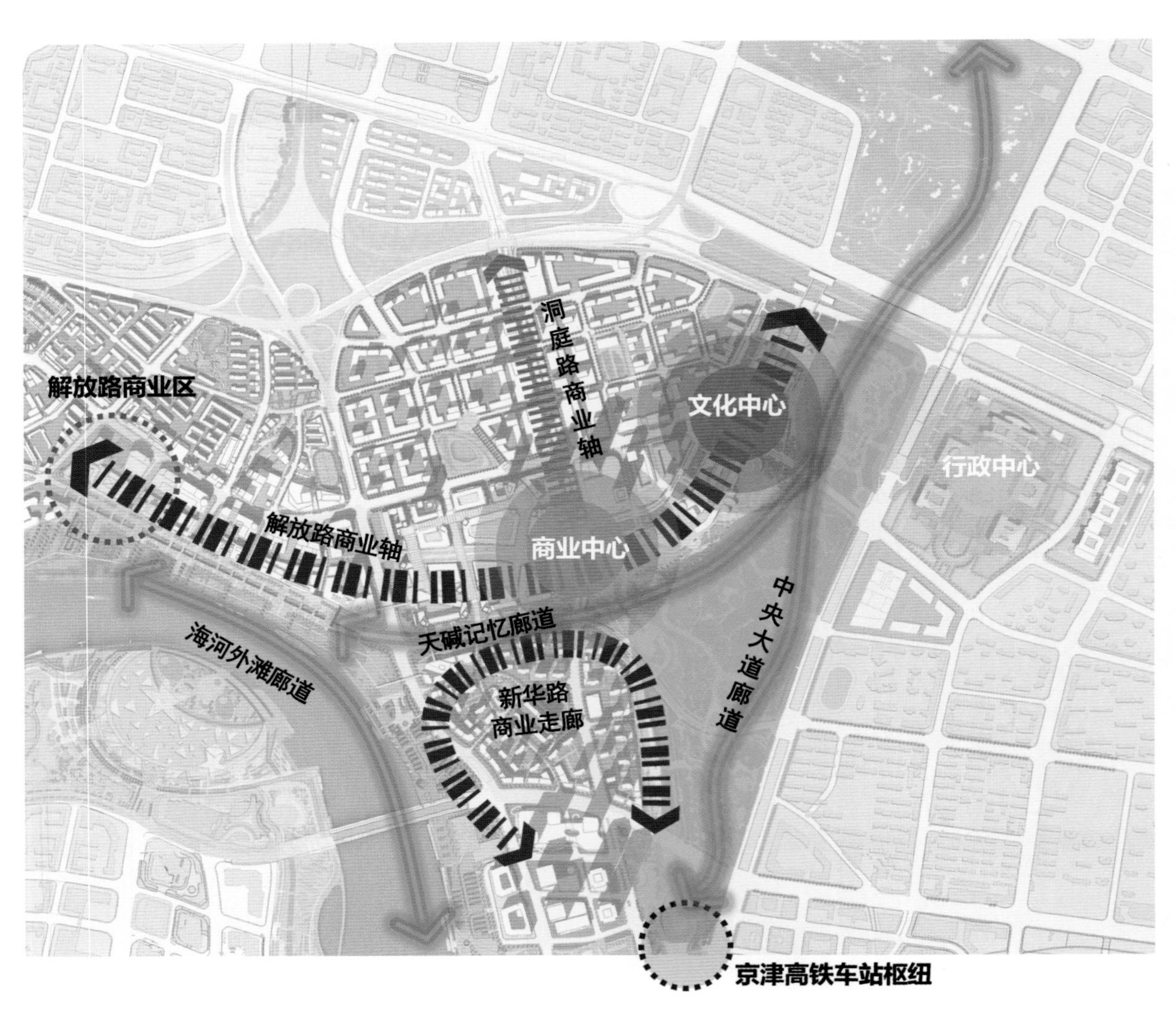

功能结构示意图
Functional Structure Sketch Map

总体布局

天碱解放路商业区规划发展形成“两心、两轴、三廊道”的总体结构。

两 心：天碱商业中心、滨海新区文化中心。

两 轴：解放路商业轴、洞庭路商业轴。

三廊道：海河外滩廊道、中央大道开放空间廊道、天碱记忆廊道。

Overall Layout

Develop the overall structure of "two centers, two axes and three corridors".

Two centers: Tianjian Commercial Center and Binhai New Area Cultural Center.

Two axes: Jiefang Road Business Axis and Dongting Road Business Axis.

Three corridors: Haihe River outer bank corridor, Central Avenue open space corridor, Tianjian Memory corridor.

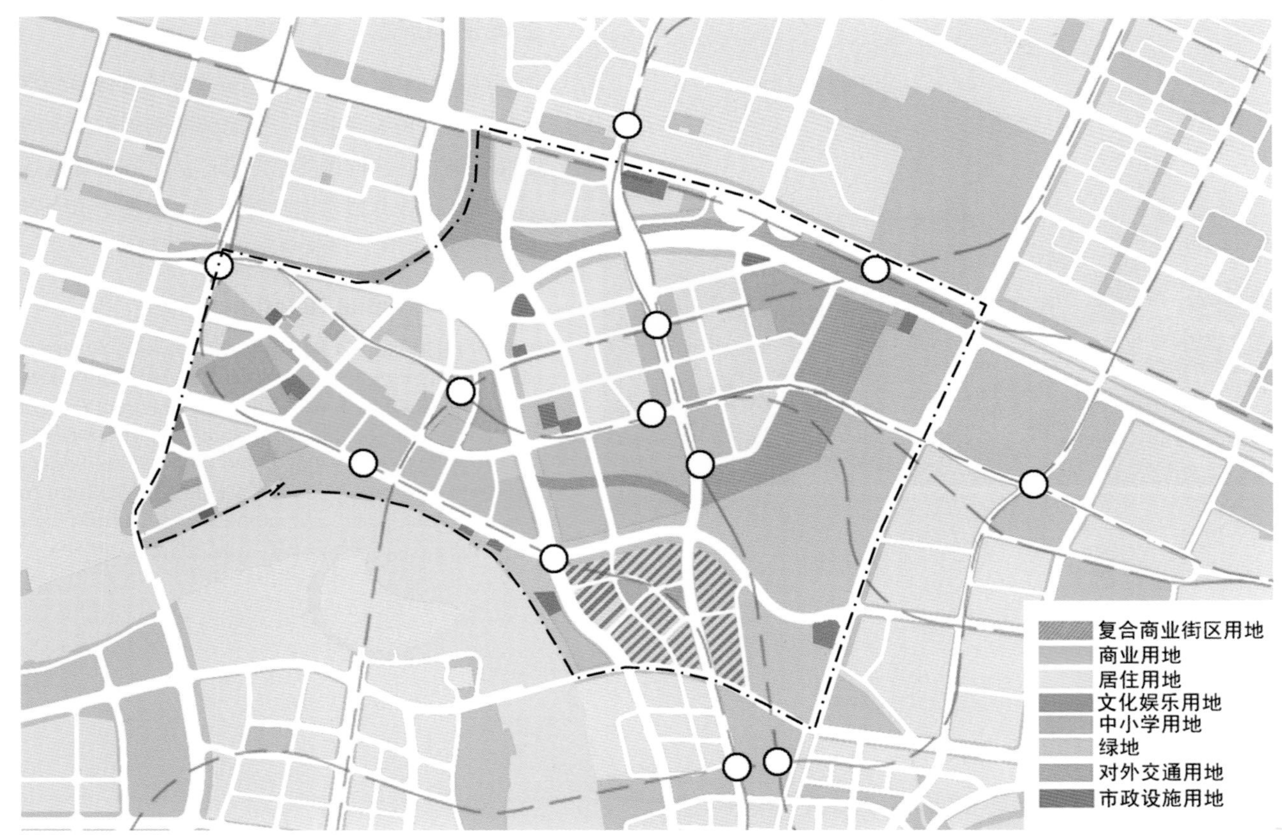

用地性质图
Land Use Map

用地布局及业态

商业集中于解放路、洞庭路两条商业轴布置。重点建设大型商业综合体、街廓商业和底层商业街区等丰富多样的商业形态，构建大型超市、精品百货、家用电器、家居用品、数码影院和餐饮娱乐等多业态互为补充、相互促进的新格局，并最终与西侧解放路传统商业街区共同形成具有一定规模、形态多样、商业与文化娱乐结合的天碱解放路商业区。

Land Use Layout and Business Model

Commerce is concentrated on the two commercial axes of Jiefang Road and Dongting Road. Focus on the construction of large-scale commercial complexes, street commerce and bottom-level commercial blocks, and other rich and diverse business forms, and build a new pattern of complementary and mutually reinforcing business forms, such as large supermarkets, high-quality department stores, home electronics, household appliances, digital cinemas and food and entertainment, and eventually form a Tianjian–Jiefang Road commercial center district that is of a certain scale, diverse in form, and a combination of business and culture and entertainment.

周边天际线控制示意图
Control Sketch Map of Peripheral Skyline

区域空间关系

规划形成以于家堡 588 米地标建筑为区域中心高度控制点，响螺湾地标、于家堡南制高点和天碱制高点与之呼应的整体空间形态。区域内建筑顶部可灵活设计并提供退台看海河，使整个沿河界面形成高低起伏、错落有致的城市天际线。

Regional Spatial Relations

Plan to build the overall spatial form with Yujiapu's 588-meter landmark building as the regional central high point, Xiangluowan landmark, southern Yujiapu high point and Tianjian high point corresponding to it. The top of the buildings in the area can be flexibly designed and provided with a set-back platform to view the Haihe River, so that the whole riverside view forms a rising and falling, well-spaced urban skyline.

现代主义风格的海河沿岸
Modernist Style Haihe Riverside

古典主义风格的海河沿岸
Classical Style Haihe Riverside

滨河商业界面

充分发挥传统商业街区地段优势，发掘和利用好海河自然景观资源，体现其景观的亲水性和休闲性，并有机衔接天碱地区东侧的中央大道绿化景观廊道、海河景观廊道，从整体景观上将其打造成为凸显本区域城市形象的一大亮点。

建筑设计在风格、体量和密度上应既统一又富有变化，使广大公众深刻感受到区域整体城市形象的提升。在整体风格上，建筑外观设计探讨了现代主义与古典主义两种方案，寻求体现新城市主义建筑特色，实现中心商务区形象的整体统一。

Riverside Business Space

Give full play to the advantages of traditional commercial blocks, excavate and make good use of Haihe River's natural landscape resources, embody the waterside and leisure qualities of its landscape, and organically connect the Central Avenue green landscape corridor and Haihe River landscape corridor on the eastern side of Tianjian area, building it into a highlight of the regional urban image in terms of the overall landscape.

The architectural design in the region should be both unified and varied in style, volume and density, so that the general public can have an understanding of the overall urban image of the region. In terms of overall style, architectural design explores the two styles of modernism and classicism, seeking to embody the architectural features of neo-urbanism and achieve the overall unity of the image of the Central Business District.

项目概况

规划借鉴新城市主义理论。新城市主义源于二战后城市的不断发展与扩张所产生的种种新问题。早期新城市主义希望通过回归传统的方式，对城市空间进行设计，以亲和行人的规划原则、回归传统的建筑风格来避免现代主义城市的千篇一律和刻板。后期以卡尔索普为代表所提出的“区域城市” 把整个都市区视为一个系统，而内中的社区都成为具有全部城市功能——包括居住、就业、娱乐、商业、服务、交通等六方面内容的完整城市。新城市主义的建设需要达到功能混合、空间宜人、生活和谐的要求，这也是规划设计实践中所探寻的方向。

Project Overview

The emergence of New Urbanism stems from the various new problems that rose from the continuous development and expansion of cities after World War II. Early New Urbanism aimed to design urban space through returning to the traditions, avoiding the uniformity and rigidity of modernism by adhering to the principle of prioritizing pedestrians and returning to traditional architectural styles. In the period that follows, the concept of "regional city" proposed by Calthorpe regards the whole metropolitan area as a system, and the communities in it become complete cities with all urban functions, including residence, employment, entertainment, commerce, services, and transportation. It is evident that the construction of cities under the guidance of New Urbanism needs to meet the requirements of mixed functions, pleasant space and harmonious life, which is also the direction we seek in the design practice.

总平面图
Master Plan

混合多样化的城市功能支持标志性建筑

国家海洋博物馆是展现国家海洋战略的标志性建筑，是我国首座国家级、综合性、公益性的海洋博物馆。有别于传统博物馆坐落于城市建成区中，国家海洋博物馆选址位于完全没有任何城市基础的填海区域中，紧邻南湾。一方面其建设必将成为滨海旅游区发展建设的标杆性工程，有利于推动开发建设；但另一方面我们也深刻地认识到，需要在海博周边地区规划建设功能混合多样化的综合片区来为海博提供全天候的城市活力，避免大型城市公共空间成为被社会遗忘的角落。

以此为出发点，本次城市设计希望为国家海洋博物馆营造一个激动人心的场所，为人们提供丰富的旅游城市体验，让人与自然和谐共处，其完整的步行系统和便捷的公交配套体系，创造了多样的、可步行的、紧凑混合社区，使其成为宜居性与可持续性兼具的旅游城区。

Mixed and Diverse Urban Functions Supporting Landmarks

National Maritime Museum of China is a landmark building demonstrating the national marine strategy, and is the first national, comprehensive and public-spirited marine museum in China. Unlike traditional museums, which are situated in existing urban areas, the National Marine Museum is located in reclamation areas without any urban foundation, close to the Nanwan Bay. On the one hand, its construction will certainly become a benchmark project for the development and construction of Binhai tourist areas, which is conducive to promoting the development and construction; on the other hand, we also fully recognize that it is necessary to plan and construct a general area with mixed and diverse functions around the museum to provide day-long urban vitality for the museum, so as to avoid a large-scale urban public space becoming a forgotten corner.

Taking this as a starting point, this urban design hopes to create an exciting companion environment for the National Marine Museum, providing the rich experience of a tourist city, harmonious coexistence of nature and man-made, complete walking system and convenient supporting public transport system, creating a diverse, walkable and compact mixed community, making it a tourist city with both habitability and sustainability.

国家海洋博物馆
National Maritime Museum of China

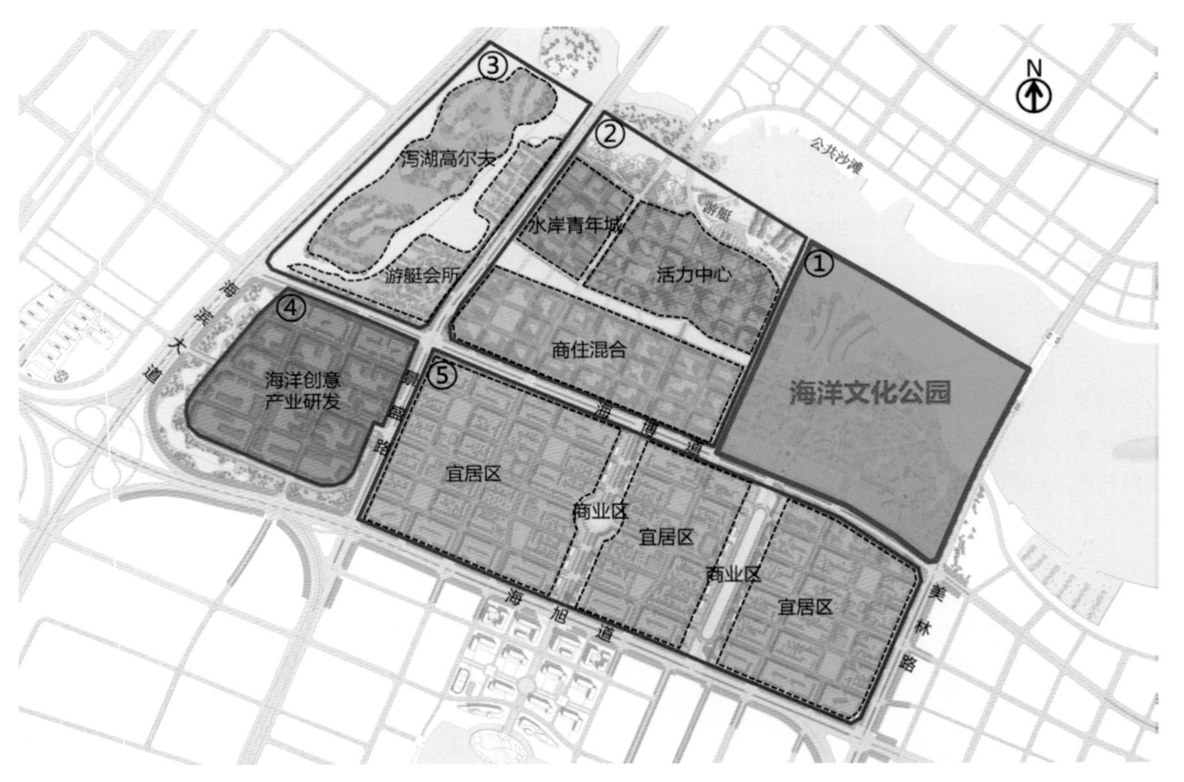

功能分区图
Functional Zoning Map

国家海洋博物馆周边片区功能研究

国家海洋博物馆周边片区定位为中国海洋文化博览产业基地和生态宜居旅游城区。功能构成上，海博周边片区内需统筹并体现滨海旅游区主题公园、休闲总部、生态宜居、游艇总会为核心的主要功能。围绕海博周边区域，形成居住、办公、消费、生产、娱乐、休闲一体化的城市功能体系，尽量延长城市一天中的活动时间，形成 7×24 小时的综合活力。

Study on the Functions of the Surrounding Areas of National Maritime Museum

The surrounding area of National Maritime Museum is positioned as a China Marine Culture Exposition Industrial Base and an ecologically livable tourist area. In terms of functional composition, the Marine Museum surrounding areas need to coordinate and embody the main functions of Binhai tourist area, which are theme park, leisure center, ecological livability and yacht club. Around the surrounding area of the museum, an integrated urban function system of residence, office, consumption, production, entertainment and leisure will be formed, which will extend the time of urban activities in one day as much as possible and form 24/7 urban vitality.

小街廓、密路网
Small Street and Dense Road Network

小街廓、密路网城市空间构造宜居城市

恬静美好的旅游区不应该有交通拥堵等出行问题，也不应该有同质化的建筑立面造成城市空间的迷失。对此新城市主义提出的解决方式是采用小街廓、密路网的城市格局。一方面，更小的街廓提供更多的城市建设地块，有更多的开发建设可能，有利于居民和游客获得更加人性化的城市体验。另一方面，更密的路网提供更多步行路径选择，有效降低两点之间的步行距离，吸引人们采用步行出行方式，从而形成慢节奏、有情调的街道空间。

Small Blocks and Dense Road Network Urban Space Supports a Livable City

Tranquil and beautiful tourist areas should not have traffic congestion or other travel inconvenience, nor should they have homogeneous architectural facades that lead to people being lost in urban space. The solution proposed by New Urbanism is to adopt the urban pattern of small blocks with a dense road network. On the one hand, smaller blocks provide more urban construction zones and more development and construction possibilities, which is conducive to the formation of a more human-centered urban experience for residents and tourists. On the other hand, a denser road network provides more choices of walking paths, effectively reduces the walking distance between any two points, and attracts people to choose to travel on foot, thus forming a slow-paced, romantic street space.

围合式的居住形态

海博周边片区的旅游城区特质决定了其应具有较高的居住品质。这不仅仅是满足基本的日照采光、居住安全等要求，还应更多地注重社区环境品质的塑造。围合式布局的住宅有利于承载社区内部生活，满足居住者心理安全需求以及社会交往的需求。相比较于千篇一律的板式高层同质化空间，围合式住宅更有利于形成本土化的归属感，突出旅游区宜居的城市品质。

Enclosed Residential Form

The characteristics of the Maritime Museum's surrounding tourist area determine that it should provide a high living quality. This is not only to meet the basic requirements of lighting and living safety, etc., but also to focus more on the shaping of the quality of the community's environment. Enclosed housing is conducive to community life, satisfying the residents' needs for psychological security and social interactions. Compared with the uniform homogeneous space formed by wide high-rise apartment buildings, we believe that enclosed housing is more conducive to forming a sense of local belonging, highlighting the tourist area's livable quality.

围合的居住区
Enclosed Residential Area

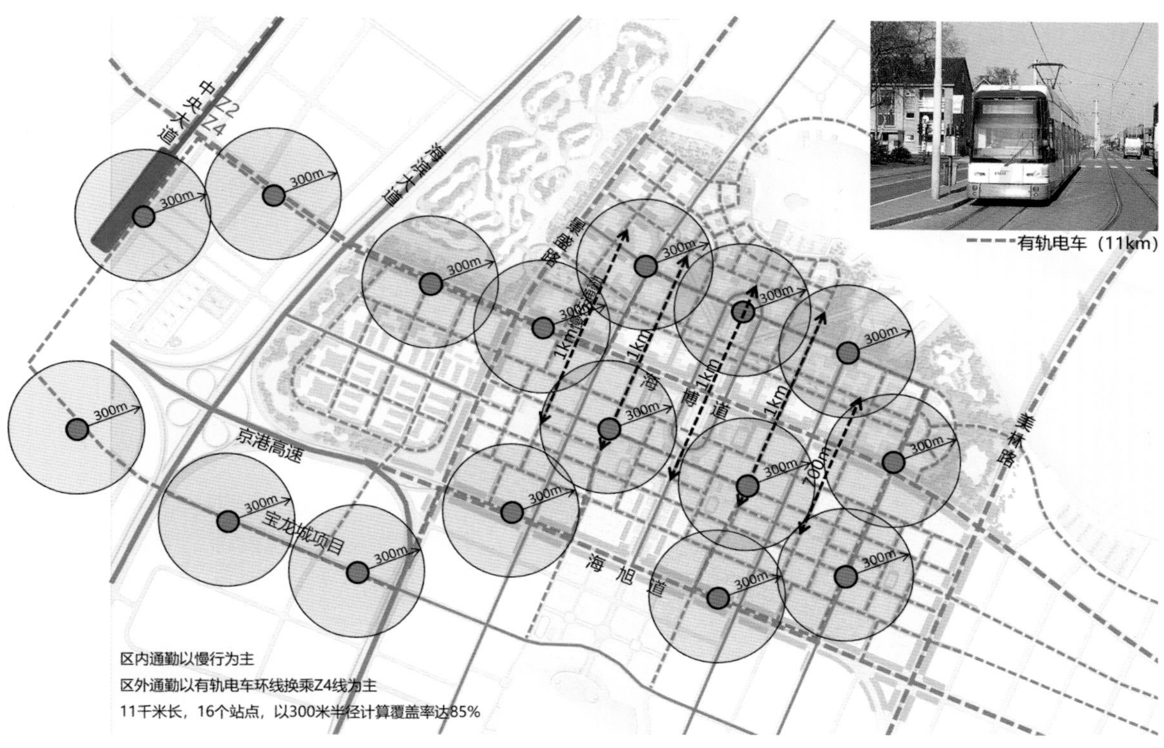

轨道交通示意图
Rail Transit System Map

步行公交相结合的出行方式

旅游区的特质决定了外来人口多，随着快捷高效的轨道交通的建成，本地区出行方式将会以公共交通为主。这也成为小街廓、密路网模式可以在本地区实践的重要保证。

引入与轨道交通相接驳的快速公交系统，形成大运量公交体系，使 5 分钟步行出行圈基本覆盖城市生活区域，并重点与海博园区公交站点进行接驳。

A Combination of Walking and Public Transit

The characteristics of the tourist area determine that it will have a large mobile population. With the construction of rapid and efficient rail transit, we believe that public transit will be the main mode of travel in this region. This has also become an important guarantee for the practice of the small blocks and dense road network model in the area.

We plan to introduce the BRT (Bus Rapid Transit) system, which connects with rail transit, to form a mass transit system, allowing a 5-minute walking circle to mostly cover the urban living area, and emphasizes on connecting with the bus stations of the Marine Museum.

国家视角：转型时期对城市空间的品质与精细化管理提出更高要求

随着新常态下经济发展方式转变，城市建设进入从“量”到“质”转变的阶段，对城市空间的品质与精细化管理提出了更高的要求。需要通过城市设计的手段，提升城市规划管理水平与城市空间质量。

National Perspective—Higher Requirements for Quality and Refined Management of Urban Space in Transitional Period

With the transformation of economic development mode under the “New Normal”, urban construction has entered the stage of “quality” from “quantity”, which puts forward higher requirements for the quality and refined management of urban space. It is necessary to improve the level of urban planning and management and the quality of urban space by the means of urban design.

新八大里地区城市设计鸟瞰
Aerial View of Urban Design in the New Badali Area

编制体系：与现行城乡规划法定体系相对应的城市设计编制体系

城市设计贯穿于城市规划的各个阶段，与城市规划的阶段划分相对应，可分为：总体城市设计、详细城市设计、专项城市设计三个层次。

Compilation System: Urban Design Compilation System Corresponding to the Current Legal System of Urban and Rural Planning

Urban design runs through all stages of urban planning. Corresponding to the stage division of urban planning, it can be divided into three levels: overall urban design, detailed urban design and special urban design.

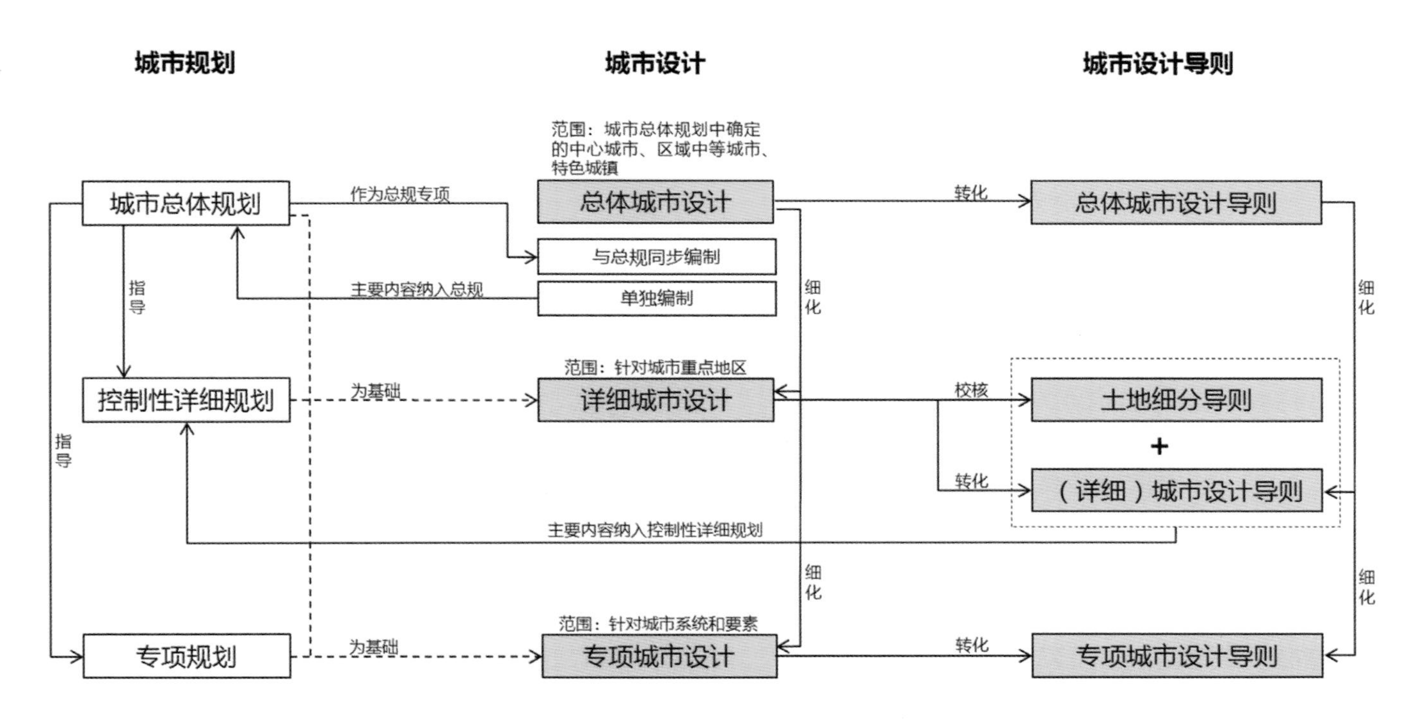

城市规划编制体系
Urban Planning Compilation System

规范总体城市设计编制要求

1. 总体城市设计编制范围。

城市总体规划中确定的城市集中建设区。

2. 总体城市设计编制内容。

从宏观层面研究确定城市空间的总体形态，提出改善城市景观形象和空间环境质量的总体目标，构建富有特色的城市空间形态格局与人文活动场所的总体框架。

3. 总体城市设计的成果要求。

总体城市设计综合报告，包括总体城市设计图纸和文字说明。

Standardizing the Requirements for the Compilation of Overall Urban Design

1.Scope of Compilation of Overall Urban Design

The urban centralized construction area determined in the urban master plan.

2.Compilation Content of Overall Urban Design

From the macro level, the overall form of urban space is determined, and the overall goal of improving the image of urban landscape and the quality of spatial environment is put forward, and the overall framework of urban spatial form pattern and human activities venues with distinctive features is constructed.

3.Achievement Requirements of Overall Urban Design

Comprehensive report of overall urban design is including general urban design drawings and text description.

制定城市设计目标与总体思路

研究和塑造城市整体风貌特色

确定城市空间形态与结构

城市综合环境特色指引

城市重点地区设计指引

城市主要功能区特色指引

城市设计实施的措施和建议

总体城市设计编制内容
Compilation of Overall Urban Design

详细城市设计编制内容

基本内容：详细城市设计编制必须具备的内容。

拓展内容：在基本内容的基础上，须根据不同类型重点地区的特征和需求，额外编制的内容，以此方式对重点地区详细城市设计进行有针对性的扩展。

Content of Detailed Compilation of Urban Design

Basic content: The content necessary for detailed urban design compilation.

Expanded content: On the basis of the basic content, additional content should be compiled according to the characteristics and needs of different types of key areas, in order to expand the detailed urban design of key areas in this way.

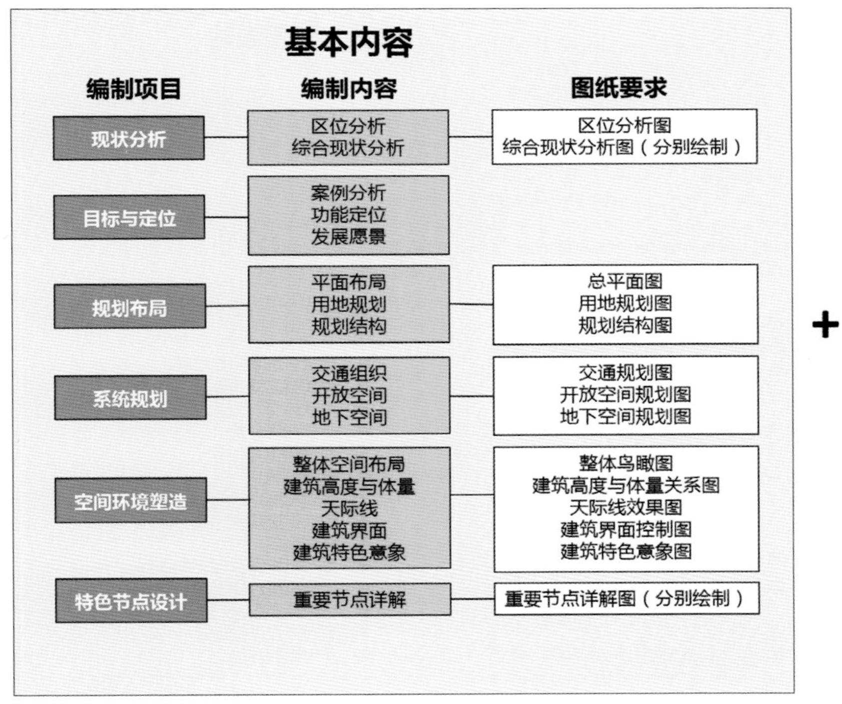

+

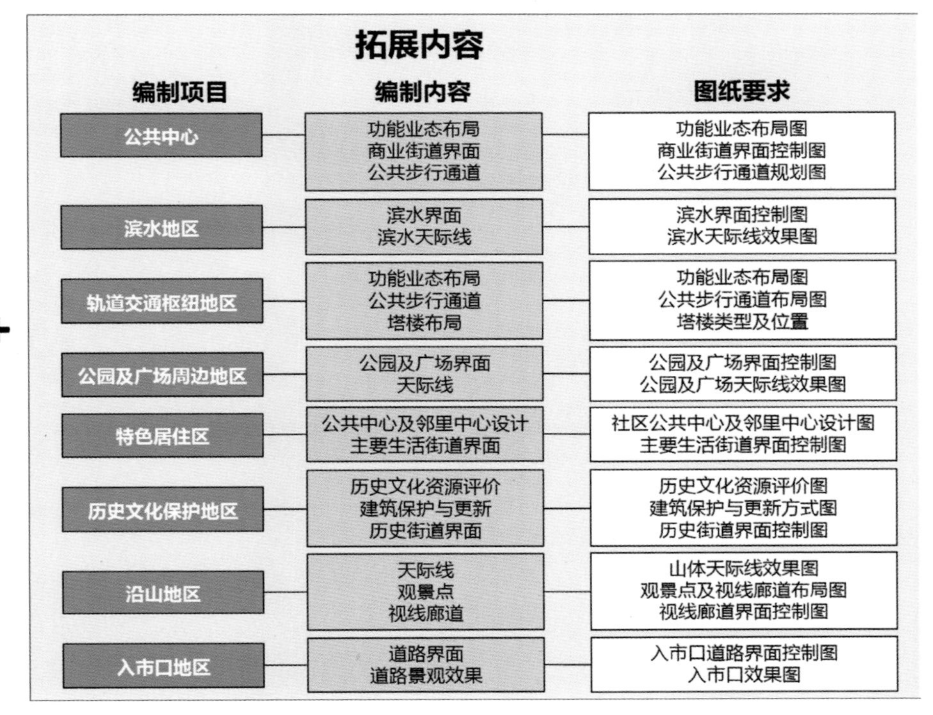

详细城市设计编制内容
Content of Detailed Compilation of Urban Design

详细城市设计成果要求

详细城市设计的成果通常为城市设计图册和模型。

Requirements for Detailed Urban Design Achievements

The results of detailed urban design are usually urban design atlas and models.

详细城市设计成果要求
Requirements for Detailed Urban Design Achievements

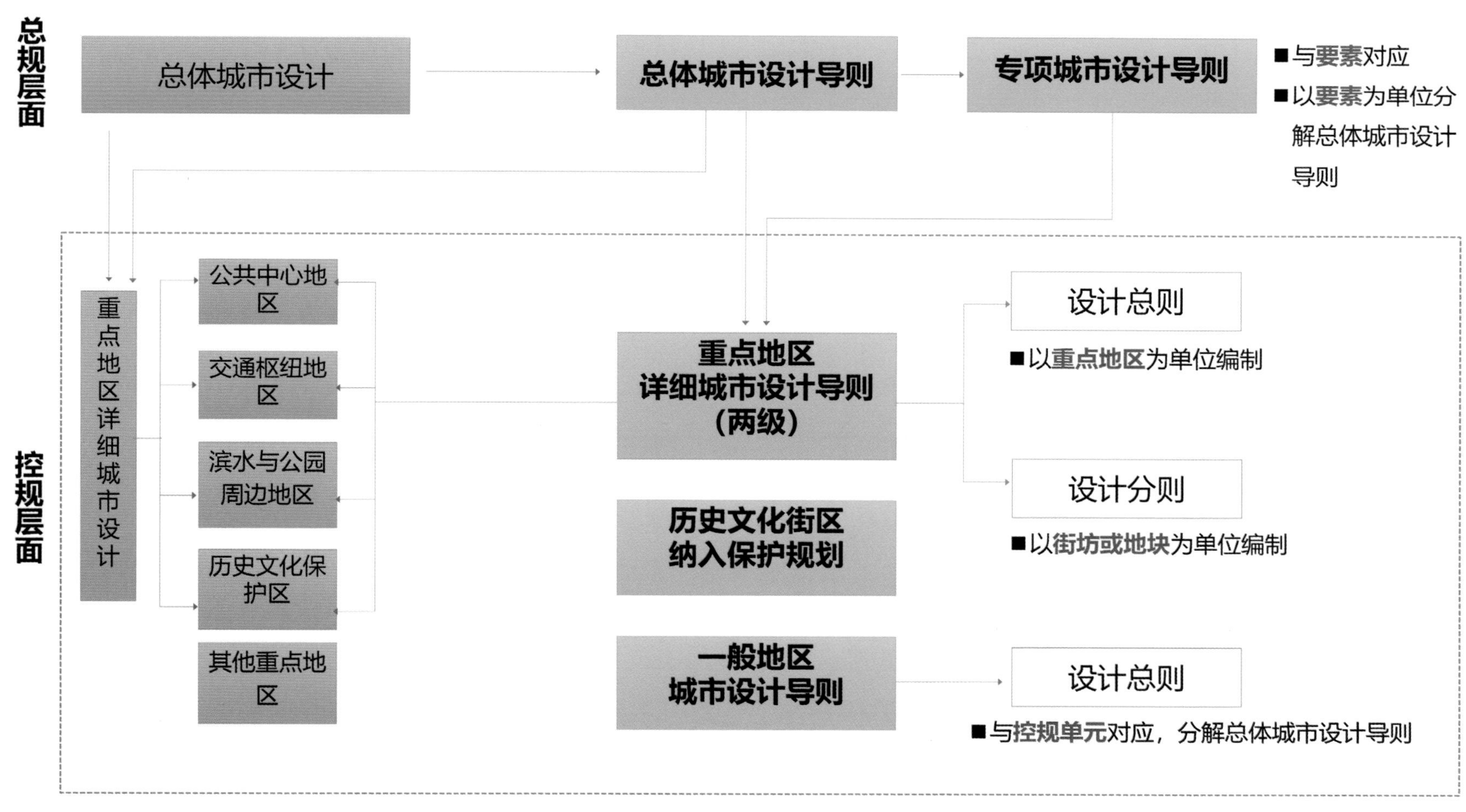

详细城市设计导则编制体系的完善思路
Thoughts on Improving the Compilation System of Detailed Urban Design Guidelines

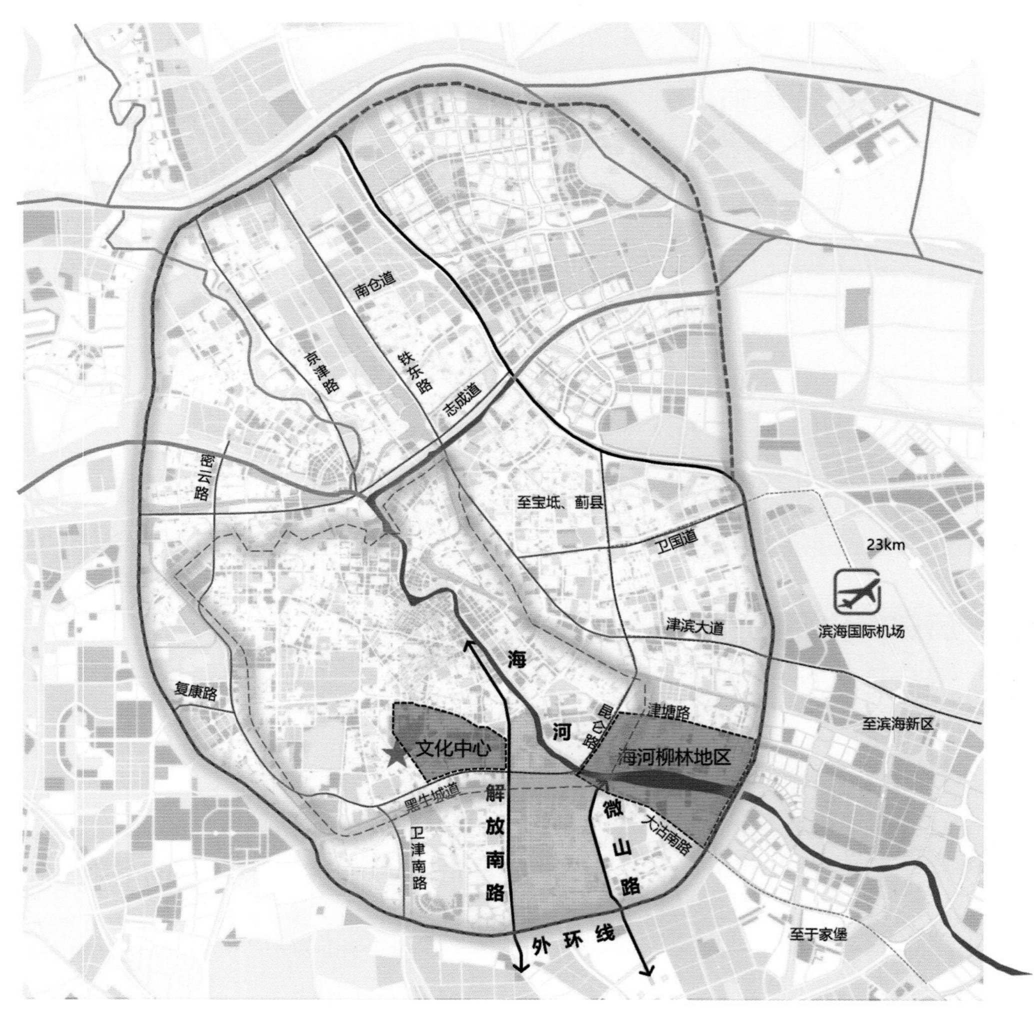

项目区位
Project Overview

项目概况

天津解放南路地区位于天津市中心城区南部，是天津市“十二五”规划建设的重点地区，总占地面积 16.67 平方千米。西侧紧邻天津市文化中心，东侧紧邻海河柳林地区，是连接两个城市重要功能区的重点区域。

天津规划的两条重要的环城生态廊道——外环线“一环十一园”绿化廊道和环城铁路绿道公园，从该地区穿过。

Project Overview

The area around South Jiefang Road in Tianjin is located in the south of Tianjin downtown area. It is the key area for the “Twelfth Five-Year Plan” of Tianjin, with a total land area of 16.67 square kilometers. The west side is adjacent to the Tianjin Cultural Center, and the east side is adjacent to the Haihe Liulin area. It is a key area connecting the important functional areas of the two cities.

Two important urban ecological corridors planned by Tianjin—the outer ring line "One Ring and Eleventh Parks" greening corridor and the Ring Road Railway Greenway Park, pass through the area.

三个发展阶段
Three Stages of Development

历史照片
Historical Photos

天津老重工业基地的发展沿革

基地南部以水塘空地为主，闲置多年。

北部是 20 世纪 50 年代发展起来的天津老重工业基地。清代陈塘庄是海河的重要渡口，当时建成天津第一个能组装火车头的车辆工厂。第一阶段是天津沦陷时期，日本人沿海河建造了现在的棉纺四厂、天津毛织厂和天津市第一钢丝绳厂；第二个阶段是 20 世纪 50 年代初，市机械局、建材局、化工局、一轻局、劳动局、电子仪表局等 14 个系统相继在此建厂，成为当时天津大型的工业企业聚集区；第三个阶段是 21 世纪初前后，环渤海金岸集团在该地区中部建起了装饰城和汽配城，并逐渐形成家居、汽修产业集聚区。1991 年在陈塘庄建起的热电厂已于 2015 年关停。

Development History of Tianjin Old Heavy Industry Base

The southern part of the base is dominated by ponds and open spaces for many years.

The northern part is the Tianjin old heavy industry base that was developed in the 1950s. In the Qing Dynasty, Chentangzhuang was an important ferry crossing for Haihe River. At that time, the first vehicle factory able to assemble a locomotive was built in Tianjin. During the fall of Tianjin, the Japanese built the current No.4 Cotton Spinning Plant, Woolen Factory and Tianjin No.1 Wire Rope Factory along the Haihe River. The second stage was in the early 1950s, 14 systems including the Municipal Machinery Bureau, the Building Materials Bureau, the Chemical Industry Bureau, the Light Bureau, the Labor Bureau, and the Electronic Instrument Bureau successively built factories here, becoming a large industrial enterprise in Tianjin at that time. The third stage is around the 21st century. The Bohai Sea Jin'an Group built a decorative city and an auto parts city in the middle of the region, and gradually formed a home and auto repair industrial cluster. In 1991, the Chentangzhuang Thermal Power Plant was built and shut down in 2015.

2011 年批复版城市设计
Urban Design Approved in 2011

2015 年再深化版城市设计
Urban Design Deepening in 2015

2012 年深化版城市设计
Urban Design Deepening in 2012

规划沿革及设计初衷

自2010年开展城市设计国际方案征集开始，解放南路地区在这九年期间，一直在进行城市规划的动态更新与提升。为将解放南路地区建设成为新时代社区的范本，于规划之初制定了一些发展原则与目标，希望可以通过挖掘工业遗存的场所精神、生态优先构建区域生态骨架、以人为本组织居住社区、实验性建构开放式街区等新思路新模式，最终将该地区建设成为高生态品质、高文化品质、便捷高品质配套、具有持久产业动力、实现职住在地平衡的魅力活力城市片区，这也是整个解放南路地区规划的出发点和九年建设的回归点。

从近两年出台的相关政策和行业规范来看，当时提出的很多思想和原则是极具前瞻性的。

Planning History and Design Intention

Since the launch of the International Design for Urban Design in 2010, the South Jiefang Road area has been undergoing dynamic updating and upgrading of urban planning during these nine years. In order to build the South Jiefang Road area into a model of the new era community, some development principles and goals were formulated at the beginning of the planning. It is hoped that the ecological spirit of the place where the industrial remains will be built, the ecological framework will be built preferentially, the people-oriented organization will live in the community, and the experiment will be Constructing new ideas and new models such as open-plan neighborhoods, and finally building the area into a glamorous and energetic urban area with high ecological quality, high cultural quality, convenient and high-quality support, sustainable industrial power, and balanced occupation and living. This is also the entire liberation, the starting point of the planning of the South Road area and the return point of the nine-year construction.

Judging from the relevant policies and industry norms promulgated in the past two years, many of the ideas and principles put forward at that time were extremely forward-looking.

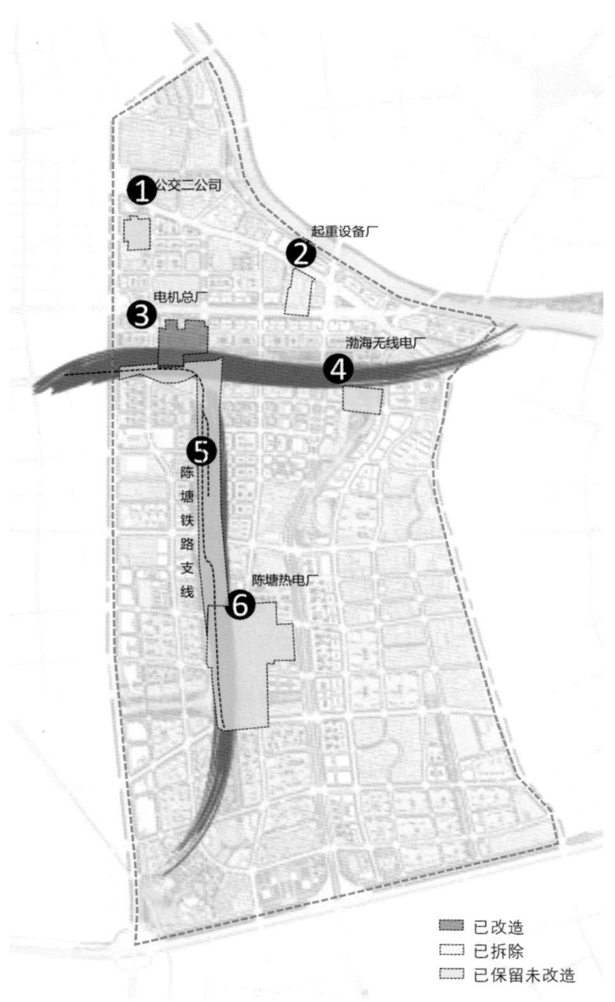

场地内工业遗存
Industrial Heritage in the Site

挖掘地区工业遗存的场所精神

挖掘场所精神并非简单保留单体工业遗存，而是以工业遗存作为精神和空间的脉络，串起整个地区的规划。

结合复兴河、陈塘铁路支线、陈塘热电厂冷凝塔、老厂房等工业遗存，打造T形开放空间作为规划结构主骨架，形成工业文化记忆之轴。

Excavate the Spirit of the Regional Place

The spirit of the excavation site is not simply to preserve the remains of the single industry, but to use the industrial heritage as the context of the spirit and space to string the planning of the entire region.

Combined with industrial relics such as Fuxing River, Chentang Railway Branch Line, Chentangzhuang Thermal Power Plant Condensation Tower and old factory building, the T-shaped open space is built as the main skeleton of the planning structure, forming the axis of industrial cultural memory.

生态骨架的要素分布
Distribution of Elements of Ecological Skeleton

生态优先建构区域生态骨架

以生态空间而不是干道系统作为规划骨架，规划长 5 千米、宽 200 米的 T 形开放空间，接入两个环城绿道，并完善中心城区整体生态系统。T 形开放空间是地区生态环境塑造的核心，也是整个解放南路地区“自然积存、自然渗透、自然净化”的海绵体系。

T 形开放空间串联四个产业带。通过升级现有汽配、家居等产业，形成地区活力引擎，促进职住在地平衡。

Ecological Priority Construction of Regional Skeleton

Taking the ecological space instead of the trunk road system as the planning skeleton, we plan a T-shaped open space of 5 kilometers long and 200 meters wide, access two green roads around the city, and improve the overall ecosystem of the central city. The T-shaped open space is the core of the regional ecological environment, and it is also the sponge system of “natural accumulation, natural penetration and natural purification” in the entire South Jiefang Road area.

The T-shaped open space is connected in series with four industrial belts. By upgrading the existing auto parts, home and other industries, the regional vitality engine will be formed to promote the balance of occupation and residence.

中心城区生态系统
Central Urban Ecosystem

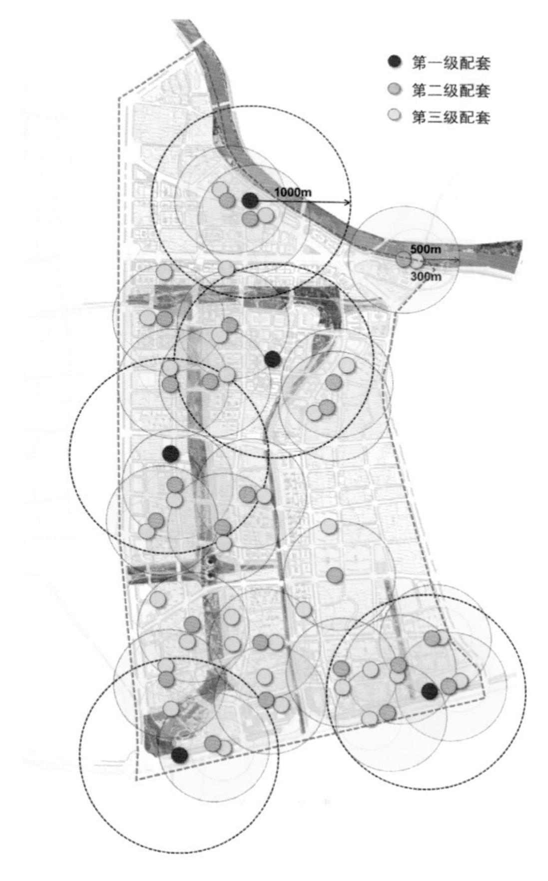

配套分布
Function Allocation Map

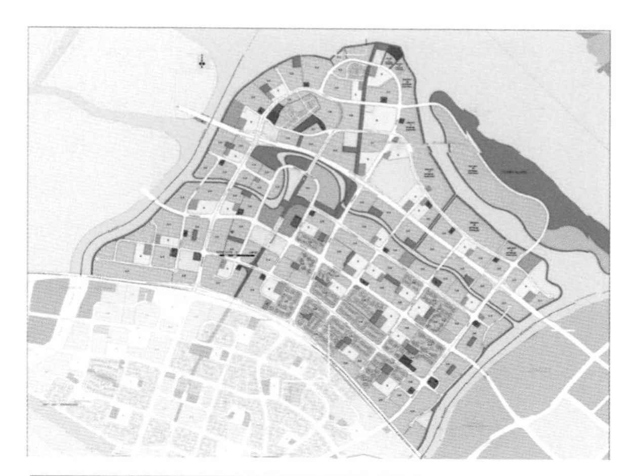

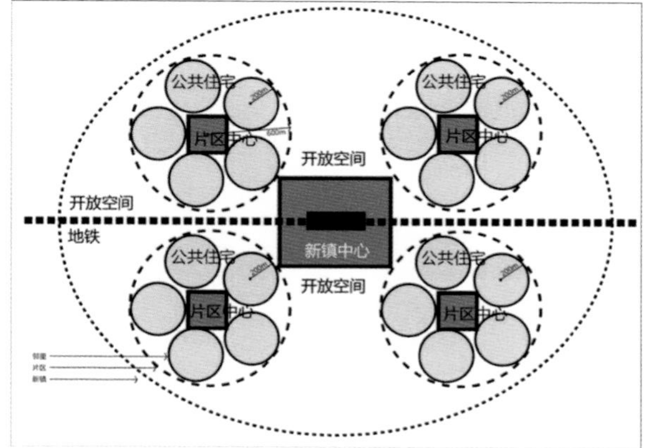

新加坡榜鹅新城的新镇组织模式
New Town Organization Model of Singapore Punggol New Town

以人为本组织居住社区

居住区配套借鉴新加坡邻里中心模式，三级社区服务中心紧密布局在开放空间的两侧，与开放空间内的休憩场所互为补充，构建生活配套服务网络。

街道空间的设计突破了传统道路设计的做法，将绿化带从道路红线外侧置换到红线以内，在机动车道、非机动车道、人行道之间设置绿化带，拉近了行人与建筑的距离，活跃了街道氛围。

People-oriented Organization Living Community

The residential area is based on the Singapore Neighborhood Center model, and the third-level community service center is closely arranged on both sides of the open space, complementing the rest places in the open space, and building a life supporting service network.

The design of the street space broke through the traditional road design practice. The green belt is replaced from the outside of the red line of the road to the red line. Green belts are set up between the motor vehicles, non-motor vehicle lanes and sidewalks, which narrowed the distance between pedestrians and buildings to active street atmosphere.

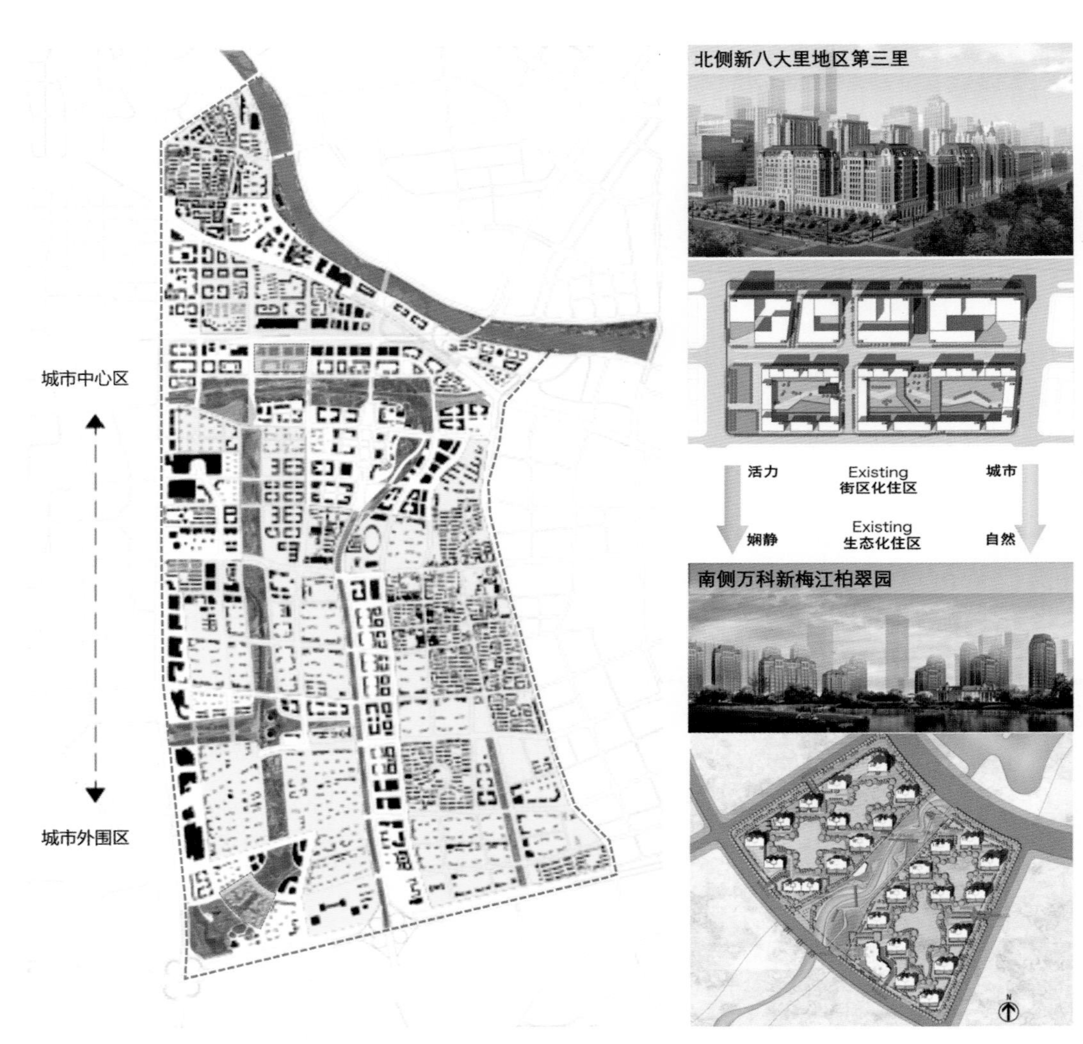

肌理演变
Urban Fabric Evolution

实验性建构开放式街区

居住布局形态南疏北密，在北部新八大里地区，实验性地建构了“窄路密网”的开放式街区，地块内部采取院落式布局，外侧底商为街道提供更多的商机和活力，内部围合院落进行封闭式管理，使居住环境安全私密。

Experimental Construction of Open Blocks

The layout of the residential blocks is of high density in the north and low density in the south. In the northern Badaili area, an open block of narrow road dense nets is experimentally constructed. The interior of the plot adopts a courtyard layout, and the outside base provides more business opportunities and vitality for the street. The enclosed courtyard is closely managed to make the living environment safe and private.

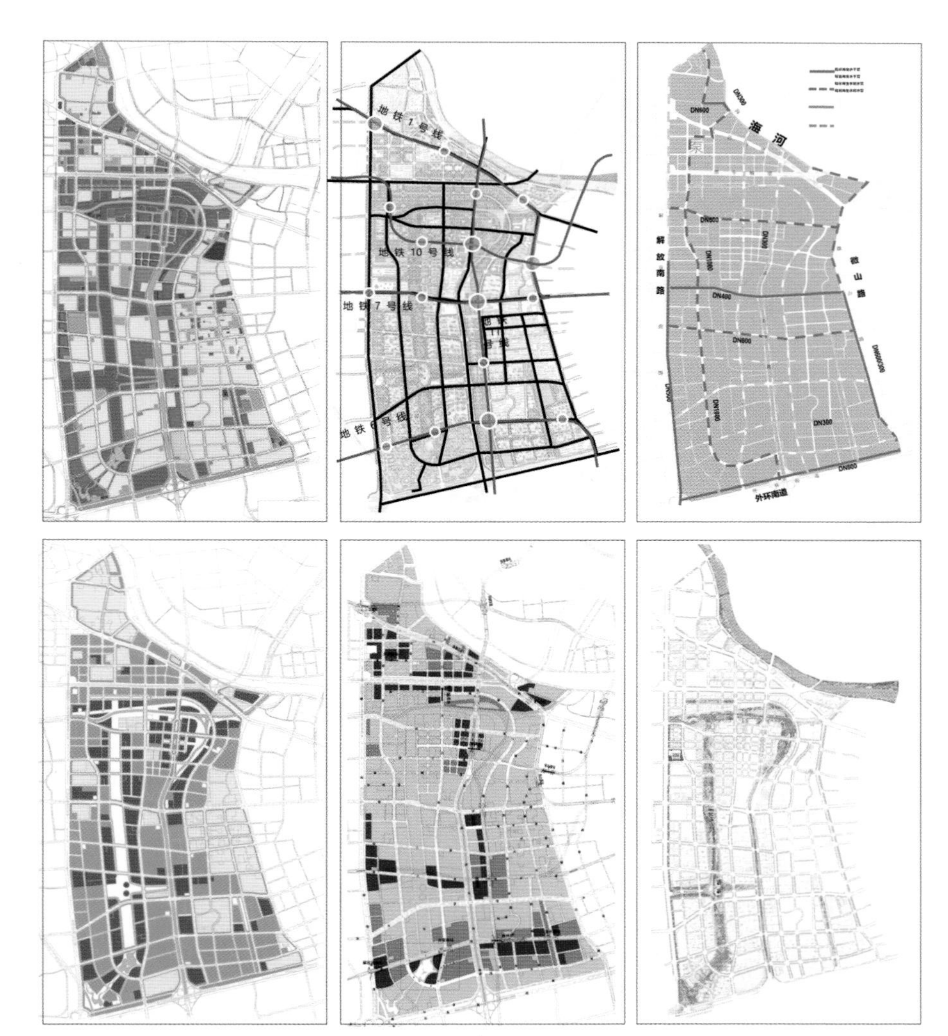

控制性详细规划和交通、市政、生态、地下空间、景观专项规划
Controlled Detailed Planning and Transportation, Municipal, Ecological, Underground Space, Landscape Special Planning

编制控规和专项规划

为将规划理念转化为对建设开发的有效管控，在城市设计基础上编制了设计导则，从整体层面对地区的空间形态、开发形态等方面进行全方位的把控，并对“两核三轴四带”的重点地块进行了分图则的详细编制。

在指挥部规划前期组的组织下，一并编制了控制性详细规划和交通、市政、生态、地下空间、景观等专项规划，从定性到定量，把城市设计转化为法定规划，有效控制建设实施。

Prepare Control Regulations and Special Plans

In order to transform the planning concept into effective management and control of construction and development, design guidelines were prepared on the basis of urban design, and all aspects were controlled from the overall level of the regional spatial form and development form, and the two cores were The key blocks of the "Two cores, three axises, four belts" were compiled in detail.

Under the organization of the pre-planning group of the headquarters, a detailed plan for control and special plans for transportation, municipal, ecological, underground space, and landscape were compiled. From qualitative to quantitative, urban design was transformed into statutory planning, and construction was effectively controlled.

用地建设情况
Land Use Construction

实施评估

经九年建设，整个解放南路地区已建设用地、已出让未建、在建用地合计达 200 公顷。

在规划指导下，南侧起步区和北侧新八大里地区已成规模：空间形态得到有效控制；多条道路已完成了设计与施工；多个居住地块已入住；环城绿道公园解放南路段和起步区两个大型生态公园已完成施工并投入使用，中央绿洲公园作为天津市海绵城市建设试点的重点建设内容正在推进中。

Implementation Evaluation

After nine years of construction, the entire land of South Jiefang Road has been constructed, and the total land that has been transferred and not built is 200 hectares.

Under the guidance of the plan, the starting area on the south side and the New Badali area on the north side have become scaled, and the spatial form has been effectively controlled; many roads have been designed and constructed; multiple residential plots have been occupied; the greenway park in the city has been liberated two large-scale ecological parks in the southern section and the starting area have been completed and put into use. The central oasis park is being promoted as the key construction content of the pilot construction of the sponge city in Tianjin.

解放南路地区起步区航拍实景（摄于 2019 年 7 月）
Aerial Map of the Starting Area of South Jiefang Road (Photo Taken in July 2019)

解放南路地区起步区规划效果图
Rendering of the Starting Area in South Jiefang Road Area

新八大里地区航拍实景（摄于 2019 年 7 月）
Aerial Picture of the New Badali Area (Photo Taken in July 2019)

新八大里地区规划效果图
New Badali District Planning Rendering

公园实景
Park Scene

生态建设评估

生态骨架首尾的绿道公园起步段、卫津河公园和太湖路公园已完成建设并投入使用。中段的中央绿洲公园尚未实施，目前作为天津市海绵城市建设试点的重点建设内容正在推进中。

公园没有先于周边地块实现建设，生态骨架未能南北贯通，对整个地区而言比较遗憾，无法体现生态品质对区域开发的带动效用。

Ecological Construction Assessment

The Green Road Park starting section, the Weijin River Park and the Taihu Road Park, which are the first end of the ecological skeleton, have been completed and put into use. The central oasis park in the middle section has not yet been implemented, and the key construction content of the pilot construction of the sponge city in Tianjin is currently being promoted.

The park did not build above the surrounding plots, and the ecological skeleton failed to penetrate north and south. It is regrettable for the whole region and cannot reflect the driving effect of ecological quality on regional development.

改造后的电机总厂
Transformed Motor Factory

现状陈塘热电厂
Current Situation of Chentang Thermal Power Plant

文化重塑评估

规划希望挖掘地区工业文化的历史特质，对于区域内各大厂区都做过大量的调研和设计工作，经多年努力保留下几处具有时代特色的工业建（构）筑物，其中电机总厂经改造现已作为售楼处投入使用，公交二公司、渤海无线电厂、陈塘热电厂的两个冷却塔因规划得以保留，为未来提升区域亮点预留了机会。陈塘铁路支线复兴河段结合环城绿道公园的建设，公园已建成并对公众开放，位于规划中的中央绿洲公园范围内的区段现也大部分得以保留。

比较遗憾的是，位于新八大里地区的起重设备厂因诸方原因于 2015 年随建设拆除，中部地区的陈塘热电厂当年因平账等原因只确定保留下来两座冷却塔，厂区内还有很多特色工业建（构）筑物，面临被拆除的命运。

Cultural Remodeling Assessment

The plan hopes to explore the historical characteristics of regional industrial culture. It has done a lot of research and design work for all major factories in the region. After years of efforts, it has retained several industrial structures with characteristics of the times. The sales office transformed from the moter plant was put into use, and the No.2 Bus Company, Bohai Radio Plant and two cooling towers of Chentangzhuang Thermal Power Plant were retained due to the plan, which provided an opportunity for future highlights. The construction of the Fuxing River section of the Chentang Railway branch line and the construction of the Ring Road Greenway Park has been completed and opened to the public. Most of the sections within the planned central oasis park are now preserved.

It is a pity that the lifting equipment factory located in the New Badaili area was demolished with the construction in 2015. The Chentangzhuang Thermal Power Plant in the central area was only determined to keep two cooling towers for the reasons of the account and other reasons. There are many characteristic industrial buildings and structures that face the fate of being demolished.

陈塘自主创新示范区
Chentang Independent Innovation Demonstration Zone

产业提升评估

原陈塘庄工业区经整体改造提升为“陈塘自主创新示范区”，是“天津国家自主创新示范区一区二十一园”之一，由原以纺织、化工、机械为主的高污染制造业成功转型为发展智能制造、文化、环保的都市型工业。

原规划的家居公园和汽车公园产业带未能实现出让及建设，现状仍为低产能的建材市场和汽修市场，因为无法实现职住在地平衡，原设想的跃迁式产业升级未能实现。

Industry Improvement Assessment

The original Chentangzhuang Industrial Zone was upgraded to “Chentang Independent Innovation Demonstration Zone” and was one of the “Twenty-one Districts of Tianjin National Independent Innovation Demonstration Zone”. It was highly polluted by the original textile, chemical and machinery industries. The manufacturing industry has successfully transformed into an urban industry that develops intelligent manufacturing, culture and environmental protection.

The original planned home park and automobile park industry belt failed to realize the transfer and construction. The current situation is still the low-capacity building materials market and the auto repair market. The original like transitional industry upgrade has not been realized because it is impossible to achieve the balance of occupation and living.

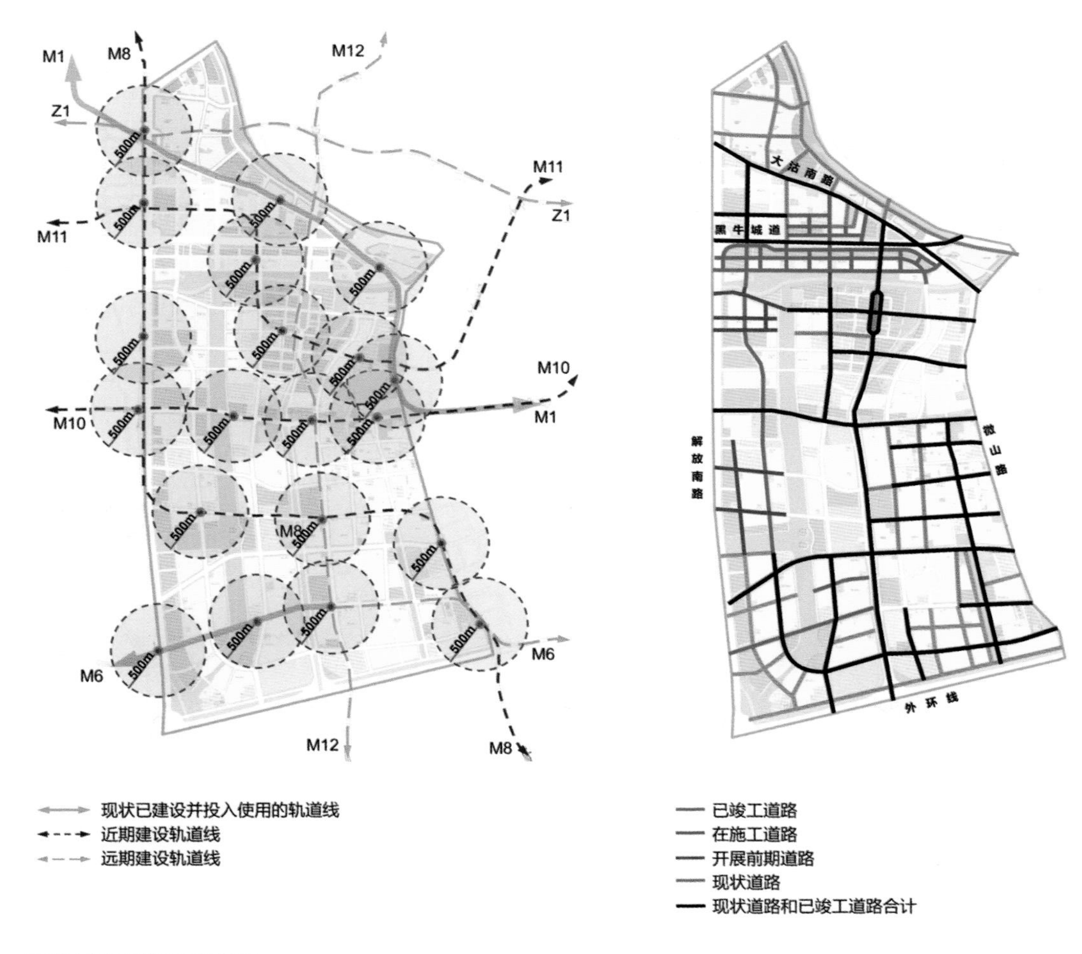

轨道交通及道路建设情况
Rail Transit and Road Construction

交通配套建设评估

交通配套方面，规划五横两纵共 7 条轨道线，其中 2 条现营运中，1 条在建，2 条计划于 2019 年开工建设。现状及近期建设的线站基本覆盖新八大里地区、陈塘科技园、起步区等人口较集中区域。已实施道路也集中在这三个集中建设的区域，已竣工、在施工和开展前期的道路合计 43 千米。近期将打通跨越复兴河的太湖路、内江路这两条重要的南北向连通道路。

Supporting Construction Assessment

In terms of transportation support, there are 7 track lines for five horizontal and two vertical lines, two in operation, one in construction, and two plans to start construction in 2019. The current situation and the recently constructed line stations basically cover areas such as the New Badali area, Chentang Science Park, and the starting area. The implemented roads are also concentrated in the three concentrated construction areas, with a total of 43 kilometers completed, construction and pre-development roads. In the near future, the two important north-south roads connecting Taihu Road and Neijiang Road across the Fuxing River will be opened.

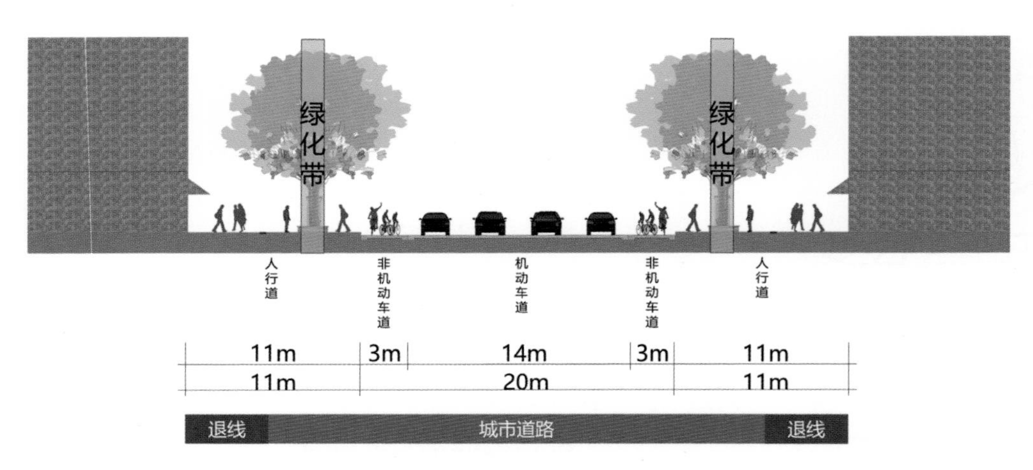

新八大里地区商业大街实施断面
Implementation Section of the Commercial Street in the New Badali Area

起步区创新断面道路实景
Innovative Section and Road Scene in Starting Area

交通配套建设评估

创新断面达到了规划预期，形成了良好的交通氛围和生活氛围，但在实施过程中也遇到了一些问题，如一些路段绿化部分已移交，日常有专人养护，但是道路部分还未移交，路面较破乱。

新八大里地区吸取了经验与教训，在次干道和支路设计上取消绿线，例如将商业大街人行道加宽至 11 米，其中 5 米作为建筑退让红线的距离，由市政代建后移交开发商管理，6 米为城市道路的组成部分，由市政建设管理。

Supporting Construction Assessment

The innovative section has reached the planning expectations and formed a good traffic atmosphere and living atmosphere. However, some problems have also been encountered in the implementation process. For example, some green sections of the road sections have been handed over, and there are special personnel to maintain daily, but the road parts have not been handed over.

The New Badali area has learned from the experience and lessons, and canceled the green line on the secondary trunk road and branch road design. For example, the commercial street sidewalk will be widened to 11 meters, of which 5meters will be used as the distance for the red line of the building, and will be handed over to the developer after the municipal construction. The part of 6 meters is an integral part of urban roads and is managed by municipal construction.

天津市第二新华中学和河西中心小学实景
Tianjin Second Xinhua Middle School and Hexi Central Primary School

邻里中心和文体中心实景
Neighborhood Center and Cultural Center

文化教育配套建设评估

其他配套方面，起步区的河西中心小学、天津市第二新华中学、河西锦绣幼儿园均已建或在洽谈中，邻里中心和文体中心已建成并移交使用。新八大里地区的上海道小学私立分校、师大二附小已建成。

Supporting Construction Assessment

In other aspects, the Hexi Central Primary School, Tianjin Second Xinhua Middle School and Hexi Jinxiu Kindergarten in the starting area have been built or under negotiation. The neighborhood center and the cultural and sports center have been completed and handed over. The Shanghai Dao Primary School Private Branch School and the Second Normal University of the New Badali area have been completed.

新八大里地区居住社区环境实景
Real Environment of New Badali Area Community

起步区全运村实景
Starting Area of Quanyun Village

小区建设评估

起步区与新八大里地区集中建设区已有大量还迁房、商品房入住或即将交房，已建成的小区具有优良的建设品质与环境效果，入住率较高。

Community Construction Assessment

There are a large number of relocated houses, commercial houses or upcoming houses in the starting area and the newly built area of the New Badali area. The completed community has excellent construction quality and environmental effects, and the occupancy rate is high.

实施总结

解放南路地区东西两翼均已形成以居住功能为主的成熟社区，该地区实施的建设因其中央绿洲公园、特色工业文脉与优良的人居环境等资源优势，成为区域提升的最佳承载地，为天津南部地区打造了具有文化、生态、活力特色的新城区。

Implementation Summary

Both the east and west wings of South Jiefang Road have formed a mature community with residential functions. The implementation of this area has become the best place for regional improvement due to its central oasis park, characteristic industrial context and excellent living environment. It has created a new urban area with cultural, ecological and vital characteristics for the southern part of Tianjin.

解放南路地区鸟瞰
Aerial View

以人为本的城市设计
People-Oriented Urban Design

当代中国城市设计体现出新的方向：生态文明、以人为本等方向

生态文明：现代城市规划主要是解决工业化后城市建设扩张带来的城市问题。1791 年，在朗方的华盛顿规划中，城市与自然浑然一体。1824 年，英国空想社会主义者罗伯特·欧文，在美国印第安纳规划建设了自给自足的“新和谐村”，探索了有组织的规划，避免工业化对农村地区的肆意侵害。1860 年，被称为“美国农民”的奥姆斯特德用自然的方式设计了纽约中央公园以及波士顿“翡翠项链”绿地系统，使人工化的城市中融入了自然的气息。1898 年，埃比尼泽·霍华德提出“城市与乡村磁铁”的田园城市理论，开启了现代城市规划的新篇章。1915 年，盖迪斯《进化中的城市》出版，认为城市规划应该是城市和乡村结合在一起的区域规划。可见，现代城市规划诞生的目的就是使城市与自然乡村融合。再反观我们的城市规划和城市设计，却主要以城市规模的扩张为目的，过度考虑设计实体的城市和建筑，而较少考虑生态环境和大尺度的公园绿地系统。实际上，中国传统的城市规划一直强调天人合一，道法自然，即使严格按照《周礼·考工记》形制规划的北京城，依然保留了自然水系和北海、中海、南海等三海系统，自然灵活的三海与严谨的中轴线交相辉映。

以人为本：与“人”尺度接近的细节决定城市设计的成败，要支持城市中心商业的复兴，就必须进一步完善步行系统，这是世界各国的成功经验。城市的步行化建设在天津大部分地区，特别是中心城区，还与国外许多案例存在较大差距，整体上不宜于人的步行。步行道被占用、景观缺乏设计的情况比比皆是。公园建得太少，无法满足人们的多种需求。外环线周边规划了十一个公园，但只建成了四个，还有七个未建。另一个是居住问题，发展的最终目的是提高人民的生活水平，城市设计要面对量大面广的居住和社区问题。但以往的城市设计在这方面的考虑不多，一直沿用计划经济的居住区规划方法，建设了大量大规模封闭的小区和没有个性的住宅，不适应人民对美好生活的需要。

Contemporary Chinese Urban Design Embodies New Directions: Ecological Civilization, People-oriented and Other Directions

Eco-civilization: The emergence of modern urban planning is mainly to solve the urban problems caused by the expansion of urban construction after industrialization.In 1824, Robert Owen, a British utopian socialist, planned to build a new harmonious village of self-sufficiency in Indiana, USA. Although the result was unsuccessful, he explored organized planning to avoid the wanton infringement of industrialization on rural areas. In 1860, Olmsted, known as the "American Farmer", designed the New York Central Park and Boston's "Jade Necklace" green system in a natural way, bringing a natural flavor into the artificial city. In 1898, Ebenezer Howard proposed the rural and urban magnet as a pastoral city, which opened a new chapter of modern urban. In 1915, Gaddis published *The Evolutionary City*, which argues that urban planning should be a regional planning that combines urban and rural areas. Therefore, the purpose of the birth of modern urban planning is to integrate the city with the natural countryside. On the other hand, our urban planning and design are mainly aimed at expanding the scale of the city, excessive one-sided consideration of the design of physical cities and buildings, and less consideration of the ecological environment and large-scale park green space system. In fact, China's traditional urban planning has always emphasized the unity of heaven and man, Taoism and nature. Even if Beijing City is strictly planned according to the form of *Zhou Li Kao Gong Ji*, it still retains the natural river system. The natural and flexible three seas intersect with the rigorous central axis.

People-oriented: Details close to human scale determine the success or failure of urban design. To support the revival of urban central commerce, it is necessary to further improve the pedestrian system, which is the successful experience of all countries in the world. In most areas of Tianjin, especially in the central urban area, there is still a big gap in the construction of urban pedestrianization, which is not suitable for people to walk on the whole. Walkways are occupied and landscape design is lacking everywhere. Also, there are too few parks and can't meet the various needs of people. We have planned eleven parks around the outer ring road, but only four have been built, and seven have not yet been built. The other is the problem of housing. The ultimate goal of our development is to improve people's living standards. Urban design is to face a large number of housing and community problems. However, our urban design has not given much consideration in this respect. We have been using the planned economy residential area planning method to build a large number of large-scale enclosed areas and houses without personality, which is not suitable for the people's desire for a better life in the future.

海河区域的水系统

每一个城市都是其所在流域的结晶。城市通过水系网络影响流域内功能区域，增强流域内部联系，维持流域生态承载力。

海河干流所在区域形成了以海河为主干，各支流为支干，小型河道、沟渠为“毛细血管”，各湖泊、水库、鱼池、坑塘为蓄滞洪区的水网系统，称为“九河下梢，水乡泽国”。

随着突飞猛进的城市化建设，一个个大型建设单元落位海河干流区域，各坑塘沟渠相继被填平，原有致密的水系结构遭到破坏，水网密度与水域面积降低，由此引发城市水环境的生态问题。

Water System of Haihe River Area

Every city is the crystallization of its basin. The city affects the functional areas within the basin through the water network, enhances the internal links of the basin, and maintains the ecological carrying capacity of the basin.

The area where the main stream of the Haihe River is located formed a water network system , the Haihe River is the main part, the tributaries are the branches, the small rivers and ditches are the capillary, and the lakes, reservoirs, fish ponds and pit ponds are of the storage area, called "nine rivers downstream, water town and swamp country".

With the rapid development of urbanization, one large construction unit locate in Haihe main stream basin by another. The pits and ditches have been filled successively, the original dense water system structure has been destroyed, and the water network density and water area have been reduced, thus causing ecological problems of urban water environment.

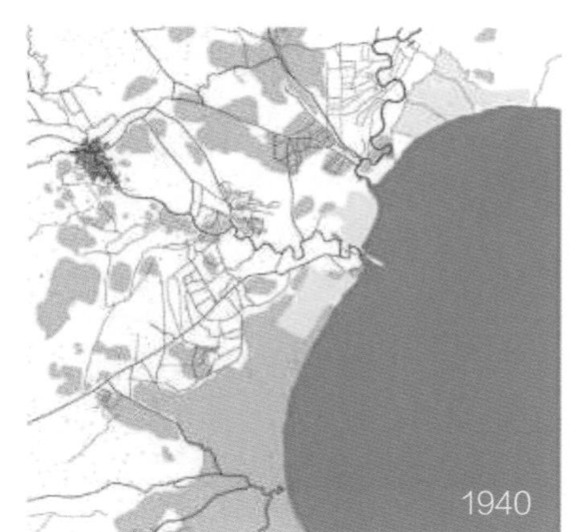

天津城市与水系统演变
Tianjin City and Water System Evolution

海河区域的水环境问题

水系统破坏引发环境问题，例如雨季暴雨引发城市内涝，旱季水资源不足影响农业生产等。区域用水对地下水依赖过大，天津所在的华北地区已经成为世界上最大的地下水漏斗区域，地下水超采会引发地下水污染和地面沉降等环境问题。

Water Environment Problems in Haihe River Area

Water system damage causes environmental problems, such as rainstorms in the rainy season and urban floods, and insufficient water resources in the dry season affect agricultural production. Regional water use is too dependent on groundwater. The north China region where Tianjin is located has become the world's largest underground funnel area. Groundwater over-exploitation causes environmental problems such as groundwater pollution and land subsidence.

海河区域的水环境问题
Water Environment Problems in Haihe River Area

景观生态规划理念

景观生态规划从 20 世纪初期，经历自帕特里克·盖迪斯的“区域规划”理论，到麦克哈格的“设计结合自然”，再到近年瓦尔德海姆的“景观都市主义”理念的演变，景观生态规划已成为解决城市环境问题切实有效的方法。景观已不只是游览与展示的“景”，更是切实解决生态问题的功能区域，是维护城市生态安全的基础设施。城市规划的对象将不只包括城市的功能区域，更包含整个生态片区的自然生态结构。因此规划应当是一种景观优先，在大的区域视角协调整个片区，构建区域生态格局的区域协调策略，应优先确定对区域生态安全具有关键意义的片区、廊道、节点，并将这些区域组合成维持地区生态安全的景观系统，这个系统应当具有明确的边界和严格的管控措施，包含多样的生态系统和一定的活力片区。

Landscape Ecological Planning Concept

From the early 20th century, the landscape ecological planning experienced the theory of "Regional Planning" from Patrick Geddes, to "Design With Nature" of Mcharg, and then to the "Landscape Urbanism" concept of Waldheim in recent years. It has become an effective way to solve urban environmental problems. The landscape is not only the "view" of the tour and display, but also the functional area to effectively solve the ecological problems of the district and the infrastructure to maintain the ecological security of the city. The object of urban planning will include not only the functional areas of the city, but also the natural ecological structure of the entire ecological area. Therefore, planning should be a landscape priority. Coordinating the entire area in a large regional perspective, and constructing a regional coordination strategy for the ecological pattern of the region, priority should be given to identifying areas, corridors, and nodes that are critical to regional ecological security. Synthesize a landscape system that maintains regional ecological security. This system should have clear boundaries and strict control measures, including diverse ecosystems and certain dynamic areas.

帕特里克·盖迪斯（1854—1932）

他认为应将自然地域作为规划的基本骨架，城市规划应该成为城市和乡村结合在一起的区域规划。

Patrick Geddes (1854—1932)

Taking natural areas as the basic skeleton of planning, urban planning should become a regional plan combining urban and rural areas.

伊恩·麦克哈格（1920—2001）

他认为大都市应保留关键的自然地域，为城市提供生态服务，并构建城市开放空间体系。规划应以自然优先，多学科、多因子相叠。

Ian Mcharg (1920—2001)

Metropolises should retain key natural areas, provide ecological services to cities, and build urban open space systems. Planning should be a natural priority, multidisciplinary, multi-factor overlapping.

查尔斯·瓦尔德海姆

他认为景观是一种城市支撑结构，能够容纳各种自然过程。他提出景观应作为生态基础设施，为城市提供持续的生态服务。

Charles Waldheim

The landscape is an urban support structure that can accommodate a variety of natural processes. The landscape should be used as an ecological infrastructure to provide sustainable ecological services for the city.

景观生态规划理论先驱
Three Masters of Landscape Ecological Planning

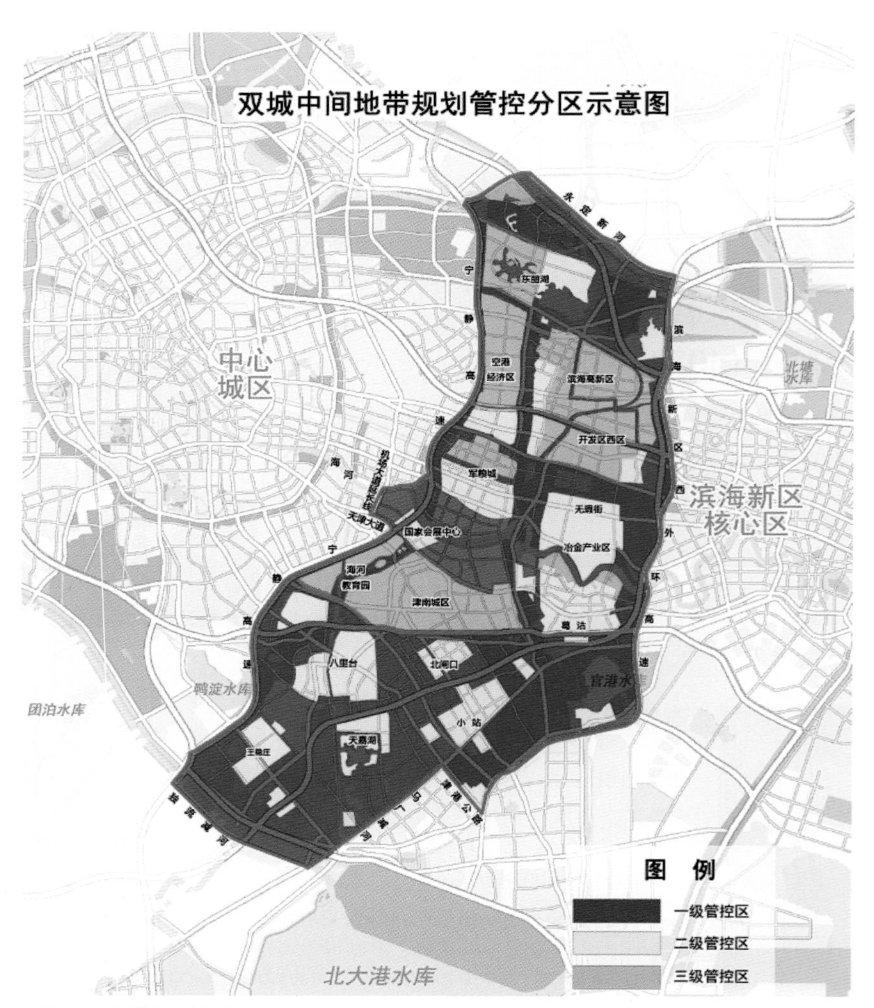

双城中间地带规划管控分区
Planning and Control Division for the Region Between Twin Cities

天津市双城间绿色生态屏障区规划

为贯彻国家生态文明建设和京津冀协同发展战略部署，保障京津冀地区和天津双城的生态安全，天津提出“滨海新区与中心城区要严格中间地带规划管控，形成绿色森林屏障”的决策部署，避免两城区持续扩张，未来连成一片，走“摊大饼”式的城市发展老路。

规划在东至滨海新区西外环，南至独流减河，西至宁静高速，北至永定新河，总计 736 平方千米的范围内，确立三个层级的管控分区，其中一级管控区为严格禁止建设区域，将成为对京津地区生态安全具有关键意义的生态区域。未来双城之间将呈现“大水、大绿、成林、成片”景观，形成“双城生态屏障、津沽绿色之洲”。

Planning of Ecological Barrier Between Twin Cities in Tianjin

In order to implement the national ecological civilization construction and the strategic deployment of the Beijing-Tianjin-Hebei region coordinated development, and to ensure the ecological security of the Beijing-Tianjin-Hebei region and the Tianjin twin cities, Tianjin proposed the decision to "strictly control the middle area of Binhai new district and the central city to form a green forest barrier". Deployment, to avoid the continued expansion of the two cities in the future will be one after another, take the "big cake" style of urban development.

It is planned to be located in the west outer ring highway of Binhai New Area in the east, to the Duliujian River in the south, to the Ningjing highway in the west, to the Yongdingxin River in the north, and within a total area of 736 square kilometers, three levels of control zones are established, of which the first-level control area is a strict ban on construction areas will become an ecological area of vital importance to the ecological security of the Beijing-Tianjin region. In the future, between the twin cities will present a landscape of "big water, big green, forest, and pieces", forming a "twin cities ecological barrier, a green continent of Tianjin".

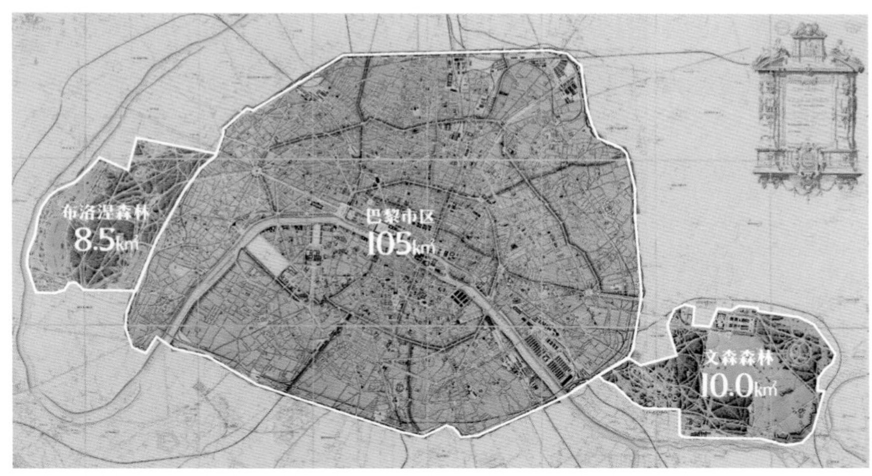

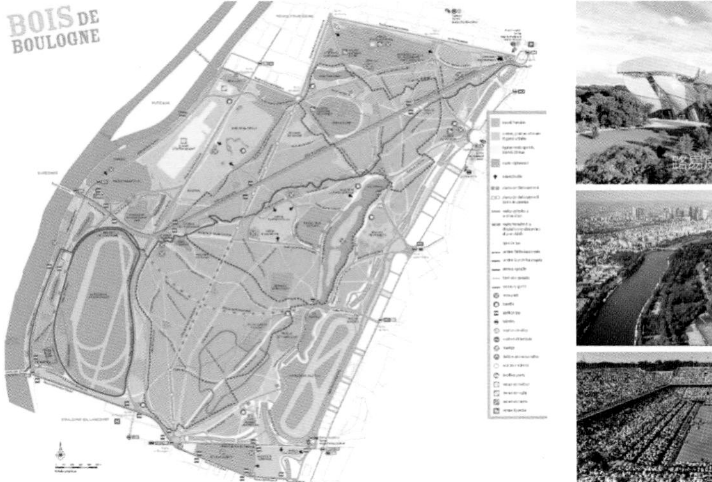

巴黎的都市绿心
Urban Green Heart of Paris

国际都市绿心研究

巴黎东西各有一个很大的森林公园，西部的布洛涅森林公园面积约 8.5 平方千米，相当于整个巴黎城区面积的十二分之一，东部的文森森林面积约 10 平方千米。两公园历史悠久，植被丰富，被视为巴黎的两叶肺，还包含动物园、游乐场、赛马场、足球场、体育馆、自行车赛场、临时展览馆、花卉公园等活力区域。

布洛涅森林公园承担多个具有国际影响力的活动，建有路易·威登基金会艺术中心、玛摩丹美术馆、罗兰·加洛斯网球中心、隆尚赛马场等。由《费加罗报》发起的“费加罗越野赛”迄今已经在布洛涅森林跑了 40 多年，参赛人数累计 120 万，是世界上规模最大的群众性越野障碍赛。

International Urban Green Heart Research

There is a large forest park on the east and west of Paris. The Boulogne Forest Park in the west is about 8.5 square kilometers, which is equivalent to one-twelfth of the area of the entire Paris. The Vincennes Forest in the east is about 10 square kilometers. The two parks have a long history and rich vegetation. They are regarded as the two lungs of Paris. They also include the zoo, playground, racecourse, football field, gymnasium, cycling stadium, temporary exhibition hall, flower park and other vital areas.

Boulogne Forest Park undertakes a number of internationally influential events, including the Louis Vuitton Foundation Art Center, the Mamodan Art Museum, the Roland Garros Tennis Center, and the Longchamp Racecourse. The "Figaro Cross Country" initiated by The Figaro has been running in Boulogne Forest for more than 40 years and has a total of 1.2 million participants. It is the largest mass off-road obstacle course in the world.

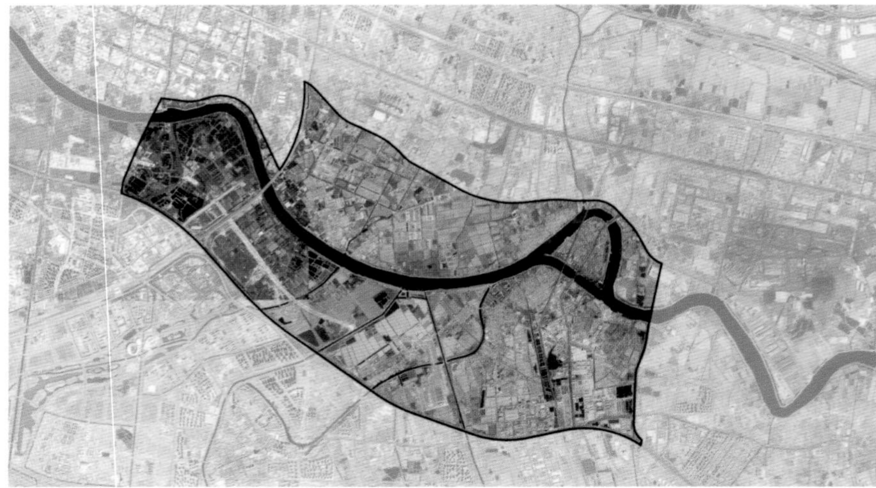

海河中游绿心范围
Range of Green Heart in Haihe Middle Reaches

海河中游绿心

海河中游区域一方面成为生态屏障区内“一轴三带五湖多廊”空间结构中的“海河景观带”，将成为连接东西两城、南北生态的核心区域，是维持城市生态承载力的关键性生态基础设施。

在另一方面，海河中游 23 千米形成“一带一城一区三镇多园多点”的区域结构，为区域内的津南新城、海河教育园区、葛沽民俗小镇、天钢文创小镇和军粮城宜居小镇等城市建设区域提供持续的生态供给。

海河中游葛沽以上段 12.5 千米，连同南北两岸共计超过 45 平方千米为建设区域，成为确保城市群生态安全的生态屏障，以及具有国际品质的都市绿心。

Green Heart in Haihe Middle Reaches

On the one hand, the middle reaches of the Haihe River has become the “Haihe River Landscape Belt” in the spatial structure of “one axis, three belts, five lakes and multiple corridors” in the ecological barrier zone, which will become the core area connecting the east and west cities and the north and south ecology, and the key ecological infrastructure to maintaining the urban ecological carrying capacity.

On the other hand, 23 kilometers in the middle reaches of the Haihe River formed a regional structure of “one belt, one city, one district, three towns, multiple gardens and more points”, provide a continuous ecological supply for urban construction areas such as Jinnan New City in the region, Haihe Education District, Gegu Folk Town, Tiangang Cultural Town, Junliang cheng Livable Town, .

In the middle reaches of the Haihe River, 12.5 kilometers above Gegu Town, together with the total area of more than 45 square kilometers on both sides of the Haihe River, is an ecological barrier that ensures the ecological security of city cluster and has an international quality urban green heart.

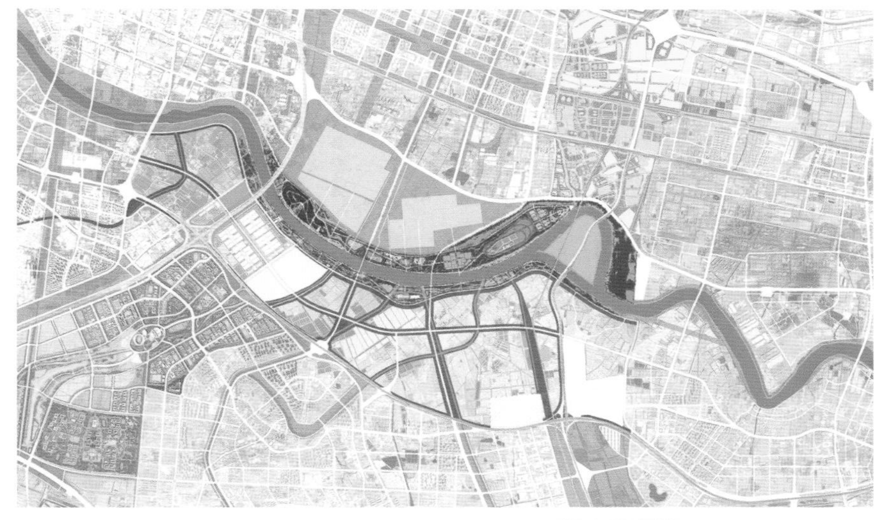

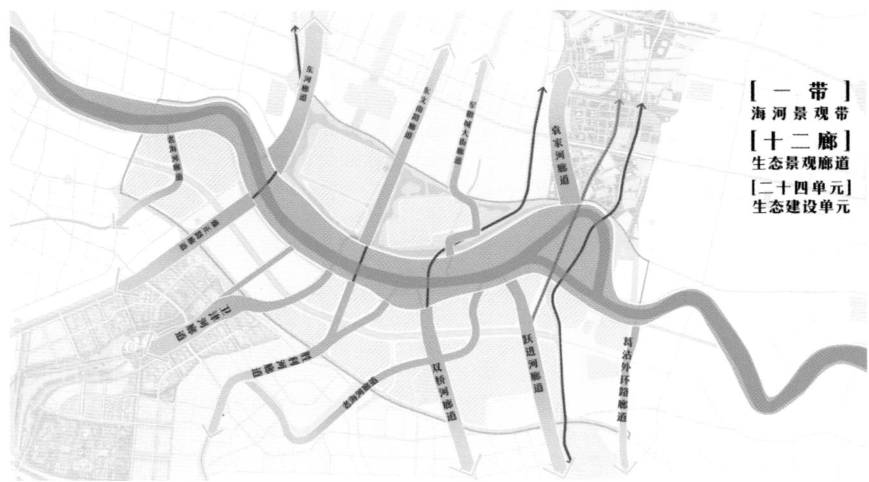

海河中游绿心规划
Planning of Green Heart in Haihe Middle Reaches

海河中游绿心规划

海河中游形成“一带、十二廊、二十四单元”的景观结构。海河中游鱼骨状生态骨架将区域划分为多个生态建设单元。

“一带”即海河景观带。未来沿海河形成长度超过 12.5 千米具有宽窄变化的景观带，总面积超过 8 平方千米。景观带除生态要素，还包含展出花园、活力公园、运动场等功能区域。景观带设置堤顶游线和水上游线，包含码头、驿站等节点。

“十二廊”即十二条生态景观廊道。是海河中游区域沿海河支脉向南北延伸的生态廊道，构建区域生态与游憩网络，使绿带向海河南北渗透。

“二十四单元”即二十四个生态建设单元。以水、路为边界，形成 24 个面积大小不一的生态岛屿。

Planning of Green Heart in Haihe Middle Reaches

The Haihe middle reaches form a landscape structure of “one belt, twelve corridors and twenty-four units”. In the middle of the Haihe River, the fish bone-shaped ecological skeleton is divided into several ecological construction units.

“One belt” is the Haihe River landscape belt. In the future, the coastal river will form a landscape belt with a wide and narrow variation of more than 12.5 kilometers, with a total area of more than 8 square kilometers. In addition to ecological elements, the landscape also includes functional areas such as exhibition gardens, vital parks, sports fields. The landscape belt is provided with a top line and a water upstream line, including nodes such as docks and stations.

“Twelve corridors” are twelve landscape ecological corridors. It is an ecological corridor extending from the coastal river branch in the middle reaches of the Haihe River to the north and south, constructing a regional ecological and recreational network, and infiltrating the green belt to the north and south of the Haihe River.

"Twenty-four units" are twenty-four ecological construction units. With water and road as the boundary, there are 24 ecological islands of different sizes.

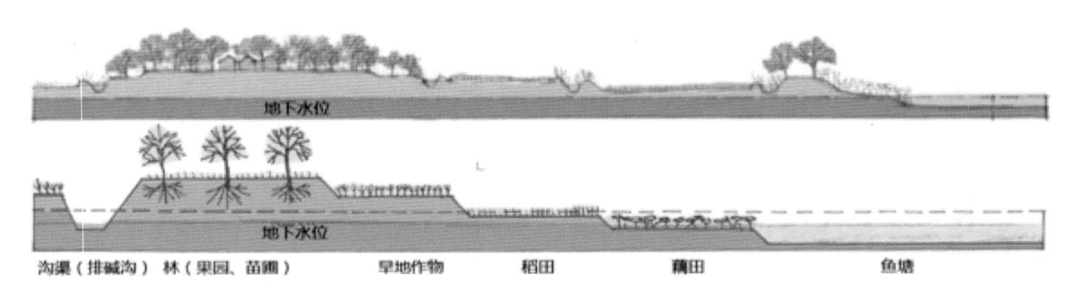

传统台田营建策略
Traditional Platform Field Construction Strategy

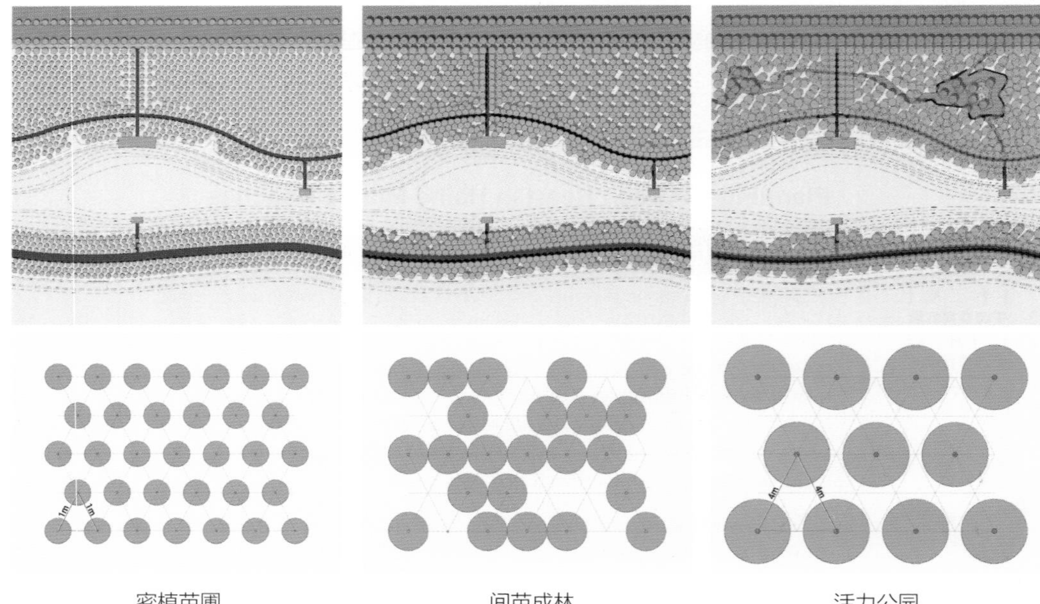

密植苗圃
Close Planting Nursery

间苗成林
Sparse Trees

活力公园
Vital Park

基于本土的设计策略
Local Based Design Strategy

海河中游绿心设计策略

天津整体地势较低，且处于盐碱地带，直接种植在地面的乔木生长数年后，根系将扎入盐碱含量较高的浅层地表水层，难以存活。借鉴天津地区传统台田经验，通过简单的土方工程，将低地的土堆到高地上，营造富有变化的地形，在高地上种植树木，保证乔木存活率，并与景观结合，形成多样的景观体系。

现阶段采用高密度种植低胸径乔木的办法，将生态涵养区建设成为储备林地和苗圃基地。随着林地乔木长大，逐步进行间苗，以园养园，营造生产型的低碳高效景观，实现公园与城市协同共生，将城市规划成为“生长的城市，变化的公园”。

Design Strategy of Green Heart in Haihe Middle Reaches

The overall terrain of Tianjin is low, and it is in the saline-alkali zone. After several years of growth of trees directly planted on the ground, the roots will be plunged into shallow surface water layers with high salinity and high content, which is difficult to survive. Drawing on the experience of traditional platform field in Tianjin, we use simple earthworks to pile up lowland mounds into highlands, create varied terrain, plant trees on highlands, ensure tree survival, and combine with landscapes to create diverse landscape system.

At this stage, we use high-density planting of low-breasted trees to build ecological conservation areas into reserve forests and nursery bases. With the growth of arbor trees in the forest, the gradual planting of seedlings will be carried out to raise the production-oriented low-carbon and high-efficiency landscape, and realize the synergy between the park and the city, and become a “growing city, changing park”.

海河中游绿心展望
Outlook of Green Heart in Haihe Middle Reaches

海河南岸的未来畅想

未来位于海河南岸的国家会展中心建成后，每年都会举办国际级别会展，南岸片区逐渐成为天津各项国际活动的承办地。国际会展中心东侧 180 公顷区域，可以承办世界博览会，将整个片区变成世博文化园，并继续举办相关国际活动。

世博文化园内西侧一个 11 公顷的岛屿规划为国际网球赛事的场所，里面新建的网球主场馆是世界顶级网球硬地赛场之一，赛场南侧一片 33 公顷的岛屿将依托国际网球中心入驻多家网球俱乐部，每到周末京津冀地区有大量的人驱车前来参加网球活动。

世博文化园东侧一片 67 公顷的岛屿群，可将其规划为举办园博会的场所，在园博会的基础上修建天津植物园和新的天津动物园，成为天津一个重要景点，每年吸引数百万的游客到这里参观游玩。

Future Imagination on the South Bank of Haihe River

After the completion of the National Convention and Exhibition Center on the south bank of the Haihe River, an international level exhibition will be held every year. The south bank area will gradually become the host of various international activities in Tianjin. The 180-hectare area on the east side of the National Convention and Exhibition Center will host the World Expo, transform the entire area into an expo cultural park, and continue to hold related international events. An 11-hectare island on the west side of the expo cultural park has become a venue for international tennis events. The new tennis stadium is the world's finest tennis hard court. The 33-hectare island on the south side of the stadium is based on the international tennis center. At the tennis club, a large number of people drive to the tennis event every weekend in the Beijing-Tianjin-Hebei region.

A 67-hectare island on the east side of the Expo Cultural Park will host the garden expo. The Tianjin Botanical Garden and the new Tianjin Zoo will be built on the foundation of the Expo. It will become an important attraction in Tianjin, attracting millions of visitors every year.

海河中游绿心展望
Outlook of Green Heart in Haihe Middle Reaches

海河北岸的未来畅想

海河北岸的城市森林将成为天津重要的生态储备林区和苗圃基地，这里如同一个绿色心脏，一直向周边城市输送苗木，成为城市行道树和公园树。东侧 145 公顷的区域将成为国际花园节的举办场所，世界各地的设计师将在这里通过设计进行艺术交流。更重要的西侧 250 公顷片区，将成为天津未来举办奥运会的场地，迎接天津百年未有的大事件。

Future Imagination on the North Bank of Haihe River

The urban forest on the north bank of the Haihe River has become an important ecological reserve forest area and nursery base in Tianjin. It is like a green heart. It has been transporting seedlings to surrounding cities and has become a city roadside tree and a park tree. The 145-hectare area on the east side will be the venue for the international garden festival, where designers from all over the world will communicate through art. While the 250-hectare area on the west side will become the venue for Tianjin to hold the Olympic Games in the future, and other big events.

美国纽约实施“抢街革命”前后对比图
Comparison Before and After The “Street Fight” Revolution in New York

世界城市发展趋势：更多步行 = 更少拥堵 = 更多活力

当前，全球有 29 个国家的 50 多个城市正在开展步行城市建设，如纽约的“抢街革命”、伦敦的特拉法加广场改造等，越来越多的城市用实践证明，建设步行城市是破解城市拥堵的重要手段和复兴城市活力的关键举措。

World Trends: More Walking = Less Congestion = More Activity

Currently, more than 50 cities in 29 countries around the world are building walkable cities, such as the "Street Fight" revolution in New York and the reconstruction of Trafalgar Square in London, etc. More and more urban practices show that the construction of walkable city is an important means to break urban congestion and a key measure to revive urban vitality.

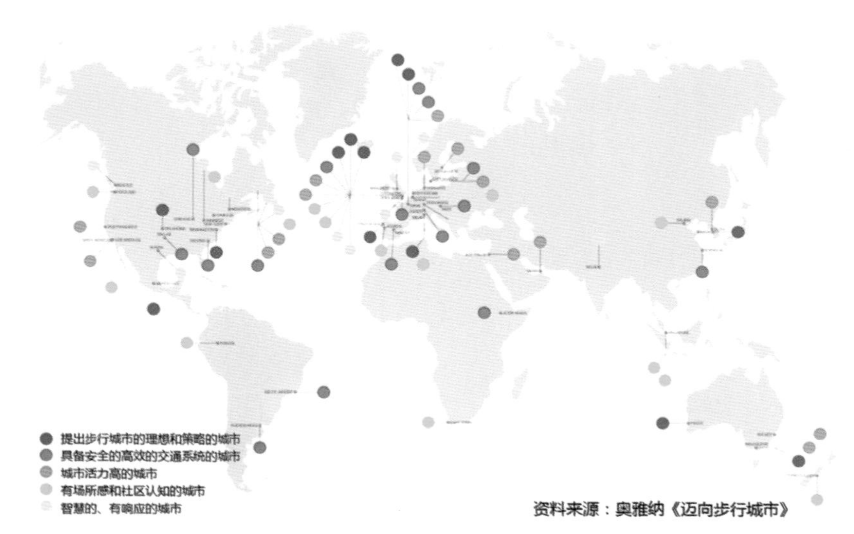

全球范围内进行步行城市建设的城市
Cities Building Walkable Cities Around the World

天津历史文化街区道路系统：窄路密网，特色鲜明

“近代百年看天津”，天津是近代北方京畿门户和金融商贸中心，1986 年被列为国家历史文化名城，拥有十四片历史文化街区。天津历史街区道路系统始建于明清，繁荣于民国时期，是西方窄路密网规划理念在近代中国最早的实践地区之一，具有路网密度高、街区尺度小，商业氛围浓厚的特点。历史街区道路系统是中心城区路网的重要组成部分和步行化人性城市建设的重要载体。

Road System of Historic Blocks in Tianjin: Narrow Road,Dense Network and Distinctive Features

As the saying goes, "If you want to see the modern China in last 100 years, go to Tianjin," Tianjin served as the gateway to capital Beijing and the cradle of modern industry and commerce during the 18th to 19th In 1986, it was listed as a national historical and cultural city, with 14 historical and cultural areas .The road system of historic blocks was built in Ming and Qing dynasties and flourished from 1912 to1949. The narrow road network was first planned in Tianjin, with the characters of high road network density, small street scale, strong business atmosphere. The historical block road system is an important part of the road network in the central city and an important carrier for the construction of walkable cities for people.

1945 年天津街道地图
Map of Tianjin in 1945

历史街区的道路肌理
The Road Texture of the Historic Areas

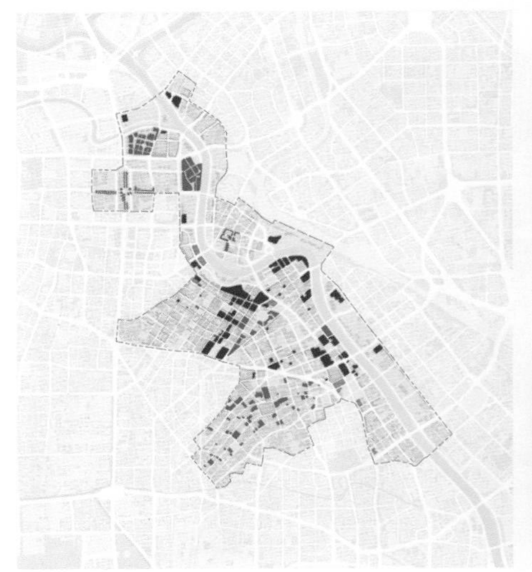

历史街区的商业设施
Commercial Facilities in the Historic Areas

和平路今昔对比
Comparisons of Heping Road Today and Yesterday

规划目标：世界一流的适宜步行的活力城区

规划提出聚焦中心，对标国际，用 3 ~ 5 年时间将天津历史文化街区提升成为世界一流的适宜步行的活力城区。

结合新时代新要求，实现街道空间三方面转变：

在空间属性上由机动车交通空间向步行化生活空间转变；在路权分配上由“以车为本”向“以人为本，兼顾车行”转变；在生活方式上由快节奏、讲效率向慢节奏、讲品质的转变。

Goal of Planning: A World-class Walkable and Vibrant City

Focus on the city center and according to international standards, we hope to promote Tianjin historical and cultural areas into a world-class walkable vibrant city in 3-5 years.

Combined with the new requirements of the new era, the street spaces will be transformed into three ways:

From motor vehicle traffic space to pedestrian living space; From "car oriented" road to "people oriented" road; From fast-paced life with high efficiency to a slow one with high quality.

海河步行桥
Pedestrian Bridge Over the Haihe River

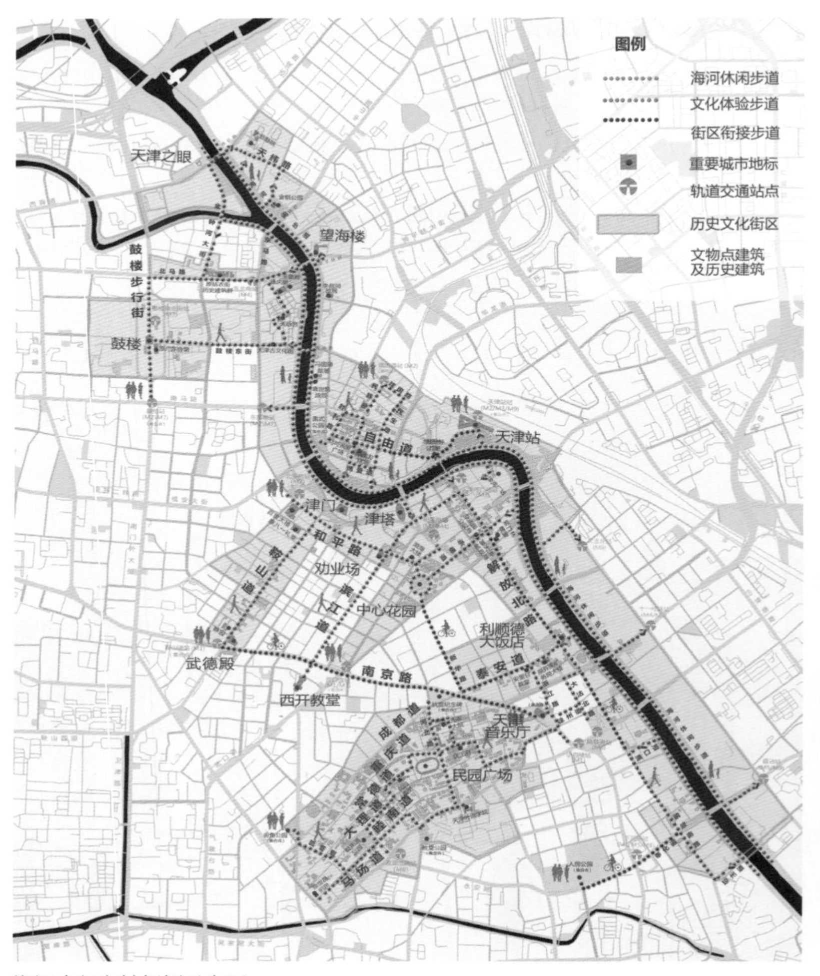

海河步行主轴规划示意图
The Schematic Diagram of Haihe Pedestrian Axis Planning

规划策略 1：强化一条海河步行主轴

重点提升海河主轴的贯通性，提出三条贯通措施。一是顺河贯通，逐一贯通两岸步行断点，建立连续无障碍的步行空间。二是垂河管控，结合道路交通规划，打通鞍山道、四平东道等道路的垂河断点，提高滨河可达性。三是跨河增桥，结合两岸桥梁建设规划，加快推动四处跨河步行通道建设，提升两岸历史文化街区可达性。

Strategy 1: Strengthen the Haihe Walking Main Axis

Focus on improving the walking continuity of Haihe axis, there are 3 strategies. Firstly, clear the walking break points to create a continuous and accessible pedestrian space. Secondly, combined with road traffic plan , connected the break point of vertical river Anshan Road and East Siping Road and improve the accessibility of riverside. Thirdly, combined with the planning of bridge construction, accelerate the construction of four cross-river pedestrian routes to improve the accessibility of historical and cultural areas on both riversides.

鞍山道垂河断点改造示意图
Create the Slow Lane to Clear the Walking Breakpoint in Anshan Road

桥梁步行断点改造示意图
Reconstruction of the Walking Breakpoint Under Bridge

规划策略 2：贯通三级城市步行网络

· 区域层面，建立外围交通保护圈，限制过境交通进入历史街区，实现“五横四纵”慢行骨架的静稳化改造。

· 街区层面，优化路权分配，对八个步行街区内贯通性强且步行流量较大的道路实施“抢街革命”，缩减车行空间，增加慢行空间。

· 邻里层面，清理和疏通街巷里弄，形成可穿越的里巷空间步行网络。

Strategy 2: Connect Three - level City Walking Network

· Regional level: Plan a peripheral traffic protection ring to restrict transit traffic to historic areas and calm “five horizontal and four vertical” slow-traffic system.

· Block level: Carry out "street-fight revolution" for roads with strong connectivity and large pedestrian in 8 characteristic walking blocks to reduce the car space and increase the slow-walking space.

· Neighborhood level: Clear and dredge streets and lanes to form a walkable space network.

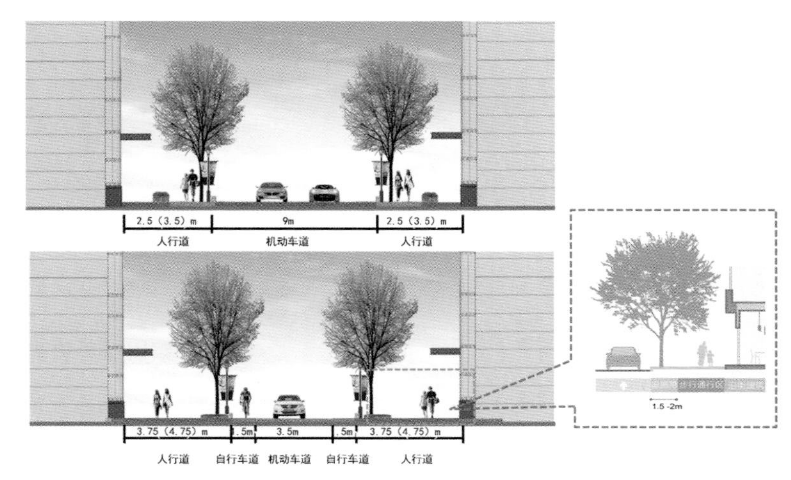

滨江道（大沽北路—解放北路段）道路断面改造设计示意图
The Design of Section Renovation on Binjiang Road.（Dagubei Road-North Jiefang Road)

滨江道（大沽北路—解放北路段）道路断面改造空间效果示意图
The Effect Picture of Section Renovation on Binjiang Road（Dagubei Road- North Jiefang Road)

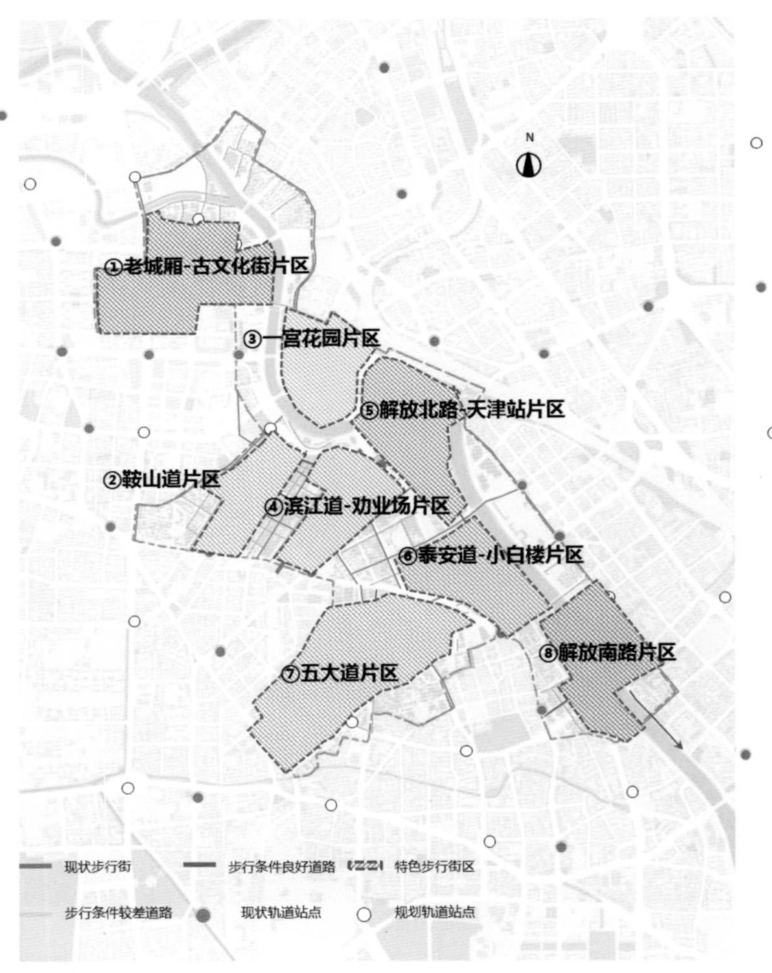

八个特色步行街区示意图
Eight Characteristic Walking Areas

规划策略 3：提升八个特色步行街区

深化八个特色步行街区城市设计，将其打造成特色鲜明、功能完善、活力十足的步行者天堂。做法包括：完善街区功能，增加商业、文化设施配套；突出鲜明场景，重点提升各街区的核心标志空间，明晰标识指引，建立智能化、特色化的标识指引系统；提升空间品质，增加休憩设施，改造碎片化空间。

Strategy 3: Upgrade 8 Characteristic Walking Blocks

Deepen the urban design of 8 walking blocks to make them pedestrian paradise with distinctive features, complete functions and full vitality. The measures include: increase the commercial and cultural facilities to improve the block function; highlight distinct scenes to improve the core sign space of each block, establish intelligent and characteristic sign guidance system; increase recreation facilities and transform the fragmented space to improve the quality of space.

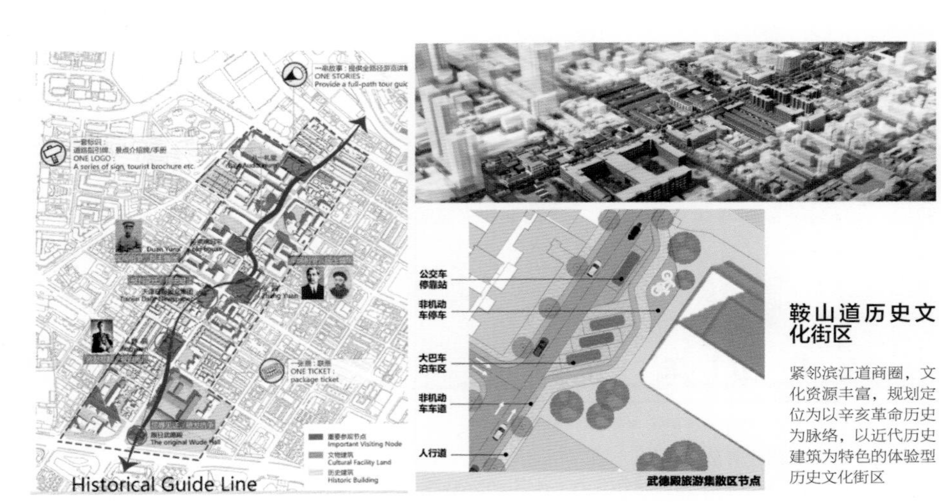

鞍山道历史文化街区城市设计
Urban Design of Anshan Road Historical and Cultural Area

规划策略 4：优化三类关键步行节点

结合步行适宜性评估结果，对地铁出入口节点、交通枢纽节点和干路交叉口节点，有针对性地优化提升，增强步行安全性和舒适性。

一是增加交通渠化设计，引导车流和行人各行其道，增强安全性；

二是改变路口转弯半径，缩小行人过街距离，降低车辆过街车速；

三是拓宽节点慢行空间，美化亮化节点环境。

Strategy 4: Optimize 3 Kinds of Key Walking Nodes

Combined with the assessment results of pedestrian suitability, we plan to improve the subway entrance node, transportation hub node and main road intersection node in a targeted way to provide safer and more comfortable walking experience.

Firstly, increase channelization to guide traffic flows and pedestrians in different ways and to enhance safety;

Secondly, change turning radii to reduce the distance of crossing the street and the speed of vehicles;

Thirdly, increase slow traffic system space and beautify the environment.

小白楼地铁站周边现状
The Current Environment of Xiaobailou Subway Station

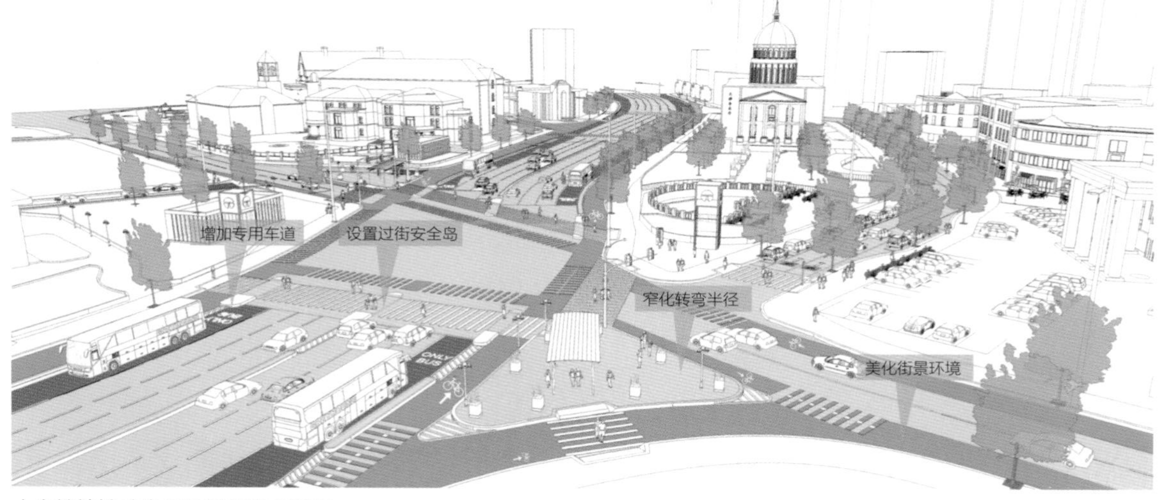

小白楼地铁站步行环境改善示意图
Walking Environment Renovation of Xiaobailou Subway Station

规划策略 5：塑造一批文旅步行线路

以“时光漫步，感知天津” 为主题，基于各街区内丰富的历史文化遗产，策划八条文化体验路径，每条线路步行时间约 2 小时，采取多元化的导览形式，通过串联景观节点、历史建筑、商业中心和轨道站点，以点带面，实现街区文化旅游融合发展。

Strategy 5: Create a Set of Cultural Walking Routes

With the theme of “Wander in history and experience Tianjing ”,we will design 8 cultural pedestrian routes based on the rich historical and cultural heritage. Each route takes about 2 hours to walk. Some landscape nodes, historic buildings, commercial centers and rail stations will be connected by routes and different guidelines to develop district cultural tourism.

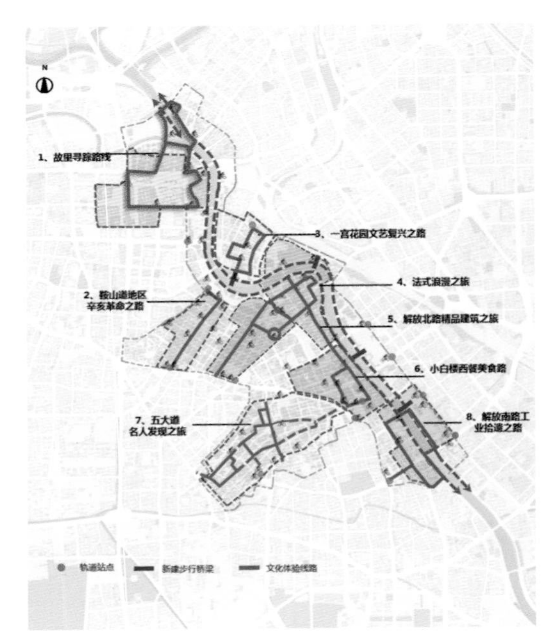

八条文化旅游体验路径示意图
Eight Cultural Walking Routes

多元化步行导览示意图
Diversified Walking Guidelines

巴黎塞纳河步道品牌建设
Case Study: Brand Building of Paris Seine Trail

保障措施

健全管理机制，统筹指挥建设：建立市级道路改造指挥部，协调相关部门和利益主体，对建设项目和开发时序统一部署安排。

鼓励多方参与，拓宽资金渠道：完善财政机制，采用政府设立专项资金和多元主体参与相结合，共同拓宽步行系统建设的资金渠道。

编制技术规程，加强规划引导：编制历史街区道路设计导则和技术规程，指导各街区、各路段的道路改造。

加强宣传推广，打造特色品牌：通过“线上＋线下”渠道广泛宣传，形成特色鲜明的街区形象，打造国内外知名的品牌步行街。

推出智能平台，发展智慧交通：加强道路信息化基础建设和大数据应用，实现步行信息归集、整理和分析，优化城市交通管理。

Safeguard Measures

To improve the management mechanism and establish a municipal road reconstruction headquarters to coordinate relevant departments and stakeholders, and make unified arrangements for the construction projects.

To improve the financial mechanism, we suggest the government set up special funds and encourage the participation of various parties to broaden the financial channels for the construction of walking system.

To publish the road design guidelines and technical regulations for historical blocks by the planning department,in order to guide the road reconstruction of each block.

To strengthen the publicity and promotion, through “online + offline” channels of wide publicity to form a distinctive street image, in order to build a well-known pedestrian street.

To develop smart transportation an intelligent platform for walking system should be established by strengthening road informatization infrastructure and big data application.

The platform can collect, organize and analyze walking information and optimize urban traffic management.

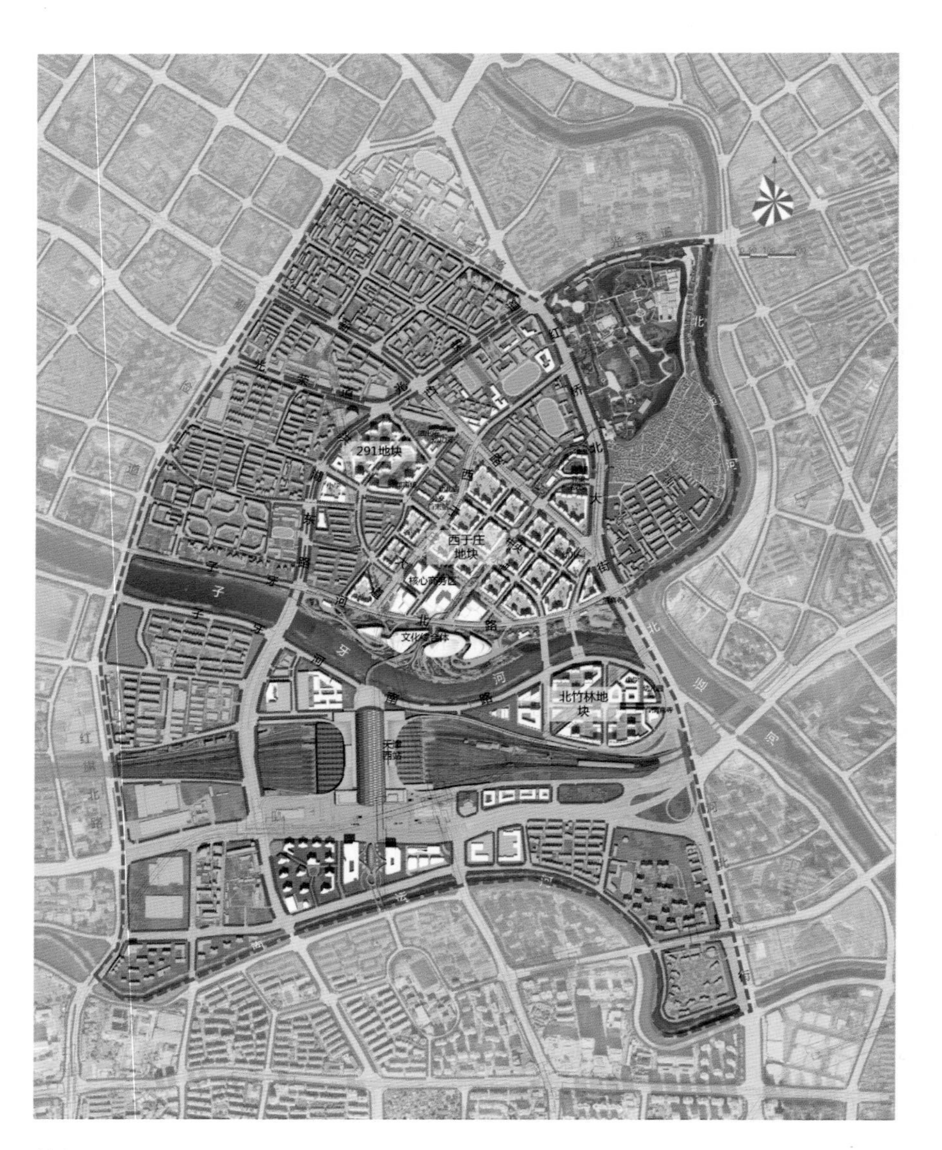

项目概况

高铁是近年快速发展的新型交通模式，高铁站影响下的大型城市综合交通枢纽作为多种现代交通方式的汇集点，势必带动周边地区的发展。从枢纽功能与区域功能一体化的角度来看，西站地区将建设成为集商务金融、商业贸易、文化休闲、居住于一体的，集中展现天津崭新城市形象的地区。规划通过完善子牙河两岸的滨水空间体系，复兴河岸公共空间，使之成为城市生活的重要载体，借此提升区域活力；挖掘历史人文资源，通过运河复航等策略，将各要素协调统一，合力发展。以大运河国家文化公园建设为契机，基于区域优良生态本底，再塑一条天津经济、文化发展脉络，提升城市品质。

Project Overview

The high-speed railway is a new type of transportation mode that has developed rapidly in recent years. The large-scale urban comprehensive transportation hub under the influence of the high-speed railway station is a gathering point of various modern modes of transportation, which is bound to drive the development of the surrounding areas. From the perspective of the integration of hub functions and regional functions, the West Station area will be built into a region that integrates business finance, commercial trade, cultural leisure, and residence, and focuses on the image of Tianjin's new city. The plan is to improve the waterfront space system on both sides of the Ziya River, revitalize the public space on the river bank, and make it an important carrier of urban life, thereby enhancing regional vitality; excavating historical human resources, and reorganizing various elements through strategies such as canal resumption , joint development. Taking advantage of the construction of the Grand Canal National Cultural Park, based on the excellent ecological background of the region, it will reshape a vein of Tianjin's economic and cultural development and enhance the quality of the city.

西站地区设计总平面图
General Plan of West Railway Station Area

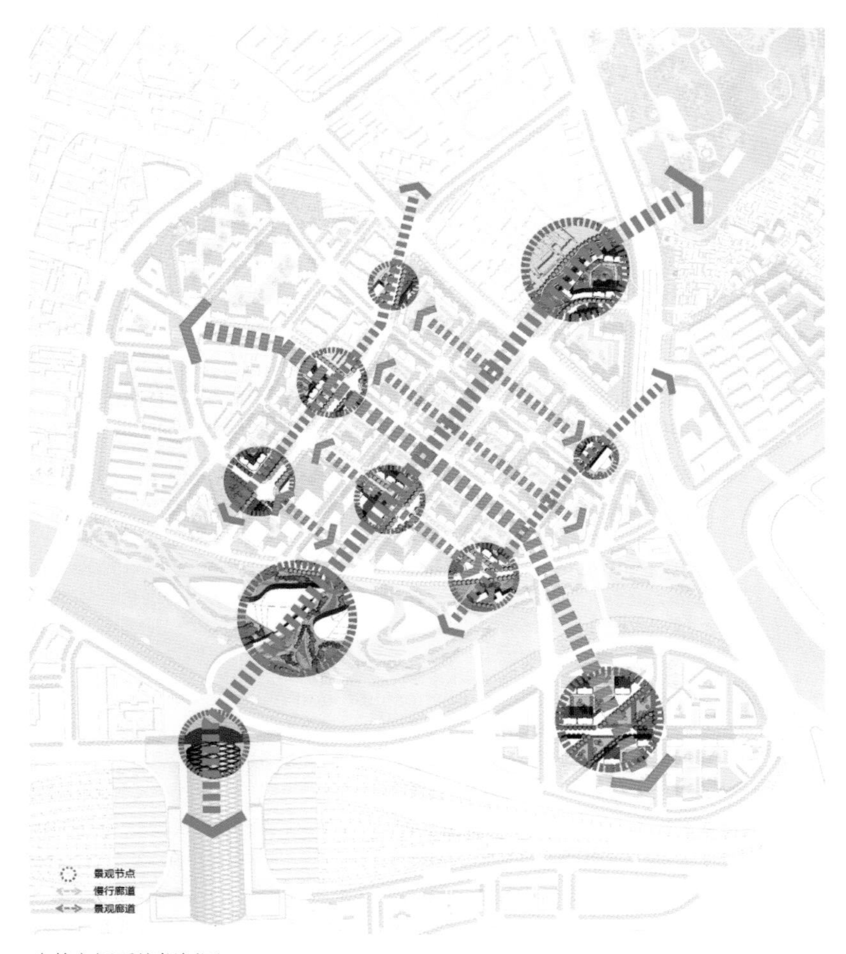

公共空间系统规划图
Public Space System Plan

复兴——依托人文资源激活滨河空间

1963 年秋季，海河流域的暴雨引发特大洪水，造成了巨大经济损失。为降低洪水威胁，子牙河沿线修建防洪堤岸，西于庄地区的堤岸比腹地高出 2 米左右，区段内的滨河空间直接由水岸过渡到居住，主要承担的是防洪功能。作为城市公共活动空间的重要载体，子牙河与城市空间分离。应具备的公共性、生态性、亲水性未得到体现，滨河空间的活力亟待挖掘提升。

规划尝试从人文、生态的角度重构公共空间体系，创造多层次的城市生活空间。节点——街角公园，为市民提供就近的活动场所，促进市民的交流。路径——街道，是连续性最强的公共空间，通过梳理步行体系，可在不同节点空间之间实现互联互通，塑造了清晰的城市空间骨架，促进了各节点空间转化成活力场所。开放空间——打造丰富多彩的滨河亲水步道空间，使之成为地区最具活力和吸引力的要素。

Revitalization: Activate Riverside Space With Human Resources

In the autumn of 1963, heavy rains in the Haihe river basin caused huge floods, causing huge economic losses. In order to reduce the threat of flooding, flood embankments were built along the Ziya River. In Xiyuzhuang area, the embankments are about 2 meters higher than the hinterland, and the riverfront space in the section is directly transferred from the water bank to the residential area, mainly bearing the flood control function. As the carrier of important urban public activity space, Ziya River has become a cutting state of urban space. The publicity, ecology and hydrophilicity that should be possessed have not been reflected, and the vitality of riverside space needs to be improved urgently.

We try to reconstruct the public space system from the perspective of humanity and ecology, and create multi-level urban living space. Node — corner park, provides the nearby activity place for the citizen, promotes the citizen's communication. Path — street is the most continuous public space. By sorting out the pedestrian system, interconnection mechanism can be established between different node spaces, shaping a clear urban space framework and promoting the transformation of node spaces into dynamic places. Open space — create a colorful riverside waterside footpath space, making it the most dynamic and attractive element in the region.

理水——结合水脉贯通区域特色要素

基地内河流沿岸有诸多特色要素，如北运河、西沽公园、子牙河、西站等，但要素之间缺乏联系，自成体系。

规划连通城中水系，再现津沽特色，并形成海绵城市的主干系统。梳理后的水系兼具生态性、景观性、人文性等特征。

生态性：使用一系列景观与工程手段使城市的排水系统遵循雨水循环规律，规划生态、可持续发展的城市。同时也为生态多样性预留出足够的空间。

公共性：规划层次丰富的景观水系，建立从西沽公园到子牙河的直接联系，更加人性化的尺度，使城市公共空间和自然处于有机融合状态。

人文性：水系沿岸树木以桃树为主，呼应和延续区域内桃花堤的特征，同时，布局与水系紧密联系的二层连廊体系——新红桥，借此激活区域活力、带动区域发展。

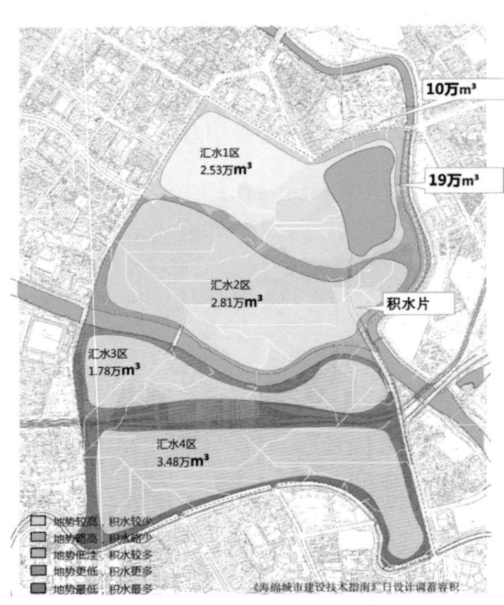

现状水文特征图
Current Hydrological Features

方案水系梳理图
Water System Carding Diagram Scheme

Water Management: Combining the Characteristic Elements of Water Veins Through the Region

There are many characteristic elements along the river in the site, such as the North Canal, Xiqiao Park, Ziya River, West Station, etc., but the elements are lacking in connection and self-contained.

The plan connects the water system in the city, reproduce the characteristics of Jingu, and forms the backbone system of the sponge city. The combed water system has the characteristics of ecology, landscape and humanity.

Ecologicality: a series of landscape and engineering methods are used to make the city's drainage system follow the law of rainwater circulation and achieve an ecological and sustainable city. It also leaves plenty of room for biodiversity.

Publicity: plan the landscape water system with rich levels, establish the direct connection from Xigu Park to Ziya River, and make the urban public space and nature in the organic integration state.

Humanism: the trees along the river system are mainly peach blossom trees, which echo and continue the characteristics of the peach blossom dyke in the region. At the same time, a two-story corridor system — new red bridge — which is closely related to the river system is distributed to activate regional vitality and drive regional development.

滨河空间现状树木保留图
Riverfront Space Status Tree Retention Map

护树——结合公共空间保留现状树木

规划范围内树木分布随机、自然生长，具备良好的生态本底。如何处理现状树木，并进行最大限度的保留，成为方案推进的重大挑战。

规划综合考虑现状树种、树龄、胸径等因素，保证大树原地保留和移植的成活率，遵循城市与自然资源共存共生的原则。

公共空间分成两个层级：一级公共空间，如滨河空间及中轴空间等；二级公共空间，如街角公园及带状绿地等。通过道路线形的灵活调整，街头公园的随机设置，以及沿河公园的集中安置等手段，使公共空间体系尽可能和成片树木结合，构建层级丰富的绿地空间系统。与此同时，避免过重的人工规划痕迹，形成一个有机生长的绿地系统。

北运河、西沽公园、子牙河、西站等重要区域之间，形成连续、开放而又亲民的公共空间纽带，为城市未来的发展打下坚实的基础。

Tree Protection: Preserving the Existing Trees in Combination with the Public Space

The trees in the planning area are randomly distributed and grow naturally, and have a good ecological background. How to deal with the existing trees and maximize the retention has become a major challenge for the program.

The plan comprehensively considers the current tree species, tree age, breast diameter and other factors to ensure the survival rate of large tree in retention and transplantation, and follows the principle of coexistence of city and natural resources.

The public space is divided into two levels: a first-level public space, such as riverside space and a central axis space; and a second-level public space, such as a street corner park and a strip of green space. Through the flexible adjustment of the road line type, the random setting of the street park, and the centralized placement of the river park, the public space system is combined with the trees as much as possible to construct a layered green space system. At the same time, avoid excessively heavy manual planning traces to form an organically grown green space system.

The North Canal, Xiqiao Park, Ziya River, West Station and other important areas form a continuous, open and intimate public space bond, laying a solid foundation for the future development of the city.

项目概况

2006 年国务院批准的《天津城市总体规划（2005—2020）》确定天津中心商务区位于海河两岸，由解放北路、小白楼和南站（六纬路）地区共同构成。占地约 1.8 平方千米。

河东南站（六纬路）片区是天津中心商务区的重要组成部分，片区东至七纬路，南至十一经路，西至海河东路，北至李公楼桥。历史上河东南站（六纬路）区域所在地位于天津俄租界。随着时间流逝，俄租界已从天津版图上消失，现仍有少数可考证的俄式建筑。比如 1919 年建厂的天津老卷烟厂、1941 年建成的天津老南站（九纬路）等；位于十一经路保存完好的原俄国驻天津领事馆，是天津特殊保护等级历史风貌建筑。项目规划总用地 57.1 公顷，项目规划建筑面积 239.55 万平方米，项目规划居住人口 2.7 万。

Project Overview

The general plan for tianjin (2005-2020) approved by the state council in 2006 determined that tianjin CBD is located on both sides of haihe river and consists of north jiefang road, xiaobailou and south railway station. It covers an area of about 1.8 Square kilometers.The area of hedong south railway station is an important part of tianjin central business district, which is as far as qiwei road in the east, shiyi road in the south, haihe east road in the west and li gonglou bridge in the north. Historically, the area of hedong south railway station was located in the russian concession of tianjin. As time goes by, the russian concession has disappeared from the map of tianjin. For example, the old cigarette factory in tianjin, which was built in 1919, and the old south railway station in tianjin, which was built in 1941. The former russian consulate in tianjin, located in shiyi road, is well preserved and is a historical building with special protection level in tianjin.The total planned land area is 57.1 Hectaresthe planned construction area of the project is 2,395,500 square metersthe project has a planned resident population of 27,000.

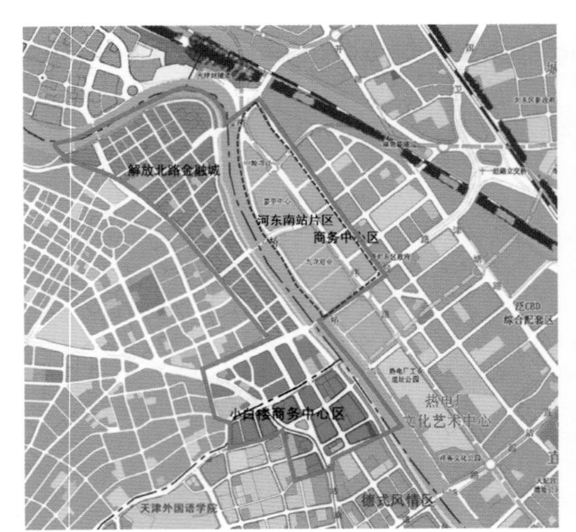

南站（六纬路）区位及范围
Location and Scope

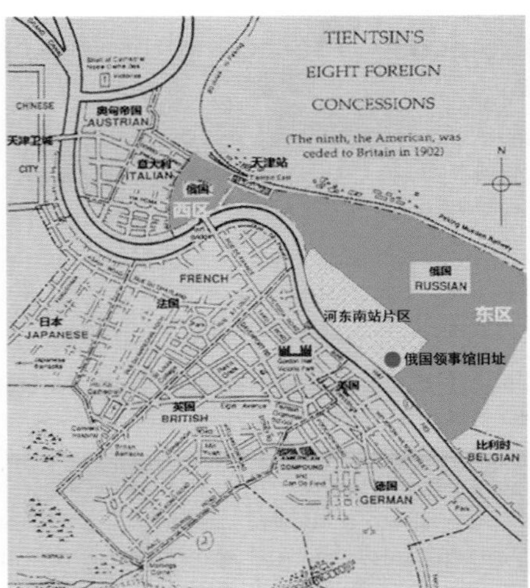

南站（六纬路）历史区位及照片
Historical Location and Photos

天津老卷烟厂（1919—）旧地位于六纬路
Tianjin Old Cigarettte Factory

天津老南站（1941—2006）旧地位于九纬路
Tianjin Old South Railway Station

新时代天津中心城区的 CBD

经过十余年的建设，南站（六纬路）地区取得了一些成绩：项目沿海河的城市界面已经基本完善，区域内规划的主要道路均已建设完成，地铁九号线在区域内设有两站，分别为大王庄站和十一经路站。综合来看，南站（六纬路）地区是天津中心城区 CBD 区域中最具发展潜力的区域。

CBD of Tianjin Downtown Area in New Era

After more than ten years of construction, the South Station area has achieved some achievements: the urban interface of the project coastal river has been basically improved, and the main roads planned in the area have been completed. Metro Line 9 has two stations in the area, respectively Dawangzhuang Station and Eleventh Road Station. On the whole, the South Station area is the most promising area in the CBD area of Tianjin's central city.

南站（六纬路）鸟瞰
Aerial View

功能复合

简·雅各布斯提出“主要用途复合之必要性”，城市发展需要两种形式的多样性：第一种多样性是首要用途，它把人群引向一个地方，承担城市功能的“抛锚地”；第二种多样性是那些回应第一种用途而发展起来的商业（商店和服务设施），主要服务于被首要用途吸引过来的人群。

当代 CBD（中心商务区）的规划建设也越发重视功能的复合，所谓 CBD，其商务办公的功能就是其“抛锚地”，但要充分发挥其商务办公的价值，相应的配套设施，比如商业设施、居住配套和生活配套也应健全发展。

规划依托海河景观带，基于项目三处跨海河的对外交通要道，规划三处超高层商务办公的组团，连接六纬路两侧现有的商业资源，整合形成 1.6 千米长的商业系统，规划形成“一带、三节点、活力轴”的复合功能架构。

Functional Composition

Jane Jacobs believes the "necessity" of main use compound, city development need two forms of diversity, the first kind of diversity is the primary purpose, it puts people into one place, to undertake urban functions "anchor", the second kind of diversity is the response to the first use and developed business (shop and service facilities), the main service is the primary purpose attract crowds.Contemporary CBD planning and construction also pay more and more attention to the complex functions, the so-called CBD, its business office function is its "anchor", but to give full play to the value of its business office, the corresponding supporting facilities, such as commercial facilities, residential facilities and living facilities should also be sound development.In project planning and design, we rely on Haihe River landscape zone, based on project three places across the Haihe River external transport, planning of three high-rise business office of the group, to connect Liuwei Road on both sides of the existing business resources, integration of the formation of 1.6 kilometers of business systems, planning to form a compound functional structure of " one belt, three points, dynamic axis" .

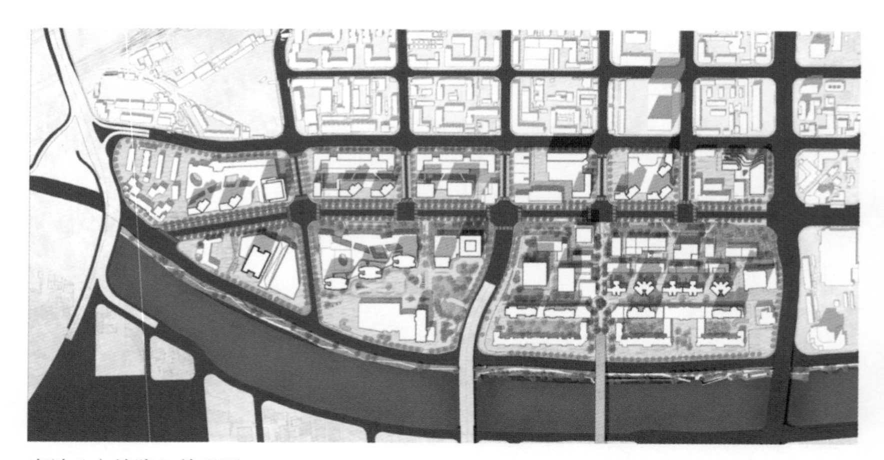

南站（六纬路）总平面
General Plan

南站（六纬路）商务办公组团
Business Group

开放、有活力

雅各布斯在其著作《美国大城市的死与生》一书中讲到“小街段之必要”，大多数的街段必须要短，也就是说，在街道上能够很容易拐弯。孤立的、互不关联的街道街区从社会的角度讲，会陷入孤立无助的处境。街道的频繁出现和街段的短小都是非常有价值的，因为它们可以让城市街区的使用者拥有内在有机的交叉使用方式。

规划基于现状的路网基础，有效地梳理区域内的慢行系统，尽量打通断头路，保证慢行系统的连续性和有序交织，激发城市活力。

Open Energy

In her book *The Death and Life of Great American Cities*, Jacobs talks about "the necessity of small streets", in other words, it is easy to turn on the street. Isolated, disconnected streets and blocks can be socially isolated and helpless. The frequency of street appearances and the short length of street segments are valuable because they allow users of city blocks to intersect with each other organically.

In the design, based on the current situation of road network effectively sort out the slow-moving system in the area, try to break through the broken road, ensure the continuity and orderly interweaving of the slow-moving system, and stimulate the vitality of the city.

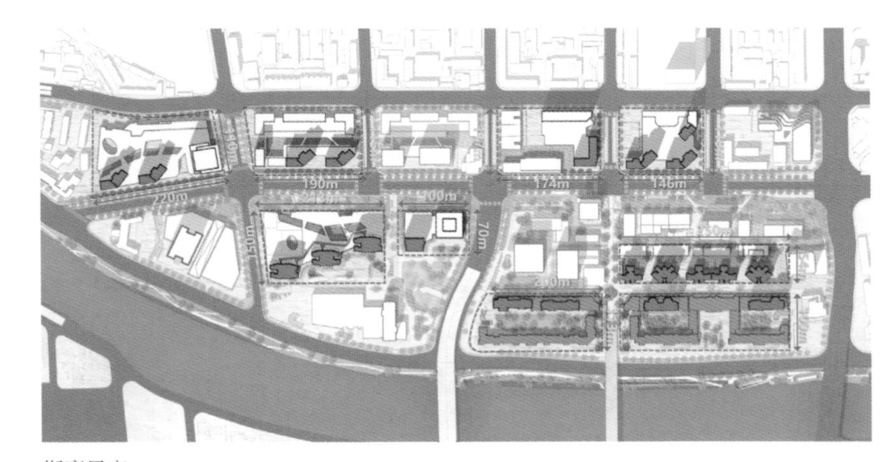

街廓尺度
Block Size

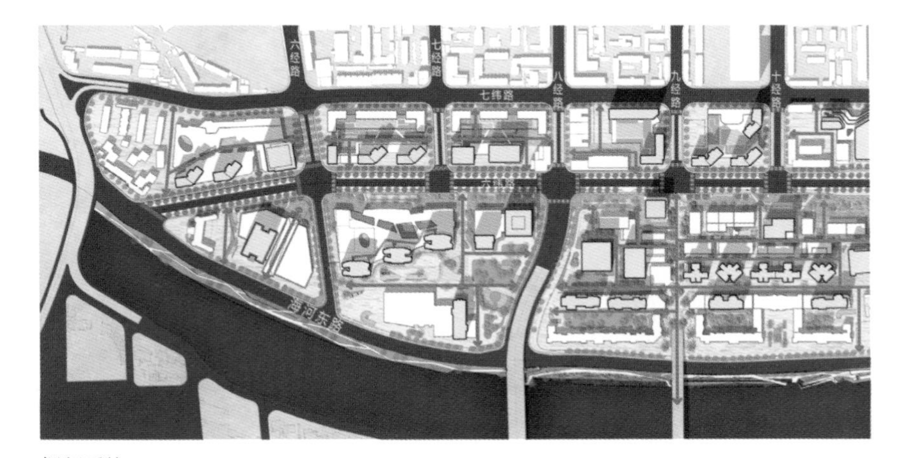

慢行系统
Non-Motorized Traffic System

风情街区
Style Street

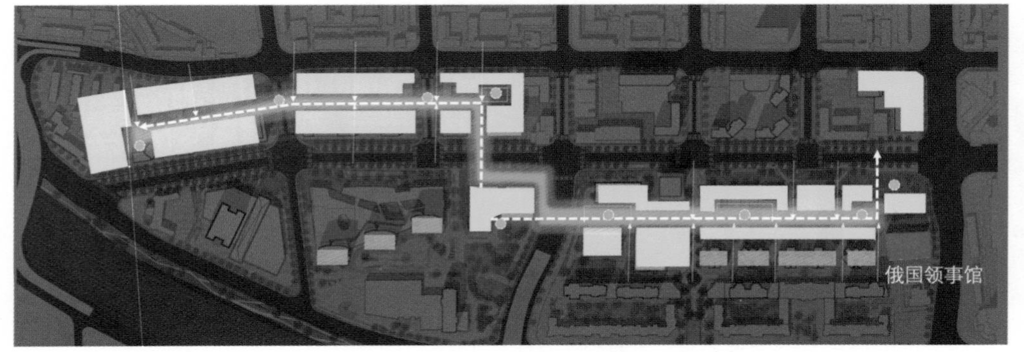

风情街区平面图
General Plan of Style Street

文化回归

老建筑对于城市是如此不可或缺，如果没有它们，街道和地区的发展就会失去活力。在大城市的街道两边，最令人赞赏和使人赏心悦目的景致之一是那些经过匠心独运的改造而形成新用途的旧建筑。

城市里新建筑的经济价值足可以由别的东西，如花费更多建设资金建成的建筑来代替。但是，旧建筑是不能随意被取代的。这种价值是由时间形成的。这种多样性需要的经济必要条件对一个充满活力的城市街区而言，只能继承，并在日后的岁月里持续下去。

规划积极保留区域内现存的风貌建筑——原俄国驻天津领事馆，并以此为“锚点”，梳理一条“叙事线索”，规划一条风情街区，从而以此为载体，注入天津的地域文化，使之成为“既有表，又有里”的建筑载体，促进区域的文化回归。

Cultural Regression

The necessity of old buildings. Old buildings are so indispensable to the city that without them, the development of streets and areas will lose their vitality. On either side of the streets of the great city, one of the most admirable and pleasing sights is the old buildings which have been cleverly transformed into new uses.

The economic value of new buildings in the city can be replaced by something else, such as more construction fees. However, old buildings cannot be replaced at will. This kind of value is formed by time. This diversity requires economic imperatives that, for a vibrant urban neighborhood, can only be inherited and sustained for years to come.

For the existing historical and stylistic architecture in the area—the former Russian Consulate in Tianjin, the original style will be kept. Regarding it as an "anchor", acarding a "narrative clue", and planning a style block will be the carrier of the regional culture of Tianjin, thus promoting regional cultural regression.

劝业场地区的“前世今生”

1900 年，劝业场地区沦为日法租界的管辖范围。

1949 年，劝业场地区整体格局形成天津的商业与娱乐中心。

20 世纪 70 年代，地铁一期工程开工，墙子河被填平，建设了如今的南京路。鞍山道周边地区陆续建设了一部分 4 ~ 6 层“筒子楼”住宅。

20 世纪 90 年代，中原百货、麦购休闲广场等大型商业综合体相继建成。

2000 年后，随着海河沿岸开发改造，津门、津塔等高层地标建筑相继建成。南京路沿线出现了现代城等超高层办公建筑，整体形成了南北高、中间低的空间格局。

Persuasion of the Past and Present Life in Quanyechang Area

In 1900, the area of the Quanyechang area became the jurisdiction of the Japan and France concession.

In 1949, the overall pattern of the concession was formed, and the area of the company was the commercial and entertainment center of Tianjin.

In the 1970s, the first phase of the subway started, and the wall river was filled in, and the Nanjing road was built. A part of the 4-6-storey “tube-shaped apartment” has been built in the surrounding areas of Anshan road.

In the 1990s, large-scale commercial complexes such as Zhongyuan Department Store and Maigou Leisure Plaza were built.

After 2000, with the development and reconstruction of the Haihe River, high-rise landmark buildings such as the Jinmen and Tianjin Tower were built. Some high-rise office buildings such as the Modern City appeared along the Nanjing Road, forming a spatial pattern of high north and south and low in the middle.

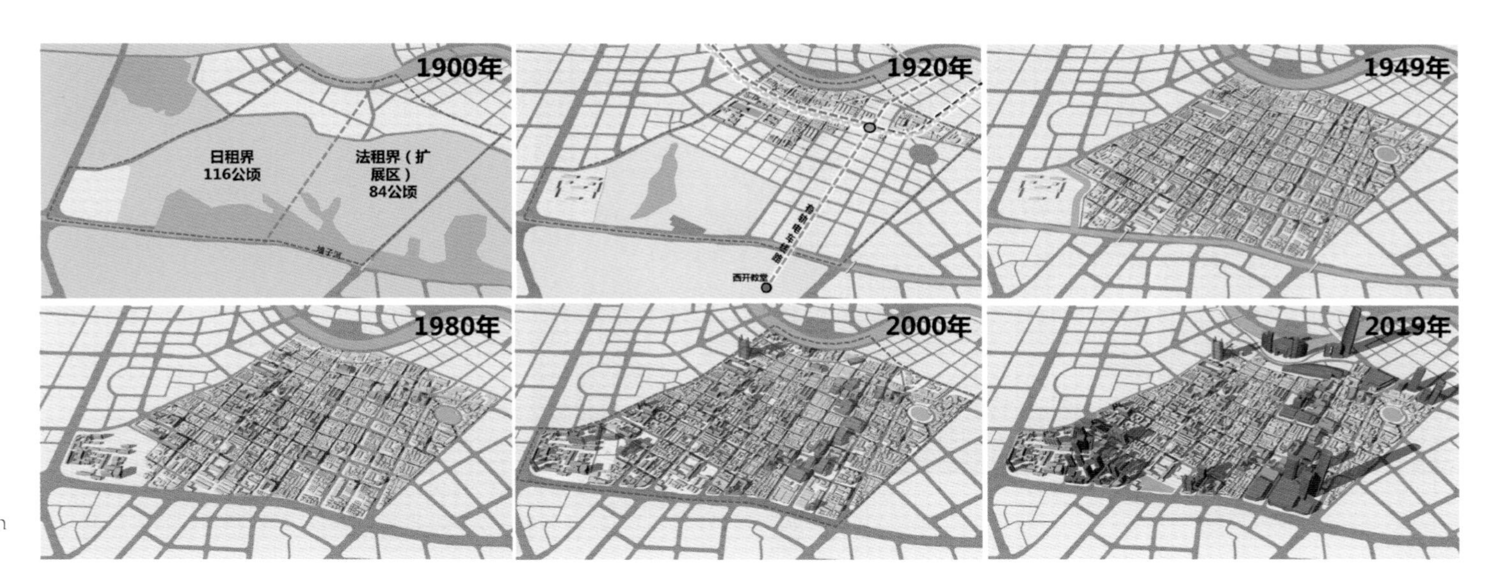

哈密道各时期空间形态
The Spatial Form of Each Period of Hami Road

街区的价值

商业经济价值：劝业场周边地区是天津现代商业的起点。

空间价值：天津“小街廓，密路网”最典型的地区。

历史文化价值：多样的居住建筑风格与丰富的历史人文资源。

Block Value

Business economic value: the surrounding area of Quanyechang is the modern commercial origin of Tianjin.

Spatial value: "small street and dense road network" in Tianjin is the most typical area.

Historical and cultural values: diverse residential architectural styles and profound historical and cultural resources.

街巷现状
Current Street photo

十四片历史文化街区的路网密度
The Road Density of 14 Historic Cultural Areas

面临的挑战与问题

随着城市综合体在中心区以外地区涌现，加之电商的冲击，劝业场地区的商业核心地位受到极大挑战。

地区人口密度高，但人户分离、老龄化、贫困化现象明显，影响了街区的整体活力。

中华人民共和国成立前老建筑数量多，多数建筑年久失修，缺少维护，存在安全隐患，同时居住配套设施不完善、陈旧老化、不堪重负。

Challenges and Problems

With the emergence of urban complexes outside the central area, coupled with the impact of e-commerce, the core position of business has been greatly challenged;

The population density of the region is high, and the separation of households, aging, and pauperization are obvious, affecting the overall vitality of the neighborhood; There were many old buildings in the period before the founding of the People's Republic of China.

Most of the buildings were in disrepair, lack of maintenance, and there were potential safety hazards. At the same time, the residential facilities were imperfect, old and aging, and overwhelmed.

现状照片 Status photos

规划历程

2007 年在劝业场地区城市设计国际竞赛中，由境外设计公司率先提出在哈密道一侧拆出一条南北长 1 200 米、东西宽近 100 米的城市绿化带，同时沿绿轴两侧高密度开发，用以平衡土地成本，更新地区整体环境的方案，这在那个整体更新、追求项目自身资金平衡为主流方式的时代，受到了多数人的认可。

之后受市规划局委托，进一步研究规划方案。出于对城市文化遗产的责任感，试图从中找到一个平衡点，既要考虑到历史街区的保护要求，体现城市文化的价值和力量，又要尽量为城市未来的发展找出足够的、合理的空间，同时还要综合协调各种因素形成富有特色的、充满活力的城市环境，为此讨论过中强度、低强度的多种可能性方案。

Planning Process

In 2007, in the international competition for urban design of the Quanyechang area, the overseas design company took the lead in proposing to remove an urban green belt with a length of 1200 meters from north to south and a width of nearly 100 meters from east to west on the side of Hami Road. Density development, which is used to balance land costs and update the overall environment of the region, has been recognized by the majority in the era of overall renewal and the pursuit of the project's own capital balance as the mainstream.

Afterwards, it was commissioned by Tianjin Planning Bureau to further study the planning plan. In the sense of responsibility for the cultural heritage of the city, trying to find a balance point, not only must consider the protection requirements of historical areas, reflect the value and strength of urban culture, but also try to find enough and reasonable space for the future development of the city. At the same time, it is necessary to comprehensively coordinate various factors to form a distinctive and dynamic urban environment. For this reason, various medium-strength and low-intensity possibilities are discussed.

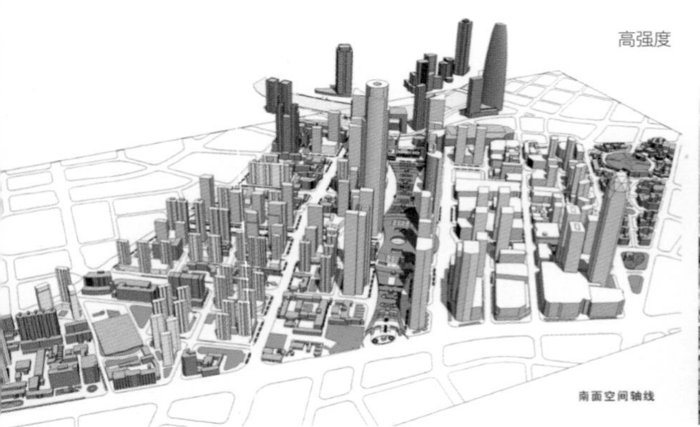

多种开发强度方案对比
Comparison of Various Open Intensity Schemes

规划目标

1. 振兴中心商业区

滨江道与和平路构成的“金街”商圈一直以来就是天津市的商业中心，但发展腹地受限，本次规划结合周边历史街区，激活发展内生动力，打造世界级商业中心。

2. 改善在地民生

在四周高楼大厦的包围下，留下约 60 公顷相对完整的城市街区，但这些建筑损坏较为严重的非历史街区，多为公产与企业产，人均居住面积 18.8 平方米，属于高楼大厦后的贫民区。此次规划通过旧城与新区互动的方式，疏解人口，改善民生。

3. 实现片区微更新

对中心商业区内的原居住建筑，在人口疏解后，结合建筑自身特点，保留一批，更新一批，通过对建筑的有机更新，引入新的业态，发展创意商业街区，作为“金街”沿线大型商业的补充。

Planning Objectives

1.Revitalization of the Central Business District

The "Golden Street" business district consisting of Binjiang Street and Heping Road has always been the commercial center of Tianjin, but the development hinterland is limited. This plan combines the surrounding historical districts to activate the development of endogenous power And create a world-class business center.

2. Improve Local People's Livelihood

Surrounded by high-rise buildings, about 60 hectares of urban fabrics remain relatively intact, but non-historical blocks with more serious building damage, mostly for public and commercial production, per capita living area is 18.8 square meters. It belongs to "the slum area behind the high-rise building". The plan will use the old city and the new district to interact to ease the population and improve people's livelihood.

3. Realize the Micro Update of the Area

For the original residential buildings in the central business district, after the population is deconstructed, in combination with the characteristics of the building, some buildings will be kept and other buildings will be updated. Through the organic renewal of the building, a new format is developed, and the creative commercial district is developed with those large commercial complements along the "Gold Street".

滨江道步行街现状
Binjiang Street Pedestrian Street Status

热河路居民区现状
Current Status of Rehe Road Residential Area

田子坊现状
Current Status of Tianzifang

振兴中心商业区

借鉴国内外先进经验，在疏解人口的同时，完成产业调整。通过发展金街“后街经济”，引入带有文艺、时尚、潮流元素的新兴业态，提升片区活力。将劝业场地区打造成为天津最时尚、最具历史感、最多元的中心商业区。

Revitalization of the Central Business District

Drawing on advanced experience at home and abroad cities, we will complete industrial adjustment while demographics are being resolved. Through the development of the "backstreet economy" of Gold Street, the introduction of new formats with literary, fashion and trend elements will enhance the vitality of the district. It will be the most fashionable, historical and pluralistic central business district in Tianjin.

文艺

时尚

潮流

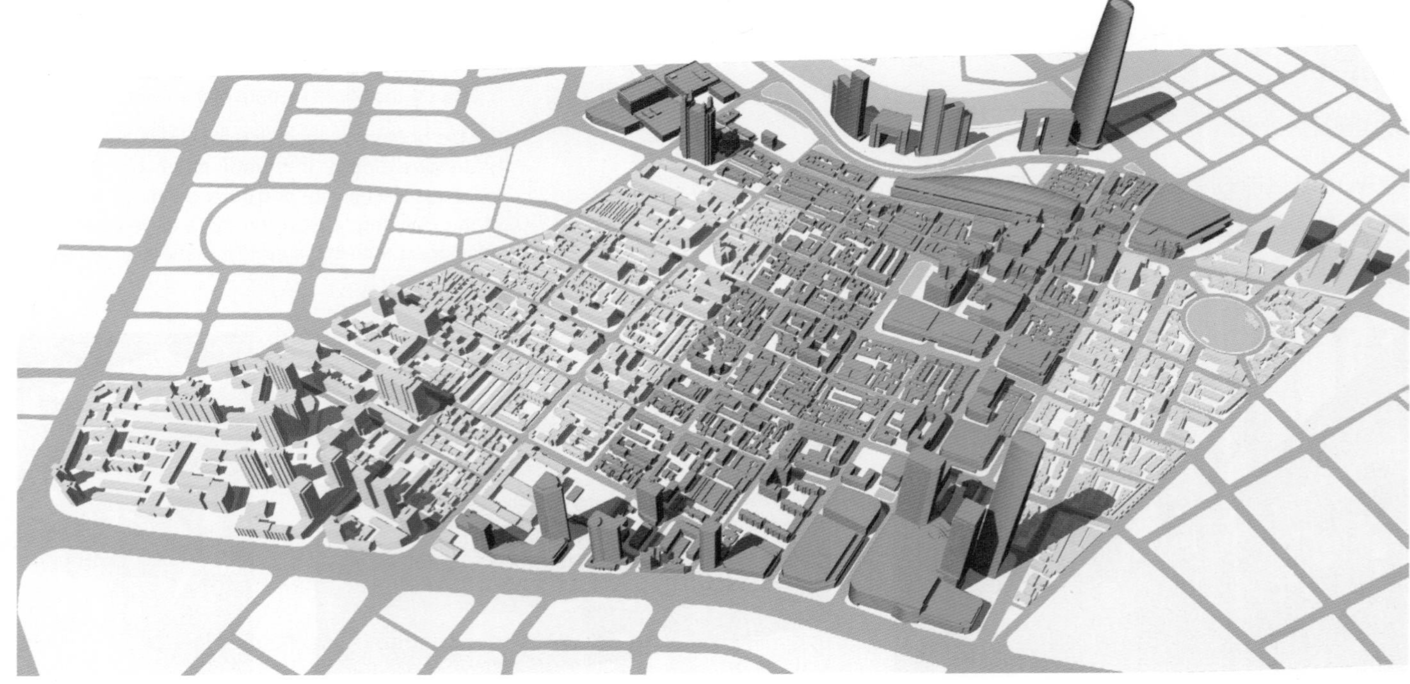

中心商业区范围图
Central Business District Area

对标国际知名商业中心——东京银座商业区

银座地区是日本最繁华的商业区，具有“东京心脏”之称，占地约0.9平方千米。布局特点是中央大道与后街群组合的商业布局模式。中央大道集中了众多大型商业综合体、品牌旗舰店，同时也由传统的“购物场所”向综合性的“生活广场”转化，不仅具有一般的购物功能，还有观光、休闲、文化、娱乐等多种功能。后街群承接了各类日本老字号与夜色经济，与主街的现代商业氛围呈现强烈反差。中央大道加后街群的这种格局，赋予游客独特的购物和文化体验，激发了沿街商业的活力。

Benchmarking Internationally Renowned Commercial Center—Tokyo Ginza Business District

The Ginza area is the most prosperous business district in Japan and is known as the "Heart of Tokyo", covering an area of about 0.9 square kilometers. The layout feature is the commercial layout mode of the combination of the central avenue and the back street group. The central avenue is home to many large commercial complexes and brand flagship stores. It also transforms from the traditional "shopping place" to the comprehensive "life square", which not only has general shopping functions, as well as sightseeing, leisure, culture, entertainment and many other functions. The back street group has undertaken all kinds of old Japanese brands and the night economy, and the modern business atmosphere of the main street is in sharp contrast. This pattern of the Central Avenue plus the backstreet group gives visitors a unique shopping and cultural experience that activates the vibrancy of the business along the street.

银座商业区范围图 Ginza Business District Scope

银座中央大道 Ginza Central Avenue

在保护中发展，在发展中保护

五大道地区作为 20 世纪初至 20 世纪中叶中国沿海开埠城市高档居住建筑最集中的区域之一，集中体现了中国由传统封闭型社会向现代开放社会转变的轨迹，集中展示了近现代中国居住建筑、生活方式的演进历史，是不可多得的活化历史书，具有潜在的世界文化遗产价值。如何承续其历史，让历史街区融入当代的生活并为当代生活助力、接续发展，成为我们这个时代的重要命题。五大道地区的保护与更新始于 20 世纪 90 年代，一方面强调整体保护，深入研究历史文化资源价值及空间格局特色，延续和保持历史文化街区的空间肌理、建筑形态、建筑高度及历史环境特征。另一方面，注重保护的同时完善街区功能及布局，优化地区环境品质和空间景观，提高历史建筑的使用价值，促进街区自我良性发展。

Develop in Protection and Protect in Development

As the most concentrated area of high-end residential buildings in China's coastal open cities from the early 20th century to the middle of the 20th century, the Wudadao area embodies the trajectory of China's transition from a traditional closed society to a modern open society, focusing on the modern Chinese residential architecture and life. The evolutionary history of the way is a rare activation history book with potential world cultural heritage values. How to continue its history, integrate historical districts into contemporary life and continue to develop for contemporary life has become an important responsibility of our time. The protection and renewal of the Wudadao area began in the 1990s. On the one hand, it emphasizes the overall protection, in-depth study of historical and cultural resources and spatial pattern characteristics, and continues and maintains the spatial texture, architectural form, architectural height and historical environment characteristics of historical and cultural blocks. On the other hand, ,to improve the function and layout of the neighborhood, optimize the regional environmental quality and spatial landscape, improve the use value of historical buildings, and promote the self-development of the neighborhood while paying attention to protection.

五大道整体鸟瞰
Overall View of Wudadao Area

社会、文化和经济生活的整合——国际宪章的反思

在当前国际普遍对城市保护展开反思的情况下，国际历史村镇委员会（CIVVIH）修订了《华盛顿宪章》和《内罗毕建议》中的一些方法和受到关注的事项，在 2011 年 11 月 28 日于巴黎举行的国际古迹遗址理事会（ICOMOS）第 17 届全体大会上通过了《关于维护与管理历史城镇与城区的瓦莱塔原则》（简称《瓦莱塔原则》）。《瓦莱塔原则》将保护的领域延展到非物质维度，它重申了诺伯舒茨的“场所精神”理论，还提出了在历史街区介入新要素的原则，将社会、文化和经济生活整合进去。此外文件特别指出，数量和时间会对历史街区的保护与更新产生重大影响。对街区的很多小的改变可以积累成为巨大的变革，变化的速度也决定了对历史街区产生影响的强度。这与五大道历史文化街区城市规划一直秉承的“整体保护，有机更新”不谋而合。

五大道街景
Street View of Wudadao Area

Integration of Social, Cultural and Economic Life: Reflections on the International Charter

In the current international community's general reflection on urban protection, CIVVIH (affiliated with ICOMOS) revised some of the methods and concerns in the Washington Charter and Nairobi Recommendations, in November 2011. At the 17th plenary session of ICOMOS hold in Paris on the 28th, the Valletta Principles for the safeguarding and management of historic cities, towns and urban areas ("Valletta Principles") was adopted. The Valletta Principle extends the field of protection to the non-material dimension. It reaffirms Christian Norberg-schulz's "Genius Loci" theory and proposes the principles of involve new elements in historical areas, such as integrating social, cultural and economic life. In addition, the document specifically states that the quantity and timing will have a major impact on the protection and renewal of historic areas. Many small changes to the neighborhood can accumulate into huge changes, and the speed of change also determines the intensity of the impact on the historic neighborhood. This is consistent with the “whole protection, organic renewal” that the urban planning of the Wudadao historic area has always been adhering to.

建筑组合类型	门院式	里弄式	院落式
抽象出的建筑原型			
与街道的关系	主要位于街道交叉口处	在城市街道上有明确入口和通道	有唯一的入口并直接伸入内院
建筑形式	有主要的临街面，另一面与周边建筑保持着整齐的界面	里弄内部有明确紧密的界面	建筑造型丰富
使用特性	目前主要用于公共机构，开放程度低	小户型居住，比较开放	混合居住，开放程度高

通过类型学发现建筑组合的原型
Discover the Prototype of the Building Portfolio Through Typology

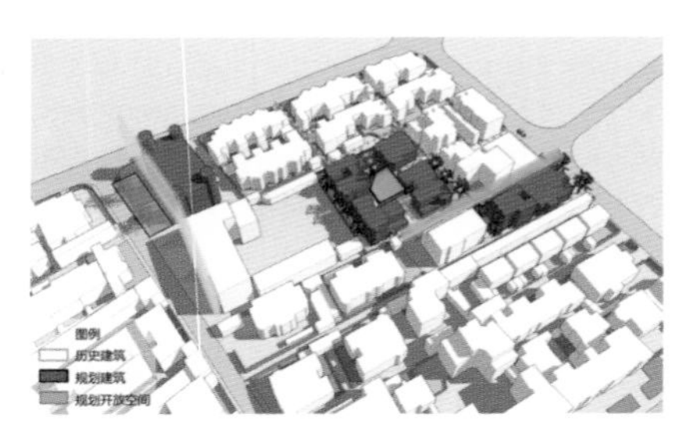

突出建筑类型本身的特质，根据环境恰当选型
Highlight the Characteristics of The Building Type Itself and Select Model According to the Environment

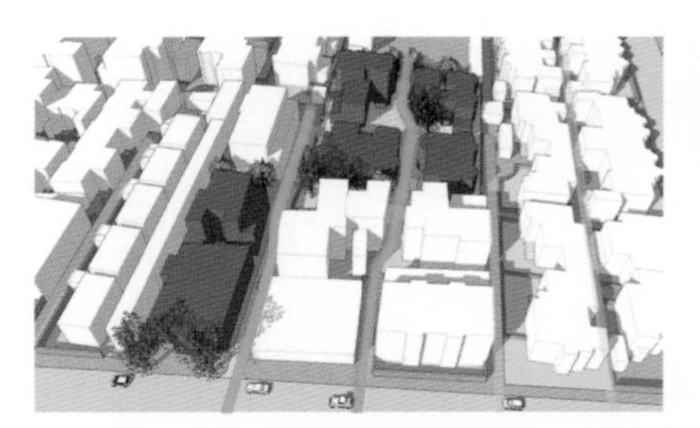

延续建筑划分并贡献街坊内部的联系通道
Continue the Division of Buildings and Contribute to the Communication Channels Within the Neighborhood

尊重城市记忆，延续场所精神

五大道是一个活着的历史街区，它引人入胜的地方，不仅是建筑本身的历史感和美学价值，还包括那些在小洋楼里居住的人，是他们的传奇故事及居民们每天的生活在书写着这些建筑的历史。当空间与时间的关系固化在人们的记忆中时，就创造出了身份认同感，于是使城市空间获得了更多的意义，形成了“场所精神”。

借由城市形态学和建筑类型学的分析方法，可以发现在空间、建筑和它们的文化及生活方式之间存在着清晰的对应关系。五大道始终是市民生活居住的空间，与日常的各种琐事息息相关。正是这些琐事构成了生活在其中的人们的共同回忆与情感联结，而它们蕴藏在空间之中，成为这个地方的“场所精神”。保留街区的肌理、氛围、现存的居民交往空间，在历史街区更新中介入新的要素时遵循原有的空间逻辑，才能使不损害历史街区核心价值的进一步发展成为可能。

Respect City Memory and Continue Genius Loci

The Wudadao area is a living historical district. Its fascinating place is not the historical and the aesthetic value of the building itself. It is the people who lived in these small houses. It is their legendary story. The life stories and new chapter of residents are writing every day. When the relationship between space and time solidifies in people's memory, it creates a sense of identity, which makes the urban space more meaningful and forms a "Genius Loci."

Through the analysis of urban morphology and architectural typology, it can be found that there is a clear correspondence between space, architecture and the culture and way of life that produces them. The Wudadao area is always a living and living space for citizens, and it is closely related to daily trivial matters. It is these trivial things that constitute the common memories and emotional connections of the people in life, and they are hidden in space and become the "place spirit" of this place. Retaining the texture, atmosphere and existing residents' communication space in the neighborhood, and following the original spatial logic when updating the new elements in the historical district, it is possible to make further development of the core values of the historical area.

五大道体验之旅的规划结构
Planning Structure of the Wudadao Experience Tour

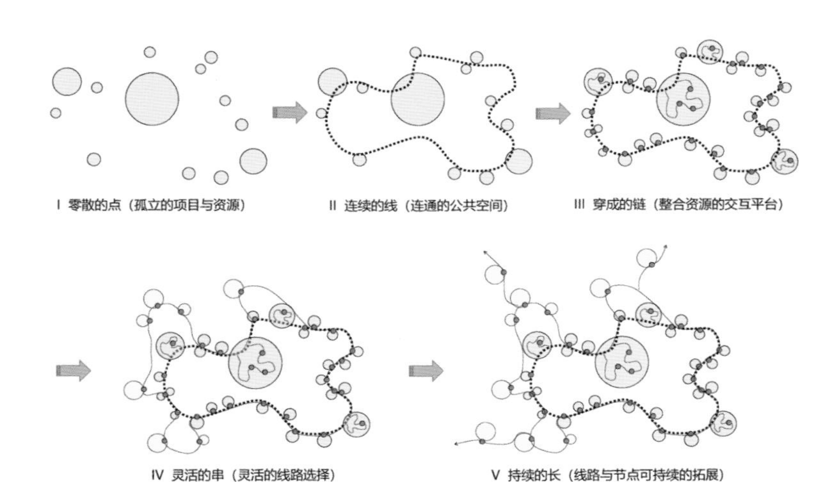

五大道体验之旅的发展模式图
Development Model of the Wudadao Experience Tour

探寻历史底蕴，展现城市魅力

为了让游客可以在行走间领略五大道历史文化街区的独特魅力，在规划游览路线时，充分考察了最能代表五大道历史文化底蕴的空间节点。将孤立的项目与历史文化资源点穿成连续的线，形成线状公共空间，再通过整合资源的互联网交互平台形成链，使之不断生长、灵活串联，形成“互联网 + 五大道体验之旅”。在这条路线上营造丰富多样的业态及多维文旅体验，依托线路持续催生一些城市事件，有序整合社群活动，成为汇聚文化精髓与城市生活互动的载体。这条线路具备未来生长的可能，随着游客增加可向外延伸，开发设计更多资源。线路的整体运营将成为五大道文旅发展的新动力，并以此推动现存历史文化资源的挖掘与利用，带动文化经济的发展，展现城市独特魅力。

Excavate the Historical Heritage and Show the Charm of City

In order to let visitors enjoy the unique charm of Wudadao historical and cultural district in the walking room, when planning the tour route, the space nodes that best represent the five historical and cultural heritages are fully examined. The isolated project , the historical and cultural resources are put into a continuous line to form a linear public space, and then the chain is formed through the internet interactive platform integrating resources, so that it grows and flexibly connects in series to form an "Internet + Wudadao Experience Tour". In this route, we will create a rich and diverse format and multi-dimensional cultural and travel experience. We will continue to generate some urban events on the basis of the route, and integrate social activities in an orderly manner, becoming a carrier for the interaction between the essence of culture and urban life. At the same time, this route has the possibility of future growth, and as the number of tourists increases, it can develop and design more resources. The overall operation of the line will become a new impetus for the development of the Wudadao cultural tourism, and promote the mining and utilization of the historical and cultural resources of the stock, promote the development of cultural economy, and show the unique charm of the city.

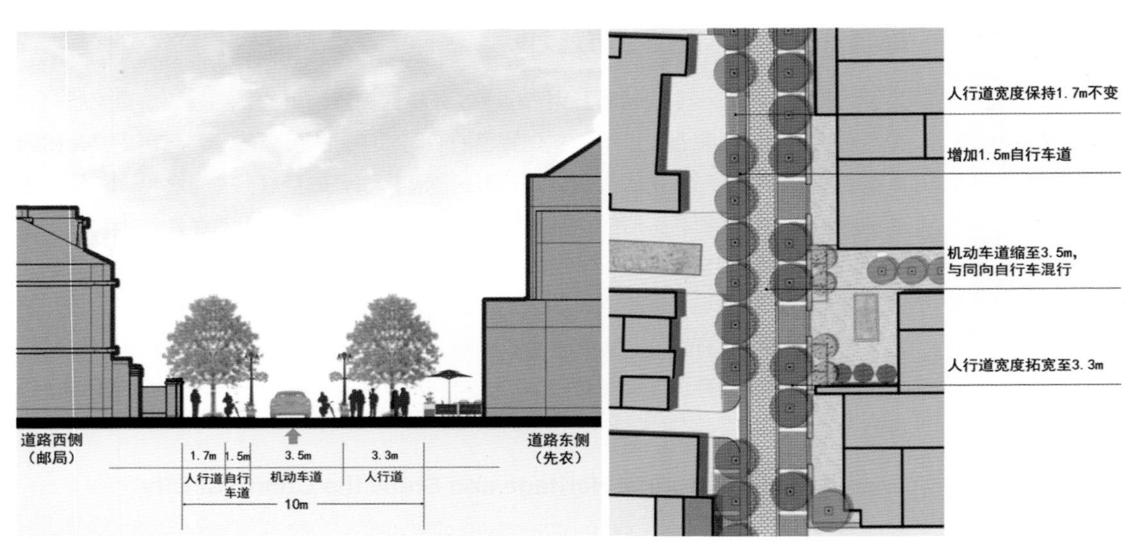

湖南路道路断面调整方案
Hunan Road Section

湖南路调整前后对比
Comparison of Hunan Road Before and After Adjustment

改善交通系统，优化慢行体验

结合天津历史城区步行化和城市中心文化复兴，将五大道体验之旅纳入全市的步行体系，赋予行人和自行车优先路权。主环线沿线街道采用适当缩窄机动车道，拓宽慢行道断面优化设计，以围墙、绿植、商业外摆、街道家具等细腻设计塑造舒适的林下漫步休闲空间。同时，对沿线历史风貌建筑进行精心的立面修缮，使其重焕光彩。

Improve Transportation System and Optimize Slow-traffic Experience

Combining the pedestrianization of Tianjin's historic city and the cultural revitalization of the city center, the Wudadao Experience Tour will be incorporated into the whole city's pedestrian system. Give priority to pedestrians and bicycles. The street along the main ring line adopts appropriate narrowing of the motor-traffic way, widening the design of the section of the slow -traffic road, and creating a comfortable under- walking leisure space with the delicate design of the wall, green plant, commercial pendulum and street furniture. At the same time, the façades of the historic buildings along the line have been carefully restored to rejuvenate them.

五大道局部鸟瞰
Partial Aerial View of Wudadao Area

保留街区本真，助力城市发展

小规模、渐进式改造的方式真实地反映了五大道内在的更新需求，唤醒了城市中心历史街区的活力，并且在规划的控制下能够很好地契合五大道固有的建筑形制和环境氛围，维系原有的社会和邻里关系，保护街区原真性。一幢幢蕴藏着历史文化积淀的建筑，被重新赋予了现代都市生活方式及生活氛围，不再仅仅是站在围墙后面供人瞻仰的文化符号，而成为拥有鲜活灵魂的人文历史街区。

Maintaining the Authenticity of the Neighborhood and Helping the City Development

The small-scale and gradual transformation method truly reflects the intrinsic renewal demand of The Wudadao area, awakens the vitality of the historical center of the city center, and under the control of the plan, it can well fit the architectural style and environment of The Wudadao area. The original social and neighborhood relationship, the original authenticity of the block is protected. The building with historical and cultural heritage has been re-emphasized to the modern urban lifestyle and living atmosphere. It is no longer just a cultural symbol that stands behind the wall for people to admire, but becomes a humanistic historical block with a fresh soul.

先农大院实施照片
Implementation Photo of Xiannong Yard

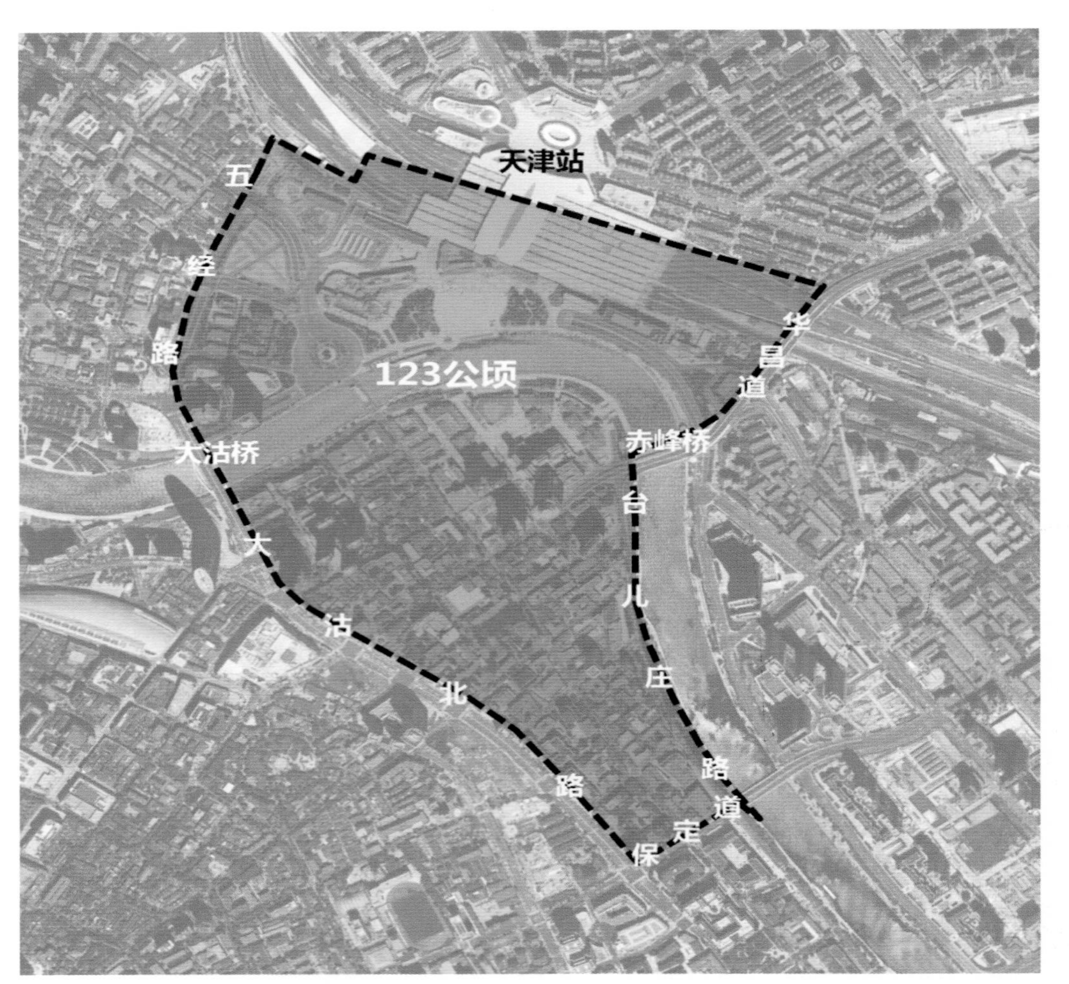

解放北路与天津站地区城市设计范围
Urban Design Scope of North Jiefang Road and Tianjin Railway Station Area

重塑步行网络，激发城市活力——以人为本的设计导向

津湾广场、天津站是天津最具代表性的城市地标之一，但凡来过天津的人，对这里都不陌生。解放北路与天津站地区形成于租界时期，一直以来都是天津的金融中心和交通枢纽。然而，随着近年来城市商业升级和旅游业的快速发展，这里出现了一些城市历史街区的共性问题：城市衰落现象明显，街道空间缺乏活力；传统的金融业态吸引力不足，整体商业氛围低迷；处于滨江道商业中心和天津站之间的津湾广场却形成了人流洼地。

Rebuilding Walking Network and Stimulating Urban Vitality: People-oriented Design Orientation

Everyone who has been to Tianjin is familiar with it. Jinwan Square and Tianjin Station are the most representative landmarks in Tianjin. North Jiefang Road and Tianjin Station area have been the financial center and transportation hub of Tianjin since the concession period of 1860. However, with the rapid development of urban commercial upgrading and tourism in recent years, there have been some common problems in urban historic districts: obvious urban decline, lack of vitality in street space; lack of attraction of traditional financial industry, depressed overall business climate; between Binjiang Street Business Center and Tianjin Station. Jinwan Square has formed a depression of people.

“东方华尔街”与“百年老龙头”

历史上的解放北路曾经云集着众多享誉国内外的金融机构，如汇丰银行、花旗银行、中央银行等，成了当时的“东方华尔街”。旧时的解放北路，即便与上海外滩相比，也是毫不逊色的。不过如今历史尘埃烟消云散，百年风云过后，繁华鼎盛早已不再，只有一幢幢造型独特的西洋建筑作为昔日繁华的见证者。

沿着解放北路向北，便来到了天津最重要的火车站——天津站。回望 1892 年，醉心洋务运动的李鸿章不会想到，当初为保障北洋海军用煤而设计的老龙头火车站，历经百年沧桑，变成了这座城市的新地标，与近在咫尺舒缓流淌的海河、优美大气的津湾广场，共同谱写着天津历史与现代交融的乐章。

"Oriental Wall Street" and "Centennial Laolongtou"

Historically, North Jiefang Road once gathered a large number of well-known financial institutions at home and abroad, such as HSBC, Citibank and the Central Bank, and became the "Oriental Wall Street" at that time. The old North Jiefang Road was no less impressive than Shanghai Bund. But now the dust of history has disappeared, and after a century of storms and clouds, the prosperity has long disappeared, leaving a few western buildings with unique shapes as witnesses of the past prosperity.

Along North Jiefang Road to the north, we arrived at Tianjin station, the most important railway station in Tianjin. Looking back on 1892, Li Hongzhang, who was fascinated by the westernization movement, could not imagine that the old leading railway station designed to guarantee the use of coal by the Beiyang Navy had gone through a century of vicissitudes, transformed into a new landmark of the city, and composed the blend of Tianjin's history and modern times with the gentle flowing seas and the beautiful atmosphere of the Jinwan Square close by the movement.

解放北路老照片
Old Photos of North Jiefang Road

连接解放北路与天津站地区的解放桥老照片
Old Photos of Jiefang Bridge Between North Jiefang Road and Tianjin Railway Station Area

改造完成的解放北路与天津站

改革开放后的解放北路又逐渐恢复了原有的金融功能，2008 年解放北路完成了一次卓有成效的整治提升，街道面貌焕然一新，历史空间得到了很好的恢复，成为天津历史街区改造的一个典范。

进入 21 世纪，经历了数次改造扩建的天津站，成为铁路、地铁、公交、出租车等多种出行方式的大型交通枢纽，除了承载着交通功能外，更变成了一座集吃、喝、住、行、玩于一体的“活力体”。

Reformed North Jiefang Road and Tianjin Railway Station

After the reform and opening, North Jiefang Road gradually restored its original financial function. In 2008, North Jiefang Road completed a fruitful renovation and upgrading, the street looks brand-new, the historical space has been well restored, and it has become a model of the transformation of Tianjin's historical areas. In the 21st century, Tianjin Station, which has undergone several renovations and extensions, has become a large-scale transportation hub of railway, subway, bus, taxi and other transfer modes. In addition to carrying traffic functions, it has become a "dynamic body" that integrates food, drink, housing, transportation and entertainment.

解放北路与天津站地区
North Jiefang Road and Tianjin Railway Station

规划目标

本次城市设计是以保护历史街区整体环境为前提，将解放北路与天津站地区作为一个整体进行统筹规划，重塑片区的慢行系统，激发津湾广场商业活力，实现天津站人性化改造，力求达到盘活存量建筑、更新商业业态、提升街道空间品质、增强城市活力的效果，打造具有天津特色的都市旅游文化。

Planning Objectives

The urban design is based on the premise of protecting the overall environment of the historical area. The North Jiefang Road and Tianjin Station area will be planned as a whole, rebuilding the slow-traffic system of the district, inspiring the commercial vitality of Jinwan Plaza, and realizing the humanized transformation of Tianjin Station. Strive to achieve the revitalization of stock buildings, update the commercial format, improve the quality of street space, enhance the vitality of the city, and create an urban tourism culture with Tianjin characteristics.

从天津站后广场上空看海河对岸
Looking at the Opposite Side of the Haihe River From the Back Square of Tianjin Railway Station

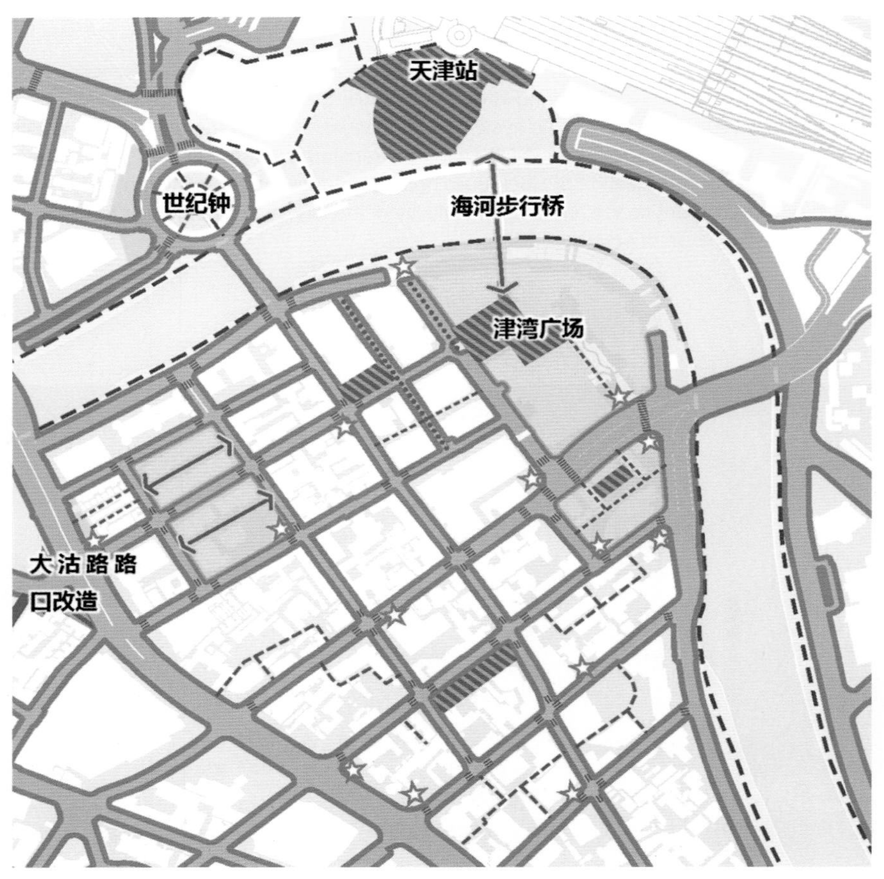

解放北路与天津站地区步行网路
Pedestrian Network of North Jiefang Road and Tianjin Railway Station Area

由机动交通向步行化设计的转变

近几十年来，城市建设大多是从机动交通发展的视角出发的，解放北路地区原有的步行网络已被割裂。根据百度热力分析，2018 年滨江道地区日人流量的高峰达到了 70 余万，天津站地区日均客流可达 20 余万，而一河之隔、一路之隔的津湾广场却成了人流洼地，日均客流不足 3 万人。由于缺少直接的步行联系，不少乘客只能在对岸看风景而不能真正走进津湾广场；大沽北路与兴安路交叉口过宽，步行穿越困难，也阻断了前往津湾广场的人流。本次规划从构建完善的步行网络出发，重点改造三项内容：建设海河步行桥、改造大沽路路口，加强津湾广场与天津站、滨江道的步行联系；改造世纪钟节点，强化天津站的步行交通枢纽作用；提升街道空间品质，展现街道特色，改善步行环境。

The Change From Motor Traffic to Walking Design

In recent decades, urban construction is mostly from the perspective of the development of motor traffic, and the original pedestrian network in the area of North Jiefang Road has been split. According to Baidu thermodynamic analysis, the peak daily passenger flow in Binjiang Street area reached more than 700,000 in 2018, and the average daily passenger flow in Tianjin Station area reached more than 200,000, while the Jinwan Square across the river became a depression of people flow. The average daily passenger flow was less than 30,000. Because of the lack of direct walking contact, many passengers could only cross the bank. The crossing of North Dagu Road and Xing'an Road is too wide, and it is difficult to walk through, which also blocked the flow of people to Jinwan Square. Starting from the construction of a perfect pedestrian network, the plan focuses on three aspects: the construction of Haihe pedestrian bridge, the transformation of Dagu Road intersection, and the strengthening of the pedestrian links between Jinwan Square and Tianjin Station and Binjiang Street. Renovate the Century Bell Node, strengthen the role of Tianjin Station's pedestrian transport hub, improve the quality of street space, show the characteristics of the street, and improve the walking environment.

街道改造效果图
Street Renovation Renderings

恢复历史风貌，提升街道活力

改造与解放北路垂直的 8 条横向街道，以“一纵八横”为骨架，形成了该地区的步行网络骨架，街道改造主要包括三方面措施：

一是改造建筑外檐，增加商业氛围。对建筑进行分类整治，文保单位和历史建筑要做到修旧如旧，恢复原有的历史风貌。

二是调整街道断面，加宽人行步道。缩窄车道以限制机动车流量，将人行步道加宽，并鼓励沿街的商业外摆，提升街道活力。

三是完善街道设施，增加绿化空间。增设休憩座椅、分类垃圾桶等街道设施，同时见缝插绿，增加街角公园。

Restoring Historic Style and Enhancing Street Vitality

Eight crosswise streets perpendicular to North Jiefang Road have been transformed into a pedestrian network structure in the area with the framework of "one vertical and eight horizontal". Street reconstruction mainly includes three measures:

The first is to rebuild the outer eaves of buildings and increase the commercial atmosphere. Classified renovation of buildings, cultural protection units and historic buildings to be as old as before, to restore the original historical style.

The second is to adjust the street section and widen the pedestrian walkway. Streams are narrowed to limit motor vehicle traffic, widen sidewalks, and encourage commercial outdoors along the streets to enhance street vitality.

The third is to improve street facilities and increase green space. Street facilities such as recreational seats and classified trash cans will be added, while green seams will be inserted and street corner parks will be added.

打造一个可亲近的城市名片

世纪钟是天津的一张城市名片，初到天津的外来游客尤其喜欢与它来一张合影，但环岛机动车交通与人流混行，大量的护栏和过窄的人行步道也使人不便近距离接触。本次改造重新梳理了交通流线，将公交车、出租车和不停靠车辆进行分流引导，增加自行车道和人行步道，同时增加了广场的活动空间，将世纪钟广场打造为一个可亲近的城市名片。

Create the Close City Business Card

Century Clock is a city card in Tianjin. First-time visitors to Tianjin especially like to take a picture with it. However, the vehicle traffic around the island mix with people. A large number of guardrails and narrow sidewalks also make it inconvenient to get close contact with people. This renovation has reorganized the traffic flow, diverted buses, taxis and no-parked vehicles, increased bicycle lanes and pedestrian walkways, and increased the activity space of the square, making Century Bell Square a close city business card.

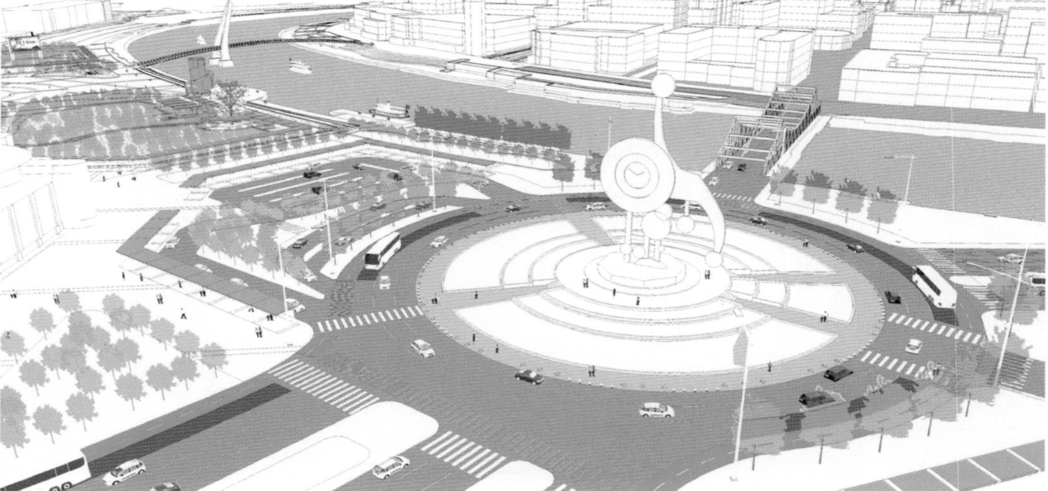

世纪钟节点设计
Design of Century Clock Node

开放、包容、人性化的天津站改造

在世纪钟环岛处规划改造出一处出租车落客区，方便车辆接送乘客。按照人流流线增加一组玻璃与膜结构的设施，整体效果如同一张浮动的巨大画布，中间凸起并向两侧延伸，以开放包容的姿态面向海河，并与海河步行桥融为一体。

进站的乘客下车后可沿着长廊进入主站房，长廊两侧设置了小型零售点、咖啡座等商业外摆，还为在室外候车的乘客提供了休憩座椅。改造后的天津站将成为展现天津国际化大都市形象的新地标。

Open, Inclusive and Humanized Reconstruction of Tianjin Station

A taxi landing area will be rebuilt at traffic island around the Century Clock in order to facilitate the transport of passengers. A group of glass and membrane structures will be added according to the flow line. The overall effect will like a floating giant canvas, which raised in the middle and extended to both sides. It is facing the Haihe River with an open inclusive attitude and integrated with the Haihe pedestrian bridge.

Inbound passengers can enter the main station along the corridor after getting off. On both sides of the corridor, there are small retail, coffee seats and other commercial outdoors. They also provide recreational seats for passengers waiting outside. The transformed Tianjin Station will become a new landmark to show the image of Tianjin as an international metropolis.

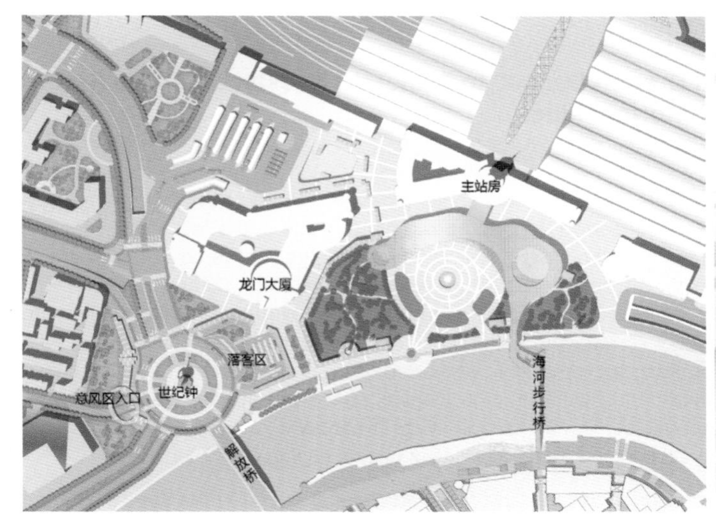

天津站前广场设计
Uban Design of Tianjin Railway Station Front Square

中心城区与环城四区现状建设格局
The Current Construction Pattern of the Central City and the Four Districts Around the City

从“金龙起舞”到“翡翠项链”——生态优先价值观的回归

1. 生态价值的褪色

曾经的九河下梢、七十二沽美景，在快速的城市空间拓展中褪色，这是上个城市发展阶段中自然价值消退的一个缩影。从 20 世纪 90 年代开始，天津进入快速发展的轨道，中心城区以典型的圈层蔓延模式扩展了 200 多平方千米，并即将填满快速环路与外环线之间的区域，再往外就与环城四区连为一体，“摊大饼”蔓延的趋势越来越明显。大面积的湿地、坑塘、农田、林地等近郊的生态性空间在这个进程中消失，曾经的沽塘绿树、鸟语花香、蝉鸣蛙叫的自然生态美景不复存在，近郊的自然生态系统遭到了很大的破坏。

今天中心城区仅存的湿地、林地生态空间几乎都集中在外环线沿线，也就是“一环十一园”地区。一个近 50 平方千米的绿色生态空间，毋庸置疑，将成为中心城区生态转型的战略区域。

From “Golden Dragon Dance" to "Emerald Necklace" – The Return of Ecological Priority Values

1.The Fading of Ecological Values

A little under the 9 rivers, with 72 scenic spots, faded in the rapid development of urban space, which is a microcosm of the fading of natural value in the last stage of urban development. Starting from the 1990 s, Tianjin entered the track of rapid development, in a typical downtown envelops spread pattern extends more than 200 square kilometers, and is about to fill quickly the area between the fast-traffic ring road and the outer ring road, farther away four surrounding areas is an organic whole repeatedly with the outer ring, the spread of booth pie trend more and more obvious, large areas of wetland, pits, farmland, forest land on the outskirts of ecological space disappeared in the process, once sell pond, green trees, flowers, cicadas frogs is natural ecological beauty ceased to exist, have done great damage to the outskirts of natural ecological system.

Today, the only remaining wetland and forest ecological space in the central city are almost concentrated along the outer ring road, namely "one ring and eleven gardens" area, a green ecological space of nearly 50 square kilometers, which will undoubtedly become the strategic area for the ecological transformation of the central city.

2. 生态优先价值观回归

今天，天津提出“生态优先”的战略，这既是对以往的反思，也是对未来的指向。自然生态不再从属、依附于城市，不再被动地为城市发展让步，而是有了一个与城市发展空间平等甚至更重要的地位。人与自然将回归麦克哈格在《设计结合自然》一书中提出的“我们”的关系。

上个三十年，海河的综合开发战略带动了津城的经济腾飞，实现了城市跨越式发展。下个三十年，中心主城“翡翠项链”格局的建立，将推动城市以生态为导向的高质量发展。

2.Return of Ecological Priority Values

Today, Tianjin put forward the strategy of "ecological priority", in a sense to set up a different priority values by economic development, which is both reflection on the past, is also a point of the future. Namely, natural ecology is no longer dependent, attached and no longer passive concessions for the city development, with an equal space with urban development is even more important position. The human and the nature will return the "our" relationship in the book *Design With Nature* by Ian L. McHarg.

In the last 30 years, the comprehensive development strategy of Haihe promoted the economic take-off of Tianjin and realized the leapfrog development of the city. In the next 30 years, the establishment of "Emerald Necklace" of the central city, will promote the city's ecologically oriented high-quality development.

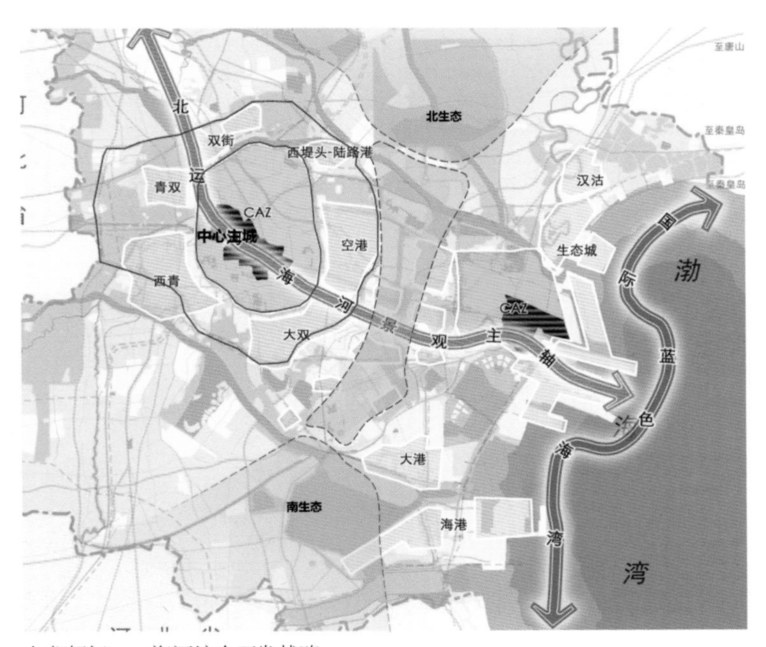

金龙起舞——海河综合开发战略
Golden Dragon Dance-Haihe River Comprehensive Development Strategy

海河综合开发与一环十一园设计
Haihe River Comprehensive Development and One Ring Eleven Gardens Design

从“伦敦绿环”到“波士顿项链”——作为城市战略的环城公园系统

1. 外环绿带得与失

20 世纪 90 年代规划建设的天津市外环线及外环 500 米绿带，最初的目的在于划定一个中心城区明确清晰的增长边界，建设中心城区的绿色屏障，是学习借鉴了伦敦绿环规划的经验，在当时具有前瞻性和现实意义，外环 500 米绿带成了中心城区城市结构的重要组成部分，也建立了基本的生态空间框架。多年来，历版天津城市总体规划延续和不断完善外环绿化带及“绿楔”规划，逐步形成了今天以外环绿带及“绿楔”共同构成的城市公园体系——“一环十一园”。

但在今天看来，也存有遗憾，外环绿带过于强调防护职能，且自成体系，作为“增长边界”的角色，始终独立于中心城区的生态与空间系统，未能发挥“增长边界”之外更大的职能。“一环”虽学习“伦敦绿环”，但空间尺度不足，难以孕育完整的生态系统。对于天津践行生态优先战略，建设生态宜居城市，“一环”应有更高的定位，发挥更大的作用。

From "London Green Ring" to "Boston Necklace" — Ring Park System as Urban Strategy

1.The Outer Green Belt Gains and Loses

In 1990s, planning and construction of Tianjin outer ring and outer ring green belt width 500 meters, the original purpose is to draw a clear central urban growth boundary. The construction of urban green barrier, it is to learn from the experience of the London Green Ring Planning prospective and realistic significance at that time, outer ring green belt width 500 meters became the important part of urban structure, city center has established the basic framework of ecological space. Over the years, the overall urban planning of Tianjin has continued and improved the planning of green belt and green wedge on the outer ring, and gradually formed the urban park system composed of green belt and green wedge on the outer ring — "one ring and eleven parks".

However, in today's view, there is also a pity that the outer green belt places too much emphasis on the protection function and has its own system. As the role of "growth boundary", it is always independent of the ecological and spatial system of the central city and fails to play a bigger role beyond the "growth boundary". Although "one ring" studies "London Green Ring", it is difficult to breed a complete ecosystem due to insufficient spatial scale. For the implement the ecological priority strategy in Tianjin and build an ecological livable city, "one ring" should be positioned higher and play a greater role.

外环绿带的规划与实施
Planning and Implementation of Outer Ring Road Greenbelt

中心主城的自然生态要素分布
Distribution of Natural and Ecological Elements

2. 中心主城的“翡翠项链”

随着外环线交通职能的调整以及双城之间绿色屏障的确立，中心城区与环城四区一体化成为必然，高速环路以内约 1 300 多平方千米的空间需要重塑结构体系，在这个更大的空间尺度上，容纳了更多的自然生态要素，既有环内的 11 个大型城市公园，也包含了环外的 5 个郊野公园以及大面积的农业生态用地，多条河流绿化走廊，为构建一个更大的城市生态系统提供了机会，也为中心城区建立一个生态系统引导的健康的空间结构体系提供了可能。

从这种意义上讲，“一环十一园”不再是“伦敦绿环”，而是“波士顿项链”，是一个生态结构更稳定、多样，与城市空间系统更加契合，且为人居环境提供有力支撑的“翡翠项链”，在这样一个绿色生态系统的支撑下，高速环路以内的国土空间将进一步整合重塑。

2.The Emerald Necklace of the Central City

Along with the adjustment of the outer ring road traffic function and the establishment of the green barrier between two cities, central city with four surrounding areas integration is inevitable, the space of about 1300 square kilometers within high speed loop need to reshape the structure system. In the larger spatial scales, more natural ecological factors contains both the ring of 11 large city park the five country park outside the ring as well as large areas of land for agricultural ecology and many river's green corridors, to build a bigger city ecosystem provides opportunity and build an eco-system for central city spatial structure of the health system.

In this sense, the "one green ring and eleven parks" is no longer "London Green Ring", but "Boston Necklace" , It is an ecological structure with more stable, diversity, and more urban space system. The living environment provide strong support for "Emerald Necklace", in such a green ecological system, under the support of high-speed ring within the territory will further integrate reshaping.

3. 以“翡翠项链”作为城市战略的环城公园系统

1895 年，奥姆斯特德设计了 800 公顷的“翡翠项链”——一个由公园、生态保护地、绿道、河流廊道、林荫道组成的都市线形公园系统，作为波士顿城市发展框架，这也是 19 世纪第二次工业革命后，美国城市以公共空间作为有效工具引导城市发展，重塑城市面貌的体现。

作为城市战略的公园系统，对城市发展的持续影响主要通过三种途径：一是开创某个地区的城市化进程，二是改变某个地区土地使用模式，三是建立塑造城市生活面貌的综合系统。

波士顿“翡翠项链”的成功主要基于以下六个要素。

位置：距离、进入方式、地形，自然特色；市场：适应人口增长趋势，提供公共设施，引发周边的私人商业活动；设计：面积充裕，设计兼顾形态可行性，地形条件及周边环境，规划可容纳多样活动，成为相互联系的统一体系；资金：公园的建设资金投入，运营成本；企业家：公园开发商，独立的公园委员会；时间：建设完成后，需要长期维持。

3.Emerald Necklace —The Ring Park System as the Urban Strategy

In 1895, Frederick Law Olmsted designed “Emerald Necklace” of 800 hectares area including a park, ecological landscape, green road, river corridor, avenue of urban park system, linear as Boston city development framework. This is a American city public space as an effective tool to guide urban development and reshape cities in the 19th century after the industrial revolution.

As a strategic urban park system, the sustainable impact on urban development can be achieved in three ways. The first way is to process the pioneering of region urbanization. The Second way is to change the pattern of land use in a certain area. The third way is to establish a comprehensive system for shaping urban life. The success of the Boston Necklace's emerald necklace is based on six factors:

Location: distance, entry mode, terrain, natural features.

Market: adapt to the trend of population growth, provide public facilities, and trigger the surrounding private commercial activities; Design: morphological feasibility, abundant area, topographic conditions and surrounding environment; a unified system that accommodates diverse activities and interconnections. Capital: park construction capital input, operating costs. Entrepreneur: park developer, independent park board. Time: after construction is completed, it needs long-term maintenance.

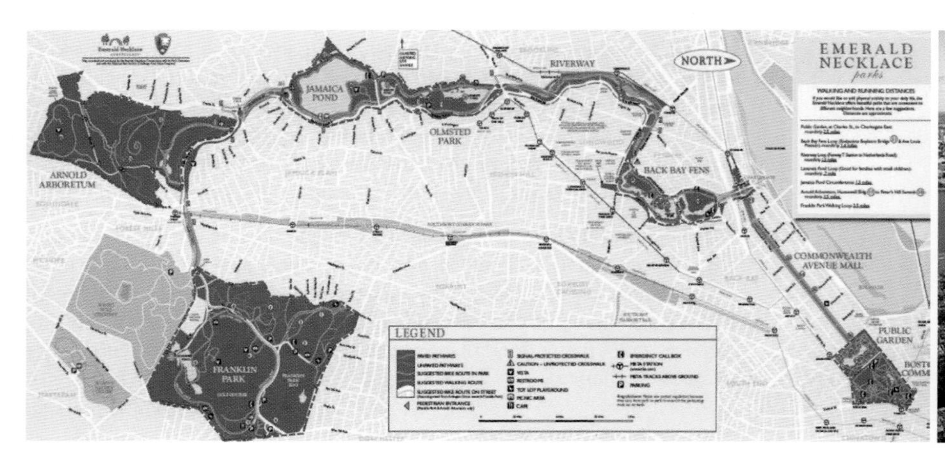

波士顿的“翡翠项链”示意图 Boston's Emerald Necklace

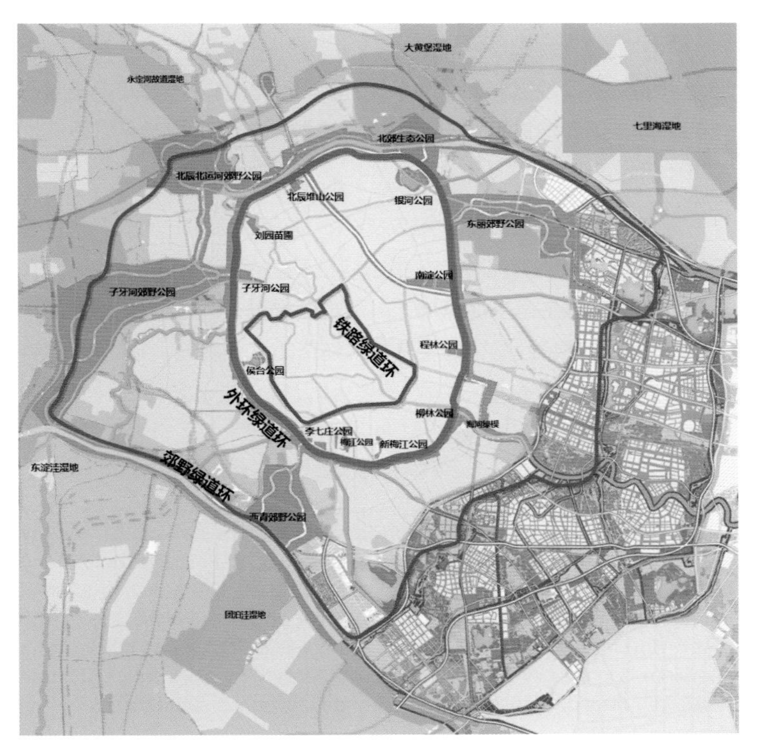

中心城区蓝绿空间结构
Blue-green Spatial Structure of Central City

“翡翠项链”——重建中心城区生态系统的完整性、多样性

1. 三环体系，蓝绿生态结构

借鉴奥姆斯特德的波士顿“翡翠项链”建构过程，首先进行自然保护地、特色自然生态选择与景观重塑，其次通过绿道、林荫道、河流绿廊进行生态景观连接，以形成整体、统一的公园系统。

在外围田园城市圈层，构建郊野绿道环 + 郊野农林生态；在生态宜居圈层，构建外环绿道环 + 十一个城市公园；在中心活力圈层，构建铁路绿道环 + 社区公园，最终形成自然生态体系与人居空间体系契合的三环体系。做到内疏外联，三环连通：一是水系连通，一环十一园及周边地区与中心城区内部水系连通，建构海绵城市的调蓄功能；二是绿道连通，依托河道、绿廊构建特色绿道与外围郊野公园、外环绿带与高速生态环相连通，强化与双城之间绿色屏障的联系。

Emerald Necklace —Restoring the Integrity and Diversity of the Central Urban Eco-system

1.Three-ring System, Blue-green Ecological Structure

Drawing on the construction process of Olmsted's Emerald Necklace in Boston, it firstly carries out Natural Conservation, featured natural ecological selection and landscape remodeling. Then it connects the ecological landscape through greenway, avenue and river veranda to form an integrated and unified park system.

There are construct countryside greenway ring and Countryside agroforestry ecology in the outer rural urban circle. In the ecological livable circle, there are the construction of the ring greenway ring and 11 city parks. In the central vitality circle, there are the railway greenway ring and community park which will be constructed. Finally, three-ring system that fits the natural ecological system and human living space system will be formed. At the same time ,outside the sparse connect three rings. The first is the connection of water system. That is the connection between the first ring 11 garden's the surrounding area and the inner water system of the central city, So as to construct the regulating and storing function of sponge city. Second, the greenway is connected. The characteristic greenway is connected to the peripheral country park, the outer green belt and the high-speed ecological ring by relying on the river channel and the greenway, So as to strengthen the connection with the green barrier between the two cities.

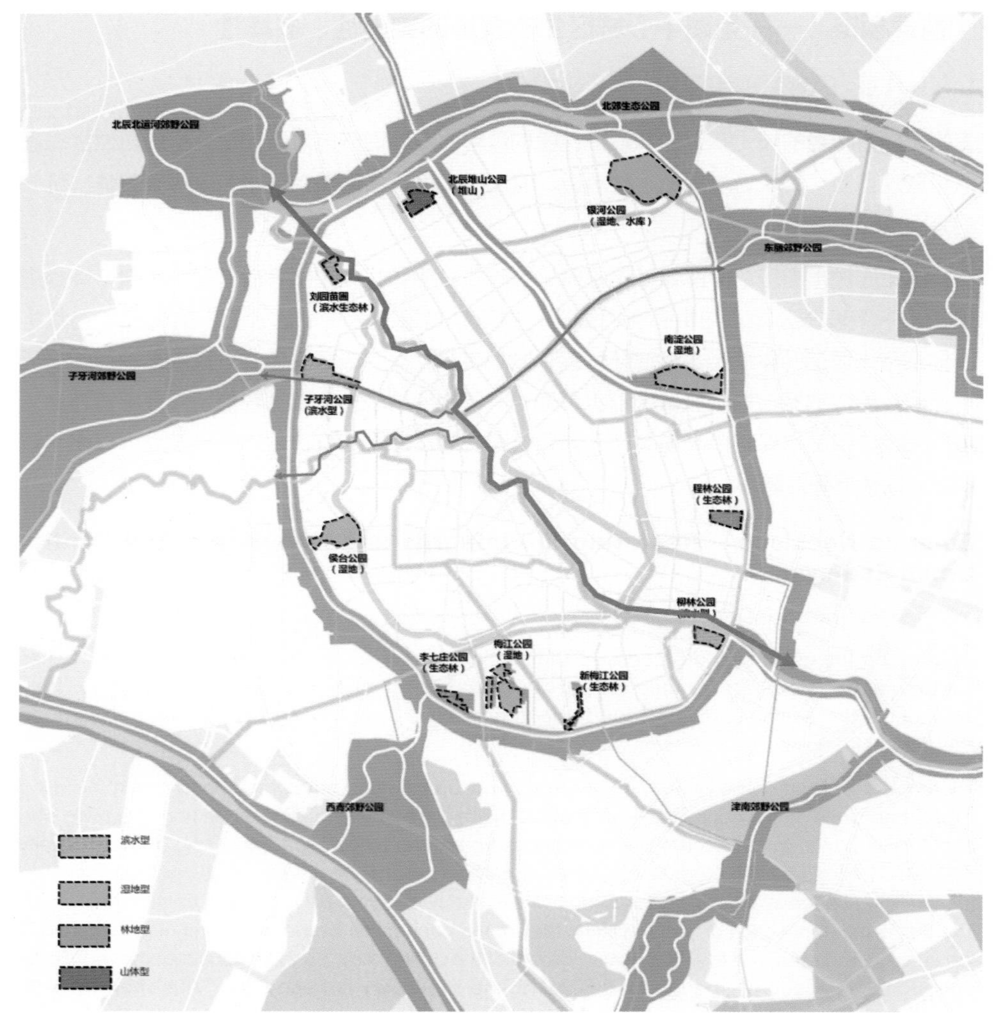

城市公园特色
Characteristics of Urban Parks

2. 特色公园

波士顿的“翡翠项链”，并非单纯的自然保护地，奥姆斯特德对已有的湿地、沼泽地进行了一系列的生态改造与景观重塑，以适应周边城市开发的需要。

十一园设计以生态为基础，融入文化，结合海绵城市功能要求，塑造一系列具有鲜明文化特色、生态特征及功能特色的城市公园。在保护现有的自然生态基底基础上，形成湿地型、滨水型、 林地型、山体型四种自然特色的公园，并融入城市的文化脉络， 赋予公园鲜明的文化特色，形成传统中式、西式以及现代文化艺术三种典型的文化特征，将城市公园打造为特色“城市客厅”。

2. Characteristic Park

The Emerald Necklace in Boston is not simply a natural protection site. Olmsted has carried out a series of ecological reconstruction and landscape reconstruction on the existing wetlands and marshes to meet the needs of the development of surrounding cities. Therefore, the design of 11 urban parks takes ecology as the basis, integrates culture and combines sponge city's functional requirements to create a series of urban parks with distinct cultural, ecological and functional characteristics.

On the basis of protecting the natural ecology of the existing basement, there are wetlands type, waterfront type, forest type and shape type of parks ,Parks are Integrated into the cultural context of the city and given distinctive cultural features to formate the traditional Chinese, Western and modern culture art three typical cultural characteristics of the city park features to build "city living room".

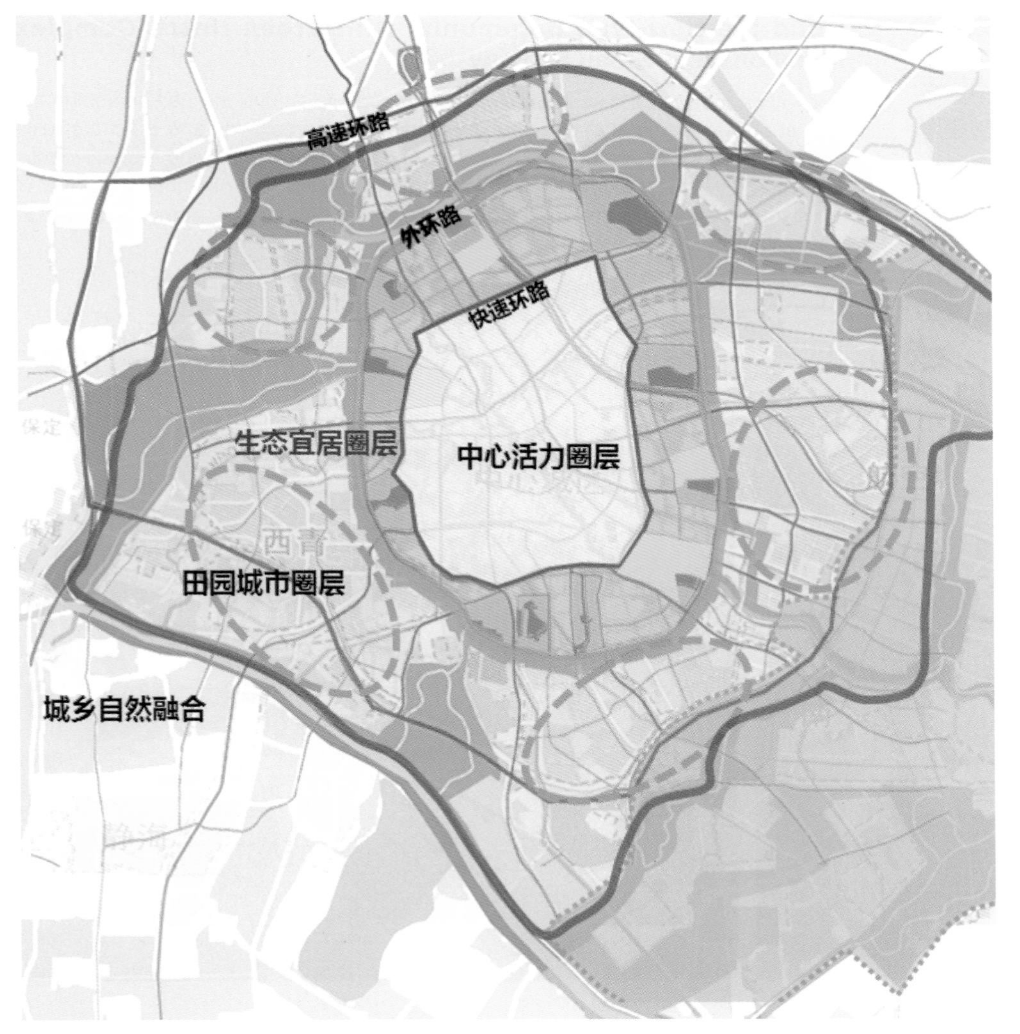

中心城区三圈层空间特色
Spatial Characteristics of Three Layers

生态导向的中心城区与环城四区空间整合与结构优化

采用适应不同生态特色导向的三圈层模式。中心活力圈层、生态宜居圈层、田园城市圈层三个圈层确立了建立在都市自然生态系统之上的城市发展空间。中心活力圈层是快速环路以内的城市建成空间，是城市活力中心区，以城市更新为主要特征，采用高强度、高密度的空间开发模式。生态宜居圈层是快速环路与外环之间的区域，突出居住空间与大型城市公园结合的特征，采用中低强度开发，TOD（以公共交通为导向的开发模式）、SOD（以社会服务设施建设为导向的开发模式）局部高强度。田园城市圈层是外环至高速环路之间的区域，以存量、减量发展为主要特征，居住空间与城郊生态空间结合，采用低强度、低密度开发，具有较高的蓝绿空间占比、森林覆盖率。

The Spatial Integration and Structural Optimization of the Ecologically Oriented Central City and Four Surrounding Areas

Adapt to Different Ecological Characteristics of the Three - Layer Model.The establishment of three circles: central activity circle, ecological livable circle and rural urban circle which are urban development spaces based on urban natural eco-system. The central activity circle layer is the urban built space within the fast-traffic ring road , and it is the central area of urban vitality. It takes urban renewal as the main feature and adopts the spatial mode of high intensity and high density. The ecological livable circle layer is the area between the fast-traffic ring road and the outer ring road , highlighting the characteristics of combining residential space with large urban parks. It adopts medium and low intensity development and partial high intensity TOD and SOD. The rural urban area, from the outer ring to the high-speed ring, is mainly characterized by the development of stock and quantity reduction. The residential space is combined with the suburban ecological space, and the development is carried out in low intensity and low density, with a high blue-green space ratio and forest coverage rate.

建设以公园为绿心、功能复合、绿色生态宜居的新型社区

强化公园周边城市空间控制要求，结合社区的空间品质、空间类型的要求，建设各种开发强度适宜、整体高度有序、多元化混合的新型邻里社区。外圈层为中强度开发的小高层、高层，内圈层为低层高密度、联排、多层住宅，站点周边为高强度开发的中高层。

Build the New-type Community with Green Heart, Complex Functions and Green Ecology

Strengthen the spatial control requirements of the city around the park, combine the spatial quality and spatial type requirements of the community, and build the new-type community with appropriate open intensity, high overall order and diversified mix. The outer layer is medium intensity, small high-rise and high-rise, the inner layer is low density, townhouse and multi-story residence, and the site is surrounded by high intensity, medium high-rise.

新型社区空间模式
The New-type Community Space Model

工作意义

城市开放空间是城市的客厅，是居民进行公共活动和工作生活的场所，是展示城市风貌特色的载体，是眺望城市轮廓线和对景的通廊，是衡量城市品质的主要指标。因此，城市开放空间是城市设计行业的重点工作内容。

Working Objective

Urban open space is the living room of a city, the place where residents carry out public activities and life, the carrier of displaying the features of the city. It also serves as the corridor overlooking the city outline and the landscape. It is the main index to measure the space quality of the city. Therefore, urban open space is the critical work of urban design specialty.

水西公园
Shuixi Park

研究对象

城市开放空间指城市内供社会公共使用和居民日常生活的室外空间，包括河湖水系、公园、广场、街道等空间。本书研究重点为独立占地的水系、公园、广场、街道等开放空间。

Research Object

Urban open space refers to the outdoor space in the city for public use and daily life, including rivers and lakes, parks, squares, streets and other spaces. The focus of the planning study are those open spaces such as water system, parks, squares, streets, and so on.

水系

公园

广场

街道

研究对象
Research Object

早期公共生活研究　　公共生活研究作为策略工作　公共生活研究成为主流

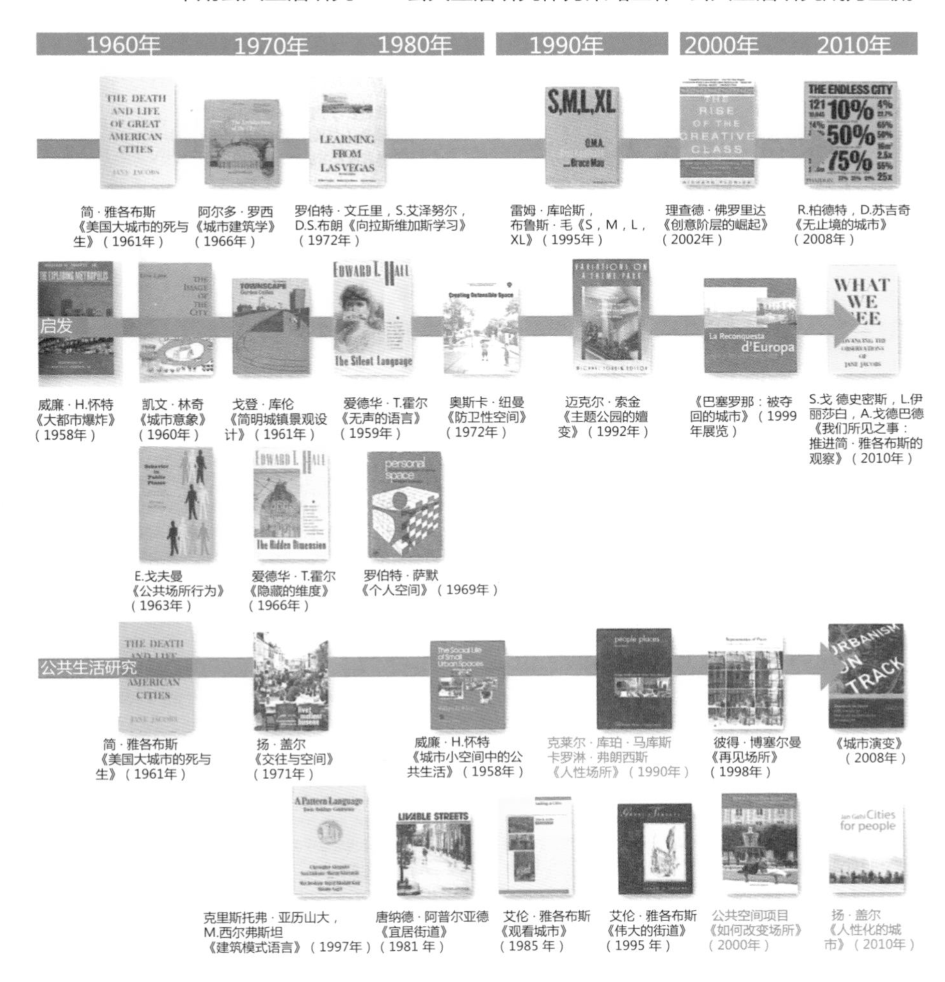

西方学术界的学术研究

Academic Research in Western Academia（ Source: Research Methods of Public Life ）

项目借鉴理论

对优秀城市开放空间的共同特质进行研究，按照公共生活研究的发展对西方学术界多年来的学术研究进行分析整理，从众多的研究团队中选取三个最具代表性和影响力的研究团队，即丹麦皇家艺术学院扬·盖尔教授、加州大学伯克利分校C.C.马库斯与C.弗朗西斯教授、“公共空间”项目，作为项目理论借鉴的重点分析对象。

The Theory

The project team studied the common characteristics of excellent examples of urban open space. According to the development of public life research, we analyzed the academic research of western academia for many years. Three most representative and influential research teams were selected as analysis objects for theoretical references. They are research projects led by: Professor Jan Gehl of the Royal Danish Academy of Art, Professor C.C. Marcus and Professor C. Francis of the University of California, Berkeley, and the team called “Project for Public Spaces” .

保护

免受交通状况与事故困扰——安全感

- 对行人的保护
- 消除行人对交通状况的畏惧

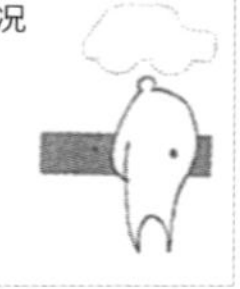

免遭犯罪与暴力——人身保护感

- 活跃的公共领域
- 街道上目光的注视
- 日间与夜间功能的重合
- 良好的照明

免于不愉快的感官体验

- 风
- 雨、雪
- 冷、热
- 污染
- 灰尘，噪声，强光

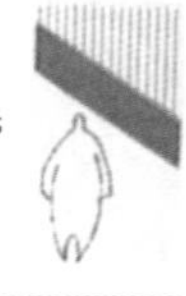

乐趣

行走的机会

- 行走的空间
- 不受阻碍
- 良好的路面
- 为所有人服务的无障碍设计
- 有意思的建筑立面

站立、停留的机会

- 边界效应
- 吸引人站立、停留的区域
- 给站立者提供的支持倚靠

坐下的机会

- 可就座的区域
- 利用优势：视野、阳光、人群
- 适合就座的场所
- 方便休息的长凳

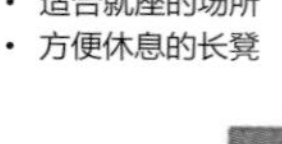

观看的机会

- 合理的观看距离
- 不受阻碍的视线
- 有趣的景观
- 照明（黑暗时）

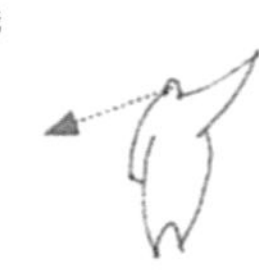

交谈和倾听的机会

- 低噪声等级
- 构成“适于谈话的景观”的街道设施

嬉戏与锻炼的机会

- 鼓励进行创造性活动、体育活动、锻炼和嬉戏
- 昼夜都可以
- 冬夏都可以

舒适

尺度

- 按照人性化尺度设计建筑和空间

享受当地气候优势的机会

- 阳光、阴凉
- 热、冷
- 微风

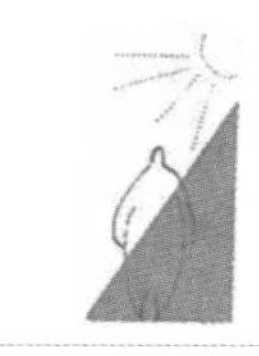

好的感官体验

- 好的建筑设计与细部设计
- 好的材料
- 精美的视觉景观
- 树、绿植、水

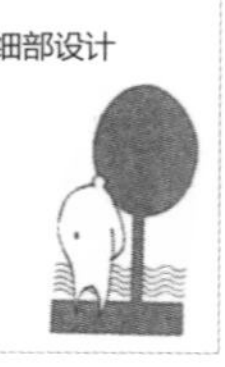

公共空间 12 项品质标准
Twelve Quality Standards of Public Space

扬 · 盖尔教授

扬 · 盖尔提出用人的尺度进行规划，将产生持久、高质量的设计。《人性化的城市》中提出公共空间的 12 项品质标准，可以归纳为保护、乐趣、舒适三个方面。

C.C. 马库斯与 C. 弗朗西斯教授

针对北美城市（以旧金山市为主）七类开放空间的建成案例，基于使用者的真实感受进行使用状况评价，在调查、统计、评价的基础上针对特定的空间类型提出操作性的而不是规范性的设计导则。

Professor Jan Gehl

Jan Gehl proposed that planning based on the scale of human beings would produce durable and high-quality designs. The 12 Quality Standards of public space mentioned in the book Cities for People can be summarized into three aspects: protection, pleasure and comfort.

Professor C.C. Marcus And Professor C. Francis

According to the seven types of open space construction cases in north American cities (mainly in San Francisco), they evaluated the use status based on users' real feelings, and put forward operational rather than normative design guidelines for specific space types on the basis of investigation and statistics.

公共空间度量标准（资料来源：“公共空间”项目）
Standard of Public Space（Source：Project For Public Spaces）

公共空间项目

成功的公共空间有四个关键属性，分别是社会交融（利于社交活动）、通达与连接（可达性）、舒适与印象（舒适性）、功能与活动（多样性，满足不同活动的需求）。从关键属性出发，将每个关键属性对标到多个无形价值，再进一步转化为度量标准。

Project for Public Spaces

A successful public place has four key attributes: social integration (social-activity friendly), accessibility/connectivity, comfort and impression, functions and activities (to fulfill diversity and to meet the needs of different activities). Starting from these key attributes, each was marked to more than one intangible values, and further be transformed into a set of evaluation standards.

实践情况

构建由主次轴线和公园、广场、滨水空间组成的开放空间系统，将其整理为轴线与节点的关系，以进一步引导设计。

The Practice

Firstly, we constructed an open space system consisted of primary and secondary axes, parks, squares and waterfront spaces. Then, the relationship between axis and nodes were refined for further design guiding.

开放空间系统引导图
Open Space System Guidance

实践情况

天津河流的不同区段具有不同的特点，依据现状与规划定位，将滨水空间划分为都市活力型、历史风貌型、自然生态型三类。

The Practice

Different sections of rivers in Tianjin have different characteristics. According to the current situation and the planning goals, the waterfront spaces were divided into three types: urban vitality type, historical style type and natural type.

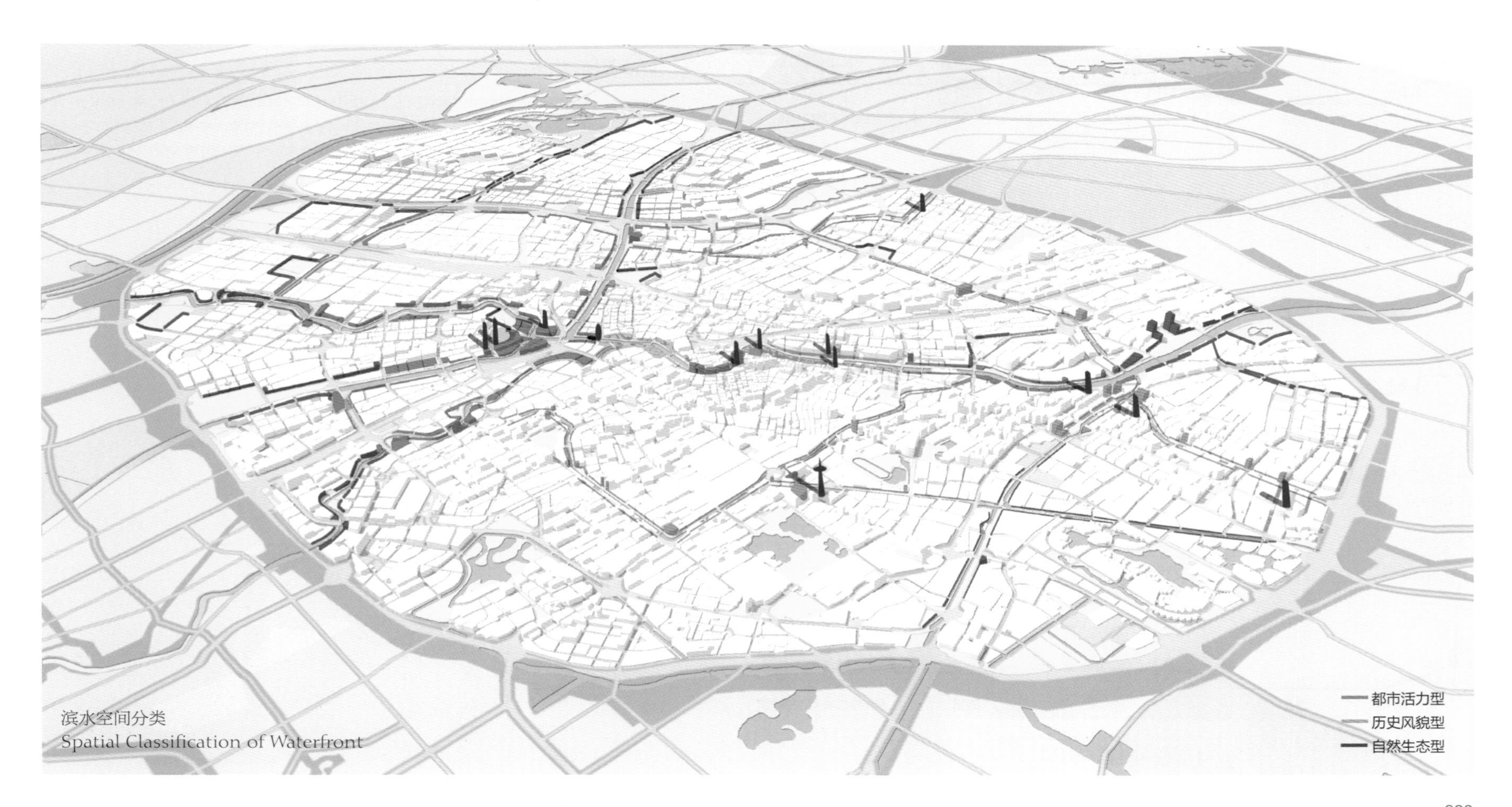

滨水空间分类
Spatial Classification of Waterfront

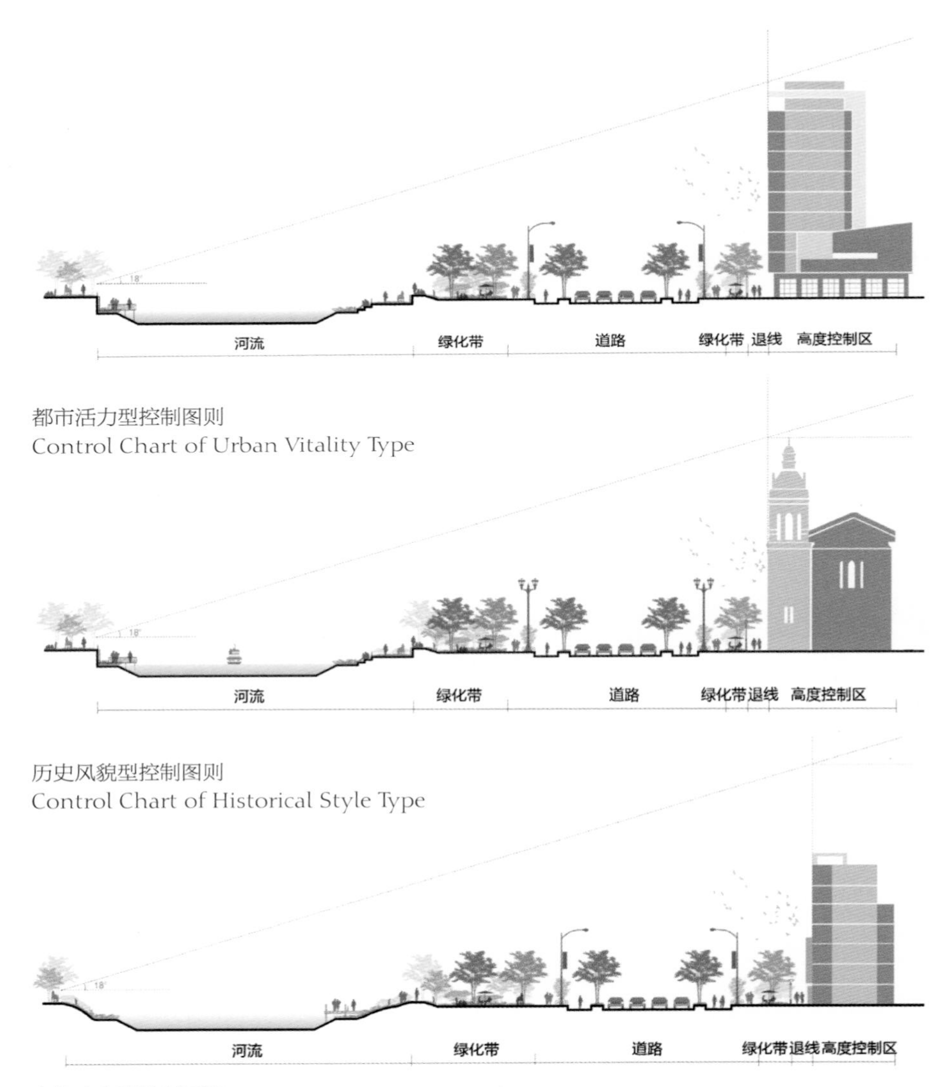

都市活力型控制图则
Control Chart of Urban Vitality Type

历史风貌型控制图则
Control Chart of Historical Style Type

自然生态型控制图则
Control Chart of Natural Type

实践情况

根据不同类型滨水空间的特点，有针对性地对空间主题、竖向设计、公共交通、慢行系统、园林植栽、景观通廊、建筑高度、建筑风格色彩等要素提出管控要求。

The Practice

According to the characteristics of different types of waterfront space, we put forward the control requirements for space theme, vertical design, public transport, pedestrian system, landscape plants, landscape corridor, building height, architectural style and color.

实践情况

滨水空间与相邻道路空间应统筹设计，重点在于：滨水空间应配合与之平行的道路，确保滨水慢行空间的连续性；呼应与之垂直的道路，在交口处设置节点。以道路作为轴线时，滨水空间的设计应达到呼应和强化轴线的效果。

The Practice

Waterfront space and adjacent road space should be designed as a whole. Firstly, waterfront space should work together with the parallel road to ensure the continuity of waterfront pedestrian space. Secondly, nodes should be placed at the intersection to mark the perpendicular path. When the road served as an axis, the design of waterfront space should be designed to strengthen the axis.

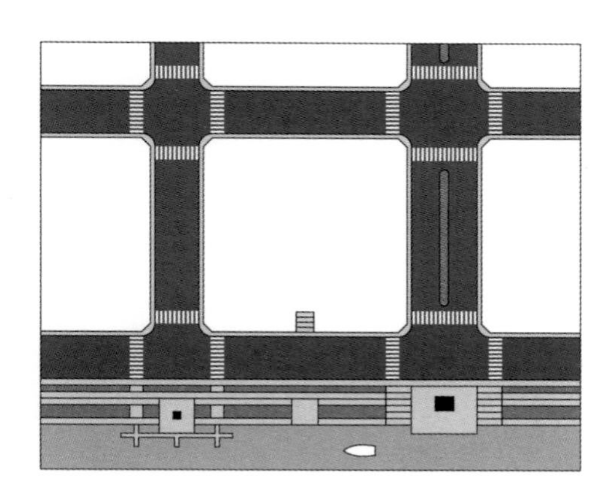

滨水堤坝与道路无高差
1. 滨水人行道植栽种植宜避免遮挡水景。
2. 临水步道可采用木栈道形式，适当挑出。
3. 滨水低级别道路可通过更换铺地材料，设置降速带等方式，优化其步行体验。
4. 垂直于滨水的道路，可在对应交叉口处设置滨水节点与标志物。
5.路口相隔过远时应考虑设过街天桥或地下通道。
6. 如纵向道路为轴线，对应滨水节点应适当扩大规模，并设标志物，有条件的增设游船码头。

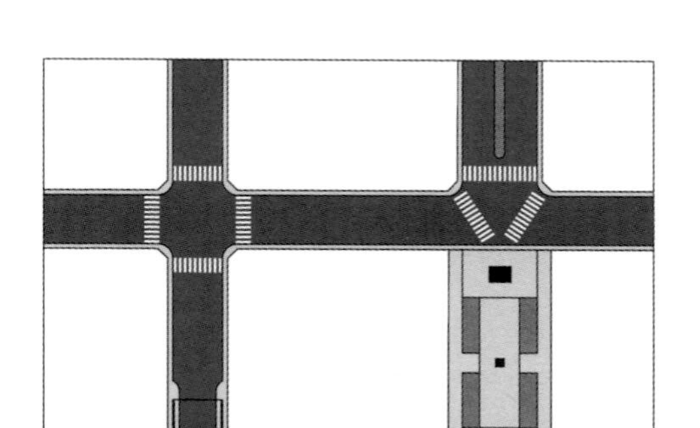

滨水堤坝低于道路
1. 滨水空间不应由临水地块业主独占，应留出一定宽度的公共空间。
2. 桥梁不应阻挡滨水空间步道的连续性。
3. 应在必要位置增建步行桥，连通两岸的开放空间。
4. 靠近滨水空间的重要交口，在不干扰建筑正常使用的情况下应开辟步行道与滨水空间相接。
5. 如纵向道路为公共空间体系轴线，对应滨水节点应适当扩大规模，并设标志物，有条件的增设游船码头。

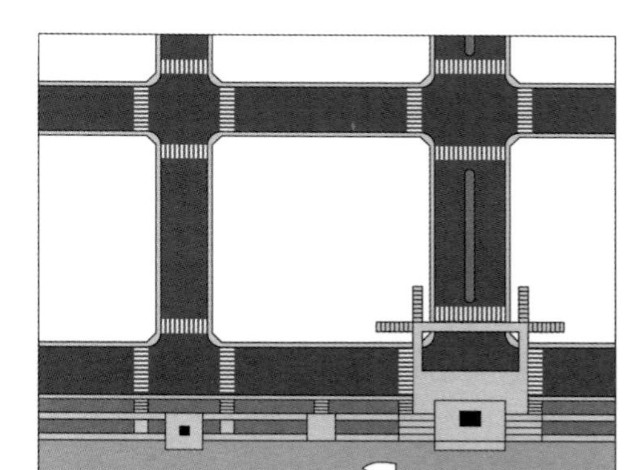

滨水堤坝高于道路
1. 水景被遮挡的情况下，人行道应同时设于堤上、堤下，堤与道路过渡可用坡道、台阶或矮墙等，视高差大小而定。
2. 道路交口与滨水空间宜用过街天桥连接，应于对应位置设亲水平台，亲水平台可与过街天桥结合设计。
3. 如纵向道路为公共空间轴线，对应滨水节点应适当扩大规模，并设标志物，有条件的增设游船码头。

滨水空间与街道的关系
The Relationship Between Waterfront Space and Street

实施效果

天津形成以水网为主的开放空间系统与高品质的滨水开放空间。以海河为代表的滨水开放空间，汇聚最高级别的城市公共中心与核心服务职能，呈现最具时间轨迹的历史人文特色资源，最大限度地体现天津城市的“时代感、地方性、人情味”。

天津建设了展现沽上美景的大型城市公园。天津地势低洼，汇水集中，历史上形成“七十二沽花共水，一般风味小江南”的津沽美景。城市建设多利用河湖湿地建设了水西公园、水上公园等以湿地水体为特色的城市公园，保护水系资源，涵养城市生态环境。

The Implementation Review

In Tianjin a water network-based open space system and a series of high-quality waterfront open spaces have been constructed. The waterfront open space represented by Haihe River brings together the highest-level urban public center and service cores, presents the most historical and humanistic resources, and reflects the "sense of the times, locality and humanity" of Tianjin to the greatest extent.

Large-scale urban parks has been built to show the beautiful scenery of ponds and lakes which used to be the most prominent landscape feature of Tianjin. Tianjin has a low-lying terrain and concentrated water catchment. In Qing Dynasty there was an antithetical couplet said: “flowers bloomed in seventy two pools, just feels like micro Jiangnan views”. In later urbanization, lakes and wetlands were often transformed to parks, such as Shuixi Park, Water Park, taking use of wetland feature and protecting water resources in the same time.

人民公园
People's Park

水西公园
Shuixi Park

实施效果

经过重点整治，塑造了主要迎宾道路，保护并传承了十四片历史文化街区内的历史风貌街道，尺度亲切宜人，建筑多姿多彩。传承天津城市的历史积淀与人文情怀，集中展示天津“中西合璧、古今交融”的历史文化名城风貌特色。

重点地区建设了高品质的城市广场，聚集了活力与人气。设计水准较高、细节处理到位，建成了文化中心、民园广场、津湾广场等城市广场，作为市民活动的新场所，展示了城市特色，聚集了城市人气，提升了城市活力。

The Implementation Review

After the renovation, the main roads of welcoming guests have been created, and the historic streets in 14 historic and cultural blocks have been well preserved. The scale is human friendly, and the buildings are in diversified figures. The historic and cultural blocks inherits the historical and humanistic characteristics of Tianjin, and focuses on displaying the features of Tianjin's combination of Chinese and western cultures and fusion of ancient and modern cultures.

High-quality urban squares were constructed in the core areas to gather vitality and popularity. The design is intricate and the details were well handled. Urban squares in the Cultural Center , Minyuan Garden Square and Jinwan area, have been built as new places for citizens' activities, displaying urban characteristics, gathering urban popularity and promoting urban vitality.

民园广场开放空间
Openspace of Minyuan Square

民园广场建筑风貌
Architecture of Minyuan Square

海河岸线
Haihe River Coastline

城市设计导则的作用与意义：加强城市意象，凸显城市特色

以城市设计导则的方式进行控制，通过精心的规划和设计，通过城市的重要场所，包括历史的和现代的，城市中心和边缘的，展示出一幅生动的城市画卷。

The Role and Significance of the Urban Design Guidelines: Strengthen Urban Imagery and Reflect Urban Characteristics

Through careful planning and design, controlled by urban design guidelines, the city's important places such as historical and modern, urban centers and edges, to show a vivid urban picture. The design of the water park the sky tower and the old town pass path are naturally connected. The citizens and tourists can feel the connection between some important spaces in the city.

旧金山鸟瞰
San Francisco Aerial View

向旧金山学习：建立一套行之有效的管理法则

使城市设计导则成为实现城市设计目标的过程中必不可少的重要环节，通过城市设计导则来完成城市设计的规范化与法律化，将其纳入城市规划管理机制中，从而形成城市良好的形体秩序并提升城市环境品质。

Learn From San Francisco: Establish a Set of Effective Management Rules

Making urban design guidelines an indispensable part of the process of achieving urban design goals, through the urban design guidelines to complete the standardization and legalization of urban design, and into the urban planning management mechanism, thus forming a good shape of the city and improve the quality of the urban environment.

向旧金山学习：传递一种城市自身的内在精神

城市设计的实践和探索并不是就事论事的，而是以人的生活及其品质的提升作为重要的依托。城市之所以美好，在于城市长期以来形成的多元性与混合性，在于它所承载的美好生活。

Learn From San Francisco: Passing on City Inner Spirit

The practice and exploration of urban design is not a matter of fact, but an important support for the life of people and citizens and the improvement of their quality. The beauty of the city lies in the long-term diversity and mixture of the city, and the beautiful life it carries.

旧金山 Misson Dolores 公园：不仅提供了丰富活动的公园场地，更为城市提供了绝佳的观赏整体风貌的视角和拍摄点
San Francisco Misson Dolores Park : not only provides a wealth of park activities, but also provides the city with an excellent view of the overall style and shooting points

旧金山十分注重历史与自然环境的保护，在总体城市规划中将保护与发展并举列为恒久的城市目标。山顶挺拔的建筑突出山的形状并保护了景观，低矮的小尺度建筑建在山坡上
San Francisco pays great attention to the protection of history and the natural environment, and protects and develops it as a permanent urban goal in overall urban planning. The tall buildings on the top of the mountain strengthen the shape of the mountain and protect the landscape. The low-rise small-scale buildings are built on the hillside

第一步：建立城市的骨架

第一个层次：中心城区总体城市设计导则。

自明永乐二年在三岔河口设卫建城至今，天津经过 600 多年的发展变迁，逐步形成了沿海河发展的城市格局。海河像一条玉带蜿蜒流过，它是物质的也是精神的，它是城市生活的起源，它是这座城市的生命与灵魂。中心城区总体城市设计导则的工作重点是建立总体空间艺术骨架，从战略视角对城市特色和山水格局进行宏观尺度上的预先控制和约束。突出海河在城市文化、景观特色上的价值，促使城市沿海河向外拓展，缓解市中心高强度开发与历史遗产保护的矛盾。

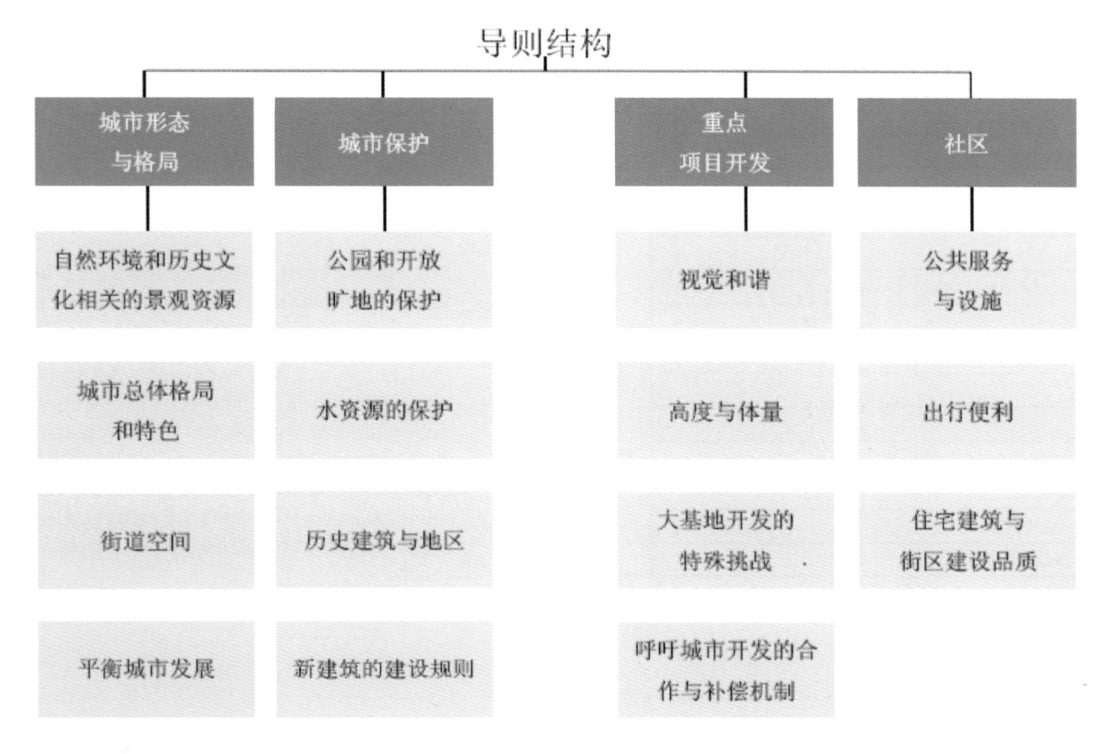

中心城区总体城市设计导则结构
Central City Overall Urban Design Guide Structure

The First Step: Building the Urban Framework

The First Level: Central City Overall Urban Design Guidelines.

Since 2nd year Yongle of the Ming Dynasty the acropolis was established in the Sancha Estuary, Tianjin has gradually formed the urban pattern of coastal river development after more than 600 years of development. Haihe River flows like a jade belt. It is both material and spiritual. It is the origin of urban life. It is the life and soul of the city. The overall urban design guidelines for the central urban area focus on the establishment of an overall spatial art framework, with a pre-control and constraint on the macro-scale of urban characteristics and landscape patterns from a strategic perspective. Highlight the value of Haihe River in urban culture and landscape features, and promote the expansion of urban coastal rivers, alleviating the contradiction between the high development intensity of the city center and the protection of historical heritage.

控制层面	系统分类	控制内容		成果表达方式
设计总则（单元层面）	整体风格	对本单元的街区特色、历史文脉、自然资源等进行总结提炼，在使用功能的基础上，提出地区风貌特色塑造的整体要求		文字说明+相应图则
	空间意象	对本单元的空间形态和城市意象进行整体描述，指出重要的特色区域、地标节点、视线通廊等主要意象元素，并提出控制指引		
	街道类型	综合考虑交通组织与街道界面性质，根据城市道路的不同使用功能将街道划分为四种类型——交通型道路、景观型道路、商业型道路及生活型道路，应指出分属各类型道路的路名及其总体控制要求。重点地区与历史文化街区应对有特色的景观型道路、商业型道路等提出道路断面建议		
	开放空间	充分考虑本单元内开放空间系统的整体组织和布局，标明各类开放空间（公共绿地、生产防护绿地、广场）的位置，提出总体控制要求		
	建筑	根据设计地段的自然和人文环境特征，对建筑群体控制、高度、体量、建筑风格、外檐材料及色彩等提出控制与引导建议		
	其他：历史文化保护	明确历史文化街区核心保护范围和建设控制地带		
	其他：商业街区特色控制要素	根据商业街区的特征，对建筑首层通透度、建筑墙体广告、建筑裙房、建筑骑楼提出控制与引导建议		
设计分则（地块层面）	街道	1 建筑退线	对某一具体道路（名称）提出建筑退让规划控制线的要求	以文字为主填写表格
		2 建筑贴线率	对某一具体道路（名称）提出建筑贴线率的要求	
		3 建筑主立面及入口门厅位置	提出建筑主立面及入口门厅的位置要求（应面向哪条街道）	
		4 机动车出入口位置	提出居住小区出入口或停车场库等机动车出入口的位置要求（应面向哪条街道）	
	开放空间	5 类型及控制要求	分类叙述公共绿地、生产防护绿地、广场的分类、等级和主要功能	
	建筑	6 建筑体量	提出地块内部建筑的三维体量控制指引	
		7 建筑高度	对高度限制（限高及限低）、高度分布、重要（地标）建筑位置及天际线控制要求	
		8 建筑风格	提出建筑风格（一般分为五类：西洋古典风格、中国传统风格、现代风格、新古典风格、新中式风格）	
		9 建筑外檐材料	提出推荐及限制使用的建筑表面材料及比例	
		10 建筑色彩	提出推荐及限制使用的建筑色彩	
	其它	1 建筑首层通透度	提出商业建筑首层通透比例的要求	
		2 建筑墙体广告	提出建筑墙体广告的风格、在建筑立面中所占比例等方面的要求	
		3 建筑裙房	对连续建筑裙房的位置、高度等提出控制要求	
		4 建筑骑楼	对建筑骑楼空间的位置及进深、净高等提出控制要求	
		5 围墙	对有围墙的地块提出其通透度和高度的控制要求	

以控规单元作为编制单位的城市设计导则编制框架
Urban Design Guidelines Framework for the Preparation of the Control Unit as the Compilation Unit

第二个层次：控制性详细规划单元层面的城市设计导则。

这一层次的导则成果重点服务于规划管理，可以直接应用到项目审批中。

在充分挖掘天津建筑文化、街道肌理等空间特征的基础上提出相关的指导性控制要求，推动已经编制完成的各层次城市设计成果转化为系统、全面的城市设计导则。

The Second Level: Urban Design Guidelines at the Level of Control Detailed Planning Units.

The results of this level of guidance are focused on planning management and can be directly applied to project approval.

On the basis of fully excavating the spatial characteristics of Tianjin architectural culture and street texture, relevant guiding requirements are proposed, and the urban design achievements of all levels that have been compiled are transformed into systematic and comprehensive urban design guidelines.

第二步：树立城市的品格

伊利尔·沙里宁曾说：“许多人士把城镇规划当作纯技术问题，在进行规划时只管就事论事，而忽视了重大的精神要求。”在编制天津市城市设计导则的过程中，管理者和规划师们始终没有忘记，正是天津的历史文化底蕴和天津市民的精神追求、生活方式，塑造了这座城市生活的整体质量和内在品格，它既是城市魅力的关键，也是城市未来发展的基础。

在认清这一问题的前提下，以中心城区总体城市设计导则确定的城市格局为基础，在第二个层次，也就是与传统控规单元的规划范围相适应的城市设计导则的控制中，将编制单元划分为三种类型：历史文化保护地区、重点地区和一般地区。与其对应，编制三种类型的城市设计导则，涵盖三个不同的方向。

The Second Step: Establish the Character of the City

Eliel Saarinen once said: "many people regard urban planning as a purely technical issue, and they only ignore the major spiritual requirements when planning." In the process of preparing Tianjin urban design guidelines, our managers and planners have never forgotten that it is Tianjin's historical and cultural heritage and the spiritual pursuit and lifestyle of Tianjin citizens that have shaped the overall quality and intrinsic character of the city's life. It is both the key to the charm of the city and the foundation for the future development of the city.

Under the premise of recognizing this problem, based on the urban pattern determined by the overall urban design guidelines of the central urban area, the second level, that is, the control of urban design guidelines that are compatible with the planning scope of the traditional control unit. The compilation units are divided into three types: historical and cultural protection areas, key areas and general areas. Corresponding to this, three types of urban design guidelines are developed covering three different directions.

海河发展轴
Haihe Development Axis

第一种类型：历史文化保护地区城市设计导则

本导则针对历史文化遗存较为丰实，能够比较完整、真实地反映天津在一定历史时期的传统风貌或地方特色，存有较多文物古迹、近现代史迹和历史建筑的地区。毫无疑问，历史街区是城市记忆保存最完整、最丰富的地区之一，历史街区内的建筑、街巷、河流和树木都是市民对家乡认知以及感情依托的重要载体。

这类导则以保护为导向，其目标是整体保护历史街区，并使其成为当代城市生活的一部分。导则的编制原则是保护各类物质及非物质文化要素，强化街区风貌特色，小规模地、缓慢地更新，强调自然真实的演进，以此来维护历史街区中真实的城市生活。

The First Type: Urban Design Guidelines for Historic Areas

For the historical and cultural relics, it is relatively rich and can reflect the traditional features of a certain historical period or the local characteristics of Tianjin, and there are many cultural relics, modern and historical sites and historical buildings.

There is no doubt that the historical block is the most complete and richest area of urban memory preservation. The buildings, streets, rivers and trees in the historical block are important carriers for the citizens' understanding of their hometown and their emotional support.

These guidelines are protection-oriented and aim to protect historical neighborhoods as a whole and make them part of contemporary urban life. The principle of the guidelines is to protect all kinds of material and non-material cultural elements, strengthen the characteristics of the neighborhood, small-scale, slow renewal, emphasizing the true evolution of nature, in order to maintain the real urban life in the historic areas.

位于五大道历史文化街区的安乐邨
Anle Village in Wudadao Historical Area

第二种类型：重点地区城市设计导则

重点地区以满足城市公共生活需求为主要功能，体现在用地性质上，即以公共服务设施用地为主。其控制对象通常是中心城区中的新建项目，对城市的经济、政治、文化、景观环境等具有重要影响，与建设实施紧密相关，故其控制深度和广度也都超越另外两类导则。

以天津文化中心周边地区为例，城市设计导则成果有数百页，从宏观、中观、微观不同层次逐级深入地提出控制与引导要求，但其根本目的并没有变——为市民提供更好的、多样化的城市公共服务。如针对街道的控制，导则在宏观层面将街墙划分为 6 种不同的类型，并通过限定街墙高度、建筑贴线率、建筑退线等指标的方式，赋予了各类街道不同的特征；中观层面则从地面层用途、出入口、停车场等方面进一步深入控制；最后所有的控制指标落实到地块层面以便于管理人员使用。其各个层次控制指引的目标是一致的：改变固有的以机动车通行效率为唯一衡量标准的道路设计方法，强调街道应集商业、休闲、交流和交通功能于一体，承载城市生活。

The Second Type: Urban Design Guidelines for Priority Areas

Priority areas meet the needs of urban public life as the main function, which is reflected in the nature of land use, that is the land for public service facilities. The control object is usually a new project in the central urban area, which has an important impact on the city's economic, political, cultural, and landscape environments. It is closely related to the construction and implementation, so its control depth and breadth also exceed the other two types of guidelines.

Take the surrounding area of the Tianjin cultural center as an example. The results of the urban design guidelines are hundreds of pages. The requirements for control and guidance are gradually and intensively drawn from the macro, meso and micro levels. However, the fundamental purpose has not changed which is providing citizens with a better and more diverse urban public life. For the control of the street, the guide divides six different types of street walls at the macro level, and assigns different character characteristics to various streets by limiting the height of the street wall, percentage of build-to-line, and setbacks line. The meso level is further deepened from the ground floor use, entrances and exits, parking lots, etc. Finally, all control indicators are implemented at the parcel level to facilitate management. The goal of each level of control guidance is the same: to change the inherent road design method that uses motor vehicle traffic efficiency as the sole measure, emphasizing that the street should integrate business, leisure, communication and traffic functions to carry urban life.

天津文化中心周边平江道效果图
Rendering of Pingjiang Road in the Surrounding Area of Tianjin Cultural Center

第三种类型：一般地区城市设计导则

一般地区即除上述两类地区以外的其他地区，以不同年代建成的居住社区为主。它们没有历史街区的厚重与沉稳，也没有重点地区的时尚与繁华，却在城市中占有绝对比例，身处这样最为普通的城市社区中，也最能真切地感受到一个城市的品格。

体院北、王顶堤、民权门……这些老社区通常是由多层条形住宅、小型街心花园以及林荫小路组成，整齐中富有韵律。城市设计导则在这类地区所起到的作用更多的是保持社区的稳定性，保障社区的生活品质，促进邻里关系的延续，并提供更优质的公共活动空间——老人在楼下乘凉调侃，孩童在草地嬉笑打闹，年轻人骑着共享单车悠悠地经过，猫咪在窗边呆望闪烁的树影，这一幕幕生活图景是多么欢愉而自在。

The Third Type: Urban Design Guidelines for General Areas

The general areas, except those in the above two types of areas, are mainly residential communities built in different years after liberation. They do not have the weight and stability of historical districts, nor the fashion and prosperity of priority areas, but they occupy an absolute proportion in the city. In such a common urban community, they can most truly feel the character of a city.

These communities, such as Wangdingdi community and Minquanmen community, also strengthen the self-identity of the community residents. The old community is usually composed of multi-story houses, small street gardens and tree-lined paths, neatly rhythm. The role of urban design guidelines in such areas is to maintain the stability of the community, to ensure the quality of life in the community, to promote the continuation of the relationship between the neighborhood, and to provide better public space for public events. The old man took a cool tune downstairs, and the children laughed and laughed in the grass. The young man rode a shared bicycle and the cat looked at the flashing tree shadow at the window. How happy and comfortable the scene was.

居住区街道
Residential Street

实施效果

城市设计导则通过建筑高度控制，使历史街区保持了优雅安静、亲切宜人的历史氛围；城市公园与周边建筑形成了相得益彰、资源共享的良好效果；海河两岸有序发展，海河成为一条具有天津特色及人文尺度的魅力之河。

“罗马不是一天建成的”，编制城市设计导则的目的不是得出最终的结果，而是保证作为生命体的城市朝着更好的方向相对稳定地生长。

Implementation Effect

The urban design guidelines maintains an elegant, quiet and intimate historical atmosphere through controlling the building height . The urban park and the surrounding buildings form a good effect that complements each other and share resources. The orderly development of the two sides of Haihe River has become a characteristic of Tianjin.

And the river of charm of the human scale." Rome was not built in a day." The purpose of our urban design guidelines is not to produce the final result, but to ensure that the city as a living body grows relatively stably in a better direction.

历史街区
Historical Area

城市公园
City Park

海河两岸
Two Sides of the Haihe River

独特的空间结构决定多元交通诉求

滨海新区“多组团、长距离、网络化” 的空间结构特性决定了其多元交通目标诉求：

（1）长距离跨区及对外客运交通：安全、畅通、快捷、高效。

（2）短距离区内客运交通：安全、便利、绿色、经济。

（3）内外交通分离、长短交通分开、快慢交通分流、体系衔接高效。

Unique Spatial Structure Determines Multiple Traffic Demands

The spatial structure characteristics of Binhai New District's "multi-group, long-distance and network" determine its multiple traffic objectives:

(1) Long-distance trans-regional and external transportation: safe, smooth, fast and efficient.

(2) Passenger Transportation in Short Distance Area: safety, convenience, green and economy.

(3) Separation of internal and external traffic, separation of long and short traffic, diversion of fast and slow traffic, efficient system convergence.

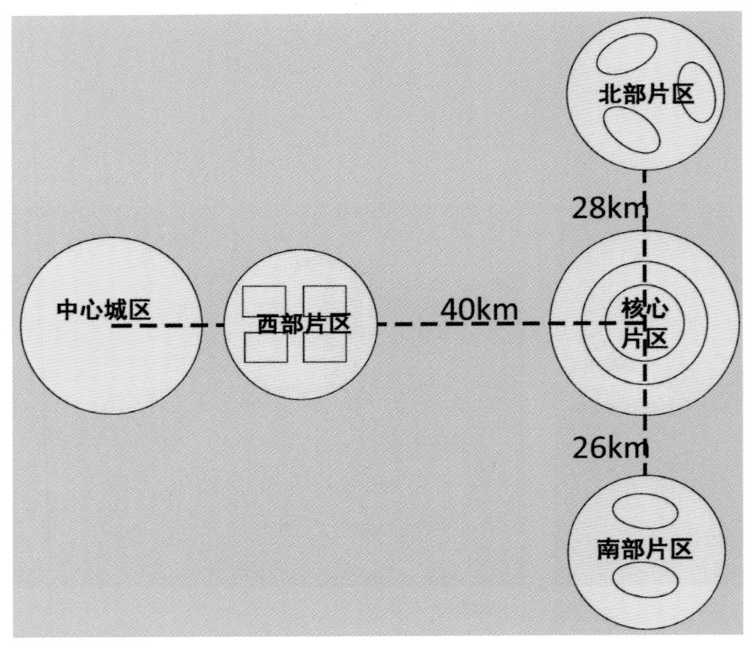

滨海新区空间结构分布概念图
Concept Map of Spatial Structure Distribution in Binhai New District

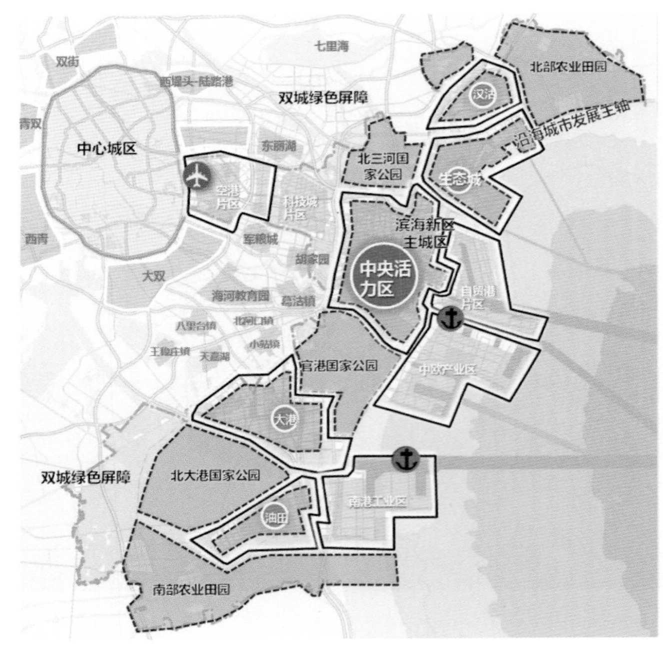

滨海新区空间发展格局示意图
Sketch Map of the Spatial Development Pattern of Binhai New District

政策导向与组织模式协同多元目标

1. 差异化的交通发展政策

公交主导发展区：如滨海核心标志区、生态城核心区，公交出行比例为 50% 以上，小汽车出行比例为 20% 以下。

并重模式发展区：公交主导发展区以外，公共交通和小汽车出行比例均控制在 35% 左右。

Policy Orientation and Organizational Model Coordinating Multiple Objectives

1.Differentiated Transportation Development Policies

Bus-oriented development zones: such as central urban area and eco-city core area, the proportion of bus trips is more than 50%, and the proportion of car trips is less than 20%.

Pay equal attention to the model development zone: outside the bus- oriented development zone, the proportion of public transport and car travel is controlled at about 35%.

交通发展政策分区示意图
Schematic Map of Transportation Development Policy Zoning

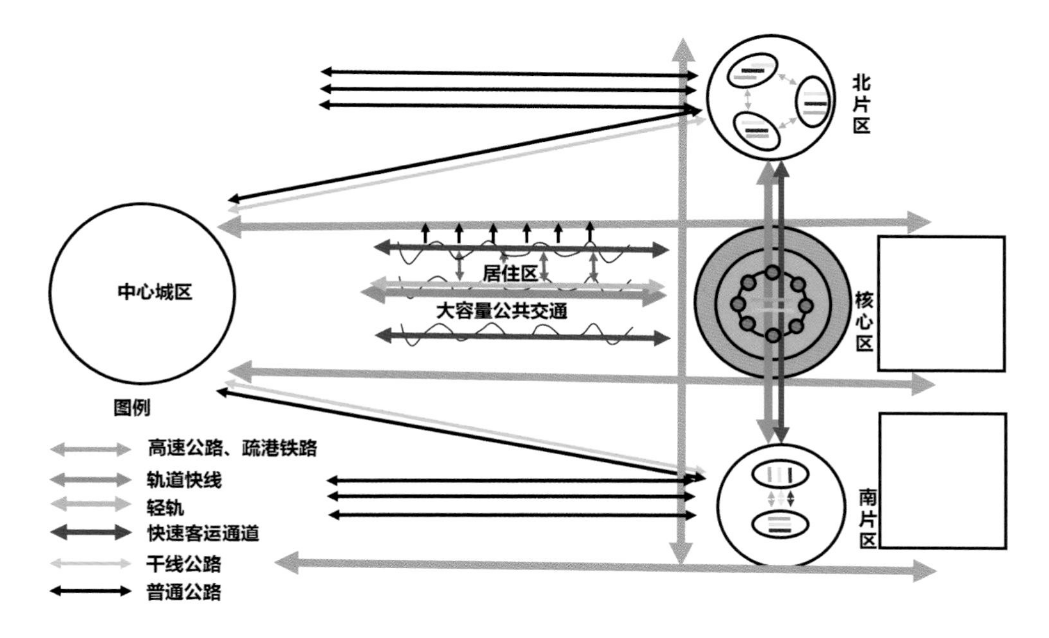

交通组织模式概念图
Concept Map of Traffic Organization Model

2. 层级化的交通组织模式

区间及对外交通：打造“双快”客运模式，兼顾机动性与绿色化。

考虑新区组团间及对外交通出行频次高、距离长的特点，采用“快速轨道”和“快速道路”连接主要组团，缩短出行时间。

在为区间公共交通提供便捷出行条件的同时，考虑区间交通对机动性的诉求。

2.Hierarchical Traffic Organization Model

Intersection and External Traffic: Creating a "Double Fast" Passenger Transport Model, Considering both Mobility and Greening Considering the characteristics of high frequencies and long distances between new districts and groups as well as for external traffic Use "fast track" and "fast road" to connect the main groups, shorten the time and Distance.

In order to provide convenient travel conditions for interval public transport, the demand for mobility of interval traffic is considered.

滨海新区轨道系统规划示意图
Schematic Map of Track System Planning for Binhai New District

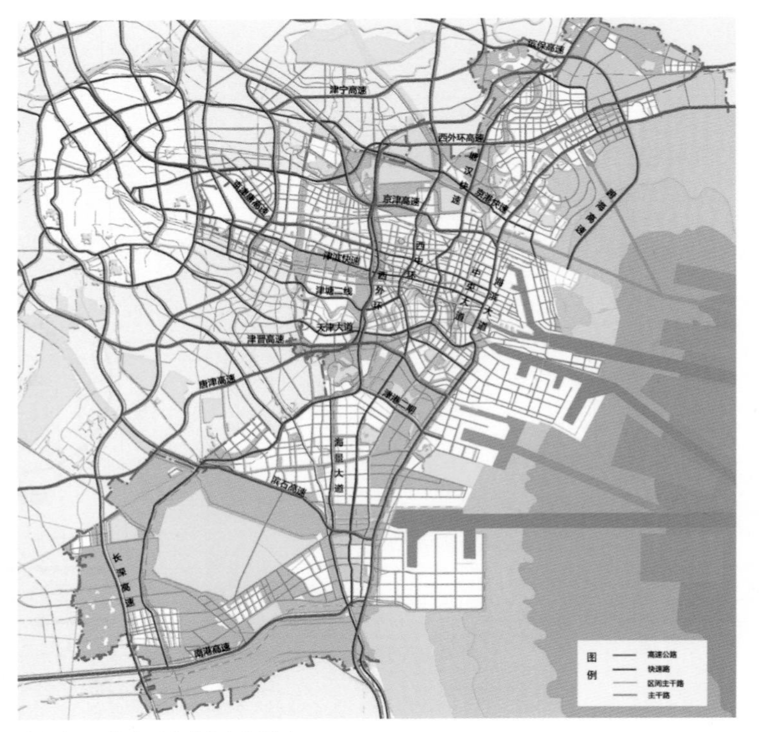

滨海新区道路系统规划示意图
Schematic Plan of Road System in Binhai New District

滨海新区客运交通规划 | PASSENGER TRANSPORTATION PLAN OF BINHAI NEW DISTRICT

区内交通规划目标为:

近期：改善步行、骑行环境，合理引导绿色出行。依托公交枢纽站点、公共活动中心，打造一体化、高品质的非小汽车出行环境，提高绿色出行方式吸引力。

远期：轨道成网后，公交主导发展区限制小汽车出行。

Regional traffic:

Recent-term: Improving the Walking and Riding Environment and Reasonably Guiding Green. Travel Relying on bus hub station and public activity center, Creating an Integrated and high quality Non-car travel environment and improving the attraction of green travel.

Long-term: After the rail network, the bus-oriented Development Zone restricts car travel.

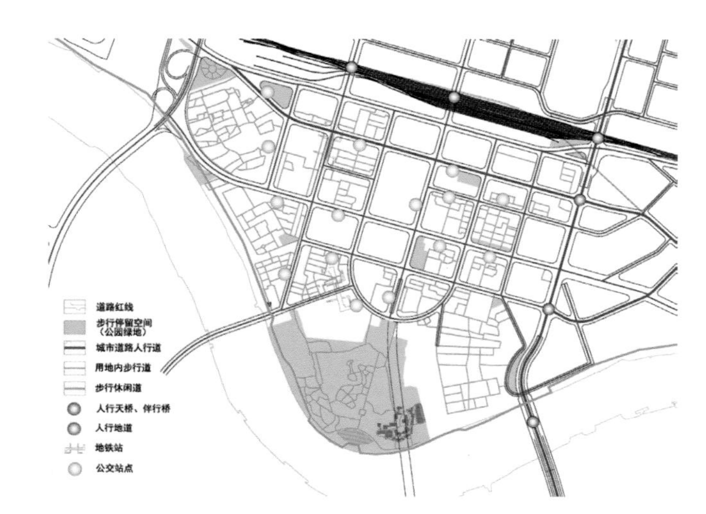

老城区三级步道示意图
Schematic Map of Third-Level Walkway in Old Town

核心标志区（新建地区）三级步道示意图
Sketch Map of Three-Level Trail in Core Landmark Area (Newly Built Area)

核心标志区（新建地区）三级自行车道示意图
Sketch Map of Three-Level Bicycle Lane in Core Sign Area (Newly Built Area)

明末至中华人民共和国成立前：五方杂处、中西合璧

明末至中华人民共和国成立前的天津五方杂处，是中西文化的熔炉。码头市井的“哏儿”生活和租界里优雅的“洋”生活，在这里交融生长。丰富多元的生活孕育出了近代天津独具特色、风格多样的住宅类型，其整体发展脉络表现为传统合院—院落式里弄—锁头式里弄—新式里弄—独立式住宅。

1. 传统合院

天津传统民居，始建于明末清初，经历数次易主、翻建之后，多建于清末民初，距今有一百多年的历史，主要集中在老城厢。其建筑形制、风格与北京四合院属同一体系，但又具有自己的地域和文化特色，称“四合套”。在院落布局上更讲究实用，更加自由化、多样化。大中型四合套，普遍同时具备串联式和并联式两种内部交通体系，并联交通系统是通过箭道来实现的。一些商号常常采用前店后宅的形式，将临街住宅作为铺面。建筑风格上，有的合院采用中西合璧式，且结合得十分自然。

Modern Chinese and Western Combination

Modern Tianjin is a mixture of five elements and a melting pot of Chinese and western cultures. Pier city's "straight" life, and concession in the elegant "foreign" life, where blend growth. The rich and diversified life gave birth to the unique and diversified living forms in modern Tianjin. The overall development of Tianjin is reflected in the traditional courtyard — court yard alley — lock alley — new alley — free standing house.

1.Traditional Courtyard Houses

Tianjin traditional residential houses were built in the late Ming and early Qing dynasties. After several changes of ownership and reconstruction, most of them were built in the late Qing and early Republic of China. Its architectural form, style and Beijing courtyard is the same system, but has its own regional and cultural characteristics, known as the "four sets". The layout of the courtyard is more practical, more liberal and diversified. Large and medium-sized quadrangle sets are generally equipped with both serial and parallel internal traffic systems, and the parallel traffic system is realized through the arrow way. Some firms often adopt the form of front point and back house, using the house facing the street as the pavement. On building style, some close a courtyard to use type of combination of Chinese architecture and western architecture, combine very natural.

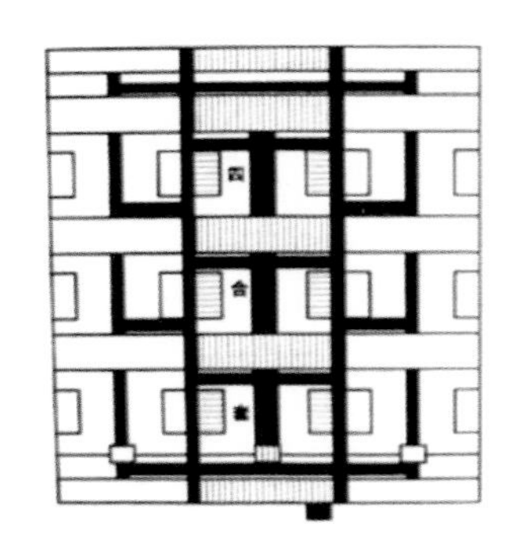
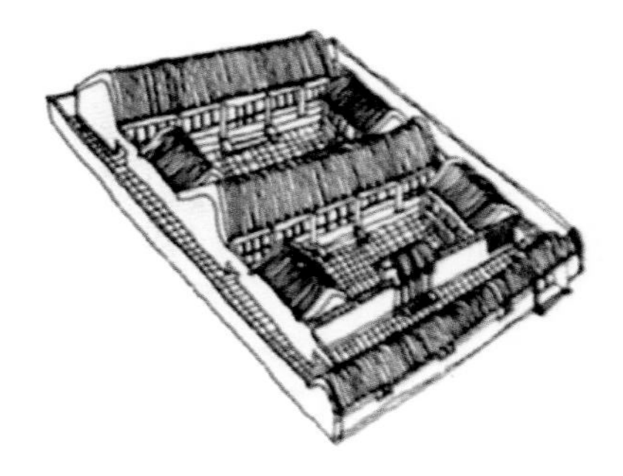

传统合院示意图
Schematic Diagram of Traditional Courtyard House

1900 年天津老城厢
Old City of Tianjin in 1900

2. 院落式、锁头式、新式里弄

开埠之初的老城厢延续着昔日的繁荣，居住形态延续了传统的合院式。随着经济的快速发展，人口的迅速增长，需要提供更高密度的住宅与之相适应。

20 世纪初河北新区崛起后，城市的经济和政治中心转移到那里，大量的院落式里弄住宅适时出现，成为当时住宅建筑比较集中的地区。院落式里弄脱胎于传统合院式住宅，通过加密单元排列、减小院落空间来达到增加住宅容积率的目的，建筑层数几乎没有改变。这种形式在丧失居住舒适性的同时只是容积率略有提高，所以没能得到广泛的应用。南市区作为城市经济向边缘的渗透地区，带来该区域住宅的规模建设，有房地产公司建设经营了一批锁头式里弄住宅。它源自上海早期石库门里弄，同院落式里弄住宅相比，是以增加建筑层数和进一步缩小院落空间来提高容积率，虽然舒适性尚可，但由于南北方气候及风俗习惯的差异而未能得到推广。随后，劝业场地区后来居上，成为寸土寸金的商业黄金地段，建设了新式里弄住宅。它源于西方的两排式住宅，通过增加建筑层数（3 层）来得到足够的使用空间，舒适性较好，从而与城市住宅的发展合拍，成为近代天津建造量最大的住宅形式。

2.Courtyard Style, Lock Style, New Style Lane

The old city at the beginning of the port continued the prosperity of the past, and the residential form continued the traditional courtyard style. With the rapid development of economy and the rapid growth of population, it is necessary to provide higher density house to adapt to it.

At the end of the 19th century, after the rise of Hebei New Area, the economic and political center of the city moved there, and a large number of courtyard-style residential buildings appeared in time, becoming the area where residential buildings were concentrated. The courtyard lane is derived from the traditional courtyard house, and the Floor-area Ratio of the house can be increased by infilling the arrangement of units and reducing the courtyard space, and the number of building floors has hardly changed. This form is not widely used because it only increases the Floor-area Ratio slightly while losing the comfort of living. As the infiltration of urban economy to the edge, the Nanshi Area still brings about the scale construction of residential buildings in this area.

Some real estate companies have built and operated a batch of locker-type residential buildings. It originated from the early Shikumen alley in Shanghai. Compared with the courtyard house, it increased the number of building floors and further reduced the courtyard space to improve the Floor-area Ratio. Although it was comfortable, it failed to be popularized due to the differences in climate and customs between the South and the North. Later, the Quanyechang area came from behind to become a golden commercial area with land and land value. It originated from the duplex house in the west, and gained enough space for use by increasing the number of building floors (3 Floors), with better comfort, so as to keep pace with the development of urban house and become the House form with the largest amount of construction in modern Tianjin.

民园西里
West Minyuan Lane

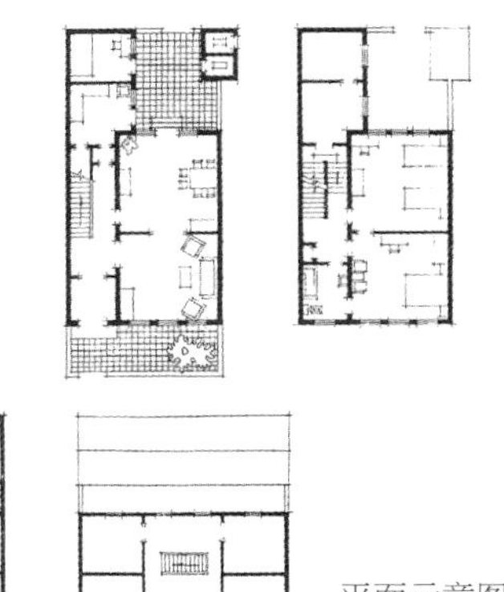

平面示意图
Schematic Diagram

3. 独立式住宅

经济相对稳定后，住宅建设不再受商业利益驱使，转而以追求居住品质，选择方便、安静、比较繁荣的地区作为住宅区。于是英租界新区、意租界等地区以追求居住的安逸舒适为目标，进行了有目的、有计划的成片住宅开发，除了新式里弄住宅外，出现了大量的独立式住宅，逐步形成了天津近代“小洋楼”风貌。

独立式住宅中西方住宅样式、设备和中国传统的生活方式交织在一起，构成复杂的综合体。以五大道地区为代表的大量独立式住宅，大多趋向“新奇”、内容繁杂、形式多样，突出反映了当时社会上层人群的生活情趣和精神追求。

3.Detached Houses

After the economy is relatively stable, residential construction is not driven by commercial interests and flow everywhere, turned to the pursuit of living quality, convenient, quiet, more prosperous region as a residential area, so the British Concession, Italian Concession, and other regions in the pursuit of living comfort as the goal, purposeful and planned residential development, in addition to the new style lane, emerged a large number of houses, gradually formed the modern Tianjin "small garden house " style.

A complex of western housing styles, equipment and traditional Chinese lifestyles are intertwined in the detached house. A large number of detached houses represented by the Wudadao area tended to be "novel", with complex contents and diverse forms, which highlighted the life interest and spiritual pursuit of the upper class of the society at that time.

马占山旧居
Former Residence of Ma Zhanshan

张学铭旧居
Former Residence of Zhang Xueming

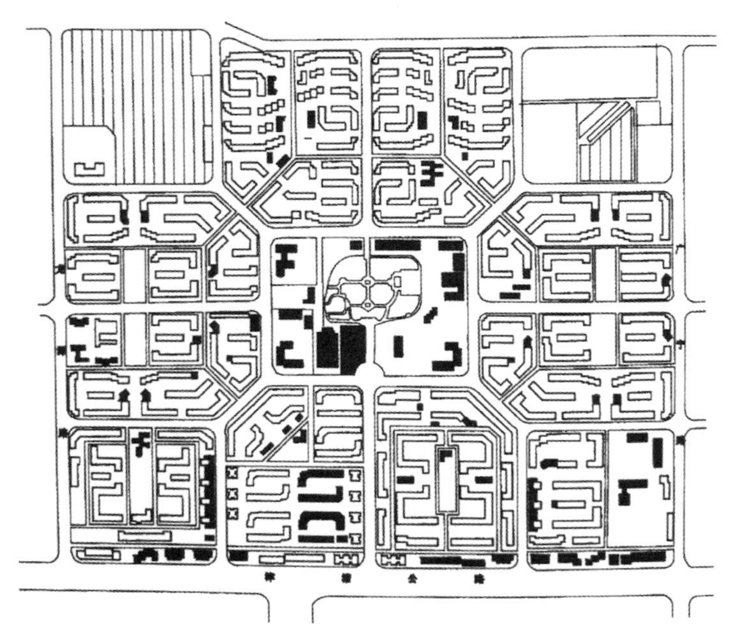

中山门工人新村
Zhongshanmen Workers' New Village

计划经济时期（1949—1956）：单位大院

中华人民共和国成立后，伴随着住宅建设的实践，在“邻里单位”和苏联“扩大街坊”的影响下，居住小区规划理论初步形成，采用“邻里单位”模式，形成了新型的、以公有制为基础的、以行政控制为主导、以“单位大院”为特征的多层行列式住宅类型。

1. 单层工人住宅

这个时期代表性的大型居住区为中山门工人新村，这是天津市第一个大规模成片建设的居住区。新村占地 92 公顷，建筑面积 16.45 万平方米，住房 1.01 万间。中山门工人新村原规划设计以邻里单位理论为依据，内部道路为八卦形，共有 12 个街坊，周边居住街坊围绕中心的公共建筑和中心公园布置。实施过程中保留了规划路网和中心的公建街坊，住宅为单层，一户一室，半间厨房，每 10 间或 12 间连成一排，均为正南北向布置，集中使用公厕和上下水道。在当时的条件下，工人新村的建设解决了一大批人民群众的燃眉之急。

Planned Economy Period (1949-1956) : Unit Compound

After 1949, along with the practice of residential construction, in the "neighborhood unit" and "expand neighborhood" under the influence of Soviet Union, the preliminary formation of residential area planning theory, adopt the "neighborhood unit" mode, formed a new, on the basis of public ownership, dominated by administrative control, the characteristics of the image of the "unit compound" living forms.

1.Single-storey Workers' Housing

The representative large residential area in this period was Zhongshanmen workers' new village, which was the first residential area built on a large scale in Tianjin. The new village covers an area of 92 hectares, with a construction area of 164,500 square meters and 10,100 houses. The original planning and design of Zhongshanmen workers' new village was based on the theory of neighborhood units. The internal roads are Bagua-shaped, with a total of 12 neighborhoods. The surrounding residential neighborhoods are arranged around the central public buildings and central parks. In the process of implementation, the road network planning and the central public buildings are retained. The houses are single-storey, with one room for each household and half room for kitchens. Every 10 or 12 rooms are arranged in a row, which are in the north-south direction. Under the current conditions, the construction of a new village for workers met the urgent needs of a large number of people.

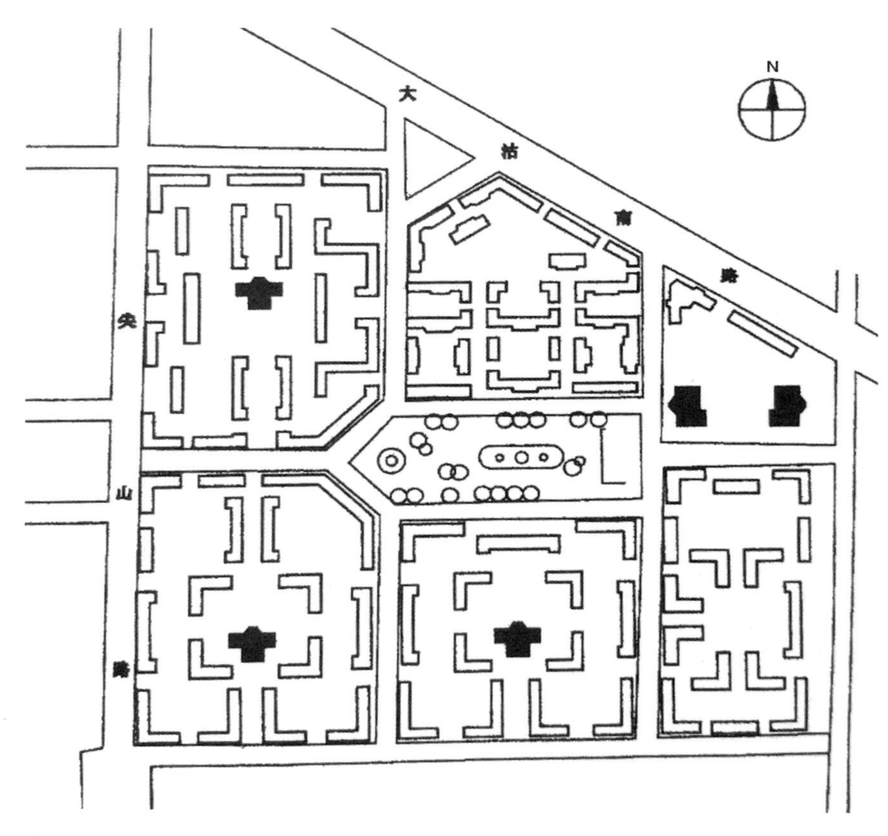

尖山居住区
Jianshan Residential Area

2. 周边式、行列式住宅

苏式居住区规划思想及单元式住宅设计引入中国后，天津住宅小区规划出现了大量的周边式、行列式布局。住宅采用单元式，设计适应了新的生活方式，取消以起居室为中心的居住模式，改为走廊式布局，增加独立房间，改善厨、卫条件。同时住宅标准化为大量建设住宅提供了条件。

这时期住宅层数以 3 ~ 4 层为主，典型的代表有尖山居住区。它是天津市第一个楼房居住区，规划划分独立街坊，并以占地 2 公顷的中央长形公园为核心，街坊之间用轴线联系，建筑群以轴线组织。住宅采用双周边的布局形式，形成了封闭或半封闭的居住结构，也产生了大量东西向住宅。

2. Surrounding Type, Determinant Houses

The planning idea of Soviet residential district and the unit residential design were introduced into China, and a large number of peripheral and determinant layout appeared in Tianjin residential district planning. The residence uses unit type, the design adapted new lifestyle, cancel living room to be the living mode of the center, instead corridor type layout, increase independent room, improve hutch, defend a condition. At the same time, housing standardization also provides conditions for the construction of a large number of houses.

During this period, the number of residential floors is mainly 3-4, and the typical representative is Jianshan residential area. It is the first residential building in Tianjin, which is planned and divided into independent neighborhoods, with the central long park covering an area of 2 hectares as the core. The neighborhoods are connected with an axis, and the buildings are organized with an axis. The house takes the form of a double-perimeter layout, forming a closed or semi-closed residential structure and producing a large number of east-west houses.

体院北居住区
Tiyuanbei Residential Area

改革开放初期（1978—1990）：大型居住区

改革开放带来了巨大的发展契机，居住区建设开始复苏，逐渐步入正轨，居住区规划结构形成了“居住区—居住小区—组团”三级结构，为以后住宅建设开拓了新的起点。

这时期代表性的大型居住区为 14 片之一的体院北居住区。体院北居住区占地 89.7 公顷，总建筑面积约 82.64 万平方米，其中住宅建筑面积 70.97 万平方米，规划人口约 5 万。它采用“居住区—街坊—组团”的规划结构，道路系统采用弧形和直线相结合，丰富了沿线的建筑景观。生活服务设施分两级设置并设有大型中心绿地。住宅层数以多层为主，低层高密度小区和高档住宅小区为 3~4 层，在居住区北部和中心为点式高层。

The Early Stage of Reform and Opening up (1978-1990) : Large Residential District

The reform and opening up has brought great development opportunities, the construction of residential areas began to recover, gradually into the track, the residential area planning structure formed a "residential area — neighbourhood — residential group" three-level structure, for the future residential construction opened up a new starting point.

During this period, the representative large residential area was the Tiyuanbei area of the sports courtyard, one of 14. The Tiyuanbei area covers an area of 89.7 hectares, with a total construction area of 826,400 square meters, including a residential construction area of 709,700 square meters and a planned population of about 50,000. It adopts the planning structure of "residential area — neighborhood block — residential group ", and the road system adopts the combination of arc and straight line, enriching the architectural landscape along the road. Life neighbourhood facilities are divided into two levels and there is no large central green space. The residential area is mainly composed of several layers, the low-rise high-density residential area and the high-grade residential area are 3-4 floors, and the north and center of the residential area are the point-type high-rise buildings.

华苑居住区
Huayuan Residential Area

经济转型时期（1990—2000）：试点小区

《城市居住区规划设计规范》（GB 50180—1993）使得我国居住区规划理论走向成熟。明确提出“居住区—居住小区—组团”的规划结构以及相应的规划控制。同时，1990 年颁布《中华人民共和国城镇国有土地使用权出让和转让条例》，1994 年发布《关于深化城镇住房制度改革的决定》，明确了城镇住房改革的基本内容，住宅市场化开始实行。

这一时期，天津住宅建设采取了以商品房形式筹措资金、单位之间联建、单位与个人联建等多种形式，包括建设部城市住宅试点小区、建设部小康住宅示范小区等。其中最具代表的是华苑居住区和梅江居住区，规划多采用 “居住区—居住小区—组团”的规划结构，“顺而不穿，通而不畅”的交通设计原则，以及“四菜一汤” 式的功能布局，住宅以多层、高层相结合，进而逐步形成“超大封闭街区”的居住形态。

Economic Transition Period (1990-2000) : Pilot Community

"The Code for Urban Residential Area Planning and Design (GB50180-1993)" makes the theory of residential area planning mature in China. The planning structure and corresponding planning control of "residential area - neighbourhood - residential group" are put forward. At the same time, "the Regulations of People's Republic of China on the Transfer and Transfer of the Right to Use State-owned Land in Urban Areas" was promulgated in 1990, and "the Decision on Deepening the Reform of the Urban Housing System" was issued in 1994, which clarified the basic contents of the urban housing reform, and the marketization of housing began.

快速城镇化时期（2000 年至今）：疯狂的“玉米地”

1998 年，国务院发布的《关于进一步深化住房制度改革加快住房建设的通知》，全面实行住房分配货币化，以此为标志，住房制度改革全面展开。由此，房地产业开始蓬勃发展，城镇化进入快速发展时期。

2000 年后，在土地财政的推动下，城市建设用地快速扩张，房价不断攀升，在强大的资本逻辑下，住宅容积率随之猛增。居住形态逐渐演变为“超大封闭街区” +“点式高层”的单一模式。不论是在市中心还是郊区，甚至连近郊乡村，都呈现千篇一律、乏味无趣的疯狂“玉米地”景象。

Rapid Urbanization Period (2000 - Present) : Crazy "Corn Field"

In 1998, the State Council issued the "Notice on Further Deepening the Reform of the Housing System and Accelerating the Housing Construction", which fully implemented the monetization of housing distribution. As a result, the real estate industry began to flourish and urbanization entered a period of rapid development.

After 2000, driven by land finance, urban construction land expanded rapidly and housing price kept rising. Under the strong capital logic, floor-area ratio soared accordingly. Residential form gradually evolved into a single mode of "super-large closed block" + "point-type high-rise". In the center of the city, in the suburbs, and even in the countryside, it's all the same, boring, crazy cornfields.

超大街廓，高层居住区
High-Density Sprawl

北美式低密度蔓延
The North American Style Low-Density Sprawl

中国式高密度蔓延
China-Style High-Density Sprawl

高密度蔓延

回顾历史，天津曾拥有独具特色、生动多样的居住形态，然而走到今天，像病毒一样蔓延的疯狂的“玉米地”，已经成为天津，乃至中国城市病的病灶之一，威胁着经济、社会和环境的可持续性，是我们不能回避的关键问题。

新都市主义代表人物，规划大师彼得·卡尔索普曾提出，地球上有三种类型的蔓延挑战城市：其一是北美式的高收入蔓延，其建筑密度低，功能隔离，以小汽车为主要交通工具；其二是拉美、非洲的低收入蔓延；第三就是中国疯狂的“玉米地”现象，这是一种特有的高密度蔓延，也具有一定的讽刺意味和悲剧性。

High-density Sprawl

In retrospect, Tianjin used to have unique and vivid living patterns. However, till now, the crazy "corn field" spreading like a virus has become the root cause of urban disease in China, threatening the sustainability of economy, society and environment. It is a key issue that we cannot avoid.

Peter Calthorpe, the master of planning and the representative of new urbanism, has proposed that there are three types of sprawl challenges on earth. The first is the spread of high income in North America, which is a characteristic low-density spread with functional isolation and motor vehicle traffic. The second is the spread of low income in Latin America and Africa. The third is China's crazy "cornfield" phenomenon, which is a characteristic high-density spread, but also ironic and tragic.

单调的建筑景观
Monotonous Architectural Landscape

过大的尺度
Oversize

乏味无生气的街道
Boring Streets

消极的公共空间
Negative Public Space

高密度蔓延

正如彼得·卡尔索普所说，“高密度的城市不能围绕汽车来设计，这是一个简单而直白的事实”。巨型主干道围绕的超大街区，本质是为小汽车服务，巨大的尺度牺牲了行人和慢行交通的舒适性和安全性。高层塔楼带来的高密度更是让小汽车的负面效应加剧。即使在低密度下，以汽车为基础的城市也无法运作。城市被拥堵和污染充斥，几百万人出行成为每天重复的“磨难”；不合理的土地财政导致房价高企与高库存；千城一面，人居环境单调乏味，缺乏特色；场所交往缺失等，这些城市病都由此引发。

High-density Sprawl

As Peter Calthorpe said, "It is a simple and straight forward fact that high-density cities cannot be designed around cars." The superblocks around the giant main roads are essentially for the car, at the expense of the comfort and safety of pedestrians and chronic traffic. The high density of high-rise towers adds an order of magnitude to the negative effects of cars. Even at low densities, car-based cities cannot function. Cities are plagued by congestion and pollution, and millions of people travel daily as a daily grind. Land finance leads to high housing prices and high inventory; one side of the city, the living environment monotonous, lack of characteristics; lack of place contacts and other urban diseases are caused by this.

城市吞噬乡村
City Devours Country

乡村荒地
Blighted Country

城乡割裂

放大到区域的尺度，城市“摊大饼”式的快速扩张，并以高密度蔓延的模式不断吞噬着周边的乡村和农田，其本质是无序的郊区化，严重破坏了区域空间的生态连续性，城乡割裂严重。城市中心密不透风，居住拥挤，绿地缺失；城市边缘违法建设、产权混乱、垃圾围城等现象普遍；城市周边生态环境退化，乡村面临危机。

Split Between Urban and Rural Areas

At the regional scale, the rapid expansion of cities in the form of "spreading the cake" and constantly engulfing the surrounding countryside and farmland in the pattern of high-density sprawl, is essentially a disorderly suburbanization, which seriously destroys the ecological continuity of regional space and severs the urban and rural areas. The city center is airtight, crowded and lack of green space; illegal construction on the edge of the city, property right chaos, garbage siege and other common phenomenon; the ecological environment around the city is degraded and the countryside is in constant decline.

社会矛盾

社区是连接居民、政府和开发商的纽带，是社会组织运行的基本环节，是城市治理的基本细胞。回溯天津居住形态的演变，每次变迁都强调自上而下的社会动员力量，对个体和局部组织的需求和利益有所忽视。特别是以单位大院为主的居住模式解体后，原有的社会关系网络也随之消失，但并没有建立起相应的新的社会网络，社区的作用变得薄弱。由于缺少了社区组织来充当缓和社会矛盾的润滑剂，居民的需求难以得到及时有效的反映，常常导致公众与开发商和政府直接站在对立的两端，难以形成有效的沟通。与发达国家已经较为成熟的社会调节机制不同，社区力量的薄弱，不利于缓解社会矛盾。

几个误区

以上的问题，原因是复杂的，其中包含两个一直以来重要的观念误区。

误区一：低层住宅不适合人多地少的国情。

许多人认为，中国人多地少，低层住宅并不适合我国具体国情。因此，居住用地分类中的一类居住用地，在实际的规划管理中，几乎被一刀切地杜绝。这也直接扼杀了低层住宅如独立式住宅、合院、联排等丰富的居住形态再出现的可能性，这也间接加剧了中国式的高密度蔓延，反而造成大量住宅空置，土地浪费。

误区二：背景建筑（住宅建筑）并不重要。

改革开放 40 年来，中国的建筑文化繁荣发展，具有国际影响力的优秀公共建筑层出不穷。我国每年商品住宅建设量约 30 亿平方米，竣工量和销售量为十几亿平方米，然而，在这巨大的建设量背后，住宅规划和设计却鲜有改革创新。

Social Contradictions

Community is the bond connecting residents, government and developers, the basic link of social organization operation and the basic cell of urban governance. Looking back on the evolution of the residential form in Tianjin, every change emphasizes the top-down social mobilization and ignores the needs and interests of individuals and local organizations. In particular, after the disintegration of the residential mode dominated by units and courtyards, the original social network was also dissolved, but no corresponding new social network was established, and the role of community was always weak. Due to the lack of community organizations to serve as a lubricant to alleviate social conflicts, it is difficult for residents to get timely and effective responses to their needs, which often leads to the public directly standing on opposite sides of the problem developers and the government, and it is difficult to form effective communication. Different from the relatively mature social adjustment mechanism in developed capitalist countries, the weak community power is more likely to occur and aggravate social conflicts at the current stage of weak economic growth. Currently, the spread of high density, urban and rural fragmentation brought by the disorder.

Several Misunderstandings

The reasons for these problems are complex and include two important misconceptions.

Myth 1: low-rise housing is not suitable for the national conditions of more people and less land

We always think, more people and less land in China, low-rise residence does not suit our country specific national condition. Therefore, type I residential land in the classification of residential land, in the actual planning and management, is almost universal to eliminate. This also directly killed the possibility of the re-emergence of low-rise houses such as detached houses, courtyard houses, townhouses and other rich living forms, which also aggravated the spread of Chinese-style high density from the side, but caused a large number of empty houses and land waste.

Myth 2: background architecture (residential architecture) doesn't matter

Since the reform and opening up 40 years ago, China's architectural culture has flourished and developed. However, China operates on approximately 3 billion square meters of commercial housing construction each year, and those on operations and sales were more than 1 billion square meters. However, behind the huge amount of construction, there are few reforms and innovations in residential area planning and design.

现在对我们影响最深的依然是苏联的居住区规划，柯布西耶理论和邻里单元等理论，以及计划经济时期的“大杂烩”，缺乏完备的理论研究和理论基础。千人配套指标、日照间距等为主的规划设计标准，没有充分考虑市场需求，对住宅多样性、人性化空间塑造缺乏全面引导。广大的中国建筑师们也多把公建设计作为一直追捧的方向，对本土住宅设计的创新探索寥寥。而吴良镛先生的“菊儿胡同”，保留建筑地域性、民族性，是中国诗意栖居的代表，却并没有在中国的住宅规划设计中得到推广，不得不令人遗憾唏嘘。

而事实是，作为背景建筑的住宅建筑比地标建筑更重要。住宅在城市中占有极大的比例，对城市形态的基本构成起着决定性的作用。城市形态和城市特征在相当程度上受背景住宅的影响，没有住宅，城市中心无法孤立存在。住宅建筑类型主导着城市街区形态，进而主导着城市整体空间形态，是决定城市品质的关键，而这正是我们重视不够的方面。

At present, the most important influence on us is still the planning of residential areas in the Soviet Union, the Utopian socialist collective residence, Corbusier's theory and Neighborhood Unit theory, as well as the hodgepodge of planned economy, lacking of theoretical research and theoretical basis. The planning and design standards, such as supporting index for thousands of people and sunshine spacing, have not fully considered the market demand and lack comprehensive guidance for the building of diverse and humanized space. The majority of Chinese architects also regard the public building design as the direction they always pursue, and seldom explore the innovation of local residential design. However, "Ju 'er Hutong" by Mr. Wu Liangyong, which retains the regional and national characteristics of architecture and represents the poetic dwelling of China, has not been promoted in China's housing planning and design, which is regrettable.

The fact is, the residential building as the background building is more important than the landmark building. Housing occupies a great proportion in the city and plays a decisive role in the basic composition of urban form. Urban form and urban characteristics are largely influenced by the residential background, without which the urban center cannot exist in isolation. The repeatability of residential building type dominates the urban block form, and then the overall spatial form of the city, which is the key to determine the quality of the city, and this is exactly what we lack the most.

中国宏村
Hongcun Village in China

美国海滨城市
Seaside City in America

在中国，古人云“宅，所托也 ”（《说文解字》）。“人以宅为家，居若安，则家代昌吉”（《释名》）。中国传统文化中，住宅和当时的生活方式、文化传统、精神追求达成了完美的结合。住宅，不仅是物质生活的载体，还是精神审美的载体，更是自我心理理疗的空间。

而在西方，住宅更被认为是生命存在的象征，道德高尚的象征。如富兰克林 · 罗斯福所说，如果一个国家的人民拥有自己的住房，并能够在自己的土地上赢得实实在在的份额，那么这个国家是不可战胜的。

In China, the ancients said, " The houses are the foundation of People's lives" (*Shuo Wen Jie Zi*). "If People lead a peaceful life in the house, the whole family will prosper" (*Shi Ming*). Under the Chinese traditional culture, the residence and then the life style, the cultural tradition, the spirit pursue have achieved the perfect union. Residence is not only the carrier of material life, but also the carrier of spiritual aesthetics, and more importantly, the space of self-psychological therapy.

In the west, the residence is regarded as the symbol of the existence of life, the symbol of moral nobility. As Franklin Roosevelt said, a nation is invincible if its people own their homes and can win a substantial share of their land.

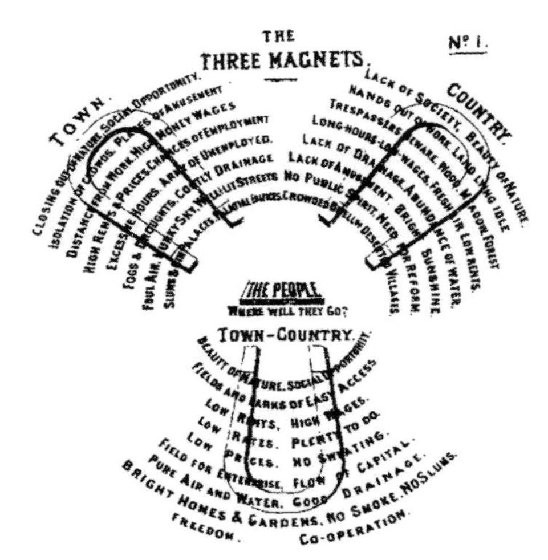

“三磁铁”学说
The Three Magnets

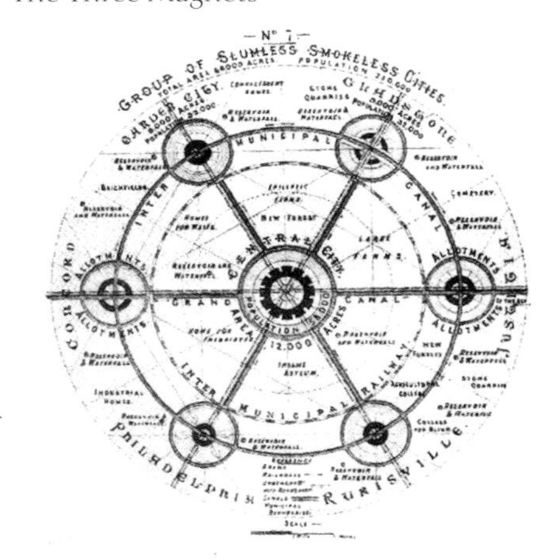

无贫民窟、无烟尘的城市群
Group of Slumless and Smokeless Cities

自工业革命以来，住宅问题一直是困扰现代城市和社会的核心问题，住宅也一直是现代规划研究的主要内容，许多思想家、政治家、建筑大师都做了深入的研究和实践，如今看来，住宅依然是解开当前许多困惑的“万能钥匙”。

百年现代城市规划理论照进现实——明日的田园城市

19 世纪末，埃比尼泽·霍华德（1850—1928）发表了《明日的田园城市》这一影响后代的著作。他当时思考着与当前中国相似的城乡割裂的问题，随着工业经济发展和资本主义扩张，英国大城市人口急剧膨胀，环境恶劣，同时出现乡村日益萧条衰败的危机。他提出“城市是人类社会的标志，乡村是上帝爱世人的标志，城市和乡村必须联结，构成一个‘城市—乡村’磁铁”。霍华德田园城市的根本出发点是为人们提供适宜的住宅，它既可享有城市社会文化成果，又可使居民身处大自然美景之中，用以社区为基础的新文化来医治城市中心区的“脑溢血”和城市边远地区的“瘫痪病”。它统筹城乡要素，形成可渗透的和自然和谐相处的城市区域，使城市与乡村在更大范围的自然环境中取得平衡。同时建立社区自治，把最自由和丰富的机会同等地提供给个人和集体，闪耀着人本主义光芒。

Since the industrial revolution, the housing problem has been the core problem troubling modern cities and society. Housing has been the main content of modern planning research. Many thinkers, politicians and architects have conducted in-depth research and practice.

One Hundred Years of Modern Urban Planning Theory Into Reality –Garden Cities of Tomorrow

The end of the 19th century Ebenezer Howard (1850-1928) published the epoch-making classic *Garden Cities of Tomorrow*. He was faced with the same rural-urban divide as China is today. With the development of industrial economy and the expansion of capitalism, the population of Britain's big cities expanded rapidly, and the environment was harsh. He proposed that "the city is the symbol of human society, and the country is the symbol of God's love for people. The city and the country must be married to form an urban-rural magnet". Howard's rural city, at its core, is about providing affordable housing that allows for both the social and cultural fruits of the city and the natural beauty of its inhabitants, as well as a new community-based culture to heal the cerebral hemorrhage of the city center and the paralysis of the urban frontier. It integrates urban and rural elements to form permeable urban areas living in harmony with nature, so as to balance urban and rural areas in a larger range of natural environment. At the same time, the establishment of community autonomy, the most free and rich opportunities equal to individual efforts and collective cooperation, shining the humanistic light of independent autonomy!

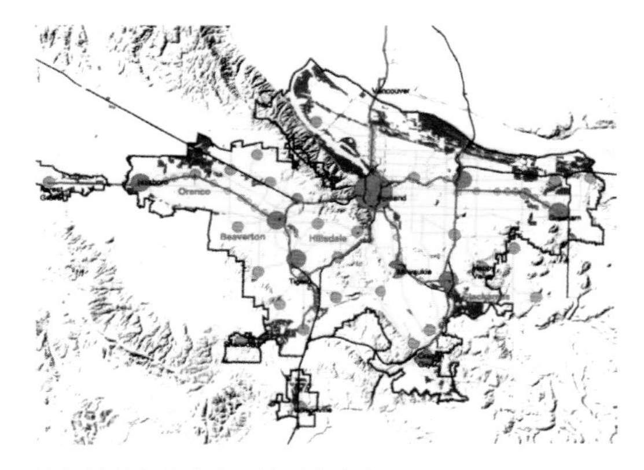

波特兰大都市中心、地区和走廊
Portland Metropolitan Center, Area And Corridor

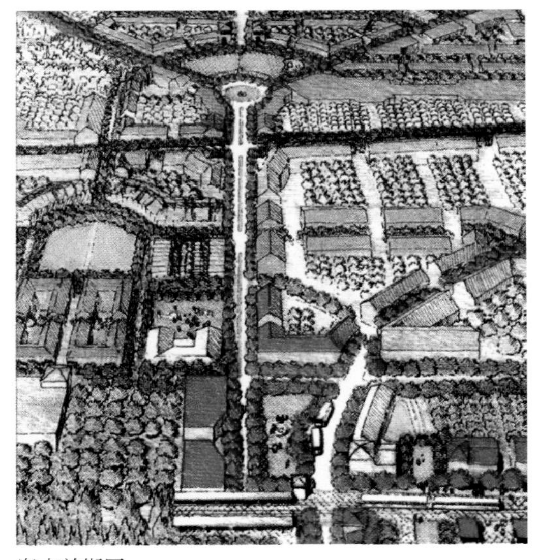

奥克兰街区
Oakland Block

百年现代城市规划理论照进现实——区域性城市

20 世纪，刘易斯 · 芒福德（1895—1990）继承了霍华德田园城市理论，并受盖迪斯区域生态思想的影响，进而提出了区域性城市的思想。而区域性城市（regional city）即指比较大的城市向周围的扩展而形成的城市。芒福德区域城市的概念是从居住出发，认为居住既能享受城市的便利和文明，同时又可以接近自然。住宅以集合多层住宅和 1、2 层独立住宅为主。

21 世纪，新都市主义代表彼得 · 卡尔索普的《区域城市——终结蔓延的规划》，是在霍华德、芒福德等人思想基础上的继承和发展，与田园城市一脉相承，是霍华德“社会城市”在新时代的诠释。他提出区域应被视为一个整体来进行设计，而设计区域就是设计街区。把区域和它的元素——城市、郊区、自然环境看作一个整体；同样，把街区和它的元素——住宅、商店、开放空间、市政机构和企业看作一个整体。两者都需要有被保护的自然环境，有须臾不可离开的中心，“人”的尺度的交通系统，公共场所和综合的多样性。开发区域就是在创造健康的街区、地方和城市环境，同样开发这样的街区就等于在创造持续发展的、综合的和有凝聚力的区域和环境。因此区域与社区在两个尺度相互配合，协调共生。

One Hundred Years Of Modern Urban Planning Theory Into Reality –Regional City

In the 20th century, Lewis Mumford (1895-1990) inherited Howard's theory of rural city and, under the influence of Geddes' regional ecological thought, put forward the idea of regional city. Regional city refers to the city formed by the expansion of larger cities to the surrounding area. The concept of Mumford regional city is based on residence, which can enjoy the convenience and civilization of the city and be close to the nature at the same time. The residence is mainly composed of multi-storey houses and independent houses on the first and second floors.

In the 21st century, the Charter of New Urbanism, Peter Calthorpe's *The Reginal City Planning for the End of Sprawl*, is the inheritance and development based on the thoughts of Howard, Mumford and others. It is consistent with the rural city and the interpretation of Howard's social city in the new era. He proposed that the area should be designed as a whole, and the design area is the design block. Consider the region and its elements – the city, the suburbs, the natural environment – as a whole. Similarly, consider the block and its elements – homes, shops, open spaces, municipal institutions and businesses – as a whole. Both require a protected natural environment, an inescapable center, humanscale transportation systems, public spaces, and integrated diversity. The development area is to create healthy blocks, places and urban environment. Similarly, the development of such blocks is to create sustainable, integrated and cohesive areas and environment. Therefore, the region and the community cooperate with each other on two scales to coordinate the symbiosis.

百年现代城市规划理论照进现实——城乡断面

新都市主义另一个杰出代表杜安尼夫妇，在区域性城市的基础上提出城乡断面，更揭示了住宅分布的空间类型，从市中心的高层高密度，逐步到多层、联排和独立住宅，这是住宅类型分布的生态规律，更包含了区域经济活动、级差地租、交通运行、城市管理和社会组织等方面的客观规律。

One Hundred Years Of Modern Urban Planning Theory Into Reality –A Typical Rural-urban Transect

Another outstanding representative of New Urbanism, Duany couples in urban and rural cross section is proposed on the basis of regional cities, more reveal the distribution of the residential space type, in the center of the city high-rise high- density, gradually to the multilayer, united and independent housing, this is residential type distribution of ecological law, more includes regional economic activity, differential rent, traffic, city management and the objective laws of social organization.

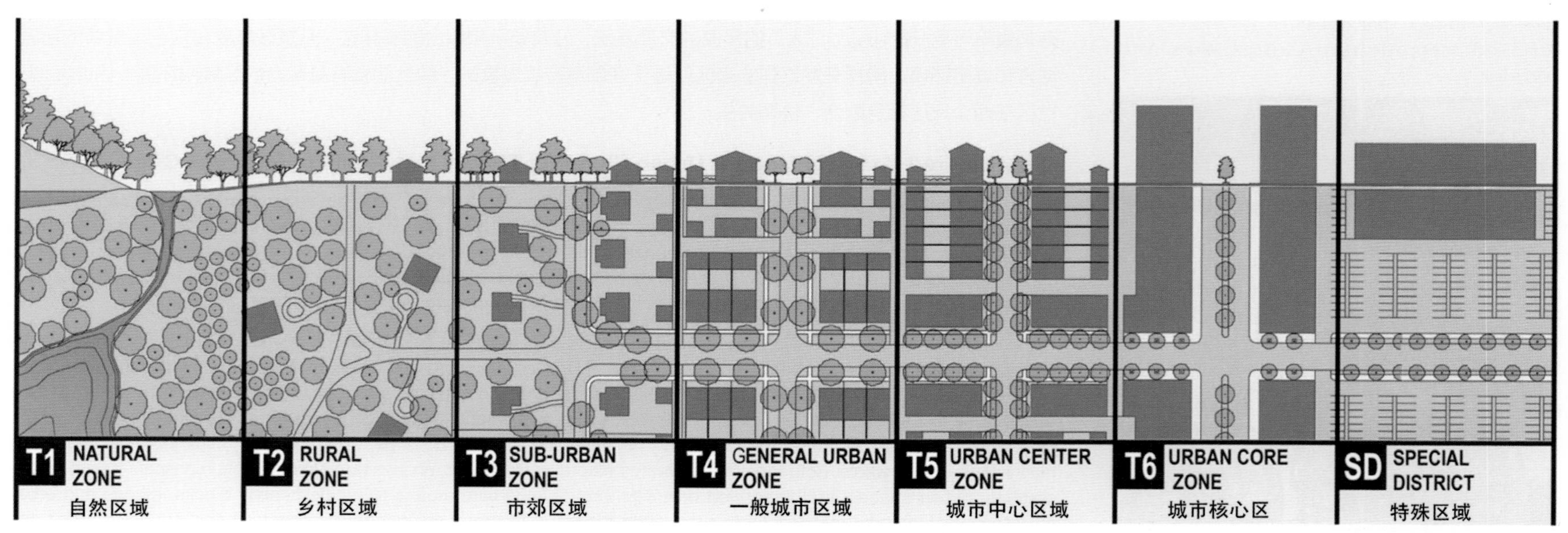

城乡断面图
Typical Rural-Urban Transect

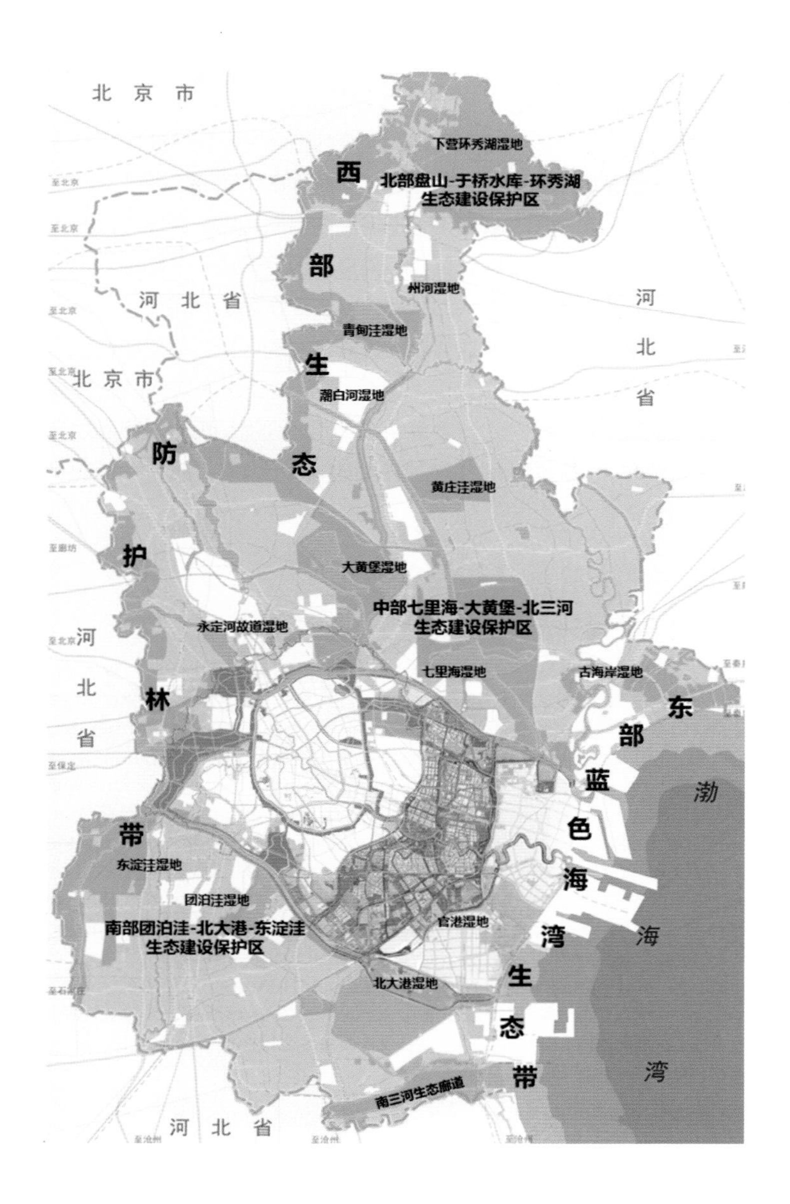

天津全域城乡生态结构
Ecological Structure of Urban and Rural Areas in Tianjin

城乡融合的区域性城市

当今，中国由住宅引发的城市问题，已经日益突出。解决好住宅问题，不仅关系到老百姓的切身利益，更关系到我国社会经济的可持续发展，是实现中华民族伟大复兴中国梦的重大课题。百年传承的理论指导我们，探索中国居住形态的特色之路，必须站在区域的高度，构建城乡融合的区域性城市。

Urban and Rural Integration of Regional Cities

At present, the urban problems caused by housing in China are already imminent. Solving the housing problem is not only related to the vital interests of the people, but also to the sustainable development of China's social economy and the realization of the great theme of the Chinese nation's great rejuvenation of the Chinese dream. In the inheritance of the theory of centuries, we will explore the characteristic road of China's living form, and must stand on the scale of the region to build a regional city with urban and rural integration.

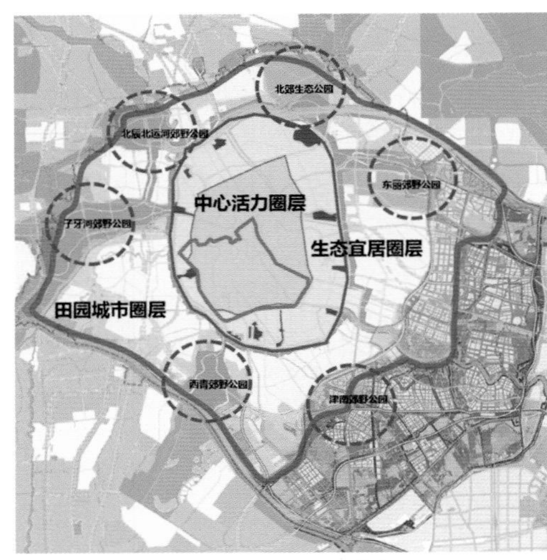

天津中心主城区城乡空间结构
Urban-Rural Spatial Structure of the Main Urban Area of Tianjin Center

天津中心主城区城乡断面分区
Urban-Rural Sections in the Main Urban Area of Tianjin Center

合理的城乡断面

天津特有的双城结构下，中心城区（市内六区）和周边四个外围郊区、滨海新区以及双城管控地区，都可以理解为一个城市区域。以中心城区（市内六区）和周边四个外围郊区形成的城市区域为例，在“一环十一园”规划下，形成了与自然和谐、城乡融合的居住空间整体格局优化。以城市快速路、外环快速路、环外环边界，形成三个圈层。中心活力圈层是快速环路以内的城市建成空间，以人口疏解、城市更新为主要特征，采用高强度、高密度的空间模式；生态宜居圈层是快速环路与外环之间的区域，突出居住空间与大型城市公园结合的特征，采用中低强度开发，TOD（以公共交通为导向的开发模式）、SOD（以社会服务设施建设为导向的开发模式）局部高强度；田园城市圈层是外环至高速环路的区域，以存量、减量发展为主要特征，居住空间与城郊生态空间结合，采用低强度、低密度开发模式。

在三个圈层的基础上，进一步细分，形成 11 个城乡断面分区。

Reasonable Urban and Rural Sections

Under the unique two-city structure of Tianjin, the central city (six districts in the city) and the surrounding four peripheral suburbs, Binhai New Area, as well as the twin cities control area, can be understood as an urban area. Taking the urban area formed by the central city (six districts in the city) and the four peripheral suburbs as an example, under the planning of "one ring and eleven gardens", the overall pattern of residential space that is harmonious with nature and integrated with urban and rural areas is optimized. The urban expressway, outer ring expressway and outer ring boundary form an ecological livable circle. The central circle is the urban built space within the fast ring, and it is the central area of urban vitality. It takes urban renewal as the main feature and adopts the spatial mode of high intensity and high density. The ecological livable circle layer is the area between the rapid ring and the outer ring, highlighting the characteristics of combining residential space with large urban parks. It adopts medium and low intensity development and partial high intensity TOD/SOD. The rural urban area, from the outer ring to the high-speed ring, is mainly characterized by the development of stock and quantity reduction, and the combination of residential space and suburban ecological space adopts the low-intensity and low-density development.

On the basis of the three circles, it is further subdivided into 11 urban and rural sections.

四个城市城乡断面比较
Cross-Sectional Comparison of Urban and Rural Areas in the Four Cities

多样化的社区

简·雅格布斯说，任何专制的、家长式的规划，试图用一种类型、一种模式解决所有的问题的规划，注定是不会成功的。不同时代，不同科技发展阶段，人们对生活场景的要求也在变化，因此城市住宅也应该是丰富多样的。在不同的人生阶段，人们的生活场景在不断发展变化。丰富多样的居住形态，代表着多种多样的生活方式，更折射出多姿多彩的精神生活。

因此，我们需要在城乡融合的区域格局下，根据不同的地区，提出不同的住宅类型和社区设计指引，为市民自由生活的表达，提供多样化的选择，让住宅回归对生命、对社会、对自然的本质意义。

Diverse Community

Jane Jacobs says any authoritarian, paternalistic program that tries to solve all problems with one type, one model, is doomed to fail. The key to urban housing is its rich diversity. In different times and different stages of technological development, people's requirements for life scenes are also changing. At different stages of life, people's life scenes are constantly evolving and changing. Rich and diverse living forms, representing a variety of ways of life, but also reflects the colorful spiritual life.

Therefore, under the regional pattern of urban-rural integration, we need to put forward different residential types and community design guidelines according to different regions, so as to provide diversified choices for the expression of citizens' free life and return the essential meaning of housing to life, society and nature.

面向 2035、2050 年的居住愿景

2035 年，人人拥有安全舒适、可负担、有尊严的住房。2050 年，建成美好、可持续发展的人居环境，实现中华民族的诗意栖居。

生活品质

1. 人均住宅建筑面积

2035 年，天津城镇人均住宅建筑面积 40 ~ 45 平方米。

2050 年，天津城镇人均住宅建筑面积 45 ~ 50 平方米。

住宅类型	人均住宅建筑面积（平方米）			
	中心活力圈层	生态宜居圈层	田园城市圈层	双城管控区
普通住宅	35	45	50	45

2035 年人均住宅建筑面积
By 2035, Per Capita Residential Area

住宅类型	人均住宅建筑面积（平方米）			
	中心活力圈层	生态宜居圈层	田园城市圈层	双城管控区
普通住宅	40	50	60	45

2050 年人均住宅建筑面积
By 2050, Per Capita Residential Area

2. 户型及面积

随着城镇家庭结构变化，城镇住房需求也发生变化，呈现高品质小户型需求量增加，改善型住房需求增加，高档住房需求上升的趋势。因此，需提供与家庭结构、住房需求对接的房型与面积，在增加面积的基础上，合理化住房功能。

需求类型	户型	住房功能	面向人群	新建商品住宅面积
高品质小户型需求	1 室 1 厅 2 室 1 厅	卧室、客厅、厨房、卫生间	单人家庭、丁克家庭、空巢家庭	户均 75 平方米
改善需求	2 室 2 厅 1 书	卧室、客厅、餐厅、书房、厨房、卫生间	核心家庭（两个孩子）	户均 120 平方米
	3 室 2 厅 1 书	卧室、客厅、餐厅、书房、儿童房、厨房、卫生间		户均 135 平方米
高档需求	多室 2 厅 1 书	卧室、客厅、餐厅、书房、儿童房、活动室、储藏室、厨房、卫生间	核心家庭（两个孩子）	户均 150 平方米以上

居室数及面积
Number of Bedrooms and Area

Housing Vision for 2035 and 2050

By 2035, everyone will have safe, comfortable, affordable and dignified housing.

By 2050, we will build a beautiful and sustainable living environment and realize the poetic dwelling of the Chinese nation.

The Quality of Life

1.Per Capita Residential Area

By 2035, the per capita housing area in Tianjin will be 40-45 square meters.

By 2050, the per capita housing area in Tianjin will be 45-50 square meters.

2.Number of Bedrooms and Area

With the change of urban family structure, urban housing demand also changes, showing a trend of high quality small family demand increase, improved housing demand increase, high-end housing demand rise. Accordingly, the bedroom number that needs to provide docking with domestic structure, housing demand and area.

住宅与社区的多样性

1. 多样的住宅类型

在城乡融合的格局下，形成与自然环境连续统一的城乡断面，在不同地区，提出不同的住宅与社区类型引导。

在区域内形成丰富多样的居住形态，承载多种多样的生活方式，折射多姿多彩的精神生活。

区位引导	社区类型	住宅类型	生活方式
中心活力圈层	高密度、高强度社区	高层Ⅱ类住宅（≤80m） 超高层住宅（>80m）	繁华都市 历史记忆
	历史社区	低层住宅（≤18m） （独立式、联排式、合院式）	
	老旧社区	多层Ⅰ类住宅（≤27m） 多层Ⅱ类住宅（≤36m）	
生态宜居圈层	公园周边社区	低层住宅（≤18m） （联排式、合院式） 多层Ⅰ类住宅（≤27m） 多层Ⅱ类住宅（≤36m） 高层Ⅰ类住宅（≤54m）	品质生活 邻里交往
	TOD 社区	多层Ⅰ类住宅（≤27m） 多层Ⅱ类住宅（≤36m） 高层Ⅰ类住宅（≤54m）	
田园城市圈层	郊区社区	低层住宅（≤18m） （独立式、联排式、合院式） 多层Ⅰ类住宅（≤27m）	亲近自然 对话自然

天津中心主城区主要住宅类型引导
Guidance of Main Housing Types in the Main Urban Areas of Tianjin Center

2. 多样的建设主体

社区内部住宅建筑类型相似，单体建筑丰富多样，风格各异。通过住宅建筑实现市民多样的审美要求、精神追求。

规划中避免单一开发商大规模建设的粗放模式，鼓励小规模建设面积、多数量建设主体的精细模式。通过产权地块切分，充分发挥各建设主体的个性特色。

Diversity of Homes and Communities

1.Diverse housing types

Under the integration of urban and rural areas, urban and rural sections are formed in a continuous and unified way with the natural environment. Different residential and community types are proposed in different areas. It forms a variety of living forms in the region, carrying a variety of life styles, reflecting the colorful spiritual life.

2.Diverse construction subjects

The residential building types inside the community are similar, and the individual buildings are rich and varied with different styles. Through the residential building to achieve a variety of aesthetic values, the expression of spiritual pursuit. In the planning, the extensive mode of large-scale construction by a single developer is avoided, and the detailed mode of small-scale construction area and multiple construction subjects is encouraged. Through the property rights of land segmentation, give full play to the individual characteristics of the main construction.

3. 多样的政策保障

倡导低端有保障、中端有供给、高端有市场的住房政策，重点增加公共住房比例和种类，形成公共住房、市场住房两个独立市场，避免对商品房市场的过度冲击。

4. 混合的住宅布局

引导不同类型居民的适度混合与和谐共处，提倡不同收入、不同年龄的人群共同居住，加强社区融合。

以“大分散、小集中”的原则布局公共住房，保障型住房宜以独立地块形式穿插在商品房中布局。适当选择地段较好的地块建设改善型限价商品房。

老年公寓以独幢形式靠近社区中心，或与社区中心结合布局，不鼓励设置大面积的老年社区。

3.Multiple policy guarantees

We will advocate a housing policy that guarantees housing at the low end, supplies at the middle end, and has a market at the high end. We will focus on increasing the proportion and types of public housing, and form separate markets for public housing and market housing, so as to avoid excessive impact on the commercial housing market.

4.Mixed residential layout

We should guide the moderate mixing and harmonious coexistence of residents of different types, encourage people of different incomes and ages to live together, and strengthen community integration. Public housing should be distributed in the principle of "large dispersion and small concentration", and low-income housing should be distributed in commercial housing in the form of independent plots. Appropriate choice of a good lot of land to build improved limited price commercial housing. Apartments for the aged are in the form of a single building near the community center, or combined with the layout of the community center, so it is not encouraged to set up a large area of community for the aged.

住宅类型		住宅类型	主力户型	面向人群	政策支持
公共住房	保障型 （公租房、还迁房、经济适用房）	多层 II 类住宅（≤ 36m） 高层 I 类住宅（≤ 54m）	两室	低收入家庭	5 年后可上市，政府优先回购或售予同等资格购房者
	限价商品房	多层 I 类住宅（≤ 27m） 多层 II 类住宅（≤ 36m） 高层 I 类住宅（≤ 54m）	三室 四室	中等收入家庭	城市居民均有一次机会 5 年后可上市，政府优先回购或售予同等资格购房者 部分可实行共有产权
市场住房	高档商品房	低层住宅（≤ 18m） （独立式、联排式、中式合院式）	四室以上	高净值家庭	
		超高层住宅（> 80m）	一室 两室 四室	高净值家庭、外籍人士、精英人士	

住房政策引导
Housing Policy Guidance

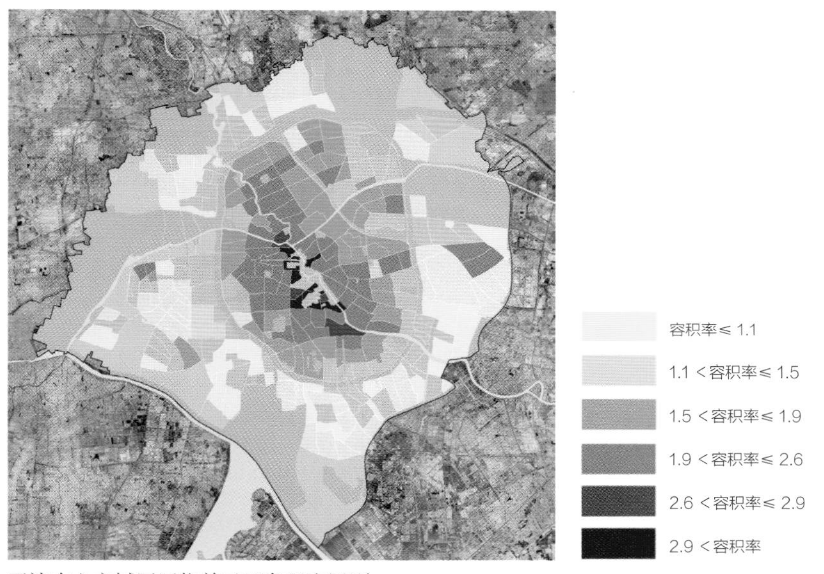

天津中心主城区居住单元开发强度导则
Guidelines for Residential Unit Development Intensity in Downtown Tianjin

区位引导	容积率建议
中心活力圈层	核心地段容积率 > 2.5 轨道站点周边 650 米范围内容积率≤ 2.5 其他地区容积率≤ 2.0
生态宜居圈层	轨道站点周边 650 米范围内容积率≤ 2.0 其他地区容积率≤ 1.6
田园城市圈层	1.0 <容积率≤ 1.2

天津中心主城区居住开发强度引导
Residential Development Intensity Guidance in the Main Urban Area of Tianjin Center

合理的开发强度和密度

控制合理的开发强度和密度，确保住宅类型的多样性，营造舒适优美的住宅环境。规划中社区内居住用地整体容积率不得低于 1.0。

除城市中心核心地段外，严格限制居住用地容积率，不得超过 2.5。

Reasonable Development Intensity and Density

Control reasonable development intensity and density, ensure the diversity of housing types, create a comfortable and beautiful housing environment. The overall Floor-area Ratio of residential land in the planned community shall not be lower than 1.0.Except for the core areas of urban centers, the Floor-area Ratio of residential land shall not exceed 2.5.

社区中心

营造邻里守望、具有归属感的社区场所。充分表达社群个性，传达丰富多样的社区场所精神。

功能构成中，保障并完善基本配套设施。根据每个社区不同的群体需求，补充符合群体生活方式，能触发公共生活、社会交往的服务功能。鼓励各功能集中布局，围绕社区中心组织各类活动。

城市设计中，社区中心选址宜毗邻轨道、公交站点。与社区公园相结合，塑造具有人性尺度的公共空间，形成聚集人气的社区场所。与幼儿园、小学相结合，实现场地共享，突出温馨家庭的生活氛围。与运动场地相结合，提供便捷多样的健身场所。

建筑设计中，避免单一模式的简单复制，要根据社区地域文化特征，形成富有个性的建筑设计。强调人与自然的融合和交流。在各功能合理组合的基础上，强化共享空间设计，满足全时段的空间活力，促进交往。强调空间的可变性，根据社群需求变化，实现不同功能的转换。

Community Center

Create a neighborhood watch and community place with a sense of belonging. Full expression of community personality, convey rich and diverse community spirit.

To ensure and improve the basic supporting facilities. According to the needs of different groups in each community, it can supplement the service functions that conform to the group's lifestyle and trigger public life and social communication. Encourage the centralized layout of various functions and organize various activities around community centers.

In urban design, the community center site adjacent to the subway, bus station. Combined with the community park, it can shape the public space with human scale and form the community place where people gather. Combined with kindergartens and primary schools, the site is shared, highlighting the warm family living atmosphere. Combined with sports venues, it provides convenient and energetic fitness venues.

In architectural design, the simple replication of a single mode is avoided, and the architectural design with rich personality is formed according to the regional and cultural characteristics of the community. Emphasize the integration and communication between human and nature. On the basis of the reasonable combination of various functions, the design of shared space is strengthened to meet the space vitality of the whole period and promote communication. It emphasizes the variability of space and realizes the transformation of different functions according to the changing needs of the community.

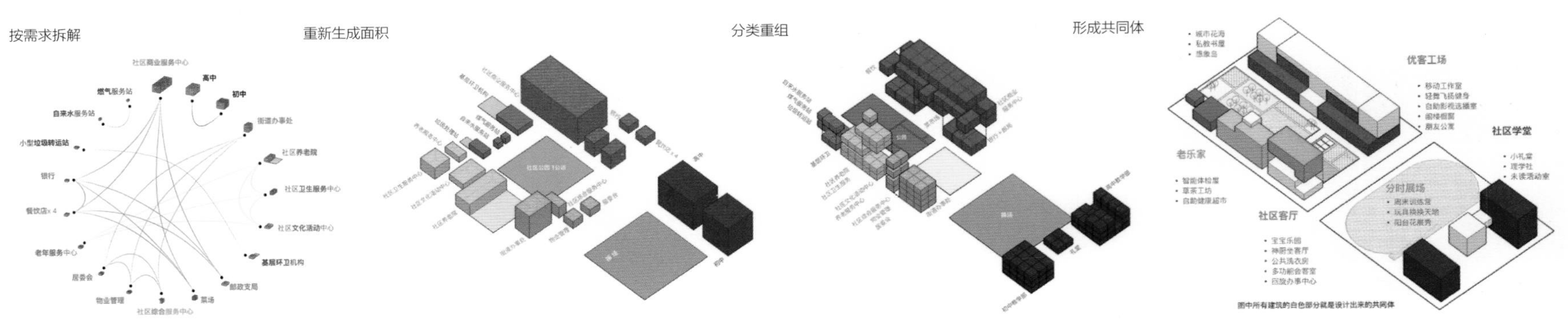

社区中心的生成
Function Composition of Community Center

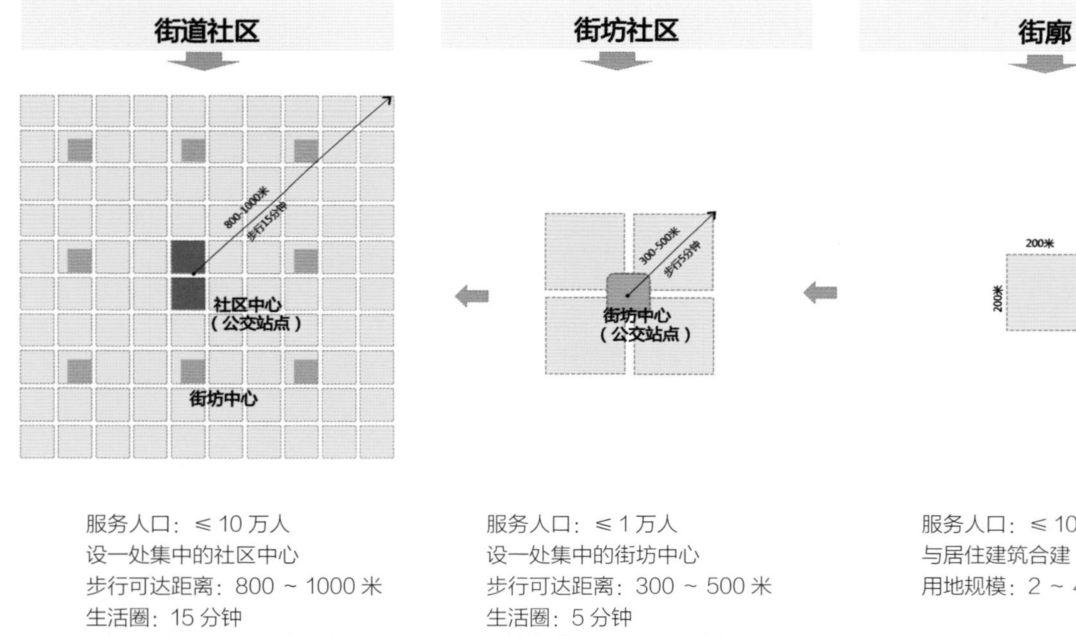

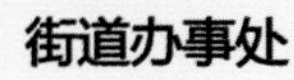

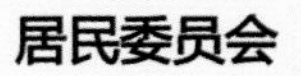

业主委员会

社区结构分析
Community Structure Analysis

社区中心层级体系

规划构成与服务半径、心理认知、社区管理相对接的社区中心体系，对应新型社区三级体系，形成三级社区中心体系：社区中心、街坊中心及街道底商及配套。

Community Center Hierarchy System

The planning constitutes a community center system which is relatively connected with service radius, psychological cognition and community management. Corresponding to the new three-level community system, there is three-level community center system: community center, neighborhood center and sub-district business and supporting facilities.

“见缝插绿”的中心活力圈层示意
"See Seam Inserted Green" Central Dynamic Circle Diagram

“留白增绿”的生态宜居圈层示意
"White Space and Green" Ecological Habitable Circle

“生态大绿”的田园城市圈层示意
"Ecological Big Green" Rural Urban Circle

生态环境

倡导社区接近自然和公园，使人类社区与自然世界形成连续的统一体。在城乡融合的格局下，形成接近自然与公园的人居环境。

提升人均公园绿地指标，建设宜居城市。根据不同地区，提出不同类型的控制指标。

Ecological Environment

Advocate for community access to nature and parks so that human habitat forms a continuum with the natural world. Under the pattern of urban and rural integration, the habitat environment close to nature and park will be formed. We will improve per capita green space in parks and build livable cities. Different types of control indexes are proposed according to different regions.

区位引导	策略
中心活力圈层	“见缝插绿” 提升公园、绿地、广场品质
生态宜居圈层	“留白增绿” 强化外环绿带与沿线公园的联系
田园城市圈层	“生态大绿” 建设大尺度生态绿化，建立区域生态安全格局

天津中心主城区社区生态环境引导
Community Ecological Environment Guidance in Downtown Tianjin

交通

1. 交通出行指标

在交通出行指标中，减少通勤交通出行距离及耗时，提升绿色交通（公交 + 慢行）出行比例，提高居民生活幸福感。中心活力圈层路网密度和公交站点密度较高，可以有效支撑慢行出行和公交出行需求。外围圈层通过增加社区内部路网密度支撑慢行出行比例。

2. 绿色交通为导向的道路系统

规划绿色交通为导向的道路系统，加强居住街坊的可达性，建立慢行交通为导向的社区内部出行模式。道路网密度一般不大于 200 米 ×200 米；轨道站点周边 300 米半径范围内路网密度宜进一步加密，不大于 150 米 ×150 米。社区内道路应采用交通稳静化措施，限制车辆行驶速度及噪声，提升居民慢行交通出行安全感和舒适感。

3. 建立便捷的公共交通出行环境

以地区轨道站点及公交设施为中心，设置相关配套设施，引导居民使用公共交通出行。轨道车站周边应结合居住区所处区位居民出行特征，合理布置与轨道换乘相关的交通设施用地，包括公共交通换乘枢纽、社会停车场库、自行车存放场、出租车候客点等。轨道车站周边 300 米半径范围内可适当提高土地使用强度。社区及公共服务设施主要出入口应与轨道车站出入口、地面公交站点充分结合，通过高品质的步行系统与公交站点衔接，方便居民出行。

4. 完善的静态交通设施

根据不同圈层居住区交通特征，提出与地区特点相适应的停车配建指标。

The traffic

1.Traffic indicators

In the traffic index, reduce the distance and time consumption of commuting traffic, increase the proportion of green traffic (bus + slow-traffic), and improve residents' happiness in life. The network density and bus station density of the central dynamic circle are high, which can effectively support the slow-traffic l and bus travel demand. The outer circle supports the proportion of slow-traffic by increasing the density of road network inside the community.

2. Green Traffic-oriented Road System

Planning the green transportation-oriented road system and strengthening the accessibility of residential neighborhoods, we will establish a slow-traffic oriented intra-community travel mode. The density of road network is generally not more than 200 X 200 meters. The network density should be further encrypted within 300 meters around the track station, no more than 150 meters X150 meters. Traffic calming measures should be adopted on the roads in the community to limit the speed and noise of vehicles, so as to enhance residents' sense of safety and comfort in slow-traffic.

3. Establish the Convenient Public Transportation Environment

Centering on regional railway stations and public transportation facilities, relevant supporting facilities are set up to guide residents to use public transportation. The surrounding areas of track stations should be combined with the travel characteristics of residents in the location of the residential area, and reasonably arrange the land for transportation facilities related to track transfer, including public transportation transfer hub, social parking garage, bicycle storage yard, taxi waiting point, etc. Land use intensity can be appropriately improved within a radius of 300 meters around track stations. The main entrances and exits of community and public service facilities should be fully integrated with the entrances and exits of track stations and ground bus stations, and connected with bus stations through high-quality walking system to facilitate residents' travel.

4.Perfect Static Traffic Facilities

According to the traffic characteristics of different residential areas, the parking allocation and construction indexes suitable to the local characteristics are proposed.

区位引导	可再生能源	规划要求
中心活力圈层	太阳能 浅层地热	新建住宅、公建可因地制宜设置分散式太阳能集中供热系统； 新建公园绿地、广场、湖面等开敞空间宜利用浅层地热为周边住宅、公建提供集中冷热源。
生态宜居圈层	太阳能 浅层地热 深层地热	新建住宅、公建可因地制宜设置分散式太阳能集中供热系统； 结合一环十一园建设可再生能源站； 深层地热资源覆盖区内的住宅建筑宜采用深层地热进行供热。
田园城市圈层	太阳能 浅层地热 深层地热 污水废热	新建住宅、公建可因地制宜设置分散式太阳能集中供热系统； 大型郊野公园、广场、湖面等开敞空间宜利用浅层地热为周边住宅、公建提供集中冷热源； 深层地热资源覆盖区内的住宅建筑宜采用深层地热进行供热； 污水处理厂周边 3 千米范围内的住宅、公建、工业厂房宜采用污水废热进行集中供冷、供热。

天津中心主城区可再生能源规划要求
Renewable Energy Planning Requirements for the Main Urban Area of Tianjin Center

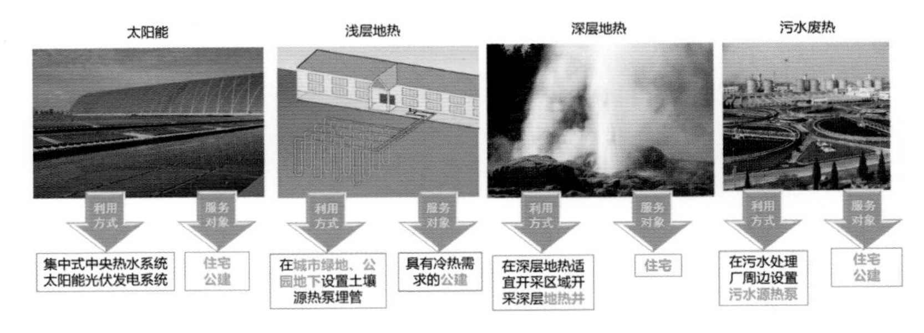

可再生能源利用方式示意
The Use of Renewable Energy

绿色基础设施

1. 可再生能源

坚持“因地制宜、绿色发展”的总体思路，按照具备条件的区域优先利用可再生能源的原则，积极推动可再生能源站建设。

2. 海绵城市

将海绵城市建设理念充分融入城市发展，最大限度地减少城市开发建设对生态环境的影响，构建源头、中端、末端全过程控制的雨水排放系统，实现绿色、生态的城市发展模式。

到 2035 年，年径流总量控制率不低于 75%，年悬浮固体总量去除率不低于 65%，雨水资源化利用率不低于 6%。城市建成区 80% 以上的面积应达到目标要求，内涝防治标准达到 100 年一遇，再生水利用率达到 62%。

Green infrastructure

1.Renewable energy

We will adhere to the general idea of "adapting measures to local conditions and pursuing green development" and actively promote the construction of renewable energy stations in accordance with the principle of giving priority to the use of renewable energy in regions where conditions permit.

2.Sponge City

The concept of sponge city construction should be fully integrated into urban development to minimize the impact of urban development and construction on the ecological environment, build a whole-process controlled rainwater drainage system at the source, middle and end, and realize a green and ecological urban development mode.

In 2035, the total annual runoff control rate shall be no less than 75%.The annual removal rate of total suspended solids shall be no less than 65%. The utilization rate of rainwater shall not be less than 6%. More than 80% of the urban built-up areas should meet the target requirements. Waterlogging control standards reaches 100 years. The utilization rate of recycled water reaches 62%.

住宅类型		断面分区（黑底 允许 / 白底 不允许）
传统中式合院	此类住宅是天津传统民居，单层，形制一是以正房为核心，以方形封闭院落为组织形式，二是轴线对称布置，三是中小型住宅沿纵轴发展，大型住宅横纵轴发展。同时，形成纵横丰富的街巷胡同肌理。此类住宅可位于乡村区域的传统聚落以及城市中心区保留的传统聚落。	T1 **T2** T3 T4a T4b T5a **T5b** T6
里弄式住宅	此类住宅是天津近代主要住宅类型之一。2~3层为主，由多个分户单元并联组成，形成尺度宜人的巷弄，各分户具有前后院，每户均有独立的出入口，此类住宅可位于市郊区域以及城市中心区的历史社区。	T1 T2 **T3** T4a T4b T5a **T5b** T6
独立式住宅	此类住宅是天津近代高级住宅。2~3层为主，独门独户，顶天立地，拥有良好的采光通风，每户有独立的花园和车库，密度较低，亲近自然。住宅风格形式突出个性，极具多样性。此类住宅可位于乡村区域、市郊区域以及城市中心区的历史社区。	T1 **T2** **T3** T4a T4b T5a **T5b** T6
院落式住宅	此类住宅由一组独立住宅单元围合形成，2~3层为主，独门独户，顶天立地，每户有独立的庭院，同时各户围合形成公共庭院，具有尺度宜人、层次丰富的院落空间，兼具私密性和邻里交往。此类住宅可位于市郊区域，一般城市区域的毗邻公园周边地区以及城市中心区历史社区。	T1 T2 **T3** **T4a** T4b T5a **T5b** T6
联排式住宅	此类住宅由三个或更多独立住宅单元并联组成，2~3层为主，有天有地，每户拥有独立的院子和车库，以跃层和复式为主，立面形式丰富多样。前院面向社区街道，形成宜人的街道空间和步行环境。此类住宅可位于市郊区域以及一般城市区域的毗邻公园周边地区。	T1 T2 **T3** **T4a** T4b T5a **T5b** T6

住宅类型		断面分区（黑底 允许 / 白底 不允许）
多层围合式住宅	此类住宅由堆叠的集合住宅组成，共享入口，4~6层为主，通过围合街廓，塑造宜人尺度的街道空间，内部形成公共院落，促进邻里交往和社区生活。水平、垂直空间业态混合，邻主街布局首层商业，激发城市活力。此类住宅可位于一般城市区域。	T1 T2 T3 **T4a** **T4b** T5a T5b T6
高层 I 类住宅	此类住宅由堆叠的集合住宅组成，共享入口，18层（54米）以下为主，由多层裙房围合街廓，塑造宜人尺度的街道空间，上部增加点式高层。此类住宅可位于一般城市区域的公交站点周边，城市中心区的地区中心以及城市核心区，以局部聚集为主，避免大面积重复。	T1 T2 T3 **T4a** T4b **T5a** T5b **T6**
高层 II 类住宅	此类住宅由堆叠的集合住宅组成，共享入口，26层（80米）以下为主，由多层裙房围合街廓，塑造宜人尺度的街道空间，上部增加点式高层。此类住宅可位于城市中心区的地区中心以及城市核心区，以局部聚集为主，避免大面积重复。	T1 T2 T3 T4a T4b **T5a** T5b **T6**
超高层住宅	此类住宅由堆叠的集合住宅组成，共享入口，26层（80米、100米）以上，由多层裙房围合街廓，塑造宜人尺度的街道空间，上部增加点式高层。拥有良好的景观视线条件，更复杂的电梯数量、消防设施、人员疏散要求，此类住宅可位于城市核心区。	T1 T2 T3 T4a T4b T5a T5b **T6**

T1 自然区域
T2 乡村区域
T3 市郊区域
T4a 一般城市区域(多层+高层)
T4b 一般城市区域(老旧社区)
T5a 城市中心区(地区中心)
T5b 城市中心区(历史社区)
T6 城市核心区

中心活力圈层
The Central Dynamic Circle

中心活力圈层内社区

市中心既是天津活力中心区，同时也聚集了14片历史街区，这里是天津城市生活最丰富的地方，也是最具人文精神的场所。这里吸引着天津最活跃的人群，他们大多在市中心工作，追逐潮流，热衷交往，有丰富的业余生活，追求生活品位。这里具有完备的步行网络，步行可达一体化的公共交通系统、高品质的公共空间、商业街区等。

Community in the Center Dynamic Circle

The city center is not only the dynamic central area of Tianjin, but also the gathering of 14 historical areas. It is the place with the most abundant urban life and the most humanistic spirit in Tianjin. It attracts the most active people in Tianjin, most people work in the city center, follow the trend, are keen on communication, have rich spare time life and pursue the taste of life. There is a complete walking network, which can reach the integrated public transportation system, high-quality public space and commercial blocks.

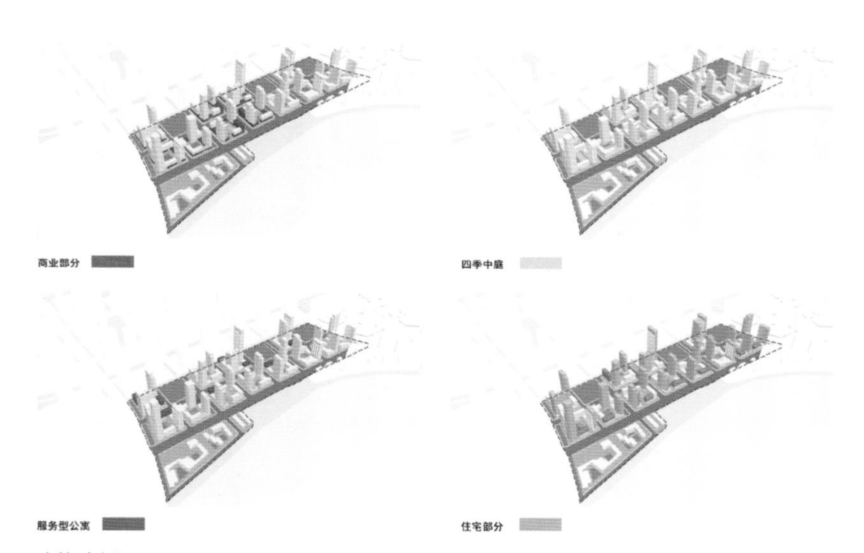

功能分析
Functional Analysis

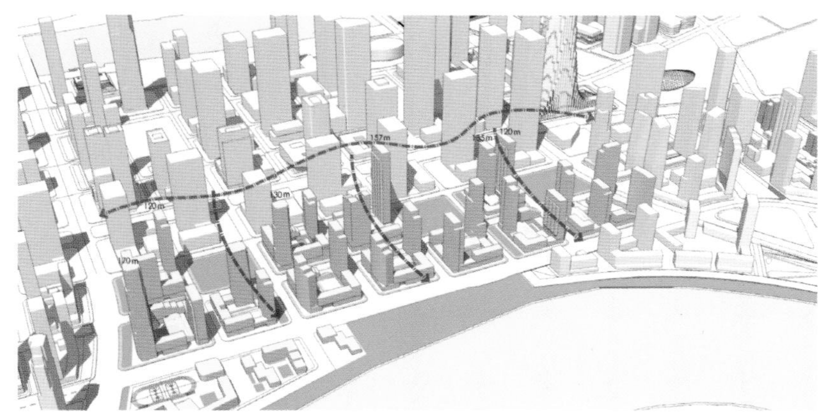

体量与天际线
Volume and Skyline

1. 高密度社区——于家堡东滨路地区

东滨路地区位于滨海新区核心区的于家堡金融区，东滨路地区重新定义城市与住宅的意义和形式，高度混合的功能，创建了一个生机勃勃、进入公共领域极为便捷的邻里社区。这里小街廓密路网，延续城市肌理模式，并与主街道相连，适于步行。错落的高度，与金融区整体城市形态融为一体，形成优美的滨水天际线。

1.High-density Community — Yujiapu Dongbin Road Area

Located in Yujiapu financial district, the core area of Binhai New Area, Dongbin Road area redefines the meaning and formation of city and residence, highly mixing functions and creating a vibrant neighborhood community with extremely convenient access to the public sphere. A network of narrow streets, extending the pattern of urban fabric, connected to the main street in the walkable neighborhood. The scattered height integrates with the overall urban form of the financial district to form a beautiful waterfront skyline.

于家堡东滨路地区效果图
Rendering of Yujiapu Dongbin Road Area

高密度、高强度开发，创造核心地段土地最大利用率，充分挖掘综合价值。社区拥有高速铁路、地铁、城市有轨电车等多种形式的公共交通，发达的公共交通减少了社区对小汽车的依赖性。营造了丰富的街道生活，亲人尺度的街墙增加了街道的活力。多样化的景观环境，包括屋顶花园、街廓庭院、社区公园以及滨河公园，具有鲜明特征，形成独一无二的景观体系。同时，社区倡导绿色经济，利用最先进的绿色技术，促进持续性城市发展。

High-density and high-intensity development creates the maximum utilization rate of land in core areas and fully excavates comprehensive value. Community has high-speed railway, subway, urban tram and other forms of transportation system, developed public transport less community car dependence. To create the plentiful street life, the family scale of the street wall to increase the vitality of the street. The diverse landscape environment, including roof garden, street courtyard, community park and riverside park, has distinctive characteristics, forming a unique landscape system. At the same time, the community advocates green economy and uses the most advanced green technology to promote sustainable cities.

2. 城市更新社区——劝业场地区

劝业场地区是天津现代商业的发祥地，是天津城市中心的顶级核心商圈，已有百余年的历史。它是天津“窄路密网”最典型的地区，街廓内建筑肌理的变化记录着近现代天津居住模式的变化，但街廓肌理一直保持不变，拥有丰富的街巷空间。其建筑类型丰富，存留了各个时期的典型建筑形式，具有丰富的文化价值，是近现代人居建筑的博物馆。

2.Urban Renewal Community — Quanyechang Area

Quanyechang area is the origin of modern commerce in Tianjin and the top core business circle in the city center of Tianjin. With a history of more than 100 years, it is the most typical area of "narrow road and dense network" in Tianjin. The changes of building texture in street width record the changes of living mode in modern Tianjin, but the street width texture remains unchanged and has abundant street space. It is a museum of modern and contemporary residential architecture with rich architectural types and typical architectural forms of various periods.

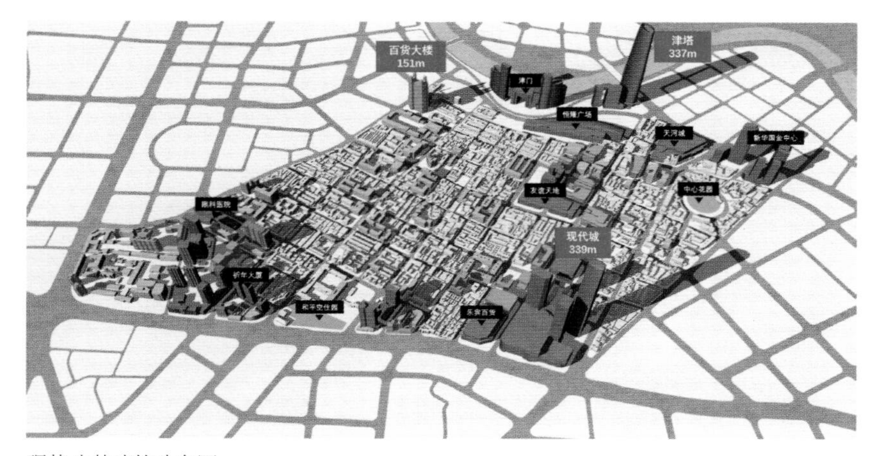

现状公共建筑分布图
Map of Present Public Buildings

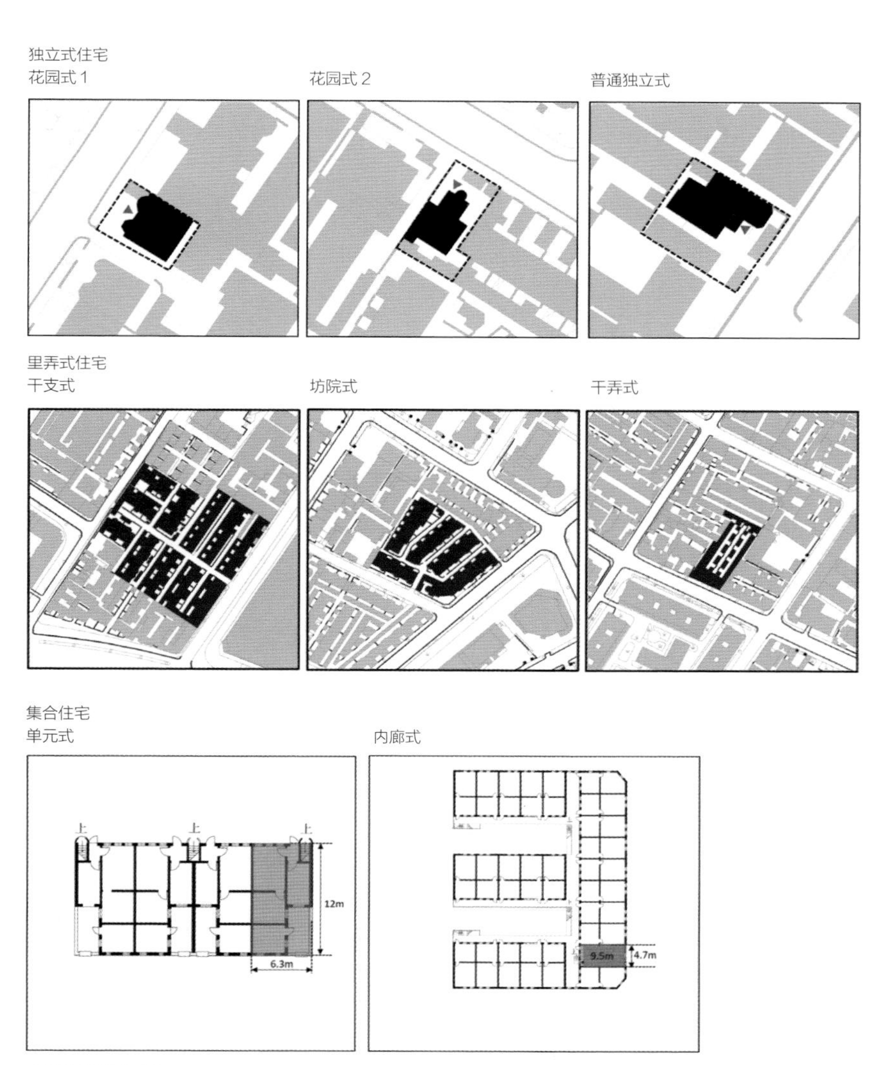

住宅类型分析
Housing Type Analysis

劝业场地区鸟瞰图
Aerial View of the Quanyechang Area

劝业场地区以“改善人民生活、传承历史文化、提升城市活力”为目标，通过城市更新，营造历史城区宜商宜旅宜居的示范区。首先，释放“不拆”信号，制定人口疏解方案，明确保护范围，总量不变情况下，调整历史街区保护范围。在采用异地还迁，明确疏解人口范围与规模（危陋房认定范围 + 现状居住地块范围 + 中华人民共和国成立前老建筑范围，最终确定疏解人口的居住建筑面积）。其次，梳理空间结构，形成一条步行街、两个活力圈层、十个微更新单元，整体保护，分级控制。第三，发掘人文历史资源，打造多层次的活力圈层，提升片区文化“软实力”。最后，基于类型学研究制定微更新单元保护与开发模式，同时编制微更新实施导则，保证微更新策略落实。

With the goal of "improving people's life, inheriting history and culture and enhancing urban vitality", the Quanyechang area aims to build a demonstration area suitable for business, travel and living in the historic district through urban renewal. First of all, we should release the signal of "no disconnection", formulate the plan of Population deconstruction, define the protection scope, and adjust the protection scope of historical blocks under the condition that the total amount remains unchanged. By relocating to other places, the scope and scale of population can be clearly alleviated (the recognized scope of dangerous houses + the superimposed scope of current living land + the scope of old buildings before the founding of the People's Republic of China, and the living building area of population can be finally determined). Secondly, the spatial structure is organized to form a pedestrian street, two dynamic circles, ten micro-renewal units, overall protection and hierarchical control. Thirdly, explore cultural and historical resources, build a multi-level circle of vitality, and enhance the cultural "soft power" of the region. Finally, the protection and development mode of micro update unit is formulated based on typology research, and the implementation guidelines of micro update are compiled to ensure the implementation of micro update strategy.

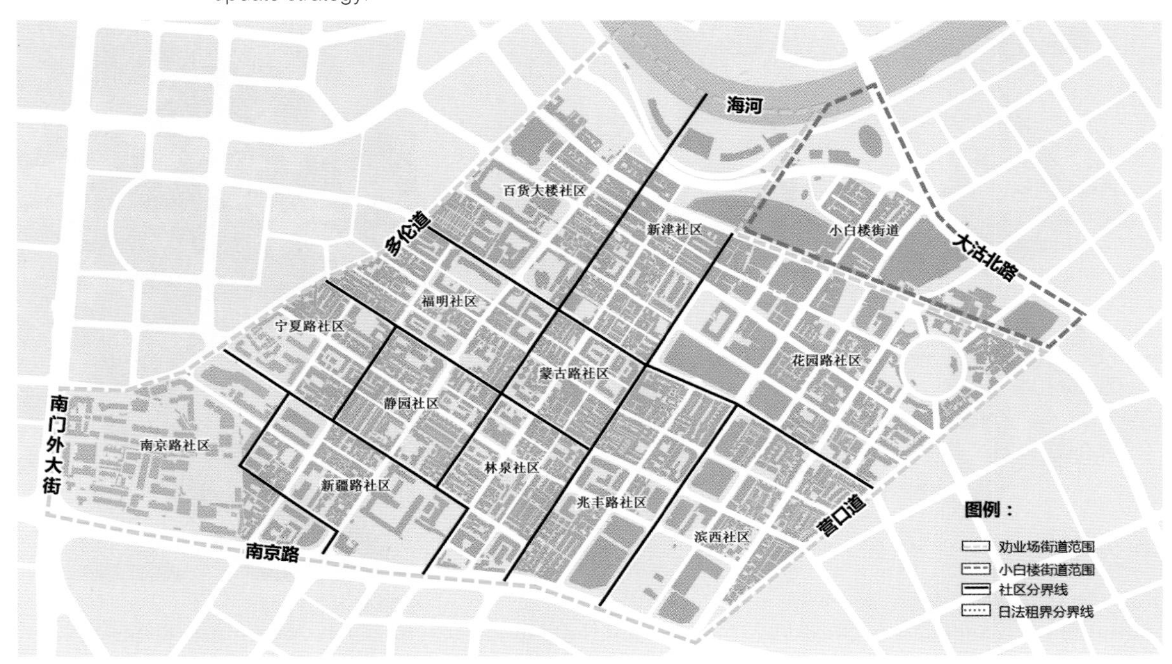

社区分布
Community Distribution

中华人民共和国成立初期、“文革”时期、1976 年地震时期三个阶段有大量居民涌入，以分配和抢占的形式，重构了原有私户的居住模式。

城市更新计划将通过市场机制引导居民恢复最初的居住形态，并满足当今的生活需求。

During the three stages of the early period of liberation, the period of the Cultural Revolution, and the period of the earthquake in 1976, a large number of residents poured in and reconstructed the living patterns of the original private households in the form of distribution and preemption.

The urban renewal plan will guide residents to restore their original living patterns through market mechanisms and meet the needs of today's life.

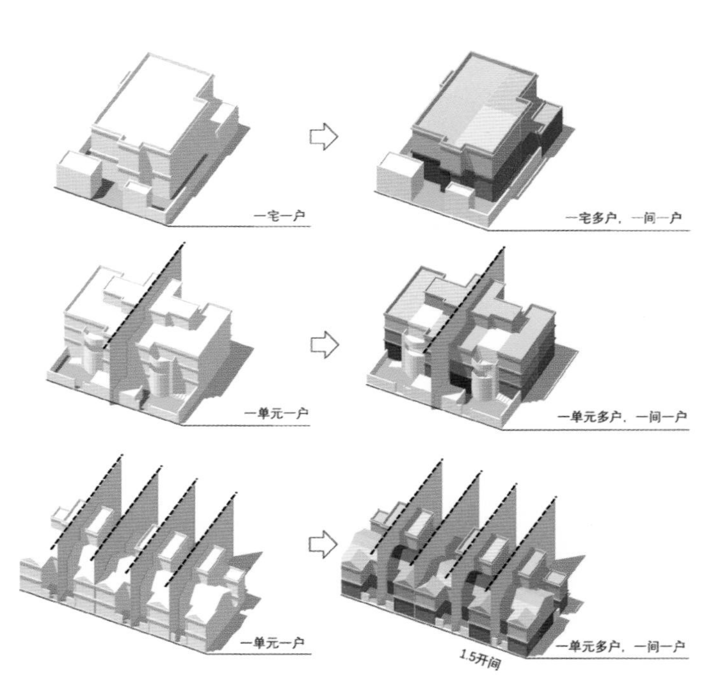

人口集聚造成的居住模式演变
Evolution of Living Patterns Caused by Population Agglomeration

静园社区更新示意
Jingyuan Community Renewal

社区分布
Community Distribution

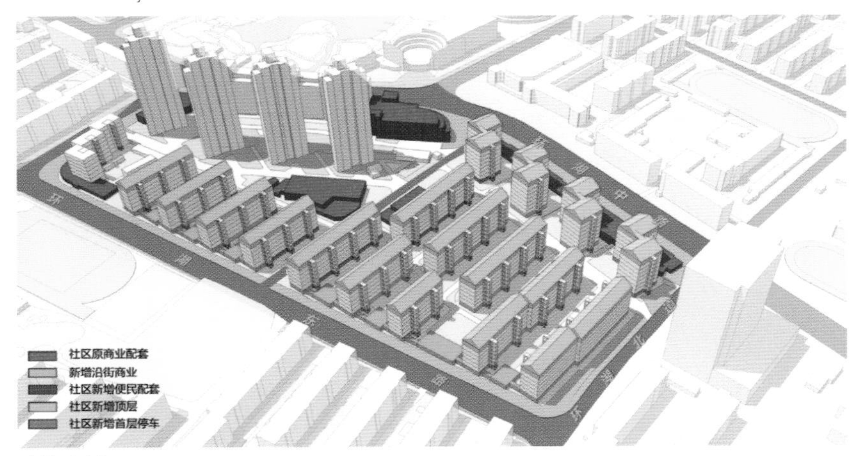

功能分析
Functional Analysis

3. 老旧社区——体院北地区

体院北是改革开放初期，天津规划兴建的“新十四片”大型居住区之一，是中心城区老旧社区更新的典型，具有高质量提升的示范意义，具有社区共同缔造模式的探索意义。重点从增加毛细支路、加密城市路网、营造社区活力中心、提升社区品质四个方面全面更新社区。

3.Old Community — Tiyuanbei Area

As one of 14 new large community planned in the early stage of reform and opening up, Tiyuanbei area is a typical example of the renewal of old communities in the central urban, with the demonstration significance of high quality improvement and the exploration significance of the community jointly creating a model. The focus is to comprehensively renew the community from the following four aspects: the capillary branch road, the urban road network, the building of community vitality center, and the improvement of community quality.

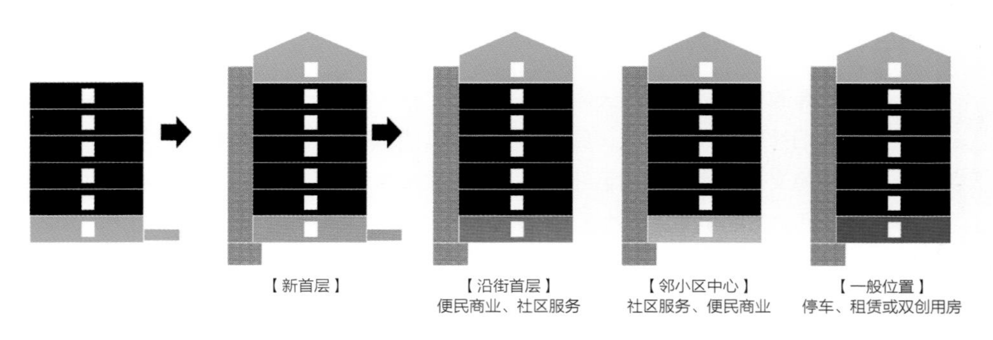

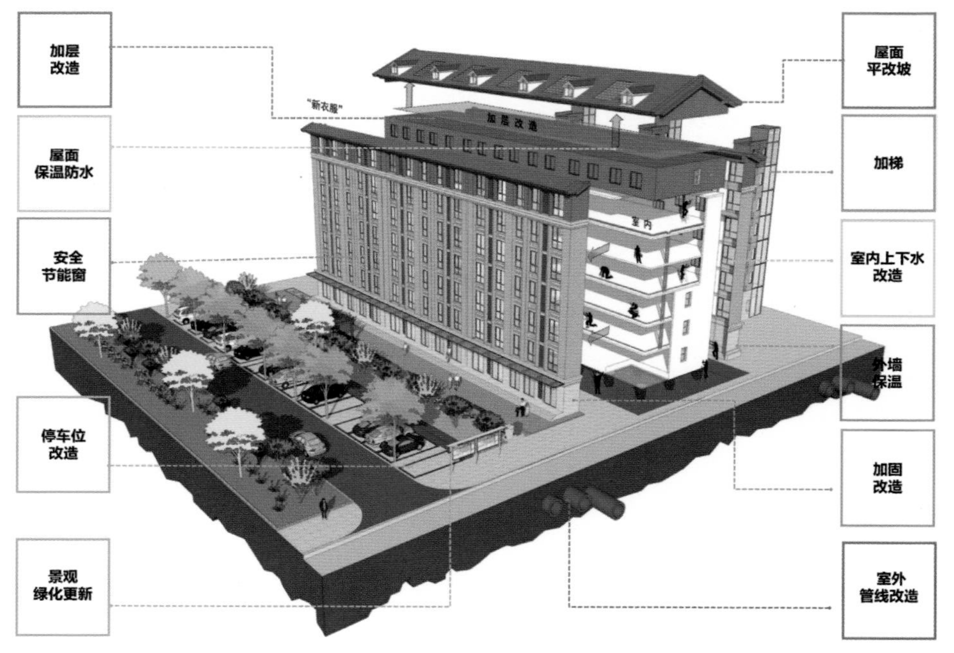

老旧住宅改造示意
Old House Renovation

采用菜单式改造模式，从住宅建筑、服务设施、开放空间三方面，加强整体改造规划，鼓励居民参与，启动示范项目，逐步推行。其中，住宅建筑采用加层、加梯方式，置换首层功能，综合解决老旧社区停车、公共服务配套不足的问题，做到增设沿街底商，不减街道红线，无损住户利益。

Adopting the menu-type renovation mode, the overall renovation plan will be strengthened from three aspects of residential buildings, service facilities and open space, encouraging residents to participate, launching demonstration projects and gradually implementing them. Among them, the residential building adopts the method of adding floors and elevator to replace the function of the first floor, comprehensively solve the problem of insufficient parking and public service in the old community, and add first floor business along the street, without reducing street red line , without damaging the resident's interests.

生态宜居圈层
Ecological Livable Circle

生态宜居圈层内社区

在快速路和外环快速路之间，结合十一个公园周边，形成宜居社区。这里紧邻公园，有良好的生态环境，完善的生活配套设施，便捷的公共交通系统，周边有优质的学校和医院，是众多有小孩，有老人的家庭的选择，也是市中心老旧社区中家庭提升改善居住环境的选择。在这里，孩子们可以在草地中自由奔跑，老人可以在林荫下漫步，一家人可以在树下野餐度假，共享时光。这里的居民有的在市中心工作，有的在社区内工作，通勤以公共交通为主，休闲出行以小汽车为主。

Community in Ecological Livable Circle

Between the expressway and the outer ring expressway, eleven parks are combined to form a livable community. It is close to the park, with a good ecological environment, complete living facilities, convenient public transportation system, and excellent schools and hospitals nearby. It is the choice of more families with children and the elderly, as well as the choice of families in the old communities in the downtown to improve their living environment. They pay more attention to family life, children can run freely in the grass, the elderly can walk in the shade, the family can have a picnic holiday under the tree, share the family. Some of them work in city center, some work in community, mainly commute by public transportation, leisure travel by car.

1. 公园周边社区——水西公园周边地区

社区以水西公园为核心，形成创新型商务商业、生态宜居的开放活力园林社区。通过街坊、胡同、院落塑造开放空间，构建层级分明的绿地系统，让城市回归自然。

1. Community Surrounding Park — Shuixi Park Area

With Shuixi Park as the core, the community has formed an open and dynamic garden community featuring innovative business and livable ecology. Create open space through neighborhoods, hutongs and courtyards and construct a clear hierarchy of green space system, so as to return the city to nature.

城市设计总平面图 Master Plan

效果图 Rendering

开放空间系统 Open Space System

临公园建筑以混合多元的建筑形态为主，采用前低后高的空间形态为周边区域留出视线通廊，塑造高活力混合多元的滨水界面。作为服务居民的生活大街以开放的姿态欢迎居民的到来，营造温暖和谐的街区形象，商业商务界面以公建为主。

The architecture of the adjacent park is dominated by the mixed and diversified architectural forms, and the front low and back high spatial forms are adopted to set aside the sight corridors for the surrounding areas, so as to shape the waterfront interface with high vitality and mixed diversity. As a living street serving residents' life, it welcomes residents with an open attitude and creates a warm and harmonious image of the block. The main interface of commerce and business are public buildings.

街道空间效果图
Street Space Rendering

2.TOD（以公共交通为导向的开发模式）社区——和谐社区

和谐社区通过简单易懂的组织结构，为全区创造了具有特色的地方感，进一步丰富了基地的场所意识。新月形的中央公园由连续的街道界面界定，塑造宏伟的城市印象。弯曲的公园及新月形街道与 Y 形的人行与自行车林荫大道相交，将子区域与主要的公共交通节点相连。Y 形的人行与自行车林荫大道为三个社区的中心。通过对朝向、采光、密度以及避免大规模平面过于单调的要求，决定了建筑高度的变化。较高的建筑位于不会遮挡临近建筑的地方，而且他们的位置强化了方案中的特殊场所。

2.TOD Community: Harmonious Community

Through a simple and easy-to-understand organizational structure, the harmonious community has created a unique local sense for the region, further enriching the base's sense of place. The Crescent Moon Park is defined by a continuous street interface that shapes the impression of the city on a magnificent street. Curved parks and crescent-shaped streets intersect with Y-shaped pedestrians and bicycle boulevards, connecting sub-areas to major public transportation nodes. The Y-shaped pedestrian and bicycle boulevards are three community centers. The change in height of the building is determined by the need to illuminate the sun, density targets, and avoid the monotony of large-scale planes. The higher buildings are located where they will not block adjacent buildings, and their location reinforces the special locations in the scheme.

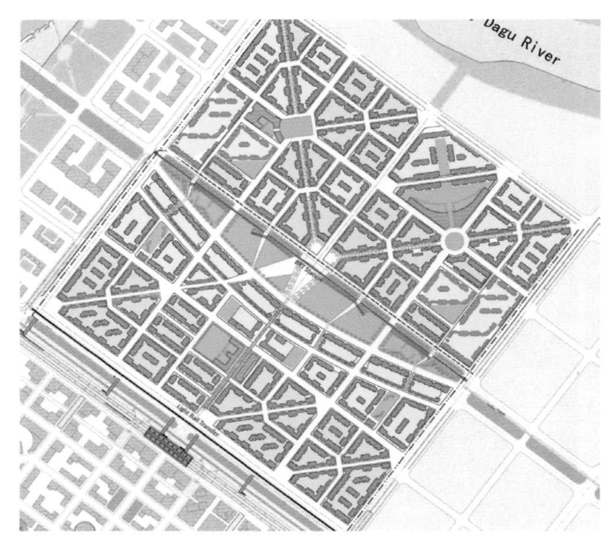

城市设计总平面图
Master Plan

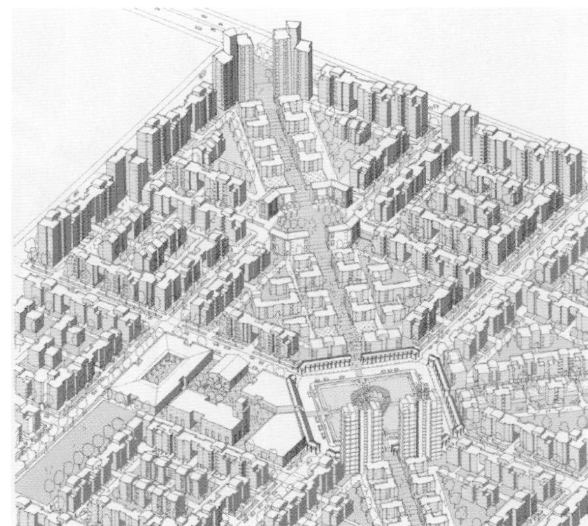

空间形态
Spatial Morphology Analysis

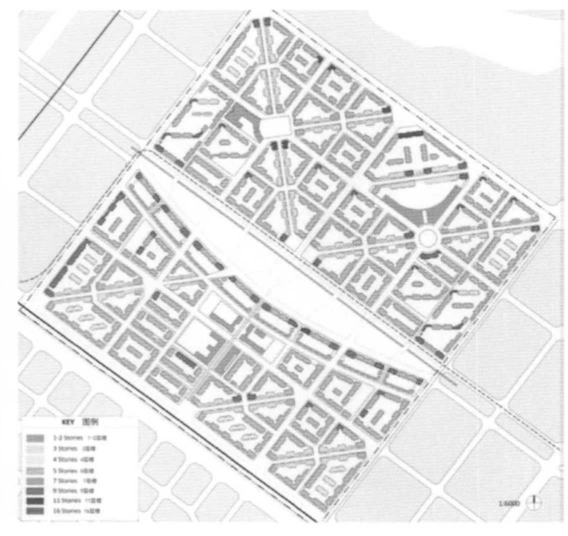

建筑层数
Building Floor Analysis

街道空间设计
Street Space Design

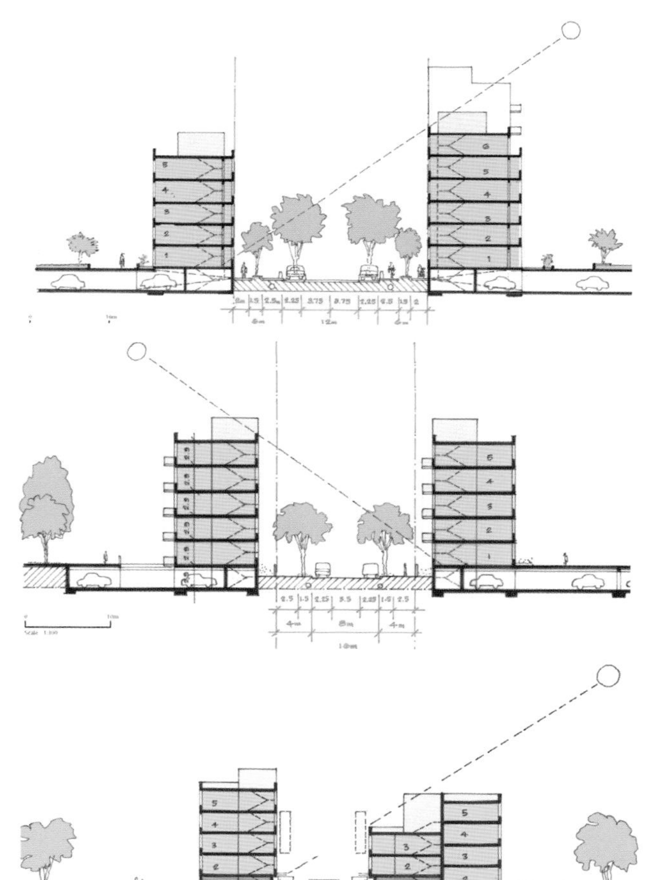

城市里的步行生活取决于街道的不同功能，有些用于机动车通行，其他用于商业活动与步行。街道不同的功能形成了不同的街道宽度。和谐社区在遵循日照管制的前提下，策略性地降低某些地方的建筑高度，依据街道性质进行景观设计，创造不同的街道特质。

Walking life in a city depends on the different roles that streets play, some facilitating motor traffic, others nourishing commerce and walking. The different roles of streets form very different street widths. Under the premise of sunshine control, the harmonious community strategically reduces the building height in some places, carries out landscape design according to the street nature, and creates different street characteristics.

城市设计总平面图 Master Plan

3. 混合多样社区——西营门社区

西营门社区是保护传承大运河文化遗产，展现运河城区段市井文化与当代时尚生活的滨水文旅混合社区。交通型街道界定基本街区尺度为600 ~ 800 米，生活型街道界定基本街廓尺度为 120 ~ 300 米。层级分明的绿地系统让城市回归自然，延展通达的慢行网络 ，精心营造通河通园的开放街区。以运河保护为出发点，降容积，控高度，并且在居住地块形成前低后高的空间形态，打造良好的运河观感。

3.Encourage Mixed and Diverse Community — Xiyingmen Community

Xiyingmen community is a mixed community of waterfront hydrological tour, which protects and inherits the Grand Canal cultural heritage and shows the urban culture of canal city section and the modern fashionable life. The basic block size of the traffic street is defined as 600-800 meters, and the basic block size of the living street is defined as 120-300 meters. The well-stratified green space system allows the city to return to nature, extends the accessible slow -traffic network, and elaborately creates an open block with river and garden. Taking Grand Canal protection as the starting point, it reduces the volume and controls the building height, and forms the spatial form before low and after high in the residential land to create a good canal appearance.

混合街区效果图 Rendering of Mixed Block

社区容纳多样化的混合街区，其中混合科创街区将主题商业、休闲、创意展示等特色功能交融，与创意孵化产业形成联动，形成“文化 + 创意 + 设计 + 艺术”复合的创意基地交易展示平台，高效促进就业、提升产能。混合乐活街区，保留现状良好的工业厂房，注入新的功能。艺术性架建的大跨距建筑，与旧厂房碰撞出新火花，达到了平衡，产生了新空间。以围合院落形态为基底，形成人性化的空间肌理。依托厂房改造提升和新建筑的融入，打造具有特色的多元混合活力街区。

The community accommodates diversified mixed blocks, among which the mixed science and innovation blocks integrate the theme business, leisure, creative display and other characteristic functions, form linkage with the creative incubation industry, and form a "culture + creativity + design + art" composite creative base trading and display platform, effectively promoting employment and productivity. The mixed happy-life block retains the industrial plant in good condition and injects new functions. The artistic long-span building colliding with the old workshop creates a new dialogue to achieve balance and create a new space. Taking the enclosed courtyard form as the base, it forms the intimate spatial texture of human nature. Relying on the transformation and upgrading of factory buildings and the integration of new buildings, to create a diverse and dynamic block with characteristics.

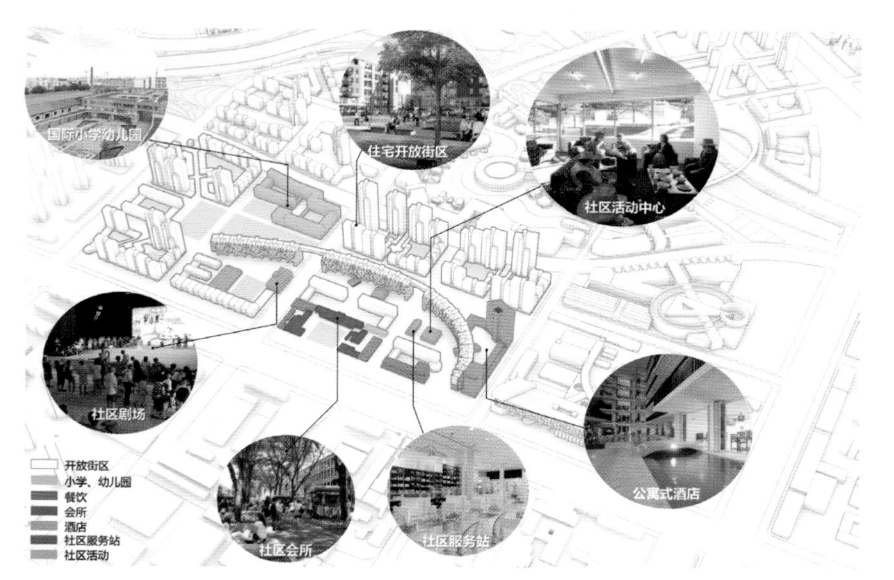

功能分析
Functional Analysis

业态构成
Business Composition

田园城市圈层内社区

外环快速路和高速环路之间，是人和自然对话，城市和乡村交融的地方。这里的人们追求宁静自由的生活，崇尚亲近自然，享受人生的生活方式。花园和庭院是每个家庭日常生活的中心，是对话自然，关照内心的场所。社区里有小镇中心、商业主街和学校，离家不远有家庭医生。出行以小汽车为主。

Community in Rural City Circle

Between the outer ring road and the highway loop, there is a dialogue between man and nature, where city and country meet. Here people pursue a quiet and free life, advocating close to nature, enjoy life style. Gardens and courtyards are the center of every family's daily life, a place to talk to nature and care for the heart. The community has town centers, main business streets and schools, and a family doctor not far from home. The trip is mainly by car.

田园城市圈层住宅
Rural City Circle

天南云智小镇

云智小镇位于海河教育园区内，本着完善海河教育园区定位，激发园区创新、研发应用能力的出发点进行设计。园区要求建立最优学、产、研、用创新平台，提升园区整体配套服务水平。最终给市民呈现出一个“智能的小镇、文化的小镇、生态的小镇、活力的小镇”，为天津塑造一张靓丽的城市新名片。

城市设计总平面图
Master Plan

Yunzhi Town in Tiannan

Located in Haihe Education Park, Yunzhi Town is designed to improve the orientation of Haihe Education Park, stimulate its innovation, research and development and application capabilities. The Park proposes to establish an optimal learning, production, research and innovation platform to enhance the overall supporting service level of the park. Finally, it presents the citizens with a smart town, a cultural town, an ecological town and a vibrant town, and creates a beautiful new business card for Tianjin.

鸟瞰效果图
Rendering of Aerial View

院落式（两进）

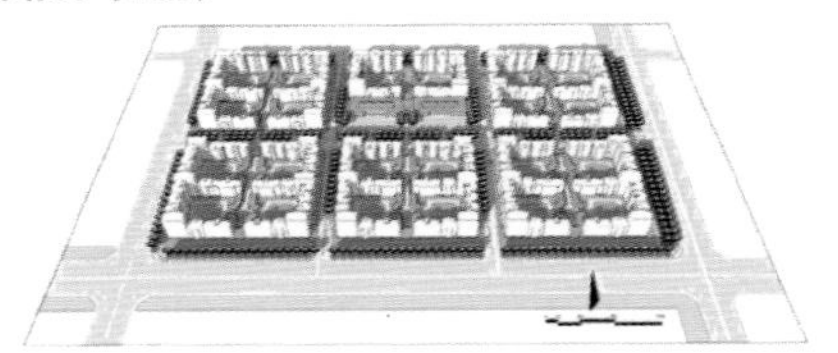

院落式（三进）

街区式

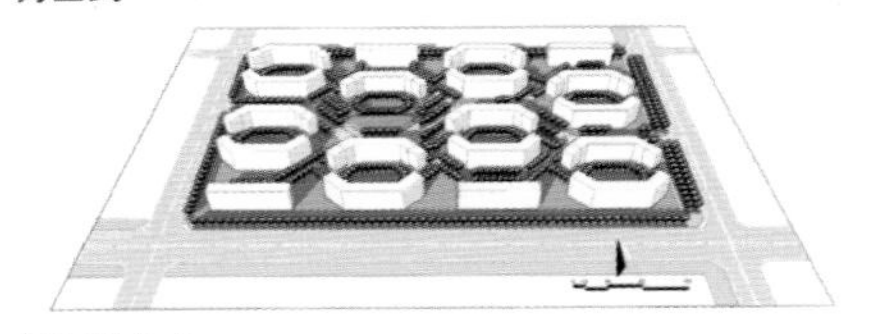

六边围合式

天南云智小镇街廓类型
Block Types in Tiannnan Yunzhi Town

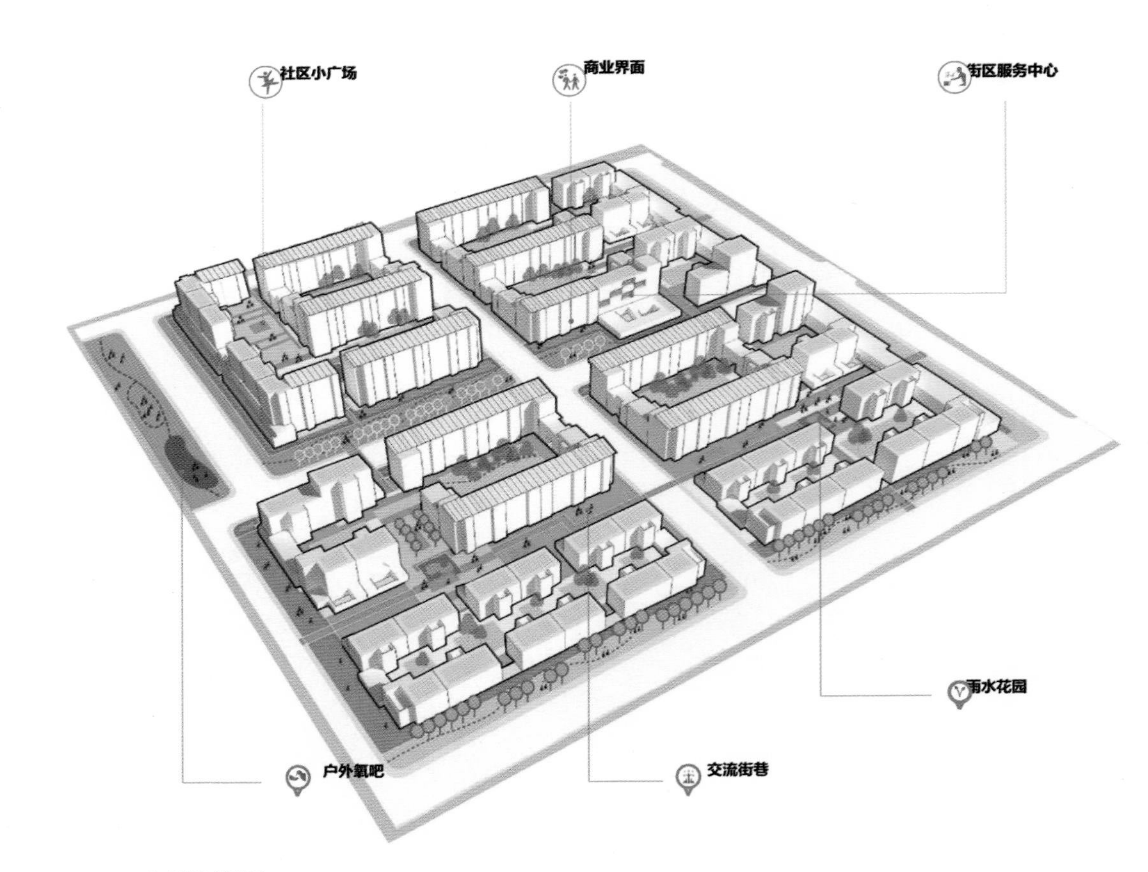

典型街廓设计
Typical Block Urban Design

新时代，随着土地财政逐渐难以为继，房地产拉动力渐渐衰弱，加上最新《城市居住区规划设计标准》等一系列新行业规范的出台，城乡融合的多样化社区将取代高密度蔓延。

In the new era, with the land finance becoming unsustainable, the real estate driving force gradually weakening, coupled with the latest " Standard for Urban Residential Area Planning and Design " and a series of new profession specification, urban and rural integration of diverse communities will replace the spread of high density.

万科城社区中心
Vanke Community Center

未来天津的城市中心不再拥挤不堪，小白楼城市中心得到强化和复兴，尽情彰显城市魅力。像劝业场地区这样的历史街区，注入更多样化的业态和人群。市民的住房选择变得自由而多样，无论是便利的公寓，环境优美，教育资源良好的洋房，医疗资源良好的养老社区，还是彰显个性的独立住宅，都可以按自己的不同需求，找到理想的住房。大家可以和自己志同道合的朋友毗邻而居，让社区更具有幸福感和归属感。城市郊区与乡村自然景观相融合，郊野和乡村休闲也成为城市居民旅游度假的绝佳选择。人与自然、人与城市、人与人构建起了和谐的关系。

In the future, the urban center of Tianjin will no longer be crowded, and the urban center of Xiaobailou will be strengthened and revitalized. In historical districts like the Quanyechang district, more diversified formats and groups of people have been injected into the district, which has been gradually stimulated. The housing choice of citizens becomes free and diverse, no matter it is convenient apartment, small family; beautiful environment, good educational resources housing; elderly care community with good medical resources; still reveal the independent residence of individual character, can press oneself of the different demand, find ideal housing. People can live next to like-minded friends, making the community more happy and belong. The city suburbs are integrated with the natural landscape of the countryside. On weekends, the countryside and leisure are also excellent destinations for urban citizens to travel and spend their holidays. People and nature, people and cities, people build the harmonious relationship.

水上公园 Water Park

文明发展的进程中有一点是始终不变的：社会要进步，人类要追求更美好的生活。回顾历史，一个民族的发展始终与美好的人居环境相伴随，人居环境建筑的最终目标是社会建设。吴良镛先生曾在《北京宪章》中提出：“美好的人居环境与美好的人类社会共同创造，建设一个美好的、可持续发展的人居环境，是人类共同的理想和目标。”

One thing remains constant in the progress of civilization: society wants progress and mankind wants a better life. In retrospect, the development of a nation has always been accompanied by a beautiful living environment. Mr. Wu Liangyong once put forward in the "Beijing Charter" : "It is the common ideal and goal of mankind to jointly create a better living environment and a better human society and build a better and sustainable living environment.

巴黎勒普莱西公益住宅
Le Piessis-Robinson Public Housing

城乡融合奠定了人与自然和谐相处的健康格局，而多样化的社区是社会的基本单元，是从微观出发，进行社会重组，通过对人的关怀，为每个人的意愿提供支持，从而激活丰富多样的生活，维护社会公平，最终实现和谐社会的理想。“人，诗意地栖居在大地上”是海德格尔的理想，更是我们一代代城市规划建设者的理想。

Urban and rural integration laid the pattern of health, harmony between man and nature and the diversity of the community is the basic unit of society, from the microcosmic view, social restructuring through the care of people, for everyone will be able to stretch and possible development and support, to activate the rich variety of life, safeguard social fairness, eventually achieve the ideal of a harmonious society. "People, poetic dwelling on the earth" is Heidegger 's ideal, but also our generation of urban planners' ideal.

滨海湖
Binhai Lake

结语
Conclusion

总结与反思：天津的城市设计起步早，取得了许多成绩，但也存在着一些问题。在本轮城市试点工作中，我们进行了全面的总结与反思。曾经的问题暴露出我们依然受现代功能主义城市规划思想方法所控制，未来必须改变居住区规划模式，结合现代住房制度改革，建立真正用城市设计的理论、方法来研究城市的历史文化，用综合统筹的城市设计措施，创造高品质的城市形式、功能和环境。

挑战与机遇：新时代，新发展。当前，国家正在转型升级，对天津城市发展也提出了新的要求。在《京津冀都市圈区域规划》中，要求天津要融入区域协同发展的战略格局，共建京津冀世界级城市群；在新一轮《天津市国土空间总体规划》中，要求天津落实“以北京、天津为中心引领京津冀城市群发展，带动环渤海地区协同发展”的国家要求，打造引领环渤海地区发展的世界名城。新时期的国家战略发展要求使天津面临新的挑战，也使天津面临前所未有的发展机遇。

未来格局与展望：天津的城市规划建设能取得今天的成绩，是天津一代又一代规划人努力的结果，是无数城市设计师一笔一笔画出来的。老一代规划师给我们留下了宝贵的传统，我们要一代一代传承下去，发扬光大。2002 年，天津编制《海河综合开发总体规划》时，定下目标，要把海河打造成世界名河。十几年过去了，海河的水变得清澈洁净，海河两岸变得美丽动人、特色鲜明。天津市新一轮国土空间总体规划提出，到 2035 年要把天津打造成世界名城。我们相信，通过努力，天津一定会实现这一目标，从世界名河到世界名城。让我们相约 2035 年，重回海河边！

Summary and Reflection. Tianjin's urban design started early and made many achievements, but there are also some problems. In this round of urban pilot work, we have made a comprehensive summary and reflection. The past problems reveal that we are still under the control of modern functionalist urban planning ideas and methods. In the future, we must change the residential area planning mode, combine with the reform of modern housing system, establish the theory and method of urban design to study the history and culture, and create high-quality urban form, function and environment by means of comprehensive and integrated urban design measures.

Challenges and Opportunities: New Age and New Development. At present, the country is undergoing transformation and upgrading, and new requirements are put forward for Tianjin's urban development.In the Outline of Collaborative Development of Beijing, Tianjin and Hebei Province, Tianjin is required to integrate into the strategic pattern of regional coordinated development and build a world-class urban agglomeration of Beijing, Tianjin and Hebei Province. In the new round of Tianjin Land and Space Master Plan, Tianjin is required to implement "Beijing and Tianjin as the center leads the development of Beijing-Tianjin-Hebei urban agglomeration and promotes the coordinated development of the Bohai Rim region. It is also the national requirement to build a world-famous city leading the development of the Bohai Rim region. The requirement of national strategic development in the new period makes Tianjin face new challenges and unprecedented development opportunities.

Future Pattern and Outlook. Tianjin's urban planning and construction can achieve today's results, is the result of the efforts of generations of planners in Tianjin, is drawn by countless urban designers. The old generation of planners has left us a precious tradition, which we should inherit and carry forward from generation to generation. In 2002, when compiling the Comprehensive Development Master Plan of Haihe River, Tianjin set a goal to build Haihe River into a world-famous river. Over the past ten years, the water of Haihe River has become clear and clean, and both sides of Haihe River have become beautiful and charming. Tianjin's new round of land and space master plan proposes that Tianjin should be built into a world-famous city by 2035. We believe that through our efforts, Tianjin will surely achieve this goal, from world-famous river to world-famous city. Let's go back to the Haihe River in 2035!

BACK TO THE HAIHE WATERFRONT

THE TALE OF URBAN DESIGN IN TIANJIN

ZHU XUE MEI

TRANSLATED BY HUO YI HAO YANG HONG

Preface

People · Tianjin City

Tianjin is a city born of a river. The 72-kilometer-long Haihe River flows slowly through the city, forming a historic and cultural accumulation of more than 600 years. Every important historical period has left its mark on the development of the city, melting into the city.

In the history of urban development around the world, China and the West are two major veins, and Tianjin is a city where the urban planning and design practices of China and the West meet. In the second year of Yongle of Ming dynasty (1404), Zhu Di ordered to set up a guard city in Tianjin, adopting a typical traditional Chinese square city layout of geometric lines pointing to due north and due south. Since the Haihe River is located in the east of the city, although the city was in the shape of a square, the construction emphasis of the city was inclined to the west bank of the Haihe River. More than 400 years later, the Second Opium War broke out and Tianjin was forced to open as a treaty port. Western powers established the Nine-Nation Concessions respectively on both sides of downstream the Haihe River in the southeast of the city, and the planning and design of the concessions adopted the neoclassical technique popular in the West at that time. Since then, Tianjin has experienced the integration of Chinese and Western urban characteristics for more than one hundred years, and has eventually formed its own urban characteristics and suitable urban planning and urban design methods.

Modern urban design was born in the 1950s in the United States, and gradually entered China after the Reform and Opening Up. The urban design work in Tianjin started early, with attempts going back to as early as in the 1980s after the earthquake reconstruction. In the 1990s, the urban design of the Haihe River and historic blocks was carried out. In 2002, the comprehensive development and renovation of the Haihe River were launched, and a large-scale international call-for-proposal for urban design were held. In 2006, urban design was widely implemented in the Binhai New Area. In 2008, the whole city began to promote urban design. After nearly 20 years of large-scale practice of urban design, Tianjin's urban appearance has undergone great changes, with a large number of successful cases and big national influences. In 2017, Tianjin was listed as one of the second batch of national pilot cities for urban design, its main tasks being to compile successful experiences of establishment and implementation of urban design management, perfect an urban design system, aiming at the current problems in Tianjin and other domestic cities, fully improve the theoretical and practical levels of urban design through demonstrative, high-level urban design pilot projects, guarantee the transformation and upgrading of cities.

This book comprehensively retraces the course of development of urban design in Tianjin. Through the perspectives of real people in the city, it vividly demonstrates Tianjin's successful urban design cases and experience, and especially takes how this urban design pilot project meets people's wish for a better tomorrow as the main content. This book revolves around the public space on both sides of the Haihe River in Tianjin, focusing on humanized urban design practices, which include: Improving the quality of public open space by constructing the pedestrian system and public transport system in the urban center, so as to increase the vitality of Tianjin urban center and revitalize the city's historic districts; Restoring the Tianjin-Tanggu urban ecosystem and achieving the harmonious co-habitation of human and nature through greatly increasing the amount of urban parks, country parks, rural suburbs and other green living spaces; designing the wonderful life scenes that Chinese people yearn for by exploring ideal housing types and new types of communities in 2035 and 2050 in Tianjin; Pointing to the future development direction of the city through the questioning of the ideal city, drawing a beautiful vision for the realization of the Chinese dream of the great rejuvenation of the Chinese nation. This book is not only a collection of practical cases of urban design in Tianjin, but also a demonstration of the key content of urban design theory in China in the new era.

The characters in the book are mostly fictional, but the stories generally are real. Old Zhang and his family represent the general public of Tianjin, Bauer represents the many overseas planners and designers who participate in Tianjin's urban planning and construction, Song Yunfei represents the young people who start up businesses in Tianjin and the Binhai New Area, and Chief Chu and ZhuGe Min are the collective portraits of the several generations of urban designers in Tianjin.

Contents

Chapter I

Two Tianjins:
The Old Town vs. The Concessions

1. People Living in Tianjin

Old Zhang

The sun rises at six in the morning. Old Zhang goes out to the Shizilin Bridge by the Haihe River as usual. He was born in the Old Town; In 2002 the Old Town was completely demolished, and he was relocated to Huayuan community and moved into an apartment unit on the fifth floor in a six storey walk-up. The flat was 60 square meters large, much more spacious and better equipped than his 16-square-meter bungalow in Old Town. The apartment came with kitchen and bath facilities, and the supporting services in the neighborhood was relatively convenient, but without the old neighbors and the familiar environment, Old Zhang often felt a sense of loss. Old Zhang's eldest son inherited the family craft and rented a shop selling clay figurines and kites at the Tianjin Ancient Cultural Street. After the 2008 Beijing Olympic Games, the Haihe River cruise started running, bringing a stream of visitors, and the shop's business was booming. With holidays and festivals coming and going the street was always crowded.The family was too busy with the business to lift their heads. Over a few years they saved some money, and in order to take care of the shop from closeby, the eldest son's family bought a new apartment in the Old Town. The son was also thinking of Old Zhang, and arranged for all the brothers and sisters to chip in to buy a new apartment for Old Zhang in the Old Town too, to satisfy Old Zhang's wish. The apartment was 90 square meters, with two bedrooms, on the 18th floor of a highrise. The apartment was bought in 2012, and the price was not cheap back then either—more than 9,000 yuan per square meter. In order to buy the new apartment, Old Zhang sold his old apartment at Huayuan, and his children paid for the price difference of 200,000 yuan together. In these years, the housing price in Tianjin had been rising fast, and now Old Zhang's house is worth almost 50,000 yuan per square meter.

Old Zhang is very pleased—not only is the apartment worth more, but more importantly, he can live in the Old Town again. Although the old neighbors are gone, the environment has changed, but it is much cleaner, the road is wider, the quality of buildings has also improved, and he can walk to the Haihe River side every morning. The Haihe River has changed so much! But Old Zhang still feels like this is the Haihe River he knows and loves; the Ancient Cultural Street, the Jade Emperor Pavilion, the Shizilin bridge, and the Wanghailou Church on the other side... These places witnessed the first half of his life. He went to the Haihe River side every morning, not only to exercise, but also as if there is an invisible force drawing him towards it; it is where he was born, where he and his childhood friends always started up trouble—it is the "Old Town" that carries his thoughts and memories. Every scene in the memory, along with feelings from his childhood, seems to be a part of his body, awakening every time he scents the smell of the Haihe River's water. He always takes a deep breath when he comes to the Haihe River side.

Mr. Bauer

Every time Mr. Bauer visits Tianjin, he stays at Lishunde Hotel. Although the newly opened Ritz-Carlton Hotel next door has more advanced facilities, he still likes Lishunde more—it was the first luxury hotel of Tianjin, built in 1863, and was once where various Presidents of the Republic of China (Sun Yat-Sen and Yuan Shikai) and celebrities (Mei Lanfang, Zhou Enlai) had stayed; even Zhang Xueliang and Miss Zhao Si had danced here. It was one of China's trendiest hotels and a popular destination for celebrities and the wealthy when it was first built, and its charm is still present to this day. By the entrance is the eldest push-and-pull gate-style elevator still in use in China; the exquisite guest rooms in the old building, the antique wooden floors, the first-class Indian waiters—each space exudes the fragrance of roses from the distant England, carrying a gentleman-like elegance and the comfort of a countryside at the same time.

Mr. And Mrs. Bauer arrive at the hotel in the evening by taxi from Beijing Daxing International Airport after a two-hour drive. They had been discussing the airport terminal building designed by Zaha Hadid, which Shirley described as "huge, bold and domineering" during the journey. After checking in the hotel and resting for a little, they are ready to go to the Minyuan Square located in Wudadao, as their local designer friends have offered to hold a reception event for them. Because of the long flight and the jet lag, the Bauers were tired, but were unable to refuse the enthusiastic offer; moreover, they have been thinking about Wudadao over this time, looking forward to seeing what has happened to it in recent years. The Bauer family has deep historical ties to Tianjin, where Bauer's father's old friend, the Austrian architect Geyling, was active in the 1930s, designing and building hundreds of small Western-style villas. Bauer's father had worked with Geyling in Tianjin for more than a decade. Bauer has been influenced by his father since he was a child, and has seen many sketches of architectural design projects in Tianjin drawn by his father. In recent years, the American architectural firm that he is working for has been taking up several projects in Tianjin, and he is excited to be able to return to the city where his father worked as a young man.

Old Zhang

Old Zhang takes an hour-long walk every morning. His usual route is walking toward the north, which he was used to since he was young. In the past, the south side of the Old Town is the area of "nobody's jurisdiction", and adults was reluctant to let children go there in fear of accidents; Further south was the concession, which had been reclaimed by then, but the people of the Old Town were generally still afraid to enter. From home, heading east along the Beicheng Street, past the archway that scribes "Gushang Yiyuan (the cultural birthplace of Tianjin)" at the north entrance of Ancient Cultural Street, to the Shizilin bridge, where more than 70 small bronze lions of different postures sit on the bridge handrails. Going under the bridge to walk on the Haihe waterfront platform adjacent to the East Haihe Road, there are many senior citizens exercising in the morning every day, as well as people sitting there all day with a small campstool, fishing. In the front is the relief sculpture "Illustration of the Supervising of Transit in Luhe River", recording the prosperous history of Tianjin's shipping ports. Beyond that are the Jingang Bridge and the North Zhongshan Road in Hebei District, which was originally the Hebei New Area built by Yuan Shikai and under Chinese rules. The Dabei Buddhist Temple is the largest buddhist temple in Tianjin, attracting many believers and has been very prosperous; The Lunar December 8th of every year is the day for worshipping ancestors and divinities such as the god of gate, the god of door, the god of house, the god of kitchen, the god of well and so on, and people rush to line up to eat the Laba Porridge crowded head. It is said that there are people offering to pay big money in exchange for being able to burn the first stick of incense every year.

Next to the Dabei Buddhist Temple is the ferris wheel known as the Eye of Tianjin, which straddles the Haihe River. According to Old Zhang's youngest son, it was designed by the famous Japanese structural designer Mamoru Kawaguchi. The Yongle Bridge is integrated with the ferris wheel. The ferris wheel is 140 meters high, and the shaft in the middle alone is 6 meters thick, made in France. Crossing the Yongle Bridge and entering Hongqiao District's boundary, this area is the "Three Stones" area made up by the Hebei Main Street and the south and the north canal, originally famous for its iron industry workshops, and is one of the birthplaces of Tianjin's national industry. The text "History of Blood and Tears at Three Stones" that Old Zhang learned in elementary school as a kid took place in this area, and the school had led them to visit the Three Stones Museum at the time. In order to commemorate this part of history, when the Haihe River was being redeveloped, a large Modern Industrial Museum of Tianjin was built by the Haihe River. The design of the museum was bold and rough, taking shape from the meaning of "three stones". For some reasons, the place is now Tianjin Jurior's Palace. Although it seems somewhat a mismatch, the building is better served being utilized rather than being abandoned. Perhaps it was in the hope that the children here would not forget about the history of blood and tears, and become entrepreneurs with conscience.

Old Zhang goes on until he reached the bend of the Ziya River and the south canal, and the statue of the water-seeking mother stands before him. This statue was built to commemorate the project of bringing clear water of Luanhe River into Tianjin. Originally, the people of Tianjin could only drink bitter and salty underground water, until they had access to the sweet water from Luanhe River—after all, one must not forget about the well-diggers when drinking from it. By this time, Old Zhang's morning routine is almost over and he heads back down the south canal. It is said that this is the place of "Tianzi Jindu (The Son of Heaven crossing river)", and the origin of the name of "Tianjin", that is, Zhu Di, emperor of the Ming dynasty, travelled down south from here to conquer Nanjing in the Jingnan Campaign. Next across the south canal is the Big Hutong small items wholesale market. In the past, vendors from Hebei and Shanxi Province all came here to restock, people coming and going, goods-carrying vehicles everywhere; back in the days, some of the small souvenirs in Old Zhang's shop were also purchased here. But now the market is empty. People are saying that the market is shut down because it was not up to fire safety standards. It is not clear what the future holds for it.

Today, Old Zhang comes home a little early. He plans to go visit the Planning Exhibition Hall as he wants to choose an apartment for his eldest grandson. Old Zhang has three children; besides the eldest son who runs the shop, there are also a daughter who has settled abroad and the youngest son, Zhang Hui, who works as a driver at the Planning and Design Institute. He has two grandsons and a granddaughter. Recently, the eldest grandson has made his mind to come back from the United States to work and get married. Old Zhang has little to do anyway, so to choose a wedding apartment for his grandson has become his top priority . His younger son said that the Tianjin Urban Design Exhibition will be held at the Tianjin Planning Exhibition Hall today, and there will be many new ideas, new technologies, and displays of newly planned communities and subwaylines. Old Zhang wants to go to see the exhibition as soon as possible, so that his grandson can choose a good location when he buys the apartment. Meanwhile, as an old Tianjiner, he loves to know about the city's affairs, and he is proud of the changes along the Haihe River in last decade. He has always been paying a lot of attention to the construction of Tianjin, and also wants to see how Tianjin will be planned out in the future.

2. Ming Chengzu Set up the Guard City

Zhang Hui

Zhang Hui's Chinese Zodiac Sign is horse, and he is 41 years old this year (2019), born and raised in Tianjin; his ancestors had lived in Tianjin since the end of the Ming Dynasty for generations, so he can truly be called an authentic Tianjiner. Although Zhang Hui not so old yet, he likes to see about old and traditional things. He would listen to radio programs like "Talking about Tianjin Wei" while driving and would also spend time to discover and collect stories about old Tianjin while driving through the country. These days, Zhang Hui has been doing a lot of extra work as a driver in preparation for the upcoming International Seminar and Exhibition on Urban Design. Zhang Hui is willing to do the extra work as he sees his own value in it, and his boss would also see his hard work. Early this morning, Zhang Hui drives the exhibition boards to the Planning Exhibition Hall, and sees the exhibition staff are still busy making final adjustments. With half an hour to spare before meeting the experts, he takes time to see the exhibition, which begins with the city's history of changes.

Tianjin, located at the lower reaches of the Haihe River Basin in the north China plain, is the place where the five branches of the Haihe Rivers meet. So there always is saying that Tianjin is at the "lower ends of nine rivers". It faces the Bohai sea in the east and Yanshan in the north, borders the Hebei Plain in the southwest, and is 137 kilometers away from Beijing in the northwest. In prehistoric times, the location was a shallow sea. 6000 years or so from now, seawater began to retreat to the direction of Bohai sea, leaving a quite thick layer of marine deposit on the ground. The old Yellow River, one of Haihe Rivers, has changed its course many times. 3000 years ago, it entered the sea near the Ninghe District in Tianjin; in the Western Han Dynasty, it entered the sea near the Huanghua County in Hebei province; in the Northern Song Dynasty, it entered the sea in the southern suburbs of Tianjin. During the Jin Dynasty, the Yellow River moved southward and captured the Huaihe River to reach the sea, and the coastal line of Tianjin was finally stable by then. The alluvial of the old Yellow River had made the surface with existing marine deposit form a large alluvial surface, gradually silted into a flat land, simultaneously pushing the coastline eastward. Tianjin was created when the Yellow River shifted its course several times, filling the sea with tens of millions of tons of silt. As the sea receded, it in turn left tens of millions of tons of salt in the soil. Salt-cured and alkali-burned, Tianjiners are warm hearted inside, and have always valued justice and loyalty. The prosperity of Tianjin was for the Haihe River: The construction of the Grand Canal between north and south in the Sui Dynasty made Tianjin an important transit center on the canal. Tianjin was increasingly prosperous and its role as a transportation hub and military base was getting increasingly prominent. In 1404, Zhu Di (Ming Chengzu) set up a guard city at the Sancha Estuary.

Looking at these old pictures and texts, Zhang Hui remembered the old story of his family. According to the elders in the family, the Zhangs came to Tianjin in late Ming Dynasty. At that time, people who had been to Tianjin and back would talk about how bustling and lively Tianjin Wei was, and the hearts of those who had not been there were always full of longing. The Zhang family originally came from Anhui Province , where they relied on nature to farm the land.One year of pest after one year of flooding after one year of drought, the family could not feed itself. One of the braver ancestors came straight to Tianjin Wei with his wife and children. He did not have any special skills, but had a lot of strength. Looking at so many street performers making money, he joined them too. Later it was said that he became rather famous in Tianjin Wei, and people gave him a nickname of Zhang Dali ("Big Strength") because of his solid build and incredible strength.

Old Zhang

Old Zhang lives in the Longting Garden in the Old Town. The Haihe River is on the east side of the Old Town and the Sancha Estuary where the north and south canal intersect naturally formed a trade center due to the transportation and trade. There were used to be many wharfs, religious establishments such as the Tianhou Palace and the Jade Emperor Pavilion, the Gongnan and Gongbei shop street. Therefore, although the Old Town is in a typical traditional Chinese square city with symmetric layout, the northeast part of city that is adjacent to the Sancha Estuary turned out to be more prosperous, creating the saying of "north gate noble, east gate rich, south gate poor, west gate cheap." Gulou (the Drum Tower) is situated in the middle of the Old Town, but for old Tianjiners, the square in front of the Tianhou(Sea Goddess) Palace is more meaningful, where two poles stands high with lamps and flags up guiding ships on the Haihe River. It is just like a big stage, where various novel characters and events are staged in turn. In 1986, the Gongnan and Gongbei Streets were renovated in the original old style and renamed as the Ancient Cultural Street, which mainly sold folk culture products of Tianjin. Old Zhang, who had just been laid off, rented a shop to make a living. But at the time, the entire neighborhood was as old as the Old Town itself, with 600 years of history.In modern times, China was long in a state of chaos, the governments had no spare to take care of houses. . The houses were in disrepair for so long time that became dilapidated. The streets were very narrow. Many alleys could not fit even two people walking side by side, and ambulances and fire trucks had no way to enter. Since 1994, Tianjin began to carry out the renovation of dilapidated bungalows, but since the Old Town, the Nanshi in the south and the Three Stones area in the north were too crowded and densely populated, reconstruction was not financially feasible, so the redevelopment did not start in these areas. By the end of the 20th century, it was said that there was not even one section chief, the lowest level carde, living in the Old Town. Everyone with power, money, or capabilities had moved out. The business of Old Zhang's shop had been slow; although he had a little savings, but with no access, he had been trapped in the Old Town.

In 2002, Tianjin started the comprehensive redevelopment of both sides of the Haihe River. The government made up its mind to raise funds and carry out the urban renewal of the Old Town, Nanshi and Three Stones areas to improve the living conditions of residents, which was also a response to the vocal wishes of local residents. The Zhang's used to have two rooms in a courtyard house, where their family of five lived. When the eldest son got married, they put up a shed in the courtyard. Whenever it rained heavily outdoors, it would rain lightly indoors; as soon as the rain stopped, they had to scoop water out of the house using an enamel basin. Those were the difficult days. Old Zhang's family got compensated with two apartment units in Huayuan. The day they received the relocation notice, the family sat together and chatted until midnight. Old zhang was so excited that he did not shut his eyes that night. He waited every day for the day to come when he could move out of the pit of the Old Town. Although it was a little bit sad to part with his old home and neighbors , it was hard to place sentimental values above practical needs. He was grateful to the Party and the government from the bottom of his heart.

Later his youngest son, Zhang Hui, told Old Zhang a sotry. Early in 1994, Tianjin Municipal Government designated 14 areas as historic and cultural districts, including the Old Town, the Ancient Cultural Streetand Wudadao, etc.. At the time the relocation decision was made, a vice mayor said: "since we have spent a lot of money to move the people out, can we keep the houses and not to tear them down for now, in order to renovate them later. Just put walls around?" This proposal was considered quite impractical then. In the city center, if a total area of more than 2 square kilometers, including the Old Town which was 1.5 square kilometers plus the Nanshi area, becomes no man's land, not only will there be huge safety risks, but also a large amount of capital investment will not be recovered. "From the point of view of today, it looks like you can renovate the Old Town slowly," Zhang Hui said, "experts call it 'organic renewal'." However, at that time, the housing situation in the Old Town, the social environment, the economic pressure and people's understanding levels were all far from what it is like now, and slow renovation meant giving up renovation.

Therefore, after residents of the Old Town moved out, the demolition of buildings was the inevitable choice. In 1994, a developer paid a lot of money to entrust the famous P&T Group in Hong Kong to make a redevelopment plan for the Old Town. A lot of investigation and analysis about the Old Town were done before the plan was made. As prerequirment, it makes clear that the layout of the Old Town as well as the historic buildings, such as Gulou(the Drum Tower), the Cantones Pavillion, the Cangmenkou Church, the Bian Family House and the Confucius Temple must be preserved in the plan. Taking the financial pressure into consideration, the final plan was to keep the cross-street layout of the Old Town and, putting 2- to 3-storey low-rise buildings around Gulou and along the cross-streets, surrounding them with high-rise residential buildings nearly 100 meters, and laying public buildings of around 150 meters high at the four corners, which were a metaphor for the turrets in the past. In practice, a large aspect of the protection of historic districts is to control the height of buildings; what was done for the Old Town was a compromise after weighing the financial pressure and historical protection as well as many other factors. It can be said to be a reduced-size Old Town.

But in Old Zhang's mind, this reduced-size Old Town is not the same as the true Old Town. In the past, it was the place where ordinary people like him lived. Although the living situation was bad, the alleyways and the courtyards were always filled with vivid air. At present, there are only fenced-off high-end residential estates. The alleys are gone, and walking on the spacious street, he seldom meets anyone he knows. Gulou Square was still at the center of the Old Town, with Gulou rebuilt and the the Cantones Pavillion retained, where Peking Opera performances are often held. It only attracts tourists with the expensive ticket price. Old Zhang has always wanted to go in and see if the place is still like the way he remembers, but do not actually enter in ever since, feeling like he is now somehow a stranger in this Old Town. The cross-streets have been transformed onto four pedestrian shopping streets on the four sides of Gulou respectively, with business establishments such as movie theaters, restaurants, spatea houses and so on, but for some reason it has never gotten as popular as the Ancient Cultural Street. The shops along the streets remain unvisited. Old Zhang doesn't know why—the place has gotten better, but there is less people visiting. What would it have been like if the Old Town was not demolished? What if this is how the Old Town is destined to be?

3. The Planning and Construction of the Nine-Nation Concessions

Mr. Bauer

The car moved smoothly and gracefully, the buildings and streets backing away—a familiar scene for the Bauers. More than 100 years have passed. In the 18th century, after the Industrial Revolution and the bourgeois revolutions, the West began to be powerful with its newly developed science and technology. China, once the center of the world, was declining by the end of the Qing Dynasty. In 1860, The Second Opium War broke out, and the Qing Dynasty was forced to sign the Beijing Treaty with Britain and France and ratified the Tianjin Treaty. Tianjin was forced to open up as a treaty port. Britain was the first to establish a concession in 1860, on the west bank of the Haihe River. Then France set up a concession in the north of the British concession, in the Zizhulin area. During this period, although the construction in the concessions were relatively slow, the scope of the concessions were constantly expanding. The German concession was not established until 1895. In 1900, in the name of suppressing the Boxer Rebellion, the Eight-Nation Alliance conquered Dagukou, aiming to capture the capital. After the Alliance captured Tianjin, the Old Town was a chaotic place of burnt down houses and looted shops. The Alliance formed a provisional government, also known as Dutong Yamen, and began to administer the city. The city wall, which was more than 500 years old, was torn down for military purposes, and the bricks from the wall were used to pave roads. The four ring roads around the city—East Road, West Road, South Road, and North Road—was built during this time. Later, a tram line was built on the ring road, which was the first tram in Tianjin's history. With the continued occupation of the Eight-Nation Alliance, the original concessions were further expanded, and Russia and other countries that had no concessionary territories before also forcibly established their own. Eventually, a total of nine countries (Britain, France, Germany, the United States, Japan, Italy, Russia, Austria, and Belgium) set up their own territories in Tianjin respectively; each had its own governmental offices, churches, hospitals, post offices and military camps. These buildings were all in the original style of the countries of their respective concessionary territories. Many foreign architects undertook design projects in the concessions. Bauer's father followed the Austrian architect Geyling to carry out design activities in the concessions. Because Geyling was very famous, he not only practiced in the Austrian concession, but also designed buildings in various concessions including the British and the French. After 1945, all the concessions were taken back, but the marks of the concessions remained. Because of historical conditions, Bauer never had the opportunity to visit the city where his father had worked as a young man, to see the houses he had designed. It wasn't until 1978 with the Reform and Opening Up that Bauer had had the opportunity to travel to Tianjin for the first time in the early 1980s. Time flies, and now 30 years have passed, the old houses in these concessions have new owners once again, and new stories are taking place in these very houses.

Bauer asked the driver to go to Minyuan Square, which his friends said was a new place for them to experience Tianjin. At the intersection of Chongqing Road and Hebei Road, a brand new building appeared in front of them—Minyuan Square. Bauer had come here six years ago, and the place was an old stadium back then. According to his friend, Tianjin Minyuan Stadium was built in 1925, when the British Municipal Council invited Eric Liddell, a Scotsman who won the 400m gold medal at the 1924 Paris Olympic Games, to design and renovate the Minyuan Stadium. Minyuan Stadium, which was finished a year later, was the most advanced comprehensive stadium in Asia at the time. After the founding of the People's Republic of China, the first national soccer match was held here, and it was the home of the Chinese White Team, which later became the "King's Team". After the Chinese Super League (CSL) was founded, the stadium was the home of Tianjin TEDA Football Club for eight years; it was lively every time there was a match on the weekends. In 2004, TEDA constructed a professional soccer stadium in Binhai and hosted the first opening ceremony of the CSL. In 2008, to host soccer matches for the Beijing Olympics, the 60,000-seat "water drop" complex (Tianjin Olympic Center Stadium) was built in what is now the Olympic City. Due to the small number of seats in Minyuan Stadium and the bad traffic and sanitary conditions from the large number of people gathering during the match, it gradually stopped hosting soccer matches. One cannot help but feel sad to see a once bustling huge building fall silent as it is now.

In 2012, in order to improve the vitality of Wudadao area and improve supporting facilities for tourism, the government of Heping District proposed to reconstruct the Minyuan Stadium.There was once before a Hong Kong developer who took a fancy to the land of Minyuan Stadium, proposed to demolish the stadium and build high-rise residential housings, and offered to build a modernized stadium for Tianjin for free. Despite of its urgent desire to attract more foreign investment, the government refused the developer's proposals in order to preserve Wudadao's overall scale and historical context. The reconstruction of Minyuan Stadium was not a pure commercial development project, but a single function sports facility transforming into a multifunctional venue opening to the public for free. Meanwhile, the supporting facilities of Wudadao had been improved, and an underground parking lot with 450 parking spots was built, allowing it to become an focal point for tourists gathering in the Wudadao.

Bauer sees that the newly built Minyuan Square is a building imitating British classical style, mostly retaining the layout and outline of the original Minyuan stadium, with a square mainly composed of turfs in the center. In the first floor of the building, there are visitor center, Wudadao Museum, Sports Museum, all kinds of restaurants, duty-free shops and other services.Today is a holiday, and there are musicians playing in the open air and the square is full of people. The Bauers and their friends emerge from the restaurant SMOKI&CO as the night falls and they can't help but join the stream of people walking counterclockwise on the running track. Bauer walks slowly and he can feel the breaths and heat of the people around him, hear them talking and whispering, scent the smell of baking bread, see the lights flickering on people's faces... It is a wonderful feeling, Bauer contemplates, and it is amazing how these moving people form a scene that resembles a religious pilgrimage. Looking at this magnificent building, Bauer recalls the words of famous architect Rossi in his book *The Architecture of the City*: The functions of many buildings change over time, but the meaning of the building itself is unparalleled. Obviously, Minyuan Square is a successful urban building.

Chief. Chu

Chief. Chu is the chief planner of Tianjin Urban Planning and Design Institute(TUPDI). She has cooperated with Mr. And Mrs. Bauer many times before, and they are already old friends. The Bauers have been to Tianjin many times, and they have eaten the Goubuli steamed buns too many times. They are able to have authentic Western-style food in Wudadao this time, which also reflects the fruits of Tianjin's efforts on internationalization. After dinner, Chief. Chu and her colleague, Luo Yun, the chief planner of the Urban Design Lab of TUPDI, accompany Bauer and his wife to their new office, "Wudadao College and Urban Design Lab of TUPDI" in the Xiao Guangming Li at Xiannong Block. The renovated small villas are suitable for small and medium-sized design firms to work in, and the interior is bright and elegant. The exhibition hall on the first floor showcases several projects of urban design: Tianjin Wudadao, North Jiefang Road, community centers and pedestrian system. It reveals that this is a planning and design firm dedicated to research-based design. In addition, the reason why they are called "Wudadao College" is because their services also include urban design vocational education, including both theoretical learning and hands-on practice, which is very helpful to the cultivation of urban designers. Chief. Chu starts from this small villa and talks about the bitter-sweetness of revitalizing the beautiful-looking historic buildings in Tianjin. Baur sees the spark in Chief. Chu's eyes. He can tell that these people love Tianjin, love Wudadao, love design, love life, and they try their best to experience and be part of this city. All this makes Bauer feel moved.

Unlike the Minyuan Xili and Qingwangfu, which are attractions for young hipsters and mainly composed of high-end restaurants and hotels, Xiannong Block, in addition to drawing food and drinks establishments such as Starbucks and restaurants like Yujian Yuntai and Najia during initial business invitation, attracts the headquarters of various companies to settle, and has a more diverse, risk-resistant, and long-lasting and vibrant business model overall. According to Chief. Chu, Tianjin is rolling out a series of policies to excavate and utilize historic buildings to develop the "Small Villa Economy", that is, to attract headquarters settling down in these villas, fully utilizing the diverse potential of historic and cultural assets of the city, truly rejuvenating historic districts.

Mr. Bauer

Back at the hotel, the Bauers are wide awake, and they take a stroll outside the hotel. Around this area is Tianjin Tai'an Road Historic and Cultural Block, once the core area of the British concession. After the founding of People's Republic of China, it became where offices of Tianjin Municipal Committee, Tianjin Municipal Government, Municipal People's Congress, CPPCC and other committees and bureaus were concentrated. Gordon Hall, built in 1890, was the Municipal Council of the British concession. It was the largest building in Tianjin in the 19th century and became the location for Tianjin Municipal Government after the founding of New China. The building was badly damaged in the 1976 earthquake and was later demolished to replace it with a new municipal government building. In 2010, after Tianjin Municipal Committee and Municipal Government moved to the new site on Youyi Road, the comprehensive development and construction of the Wudayuan(Five Big Courtyards) on Tai'an Road were implemented. The municipal government building was demolished and a Ritz-Carlton Hotel was built on the original site. Bauer has heard that in addition to the hotel, this group of new buildings also includes the complex functions of shopping, office, apartment and so on. It takes the form of an enclosed layout, forming a complete city block with square, but also matches the tradition and retains the old urban texture. Indeed, the layout is distinctive, and the Bauers can feel the nobility of the red-brick buildings flanked by trees. However, it has been seven years since the construction and opening of the Wudayuan building complex in 2012, but it seems to be somewhat unpopular. They feel a little lonely as they walk on the street.

4. Mother River of Tianjin

Every world-famous city has a famous river, such as Paris and the Seine, London and the Thames, Frankfurt and the Main, Melbourne and the Yarra, to name a few. The Haihe is the mother river of Tianjin. Tianjin is a city born of a river. 72 kilometers of the Haihe River flows through the city slowly, forming a historic and cultural accumulation of more than 600 years. Every important historical period has left its mark on the development of the city, melting into the city.

Mr. Bauer

Mr. Bauer wakes up at 7 am. Although he has only slept for less than six hours, Mr. Bauer feels recharged. He opens the curtain and comes out to the balcony. He sees the Haihe River bathing in the rising sun. On the other side of the river is the Tianjin Old South Cargo Railway Station Business District. Bauer recalls his first visit to Tianjin in the late 1980s, when he lived in this same room. The hotel clerk said the room had been used by Hoover, the 31st President of the United States. At the time, he saw the Tianjin South Railway Station across the Haihe River busy and bustling, with trains entering and leaving constantly. The site was a Russian concession a hundred years ago, but by the time Bauer saw it through the window of his hotel room in Lishunde, the only historic building that had left was the Russian consulate, then the Tianjin Cigarette Factory.

In 2002, Tianjin initiated the International Call for Urban Design Proposal of Six Nodes for the Comprehensive Development on Both Sides of the Haihe River. That was when Bauer came to Tianjin for the second time, staying in the same hotel room at Lishunde. Although he participated in the CFP for the Central Square node, he noticed several proposals for the Tianjin Old South Cargo Railway Station business district as well, one of which caught his eye from a French company that planned for a group of delicate glass-walled buildings to surround a green riverfront. Later, Tianjin put limitations for buildings with glass curtain walls In order to save energy, ensure safety and avoid light pollution. The plan was not implemented. The buildings in the Tianjin Old South Cargo Railway Station area are now built in a simple European style, which has nothing to do with Russia. Bauer thinks about how interesting it would have been to build a few "onion head" domes representative of Russian architecture—the old Russian concession would have been immediately recognizable. At present, the buildings are built in strict accordance with the requirements of Tianjin planning and design guidelines, in a regular and simple manner. It was a pity that the factory building of Tianjin Cigarette Factory was not preserved. The quality of the factory building was very good, and the corbels in the structure were distinctive. The Riverview Place was built on this site by Kerry Properties Ltd. in Hong Kong, and includes the Shangri-La Hotel, commercial complex, movie theater, swimming pool and high-rise apartments and offices. The project was originally planned to locate in the Haihe Central Square. Later, considering the memorial significance and importance of the Central Square, and fearing that it would create pressure for the Italian Style District, the Tianjin Municipal Government did a lot of work to relocate the project here.

In the past, the urban environments on the two sides of the Haihe River were very different. Although only separated by a river that was 100 meters wide, the west side of the Haihe River was fashionable and prosperous, while the east side of the Haihe River was poor and dilapidated. North Jiefang Road is a traditional financial district, historically known as the "Wall Street in the north", located in the downtown Heping District. On the other side of the east bank of the Haihe River, there were a lot of industrial plants and storage land , and the land and housing price was much lower than the west bank. At that time, the planning for the Haihe River proposed to intensify the contact between the two sides of the Haihe River, with the purpose of balancing the development of the two sides. Over the years, many new bridges have been built and rebuilt on the Haihe River, linking the two sides together, and these new and high-quality bridges have formed a beautiful scenery on the river. Bauer spends a few days in Tianjin and is impressed by the design of the bridges on the Haihe River. For example, there are four bridges in his line of sight from where he is now standing. The Chifeng Bridge to the north is a curvilinear cable-stayed bridge. This kind of structure and shape is generally used for broad rivers, but luckily Chifeng Bridge is located at one of the turning points of the Haihe River and the use of the horizontal curve lessened the sense of interruption for the environment. Next is the Baoding Bridge that does not work so well. it is also a cable-stayed bridge, , too big and stiff in the city center, like a highway bridge. From Bauer's point of view it is a failed design. Looking further south is ZhuGe Min Guangming Bridge; like the Bei'an Bridge in the upper stream, it is reconstructed by lifting an old bridge, and it is similar to the Pont Alexandre III in Paris in shape. In the distance, he can see the Jinfu Bridge designed by the famous French bridge designer and architect Mark Mimram. It is a excellent work, with its smooth modern design and lightness in shape. Last night, having just arrived at the entrance of the hotel, Bauer noticed that a new bridge was being built over the Haihe River. The hotel staff told him that it would be a pedestrian bridge which would make it easier for the North Jiefang Road and Tianjin Old South Cargo Railway Station Business District to communicate. It struck a chord with Bauer, who felt that China was growing fast not only economically but also intellectually and culturally.

Old Zhang

Old Zhang has had breakfast and changed his clothes, and goes out to visit the Planning Exhibition Hall. He is no stranger to the Planning Exhibition Hall as he has been there several times. The Planning Exhibition Hall is only two kilometers away from his home, a half-an-hour walk. He still remembers that when the museum was just opened in 2008, there was a large crowd and a continuous stream of people. Old Zhang had curiously followed the crowd into the Planning Exhibition Hall, which was the first time that he had a comprehensive and intuitive understanding of Tianjin, and connected his own subjective impression of Tianjin with models and orientations.

Old Zhang likes to walk through the Ancient Cultural Street. The street is bustling with people, all jostling with each other. Old Zhang has an ineffable sense of pleasure and satisfaction; with all these people, businesses would be booming. It is yet to be 9 o 'clock, the shop is still getting ready, and many tour groups have already arrived. The eldest son's shop is on the Gongnan Street, not far from the Tianhou Palace, with a good location which has led to the business doing well. In the past, there was only one shopping street, Gongnan and Gongbei Street, in the whole Ancient Cultural Street area. When it was redeveloped again in 2002, the neighborhood was totally demolished except for several historic buildings, and the whole area was turned into a shopping district. Many small streets and courtyards were extended from the original pedestrian street within, and modern buildings with traditional Chinese feathers were built alongside the Haihe River.In terms of atmosphere of commerce, the old street is still winning. By the door of the shop, the clerk is arranging the product displays, sees Old Zhang and cordially greets him. Since the shop is an old shop, the eldest son and his wife only show up to the shop in the holidays when there are more customers. Usually the couple seems to be very busy, and Old Zhang wonders about what keeps them so busy. Through the "Jinmen Guli (the Old Place in Tianjin)" archway at the south exit of Gongnan Street is the Shuige ("water pavilion") Street leading to the Haihe River. Old residents in this area all know that this is originally a water-selling place, next to the Shuige Hospital where the eldest son was born.

From The Shuige Street to the Haihe River onto the Jintang Bridge—this bridge can open by rotating horizontally, and is the place where Liberation Army joined forces to liberate Tianjin in the Beiping-Tianjin Campaign. Walking on the bridge, Old Zhang can see the place that has many beautiful European-style buildings lined up on other side of the Haihe River, which is the former Austrian concession. Inside, there are three old buildings that are respectively the Austro-Hungarian Consulate, Feng Guozhang's former residence, and Yuan Shikai's former residence. Before the comprehensive development and reconstruction of the Haihe River, East Haihe Road was a small road and the buildings by the roadside were in a bad shape. Old Zhang wondered at that time: Yuan Shikai planned and constructed the Hebei New District , and also built the Tianjin North Station in order to not have to pass the Lao Longtou Railway Station, not have to step into the concessions, so why did he build this house in the Austro-Hungarian concession?

Old Zhang has heard that, when the Haihe River was redeveloped, in order to preserve these houses in the Austrian concession, the East Haihe Road was redirected behind these buildings, and stone-paved pedestrian walkpaths were laid down by the Haihe River. A lot of tall and elegant Chinese parasol trees were planted on the river bank. The new buildings also invite long-last appreciation; the roofs were originally dark gray and were later painted into red, and Old Zhang feels that it looks better than before. The ambiance is good, but toward the south half of the street there used to be three bank branches with big lion statues sitting by their entrances, giving people an unsettling, chilly feeling, let alone that the statues do not blend in with the environment well. In Old Zhang's opinion, the buildings facing the Haihe River should open some coffees, small shops on the street level, set up some umbrellas and put out outdoor seats. Business will be good. Old Zhang's idea finally came true last year as the Municipal Government began to promote night economy. This section of enbankment become another one of the must-see places along Haihe River.

Further ahead, located at the entrance of Bei'an Bridge and adjacent to the Austrian concession is a small triangle-shaped land converted into a Music Park. Old Zhang gets it—Austria produces a lot of musicians. The park is not big, with statues of musicians such as Beethoven, Johann Strauss and others, not very eye-catching. The Bei'an Bridge on the Haihe River, on the other hand, is resplendent and invites attention. Built in 1939, the Bei'an Bridge was originally a wooden bridge, called "The New Bridge" or "The Japanese Bridge", because it led to the Japanese concession. It was rebuilt after the victory of the Anti-Japanese War in 1945 and renamed "Victory Bridge". In 1973, it was rebuilt into a cement structure bridge, called "Bei'an Bridge". Bei'an Bridge is a new "plate girder bridge" built after the liberation. There are two rows of pillars in the river, and the concrete slabs above are used as the bridge deck to allow traffic to pass through. To meet the new navigation standard in the Haihe River redevelopment plan, the bridge structure had to be lifted. At that time, the media covered the bridge's reconstruction, and Old Zhang also went to the scene to watch. Digitally controlled, dozens of hydraulic jacks lifted the bridge body synchronously, and then reinforced and widened after it was in place. The final result was beyond Old Zhang's expectation: the marble bridge body and the gold-infused statues shined brightly under the reflection of the blue sky, white clouds and the river. Some people say that the Bei'an Bridge is an imitation of the Pont Alexander III in Paris, France, but Old Zhang disagrees, because the sculptures on the bridge columns are the traditional Chinese four symbols—Azure Dragon, White Tiger, Vermilion Bird, and Black Turtle, meaning to wish the four directions (north, south, east and west) safety and peace. Statues on the pillars of the bridge are adornment coiled dragons, and the four sculptures of dancing deities on the bridge are elegant and life-like. How can these be copied from Paris? Even if so, it should at least be called a combination of Chinese and Western elements.

Just across the road is the Planning Exhibition Hall and the Italian Style District; it says on the Internet that it was Italy's largest overseas colony. Old Zhang still has some impression of this area. When he was young, he used to watch basketball matches with his friends in the adjacent Yigong Huili Court. During the Cultural Revolution, he participated in the parade in the Central Square, and had brought his kids to the playground in the Central Square in the 1980s. But he has always had a kind of gloomy and horrible feeling about the concession, as if terrible things are happening in those buildings, like in Cao Yu's drama Thunderstorm. As he walks along the Bo'ai Road, he feels as if someone were watching him from one of the villas. Although the roads in the Italian concession are comparatively wide, it somehow does not feel as safe and comfortable as walking in the narrow alleys of the Old Town. Old Zhang has an impression deep in his heart that most people who live in the small villas do not speak the Tianjin dialect and are not Tianjiners. The Italian Style District has become a popular tourist spot now, with all kinds of restaurants, but from Old Zhang's point of view they are too flowery and are designed to make money off of the tourists, so he rarely eats there. He looks around while walking, and looks up and sees the Planning Exhibition Hall. The square is crowded; it seems that many people have come to see the exhibition today.

5. Tianjin Binhai Area

Speaking of Tianjin, the Bohai Bay and Binhai areas are inevitable topics. Binhai New Area is located on the east side of Tianjin and on the coast of the Bohai Sea.

Historically, in the Wanli years of the Ming Dynasty, Tanggu was already a coastal military town. By the end of the Qing Dynasty, due to serious deposition in the Beijing-Hangzhou Grand Canal, the transport of grains from the south to the north moved from canals to the sea; from Dagu Kou into the Haihe River to the Sancha Estuary, Tanggu gradually became a transfer station and cargo distribution center for transportation connecting rivers and seas, and also a coastal defense fortress, as Dagu fort was an important coastal defense barrier. In 1860, during The Second Opium War, the Eight-Nation Alliance was defeated in Dagu fort and stealthily landed in Beitang, opening China's gate to the West. Tanggu became an important base of military industry and national industry development at that time. In the 11th year of Guangxu reign (1885), Li Hongzhang founded the "Beiyang Naval Shipyard" in Dagu. In the 14th year of Guangxu (1888), he extended the Tangshan-Xugezhuang railway ran by Kailuan Mining Bureau to Tanggu. In 1900, the Eight-Nation Alliance captured Dagu Kou, captured Tianjin and advanced all the way into Beijing's Forbidden City. In 1914, industrialists Fan Xudong, Li Zhuchen and others established Jiuda Refined Salt Factory and China's first pure alkali factory—Yongli Alkali Factory in Tanggu, which became the birthplace of China's nationalized industry. After the outbreak of the Anti-Japanese War, the Japanese invaders built an artificial harbor at the mouth of the Haihe River in 1939 for exploitation purposes. In the 1930s, Liang Sicheng and Zhang Rui proposed to develop the Binhai coastal area in their planning proposal for Tianjin Special City. After the founding of the People's Republic of China, Tianjin Port re-opened, which marked the beginning of the development of Tianjin's Binhai area. After the Reform and Opening Up, Tianjin, as one of the 14 coastal open cities, built the Tianjin Economic-Technological Development Area (TEDA) on the site of Tanggu Salt Farm, attracting a large number of foreign enterprises. In 1994, Tianjin set the goal of building Binhai New Area in ten years. After more than 10 years of hard work, fruitful results have been achieved. In 2006, the central government officially included Binhai New Area into the National Development Strategy.

Song Yunfei

Song Yunfei, like many talented elites who have studied abroad, obtained master's degrees, and worked in urban planning and design, architectural design and traffic planning, has chosen to pursue his career in China for better opportunities. China is the largest construction site in the world, and there is plenty of room for people to fully exercise their skills and talents. Overseas returnees with similar experiences as Song Yunfei would often choose to work at the Beijing, Shanghai, or Guangzhou branches of Hong Kong or international companies. Relying on his professional skills and the prestige of his school, Song Yunfei first chose to join the famous MVA Transportation, Planning & Management Consultancy in Hong Kong. The company was mainly targeting at the markets of Hong Kong and major cities in mainland China. At first, Song participated in projects mainly in Shenzhen and Guangzhou in The Pearl River Delta and Shanghai and Nanjing in the Yangtze River Delta. In 2006, after Tianjin Binhai New Area was incorporated into the National Development Strategy, MVA began to participate in the planning of rail transit in Tianjin and Binhai New Area, which gave Song Yunfei the opportunity to come into more contact with and to understand Tianjin, a city relevant to his own life.

After working in Hong Kong for more than a decade, Song Yunfei, now in his 40s, decided to return to mainland China to start his own business. This decision was made after careful consideration; he took into consideration firstly the potential glass ceiling in a foreign company, secondly the opportunity to put his own ideas on transportation planning in China into practice, and thirdly his wife and their kid who had been living overseas for more than a decade and the family's wish to reunite and settled down. In the choice of the city to settle, Song Yunfei and his wife had repeated discussions. In cities like Beijing, Shanghai, Guangzhou and Shenzhen, which have developed economy and abundant talents, the overall development of the city started earlier and the policy environment had relatively matured; however, the competition would be intense, living cost would be high, and the pace of living would be fast, which worried Song Yunfei's wife who had been living in North America for many years. On the other hand, big cities in central China, including his hometown, lacked urban scale and influence, which provide little room for improvement on urban layout and transportation planning. Eventually, after careful consideration, they decided to choose Binhai New Area. In the past two years, Song Yunfei had done several transportation planning projects in Binhai, and had a certain business basis. At the same time, he also had an understanding of the overall planning of Binhai New Area. In addition, Binhai New Area is quite suitable for living, with sound establishment of educational and medical facilities, good quality of houses and low housing prices. His emotional connection with his family made him determined to take root in this land.

He had heard his wife's stories of family struggles when they had met in college. His wife's grandfather, in his younger years, followed Li Zhuchen, a fellow countryman from Hunan province, to Tianjin Tanggu. Influenced by the ideal of "Saving the Country by Industry", they founded Tianjin Jiuda Refined Salt Factory and Tianjin Yongli Alkali Factory, creating nationalized industry and becoming the forerunner of the world alkali industry. During the Anti-Japanese War, in order to keep the lifeline of national industry, they moved factories to Sichuan under crisis and hardship. Almost penniless, they broke through heavy resistance, started again, and not only continued the production, but also expanded the market. After the liberation, the two factories of Jiuda and Yongli returned to Tianjin again, and Song Yunfei's wife's grandfather also moved back to Tianjin Tanggu with the factory, and afterwards Song Yunfei's father-in-law also became a technician of the alkali factory. Half a century passed in the blink of an eye, the alkali factory has stopped running and moved to the Lingang Industrial Zone. Song Yunfei still remembers the first time when he followed his father-in-law to the alkali factory's old site; in front of the huge "lime kiln", the old man told the legendary and tough enterprise experiences of their generation, recalling his experiences with the alkali factory through all these years. Under his simple, rigorous appearance, there glistens a sense of firm resolution in his eyes.

Knowing his wife had allowed Song Yunfei to forge a bond with Tianjin and Binhai New Area. Although having studied and lived overseas for many years, he is guided by the generations of entrepreneurial determination and spirit in his wife's family, which secretly draws him back to this land, inspiring him to continue his entrepreneurship story.

6. Post-earthquake Reconstruction in Tianjin

Chief Chu

Chief. Chu came to the Planning Exhibition Hall early in the morning to prepare for the international forum and seminar being held today. Her Urban Design Lab mainly works in urban design, and most of the pilot projects is headed by the Lab. She is glad to see that urban design is becoming more and more popular. Many planning and design institutions, including many other departments of TUPDI, are involved in urban design pilot projects, and the meeting room in the Lab is often full with people participating in discussions.

Chief. Chu can be said to be almost a Tianjiner. She was born in Zhejiang Province in the 1960s, when the Cultural Revolution was in full bloom. She came to Tianjin with her parents when she was 10 years old. Two years later, the Tangshan Earthquake hit Tianjin. Chief. Chu experienced the horror of the earthquake and witnessed the devastation. She chose architecture as her major in the college entrance examination, hoping to build her own city in the future.

On July 28, 1976, around 3 a.m., Chu, at her age of a pupil of elementary school, was awakened suddenly in her sleep by her mother's urgent voice, "Earthquake, run!" Chu was dragged into the courtyard by her father. The earthquake brought a devastating blow to Tangshan, 108 kilometers away from Tangshan. Tianjin was also hit hard. The worst affected Siping Road, Gansu Road, Guiyang Road areas were in ruins with houses collapsing. According to the later statistics, 64% of the houses in the six districts in the center of Tianjin were damaged, 14% of them were completely destroyed, and more than 30% of the schools, hospitals, cultural facilities and commercial outlets collapsed or were seriously damaged. 24,296 people were killed and 21,568 were seriously injured across the city, and the direct economic loss amounted to 3.92 billion yuan. All the people living in buildings and most of the people living in bungalows were afraid to go home and sleep. All the open areas, school playgrounds and street parks, had temporary shelters set up, and both sides of main streets were also full of small sheds, leaving a narrow path in the middle of the road. Chu heard her classmates saying that family members of workers in Tanggu shipyard stayed on ships in those nights.

People in Tianjin, recovering from the devastating earthquake, began the long process of reconstruction. In 1977, the municipal government designated Guiyang Road and other five worst damaged blocks as reconstruction areas. In 1978, the state council approved the "Tianjin Earthquake Damaged Housing and Supporting Facilities Restoration And Reconstruction Three-Year Plan". Under strong support from the state and all parties, the post-earthquake reconstruction of Tianjin made significant progress. The Luanhe river diversion to Tianjin project was completed; coal gasification and centralized heating in the central areas was achieved in three years; the "3 rings and 14 radiation rays" road network formed the basic skeleton; Lao Longtou Railway Station(The East Railway Station) was reconstructed; the airport was upgraded to an international airport. Urban public facilities were greatly amended, and urban traffic improved significantly. At the same time, the city had repaired some distinctive old buildings, redecorated the facades of existing buildings and built several high-rise flats along the ring roads, and constructed linear parks on both sides of the Haihe River and added planting to streets and courtyards. The urban appearance is greatly improved, and Tianjin became a model of environment renovation for domestic cities to learn from at that time. Chu always felt the rapid changes of Tianjin whenever she came back home during breaks of holidays and vacations, and she could always hear praises for Tianjin from classmates from other provinces and cities.

In 1984, the state council made the decision to open 14 coastal port cities, including Tianjin. After comparing multiple proposals, Tianjin Economic-Technological Development Zone was eventually decided to be located on the Tanggu Northeast Salt Farm, which kicked off the development and construction of the Binhai New Area. In 1985, the mayor of Tianjin presided over the revision of Tianjin's comprehensive master plan, as well as 30 corresponding specialty planning. Before the plan was submitted to the state council , the government held an urban master plan exhibition to solicit opinions widely…

At this time, Chief. Chu graduated from university and returned to her hometown to work in the Municipal Urban Planning Bureau. She vividly remembers her first few days on the job, which coincided with the 10th anniversary of the Tangshan Earthquake on July 28, 1986. On the day, she accompanied colleagues to the Huangjia Garden Plaza on Nanjing Road to participate in the commemorative event, which was also the grand unveiling of the Tianjin Earthquake Memorial. This year, Tianjin's earthquake relief work was finally completed. On August 4, 1986, the state council approved "Tianjin's Comprehensive Master Plan (1986—2000)". This is the first state-approved urban master plan in the history of Tianjin, ending the history of Tianjin's practice of implementing urban construction while simultaneously compiling the urban master plan since the founding of the People's Republic of China. Because this plan was so important, the Municipal Government decided to hold another exhibition to publicize the urban master plan of Tianjin. The exhibition was held the Tianjin Industrial Exhibition Hall on the Machang Road.

Chu and other young planners participated in the whole process of curation with several younger senior planners. In those days, there were no computers or printers, and large drawings needed to be drawn by hand. With the drawings mounted on big chart boards, they climb on it to draw and color, and were completely tired out everyday. When the exhibition opened, Chu and several other young planners also worked as narrators. In this process, she gained a preliminary understanding of Tianjin's Comprehensive Urban Plan.

After the planning exhibition, Chu returned to the Planning Bureau and was officially assigned to the chief engineer office of the Planning Bureau. In the office, there were Mr. Liu, Mr. Shen and Mr. Chen, old experts who had been working in urban planning since before the liberation. It was said that Mr. Liu graduated from Tianjin's famous Industrial and Commercial College and had worked in the government of the Kuomintang period. The old experts were warm and kind, teaching Chu patiently. From them, Chu had a deeper understanding of Tianjin's Urban Planning history.

The Tianjin's Comprehensive Master Plan (1986—2000) approved by the state council was actually formed on the basis of 21 drafts of the master plans formulated continuously since 1952. The plan determined the layout of urban and industrial development within the whole city. Considering the trends of ports transitioning from inner river ports to seaports and locating major industries by the sea, the plan put forward the development strategy of "industries moving eastward" and the overall urban pattern of "One Pole Carries Two Ends". The urban layout centered on the urban areas of Tianjin, including the Central City, Tanggu and the industrial zone of the lower reaches of the Haihe River, forming the main body of the city, and constructing a multi-level urban network system with surrounding coastal towns, suburban satellite cities, five counties and dozens of administrative towns.

The goal of the plan was to build Tianjin into a comprehensive industrial base with advanced technology, an open multi-functional economic center and a modern international port city by the end of the 20th century. The total population of the city would grow to about 9.5 million, of which 6.1 million would be permanent urban residents. By systematically spreading industry and population to coastal and satellite towns, the population in the Central City would be controlled within 3.8 million. The population of Tanggu, including the port area and TEDA, would grow to 600,000 people. The entire coastal area would have an urban population of 1.2 million. The Haihe River would become a natural axis running through the central urban area and the Binhai coastal area. The river section above the second watergate would be used mainly for clear water storage. Amusement facilities should be developed appropriately. Large public buildings should be built along the waterfront, and green space should be expanded and forms a main scenery line in the urban area. Science, culture and education as well as the tertiary industry should be developed in the 71-square-kilometers central urban area. City-tier commercial center should be appropriately expanded to cover more than the original area of Quanyechang and Heping Road.

The main road system is consisted of the inner, middle and outer ring roads and 14 radiation rays as the structure. Public transportation and bicycle systems would be prioritized. The underground railways and rapid rail transit would continue to be developed, strengthening the links between urban and suburban areas. The lower reaches of the Haihe River would mainly be used for shipping, with wharves and warehouses along the coast. North bank would develop steel and mechanical processing enterprises, south bank construction of light and textile industry and fine chemical enterprises. The Binhai coastal zone would develop ports, shipbuilding, machinery, building materials, chemicals, light and textile industries, and takes advantage of coastal ports to actively develop fishery and tertiary industries and promotes the development of marine engineering. The port city formed with Tanggu as the core would become the center of economy, culture, science and technology in the Binhai area. Hangu in the north would be built into an industrial city dominated by marine chemical industry. Dagang in the south would be built into an industrial city dominated by petroleum and chemical industry. Small towns such as Beitang, Gegu, Xiaozhan and Guangang, forming the Binhai town system, make contributions to the prosperity and development of Bohai bay economic cooperation circle by utilizing its advantageous center position in the Bohai bay.

In a word, this edition of Tianjin City Master Plan is comprehensive and with distinctive features. The proposed road system of "3 rings and 14 radiation rays" is to solve not only the traffic issue but also the problems of urban layout disorder and various facilities not being systematic caused by the rigid division of Nine-Nation Concessions. Meanwhile, a 500-meter green belt was set up on the outer ring road to avoid disorderly urban sprawl. These were domestically leading practices at the time. These details were all memorized by Chu at that time and are still often recalled for reference.

At that time, "One Pole Carries Two Ends" sounded corny and not academic enough in Chu's ears. Mr. Liu read her mind and enlightened her. Although there were plans for Old Town and different concessions, there was always no an unified plan for the overall urban development of Tianjin. This lead to a bad situation with intermingled industrial, residential landuse, heavy traffic congestion and poor living conditions. In order to reduce the heavy pressure on the central city and improve living conditions there, the plan decided to gradually move the industrial enterprises in the central city to the Binhai area and the industrial zone in the lower reaches of the Haihe River. The central city would focus on the development of residence, living, business and culture, and the Binhai area industries and ports. This was the strategy of "industries moving eastward". The strategy needs the support of transportation and other infrastructure. In this way, the Haihe River and paralleling transportation artery become "the Pole", which "carries" the central city and Binhai area at two ends. The plan put forward the urban layout structure of "One Pole Carries Two Ends", which was very vivid and easy to unify the thoughts and understandings among the masses and officials. "Industries moving eastward" is more concise than "the city's industrial layout shifting to the coastal area"; "One Pole Carrying Two Ends" is easier to understand than "The Transformation of Urban Spatial Structure from Single Core to Double Core". Perhaps it was these easy-to-understand concepts that had made Tianjin city master plan influential and feasible. So, 20 years later, people saw the Binhai New Area of Tianjin standing on the once barren salt marsh.

7. Two New Tianjins: The Central City and Binhai New Area

In the past 100 years, there have been two Tianjins: the Old Town and the concessions. Today there are two new Tianjins: the Central City and Binhai New Area. In the past 40 years, the urban appearance and functions of the central city have been greatly improved after the post-earthquake reconstruction and large-scale urban redevelopment as well as preservation of the historic districts. In recent 20 years, Binhai New Area, as the frontier of opening up and reform, with ports as the core resources , has been accelerating their development, and the proportion of the total economic volume of GDP in Tianjin that Binhai New Area held has over 50%. Although they are two areas of the same city, the urban appearances and styles are obviously different; Although only 50 kilometers away, many people in Tianjin have hardly ever been to Binhai. With the transformation of Tianjin's economy and the further improvement of the traffic conditions between the two areas, especially the emergence of new attractions in the Binhai New Area such as Binhai Cultural Center Library and National Marine Museum, the two Tianjins will be integrated in the near future.

Mr. Bauer

At 8 am, Bauer put on his formal suits. His schedule today is to attend the opening ceremony of Tianjin Urban Design Exhibition and urban design forum, and attend the urban design practice seminar in the afternoon. He has made an appointment with chief planner Zhuge at noon to continue the discussion of the Haihe River that they did not finish at breakfast. The schedule is a little tight. His wife Shirley is also getting ready to go out. She plans to visit The West Water Park and The Pillow Park this morning. The West Water Park is a large-scale Chinese park designed by a local design institute, and The Pillow Park is a Western-style garden designed by Charlie, a famous French landscape designer. Shirley wants to learn about the key how to design Chinese-style modern gardens, so that she will have the opportunity to use the knowledge in the future in Chinese projects such as the Chenglin Park. In the afternoon she is going to middle reaches of the Haihe River to attend a workshop on ecological green barrier between two cities, meeting Mr. Bauer at Yujiapu in Binhai New Area at night to attend a concert of their daughter Vivian and her students at the Juilliard School. Vivian is a violinist who teaches at the Juilliard School in Tianjin. It is a coincidence that the family happen to gather here in Tianjin. Since she would be wearing an evening dress, Shirley brings a suitcase with her. She reserves a car online, and her phone shows that the car will soon arrive at the hotel. Mr. Bauer carries the suitcase downstairs.

After saying goodbye to his wife, Bauer walks along North Jiefang Road to the Tianjin Planning Exhibition Hall in the Italian Style District. He checked the map on his mobile phone—2.5km, about 18 minutes' walk. It is the end of the morning rush hour, and although there are many cars on North Jiefang Road, the one-way traffic in good order. North Jiefang Road was first built with deep wooden stakes on the riverbank. It was the first asphalt road in Tianjin or northern China (the Great France Road in the French concession, 1918). In 2008, Tianjin carried out environmental renovation for the Olympic Games. Shirley cooperated with Chief. Chu from the TUPDI to work on the landscape design of street renovation. In order to provide convenience for pedestrians, the plan narrowed down the car lane from 15 meters to 8 meters, releasing space to widen the sidewalk, and also reduced the common 15-centimeter height difference between the car lane and the sidewalk to 5 centimeters and turned the asphalt road into stone-paved road. It fitted the historic street site of North Jiefang Road and also limited the speed of cars. In the past, the sidewalk on North Jiefang Road used to be very narrow. With the car speed so fast, pedestrians could not help but feel unsafe and nervous while walking. If it happened to meet with the steps of buildings, the sidewalk became too thin to be almost impassable. Today, walking on North Jiefang Road again, although there are still cars, the speed is slow and the sidewalk has been widened; especially with the height difference between the sidewalk and the car lane reduced, it is very comfortable to walk on, with no more sense of being cornered as in the past. The environmental renovation of North Jiefang Road is satisfactory and the construction quality is good, and therefore is able to fully reflects the designer's intention. In some irregular road intersections, the ground is paved as an oval pattern, which not only cleverly organizes the roadway and sidewalk, but also serves the function of marking. At that time, the project manager of the North Jiefang Road marveled as he saw the design plan: "I have done so many projects in my life and this is the first time I have ever done work so fine, almost like embroidery... We will build North Jiefang Road into the road of the best construction quality in Tianjin." Today, more than a decade has passed. Having rolled over by wheels and exposed to wind and rain, the stone-paved road has a color texture more layered with a sense of history, matching the historic buildings on both sides. Last night, Bauer and his wife took a short walk down North Jiefang Road, where the stone-paved road sparkled under streetlights.

North Jiefang Road was once the most important street in the British and French concessions, flanked by classical bank buildings, known as the Wall Street of northern China. Although none of the bank buildings are very large in size, the proportion of facades and details of the buildings are exquisite and interesting. After the Reform and Opening Up, due to the development of banking services, the old buildings could not meet the new functional requirements, and the traffic condition was poor, so many banks had built office buildings in new sites such as Youyi Road. As early as in the early 1990s, the People's Bank of China built a high-rise building as an extension of its existing office on North Jiefang Road, which was not in harmony with the scale of the old street with its relatively big volume. While much of the business had moved out to new locations, each bank had retained its original site, and some set up offices on the ground floor of the old sites. Tianjin plans North Jiefang Road as financial street out of the notion on historic and cultural inheritance. There was a time when the city tried to turn North Jiefang Road into a high-end commercial street, but failed; it seemed that it would be the best choice to keep the street as a financial location. As he walks along, Bauer thinks of the fact that there is also China's first modern post office on North Jiefang Road—Daqing Postal Office Tianjin Branch, which issued China's first stamp: Dalong (Big Dragon) Stamp. Now there is also a China Post Museum on North Jiefang Road. Not far away, the former French Ministry of Industry building, once used to have been converted to an art gallery, is now an office building, closed to the public. Further on, the former French club is now the Chinese Museum of Finance. In front of the museum is a new small garden, which is also the entrance of Jinwan Square. There are people gradually gathering in this area.

In front of him is the Jiefang Bridge, which was once called Wanguo Bridge. It is a well-maintained iron drawbridge. Because of its traffic function, Jiefang Bridge cannot be up for a long time, so it is only up in holidays and special occasions, for roughly two hours each time. The bridge is said to have been designed by the company that designed the Eiffel Tower in Paris. Standing on the bridge, Bauer sees the Jinmen Tower complex, designed by Skidmore, Owings and Merrill, LLP (SOM) and developed by Beijing Financial Street Company, which has become a landmark of Haihe Central Square. Bauer has been familiar with the site since he participated in the international call for urban design proposals for Haihe Central Square. Here is the first bend of the Haihe River, which is also the geometric center of Tianjin. The north bank was once the Italian concession. After the liberation, the

Central Square was the place where people gatheredand paraded, and later it was turned into a children's playground. The south bank of the Haihe is the Japanese concession and Heping Road commercial street. There were warehouses along the bank and a nine-story publishing house building that was demolished later. There were several famous design firms shortlisted for the urban design competition of the Haihe Central Square. Eventually the collaborative proposal of WRT, LLC , Bauer and TUPDI was selected. It distinctively proposed to design a 300-meter-high landmark "Jin Tower (Tianjin Tower)" by the Dagu bridge, and a concert hall that was similar to the Bilbao Museum designed by Frank Gehry in the Central Square. Another plan that was selected designed a group of large public buildings such as "Jinmen(Gate of Tianjin)" and a circular promenade by the Haihe River. The final plan combined "Jinmen" and "Jin Tower"; although not necessarily the best plan, it was quickly implemented, with high-end office buildings, hotels and apartments, attracting some high-income people to return to the city center. Later, the huge Riverside 66 Shopping Center was built at the intersection of Heping Road and Binjiang Street. Its quality of architecture and internal space was very good. However, the project was delayed for several years, and the best opportunity for brick-and-mortar shops was missed. Its business was affected to a certain degree with the fast rising of e-commerce. The Heping Road pedestrian street, where the Quanyechang and the Baihuo Dalou (Tianjin Department Store) are located, is also increasingly empty, and the large area behind it is called "the Quanyechang area", with dense population in poor living conditions. How to realize the urban renewal of Quanyechang area is also a key question of this urban design pilot project.

Bauer's gaze turns to the right, and the familiar shapes of the bell tower at Tianjin Railway Station come into view. Built in the 1980s, the building is a modern form of architecture, with a three-pronged layout that encloses several squares. In 2008, when the Beijing-Tianjin intercity high-speed railway, which was the first high-speed railway in China, was gonging to operation, the railway station was renovated and expanded in a large scale. The overall plan retained part of the passenger building and its trademark facade that faced the Haihe River, which was an important landmark of Tianjin. Across the Haihe River from Tianjin Railway Station is Jinwan Plaza—a group of multi-storey and super-high-rise buildings in set-back form. The multi-storey buildings, that are closer to Haihe River, imitate traditional European architectural style, and the high-rise buildings are modern in general, combined with new techniques and classical symbols. Jinwan Plaza is located in the second bend of the Haihe River, which is at the due opposite of Tianjin Railway Station. The opening of Jinwan Plaza, with Tianjin's first IMAX cinema, coincided with the release of James Cameron's film Avatar and it was extremely popular, with theaters crowded and tickets hard to get. However, it did not last long; the business soon declined, which perhaps had something to do with the detour from Jinwan Plaza to Tianjin Railway Station being too far. Today, Bauer makes a new discovery: On the Haihe River between Tianjin Railway Station and Jinwan Plaza, he sees a novel-looking pedestrian bridge. The bridge has an undulating roof which extends out connecting as a whole with the newly-added corridor in front of the existing railway station. Many people are walking and take pictures on the bridge, and the Jinwan Plaza seems lively again. It seems that the planning and construction of the pedestrian system in central Tianjin really have practical effects on the city.

Bauer moves on, and appears in front of him the elegant Century Clock, which is a gargantuan steel-structured sculpture constructed in order to greet the 21th century. Its pendulum frame is shaped like an "S", drawn from the division line in Tai Chi chart, referring to the transition of Yin and Yang, meaning ending and beginning. The appearance and material of Century Clock match with the ones of Jiefang Bridge, and also with the characteristics of Tianjin as a modern industrial city. Bauer finds that the road around the plaza under the Century Clock has also been renovated. In the past, the car lanes around the Century Clock was wide and the traffic was chaotic; tourists were hard to cross the road to take pictures in front of the clock. Today, road calming measures have been adopted to stabilize and tranquillize the traffic: the car lanes were narrowed down and the sidewalk widened. Now it is much easier to reach the Century Clock Square through the pedestrian crossing. Entering the Century Clock Square, Bauer walks west and sees the entrance square of the Italian Style District. The signs are very clear, and Bauer can testify himself that people coming out of Tianjin Railway Station can easily be guided here. Bauer looks up and finds that he has arrived at the Tianjin Planning Exhibition Hall. He feels a little excited.

Song Yunfei

Today, Song Yunfei is going to attend the Tianjin Urban Design Seminar in the downtown of central Tianjin, and he is taking the train from the Yujiapu High-Speed Railway Station. The railway station is 16 kilometers away from his home, and it takes 30 minutes by car and 1 hour 20 minutes by bus and walk respectively. He usually goes to work at 9 o 'clock by bus and listens to audio books on the bus. Today, his wife tells him to drive. He leaves the house early in the morning to avoid traffic jams.

He leaves at 7:30 and arrives at the underground parking lot of Yujiapu High-Speed Railway Station at 8:00. He buys a sandwich and a latte at the underground cafe. There are still a few minutes left until the ticket-checking begins and out of habit Song Yunfei looks up to see the dome of the Railway Station. Yujiapu High-Speed Railway Station is the first fully underground high-speed railway station in China, and is planned to have three transfer subway lines. It is the transportation hub of Binhai New Area's central business district. The architectural design scheme of the above-ground parts of Yujiapu Station use the concept of "shell"; inspired by the helixes of nautilus and sunflowers, it constructs double-end helix grids on the basis of a circle and stretches them to form an initial two-dimensional shape, and forms an initial three-dimensional shape by "hanging" vertically, then reverses to reach a shell shape. It forms an open, bright, modest and innovative architectural space, achieves the perfect unification of structure and architecture, and has become a Binhai landmark.

After eating breakfast on the 8:15 high-speed train, he arrives at Tianjin Railway Station 20 minutes later. Song Yunfei quietly praises the convenience of high-speed rail transit between the two cities; this way of travel in urban areas is convenient and efficient. He calculates in his mind: from Tianjin East Railway Station to the Urban Planning Exhibition Hall is a 20-minute walk; he will be just on time. He walks out of the station, comes to the front square, and sees the square that used to be empty now has a newly-built long glass corridor. Passengers now finally have a place to shelter from rain and snow. At the entrance and safty checkpoint, passengers line up to get into the station in an orderly fashion. The pedestrian bridge connecting Tianjin Railway Station to Jinwan Plaza that has created a hype on the Internet is southeast of the front square, where many people are taking photos. This scene makes Song Yunfei feel like the city has began to wake up and is full of hope.

Song Yunfei comes to the Century Clock Square, following the guide system of the Italian Style District, and finds that a traffic re-organization has also taken place here. There is now a designated taxi pick-up area, and traffic calming measures greatly improving the pedestrian environment. Song Yunfei thinks: urban designers and planners have been calling on this for many years, and now Tianjin really seems to be pushing walkable city campaign in the urban center. He recalls Sadik-Khan, the former New York City transportation commissioner and author of *Street Fight*, visiting Tianjin recently; in the very planning exhibition hall, she gave a brief introduction to New York's experience of implementing a pedestrian system through three steps: mapping a common vision, acting quickly and speaking with data. Sadik-Khan described a few "fights" through stories, big and small, such as changing streets flexibly according to weather conditions, starting from a temporary walking area in certain parts over the weekend, collecting feedback and immediately presenting data, and so on, to change the pursuit of profit deeply rooted in people's minds and change people's cognitive inertia. The title of the book, "Street Fight", is also very interesting. At first glance, "Street Fight" easily reminds people of video games and hip-hop dance battles, which is very eye-catching. Although the book's content may not be as hot as the title suggests, it is also full of dramatic tension, which is enough to attract readers other than professionals.

Song Yunfei clearly remembers that he asked Sadik-Khan a question that had been bothering him for years: Tianjin wanted to promote pedestrian, and the street space between buildings needed to be designed and reconstructed as a whole, which involved many departments, including municipal administration, transportation, planning... Due to the different responsibilities of different administrative departments, disciplinary barriers made it difficult to integrate and design for street space. How does New York do this? What is the role of New York City's Department of Transportation? Where do you stand when communicating with other departments? Sadik-Khan briefly answered the question: every job requires cooperation. When I was New York's transportation commissioner, there were 11 departments involved in street management, each with its own standards. It was very difficult and time-consuming to gather the 11 departments for discussions and for us to reach a common standard. We had weekly meetings, three years in a row, and the result of our hard work was a unified design guide for New York's streets that all 11 departments agreed on, so that whoever became mayor or traffic commissioner later on would have the guide. Mayor Bloomberg has been a big supporter of the program and has made a point of bridging the gap between departments, like the fire department, which traditionally has not been very supportive of slow streets because of the difficulty of dealing with large hydrants and bicycle lanes. Under the leadership of the mayor, in order to implement the slow street plan, the chiefs sat together and reached a consensus through discussion and negotiation.

Song Yunfei thinks that the situation in China is similar to that in New York. The process of gaining consensus requires professionals to conduct long-term and unremitting research and exploration in their own fields, and to present convincing results and visions, including mobilizing citizens' sensory experiences to endorse good implementation effects of designs. They also need to be good at communication; one highlight of the experience from New York is that professionals pushed policy-makers to think about how to gather more social resources to promote a common value and the implementation of a vision in a very skilled way, and they need to know about the stakeholders mostly likely to support the realization of this vision, the infrastructure construction that would support the vision, and how to integrate them into the project. He admires the Sadik-Khans for their courage, their discipline and their ability to communicate across a range of voices.

Song Yunfei thinks while walking, and when he looks up, he sees the entrance square of the Italian Style District. Suddenly he spots a familiar figure in the crowd—Mr. Bauer. He hurries up.

Chapter II

People · Housing · Parks

8. The Urban Planning Exhibition Hall

Tianjin Planning Exhibition Hall is a four-storey modern building facing the main entrance of the square in Italian Style District, with colonnades on the front facade and irregular patterns on the ivory sand travertine, avoiding rigidness in appearance. The building was originally intended to be a showcase center for Italian merchandises, but it never came into operation. In 2008, Tianjin carried out successfully a key planning campaign, and a lot of planning and urban design projects needed a place to be displayed. According to the original plan, the planning exhibition hall would be built on the land to the north of the Central Square, and architectural design proposals were collected for this site. But the planning exhibition was requested to taking place in very limited time, so this ready-made building was chosen and transformed into the Tianjin Planning Exhibition Hall. It took only half a year for the exhibition to be arranged and officially opened in January 2009. The exhibition hall covers an floor area of 15,000 square meters and the exhibition area 10,000 square meters. Although the space is not large, it shows the history and future blueprint of Tianjin's urban planning to the point, both comprehensive and detailed, and has received praise from various sources. After the opening, the crowd was huge. In the first five years of opening, the number of visitors reached one million each year.

This Tianjin Urban Design Exhibition is arranged in the temporary exhibition area on the first floor of the Planning Exhibition Hall. The exhibition content is rich and well-knit, from the historical review of urban design development in Tianjin, the summary of urban design implementations in recent years to the showcases of various new urban design projects. The Evaluation of the Comprehensive Development and Reconstruction Plan of Haihe River is a key point, serving as the climax of the first half of the exhibition. The second half exhibition is divided into different sections by themes, including "The Haihe River—Waterfront and Public Space", "Humanized City—Pedestrian and Revival of the Urban Center", "The Urban and Rural Magnet—New Urbanism and Suburbanization", and "Livable City—Major Urban Parks and New Communities". The four themes embody Tianjin urban design's new ideas of people-centered and ecology-oriented. Urban design is based on the feelings of residents, to meet people's needs for housing, education, health care, sport, to create beautiful and authentic living places, to create public and open spaces, to improve ecological environment, aiming to further increase the urban vitality of Tianjin. This exhibition is not only a display of urban design projects, but also an invitation for all residents, stakeholders and communities to involve in. A lot of new thought was put into the exhibition. As it should not only reflect the theoretical height of urban design project, but also be easy for the audience to comprehend, combining the highbrow with the lives of ordinary people. Each of the four themes of exhibition is divided into three sections: Yesterday, Today and Tomorrow. The most attractive is the Tomorrow section, divided into "How will you travel tomorrow?", "Which park are you going to tomorrow? ", "What kind of house will you live in tomorrow?", fascinating and entertaining. This is the first time in decades that Tianjin urban planners have tried to look at the real influences of urban design on residents from the perspective of ordinary people.

There is still a quarter of an hour to go before the opening of the exhibition, and Old Zhang has already sneaked in and started browsing. Mr. Bauer and Song Yunfei enter the lobby of the Planning Exhibition Hall together. Mr. Song, director of the Tianjin Urban Planning and Design Institute(TUPDI), greets them and leads them into the VIP room. There are many guests seated inside and Mr. Bauer sees many familiar faces, including Director Liu and Chief Planner Zhuge of Tianjin Planning Bureau, several academicians and professors from Tsinghua University, Tongji University, Southeast University and Tianjin University as well as old friends from Tianjin Urban Planning Academy, Tianjin Urban Planning Association, TUPDI In the front row, Mr. Bauer meets some of the expected international gurus. Prof. Jonathan Barnett, an octogenarian, is the internationally renowned urban design theorist, whose book *Urban Design As Public Policy* has become a textbook for universities around the world. Since 2000, he has participated in several key urban design projects of Tianjin, e.g. the French Concession along the Haihe River, the central business district of Binhai New Area, etc.. Mr. Bernard Tschumi is a world-renowned architect and architectural theorist, who designed the Parc de la Villette in Paris and the New Acropolis Museum in Athens as well as the Science and Industrial Museum in the Cultural Center of Tianjin Binhai New Area. Mr. Dan Solomon is renowned architect and a former professor at the University of California, Berkeley, who is actively involved in New Urbanism and has been engaged in affordable housing for a long time. He is the author of *Global City Blues* and other books. He has been involved in the urban design of the New Harmony Residential Community in Binhai New Area since 2011. Having met all these great figures, Mr. Bauer is excited and looking forward to this seminar, which should be a feast of ideas on cities and urban design.

The opening ceremony of Tianjin Urban Design Exhibition is simple and creative. First, Director Liu gives a speech: "In the past, most of our planning and design were dogmatic and paternalistic. The planners took their pens, looking at the city from high above, to engage in one-sided planning and design from their own. Today, our planners should step onto the ground and conduct planning and urban design based on real people's experiences and feelings, not only with rational physical analysis, but also with mental perception and empathy with citizens. Of course, planners are indeed different from ordinary people. They are professionally trained and have certain planning and design knowledge and skills, so that they can depict the future of a beautiful city. Let's look forward to a better tomorrow." Later, Director Liu invites several international experts and planners who have participated in these urban design projects to come on stage, to scan a QR code simultaneously and enter their short phrases about visions and expectations for tomorrow, to open the symbolic door in the middle of exhibition.

9. How will you travel tomorrow?

Song Yunfei

Song Yunfei is invited by Director Liu onto the stage as a young planner representative. He enters his expectations for tomorrow: No more traffic jam; to enjoy the happy life of driverless car. People are active animals. Modern people sit for too long. They need to date, meet, communicate, work, go to the countryside, walk, run, ride, drive, take buses, taxis, subways, high-speed trains, planes and ships, and they need to choose from a variety of transportation modes. Song Yunfei feels that Tianjin does a good job in terms of seamless transfer between different means of transportation. Tianjin East Railway Station and West Railway Station are old stations with long histories, and have been modified to become high-speed inter-city railway stations and comprehensive transportation hubs that connect to subways, buses and taxis. The Yujiapu High-Speed Railway Station in Binhai New Area is also such a hub, as it is located in the center of the central business district, with more than three subway lines. Song Yunfei believes that in domestic cities, objectively, the planning and design of transportation hubs in Tianjin are relatively good, although it is certainly no match to advanced cities such as Tokyo, London, and Hong Kong. The superiority of these cities' systems is mainly reflected in the comprehensive development and compound functions of stations, which is also an issue Song Yunfei has been thinking about for a long time. His company plans to invest efforts in this research direction, and is now in a stage of research and collecting basic data, combining with the specifics of the project to do some preliminary analysis work.

Coming into the central city for the seminar this time, he also wants to take the opportunity to go to the West Railway Station to investigate the feasibility of the renovation of the West Station area. In terms of subway construction, although Tianjin began to construct subways using the Qiangzi River at Nanjing Road in as early as the 1970s, and is the first city in China to build a subway aside from Beijing, at present, the subway construction speed is slow. The total running mileage is 215 kilometers, and the total length of its six lines ranks the tenth in China. Shanghai has 705 kilometers with 16 lines and Beijing 640 kilometers with 22 lines. Tianjin now ranks 13th in passenger volume with up to 1.5 million passengers per day, while Beijing, Shanghai and Guangzhou have all exceeded 10 million passengers per day. Is the greater the mileage, the better? Song Yunfei disagrees. For example, Hong Kong subway has six lines and a total running mileage of 260 kilometers, but the maximum passenger volume can reach 5.6 million passengers per day, averaging to 21,500 passengers per kilometer. The level of services is also very high. Tianjin only has an average of 7,000 passengers per kilometer, with the utilization rate of subway being low at present. Beijing, on the other hand, has an average of 22,400 passengers per kilometer, higher than Hong Kong, and the subway is too crowded during rush hours, leading to bad riding experiences. In the past, some people said that Tianjin subway has a low passenger volume because of the lack of a network. At present, Tianjin subway lines have mostly formed a network, but the increase in passenger volume is not as large as anticipated. Others say Tianjin is small and suitable for walking and non-motorized travel, which Song Yunfei feels may be the real reason. And traffic congestion in Tianjin is at an acceptable level, much better than in Beijing. Transportation planning does depend on the individual city.

Walkability, or the return to walkable and humanized cities, is the latest trend in today's world. Jan Gehl, a Danish architect who has drawn on the work of many academics to write *Life Between Buidings* in 1971, has begun to focus on humanized cities from very early on. He has been engaged in the research and planning of urban walkability for a long time. Many cities, such as Copenhagen and Melbourne, have made great breakthroughs under his guidance. In the early 20th century, faced with global competition, London invited renowned architect Sir Rogers to lead the development of urban walkability strategy in order to revitalize the downtown area. The high level of implementation of the pedestrian system has played a very important role in the revitalization of London. The aforementioned former mayor Michael Bloomberg and traffic director Sadik-Khan of New York City were also influenced by Jan Gale to carry out the walkablizing "Street Fight" project in New York and received great success. The photos that compare and contrast the Times Square before and after the walkability project render the truest, most compelling visual effect of the improvement of street conditions. Today, the renovation and reconstruction of Tianjin's urban center has been mostly completed after more than 20 years. After the renovation, many large commercial facilities are concentrated in the downtown area, but some new establishments are good in architectural qualities yet always lack popularity and have a hard time making profit. The rapid development of e-commerce impacting brick-and mortar stores and other factors play a role, but the city's imperfect pedestrian system, low level of pedestrian facilities, inconvenience in walking, and the lack of direct pedestrian links between commercial and historic districts are also important reasons. Song Yunfei has a deep understanding of this. In the past, if he goes to the Italian Style District, Jinwan Plaza or Quanyechang from Tianjin Railway Station, he has to take detours even though the crow-fly distances between these locations are not so long, and the walking experiences are neither safe nor comfortable. Through the cheerful experience of this morning, Song Yunfei feels that Tianjin's urban design pilot project accurately has captured the theme of urban walkability. At present, Shanghai and other cities in China have formulated "Guidelines For Street Design", and Tianjin can be more systematic and comprehensive.

Next, the eye-catching title on the display board comes into his view: "Guarantee Two Parking Spaces for Each Household." Song Yunfei is shocked. Garages that can fit two cars are standard in American suburbs, but in China, the lack of parking spaces has always been a big problem. Moreover, the ground traffic is congested in the urban center and some areas. Cities such as Beijing have implemented methods such as limiting the purchase and registration of automobiles and promoting public transportation with an emphasis on the subways, but it cannot be denied that the automobile industry is one of China's pillars of economy, and this goes for metropolises such as Beijing, Shanghai, Tianjin and Guangzhou. The problems of pollution and traffic congestion will gradually be addressed with the development of technologies such as new energy vehicles and driverless cars, but is it appropriate to bring up "Two Parking Spaces" this early? Song Yunfei, who specializes in traffic parking policies, isn't sure either. He is about to read the detailed explanations on the board when the broadcast announces that the seminar is about to begin. Song Yunfei walks toward the conference hall with questions.

10. Where will you live tomorrow?

Old Zhang

From the arrangement and vocabulary of the exhibition, it can be clearly seen that the purpose of this exhibition is for ordinary citizens to understand the values conveyed by these urban design pilot projects. Old Zhang looks at the exhibition carefully for a long time; the words are very simple, unlike the confusing and deceptive advertising slogans at real estate fairs or the kind of professional jargons only experts understand. Old Zhang takes a good look and some of the words really touches his heart. For example, "people are walking animals and can't sit for a long time," which Old Zhang agrees with very much. His wife used to be sick in bed, which is a living example. Not being able to move is too painful, and he feels good every morning when he goes out to take a walk. But for some titles, such as "Humanized City", he doesn't quite understand what it means as it feels a little vague. "Walkable City"? Is it necessary to use this kind of terms? Isn't it just walking? Also, "limiting cars in urban center"—businesses are not doing well, and now they want to limit cars; without traffic, how many more shops will close their doors? And the proposal of building three more bridges over the Haihe River to connect several scenic spots —Will it be too dense, almost like putting a lid on top of the Haihe River? Old Zhang's whispering to himself draws the attention of a nearby volunteer, who explains: "These bridges are pedestrian bridges, and will be designed to be very light, so they will not affect the landscape of the Haihe River. Who knows, maybe they will even go viral on the Internet.""Go viral on the Internet"? Looking at the young face in front of him, Old Zhang secretly thinks: "Downy lips make thoughtless slips." He walks straight ahead.

These years, Old Zhang's life is really a far cry from the past, not having to worry about making a living anymore at all. But sometimes, a sense of melancholy will stem out of his heart, bringing in an emptiness. He often recalls the days when his family of seven, across three generations, lived in the two rooms of the tattered bungalow in the Old Town. Many of the houses were built with bricks mixed with mud; a bed would occupy half of a room; it was chaotic every time it rained; and every time they went to the grocery store to buy some meat, there wouldn't be any pieces that was worth more than 20 cents and it would all be fat. But Old Zhang incredibly misses that kind of life. Back then, neighbors often met and talked, and everyone knew which household was eating lo mein today and which family was quarreling. Grandma Wang was so nosy and high-pitched voice that she could be heard half the way down the alley, and was often scolded by her daughter. After Doudou's mom had given birth to her little brother, she always sat in the middle of the courtyard, passing on gossip while nursing the baby. Old Liu rushes to the bathroom with his back lumped every morning, and Old Zhang—who was Little Zhang at the time—and a few childhood friends timed his habit and deliberately occupy the stalls, making Old Liu so anxious that he kept walking in circles; they called this "holding the old man" as they liked to torture Old Liu Tou for fun. Every day when Little Zhang came home from school, he could see his grandma cooking in the small makeshift kitchen. There was a cast iron pot on the stove at the corner of the kitchen; if it happened to be the day of steaming cornbread cones, a hot steam would rise up as the steamer's lid was lifted, along with the delicious smell of corn. Grandma would pick and pull the cornbread cones one by one, and then mumble "hot" as she tore off a small piece and put it into Little Zhang's mouth. That was his happiest moment; In his memory, the smell of hot steam and corn was the smell of home... Now, he rarely eat cornbread cones, and when he occasionally takes a bite, he can no longer taste the flavor from his childhood.

Living conditions have indeed really improved. His parents have passed away, and the children have grown up. He lives on his own, with no grandchildren around, only a few family meals on holidays. The old practices in the past are also gone, like visiting relatives before holidays to pass out holiday gifts, thanking them for a year of care, or paying New Year's greetings to each other by the doorsteps with auspicious wishes, the junior bowing to the senior and the senior giving the junior red packets. Also, for example, cleaning the house till it is spotless before the New Year, putting up the Fu character and New Year's decorations, worshiping the kitchen god, setting off firecrackers and fireworks and making dumplings on New Year's Eve, and wearing new clothes and eating dumplings on New Year's Day. There are conventions on what should be eaten every day during the New Year's, and every day is full of things to do. It was tiring but lively, creating a feeling of steady and safe. Now, all of these old practices are gradually getting forgotten, not only in terms of celebrating the Chinese New Year, but also in daily life, which is completely different from the life in the bungalow in the Old Town in the past. There are fewer opportunities to meet friends and relatives, even to quarrel with or yell at each other, not to mention all sorts of crowd performances, consolation activities for workers and the elderly, and the group study sessions of important government documents organized by unites and communities. There would always be people coming out to mediate whenever a couple fought, and if a household bought a good item and put it in the courtyard, it would attract a bunch of people to look at and comment at it. It felt that like with everything there would be people talking, asking, putting their hands on. Now in the apartment unit, Old Zhang has lived there for several years and has not talked with the neighbor next door more than a handful of times. In the past, the bungalows were not as clean, and neighbors and relatives just causally dropping by and wearing shoes into the house was a regular thing; now the unit is too private, and whenever neighbors visit it feels like a prying of privacy, making everyone uncomfortable, so it results in casual greetings and not actual interactions or deep relationships.

When the old people in the community bask in the sun and chat on the small green square, what they talk about the most are the big affairs of the country and the city and the trivial matters of family, and also about going to the farmer's market and going to the hospital. Old Zhang's wife was very active and quick-tempered. She liked to be busy until she had a stroke a few years ago. Luckily it was treated in time and she did not suffer from serious consequences, but it damaged her motor functions on the right side and made it difficult for her to go about, which has created a constant bad mood for her. Old Zhang tries his best to take care of her and have her eat well, but his wife is unavoidably depressed and everyone is consequently in a bad mood. The children are very busy now and can only get together with them during holidays; usually they just make occasional phone calls and forward some jokes in the WeChat group. During the recent Labor Day holiday, the eldest son and his wife went out to eat with the old couple, and told them that their grandson and his wife were planning to go back to China, which the old couple were happy to hear. The daughter-in-law also said that business was slow in the shop, with labor costs rising fast, and it was getting difficult to maintain the business. She said that the big shops in the Ancient Cultural Street, like "Zhang's Clay Figurines Family", and a lot of other shops were building brands. They also wanted to try to do brand shops, set up studios and open new shops in new locations. Hearing the daughter-in-law saying that the business was rough, Old Zhang disagreed. He believed that the location of the store was the most important, and as long as the location of the store was good, all the other problems were caused by the operation. Hearing Lao Zhang saying this, everyone fell into silence; the son looked at his phone and the daughter-in-law went out to make a phone call. Old Zhang knew that he had said something that the others disliked and knowingly stopped talking. He could not help but feel a little annoyed. He had no control over the children, and saying things wouldn't make any differences. Walking by the Haihe River early every morning is the happiest thing for Old Zhang. In addition, helping look for a house for his grandson is also something happy to do for himself. After all, he has lived a long life, and there is no point in cutting off one's nose to spite one's face and bringing troubles on himself. Old Zhang knows this truth.

Recently, there is another thing in Old Zhang's mind. A while ago his cousin came to his home, brought a lot of top grade fruit, expressed concern for Old Zhang's wife's health, praised Old Zhang's house, and then started talking about his own troubles. The cousin's son was about to get married and was waiting for the marriage house. Originally, they had been looking forward to the government demolition of their family's old house and to buy a new house with the compensation from the demolition, but there has been no words on the demolition recently, and the cousin's son was getting impatient. The cousin and his wife talked about selling the old house, using the payment plus their own savings for the down payment to buy a house for their son to get married, and they would live in a rental house. Hearing this, Old Zhang felt bad and didn't know what to say. Old Zhang's father had three brothers, all of whom were born in the Old Town; after growing up, each brother started his own family and lived close to each other. Old Zhang and his cousin used to often play together with the children in the alley and were close to each other. After the liberation, Old Zhang's second uncle got assigned a big room within a two-storey building, located on the edge of Quanyechang , Because he used to work as a underground Party member and later worked at a government office. Although he had to share the house with a few other families, but they had a common kitchen and bathroom, inspiring feelings of envy from the Zhang's. Old Zhang went to school in the Old Town. Although Zhongying Elementary School was the earliest modern school in Tianjin, all the schoolhouses were single-storey houses; but his cousin's school was full of tall buildings made of cement. Later, the cousin went to the famous Huiwen Middle School. Old Zhang lived in the courtyard house in the Old Town, which became a shattered bungalows as the residents increased, but the living conditions and facilities were not improved along with the amount of people. In the winter, he had to go outside to the public restroom at the entrance of the alley, the water tap would be frozen and had to be defrosted by pouring boiling water on it, and he could only shower once every one or two weeks when he went to the public bathhouse. But the cousin's house had a bathroom, a Western toilet, and he could take showers at home. Old Zhang dreamed about living in such a great small villa one day. However, things changed quickly, and had not been in his second uncle's favor ever since the "Three-anti and Five-anti Campaign" began. During the Cultural Revolution, his second uncle and aunt were sent to the countryside, leaving the cousin at home by himself. Although he had been doing very well on his studies and had taken the college entrance examination, the Cultural Revolution came suddenly and he didn't get to go to college. He went to the countryside to live and work in a production team in Inner Mongolia.

Although the cousin received a few blows, he was a proud man, and yet with his mediocre family background he had not been able to find someone to marry. At the beginning of the Reform and Opening Up, it was very difficult to find a room in the city, and the cousin who had a spacious room became popular and finally found a sweetheart and got married. When his son was born one year later, he was already nearly middle-aged. The company the cousin worked at had general business benefit, and he did not have access to any backdoors, so he kept missing out on the assigned housing. After the reform of the housing system, housing was no longer directly allocated, and he only received a dozen thousands of yuan of monetary compensation. It can only be said that he had a bad luck and had missed the proper timing several times. With the rising of prices of commercial and second-hand housing, the cousin's chance of getting a good house was getting more and more slim. Their old house was getting more and more shabby, and with living quality in other dimensions having improved, housing became their biggest concern. On the other hand, Old Zhang, who had usually lived in a worse house than his cousin, was able to live in an apartment unit because of the demolition of the Old Town, and then bought a new apartment flat with the increasing amount of income from the small shop. Old Zhang occasionally went to his cousin's home to visit, and felt the old Japanese concession was becoming more and more similar to the Old Town in the past: cars parked everywhere on the narrow streets, electrical wires everywhere, illegal sheds and add-ons all over the place. The old appearances of the small villas had disappeared. Housing management station occasionally sent workers to do some repairing, but the houses and the pipelines were so old, simple repairs could not solve the problem. Therefore, the demolition was the cousin's only hope. But his old house was in the downtown area and densely populated, and people said there was not enough money for the demolition to take place, and there were also experts proposing to put this area under historic and cultural block protection. Whatever they had said, it resulted in the demolition not taking place.

A few years ago, the State Council pushed for the reconstruction of shantytowns and put in a lot of money, so some long-standing areas finally got to be demolished. But the buildings around Quanyechang were all above two or three stores and could not be counted as shantytowns. Moreover, the situation is totally different from before; buildings were possible to be demolished only if it was making way for "major infrastructure" such as subways. The demolition compensation amounted to almost 100 thousand yuan per square meter with the monetary reward for early movers. Although the cousin's family only had one room, it was a large room that was more than 20 square meters; plus the shared kitchen, bathroom and stairs whose floor area would be divided among residents, it should come to more than 30 square meters in total. With the demolition policy now, it would come down to nearly 3 million yuan of compensation, and the cousin was waiting for this money to buy a house.

But man proposes and God disposes, there had been no subway lines planned around this area so it could not be included in the infrastructure demolition. If simply selling it as an old house, as it was not school district housing, it could only sell for 1 million-something yuan tops. The cousin was really unwilling. Old Zhang had advised his cousin for many times to just let it go, but the cousin was stubborn. Looking at his cousin, his childhood friend who used to ride so high and who was now like this, Old Zhang felt bad. One should not be too honest and timid. This is Old Zhang's summary of his life experience.

11. What kind of house will you live in tomorrow?

Old Zhang

As Old Zhang looks at the exhibition, things flash by in his mind. He is familiar with the part of the exhibition about the past and present of Tianjin. Looking at the old photos from the past as well as the contrasting new photos from today, it feels like that Tianjin really has undergone earth-shaking changes. Slowly, he comes to the third part of the exhibition: Where will you live tomorrow? What kind of house will you live in? Under the title reads: Living is one of the main functions of the city. With the progress of human civilization, the forms of housing are also evolving, with the facilities more and more developed and the forms more and more diverse. Different places have different housings. What will your house be like tomorrow? Reading this question, Old Zhang has some doubts: aren't all housing apartment units? Only the few and wealthy can afford to live in villas; and according to the government, China has more population than land, so building villas are not allowed. But when Old Zhang visited his daughter's home abroad, he really realized the beauty of big houses. In his life, Old Zhang has lived in a bungalow in a courtyard house, an apartment unit on the fifth floor with no elevator, an apartment unit on the 18th floor with an elevator, and the big American house of his daughter's. Each house is closely related to the different stages of Old Zhang and becomes an integral part of his life. Old Zhang is glad that he has bought an apartment with an elevator, and the price was also not too high back then. After his wife's stroke, it has become inconvenient for her to move about, and Old Zhang has been helping her walk downstairs to take a walk every day. Good thing that there is the elevator, otherwise if the couple still live in the 5th-floor apartment in Huayuan, it would be very difficult for Old Zhang's wife to go downstairs. Old Zhang has heard that some old people live buildings with no elevators and it feels like being in jail for them after they have aged. Now there is rumor that Huayuan plans to install elevators for the buildings, but residents have different opinions on the matter—people living on the 6th floor insist on installing elevators while residents of the 1st floor are reluctant. It is hard to meet everyone's need. It costs 400 thousand yuan to install an elevator, and the operation, repair and maintenance all cost money. Especially for some older communities, property management fees are already cheap (usually a few yuan every month for every household), and there are also residents unwilling to pay, causing property management companies to rotate like crazy. Sometimes there isn't anyone there to collect the garbage and the residents can only rely on the government to solve the problem. Now the government is pushing for the renovation of old communities, re-painting the outer walls, replacing pipelines, installing security doors, replacing the courtyard pavement, repairing the community roads and so on. But some of the replacements are also of mediocre quality and break after a while. This can't solve the problem from the source.

Old Zhang sees a picture on the exhibition wall—The Evolution of Tianjin's Types of Residential Housing, starting from the northern Chinese courtyard house in the Old Town with one, two, three courtyards or more, to the colorful small villas during the time of concessions, more than diverse in form than courtyard house, and then the new workers' villages, neighborhood unit communities, Soviet-style enclosed big courtyards; later there were the rows upon rows of barracks multi-storey residences and large terraced residential communities, and by the 1990's , high-rise residences had been on the rise and there were also more forms to the exterior appearances of the houses. But for some reason, the houses built in recent years have all become high-rise residences, the same 30-storey high-rise buildings no matter in the city, outside the city, in the city center, or in the suburbs, and even in many counties. Old Zhang doesn't like it. Why do you live in the same house everywhere? There is a cartoon drawing on the exhibition board that depicts high-rise buildings aggregated together to become a boundless concrete jungle, and every house is like a pigeonhole in the sky. "Pigeonhole"—Old Zhang feels this word is quite adequate. Indeed, there is a sense of unexplainable loneliness and anxiety that comes with living in a high-rise flat, as if one is disconnected from the land and the society. He always worries about the possibility of fire and earthquake. While it feels lonely on the one hand, on the other hands it feels inconvenient to be influenced by other households—right after the move-in, every household was furnishing, and it was still not completely finished 2-3 years later. Finally, there were a few years without noises, and then someone sold their apartment and the new owners started re-furnishing. By this pattern, this is no end in sight. In this way, it seems better to have individual houses and avoid interfering with each other.

The narrator says that according to the reality of the large amount and low quality of existing housing in Tianjin, the new planned housing in Tianjin in the future expects to highlight diversity to meet the increasing demand of residents. The newly planned communities in the central city are mostly located by the outer ring road and adjacent to the large city parks and the green belt, providing not only good environments but also complementing facilities for education, health, and sports. The types of the housing are planned to be mostly low-rise parallel row houses, multi-level townhouses and lower high-rises, with a small amount of improved high-rise residence with higher standards than average. For the old apartments around the middle ring road that were gradually built since the 1980's and are now more than 30 years old, they should be gradually transformed through urban organic renewal into houses for migrants to settle down in and for young people to buy for the first time. In this way, the general housing level and diversity of Tianjin can be improved. As to the planned new communities around the eleven large parks on the outer ring road, Old Zhang knows a little. Meijiang area was built earlier and there is no brand-new apartment there now. Living there is not as convenient as in the downtown area, but the environment is good and the supporting facilities are relatively mature, but the housing price is relatively high. South Liberation Road Area has been built for several years. On its south end, it forms a group of residential estates with the surrounding of National Games Village. The park and most of the land in the middle have not been built yet, creating a sense of emptiness and marginalization. The area around West Water Park, which used to be called Houtai Park, is close to the expressway, the location is good, the park is built, and there is a large-scale modernized hospital. Currently the government is putting restrictions on housing prices, so these houses cost only 32,000 yuan per square meter or so, causing a shortage on housing supply; every time developers launch a batch of houses, they would immediately be snapped up, with hundreds of people signing up for a house and still with no way to buy.

Old Zhang mainly looks at the newly planned communities and writes them down one by one: The Pillow Park area along Haihe River is located at the junction of Jinnan District and Dongli District; Dongli District has Chenglin Park area and Nandian Park area. Although these new communities are inside the outer ring road, but Old Zhang who lives in the Old Town still feels like they are rather far away. The farthest is the Galaxy Scenic Area, outside old outer ring road and inside new outer ring road, adjacent to the Beijing-Binhai High-Speed Railway Beichen Station, takes an hour-long car ride and feels like basically the countryside. The narrator says that these parks and the new types of ecological communities can increase more than 10 square kilometers of large park green space, and with all levels of community green space, the total park green space can be up to 12 square meters per capita. The total floor area of newly built housing will be 30 million square meters and nearly 300 thousand sets, meeting the residence-improving demand of 1 million people by 2035. Therefore the next 15 years are still the prime residential construction period of Tianjin. What does this mean? Old Zhang needs to chew on it more, but he has a vague feeling that Tianjin will have many good parks and good houses next to the good parks around the outer ring road.

Next, Old Zhang sees a "Map of Housing Type Distribution". The narrator explains that just like commercial buildings and high-rise business buildings have their own distribution patterns, different types of residential buildings also have their own distribution patterns. High-rise residences are generally located in the center of cities or districts, lower high-rises and multi-storey residences in urban areas surrounding the city center, and low-rise housing on the outskirts of cities. This had to do with the urban land price distribution, population density, job density, traffic pattern and so on. Old Zhang recalls that he realized when he visited his daughter in the U.S. that there were only high-rises in the center of the cities, with multi-storey buildings and row houses on the outer areas and regular houses in the suburbs. It got lower and lower from inside to outside and higher and higher from outside to inside, and the downtown area was immediately recognizable. But it doesn't seem like this in China, with the inside and outside both of the same height, all 100 meters high. Isn't that a violation of the rules?

The narrator also says that people like to be close to the natural world by nature. With the widespread of railways and trams, the phenomenon of urban suburbanization has emerged, and the popularity of private cars has promoted the sprawl of suburbanization. We want to avoid the unorganized urban sprawl of suburbanization, and also satisfy people's yearning for nature at the same time. Here we have referred to the concept of New Urbanism and planned the city using a relatively compact layout, ensuring the balance between the city and the ecological environment. In the planning of new types of communities around the outer ring road, not only a lot of parks were planned, but a floor area ratio of 1.2 or so was also ensured to avoid overcrowding. Also, for the suburbs outside the outer ring, different from the low-density sprawl in countries like the U.S., the problems in our current situation are too large of a planned expansion, too high of buildings, and too intense the development. If this continued development is allowed, Tianjin's central urban area will form a development tendency of expanding over the 500-meter wide outer ring green belt, similar to the process of spreading pancake dough. This is not the "crepe" in the U.S., but a thick pancake, like the high-intensity sprawl of urban areas in Beijing.

12. People's Parks

Mr. Bauer

After the opening ceremony of the Urban Design Exhibition, Mr. Bauer finally has the opportunity to say hello to the busy Chief Planner ZhuGe and confirm the time and place of their meeting at noon. Bauer knows all about Tianjin's history and urban design. He is more concerned about the content of the Tianjin urban design pilot projects. "How will you travel tomorrow?" Walkable city is a worldwide trend, and Bauer thinks Tianjin is on the right track, and he has already seen the good results this morning. Bauer also approves the notion of "Two Parking Spaces", but only if the floor area ratio(FAR) of the land can be decreased. As an architect, his views on cars are consistent with those of the great masters such as Le Corbusier and Frank Lloyd Wright, and They are not against cars. The key is how to do urban design more intelligently.

The next section title is "Which park do you choose to go to?" This is also an attractive topic. Bauer learns from his wife Shirley and Chief Chu that Tianjin will build a lot of parks in the coming years—seven parks just in the urban center, and that a huge ecological barrier is planned between the central city and Binhai New Area. The plan is intended that each park will have its own characteristics, forming a Green Necklace. Bauer associates it with the Emerald Necklace designed by Olmsted, a pioneer in American urban planning and landscape architecture, in Boston. There is no doubt about the importance of parks to people; they are crucial to a city. In feudal society, the royal families had hunting gardens, which gradually evolved into royal gardens. After the Industrial Revolution, cities rapidly expanded and parks became an important means to cure "urban diseases". By the end of the 20th century, people all over the world had become aware of the importance of ecological sustainability, and parks became the main medium to cure pollution, improve ecology, and for harmonious coexistence between human and nature. Among the five strategies of "A Region at Risk", the fourth tri-state regional plan of New York in 1996, Greenwards strategy was also ranked first among the other five.

Nowadays, the ecological environments of western developed countries have been greatly improved. However, in order to achieve the goals of economic transformation and upgrading, fundamental improvement of ecological environment and easing the mental pressure of citizens in China, park construction is a key point, and it may be more difficult to design and build high-level parks than architecture. Then comes the questions, "What kind of housing will you live in? Where will you live?" Bauer feels like it is just like the philosophical questions of "Who are you? Where are you going to?"—they directly pinpoint the core issues of urban planning and urban design in China. ZhuGe realizes that the urban design in Tianjin has entered a new era of giving weight to people and their psychological needs, from the initial physical and functional design, from the action plan of the Haihe comprehensive development to the emphasis on green ecology and smart technology.

13. The Renewal and Revival of Quanyechang Area

Old Zhang

Coming to the "Renaissance of Urban Centers" section, Old Zhang sees a new term "Urban Renewal". He understands what the "three-level pit" reconstruction, post-earthquake reconstruction, dilapidated bungalow reconstruction, shantytown reconstruction, subway construction demolition all means, since those are all things that his relatives and friends have experienced and understood the policies, but what exactly does "renewal" mean? What benefits can the common people involved in urban renewal get? The narrator says that there are 70 million square meters of multi-storey houses without elevators in Tianjin, which were built from the 1970s to 2000. They are mainly distributed around the middle ring road, with large quantity and widely scattered. The houses are all more than 20 or 30 years old, and are in need of renewal because the pipelines have deteriorated and the outer walls are not up to the newer energy standards. Because the majority of the buildings are brick-concrete structure, the indoor dividing walls, supporting walls at same time, are difficult to alter, and the units are too small to have independent living room or dining room. Nearly 3 million urban residents and migrants live in these kind of houses, and parking is a big problem. However, the locations of these multi-storey houses are relatively good, with good environments and complementing facilities and bus services. From the experience of European and other countries, urban renewal and reconstruction, besides installing elevators and repairing the pipelines, need to reduce the units and increase floor area of each unit. To put it simply, it means to merge three units of housing into two units of housing. Aside from technical problems such as building structures, how to realize the economic balance, how to formulate a specific policy, how the negotiation mechanism works—these are also important questions to think about. "It's not easy! Old Zhang blurts out. He has lived in Huayuan for five years and knows the difficulty of it. The problem is who leaves and who stays. Why should the ones move out, what benefits can they get and what kind of compensation can they receive? The ones who stay will have a bigger apartment and elevators, and the house could possibly be in a future school district.

But these cannot be gained without a price; the ones who stay have to pay for... It's too complicated.

The title of "Urban Renewal of the Quanyechang Area in the Central Commercial District of Tianjin" is very long, but Old Zhang catches the word "Quanyechang" at the first glance. He is here to look at houses for his grandson, but in fact his grandson is doing well on his own and does not need Old Zhang to worry after him. It is his cousin's housing difficulties indeed that are making Old Zhang's feel troubled. Although old, Old Zhang still has the loyalty of Tianjin people in his bones, and hopes the best for his whole family. He does not want the situation of some people being filthy rich while some people struggle to make ends meet. He recalls the famous line of Mr. Ma Sanli, Tianjin's reputed Xiangsheng (comic talkshow) performer: "What is this all for? Just have a laugh." The urban design scope of the surrounding areas of Quanyechang is south to Nanjing Road, north to the Haihe River, east to Chifeng Road, and west to Duolun Road. The enclosed area is 1.5 square kilometers. Now it is the central commercial district of Tianjin, with Heping Road and Binjiang Avenue's pedestrian shopping street still attracting constant flow of people. Historically, it was once the concession of France, Britain and Japan. At that time, it was also the most prosperous and fashionable commercial and entertainment district in the north. From the former state prime minister, warlords and bureaucrats to wealthy ladies and company staff, everyone can find something for themselves in Quanyichang, Hendry, Shengxifu... The latest and most fashionable items in Europe and the United States can soon be found here. All of the country had heard about the exciting modernity in Tianjin, and a lot of famous Beijing opera performers had come out of the Chinese Opera Theater not far from Quanyechang. After the liberation, it was still a fashion center known across the three northern regions of China. At that time, even people from Beijing thought that Tianjiners knew how to dress up and eat well. In the later period of the Cultural Revolution, with the decline of the national economy, the scenery of Heping Road and Binjiang Avenue was no more. In 1976, Tangshan Earthquake had impacted here, and temporary earthquake shelters had filled the street side of Nanjing Road and the Haihe River bank. With the Reform and Opening Up in 1978, Tianjin carried out the post-earthquake reconstruction, and the urban construction and development were robust, which can be said to be at the forefront of the country. Many cities sent people to Tianjin to learn from. The Old Longtou Railway Station by the Haihe River was reconstructed, and had a naked oil painting depicting the myth of Jingwei filling the sea with pebbles painted on the dome of the hall; it was unprecedented and caused wide discussions. The International Mall was built in front of the Xikai Church at the end of Binjiang Avenue, showcasing and selling exotic merchandises from developed countries; then Binjiang Avenue also became a quite busy pedestrian street with vendors, selling the trendiest merchandises from Guangzhou at the time. In the following 30 years, many large shopping malls were gradually built alongside Binjiang Avenue and Heping Road, such as Binjiang Shopping Mall, Youyi Shopping Mall, New Quanyechang Building, New Baihuo Dalou Building, Lebin Department Store, ISETAN and so on. After the Comprehensive Development and Reconstruction of Haihe River in 2002, Jinmen, Tianjin Tower, Riverside 66 Shopping Center, and the recently opened Tee Mall were built between Haihe River and Heping Road. The old Nanshi "nobody's jurisdiction" area to the north of Quanyechange was also comprehensively transformed and high-rises were built up, leaving the Quanyechang area in the middle having not been renovated for years.

The narrator explains that the Quanyechang area is located in the center of the city. In the 1950s, Soviet expert Mochin helped guide the compilation of Tianjin city master plan. He tried to adopt the popular classical urban design technique at that time by planning a green axis from the Central Square through the Quanyechang area to Tianjin University, Nankai University and Water Park, which was used as an order to integrate the collage and chaos of urban layout caused by the separate administration of each concessions. In 1986, the state council approved the "Tianjin's Comprehensive City Master Plan" and put forward the "3 rings and 14 radiation rays" road framework, and Tianjin became one of the second batch of famous historic and cultural cities in the country. In 1994, the 1986th citymaster plan began to be revised, and the new version highlighted the important role of the Haihe River, proposed the Hami Road Green Axis, and also drew nine historic and cultural protection areas and five historic and cultural areas known as "the 14 areas", including both sides of the Haihe River, Central Garden, Chifeng Road, and Anshan Road. The role of Heping Road and Binjiang Avenue as the city's central commercial district and historic and cultural protection area was also emphasized again. After that, only 60 hectares of mainly residential land in the Quanyechang area were not defined as special districts. Over the years, many attempts have been made to carry out redevelopment. However, due to the high population density and relatively small land mass, the technical regulations on sunlight and parking were difficult to meet, and an economic balance was hard to reach too. More than 20 years have passed, and the large-scale demolition and reconstruction in Tianjin has almost come to an end. Surrounded by high-rise buildings, this non-historic protection area with relatively intact urban texture and severe building damage is left. Although the residents live near the city center, the living cost is expensive and it is not convenient to buy groceries or park cars. Some of the owners are no longer living here, renting out the houses to migrant workers or the families who are accompanying patients who are hospitalized in the Blood Institute and Hospital nearby.

After several in-person visits, the secretary of the municipal committee pointed out that there should be no slums behind high-rise buildings. The district committee and district government were determined to change this situation. They went to study from Shanghai, Guangzhou and other places to prepare for urban renewal. After studying and analyzing experience from domestic and foreign cities, the TUPDI put forward an urban design scheme for the interaction between the old area and the new district based on the actual situation. The content consisted of two parts. First, urban renewal in the area of Quanyechang to reduce the population density, improve living conditions and environment, and develop creative commercial blocks, which will supplement the linear commercial streets and large commercial streets on Heping Road and Binjiang Avenue, and form a city-level central commercial block covering an area of 1 square kilometer. Second, site selection and planning for an around-2-square-kilometer "Heping Quanye New Town" in areas near the outer ring road. In terms of layout Heping Quanye New Town, considering restoration of the spatial scale and neighborhood relations of the old Quanyechang area and drawing the good education and health resources in Heping District. It would serve not only as a population density relief for the Quanyechang area, but also aim to realize a higher quality of life for resident through the development of tourism, elderly care and other related industries, ensuring the employment of residents and becoming the "urban life engine" driving the development of the region.

As a new community, the housing types here are abundant, forming a diverse community life. In addition to continuing some of the traditional residential building types in the Quanyechang area, the planning and design of public rental housing, "well-off" housing, high-end housing and other types of housing are also carried out. The area around the Quanyechang has a population of more than 40,000, and the plan is to resettle about 20,000 people elsewhere. At present, most of the residents live in public rental housing, and each household area is very small, less than 20 square meters on average. For the families with real housing difficulties, even if they are given monetary compensation, the amount would make it difficult for them to purchase appropriate housing. Due to a series of subsidy and incentive policies of "Heping Quanye New Town" and the cheap land price adjacent to the outer ring road, it is relatively easy for residents to apply for public rental housing that meets the national housing standards. For the vacated old houses, the district government can choose some of them to continue to use as public rental houses, and the other part can be used for a mixuse of functions such as creative and services according to the planning, which can be either rented or sold. The part sold can retrieve some funds to relieve the financial pressure of the construction of "Heping Quanye New Town". The total investment of the 2-square-kilometer "Heping Quanye New Town " is estimated to be more than 30 billion yuan, and the district government's investment and debt should be about 10 billion yuan through the operation of the platform company, including land acquisition fee of 3.0 billion yuan, infrastructure construction cost of 3 billion yuan, construction cost of public rental housing of 3 billion yuan, and other unforeseeable expenses of 1.0 billion yuan. "Heping Quanye New Town" has an available-for-sale building floor area of 1 million square meters, equating to a fund of 8 billion yuan according to the land price of 8,000 yuan per square meter floor area, which can balance the costs of land acquisition, infrastructure and park construction. Implementing small-scale and micro-updates to the Quanyechang area based on residents' willingness; the ones who are in dire need to improve living conditions can choose public rental housing in Heping Quanye New Town, with a unit area no less than 20 square meters per person and a rent level is slightly higher than the past, with a certain property management fee. After the residents move away, the area would be remodeled, repaired or reconstructed based on the actual situation, no widening the streets, nor dramatically increasing building heights, preserving the traditional pleasant street scale and urban texture.

So much content and all new concepts, different from the old practices of demolition compensation, Old Zhang is having trouble understanding in this short amount of time. The Quanyechang area on the illustration is still the way it is now, not with high-rises, and the plan for the "Heping Quanye New Town" to the next is very similar to Quanyechang area, with a central park, the Huiwen Middle School, Anshan Road Elementary School, and the Hami Road Green Axis, as if the two areas are blood brothers, both alike and not exactly the same. "Old uncle, the reason why Heping

Quanye New Town is designed like Quanyechang is to make the residents there feel familiar," a volunteer comes up and explains to Old Zhang. Old Zhang nods vaguely. He really does not like it when people call him old uncle.

"How is this old house to be valued, young lady? Would there be compensation for those without an ownership certificate? How much is it per square meter?"

"Old uncle, this has nothing to do with money."

"How come? I know all about it. It's either cash or a deposit book, or even a card, the money is all transferred into the card and you can have it once you sign the demolition compensation agreement. Whoever moves first would get a reward of a hundred-something thousand bucks!"

"You are talking about the monetary-compensation-based relocation. Now the area of Quanyechang follows the policy of urban renewal. Based on the existing housing and family conditions, we issue vouchers to the families for renting units in Heping Quanye New Town. Of course, they can also choose to apply to live elsewhere with public rental housings."

"But the Heping Quanye New Town is near outer ring road! Won't you give me any compensation for moving me from downtown to the outer ring road?"

"Old uncle, urban renewal is voluntary; you can move if you want to live better, and you can also stay in the old place. It's just that in that case if you want to improve living conditions, you have to wait until there are neighbors moving out. After the housing authority has renovated the empty old units, there would be a lottery to decide who gets to live in it first. "

"Will that do?"

"Old uncle, this is just a pilot project. They have proposed a lot of different ideas, you know that this area has not been demolished over so many years because of the cost. There is not so much money just from government finance. The government did its best to build Heping Quanye New Town. Once the planning and design are done you will understand what you have gained. Look at this room layout—better than an apartment unit. Even I want to go live there, but I don't meet the conditions. The houses are limited in number and priority is given to residents who have housing difficulties around the Quanyechang."

"If there is an older child in the family who is getting married, can they family get two units?" Old Zhang thinks of his cousin's situation.

"It should work, but we only do the planning and design. About the specific policies and management methods you need to learn from the District Housing Management Department. The design has just come out and is on display starting from today. After that, residents and other parties will be asked for their opinions and approval before implementation."

"Oh, thank you, miss!"

Old Zhang thinks that his cousin and his cousin's wife should have no problem with getting two houses from this plan, but he doesn't know if the nephew and his future wife would agree to it. Why? Because this way their marriage house would be public rental housing with no property rights for their own.

14. Entrepreneurs

Developers

Old Zhang has seen what he wants to see, and the questions he wants to know are answered, but the answers make Old Zhang have more questions in his mind. Old Zhang is anxious. He sees the crowd at the door of the conference hall and goes over. "Old uncle, the exhibition is over there." The staff at the door stops Old Zhang. "I've seen it. I'm going in.""Old uncle, this is an international seminar, you may not understand.""That's why I'm going in to learn." Seeing Old Zhang getting anxious, someone next to them who seems like an official figure says: "Don't worry, old uncle, let's come here and talk it out." Then a middle-aged fat man comes over: "Director Liu, I'll take it from here. Uncle, tell me what you're looking for. You want to buy a house, right?""How do you know? "

"I am from a real estate development company. What kind of house do you want to buy?"

Old Zhang has looked at the exhibition for a long time, and it is just like his youngest son has reminded him—the content of the exhibition is too professional, and although they have used a lot of simple and straightforward language, he still does not quite understand, feeling like it's not down to earth. There is also not specific house-selling information. Now having met a developer, he wants to ask his questions.

"The exhibition says there are new communities around the eleven outer ring parks. Are there any for sale?"

"Yes, a lot! There are a few properties recently started selling around The Pillow Park. You can go take a look; it's very good. Let me give you Miss Wang's business card. She's a real estate agent in that area." Old Zhang takes the card. "Uncle, if that's everything, I'll go to the seminar.""Wait a minute. What does "the urban renewal of the Quanyechang area" mean in the exhibition? Where exactly is Heping Quanye New Town? Has construction begun?""The city renewal of Quyechang is a new thing and I am also learning. The specific location is by the outer ring road, inside the new ring road, a 40-minute drive from the expressway if avoiding the morning and evening rushes. The Park has already started construction.

Old Zhang looks at his phone. It is already 11 o 'clock, and he has to go home and cook for his wife. He sends a WeChat message to his eldest son: "Come and look a house with me at 2:00 PM." He is putting the mobile phone in his bag when it rings with a received WeChat message, his eldest son replying with "OK, yes sir."

Entrepreneurs

"High-Intensity Urban Sprawl" is how Carl Thorpe, the founder of the American New Urbanism, describes Beijing. He opposes the sprawl of suburbanization, but he is more concerned about the construction of high-rise housing blocks in urban suburbs. There are 1000 square kilometers of land between Tianjin's outer ring and the outer freeway loop, more than 60% of which is planned urban construction land, around 600 square kilometers, 100 square kilometers of which is undeveloped residential land. If calculated using the floor area ratio of 1.0, a total residential space of 100 million square meters can be built, and if calculated by the standard of 100 square meters per household, there will be 1 million houses with 2.8 million residents. In this way, the main urban area of Tianjin will become a big "pancake" with a population of 10 million and an area of more than 1000 square kilometers. The traffic congestion and environmental pollution problems will be several times more serious than Beijing today, and there will be no need or possibility for Binhai New Area to be built. So the development intensity must be controlled, building heights reduced, amount of construction land decreased, water area and green space increased. Like the plan for the ecological barrier between the central city and Binhai New Area, The ecological barrier is spread across four districts. The center of the district are multi-storey residences, and the surrounding area are low-rise row houses or detached houses or courtyard houses. Under the shade of green trees, it perfectly integrates with nature.

In addition to experts and officials, most of the participants in this international urban design forum and seminar are planners and designers from various planning and design institutes as well as university teachers and students, and there are also a small number of developers. It has become a common practice to solicit opinions from development companies when compiling urban design. In addition to developers, the forum has also specifically invited several entrepreneurs, and large investors to attend. To be successful in urban design, apart from the support and guidance of the government, the key lies in the response of the market. Developers are actual operators who understand the needs of the market. But developers are generally short-sighted and profit-driven, so entrepreneurs with a strategic vision are needed. Many of the pilot urban design projects in the exhibition require large, long-term investments. Several of the entrepreneurs at the conference looks at the data above and considers it carefully. Whether these data analysis is accurate or not, they intuitively believe that the thinking and direction are right.

According to the plan, there will be a 1-million population increase inside the outer ring, 1 million inside the new-four-district ring, 2 million in the Binhai New Area, and nearly 1 million in the five outer suburban districts, a total of 5 million. Anyone who knows Tianjin's population will know how hard it is to fill the houses with 300 thousand new people a year over the next 15 years. However, if only a large number of migrants are drawn here without industrial development and significant increase in other taxes, the city can only rely on the real estate sales as the main tax revenue, which cannot cover the education, health and other expenses brought by the additional 5 million people. Therefore, Tianjin must develop internally while absorbing migrant population.

Chapter III

Master Plan of the Redevelopment of the Haihe River in 1990s

15. Rediscover the Haihe River

Chief Planner ZhuGe Min

Mr. ZhuGe, who is always the chief planner of Tianjin Urban Planning Bureau, comes to the waterfront in the early morning. Today, he will have breakfast with Mr. Bauer. He is nearly 60 years old and was born in Jiangsu Province and came to work in Tianjin after his Doctor's degree. He has been working diligently in Tianjin for 30 years, and has witnessed the whole process of urban planning and development of Tianjin since the Reform and Opening Up. He is a disciple of Wu Liangyong, the master of urban planning in China, who given him a message upon graduation: To plan a city well, you must plant your roots in the city. This sentence has been with him for 30 years, and half of his life's efforts and struggles have melted into the city of Tianjin. He is using his own efforts to push for the inheritance of Tianjin's urban design tradition.

Mr. Bauer

At 7 a.m., Mr. Bauer arrives in the dining room downstairs and sees a familiar figure. He has invited an old friend for breakfast, Chief Planner ZhuGe of Tianjin Urban Planning Bureau, to talk about Tianjin and the Haihe River. Bauer is currently writing a book on Tianjin's urban planning, which can't go without the Haihe River. "Chief ZhuGe, we have met many times, but there are still things that I have not been able to ask you about in depth and that you have not been able to talk about in detail because of limited time. This time I want to ask you some important questions, such as about the Haihe River, the mother river of Tianjin. It runs through the city, and had a huge influence on Tianjin, and I want to know when did the Haihe River planning began, and what kind of evolution has it experienced?"

ZhuGe Min

Mr. Bauer's questions prompt Mr. ZhuGe's recollection. In the early 1990s, ZhuGe Min had graduated from Tsinghua University with his doctoral degree and came to Tianjin Urban Planning Bureau to work. He remembers that soon after he was reassigned to the chief engineer office of Tianjin Urban Planning Bureau along with several young colleagues. Under the direction of Liu Yuxiang, Shen Yingchu, Chen Yuebo and other chief engineers, they began compiling the master plan of the Haihe River according to the instructions of the Municipal Government. ZhuGe Min once accompanied several high level city officials on a boat tour on the Haihe River. In the golden autumn of October, they spent more than two hours traveling from Sancha Estuary to the outer ring road. At that time the flood season had passed and the water level was not too high, so they chose a shallower boat so that they could cross the dozen of bridges on Haihe River. The city officials, all in their 60s and with amiable faces, watched both sides of the river, admiring its graceful shape and appropriate width. He also heard and saw the expressions of regret on the faces of the officials when they talked about how there were so many bungalows, factories and warehouses scattered on both sides of the Haihe River, which had great potential to be transformed yet with no funds to realize it. Before disembarking from the boat, the city officials asked the planning bureau's leadership to do a good job in the planning of both sides of the Haihe River, and to strictly control to leave enough space for future developments. From the calmness in the officials' tones, ZhuGe Min felt a state of open-minded detachment. At that time, Tianjin had just completed the post- earthquake reconstruction, and the state support had expired while the urban economic transformation and upgrading were in a difficult period. The operation of the city was already very difficult, not to mention making large-scale investments in the Haihe River's development and renovation.

Without urgent need for development and renovation, there was no time pressure. This round of the Haihe River planning was carried out calmly and in-depth, taking more than two years. ZhuGe Min had looked at an abundant amount of historical documents and went to the site over and over again to survey. At that time, he only found an article by Chen Yongyang, deputy chief engineer, on the detailed construction plan of the Haihe River central urban area of "Tianjin Planning Newsletter". The plan was demonstrative and the attached detailed plan was relatively simple. The content was mainly to arrange some cultural buildings on both sides of the Haihe River. ZhuGe Min's graduation thesis is "Preliminary Research on Comparative Urban Planning", and he collected the planning schemes and experience in development of many waterfront cities in different countries. However, he had never visited other countries before graduation, and had only had opportunities to go abroad after he started working. The first time he went abroad, he visited Tailand, Malaysia and Singapore for survey purposes via Hong Kong, so at that time, he had not seen many of the world-famous waterfront city like Paris or London, not to mention an intuitive feeling and perception of them, and could only look at drawings and pictures and rely on his imagination.

Tianjin was prosperous because of the Haihe River. The construction and running of the Grand Canal between north and south during the Sui Dynasty made Tianjin an important transit station on the canal. In the Yuan Dynasty, the Grand Canal was silted up, and the water transportation was changed to sea transportation. From Dagu Kou into the Haihe River to Sancha Estuary, Tianjin's strategic position became increasingly prominent. In 1404, Ming Dynasty Chengzu set up the guard city, and Dagu Kou later became an important military fortress. The 72-kilometer winding the Haihe River from Sancha Estuary has recorded the evolution of Tianjin in the past six centuries. Compared with many coastal and riverside cities in the world, Tianjin not only follows the common pattern of urban development, but also has its own distinctive characteristics. In terms of transportation, with the increase in the volume of ships, the water transport in Tianjin's canal and the Haihe River gradually shifted to the seaport at the estuary. This is a common characteristic of many waterfront cities, so in 1986, the strategy of "industries strategically moving eastward and developing coastal area" from the "Tianjin Comprehensive Master Plan" approved by the state council was consistent with the pattern of urban development. On the other hand, Tianjin's Bohai Bay and the Haihe River it is nested by have many distinct characteristics. The Bohai Sea is a 70,000-square-kilometer inland sea that reaches deep into the land, so the chances of Tianjin suffering from typhoons and storm surges are small. But the self-purification ability of Bohai Sea is weak; it takes ten years for the seawater to circulate, and the coastal area of Tianjin is geologically a land formed by the recession of the sea into the land; the silt or the beach has a slope of only one of a thousandth, and when the tide falls it is several kilometers of mud beach. The seawater carries a high percentage of sand and is muddy. Tianjin is the most typical case of the saying "seeing no sea by the sea". Tianjin port is the largest artificial port in China excavated from a estuary. For many years in the past, the Tianjin Port waterway has been silted up continuously and required constant desilting.

The Haihe River is one of the seven major water systems in China. Tianjin is located in the lower reaches of nine rivers. Although it enjoys the benefits of being waterfront, it cannot avoid the harm of being waterfront either. Historically, Tianjin has often suffered from floods. Historical photos of the flooded Quanyechang area during a flood in the 1930s are unforgettable. After the founding of the People's Republic of China, there was a flood. The people of Hebei and Tianjin fought against the flood and saved the Tianjin and Tianjin-Pukou Railway from being flooded. In 1954, Chairman Mao gave an inscription: The problem with the Haihe River must be eradicated, fighting a people's war. So a large number of reservoirs were built in the upper reaches of the Haihe River, and the river course was dug deep and river bank raised. With all this, the floods were prevented, but the water from the upper reaches was much less available, and because of the lack of running water from upstream, the seawater rose along the river course at high tide, resulting in salinization of the water and salinization and alkalization of the earth on both sides. A ship lock was built at the mouth of the Haihe River to avoid rising tides. After the decrease of upstream inflow, the drinking water supply mainly relied on underground water, but the fluoride content was high, so the children who grew up in some areas of Tianjin (such as Donggu and Xigu in Tanggu) had yellow front teeth. In 1981, to solve the problem of drinking water, the state Council decided to implement the project of bringing Luanhe River into Tianjin and built a more than 200-kilometer-long canal to draw the Luanhe water from the Panjiakou Reservoir in Tangshan, reaching Tianjin through mountains and hills. By then, people in Tianjin could finally drink sweet water, and the Haihe River became the only alternate water source in Tianjin apart from the Yuqiao Water Reservoir. To avoid salinization, the second water gate were built in the middle reaches of the Haihe River. The 1986 version of the city's comprehensive master plan stated clearly that the Haihe River should "reserve water above the second water gate and run ships under the water gate." As a result, the upper reaches of the Haihe River were under strict environmental protection and various activities such as cruise ships were strictly controlled to prevent pollution of the water quality of the river.

When ZhuGe Min studied the role of the Haihe River, he saw that it was clearly stated in the 1986 Urban Master Plan that "The Haihe River is the scenic axis of Tianjin central city." At that time, he didn't quite understand—the Haihe was so important to Tianjin and was the mother river of Tianjin; Why was it only defined as the scenic axis? With the in-depth study of the complex functions of the River, he gradually understood the deep meaning behind it. Although there had been no major floods since the radical reconstruction of the Haihe River, they still had to be prepared for as there would be serious consequences by the silm chance it happened. Whenever the flood season came, the whole city prepared nervously as if facing the enemy.

The biggest headache for municipal leaders and water conservancy departments in charge of agriculture and water conservancy at that time was: should they let out the water in the Haihe River and other major channels, waiting for the upstream flood, or bet on the upstream not flooding and keep the precious water in the river? They dreamed of avoiding the awkward situation of letting out of the water and not having a flood and not having enough water for the next year. According to the flood control plan of the Haihe River at the time, the flood discharge capacity of the Haihe River should be 2,000 cubic meters per second. In other words, the height of the dikes on both sides of the Haihe River in the downtown should be 6 meters, up from 4 meters from the Dagu elevation line now. It was said that the top city officials and a professor of the Department of Water Conservancy of Tianjin University had an intense argument for this matter. If flood control walls of 6 meters high were built on both sides of the Haihe River, more than 2 meters above the ground, people's views would be blocked under these walls, and what landscape was there to speak of? What's more, with the years of siltation, the riverbed of the Haihe River kept rising, and the embankment also kept rising with the water level. At this time, the embankment was already 2 meters higher than the city's general ground level. If the embankment was to be more than 2 meters higher, it was no different from the river hanging over head; if there was an accident, the harm to the city was unimaginable.

Thanks to the abundant planning time for the Haihe River, ZhuGe Min was sent back to Tsinghua University by his advisor for half a year to work on the comparative study on the structures and forms of China's megacities. He still vividly remembers that there were only 13 megacities with a population of more than one million, and Tangshan ranked last with just over one million. According to the requirements of his advisor, he used the digitizer to enter the two-dimensional forms of Chinese mega-city built-up areas with more than a million population as well as major U.S. cities built-up areas into the computer, overlaying them under the same scale to compare, and found that China's mega-cities city were much smaller than that of the United States. His explanation was that American cities had entered the post-industrial era after suburbanization, and China was still in the early stage of industrialization and urbanization. However, this comparative finding made ZhuGe Min ponder over whether China's basic national policy of "Strictly controlling the size of big cities, reasonably developing medium cities and vigorously developing small towns" was correct or not. During this period, some economists proposed to give full play to the agglomeration effect of big cities and not limit the development of big cities in China. However, the planning circles ignored all these arguments, and believed that the absolute truth of the basic national policy should not be doubted or discussed. Seeing from the situation today, the strict control on the sizes of China's megacities, especially the capital, Beijing, was not only ineffective, but these cities further expanded and showed the tendency of "spreading pancake dough". This is different from the United States low-rise sprawl of suburbanization, but a development with the expansion of high-rise buildings, which Carlthorpe called "high-intensity urban sprawl", causing serious traffic congestion, environmental pollution and other "urban diseases". China's rural areas used to be very good living environments, but now a lot of people opposed suburbanization in a politically correct sense, and then the suburbs are all built into city-like high-rise buildings—a sad fact.

Choosing his thesis topic to be comparative urban planning, ZhuGe Min's original intention was to learn from advanced foreign experience and avoid the detours that Western developed countries had taken; however, in the end, China not only failed to avoid the West's detours and "urban diseases", but was also more seriously affected by the diseases. In a course of political economy, ZhuGe Min analyzed a large amount of data and wrote a paper suggesting that China's one-child policy should be changed immediately to allow one couple to have two children, or the country would soon face a severe aging population problem. It was funny to think of; at that time, the teacher gave his paper 85 points, not high or low, and ZhuGe Min did not know if the teacher had read it carefully. At that time, family planning was still a national policy, a taboo that could not be touched, and he was lucky that his teacher did not take it seriously. Today, it seems that China really should have changed the family planning policy earlier. This small antecdote shows that the reform of our country has entered a difficult phase, and it is necessary to reform some doctrines that were considered as unmovable as the Bible, especially in the field of urban planning. For example, by all accounts, there are more people than land in China, and China has adopted a very strict land management policy and required for high-intensity development, but the fact is there is too much wasted and inefficiently used land, and the amount of waste it has caused is so much worse than low-intensity development. Yet no one dares to "hope" for the rational reduction of development intensity. The notions of "too many people, too little land" and "high-intensity development" are hypnotic, which have blinded our minds and limited our imagination, creativity and motivation and vitality for reform. For these topics, he had not been able to conduct in-depth research, and also lacked practical experience at the time, and just often had ideas in mind. So he still put more energy into the study of world famous riverside cities such as Paris and London in his spare time. With the deepening of the study, his thoughts become clearer and more firm.

Back in Tianjin, ZhuGe Min began to organize the results of the Haihe River's planning. It was a thin mimeographed booklet with few illustrations; computers and printers were not widely popularized at the time, and text was printed from typewriters and illustrations hand-drawn with fine-point pens, with the layout also manually arranged. The title of the plan was very simple: Role and Plan of Both Sides of the Haihe River. The text was only 5000 words and was like a planning outline, with a focus on the role of the Haihe River and hoping to increase people's recognition of the importance of the Haihe River for Tianjin. At that time, the planning scope was only to the second water gate of the middle reaches of the Haihe River, a total length of 31.8 kilometers. First of all, the plan pointed out that: Tianjin's comprehensive master plan determines that Haihe River is a multi-functional river, and it should reserve water above the water gate and run ships under the water gate during non-flood seasons after the completion of the second gate; during flood seasons, it will serve the functions of intaking flood from upstream and rainwater in urban area. Then it pointed out that according to the present situation of the Haihe River and the arrangement of the comprehensive planning, the Haihe River mainly stores water and has functions such as flood discharge, rainwater discharge, water supply, shipping, traveling and hosting cultural and sports activities in flood seasons. For non-flood seasons, in addition to the above-mentioned functions, no rainwater discharge was allowed. These texts were carefully written under the supervision of several chief and senior engineers, slightly different from the city master plan, and were intended to add the functions of tourism and cultural activities. The crucial consideration behind not discharging rainwater in non-flood season was to protect the water quality of the Haihe River, as the first few rains in northern cities in spring were usually very dirty and polluted. Although there was little change in function, there were significant changes in the nature of the Haihe River's role compared to in the master plan. The plan clearly put forward for the first time that the Haihe River is the central axis of Tianjin, the central line running through the city's politics, economy, history and culture, and the lifeline of revitalizing Tianjin's economy and activating Tianjin with economic development in the center.

It should be noted here that in the past, it was generally agreed that the Haihe River was only a scenic axis and a secondary axis, because the development of Tianjin gravitated toward the southwest of the south bank of the Haihe River, so the axis from the Central Square to the Water Park in people's mind, or from the Old Town to the due south is the main axis of Tianjin. Over the years, due to the cut of the Haihe River and other reasons, the development of the north bank of Haihe River had been relatively slow. Meanwhile, historically there had been wharves, warehouses, and factories by the Haihe River. These functions were developed first with water freight, while urban life gradually deviated from the Haihe River and developed in the opposite direction. This planning of the Haihe River hoped to convert the factories and warehouses on both sides to commercial, financial and cultural service facilities, setting the Haihe River as the city's main axis, and the street perpendicular to the Haihe River, including the Harbin Road green axis, as the city's secondary axis, forming a balanced development in a "fishbone-shaped" structure with both sides of the Haihe River using it as an axis. In 1994, ZhuGe Min and several planners went to Australia to study the planning of the riverside, especially the Yarra River in Melbourne, its width and shape very similar to the Haihe River. Local planning officials said that in order to balance the development of both sides of the Yarra River, their plan proposed that the city of Melbourne should develop facing the Yarra River. ZhuGe Min immediately felt a connection across thousands of miles.

Mr. Bauer

Upon hearing this, Mr. Bauer asks, "Can all parties reach a consensus on the great changes in the understanding and role of the Haihe River?" The reason he asks this question is that Bauer has visited China many times. His wife is a Chinese-American born in Taiwan, so he knows China and Tianjin well. Chief ZhuGe has been a visiting scholar in the United States. His English pronunciation is not native-level, but his expression clear and accurate, so there is no communication barrier. "Your question is very important. Mr. Wu Liangyong once told me that a good plan requires consensus in three aspects: academic consensus, social consensus and leadership consensus . These words have been engraved on my mind, and I have realized more and more the importance of the three consensuses in 30 years of planning practices. The reason why the comprehensive development and renovation of the Haihe River was able to be successfully implemented in the end was because of the three consensuses, which took ten years." Chief ZhuGe especially put emphasis on the last sentence. "Ten years is not a long time for a planning project of this size," Bauer says. "It's a reasonable period. It took us 16 years from 1946 to 1962 to reach consensus to build the rapid transit system BART in the Bay Area.""That was in the United States, but China was at the height of Reform and Opening Up, and development was accelerating everywhere."

ZhuGe Min

ZhuGe Min goes on to explain. The experts all agreed on the elevation of the role of the Haihe River. Truthfully, at that time, few people paid attention to Haihe River, and since the plan did not involve concrete implementation yet, the other departments did not have too many opinions. In those years, Tianjin encountered great difficulties in its economy. For the post-earthquake reconstruction, the state gave Tianjin over 2 billion yuan of financial support, allocating several hundred million yuan every year. You know, several hundred million yuan was quite a large amount at that time. These funds guaranteed the construction of essentially road system and traffic facilities such as the "3 rings and 14 radiation rays" and housing as well as improvement in environment. When reconstruction ended, that money was gone. However, Tianjin's actual situation was that industries were generally backward, and factories had very poor profit, with many equipments and technologies from before the liberation. Companies was having a hard time and a large number of workers were laid off. At that time, there were many complaints from the society: some people said that all the money in the post-earthquake reconstruction was invested in urban construction, but not in industry, resulting in the lack of strength and sustainability in urban development; some said that Tianjin was too close to the capital of Beijing, and many big projects were snatched away by Beijing. For example, the state wanted to build a 300,000-ton large vinyl plant in northern China. Every city fought for it, and the result was that Beijing and Tianjin each built a small 140,000-ton vinyl plant, neither reaching a large scale. By 1992, there were different voices in China, one saying to continue Reform and Opening Up, the other saying to go back. In January, Deng Xiaoping visited Shenzhen and other cities and delivered an important speech on his southern tour, pointing out that China's Reform and Opening Up should continue, following the path of socialism with Chinese characteristics and developing market economy and bringing China's Reform and Opening Up to a new stage, a new climax.

Time passes quickly. Chief ZhuGe stands up: "Mr. Bauer, it's getting late. I have to go to the planning exhibition hall to prepare for today's meeting. We'll talk at noon."

16. One Pole Carries Two Ends

Song Yunfei

After deciding to return to China, Song Yunfei has compared Tianjin's downtown area with Binhai New Area, two cities carried by the pole that is the Haihe River. Although the central city has better supporting facilities and is more convenient, it always gives out a sense of lack of vitality. By contrast, he prefers the young, vibrant Binhai New

Area. After deciding to develop his career in Binhai New Area, Song Yunfei decides to place his office at Yujiapu. It is a Free Trade Zone with various preferential policies for start-up companies and convenient internal and external transportation. He has once participated in the transportation planning of Yujiapu Financial District, where a large transportation framework has been formed, and there are direct high-speed trains from Yujiapu to Tianjin Railway station and Beijing South Railway Station. The current office space and surrounding environment of Yujiapu are first-class, situated next by the Haihe Park with open views, pleasant pedestrian streets and established underground space, which can almost rival the business districts of domestic and even many foreign metropolises. Currently supporting residential facilities are not yet fully established. Originally he has wanted to buy a apartment at Yujiapu or Tianjian, where there are many parks, the Binhai Cultural and Art Center, with good house layouts and environment, but the schools there right now are mediocre. He has heard that the relevant departments are talking with some famous universities about setting up campuses here, but there is no word on a timeline. Not far away from here, the living environment of TEDA has already matured and there is an international school, but the children who are used to living in big houses still don't like it. Finally, they have chosen an ecological townhouse by the lake of Sino-Singapore Tianjin Eco-City. The environment, the schools and the housing here all give out the feel of an American suburb, and while you can see some high-rise residences, it's not too crowded yet.

Song Yunfei does traffic planning and is attentive to traffic problems. The Sino-Singapore Tianjin Eco-City has designed a set of targets for eco-city in its planning, and most of them have been realized, except for the large gap in transportation. The original technical target of the plan was that green travel should account for 80% of the total travel, but due to the long-term lag of subway construction, this target is far from being completed. When he was abroad, Song Yunfei and his family lived in a suburban house, just like Americans. They had two cars. He usually drove a Toyota Corolla Hybrid, and his wife drove a multi-purpose van for shopping and driving their kid. After arriving in Hong Kong, Song Yunfei mainly relied on public transportation and walking. Back in the mainland, he has wanted to experience green travel in all its aspects, like Jane Jacobs, but it seems too idealistic. Subways, buses and walking are not as convenient as cars, and living without a car in the Binhai New Area is impossible. He has discussed with his wife and bought a Tesla S3, which is not only environmentally friendly but also has a large internal space. His wife has originally wanted to follow her habit in the United States and buy a van, but parking in China is really a problem, inconvenient no matter in their own home or in public, so she gave it up.

17. The Century Renovation of Dilapidated Bungalows

ZhuGe Min

ZhuGe Min steps out of the hotel and into the car. Looking out of the window at the scene of the Haihe River bank, his thoughts are still going back. After the

Shenzhen special zone was set up, the setting up of Shanghai Pudong New Area was approved by the state council in 1992, and the Bohai Rim region, where Tianjin is located, was also accelerating its development. In March 1992, Motolora Tianjin factory was set up in TEDA, producing mobile phones and other products, along with products such as Xiali(Charade) automobiles and Master Kang instant noodles, which became the nationally renowned "one phone, one bowl of noodles and one car". Although great progress had been made in the Opening Up, it could not reverse the overall decline of Tianjin's industry. In order to implement the strategy of "industries moving eastward", the central urban area had implemented the "Double Optimization" project, which optimizes the urban land use structure by optimizing the industrial structure. Some industrial land, such as Friendship Cannery Factory and Renli Wool Mill, was converted into land for real estate development, but the range was limited. On the other hand, despite the post-earthquake reconstruction greatly improving Tianjin's urban functions, road traffic and infrastructure, due to years of war and disputes since the end of the Qing Dynasty and the beginning of the Republic of China, the urban construction has been stagnated for a long time. After the founding of new China, and the implementation of the "produce first and live later" policy, there was a lot left undone in terms of housing construction and other aspects. With the continuous growth of urban population, the problems of housing shortages and backward supporting facilities became more and more prominent. Chief ZhuGe clearly remembers one scene at the time: he accompanied officials of Planning Bureau to participate in a municipal government meeting. It was early spring and yet was still cold. At the beginning of the meeting, the mayor walked into the conference room wearing his green military quilted coat, looking like in a very bad mood. His eyes swept to the comrades from the Municipal Party Committee's Propaganda Department and immediately got into a mood: "Where have you people from the Propaganda Department been? Last night it rained heavily and many residential areas were flooded. I did not see a reporter from a TV station or a daily newspaper on the spot. The newspaper this morning said nothing about people being flooded and was all just some big talk. What's the use?"

At that time, most of the pipe network in Tianjin city was aging, especially in some low-lying bungalow areas, which would be flooded every time it rained. The residents had been complaining, but due to the lack of funds, it had not been solved for a long time. Tianjin learned from the experience of the Shenzhen, Guangzhou and Shanghai, and also wanted to introduce foreign capital to solve the problem of reconstruction fund shortage, but the housing market was not yet open to the public by then, so the city took out some land that was adjacent to Quanyechang and International Mall to attract foreign investment. The land was planned to be used for commercial and office projects, but only attracted a few foreign Chinese investors. The city originally hoped to build a office tower on both sides of the International Mall and make use of the advantages of Tianjin's metallurgy industry to create a national or regional market of steel raw materials, but the construction had not started for a long time. Chief ZhuGe once led a team to learn the experience of the south. After coming back, he compiled the controlled detailed plan of these land, and put forward the planning indexes such as the land use nature, floor area ratio(FAR) and building height for investment attraction. But the overall economy of the city was not good, and all kinds of work were affected. Worse, in February 1993, Tan Shaowen, secretary of Tianjin Municipal Party Committee and member of the Political Bureau of the Central Committee, died of illness at the age of 64. It was a rare case of main leadership figure passing away while still in position in the history of Tianjin.

Fast forward to 1994, when China's economy began to overheat as Deng Xiaoping's southern tour speech led to accelerated development activities in all cities. In 1992 and 1993, the growth rate of fixed asset investment was 42.5 percent and 58.6 percent respectively, much higher than that of previous years. Investment demand drove consumption demand, causing serious inflation and posing a huge risk to social stability. Since 1994, the central government had put forward a "moderately tight" fiscal policy to regulate and control the overheated economy. At this time, many Tianjiners began to complain that Tianjin was always unable to grasp the opportunity and find the right time; other cities had accelerated developments and Tianjiners had not yet woken up, and when they had woken up and wanted to develop, the macro-control happened. There were a lot of negative emotions in society. In fact, at the beginning of 1994, the new Tianjin Municipal Party Committee and Municipal Government analyzed the situation at home and abroad, and put forward and began to implement the goal of "Three, Five, Eight and Ten", especially from the point of view of Tianjin's reality, the requirements of the times, the historical mission and the aspirations of the people. "Three" was to quadruple GDP three years ahead of schedule; "Five" was to use five to seven years to complete the reconstruction of dilapidated bungalows in urban area; "Eight" was to use eight years to adjust, graft and transform state-owned large and medium-sized enterprises; "Ten" was to use ten years to mostly finish building Binhai New Area. It could be said that these four goals accurately grasped the key issues of Tianjin. But at that time, many people did not understand, including the planning community; the colleagues at the Planning Institute thought that the term "Three, Five, Eight and Ten" is too simple, and would be hard for outsiders to understand without explanations. The new Party Committee secretary attached great importance to the planning work, and went to the Urban Planning and Design Institute to study the issue of layout of harbor industrial planning in person. After learning that the Planning Institute was a department level institution, he instructed the Planning Institute to be upgraded to the deputy bureau level institution. On hearing this, the whole Planning Institute was so excited, so they no longer talked about their oppositions. The cardres of the Urban Planning Bureau were mainly opposed to the "use 5 to 7 years to complete the reconstruction of dilapidated bungalows in urban area" part; they argued that city development was a long-term process, and to use 5 to 7 years to finish transforming dilapidated bungalows, then what was left to do after that? The point was that future developments could not be predicted based on the current understanding and development level, so room needed to be left for the future. Indeed, there were many problems with the initial reconstruction of dilapidated bungalows. Due to the lack of funds from the city and the state, the project could only rely on the self-balance of land demolition and construction ratio, resulting in the design and construction standards of relocation housings to be economic. Some relocation housing had small floor areas and had less than one-square-meter bathrooms, where the doors could not even be closed after a person went in. Problems did come up but were also continuously being solved.

ZhuGe Min also had some opinions on these issues. He remembers going to Hong Kong's Housing Authority and learning about their housing crisis in the 1950s. Many people had nowhere to live and most could only live in shantytowns, where safety hazards like fires were very likely to happen. Therefore, the Hong Kong and British Government had implemented the renovation of shantytowns and new housing constructions. The newly built apartments were relatively dense seven-storey walk-ups, each with only a kitchen and bathrooms were shared by more than a dozen families on each floor. Due to low living standards and poor construction quality, the Hong Kong government began to demolish these multi-storey buildings with only 30 years of use in the 1980s. In order to balance the fund, they had to build taller super high-rises. Looking at Tianjin's implementation of the reconstruction—most of the new buildings were six-storey brick and concrete residences, with no elevator and small unit sizes; because of the brick and concrete structure, most of the walls were load-bearing and could not be removed or altered. ZhuGe Min always had a kind of worry: how would this large amount of buildings be upgraded 30, 40 years later? It's not like they could just copy Hong Kong's way. Hong Kong was limited by the amount of usable land and had rather low living standards in the world. On the other hand, ZhuGe Min also deeply empathized with the sufferings of the residents and realized the necessity and urgency of implementing housing reform. As a young planner, he was willing to stay in a city and get to know it. He had a hobby, which was to wander around the city whenever he had time. Not only did he often go to both sides of the Haihe River and the city's commercial centers and the old concession areas, but he also visited the Old Town, Nanshi, Qiandezhuang and other areas by bike. At that time, the living conditions of the Old Town and Nanshi was already very bad, with the main building materials of the original houses being bricks and mud, but they could to compare to Qiandezhuang as the real shantytown! The houses there could hardly be called houses. ZhuGe Min remembers that the city officials in promoting the reconstruction of dilapidated bungalows often said "to do the reconstruction with emotions", and stressed that "it has been more than 40 years since the revolution under the leadership of the Communist Party and the founding of new China, and yet the people still live in such dilapidated houses and such bad conditions. We as leaders and officials should feel ashamed!" Under the strong advocation of the Municipal Party Committee and Government and the collective efforts of the district committees, district governments, all levels of officials, urban development enterprises and the masses of residents, by 1999, the reconstruction of dilapidated bungalow areas was mostly completed. According to statistics, a total of 10 million square meters of dilapidated bungalows were demolished, benefiting 800,000 residents, accounting for about one third of the total urban population at that time. It was a huge accomplishment. Since 1999 was the last year of the 20th century, some people named it "the Century Renovation". The completion of the Century Renovation was of great significance to the future development of Tianjin. ZhuGe Min remembers in 2002, when preparing to implement the Comprehensive Development and Renovation of Both Sides of the Haihe River, the party secretary spoke meaningful words: "if we had implemented the Comprehensive Development and Renovation of the Haihe River ten year or five years earlier, some must would have said that we are doing vanity projects, because the people were still living in dilapidated bungalows, the city's road system was not perfect, the municipal ability insufficient, the facilities aged, and it often flooded when raining. Only today, when the reconstruction of dilapidated bungalows has been mostly completed and the urban infrastructure has been improved, can the necessary conditions be met for the Comprehensive Development and Renovation of the Haihe River."

The reconstruction of dilapidated bungalow areas had greatly improved the living conditions of urban residents, but some other dilapidated areas such as the Old Town, Nanshi and West Yuzhuang were not reconstructed due to funding issues. The reconstruction plans mostly retained the urban road pattern and street texture. Compared with other big cities in China, Tianjin led the country in the planning concepts of dilapidated housing reconstruction. As the ancient capital for thousands of years, Beijing had a large number of courtyard houses in its Old City area. Save besides a few were preserved better as orgnization units, most of the residential courtyard houses had bad living conditions because of age, disrepair, and the dwellers building and adding facilities illegally due to increased family population. However, because it was about the preservation of Old City areas, all sides of society paid a lot of attention, and it was difficult to carry out large-scale renovations. Through years of research, Mr. Wu Liangyong promoted the implementation of the renovation of Ju'er Hutong in the late 1990s, and successfully carried out a pilot for multi-storey courtyard houses, which was well received, but unfortunately it was not widely popularized. Most of the old part of Shanghai consisted of two-storey shikumen (stone houses) and lilong (back alley) buildings, which were of better quality than the average Chinese urban compound, but residents had go to the public bathrooms every morning to clean the toilets. The Old Town areas in Guangzhou were more densely located and with the rapid development of the city, a large number of urban villages were developing out of order. The living conditions of urban and rural residents and migrants were very chaotic and crowded. It could be said that at that time, Tianjin led the country in the reconstruction of dilapidated urban bungalows, 15 years earlier than the large-scale reconstructions of shantytowns in China after 2010.

The renovation of dilapidated bungalow areas is an important means to drive urban economic development. In urban economics, real estate is a very important component. Historically, Tianjin was a commercial circulation city. Some people say that Tianjin supported 3 million people by commercial circulation. At the beginning of the post-liberation period, Tianjin's light industrial products were relatively developed. Later, with the blockade of China by the imperialists and the policy of transforming consumption cities into production cities, Tianjin, as a port city of foreign trade, had neither raw materials nor water, hit a bottleneck with its industrial development. In this case, the government had no money to upgrade old enterprises, and no money to invest in large-scale projects. As a result, the economic growth rate had slowed down, the income of employees had decreased, and the service industry had not developed. The new city leadership of Tianjin thought hard and finally found a way to drive the overall economic development by advocating for the reconstruction of dilapidated bungalows. Since there was no reserved fund, the government could only rely on the market-oriented means of increasing the demolition and construction ratio, and using the sales revenue of the newly built commercial housings to cover the investment in reconstruction. The government provided policy support in the land taxes, fees and other aspects of the reduction of supporting fees. With the gradual spread of the reconstruction, Tianjin gradually formed a sizable real estate market—housing construction drove the development of construction materials and construction and other engineering enterprises; people moved into new homes and drove the sales of decoration, furnishing, furniture and other goods. The real estate industry chain drove the recovery of Tianjin's overall economy.

The success of the reconstruction of dilapidated bungalows is not only reflected in the improvement of residents' living conditions, the change of urban features and functions, and the promotion of urban economic development, but more importantly, the transformation of urban ethos and the liberation and progress of ideas. Before the liberation, Tianjin was a city open to the outside world, leading the fashion trend in northern China. However, from the founding of new China till the Cultural Revolution, Tianjin became a closed and stagnant city. Some people say that there are two kinds of Tianjiners, one is the open Tianjiner, and the other is the backward Tianjiner. Easily satisfied, unmotivated, relying on connections and valuing personal loyalty—these are typical characteristics of the Tianjiner. Feng Jicai called it the quay culture. But when urban development is in crisis and the city gets derided, Tianjiners are also bursting with strength. The reconstruction of dilapidated bungalows pushed down not only the old decayed bungalows, but also the old notion of holding on to the broken and being easily satisfied, and the psychological inertia of relying on planned economy and housing assignments from working institutes. In the late period of the reconstruction, the state implemented the reform plan of deepening the housing system, and no longer carried out physical housing distribution. By this time, Tianjin was already prepared. In order to promote the renovation and real estate development, the party secretary at that time always repeatedly told the story of American and Chinese old ladies buying houses: the American old lady bought a house on mortgage, died many years later, having paid off the mortgage and enjoyed a lifetime of good house; the Chinese old lady lived in a very poor house, saved up all her life to buy her own house, was finally able to afford it near the end of her life but died before she got to enjoy the new house. With this vivid example, the party secretary told people to have expectations for the future and to spend money unearned yet. This was not only for the officials and the people, but also for the whole city of Tianjin to hear. The party secretary had been engaged in industrial work for a long time, and had been a top manager of state-owned enterprises. Supposedly, he had been doing planned economy all this time, but he could understand the patterns of the market economy both from his personal experience and for the necessity of the times. Through the reconstruction of dilapidated houses, which touches the vital interests and hearts of the residents, the old Tianjiners concept began to change again. Although people still had complaints about the city, but the momentum of their desire to engage in work and entrepreneurship doubled.

18. Central Business District · Central Commercial District

ZhuGe Min

Six years of large-scale reconstruction can be said to be the prelude to the Comprehensive Development and Renovation of the Haihe River. During this period, ZhuGe Min conducted several important work closely related to the planning of the Comprehensive Development and Renovation of the Haihe River in future. In 1994, he cooperated with several young people from Tianjin Urban Planning and Design Institute (TUPDI) and completed the research project of "Planning of Central Business District (CBD) in Tianjin". The research summarized the basic theories about the CBD, and compared the patterns and individual characteristics of the CBDs of major domestic and international cities, and finally, according to the actual situation of Tianjin, proposed a scheme for the planning and site selection for Tianjin's CBD. It was to be located in the city center inside the inner ring road by the two sides of the Haihe River, covering an area of 1.8 square kilometers, relying on the historically financial function of the North Jiefang Road and the traditional commercial and trade function of Xiaobailou, and using warehouses such as Tianjin Old South Cargo Railway Station as a new development space on the other side of the Haihe River at the same time. This selected site continued and grew from the historical context of the city, and highlighted the Haihe River as the urban axis and the characteristics of its natural environment, combining the old with the new, and had been recognized and agreed by all parties. At that time, many domestic cities were planning and constructing CBD, represented by the Pudong Lujiazui Financial District in Shanghai. The vast majority learned from the experience of La Defense of Paris and the Docklands of London, leaving the old center for a "frog-leaping" planning and construction of the new CBD. The construction of La Defense in Paris was largely driven by the historic preservation requirements, which posed a ban on high-rise buildings, so a new modern CBD was planned outside the old city area. But ZhuGe Min knew that building a new CBD required huge investment and took a long time. The La Defense in Paris suffered a decade of stagnation, while Docklands in London had also bankrupted the famous development company Olympia and York. At that time, some people in Tianjin proposed to plan and locate new CBD in Youyi Road and other areas in the central city, but after repeated research, ZhuGe Min's team thought it was not suitable. For Tianjin was a a city with "One Pole Carries Two Ends" structure, a new CBD needed to be planned for Binhai New Area; at the same time, looking from the regional perspective, Beijing was northern China's largest commercial, financial and international institution center, and Tianjin did not need nor had the capability to plan another new CBD, and therefore the selected site on both sides of the Haihe River in urban center was the best choice. At the time, North Jiefang Road was where Tianjin's Municipal Party Committee, Municipal Government, the People's Congress, CPPCC Committee and other main governmental establishments were located, adjacent to the CBD, and the selected site also matched well with the Haihe River's functions specified in the "Role and Plan for Both Sides of the Haihe River" whose compilation ZhuGe Min had led in 1991. This again confirmed that the previous direction of in-depth development for the

Haihe River's role was realistic and pragmatic, which further solidified ZhuGe Min's confidence.

19. Revision of Urban Master Plan

Also in 1994, in order to welcome the new century, Tianjin Municipal Party Committee and Government decided to start the revision of the last urban master plan (1986-2000) approved by the state council, and to work out a new cross-century master plan for Tianjin (1994-2010). ZhuGe Min was only 31 years old at that time, having worked for 5 years after graduation. Under the trust of the leadership of TUPDI, he was entrusted with the important task as the leader of the working group for the revision of the master plan. There is a saying in the planning circle: to become a real planner, one must have done an urban master plan. Only through doing the master plan can one understand the breadth of urban planning and also have a comprehensive and in-depth understanding of a city. ZhuGe Min led a group of young people in the planning team to devote themselves wholly to their work. First of all, they must agree that the city master plan across the century should be advanced and innovative in theory and method, and also learning from the sustainable development agenda proposed by the United Nations conference on environmental development, should propose sustainable development in the guiding ideology of the urban master plan: "development that can meet the needs of the present without jeopardizing the ability of future generations to meet their needs." The leadership figures of the planning bureau thought the idea was good upon hearing this new concept, but was also uncertain, and asked: "now all domestic cities are accelerating development. Is it appropriate for us to propose sustainable development at this point?" It was not until 1997 that the 15th National Congress of the Communist Party of China explicitly took sustainable development as the long-term principle of China's economic and social construction that everyone realized the advanced nature of this concept. However, at that time, it was just that. The planners who put forward this concept did not have a mature way to implement the concept of sustainable development in the master plan, but only paid more attention to ecological environment and environmental protection, which was already advanced in China at that time.

Another relatively advanced approach was the use of computer-aided planning and design. Prior to this, the maps for the city master plans were made by planners by hand. A very large drawing plate needed to be customized, and the official planning map was made from having land map mounted on the drawing plate, rendered and colored with watercolor, and then cutting the drawing from the drawing plate after drying, so as to ensure the flatness of the map. At that time, the rapid development of personal computers led to a significant increase in computing speed, and the 586 processor and other good digital input and output equipments emerged, with an increasing amount of Local Area Network and application software. ZhuGe Min applied for TUPDI to purchase a computer aided system for the master plan and the leadership supported it, spent 170,000 yuan to purchase five computers with 586 processors and monitors with good graphic performances, a Tablet digitizer and a large color printer, connected to Wireless Local Area Network, and purchased the Geographic Information System software Automap, Office Powerpoint and animation software 3D Station, a LCD screen and a projector. At the time, the projector was so dim that simply a transparent LCD screen connected to a computer placed on a normal thin-film projector, with the so-called enhanced screen, was not able to be clearly projected. With this kind of equipments, the contents of the master plan for the first time in history were made into animation and powerpoint and presented to leadership at all levels including the People's Congress. All the master plan illustrations were drawn printed with computers, greatly improving the result and quality of the presentation. It was not only the first in the history of Tianjin's city master plan compilation, but also among all cities in China. Great achievements were made, but behind the success were also great efforts. To adopt new technologies, they must first learn new software and hardware and input a large amount of data. In the work, they also encountered many unexpected problems that needed to be solved immediately, and it was common to work overtime and stayed up all night to work. Once, in order to rush out the illustrations and make the presentation material, ZhuGe Min did not go home for four consecutive days. To think how young he was at that time! ZhuGe Min 's daughter was born in 1994; he didn't get to accompany or care for his wife when she was pregnant, and did not have many opportunities to meet his daughter after she was born. His wife sometimes took the daughter who was hobbling to walk to the office to visit; his daughter sat on the big picture board that used to be used for manual drawing, learning to speak, making everybody laugh. Looking back on those days, ZhuGe Min felt guilty to his wife and daughter that he could not be there for them at the time, even if it was for the sake of his career, but he could have been more considerate as a husband and father.

In addition to the innovation in theory and method, ZhuGe Min led the revision team to consider how to inherit the tradition and make some innovations in the planning content, especially on including the preliminary research results of Haihe River and the CBD into the formal statutory master plan. According to the 1986 city master plan of Tianjin, the general layout of "industries moving eastward" and "one pole carries two ends" was in line with the law of urban development in the world and the characteristics and reality of Tianjin. This revision of the master plan shall inherit all the aforementioned content. Considering that the saying of "one pole carries two ends" was accurate but not formal enough, "Double Centers with an Axis" was chosen as the official name. "Double center" referred to the central urban area and the core area of Binhai New Area, and "axis" referred to the urban development axis between the twin cities composed of the Haihe River, Beijing-Tianjin-Tanggu Highway and Beijing-Shanhaiguan Railway. This plan put forward the concept of a metropolitan area for the first time, which included the central city and the core area of Binhai New Area, to Yongding New River in the north, the Duliujian River in the south and the coastline in the east, covering a total area of more than 3,000 square kilometers, and was the key area of urbanization in Tianjin. The plan started from the concept of sustainable development, and expanded the concept of a 500-meter green belt on the outer ring road of the central city determined in the last version of the plan to the whole range of the 3,000-square-kilometers metropolitan area. At the same time, the safety of the city was to be strengthened. The dikes built around the river and along the coast in the metropolitan area formed the flood control circle in Tianjin. By dredging Yongding New River and Duliujian River, the dikes were raised to improve flood control and drainage capacity and reduce the pressure of the Haihe River to discharge the upperstream flood. This was very important! After Tianjin's city master plan was approved by the state council, the state commission for the control of the Haihe River basin formulated a new flood control plan for the Haihe River basin, increasing the flow of Yongding New River and Duliujian River to receive the flow transferred from the Haihe River. The flood discharge in the main stream of the Haihe River had been reduced from 2,000 cubic meters per second to 800 cubic meters per second, and the height of the embankments on both sides of the river had been greatly reduced. This was illustrative of the fact that the city master plan involved a wide range of aspects, and had a decisive influence on urban form and urban design in many aspects.

20. Historic and Cultural Blocks

This master plan had clearly promoted the status of the Haihe River. According to the content of the Haihe River plan in 1991, the Haihe River was defined as the main axis of the city, which was the central line running through the city's history, economy, politics and culture. In addition to its functions of water conservation, navigation and flood discharge, it was also the focus of developing the city's economy, displaying the nature, history and culture of Tianjin and developing tourism. According to the content of the research of "Planning of Central Business District (CBD) in Tianjin" in 1994, it was made clear that Tianjin's CBD was located on both sides of the Haihe River, consisting of North Jiefang Road, Xiaobailou and Tianjin Old South Cargo Railway Station Business District, with an area of 1.8 square kilometers, and Binjiang Avenue and Heping Road were defined as the central commercial districts of the city. At the same time, taking into consideration Tianjin's role as one of the second batch of national historic and cultural cities, nine historic and cultural protection areas and five historic and cultural scenic areas were put forward and drawn out, including Wudadao, Tai'an Road, North Jiefang Road, South Jiefang Road, Central Garden, Quanyechang, Chifeng Road, Yigong Garden, Anshan Road, the Ancient Cultural Street, the Old Town, Guyi Street and both sides of the Haihe River. Most of these historic and cultural blocks were located on the two sides of the Haihe River, and the master plan had put forward the requirements for preservation. In this way, by improving the status of the Haihe River and clarifying the functions and preservation scope of both sides, the strategic direction of planning on both sides of the Haihe River were very clear. There were many industrial and warehouse land on both sides of the Haihe River in the last version of the master plan, most of which was converted into commercial land in this round of planning. Next, the key to the plan of both sides of the Haihe River was detailed planning and high-level urban design.

Chapter IV

Travel Thousands of Miles and See the World

21. The Thames · London, England

Mr. Bauer

This morning's forum and discussion is over, but everyone is still immersed in the mood. Chief ZhuGe finally has time to meet with Mr. Bauer separately, and the two of them walks and talks, arriving at an Italian restaurant on the edge of the square before the Planning Exhibition Hall. It is lunch time, and many guests are having lunch in the terrace under the shade of parasols, with breeze blowing and tasting Western cuisine, a very comfortable arrangement. In order to talk, the two sit down in a quiet corner of the restaurant, and Mr. Bauer starts asking before they get to order: "Congratulations once again, Chief ZhuGe, what a wonderful speech. The urban design pilot projects in Tianjin this time, including the exhibition, does urban design from the perspective of the citizens and solves the problems of walkability, planning and construction of parks, and living conditions, all closely related to the lives of citizens, very good! This is also the current international trend. I'm sorry, but due to time constraints, I'd like to ask another question. I would like to know how the plan of the Haihe River itself was drawn up and what measures were taken to realize a high-quality plan?"

Chief Planner ZhuGe

"Over breakfast I introduced you to the changes in Haihe's role and status, which is a premise. Later, we revised a new cross-century city master plan. It took nearly five years to implement the role of Haihe river and related contents in the statutory planning, laying a solid legal planning foundation for the compilation of Haihe River's new plan. In this process, we were also carrying out urban design projects related to Haihe River. Generally speaking, the urban design at this stage was exploratory. In fact, during this time we had opportunities to study abroad extensively. During the four years from 1993 to 1996, I went to the United Kingdom, Australia, Germany, Japan, visited nearly 20 cities an especially some famous waterfront cities in the world, increased knowledge and broadened horizons. The comparison between what I had learned in the past and the actual cities had laid a knowledge foundation for the high-level Haihe River planning. It happened to confirm the traditional Chinese value: Travel thousands of miles and read thousands of books."

During the four busy years of revising the master plan, ZhuGe Min had the opportunity to go abroad, and also started the urban design of the Haihe River. In 1994, in order to do a good job on the master plan revision, together with a young division chief, the director and the chief engineer of the Planning Bureau, ZhuGe Min surveyed Britain for more than ten days. This was the first time he went to a Western developed country. In Britain, the birthplace of industry and urbanization, and the birthplace of modern urban planning too, the British side of the receiving institutions were relatively high-class, and the schedule was very tight. There was no local guide, no full-time interpreter, and ZhuGe Min was responsible for the translation work. They had to find the survey routes by themselves, and take airplanes, trains, subways and taxis to arrive at the appointed government departments, universities and research destinations on time. They were too busy to take a break. There was also a native Muslim who stopped at the smell of certain restaurants, let alone eating there, so choosing where and how to eat became a problem. However, at that time, everyone were used to living a hard life. They went abroad with instant noodles, electric kettles and dry food. Due to financial constraints, they spent most of their time in small hotels, and they ate simple as well.

The group surveyed, or to say actually saw, the famous Thames, Westminster Church and Big Ben, the Windsor Castle and Tower Bridge, the City of London, St. Paul's Cathedral and the Docklands Financial District under construction; They went to the National Theatre on the south bank of the Thames, and to Greenwich, the meridian of the world. In London, they visited the Department of Environment, which was responsible for urban planning at the time; they visited their Thames East Corridor Panning Branch, which had been preparing a strategic plan for the Thames east corridor. They listened to the introduction that, with the upsizing of shipping, the main port of London was also transitioning from inland port to sea port, which was very similar to Tianjin's situation. At that time, in order to prevent storm surge, Britain was building a turning-style tidal gate at the Thames's estuary to "build a city facing climate change". It was designed by Norman Foster, a British High-Tech architect, whose technical, technological and aesthetic achievements were eye-opening and breathtaking. On a side note, Forster is also the architect of the Berlin parliament dome, the Swiss Re Tower in the City of London, and Apple's new headquarters.

The group arrived at the London City Council. It was in the heyday of the Iron Lady Margaret Thatcher, who pushed for neoliberalism and deregulation, reducing government interference in the economy and market. In 1986 she abolished the Greater London Council, a regional government that included the city of London and 32 surrounding cities. An official pointed to a building on the other side of the street and introduced to them that it was where the Greater London Council once worked, now an upscale hotel. The administrative system and laws of Britain's city planning are in constant change. More than a decade later, when ZhuGe Min visited London again, not far from the Thames and Tower Bridge, the egg-shaped glass building designed by Sir Foster, perched askew on the ground, glinted in the last rays of the sun. It was the restored new office building of the Greater London Government. Just then, a long column of smoke rose in the distance, and people in the office buildings ran to the balcony to watch. It must have been a fire. It seemed that the city was getting bigger and bigger along with its problems.

The group also visited Britain's New Town Development Company (Britain was the birthplace of new towns/satellite cities). It was located in a modest building in inner London. A gentlemen of the company told that, according to the "New Towns Act", the company was entrusted by the government to develop a new town and was responsible for the whole process of planning, design, financing and construction, including the operation and management of the construction period. When the construction was completed, the new town was handed over to the local government in its entirety. During the construction period, the company actually assumed some functions of local government, which greatly improved the administrative efficiency. By the end of the 20th century, the construction and transfer of new towns in Britain were basically finished, and the central government decided to dissolve the Company. New

Town Development Company, the most influential planning model in the 20th century, had come to the end of its historical mission. The company were in the stage of settling accounts at the time. In the planning of China's big cities after the Reform and Opening Up, the idea of strictly controlling the size of central cities and building satellite cities on the periphery was commonly raised, but it had never been successful. Even today, some leaders would put forward the idea of building satellite cities. It seems that they need to learn some basic knowledge about new town.

The last stop on the trip was to visit the prestigious London School of Economics (LSE). Located in the center of the city, the LSE had a semi-open campus. The old buildings were shaded by trees, which seemed to tell the history of the city and the school. They came to the Department of Economic Geography, where they were received by a professor. The British visit was arranged by a fellow Tianjiner who graduated from the school and was a relative of a division chief of the Planning Bureau. The professor told them about the history of the city of London. He said that the city of London was like an onion as it was developed layer by layer. "Onion" was a suitable word.

ZhuGe Min was also impressed by the professor's description of the interconnectedness of the various functions of the city center: central London historically had two cities, the royal Westminster in the west and the merchant city of London in the east, which was the one-square-mile City of London today. Merchants paid taxes to the royal family, and the amount often had disputes that had to be decided by the courts, so there was a court district between the two cities. The court district spawned a boom in law firms and a range of services to make wigs and costumes for judges and lawyers. As the court was the most conflicted place, it was also the source of news, so newspaper presses clustered on the nearby Fleet Street.

At first, ZhuGe Min was taking notes while translating, and scrambled for a bit, unable to hear all of the professor's words, and could only understand the general idea. But upon hearing this, he felt a sense of understanding. Around this time he was doing research on the planning of Tianjin's CBD. The professor's talk on the correlation of various functions in the city made him clearly realize further that urban planning must follow the inherent logic and development law of the city rather than being presumptuous.

Finally, the professor said that the transformation of the city was very difficult, because the inner city of London was 30 square kilometers large and was the extension of the central business district of London, and the areas outside the core of the City of London were limited by the strict requirements of historic city preservation and more and more management items, with no new development space. So in the early 1980s, the British government set up the Docklands Development Company, LPDC, and decided to use the once prosperous Canary Wharf District in Docklands to plan a new CBD in London. SOM, the famous design company in the United States, completed the overall plan. Olympia and York Development Company, which was very established in Canada, began to build the 200-meter high landmark building for Docklands, which was also the tallest building in Europe at that time. The development of Docklands stalled amid the economic crisis, and Olympia and York went bankrupt over the project. By the time ZhuGe Min set foot on Docklands pier again, more than a decade later, the economy had recovered considerably, and the big newspaper presses and printing houses on Fleet Street in the city center had moved to Docklands, as their equipments needed to be upgraded and the old town could not accommodate it.

Some put it down to the fact that Docklands's tracks didn't run to the Old Town center, so it couldn't develop. The DLR, the Docklands Light Rail, was completed and opened to traffic in 1987, but in order to get the London city center, one needed to transfer at least once halfway, and it was inconvenient. Hence the government planned to extend the Jubilee Line in the city center directly to Canary Wharf and Isle of Dogs, a ten-year project. By the time ZhuGe Min visited Docklands for the second time in 2010, the Jubilei Line extension was open to traffic. The Docklands financial district was revitalized and in full swing. After the professor finished, everyone discussed a little, then the dean formally met with the delegation.

After the meeting, in the dining room next to the dean's office, the dean hosted the group members to dinner, which was said to be the highest courtesy of the college. When ordering, ZhuGe Min did not know what the names of the dishes stood for, only that the ingredients were probably beef or fish. The division chief of the group had received three months of training in England and knew some about British food, and had ordered beef, which proved to be a wise choice. ZhuGe Min ordered fish for the two heads as the main course, only to realize after it was served that it was like the Chinese Shengjianbao (pan-fried stuffed buns), stuffing cod pieces in bread and baking it, creating a heavy butter taste and did not suit the three people's appetites. But in order to show respect, they finished the food and praised the chef's skills, "delicious!" During the dinner, they learned that the dean had been the mentor of the Tianjin fellow. The dean said that he had never heard of Tianjin until he advised this student from Tianjin. Then he went to Tianjin and learned that there was such a big city beside Beijing. Everyone was a little embarrassed to hear the dean saying it. The group saw a map of the world in the corridor of the economics department building, and the Tianjin marked on it was still named Teintsin.

Coming out from the college, the group was scheduled to visit Milton Keynes new town. It was 70 kilometers from London, and one of the professors who accompanied them for the dinner said that he happened to live in Milton Keynes and was willing to accompany the group to the railway station. After a short walk to Euston station, the interchange station for the tube and railway, they boarded a commuter train to Milton Keynes. There were many people on the train, all of whom seemed to be commuters. The professor said goodbye to us, went to one end of the carriage and stood there. From his calm and slightly tired face, ZhuGe Min saw the life of the English middle-class. One hour and twenty minutes later, they arrived at Milton Keynes; it was already dusk when they exited the station.

This British survey tour lasted for more than ten days, mainly focused on London and radiated outward from London. Therefore, there was plenty of time for surveying, and each project could be learned in detail. Unlike the later surveys where they went to four or five European countries at a time, and could only stay for one or two days in each city, spending most of the time on the way. Coming to the UK, one cannot fail to mention satellite cities. As early as in the 1940s, Sir Abercrombie proposed to build satellite cities outside London in the Greater London Plan. ZhuGe Min had learned about satellite cities since he was in college. He knew all the illustrations and maps ranging from Harlow new town, a first-generation satellite city, to Milton Keynes new town, one of the fourth generation. A few days ago, the group took time to go to Harlow new town and University of Cambridge by car, all the way under a light rain, full of greens outside the car windows. Only then did he experience the 35-mile width of London's green belt, which was equal to 56 kilometers, while the planned width of Tianjin's green belt was 500 meters, which was less than 1/100 of London's green belt, and had not been fully implemented. This made him directly understand the gap between China's urban ecological environment and that of developed countries.

There were many cars on the road and many white-haired elderly people were sitting in the driver's seats. A driver was hired for the whole day, and he said in his conversation that there were more and more seniors living independently in Britain at present. They arrived at Harlow new town. The public buildings in the center and the surrounding houses were modern square boxes that look like northern or eastern European cities. Some of the buildings in the center of the city were 20-storey towers. The whole city, including buildings and squares, seemed to have little attraction. The satellite cities were supposed to absorb people from London, but the reality was that most of the people who lived there were from elsewhere. Due to its small size and lack of employment, the satellite cities were in fact commuter towns. Two million people commute during the morning and evening rush hours every day, worsening to London's traffic.

Further north from Harlow new town was the fascinating University of Cambridge. The cobbled streets of Cambridge were lined with bars and cafes where many students were studying. The University of Cambridge was also known as the University of the City. The University of Cambridge had many colleges and many classic buildings. Over the River Cam, which winded its way through the university, there were more than 20 bridges of exquisite designs and unique shapes, hence the name of Cambridge. "Cambridge's spirit is all in a river; The River Cam, I dare say, is the most beautiful water in the world." Xu Zhimo wrote in *The Cambridge I Know*, "It was Cambridge who taught me to open my eyes, Cambridge who moved my curiosity, Cambridge who gave me the embryo of my self-awareness,". He was generous in his praise of Cambridge. And "softly I went, as softly as I came; I flick my sleeves not even a wisp of cloud will I carry away."Farewell to Cambridge.

It was a good weather the next morning. ZhuGe Min thought he knew all about Milton Keynes from his knowlege. This was a fourth-generation satellite town, the scale was larger, planned to be 400000 people, as to achieve balance of working and living populations. Different from the traditional neighborhood layout where community centers were located inside the neighborhood centers, the planning here put community centers, primary schools on the periphery of the superblocks, in order to increase people's freedom of choice. The road network was a 1×1 kilometer grid, and the shape of the road changed with the ups and downs of the terrain; the new town clearly established the priority of cars in terms of internal traffic. But what ZhuGe Min saw was still unexpected. Out of the neat downtown, what immediately came into view was an idyllic landscape. Many rivers, ponds and wetlands were preserved in the plan, and water birds and ducks were playing in the water.

What surprised him the most was that most of the houses in Milton Keynes were typical British semi-detached houses and detached houses, with red bricks and red tiles, full of the warm feeling of home. There were a few multi-storey apartments around the city center. There were no high-rise buildings in Milton Keynes! The poetic scene helped ZhuGe Min understand why Londoners commute 70 kilometers a day to live in Milton Keynes, which was more like Howard's vision of the Garden City than Letchworth. He had bought a set of Milton Keynes's maps, which he wanted to take home and study. In 2009, he visited the UK again and specially came to Milton Keynes again. He found that a huge arched shopping mall was built in the center of the city, which was full of vitality, and the high-tech industry was developing very well. Milton Keynes was becoming a rising new high-tech city in Britian.

Before arriving back in London and returning home, the group went to visit the cultural counselor's office of the Chinese embassy in the UK. The cultural counselor's office was located in the center of the city, not far from where the group was staying. They walked to the destination and passed the famous Hyde Park; there was a crowd in the distance listening to a speech, as London still maintained its traditions. On the way they also saw someone riding a motorcycle with a map under the windshield. A warmhearted Londoner explained that in order to get a taxi license, these people needed to remember the roads of London. London's iconic black cab had a spacious carriage in which passengers could sit opposite to each other and the drivers were polite in their neat black uniforms. London's strict regulation of taxis could be seen from the licensing of taxi drivers.

Taxis are not only vital to a city's public transport, but also have an important impact on its image. Generally speaking, a city has tens of thousands of taxi drivers, year after year, accumulated over time, they receive a very large number of tourists, and have a huge impact on the city's reputation. Domestic and overseas cities pay careful attention to the management of taxi driver groups. Shanghai is doing well among domestic cities, setting up several major taxi companies to coordinate and regulate their management. On the other hand, Tianjin has dozens of small companies, and has been lacking effective management of taxi drivers.

The cultural counselor's office of the Chinese embassy in the UK was in a three-story building, a classic old building. When they entered the room, it was the first time after all these days when they did not need translation. Everyone spoke Chinese. Although the decoration in the room was British, ZhuGe Min felt as if he was back home. At lunchtime, the counselor knew that everyone had been out for more than ten days and must miss home very much, so he invited everyone to eat in the canteen. The canteen was in the basement; as soon as they went down the stairs, they smelled a familiar smell. After they went in, they found out that it was complete a Chinese canteen, with the tableware, spices, and the cooks all from China. There was scrambled egg with tomato and beef and potato stew in the pot. ZhuGe Min felt like he was in a "red revolutionary base" in the British empire. It seemed that people's ideas and habits were extremely difficult to change, but to do urban planning was to change people's ideas.

After the visit of the counselor's office, the business activities of this survey trip ended successfully. In the evening ZhuGe Min and the young division chief wanted to go see a movie, so they went to the leader to ask for permission to leave. Because the survey trip was very successful, the leader was very satisfied with everyone's performance, so they happily approved it and specially told them to be safe and come back early. The two young men arrived at Trafalgar Square, London's cultural and entertainment hub, with its many cinemas. It happened to be the weekend and there was a long queue at the cinema. They chose Fugitive, a suspense film starring Harrison Ford. The viewing hall of the cinema was very large, a hall that accommodated nearly one thousand people. To ZhuGe Min's surprise, the audience were all dressed in formal clothes, and the male ushers were all dressed in exquisite uniforms, with silver laces on the cuffs and collars, like the attendants of the royal court. The images and sound effects of the film were truly impressive, an audio-visual feast. ZhuGe Min asked himself, when can there be such a cinema, such a film and such an audience in China.

Mr. Bauer

"Am I getting too far away from the topic?""Never mind, it's very interesting." Bauer was listening with interest. "These things have something to do with the Haihe River plan. I understand that today's comprehensive development and renovation of the Haihe River would not have been possible without the improvement of the role and status of the Haihe River in the city master plan and the reduction of the discharge of the Haihe River by setting up urban flood control circles. Without the designation of Tianjin's CBD and 14 historic and cultural blocks, there would not be so many historic buildings and the Haihe River's features would not be so distinct today. Also, if you had not gone to Britain to survey, your understanding of waterfront cities would not be so deep. Your talk is too rich for me to remember. May I record it?" With ZhuGe Min's approval, Bauer turned on his phone's recording function.

"What you have summarized is very accurate, and it also shows that the urban design of this scale, with something as important as the both sides of the Haihe River, needs to consider various aspects. The positioning of the upper planning is the premise, and we must plan well in advance to pave the way; Relevant special plans and opinions of relevant departments need to be integrated, all parties need to reach consensus, and finally, exquisite skills and superb techniques are required. All of these require comprehensive knowledge and the accumulation of long hours." Although ZhuGe Min was busy presiding over the city master plan revision, he still squeezed time out to visit abroad. During the several years after the visit to Britain, ZhuGe Min went to many countries and cities to survey, although each trip was not long. The Comparative Urban Planning Theory and Method, which he wrote as his master thesis, started to be of use.

22. Victoria Bay · Hong Kong, China

Chief Planner. ZhuGe

In August 1994, in order to do a good job on planning of the Old Town's renovation,

ZhuGe Min stayed with the TUPDI's working group in Hong Kong for half a month, together with the reputable Hong Kong design company P&T Group to complete the renovation planning for the Old Town, which was in fact an urban design project. In addition to their design work, they were accompanied by Lin Yunfeng, the young director of the P&T Group and vice president of the Hong Kong Institute of Architects, to visit many places. Hong Kong's Central is its CBD and is famous for its elevated pedestrian system. During lunchtime on a weekday, Mr. Lin led ZhuGe Min's group to the Central. There were crowds on the elevated pedestrian platform, all workers in the office buildings coming down to have lunch. The walkway was lined with restaurants of all kinds, where people ate and chatted.

It seemed that lunchtime at CBDs was a very important social time! ZhuGe Min thought of most working places in China having their own canteen, and then thought of the canteen in the counselor's office of the Chinese embassy in London. He couldn't help but laughed. If CBDs were to be truly built in China, lunchtime could not take place in canteens. It needed to be more open, efficient and international, but it was not easy to change. Now in the Modern Service District(MSD) of TEDA and the initial area of Yujiapu Financial District in Binhai New Area, there were some signs that young white-collars were more willing to go out of the office to have meals and meet friends at noon.

The P&T Group with 80 years of history was originally located in the Central. The fast rising of Central led to climbing rents, and a design firm like P&T could not afford it, so they moved to the Wan Chai business district in the east. A lot of design, law, and accounting agencies and companies gathered here. The CBDs in Hong Kong gave ZhuGe Min a deeper understanding of the planning and design of Tianjin's CBDs. Aside from visiting the buildings of Hong Kong University of Science and Technology and Hong Kong Science Museum, Mr. Lin also took the group to visit Hong Kong's housing. The last time ZhuGe Min came to Hong Kong in 1991, he visited the Hong Kong Housing Authority and saw the housing exhibition inside the building. He also visited the housing in the new town in the New Territories. It was located upon a subway station. Its lower level was a large shopping mall and above the shopping mall were all residences. He did not have the chance to visit the inside of the housing.

This time Mr. Lin took them to visit two projects in downtown. One was a private high-rise housing project, with a layout of a few high-rise residences enclosing a central garden. The greening was very good and the standards for the decoration and furnishings indoor were all high end. There were two to four units each floor and a hanging garden every four floors. Another was a public housing. It was a typical small-unit layout, with around a dozen units each floor and indoor corridors. ZhuGe Min entered the corridor, and found that every household had their front door open with the steel bar gate closed for the need to ventilate in the hot weather. It lacked privacy and looked crude. But Hong Kong people needed to wait in line for several years to apply for this kind of housing.

ZhuGe Min felt that Hong Kong's housing was not very successful in both the housing system and the architectural design of public housing. Compared with western countries, the standards were too low. There were a lot of factors behind this phenomenon. Hong Kong encountered housing shortage in the 1950s, which led to many tin houses and sheds with very poor living conditions, and fire and other accidents often occurred. To solve this problem, the Hong Kong and British government began large-scale housing constructions, demolishing the old tin houses and sheds and building multi-storey residences without elevators in a large amount. Each family only had a kitchen and no individual bathroom, and dozens of families shared one bathroom on each floor. This kind of housing had very low standards, was to be used as transitional housing for a period of time, and faced the fate of having to be demolished in the 1980s. In order to be economically feasible and profitable, the new buildings after demolition were high-density high-rise housing groups, which made people afraid to walk in.

ZhuGe Min thought of Tianjin–the housing built during the reconstruction of dilapidated bungalow areas were mostly six-storey buildings without elevators and very dense; what to do with them in the future? It would be horrible to simply demolish them and build high-rise buildings like Hong Kong. There were some truth in people's worries about the reconstruction of Tianjin's dilapidated bungalow areas. Today, it seemed a large number of multi-storey residential buildings built in Tianjin since the 1980s would become the focus and main difficulty in the next step of urban renewal. Walking in Hong Kong's blocks such as the Central and Wan Chai, along the street were mostly a dozen stories of large, wide or U-shaped, enclosed old residential slabs, with each unit being tiny. Drying poles stuck out of every window and colorful clothes hanged in the air, very messy. Each unit seemed like a pigeonhole. ZhuGe Min believed that Hong Kong's housing policy and residential planning were by no means a model for most Chinese cities.

23.Yarra River · Melbourne, Australia

Chief Planner ZhuGe

With the increase of Tianjin's international exchanges, the planning of the Haihe River has also become a cooperative exchange project. Tianjin and Melbourne are sister cities. In 1994, the Australian side sent experts to assist with the histrioc preservation planning of Tianjin, and ZhuGe Min recommended the old south railway cargo station business district on the east bank of the Haihe River as the field . The Australian experts choosed Tianjin Cigarette Factory located in the area as target and spent two weeks on site conduct survey and research, and finally formulated a protection and reuse plan scheme. The quality of the cigarette factory was very high of both the design and construction and it was well preserved; the concrete corbel structure indoor not only increased the span of the factory, but also reflected the aesthetics of power for the industrial building. The cigarette factory was the first industrial heritage protection and renovation project in Tianjin, but sadly it was later demolished with the construction of the Old South Railway Cargo Station business district and the Riverview Place was built on the site. ZhuGe Min thought resignedly that perhaps this was a dilemma.

In August, Tianjin organized a group to pay a return visit to Australia, with the theme of learning from Australia's experience in the development and renovation of waterfront cities. The group six members of visited Sydney, Melbourne, Canberra, the Gold Coast, Brisbane and other cities. To save time and money, the group rented a van and traveled north along the east coast, experiencing the vastness of Australia along the way. They would usually choose a town by the road to rest for a while. In the middle of the town, there was usually a short, clean and beautiful commercial street. There would be one or two floors of commercial buildings on both sides of the street, such as supermarkets, grocery stores, interior decoration stores, gardening and flower stores, etc. They also saw a Chinese restaurant with very traditional and elegant furnishings. The Chinese restaurant's interior decoration was very Chinese, not a new Chinese style, but a colorful old Chinese style. A Chinese couple ran the restaurant, both as the bosses, but also as chefs and waiters, and their relationship with the guests were very good. The guests also seemed to be acquaintances of the couple in the small town. The business was booming and they needed to make reservations to eat there. Unlike the hustle and bustle of a typical Chinese restaurant, it gave out a sense of calm amid the bustle. ZhuGe Min wondered that Chinese people are indeed everywhere. They lived with dignity in white communities, running their own businesses and offering locals a different culinary and cultural experience.

Sydney is the largest city in Australia and a city with a sense of modernity. Located on the edge of Sydney Bay are the Sydney Opera House and the Sydney Bridge, which have become the unique symbols of the city with countless visitors. Sydney Bay centre is lively with yachts, sailboats and fireworks over the Sydney bridge in the New Year eve. The CBD of Sydney is located at the southern end of the bay, between the Sydney Bridge and the Sydney Opera House–a group of high-rise office buildings in the a one-square-kilometer range. To ensure that each office building has a good view, the planned high-rise complex is perpendicular to the bay, a very clever way of ensuring views as more more possible and avoiding forming a high wall along the waterfront. The CBD is flanked by the famous Darling Harbor, which used to be the wharf warehouses of Sydney Harbor. With the outward relocation of the harbor, Sydney has focused on the transformation of this area and the construction of a financial center. The public buildings include Maritime Museum, Convection Center and a Chinese garden. The hotel, shopping mall and dining facilities were converted from preserved warehouses, creating a lively waterfront in central Sydney that contrasts with the quiet financial district next to it.

ZhuGe Min and the group came to visit the Sydney city government. On the top floor they saw the planning model of Sydney city center, in a 1:500 scale, made with cork wood and very elegant. From the introduction, they knew that in order to encourage public transport and limit car traffic, Sydney strictly controlled the number of parking spaces within the CBD, and the total number of parking spaces was limited to no more than 5,000. Additional parking spaces were not allowed in new projects. ZhuGe Min found it strange–it was unlike the way Chinese cities handled parking in general, and parking standards in China, in particular, were subject to lower limits no matter where they are located. Sydney's policy of controlling parking spaces in the CBD was very effective. Walk in the streets of the CBD, it felt very safe and quiet, with no motor traffic noises common in regular cities' CBDs. For example, in Hong Kong's Central, even when walking on elevated pedestrians, one could see the traffic in the car lanes on the ground below and hear the giant noises created by motors and tires chafing on the road.

It was also interesting to see the elevated monorails passing through the buildings at the southern end of Sydney's CBD. Although the monorail was public transport, the carriage's volume seemed to be limited judging by its appearance. It was later heard that Sydney's elevated monorail had been removed to improve the city's landscape. Several investors wanted to build a similar elevated monorail along the Haihe River during the comprehensive development and renovation of the Haihe River. Considering the impact on the landscape of the river, the plan was never approved of. In order to promote the so-called new form of public transportation, some domestic manufacturers would exaggerate the role and effect of elevated monorail, so decision-makers should be cautious.

ZhuGe Min and the group were met by Zeng Youlong, Sydney's deputy mayor, who was a Chinese and an architect who helped build the Chinese Garden at Darling Harbor. Through his own hard work, he ran for deputy mayor and became the first Chinese deputy mayor of Sydney in 150 years. At noon, Mayor Zeng invited them to lunch. Before lunch, they took photos of everyone holding up the lobster king the size of a rugby ball together. During the meal, Mayor Zeng kept using two spoons to pick up food for everyone, looking very skilled. "I know you guys never worked this kind of jobs," he joked, "I used to work at a restaurant in Chinatown and washed enough dishes for me to use for my whole life." Mayor Zeng spoke Mandarin very well and often spoke slang, making everyone admire him.

In Sydney, they also visited some public buildings and the new Olympic sports center built for the 2000 Sydney Olympics. The sports centre was designed by COX, a leading Australian architect firm, who also designed the Maritime Museum at Darling Harbor. COX's design made use of the extensibility of steel and tie rod technique, making the structure of the large-span natatorium very light. The game pool next to the diving pool adopted a undulating shape and integrated with the outdoor environment through large areas of glass. Then they went to visit Mr. Cox at the COX office in the center of Sydney. He must have been under the age of 60 at that time, and he was full of energy. He was accompanied by a very handsome guy who was from Beijing Forestry College and worked in COX office after studying in Australia. He was an assistant of Mr. Cox and translated for the group. Twenty years later, when ZhuGe Min presided over the international CFP for the design scheme of the National Marine Museum in Binhai New Area, Mr. Cox's scheme won the first place in both rounds.

When ZhuGe Min saw Mr. Cox again, he was in his seventies, seemingly transformed from middle aged into a kind old man, but still creative. Mr. Cox had no recollection of the visit back in the day, and the handsome Chinese assistant beside him was nowhere to be found. Today, the National Marine Museum designed by Mr. Cox has been completed in the bay of China-Singapore Tianjin Eco-City in Binhai New Area. At that time, Mr. Cox introduced his design idea, saying that the museum is a harbor to stop by the sea. From the natural transformation of the building, many metaphors can be interpreted, such as animals climbing onto the land from the sea, people further communicating with the sea and so on.

Whether as a planner or an architect, in addition to paying attention to urban centers and public buildings, we must also look at the housing of a country or a city, and learn about housing policies, community planning and architectural design of houses when surveying. Housing is a large number of buildings, and it reflects the overall level of a country's buildings and people's living standards. ZhuGe Min and the group visited an detached house in the suburbs of Sydney, 30 kilometers away from the downtown, with a floor area of more than 200 square meters. The house cost 170,000 Australian dollars, or about 800,000 Yuan, which was a high price in China at that time, equivalent to 4,000 yuan per square meter, but was affordable to many new Chinese in Australia.

With the suburbanization, the hollowing phenomenon happened to the downtown of Sydney, but Australia did not have serious racial problems as the United States, so the downtown did not decline dramatically. In recent years, many young people liked to live in the downtown area, especially rich people from Japan and Southeast Asia. To meet their needs, many high-rise apartments with steel and glass as the main building materials had been built in the downtown area, which was gentrification.

The next stop was Canberra, the capital of Australia, whose planning had a special place in the world's urban planning history. At the beginning of 20th century, to solve the dilemma of Sydney and Melbourne striving to be the federal capital, the Australian authorities decided in 1911 to create a third place, and finally chose Canberra with its beautiful scenery as the new capital. In 1912, a competition was held for the plan of the new capital. Griffin, a 36-year-old Chicago landscape architect, stood out. His multi-hexagonal composition scheme was very consistent with the rolling hilly terrain of Canberra. Construction of the city began in 1913 and was halted during the First World War. By 1927, when Canberra officially became the new capital of Australia, it had taken 14 years.

The name of the new capital was also long debated, with Canberra, meaning "meeting place", the traditional name of the local population, eventually chosen. Canberra is more than 230 kilometers from Sydney. When the group arrived in Canberra by bus, it was dusk. The local guide hurried them to a Chinese restaurant for dinner. There were only a few hundred thousand people in new Canberra. After 8 PM, all hotels, restaurants, shops and entertainment venues were closed. The restaurant was owned by Song Xiaobo, a former Chinese women's basketball player, and the business seemed to be doing well. On the way back to the hotel, there were few people and it was quiet all around.

The next morning, the group went to Capitol Hill. They visited the congress first, and after listening to the debate between the congressmen for a while everyone was confused, so they got up and went out of the building. The congress hall is located in the center of the city axis–the congress hill, and is an earth sheltered architecture, the building integrating into the mountain. Looking north from the colonnade, they could see the vast green areas and the deep blue Lake Burley Griffin as well as the magnificent Captain Cook Memorial Jet, which could spray water up to 137 meters high. The scenery was heavenly.

Because Australia is south of the equator, most of the buildings face north. Canberra has no industry, but its total area is 2,400 square kilometers, which is the same as Tianjin's Binhai New Area, but its population is only one seventh of Binhai. The city area is similar to Shenzhen's, but the population is only 25 % of the permanent population of Shenzhen. Griffin's plan for Canberra, as it went back to as early as the beginning of the 20th century, was a baroque classical plan characterized by radiation, symmetry and axes, but he creatively retained and designed large areas of greenery and water, so some people called his plan ecological baroque or landscape baroque. Is there any need for economic development in this heavenly, pastoral city?

Next stop is the Gold Coast, a legendary holiday paradise. To experience the Australian road trip, the group took the van north along the famous coastal highway 1,000 kilometers from Canberra to the Gold Coast. The seat of van was rather capacious, but an official from the Commission of Housing and Development and the chief architect of the Architectural Design Institute were more burly, and could not stretch on the car, so they felt very tired after a more than ten-hour drive. Fortunately, the beautiful scenery outside the window attracted people's attention. Every time they stopped to take a midway rest, they got off to wander in the beautiful scenery and breathe the fresh sea breeze, feeling their fatigue mostly dissipated. Seeing the bustling traffic on the coastal highway, ZhuGe Min has a feeling that with the improvement of living standards and the popularity of private cars, self-driving travel in China would surely rise, which is also an important direction to boost domestic demand and develop the economy, and urban planning needs to deal with it as soon as possible.

They left Canberra at 5am and reached the Gold Coast at dusk. The kitchen of the reserved holiday apartment was equipped with all kinds of utensils and tableware. The apartment was next to the beach. ZhuGe Min stood on the balcony looking out; in the sunset, the sea was sparkling, and it was a peaceful scene. After a while, the fatigue and sleepiness of the journey hit, and he soon fell asleep. The next morning, the sound of waves woke everyone up. With the blue sea and blue sky, a few young people were running on in the beach, looking very happy. After breakfast, the first item on the itinerary was riding the tourist helicopter. The cabin was too noisy so that everyone was wearing headphones and had to talk loudly through microphones.

The Gold Coast had more than a dozen quality beaches in a row, and the coastline stretched for 42 kilometers as far as the eye could see. But the whole city could be viewed from the air, with a wall of holiday apartment blocks just behind the beach nearby. Then there was the water grid; the helicopter flew low over it, and what appeared was a group of villas. Contrary to the common villa areas, the villa areas here were composed of waterway and land roads, forming a double-way network pattern, similar to the double chessboard pattern of the ancient city of Suzhou in China. Suzhou's architectures are traditional courtyard houses with white walls and red tiles, and here were all two, three-storey detached house, with a yacht for each home, typical of waterfront holiday villas, also known as the "second homes".

ZhuGe Min thoughts of in 2001, when the economy professor of Peking University, Li Yining, proposed to boost domestic demand by encouraging citizens to buy second homes, as in well-off societies families are to have second houses, and was attacked by many "justice warriors", saying that there were still a lot of families with housing difficulties in China and encouraging second homes deviated from the purpose of serving the people. It was apples and oranges, and rendered him speechless. According to the logic of these people, if everyone is poor, no one can get rich first. Deng Xiaoping said that we should let some people get rich first and get on the road of universal prosperity eventually, and maybe he had said it for these people to hear. As a matter of fact, with the improvement of Chinese people's living standards and mobility, the rise of the second homes is inevitable and not dependent on the will of some people. Urban planning needs to be prepared for future developments. Of course, the basic housing needs of the public must also be addressed at the same time.

Gold coast is a famous tourist city. In addition to beaches, holiday yachts, villas, apartments, hotels, a large number of theme parks and other tourist facilities were also built. ZhuGe Min and the group visited the Universal Studios and Water World and saw huge crowds. At that time, he wondered when Tianjin could develop such a tourism industry. Soon after they returned home, three major Chinese cities, Beijing, Shanghai and Tianjin, began planning to introduce theme parks. Perhaps it was due to the opening of the country at that time. Many people saw the thrilling rides in theme parks abroad and thought that Chinese people should also experience and enjoy roller coasters. Another possibility is that foreigners think with the population of more than a billion, it would be very profitable to build theme parks in China. It's been 40 years since Disneyland opened its doors in 1955, and it's still packed with people to this day, making an incredible amount of money.

At that time, Beijing intended to introduce Universal Studios into China. In combination with the relocation and transformation of the Capital Steel Company, the site was located in Shijingshan. The smart Shanghainese had chosen Disneyland, believing that Disney's content was mainly animation, involving little ideological problems and would be easy to operate. Tianjin was a late addition, choosing the Paramount Park. Although Paramount Pictures is on par with Universal and Disney in film production, it is far from them in terms of theme parks. Paramount's theme parks, most of which are in second-tier cities in the United States, typically buy traditional amusement parks and add a few movie-related rides to them, such as Tomb Raider.

As a theme park would cover an area of at least 1 square kilometer, plus the surrounding supporting facilities, the investment and risk are large, so the state strictly controlled the construction of them. A theme park project is subject to the National Development and Reform Commission and the state council's approval, and is required not to use domestic funds and so on. With these requirements, the application process was very difficult, so every city formed a strong team to participate in the preparation of the feasibility report and other preliminary work. The Paramount theme park in Tianjin was initially planned to be located in Yangcun, Wuqing District, relying on the film and television base where the Tianjin TV station shot TV series, and inviting a famous design company, the former vice president of the world bank, a prince of Saudi Arabia who was planning to invest in the park and so on, a very strong lineup. Considering ideological issues, it was also considered that Chinese elements such as Journey to the West would be added.

ZhuGe Min also participated in the preliminary work of Paramount theme park. In 2002, he went to the United States and Japan as part of a team to negotiate with Paramount and visit theme parks. This was his first trip to the United States and Japan. However, due to a tight schedule, they only briefly visited the three cities of Los Angeles, Cincinnati and Tokyo. They stayed at the Beverly Hills Hotel in Los Angeles, and went to Paramount for talks on the second day and visited Universal and other studios and filming sites, and experienced Universal Studios and Disneyland; Then they visited the Paramount Park in Cincinnati, Ohio. It was the weekend and there were a lot of tourists, and it was said to be residents of surrounding cities driving themselves here.

In order to feel the attitude of Asian people towards American theme parks under the background of Asian culture, the delegation turned to Japan before returning from the United States and visited Universal Studios Japan and Tokyo Disneyland. They found that Japanese people loved American theme parks and there were so many people in the parks. A lot of tourist hotel facilities had been built around the theme parks. People said that the theme parks couldn't be experienced in only one day, so they needed to stay for several days. Each ride in the park had long lines, but ZhuGe Min and the group had VIP passes so they did not have to wait in line. On top of that, he was young and in good physical condition back then and almost got on all the main rides, ending up having a blast. Finally, they went to the Astro Boy Park in Hyogo Prefecture to find out about the local theme parks in Japan. Japan's economy and culture are very developed, especially the animation industry, but the Astro Boy Park did not have many visitors.

Years passed. Hong Kong Disneyland opened in 2005 and Shanghai Disneyland opened in 2015. Shanghai was the first city to open a world class theme park in the competition; it seemed that the Shanghainese were truly both smart and competent. Beijing's Universal Studios finally settled in Tongzhou after several twists and turns, have started construction for many years and there is no word about the time of the opening. And Tianjin? There was no further development. Later, the location of the Paramount theme park was changed to next to the carrier theme park, near the sea. Although Binhai New Area was incorporated into the national development strategy in 2006, ushering into a decade of rapid development, there is still no news about the Paramount theme park project. In the China-Singapore Tianjin Eco-City, small-scale theme parks independently developed and constructed by Chinese investors, such as Fonte Happy World and Water Cube, have been built in the former Binhai tourism area, attracting some tourists. And this may suggest that Tianjin and Binhai New Area should be determined to come up with the Chinese's own world-level theme parks with completely independent intellectual property rights and brands, become an engine that drives industry, promotes the tourism upgrade of Tianjin and Binhai New Area, and serves the 100 million people of the Beijing-Tianjin-Hebei region, and construct the "Gold Coast" of the Tianjin Bohai Bay.

After the Gold Coast trip, the team drove an hour to Brisbane, Australia's third-largest city and the capital of Queensland, a state of 2.1 million people with an area of about 1,200 square kilometers that hosted the 1988 World's Fair. The group visited the city because there was a river, the Brisbane river, which zigzagged through the city like the Haihe River. The Brisbane River was relatively wide, a little like the Thames in London. There were many urban viaducts and bridges passing by the river, which were relatively large in scale. In Brisbane's CBD, which was along the river, he found the design of the floating dock on the riverbank to be unique. The planning department said that it was to prevent the shadows of the CBD buildings from affecting the river in the long run; it was required that the shadow of high-rise buildings could not cast more than 1/3 of their total shadow area on the river, which was an effective way.

The last stop of the survey trip to Australia was Melbourne. ZhuGe Min had been thinking of Melbourne during the first dozen days of the trip. In order to save time, they flew directly from Brisbane to Melbourne. Sitting on the plane, he replayed the experience of the previous few days like a movie in his mind: the waterfront in central Sydney was full of character and vitality; the 42-kilometer long beach of the Gold Coast, making everyone cannot help but linger; it was nice to sit on a big cruise ship tasting seafood with the Brisbane River flowing... No matter how good the impressions of the first few cities were, he was still looking forward to arriving in Melbourne as soon as possible, because it was the most important city for this waterfront-themed survey.

As the second largest city in Australia, Melbourne is a famous industrial and commercial city, once the capital of Australia, and also a well-known livable city in the world. The Yarra River, which begins in the eastern mountains, is more than 240 kilometers long. It flows through the city center of Melbourne and eventually into the Hobsons Bay by the Pacific. Yarra River is about 100 meters wide in the city center, which is very similar to the Haihe River in the central section of Tianjin. Yarra river is also known as Melbourne's mother river, which is also the same as the Haihe River in Tianjin. However, the Yarra River flows from east to west, and the traditional urban center is located on the north bank of the river, which is the opposite of the Haihe River. Meanwhile, Yarra River and the Haihe River also have many similarities. For example, historically, like the Haihe River, the Yarra River has been flanked by wharves, warehouses, factories, railway stations and freight yards. Because of the river division, the city of Melbourne is better developed on the north side of the Yarra River and away from the Yarra River, and the south side of the Yarra River has not developed very much. Although the Yarra River is only 100 meters wide, its blocking effect on the city is still obvious.

Melbourne has a long history. Founded in 1849, it has always been an Australian city of industry, commerce and culture. It was the capital of the Commonwealth of Australia from 1901 to 1927. Due to the unique geographical location of Oceania, the two World Wars in the 20th century had little impact on Australia. Melbourne hosted the summer Olympics in 1956 as the first southern hemisphere city to host the Olympic Games. Melbourne has always attached great importance to urban green environment and livable city build. In 1971, Melbourne Metropolitan Development Plan was formally issued to adapt to the economic recovery and further expansion of the city size after World War II. Due to the continuous expansion of city scale and the suburbs, Melbourne's urban center was rapidly declining due to traffic congestion and environmental degradation. With the rapid growth of the number of motor vehicles, Hoddle Grid, the "narrow road, dense network"-style layout designed by Hoddle in the 19th century could not bear the large traffic pressure, leading to more and more serious conflicts between motor vehicles and pedestrians, and the attraction of Melbourne declined greatly.

In response to the above problems, the Melbourne Urban Design Act enacted by the Planning Department of Victoria in 1982 made a decision that would have an important impact on Melbourne in the future: Melbourne needs to continue to maintain a walkable, safe and attractive urban environment for business and leisure. The act proposed specific design guidelines from three aspects: city image, building and land development, transportation and walking. Based on this act, Melbourne formulated the "Streets for people: a pedestrian strategy for the central district of Melbourne" in 1985. Even from the perspective of today, the strategy more than 30 years ago had the foresight to lay the foundation for Melbourne as a humanized city: the first is shifting focus of street design from motor vehicles to people, pointing out the important value of walking for improving urban culture and business vitality, making streets a key part of urban design; the second is shifting focus of urban design from the buildings to creating a more comfortable and pleasant urban environment so as to attract and enhance walkability, which is the right direction of sustainable development that has been widely recognized; the third is that it was the first time an urban street design strategy was formulated from the a human perspective.

It had been ten years since the promulgation of the act when ZhuGe Min and the group visited Melbourne. Melbourne had made remarkable achievements in building a humanized city, which he only learned about in recent years. In the year of their visit, for the first time, Melbourne collaborated with the world renowned Jan Gehl of Denmark, to develop the Public Space Public Life survey for the city. Besides streets, Melbourne had carried out another round of humanizing transformation of city parks, squares, public green spaces and public buildings. After that, Melbourne cooperated with Jan Gehl again to create a new round of humanized space plan in 2004, which had been a decade-long routine since then. In 2015, the third round of humanized space plan was created in Melbourne, which expanded the survey scope to south coast district and waterfront new town, breaking through the more than 2-square-kilometer traditional Hoddle Grid in the city center.

Mr. Bauer

"Mr. Bauer, our domestic cities, including Tianjin, have only started to pay attention to humanized cities and pedestrian systems in the past two years, which is 30 years later than Australia. Does it mean that our urban design is 30 years later than the overall level of developed countries?" Bauer hasn't fully recovered from He's previous long speech. "It's an interesting question. But I don't think so. Thirty years ago, China was very open and had a lot of exchanges with other countries. According to the theory of comparative study, the influences between cities are diachronic and synchronic, and diachronicity is sequential, but synchronicity is mutual. Now that China is on the world stage, I don't think there will be an gap of 30 years. I do think it will be very difficult to shift from focusing on urban expansion to improving the quality of cities. If this transformation can be achieved, both conceptually and practically,

I think China's urban design can keep pace with that of developed countries. Mr. ZhuGe, as an expert on comparative urban planning, you have more say than I do."

ZhuGe Min agrees with Bauer's comment, and he's been thinking lately about the importance of Chinese urban planning and design going in the right direction in order to achieve the great rejuvenation of the Chinese nation and reach the level of a moderately developed country by 2050. "Mr. Bauer," he says modestly, "the gap between China and developed countries in urban design is still very big. We need more practice to learn and improve."

Chief Planner ZhuGe

During the survey in Melbourne, ZhuGe Min set his mind on learning from the Yarra River planning that could be directly used for the Haihe River. In addition to the Yarra River plan, the peers in Melbourne should have mentioned the collaboration with Jan Gehl on the humanization of the city, but he didn't know about Jan Gehl at that time and left no memory. However, the strategic idea proposed by the Melbourne peers to change the trend of the city's back-against-water development to one that is towards Yarra River struck a chord with ZhuGe Min and impressed him so much that he blurted out: our Haihe River plan is the same!

Melbourne also faced the same problem as Tianjin, which was how to cross the Yarra River cut and promote the development of the south bank. ZhuGe Min and the group went to survey the south bank and found that Melbourne had placed their only casino, Crown Casino on the south bank along with other public buildings such as the aquarium. When they were there, the construction of the Crown Casino was completed, but had not yet opened, so ZhuGe Min and the group had not been able to see the casino's fire show with the eight big fire-spitting columns by the Yarra River. He speculated that they might have learned from Las Vegas casinos. It attracted a large number of tourists to watch and gained popularity for Yarra River.

Later, in the planning of Haihe River, deputy director Lian of TUPDI in the planning team, for he had trained for three months in Melbourne and experienced the exciting of the fire show, suggested for many times for this kind of fire-spitting devices to be built in the Central Square by the Haihe River, but it was not implemented due to various reasons. Eventually, a row of landscape lamps that were shaped like fire-spitting columns were installed. If fire-breathing devices can actually be installed one day, the attraction of the Haihe River will surely be further enhanced.

In addition to fire shows, there are many activities on the Yarra River, such as cruise and rowing boats. There are also many parks on both sides of Yarra River, such as the Olympic Park and botanical garden, forming a complete green space system together with Yarra River. On both sides of the dyke of Yarra River, promenades have been built where people can stroll by the water. A total of more than 20 bridges sit upon the river, all in various shapes, including one uniquely shaped pedestrian bridge slanting across the river; from south bank, the bridge points in the direction of the Melbourne CBD–a group of modern high-rise buildings. Buildings across the whole city of Melbourne, including both sides of the Yarra River, feel modern and vibrant.

At the same time, Melbourne has spared no effort to protect its historic buildings and districts. The group strolled through Melbourne's Central Urban Area. Each block was a modern shopping mall, but the planning followed the city's historical road network and texture in their entirety. ZhuGe Min and his companions walked past several large department stores. When they entered the next building, they found a huge conical hall, beneath which was a historic red-brick building with a tall tower. They learned that a hall was specifically designed to preserve the historic building.

ZhuGe Min later learned that the modern building, Melbourne Central, was designed by renowned Japanese architect Kisho Kurokawa, and was built in 1991. The historic building in the middle is Coop's Shot Tower, a century-old 50-meter high tower built in 1880 to make aluminum bullets. With more than 300 steps, the tower was Melbourne's tallest building in the 1940s and the Coop Factory was once the largest bullet factory in the southern hemisphere. After World War II, lead bullets were phased out and the bullet factory was finally closed in 1961. In 1973, it was listed by the Australian government as a "building of historic value" and protected. At the time, the City Loop was under construction, and when Melbourne Central Station was built in 1981, the tower was deliberately preserved. The completion of the railway and the station had increased the value of the surrounding land, and the Melbourne government decided to build the surroundings of the station into a commercial center.

Having taken five years since 1986, the $1.2 billion Melbourne center commercial building was built. Kurokawa designed the world's largest glass conical dome, and completely engulfed the Coop's Shot Tower. A third of the center of Melbourne was occupied by the six-storey Daimaru department store, a Japanese company. It attracted huge crowds when it opened, but in 2002 the Japanese company withdrew due to poor management and other reasons. Now the building hosted a native Australian department store.

There was another thing that left an impression on him, and he remembered it years later. At the time, he saw two modern 20- storey residential slabs by the Yarra River in the city center, which his Melbourne counterpart said were proposed to be demolished for a better view of the Yarra River. ZhuGe Min was surprised. He looked at the two buildings from different angles along the Yarra River and found that they were not ugly. They in fact looked good aside from the slightly distance. He doubted the buildings could be demolished, so after that, whenever he heard that someone had gone to Australia, he would ask them if the buildings had been torn down, but many people didn't know about it.

Until 2007, when a famous architect of the Australian LAB Studio was working on a project in Binhai New Area and introduced his own design of the Melbourne Federation Square, ZhuGe Min suddenly realized that this Federation Square was built on the original site of the two residential slabs. Due to the lack of planning after the liberation, some high-rise buildings were built haphazardly by the Haihe River, which had a negative impact on the landscape of the Haihe River. However, for Tianjin at that time, there were not many high-rises in the whole city, and it was a dream to demolish high-rises for the sake of the landscape. It was not unthinkable, but seemed impossible. How difficult would it be! But after studying in Melbourne, he started to have a little courage. In 2002, when Tianjin began the comprehensive development and renovation, a 9-storey building–the Tianjin Publishing Building–was in fact demolished, which was located on the other side of the Central Square and on the south bank of the Haihe River. It was the highest building demolished in Tianjin at that time. ZhuGe Min felt that there was no construction without destruction, and correcting mistakes might come with a cost.

Mr. Bauer

"What do you mean, exactly, by 'no construction without destruction'?" Bauer was not quite sure about this idiom. "Oh, it translates from a traditional Chinese idiom. It literally means you can't build a new building without demolishing the old one, but it has broader meanings, including philosophical ones. I think it means it's hard to accept new ideas if you don't break the old ones in your head. Of course, there is a certain balance here, which is not easy to grasp. Mao Zedong said, "no construction without destruction". He launched a number of campaigns such as "breaking down the four olds" and the Cultural Revolution, which destroyed China's traditional culture, including cities and ancient buildings. This may be the price of development. Sometimes a happy medium can be found, such as the case of the Coop's Shot Tower in central Melbourne, which coexists well with the new buildings outside, but at other times they are in opposition. In the case of the Federation Square in Melbourne, it wouldn't have worked so well if two tall buildings on the Yarra River hadn't been demolished.""It seems that you gained a lot from your survey in Australia, especially the survey of Yarra River in Melbourne. You learned about not only the specific design, but also the implementation of planning and design, historic blocks protection and many other aspects. Is there anything else that left an impression on you?""Yes. It has strengthened my confidence to build Tianjin into an international city."

Although in 1994, Tianjin was relatively shabby on the whole and seemed to be far away from catching up with London, Sydney, Melbourne and other international cities, Tianjin had the natural conditions for the development of the Haihe River, a long history and deep-rooted culture, and it was because of the large scale of dilapidated bungalows, warehouses and factories that had a greater potential. Australia was a young country that became a British colony in 1770. Captain James Cook discovered the east coast and named it New South Wales, and it was initially a place for prisoners to go in exile. The first 736 prisoners landed in 1788, and the first free settlers arrived in 1790, so the country only has a history of around 200 years. In 1901, the Commonwealth of Australia and its first constitution were established. It has only been 100 years since then. The rapid development of Sydney and Melbourne makes us believe that Tianjin, with its 600-year history and the Haihe River, is sure to become a world-renowned international metropolis.

24. The Main · Frankfurt, Germany

Chief Planner ZhuGe

It is true that every famous city has a famous river. In the years that followed, ZhuGe

Min had the opportunity to visit cities in other countries, many of which had rivers running through them. In 1995, he went to Germany with a group from Tianjin Urban Planning Bureau. The first stop was Berlin. When the Berlin Wall fell in 1989, east and west Germany merged, ending more than 40 years of division and confrontation between east and west Berlin. When ZhuGe Min and the group visited, it was the time of large-scale construction in Berlin. In order to move the federal government in Bern to Berlin as soon as possible, the renovation and construction of the federal parliament and the presidential palace were being carried out, and the planning and construction of new public buildings were also being pushed forward in Berlin.

ZhuGe Min and the group went to visit the Potsdamer Platz in Berlin. The site was a construction ground with many cranes. In a temporary hall, the head of the project from Daimler-Benz briefed them on the status of the planning of the Potsdamer Platz, the Mercedes-Benz project, and the Sydney plaza project. Potsdamer Platz was historically the commercial center of Berlin. Before World War II, it was the largest square in Europe and the cultural and business center of Germany and even Europe. After the war, it became the dividing line between east and west Berlin. The Berlin wall was built, and the bustling urban center was reduced to an empty isolated zone. Many people who fled to west Berlin were killed here by machine-gun fire from sentry posts. After reunification, the Berlin wall was torn down and Berlin became the new capital.

In order to stitch together the traumatized east and west of Berlin, the German government decided to do something about the old dividing line, and building Potsdamer Platz was the best option. It was Germany's biggest post-World War II project, with 1.2 million square meters of floor space, and the land was auctioned off to multinationals including Mercedes-Benzes, Chrysler and Sony of Japan. Prior to this, there was an international CFP for the entire district, involving well-known architects and design firms including Sir Norman Foster and the then relatively unknown Daniel Liberskind. Due to the different styles of the proposals, some of the proposals were strictly cut out due to the influences and efforts of the competition organizers based on considerations about the coherence of the city's culture and history. Eventually the proposals of Munich architects Heiz Hilmer and Chrisoper Sathermin were selected, and one of the judges, Rem Koolhaas, was so angry that he walked away.

The winning plan restored the famous octagonal shape of Leipzig square, which was adjacent to Potsdamer Platz. It adopted the form of neat and uniform traditional block, with no high-rise buildings. The overall height was no more than 35 meters, which was reduced to 28 meters when implemented. The size of the land parcel was 50×50 meters. After the overall plan was finalized, an urban design proposal bidding was held for the developers. The architect of the winning bid for the Mercedes lot was Renzo Piano, while for the Sony lot it was the Helmut Jahn Firm. They could only see the model and renderings of the project at that time. The oval plane of the Sony center and the neat and uniform shape of the Benz center left a deep impression on ZhuGe Min, and is still fresh in his mind, but one could not really understand the essence of designs only by looking at models.

It was not until five years later, when he visited Berlin again, that he realized the superb level of the elaborate design of the two groups of buildings in Potsdamer Platz. The Sony Center as a whole was designed by the Helmut Jahn Firm, and its most prominent feature was that it was enclosed by a few buildings, the tensioned membrane structure covering a huge, 11-storey high semi-outdoor space, which included stores, theaters, offices, IMAX theaters and other functions. The open-air restaurant was on top of the stairs and became a very attractive public space. it was said that the design was rather different from the original plan.

On the other hand, Renzo Piano's Mercedes Benz Center strictly followed the master plan in its design, with an indoor pedestrian shopping street connecting the various areas, each with multiple functions: offices, residences, restaurants, theaters, concert halls, casinos and so on. The Mercedes Benz Center, under the leadership of Renzo Piano, was designed by a number of famous architects and firms, including Richard Rogers, Shigeru Ishizaki and Aldo Rossi, and was a unified yet varied ideal architectural complex. Under the constraint of the guidelines on general building height, street scale and functions, each architect used metal, glass and brown pottery plates as materials for the facade in their own design, and determined the theme and details of the design individually. None of the buildings was not self-centered, but they all tried to participate in the construction of public space as much as possible, to make contribution to the public space.

Potsdamer Platz was an outstanding case that reflected the whole process of the implementation of urban design. From the overall urban design to detailed urban design, and then to the final architectural design, in different design stages, there were both division of labor and cooperation; they plan for different areas together and carried out construction gradually, respected and communicated with each other. Not only internal communication, but a lot of work had also been put into communication with the public, and it had been widely recognized by the society. During the five years of construction, hundreds of people came to visit. After completion, Potsdamer Platz lived up to expectations as a city center with cultural and entertainment characteristics as well as a highly complex and transforming CBD featuring cultural and entertainment functions.

Later, in the planning and design of Yujiapu Financial District and Binhai Cultural Center in Binhai New Area, ZhuGe Min always had a strong desire to apply the organized planning and design method of Potsdamer Platz to it. Although he tried very hard to push for it, he thought he had only scratched its surface. For example, the SOM was in charge of the overall urban design plan and guidelines for the initial area of Yujiapu, and nine young domestic architects were responsible for the architectural designs of the nine blocks, respectively. It was relatively successful speaking from the aspects of form combination and architectural design, but the functions were not complex enough, and publicity and public participation were not enough, plus the influences of location and overall environment, the popularity was still lacking as of now.

In contrast, Binhai Cultural Center was slightly better. A lot of proposal comparison work was done in the urban design stage, and masters such as Bernard Tschumi and Zaha Hadid and new cutting-edge agencies such as MVRDV were invited to participate in the urban design stage, to increase the vitality of the cultural center.

In order to really learn from Potsdamer Platz's experience, ZhuGe Min had also invited Piano, Rogers and Murphy Yang and other masters who had participated in the Potsdamer Platz project, as the industrial-style architecture they designed fit the urban characteristics of Binhai New Area perfectly. In the end, due to various reasons, only Yang participated in the design of the Cultural Exchange Tower. He devoted a lot of effort, and it was a pity that the project was not implemented. If it were to be built, Yang's design of the highly industrial circular steel and glass tower would without a doubt add splendor to the Binhai Cultural Center complex.

The Potsdamer Platz complex not only continues the coherence of urban context in terms of spatial texture and stitches together east and west Berlin, but also plays an innovative role in urban function. At the time, ZhuGe Min and colleagues also surveyed the original east Berlin Alexanderplatz buildings. After 40 years of socialist construction, the buildings were branded with the distinctive architectural style of the Soviet Union and Eastern Europe. In the center of the square was the East Berlin TV Tower built in 1959, 369 meters high, 45 meters higher than the Eiffel Tower in Paris and 220 meters higher than the West Berlin TV Tower at the time. In many cases, architecture also plays a political role. 200 meters above the TV tower is a seven-storey spheric building with a revolving restaurant and a viewing hall overlooking Berlin. Of course, from west Berlin you can see the TV Tower towering over the sky. It is also a form of ideological propaganda.

After the reunification of east and west Germany, the public clamour to tear down the TV Tower became stronger, but it was decided to keep the tower as a historical symbol of the division, which might have been more uncomfortable for some people than the remaining Berlin Wall. In addition to the obvious differences in architectural styles between east and west Berlin, the differences between east and west Berliners are even more obvious, whether in clothing, language or behavior. True integration has a long way to go.

As the culprit and defeated country in both World Wars, Germany attaches great importance to the preservation of urban history and ancient buildings while carrying out historical reflection. A large number of ancient buildings were destroyed during the World War II, including the Kaiser Wilhelm Memorial Church on the Breitscheidplatz, a busy commercial street in west Berlin. The church was built in the 1890s and was then bombed to become known as the "beheaded church". To warn future generations against war, the church in the center of the city has not been restored, which is very striking. In 1957, a design competition was held for the reconstruction of the church. The winning plan was to demolish all the remnants of the church and build a brand new church, which caused great controversy. The final compromise preserved the 68-metre-high "severed head", the remains of the main clock tower, and surrounded it with four new buildings that were built according to the winning proposal. The building is grey, but at night the blue light from the latticed walls, which are made up of more than 30,000 glass windows, makes the church look beautiful.

A similar project is the German parliament building, which was built in 1884. It adopted the classical style and integrated the architectural styles of Gothic, Renaissance and Baroque. It was a symbol of the unification of Germany, and was initially the parliament for Deutsches Reich, and became the parliament for the republic in 1918. On February 27, 1933, the building caught fire and part of the building was destroyed. The Reichstag fire became an excuse for Nazi rulers to persecute political opponents. On April 30, 1945, the Soviet Red Army placed a red flag on the roof of the parliament building, declaring victory in World War II. Between 1961 and 1971, the building was rebuilt with a simplification that removed the dome that had been blown up in 1945. After the unification of east and west Germany, the building became the location for the Bundestag.

During ZhuGe Min and his colleagues' visit to Berlin, the building was being rebuilt. It was designed by the famous British architect Sir Foster. He retained the appearance of the building and built a new glass dome with various functions including sightseeing, the same size as the old dome that had been bombed. On his next visit to Berlin in 2001, ZhuGe Min visited the new parliament building, which had an alluring exterior, a mix of classical and modern. ZhuGe Min wanted to enter the glass dome to have a look inside, but an advanced appointment was needed, so his wish was not fulfilled. A few years later, Tianjin planned to build an city planning exhibition hall in the central square. In order to echo the Italian historic district behind, a building that combined the new with the old was designed, which was similar to the German parliament building with a glass dome on top. Now it seems that the Central Square should have be left blank after all, so it was a good thing that it didn't get built.

In addition to cherishing historic buildings, Germany also has great respect for modern buildings built after World War II. Next to the west side of Potsdamer Platz is the Kultur Forum, which was built in the 1960s. It was planned by Hans Scharoun, a famous German architect, and a group of buildings including the famous German Philharmonic Hall were designed. When Piano was in charge of the Mercedes Benz Center in Potsdamer Platz, it echoed well with the Berlin Library which was part of the group of buildings. On the south side of the Kultur Forum is the New German art Museum (Neue Nationalgallerie) designed by Mies Van der Rohe. It is well known in the architectural field that Mies is one of the four masters of modernism, and this work is also his masterpiece, which is listed as one of the most influential buildings in the 20th century.

ZhuGe Min and colleagues visited the new art museum in a spirit of worship. Its classic simplicity and powerful steel and glass construction were truly impressive. This building was completed in 1968 after Mies's death. It was Mies's swan song in his lifetime exploration of the pure architectural art style of steel and glass. It is known as the Parthenon of steel and glass, and the industrial construction skills and rigorous national spirit of the German people could be seen from it. Northwest of the Brandenburg Gate, they found a building encased in orange canvas with a building's facade printed on it. According to the local architect, this was a volume model built on site to see whether the volume of a building was in harmony with the surrounding environment. It showed that the Germans are truly meticulous in doing things.

Berlin is one of the most famous cities in the world, with a large green botanical garden in the center of the city. The buildings of the whole city are low in height and moderate in density, blending well into the surrounding natural environment.

But the two sides of the Spree River were indeed not as impressive. Talking about German famous riverside city, it would be Frankfurt, as the full name of Frankfurt is "Frankfurt am Main". Frankfurt is the fifth largest city in Germany, but it is home to many important business, financial, exhibition, cultural and transportation centers in Germany and even Europe and around the world.

Frankfurt's city center and the old inner city are located on the north bank of the Main River. The Main River is a tributary of the Rhine River. There are many bridges over the river, connecting the inner city with the south bank. There are many museums in the inner city and the south bank, many of them are outstanding buildings designed by famous architects, and there are many cultural buildings in the inner city, such as the former residence of Goethe. Frankfurt's CBD, just outside the city, is a group of modern high-rise buildings, including Europe's tallest and the world's first high-rise eco-building, the Commerz Bank Headquaters in Frankfurt, designed by Sir Foster, which was under construction at the time. Frankfurt is home to 80 percent of Europe's new high-rise buildings and has managed to maintain their relationship with the Old Town and also takes the Main River into consideration to avoid overshadowing the river.

ZhuGe Min and several young guys in the group lived in an apartment on the edge of Frankfurt's Old Town. The room was a mezzanine under a pitched roof, a steel spiral staircase and skylights filling the interior with variety and spirit. This was the first time ZhuGe Min had lived in a modern European residence; it was equipped with a variety of indoor facilities, the quality and sealing of doors and windows was very good, and the indoor temperature could be adjusted through central heating, allowing them to truly experience the quality of German residences. China's residential construction quality dwarfed in comparison. We have put too much emphasis on the economy, without paying attention to quality whereas quality is in fact what would last. ZhuGe Min also found time to visit some of the new residential estates in the downtown. They were all in small blocks with enclosed layouts, and the buildings were generally three floors, each yard and building different in shape, showing great richness and diversity. ZhuGe Min walked into an inner courtyard enclosed by a three-storey house at random, an about 30×30 meter square, and inside of the courtyard was completely a sand pit for children to play in. He did not see a child as perhaps it was not the right time.

The group also visited Munich, the capital of Bavaria and German third-largest city. With a long history of 850 years, Munich has a profound history of German traditional culture. It was also once a stronghold of the Nazis. It is an important industrial and commercial city in Germany, and the automobile, mechanical and chemical industries of Munich play important roles in German economy. The group tasted the traditional beer of Munich and enjoyed folk songs and dances, and visited the BMW museum and the BMW headquarters in the shape of "four cylinders", which showed all kinds of high-quality cars and products.

Not far from BMW's headquarters is the famous Munich Olympic Center Stadium. Its unique mesh and transparent acrylic materials cover an area of 75,000 square meters and is the largest roof in the world, equivalent to 10 football fields. Due to its good light transmittance, all kinds of plants grow vigorously, making people feel as if they are in a crystal clear green world of art. The plan embodies the creativity of technology, but also the rarely demonstrated imagination and sense of romance of the Germans. The 20th Olympic Games were held here in 1972, and there was a shocking terrorist attack on the Olympic village during the game.

During the visit, the group experienced the highly developed traffic planning in Germany. Frankfurt airport is one of the world's top ten airports, and the country also has a dense railway network. ZhuGe Min and the group rode the high-speed rail from Berlin to Munich, and the newly running trains, with their opulent interiors, leather seats and speeds of more than 300 kilometers per hour, became the initiating point of their later proposal to build intercity high-speed trains between Beijing and Tianjin. In terms of highway construction, Germany is even more admirable; many of their highways were built with reinforced concrete during the war. And in ZhuGe Min's impression Germany might be the only country without a highway speed limit.

German urban rail transit is also very developed, with the S-bahn fast rail, most of which runs on the ground. There are also the U-bahn metro, Stadtbahn light rail, tram and region trains. No matter what kind of rail it is, the carriages are always clean and tidy, and the seats are very comfortable, except that every now and then cyclists take their bicycles onto the car, and occasionally very large dogs are also brought into the carriages. The general sense was that the number of passengers in the German subway was less than London, and it was relatively quiet and comfortable. On the whole, compared with the trip to Britain and Australia more than a year ago, they had surveyed a broader scope of planning and design and learned more. However, they had not learned much in terms of waterfront cities.

The trip was prompted by a former planning official in Berlin. Since the 1980s, the Institute of Ecology of the Chinese Academy of Sciences had been working with the German side to research urban ecology in Tianjin as an example. As it was part of UNESCO's Man and Biosphere program (MAB), the Tianjin Environmental Protection Bureau and the Planning Bureau had both been involved in the relevant work, including this German planning official. After his retirement, he wanted to further cooperate with Tianjin in its development. The group visited some urban renewal projects in Germany, and learned about the German Land Order, which literally meaned land acquisition–that governmental organizations or trust companies invest in a region's road traffic and municipal infrastructure and improve the greenery environment, and then relevant beneficiaries of land pay a fee according to the land price growth. If they do not pay, the government has the right to purchase land and housing at pre-implementation prices.

After comprehensive consideration, Tianjin selected the Italian Style District as the research subject to explore feasible paths. A tall young woman, whose German name was pronounced "Schwarz" and was jokingly called "Shiwazi (wet socks)" for memorization, worked for a while at the Planning Bureau. The project was not finished in the end due to various factors. But later, with the implementation of the comprehensive development and renovation of the Haihe River, the Italian Style District was renovated through another way, that is, the area where ZhuGe Min and Bauer are now sitting and chatting. Throughout the whole process, ZhuGe Min and his partners have learned from many experiences, combined with their own reality in practice, and continued to explore and innovate.

Mr. Bauer

Upon hearing this, Mr. Bauer concludes: "although you did not gain much experience in planning waterfront cities in Germany, but the important thing is that you learned about the planning and design organization mode of large-scale projects like Potsdamer Platz and urban operation mechanisms such as the Land Order in Germany.""Exactly! In fact, riverside city is just one kind of city. All cities in the world are common in many aspects. To do a good job in the comprehensive redevelopment of the Haihe River requires knowledge and ability in all aspects. In fact, the protection of historic and cultural blocks and the renovation of historic buildings are among the 'Ten Projects' of the Haihe River that we implemented later. We have a completely different situation of historic building use and property rights than Germany. The government has invested in improving the Haihe River's environment, but it is unable to collect value-added taxes from land and housing owners on both sides.

"Why is that? One is that there are no laws regarding such situations yet; Second, different from the private ownership of land in Germany, our land is owned by the state, and the property rights of most houses are also owned by the government. It is meaningless to pay value-added taxes to yourself. After the founding of new China, some of the historic buildings were used as offices, and most were still used for living. Each room in a small villa was assigned to a household, and a building could accomodate more than a dozen families. The most urgent need for these residents was to improve living conditions. Therefore, according to the plan, the government put these buildings under protection, paid the residents living in small villas to move out, and improved living conditions. Because there were no specific laws and regulations for this kind of situation, in accordance with the general demolition compensation policy, agreements were signed with the residents so that they enjoyed the same treatment as residents who lived in dangerous bungalow.

"However, after the residents had moved out and the platform company had just begun to repair these small buildings, the residents came back, saying that we signed a demolition agreement and yet why aren't you demolishing the buildings? Later, in response to this new situation, in 2005, Tianjin People's Congress approved and enacted the "Tianjin Historic Buildings Protection Ordinance", categorizing small villas or similar buildings of protection values as historic buildings, and preserve, move, maintain, renovate, and utilize them according to the regulations, and to "empty and move" rather than "demolish and move" in terms of relocation of residents living in historic buildings. Of course, in addition to the rental of public housing, there are many other forms of property rights. Some houses were built by churches, and no one applied for the ownership when they were used as tenements to live in, but when the government moved the residents out, people from the church came, took out the ownership certificate and said the property rights of the house belonged to the church. There were a variety of unexpected problems. After hard work, the Italian Style District, Wudadao and other historic districts were finally revitalized, adding colors for both sides of the Haihe River, and they had also become major attractions of Tianjin.

"Tianjin has a total of 14 historic and cultural blocks, divided into two types: traditional Chinese and Nine-Nation Concessions. Most of them are well protected, such as Wudadao, Italian Style District, North Jiefang Road, etc., but a few are not ideal, such as Anshan Road, which is also the focus of our urban design pilot projects. Additionally, we plan to to further increase the vitality of Wudadao, North Jiefang Road, Central Garden and other areas by combining urban design with historic district preservation. We regard the improvement of the pedestrian system of these historic districts as an important means to increase the vitality of the city and revive the urban center. Feel free to give us any suggestion, Mr. Bauer, and we will value it greatly.

"We also want to further improve the connectivity of the pedestrians on both sides of the Haihe River, by building three pedestrian bridges on the Haihe River, connecting the Italian Style District to the original Japanese concession of West Anshan Road, the Jinwan Plaza to the Tianjin Railway Station transport hub, and the North Jiefang Road, Taian Road to Old South Railway Station business district, forming a complete pedestrian system. One of the three bridges has already in operation and become a hype in internet; the other two are under construction. "It is a good thing, and I think you must keep doing it. You talked about the arduous process of moving the residents and repairing the old historic districts such as the Italian Style District. I can recognize it from the comparison between some old photos and today's scenes. On the one hand, the comprehensive development and renovation of the Haihe River has transformed the waterfront of the Haihe River; on the other hand, it is more important to protect, restore and revive the historic blocks."

25. Tokyo Bay · Tokyo, Japan

Chief Planner ZhuGe

In 1996, ZhuGe Min was still working at TUPDI, which organized a surveying trip to Japan, including Tokyo, Osaka, Nagoya, etc. They visited planning departments such as the Ministry of Land and Resources of Japan, visited old urban centers such as Akihabara and Ginza, and visited the new constructions in Shinjuku, the Bay and the Osaka Business Park. Japan's economic bubble burst in 1990 and the economy entered a long period of stagflation, and they saw in their trip that the ambitious plans back in the days, including the plan to move the capital that had already passed the legislation, stalled. The old plan to build a new capital by selling the land of government agencies had become difficult because of the slump in land and property prices.

ZhuGe Min wondered why land prices and housing prices had risen so high in Japan, and he was even more worried about the housing bubble when he looked at the soaring prices in many Chinese cities today. In Osaka, they looked at reconstruction efforts, and although it was only a few months after the severe Hanshin earthquake, the city was largely back to normal. ZhuGe Min believed that if an earthquake of this magnitude had struck China, the death toll would not have been just 6,000. Japanese urban planning attaches great importance to urban disaster prevention and reduction, and national education and daily exercises and training play an important role in disaster response, which is worth learning from for China.

In Japan, they also looked specifically at the design, construction and distribution of public housing. They visited the design department of a big housing company, saw the construction drawings of the residences, and it was as accurate as mechanical drawings, and construction errors were calculated by millimeter! Led by their Japanese peers, they entered a newly built 50-storey public housing, whose interior design was a combination of Western and Japanese style, with an open kitchen and dining room, which was rigorous and exquisite. Japanese regulations did not require dining rooms to have separate lighting and ventilation. They also visited a residential project in Osaka, which was mainly composed of small high-rise slabs with corridors hanging outside, and high-rise towers. The layout, construction and installation of the pipelines of the whole building were orderly and precise, and the level of precision was just like the accurate construction drawings seen before. The garbage was collected in the basement, the first underground level was used for collection and compression, and the second underground level had a special garbage truck to transport the compressed garbage away. The whole process was clean, tidy, no smells and environmentally friendly.

They also visited the sales office and showroom of a corporate residence. They heard that Japan's public housing policy was similar to that of Hong Kong. 30 percent of public housing was provided by the government and applicants had to wait in line for a certain period of time. In addition to public housing and corporate housing, there was always a large area of low-rise detached houses on the outskirts of the city. Although the area was not large, but it was always one courtyard per household. They learned that this kind of house called "one-household build" in Tokyo also accounted for 40 percent of the total number of residential units, which surprised ZhuGe Min and the group. Japan's population density is 338 people per square kilometer, 2.5 times that of China's population density of 138 people per square kilometer. The population density of Japan's megacities, especially Tokyo, is much higher than that of China, but they still have a large number of low-rise single-family houses, and even half of the total national housings are single-family houses. Why does our country simply say that the residential area FAR cannot be lower than 1?

"What do you think of our food?" The restaurant's Italian owner interruptes ZhuGe's excited speech by offering him a lava cake. This Italian restaurant has been in the Italian Style District for more than three years, and the owner is very familiar with ZhuGe Min. After academic activities in the Planning Exhibition Hall, there are always planners who like to meet and chat here with ZhuGe Min. "You have just told the story up till 1995, but there are still seven years between then and the comprehensive development and renovation of the Haihe River in 2002. What happened during this period?" Mr. Bauer asked. "I had not gone abroad in those years. Some of the things I did during this period seem to have no direct relationship with the Haihe River planning, but in retrospect, they were all preparation for the planning of the Haihe River in 2002."

ZhuGe Min paused and took a bite of the cake. He liked the sweet taste. "I'll tell you a little story. What is the most famous waterfront city in the world? Right! Paris on the Seine, no doubt. Over the years, I have collected a lot of information about the Seine in Paris, but I have never had the opportunity to visit the Seine. Perhaps it was God's will that during the trip to Germany, we unexpectedly came to Paris before landing in Germany and saw the Seine, but it was at night and the lights were dim, so we did not clearly see the Seine."

26. Paris in the Shades of Evening, France

At the time, Europe's Schengen agreement was just beginning to take effect, and the group, who had all prepared their German visas, planned to enter Germany by changing flights at Charles De Gaulle airport in France. They needed to move from one terminal to another to get on the next flight. At the Customs they experienced the bureaucracy of French border officials. 'You have to have a French visa, and you're not allowed to pass without a French visa,' the border official said. Schengen or no Schengen, he said a visa for the country of landing was required. After several rounds of negotiations, nothing came of it. The members saw another terminal within sight but could not get through. As the hours passed, they could only watch as their plane flew away. Finally, each member of the group had to pay 200 francs to apply for a temporary visa and book a new flight.

It was dusk when we left the airport and it was dark when they reached the hotel. The group had booked an early flight the next morning, and it was impossible to see Paris during the day. After discussion, the leader said: since we are already in Paris, we can survey it overnight. The members immediately called two cabs for a night tour of Paris. The taxi drivers looked Asian, but could not understand Chinese; turned out they were Vietnamese. It was late at night, Paris was asleep, the streets of the Champs-Elysees were deserted with only streetlamps still standing. The banks of the Seine were almost completely dark, quite different from the scenes of bright lights by the Seine in everyone's imagination. When ZhuGe Min visited the Seine again in 2001, he found that the lights on the buildings were from the searchlights on the cruise ships. The tourists on the ships could clearly see the brightly lit buildings on both sides of the river. After the cruise ships passed by, the buildings disappeared into darkness.

They were trying to get the driver to Montmartre to get a good view of Paris from the top when sirens went off and a police car followed. After stopping their taxis, the policemen checked the id of the two taxi drivers and said something. When the policemen were done, the night was getting dark, and the members were getting sleepy. Considering that they had to get up early tomorrow to catch the plane, they had to give up. This was ZhuGe Min's first trip to France, catching a glimpse of Paris under the night for just a few hours. ZhuGe Min never thought that it would be six years later when he finally got to see the true colors of the Seine!

Chapter V

The Golden Dragon Dancing · The Haihe River

27. The Prelude of the Haihe River Planning

In 1998, the Urban Planning Bureau changed for a new director who has construction background. Because always suffered bullying from the "little bastards" of the planning bureau, he had already vowed that he would determin to give them the hell if he became the director of the bureau. As soon as becoming new director, he proposed that the Urban Planning Bureau should not only be busy approving projects, but also give advices to the municipal government. Therefore, Tianjin Urban Planning and Design Institute was assigned to engage in series of planning and design projects. Chief Planner. ZhuGe was responsible for organizing the urban design of Xiaobailou of the Central Business District, Old South Station and North Jiefang Road on both sides of the Haihe River. The plan of Old South Station was based on the plan by Australian experts in 1997. In order to elaborate the plan, a large working model for the entire area inside Inner Ring Road of Tianjin was made. New director saw it and considered that the effect of the model was very intuitive, also very novel, so he asked what kind of plan was for. When he knew it was called urban design, after some reflection, asked Chief Planner. ZhuGe to give a lecture in the Planning bureau, to introduce what the "urban design" is. In the absence of available lecture materials, he spent two hours to introduce ten classics on urban design: Design of Cities (Bacon, 1967), Concise Townscape (Cullen, 1995), The Image of the City (Lynch, 1960), The Design of Outer Space (Yoshihiro Ashihara, 1975), Finding Losing Space (Trancik, 1986), etc. Chief Planner. ZhuGe felt his speech was boring when he spoke in front, and he can see all audiences have no reactions. It was his first academic lecture at the Planning bureau and also the most unsuccessful one in his life. When he thinks about it now, he always believes that it would be more acceptable to introduce a specific urban design project than to talk about theory only.

In August of 1999, the state council formally approved the revised city master plan of Tianjin (1999-2010), which was initiated by Chief ZhuGe in 1994. The revision work was basically completed in 1996, but it was officially approved in 1999. That was the common condition of China's mater plan which always took five or more years from the revising to the formal approval. In order to implement the very top and long-term master plan and make regulatory plan for real projects at the same time, the Urban Planning Bureau decided to carry out the full coverage of regulatory plan for the central city. In the beginning, most planners from TUPDI were afraid of difficulties, and Chief ZhuGe always worked with colleagues on making work plan, definiting regulatory units and formulating technical standards. With the joint efforts by all departments in two years, the regulatory plan for the central city, of more than 180 units, 334 square kilometers, was completed successfully and approved by the municipal government, which became the first full-coverage statutory plan in the Central city of Tianjin in history, ranking in the top all over the country in that time. Even now, 20 years later, when we travel around the country, we still often found that most big cities do not have a complete road alignment map of the central city. While Tianjin had completed the basic work as early as 20 years ago. Since then, only dynamic maintenance, constant adjustment and improvement need to be taken on.

At that time, some people doubted the level of the planning, but practice has proved that this version of regulatory plan is successful. Although there were, more or less, some problems, it did play an important role in guiding and controlling the urban construction of Tianjin and made the urban planning step by step on the track of standardization and legalization.

In the late 1990s, in the process of promoting the full coverage of urban design and regulatory plan in Tianjin, the planning on both sides of the Haihe River gradually matured and took shape. In 1996, Chinese Ministry of Construction and Australian Ministry of Industry Export and Tourism carried out a riverside city cooperation project funded by Australian government. Australian exports, cooperating with Tianjin side, carried out conceptual planning for Old South Station Area of Central Business District on the east bank of the Haihe River, which was completed in 1997. In 1998, combining with North Jiefang Road, Xiaobailou and Old South Station in Central Business District, the plan for both sides of the Haihe River continued to be deepened, of which general structure showed basically stable in this period. At this time, in 1996, Deng Jiaju, a painter from Tianjin, showed the Water Transport Map of the Haihe River at the United Nations, based on the water transport history of the Haihe River, creating impact. UNESCO sent a letter to Tianjin Municipal Government to promote the Haihe Plan. Therefore, in 1999, a delegation from TUPDI went to Paris to study the Seine, contacting with the famous airport design company and Paul Andrew about the planning of the Haihe River, but failed to cooperate due to various reasons. In 2001, with the push by the Municipal Introducing Foreign Intelligence Leading Group Office and French embassy, France CONCEPETAL Design Company came to Tianjin for inspection, to prepare the Haihe River concept planning.

28. The Golden Dragon Dancing: The Comprehensive Development and Innovation Plan of the Haihe River

In 2000, when the world greeted the new century with jubilation amid fears of Y2K in the computer operating system, Tianjin entered a new historical period. Thanks to the hard work of the previous stage, the goals of "3-5-8-10" have been accomplished. In particular, the reconstruction of dilapidated shanty dwellings in urban areas had been basically completed, greatly improving the appearance and functions of the city, which indicated that Tianjin had entered the new century. How to develop in the future, became the first urgent problem in front of the whole municipal government. Under the arrangement of the Urban Planning Bureau, while cooperating with foreign countries, the TUPDI, basing on itself, began to work out a new round of the Haihe River plan. Chief ZhuGe was in charge of the project and set up a working team inside the institute. The plan was open-ended and need to answer three key questions: Why does the Haihe River need to be developed? If being developed, what content do we need, what standards should it meet? How can the plan be implemented? It should be said that the Haihe planning has been accumulated for nearly ten years: it had completed the city master plan, the full coverage of the regulatory plan, the urban design for key areas, laying a solid foundation for the preparation of a good plan in the next step. But that was not enough. It's impossible to make a high-level plan if you just follow the script. The team adopted an open planning method, often sitting together to brainstorm randomly, stimulating each other. At the same time, Chief ZhuGe organized relevant departments to compile data collections like World-famous Rivers, History of Haihe River, History of the Haihe Riverbanks and Architecture of the River, etc. Through continuous discussion, people realized that the key to answer the above three key questions is to solve several key issues affecting the Haihe River plan. The Haihe River planning is a project that can affect the whole, so first of all, the planning need to be given a resounding name: The Comprehensive Development and Renovation of Tianjin Haihe River. As the saying goes, Name is not right, words are not right. People all approved this name but also think it still too long to be readable. The deputy director Lian of TUPDI, and Mr. Qi, the head of the team, two talented people together, proposed a resounding and picturesque name: "Golden Dragon Dancing". It's so catchy that everyone said it great but felt a little too much. However, for this kind of motivational "Social Mobilization" type planning, the name should be loud. The meaning of "Dragon Dancing"is also very clear: the Haihe River winds through the city of Tianjin, stretching 72 kilometers to the sea, like a sleeping dragon lying prone on the JinGu (Tianjin) land. To implement the Comprehensive Development and Renovation of Tianjin Haihe River is to drive Tianjin's urban economy, society and culture, like a vibrant golden dragon getting up and dancing. Some people asked: where is the dragon's head? Traditionally, the head is the source of the Haihe River at the Sancha Estuary in the central city. Others asked: with the coming of the new era and China's participation in the maritime order, should the dragon's head be reversed at Tianjin Port at the estuary of the Haihe River? Others still said, In history, Tianjin's earliest railway station located on the Haihe River's bank was called "Old Dragon Head". But there didn't seem to be much response from the outside. The plan of the Haihe River divides 72 kilometers into upper, middle and lower three sections. The upper section is 19 kilometers, from the junction of three rivers to the outer ring road of the central city, which is the focus of Comprehensive Development and Renovation of Tianjin Haihe River.

As for the orientation and function of the Haihe River, there is not much dispute due to the basis of consensus formed by previous studies and discussions, including the support from Tianjin city master plan approved by the state council in 1999. In the beginning, for the orientation of the Haihe River with a lot of descriptive language, such as: "The Haihe River, is the mother river, the creation river, hope river of Tianjin. It is the blue axis of urban layout, is the hope and the lifeline of Tianjin. It is the center line of Tianjin's politics, economy, history and culture, the axis of greening, tourism and scenic landscape. It is a service-oriented economy belt and landscape belt. It is the core zone representing Tianjin as Chinese northern economic center and international port metropolis, etc. "

From these expressions, it can be seen that planners have deep feelings and infinite expectations for the Haihe River, so they use words to the utmost in fear of missing anything. Later, in the several rounds of presentation and consultation for the Haihe River plan, especially with the help by someone fromthe Municipal Research Office, chose the sentence that "The Haihe River is a service-oriented economy belt and landscape belt" as the foundation, considering all options and factors, and combined with "culture belt", forming a highly generalization and clear orientation of the Haihe River.

Mr. Bauer

"After 20 years of practice, I have come to realize the importance of writing for planning and design. Chinese, unlike English, has its own power and charm. What do you think, Mr. Bauer?""I agree with you that we also have many short and strong slogans, which are very important."

Chief Planner. ZhuGe

Compared with the orientation of the Haihe River, the functions of the Haihe River is a scientific problem that cannot be ambiguous. There are contradictions between the functions of flood discharging and water conservation, and the project of comprehensive development and renovation. Fortunately, after many years of hard work and communication, with a variety of technical means, all aspects had reached a basic consensus, to be implemented into Tianjin City Master Plan which had been approved by the state council in 1999. That is how the Haihe River Planning has superior legal planning support, how the Haihe River comprehensive development got the basis for determination of function, how the comprehensive development and renovation set the foundation of success, which was always taken delight in talking about by Chief ZhuGe. He began to be in charge of revising the city master plan from 1994. Even if he was no longer in charge of the work, however, Chief ZhuGe still corrected the plan text of the city master plan word by word before report to the state concil, ensuring without any missing and error. Because city master plan has to go through the review of the minister-level joint meeting organized by the Ministry of Construction (Minstry of Housing and Town-Country development now), the drawings and relevant texts must all pass the strict review of the relevant ministries and commissions of the state. The state council will finally approve the plan if the ministries and commissions have no opinions. It is a great deal to reduce the discharge standard of the Haihe River within the urban flood control circle of built-up area in Tianjin. If it was clear in the master plan, it is equivalent to the approval of the Ministry of Water Resources. Therefore, after the state council approved Tianjin Master Plan (1999-2010), the National Haihe Watershed Management Commission organized the formulation of the Haihe Riverflood control basin plan, submitted it and get approval, which increasing the flood discharge capacity of two longitudinal rivers, the Yongdingxin River and the Duliujian River, raising the corresponding levee elevation, reducing the runoff of the Haihe River trunk stream, from 1200 to 800 cubic meters per second, dropped by a third. At the same time, the height of the dikes on both sides of the Haihe River has been clarified, which is more than 1 meter lower than the original plan. The implementation of upper planning adjustment provided guarantee for the Haihe River Plan, making Chief ZhuGe's team members also took a reassuring.

After the orientation and function of the Haihe are determined, the next major work is specific planning and urban design. How can both bunds transform into waterfront platforms? What will the section and the elevation of the banks be? How can we form continuous, uninterrupted promenades on both sides of the Haihe River? Planning need to solve many specific problems. To develop tourism, the Haihe River must be navigable for cruise ship. So what should the height for cruise ships be? There are many bridges built in different periods in the upstream of the Haihe River. How can it be coordinated? Another key issue is road traffic on both sides of the Haihe River. To carry out comprehensive development and renovation on both sides of the Haihe River, and to develop the service industry, the transportation problem must be solved first, otherwise nothing else can be done. People can't reach the riversides, or it is very inconvenient to reach, how can tourism be developed? However, the situation at that time was that there were few bridges on the Haihe River with narrow width, being very congested during rush hour. Moreover, the roads along the Haihe River were all at the same level with the bridges, where traffic lights make traffic more congested in the conjunctions. In order to promote the development of the city towards the Haihe River and strengthen the connection between the two banks, the plan must increase the number of bridges. If the bridge is still level crossing, the traffic jam will be further increased. If the viaduct is built, it will damage the Haihe River landscape. Chief ZhuGe always remembered that a few years ago, in order to ease the traffic pressure on Zhongshan Road, the Old Jingang Bridge was dismantled and a double-deck bridge was built. All circles in the society reacted strongly. Most people focus on two facts. On one hand, many old citizens can't accept it emotionally. On the other hand, someone considered the new bridge too high to preserve the Haihe River landscape. After careful analysis and selection of multiple schemes, the working team proposed at the early stage that the elevation of the bridge should be raised according to the improvement of navigation standards of the Haihe River, and the road along the river should be crossed under the bridge head. This is a bold and ingenious scheme. It is understood that in order to solve the traffic problems in the old urban area of Paris, one side of the Seine was occupied to build an urban expressway, while some technical details is unknown. The river course of the Haihe is not wide enough and impossible to be occupied, while the traffic problem must be solved.

Another key issue is the goal of the Comprehensive Development and Renovation Plan. The new city master plan approved by the state council in 1999, has put forward the goal as "Using 10-15 years, to build both sides of both sides of the Haihe River into the center line of Tianjin's politics, economy, history and culture, the axis of greening, tourism and scenic landscape running through the central area of Tianjin, to become the core zone representing Tianjin as Chinese northern economic center and international port metropolis". In the implementation process, the leader required to adjust the period to 3-5 years. Although the time is too short to complete, but as a kind of social mobilization, it is also understandable. Things must be done one by one.

The more uncertain is the idea of creating a world -famous river. There was a variety of statements in the beginning which is "Within 3-5 years, to build the Haihe River into a unique, world-class service-based economic belt, landscape belt. By carrying forward the Haihe culture, to create a world-famous river." After that, it was adjusted to "Within 3-5 years, to build the Haihe River into a unique and world-class service-based economic belt, landscape belt and cultural belt. With a long-term effort, to create a world-famous river.", which was still objected and scoffed by many people.

Indeed, from the situation on both sides of the Haihe River at that time, we could not see any signs of the world-famous river.

While Chief ZhuGe was very sure. After ten years of planning research and the understanding and accumulation of domestic and overseas inspection, he firmly believes that Haihe River has great potential and Tianjin has great potential. This belief has never been shaken. In the early 1990s, many people were bearish on Tianjin. When there were many complaints in the society, he wrote an article Tianjin's Strategic Position and Self -confidence, which was published in *Tianjin Daily*. He believed that from the objective law of city development in the world and the future of of Chinese urbanization, industrialization and coordinated development for the Beijing-Tianjin-Hebei region, Tianjin will surely revitalize the northern economic center. He had great confidence in making the Haihe River a world-famous river. In his heart there was still an unfinished desire –visit to the world's most charming waterfront city – Paris. He would like to go to the Seine by himself, to give him a final confirmation.

29. The Seine, Paris, France

Just on May 2001, according to the instruction of Tianjin Municipal Government, the Urban Planning Bureau organized an investigation group to visit Europe, also to discuss planning scheme with Concept Firm from France who was now in charge of the Haihe River conceptual plan. The delegation, led by the head, six people in 15 days travelled eight European countries. Though it was called eight countries, for including small ones like the Vatican and Luxembourg, the main destination is Paris, France. The Concept Firm was just located in the building designed by themselves in La Defense. After discussing the Haihe River concept plan, the firm's chief designer took them on a tour of LaDefense, a new planned, modern central business district. Maybe French people are limited too much for protecting Paris, and Parisians' enthusiasm for construction has been suppressed for too long time. Therefore, when they planned to develop a new central business district, which is not subject to the constraints of height and scale, it is obvious that the scale is too large.

The core area of La Defense Business District covers an area of 1.6 square kilometers, with a development capacity of nearly 3 million square meters. It was a multi-functional business district, including office, business and commerce, etc. As a transportation hub in the northwest of the city's main axis, the plan is organized in a completely three-dimensional way: all the buildings are built on a large platform. They were standing on the platform of 40 hectares, catching up in the clear sky and the direct sunlight. They had no place to hide in the hard-paved square, so they had to come under the big arch. The arch echoes the Arc de Triomphe, in 110 meters long, wide and high, covering an area of 5.5 hectares. The big step is very high, a group of people finally climbed the big step, barely hiding under the tensioning membrane. On the top floor of the great arch, which is the horizontal sky corridor of the building, there was an exhibition that told the story of the planning, designing and construction process of the great arch, which they watched with great interest. But the bad feeling of having nowhere to hide in the sun in the square, left in his mind for the whole day. This experience made La Defense a negative image to Chief ZhuGe . In the following years, during the planning management, architecture and urban design review, he often mentioned it to remind the designers to compare proportion with it, in order to find the appropriate scale.

The Seine water flows. Standing by the Seine, overlooking the Notre Dame de Paris on Cite Island, Chief ZhuGe's excited mood gradually calmed down. More than 2000 years ago, hundreds of residents lived in less than half a square kilometer on Cite Island. In the 4th century BC, a Roman tribe forcibly occupied the Gaul's village on the island, establishing the Parisii people's capital, where Paris got its name. by the 6th century AD, Paris became the capital of Kingdom of France. For more than 1,400 years, Paris has been the capital of all dynasties. On July 14, 1789, the people of Paris destroyed the Bastille, which led to the outbreak of the bourgeois revolution, the declaration of human rights, and the abolition of the monarchy. Since then, Paris has entered a period of rapid development, especially when Napoleon III appointed Paris administrator Hausmann to carry out the famous large-scale reconstruction of Paris, which made Paris what it looks today.

Although demolished a large number of old buildings and widened the streets, which caused great controversy, Hausmann laid down a strict urban pattern for Paris today. He planned important streets, squares and a large number of public buildings, determined the street trees and lamp posts, and determined the height and shape of buildings on both sides. In addition, he creatively upgraded the water and sewer system in Paris, creating lots of open spaces and parklands. It can be said that without Hausmann's transformation, there would be no Paris today.

The Tianjin delegation investigated both sides of the Seine from multiple perspectives. Firstly, they visited the waterfront path. The Banks of The Seine in Paris was added to the UNESCO World Heritage list in 1911. The evaluation from the UNESCO World Heritage Committee is, "From the Louvre to the Eiffel Tower, from the Place de la Concorde to the Grand and Petit Palais, the evolution of Paris and its history can be seen from the Seine. The Cathedral of Notre-Dame and the Sainte Chapelle are architectural masterpieces while Haussmann's wide squares and boulevards influenced late 19th- and 20th-century town planning the world over." Historic buildings, such as the Notre Dame Cathedral, the Sacre-Coeur Cathedral and the Louvre Museum, represents buildings of different artistic styles in different periods. The beauty of Paris is inseparable from the Seine. With this river, which flows slowly, Paris is tender and charming. The Seine is the mother river of Paris, the soul of Paris, and a flowing picture.

For tourists with limited time, they usually choose to take a cruise to enjoy the scenery of the Seine, to roughly browse the important buildings and landscape on both sides of the Seine. While if you want to really know Paris, you should walk along the Seine, to walk to stop, to enter famous attractions or stroll on the shore, to watch the painters drawing portrait for tourists, to look for treasure and souvenirs in more than 200 famous "Green Book" old bookstalls, or to walk down to the waterfront platform and sit on the floor, watching boats coming and going and deck visitors people greeting each other with the shore.

The Seine is more than 700 kilometers long, with a basin area of 78,000 square kilometers. Its length is more than 10 times of the mainstream of the Haihe River, and its basin area is one quarter of the Haihe River. Rising from Langle Plateau in Burgundy where the peak is 470 meters, the Seine has a drop of only 250 meters over 40 kilometers from its source. Paris is altitude 24 meters, located in the middle of the Seine, 365 kilometers to the estuary. Because the Seine has a gentle flow, it's suitable for shipping. Moreover, it is more seaworthy because there is no special water quality requirement and it has a regular fluctuation. The highland of Paris, where the Seine flow through, are France's most fertile agricultural region. The Seine flows from the southeast to the northwest, to the flat central part of the basin, where it slows to form a meander, crosses the center of Paris, finally at the bottom of Mantes La Jolie passing Normandy to the estuary of the English Channel.

In downtown Paris, the Seine has been renovated and the channel between its docks has been narrowed. Downtown along the river more than 10 kilometers are stone docks and embankment. Due to the water flow and flood level having a height difference of about 6 meters, so the height difference between docks, waterfront embankment and city roads was 5 meters. Generally speaking, the waterfront embankment looks very natural and pristine. The bridge is connected under the bridge, and the upper embankment is connected by steps or large ramps. The waterfront paths and embankments of the Seine are the most poetic places in Paris.

Parisians called north of the Seine the Right Bank, where Champs-Elysees, Arc DE Triomphe, the Louvre and many high-end department stores, boutiques and restaurants locate on; South of the river is known as Left Bank, where there are many universities, colleges and other cultural and educational institutions. As literary and art circles gather on Left Bank, various bookstores, publishing houses, small theaters, art galleries and museums are gradually established. Cafes and beer bars also emerged, becoming important meeting places for intellectuals on Left Bank. When you walk into a coffee shop, you will sit in the chair that Hemingway used before, under the lamp made by Sartre, next to the window where Picasso in a daze. Parisians tell jokes, "Left Bank uses brains, Right Bank uses money."Walking on Left Bank, professors and college students brush; and on Right Bank, you should be careful not to step on other people's feet –ten to one, those of bankers in fine leather shoes.

There are more than 30 bridges on the Seine, built in different historical periods, in various styles, with many stories. The balustrades of more than one bridge are covered with lover locks, densely, large or small, of various of shapes and sizes. Lovers come to Paris, write down names of two people on a lock, then lock it on the bridge, embrace and kiss, throw the same key into the Seine, on behalf of permanent love. The first bridge to be covered with locks was Pont des Arts, also known as the Bridge of Love. In 1802, Napoleon ordered that the bridge should be designed in English style. It is the earliest metal building in Paris. In 2014, the government removed more than 100 don of locks to ease the load on the bridge. Another bridge over the Seine, Pont de l'Alma, witness romance of royalty.

To see the relationship between bridges and roads over the Seine, the group invited a French architect to accompany them to take a tour of the site of an interchange.

The architect immediately thought they were going to see the scene of princess Diana's car crash. He drove them to the tunnel under Pont de l'Alma on Right Bank, slowed down, pointed to a pillar in the middle of the road, and said, "There it is." The short tunnel runs two-way, two-lane on each side. Princess Diana and her boyfriend Dodi, the son of the owner of the famous Harrods Department Store, died in a car crash on August 31, 1997 as they tried to escape the paparazzi. The car hit the central pillar at high speed, and the marks were still visible. Since then, tabloid journalists came here gathering materials to piece together of royal conspiracy theories, while fans mourned and left flowers at the bridge.

The bridge, connecting Island with the left bank and the right bank is the oldest existing bridge in Paris, while whose name is "the New Bridge". The bridge founded in late 16 century, completed nearly 30 years later. It is 232 meters long and 22 meters wide. It was completed in 1607. It became the first bridge in Paris without building any houses, so called "the New Bridge". The most spectacular and magnificent bridge on the Seine is the Alexander III bridge, built in 1900 to celebrate the alliance between Russia and France. Russia and France two countries were still feud a hundred years before, Napoleon led the attack on Moscow, massive killing. after two hundred years, the two countries solved problems with a smile. This is a constant drama across the continent of Europe."The Hundred Years' French British War", which took place earlier, lasted 116 years and was the longest war in human history.

The Alexander III Bridge, a single-arch bridge with a span of 107 meters, connects the Champs-Elysees with Hotel des Invalides on the other side, where Houses Napoleon's Tomb. There are two 17-meter-high bridge heads at each end of the bridge. On the top of the bridgeheads are golden sculptures of the goddess reining flying horses. These statues, even in gloomy weather, are glittering and sparkling.

Tourists to Paris finally have to take the cruise in the Seine. In the tourist season, the cruise is very dense, every 20 minutes. The main route concentrate in the most exciting part of the Seine, about an hour for a round trip, 10 francs per person, equivalent of about 100 yuan. It is said that there are night cruise ships, with dinner, requiring formal dressed, about 1000 yuan. Because members thought that the price was a little expensive, so they gave up. Cruise ships are generally two stories, the top floor is open for standing on to enjoy the scenery of the two sides. The members were surprised to find that the cockpit could be descended and lifted when passing under the bridge due to its long hull and high altitude. The headroom of the bridge is a problem of great concern to all of them. After many inquiries, it is learned that the headroom of bridges on the Seine is 9 meters to ensure the passage of double-deck cruise ships, while the headroom of bridges on the Haihe River is relatively low, only about 3 meters, which is a subject that must be studied.

The group climbed the Eiffel Tower and took in a stunning view of the city, which is both artsy and orderly. The combination of strict planning control and exquisite design have made Paris the fashion capital of the world. The delegation came to Louvre Plaza, saw the glass pyramid designed by I. M. Pei, felt the master's exquisite attainments at close range, and understood why President Mitterrand insisted I. M. Pei do the renovation design of the Louvre. Through the galleries, they rushed to browse paintings by masters, pushed their way to the front of a throng of people to see the Mona Lisa, the treasure of the Louvre. It turns out that this famous painting is only about the size of octavo paper, which is not the large size they thought at all. Maybe the smaller, the more valuable it is. Chief ZhuGe began to understand, the reason why Paris has become a world -famous city and why the Seine has become a world-famous river, Besides the river itself, the beautiful buildings and urban environment on both sides, the cultural and artistic connotation of Paris plays a crucial role.

In addition to visiting the Seine and the historic buildings of Paris, the group paid a special visit to a new building in downtown, the famous Pompidou Cultural Center, which also caused a great deal of excitement at the beginning of its construction. The Pompidou Cultural Center's unique and pioneering steel structure and exposed shapes of various pipelines are easy to see in various magazines and books. When you approached it, you still feel very familiar and intimate. Although its shape is different from the surrounding historic buildings, its volume and scale are quite coordinated with the surrounding environment, as to announce to the world that Paris has a long history, but also fashionable, trendy and pioneering. The experience of Pompidou sets a unique example for the protection of historic cities and neighborhoods in China.

The group made a special trip to La Villette located in the northeast of downtown Paris, covering an area of 35-hectare. It was a former slaughter house and wholesale market, but now it is one of the city's largest parks. When Chief ZhuGe was in the park, he saw the City of Science and Industry on the north bank of the canal and the red structures scattered in the park, as well as the huge bicycle tires and handlebars inserted diagonally in the green space, Chief ZhuGe realized the image of the 21st century city park in the mind of the master designer Bernard Tschumi. It is said by the industry that La Villette Park have started a new era of urban design. In 1982, President Mitterrand proposed to hold a competition for the park plan, with the title of "Urban Parks in the 21st Century". Rem Koolhaas, Zaha Hadid and other internationally renowned architects were invited to participate. American architect Bernard Tschumi won in the last. He used a chic, unusual way to express the concept that the park is a continuation of city space, through point — red structures (folly), line — irregular road, face — green landscape area of grid system and deconstruction, to create a space interacted with the surrounding urban area, public buildings and communities.

The group arrived at the National Library of France on the left bank of the Seine in the southeast of Paris. The sculpt of library is just like four opened books on a huge wooden platform, enclosing a sink yard. This project was also promoted by President Mitterrand. In 1989, France held an architectural design competition of national library. 244 architects from all over the world participated in bidding. Finally, the surprising 36-year-old French architect, Dominique Perot, won the competition. The project was completed in 1995 and won a series of awards including the European Ludwig Mies van der Rohe Prize of 1996. The National Library of France is one of the few large libraries in the world. Perot's design overturns the traditional design concept and method of libraries, making the library a new cultural landmark on the

Seine.

On the evening of their departure from Paris, the group returned to the Champs-Elysees, with mottled trees, tourist-filled streets, brightly lit mansions, wide walkways and plane-trees perfectly separated from the noisy motorway. Overlooking Arc de Triomphe and the new high-rise skyline of distant La Defense, Chief ZhuGe was filled with emotion. In the past few days, he had paid intensive visits to Paris and the Seine. He has seen many great experiences and practices in Paris at close range, realized the similarities and huge gaps between the Haihe River and the Seine, Tianjin and Paris. Because of these similarities and huge gaps, Chief ZhuGe had further seen the potential and huge space of the Haihe and Tianjin in development, which confirmed Chief ZhuGe's determination to build the Haihe River into a world-famous river.

After leaving France, the delegation visited Germany, the Netherlands, Italy and other countries. They visited six major cities, including Frankfurt, Amsterdam, Venice, Florence and Rome, and several towns. Because of previous trip to the Seine in Paris, in Frankfurt, Chief ZhuGe came to the Mein River for the second time, which he felt different from the first time. The River in the central part of the city could not only be ecological, but also be more urbanized. Chief ZhuGe was impressed by the canals in Amsterdam and the water city in Venice. The city and water can coexist in harmony. The classic St. Mark's Square is a haunting place.

Florence, the birthplace of the Renaissance 500 years ago, is the center of European culture and art. An Arno River runs through the city. There are seven bridges in the center of the city, of which Vasari Corridor built in 1345 is the most famous. It has two layers. The lower layer is a shop, and the upper layer is a long corridor connecting the Vecchio Palace and the Pitti Place on both sides. Vasari Corridor is also called the Ancient Vecchio (PONTE VECCHIO). VECCHIO means ancient. It is a historic arch bridge in Florence. Historically, there are commercial functions on all bridges. The Vasari Corridor in Florence is very similar to those in some parts of China, such as the Hongqiao Bridge in the ancient city of Fenghuang in Hunan Province. Looking at the Florence Vasari Corridor, Chief ZhuGe began to consider where to build such a bridge in Tianjin. Florence is only 100 square kilometers, with a population of more than 400,000. They came to the high Michelangelo Square, where the statue of David stood. Chief ZhuGe looked back at the ancient city of Florence, which is well preserved. Against the setting sun, the red roofs of the buildings in ancient city and the beige walls sent out a warm feeling. The water of the Arno River passing through the city was also tinged with golden sunlight. Bridges on the river leave golden silhouettes, picturesque.

The last stop was Rome, a world-famous cultural city with a history of 2500 years. As the capital and largest city of Italy, Rome has only 200 square kilometers of urban area and a population of nearly 3 million. The city of Rome was built on seven hills in the lower reaches of Tiber River, so called "the City of Seven Hills". It is the birthplace of the ancient Roman Empire. Because of its long history, Rome has the name of "the Eternal City" and "the City of World". The Vatican is the home of the Catholic Pope and the Holy See. It has more than 700 churches and monasteries, 7 Catholic universities. Compared with the numerous cultural heritage and monuments in the

Old Town, the Tiber River, less than 100 meters wide, is slightly inferior. In 1980, the historic city of Rome was listed as a UNESCO World Heritage. Walking in the middle of the historic monuments, whether it is the Pantheon or the Colosseum, the voice of history echoed in his ears, and he experience the depth of history in his hearts.

30. Technical Report of Abroad Investigation

After returning to Tianjin, Chief ZhuGe began to write investigation report. After changing several versions, the final title was confirmed as Idea Innovation—Realizing Leap-forward Development of Urban Planning and Construction in Tianjin. Since the beginning of his writing, Chief ZhuGe had decided not to write academic articles, but to write an easy-to-understand technical report. The article was divided into three parts with more than 10,000 words.

At the beginning, Chief ZhuGe wrote, "During the investigation, we often changed our position and thought about what would happen if we planned and managed today's cities in Europe, or in turn let Europeans plan and build our cities. Although this may not exist, through transpositional thinking, we realized that some basic laws and principles of urban planning and construction are common. We need to look at the experience and lessons of urban development in Europe from a historical perspective. At the same time, we have to measure our urban planning and construction work by the same standards as those in the West today."

Drawing lessons from European experience, we propose to establish new ideas: e.g. "Landscape Economy", "Urban Characteristics", "Urban Culture", "Zone Management". In this investigation, we are most impressed that all the advanced cities have put the landscape construction and preservation in a very important position. This is the general trend of human civilization in the 21st century. First of all, only by creating a "bluer sky, greener land and clearer water", can citizens enjoy a relaxed and happy life, can talents be widely attracted to live and work in peace and contentment and greatly promote economic development. Secondly, the environmental landscape itself can directly create huge economic benefits.

We can see this in Paris, Rome, Venice, Florence and other famous cities around the world, there tourists are endless. Paris's annual tourism revenue is said to be as high as $4 billion. The Eiffel Tower, the Seine, the Louvre and the Versailles are crowded with tourists from all over the world. Beautiful city scenery and famous scenic spots bring constant wealth to Paris like a full-powered "Banknote Printing Machine". Paris's efforts to protect the urban landscape and environment have been greatly rewarded, leaving a treasure trough for future generations. Furthermore, landscape is an important medium to improve the taste of citizens.

Churchill said, "We shape our environment and then our environment shapes." Besides various cultural facilities, the beautiful, clean urban environment can also restrain and regulate people's behavior, cultivate people's sentiment and enhance residents' cultural taste. Purely pursuing the immediate economic benefits by destroying the landscape will cause greater damage to us in long term. If Paris,

Rome, Venice and other cities do not pay attention to the protection of urban history and landscape, they surely won't have today's brilliance.

Emphasizing individuality and characteristics has always been the goal of human society. Since the 20th century, with the development of industrialization, there has been a phenomenon of urban and cultural convergence. Especially in the 21st century, with the continuous development of knowledge economy and global economic integration, this problem is more prominent. The unique beauty of European cities once again makes us realize the importance of individuality and characteristics to the urban landscape. The Old Town of Paris and the Seine, the Roman monuments, the water city of Venice, the bridge of Florence, the music capital of Vienna and the canal of Amsterdam have their unique and ingenious urban layout and distinctive, diverse and unified architectural style, which make us nostalgic and admire.

A country, a nation, or even a city emphasizes individuality and characteristics as its capital on the global stage, which is an important magic weapon in today's global competition. Let their own characteristics become a gold-lettered signboard, a strong cohesion and appeal of the city. Urban characteristics are composed of many factors, such as natural, artificial and historical. Among them, the historic and cultural heritage of a city is the most important components of urban characteristics, which is also a very valuable asset. Every city in Europe has placed the protection of cultural heritage and monuments in a very important position. Urban cultural heritage and monuments are an important part of urban culture, reflecting the city's cultural taste.

The excavation, arrangement and exploitation of urban cultural relics and historical heritage can't only bring tourism benefits, but also an important means of educating citizens. Tianjin, as the most important city in the modern history of China, has left a large number of historical heritages, such as the humanistic landscape of Wudadao, the Italian Concession, North Jiefang Road, and the unique natural landscape of the Haihe River. We should vigorously excavate and carry forward Tianjin's urban historic and cultural tradition, making it an important means to develop tourism, to create urban characteristics and to enhance the city's cultural taste.

Investigating European cities, reviewing the history, we can see that urban planning and construction have always been an important part of world civilization. The city is the history written in stone. Architecture is like solidified music. Let's read, appreciate and listen to it. Sometimes we forget this because of the problems we are facing. In the 20th century, the rapid expansion of cities has brought many urban problems, and the development of modern industrial technology has made many fantasies possible. Complex social problems and colorful modern technological means make us doubt the eternal proposition that "Architecture is both technology and art." Thus, the trend and phenomena of "Functionalism", "Housing is a living machine" and "the International Architectural Style" have emerged.

In the 21st century, man' understanding has returned, and the urban Renaissance movement has sprung up again in Europe. History once again proves that urban planning and construction is not only a simple material work, but also the spiritual pursuit and creation of mankind. Today, if a city only pursued economic development without cultural exploration, there wouldn't be real long-term economic development. Similarly, if a city only takes economic benefits as the highest goal for its planning and construction, it will not be a good city. Urban planning and architecture have a profound impact on urban culture, and the impact of large infrastructure such as road system on the long-term development of cities is also enormous, which often becomes a part of history and culture.

We can see this impact from the changing relationship between urban layout, open space and large infrastructure in Europe. Therefore, the planning and construction of urban roads, bridges and infrastructure must also be considered in depth from the perspective of the impact of urban culture. Urban planning is not only about shape and function, not only about building a house or a road, but also an exploration of urban culture. To improve the level of urban planning and construction in an overall way, it is necessary to take urban planning and construction as an important part of urban cultural construction, and to raise urban construction to the height of cultural construction.

In order to maintain the long-term prosperity and high construction level of a mega-city, it is necessary to establish the concept of urban management and take the overall consideration of urban development. Traffic congestion, environmental pollution, housing shortage and high crime rate are the "Urban Diseases" encountered by western big cities, which we have been trying to avoid in urban planning for decades. Today, we visit Europe again and find that these urban problems have gradually been solved.

Traffic congestion was once the most serious chronic disease in major European cities, and it also brought air pollution. Today, with social and economic development and technological progress, major European cities have established integrated urban transport systems. By means of planning and layout adjustment, vigorous development of public transport, establishment of modern traffic management system and technological progress (such as the establishment of automobile emission standards), the problem of traffic congestion has been solved basicaly. Moreover, with the integration process of European unification, European countries canceled their borders. The highway network across the continent made cross-border traffic very convenient, greatly promoting personnel exchanges and economic exchanges among countries, shortening the space-time distance of Europe. Convenient transportation has become a major advantage of European economic development.

Therefore, facing urban problems, we can't adopt the"to treat only where the pain is" single-play method, but must base on social and economic development, scientific and technological innovation, to solve it by comprehensive means. The operation of a city should not only be limited to the built-up area, but also including the land and environmental resources around the city. The concept of land resources operation and environmental resources operation must be established. In Europe, where we visited, whether in cities or suburbs, full of eyeful green, blue water and blue sky. In order to achieve a high level of urban planning and construction, it is not enough for a city full of dusk to without clear sky. The regional environment around the city must be taken into account. Verdun, a small town in Austria, not only has beautiful buildings, but also shows respect for the land and the natural environment, making people feel the paradise on earth, which is the basis of victory.

Zoning is the inevitable outcome of urban evolution. Within the city, different parts built in different periods have formed different historic zones, and the relative agglomeration of various urban land use have formed different urban functional zones, each of which has its own distinct features. In Europe, urban planning and building management is basically based on different zones, to formulate corresponding regulations, unlike our current simple "one size fits all" method. Rome has strict protection and control over the ancient city. New buildings must respect the historical environment around them, so they basically maintain the original appearance of the ancient city.

Generally speaking, Paris adopts the planning layout of strictly protecting the old urban area and creating a new area of La Defense. The new area and the old area adopt completely different planning and building management methods. The Old Town has formulated strict regulations for planning and building management in order to maintain its traditional style and pattern. Building height is strictly limited. New buildings are not allowed to exceed height limits. Street buildings have the same setback, eaves have the same height, and the building shape and style should be the same. After more than a hundred years of strict management, Paris has become a world-famous city with a strong sense of integrity and beautiful. The French nation is a romantic nation, but after arriving in Paris, we can see the beauty of order within the nation through the appearance of the city. The only super high-rise building in the Old Town of Paris, the Post Office of Paris, has become a typical negative example for destroying the style and features of the Old Town of Paris.

The result of district management reflects the integrity and unity of urban blocks and the diversified changes of urban landscape, which reflects the seriousness, continuity and rigidity of planning management regulations. With the change of the times, the function, property and modernization of buildings can be changed accordingly, and the form of buildings can be continuously developed and innovated on the premise of strictly following the historical context. However, the stringent architectural height and the uniform architectural setback and alignment can never be surmounted, even if it is like the "most avant garde" building in Paris, France. The Pompidou Cultural Center has changed everything, but its alignment and height remain unchanged from the surrounding buildings.

"Let a hundred flowers blossom and a hundred schools of thought contend" is the consistent policy of our literary and artistic creation. In architectural creation, we have long quoted such a policy. Generally speaking, it is beneficial to encourage the prosperity of architectural creation. However, architecture has its own particularity, it must be in a specific place in the city, must consider the coordination with the surrounding environment. Otherwise, it can only produce chaos. We often see a group of buildings, in styles of Chinese, European and modern, stacked together without any correlation. Experience in Europe has proved that district planning and building management is an effective method. Its essence reflects the order of the city and the aesthetic principles of diversity and unity. It emphasizes the importance of historical context to the city and architecture, which is very important to create a distinctive image of the city as a whole. In the history of Tianjin, there were nine-nation concessions, each of which had its own characteristics. Therefore, it is particularly important for us to advocate zoning planning and building management.

On the basis of summing up European experience and setting up new ideas, we propose that the leap-forward development of Tianjin is a leap-forward development of economy, urban space and urban culture. Urban planning and construction should be integrated into the trend of urbanization in the world to enhance their cultural taste. Therefore, it is proposed to construct "the Platinum Necklace" of Tianjin's historic and cultural heritage, "the Golden Passage" of the high-speed viaduct in the central urban area, to implement the building zoning management, and to build "the City Of Nations", to form a new mechanism of "Public-Private Joint Venture" in urban construction and other proposals.

When we appreciate the historic and cultural heritage of famous cities in Europe, we always associate Tianjin with it. Tianjin history can be traced back thousands of years. The city has been built for nearly 600 years. It is a city integrating modern Chinese and Western architectural culture. It has a large number of historic and cultural heritage. Learning from the experience of European cities, we suggest linking the famous scenic spots in the old urban area of Tianjin with the planned roads and rivers to form "the Platinum Necklace" of the historic and cultural heritage of Tianjin. It is not only the main line of urban tourism development, but also the window to display Tianjin's historical heritage, cultural taste and city image.

Specifically, two lines will be formed, one is the water line, highlighting the natural characteristics of the northern water capital of Tianjin. With the Haihe River as the main axis, it extends to the North Canal, the South Canal, the Ziya River and the Xinkai River, connecting the landscape along the rivers to form a blue pearl necklace. The second line, as a major tour route, is mainly composed of Tianjin core area, total length of 15 kilometers. It will serious famous scenic spots such as Old Town, South Market, Heping Road, Binjiang Avenue, Wudadao, Xiaobailou, Jiefang Road and the Italian Concession and Tianjin Railway Station to form "the Platinum Necklace". Along the route, special tourist bus lines will be operated. Recent work has focused mainly on the renovation of both sides of the Haihe River, with the theme of highlighting the architectural features of Old Town and nine-nation concessions along the river, forming a unique Riverside Landscape in Tianjin. Through the excavation, protection and reuse of Tianjin's historical heritage, Tianjin's historic and cultural heritage and cultural tastes are fully displayed, and the comprehensive functions of the old urban areas are enhanced, so that Tianjin can become a renowned international tourist city, forming our own "Banknote Printing Machine".

Traffic congestion in urban areas is a more and more urgent problem at present. In recent years, with the reconstruction of shanty dwellings, more than 20 roads have been widened, which has greatly improved the road traffic conditions in the urban area. However, with the growth of urban social and economic activities and the increasing number of motor vehicles, and the increasing intensity of development, the problem of traffic congestion is still very prominent, and there is a trend of further deterioration. From the experience of foreign countries, there are three main modes to solve the problem of urban transportation. One is to restrict the development of private cars and vigorously develop public transport, such as Hong Kong; the other is to encourage the development of private cars, such as Los Angeles; the third is to vigorously develop public transport, while reasonably developing private cars, such as most of European cities.

In the past decades, like other cities in China, our city has adopted the first mode, that is, to strictly restrict the development of private cars and vigorously promote public transport. In recent years, through a variety of reform measures, our city has vigorously developed public transport and made great progress. At present, the construction of Metro Line 1 will further promote the role of public transport, and the problem of excessive bicycles will be gradually solved. Meanwhile car transportation has become the most important problem we are facing. From the current reality, it is obviously out of time to restrict the development of private cars. We should proceed from reality and adjust our thinking to meet the needs of the reasonable development of private cars while vigorously developing public transport.

From the perspective of European cities, on the premise of adopting different traffic policies according to different urban districts, the problem of car development is mainly solved by adopting modern traffic management and improving the speed of motor vehicles. Because of the narrow roads and tight parking, the entry of cars is objectively restricted in the old urban areas. The problems in the old urban areas can be basically solved through traffic management and modern signal system. In the new areas around the city, the construction of expressways is the main method. The improvement of speed not only improves the efficiency of the city, but also reduces congestion and vehicle exhaust pollution. Tianjin must establish an expressway system.

Through research and analysis, we believe that the existing functions of Middle Ring Road should be maintained in terms of its location, current situation and importance. At present, the two semi-ring roads in the southeast and southwest are planned as quasi-expressways, which are not closed to the existing city roads. Although some of the pressures of Middle Ring Road can be shared after completion, their effects are not obvious. We propose to connect the southeast and northwest half-rings to build an elevated and fully enclosed expressway system. The elevated ring road is initially planned to be 40 kilometers long, with a design speed of 80-100 kilometers per hour and two-way six lanes. The cost is estimated to be 4 billion yuan. On the basis of maintaining the original "3 rings and 14 radiations" road network, the construction of elevated expressway will superimpose a layer of Expressway system, which will greatly complement the existing road system, improving road traffic conditions and forming a leading road network system with Tianjin characteristics.

At the end of the report, He concluded with "careful planning, scientific decision-making, looking forward to the future" the twelve words. Urban planning is a complicated work, which has a wide influence. It is an important government action, no matter how big or small it is. On one hand, we should try our best to meet the actual needs of the citizens and implement the Party's purpose. On the other hand, we should pave the way for future development and grasp the direction so as to make Tianjin moving towards an international city.

As the guide of urban development and construction, urban planning must constantly innovate and improve its concept, ideas and methods. Although we have always put urban planning in an extremely important position, despite the continuous exploration and innovation of the planning work and some achievements, there is still a big gap from the requirements of achieving leapfrog development and building an international city.

Our traditional city master plan has too many marks of planned economy. Closed planning method requires accurate prediction of the development speed and proportion of various functions of the city. It overemphasizes the time limit of planning and the control of the planned population and land scale. The master planning is "too detailed to die", while detailed planning is not enough. It generally takes several years for the city master plan to be compiled and approved. The effective planning period of 10 to 15 years is quite restrictive to the long-term development of the city. Structural planning, strategic planning and conceptual planning are modern planning methods widely adopted by many western countries since the 1960s as innovative planning practices. They jumped out of the traditional planning mode dominated by time limit and the selectivity and flexibility of planning are emphasized.

To achieve the goal of leapfrog development of the city, strengthening urban design is an effective measure. Modern urban design is formed in 1950's and wide spreaded. Its main objective is to create pleasant urban environment. The quality of urban design is related to the quality of urban image and construction level. Our urban design has just started in a decade. Basing on the current situation, we should focus upon the major urban design projects, such as Both Banks of the Haihe River, the Central Business District, the Central Commercial District and the historic districts, etc., in order to provide the design guideliness for the planning and construction management of these districts and key areas. Organizational work should be greatly strengthened, and a number of efforts should be concentrated to strengthen the compilation, implementation and control of urban design. Urban design adheres to the principle of people-oriented, and deals with the relationship between urban landscape, urban space and function in an all-round way, especially in detail to meet the needs of people's activities.

At the end of the survey report, Chief ZhuGe wrote: With the continuous improvement of our city's economic capacity, social civilization, planning and management level and construction quality, we have the abilities to face the world level and draw directly on the advanced international experience. Through investigation, we have improved our understanding, found some ideas, and strengthened our confidence in building Tianjin into a world-wide city. As long as we work hard and make unremitting efforts, we will surely build Tianjin better and better, and achieve a new leap forward in the new century.

Mr. Bauer

"Chief ZhuGe , your report was written in 2002. Now 17 years have passed. You still remember so clear, which means the report condenses many of your personal experiences. From today's perspective, what do you think of the report? Did the report work at that time? In addition, this report is so long. Can you send me the original? "Well, I'll send it to you in WeChat. This survey report is not only an investigation of eight European countries, but also contains the achievements of my research on western developed countries over the years, as well as my thinking on Tianjin's planning over the years. In order to attract attention, there are indeed some radical suggestions. That year, I was 38 years old, and relatively young and vigorous. Some problems were not fully considered. From today's perspective, many points of view need to be further discussed. However, the report did attract enough attention at that time."

Chief ZhuGe

Shortly after the report was handed in, one day the director of the bureau met Chief ZhuGe in the corridor and said, "The leader has seen the report." From the mysterious expression of the director, he stopped asking the leaders' opinion. Since the return of the European investigation, the pace of comprehensive development and renovation planning of the Haihe River is getting faster and faster. The city leaders listened to the reports for six times, and made instructions and specific requests on planning thoughts, design ideas, market operation and implement strategy. By July 2002, the plan had been basically mature. The municipal leaders held large-scale meetings in the Urban Planning Bureau to listen to the presentation and review the plan. Besides leaders of relevant committees and departments, many leaders of municipal-administered, state-owned development companies and construction enterprises participated in the meeting. A leader of a big development company who maybe had not recognized the importance of comprehensive development of the Haihe River, or maybe still was lack of confidence in the market of the Haihe River development, didn't come but sent deputy leaders to attend the meeting. The municipal leader was very unhappy and said frankly, "Call him to come here". At the end of the meeting, the municipal leader required the bureau to do an extensive consultation on the plan.

31. Colorful River and Integrated Planning

Public participation and listening to the opinions of all parties had been an usual measure that time, but the scope of this time was unprecedented in Tianjin urban planning history. The planning team reported the plan to the leaders nad members of the Municipal People's Congress and the Municipal Committee of CPPCC, to part deputies and members, and came to relevant district governments for consultation, and listened to the cultural, historical, business and financial circles on a number of occasions. Famous academicians in Tianjin were also invited to comment and argue. Statistically, face-to-face reports amounted to not less than one hundred times and 2,000 people. Mr. Qi, the head of the team, who is handsome and eloquent, has become a export reporting the Haihe River Planning. With continuous reporting, extensively absorbs opinions from all sides, The Comprehensive Development and Renovation Plan on the Haihe River turn more and more perfect, and the writing turn more and more fluent and graceful. The plan depicts the Haihe River as a colorful river.

> The Haihe River is red, it is the ebullient blood of our city, feeding Tianjin from birth to growth, from childhood to maturity.

The Haihe River is orange, it is gestated and preserved with inexhaustible and precious resources.

The Haihe River is yellow, like an original historical book, which records the vicissitudes of Tianjin's development from a waterway terminal to a modern metropolis in China.

The Haihe River is green, it injects fresh natural flavor into our huge city island.

The Haihe River is cyan, she blends into the spirit of the sea and forms a unique regional culture in Tianjin.

The Haihe River is blue, it gives birth to our hope of building an ecological city in the future.

The Haihe is purple. She has bred many leading artists, showing the artistic charm of Tianjin people.

In the near future, when the both sides of the Haihe River stand up a city that attracts worldwide attentions, the Haihe River will also be happy to smile, and this crystal city will reflect the more brilliant seven-color rainbow of the Haihe River!

From the aspects of history and culture, industrial economy, landscape environment, ecological construction, road traffic, tourism and leisure, we determine six thematic objectives of Comprehensive Development and Renovation on Both Sides of the Haihe River:

Through the development and Renovation of the Haihe River, to show the long history and culture—Life is like a song, and the river is full of charm.

Through the development and Renovation of the Haihe River, to develop the Riverside Service industry—Economy will take-off, and the river will be its power.

Through the development and Renovation of the Haihe River, to highlight the waterfront city image—City and river will fuse into one, and the river will become the symbol of city.

Through the development and Renovation of the Haihe River, to build the ecological city foundation—Green will infiltrate into city, and the river will return its natural beauty.

Through the development and Renovation of the Haihe River, to improve the road traffic system—Both banks will be linked, and the river will form a cohesive one.

Through the development and Renovation of the Haihe River, to develop tourism and leisure resources—People will share the space, and the river will full of vitality.

According to the historical development and construction along the Haihe River, the central city part of the Haihe River within Outer Ring line is divided into four sections: the section from the Beiyang Bridge to Nanma Road is the Central Historical District (CHD), the section from Nanma Road to Chifeng Bridge is Central Recreation District (CRD), the section from the Chifeng Birdge to the Fenghua Bridge is Central Business District (CBD). The name for the section from the FenghuaBridge to Outer Ring Road bothers us a lot. At this moment, American EDAW Company begins to participate in the master planning of Both Sides of the Haihe River. They propose to define this section as Smart City (STD), which everyone finally agree with.

Central Historical District (CHD) is the birthplace of Tianjin, including the Sancha River Estuary, the Old Town, Guanshang, Dahutong, the Dabei Temple, the Wanghai Tower Church and Ancient Cultural Street. It shows the historical features of Tianjin's origin and early development. The plan is vigorously to promote tourism and commercial activities, relying on traditional cultural resources, giving the region active vitality.

Central Recreation District (CRD), including the central commercial area with Heping Road and Binjiang Avenue, Central Plaza and Tianjin Railway Station, embodies the consumption and entertainment functions of modern cities. Based on the existing facilities, the plan aims to increase culture and entertainment facilities, to enhance its vitality and form a unique central leisure area. As this section is also the most varied section of the river waterline, large-scale green space and square is planned, creating a image commensurate with the city center, making this area a symbol of the city's image.

Central Business District (CBD), includes the Financial District of North Jiefang Road, Old South Station and the Xiaobailou Business District, forming the core part of CBD. Plan focuses on the construction of business, office and exhibition facilities, creating a good business environment to attract international enterprises, and becoming a prominent symbol of the modern economic center. In this area, many important historical relics will be excavated, to form a small museum group, showing the rich history of modern urban development of Tianjin.

Smart City (STD), starting from Fenghua Road, there are a large number of land along the Haihe River, which is also valuable resource for the future construction of eco-city in the central city. The plan takes sustainable and ecological development as its theme, to construct a new urban form in this area. It industry focuses on advanced network, intelligent technology as its core, creating high output and high added value in loose urban form. The whole area will show the high environmental and living quality of an international metropolis.

In the process of wide consultation, the members of the project team were educated again and again, and gradually reached consensus with all sectors of society in many aspects. Teng Yun, a well-known literary critic in Tianjin, said, "Economy is the blood of the Haihe River, landscape is the appearance of the Haihe River, and culture is the soul of the Haihe River." In the eyes of Feng Jicai, chairman of Tianjin Literary Federation, the Haihe River is "the soul" of Tianjin. Tianjin people all attach great importance to the Haihe River. Over these years, with the care and common efforts in all aspects, great achievements have been made in urban construction on both sides of the Haihe River. However, compared with the world's famous rivers, the Haihe River still has more opportunities for development. Through repeated studies of the situation and problems faced by the development and renovation plan on both sides of the Haihe River, we deeply understand that this major project is actually an important step in the grand development strategy of Tianjin. Since the reform and opening up, a series of important projects in Tianjin have laid a solid foundation for the development and renovation of the Haihe River.

Firstly, the construction of flood control circle, as the "Lifeline Project" of the central area of the city, coordinates the flood control and flood discharge functions of the Haihe River from the regional scope, so that the Haihe River can become a living landscape river. Secondly, the shanty dwellings reconstruction project has solved the most pressing problem of living conditions of the citizens, creating a solid mass base for the Haihe River development and renovation. Thirdly, the eastward shift of industrial strategy has given a large number of industrial enterprises on both sides of the Haihe River new development space, freeing up land and space resources for development and renovation. Fourthly, the renovation of urban underground pipeline network, especially drainage network, has improved drainage system, lightened the burden of the Haihe River, and greatly reduced the investment cost of the Haihe River development and renovation. Fifth, the renovation of secondary rivers such as Jin River and the Weijin River has improved the urban landscape, solved the problem of water system connectivity, accumulated experience for the development and renovation of the Haihe River, and prepared for public opinion. Sixthly, after years of rapid economic development, Tianjin has gradually shifted the focus of industrial development to the tertiary industry. The Haihe River is the main line of urban tertiary industry development. The time for the development has also been ripe, and landscape construction and economic development has also organically linked. The above projects have prepared for the project in all aspects, and made the development and renovation of the Haihe River natural. Without these work as the basis, the Haihe River project will face many contradictions and problems that are difficult to solve, to transfer the plan from conception to reality. Through continuous propaganda and communication with all sectors of society, the establishment of higher standards and the creation of world -famous river has been recognized.

32. Ten Major Projects: Market-oriented Construction

Chief ZhuGe

After a long period of discussion, it is determined that the development and revonation of the Haihe River should be carried out in the way of comprehensive development, government guidance and market operation. Firstly, the goal of Comprehensive Development and Renovation is defined through overall planning. In order to achieve the overall goal of building world-famous rivers and the above six thematic goals, starting from comprehensiveness and systematicness, the plan determind "Ten Major Projects", which include water treatment, embankment reconstruction, road traffic, bridges and tunnels, navigation, green&plaza, environmental landscape, lighting and night scene, public buildings, renovation & replacement. Ten Major Projects should be implemented in advance, so that the surrounding environment and infrastructure can be improved, so as to enhance and show the land value on both sides of the Haihe River, to attract investment and promote economic development. The land, cultural, landscape, brand and management resources on both sides of the Haihe River should be fully tapped. Service industry such as finance, tourism, transportation, commerce, trade, entertainment, etc. should be vigorously developed, so as to further accelerate economic development and overall social progress. Development schedule, start-up projects, financing mechanism, management mode should be renovated. At the same time, through strict planning and management, high-level planning and design, and high-standard construction, the Haihe River project will reach the world's leading level, and Tianjin is to undergo tremendous changes.

The development and renovation of both sides of the Haihe River is a landmark project for the city to achieve leapfrog development. We must plan the construction contents in accordance with the principle of "think big and do big". Especially as early start infrastructure project, it must have a high level in order to ensure the high quality of the overall construction of the Haihe River. In fact, "Ten Major Projects" implemented in advance are composed of many contents, led by the Urban Planning Bureau and Tianjin Urban Planning and Design Institute, and completed in cooperation with the relevant bureaus, professional planning and design institutes. In order to be easy to remember, the tremendous projects are summed up as "Ten Major Projects". In fact, many sub-projects are included in each project.

The first of Ten Major Projects: Water Treatment Project. This project actually includes several sub-projects: water resources and river system planning, water environmental protection, drainage construction and desilting construction. The Haihe River is the backbone River in the central city of Tianjin, and its adjustment and storage function should be fully developed. At present, the tributaries of the upper reaches of the Haihe River only pass through the Qujiadian Gate to supplement the Luanhe River's water to the Haihe River twice a year after spring irrigation and flood, each time about 20 million cubic meters. Floods flow through the mainstream of the Haihe River in the year of abundant water. The Haihe River is a reservoir-type channel for most of the time every year, and there is basically no discharge. It is planned that above the second gate of the Haihe River will become a backup reservoir for the water source by diverting the Luanhe River into Tianjin and a water conveyance channel for the water source by diverting the Yellow River to Tianjin. After the South-to-North Water Transfer is realized, the Haihe River will be used as a backup reservoir for the water source by diverting water from the Yangtze River. In order to meet the landscape needs of secondary river and make water alive, it is necessary to increase the water storage of the river every year. At the same time, in order to ensure the water level of the Haihe River kept at a high water level (1.5 meters) all year and meet the water-close landscape needs of the Haihe River, it is planned to increase the water storage to the Haihe River every year.

At present, the water quality is organic pollution and saline water. The rainfall and sewage discharged in flood season are the main pollution sources of the Haihe River. The buffer capacity of Haihe River to external pollution is small, and the water quality of the Haihe River will be seriously polluted in a short time after each large rainfall, among which ammonia nitrogen and permanganate index are the most serious. The Haihe River has little inbound water. In peacetime, the river water stays in the riverbed. Various pollutants accumulate gradually in the sediments through adsorption and sedimentation. As time goes by, the concentration of various pollutants in sediments increases gradually, and through biological, chemical and physical effects, the pollutants gradually diffuse into the water body, which is the main source of pollution in non-flood season. In order to achieve the goal of river ecological restoration and water quality maintenance, all pumping stations in urban areas are strictly prohibited from discharging sewage to the Haihe River in non-flood season, and all factories along the river are strictly prohibited from discharging sewage to the Haihe River. Target water quality standards for water environmental protection are implemented in accordance with National Surface Water Environmental Quality Standards (GB3838-88). For the part above the Second Gate will be implemented in Type III of Non-Flood Season Standard and Type IV of Flood Season Standard.; for the part below the Second Gate will be implemented in Type V Standard.

The old urban areas of the central section of the Haihe River, various facilities and pipelines are mostly out of repair, which can't meet developing requirements. Most of the drainage systems on both sides of the Haihe River are combined rainwater and sewage systems, and part of them are drainage blank areas. Combined drainage system, rainwater and sewage discharge into the existing sewage treatment plants and sewage rivers in each district in dry season, while discharge into the Haihe River in flood season, resulting in pollution of the Haihe River. In order to protect the water quality of the Haihe River and strengthen the construction of drainage facilities, the combined system of rainwater and sewage was reformed into the component system according to the master plan of Tianjin. Rainwater is lifted by the current and planned nearby rainwater pumping stations and discharged into the Haihe River. Sewage and initial rainwater, according to the sewage drainage system planning, are discharged into the current and planned sewage treatment plants. All sewage pipelines on both sides of the Haihe River should be cut off to ensure the water quality of the central city water system fundamentally.

Due to the lack of regulation of the mainstream of the Haihe River itself, and the large-scale subsidence of the land on both sides of the river, the river course is generally silted up, especially in the estuary section, which results in the sudden reduction of the flood discharge capacity of the river course. Therefore, it is necessary to regularly clear the silt and keep the river course unblocked.

The second of Ten Major Projects: Modification of Embankments. The present situation of the Haihe River revetment and embankment was mostly built in modern times. It was gradually built with the development of the city. Some parts have 100 years' history and need to be treated and reformed. The part above the Jintang Bridge is slope protection; the part following the Jintang Bridge is vertical high pile cap and sheet pile (with anchor at the back). After liberation, some old wharfs were strengthened. In 1996, the right bank embankment of the Haihe River from Yingkou

Road to Nanjing Road was strengthened and rebuilt, and the revetment of following the Liuzhuang Bridge was built. At present, some embankment revetments still need to be renovated because of the old disrepair. This plan integrates the road construction along the river, the construction of waterfront, tourism and navigation wharf, reinforces and rebuilds the old Haihe River embankment, and removes the road pipelines along the river. Through measures of partial revamping of the Haihe River and adopting moving back and heightening embankment level, planning attempts to fundamentally solve the problem of flood control standard in the urban section of the Haihe and solve the contradiction between flood control and urban landscape.

The third of Ten Major Projects: Road Traffic Project. There are no connected parallel roads along the Haihe River. Planning adjusts riverside roads on both sides of the Haihe River, and builds riverside expressways, improving the accessibility of the areas along the Haihe River. The alignment and vertical elevation of the riverside roads on both sides of the Haihe Riverare planned to adjust appropriately, so to change both sides in mixed traffic flow to vehicle-only road on one side, and the other side is for green areas, walking and bicycle paths, in which the core section of the Haihe River should be freed as much as possible for waterfront space. On the east bank of the Haihe River, the section from the Beiyang Bridge to the Xinkai River, along the river, is a heightened protective green belt, followed by riverside road. The section from the Xinkai River to the Bei'an Bridge, is planned as a sunken vehicle-only road. The section from Bei'an Bridge to the Daguangming Bridge, is planned as an underground (or cut-type) vehicle-only road, to completely resolve the ground traffic tension on the front square of Tianjin Railway Station. The section from the Daguangming Bridge to the Haihe Bridge is planned as a sunken vehicle-only road. The section from the Haihe Bridge to Outer Ring Road is a 40 meters wide vehicle-only road. A green belt with a width of 10 meters is set in the middle of the road, and green belts of 30-50 meters wide on both sides of the road are set to enhance the landscape effect. On the West Bank of the Haihe River, the section from the Jingang Bridge to South Road is planned to be a sunken road, while the section from South Road to North Dagu Road is planned to adjust the road alignment, merging Zhangzizhong Road into Xing'an Road alignment, with a planned 30 meters wide red line. A 100 meters wide green belt is reserved between the road and the riverbank. The road connects south to link North Dagu Road to Qufu Avenue. The red line of North Dagu Road is planned tobe widened from 30 meters to 40 meters to relieve the traffic pressure in the area of North Jiefang Road. The section from the Chifeng Bridge to the Guanghua Bridge, along the river, is an integrated design of waterfront platform, walking and greening platform, semi-underground bicycle parking lot, bar corridor and byway. The section from the Guanghua Bridge to the Outer Ring Road is planned to be a tourist road and connected with the road outside the Outer Ring. The flood control embankment along the sunken vehicle-only road is set back. The overpasses are planned at the intersections of bridge. The return traffic is realized by underground transportation hub such as Old South Station and some overpass bridges. In the Riverside Square along the river road to the central area, combined with parking facilities and Metro stations, transport hub is set up to provide transfer services. Planning establish a number of large-scale transport hub near the Dongbeijiao Triangle, Xing'an Square, the East Station, Old South Station and Tianjin Steel Plant. Among them, a large underground parking lot with 5000 berths is planned in Old South Station area, which can relieve most of the motor vehicle flow reaching the central business district quickly through the East Bank of the Haihe River. At the same time, we should pay attention to road traffic management, such as establishing pedestrian priority areas and opening up bus lanes (special lanes) system and other special roads or lanes, and rationally organizing one-way traffic.

The fourth of Ten Major Projects: the Bridge and Tunnel Project. At present, there are too few bridges on the Haihe River. There are 12 existing bridges with an average distance of 1.6 kilometers. Most of them are mixed-traffic bridges, and the traffic of the Haihe River is over-concentrated on a limited number of bridges. The traffic volume of bridges across the Haihe River is 355,300 vehicles per day, with an average of 2963,000 vehicles per bridge per day. Among them, the Daguangming Bridge is as high as 705,000 vehicles per day. The Daguangming Bridge, the Guanghua Bridge, the Shizilin Bridge and the Bei'an Bridge concentrate 57% of the traffic across the river. At the same time, most of the intersections between bridgehead and riverside roads are plane intersections, which often seriously hinder cross-river traffic on both sides of the river.

In order to ensure the close communication between the two sides and divert the flow of bridges with special concentration of traffic, such as the Daguangming Bridge's flow, planning increases the main river-crossing channels, by building and rebuilding bridges, increases overpasses at bridgeheads, reduces the interference with traffic along rivers, to ensure the rapid passage of cross-river traffic. There will be 28 planned bridges and tunnels with an average spacing of 0.68 km, 16 among which are new bridges, including 6 non-motorized vehicles and pedestrian bridges as needed.

Bridge is the symbol of a city. Tianjin has been known as the "World Bridge Exposition" because many concessions have built bridges in their concessions in history, and the types of bridges are abundant. In the future, every new bridge on the Haihe River will be designed and constructed into a high-quality bridge. Bridge design should integrate with the surrounding environment of the city, playing creativity to form a unique scenic line above the Haihe River.

The fifth of Ten Major Projects: Navigation Project. At present, the Haihe River already has tourist routes, but because of the low center elevation of bridges like the Jintang Bridge, the Jiefang Bridge, only 4.47 meters, which can't meet the navigation standard of the Haihe River. At high water level, navigation can only be interrupted. At the same time, due to the low standard of waterway, the short hull of cruise ship, the sense of sightseeing is poor. Planning improves the Haihe River tour routes, renews the Haihe River cruise ships, and builds tourist terminals and supporting facilities. The water bus will be opened, west to Yangliuqing, north to Xigu Park and the Taohua Embankment, setting passenger wharf in Second Gate, then east to the Tanggu Estuary. Passenger wharfs will also be built in the Dabei Temple, Ancient Cultural Street, Heping Road, Central Plaza, Tianjin Railway Station, Old South Station, Dazhigu, the Pillow Scenic Spot and other places, connecting with other modes of transportation and relieving the pressure of ground traffic.

According to the navigation standard, the section from the Jingang Bridge to Guanghua Bridge is planned as Grade VI waterway with a headroom of 5.0 meters under the bridge; the section from Guanghua Bridge to Outer Ring Road is Grade IV-V waterway with a headroom of 5.5 meters under the bridge. According to the above navigation standards, most of the bridges between the Jingang Bridge and Guanghua Bridge need to be renovated. The Jintang Bridge and the Jiefang Bridge are used as pedestrian sightseeing bridges after lifting 2.00 meters, and the headroom under the bridge after lifting reaches 5.0 meters. The headroom requirements for new and rebuilt bridges in the upstream of Guanghua Bridge should not be less than 5.0 meters, and that for new bridges in the downstream from Guanghua Bridge to Outer Ring Road should not be less than 5.5 meters (7.0 meters elevation at the bottom of the beam).

The sixth of Ten Major Projects: Green Square Project. Green space on both sides of the Haihe River is mainly open space, and Haihe Belt Park is the main part of it. In the reconstruction of Tianjin after the earthquake in 1980s, through arduous efforts, a large number of dilapidated houses and adjacent buildings were demolished, and Haihe Belt Park was built, which changed the landscape of the riverside in the center of Tianjin. On the basis of the original Haihe Belt Park, a relatively concentrated green space is planned at several main nodes, forming a riverside greening system combining point, line and surface, and planning additional greening area of 2.1 million square meters. At the junction of the Ziya River and South Canal, an estuary park is planned. On the other side of Central Plaza, a large strip of green space is planned. Between North Dagu Road and Jilin Road, the centralized green space is planned. In the front of Old South Station Area and Tianjin Railway Station, the more centralized green space and square are planned to increase the external space for people's communication and activities. In the central section of the Haihe River, planning combines with the surrounding environment and greening, focuses on the development of theme squares and green parks with different characteristics, through the construction of squares, to improve the surrounding environment and enhance the value of surrounding land.

Siyuan Square in the Sancha River Estuary, Victory Square linked the Jintang Bridge, Central Plaza, Peace Square and Jinmen Square towards Tianjin Railway Station are planned, whose total area is 200,000 square meters. At the same time, combined with the construction of walkway and tour road system along the river and waterfront facilities, a perfect open space system along the river is formed.

Greening design on both sides of the Haihe River consists of three parts: River embankment, ground and vertical greening. These three parts of greening are displayed on three different elevation levels, forming a layer-by-layer upward greening landscape. The green belt of the Haihe River and the buildings along the river reflect each other and become a magnificent landscape for Tianjin's sightseeing.

The seventh of Ten Major Projects: Environmental Landscape Project. Advertisements, sculptures and sketches on both sides of the Haihe River, including street lamps, telephone booths, newspaper booths, bus stops, taxi stations and public toilets, should have local, historic and cultural characteristics, and should be coordinated with waterfront facilities, space, landmark buildings and landscape construction along the river as a whole. Among them, the focus is combined with historical heritage along the bank, setting a series of environmental art works, such as reliefs reflecting modern historical facts and group sculptures representing the success of liberation of Tianjin. In addition, the artistic works created by modern artists reflecting the life of the new century are supplemented to make the bank of the Haihe River a gallery of artistic works. In order to provide convenience for tourists and citizens, the bank areas of the Haihe River should systematically construct signs and guiding facilities in line with the needs of modern cities. The Haihe River is the symbol of Tianjin. The landscape environment along the Haihe River directly reflects the image and quality of Tianjin. Construction on both sides of the Haihe River, from revetment, bridges, greening, sketches to buildings on both sides, is an important part of Tianjin's urban humanistic landscape. Its content reflects the history and culture of the city, which should fully reflect the unique beauty of Tianjin.

The eighth of Ten Major Projects: Lighting Nightscape Project. A special lighting plan need to be formulated for the Haihe River. The Haihe Lighting Planning, combining the important nodes and landmark buildings on both sides of the Haihe River, determines the key lighting areas, light sections and levels, and determines the overall color, style and form of lighting, so as to achieve both warm, prosperous and orderly. The multi-level lighting night scene is the character of the Haihe River night scene. According to the planning of both sides of the Haihe River, it is divided into four types of night scene lights: buildings, squares, roads, bridges and garden green space. Generally speaking, bridges are the main landscape in the night scenery of the Haihe River. Bridges should be lightened to highlight their shape. Riverside greening is mainly floodlight, emphasizing integrity and privacy. Square lighting is the essence of floor lighting, which needs to show gorgeous, jumping color. Building lighting highlights the overall outline of the building, especially the roof part. The width of the Haihe River is about 100 meters. The reflection of building lights on both sides of the river can be reflected on the river surface. The river surface should almost not be illuminated. The illumination of cruise boats is an important part of theHaihe River lighting.

The ninth of Ten Major Projects: Public Building Project. In the central urban section of the Haihe River, plan will focus on the construction of several landmark public buildings. Through the construction of public facilities, we will improve the urban functions, enrich the Haihe River landscape, enhance the value of surrounding land, and promote the development and construction of the main functional nodes along the Haihe River. Ten landmark public buildings are planned: Tianjin Urban History Museum in the Sancha River Estuary, etc. The total planned building area is about 300,000 square meters. Combining with the surrounding environment, various landmark buildings along the Haihe River are formed the most beautiful notes in the Haihe River landscape.

The tenth of Ten Major Projects: Renovation and Replacement Project. The renovation of buildings and environment along the Haihe River is an important aspect of the comprehensive development of the Haihe River. The renovation project should be carried out in accordance with the unified landscape planning of the Haihe River to ensure a high level and high standard of construction. Renovation includes renovation of historic buildings basing on maintenance, restoring of the original historical style, especially relying on the original concession building areas to form exotic architectural style areas with different styles, to overall reappear of the unique landscape of Tianjin "World Architecture Expo". Shanty dwellings will be removed and transformed to coordinate the overall architectural image. For buildings in important areas along the river, the function should also be replaced and adjusted. A large number of business -oriented cultural and service facilities, such as cafes and bars, will be added for the vitality of the riverbank area.

Chief ZhuGe

"Mr. Bauer, with above detail introduction, I want to show you that the planning and design of the Haihe River is solid, and the result is a joint work by all relevant departments and professional planning and design institutes. With such a detailed project catalogue, we can have a clear understanding of the scale of the project, and can more accurately measure the scale of investment, including future income of land transfer."

The overall construction projects on both sides of the Haihe River are magnificent and need huge investment. According to the calculation, the total investment of the ten major projects is about 19.6 billion yuan (including demolition compensation fee), including 6 billion yuan for public welfare facilities, 8 billion yuan for roads and bridges, 4.5 billion yuan for municipal projects and 1 billion yuan for unforeseen expenses. The cost of capital interest is 3 billion yuan. This figure is astronomical for young planners fromTUPDI at that time. For them, it is the first time to figure out such a large investment for a project. The total area of land used in the comprehensive development plan for both sides of the Haihe River is 42 square kilometers. Preliminary estimates indicate that the full realization of the planning for both sides will require investment of 180 billion yuan in construction funds.

According to the idea of "Raising funds through various channels as market mechanism and further accelerating infrastructure construction" by municipal leaders, learning from the valuable experience of the Shanty Dwelling Reconstruction Project, they seek the way of "promoting development by infrastructure construction", that is, improving investment environment through the construction of infrastructure such as roads, pipelines and networks on both sides, improving land value, further attracting market forces, so as to jointly build the Haihe River. The operation strategy of raising start-up funds by market-oriented means, not entirely depend on finance is determined. Municipal leaders have pointed out many times that we should change the concept of using money. "In the past, we saved enough money before start, and now we will do it first, then pay back the money bit by bit later." The development and construction on both sides of the Haihe River have very good prospects, and it is entirely conditional to attract funds from all aspects to participate in. After a period of negotiations with development banks and commercial banks, banks generally show great interest and enthusiasm. Although they have not found a specific and feasible way, they have strengthened their confidence and determined the idea of market-oriented development.

The general idea of market-oriented development is that the development and renovation projects on both sides of the Haihe River can be divided into two stages. The first stage uses 3-5 years (2003-2007) to construct basic projects and cultural facilities using bank loans, to improve the investment environment and make the Haihe River change its appearance. This stage is mainly the input period. For the next 15 years (2007-2022), the comprehensive development of both sides were completed through market input, and bank loans were gradually repaid through multi-channel income. This stage is mainly the payback period.

The specific mode and steps of operation are as follows: first, to establish an organizational framework and a leading group, which is headed by the main municipal leaders and whose members are the main responsible directors of the relevant bureaus. Under the leadership, the Haihe Development and Reconstruction Project Office was established. Under it, there is a Haihe Development and Investment Company, which operates according to the market and acts as a carrier of attracting investment, responsible for financing and investment promotion within the scope of planning. The Haihe Development Investment Company adopts the mode of obtaining construction funds through land development, which are handed over to various construction departments to implement infrastructure construction on both sides of the Haihe River.

Within the defined 42 square kilometers of the Haihe River Project, there are 15 square kilometers of transferable land. Through consolidation and market operation, the total income of exploitable land will be 25.6 billion yuan. According to the static input-output analysis, the net income of comprehensive development in the upper section of the Haihe River is 1.04 billion yuan. According to 90% discount coefficient of various risks to income, the total income level of the project is 936 million yuan. In terms of contribution to national economic growth, the annual investment of 3.9 billion yuan will stimulate economic growth by 1.5% and increase employment by about 300,000 people.

33. Persistence and Regularized Management

As the saying goes: Three construction, seven management. A good city must be in a good position in urban management. Elder city leaders have more experience than young planners in this respect. Therefore, when they rejoiced that the comprehensive development and renovation plan of both sides of the Haihe River had finally formed the structure of the two major parts of overall planning and market-oriented development, the city leaders clearly demanded the addition of regulatory management chapter, especially the formulation of planning management regulations of the development of the Haihe River. On the basis of several rounds of planning already done for the Haihe River, drawing lessons from abroad good experience of planning management, through compiling detailed blueprints, deepening regulatory detailed planning, landscape detailed planning, etc., the content of planning management is clear, includes not only the control of land use, building height and building form on both sides of the Haihe River, but also the planning and design requirements of greening, water body, bridge, landscape, riverbank and lighting. The main control contents include building style, building height, navigation standard, bridge construction standard and sea level standard, etc.

Architectural styles on both sides of the Haihe River can roughly form a sequence, ranging from buildings with traditional Chinese characteristics at the Sancha River Estuary to modern western architecture in the First Workers' cultural Palace area, the North Jiefang Road area, and contemporary architecture in the downstream new area. There should be a transition between new buildings and protected historic buildings, reflecting the continuity of development.

The Haihe River is relatively narrow, so the height of buildings on both sides of the river should be limited. According to the analysis of the present situation and the effect of space landscape, the height regulatory of buildings on both sides of the Haihe River need to be formulated, to determine the landmark buildings, landscape contours, building setback lines and slope control lines.

Navigation will be the main result of comprehensive development of the Haihe River, and channel headroom must be ensured. The section from the Jingang Bridge to the Guanghua Bridge is planned as Grade VI waterway with headroom of 5.0 meters under the bridge, and the section from Guanghua Bridge to OuterRing Line is Grade IV-V waterway with a headroom of 5.5 meters under the bridge.

Bridge construction must meet the standards of navigation and others. In the future, the headroom requirements of newly built and rebuilt bridges in the upstream of Guanghua Bridge shall not be less than 5.0 meters (6.5 meters at the bottom of the beam), and that of new bridges in the downstream to Outer Ring of Guanghua Bridge shall not be less than 5.5 meters, so as to ensure navigation requirements.

The water level control of the Haihe River is very important in navigation, tourism and landscape. After the urban secondary rivers are successively renovated, they are connected with the primary rivers to form the landscape of Tianjin water system. In order to ensure the second-class river landscape and the need of cruise ship traffic, and to make the water live, it is necessary to increase the water storage of the river annually, and at the same time to ensure that the water level of the Haihe River is kept at a high water storage level (1.5 meters) all the year round to meet the waterfront requirements of the Haihe River landscape.

The above planning control content is extracted from the comprehensive development and renovation plan of both sides of the Haihe River, which is considered as the most important content. Although there are no formal documents or laws and regulations in the above-mentioned planning management provisions, they have really become the "basic law" of the Haihe River development and planning management in the implementation of comprehensive development and renovation. Today, we realize that special planning and management regulations should be formulated for important urban areas like the Haihe River.

Mr. Bauer

"Chief ZhuGe , I quite agree with you on this point of view. In this respect, our laws and regulations are relatively perfect. Of course, more rules will limit the development or reduce the efficiency of administrative approval, which is actually a double-edged sword. Listening to your systematic and detailed introduction of Comprehensive Development and Renovation Plan on Both Sides of the Haihe River, I feel deeply touched. This is indeed a highly operational plan. What kind of plan does this plan belong to?""I was thinking about this question at that time. The Comprehensive Development and Renovation Plan on Both Sides of the Haihe River does not belong to any existing planning type. From its role, it is a development plan including master plan, specialty planning, development scheme, urban design, transportation planning, civic engineering planning and tourism planning. We can also understand that this is a "High-dimensional urban design", that provides a top-level urban design for development and construction. More than a decade, I have been wondering why such a magnificent and leading urban design has only emerged once. Tianjin has created a new type of urban design, which takes space as the carrier, to coordinate all factors as economic, social and cultural. However, it is a pity that Tianjin has not developed it into Tianjin mode. Nevertheless, gold always shines. Today, the actual effect on both sides of the Haihe River proves that the urban design is great importance and in high level. In such a new era that emphasizes synergistic development, and in the context of institutional reform, perhaps the Haihe's innovations will become the direction of urban design. I think that's why Mr. Bauer is so interested in the HaiheRiver's planning. It is also the reason why I tirelessly narrate the urban design experience more than ten years ago.

34. The First Phase of Reconstruction of the Haihe River · "4-2-1-6" Projects

After a long period of planning and careful preparation, on October 29, 2002, the Municipal Standing Committee listened to the report of The Comprehensive Development and Renovation Plan on Both Sides of the Haihe River by the Urban Planning Bureau, and made the strategic decision of "comprehensive development and renovation of the Haihe River": In three to five years, to build the Haihe River into a unique and world-class service-oriented economic belt and landscape belt, carry forward the Haihe culture and create world-famous rivers, making the Haihe River a symbol of displaying Tianjin's modern port metropolis. The decision also pointed out that this is an urgent need for Tianjin to build a modern international port metropolis and an important measure to realize the leapfrog development. Through the development of service industry, we will promote the comprehensive development of both sides of the Haihe River, gathering financial and insurance industries, promoting the development of trade, transportation, tourism, entertainment, catering and intermediary services, effectively stimulating the market and attracting more people, logistics, capital and information flows. Its role is global, strategic and historical. It is another historical project of Tianjin after the basic realization of the four major objectives of the "3-5-8-10" struggle. It is also an inevitable choice for Tianjin to further accelerate its development and march toward higher goals.

To prepare this important report and presentation, planners forgot to eat and sleep. Originally, they thought they could relax after the adoption of the master plan. Unexpectedly, the resolution of the Municipal Committee calls for the immediate start of construction. Chief ZhuGe and his colleagues in TUPDI have been speeding up the preparation of the planning and design of the first phase of the project. In line with the principles of "emancipating the mind, exploring and innovating, taking the road of marketization, giving full play to the role of market mechanism" and "synchronous construction of infrastructure and development blocks", we adopt policies as market operation, unified planning and step-by-step, area-by-area implementation. After studying, repeatedly discussing and soliciting opinions with relevant bureaus and six district governments, the "4-2-1-6" project was make sure to be launched in the first phase of the development and renovation of the Haihe River. Among them, "4" represents four urban roads, including East Haihe Road, West Haihe Road, etc., and "2" represents the Yongle Bridge and the Dagu Bridge, "1" represents a light rail line, and "6" represents six key nodes along the Haihe River, including the Canal Economic Zone, Cultural and Commercial Zone, the Dabei Temple Commercial Zone, the Tianhou Palace Commercial Zone, Haihe Plaza Commercial and Entertainment Zone, Old South Station Central Business District and the Haihe Five Rings Sports City. In accordance with the requirements of the Municipal Party Committee and the Municipal Government of "keep pace with the times, strive for the upstream, seize opportunities and leap over development", the Urban Planning has organized TUPDI, Tianjin Architectural Design Institute and Tianjin Surveying and Mapping Institute, to carry out a lot of in-depth and meticulous work, making concept plan, design and economic calculation for the recently launched six nodes projects.

In order to fully publicize the comprehensive development and renovation of the Haihe River, Service-oriented Economic Belt, we also carried out propaganda and reports with the help of a number of news media units. More than 200 news reports were published in *Tianjin Daily* and Tonight News Paper, and more than 60 news topics and special interviews were conducted by Tianjin Radio and Television Station. In addition, we used modern information technology, to disseminate information and advertise widely through the Internet, which has got a strong social response. Tianjin Teda Company, Binhai Rapid Transportation Development Company, Tianjin Real Estate Development Group and Tianjin Real Estate Corporation have put forward investment intentions. Some domestic and foreign investment companies, well-known enterprises continued to come to Tianjin to discuss about cooperation. Bank of China, China Development Bank, Shenzhen Development Bank and other major banks have expressed their support for the project.

After the meeting of October 29, 2002, the demolition and relocation of some projects approved in advance have begun to accelerate. On December 17, 2002, on the eve of the Third Plenary Session of the Eighth Session of Tianjin Municipal Committee, the municipal leader inspected the start-up site of six districts in the central urban area, and again listened to the report of Comprehensive Development and Renovation Plan on Both Sides of the Haihe River and the recent construction work by the Urban Planning Bureau. The municipal leader agreed in principle with the development of six nodes along the Haihe River through the implementation of infrastructure construction such as "Four Roads, Two Bridges and One Light Rail", thus further promoting the comprehensive development of the Haihe River and forming a virtuous circle.

35. International Call for Urban Design Proposals of Six Nodes of the Haihe River

To build the Haihe River into a world-famous river, the precondition is to improve the planning and design level of both sides of the Haihe River. How to improve the level, is to open up the planning and design market, attracting world-class planners and designers to participate in and do a good job in the Haihe development planning and design from a high starting point. In the early stage of the comprehensive development and construction of the Haihe River, we focus on the first-class international scheme, plan and design. We invite internationally renowned experts to draw a blueprint for the development of the Haihe River, paying attention to utilize the world-class intellectual resources and the international advanced planning and design concepts, and to start the comprehensive development and construction of the Haihe River with high starting point and high standards. In the process of overall urban design, the Tianjin Haihe River Planning Workshop was established by TUPDI in cooperation with EDAW Company of the United States, applying advanced planning ideas and design concepts, to continuously promote the deepening and improvement of the planning scheme. In view of the six nodes and a bridge in the first phase of the start-up project, we request the consent of the Municipal Committee and the Municipal Government to organize an high-level international call for urban design of conceptual plan and design. The registration information was released on December 15, 2002. On December 18, 2002, Tianjin Municipal Government held a special press conference to announce the relevant contents of International Urban Design Collection for Six Node of the Haihe River. The response from all aspects was strong. Planning and design companies at home and abroad were enthusiastic to register. By the deadline, there are 56 well-known planning and design institutions, including 42 overseas companies, from more than a dozen countries and regions such as United States, France, Germany, Canada, Japan, South Korea, Hong Kong and so on. It is the first time in history that so many overseas planning and design companies have participated in the planning and design of Tianjin.

On December 24, 2002, the eighth Tianjin Municipal Committee held the 3rd plenary sessions, put forward the "three-step" strategy and five strategic initiatives, and drew a grand blueprint for the development of Tianjin. First, in 2003, the per capita gross domestic product reached US$3000 to achieve the main economic target of building a well-off society in an all-round way; second, three to four years ahead of schedule, GDP, per capita disposable income of urban residents and per capita net income of peasants, are respectively doubled compared with 2000, so that the total economic output and the level of people's income can reach a new step. The third step is to build Tianjin into a modern international port metropolis and an important economic center in the north of China, and establish a relatively perfect socialist market economic system, which will be one of the first areas in China to basically realize modernization. Five strategic measures are to vigorously develop the Haihe economy, marine economy, advantageous industries, district and county economy, small and medium-sized enterprises and private economy. Develop the Haihe River economy vigorously was confirmed as the top of five strategic initiatives.

After the deadline for registration on December 31, the Urban Planning Bureau conducted the pre-qualification immediately, and the results came out on January 2, 2003. Then, on January 7, the district governments determined the participating planning and design units of each node. 31 participants were selected from 56 registered institutions. There should have been 35 design units, but when the design units were determined in each district, several districts selected the same design unit. After negotiation, four units participated in two nodes at the same time. On January 10, the Urban Planning bureau issued a planning and design mission letter, organizing the planning and design units to study the sites, formally stating the collection. For saving time, the collection time was determined to be from January 10, 2003 to March 5, 2003, about 2 months. Then the expert review will be organized. Due to the time constraints, as well as the competitiveness and difficulties of the project, the planning and design work is very challenging.

The first node is Canal Economic, Cultural and Commercial Zone, which is located at the junction of the Ziya River, South Canal and the Haihe River in Hongqiao District, the Sancha Estuary, with a planned area of 64 hectares. This is the starting point of the Haihe River and the birthplace of Tianjin City. The plan relies on traditional cultural resources, through the construction of Chinese Modern Industrial Museum and Tianjin Urban History Museum, fully excavating the canal culture, to create the first landscape of the Haihe River by combining modern and tradition.

The second node is the Dabei Temple Commercial Zone, located in Hebei District, at the Sancha River Estuary, where Ziya River, South Canal and the Haihe River meet. The node aims at forming the unique advantages, injecting new contents into Tianjin's tourism and commercial activities, and building a traditional business zone of "Temple and city" as one.

The third node is the Tianhou Palace Commercial Zone, located in the north of Nankai District, only 500 meters away from the Sancha River Estuary. It is one of the earliest water transport terminals in Tianjin, with a planned land area of about 14 hectares. The plan calls for utilizing the influence and popularity of the area in the whole country, preserving the cultural heritage such as the Yuhuang Pavilion and the Tianhou Palace, rebuilding the Haihe Tower and restoring the Water Pavilion, so that the Tianhou Palace area can become a traditional culture commercial zone fully reflecting Tianjin's features, beautiful scenery, unique architectural style and economic prosperity, and become a beautiful landscape node along the Haihe River.

The fourth node is Haihe Plaza Commercial and Entertainment Zone. Haihe Plaza is the geometric center of Tianjin, with a planned land area of 72 hectares. The plan calls for making full use of the characteristics of spanning the Haihe River and vast water surface, giving full play to the location advantages, making use of the existing commercial grounds, combining with the overall construction of Tianjin commercial center area, building facilities like the Haihe Concert Hall on the north bank, creating prosperous scenery of the central commercial district, and fully reflect the international metropolis style of Tianjin.

The fifth node is Old South Station Central Business District, which is an important part of Tianjin Central Business District, with a planned land area of 30 hectares. Planning through the rational organization of waterfront space, strengthens the natural and waterfront characteristics, to create a pleasant, unique landscape, convenient transportation, advanced facilities of the first-class international central business district, with business office, exhibitions, conferences, commerce, entertainment, catering, leisure facilities, high-level apartments, to build an important economic center of an international metropolis.

The sixth node is the Haihe Five Rings Sports City, located in the Guajia Temple area, Hexi District, with a planned land area of 42 hectares. Through the construction of the Water World and the Five Rings Sports City, this area will be built into a new waterfront tourism resort fully reflecting its characteristics, with sports, commerce, entertainment, catering, leisure facilities, and high-grade apartments.

The first bridge is the Yongle Bridge, located on the Ziya River at the junction of Hebei District and Hongqiao District, connecting the Dabei Temple Commercial Zone. The width of the bridge is required to be six lanes in both directions, one span, witha 4.5 meters headroom under the bridge. The planning and design require that the bridge itself not only satisfies the traffic function, but also has the functions of business and sightseeing and leisure. Through the unique architectural style of the bridge body, it can add luster to the Haihe River.

On the New Year's Day holiday in 2003, the main leaders of the Municipal Committee and the Municipal Government inspected the start-up site of six districts on both sides of the Haihe River once again, and came to the Urban Planning Bureau to listen to the report of the recent construction work on both sides of the Haihe River by the Urban Planning Bureau. In accordance with the instructions of the municipal leaders to "take out the second phase implementation plan as soon as possible and make every effort to expand the construction scope", the Urban Planning Bureau had planned and completed the plan for the second phase start-up projects in advance, including 13 roads, 6 squares, 2 bridges, 4 bridge renovations, 10 cultural buildings and 14 blocks. Judging from the frequency of actions of every a few days, the leaders of the Municipal Committee and the Municipal Government has really made their minds to engage in the comprehensive development and renovation of Haihe River.

On January 24, 2003, at the first meeting of the 14th Municipal People's Congress, the newly elected mayor put forward in the government report that we should vigorously develop the Haihe economy. We should develop comprehensively both sides of the Haihe River, improve infrastructure, optimize the overall environment, promote service industry and related industries development, promote the transformation of old urban areas, format unique service-oriented economic, cultural and landscape belts, gradually making the Haihe River a world-famous river, and driving Tianjin to become a world-famous city. Tianjin People's Congress Conference and the Political Association Conference were all held in Tianjin Auditorium, which was built in the 1950s and had a similar architectural style to the Great Hall of the People in Tiananmen Square in Beijing. When the agenda of the Congress was over, the Municipal Committee and the Municipal Government, immediately convened a special enlarged meeting, the newly elected city leaders by the National People's Congress, to hear the presentation of the Comprehensive Development and Renovation Plan on Both Sides of the Haihe River by the Urban Planning bureau. More than 200 people were filled in the conference room, and everyone was inspired by the beautiful vision of comprehensive development and Revonation of the Haihe River.

February 1, 2003 is the Spring Festival, a traditional Chinese festival. This year is the year of sheep. The early Spring Festival heralds a bumper harvest. Chinese people have a long holiday in the Lunar New Year. They are reunited with their families to celebrate the Spring Festival. But presumably at this time, foreign design companies participating in the Urban Design Collection of the Six Nodes on Haihe River were working overtime. By March 5, participants submitted a total of 35 planning and design schemes for six nodes and a bridge. Each district government organize to carry on appraisal. Through experts' review and public announcement, these design schemes, with their own characteristics, competing to bloom with each other, reaching a higher level, has been recognized by leaders at all levels and all sectors of society.

Mr. Bauer

Mr. Bauer has been listening to Chief ZhuGe's narrative carefully and painstakingly. To be honest, listening to a non-English native speaker's lengthy remarks, especially including a lot of investment data and related analysis, much more attention is needed. When Chief ZhuGe talked about the Urban Design Collection, his expression relaxed, because Mr. Bauer has participated in the Urban Design Collection of Haihe Central Plaza, and also participated in other planning and design projects in Tianjin. It was from then on that Mr. Bauer began to pay attention to Tianjin and had a better understanding of what happened in Tianjin afterwards. Over the years, he praised Tianjin's achievements in urban planning and design. At the same time, he had some opinions on some practices, stuck in his throat. He decided to say: "Chief ZhuGe, the form of urban design collection of the six nodes is very good, so that many internationally renowned design firms can know and understand Tianjin, participate in Tianjin's urban design. However, many very creative schemes were not included in the evaluation, and some of the final comprehensive schemes were not very ideal. What do you think of this?"

Chief ZhuGe

Chief ZhuGe seems to have been prepared for Mr. Bauer's question. It seems that this is something he has already thought about. "Mr. Bauer, your question is very pertinent. Indeed, many innovative schemes failed to qualify, which shows that the jury is still relatively conservative. Generally speaking, the biggest difference between design collection and design competition lies in this. The avant garde plan failed to achieve the best results, which I think may be related to the development stage of Tianjin at that time, having its inevitability. Entrusted by Tianjin Municipal

Government, the Urban Planning is responsible for organizing, managing and supervising the collection of design schemes for comprehensive development on both sides of the Haihe River. The district governments and investors of each node are the contractors, which are responsible for the collection of design schemes for each node. By this time, ZhuGe Min had been transferred from TUPDI to the Urban Planning bureau. In order to ensure the level of planning and design, the Urban Planning Bureau urged the district governments to clarify the amount of compensation fees and bonuses for the urban design collection. The districts with good economic conditions agreed very brightly. However, the economically disadvantaged areas are reluctant to pay for the design collection. After being persuaded continuously, finally all district had identified the amount and the source of funds. The compensation fees and bonuses for six nodes and one bridge project totaled 10 million yuan, which was the first time in Tianjin's history. "It is said that these funds mobilize the enthusiasm of units participating the design collection, but in fact they are not. Ten million yuan, for a total of 35 schemes, means an average of less than 290,000 yuan per design scheme. Many design units have invested heavily, with overall model and main single model, including animation, plus international travel costs, the number should be just only enough for cost. The main purpose of each participating unit is to undertake the actual engineering design after the collection. Mr. Bauer, you have personally experienced on this, don't you?

"In order to ensure the fairness and scientificity of design collection activities, we request that 7-11 renowned international and domestic experts on planning and architecture, representatives of relevant departments of the municipal government, representatives of district governments and investors were invited to form a evaluation panel for each node. They reviewed the design collection, voted to determine the optimal scheme. After reporting to the municipal government, the results of the design collection were announced and publicized. This is the key. Because the experts themselves are more concerned about the feasibility of implementation, the recommend schemes were then reported to the municipal government, so most of the creative proposals can't be included. The Yongle Bridge is an exception, which is now "Tianjin Eye". This is also the result of a municipal leader overcoming objections. He asked the Planning bureau for its opinion. Everyone's opinion was not in favor of the plan, but it was difficult to speak out when they saw the tendency of the leader. What do you think, ZhuGe? The leader asked me, I said that bridge is still a good bridge. Then the leader reprimanded of me, "the planning bureau was too conservative".

In athletics, the only thing people remember is the champion. The non-finalist schemes had been archived, but mostly the final implementated comprehensive plan would go on. These two situations seem similar, but they are fundamentally different. The champion is formed by the process of defeating other opponents, and the final comprehensive plan is often evolved from the integration and evolution of many schemes. Therefore, when we look at the planning and construction of a region, it is not enough to only look at the final implementation of the comprehensive plan and construction effect, but also to clearly grasp the context of its program development and the causes of change in the time dimension. The follow-up deepening scheme of six nodes, combined with project investment, development and construction difficulties and other factors, continue to deepen the technical position, to form the implementation plan ultimately. As a result, the Dabei Temple Commercial Zone, the Tianhou Palace Commercial Zone, the Haihe Plaza Commercial and Entertainment Zone and the HaiheFive Rings Sports City (later renamed as the Haihe Water World) have achieved the expected goals according to the comprehensive plan. The Canal Economic Zone, Cultural and Commercial Zone and Old South Station Central Business District had changed greatly in late stage, which have not been fully implemented yet. Overall, the international urban design collection is successful.

In order to do a good job in collection, we organized the design institutes of TUPDI, Tianjin Architecture Design Institute, Huahui Design Company and Bofeng Design Company to carry out the conceptual plan and design of six major nodes of the Haihe River in advance. Conceptual plan and design scheme solicited opinions and suggestions from all factors of society extensively, demonstrating thoroughly the road traffic and the scale of planning and construction, and making economic calculation, which made preparations for international design collection. In the design assignment, we required design firms should learn from the successful experience both at home and abroad, while highlight the local characteristics of Tianjin, fully embodying the details of Tianjin's history and culture, absorbing the essence of world culture, displaying the style of urban culture, integrating functions, artistry and appreciation into one body and embodying the world level. These requirements are reasonable and appropriate. But then we found that some requirements were too detailed, which limited the imagination of designers. For example, in the design conditions of the Yongle Bridge, two forms of tower arch bridge and cable-stayed bridge are required. The planning and design required that large-scale sculptures can be arranged around the elevation of the tower base, reflecting the development history of modern national industry in Tianjin, the development history of Tianjin water transport, and the performance of Tianjin's Shanty Dwelling Renovation. If we look back now, these too detailed requirements are superfluous.

36. Reform & Innovation · Perseverance · Details Determine Success or Failure

Mr. Bauer

Mr. Bauer: "Chief ZhuGe, thank you for telling me frankly the touching story of the Haihe River planning, especially many of your real ideas. From this, I have a deeper understanding of the Haihe River and Tianjin now. I still have a final question. What do you think is the most important thing for such a high-level and grand plan as the Haihe River to be fully implemented in the end?"

Chief ZhuGe

In my opinion, people are the most important factor for a city to succeed. People here are a wide range of concepts, including all aspects of people: Municipal leaders, planning managers, planning designers, engineers, resident, extraneous participant, etc. To become a world-famous city, we must first have confidence, perseverance, constant exploration of laws and timely decision-making; secondly, we should learn extensively and draw lessons from others, make use of various resources, make clear the planning and design with its own characteristics; train our own planning and design talents; thirdly, we must rely on reform and innovation to crack technical and financial difficulties in the planning, design and construction, to find the way to cross the seemingly impossible gap. Finally, details determine success or failure.

The Haihe River planning is the result of perseverance, fortitude and continuous efforts of several generations in Tianjin, and the result of concerted efforts and common struggle in all aspects, from top to bottom. Tianjin people regard the Haihe River as their mother river and cherish it very much. The past plannings were all around the Haihe River, gradually adjusting the functional orientation of the Haihe River, gradually laying the foundation for the transformation of the Haihe River. When the Haihe River didn't have condition for large-scale development and renovation, we maintained its strength, controlled and protected both sides of the Haihe River well. We carried out continuous analysis and research, to build up a reserve force for the comprehensive development and renovation of the Haihe River through shanty dwelling reconstruction. When it was qualified, we made every effort to do a good job in planning and design, setting high standards, learning from advanced experience at home and abroad extensively, and striving to achieve a higher level. We made full discussion and then made scientific decision. After the goal was clear, we went ahead bravely to overcome difficulties. Through reform and innovation, solving seemingly unsolvable problems in technology, capital and other aspects, we can accomplish impossible tasks. In the implementation process of comprehensive development and renovation, attention should be paid to details, excellence and meticulousness, so as to ensure that planning, design and construction reach the first-class level.

Tianjin is thriving because of the Haihe River. From the development history of the world's cities, although riverside cities are different in geographical and cultural aspects, the objective law of the evolution of urban structure is the same. Most of the waterfront cities are developed because of the transport function of rivers. At first, along rivers there are always wharfs, warehouses and factories. With the development of industry and the large-scale ship transportation, the inland waterway shipping function has gradually declined, and the inland waterway ports have shifted to estuaries to form seaports. Wharf, warehouse and factory land along the river have been gradually adjusted to tertiary industry, which has become the central area to show the image and characteristics of the city. Of course, this evolution must be based on urban economic development. Anyway, the overall layout of "industry strategically moving eastward" and "one pole shouldering two ends" determined in 1986 edition of Tianjin City Master Plan conformed with the objective law of riverside cities development. The subsequent editions of City Master Plan have adhered to such a layout structure. With the accomplishment of the "3-5-8-10" phased goals, Tianjin's economic development has reached a new level. The trend of urban industrial restructuring and the increasing proportion of tertiary industry have provided the necessary conditions for comprehensive development and transformation on both sides of the Haihe River.

Planning on both sides of the Haihe River has lasted for ten years. We have deeply studied the historical evolution of the Haihe River and Tianjin City, analyzed the current situation and existing problems, learned from foreign advanced experience, clarified the key issues of the Haihe River's function orientation, land use layout, road traffic, and dealt with the relationship between the Haihe River tourism development, flood control and water conservation. In the Overall Urban Design on Both Sides of the Haihe River, TUPDI cooperated with theEDAW Company to host the workshop and invite relevant professional companies to participate. The international design collection for starting six nodes was organized. In the design of the Yongle Bridge and the Dagu Bridge, the famous Japanese "Michimasa Kawaguchi" office and the world-renowned bridge designer, Academician Deng Wenzhong, were invited to preside over the design. At the same time, give full play to the strength of local planning and design units. Overall, adhering to standards and opening the planning and design market is the direction. Nearly 100 planning and design units have participated in the planning and design of comprehensive development and renovation on both sides of the Haihe River. TUPDI, Tianjin Architectural Design Institute, Tianjin Municipal Engineering Design Institute, French Concept Design Firm, American EDAW Company and other domestic and foreign planning and design institutions conducted in-depth preliminary research, conceptual planning and overall urban design, which laid a solid foundation for the follow-up planning and design work. The majority of the participants in the development and construction are Tianjin State-owned Development Corporation, which is Tianjin's own company. From the effect of the construction, it shows that Tianjin's own company can also do well.

If these are the necessary conditions for the success of comprehensive development and renovation of Haihe River, then reform and innovation are the sufficient conditions for the success. If we stuck to the rules and do not reform and innovate, there would not the Haihe River today. Firstly, on planning and design, if we continued the past one-sided and single approach, mechanically regarded flood control and water conservation as the untouchable red line, didn't solve flood control problems from the perspective of basins and regions, and opposed the relationship between environmental protection and tourism, we would not be able to build the Haihe River into a unique, first-class international service-oriented economic and landscape belt, and to carry forward the Haihe River culture and establish the development goal of the world-famous river.

After the plan has been determined, the implementation of the plan must be supported by funds. If we continued to use the traditional method of relying on financial investment, there would not be such type and such a large-scale fund at all. We must reform and innovate and rely on market-oriented means. Although The Comprehensive Development and Renovation Plan of the Haihe River puts forward a market-oriented approach, which the broad direction is feasible, but the specific path proposed is not, including land pre-acquisition, issuance of blank land certificates, etc., are illegal and can't be operated. Loan repayment mainly depends on income from land value increment, which is guaranteed. However, the loan needs collateral, though using land as mortgage, the land still needs to be requisitioned with fund in advance, while there is no fund for requisitioning land. Finally, the experts of the

China Development Bank in financial innovation and experts of Tianjin in land market cooperated to innovatively put forward a new way of "lending by means of pledging the expected revenue of the government through land leasing". In accordance with this path, China Development Bank has loaned 50 billion yuan to Tianjin for a long-term loan of 15-20 years. Besides 19.6 billion yuan for comprehensive development and renovation of the Haihe River, Tianjin has also used this funds for urban subway construction, expressway construction and park construction. At that time, these four projects were given a good name. The comprehensive development and renovation of the Haihe River is "Golden Dragon Dancing", and the subway construction is "Space-time Connect". The expressway construction is "the Silvery Chain Around the City", and the park construction is "Emerald Layers of Green". This new mode of government financing has laid a solid foundation for the development of Tianjin, becoming the model that other cities are competing to emulate.

With planning and capital, how to achieve a high level of comprehensive development and renovation of the Haihe River depends on details, which determine success or failure. Since the Spring Festival of 2003, we have entered the headquarters of comprehensive development and renovation of the Haihe River. The headquarters was originally the office building of the Hebei District Housing Management Bureau, located next to Yuan Shikai's former residence, which is a four-storey building. Every day, we can see the heat of work scenes and project progress along the Haihe River. One day, after lunch, Chief ZhuGe looked out of the window as usual. Suddenly, he found that the embankment of the Haihe waterfront platform under construction was not on the same elevation as the sinken road of the Haihe River under implementation. When he arrived at the construction site across the river, did he realize that the elevation used by Tianjin Water Conservancy and Design Institute in designing the waterfront platform was the Huanghai elevation, while that used by Tianjin Municipal Engineering Design Institute in road design was the Dagu elevation. The difference between the two is nearly one meter. Fortunately, the discovery was in time and did not cause great damage.

EDAW Company has carried out master plan and landscape design of the embankment in the upstream starting section of Haihe River. To coordinate well, EDAW and TUPDI have deployed a young designer on the site respectively, who, the two guys, a French and a Chinese, were very conscientious and responsible for the work. At that time, the Old Town and other areas were being demolished. The French designers found many old ready-made bricks and stones then used them on the riverside footpath before the Dabei Temple. Grey bricks with a sense of historical vicissitudes fi the historical orientation and tone of the whole landscape design. The above two examples fully illustrate what finally make sure the success of comprehensive development and renovation on both sides of the Haihe River in Tianjin.

Mr. Bauer

From 12:30 to 1:30, Chief ZhuGe told Mr. Bauer the story behind the main contents of the Haihe River Planning and answered Mr. Bauer's concerns, which were also great concern to all, while he never had the opportunity to share with anybody. For the first time, Chief ZhuGe freely told the story behind the Haihe River Planning. He relaxed and took the bill from Mr. Bauer. "Mr. Bauer, let's learn from the young people of today, go Dutch.""Good! It's on you." Mr. Bauer said. The Italian restaurant owner came over and saw that there was still part of the food on the table. He asked, "Are you not satisfied with our food?""The food is very good, but we're so busy on talking. Next time we'll come here especially." After saying goodbye to the boss, they got up and walked to the Planning Exhibition Hall, which is less than 100 meters away. At the door, Chief ZhuGe said to Mr. Bauer, "I heard that you will teach at Tianjin University next semester. You are welcome to give us more lectures then." "It's a pleasure to be invited, Thank you. Chief ZhuGe, besides teaching in Tianjin University, I also want to participate in some scientific research and planning and design projects in Tianjin. My wife Shirley also plans to come to Tianjin. She is an excellent landscape architect. We want to start new careers in Tianjin." "Great! Tianjin is just lack of good designers!"

Chapter VI

No Urban Design, No Construction

37. City Beautification and Renovation for 2008 Olympic Games · Key Planning Headquarter

Tianjin International Urban Design Forum began with key-note speeches by two world-renowned masters. Prof. Jonathan Barnett introduced the latest theories and methods of urban design and his suggestions for urban design guidelines in China. Mr. Dan Solomon introduced the latest trends in residential community planning in the United States as well as in some countries in Europe. Their reports were brilliant, receiving applause continued. During the tea break, in the lounge outside the conference hall, people drank coffee, tea and greeted each other. Naturally, two groups of people formed. A group of young planners and urban designers around Prof. Barnett continued to consult important urban design issues: when and where to build a smart city and how to build it. Another group, mainly architects, as well as management and technical staffs from some development companies, surrounded Mr. Dan Solomon to ask questions about the new community.

Time passed quickly, and the hour pointed to eleven o'clock. The host invited everyone to take a seat. "Guests, just now, we invited Pro. Jonathan Barnett and Mr. Dan Solomon to give keynote speeches respectively, which are very wonderful. Now, when we enter the Tianjin Session of this forum. we first invite Chief Planner ZhuGe Min, from Tianjin Urban Planning, to introduce the pilot project of urban design in Tianjin. Then we invited Academician. Zhou from School of Architecture of Tsinghua University, Academician. Liu from School of Architecture of Southeast University and Academician Zou from China Architecture Design & Research Group to comment on it. First of all, Chief ZhuGe, please!"

Chief ZhuGe

Chief ZhuGe stepped onto the rostrum, feeling a little excited, and even a little nervous. In his mind, the scene that he told the staffs of Tianjin Urban Planning Bureau about urban design 30 years ago just came out. At that time, the staff of the bureau were relatively unfamiliar with urban design. Today, 30 years later, urban design has become very popular in Tianjin, and the implementation of urban design projects in Tianjin is very high, which has helped the city to greatly improve its spatial quality and urban feature. Chief ZhuGe has also reached the age of over 50 from 30-year-old young people. Today, the audience is totally different from 30 years ago. There are international urban design experts and top professionals on urban design in China. In front of them, Chief ZhuGe intended to conclude the 30-year experience of urban design of Tianjin. What is more challenging is that, combined with the focus of the pilot project of Tianjin urban design, Chief ZhuGe will put forward for the first time some new viewpoints on Chinese urban design in China, which are contrary to the viewpoints generally regarded as classics in the past decades in China and may cause intense reaction. However, doesn't the significance of urban design pilot lie in the exploration of reform and innovation? when he thought of this, Chief ZhuGe calmed down a lot.

"Urban design is as old as cities. We can see this clearly from the ancient history of

Chinese and foreign urban planning, such as the ancient Egyptian city of Kahun, ancient Rome, including Beijing in Ming and Qing dynasties in China. These cities are man-made cities that were planned and constructed at as a whole. Most cities were not built at once, but natural cities that are constantly changing and growing. Over the past 600 years, Tianjin has experienced four main stages of urban development. Each stage can see the important role played by urban design, which has a profound impression and far-reaching impact on the urban spatial form. When it was set up in the second year of Yongle Period of Ming Dynasty, the old town adopted the layout of typical ancient Chinese counties, rectangular city outline and North-South layout, which lasted for 450 years. Because the Haihe River is in the northeast of the town, the development inside the town also tends to the eastern and northern part. After the Second Opium War in 1860, Tianjin was opened as a trading port, and imperialism established concessions in the downstream of the Old Town along the Haihe River. Large-scale development of concessions began in the end of the 19th century. The plans of concessions were independent, and their urban designs adopted the prevalent methods in western countries at that time. Now we can also appreciate the influence of Hausman's Paris reconstruction plan and Howard's Garden City ideas afterwards. The roads in concessions are mainly grid, with a small number of circular squares, gardens and radiation roads. A common feature of these concessions is that the road network layout is parallel or vertical to the direction of the Haihe River, thus forming a Tianjin on the north-south axis of the Sancha Estuary and another Tianjin on both sides of the Haihe River, which is vertical and parallel to the direction of the Haihe River. Although the roads in Hebei New Area built by Yuan Shikai are named "Heaven, Earth, Yuan and Huang" from *Qian Zi Wen*, the grid road network there is also vertical and parallel to the Haihe River.

"By the beginning of the 20th century, the total area of the concession formed by three stages of expansion reached 15 square kilometers, equivalent to 3.5 times that of the built-up area of Tianjin in 1860. By 1949, when Tianjin was liberated, the built-up area was about 70 square kilometers. How to integrate Chinese-style urban areas mainly with old town and Western-style urban areas with concessions is the main problem needed to face in urban spatial form design. In 1938, Liang Sicheng and Zhang Rui compiled Tianjin Special City Material Construction Program, which used the north-south axis of the Old Town of Tianjin to guide the overall structure of the city. By adding some circular squares and radiation roads, the concession area was integrated into the city. Since the founding of the People's Republic of China until now, in general, these 70 years is a rapid developing stage of Tianjin, which can be roughly divided into five periods. The first 30 years are the post-war recovery period. Tianjin has transformed from a consumer-oriented city to a production-oriented city. Ten industrial zones and supporting residential areas were planned and constructed on the outskirts of the city. At the same time, various political movements made the city suffer greatly. During this period, 21 rounds of city master plan were compiled, which basically are not out of the idea from Liang Sicheng and Zhang Rui more than 20 years ago. In the 1950s, Soviet planning expert Mukhin helped Tianjin to compile master plan. He took Central Square as the city center and planned an axis vertically to the Haihe River and pointing to Water Park in the southwest. He tried to use this axis to integrate the Chinese-style Tianjin to the west of the axis with the Western-style Tianjin to the east of the axis, in order to achieve a balance, which is similar to the role of Market Street by I. M. Pei in the planning of San Francisco. It's a pity that this axis has not been implemented, so that there is no chance in the following time."

"Tianjin was seriously damaged by the 1976's Tangshan Earthquake, which aggravated the already unsustainable city. Besides the central government's financial support for post-earthquake reconstruction in Tianjin, the Ministry of Construction has also sent a team of experts to assist Tianjin in compiling post-earthquake reconstruction plan. The 1980s was an important period for the shaped of Tianjin's urban form. With the reform and opening up, great progress was made in the reconstruction of Tianjin after the earthquake. Urban water diversion and water supply projects, gas project and road construction greatly improved urban functions. A number of public buildings such as Tianjin Railway Station, Food Street, Ancient Culture Street and International Mall have been constructed, and comprehensive renovation of urban environment were carried out. Old buildings were renovated and beautified. Solid walls were broken and green showed. urban appearance and urban management level were greatly improved, which became an model to follow across the country. At the same time, Tianjin's urban planning has been continuously improved and gradually stabilized. Basing on the previous 21 editions of plans, a new edition of the city master plan (1986-2000) has been formulated. With the approval of the State Council in 1986, this edition of the city master plan clearly put forward the overall layout of the city as "industry strategically moving eastward" and "one pole shouldering two ends". The central city was clearly teased out the road skeleton of "three rings and fourteen radiations". The inner and middle rings are mainly composed of the existing road network, which connects the road network of the Old Town and the concessions, but also further loses the sense of direction. This is why many people say that Tianjin is easy to get lost. Outer Ring Road is almost completely new in planning. The relationship between Outer Ring Road and the trend of the Haihe River had been taken into account properly, and a pear shape ring road with the Haihe River as its axis has been formed. Indeed, most of the newly planned road networks between the middle and Outer Ring Roads were planned as north-south or East-West as far as possible. Objectively speaking, the planning of this stage was greatly influenced by the functionalist planning idea, and the perspective of urban design is less considered. It hardly involves urban axis, public space, skyline and other contents.

"Until the 1990s, Tianjin implemented a large-scale reconstruction of shanty dwellings in the central city, which is the repayment of debts owed over the past half century. At the same time, the development of Binhai New Area started. During this period, the urban design of Tianjin began to emerge, and the urban structure and form were thoroughly considered and discussed. Firstly, in the early 1990s, the function and orientation of the Haihe River were studied, and it was clear that the Haihe River is the main axis of the city. This is in fact the inheritance and deepening of editions by Liang Sicheng, Zhang Rui and Mukhin. It is proposed that the development of the city should be changed from back to the Haihe River to facing to, and that the unbalanced development of the South and north areas of the Haihe River should be adjusted to a balanced one. By 1994, the new round of city master plan revision made it clear that Tianjin Central Business District is located in North Jiefang Road,

Xiaobailou and Old South Station in Hedong District on both sides of the Haihe River; Central Commercial District is located in Heping Road and Binjiang Avenue on the South Bank of the Haihe River, and 14 historic and cultural blocks on both sides of the Haihe River are delimited for overall protection. Although the overall urban design of the central city had not been compiled in this period, the its overall artistic framework had been determined. As for the transitional zone between the middle and Outer Ring Road, the revised master plan initially proposes the constellation of 10 distinctive functional communities in job-housing balance, such as the Olympic Sports Center Community, the Exhibition Area Community and so on.

"By the 21st century, we have worked out a comprehensive development and renovation plan for both sides of the Haihe River and started to implement it. The comprehensive development and renovation planning on both sides of the Haihe River is actually an urban design, which is an implementational urban design. In the first stage of the overall urban design, it includes not only the aim, objectives, strategies and the implemented "Ten Major Projects", but also comprehensive planning, market-oriented development and statutory management. After the approval of the overall urban design of the Haihe comprehensive development and renovation, the international urban design collection of six nodes was held, which eventually formed as implementable schemes. At the same time, the landscape design scheme of the initial stage were compiled. With the progress of development and renovation of the Haihe River, we have organized the second round of urban design collection for key nodes, including the British and the French Concession, the German Concession, the Italian Concession, the Xigu Area and so on.

"For 2008 Beijing Olympic Games, Tianjin, as a co-host city, launched the construction of the Olympic Sports Center. Because the surrounding areas on both sides of the Haihe River did not have such space, the Olympic Sports Center was never expected to be arranged around the axis of the Haihe River. In order to strengthen the coordinated development of the Beijing-Tianjin-Hebei region, Tianjin, together with Beijing and the Ministry of Railway, launched the construction of the first high-speed railway in China, the Beijing-Tianjin Intercity Railway. the Tianjin Railway Station by the Haihe River , close to the city center, was fixed as the terminal of the Intercity express railway, which is a smart choice. With the reconstruction of the station, Tianjin Railway Station has become a comprehensive transportation hub with three subway lines. With the experience of successful urban design of Haihe River, after the Binhai New Area was incorporated into the national development strategy in 2006, we launched the urban design collection for five major functional zones. In 2007, we also organized an international seminar on urban design on both sides of the Haihe River inside Binhai Central Business District.

Chief ZhuGe

"In 2008, the Summer Davos Forum was going to be held in Tianjin. In order to greet the Beijing Olympic Games and the Davos Forum with a beautiful urban outlook, Tianjin implemented comprehensive urban environmental renovation, focusing on "One Belt, Three Districts and Five Lines". The Haihe River is a belt with three districts: Olympic Sports Center, Tianjin Station and Xiaobailou Area. The fifth lines are mainly the welcome routes, including Fukang Road-Weiguo Avenue, Weijin Road,

Nanjing Road, Jiefang Road, and Youyi Road-Southeast Half Ring Expressway. These routes in series link the urban transportation hub, the historical areas, the commercial center, the Olympic Sports Center and other important areas, which constitute the two axes spanning east-west and north-south, forming the landscape skeleton of the central city of Tianjin. The Comprehensive Renovation Project of City Appearance and Environment for Olympic Games is actually the beautification of the city. At some extent, Tianjin's comprehensive renovation of city appearance and environment in 21 century is similar to Chicago's beautiful movement in the end of 19 century. The municipal government seek effective ways and choose operable methods under the condition of short time and heavy tasks, which ensures the overall implementation and final effect of the project."

"The comprehensive renovation of city appearance and environment is a systematic project. Urban design integrates all the elements of street space into an entirety, such as building elevation renovation, green space and landscape, street furnitures, environmental facilities, road signages, advertisement, lamp and lighting, and properly handles the relationship between part and whole, form and function, environmental renovation and business upgrading, city appearance and prosperity. Starting from the details, the urban design is carefully carved, which ensures that the urban landscape has changed greatly in a short period of time, improves the living environment of the citizens, and enhances the attractiveness of the city. It preliminarily shows the urban features of Tianjin, which are style and exotic, fresh and beautiful, the combination of Chinese and Western, and the blending of ancient and modern."

"The Comprehensive Renovation of City Appearance and Environment for the Olympic Games was citywide campaign. Guidelines for public facilities inside roads, such shop signs, street furniture, were compiled to ensure the coordination of the overall effect and the consistency of the level. The human, material and financial resources invested in the renovation were enormous, and the effect is obvious. For example, it demolished the sloppy advertisements on the roofs of highrise buildings, eliminated the bad advertisements on buses, unified the appearance of taxis in the city, and removed a large number of garbage and sanitary dead corners. Many projects strived for excellence in landscape greening, and some even rework many times in order to achieve ideal results. The landscape renovation of North Jiefang Road and Nanjing Road is based on the principle of pedestrian priority, which has reached a higher level after implementation. We do not deny that comprehensive renovation has also aroused some criticism. For example, in some sections, the shop signboards are standardized with uniform size and style, which is lack of characteristics; lampposts are uniform in style and lack of personality; the original lampposts with Tianjin industry characteristics were almost replaced by uniformed posts. Some of the dark green railings along the Haihe River have been painted white, and the grey roofs of some buildings have been uniformly painted red."

"Some renovation seems to be overcorrected, which may also be the inevitable result of setting things right. Before the renovation, the urban environment was really too chaotic. All kinds of practices were quite a mixed bag like "Eight Immortals crossing the sea". Through unified renovation, the overall level has been guaranteed, while some excellent innovative designs may be buried. To change this situation requires opening up the design market step by step in the future, providing an environment for innovation and personality promotion. The improvement of urban environment must be a steady process. In fact, the practice of environmental renovation in Tianjin is inevitable, and there already has international precedents. At the end of the 19th century, in order to change the dirty and messy urban environment, the city of Chicago in the United States launched a famous City Beautiful Movement by using the opportunity of the Chicago World Expo. At that time, all buildings, railings and lampposts were painted white in order to make the city neater in the polluted air. Tianjin started the comprehensive development and renovation of the Haihe River in 2003. In the six years from 2003 to 2008, the infrastructure and landscape construction on both sides of the Haihe River were basically completed, and some major public buildings had been established. But at that time, the overall situation had not been formed, and the construction quality of other parts of the city was uneven. The comprehensive renovation of city environment is a systematic combing and upgrading of Tianjin's urban outlook and overall environment."

"In the same year of 2008, Tianjin established a key planning headquarters, concentrating on the making of 119 plans in total, nearly half of which are urban design, from the overall urban design of the central city, the overall urban design of the core area of Binhai New Area, the overall urban design of each district, to the urban design of key areas and functional zones, even to the detailed urban design of building complex. For example, two rounds of international design collections were held for urban design and architectural design of The Cultural Center in the central city; urban design guidelines were compiled for guiding architectural design in the starting area of Yujiapu Financial District; urban design was compiled for 14 historic and cultural blocks in the central city, which enabled protection guidelines to be carried out under the guidance of urban design and promoted the overall protection and orderly renewal of historic and cultural blocks. A number of high-quality projects have been erected under the guidance of urban design guidelines, such as the Cultural Center Complex in the central city, the starting area of Yujiapu Financial District, Jinwan Square, Minyuan Square, the Five Big Courtyards and Binhai Cultural Center, as well as West Railway Station, Yujiapu High-speed Railway Station, the Cruise Port Terminal and the Second Terminal Building of Tianjin Airport, making the overall urban function and environment to a new stage, showing the strength and role of urban design."

38. Urban Design of Historic and Cultural Blocks—Wudadao

Chief ZhuGe

"A good city must be a beautiful city. Urban special features are important characteristics of urban beauty. Tianjin is the second batch of national famous historic and cultural cities. It has the natural characteristics of mountains, rivers, lakes and seas, also has an important position in the modern history of China. Tianjin historical urban area has not only the remains of Chinese traditional architecture, but also the nine-nation concessions and western architecture with exotic features. The urban form is diverse, the scale of road network is delightful, and the urban pattern and features are prominent. In order to protect the historical context and urban characteristics of Tianjin and highlight the charm of the city, we carried out urban design of historic and cultural blocks at first, then transformed it into statutory protection plans (equivalent to the Regulatory Detailed Planning). By 2012, 14 historic and cultural blocks, including Historic and cultural Block on both sides of the Haihe River, had compiled urban design and protection plans."

"Wudadao is the most complete preserved historic and cultural block in our city. As early as the 1990s, Tianjin Planning Bureau compiled the construction management regulation and organized the protection plan for the Wudadao area, defining the historic and cultural relics and historical buildings to be protected, determining the modifiable land plots, regulating that the new developments should be coordinated with the surrounding environment, and the eaves height of new buildings should not exceed 12 meters. These regulations are concise and clear, ensuring that the overall spatial form of the Wudadao area is not damaged by the new construction. In fact, although the architectural form of the Wudadao area remained unchanged after liberation, the residents living there have changed mostly. Especially after the earthquake in 1976, in the Wudadao area, a lot of buildings were built illegally, as a result, the population was overcrowded, and the environment worsened. In 1999, the municipal and Heping district' governments began to carry out comprehensive environmental renovation, demolishing illegal buildings. In 2003, combined with relocation in comprehensive development and renovation of the Haihe River, Municipal Party committee and Municipal Government proposed to build a brand of "Understanding Modern China Through Tianjin", formulated regulations for the protection and utilization of historical buildings, and set up Tianjin Historical Architecture Restoration and Development Company implementing relocation and reorganization for the historical buildings inhabited by multi-households, and renovated these historical buildings according to repair old as the old principle. At the same time, promote relevant planning."

"In 2007, the urban design for the Wudadao area began to be compiled. When we surveyed the buildings in the whole area, we were surprised to find that the Wudadao started large-scale construction since 1901, and basically completed in 1946, which is already more than 100 years until now. However, only 1037 of the 2159 houses in the historic and cultural block were built before liberation, accounting for 47% of the total. The remaining 53% were built or infilled after liberation. We didn't feel so many new buildings in the Wudadao area and always feel that the old buildings in the Wudadao should account for the vast majority. This is mainly because the size, volume and height of these new buildings followed the old pattern of the Wudadao area, and the overall environment is well maintained. Before the reform and opening up, the construction scale of the Wudadao was relatively small. In the 1950s, Soviet-style residential areas such as Youhao Li and Tuanyuan Li were planned and constructed. These residential buildings have only three floors, using small enclosed courtyard layout, retaining the original road network of the Wudadao area. The scales and overall environment are also harmonious."

"After the reform and opening up, Tianjin has less foreign investment and less pressure on the redevelopment of the Wudadao area. More importantly, Tianjin has long been aware of the need to protect this area as a whole. In particular, in the 1990s, Tianjin Urban Planning Bureau formulated a strict protection plan for construction management, requiring that the height of the eaves should not exceed 12 meters, thus ensuring the overall harmony of the entire area. At the same time, through the in-depth analysis of the physical environment of the Wudadao area, we find that although some new buildings meet the requirements of building height control, the layout and shape of some buildings are more arbitrary, which destroyed the internal space and texture of the Wudadao area. In order to ensure that the new buildings do not destroy the historical heritage and coordinate with the overall environment, it is necessary to further improve the original planning and management regulations, which first requires a good urban design."

"In order to do a good job in the urban design of the Wudadao, we had adopted the analysis methods of urban morphology and architectural typology. Through the analysis of all the neighborhoods, we found that the buildings of the Wudadao can be classified into three main types: Gate-Courtyard Type, Courtyard Type and Lane Type, and found out their distribution law in the planning layout. These studies enable us to have a more scientific and rational understanding of the Wudadao area and provide a guarantee for high-level urban design."

"As the Wudadao area used to be a high-end residential area in the British Concession, most buildings have relatively high walls to ensure privacy and security of residence, which also express the identity and preferences of the owners, so the style and quality of the walls are very exquisite. Except Munan Garden and the Folk Garden Stadium, the most prominent urban public space is the street and the lane space. They can be grouped into five types. In addition, based on the detailed analysis of each building, block and neighborhood of the Wudadao, we compiled the detailed urban design of the Wudadao area, and built up the three-dimensional model of the buildings and street space of the Wudadao. Through this urban design, all parties reached an agreement to narrow the red line of the road back to the width of the current road, which was planned to be broadened before. At the same time, we put forward the urban design guidelines, and put forward the detailed requirements for the control of the new buildings involved in the historical environment for each block, so as to have a more targeted management. Its impact is also significant: first, making time and space correspondence in urban management; second, rediscovering the location of streets and blocks; third, a small amount of prototype research makes management more concise."

"While making the urban design, Tianjin Historical Architecture Restoration and Development Company has implemented the first phase of renovation of the Qingwangfu and the Xiannong Dayuan in the Wudadao area. During the implementation process, they listened to the opinions and suggestions of urban design team. The Qingwangfu was originally built by Xiao Dezhang, a court eunuch, and was bought by Zaizhen, the fourth generation of Qing Dynasty Prince. It has been using as office building of the Foreign Affairs Office of the Municipal Government since liberation. After the relocation of the Foreign Affairs Office, the

Qingwangfu is turned into a showroom. Adjacent Shan Yi Li is transformed into a small characteristic boutique hotel. The Xiannong Dayuan is a courtyard-style joint-row worker residence developed by the former Xiannong Real Estate Company in 1930's. After the relocation and renovation, Phase I of the Xiannong Dayuan is the block mainly for catering. In order to adapt to modern office function, Phase II of the Xiannong Dayuan has made further attempts to optimize the interior of some buildings without changing their appearance. At the same time, some new office buildings have been built by keeping in harmony with the original texture in volume. These renovation and renewal projects are not in large scale, but investment operators, planners, architects and planning management departments have spent a lot of effort. In 2012, close to the Qingwangfu and Phase II of the Xiannong Dayuan, Tianjin Heping District Investment and Development Co. Ltd. implemented the renovation of the Minyuan Stadium. After a lot of argumentation, the new building doesn't increase the scale of the building too much, but continues to use the type of the old building. In function, only three plastic runways were designed, which retained the intention of sports, added many functions of catering, exhibition and tourist service, and formed an open colonnade along Hebei Road, which works very well after implementation."

Chief ZhuGe

Chief ZhuGe paused and turned his eyes from the projected screen back to the audience. "Every time I talk about Minyuan Square, I think of the famous Italian architect Aldo Rossi's famous exposition in the book *The Architecture of the City*. He pointed out a building doesn't exist in isolation. A Building is the building of a city. The form of the building is very important. It is the collective memory of people. Although, as time goes by, the function of the building may change, the form of the building doesn't change. Rossi analyzed the essence of architecture with semiotics, phenomenology and other cultural and philosophical ideas. He went beyond the simple emphasis on function of modern architecture, following the practice of "forms follow functions", and put forward the objective reality of architecture as an independent artistic form. There are many such cases in Tianjin. Besides the historical block of Wudadao, the former Italian concession, which used to be mainly residential, is now called Italian Style District. Today, a large number of garden residential buildings are used as restaurants, exhibition halls, offices, becoming famous tourist destinations in Tianjin. The urban design of Italian Style District has reached a high level too, which provides planning support for the formation of a historical block with mix uses."

"The Wudadao area has a solid foundation of urban design. With theoretical and practical support, it reached a higher professional level, and won the first prize of National Excellent Urban and Rural Planning and Design. The Urban Design Lab of TUPDI has been carrying out in-depth research for a long time. Several years ago, it cooperated with the College of Environmental Design (CED) of University of California, Berkeley, to design student courses on the Construction Control Zone around the Wudadao core area to explore new possibilities for urban renewal. They also compiled the fruitful results of many years researches and practices into a book entitled *Wudadao, Tianjin, China- Conservation and Regeneration of Historic Area*, which is equally important to the protection of historical block entities. Although the level of urban design of the Wudadao area is relatively high and its achievements are obvious to all, there are still some problems that have not been solved properly. The Wudadao area is historically high-end residential communities inhabited by high-ranking officials and dignitaries. The courtyard walls are relatively high and enclosed. After liberation, although a large number of small foreign buildings were used as offices, and the people living in them also changed, the area was still relatively quiet. Although urban design adds mix uses such as catering and hotel by considering the development of tourism, it still takes mix used residential community as its development goal, so the contradiction between tourism and residential community has not been properly solved. In addition, there are about 40,000 inhabitants in the Wudadao area, with a relatively high population density. There are also some buildings with non-historical features that have low building quality and living conditions. How to form a more feasible market mechanism to achieve local renewal? We haven't got it right.

"In recent years, with the improvement of the supporting facilities and popularity of the Wudadao, the number of tourists to the Wudadao area has been ever-increasing. However, tourists can only walk on the road to watch the appearance of the Western-style villas but not able to tour inside. For this, the Urban Design Lab has continued to actively carry out relevant studies. They drew lessons from Boston's "the Freedom Trail" in the United States and designed a complete experience route within the Wudadao area, which links together main tourist facilities, former residences of celebrities, museums and unique villas and spaces in the Wudadao area to promote tourism development. At the same time, they have been exploring the new growth structure of the Wudadao area."

"As a national pilot city of urban design, Tianjin once again lists the urban design of the Wudadao area as one of the pilot projects. Combining with the current situation as the all-for-one tourism development, the application of the Wudadao area for the national AAAAA level scenic spots, and the activities of "Western-style villa Economy" and "Night Economy" promoted by Tianjin Municipal Government, this urban design defines the Wudadao area as following: the protection model of historic and cultural blocks in Tianjin, the demonstration area of all-for-one tourism and Western-style Villas Economy, and the candidate of UNESCO Heritages. Combined with the pedestrianization of Tianjin Historic Urban Area and the renaissance of the urban center, the experience tour route of the Wudadao area is brought into the pedestrian system of the whole city. Considering the increasing number of tourists, we should further improve tourist service facilities and improve public transportation. Besides Xiaobailou Metro Station currently in use, two new planned Metro Stations have been added, and the feasibility of restoring Xinhua Road Metro Station on Line M1 has been reserved. Finally, inspired by the urban design of the surrounding areas of Quanyechang in Heping District, we suggest that we should alleviate the overcrowded population of the Wudadao area and improve the living condition of some residents in urgent need by relocation, combining the construction of the Heping-Quanye New Town on the outskirts of the city. From the experience of the evolution of Paris's Old Town and London's Westminster District, it can be seen that the Wudadao area will have similar effects in the future and become a part of the urban complex center."

39. Urban Design of Tianjin Cultural Center and the Surrounding Area

Chief ZhuGe

"Cultural Center complex usually has a decisive influence on the overall spatial structure of the city. It is not only the venues of opera house, concert hall, library, art gallery, museum and so on, but also, more importantly, the public open space such as square, park and so on, which are surrounded by above venues, to form the activity center of people, and the focus of city images."

"For many years, Tianjin has not formed a centralized cultural center complex. The original plan did not prereserve a site for construction, and the city's finance is limited at that time, every public building will be strived for by many districts. In order to balance the demands of each districts and promote the development of backward districts, the only few new cultural buildings in Tianjin are scattered in many districts. Tianjin Library and Zhou-Deng Memorial were built in Nankai District, and the Science and Technology Museum is located in Hexi District. The Pingjin Campaign Memorial is located in Hongqiao District, and its location has little to do with the place where the historical events took place. The master plan is just like building a public cultural building in a unit's own courtyard, which has nothing to do with urban public space."

In 2000, Tianjin planned to build a large museum to commemorate the 600th anniversary of the founding of the guard city. The site was chosen at an open space between Youyi Road and Children's Happy Park in Hexi District. Across the road is Tianjin Auditorium built in the 1950s and the International Exhibition Center built in the 1980s. The museum's architectural design adopted the way of international design competition, and finally the famous Japanese structural architects Kawaguchi and Takamatsu won the prize. Kawaguchi has a deep knowledge of structure. He has completed the structural design of the famous Yoyogi Gymnasium in Tokyo with Tango Kenzo. It is also he who later designed the symbolic Ferris Wheel and the Yongle Bridge over the Tianjin Sancha Estuary. The shape of Tianjin Museum is like a flying swan, which indicates the new take-off of Tianjin. Later, it was said that Tianjin Museum was the most perfect long-span shell structure in the world, with the largest span but the lowest height. However, as a museum building, especially the Tianjin Museum, it does not fit very well in function and shape. Therefore, then in the overall design of Tianjin Cultural Center, it was changed into the Natural Museum. At the same time, a huge Galaxy Square covering 170,000 square meters was built at the same time. The name of Galaxy Square originates from "Tianjin", which is an alias of the Galaxy. In Qu Yuan's Lisao, it is written that "Morning begins in Tianjin". At the same time, the name has the spirit of the times, which complements the design concept of the sun, moon and stars of the whole square. However, from the actual effect, the Museum had limited effects and little contributions to the improvement of Tianjin's urban structure and form.

"In 2002, when the Haihe River was comprehensively developed, ten public buildings were initially drawn up. Three-Stone modern industrial museum in Hongqiao District, the Haihe Library in Hedong District and the Children's Theatre in Hexi District are planned and constructed. Originally, the Haihe Concert Hall was planned in Central Plaza. The name sounds good, but there was no suitable solution. Later, the original Xiaobailou Cinema was transformed into a pure European-style Tianjin Concert Hall with a big dome. Generally speaking, due to the narrow section of the Haihe River and the limitations of land use on both sides, there had been no concentrated cultural complex. In 2008, after the Beijing Olympic Games, Tianjin planned a new round of development. It proposed to build a new cultural center on the east side of Youyi Road, which has a certain foundation, including the existing Tianjin Science and Technology Museum, the Youth Activity Center and Chinese Opera Theater, with a total land area of 90 hectares. It is very rare to have such a large orderly land lot in the built-up area, which mainly occupies 30 hectares of land through the relocation and demolition of the Children's Happy Park, making many people who have many good childhood memories in the park complains."

"Two rounds of International Collection of high-level overall planning and architectural design were held for Tianjin Cultural Center, which was praised as having reached the "Star Level". The overall layout of the master plan combines the Chinese garden layout of "mountains, waters and towers" and the Western Garden techniques of "big axis and avenue". Looking from the public platform of the Grand Theater to the west, the greeting tower with a height of 60 meters is the mid-range landmark, and the Tianta TV Tower with a height of 350 meters is the foreground landmark, forming a multi-level landscape of near, medium and far, which further extends the east-west axis of the site to connect with the open space system of the city. Urban design determines the location, height, interface, primary and secondary relationship, old and new relationship and space treatment requirements of each building in the complex, forming a harmonious and orderly space experience. The whole building group highlights the overall coordination of architectural form. In the whole process of "design-construction", we should give full play to the overall planning and guiding role of urban design, to face up to disputes, respect personality, actively coordinate, seek consensus, and gather a lot of wisdom. Ultimately, the original conception of urban design is basically realized, which is a model of urban design from blueprint to the end. Tianjin Cultural Center complex and its surrounding of urban park, square and other public spaces have become the most important public space in the southern part of the city, which plays an key role in the configuration of spatial forms of this area."

"Tianjin Cultural Center has been put into use for seven years, and the overall effect is good. At the same time, there are also some problems: first, the function of the complex is not mixused enough. Although the Galaxy Shopping Center has been planned and built synchronously, the commercial types are not rich enough, and there is a lack of hotels, offices and other forms of business. Second, the overall layout lacks careful consideration of people, such as the 800-meter walk from the library to the Galaxy Shopping Mall, which may be related to management, but also unavoidable by the grand style of layout itself. Sometimes, you can't have your cake and eat it too. Despite the above regrets, the Cultural Center plays an important role in the spatial restructure of southeast Tianjin."

"As mentioned earlier, the city master plan of Tianjin in 1986 didn't give much consideration to the urban design of the central city proper and lacked a description of the urban spatial structure, but mainly clarified the road skeleton of the "three rings and fourteen radiations". By the 1999 edition of Tianjin City Master Plan, the Haihe River has been defined as the main axis of the city, meanwhile, the location and form of the Central Business District, the Central Commercial District and 14 Historic and cultural Blocks were fixed too. The spatial structure and overall form of the Historical Urban Area have been basically determined, but the spatial structure and the overall form and sharp of the areas outside Central Ring Road of the city have not been determined clearly. In the overall urban design of each district in 2008, this part is also not considered much. In 2008, Tianjin Spatial Development Strategy put forward the urban center structure of "One Main and Two Deputy" in the central city proper. One Main is to connect the traditional urban center of Xiaobailou with the surrounding areas of the Cultural Center to form a center in the central urban area of Tianjin, while in fact the two areas are far apart, so it can't really form a center. According to the spatial development strategy and the urban design of the Cultural Center, the Municipal Planning Bureau invited SOM to compile the Urban Design of Tianjin Cultural Center and the Surrounding Area."

"The scope surrounding the Cultural Center covers an area of about 2.41 square kilometers, with a total floor area development of 6.5 million square meters planned, which is to build the most important Central Business District in the central urban area. Supported by SOM, the TUPDI and Shanghai Urban Construction Design & Research Institute cooperated with other units in many specialties to form a high level of urban design, and finally formed detailed urban design guidelines and regulatory detailed planning approved by the municipal government. After the grand opening of the Cultural Center in 2012, Tianjin Infrastructure Investment Group turned its attention to the surrounding area of the Cultural Center, to begin to implement the urban design of the surrounding area of the Cultural Center and launch the demolition of the Old Badali. The Old Badali is a area with 3-5-storey residential buildings built in the 1950s and 1960s. Because of the good location and the difficulty of demolition, only two neighborhoods have been erased to ground and the demolition task has been stagnated. Today, the Cultural Center is the unique and most concentrated group of cultural buildings in Tianjin, which has played an integrated role in combing Youyi road and the surrounding space structure. However, from the current economic situation, it is difficult for the newly planned central business district surrounding the Cultural Center to be realized according to the plan. Firstly, there is no more demand in Tianjin's office market at present; secondly, it is very difficult to demolish and relocate, which is almost impossible by now. Both the density of multi-storey residential buildings and the price of second-hand housing in this area is very high, which means that the removal needs a lot of money. If there is no real estate market that can balance the huge amount of demolition funds, this good urban design will probably be put on the shelf for quite a long time."

40. Urban Design of Yujiapu Financial District

Chief ZhuGe

"Historians say that cities are the greatest invention in human history and the birthplace of civilization. Today, with the highly developed information technology in the 21st century, the function of urban agglomeration is still very important, especially in highly dense urban centers. Lujiazui Financial Zone, Luohu and Futian Central Area have played respectively a vital role in the rapid development of Shanghai Pudong New Area and Shenzhen Special Economic Zone. Since 2006, when Binhai New Area was put into the national development strategy, it began to consider how to locate and plan the Central Business District, which is the core of the New Aera. This is an urgent issue to be determined, but the difficulty is that Binhai New Area is not a blank paper, actually an old area after more than 100 years of development. Throughout in-depth study and multi-proposal comparison, the CBD of Binhai New Area is finally determined to be built along the lower reaches of the Haihe River. This area consists of existing docks, warehouses, oil depots, factories, villages, wasteland and some low-quality multi-storey residential buildings. At that time, a city-level official went to inspect the selected site of the CBD and saw such a dilapidated scene, then just asked, "Are we on the wrong place?" Yujiapu is planned to be the world's high-level central business district, which is always be doubts and comments before its really built. It's just like the criticism that we received when we planned to build the Haihe River into a world-famous river more than ten years ago. Planning needs foresight and insight, also more in-depth work. The planning of Binhai Central Business District at first place defines its functional orientation in the region and its relationship with the Central Business District of Tianjin Central City and Beijing Financial Center as well. By comparing the experience of urban central business districts at home and abroad, the planning scope and construction scale of the new central business district of Baihai were confirmed. We found that the Yujiapu Peninsula is very similar to the Dockland Financial District on the Thames River in London, which reveals the common law of the development of riverside cities."

"In order to improve the planning and design level of the Haihe River and Yujiapu Financial District, in 2007, we invited four Chinese academicians, Prof. Wu Liangyong, Qi Kang, Peng Yigang and Zou Deci, who are top experts in China, as well as international urban designer, Prof. Jonathan Barnett of Pennsylvania University, etc. as consultants to give advices on the planning. SOM, EDAW, Tsinghua University and Waterman Inc. Ltd. of the United Kingdom were hired to carry out two workshops and hold four consultation and demonstration meetings on major issues. The event lasted more year one year. The International Workshop of Urban Design on both sides of the Haihe River in Binhai Central Business District is the practice and application of urban design as a design and decision-making process in Tianjin Binhai New Area. Through the workshop, the location of High-Speed Railway Station, the Flood Control Level and the base height, the location of the starting area and other major issues were resolved. At the same time an international urban design competition of Yujiapu Financial District was held with the sponsorship of International Federation of Architects. The overall urban design of

Yujiapu District, drawing on the successful experience of Manhattan, the Magnifitiant Mile in Chicago, Lujiazui in Pudong, Shanghai and other areas, through the joint participation and collective efforts of many planning and design units, through multi-proposal comparison and selection, finally adopted a narrow street, dense road network profile and three-dimensional planning layout, extending the Beijing-Tianjin Intercity Railway to the gate location of the financial district, forming a transportation hub with the high-speed rail, subways, buses, taxies. For people's convenience and good feelings, the planning has formed a perfect ground and underground pedestrian system, based on building the cross-Haihe River tunnel of Central Avenue, underground traffic road, and municipal utility tunnel. A green belt along the Haihe River is planned to form a perfect waterfront landscape and urban skyline. The planning and design of Yujiapu fully embodies the concepts of combination of functions, humanities, ecology and technology, to achieve a high quality and the characteristics of the time, to lay a solid spatial foundation for the development of the vibrant financial innovation center, creating a beautiful place, which hopes that it will become the "Binhai Core" driving the development of the new area."

"More than a decade has passed since the site was decided in 2006, and great changes have taken place in Yujiapu, which was used to a small fishing village a hundred years ago. The ever dilapidated, lackluster peninsula became a modern business district. This change can indeed be described by the words "vicissitudes of time". At the end of 2008, after the urban design scheme was approved by the Municipal Government, the demolition began. Within a short period of time, the demolition of Yujiapu Village, Guozhuangzi Village, Laowan Avenue, Erfu Street, the wharf in the third operation area of the Tianjin Port Bureau, including Donggu and Xigu shantytowns on the South bank, was completed. In 2009, the Yujiapu Starting Area was started the construction, marked by the beginning of the Beijing-Tianjin Intercity High-speed Railway Extension Line and the new underground railway station construction. In 2015, marked by the opening of the Beijing-Tianjin Intercity High-speed Railway Extension Line at the Yujiapu High-speed Railway Station, the Yujiapu Starting Area in 0.8 square kilometers was basically completed. 14 high end high-rise office buildings have been erected from the scratch, some of them had been put into operation. The underground commercial streets Global Mall had a grand opening, and the InterContinental Hotels of the International Financial Conference Center has been put into use. High-speed Railway Station Park, the main axis of the Central Avenue landscape, Riverside Park and the North Park in the starting area have finished construction. With the opening of Haihe Central Avenue Tunnel, the traffic to and from Yujiapu was greatly improved. A financial district with beautiful landscape, pleasant scale, high efficiency and great ecological environment has taken shape. The image of Yujiapu Financial District stands on the Bank of the Haihe River."

Chief ZhuGe

"Summarizing the successful experience of planning and design of Yujiapu Financial District, the most important thing is the systematic methods of urban design and the strict implementation and management of urban design guidelines. The planning and design of Yujiapu Financial District has been continuously deepened under the overall control of the urban layout structure and spatial form of "narrow street and dense network". It mainly highlights four characteristics. First, it is to highlight the characteristics of waterfront, humanities and ecology, and focusing on shaping the urban layout and urban form in line with the orientation of the financial district. Second, it is to start with improving functions, intensive land use and optimizing land structure, vigorously develop public transport, establish a public transport system consisting of high-speed rail, subway and normal bus, and advocate green travel. Third, it is the development of underground space system, which integrates metro, underground vehicle, underground pedestrian, commercial and municipal utility tunnel. Fourth, it is the research which is carried out from the aspects of urban layout, transportation, construction, environment, energy and so on to explore the technological route of low-carbon urban development. At the same time, Yujiapu Financial District pays attention to the continuity of historic and cultural landscape, and preserves cultural relics and historical buildings, such as Tanggu South Station, *Tanggu Agreement* Site, Dagu Shipping Company Site, Mitsubishi Oil Depot Site, etc., to show the historical heritage of Yujiapu. The beautiful natural environment, superior location conditions, relatively complete land use conditions and profound historical accumulation provide a solid foundation for the urban design of Yujiapu Financial District."

"In accordance with the requirement of urban design, the Central Business District Management Committee has organized SOM and Bohai Urban Planning and Design Institute compiling the Yujiapu Financial District Regulation and Urban Design Guidelines for the Starting Area as the legal basis for planning management. The urban design guidelines of Yujiapu Financial District's Starting Area are very strict. Besides the general guidelines for the control of the Setback, Near-line Rate of street walls, the orientation of entrances and exits, and the utilization of underground space, the bulk as well as location of towers and podiums are also strictly controlled, as the so-called "Envelop Control" in foreign countries, which means that the designed building must be able to put into the envelope.

"Nine well-known young and middle-aged architects, Cui Kai, Zhou Kai, Hu Yue, Qi Xin, Zhang Qi, Cui Tong, Wang Hui, Zhang Lei and Yao Renxi, are assembled in the design of nine office blocks in Yujiapu Starting Area, providing a creative stage for outstanding Chinese architects. Guided by the unified control of urban design guidelines, the whole building group has integrated design from the aspects of height, volume, material, color, underground space, traffic organization, and so on. At first, some architects did not understand and accustom to follow the requirements of the urban design guidelines. They made many buildings with unique ideas and strong personality. They did not take into account the overall effect of the complex and the urban environment. During the design process, Philip Enquist, the partner of SOM who is in charge of the urban design and urban design guidelines in Yujiapu, worked together with the architectural design team. After persuasion and communication, the city leaders, urban designers, and the architects finally reached an agreement. But afterward, when a senior architect from SOM designed the International Financial Conference Center, the earliest scheme was also quite strange, and finally the present one came into being. It not only guarantees the high level of architectural design of single architectural scheme, but also guarantees the overall image of city streets and squares and the quality of public space such as green space and parks. Architecture really becomes a city building." Finally, the first case of architectural design strictly following urban design guidelines and a so large scale integrated underground space implementation in China has been realized, which is a bold innovation and experiment.

"Nine office buildings in the Yujiapu Starting Area, together with the subsequent International Financial Conference Center and two other office and apartment buildings, formed a 9+3 building complex in the starting area, covering 0.2 square kilometers, with a total floor area of 1.8 million square meters. The building heights of 12 towers ranges from 120 to 300 meters, and the International Financial Conference Center appeared as a horizontal landmark building. In accordance with the urban design guidelines, architects carefully design, which reflects in their own architectural style, but also ensures the overall coordination and unity of the building complex, making the final effect is very convincing. The architectural style of the building complex in the Yujiapu Starting Area is unified and diversified, which matches the temperament of the financial center, and can even be comparable to that of Potsdam Square complex in Berlin. As financial buildings, the architectural style is "classic and new". The overall building is moderate in volume, uniform in height and appropriate in density. The building materials are relatively common, the building colors are moderate and stable, together forming its own distinct architectural features. The architectural design of the starting area of Yujiapu Financial District represents the highest level and direction of the current domestic architectural complex design."

"In the process of planning implementation, Yujiapu's planning managers and builders were constantly encountering new problems and challenges. Conflicts between advanced planning concepts, design methods and relevant relatively backward domestic regulations often occur; contradictions between mixed use of land and approval of planning management are numerous; dialogue between advanced planning ideas and relatively lagging urban management means is difficult. Therefore, we have carried out the standardization and legalization reform of urban design in Binhai New Area, to formulate the reform plan and corresponding management regulations. Through continuous reform and innovation, the above problems have been gradually solved."

Chief ZhuGe

"From the point of view of urban design and architectural design, Yujiapu has reached a high level, but from the point of view of market operation, it is facing many difficulties for various reasons. I think there are three main reasons for the urban planning and construction. First, the construction of a business districts in Binhai New Are is too decentralized and competing with each other. The Modern Service District (MSD) of TEDA has built more than one million square meters of office buildings, including the 530-meter-high Zhoudafu Center. On the other side of Yujiapu Starting Area, Xiangluowan Business District has been planned and constructed with 40 buildings with a total floor area of 5 million square meters, one third of which are office buildings and one third are apartments. MSD, Yujiapu Financial District and Xiangluowan Business District are not more than 3 kilometers apart. Although the orientations are slightly different, the buildings constructed are all office buildings.

Millions of square meters of office buildings are put into the market together, which is undoubtedly disastrous for Binhai New Area dominated by the secondary industry. At that time, the three districts belonged to different governments or management committees, so they did not recognize the limited capacity of the market. The cost of land collected in Xiangluowan was relatively low. At that time, it was positioned as the business district for other provinces and municipalities' business offices in Tianjin. In order to invite business, a very preferential policy was given. However, the invited enterprises were mixed of good and bad, so the overall quality of the buildings was not high. At that time, a leader of the new district reported that Xiangluowan could be done experiments, and if it was not done properly, we still had Yujiapu, which made the city's main leader be furious, how can such a large development be experimented with? If Xiangluowan had been planned as a living supporting area, the central business district of Binhai would be much better today."

"Second, the central business district is inconvenient in transportation and lacks mutual support. Central Business District of Binhai New Area consists of MSD, Yujiapu Financial District, Xiangluowan Business District and Tianjian Cultural Business District. Due to the separation of the Haihe River and Railways, it is lack of effective traffic links between several districts. At first, experts proposed to dismantle the second-line railway into the port, but the final result was not only non-dismantled, but constructing the double-track. Despite the construction of the Central Avenue Tunnel, Chunguang Road Tunnel, Beihai Road Tunnel and Dongting Road Tunnel, the number of tunnels is still relatively small compared with traffic demand. Due to the problem of navigation standards, the construction of bridges over the Haihe River has been delayed, and the traffic across the river has been a headache. As the lack of convenient links among the districts in Central Business District of Binhai New Area, it is lack of mutual support and interaction with each other, which does not reflect the benefits of agglomeration and scale. Although the navigation standards were approved recently, the bridges under construction have not been closed more six years and played their due role, which has a great impact on the development of Central Business District of Binhai New Area."

"The third is lack of living supporting facilities and the slow construction of the subway. The functions of the contemporary central business district have become more complex and the supporting facilities are necessary. But in Yujiapu Financial District and Xiangluowan Business District, the starting projects are all office buildings, lacking residential projects and supporting educational and medical facilities, which can't form a aggregation effect of popularity. By contrast, MSD in TEDA is better. After 20 years of construction, there has been a perfect residential community and supporting facilities, such as Cardiovascular Hospital, Teda Hospital, Nankai Teda College and so on. In addition, public transport in the new district is underdeveloped. As early as 2009, the new district has formulated rail transit network planning and reported it to integrate into Tianjin Rail Transit Network. In 2015, the state council approved the recent construction plan of Tianjin Metro, including three lines B1, Z4 and Z2 in Binhai New Area. B1 and Z4 have been started construction, but over the years, progress has been slow, recently B1 was opened to traffic, Z4 is still under construction. Without rapid rail transit, it is impossible to make Yujiapu accessible for passengers from the original Tanggu District and the Eco-City, to play its role of the center, and to provide flow support to it".

"Despite these problems, Central Business District of Binhai New Area is gradually forming seven functions: administration, business, commerce, culture, entertainment, residence and transportation. The Binhai New Area Government is located beside Central Avenue. In the opposite Tianjian Area, where Binhai Cultural Center and Wanda Mall have been operated. In Xiangluowan, more than 20 administrative bureaus of Binhai New Area have been stationed, and the Immigration Service Hall of Binhai New Area has been in service. The average daily flow of people in Xiangluowan Polar Ocean Hall and Membrane Structured Stadium is 10,000. More than 3,000 apartments have been rented and sold. In Yujiapu, Baolong International Center complex, Global Mall Shopping Street and CGV Cinema have been in operation for many years. With the administrative bureaus of the new district, the daily flow of work and personnel can reach 10,000, and the flow of businessmen in enterprise offices in the district can reach 50,000. Yujiapu High-speed Railway Station makes passengers of Beijing and Tianjin convenient and smooth, with an average daily flow of nearly 10,000 people."

"More and more policies are focused on the convergence of Yujiapu Financial District. In December 2014, China (Tianjin) Pilot Free Trade Zone was established. Yujiapu Financial District is one of the three parts. Tianjin Municipal Government's call for "mass entrepreneurship and innovation" lies in the establishment of Double-Creation Special Zone in Yujiapu. The policy and functional advantages of "Financial Innovation and Operation Demonstration Zone + Pilot Free Trade Zone + Double-Creation Special Zone " are constantly emerging. At the same time, Tianjin simplified its administration and decentralized its power, implemented the "One-Seal Management and Approval" and other reforms, which made trade and investment services more convenient and made the number of registered enterprises in Yujiapu grow exponentially. According to the relevant information, in 2017, there were 8452 new market subjects in Yujiapu, with a total registered capital of 238.3 billion yuan, and the total number of market participants exceeded 338,000, of which 2433 were registered capital of over 50 million yuan. It has gathered nearly 2,000 financial and similar financial institutions, covering almost all financial subdivisions, and has managed assets of more than 2.6 trillion yuan. Of the new market subjects in the past three years, 30% came from Beijing. At the same time, more resources lie Yujiapu. The world-renowned The Juilliard School, New York in the United States has put into use its Binhai Yujiapu Campus. TEDA Administrative Commission has also moved into Yujiapu. The opening of Metro Line B1 has a great impact on Yujiapu, and we have seen the effect now. If Binhai Metro Line Z4 can be opened in the near future, the role of connecting the Eco-city, Beitang and TEDA with Yujiapu High-speed Railway Station will be more obvious. Line B1 connects the Binhai High-speed Railway Station and the second high-speed railway between Beijing and Tianjin, which has already been in construction. In the future, it will be able to reach Tianjin Binhai International Airport directly from Yujiapu Starting Area and High-speed Railway Station. It will be convenient to connect with Tongzhou New Administrative Center and of Central Business District of Beijing to further enhance the radiation level of Binhai New Area."

"Despite the good news, the difficulties faced by Yujiapu are still great. The economy is greatly affected by the envisaged deterioration of the current world economic environment. General speaking, the construction of office, commercial facilities and public service facilities in the starting area is ahead of schedule, and living facilities are seriously lagging behind. Even with the increasing number of market subjects and office workers, the crowds surging in Yujiapu Station and underground commercial street, in the large business district, these office workers are quickly submerged in the buildings, especially on weekends, there are not many people, so it is unavoidable to be questioned by "Ghost City" and will continue for some time. "

"The emergence and development of a city is a historic process, and urban design is also produced in a series of decision-making processes that are carried out every day. Prof. Jonathan Barnett put forward the concept of modern urban design as a "Public Policy". That is to say, "Real urban design is a series of administrative decision-making process and urban form shaping process. It is a continuous process of change with creativity and development flexibility. It should make design more flexible, with great freedom and flexibility, rather than creating the perfect ultimate environment or providing an ideal blueprint. If traditional urban design pays more attention to "Goal Orientation", then modern urban design should be a combination of "Goal Orientation" and "Process Orientation". Urban designers should not only hand over beautifully packaged products, but also make a flexible and repairable rule with the change of demand and time."

"Like La Defense in Paris and Dockland in London, the Yujiapu Financial District is now facing difficulties and setbacks in its development, which seems to be fate. Regardless of economic or other issues, we are also thinking about urban planning and urban design to find out the reasons. Our urban planning, including urban design, is still the ultimate blueprint, or like the construction plan, hoping to make a contribution to win the battle. In fact, the development of cities and the construction of business districts depend on the market, which is a process of gradual development and formation. Famous architect Koolhaas is familiar and criticized by the Chinese people for his design of the "Big Pants" of CCTV's new building. However, as the director of Architecture School of Harvard University, he has studied Manhattan of New York for a long time and published a series of books with great accomplishments. In *Delirious New York*, Koolhaas vividly recorded the development of Manhattan. From the earliest and simplest grid planning in New York to the promulgation of the Zoning Ordinance in 1919, the development of New York has been relying on the strength of the market, full of vitality and vigor, becoming the center of the world. In fact, every building on Manhattan's land has been rebuilt more than once. It is constantly rebuilt and rebuilt in accordance with the development changing of the market that makes Manhattan look like today. The World Trade Center was built in 1975 after the old buildings on the site was demolished in 1966. It was completely destroyed and rebuilt again in less than 30 years because of the incident of September 11, 2001. Of course, this is an extreme case. Although Manhattan's building is constantly changing, the basic road network skeleton of Manhattan has not changed fundamentally. New York's experience has proved that planning should be adjusted according to market changing, but the basic planning structure, road skeleton and infrastructure can't be changed at will."

The "Narrow Road and Dense Network" plan of Yujiapu Financial District has such flexibility. But there is also a big problem in planning, that is, the highest buildings in Yujiapu are all planned around the high-speed railway station. This is reasonable from the perspective of traffic accessibility and land price level. However, in terms of the process of market development, the buildings in starting phase should be low-cost, and the current market does not support anymore the construction of super high-rise buildings. As a result, if the surrounding area of the high-speed railway station is designed according to the ultimate plan, it will be empty space for a considerable period of time, which is harmful to the development of the business district, or even to the development of Binhai New Area. Therefore, the transitional construction scheme should be adopted in the surrounding area of Yujiapu High-speed Railway Station. In the near future, a group of buildings with market response and reasonable cost should be constructed. Together with the promenade along Haihe River near Binhai High-speed Railway Station, it will form a punch-in destination in Beijing, Tianjin and Hebei, which has an impact even on the whole country.

"Rome was not built in a day. Urban development takes time. It has been 30 years since the development of Shanghai Lujiazui Financial Zone. Both La Defense in Paris and Dockland in London have gone through painful times and are now back from death. With the progress of modern technology, one-time planning can also adapt to the changes of future development, which is important to adhere to. The 10-year urban design implementation of Yujiapu Financial District is the long-term perseverance of urban planning management departments and urban designers. We are confident about the future of the Yujiapu Financial District in Binhai New Area."

"At present, according to the implementation of the plan of Yujiapu in the past ten years and the changes of economic and social environment, and the public's continuous attention to the supporting living environment of the financial district, the decision makers, planners, managers and builders of Yujiapu are carrying out a new round of planning optimization adjustment and upgrading. The overall layout structure and road network of the urban design of Central Business District can't be changed. The transitional scheme around the Yujiapu High-speed Railway Station can be got through holding international solicitation and finding creative schemes to attract attention. In order to start the construction of high-level educational and medical facilities, the contradiction between housing and the Standard for Urban Residential Area Planning and Design should be solved through reform and innovation. The amount of high-end housing can't be large, because the preciousness of high-grade is precise and scarce. These plans will be upgraded to prepare for the next ten year development of Yujiapu Financial District."

41. Urban Design of the Cultural Center in Binhai New Area

Chief ZhuGe

"A city's cultural center is like a city's living room, usually located in the core of the city, elegant and generous, inviting city residents and tourists to gathering, reflecting the cultural heritage and characteristics of a city. The cultural center of a city at home and overseas is the main space of the city, producing many good works of art, displaying the style and taste of the city, cultivating the sentiment of the residents, and continuing the spirit of the city. The planning and design of the Cultural Center of Tianjin Binhai New Area, learning from the advanced experience at home and abroad, is intended to be distinct from the Cultural Center in the central city. Through innovative planning and design, we try to cultivate a cultural center of the 21st century with complex functions, pleasant scale, vitality, diversity and international standards."

"Beginning in December 2010, on the basis of learning from the successful experience of planning and construction of Tianjin Cultural Center, after full preparation, we organized the Conceptual Design Collection of Binhai Cultural Center complex in sponsorship with Binhai Cultural Radio and Television Bureau and Binhai Central Business District Management Committee, and invited Zaha Hadid Architects of the United Kingdom with Tianjin Urban Planning and Design Institute Architecture Branch, Academician He Jingtang Studio of Architectural & Design Research Institute of South China University of Technology, Bernard Tschumi Architects Firm with KDG Architecture Design Co., Ltd. of the United States, MVRDV of the Netherlands with Beijing Institute of Architectural Design ,four groups of top-ranking design units at home and abroad, to participate in. According to the expertise and willingness of the design masters, each master was responsible for the architectural design of a major building, at the same time, proposed the concept master plan of the building complex. Zaha Hadid's design of streamlined, undulating Binhai Theatre and the master plan of the complex is very fashionable; Bernard Tschumi' design of the Modern Industry Museum reflects the industrial history and the sense of site used to be Tianjin Soda Factory; Academician He Jingtang' design of earth-sheltered Art Gallery integrates the Cultural Center complex with the park; and MVRDV's design of the Aerospace Museum with futuristic is impressive. In February 2011, we invited experts at home, such as Academician Ma Guoxin, Li Daozheng, and Design Master Xing Tonghe to review the proposals. This conceptual design collection had laid a solid foundation for the further planning and design of the Cultural Center."

"The schemes by the four masters, whether the architecture design or the master plan, all have their own brand characteristics. How to synthesize the four Masters' schemes confounded the young designer of Tianjin Urban Planning and Design Institute. Finally, we use Academician He Jingtang's master plan, and barely integrated the other three Masters' architectural design into one piece. Although you may think it all right by appearance, it is a hodgepodge, which is actually the Master Tschumi's master plan of a platter with four geometric shapes. It seems that the master's foresight is clearly demonstrated. Later, the situation changed. Binhai New Area Party Committee and Government decided to build the urgently needed cultural venues first. The Urban Design Lab of TUPDI began to work out the urban design plan for the cultural center and the starting area. Never thought, they did 12 editions for the program during the following time. By the end of 2012, Tianjin Cultural Center was built. In order to be different from it, also absorb its advantages and avoid the same problems, the master concept plan of Cultural Corridor was formed with the efforts of planners and architects of the Urban Design Lab and Architecture Branch of TUPDI. The plan of the Cultural Corridor changed the traditional layout of cultural centers around the central square and the central axis, concentrating cultural venues on one side and forming a large area of cultural parks on the other. At the same time, in urban design, we try to form a cultural center with vitality, creativity and complex functions, combining the venues with the development of cultural undertakings as cultural industry . Ultimately, this novel scheme had been approved by all parties."

"The planning and layout of Binhai Cultural Center has been constantly deepening and improved for two years and gradually stabilized. At the same time, the construction of Binhai Cultural Center was included in the "Ten People's Livelihood Projects" of Binhai New Area. The construction was ready to begin formally. Therefore, starting in July 2013, according to the latest requirements of urban design and the master plan of the cultural corridor, the second round collection of the architectural design for Binhai Cultural Center (Phase I) was carried out. We invited Bernard Tschumi Architects and MVRDV, which had participated in the first round of design collection. Also we purposely invited some well-known overseas or domestic architects and companies, such as Helmut Jahn of USA, GMP of Germany and Huahui Architectural Design Co.,Tan Bingrong of Canada, to design five cultural venues and the cultural corridor respectively as a team. If the first round of architectural design collection is free play, this round is propositional composition, which should be designed in strict accordance with the master plan of the cultural corridor. Therefore, we have also made some considerations in the selection of designers."

"We had haven two choices for the chief designer of the cultural corridor. One is Zaha Hadid Architects who participated in the first round of design collection, the other is GMP of Germany who designed the opera house of Tianjin Cultural Center. In the first collection of Binhai Cultural Center, Zaha Hadid Architects made tremendous efforts to demonstrate their capabilities, strength, enthusiasms and completed a high-level master plan and conceptual design of Binhai Grand Theatre. They invited professional companies to do the acoustical design of Binhai Grand Theatre. While judging from the completed drawings, Zaha Hadid Architects spent tremendous human and material resources, although we only paid a $100,000 premium, which is far from enough. Although Zaha Hadid herself failed to attend the first review meeting, she later came to Binhai New Area and visited Yujiapu and the site specially. Her dedication was admirable. But this is a propositional composition, and several master architects and well-known design firms to cooperate with each. We were worried that the style of Zaha Hadid Architects is difficult to coordinate with other designers. In addition, we also realized that Binhai New Area does not have the ability to bear Zaha Hadid architects, mainly because of funds, but also human resources in management, construction and supporting. In the end, we can only reluctantly give up. Facts have proved that it is correct to choose GMP of Germany as the chief designer. The work style as well as design style of GMP is the same, serious and rigorous. They have a branch office of hundreds of people in Beijing and can be on the site at any time. In the design process, together with

Tianjin local design institutes, they played a good role in coordinating, ensuring seamless connection between the designs of the Cultural Corridor and the other five cultural venues. The cultural corridor is exquisite in design, showing the beauty of industrialization, and also in line with the overall style of Binhai New Area."

"The collective work lasted ten weeks, during which design units went to Tianjin three times to carry out the collaborative design of the "Workshop" type, and held many video conferences, in-depth study of various issues. The plan of each venue should have its own characteristics and must be coordinated with each other and unified into the cultural corridor as an entirety. In September, we invited Academician Ma Guoxin and other domestic professional authorities to review the design project with the international design masters participating in the project, and to evaluate and put forward suggestions for the architectural design scheme of Binhai Cultural Center (Phase I). The idea of building a cultural complex with the cultural corridor has been fully affirmed by the expert group headed by Academician Ma Guoxin. This collection is not a normal collection, but a collaborative design. Besides the well-known designers and companies mentioned above, there are many companies that act as consultants in various designs, including Nikken Sekkei Ltd of Japan, MVA Transport Consultants Co. Ltd. of Hong Kong, Huahui Landscape Design Co., Ltd of Tianjin☒Savills☒Tianjin Urban Planning and Design Institute, Tianjin Architectural Design Institute, Tianjin Bohai Urban Planning and Design Institute, Tianjin Municipal Planning Institute and so on, have provided various specialty consultancy services and technical support in the design process, ensuring the operability of the planning and design scheme."

"By comparison, Binhai Cultural Center project is more like the design and organization mode of Potsdam Square than the design of Yujiapu Starting Area project. The 470-meter corridor of the Binhai Cultural Center unifies five different venues, embracing changes in the unification. The Cultural Corridor is the highlight of this design, focusing on people's use and feelings, which represents "people-oriented" design idea. In the process of collection, including later design work, German GMP designers have repeatedly proposed that the corridor should be open air and not closed. Considering the cold climate in winter and summer in northern China, we insisted on making a enclosed city space. From the present effect, the enclosed corridor inside is more comfortable than outside ."

"After the grand opening of the Binhai Cultural Center, the Binhai Library has become an internet celebrity building, attracting a large number of eyeballs and a huge flow. There are always people queuing in front of the gate of the library in the Cultural Corridor. Unfortunately, due to various reasons, the Cultural Exchange Building designed by Helmut Jahn was not built synchronously, lacking the complex functions of hotel, apartments, offices, conference facilities and so on. If we want to take advantage of the influence of Binhai Cultural Center, it will be a good choice to construct Cultural Exchange Building by market-oriented means as soon as possible and improve the function of Binhai Cultural Center. In addition, the master plan of Binhai Cultural Center is to give full play to the role of suturing the core area of Binhai New Area. We should make up our minds to open up the links between Dongting Road, Yujiapu and TEDA, Jiefang Road and Haihe Bund, build the Line Z4 of Binhai

Metro as soon as possible, and build the walking system of Binhai Central Area, which will link Yujiapu, Tianjian Area, Binhai Cultural Center and TEDA MSD into a system. To speed up the construction of Tianjian Residential Community, we need to improve the supporting facilities as high-level schools and hospitals. It is necessary to plan ahead the feasibility study of Phase II and Phase III of market-oriented construction of Binhai Cultural Center."

"Two years after the completion of the first phase of Binhai Cultural Center, Binhai Cultural Park next to the Binhai Cultural Center was finally built. Although a design competition was held and eventually designed by an internationally famous landscape firm, the level of design was not good. The technique is too fancy. It competes with the main building of Binhai Cultural Center, also is not very convenient in function. This 20-hectare Park in the center of the city, which is not quiet and deep, too simple and straightforward, lacks many pleasant spaces for people to stay, and can't attract enough popularity, which seems to deviate from the original intention of the park. If there is an opportunity in the future, it should be revised and upgraded."

42. Urban Design and Management System of Tianjin

Chief ZhuGe

Above all, I reviewed the development of Tianjin urban design since 2008, introduced several key projects of urban design, and made a simple evaluation. There are many reasons for such achievements, but the key is that we have established a complete system of urban design making, legal and management. Since 2008, Tianjin has set up the Key Planning Headquarters, concentrating on making a large number of urban designs of various types. The implementation of a number of high-quality projects has pushed the urban function and the urban environment to a higher level as a whole, demonstrating the strength of urban design. In this process, Tianjin has initially formed a relatively perfect system of urban design laws and regulations and management mechanism."

As early as 2008, Article 37 of Tianjin Urban Planning Regulations made it clear that, "For the key areas and projects designated by the Municipal People's Government, the urban and rural planning authorities should organize to make urban design in accordance with urban and rural planning and relevant regulations, and formulate urban design guidelines; in the preceding paragraph: in other areas except those key areas, the urban and rural planning authorities of districts and counties, should organize to make urban design and formulate the urban design guidelines". In the same year, as one of the reforms in the overall scheme of comprehensive supporting reform experiment of Binhai New Area, the reform was carried out of "Exploring the standardization, statutory making and approval mode of urban design, and doing well the urban design of key areas and projects". The reform experiment lasted for more than three years and mainly completed three aspects of work. First, the key areas have been selected. Second, in 2010, the Municipal Urban Planning Bureau issued the *Interim Measures for the Management of Urban Design Guidelines for*

Key Areas of Tianjin Binhai New Area. Third, a lot of urban design and urban design guidelines for key areas have been affirmed, and construction project planning approval have been carried out according to the urban design guidelines, which has achieved remarkable results. It also changed the past construction project approval mode of "One Size Fits All" according to the technical planning management regulation. The urban design guidelines define specific regulations including the road intersection red line, green space ratio of block balance, the adjustment of the building setback line, and so on."

"On the basis of summing up practical work experience, Tianjin Urban Planning and Design Institute had formulated the technical regulation of urban design and urban design guidelines, which effectively links urban design with regulatory detailed planning in order to ensure the landing of the urban design. On this basis, the management system and mechanism of the "One Regulatory Plan, Two Guidelines" has been established. As the legal basis of planning management, the regulatory detailed planning should be flexible and compatible to avoid frequent adjustment and ensure its seriousness. The regulatory plan mainly controls the municipal infrastructure of the main roads, structural land use layout and overall development intensity in the city. The land subdivision guideline is the implementation plan of the regulatory plan. This uses the subdivision method of the United States for reference to determine the secondary road branch network of land use, and the land use type and development intensity of specific plots can be adjusted without changing the principle of the regulatory plan. Urban design guidelines strengthen the control and guidance of urban space environment, such as urban style, regional architectural characteristics, public space and environmental landscape, through the control of space environment and buildings or architectural groups. In order to ensure the high quality of buildings, we strengthen the management and control from the horizontal and vertical dimensions. On the horizontal level, urban design should highlight its leading role, coordinate the specialty design as a whole, such as architectural design, landscape design, traffic organization, underground space and green ecology. On the vertical level, the practical implementation of urban design should be emphasized, from project programming, architecture design, construction drawing design, material selection for construction, from beginning to end, following up the whole process, to ensure that the project be implemented in accordance with urban design guidelines"

"Specifically, at every stage of urban planning, we attach great importance to the inheritance and development of urban cultural characteristics. We have formed an urban design making and management system covering all levels and types, using the overall urban design as the general principle, using the urban design of key areas and key nodes as the precursor, and using the "One Regulatory Plan, Two Guidelines" as implementation basis, which provides strong support and guarantee for the city's high-quality urban planning and construction. Urban design plays an important role in shaping urban spatial image, features and creating pleasant environment because of its integrity and intuitive expression in the design of urban environment and spatial form. We recognize that the legalization of urban design is the key to give full play to the control and guidance of urban design on urban construction. We pay attention to the connection of urban design in the process of concept, management and implementation. "Integration" means that in the early planning stage, the overall urban design combines market demand and architectural layout to carry out conceptual design; in the statutory regulatory stage, urban design in key areas will implement the contents of underground space, ecology, transportation, municipal engineering, landscape, architecture scheme into regulatory plan and, land subdivision guideline and urban design guideline. Finally, in the construction and implementation stage, the urban design guidelines will be used as an annex to the land transfer contract, so the implementation will be ensured throughout planning management."

"At the same time, in addition to the urban design guidelines for key areas, special control guidelines will be compiled to form technical regulations for management. For example, "Tianjin Planning and Design Guidelines" put forward control requirements for residential, commercial, office and other architectural design from eight aspects: architectural layout, architectural height, architectural style and architectural color, to standardize and guide architectural design. The "Three-Side Management Regulations" for bank areas along main rivers, parks and the historic and cultural blocks has been formulated to strictly control the building height of new development. The intensity of residential areas of the general FAR inside expressway ring road can't exceed 2.5, and the outside can't exceed 2.0. The control requirements for building shape and depth-width ratio are put forward; and the control requirements for building style are also put forward, which should be in line with the surrounding areas. In order to avoid the light pollution of curtain wall, the requirement of window to wall ratio is put forward. In addition, urban design are transformed into three-dimensional digital model. The whole coverage of building block model is realized in the central urban area, and the fine model is realized in the key areas, which can be used as an additional means of individual construction project review. Through continuous practice and exploration, we realize that good urban features and high-quality urban environment can't be achieved by relying solely on the concept of planning, but must be based on the overall management and control at all levels, as well as the detailed design and substance work."

Chief ZhuGe

"Tianjin's urban design started early and made many achievements, but there are also many problems." Speaking of this, Chief ZhuGe raised his voice. "In those pilot projects of urban design, we have a comprehensive reflection, as the focus of our study for the pilot projects of urban design. Through the analysis, we find that some problems are unique to Tianjin, and some have commonalities, so solving these problems has a certain exemplary role. We summarize the problems of commonality into five aspects: The first is the problem of market. Most of our urban designers and planning managers are educated in functional urban planning, and the Planned Economy is still deeply impressed in the institutional mechanism. Therefore, most of our urban design is top-down and paternalistic. Although we always talk about market orientation and market playing a decisive role in resource allocation, there is still a lack of real attention to the market, respect for the market and response to the market. The importance of the market is reflected in many aspects, especially in the inadequate consideration and attention to human needs, improving the quality of life and the real estate market. For example, many projects built according to the urban design, the overall environment is still acceptable, the building shape is beautiful, but the market acceptance is poor, lack of vitality, such as: the Five Big Courtyards, Jinwan Square, Yujiapu Starting Area and so on. Of course, we can explain that there may be some time problems, such as the recovery of La Defense in France and Dockland in London after the recession, but in fact the decisive force of the market is what we have to pay attention to and learn from."

"The second aspect is the lack of consideration for people. Although we always talk about people-oriented, in fact, people in our eyes are only things, figures without personality. Details close to human scale determine the success or failure of urban design. To support the revival of urban center, it is necessary to further improve the pedestrian system, which is the successful experience of all countries in the world. In most areas of Tianjin, especially in the central urban area, there is still a big gap in the construction of urban pedestrianization, which is not suitable for people to walk on the whole. Walkways are occupied and landscape is lack of design everywhere, also there are too few parks. Since Water Park was built in the 1950s, there have been no large parks in the decades. In recent decades, Meijiang Park has been constructed, and West Water Park has just been completed in recent years. However, compared with the speed and total amount of urban development and construction, the park has not increased significantly and can't meet the various needs of people. We have planned eleven parks around the Outer Ring Road, but only four have been built, and seven have not yet been started. The other is the problem of housing. The ultimate goal of our development is to improve people's living standards. Besides improving the quality of urban central area and public space, the purpose of urban design is to face a large number of housing and community problems. However, our urban design has not given much consideration in this respect. We have been using the Planned Economy method of residential area planning to build a large number of large-scale enclosed housing estates without personality, which is not suitable for the people's desire for a better life in the future."

"The third aspect is that we are not familiar with the ideas of ecological civilization and the methods of landscape design, and lack of control over the large space artistic skeleton and the large ecosystem of the city. The emergence of modern urban planning is mainly to solve the urban problems caused by the rapid urban expansion after industrialization. In 1791, in Pierre Le Enfant's Washington plan, the city and nature were integrated. In 1824, Robert Owen, a British utopian socialist, planned to build a new harmony village of self-sufficiency in Indiana, USA. Although the result was unsuccessful, he explored organized planning to avoid the wanton infringement of industrialization on rural areas, so some people called Owen "the Father of modern urban planning". In the 1930s, with the operation of commercial trains, suburbanization appeared in the United States, and people yearned for the freedom and beauty of the countryside. In 1860, Frederic Olmsted, known as the "American Farmer", designed New York Central Park and Boston's "Necklace of Emeralds" green system in a natural way, bringing a natural flavor into the artificial city. In 1898, Ebenezer Howard proposed the Garden City of town-country magnet, which opened a new era of modern urban planning and showed the world the way to peace in the future. In 1915, Patrick Geddes published *The Evolution of the City*, which argues that urban planning should be a regional planning that combines urban and rural areas. Therefore, the purpose of the birth of modern urban planning is to integrate the city with the natural countryside. On the other hand, our urban planning and design are mainly aimed at expanding the scale of the city, excessive consideration of the design of physical cities and buildings, and less consideration of the ecological environment and large-scale park green space system. In fact, China's traditional urban planning has always emphasized the unity of heaven and man, learning from nature. Even Old Beijing City which is strictly planned according to the form of Zhou Li, Kao Gong Ji, it still retains the natural river system and the Three-Seas of Beihai, Zhonghai and Nanhai, intersecting with the rigorous Central Axis naturally and flexibly."

"The fourth aspect is the lack of attention paid to the impact of roads and infrastructure on planning. Urban agglomeration promotes the explosive progress of science and technology, and road traffic and civic infrastructure in turn promote the progress of urban civilization. Urban road traffic and civic infrastructure are inseparable. Our urban planners and designers are accustomed to think that road traffic and civic engineering are supporting specialties, and they do not give due attention to them ideologically. At the same time, road traffic planners and civic engineers also think that they are supporting specialties. They think that as long as they meet the specifications and standards, they can ignore important influence of the overall planning and the fact that basic design also has the problem of beauty and culture. Many people are accustomed to using some slogans to deal with a specific traffic problem, thinking that shouting "vigorously develop public transport, restrict private cars" will make everything fine, and make the world peaceful, ignoring the current situation of urban traffic congestion, ignoring the actual needs of people to travel. In fact, computer-aided planning and modeling, including the application of big data, is the most mature in the field of road traffic, while not widely used in urban design. We haven't really established the concept that urban road and civic infrastructure is the urban culture."

"The fifth aspect is the too much simplification of residential buildings. Some residential areas are said to be doing urban design. In fact, they have been still using the method of Planned Economy and Modernism in residential area planning, to make large-scale repetitive design. These are detested strongly and criticized severely by Jane Jacobs in her book *The Death and Life of Great American Cities* published in 1961. We praise Jane Jacobs' wisdom verbally on one hand. On the other hand, we do what Jane Jacobs objected to brazenly and noisily. We are accustomed to the large-scale enclosed residential area planning and development with numerous independent high-rise towers or multi-storey slabs, stifling the richness and diversity of residential buildings, and the consequence is stifling the personality of all residents."

"Coincidentally, these five problems correspond to the four functions of work, recreation, transportation and residence proposed by Le Corbusier in Athens Charter. What we are doing now only shows that we are still under the control of modern functional urban planning ideas and methods. If we don't break away from modernism in ideology and methodology, most of our so-called urban design is a wolf in sheep's clothing. Therefore, it is necessary to change the planning mode of residential areas, combined with the modern housing system, to establish the theory and method of urban design to study the history and culture of the city, and to create high-quality urban form, function and environment with comprehensive and integrated urban design measures. Tianjin' pilot projects of urban design are combining with the practice of specific projects, to explore solutions and schemes to the four aspects in market, housing, ecology and transportation."

Chief ZhuGe

"The above is my review and summary of Tianjin's urban design work. In the past, we said, "No Planning, No Construction", but now we say, "No Urban Design, No Construction". The coverage rate of urban design in Tianjin has been relatively high, and the level of urban construction has indeed been greatly improved under the guidance of urban design. Through examples, we try to objectively evaluate the effect of the implementation of urban design and urban design guidelines. While affirming the achievements, we also found the above five main problems as the key content to be solved in the pilot work. Because of the time, I would like to briefly introduce the main points of the pilot work and invite the guests to comment and give guidance. On the sub-forum in the afternoon, our designers will then give you a detailed introduction, please feel free to discuss."

Chapter VII

A More Ecological City

43. Territorial Spatial Master Plan: The Artistic Skeleton of the Metropolitan Area

In the afternoon, there are four sub-forums: Urban Design Theory and Practice, Ecological Urban Design, the Protection of Historic Cities and Urban Design, and Urban Design Education. Mr. Barnet and Chief. ZhuGe both concern urban design education and they come to the conference room of the sub-forum of Urban Design Education. The forum is chaired by Professor Zhang Zeping, Professor and Doctoral Advisor of Tianjin University and Academician of Chinese Academy of Sciences. Many faculty members from Colleges of Urban Planning and Colleges of Architecture of famous universities are present. The focus of the discussion is on the theoretical perspective of the relationship across the territorial spatial master plan, the regulatory detailed plan and the urban design. With the national emphasis on "multiple-plan integration" and the establishment of territorial spatial planning system, there is a sense of unsettlement in the Colleges of Urban Planning in universities. Professors and students are concerned about the future development of their majors and careers as well.

Academician Zhang Zeping

Professor Zhang is in his sixties, but he is relatively young among the academicians of the Chinese Academy of Sciences, especially in the field of spatial planning. He has studied and worked in the United States for nearly 20 years, mainly engaged in international urban planning research. In 2006, he was invited by Tianjin University to teach in China. After returning home, he has been engaged in the research of European Union -based spatial planning for more than 10 years, and has become an expert in this field. Since he is a professional in architecture and urban planning, he pays more attention to urban and spatial environment. As you all know, city master plan is the blueprint of urban development. Apart from guiding urban social and economic development and urban construction, its main role is to guide the shaping of beautiful urban form and create a livable environment for people. From the birth of urban planning, building an aesthetical city has been the highest pursuit and level of urban planning. Although with the rapid development, cities are becoming more and more complex, and there are more and more planning issues that need to be considered, the most fundamental purpose and content of city master plan cannot be ignored. In our planning work, especially in the city master plan as well as urban design, we must focus upon the creation of aesthetics of living environment.

Thirty years ago, Mr. Wu Liangyong published the article "Creation of Urban Aesthetics" in response to the domestic discussion on urban environmental renovation in Tianjin at that time. He pointed out:

> A good city should have such elements as comfort, clarity, accessibility, diversity, selectivity, flexibility and hygiene. People should live in it with a sense of privacy, neighborhood, nature and prosperity. The urban aesthetics include the aesthetics of all aspects including natural environment, the history and culture and environment of the city, modern architecture,

> landscape and gardens, architecture, sculpture, murals and crafts. The artistic laws of urban aesthetics include the aesthetic of the whole, the aesthetic of characteristics, the aesthetic of development and change, and the aesthetic of the rhythm of the spatial scale.(Wu Liangyong, 1989)

"Aesthetics is the ultimate goal and means of the construction of human settlements. Aesthetics is the highest spiritual experience of human beings, and human settlements are the material and spatial manifestations of aesthetics. People build cities, and the urban environment changes people. In order to create a better living environment, besides economic development, improvement of living standards, social progress and environmental improvement, we also need to design and create the aesthetics of the living environment. Some people say that our planning is the planning in traditional arts and crafts, focusing too much on the creation of physical space. In fact, international experience and the latest theoretical research show that the creation of physical environment in cities and regions is becoming more and more important, which is also the conclusion drawn from the profound experience and lessons of urban and regional planning and construction in recent decades in China. City master plan should, and must, take the shaping of urban aesthetics as one of its key elements.

"At present, the state has established the territorial spatial planning system, and one of its purposes is to solve the problems of too many types and levels of planning and the inconsistency of various policies; the other is to improve the level of urbanization construction in China and solve 'urban diseases'. Many of our colleagues engaged in urban design are confused about the relationship between territorial spatial planning and urban planning and urban design. They feel that the so-called 'three districts and three lines', 'parcels of land', 'double evaluation' and other things will replace urban planning. In fact, spatial planning emerged in Europe in the 1980s and evolved from regional planning as the spatial orientation of various EU policies. Professor Stephen M. Wheeler of the University of California, Berkeley, an advocate of new regionalism, has done a lot of research in this area. When I was a visiting scholar at the University of California, Berkeley, I talked to him. He believed:

> In recent decades, regional planning has paid too much attention to economic geography and economic development, at the expense of ignoring other connotations of regional science, which has caused great losses. In the 21st century, with the rapid evolution of the physical form of urban areas, regions are facing greater challenges in terms of habitability, sustainability and social equity. Future regional planning therefore requires holistic perspectives and new regionalism, which should include urban design, physical planning, place-making and equity, in addition to economic development as the focus of research. It needs not only quantitative analysis, but also qualitative analysis, and they should be based on more direct regional observation and regional experience.

The most fundamental motivation for this is to reevaluate the important objectives of regional development and find a balance across economic development goals, social development goals and a beautiful living environment.

"After more than a hundred years of urban and regional planning in the United States, the mechanism and methods of spatial planning have relatively matured, although they call them smart growth instead of spatial planning. It has solved the problem of the incompatibility between national will and local development, and each state and region is relatively independent. In the choice of space planning objectives, they continue to adopt relatively simple methods, such as 'quality of life', which basically continues the concept of 'American Dream'. The quality of life is determined by society, environment and economy, emphasizing social equity, environmental protection and economic development, which is also consistent with the new outlook on development. The planning objectives of "three primary color circle" proposed by New York Regional Planning Association (RPA) in 1996 in the third edition of regional planning 'New York: A Region at Risk' were very representative, where the economy is deliberately placed at the bottom. In fact, the quality of life is consistent with the concept of livable environment we have introduced, including the connotation of social equity. Despite the different titles, other states in the United States have basically adopted the same master planning objectives.

"Unlike the background of the smart growth plan of the United States, the main purpose of the European Union's spatial planning is to unite Europe, achieve territorial integration and social cohesion, create joint efforts and develop together. In the target triangle of the European Spatial Development Prospect (ESDP) in 1999, the overall goal was to pursue balanced and sustainable spatial development, and the 'three primary color circle' emphasized the unity of society, economy and environment. In documents such as the European Charter of Spatial Planning, the basic purpose of spatial planning is to seek balanced regional socio-economic development, improve the quality of life of residents, manage natural resources and protect the environment, and make rational use of land. It can be seen from this that although the quality of life is also the goal pursued by the planning, the pursuit of balanced regional socio-economic development is more important and placed at the top, which is the basis for the existence of the EU. The same trend is reflected in other levels of spatial planning in the European Union. Therefore, we say that the ultimate goal of territorial spatial planning is to create a more equal and beautiful living environment on a larger spatial scale, despite the different situations in different countries.

"How to create a beautiful living environment in the territorial spatial master plan? Ultimately, traditional urban planning, especially urban design, is indispensable. Of course, there is a need for large-scale urban design, or regional urban design. In addition to economic development and social progress, the focus of the territorial spatial master planning is the shaping of regional artistic skeleton of space. The establishment of a large-scale regional artistic skeleton of space actually involves two systems and the overall design of the vast rural areas. One is an artificial system consisting of regional urban areas and roads and civic infrastructure, the other is an ecosystem consisting of regional natural ecological reserves and ecological corridors such as rivers, mountains, trails, etc..The vast rural areas and numerous villages are Biological background. A good artistic skeleton of a region is to have a good ecological environment, a rich rural area as the basis, and beautiful cities, towns, convenient transportation and efficient civic infrastructure; to realize an ideal state of harmonious coexistence between artificial system and natural system. To achieve this, we need a variety of knowledge, including an understanding of the natural ecosystem, the transportation and civic infrastructure system, and the regional urban network system as well as related public policies. Therefore, our planners and urban designers are required to have a broader knowledge system, which also puts forward new requirements for our education system.

"To create a beautiful residential environment on the scale of the territorial spatial master plan, we should pay attention to regional characteristics. How to avoid one side of a thousand cities is a key issue. We regard large-scale urban design as the main means to enhance territorial space and urban aesthetics. In the process of making the territorial spatial master plan, it is necessary to carry out the research of the overall urban design and explore the overall spatial form and urban characteristics of the whole region. At the same time, a series of planning work, such as district overall urban design and urban design of key areas, should be carried out simultaneously, forming a situation of top-down, bottom-up, mutually blending and reflecting. Planning work should be carried out synchronously in depth and breadth, and the topic of urban aesthetics should be discussed together.

"If we still make and implement the plans according to the traditional detailed planning method of city master plan and control and manage regional urban development, we can foresee that the final result will be the same, and a monotonous, featureless city and chaotic suburbs will be formed. We really need a change, a change in the traditional planning, making and management mode, a change in the concept of architecture and space that we have become accustomed to, and carry out the overall urban design of urban areas. To establish the artistic skeleton of city region, we must reform and innovate the means and concepts of city master plan and urban design."

Prof. Barnett, American Master of Urban Design

Prof. Barnett agrees with Prof. Zhang very much. "In 2015, I published a new book, *Urban Design: Modernist, Traditional, Green and Systematic Perspectives*," which mentions:

> So far, we can adopt at least four methods to treat every urban design problem, namely, modernist urban design, traditional urban design, green urban design and systematic urban design. Designers can freely choose the most suitable urban design method or the synthesis of various methods. But modern cities have become so vast that the existing four methods are not enough. There should be a fifth method to cover such rapidly changing and complex socio-economic conditions. The new methods include: starting with natural landscapes, strengthening interchange transportation systems, understanding the design implications of the new economy, promoting compact business centers and walkable neighborhood communities, designing to meet modern building types and parking needs, formulating large-scale public design decisions through public procedures, and so on. (Barnet, 2015)

"I don't think there is a need for a general new concept of urban design. Rather, we should integrate urban design into the process of social and economic change, while maintaining a sustainable relationship with nature. The shaping of urban spatial quality has exerted more and more profound influence on people's lives through public policies. Different from Europe, the United States has no unified spatial planning system and mechanism. But some states in the United States have worked out smart growth plans, which are actually spatial planning. The focus is to grasp the larger spatial structure, which is the artistic skeleton of space. Here we first need to consider the protection of natural ecosystems and the shaping of good urban spatial form."

Chief Planner ZhuGe

Chief. ZhuGe also agrees with Prof. Zhang, adding: "Mr. Wu Liangyong's General Theory of Architecture and Introduction to Sciences of Human Settlements both emphasize the trinity of urban planning, architectural design and landscape design, which is the core of spatial planning. The spatial planning of developed countries we have studies, such as the regional planning of the three states of New York, also reflects the consistency and coordination of large-scale planning with urban and architectural patterns. In order to improve the spatial planning system and build a beautiful living environment, whether as a system or as the content and method of planning and design, spatial planning, urban design and architectural design must be an inseparable trinity. Even the national territorial spatial planning, or the large-scale spatial planning across provinces and cities, must also clearly define the housing situation and transportation mode of the vast majority of Chinese people as well as urban spatial forms, including streets, squares and parks, especially the features of the city, the beautiful skyline and the natural scenery of the countryside. Only in this way can the Chinese dream of the great rejuvenation of the Chinese nation become more visible, clearer and more approachable."

"Tianjin is an important fulcrum of the Beijing-Tianjin-Hebei world-class urban agglomeration, a national advanced manufacturing research and development base, a core area of international shipping in the north, a demonstration area of financial innovation and operation, a pioneer area of Reform and Opening Up, and an important ecological function area of Beijing-Tianjin-Hebei. In the Tianjin's territorial spatial master plan currently being edited, the optimization of ecological pattern guides the determining of the artistic skeleton of urban space. It is proposed that the spatial structure of Tianjin should be 'three districts, two belts, and a barrier in the middle; two cities, two corridors and many nodes'. The 'three districts, two belts, and a barrier in the middle' are all key ecological areas in Tianjin. The 'three districts' refer to the Jizhou ecological mountain area in the north, the Dahuangbao ecological wetland in the Qilihai wetland in the middle and the Beidaguang Tuanbo ecological wetland area in the south. The 'two belts' refer to the eastern coastal ecological barrier belt and the western shelter forest belt, and the 'barrier in the middle' refers to the green ecological barrier between the central city of Tianjin and the core area of Binhai New Area. In order to follow the strategy of territorial spatial master plan and optimize the spatial layout of Tianjin's city area, we defined the urban design ' The Necklace of Emerald and Its Surrounding New Communities 'as one of the most important urban design pilot projects , further exploring the possibility of better urban design of Tianjin City. This urban design takes the area along the outer ring road and city parks as the strategic space for the transformation, upgrading and development of Tianjin's central city, linking up the green barrier of two cities and connecting the outer ring green belts, urban parks and country parks, forming a green skeleton and creating a living environment that integrates the city and parks."

44. The Ecological Barrier between Two Cities: Integrating into Beijing, Tianjin and Hebei Province

Senior Planner Guo Yuan

While the "Urban Design Education" sub-forum is ongoing, Guo Yuan, a senior planner from TUPDI, is introducing the overall planning of the ecological barrier between the two cities at the sub-forum of "Ecological Urban Design". Guo Yuan is born in the 90s and looks energetic. She said, "The Party and the State attach great importance to the construction of ecological civilization and the coordinated development of the region. The Eighteenth National Congress of the Party clearly proposes to vigorously promote the construction of ecological civilization, strive to build a beautiful China and realize the sustainable development of the Chinese nation. In 2015, the central government promulgated the Outline of Beijing-Tianjin-Hebei Cooperative Development Planning, which clearly defined the spatial structure and development direction of Beijing-Tianjin -Hebei, and proposed to take the lead in making breakthroughs in ecology, industry and transportation. At the same time, it was clearly put forward that the regional ecological system of Beijing, Tianjin and Hebei should be constructed to realize the coordinated development of the regional ecological system of Beijing, Tianjin and Hebei. In order to implement the national strategy of coordinated ecological development between Beijing, Tianjin and Hebei, the Tianjin Municipal Party Committee and Municipal Government have carried out a series of fruitful work, actively implemented the transition zone of ecological wetland in Beijing, Tianjin and Hebei, and planned to construct the ecological barrier belt around the southeastern part of the capital.

"The Eleventh Party Congress of Tianjin made a strategic decision to strengthen the planning, management and control of the land area of 736 square kilometers between Binhai New Area and the central city and to build a green ecological barrier from the perspective of implementing the national strategy of coordinated ecological development between Beijing, Tianjin and Hebei. On May 28 last year, the Standing Committee of the Municipal People's Congress deliberated on and passed the Decision on Strengthening the Planning, Control and Construction of Green Ecological Barrier between Binhai New Area and Central City, bringing the planning and construction of Green Ecological Barrier into the legal realm. The Municipal Government has compiled the 'Planning of Green Ecological Barrier Area in the Middle of the Two Cities of Tianjin (2018-2035)'. The planning proposed that by 2035, the proportion of blue and green space in the green ecological barrier area should not be less than 70%, forming a 'Two-City Ecological Barrier, Tianjin-Tanggu

Green Island'. Through the green ecological barrier, the Qilihai and Dahuangbao Wetland Reserve in the north of Tianjin is connected with Panshan and Yuqiao Reservoir ecological protection areas, and then with Tongzhou Ecological Park and Wetland Park in Beijing; and the Tuanbo Ecological Wetland Reserve in the south is connected with the ecological park and Wetland Park in Xiongan New Area. Actively integrated into the regional ecological environment system of Beijing, Tianjin and Hebei, it will become an important part of the ecological barrier belt around the southeastern part of the capital.

"The ecological barrier between the two cities covers five districts: Dongli District, Jinnan District, Xiqing District, Binhai New Area and Ninghe District. By 2017, the total population of the region is 1.15 million, including 550,000 registered residents. The current land for construction is 198 square kilometers, accounting for 28% of the total land area; the leveled land for construction is 77 square meters, accounting for 10% of the total land; the remaining land is for blue and green spaces, including water and agricultural land, with a total area of 462 square kilometers including urban green space, accounting for 62% of the total land.

"The development of this region has gone through ups and downs. The 1986 edition of Tianjin City master plan put forward the spatial development strategy of 'industry moving eastward' and the spatial layout of 'one pole carries two ends'. With the development of the city and the past adjustment of the city master plan, the area between the two cities is also constantly changing. From 1986 to 1996, the region was in a period of self-development. According to the overall plan, a 500-meter green belt is planned outside the outer ring of the central city to avoid urban sprawl. At the same time, satellite towns such as Junliang City and Xianshui Gu were planned, but there was little development. During this period, the land between the two cities was mainly agricultural, with many villages scattered. The biggest spatial change was the implementation of the eastward movement of Tianjin steel industry and the construction of the factory of Tianjin Seamless Pipe Co. in the middle reaches of the Haihe River, and projects such as the factory of Gloria Material Steel Cooperation have also begun to be built. These projects had a tremendous impact on the ecological environment of the middle reaches of the Haihe River.

"From 1996 to 2006 was a period of gradual transformation from slow to fast development in Tianjin. Following Deng Xiaoping's speech on the South Tour in 1992, the whole country entered a new upsurge of development. In 1994, the City Master Plan began to be revised and the concept of sustainable development was put forward. Combining with the construction of Tianjin-Binhai Light Railway, Tianjin-Binhai Highway and the main axis of Beijing-Tianjin-Tanggu, the layout of 'Double Centers Axial' was put forward for the whole city and continued to implement the strategy of industrial move eastward. Considering the development of Dongli, Jinnan districts and other areas and the continuous development trend along the Jintang Highway, combined with the connectivity of the city's ecological circle, a number of new ecological corridors were planned between the two cities to connect the northern and southern ecological areas.

"After 2006, combined with the development and opening up of Binhai New Area, the airport high-tech zone, the west zone of TEDA and the Binhai high-tech zone were established along the Beijing-Tianjin-Tanggu high-tech corridor, and major projects such as Airbus A320 assembly line and large rocket had set up to promote the upgrading of regional industries. In 2008, the urban spatial development strategy put forward the layout idea of 'two cities, two ports, mutual expansion'. The middle zone of the two cities was planned and constructed as an urban expansion area, and the urban development between the two cities was accelerated. During this period, Haihe Education Park was built between the two cities, and a center for innovation and research and development was developed. National Convention and Exhibition Center was planned, and land for large projects such as sports center were reserved. The construction of small towns represented by Huaming Model Town had been implemented, and the process of urbanization had been accelerated. At the same time, with the spillover of urban functions, the construction of the 'two lakes' of Dongli Lake and Tianjia Lake and demonstration towns such as Xiaozhan and Balitai, Tianjin entered a phase of sprawl of development. Because financial balance was achieved through land sales, the urban construction land in the region had increased, and the development intensity is too high, and the building height became higher and higher.

"At present, the region has key national and municipal areas such as Binhai high-tech zone and airport economic zone, with a total industrial output value of about 280 billion yuan. With Haihe River as the boundary, the northern part of the country has formed a major project gathering area represented by large aircraft and rockets. In the south of Haihe River, the cluster characteristics of small and medium-sized enterprises and private economy are obvious. The urban space mainly concentrates in the main urban area of Jinnan, the west of TEDA, and Junliangcheng, Gegu, Xiaozhan demonstration towns, etc. There is Haihe Education Park, a municipal functional area that mainly focuses on higher education and vocational education, and other high-tech industries such as aerospace, new energy and new materials, biomedicine and so on, and it has a good foundation for scientific and technological innovation. There are abundant historic and cultural resources in the area, such as the national historic and cultural town Gegu, Tianjin historic and cultural town Xiaozhan, and Xianshuigu, the first town in east Tianjin, and so on."

"Although some achievements have been made in economic development and urban construction, the problems are serious from the perspective of regional ecology. This area is the tail of the Haihe River Basin. It has four first-class rivers, 20 second-class rivers and five lake reservoirs. The water area is about 189 square kilometers. At present, the water quality of the river is poor. Except for the Haihe River which is IV-type water, the rest are V-type and inferior V-type water. The ecological space in the area is mainly agricultural land, covering 166 square kilometers, 45 square kilometers of forest and grass land, and 56 square kilometers of green space in the construction land. Overall, the blue and green space in the region is squeezed and fragmented. There are serious water shortages in terms of resources and water quality; the water quality is poor, and there is a lack of ecological water and poor communication of water system and lack of circulation. There is extensive land subsidence, resulting in ecological degradation and serious soil salinization pollution; 60% of the area are mild salinization areas and 40% are moderate and severe salinization areas. The ecological level is low and forest land accounts for only 7%. Crop cultivation is large and the types are few, planting time is short, and the biodiversity is reduced. There are still a large number of iron and steel metallurgical enterprises in the area, causing air pollution. Many manufacturing enterprises occupy a large amount of land, with low output. The level of urban and rural planning and construction is uneven. Many cities and towns have low construction intensity, high building heights and lack characteristics. Large-scale road traffic and municipal infrastructure divide ecological space at will. In addition, there are still traditional high-impact facilities such as Junliangcheng Thermal Power Plant, with low level of village construction and a lack of pastoral features.

"Through the analysis of the current situation, we further realize that strengthening the management and control of the intermediate area of the two cities and building a green ecological barrier are strategic decisions to restore the ecological environment of our city, avoid the blind expansion of the city and expand the green space. At the same time, it is also an important measure for our city to implement the requirements of ecological civilization, shoulder the responsibility and mission of Tianjin in the coordinated development of Beijing, Tianjin and Hebei, actively integrate into the regional ecological system of Beijing, Tianjin and Hebei, construct a complete regional ecological spatial layout, and create a good 'ecological moat' for the capital. It is also an important measure to optimize the layout of Tianjin's cities and towns and the ecological environment system. In order to carry out high-quality planning, we have organized expert workshops and carried out seven thematic research projects on ecology and water system, etc...

"The green ecological barrier planning forms the overall ecological spatial pattern of 'one axis, two corridors, two belts, five lakes and multiple lines', showing the landscape characteristics of 'water abundance, green lush, forestation and patches'. Among them, 'one axis' is the Haihe ecological corridor, 'two corridors' are the two Guhai'an coasts and wetland corridors in the east and west, 'two zones' are the Yongding New River ecological belt and Duliujian River ecological belt, 'five lakes' refer to Dongli Lake, Huanggang Reservoir, Guangang Reservoir, Tianjiahu Lake, Dasunzhuang Wetland, and 'multiple lines' refer to the network corridor formed by dense rivers and green belts of traffic protection lines in the planning area. The planning focuses on eight aspects. The first is to rebuild the ecological environment of Jingu waters, implementing the concept of sponge city, establishing a multi-source water security system to guarantee ecological and agricultural water consumption, improving surface water quality, reaching the main target of IV-type water standard from the current inferior V-type category, strengthening the connection of water systems and realizing the network of North-South water systems. Secondly, to reconstruct the green forest barrier, strengthen afforestation, and greatly improve forest coverage; using the traditional 'raised field' method to concentrate afforestation and greening in the planned ecological corridor; to strengthen forest design to suit local conditions and suitable habitats, and select willow, elm, locust, and other local trunk trees. Thirdly, to reproduce the local 'town of fish and rice' tradition, keeping the existing basic farmland scale unchanged in the barrier area, actively developing agricultural economy, promoting small-scale rice growing, and encouraging the growing of cash crops. Fourthly, to reconstruct the nurturing function of lakes and wetlands, to give full play to the ecological purification function of wetlands, reduce organic pollutants, and add wetlands such as Jinzhong River Wetland, Guhai'an coastal ecological wetland corridor and Weinanwa wetland with a total land area of about 29 square kilometers. Fifthly, to reproduce biodiversity, actively create habitats suitable for animals and plants, construct waterfront habitats, retain and build new ecological corridors, and provide living space for organisms. Sixthly, to return to the poetic pastoral life, ensuring the formation of ecological corridors, and reclaim cultivated land and greening. For villages that have not yet been included in the construction of urbanization, to implement the strategy of Rural Revitalization and retain the sense of rural quaint. Seventhly, to build green valleys with high quality and development, ban scattered and dirty factories and inefficient industrial parks, implement ecological restoration projects, lead regional innovation and transformation development with green industry development, and implement the transfer of inefficient manufacturing industries. Eighthly, to build infrastructures with low impact, create an environment that is close to nature and facilitates green travel, build a greenway system, and build a five-horizontal and two-vertical parkway framework surrounding the green ecological landscape. At the same time, in order to ensure the scale of blue and green space, to optimize the original planning trunk road system by reducing the scale and adjusting the section, and construct small-scale low-impact infrastructures to achieve low-carbon emissions.

"In order to speed up the planning and implementation of the ecological barrier belt in the southeastern part of the capital, it is planned to take three years to implement the ten projects of the afforestation of the green ecological barrier, water environment control, prevention and control of air and soil pollution, ecological restoration, improvement of human settlements and rural revitalization, high-standard farmland construction, and road traffic and infrastructure construction. The project will focus on the construction of the Haihe Eco-Core, Binhe Green Corridor, Guhai'an Eco-Green Corridor, Jinnan Weinanwa Wetland Area, Binhai Guangang Eco-Forest, Jinnan Green Core, Jinzhong River Wetland Area and Xiqing Eco-Green Corridor. After the implementation of the above projects, the proportion of blue and green space in the green ecological barrier area will reach 65%. As an important part, it plays an important role in constructing the ecological barrier belt around the capital and improving the ecological environment of the Beijing-Tianjin-Hebei region."

Chief. Tu, Landscape Planner

Next, Chief. Tu, a landscape planner, makes a speech. He further introduces the ecological landscape design on both sides of the Haihe River in the ecological barrier area between the two cities. "Generally speaking, the main stream of the Haihe River is a relatively independent enclosed area in the Great Haihe River Basin. From 72 kilometers below the Jingang Bridge, it receives precipitation from about north to Yongding New River and south to Duliujian River. For a long time, this region has formed the water network system of the main stream of the Haihe River as the main branch, with side river as branches, small rivers and ditches as capillaries, and lakes, reservoirs, fish ponds and pits as storage areas. Since the Reform and Opening Up, the urbanization has made great progress. As large construction units gradually settled in the areas of main stream of the Haihe River, the pits, ponds and ditches have been filled up one after another, and the original dense water system structure has been partially destroyed. Because of water shortage, various ecological problems have come up one after another, and the ecological development of the region has reached a critical point.

"The Planning of Green Ecological Barrier between Two Cities in Tianjin defines the layout structure of large-scale ecology and urban construction, and sets up the overall ecological spatial pattern of 'one axis, two corridors, two belts, five lakes and multiple lines'. Combining with the making of the ecological barrier planning, we did the landscape design of the green center on both sides of the Haihe River axis. The length of the Haihe River that is located within the green ecological barrier area between the two cities is about 23 kilometers long, which is the middle reaches of the Haihe River. The main vein of the area is the Haihe River, which forms the main axis of ecological landscape with wide and narrow changes. Combining with the tributaries of the Haihe River, the main veins of the landscape are determined to form a fishbone-like spatial structure as a whole.

"What is the future of the middle reaches of the Haihe River? Here we use Paris as a case for reference. Paris has a large forest park on both sides of the east and west. To the west is the Bois de Boulogne Park. It is located between Neuilly-sur-Seine and Boulogne-Billancourt. Its longest part from north to south is 3.5 kilometers and widest part from east to west is 2.6 kilometers. Its area is 8.46 square kilometers, which is equivalent to one twelfth of the total area of Paris. The eastern part is the Bois de Vincennes, with its longest part five kilometers from east to west, and the widest part three kilometers from north to south. It covers an area of 9.95 square kilometers. The two forest parks have various intersecting paths, many trees and many ecological functions, such as water and soil conservation, air purification, urban heat island effect reduction, and biodiversity maintenance. They play a good role in ecological regulation and are regarded as the two lungs of Paris. The parks include zoo, amusement park, racetrack, football field, gymnasium, bicycle racetrack, temporary exhibition hall, flower park and other functional areas, covering leisure, tourism, sports, art and other activities. In addition to the basic urban functions, the parks also hold activities with international influence, having the Louis Vuitton Foundation Art Center designed by Frank Gehry, the Mamordan Art Museum, the Roland Garros Tennis Center, Longshang Racecourse and so on.

"The Figaro Cross-Country Race, launched by the newspaper Le Figaro, has been running in the Bois de Boulogne for more than 40 years. Over the past 40 years, the total number of participants has reached 1.2 million, and it is the largest mass cross-country obstacle race in the world. Speaking of this, Chief. Tu looks up from the stage, his face showing pride and excitement. He used to study abroad in France and likes extreme sports. He has experienced the Figaro Cross-Country Race firsthand. After returning home, he often participates in cross-country races all over the country and maintains the long and elegant physique of long-distance runners. His major in France was landscape design, and his mentor advocated landscape design in harmony with nature.

"The middle reaches of the Haihe River is the core function area of the ecological barrier between two cities. There are about 12.5 kilometers from the western boundary of the control area to Gegu. The 12.5 kilometers section of the Haihe

River stretches from north to East Xiangfeng Road and south to Tianjin Avenue. The areas over 4 kilometers wide are currently undeveloped and leave infinite possibilities for future development. Compared with the length of 4.1 kilometers of Central Park , New York in the United States, 12.5 kilometers of the Haihe River has the volume for at least three Central Parks. In the future, the 12.5 kilometers of the Haihe River will rely on the ecological barrier to become an great park to ensure urban ecological security and an international quality urban green core. Based on the ecological pattern and guided by the ecological principal, we should strengthen the construction of ecological infrastructure, carry out multi-directional ecological restoration, and form a clear landscape structure. That is to say, in the landscape unit divided by fishbone spatial structure, according to the needs of urban development and function, we should set up ecological parks with different functions.

"The Haihe River landscape belt which serves as the core will form a 200-meter to 700-meter wide, unilateral riverside landscape belt with changes in width, and become the urban green core between the two cities. The north bank covers an area of 3.91 square kilometers, the south bank 1.97 square kilometers and the river channel area 2.12 square kilometers, forming a total of more than 8 square kilometers of green core, including waters, woodlands, fields and gardens for sightseeing and exhibition activities, parks for recreation and sports, football, tennis, and basketball courts, racetracks for horse racing and horse riding and other recreational areas. The area strengthens the connection between the north and south shores through tourist paths, and becomes a comprehensive landscape belt with unified ecology and vitality. Combining with the 7.7-kilometer road on the top of dike on the north bank and the 7.3-kilometer road on the top of dike on the south bank, we will construct the top dike line to meet the needs of cycling and walking activities; taking the river course into consideration, we will construct a water tourist route and a land tourist route, combine the two routes through wharfs and tourist stations, and combine the routes with the surrounding activity venues and nodes, covering the whole middle reaches of the Haihe River and expanding to the downstream and other areas on both sides of the Haihe River. The twelve landscape corridors are the ecological corridors extending north and south along riversides in the middle reaches of the Haihe River. By upgrading the landscape of the river branches, the regional ecological and recreational network is constructed to achieve the infiltration of the green belt into the north of the Haihe River and increase the regional ecological carrying capacity. Relying on the Haihe River, its tributaries and main roads, the area has been divided into 24 relatively independent ecological units, with rivers and roads as boundaries, forming ecological islands of different sizes. According to the current plan, the National Convention and Exhibition Center, located at the western end of the south bank of the middle reaches of the Haihe River, is an important independent ecological unit, which will strongly promote the construction of other units. On the riverside of the south bank of the Haihe River, the ecological units will be the areas that undertake various activities at international or city levels; on the riverside of the north bank of the Haihe River, the ecological units will be the reserve for future major events in Tianjin. In the recent future, as ecological conservation areas, the ecological units will undergo large-scale greening and become a city nursery.

"There are infinite possibilities for the north and south bank of the middle reaches of the Haihe River in the future. The south bank of the Haihe River can rely on the National Convention and Exhibition Center to become an area for holding small and medium-sized international-level activities and a core area for Tianjin's foreign exchange. A large ecological unit group with a total area of 180 hectares, which can be compared with the 187.7 hectares occupied by the Shanghai Expo Cultural Park, can become a possible venue for holding the Expo in the future. The medium-sized ecological unit group with a total area of 33 hectares can be analogous to the Tennis Center of Qizhong Forest Sports City, Shanghai, and the small island with a total area of 11 hectares can be analogous to the Roland Garros Stadium in Paris, becoming the venue for future tennis tournaments.

"The south bank of the Haihe River can also rely on its location between the two cities to become the city activity center of the two cities of Tianjin in the future. The ecological units can flexibly deal with future possibilities through the integration of different adjacent islands. For example, analogous to the 14.5-hectare Paris Zoo, the 15-hectare ecological unit can become the future Tianjin Zoo. Or analogous to the current 63-hectare Tianjin Zoo, the 67-hectare ecological unit group can become the new site of Tianjin Zoo in the future. Similar to the area of 20 to 80 hectares of a single island in Xiamen Garden Expo Garden, 67-hectare of ecological unit group can also become the future area for garden exhibitions and art exhibitions. Analogous to the 217-hectare Shanghai Chenshan Botanical Garden, the 118-hectare ecological unit group can become the future Tianjin Botanical Garden.

"The north coast of the Haihe River will be the reserve for future major events in Tianjin, and the area where Tianjin may host major international events such as the Olympic Games and the Garden Expo in the future. For example, analogous to the 680-hectare Beijing Olympic Forest Park, the ecological unit on the north bank of the Haihe River covers an area of about 770 hectares, and can become the Olympic Forest Park of Tianjin. Analogous to the 275-hectare Olympic venue in London, the 250 hectares in the west of the north bank of the Haihe River can become the venue for the future Tianjin Olympic Games. Analogous to the 66 hectares occupied by the International Garden Exhibition, the 145 hectares in the east of the north bank of the Haihe River can become the venue for the future Tianjin Garden Exposition.

"In the near future, we will adopt an ecological strategy of gradual restoration. The north bank of the Haihe River is an area of ecological conservation, afforestation on a large scale will be implemented and it will become the core ecological area of the ecological barrier area between the two cities. Because Tianjin generally has a low-lying terrain and is located in a saline-alkali zone, the roots of directly planted trees will penetrate into the shallow surface water layer with high saline-alkali content after several years' growth, which makes them difficult to survive. Therefore, we draw lessons from the traditional 'raised field' practices in Tianjin, heaping the soil of the lowland on the high ground, turning the lowland lower into fish ponds, turning the highland higher into dry land platform fields, planting trees on the platform fields, and ensuring the survival rate of trees. At the present stage, we adopt the method of planting low-diameter trees with high density to build the ecological conservation area into a reserve forest land. With the growth of trees in the forest land, seedling thinning will be gradually carried out. The region will be built into a plant nursery base of the barrier area, ensuring the supply of seedlings in other areas, so as to maintain the garden and create a low-carbon and high-efficiency landscape of production type and to realize the symbiosis of parks and the city. In this way, we refer to the practical experience in Haihe Education Park area, and arrive at the calculation that the cost of a single square meter would be about 174 yuan, and the total investment in the ecological units of nearly 15 square kilometers on both sides of the Haihe River will not be more than 2.6 billion yuan.

"The Ecological Barrier Area between Two Cities adheres to global vision, international standards, high positioning and Tianjin characteristics, integrates green development and innovation drive into one, takes ecological theory and people-oriented concept as the basis of planning and design, and forms a high-level 'Two-City Ecological Barrier, Tianjin-Tanggu Green Island' with Tianjin characteristics."

45. Design with Nature: People's Parks

Landscape planner Tan Ming

Tan Ming, a landscape planner of TUPDI, is introducing the urban design of the "One Green Ring and Eleven Parks and Its Surrounding Areas" at the forum of "Urban Design Theory and Practice". This content also happens to be about ecological planning, as if this were an exchange gala for landscape planners. Tan Ming was born in the eighties and has been engaged in landscape design since he graduated from Tianjin University. He goes out to the field all year round and is tanned. In recent years, after doing a lot of practical projects, he began to think about the future direction of Chinese landscape design, just as Chinese architects kept asking themselves what Chinese contemporary architecture is.

"The 'One Green Ring and Eleven Parks and Its Surrounding Areas' is the key project among the urban design pilot projects in Tianjin, and we are honored to be able to undertake this work. The so-called 'One Green Ring and Eleven Parks' refers to eleven urban parks inside and along the outer ring greenbelt of the central city. In fact, we have been working on this plan for five years, from the initial construction of the outer green belt and walk path to the urban design of eleven parks and surrounding areas, including the input calculation of the implementation of parks and relevant infrastructure. However, the purpose and significance of this urban design pilot project goes beyond simple greening and the construction of footpaths, and also beyond the solely urban design of the surrounding areas of parks. It should be launched from the perspective of urban structure remodeling in the central city in the light of the making of the territorial spatial master plan of Tianjin. In order to do a good job on this plan, we have further studied historical materials, re-read a series of classical works of masters, conducted a large number of field surveys, interviewed local people, and visited the old planners who participated in the planning and construction of the outer green belt.

"Tianjin is located at the lower reaches of nine rivers. Although the growth of trees is generally poor in saline-alkali land, there are many pits, ponds and lakes, so it has the reputation of 72 Gu (bodies of water). It is the land where the city and water blend. In the past, because of the slow development, many people said that Tianjin was not like a municipality directly under the Central Government, but more like a large rural area. In the past 20 years, in the process of transforming "large rural area" into a metropolis, the natural beauty of Tianjin has gradually faded; the city has become larger and larger, and the green paradise carrying happy childhood has become more and more distant. In the field research, we interviewed Brother Er. Now let's start with the story of Brother Er's family."

Brother Er

"Brother Er is a native of Tianjin, born in the suburbs of the city, which is a village on the edge of today's express ring road. Because of the construction of the outer ring road in the late 1980s, Brother Er's village became an urban village. At that time, urban villages could be seen everywhere, which might be one of the reasons why Tianjin was nicknamed "large rural area" at that time. Although it was the suburbs, it was not far from Tianjin's urban commercial center, and not far from his backyard was a vegetable field with ponds, trees, cicadas and frogs, and in the distance were rivers, wetlands, farmland and woodland. Talking about the green paradise in front of his home at that time, Brother Er was full of happy memories. Later, due to the demolition, his family moved into a flat in a highrise building; although living conditions improved, the fun of going out of the house into the vegetable field was no more.

"The change Brother Er's family experienced is actually a microcosm of the changes of living and natural environment in Tianjin after large-scale urban expansion, transformation of urban villages and urban renewal. The vegetable fields and ponds submerged in urban concrete forests represent the gradually increasing distance between cities and ecological spaces after large-scale urban construction. Today, the construction of the central city has already overflowed out of the expressway ring road, and will fill the area between the expressway ring road and the outer ring road, and then connect with the four districts around the central city if going further outward. The trend of the 'spreading pancake' style of expansion is more and more evident. Inside the outer ring, the original urban ecological space, such as pit wetlands, rivers and beaches, has been further compressed, and the integrity, stability and diversity of the ecosystem have been greatly damaged. At the same time, after the new round of urban renewal, the urban construction density and development intensity are increasing day by day, approaching the level of 30,000 people per square kilometer, while the newly built urban parks is rare during the same period. The per capita urban public green space is less than 4 square meters, far below the national standard, and the gap is wider compared to developed countries abroad. Moreover, the road protection green belt occupies a large proportion in the newly built green space, and the protection green belt is generally too narrow to be considered as suitable green space for human use, which further exacerbates the contradiction of more people and less green space.

"More people, less space, overcrowded population and aging communities have become realistic problems that need to be solved urgently. After large-scale urban construction, the reduction of ecological space has become an important factor restricting the high-quality development of the central city. The spatial structure of the central city needs to be adjusted, and the living environment of the city needs to be improved urgently. Under the background of Tianjin's green ecological development strategy for the next 30 years, it will become an important strategy for urban spatial development to remold the urban ecosystem and restore the beautiful scenery of Tianjin in the past–the urban characteristics of the lower reaches of nine rivers–by integrating and upgrading the existing ecological space in the central city, and connecting the ecological system inside and outside the outer green ring belt.

"In the new century, the comprehensive development strategy of the Haihe River has led to the economic and social development of Tianjin; in the new era, by 2050, in the next 30 years, the establishment of the natural ecological system in Tianjin's urban area, which is based on the 'one green ring and eleven parks', will promote the transformation and high-quality development of the city. At the beginning of the 'one green ring and eleven parks' project, senior expert B who participated in the making of the 1986 edition of Tianjin City Master Plan and the revision of the Master Plan of the New Century brought up the view that the 'one green ring and eleven parks' should be the Emerald Necklace of Boston rather than the green belts of London. Cities and natural ecology should be a mutually supportive whole, not a division or an opposition."

Old Senior Expert B

In reviewing the making of the 1986 edition of the City Master Plan, B said, "The original planning of the outer ring road and the 500-meter green belt outside the outer ring road was based on the experience of the green belts planning in London. The original purpose was to delimit a clear growth boundary of the central city and build a green barrier for the central city. At that time, it was advanced and practical, and Tianjin was the first city in China to do so. The 500-meter green belt outside the outer ring road and the green wedges along the main rivers have become an important part of the urban structure of the central city, and have also established an essential ecological spatial framework. Over the years, the previous editions of Tianjin City Master Plan have continued and improved the outer green belt and green wedge planning, and gradually formed today's urban park system consisting of the outer green belt and green wedge–'one green ring and eleven parks'. Thirty years later, although the construction of the outer ring green belt is not ideal, most sections have only built 200-meter-wide forest belts, there are still many villages, industrial buildings and illegal construction within the 500 meter range, and only four out of the eleven parks have been built, but the basic functions of the plan have been realized and played its role.

"In today's view, there are also inadequacies in the planning. The outer green belt emphasizes too much on the protection function and has its own system. As the 'growth boundary', the role of the outer green belt is always independent of the ecological and spatial system of the central urban area, and fails to play a greater role beyond the "growth boundary". Although the "one green ring" learns from the 'green belts' in London, it is only 500 meters wide, which is only one hundredth of the width of London green belt. Its spatial scale is insufficient and is difficult to breed a complete ecosystem. For Tianjin's green ecological strategy and the construction of ecologically livable cities, 'one green ring' should have had been positioned higher and play a greater role.

"With the adjustment of the traffic functions of the outer ring road and the establishment of the green barrier between the two cities, the integration of the central city and the four surrounding districts becomes inevitable. More than 1300 square kilometers of space within the highway ring road needs to be reshaped. This larger spatial scale contains more natural and ecological elements, including 11 large-scale urban parks within the outer ring road, 5 suburban parks outside the outer ring road, large-scale agricultural and ecological land, and many rivers as green corridors, which provide opportunities for building a larger urban ecosystem as well as for establishing an ecosystem-guided healthy spatial structure system in central urban areas.

"In this sense, 'one green ring and eleven parks' is no longer London's 'green belts', but Boston's 'Emerald Necklace'. It is a more stable and diverse ecological structure, more in line with the urban spatial system, and provides strong support for the living environment. Supported by such a green ecosystem, the spatial structure of the central urban area within the highway ring road will be further integrated and remodeled."

Tan Ming

"In order to create a good urban design, we have referred to the experience of many developed countries and review several classic works, including Ian McHarg's *Design with Nature* and Robert Yaro's *A Region at Risk–the Fourth Tri-state Regional Plan of New York*. We recognize the importance of rebuilding the harmonious coexistence between man and nature, cities and nature, and of improving the quality of life of residents in parks and sports venues. Under the background of Tianjin's territorial spatial master planning, the optimization of urban spatial structure is first of all the reconstruction and optimization of ecosystem structure, putting ecological space in a position equal to or even more important than urban construction space. In fact, the urban design of the 'one green ring and eleven parks' and its surrounding areas is an ecological system reconstruction based on the main urban area of the city. It is a way to establish the ecological-oriented spatial integration and structural optimization between the central urban area and the four districts around the city. It is also a planning practice to reshape the urban-green relationship and the Town-Country relationship.

"McHarg wrote in *Design with Nature*:

> Don't ask me about your garden, and don't ask me about your flowers or your dying tree. You can treat these questions carelessly. We're going to tell you about survival. We're here to tell you the way of the world. We're here to tell you how to act wisely in front of nature. We don't care about rocks or plants themselves, because they can take good care of themselves. We care about people. People are nature's as well as their own most dangerous enemy.

We must not only recognize that human beings are an integral part of nature, but also respect the uniqueness of human beings so as to endow them with special survival values, responsibilities and obligations. And that's the idea of ecology, or human ecological planning. ... These natural elements, together with human beings, are now cohabitants in the universe, participating in the endless process of exploration and evolution, vividly expressing the loss of time. They are necessary partners for human survival, and now they work with us to create the future of the world. We should not separate human beings from the world, but combine human beings with the world to look at the problems. May this be the truth.

Let's abandon the simplistic and divisive approach to problems and give it the unity it deserves. May people abandon the self-destructive habits they have formed and show the potential harmony between man and nature. The world is rich. To satisfy human's hope, we only need to understand and respect nature. Man is the only conscious creature capable of understanding and expressing. He must be the administrator of nature. To do this, design must be combined with nature.

The relationship between man and nature: man is not only a unique species, but also a gifted consciousness. Such a person knows his past, lives in harmony with everything, continues to respect them through understanding, and seeks his own creative role. ... We should believe that nature is evolutionary, that all kinds of factors in nature interact with each other in some patterns, and that there are certain restrictions on the value and possibility of human utilizing nature, and some aspects should even be prohibited. Perhaps the most reasonable way is to investigate different environments and find out the possible and unsustainable extent of different environments as people put them to general and special uses.

Nature is like a big net, all-encompassing, and its various components are interacting, but also predictable and mutually restrictive. It forms a value system, which itself has many possibilities and limitations available to human beings. We should quantify social factors, aesthetic factors, biological factors and so on in a comprehensive way. ... From the eco-value system, the transition from 'I-it' to 'I-You' is a great progress, but 'we' seems to be a more suitable term to describe the relationship between human and the ecosystem. The economic value system must be expanded to include all the evolutionary processes of biophysics and the related systems of human desire.

Human beings should become enzymes in the biological world, that is, managers in the biological world, improving the creative adaptability between human beings and the environment, and realizing the combination of human design and nature. ... Natural imagination is an interactive and dynamic process of development, reflecting various laws of nature, and these natural phenomena provide opportunities and constraints for human use. Human adaptation to nature not only brings benefits, but also costs, but the natural evolution process does not always have value attributes; there is no comprehensive calculation system to reflect all costs and benefits. Natural evolution is holistic, and human intervention is local and increasing. Despising natural evolution is unlikely to lead to long-term revenue growth, but it can be confirmed and proved that the result of despising nature must ultimately bear a lot of costs.

City is a form, which is evolved from address and biology. It is a comprehensive product of natural evolution and artificial transformation to adapt to nature. It is also necessary to regard the historical development of cities as a series of cultural adaptations reflected in urban planning and the individual and group buildings constituting the city. Some of these adaptations are successful and thus preserved, and others are not. ... Search for the basis of urban characteristics–from the characteristics of nature and man-made cities, select expressive and valuable elements that limit and provide opportunities for new development, which reveals the basis of urban form.

"In the urban design of the 'one green ring and eleven parks and its surrounding areas', we focused on rebuilding the integrity and diversity of the ecosystem in the central city and main urban areas. According to the existing ecological factors, the selection of natural protection sites and protection factors was carried out, and the double green ring and linkage corridor system is constructed to promote the integration of a green city. First, the eleven gardens around the first ring and the surrounding areas are connected with the inner river and water system of the central city to play the storage function of a sponge city; second is to connect the one green ring and eleven parks and the surrounding areas with the urban greenways and green corridors. Through the connection of rivers, green corridors and surrounding country parks as well as rural agroforestry, outer green belt and highway ecological ring, the connection between the green barrier between the two cities is strengthened, and the ecological pattern of internal and external connection and green infiltration is formed. In the future, three 300-kilometer-long greenways around the city, the 500-meter green belt, 8 country parks and 11 city parks by the outer ring will be built. The connection with the outer suburban parks will be realized through 12 greening corridors.

"The 'one green ring and eleven parks' area will be an important supplement to the construction of green barrier between two cities, and one of the key areas for the construction of urban ecology and the improvement of the quality of urban human settlements. The outer ring green belt and urban parks are the green artistic skeleton connecting the outer ring suburban parks and optimizing the layout of the central city and main urban areas. They are the "emerald necklace" of the central urban area of Tianjin and the Green Chain of Tianjin.

"At present, the construction level of the 'one green ring and eleven parks' is relatively low. As the 'one green ring', the greening level of the outer green belt is not high, and there is a lack of human activity space. Only four of the 11 parks have been built, of which The Pillow Park has just been built and put into use. There are seven parks that need to be built. They are currently still wasteland or land to be demolished and reclaimed. In the long run, the outer green belt should also be a city park. We should speed up the construction of green belt parks and the seven other parks by the outer green belt, and promote the construction of new residential communities by implementing the strategy of greenery investments. These new communities provide green open spaces for housing and parks to meet people's needs for a better life. Alexander Garvin pointed out in his book *The American City: What Works and What Doesn't* that urban parks play an important role for people and cities, and they should be built into people's parks.(Garvin, 1995)"

46. Town-Country Magnet: Suburbanization and Town-Country Symbiosis

Tan Ming

"The main urban area of Tianjin consists of six districts in the city and four districts around the city not counting the ecological barrier between two cities, covering an area of 1300 square kilometers. The current population is 9 million, accounting for nearly 60% of the total population of Tianjin. It can be divided into three circles: central circle, ecological livable circle and garden city circle, which correspond to the natural ecosystem. The central circle is the built-up area of the central city within the expressway ring road and the most dynamic central area of the city. It is characterized by urban renewal and is of a high-intensity and high-density spatial form. The ecological livable circle is the area between the expressway ring road and the outer ring road. At present, there is still much room for development. Here, the combination of living space and the large-scale urban parks are highlighted. It will generally undergo moderate-intensity development through the transit-oriented development (TOD), partly with high intensity development. The garden city circle is the area between the outer ring road and the highway ring road, reducing development as its main characteristics. This vast area, using low-intensity and low-density development mode and combining residential space and suburban ecological space, is the main area of suburbanization and also the way out to solve the existing problems in Tianjin's inner city.

The father of modern urban planning, Sir Ebenezer Howard, wrote in *Garden Cities of Tomorrow*:

> Our beautiful land, covered by the sky, blown by the breeze, warmed by the sun and moistened by the rain and dew, embodies God's love for mankind. The solution for people to return to the land must be a universal key, because it can open an entrance. Even if the entrance is only slightly open, you can see a bright future. All causes of population concentration can be summed up as "gravity", and a "new gravity" must be established to overcome the "old gravity". Every city can be regarded as a magnet, and every person a magnetic needle. Only by finding a way to construct a magnet whose gravity is greater than that of the existing city, can we effectively, naturally and healthily redistribute the population.

The Town-Country magnet is the new gravity. We need to break the polarity between urban and rural areas. Besides 'urban life' and 'rural life', there is a third choice of "Town-Country life". It can bring all the advantages of the liveliest city life and the beauty and pleasure of rural environments together. The reality of this kind of life will be a 'magnet'.

The city is the symbol of human society, and the rural area is the symbol of God's love for the world. The urban and the rural must marry to form a Town-Country magnet. It can enjoy the same or even more social opportunities as the city; it can make residents live in the beauty of nature; it can combine high wages with low rents and taxes; it can ensure that all people enjoy rich employment opportunities and bright prospects; it can attract investment and create wealth; it can ensure health conditions; it has beautiful houses and gardens everywhere; it can expand the scope of freedom, so that happy people enjoy all the best results of concerted efforts. (Howard, 1902)

"Our country's urban planning allegedly stemmed from Howard's idea of garden city, but in the past decades of practice, we have only focused on the city, neglected the rural areas, and split the relationship between urban and rural areas. There are many prejudices against suburbanization, which may be one of the reasons for the prominent urban problems in China. In the new period of national transformation and development, we must change the old urban planning concepts and methods, and pay attention to the Town-Country magnet. As a all-in plan, the territorial spatial master planning should be carried out based on the idea of the integration of urban and rural areas."

Suburbanization is the main way to promote urban-green integration and Town-Country symbiosis and to improve the living quality of Town-Country human settlements. In the past, our thinking was rigid, and we simply thought that suburbanization means a disorderly sprawl, occupying a large amount of land and not ecologically and environmental friendly. In fact, the suburbanization sprawl problem is only present in the United States, and suburbanization is an inevitable choice for people who yearn for the natural environment. Kenneth Jackson, a famous American urban historian, wrote in his book *Crabgrass Frontier*:

As a place to settle down, as a residential and business district outside the city, the suburb has a history almost as long as the human civilization, and suburbs are an important part of the city. However, suburbanization is not the same as the suburb. It originated in the United States and Britain and can be traced back to around 1815. It refers to a process. Peripheral areas have experienced systematic growth with speed far faster than central cities. When wind, manpower and water are still the main driving forces of civilization, the social status and economic strength of the suburbs are far below the cities. At that time, the 'suburbs' represented inferior traditions, narrow visions and dirty bodies. When railways, metros and cars are highly developed, suburbs become a beautiful home for people to fulfill their yearning for the natural life. When looking at the increasingly stratified and decentralized community geography of American big cities, we should not only see the transportation technology and powerful mechanical force released by the Industrial Revolution, but also see the development of new cultural values. (Kenneth Jackson, 1985)

Suburbanization is the objective law of urban development. In the past, we over-controlled the suburbanization, using the same planning and design model as the urban center, resulting in the suburbs being also full of high-rise buildings and the situation of 'neither urban nor rural'. From the perspective of spatial planning, suburbanization is not only about suburbanization itself, but also a major strategy to optimize the urban spatial structure and solve 'urban diseases' in the main urban area of Tianjin.

The main urban area of Tianjin can be seen as composed of three circles, each of which has its own problems, but cannot be solved by itself. The central circle within the express ring road covers an area of 145 square kilometers, with a current population of 3.4 million, a population density of 23.4 thousand people per square kilometer, and a per capita public green space of 3 square meters. It includes 14 historic and cultural blocks such as Wudadao and major functional areas such as CBD, central commercial district, cultural center and sports center, etc.. After the renovation of dilapidated bungalows, the overall quality of urban buildings has been greatly improved. Apart from all kinds of urban public buildings, most of them are multi-storey residences (mostly six story walk-ups) and high-rise residences under 100 meters high, which have been built since the 1980s. Among them, the old residences built before the monetization reform of the housing system in the late 1990s have serious problems. The total area of these houses is about 70 million square meters, and there are 3 million people living in them, accounting for more than 80% of the population in the region, which is half of the total population in the central city. These old residential communities are over 30 years old, with aging houses, old buildings and damaged pipelines; second, most multi-storey residential buildings do not have elevators, and with the increase of the proportion of the elderly population, its gets inconvenient for them; third, there is a serious shortage of parking spaces in the residential areas. In order to solve the aging problem of old residential communities, Tianjin has carried out two waves of large-scale renovation of old residential communities since 2012, which has improved living conditions and community environments. Recently, the State Council also issued a notice requiring all regions to carry out the renovation of old communities, including replacing pipes, increasing external insulation, adding elevators and so on. These measures will greatly improve the quality of life for the residents in the old residential communities. It is a project in the public's interest.

However, this kind of transformation cannot change the structure and type of the housing. Because they were constructed early, the unit floor area of these residences is relatively small, and the design of the unit type only meets the basic functional needs. Generally, there is no living room, only a small hall, and the kitchen and bathroom are relatively cramped, which is not suitable for the requirements of improving living standards. In this type of residential communities, the per capita residential building floor area is little and the population density is relatively high. In the future, according to the city master plan, by 2035 to 2050, the per capita residential building floor area needs to be increased and the living conditions need to be improved. Our preliminary estimate is that about 900,000 people will be relocated in order to increase the per capita residential building floor area from 24 square meters to 33 square meters. If we learn from the experience of Europe and carry out structural renovation of this kind of multi-storey residential buildings, we will improve the layout and increase floor area of each unit, mainly by merging two units into one or three units into two, thus to completely upgrade the living conditions. This is the biggest challenge faced by Tianjin's central city in the next 15 to 30 years. It is also a key battle for the transformation and upgrading of the city. To achieve this goal, we must solve two key problems: one is the technical feasibility of the multi-storey brick-concrete housing transformation; the other is the motivational mechanism of population dispersion, which must rely on the market forces and the government's policy support. There are old community renewal projects in this urban design pilot project. The Tiyuanbei project team has done a lot of in-depth research, and they will make presentation. You are welcome to learn more about it if you are interested.

The ecologically livable circle between the expressway ring and the outer ring road covers an area of 255 square kilometers, with a population density of more than 10,000 people per square kilometer. This area is next to the planned new seven large-scale urban parks, and there are more land left for development. About 45 million square meters of residential buildings can be built, which can accommodate one million people according to the high standard of 45 square meters per capita. These people may partially be the original residents living in the old communities in the central circle who purchase new units to improve living conditions, and may partially be immigrants, hoping to have better living conditions. The key to the success of this circle planning is to achieve a high level of planning and construction in accordance with the standards of 2035 and 2050. It is not only necessary to build a new types of housing to meet future needs, but also perfect supporting facilities like education, medical treatment and senior care, with good quality. The average household floor area of newly built commercial housings is at least 120 square meters, mostly four-bedroom layouts. New communities and parks are integrated; with large green space as the core, a community open space system is organized to form pedestrian blocks; open space network is also used as the plaform for the sponge city. Surrounding the green core of the parks, various new types of community with suitable intensity, density, diversity, order and green ecological livability should be arranged. The inner blocks near the parks are low-rise, high-density, multi-storey residential row houses or townhouses. Metro stations are surrounded by appropriate high-intensity, middle and high-rise residential buildings. The outer blocks are medium-intensity, small high-rise and high-rise residential buildings. We will strengthen the requirements of urban space control around the park, combined with the entrances and exits of the parks, rationally arrange public service facilities, and form a living circle.

For the 900 square kilometers of garden city circle outside the outer ring road, the current population is 3 million, mainly concentrated in the districts where the four suburban governments are located and large residential areas close to the outer ring road. In the future, it is planned for the population to grow to 4.1 million, with a newly added 1.1 million. This region has prominent ecological functions, with

50% of the land being blue and green space. Five country parks, a large area of agricultural ecological land and many rivers as greening corridors provide an opportunity for building a larger urban ecosystem, and also provide a possibility for building a healthy, ecosystem-guided spatial structure system in the main urban area. The planning strictly protects the open space system in accordance with the natural evolution process, arranges urban land use, carries out the overall design of residential space integrated with nature, forms a smooth transition between urban and rural areas, and strengthens the development intensity and building density control of Town-Country and natural integration. According to the suitability analysis of conservation-recreation-urbanization area, the total amount and distribution are controlled, and the low-rise and low-density residential groups or clusters distributed are integrated with natural ecology. The garden city circle is our suburb. We should make people have a broad understanding of the suburbs and experience the beauty of the integration of living environment and natural environment, so that a large number of central circle residents who want to improve living conditions and be close to the nature will naturally choose the suburbs, so as to realize the population dispersion of the central circle. According to the principles of intensive and economical use of land, no waste of land and high-quality human settlements, and drawing lessons from the planning and design methods of new urbanism, suburban garden housing with Chinese characteristics will be constructed. The core of the design is a town center, and the periphery is a rural courtyard houses with traditional Chinese characteristics. The buildings are covered in the tree leaves, and people and nature and land coexist in harmony.

McHarg wrote in *Design with Nature*:

> Travel from the rural to the city represents the evolution from land-dependent life to community life, and after a certain period of time, in turn, people travel from the city to the rural, which reflects the most original relationship between people and land. People are deeply aware of the need for the rural, and the strongest evidence of this need is the rush to the suburbs, which is the largest immigration activity in history. There are many valuable things in the city, such as extensive interpersonal communication, powerful social organizations, competition, stimulation, diversity and opportunities, but people's ancient memory makes them insist on going back to the land and nature. This alternation is necessary and beneficial. (McHarg,1969)

Many of us come from rural areas, especially our parents, most of who grew up in the countryside. Although we have lived in cities for a long time, we have deep feelings for the countryside. Our yearning for a better life is not only embodied in material space, but also in the yearning for land and nature. Suburbs are indispensable when it comes to meeting the growing aspirations of the people for a better life. Moreover, through the exploration of suburbanization, we can provide technical and appropriate product support for the planning and construction of villages and the improvement of human settlements in the vaster rural areas in the suburbs, and truly realize the Town-Country symbiosis.

Suburbanization needs certain supporting conditions, and transportation is the foundation. Historical experience has proved that the development of transportation is the premise of suburbanization. Urban rail network planning needs to extend outward reasonably and forms a grape-string layout along the rail line. Besides rail and public transport, the main mode of transportation in suburbs is car, which has door-to-door flexibility and convenience. Suburb layout should be closely integrated with urban traffic planning. Different circles in Tianjin's main urban area need to adopt different planning strategies, travel modes and road traffic layout patterns.

Section Chief Ma, Sino-Singapore Tianjin Eco-city Management Committee

The eco-city's Section Chief Ma specially takes time to attend the forum. Before the forum, she thought that she had been working in eco-city for 10 years and knew enough about ecological cities. She was most concerned about how to make a city vibrant, so she signed up for the sub-forum of "Urban Design Theory and Practice". Who knew the first speech she hears was about eco-city. She cannot help raising her hand to speak a question.

As you all know, Sino-Singapore Tianjin Eco-city is a strategic cooperative project between the governments of China and Singapore. Eco-city is located in the north of Yongding New River in Binhai New Area, adjacent to Beitang and TEDA, with convenient transportation. It is 16 kilometers away from the core area of Binhai New Area, 45 kilometers away from Tianjin's central city, 150 kilometers away from Beijing, with a total area of 31 square kilometers and a planned residential population of 350,000. Eco-city is an important living area serving the functional zones of Binhai New Area.

From the land use point of view, the eco-city land is salt fields and abandoned land, including a sewage reservoir. In terms of location and scale, the eco-city is located at the edge of the core of Binhai New Area, and it should be counted as a suburb. According to traditional planning theory, it is a large satellite city. In the overall planning of Binhai New Area, it is characterized as one of the city clusters in the network layout of city region, equivalent to a small city. The initial location was at the junction of Hangu and Tangshan, and then moved southward gradually. It is near TEDA, Tianjin Port and the core of Binhai New Area, with industries as support.

The master plan of the eco-city was done jointly by the Chinese Academy of Urban Planning and Design, the TUPDI and the Singapore side. In order to do a good job, we first determined the index system of Sino-Singapore Tianjin Eco-city, including 22 control indicators and four guiding indicators of regional coordination and integration from three aspects: ecological environment health, social harmony and progress, and economic prosperity and efficiency. The master plan is based on the KPI as well as the current situation of resources, environment and human settlements in the selected area, and highlights the concept of people-oriented. The master plan was used to guide the overall development and construction of the ecological city, and provided technical support and construction viability for replication, implementation and promotion.

Corresponding to the irregular topographical features and ecological background, the master plan of the eco-city retains the old river course and sewage reservoir of Ji canal, transforms the sewage reservoir into a clean lake, and plans the ecological valley that goes through the city, connecting ecological cells in series, which are residential groups. A residential group consists of 400×400-meter residential units with a cross-shaped walking system in the middle. The eco-city learns from Singapore's experience and plans to build a centralized community center. National Animation Park, High-tech Industrial Park and Eco-environmental Protection Industrial Park are planned to achieve economic development and industrial integration. New green concepts and technologies have also been widely used in road traffic and municipal civic engineering.

The location for Sino-Singapore Tianjin Eco-city was selected in November 2007 and its construction started in 2008. Ten years later, great achievements have been made in the construction of the eco-city. There is a lot of publicity in the media. Here I will not take your time to elaborate. I would like to summarize the implementation of the previous stage of planning. We have a lot of experience of success, and also a lot of confusion and problems. The first problem is spatial form planning, that is, urban design. Objectively speaking, although the master plan of eco-city used many new terms and the technical standard is high, the actual planning is still relatively traditional, which is the traditional method of residential area planning and satellite city planning. Although urban design and regulation have been completed later, little consideration has been given to the urban form. For example, the eco-city streets are well greened, but the streets do not carry much daily living functions. The original planned walkway system failed to connect, and eventually formed a large enclosed residential area of 400 meters by 400 meters. The height of residential buildings is generally high and the form is repetitive, which may be caused by the excessive FAR of the residential land. In fact, the low-rise and multi-storey buildings here are more popular. People choose to live in the eco-city mainly in the hope of having a livable environment. I think it is mainly related to the positioning of eco-city. Is it an independent medium-sized city? a satellite city? a suburban garden city?

The second major issue is regional coordination and Town-Country coexistence, Unlike Singapore's complete urbanization, Binhai New Area also has vast rural areas, and they are also present around the eco-city. In order to speed up development and reduce conflicts with the surrounding areas, the eco-city closed its 30-square-kilometer itself up, which had its own rationale at that time. The eastern part of the eco-city is the original Binhai Tourist Area, the southern part is Beitang, and the northern part is the central fishing port. Today, it is all under the unified management of the eco-city. The west and north of the eco-city are large areas of rural areas. In fact, cities do not exist in isolation. Regional coordination and Town-Country symbiosis are advantages, but we have not made use of them, in terms of the large ecological environment and industry. Eco-city is actually a commuter town now, although we don't want to admit it and strive for balance between employment and housing. However, the vast majority of the 80,000 people currently live in eco-city work in TEDA and core area of Binhai. With the development of the tertiary industry, eco-environmental protection and all-for-one tourism are becoming big industries. At present, the Binhai Aircraft Carrier Theme Park and the National Marine Museum in the Binhai Tourist Area are the popular tourist destinations. In addition to these locations, we need to figure out a complete chain so that tourists can stay in the eco-city for a period of time, which would include residential accommodations in agricultural areas. We now realize that small towns and satellite cities located in the suburbs and urban fringes should pay attention to the coordination of natural environment, social culture, economy and policies with surrounding areas, including the coordination of architectural forms, so as to achieve regional coordination and integration.

Finally, transportation is another major issue that we often think about. Connected to Tianjin urban area and Binhai New Area, the Sino-Singapore Eco-city has comparative advantages in external transportation. The intercity high-speed railway connecting to Beijing, Tianjin and other provinces and cities provides convenient means of travel. There are several highways connecting to the eco-city from all sides, including the Beijing-Tianjin Highway) to the south of the eco-city, the Tianjin-Hangu Expressway to the north of the eco-city, and the Central Avenue and the Coastal Avenue to the east of the eco-city. However, it is not as convenient to travel between the eco-city and the Binhai core area and the Tianjin central urban area. At the beginning of location selection of eco-city, it was proposed to plan and construct rail transit, and preliminarily considered to extend the Tianjin-Binhai light rail to eco-city. However, the extended length of the Tianjin-Binhai light rail would exceed the reasonable scope, and the operation time would be too long, so it is not very suitable. In 2009, Binhai New Area planned a rail transit network: the city Z4 line along the Central Avenue, with three stations within the eco-city, reaching the Yujiapu High-speed Railway Station directly; and trams would be set up along the eco-valley in the eco-city. The plan was eventually approved by the Singapore side and incorporated into Tianjin's rail network planning. However, it was not until 2015 that the National Development and Reform Commission approved the recent construction plan of Tianjin railway. Construction has been slow for various reasons since the beginning, and has not yet opened. So now the traffic congestion problem of eco-city is relatively prominent. With the efforts of all parties, Binhai Metro Z4 line will be opened soon. We have learned that the construction of the Z2 line connecting to Tianjin International Airport, Airport Economic Zone, High-tech Zone and the west area of TEDA is also under construction. But I wonder if the traffic congestion problem in eco-city can be completely solved after the railways are running. I think there will certainly be an improvement, but I'm afraid it can't be fundamentally solved. Why? On the one hand, the slow railway construction has resulted in people forming the habit of driving; on the other hand, there is also the direct reason of the imbalance between employment and housing. Fundamentally speaking, we still haven't accepted the objective fact of the popularity of automobile and the planning is too idealistic. Therefore, I think that we should not only vigorously develop public transport, but also face the seriousness of the popularity of cars. With the maturity of new energy vehicle technologies, the problem of pollution of vehicle emissions has been basically solved. The key is how to respond wisely, and modify the road traffic policy and do a smart and intelligent transportation plan.

Chapter VIII

A More Humanized City

47. Beyond Mobility: The Transit Metropolis + Car

Director Zhong of Traffic Planning Department of Bohai Urban Planning and Design Institute

The impromptu speech of Section Chief Ma of Eco-City evoked the resonance of Director Zhong. In accordance with the agenda of the meeting, he was the next speaker, and he had been preparing silently in his mind before he spoke. Director Zhong is a professional in traffic planning. He has participated in many major plans of Binhai New Area for more than ten years and has accumulated rich experience. He likes to read and has a wide range of readings. In fact, he has read more books on urban planning than on transportation planning. Some time ago, he made an important decision to ask for leave from the institution leaders, to suspend salary and stay in office, to go to UCLA at his own expense as a visiting scholar for a academic year. He spent ten months investigating the actual traffic situation in many cities and regions in the United States, and read many books with great results. At the same time, after careful consideration, he put forward his own ideas for China's current transportation planning. Before the meeting, he had exchanged views with Dr. Song Yunfei of "Yunfei Transportation Consulting Center" in Binhai New Area, and they were not identical on some key issues. Therefore, Director Zhong has been a little worried about his speech. Just after hearing Section Chief Ma's speech, he feels a little more confident. He wanted to introduce his point of view in more vivid language.

"At present, the ranking of urban traffic congestion is very popular in China. Congestion index has become the criterion for evaluating the function or mode of urban traffic. Some cities also regard the result of congestion ranking as a reference for the level of urban governance. In the same period, many foreign cities have made great strides in the construction of humanized cities, such as the "Street fight" campaign in New York. Last year in this lecture hall, Ms. Janette Sadik-Khan, a former director of the New York Transportation Department and author of Street Fight, gave a lively lecture. This time I went to the United States to study and had a special experience in Times Square, New York. It felt very different. I wonder how you think of these two phenomena."

"Transportation is very important for urban development. From the the history of the contemporary urban transport development in the western developed countries represented by the United States, it has gone through the process from public transport to cars, and then to pedestrianization. We can clearly see this law of development. Nevertheless, it seems that we are always unwilling to acknowledge the role of cars, or to solve our current problems blindly with the current concept of transportation planning in the United States. Traffic in the United States and other western developed countries has passed the era of rapid development of cars, entering a stable period, while car traffic in China is still in a period of rapid development. China's transportation planning has been like the "emperor's new clothes" for many years. It has been shouting for vigorous development of public transport, restricting cars and ignoring the existence of cars. But the actual result is that China's car production and sales continue to maintain the world's No.1. This inconsistency of words and deeds leads to the deviation of urban planning, traffic planning and actual traffic development in our cities, especially in big cities. Traffic congestion is very prominent in some cities.".

"At present, the only way to solve the urban and rural traffic problems in China is to combine the public transport strategy, car strategy and pedestrian strategy, and put equal emphasis on the three at the same time. Urban agglomeration provides fertile soil for the progress of transportation technology, and the breakthrough of modern transportation technology promotes the great development of the city and promotes the formation of a specific spatial form of the city. Under the influence of transportation system, urban spatial form has undergone a process from "pedestrian city" to "rail city" to "automobile city". Some scholars divide urban spatial form into three stages: traditional pedestrian city, bus city and automobile city. In fact, there are corresponding problems in different stages. Today, most cities have formed a state of blending of three forms, so we need a comprehensive transportation development strategy."

"First of all, public transport strategy. We have always advocated a modern transportation system dominated by public transport, but the planning and design of public transport is not thorough and meticulous. We can briefly review the history of public transport development. Modern public transport appeared in the 19th century, and its origin can be traced back at least to 1826. A four-wheeled carriage service connecting the city centre appeared in a bathhouse in the suburb of Nantes in Northwest France. In 1829, the British George Shillibeer's Omnibus appeared on the streets of London. In 1831, the world's first steam bus began operating. Soon after, the German company Mercedes-Benz produced a bus powered by a gasoline engine instead of a steam-engine. In less than a decade, public transport services have become popular in France, Britain and major cities on the east coast of the United States, such as Paris, Lyon, London and New York. Buses enable citizens to experience unprecedented proximity to each other and shorten the distance between cities and neighbouring villages and towns."

"In 1879, German Engineer Werner von Siemens first tried to use electricity to drive rail vehicles at the Berlin Expo. Since then, St. Petersburg in Russia and Toronto in Canada have made commercial attempts to open trams. Budapest, in Hungary, founded the first tram system in 1887. In 1888, Richmond, in Virginia of the United States, opened a tram. Trams were popular in some cities of Europe, America, Oceania and Asia in the early 20th century. Tianjin is the first city in China to have tram operation. In 1904, Meyer & Co., Eduard from Belgium was allowed to invest in Tianjin and the first tram route in China. After more than 20 years of development, by the end of 1927, there were six tram tracks in Tianjin, which was about 22 kilometers long. The operation area covered Chinese Controlled Districts in Tianjin, Austrian Concession, Italian Concession, Japanese Concession, French Concession and Russian Concession."

"In 1863, the world's first underground railway was built and opened in London. Its successful operation has provided valuable experience for the development of public transport in densely populated metropolitan areas. From the 1860s to the 1930s, urban rail transit in Europe and the United States developed rapidly, and 13 cities built metros one after another. At that time, the old trams were still the main public transport facilities. However, compared with metro, its shortcomings such as low speed, high noise and low punctuality rate have been revealed. The development of urban rail transit is tortuous. From the 1930s to the 1950s, the Second World War led to the stagnation of urban rail transit, and automobiles developed rapidly with its convenience and flexibility. During this period, only five cities in the world developed urban subway. Tramways are gradually eliminated, mainly because of the increase of road traffic and the intensification of contradictions with tramways. Los Angeles is a typical example. It was originally a city with railways and trams as its skeleton. Later, with the development of cars, it demolished the trams and became the world's largest car city area. From 1964 to 1973, the tram lines in Tianjin were completely dismantled for various reasons."

"From 1950s to 1970s, due to the rapid development of automobile manufacturing industry, the popularity of cars, access to families, and the sharp increase in urban traffic volume, urban traffic has become increasingly crowded, which can lead to traffic tie-up in serious cases. In addition, air pollution, noise and other issues make people re-recognize the importance of public transport and rail transit. Paris, London and other European cities, including New York, Tokyo, Osaka, have built the most perfect rail network. Since then, cities have entered the era of rail transit, strictly speaking, it has entered the era of public transport with rail transit as the main part. In 1970, Tianjin decided to use the modification of Qiangzi River to construct the subway synchronously, which is called '7047 Project'. During this period, due to the policy of suspension of construction and the impact of the Tangshan Earthquake, coupled with financial constraints, '7047 Project' was forced to stop. Until 1984, after the third expansion, Tianjin Metro was officially opened to traffic, which was called Tianjin Metro Line 1. Tianjin became the second Metro City in China after Beijing."

"Although we have apparently also participated in the whole process of the development and evolution of public transport, our understanding of public transport is still relatively limited, and public transport as a whole is still lagging behind. Prof. Robert Cervero from University of California, Berkeley, said in his famous book *The Transit Metropolis* published in 1998.

> The term transit is used to refer to all public transport services transporting passengers in a city or region, from multiple starting points to multiple destinations, to buses and cars with no specified routes. From the types of vehicles, the ability to transport passengers, and the operating environment, the various types of public transport constitute a continuum. Common forms of public transport are well known to all, mainly including the following main types: paratransit, also known as bus-like, including vans, jitneys, shuttles, microbuses, minibuses and other taxi-like buses, bus transit, trams and light rail transit, heavy rail and metros, commuter and suburban railways, etc. In the long run, small business or public aircraft, high-speed trains, high-speed ships, high-speed long-distance highway passenger transport and other forms of public transport in urban areas will be called. (Robert Cervero,1998)

At present, in developed countries and cities, a relatively perfect public transport system has been formed. These different types of public transport are planned and used according to their respective characteristics, cooperated with each other. They are organically integrated with the urban layout and play an increasingly important role."

"For decades, our country has verbally adhered to the urban planning model based on public transport, but in the actual planning work, it has been the road-based planning layout. As for what kind of car to take on the road, it seems that the planning is not the matter to be governed. Bus-oriented is more like a slogan, and there has never been an experimental urban planning and construction based on public transport. With the rapid growth of car ownership and the deterioration of urban traffic, almost all cities continue to develop public transport as the only way to alleviate traffic congestion in the revision of city master plan. The state has also issued policies to support the development of public transport. At present, the rapid development of China's inter-city high-speed railway has begun to play an important role in inter-city links. At the same time, many cities are strengthening the construction of Metro and transit cities as well. Bus lanes and new energy buses are widely used, and some measures such as "Palm Travel" are issued to improve the punctuality and service level of bus operation, and a series of achievements have been achieved. However, there is still a lack of in-depth consideration on the key issues such as the development strategy of public transport, the relationship between urban public transport and land use. The detailed treatment of public transport services and transfer is still extensive and not people-oriented. If there is no further effective way to play the role of public transport and no new paradigm for overall planning of transport development strategy, the development of public transport is doomed to be 'times work and half success'."

"Summarizing the experience of cities at home and abroad, we can see that one of the key factors for the success of public transport strategy is to deal with the relationship between total public transport supply and demand, and the proportion of various public transport. At present, in the domestic traffic planning, the types of public transport are not rich enough, and the use of various public transport ideas are not clear. Although the difference between the express line and the urban line is clear, some urban lines extend blindly, which exceeds the reasonable operation distance. At the same time, there is not enough research on the cooperation between different public transport means, lack of detailed treatment, and can't form a rich and diverse public transport network."

The second key factor is to deal with the relationship between public transport routes and urban spatial layout, in order to build a transit -oriented metropolis. The relationship between public transport and urban form is just like the relationship between hands and gloves. It must be matching. In his book *The Transit Metropolis*, Professor Cervero divides Mass Rapid Transit (MRT) metropolises into four categories: adaptive cities, adaptive transit, strong-core cities and hybrid cities.

> Adaptive cities are bus-oriented metropolis, which guide urban growth by investing in rail transit systems and meet larger social goals, such as protecting open space and providing low-income housing for rail transit service communities. All adaptive cities are characterized by compact,

mixed suburban communities and new towns centered around bus nodes. Adaptive transit refers to a city with spreading and low-density development mode. By seeking suitable bus services and new technologies, it can best serve the whole urban area. Strong nuclear city successfully integrates public transport and urban development into a more limited central city by providing comprehensive public transport services centered on hybrid tram and light rail systems. Trams arranged on roads coexist with pedestrians and bicycles. The priority importance and health of these cities is the basis for the successful integration of urban renewal and tram renewal. Hybrid cities have achieved a viable balance between centralized development along major bus corridors and suburban and remote suburban areas where public transport services are effective. Munich combines heavy rail trunk services with light rail and conventional bus services. The two types of buses are coordinated by a regional bus authority, which strengthens the central city and serves the growth axis of the suburbs. Both Ottawa and Curitiba introduced flexible dedicated lane buses, while increasing the proportion of regional business growth around major bus stops. The combination of flexible bus-based services and the development of mixed use along bus corridors has resulted in unusually high per capita bus travel. (Robert Cervero,1998)

"From the above four types of cities, whether land use follows public transport or traffic adapts to land use, the relationship between traffic and land use is still one of the key problems, although it is not the whole problem. To deal with the relationship between public transport and land use in spatial planning, is not only reflected in the large-scale planning structure, but also in the specific urban design. Transit Oriented Development (TOD) is often talked about by urban planners and traffic planners nowadays, but our approach is too simple and conceptual. In fact, TOD includes two main types: MRT-oriented residential development and MRT-oriented commercial and employment center development. Moreover, the specific development intensity and layout of various urban locations are different. It is not only centered on public transport stations, but also surrounded by high-density development. "

"Today, the public transport system in developed countries is relatively perfect, but public transport is not a panacea for all diseases. Public transport is very important, but a single public transport strategy can't completely solve the urban traffic problem. Car traffic is still growing. London has always emphasized the public transport strategy, but today public transport travel accounts for only about 10% of the total travel, and 80% of the total travel is car transport. In many mega-cities of China, rail transit has been vigorously developed, and the speed of subway construction is unparalleled in the world, but the traffic problems are still outstanding. Moreover, the travel conditions of rail transit are bad, over-crowded at rush hour. Therefore, we can't consider that public transport is of one strike of winning and rest-easy. In the era of carriage-based public transport, a lot of horse manure once killed New York. Subsidies for public transport today are no less problematic than horse manure. If government subsidies are required, too much public transport will be unsustainable for the government. Therefore, there is a consensus in the western academic circles that the development strategy of public transport is mainly to optimize the existing public transport routes and coordinate with the layout of the city, rather than simply increase the size of the routes and network."

"Details determine success or failure. The success or failure of the planning and design of public transport stations in various forms and transfer hubs of various scales directly determines the success or failure of the urban public transport system. Urban transfer hub is usually located in the city center or sub-center, which is a combination of commercial activities and transportation hub. It is the basic function of modern metropolitan center. With the development of suburbanization, new urban transfer hubs have emerged in the near and far suburbs. These hubs are the stations of urban rapid rail transit, located on the axis of urban regional development, the transfer point of rapid rail transit with private cars and local buses, and also a center of local life. Unlike the urban central transfer hub building, the new urban transfer hub building is relatively simple, convenient, humanized, low-cost operation, is a new urban regional node, becoming the center of the community and neighborhood."

"To give full play to the role of public transport, developed countries have relatively mature "3D" experience, which is, density, diversity and urban design. Increasing the density is the best and most effective planning method to improve bus travel, but the desity should be moderate reasonable, not the higher the better. The diversity of mixed land use can also encourage bus travel and promotes the efficient use of resources, such as parking lot sharing. Therefore, apart from rational density and mixed use, urban design is a more effective means."

"At present, it is generally believed that urban design is also an effective way to improve the urban traffic function. There are many compact and mixed land-use cities in the world, such as Bangkok, Jakarta and Sao Paulo. Although they have active street life, they are not models of sustainable development and generally lack high-quality urban design. North America's new urbanism advocates compact layout and mixed land use. By using traditional design techniques such as grid road network, public places, roadside parking, open space and commercial facilities within walking range, it strives to transform suburban areas dominated by private cars into pedestrian-friendly and bus-supported cities and communities."

"To develop public transport strategy, we must form a systematic planning strategy and operable methods, and coordinate with car strategy and pedestrian strategy. Therefore, traffic planning and design is regarded as a key research content in our pilot project of urban design."

48. Vote on Wheels: Adaptive to Car

Director Zhong

"Now I would like to focus on the car strategy. In fact, car transportation and urban design are more closely related. Let's also briefly review the development of cars. In 1879, Karl Benz, a German engineer, successfully experimented with a two-stroke experimental engine for the first time. Daimler and Maybach invented the gasoline internal combustion engine in 1883. In 1885, he built the first motor tricycle in Mannheim. At the end of 1885, Daimler converted the carriage and invented the first four-wheeled car. At the end of the 19th century, when the car was introduced to the United States, only the wealthy people in a few big cities, such as New York and Philadelphia, were eligible to enjoy it. Henry Ford trialed and produced his first car in 1896. Ford Motor Company was founded in 1903, aiming to make affordable cars to all Americans. In 1908, Ford and its partners combined the design and manufacturing ideas of Alz, Lilan and others into a new type of car, the Model T, which is an undecorated, rugged, low-cost car, easy to drive and repair, viable for country roads, meeting the needs of the mass market. The body of the car changed from open to close. Its comfort and safety are greatly improved. The success of mass production not only makes the Model T the most popular car in history, but also makes the myth of family car come true. The success of Ford's assembly line mode of production has greatly reduced the cost of automobiles, expanding the scale of automobile production and creating a huge automobile industry."

"The greatest changes in the earthscapes of the 20th century are the popularity of private cars, the growth of traffic and the construction of highway networks. The development and popularization of car is a historic progress of human civilization. It not only promotes economic development, technological invention and technological progress, but also greatly increases the mobility of human beings. It can be said that car is a humanized means of transportation."

"The development of automobile and its related industries constitutes a very important and indispensable aspect of the global economy. We can see from the industrial chain of petrochemical industry that petroleum products, such as gasoline, diesel and automotive parts, need automobile consumption. The same is true for many other industries, such as steel and rubber. According to statistics, American automobiles consume 50% of American oil production, 50% of steel production, 30% of rubber production and 20% of leather production. Of the total employed population in the United States, 30% of employment is related to automobiles. American automobile manufacturers and sales companies spend $11 billion a year on advertising sales vehicles, compared with $2 billion for GM alone. The development of private cars has greatly promoted the development of the tertiary industry and created many new economic forms. The development of automobile and related industries has promoted the development of American economy and the enhancement of national strength. The automobile industry promotes the progress of a series of industries related to automobiles, promotes the growth of the national economy, solves the employment problem, and promotes consumption and innovation."

"The development of cars is the progress of civilization. Looking back on history, the development of human civilization is accompanied by the progress of transportation methods and means. Ships, trains, automobiles, airplanes and urban rail transit, which have emerged since the industrial revolution, have greatly promoted social progress and economic development. Along with the development of other technology, modern transportation has become an important part of contemporary human civilization, in which the popularity of private cars is a beautiful landscape.

For individuals, private car is the most convenient, flexible, free and comfortable way of transportation, which realizes the ideal of free travel. People call the United States a society on wheels, which is not only literal, but also has a deeper meaning. The "American Dream" formed by "Garden House, Car and Decent Work" represents the so-called democracy, freedom, equity and individuality, and is one of the boosters of the rapid development and growing strength of the United States. One of the important aspects of the specific indicators of building a well-off society in an all-round way in China is the improvement of mobility and travel ability of human and society. Aviation, railways, highways, waterways, pipelines and information transmission play an important role in economic development and the improvement of people's living standards. Private cars are also an indispensable key indicator, which represents the development and prosperity of the economy."

"There is no free lunch. Everything comes at a price. Cars bring convenience and problems to human life at the same time. These problems are mainly manifested in three aspects: energy, transportation and pollution. First, the automotive industry consumes a lot of natural resources: in addition to using steel, modern automobiles also need to use highly energy-consuming aluminum and hard-to-recycle plastics. In addition, automobiles consume a lot of oil. More than half of the world's oil is used for transportation, and one third of the fuel is used to drive the internal combustion engines of automobiles and trucks. Secondly, automobiles lead to traffic congestion and frequent traffic accidents. Traffic accidents caused by automobiles are one of the reasons leading to the greatest number of human deaths and injuries in the world today. Millions of people suffer from traffic accidents every year. At the same time, the parking of a large number of vehicles has increasingly compressed people's living space. Third, the production and use of automobiles pollute the environment. Automobile exhaust emission is very serious to urban air pollution. Photochemical smog caused by exhaust is a common problem faced by many big cities in the world. In addition, automobiles cause noise pollution."

"The impact of automobiles on cities is enormous, especially in the United States. In order to solve the problem of urban traffic congestion, some expressways go deep into the city center, causing serious damage to the urban texture and landscape. At the same time, the development of transportation has contributed to suburbanization, promoted the outmigration of population and commercial centers in old cities, and eventually led to the decline of urban centers and urban sprawl."

"In the 1970s, the oil crisis made people begin to reflect on cars, including the economic, social and urban sprawl, especially the ecological problems. Americans began a long war on cars. Apart from nuclear weapons, the most controversial issue in the 20th century is the development of private cars. Opponents abhor private cars and regard them as the main causes of urban sprawl disorderly, urban quality of life decline, environmental pollution, waste of resources, social problems and so on. Jane Kay, author of *Asphalt Nation*, points out that, 'This is a road to environmental destruction.'"

"Today there are 200 million cars in the United States, 'American cars are driving the whole country of the United States'. Since the suburbanization of cars in 1920s and the Interstate Highway Act in 1956, a series of policies have further accelerated the development of private cars. Marked by the completion of the 35-year Interstate Highway Project in 1991, the United States has entered the so-called "Post Auto Era". It was at this time that Jane Kay, a former newspaper reporter, came to live downtown with her three children from the outskirts of Boston and decided to sell her car and start writing books about cars. She spent six years publishing *Asphalt Nation* in 1997, a declaration against cars."

> Standing on the horizon of American history, the emergence of Ford T and 'the planned city' is like a silhouette against the original vision of the twentieth century. No matter how we reflect on the post-automobile era today, automobile culture has permanently eroded the American public transport system, their civic awareness and aesthetic accomplishment in the past century. Urban development has come to a huge watershed in the 1920s. What separates tomorrow's city from the beautiful homeland created by our ancestors is the gap between people-oriented and car-oriented values.
>
> On the highway, Americans seem to have the hope of exploring a new continent that has disappeared. Once again, it stimulates the imagination and enterprising spirit of Americans—these American personalities that go deep into the marrow. Roosevelt and his New Deal drew the United States out of the abyss of economic and social disasters and had a landmark impact on the construction of national parks and highways. However, in this process, roads have spread all over American cities and villages. Housing policies have encouraged the further spread of the United States. The rights of the Public Road Authority are more deeply rooted, which is like a perpetual motor, constantly pulling the United States on the road to continue asphaltization.
>
> Cars are the tools to save us and the root cause to destroy us. Despite our foolish mistakes in the automotive industry, it still has a deep-rooted existence. For three quarters of the century, our life and landscape have been shaped to meet the requirements of automobiles. In this mountain, we naturally can't see the plight of individuals and even the world, or perceive that we should start to Run away from it. We have to start rescue operations. There are many topics to end the era of automobile domination, which can't be achieved only by a declaration, a plan or a project to change automobile dependence.
>
> Abandoning automobiles, shaping a human-scale space of activity and encouraging walking urban structure, like the environmental protection movement, requires work on a small scale, with full respect for local, traditional and individual premises. This work and its objectives may not be as exciting as the lunar landing plan or the occupation of the European continent, nor the mentality of the Great Leap Forward in history, but it certainly has more sustainable prospects.(Jane Kay,1997)

"Because Jane is also a woman, and her works are equally important, Jane Kay is compared to Jane Jacobs, who wrote the world-famous book The *Death and Life of*

Great American Cities. Ten years after Jane Kay published *Asphalt Nation*, another woman named Janette Sadik-Khan launched a 'Street Fight' from a car in New York. The car culture vane began to turn slightly. All three women have made great contributions to urban planning."

"From Jane's book, we can see the breadth and depth of deep-seated problems caused by cars in the United States. In today's United States, a large number of highways, ramps and parking lots occupy a large amount of urban land, resulting in a very monotonous and spacious landscape. In the suburbs of the city, the super-large buildings based on automobiles destroy the pleasant scale. The United States is a country with heavy traffic. Local traffic congestion has become commonplace, traffic congestion in the whole region, the so-called "grid lock" has become a common phenomenon. Although roads and parking lots account for about 40% of built-up urban areas in the United States, parking spaces are often unavailable and traffic conditions are not fundamentally improved. Traffic jams in the United States cause 8 billion extra hours for commuters a year. Coastal areas with two-thirds of the population of the United States, such as the old BOS-WASH and the new LOS-DIEGOS, are the main traffic congestion routes."

"The United States is a country locked in by cars. Depending on the lifestyle of automobile and the 'travel chain' composed of private cars as the basic means of travel, Americans' lives are tightly locked with cars. As cities spread outward, the population of suburbs tripled, and the number of cities with dense, pedestrian and public transport services shrank. The spread of cities separates space from distance. They live here, shop there and work elsewhere. It is impossible to work and survive without a car, and it makes every trip independent. The inconvenience, mileage and traffic increased exponentially. In the United States, 91% of households own cars. On average, every family travels six times a day. According to the survey statistics, more than 60% of the average 10,000 to 12,000 miles per car per year are related to life, so called life is locked in cars."

"The development of cars has aggravated the problem of environmental pollution, and automobile exhaust has become the most serious source of urban air pollution. Traffic congestion and pollution are serious in the inner city. People run out of the city with their noses covered and move to the suburbs, further increases the traffic volume and the emission of exhaust gas. The per capita consumption of gasoline in the United States is five times of that in Europe. At the same time, automobile pollution in other areas afflicts some poor communities and neighborhoods in the United States. Abandoned tires, leaked batteries and abandoned cars, etc., a large number of untreated automobile waste pollutes the environment of the United States."

"In Jane Kay's view, the evils of cars are beyond description. Perhaps the above explanation of the private car problem is not comprehensive. In fact, objectively speaking, many problems are not the problems of private cars themselves, but the long-term social, political and economic problems faced during the development of human society. Economic development needs to consume resources, and production will produce pollution. The ultimate way to solve the problem is technological progress. Nowadays, new energy vehicles (NEVs) are in full swing. Therefore, despite so many problems, the development of private cars is still a trend. The traffic is unstoppable rolling. The great driving force behind this trend is that private cars meet people's needs and promote the rapid development of social economy. This is the ultimate reason why people choose to vote on wheels."

"Practice has proved that in large urban areas, a better public transport system can play a greater role, including the optimization of urban layout, which can reduce the proportion of car commuting trips. But in the weekend supermarket shopping, suburban leisure hiking and other travel needs, private cars can't be replaced by public transport. Therefore, in large urban areas, private cars, public transport systems and slow walking systems, can form complementary urban transport systems. Today, in Tokyo, Japan, urban public transport is very developed, but the ownership of private cars has reached a surprising level, with one car for every two people. People commute by bus and use their own cars for leisure. Therefore, the car is not a monster of floods, not a moral issue, but we need to use it reasonably."

"A good city must have a good traffic system, and a good traffic system needs good transportation planning to guide. As early as 1977, J. M. Thomson, a British transport expert, put forward in his *Great Cities and Their Traffic* that the struggle between public transport and car transport, track and road are the two main competing aspects of transport planning. A good traffic planning needs to coordinate the relationship between bus and car by combining its own urban structure, transportation system and economic development level. J.M. Thomson, based on his experience of dealing with urban traffic problems in five continents and 30 different cities in the world, puts forward five strategies for solving urban traffic problems and their appropriate layout forms as well, emphasizing that each strategy is not perfect, and needs to be flexibly applied in accordance with regional conditions and development stages."

"The first form of layout: fully develop car. The design idea is that cars can be freely accessible throughout the city and suitable for small towns with less population. This form of layout is widely used in New towns in Britain and the United States. There are few problems such as traffic congestion, parking and environment in small towns with this layout. Their real traffic problem is that people who can't drive have limited travel activities without cars. After husbands drive to work, the elderly and children can only stay at home in the suburbs. Another problem is that the size of cities that adopt this strategy cannot grow very large. Otherwise, traffic problems will be serious and difficult to remedy, such as Los Angeles."

"The second form of layout: restricting the city center. The city centre is small in scale and has a radial road network to serve the city centre. Most of the jobs in the city are located in the suburbs and fringe areas. Traffic mainly depends on cars, which are served by circle roads with large capacity. In addition, there is a modest public transport system, with simple radial tracks for commuters entering and leaving the city centre. This layout structure is both unstable and expensive, and too many circle-type fast loops is the root cause. The original intention of urban expressway construction is to separate the transit traffic through the city centre and alleviate the traffic congestion in the city centre. In the early stage of the opening of the expressway, it has a certain effect on traffic improvement. However, from the perspective of the urban system, this improvement is only temporary. The improvement of traffic along the Ring Road improves the value of land and stimulates the development of real estate along the Ring Road. Over time, the surrounding and central areas of the ring road have been filled and developed, the former suburbs have been merged into the city center, the former peripheral ring road has been transformed into the inner city road, the city center has not been restricted, but unorderly expanded."

"The enlargement of the size of the city centre will bring about a sharp increase in traffic volume, and the dense central area will also require an increase in the import and export of the loop. The increase of the entrance and exit will make it difficult to enter and exit the express loop, and the increase of congestion will become inevitable. The inappropriate recognition of the functions of urban ring roads and the original intention of their design have exacerbated urban traffic congestion to a certain extent. Paris, London and Copenhagen have experienced it, but they have taken remedial measures in time. London has stopped building new ring roads and Copenhagen has taken strong regulatory measures. Beijing, Shanghai, Wuhan, Zhengzhou and other cities are also experiencing congestion. The problem is that we did not summarize the experience and lessons in time, but rather expected to build more loops to alleviate congestion and fall into the vicious circle of overbuilding loops. It is noteworthy that the construction boom of ring roads is spreading towards second and third-tier cities."

"The third form of layout: to maintain a strong city centre. There is a strong city center, where the road and track form a radial traffic network, where the sub-center is set on the rail network. The rail network has a large capacity. Most people who go to work in the city center take public transport. On the transport line, public transport and private cars are balanced in the competition. This layout pattern is widely used in the top cities in Europe and America, such as Paris, Tokyo, New York, Sydney, Toronto, Athens and so on."

"The fourth form of layout: to spend less. The characteristics of this layout are: there is no high-speed road and rail network, the density of the city is very high, and the sub-center of the city is concentrated in the radiation public transport corridor and keeps a proper distance from the city center. The core of the strategy is based on the management of existing roads. Buses are responsible for most of the transport functions. In addition, land use planning, management and coordination are needed. Bogota, Curitiba and other cities adopt this layout."

"The fifth form of layout: to restrict car traffic. This strategic urban structure has a strong urban centre and a good public transport system to serve the urban centre. The track network is circular, and the sub-centers are located at the intersection points of track lines, which connect the sub-centers. Roads are also circular structure. The capacity of radial roads leading to the city center gradually decreases. When approaching the city center, through reducing the number of lanes, or changing to toll roads, cars are guided to other circular trunk roads. The most ideal way is to direct vehicles to circular roads only around the city center. There is also a transit loop outside the city to intercept transit traffic at the edge of the city. This strategy is suitable for the urban structure established in the pre-car era. It also needs to be used in conjunction with the car restriction plan, including the implementation of parking fees, the establishment of car-banned streets, public transport, walking and cycling priority, etc. London, Hong Kong, Singapore and other British background cities mostly adopt this kind of transportation development strategy."

"The five types of urban layout are different, but they contain some basic principles. First, the city activity center must have good accessibility, whether at the intersection of roads or bus corridor hub. Secondly, no matter how big or small a city center is, it must be equipped with a certain level of public transport facilities. Thirdly, a strong city center needs a large capacity of public transport to support, and the rail network connecting the city center and the sub-center is the basis for maintaining a strong city center. Fourthly, the circle-type fast loop is easy to lead to the spread and urban sprawl disorderly, which has a great negative impact on the construction of a good urban multi-center system, and the construction of the circle-type fast loop needs to be treated with caution."

"Today, in the field of urban planning, land planning and environmental protection, and spatial planning in our country, there is a strange phenomenon, a general opposition to the development of private cars and the traffic mode mainly consisting of private cars. But in fact, we are on this road. Simply based on public transport, compact urban and regional spatial planning layout structure and form can't meet the rapid development of private cars for transport needs. On the one hand, it causes the driver to drive for a long time and wait in traffic jam, on the other hand, it causes the double perplexity of passengers waiting for a long time at public transport stations. In order to avoid such a situation, we must change our concepts, to seek truth from facts, adopting the strategy of laying equal stress on public transport, car transport and pedestrian transport, and adopt the planning layout and transportation strategy suitable for the characteristics of different regions in different areas such as urban central areas, urban fringe areas and suburbs."

"Since the end of the 20th century, new urbanism has been active. It emphasizes the coexistence of bus, pedestrian, bicycle and car, which is a more successful planning and design method. It should be noted that many international masters of modern architectural design and urban planning are not entirely against private car transportation, such as Frank Lloyd Wright, Le Corbusier, etc. They are trying to find new spatial planning models in order to find solutions to the problem. Practice has proved that social, political and economic models and lifestyles are closely related to urban spatial models. Today, our country has chosen the private car as a way of economic development, which also determines our space development model. At the beginning of the automobile era, if we can't face the reality, but ignore the existence of private cars like looking at the emperor's new clothes, and try to solve the urban traffic problems simply by developing public transport, we will make historical mistakes. Therefore, we need to look at the development of private cars objectively and rationally, and to plan and design the urban and regional spatial models that are suitable for public transport, walking and private car development as well."

Director Zhong

"Foreign experts in the field of transportation have basically reached a consensus that the traffic problem is ultimately an economic issue. The choice of mode of transportation depends on how much people are willing to pay. Traffic supply is different from other service facilities such as water, electricity and gas. It has two different characteristics: urban traffic not only serves the city, but also is one of the components of the city. Traffic serves all kinds of activities, but traffic, buildings and activities inside buildings are interdependent and inseparable."

"Urban transport facilities connect the various activities that make up urban life, which must depend on transport facilities. The structure of the city, the size and expansion of the city, the way and characteristics of urban life are all determined by the property and the service quality of the urban traffic system. As a result, it is faced with the choice of what transportation mode and transportation system, and this choice can affect the function and layout of the city. The interaction between traffic and urban layout complicates the task of determining traffic functions."

"The goals of a city are diverse, and some of them are incompatible with each other. For example, people want to have spacious houses, small gardens and parks. At the same time, they want to effectively centralize industry and commerce and shorten the working distance. But these goals are incompatible with each other. People want to enjoy the comfort and convenience of cars and also many pleasures in the center of big cities. These two goals are incompatible, because the latter requires a large number of people to be transported to a small area. Therefore, there must be a compromise between convenient, fast and cheap transportation and the quality of urban life. It is of no practical significance to compare one aspect of life or traffic in different cities in isolation without comparing other relevant aspects. For example, comparing the speed of a car without comparing the distance people need to travel at that speed; comparing the density of their dwellings without comparing the time and money people spend on the distance to their workplace or local shops; or comparing the convenience of going to a supermarket without comparing it with the convenience of going to downtown and suburban areas, etc. There is no practical significance."

"People are accustomed to using a set of criteria to evaluate the quality of things. The evaluation results may directly affect the follow-up actions, as well as urban traffic. Simple recognition of traffic function leads to the selection of single evaluation index and one-sided guidance of action. A more perfect standard should be adopted to evaluate the quality of a city traffic mode. The interaction features of traffic and urban activities, urban buildings and the city itself determines the quality of a city traffic system. We can't only see whether the road is congested or not, whether the speed is fast enough, but also insert the factors of social activities to make a systematic evaluation. For example, to what extent can transportation satisfy people to participate in various activities? For example, is it because of traffic difficulties or too expensive to prevent people from going to their favorite schools, shops or work units? Or even hinder other social activities? To what extent can economic conditions permit people to live in preferred areas? People's social activities and family life are interrelated, and people must weigh the pros and cons and make their choices. For example, in the current traffic conditions, some people prefer to live worse for the convenience of going to work, but close to various social activities; others prefer to live farther or less convenient places for the sake of spacious living and better environment. For the vast majority of urban residents, the transportation related to social activities, especially to work, school, shopping, transportation to the city centre and to do business, how to be convenient, low-cost, comfortable and fast (not fast), is the core of traffic planning, and is what the public transport subsidized by the government need to achieve. At the same time, the government can provide road traffic conditions for personal travel mainly by car, and many aspects need to be solved by means of market means such as toll collection."

"What are the environmental impacts of the transportation system on people's living, working, shopping and entertainment? These require new traffic models to analyze and judge. The factors affecting these contents are not as easy to collect and quantify as those based on traffic engineering, emphasizing the indicators of motor vehicle travel and speed, but it is helpful to objectively evaluate the suitability of urban traffic system, which should be an improvement direction in the future. After car-oriented, public transport-oriented and mobility-oriented, many foreign urban traffic values are evolving to place-oriented. Professor Cervero of the University of California, Berkeley, published a new book, *Beyond Mobility: Planning Cities for People and Places*, last year. There is no Chinese version yet. As we know, architects and urban designers often talk about places, and now foreign traffic experts are also talking about places. Location more reflects the interdependence of traffic, urban activities and buildings. Different cities of different nature, or different areas of the same city, the emphasis of evaluation criteria should be different. We can't simply solve all these problems by vigorously developing public transport and restricting private cars. In the place where need to develop public transport, it should be vigorously developed. In the place where cars need to be restricted it should be strictly restricted. In the place where cars need to be encouraged it should be greatly encouraged."

Director Sun, Planning Department, Municipal Bureau of Industry and Information Technology

On hearing this, Director Sun of the Planning Department of Tianjin Municipal Bureau of Industry and Information Technology took the lead in applauding. Because he kept applauding, the people around him began to echo. When the applause fell, everyone gazed at him. "Hello, everyone. I'm Sun Chenggang, deputy director of the Planning Department of Tianjin Municipal Bureau of Industry and Information Technology. It's a great honor to participate in such a professional urban design forum today. I very much agree with Director Zhong's view that we should treat the development of cars objectively and fairly.

"In the field of urban planning in China, drawing lessons from the serious urban problems caused by the development of private cars in western developed countries, we have been restrained the excessive development of private cars for decades and firmly advocating the urban development model dominated by public transport. After the Asian financial crisis in 1997, on how to choose the economic growth point to stimulate domestic demand, the competition between commercial housing and private cars broke out. On the surface, commercial housing has gained the upper hand, but in fact, the automobile industry has developed courageously.

Today in the 21st century, the automobile industry has become one of the most important leading industries to promote the rapid development of our national economy, which has stimulated the rapid growth of many related industries and services. During this period, the discussion of environmental protection, sustainable development, circular economy and other viewpoints is the most heated topic, but China has reached the world's largest automobile production and sales volume, with an annual output and sales of 28 million vehicles, and the production and sales of new energy vehicles is also the world's first.

"The fact shows that although car has serious problems in environmental protection and resources consumption, and these problems are more and more recognized by everyone, the government and the public still choose to vote on wheels. The rapid growth of private car ownership reflects the strong demand for mobility after the improvement of people's living standards. With the emergence of a large number of private cars in cities, the traditional urban layout model based on public transport has been crushed by rolling wheels, and traffic congestion has become one of the most serious urban problems. In the case of rolling traffic, urban planning still ignores the development of cars. Almost all the city master plan and comprehensive transportation planning still follow the urban layout and traffic planning model which mainly focuses on public transport. This approach is justifiable in theory, but in real practice, we must also face the reality of the continuous development of private cars and the problem of urban traffic deterioration. We responsible planners, together with our departments in charge of the automobile industry, must face the situation objectively and find practical solutions to the problem.

"To develop cars, we must solve the problem of traffic congestion. Director Zhong, I want to ask a question. For a long time, urban centers have been plagued by traffic problems. What do you think is the most fundamental cause of traffic problems and what is the reasonable way to deal with them?

Director Zhong

"I mentioned that public transport, cars and walking should be taken as equally important transportation strategies, and different regions should adopt different strategies. I have been thinking about how to solve the problem of traffic congestion in the city center and Binhai core area. Is it necessary to further widen the road? Will viaduct overpasses be built again? Whether from the protection of historic cities or from the economic point of view, it is infeasible. At present, the trend in developed countries is that many cities have demolished the viaducts and overpasses in the city centre. We can't repeat the same mistakes. But the traffic volume is still increasing. What shall we do? Recent reading of Street Fight and several classic works of traffic planning, I felt suddenly open-minded. Downs has a famous saying, 'To solve traffic congestion by constantly widening the road is like relaxing the belt of a weight-loss person!' To solve the problem of traffic congestion in the city centre, only on the basis of further development of public transport such as track, implementation of pedestrian and humanized city strategy, through reasonable charges to limit car traffic into the city centre, this is the only and correct way out.

"Urban traffic problems are mainly economic problems because of improper charges for road use, parking, public transport and environmental pollution caused by the lack of control. For a long time, improper charges have affected the distribution of various activity sites and what kind of new transportation facilities are provided, whether to build roads or rails. After World War II, most cities in the West spent money on road construction, constantly trying to ease congestion, while the track has been stagnating and declining. The public transport company has persevered in a very difficult situation, competing with cars. As more and more people do not take public transport, but take private cars, the government is under more and more pressure. On one hand, more cars require road construction, which requires money. On the other hand, fewer public transport passengers, less income which also requires money to prevent the bankruptcy of public transport. The economic pressure is great. As a result, many cities are in despair: on one hand, roads are often congested, on the other hand, public transport is on the verge of collapse.

"The root of urban traffic problems can be said to be due to the long-term problem that urban traffic management and development didn't act in accordance with economic laws, leading to the overall failure of public transport in the competition with car traffic. The increase in population and income has exacerbated the problem. Over the past 40 years, there have been technical reason and reason of failing to reach agreement, but the situation is slowly changing. For example, city centers of London, Singapore and other cities have used mature technology to collect road tolls, environmental pollution fees, and use economic and technological means to regulate urban traffic, improve the competitiveness of public transport, with good results in practice. It should be noted that the economic means are the measures taken to achieve sustainable development after urban land is mature and highly civilized. For the rapid development of urban areas, it is necessary to coordinate the relationship between urban traffic mode, urban spatial structure and land use layout in the planning stage, and use a short 'journey' to solve the problem of people participating in activities, so as to guide the formation of a healthy and sustainable urban activity mode."

Director Sun

"The past decade has witnessed a rapid development of urban construction and motorization in Binhai New Area. Planning has played an important role in guiding urban construction. How is it considered in the transportation level?"

Director Zhong

"From the experience of transportation development in big cities in the world, there are three different modes of transportation development and five different forms of urban layout. On the basis of drawing lessons from the advanced experience, Binhai New Area has carried out corresponding innovative exploration combined with its own characteristics. In terms of spatial layout characteristics, Binhai New Area is a typical super-large city scale, multi-polar nucleus growth condition, which is different from the general urban area. It is a city region with many areas. The distance between the areas is generally about 30 kilometers, which has a high demand for mobility and exceeds the suitable service range of normal public transport. Traditional forms of urban traffic organization are difficult to adapt to this super-large-scale spatial layout."

"The spatial layout inside the area is also diversified and has obvious differences. There is not only urban area with centralized development, but also Eco-City enclosed by groups. Different types of spatial layout in the area have great differences in the requirements of traffic organization mode and traffic network layout."

"Because of its special spatial layout and development characteristics, the traffic mode of Binhai New Area is not limited to one mode, but should reflect different traffic mode at different levels. In terms of overall development ideas, Binhai New Area advocates taking public transport as the main factor, giving full consideration to the development of cars and encouraging pedestrian travel mode. Moreover, there are differences in different zones and at different levels, which fully reflects the layout characteristics of Binhai New Area. Among Binhai New Area, we adopt the strategy of "double-fast", mainly relying on fast and large-capacity rail transit and high-speed road to undertake external traffic, laying equal stress on both bus and car modes. Inside each area, green transportation mode dominated by public transport, bicycle and walking is encouraged as a whole, but there are differences among different districts. For example, in Yujiapu, Tianjian and other areas in the core of Binhai New Area, we should implement the concepts of 'small block, narrow road, dense network' and 'complete street', and focus on building an integrated green pedestrian system around the bus hub. The proportion of car trips will be very low. In industrial parks, the proportion of motorized travel will be significantly higher than that of public transport travel. For the vast urban periphery and suburbs, not only in Binhai New Area, but also in the main urban area, the density of subway and bus lines are greatly less than central urban area, resulting the car being the main mode of travel. Two new energy vehicles, NEVs, may become the standard for families living in suburbs. Director Ma's eyes swept around and did not see Dr. Song Yunfei. He worried a little about speaking openly in the forum against Dr. Song's point of view, without knowing how he felt.

Of course, the most popular strategy in the world today is the pedestrian and humanized city strategy. Jane Kay said in *Asphalt Nation*:

> we can find new ways to solve the car problem from the revival of the city and the restoration of walking scale instead of following the highway all the time. First, we must define the value of the 'place' and the role that transportation plays— making it easier for us to get there. Our aim is to transport people, not cars themselves. This understanding enables us to stand higher and see farther. (Jane Kay, 1997)

In many developed countries, especially in urban centers, the most used transportation means are walking and cycling. These simple and inexpensive transportation means are the healthiest under different land use patterns. The pedestrianization plays a huge role in the revival of urban centers, which is very important and should be carefully considered. Of course, this is guaranteed by the powerful public transport in the city centre and the restrictions on private cars, to be exact, congestion charges."

49. Pedestrian City: The Revitalization of the Downtown

Director He, Department of the Historic City of TUPDI

In the sub-forum of "Historic City Conservation and Urban Design", Director He of Department of the Historic City of TUPDI just started to report on the pedestrianization and urban revival of the old urban area, which seems to connect in mind with the meeting room next door. After graduating from Tongji University, Director He has been engaged in urban design, paying special attention to the protection and regeneration of historic blocks. Tianjin was a relatively mild city in history. The scales of buildings and streets in the traditional Old Town, in the Hebei New District and in the concessions of foreign countries are friendly and pleasant, which makes the old urban area of Tianjin have a good architectural and urban space tradition. In recent years, facing the problem of insufficient vitality in the old urban areas, Director He has been thinking about what he can do. Last year, he also attended and presided over Jane Sadik-Khan's "Street fight" lecture. The atmosphere of the hall bursts, and he learned a lot. However, due to lack of experience, there is a slight omission in the host. After introducing Ms. Sadik-Khan's rich resume, he forgot to invite her to the rostrum, which made it a little embarrassing. When Director He thought of this, he smiled to himself.

"In recent years, the protection and renewal of Tianjin's historic blocks has made remarkable achievements. However, due to the diversion of big malls, especially the impact of e-commerce, there are signs of depression in the old commercial area. New commercial facilities such as Jinwan Square and the Five Big Courtyards are also declining. How to realize the revival of the old urban area? We put forward the strategy of pedestrian city. Some may wonder whether pedestrianization can revive old urban areas. In order to answer this question, I will start from three aspects: first, an overview of the current condition of commerce in the old urban area of Tianjin; second, the history of the relationship between the old urban area of Tianjin and traffic development; and third, the successful experience of pedestrian cities at home and abroad."

"The old urban area of Tianjin is a historic urban area, which consists of 14 historic and cultural blocks. It is concentrated on both sides of the upper reaches of the Haihe River, covering an area of 10.6 square kilometers. This section of the Haihe River is 6.7 kilometers long. The area is 5.3 kilometers long from north to south and 5.3 kilometers wide from east to west. In recent years, the Municipal Bureau of Commerce has organized and approved the development plan of Haihe International Commercial Center (2010-2035). The planning statistics, in this area, the annual sales of retail goods are 35 billion yuan, and the total business floor area is 2.36 million square meters. Among the district, there are 9 shopping malls, accounting for one quarter of the city; 26 Department stores, accounting for one third of the city. Internet data further shows that there are 9001 businesses, including 3525 convenience stores and shops, 5332 restaurants and 144 hotels. Binjiang Avenue is the highest click-through area of all kinds of APP, the others are Dongma Road, Joy-City and Xiaobailou business district. The intersection of Binjiang Avenue and

Nanjing Road has the most comments on Dazhongdianping website. It can be said that this area is still the most concentrated commercial area in Tianjin. At present, there are Metro Line 1, 2, 3 and Metro line 9, four lines, 11 stations and 29 entrances and exits in the district. The scope by walking no more than 10 minutes can cover 65% of the area. There are 97 bus stops. It is also the most developed area for public transport. However, with the development of e-commerce, the traditional business format has been greatly affected. The sales volume of traditional business such as Quanyechang Plaza has been declining. The Baisheng Department Stores, Wanda Plaza and Suning Electronics Mall at the northern end of Heping Road have been closed one after another. Of course, the long-term closure of road traffic during subway construction is also one of the reasons. At the same time, with the development of tourism and experience economy, there has been a good momentum in the Wudadao, Italian Style District and other historic and cultural blocks in the region, where the flow of people continues to grow. But this phenomenon is not enough to reverse the overall decline. How to realize the revival of Tianjin's old urban area? Our view is to implement the pedestrianization."

"Looking back on history, we can find that the prosperity of Tianjin urban center is closely related to the development of transportation. Before the eighteenth century, the city center of Tianjin was very prosperous around the Sancha Estuary, relying on the convenience of river transportation. After the Second Opium War, concessions were delimited and the city centre of Tianjin moved southward. By the 1930s, Tianjin was a well-known commercial center in North China. In 1904, Meyer & Co., Eduard from Belgium was allowed to invest in and operate public transport companies in Tianjin, to invest in the construction of Tianjin's first tram route. Later, three tram lines, red, yellow and blue lines, were opened one after another. Among them, the Red Line passed North Road, the Northeast Corner, along the river road, through Jintang Bridge, Jianguo Avenue to Tianjin Railway Station. The Yellow Line went to Tianjin Customs through North Road, Northeast Corner, East Road, Southeast Corner, Zhongyuan Department Store, Four-Face Bell and Quanyechang Plaza. The starting section of the Blue line was operated in line with the Yellow Line, turning to Binjiang Avenue after the Quanyechang Plaza, and crossing Wanguo Bridge to Tianjin Railway Station. In 1921, Tianjin Tram added the Green Line, from the Xikai Church in the French concession along the Binjiang Avenue, across Wanguo Brigde to Tianjin Railway Station. Until 1921, there were five tram lines in Tianjin. At the end of 1927, the Colorful Line was added, from the Northeast Corner to Tianjin Customs. So far, Tianjin Tram has six lines, which covered Chinese controlled district in Tianjin, Austrian Concession, Italian Concession, Japanese Concession, French Concession and part of Russian Concession. Most of them are connected with Quanyechang Plaza and Tianjin Railway Station. Tram played a great role in promoting the commercial center."

"Here's a story. The first tramcar in Tianjin had a poor business at the beginning of its operation. Because the Chinese did not know much about electricity at that time and feared being electrocuted on tramcar, so no one dared to try this new thing. Later, Tianjin Tramway and Electric Lights Co., Ltd. adopted means of free trial rides, reducing fares to attract passengers, and added mattress seats in the first-class carriage of the tram, also equipped with carpets, spittoons, fans and other public facilities. Some Tianjin citizens have found that trams are not only safe and convenient, but also more comfortable than carriages and sedan chairs. Thereafter, with the increase of passenger flow, the tram business was booming day by day. The company cancelled the car class and the fare began to rise. By 1912, the company had recovered all its previous investments. From 1916 to 1928, Tianjin Tramway and Electric Lights Co., Ltd. earned 25 million silver yuan from tram lines and remitted high profits to Brussels, Belgium, every week. In the 1930s, 'tram shopping' became a popular urban lifestyle in Tianjin at that time. Trams have also promoted the development and prosperity of Tianjin's urban centre."

"For various reasons, tram lines were completely dismantled between 1964 and 1973. At the same time, the impact of the Cultural Revolution on Tianjin's urban commerce is also enormous. After the reform and opening up, Tianjin's commercial center began to gradually update. The opening of Metro Line 1 and the construction of the International Mall have made Binjiang Avenue flourishing, and the center has moved southward from Quanyechang Plaza. By 2002, the implementation of comprehensive development of the Haihe River has broadened some roads and improved the road traffic conditions in the urban center. Large commercial facilities such as Hang Lung Plaza have been built. The opening of Metro Line 3 further cements the advantages of this district."

"For many years, Tianjin has been working hard on the protection of historical and cultural blocks. Besides protecting cultural relics and historic buildings, Tianjin has also maintained the spatial scale and integrity of historic blocks. With the recent advancement of comprehensive environmental management and tourism, historic blocks are revitalized. More and more people recognize and love the pleasant architecture and urban environment, which has become a major feature of Tianjin. Urban design of pedestrian system in historic and cultural blocks is the key project among pilot urban design projects. We planned to explore the combination of pedestrianization and revitalization of the city center. Through the pedestrian network, the protection, inheritance and rational utilization of cultural heritage in series can better continue the historical context, show the city style of Tianjin, and promote the activation of historic and cultural blocks and the revival of the city center."

"In order to do this work well, we have carried out a lot of study and research. In recent years, many famous foreign cities have adopted the pedestrian strategy and achieved great success. In terms of planning theory, the theory and method of pedestrian city has been relatively mature. The famous Danish architect Jan Gehl has been engaged in the research and practice of humanized cities for many years. He has published several monographs and participated in the practices of pedestrian cities in London, New York, Melbourne, Moscow and so on, with rich experience. In his book *Cities for People*, he wrote:

> Creating walkable cities. The day when a baby can take the first step is an important day in his life. Life begins with your feet! But walking is not only walking, there are many purposes of walking, both fast and targeted progress, but also wandering. Whatever the purpose, people walking in the city will be integrated with various social activities around them, thus producing a square effect. Walking is not only a way of transportation, but

also a potential starting point and occasion for many other activities.

A city suitable for staying. There are many pedestrians in a city, which does not necessarily mean that the quality of the city is high—there are many pedestrians who may be forced to walk. However, if many people in a city stay at ease, it usually means that the city has good quality. If People choose to stay, they usually choose the boundary zone. This phenomenon is called 'boundary effect'. People standing at the boundary feel quiet and cautious there. Another phenomenon of staying is called the piano effect, which means that people like to find a place to rely on. Back to the wall makes people feel low-key and safe. William H. Whyte once called Venice 'the whole city is seatable' in *The Social Life of Small Urban Spaces*.

Encouraging people to express themselves, play and exercise in urban space helps to create healthy and vibrant cities. Without good places and good human scale in cities, there will be a lack of critical quality. Urban planners should deal with the macro and local climate very carefully, and create the necessary environmental conditions in the humanized dimension of the city for the micro-climate. The charm of urban space is not only the city life itself, but also the ingenious collection of various sensory impressions. The most important theme of urban life is people themselves. 'Man is the greatest pleasure of man', and the greatest attraction of a city is man. The core goal of humanized cities is that we need vibrant, safe, sustainable and healthy cities.

Vigorous cities require different and complex urban life. Urban life is a potentially self-reinforcing process, 'Where there are people, there will be people', and people spontaneously feel the inspiration, drive and attraction of activities and other people's existence. Urban vitality includes two dimensions: quantity and quality. Quantity can be reflected in density, but more importantly in quality, what a vibrant city really needs is attractive urban space, which is combined with the people who use it. To enhance the vitality of a city in a given environment, more people will be attracted quantitatively, and more people will be attracted to stay for a long time and speed up traffic in essence. Comparing above two, the simpler and more efficient solution is to improve quality so that people are willing to spend time.

Another important factor of urban vitality is the treatment of urban boundaries. What should we do when facing the street? That is to say, we should make the boundaries of the areas where we want to stay shortly into 'the flexible boundary'. Any single theme is inferior to an active, open and dynamic boundary, which has a greater attraction to the life and attraction of urban space. Vigorous cities are the product of elaborate planning, of which time factors and attractive flexible boundaries are the key words.

Safety includes both traffic safety and social safety. If more space and better conditions are provided for cars, the result is that walking becomes more difficult and less attractive. When designing street types and traffic schemes, the most important thing is from the human dimension. Safety is embodied in the design: Walking must have priority in all kinds of traffic. At the same time, the experience of safety and the sense of safety are of vital importance to life in cities. The presence of others in urban space means greater security. Buildings on both sides of the street also play an important role in safety. The light from windows in residential areas at night sends comfortable and peaceful signals to people nearby. At the same time, if the bottom of the building is flexible boundary, it will also make people feel safer. In addition, a clear urban structure also means security, and it is easier for us to find our way around.

Sustainability includes two perspectives: material environment and social environment. Good urban space and urban landscape play a very important role in public transportation. The station should ensure comfort and safety in both day and night. Social sustainability is an equal opportunity for different groups of people to enter public urban space and reach all parts of society. Its premise is to have accessible, enterable and attractive public space. At the same time, attention should be paid to the problem of wealth gap, and opportunities should be given to marginalized groups limited by poverty. New urban planning must begin with the design of the shortest and most attractive pedestrian and cycling links, and then elaborate on the needs of other transportation. (Jan Gehl,2006)

"This urban design takes the whole historic and cultural block as the object and scope of study. Through the careful analysis of the present situation of Tianjin historical area, especially the present situation of pedestrian system, it takes the goal and problems as the orientation, to improve the pedestrian system of historic and cultural blocks, and to improve the public space of the city through the construction of high-standard pedestrian network system. The quality and pace between them makes people return to the city centre, and makes city life return to the city centre, and makes the city centre full of vitality."

"We have adopted four strategies. The first strategy is called 'Quiet Street', which restricts car traffic into the city center by collecting congestion fees, controlling parking space numbers and increasing parking fees, and creates pedestrian-friend district by means of traffic calming measures. There are 9 social parking lots and 3000 parking spaces in this area. In addition, there are about 5000 parking spaces in the shopping mall. There are 7,000 parking spaces in shortage every day and scarce supply is adopted. The second strategy is called 'Smooth Route', which is to make clear that the pedestrian system on both sides of the Haihe River is the skeleton of the pedestrian system, to improve the main fish-bone pedestrian road and increase the cross-river pedestrian bridge, to strengthen the links between the pedestrian system on both sides, ensuring the smooth pedestrian access and networking in the old urban areas. We hope to learn from the footpaths on both sides of the Seine as world cultural heritage and brand. We should strive to build the footpaths on both sides of the Haihe River in the direction of world cultural heritage. The third strategy is 'Using Heritage', which means making further use of cultural heritage and revitalizing stock assets to form new formats. The fourth strategy is called 'Increasing

Quality', which means to improve the quality of public space carefully, to build the world's first-class pedestrian urban areas, and to realize the comprehensive revival of urban centers driven by pedestrian."

"Guided by the urban design of the pedestrian system of old urban area, the urban design for the construction and renovation of historic and cultural blocks is carried out in the areas of Wudadao, North Jiefang Road, Central Garden and Anshan Avenue. According to the overall system design, the pedestrian system in historic blocks is connected into a network. In Wudadao Historic Block, we protect the spatial relationship of architecture according to the types of buildings, and deepen the detailed protection plan, making full use of historic and cultural resources, planning a tourist route – 'Internet +' Five Avenue Experience Tour, establishing a three-level walking system to link up the walking network. At the same time, it designates Minyuan Square as the center, expanding southward to form the replacement and renewal of Heping Hotel and First Worker Sanatorium; expanding westward to transform Henan Road and reconstructing the 61 Middle School Block into a New Era Building with Chinese character and high quality, which coordinates with the overall context of the Wudadao, and explores the modernization of concession architectural style. We will strive to build the Wudadao into a model area for the protection of historic blocks in Tianjin, a demonstration area for western-style mansion economy and tourism, and an canditate for world cultural heritage."

"The North Jiefang Road Historic Block urban design concentrates on the linkage between Jinwan Square complex and Tianjin Railway Station across the Haihe River, and improves the pedestrian system of this area and stimulates the revitality of Jinwan Square. The humanization of Tianjin Railway Station is realized by the humanized renovation of Tianjin Railway Station, its Plaza and Century Bell roundabout, improving the pedestrian network. It is planned to activate Jinwan Square through the construction of Haihe Pedestrian Bridge to attract passengers. The project has been implemented and the effect is obvious. As a window of Tianjin, Tianjin Railway Station reflects its matching level. The number of visitors to Jinwan Square has surged from 30,000 per day in the past to 100,000 per day now."

"On the basis of abiding by the protection principle, the Central Garden Historic Block are carried out by means of gradual micro-renewal mode , optimizing bus routes and slow traffic space, shaping multi-level landscape, and building multi-dimensional mixed use urban space. The central garden has been restored to its original appearance. As a World Bank loan project, Chifeng Avenue and other streets have implemented pedestrian priority and traffic calming transformation, with obvious effect."

"Anshan Avenue Historic Block maintains the unique historic style of the street, further refining the elements of historic and cultural resources, introducing multiple mixed use formats, and building cultural and tourism block characterized by the Revolutionary Road of 1911. Combining with the main development nodes of historic buildings, a three-level space landscape system is formed with community parks and neighborhood green space, which especially highlights the section and traffic calming design of Anshan Avenue reconstruction."

"I want to explain 'Traffic Calming'. *Civilized Streets - A Guide to Traffic Calming* was published in 1992, in which Dr. Hass-Klau translated the German term into 'Traffic Calming'. It was Dutch urban planners and traffic engineers who first put forward the idea, noting that people's quality of life is not only related to housing conditions, but also affected by the surrounding streets. In the late 1960s, Dutch planners put forward a design idea called 'woonerf'. Instead of separating the roadway from the sidewalk, they merged the two into a single road. On such a road, pedestrians could use the roadway while motor vehicles had to travel at the speed of a horse's stroll. This design looks very much like a residential courtyard, and has been translated as 'home courtyard design'. In 1976, the design was approved by Dutch law, and quickly became popular in Germany, Britain and other European countries.

> Traffic Calming is not a sharp tool to deal with road congestion, but a tool with limitations. Only when combined with other traffic engineering methods, traffic restriction strategies and urban planning, it can achieve effect. Traffic Calming is an important part of traffic development strategy. The main task is to restrict road space mainly for motor vehicle driving and parking services, to give priority to public transport, bicycle and walking. Only when private cars are restricted can traffic calming be effectively realized."
>
> "Traffic Calming is not only a safety issue, although it is extremely important. There are limitations in dealing with traffic problems in an isolated and punitive way. Traffic Calming must be flexible. It is a new method of examining street design, not static, but formed in constantly evolvement and development. Traffic Calming design is mostly carried out in new areas, which is more expedite than that without design. But in the old urban areas, traffic calming is expected to improve the urban environment, but it conflicts with the parking need of the wealthy people who return to the city centre. (Hass-Klau, 1992)"

"In these urban designs, we attach great importance to the shaping of pleasant public spaces and great streets, while also emphasizing the overall protection and organic renewal of historic and cultural blocks. Jan Gehl wrote in *Lives Between Buildings*:

> A more rational urban development model is imperative: instead of relying too heavily on large-scale public development, highly centralized property rights and specialized planning and design models, the mechanism of building and dialogue over the material environment should be regarded as legitimate and reasonable, in order to change the development mode of defoliating existing building environment. This development attitude is based on acceptance, love and expectation of urban life, and encourages people to live in a healthy public environment.(Jan Gehl, 2006)

The pedestrian system in the city centre is a prerequisite.

50. Urban Design Guidelines for Public Open Space

Director Chen, Urban Design Lab of TUPDI

In the sub-forum of "Historic City Conservation and Urban Design", Director Chen of Urban Design Lab of TUPDI reported presenting the project team. Director Chen is the high school alumni of Director He. He has been engaged in urban design since graduated from university. He not only pays special attention to the protection and renewal of historic blocks, but also pays great attention to the shaping of urban public space. He has participated in the joint urban design of comprehensive development and renovation of the Haihe River, and also participated in the landscape design of North Jiefang Road in the comprehensive environment renovation for Beijing Olympic Games in 2008, so he understands the importance of the details of urban public space. In recent years, he has been working on urban design of public space, from the urban design of open space of five rivers in Tianjin to the urban design of open space of the whole city. In this pilot project, he presided over the making of urban design guidelines for public open spaces. Last year, he learned from his classmates working in the World Bank that Ms. Sadik-Khan was making a tour speech on "Street fight" in Shanghai, Guangzhou and other places. He immediately contacted all sides and made joint efforts with his classmates, which helped Ms. Sadik-Khan add a station in Tianjin for her Chinese trip. The response was enthusiastic.

"Looking back on human history, whether in the west or in the east, cities have always had a good relationship between architecture and public open space, including rivers, streets, squares, parks and so on. Modernism movement in the 20th century and the spread of suburbanization in the United States broke this long tradition. Le Corbusier's technical supreme garden high-rise residential building makes the city a parking lot for lonely towers and a paradise for cars, while the low-density housing and 'lollipop' road system in the suburbs of the United States have maken urban space lost. The city scenes described by Trancik in his book *Finding Lost Space* published in 1986 has become a true portrayal of urban development in the 20th century. At the same time, Trancik's Finding Lost Space puts forward three methods of urban design, namely, the figure-ground method, the linkage method and the place method. All three methods emphasize the design of urban public open space."

"In the 21st century, with the emergence of new and exotic buildings, new urbanism has emerged in the United States, trying to find the beautiful urban memory of traditional European towns from the sprawling suburbs. China, after being taunted as the World Architectural Laboratory, began to seek new architectural directions and explore new architectural and urban design patterns that focus on urban space, people's feelings. With the gradual widespread of urban design, these concepts are gradually brought into the urban design of various cities, gradually forming a space design mode of small street, dense road network, focusing on the urban street, square, park and other public space, and beginning to emphasize human interaction and humanized design."

Jan Gehl wrote in *Lives Between Buildings*:

> Urban design and urban planning involve several different scales. The large scale is the planning of the overall urban layout, which is seen from the perspective of aerial photography. Medium-scale, also known as development scale, is the design of each part of the city, the organizational principle of architecture and urban space, which is viewed urban planning from the perspective of low-altitude helicopters. The small scale is also the planning of humanized landscape, so that people can see the humanized landscape quality intuitively while walking and staying, which is an eye level city. These three levels have different rules and quality criteria. Ideally, the three scales should be integrated into an organic whole. Skyline, architectural layout, spatial proportion and so on should be combined with the spatial sequence, architectural details and urban facilities at the eye level to design and plan. There is only one successful way to design a humanized city, that is, to take urban life and urban space as the starting point. We should start from the horizon, stop from the bird's-eye view, and pay attention to three scales at the same time.
>
> The quality of cities at the eye level lies in the small scale. Enjoying good urban quality on an eye level is a basic human right. The most intimate contact between people and cities exists in a small scale, in the urban landscape seen when walking 5 km/h. Humanization is the most common human activity in planning. It must provide good conditions for people to walk, stand, sit down, watch, listen and talk.
>
> Opportunities for meeting and daily activities in urban public spaces or residential areas create conditions for interaction among residents, enabling people to be in the midst of all living beings, to hear and witness all phenomena in the world, and to experience the performance of others in various occasions. The communication in outdoor space is a form of contact, a possible way to deepen communication, an opportunity to maintain established contacts. Through these contacts, information about the social environment can be obtained, inspired, and a variety of feelings can be generated. Among them, the activity is the factor of attraction. As long as someone exists, he will attract others and seek place to close others. New activities germinate near ongoing events. We should pay high attention to the habits of activities and games, activities and seat selection. People is more interested in the activities of various people in the street itself, and the life inside and outside the building is more fundamental and meaningful than the space and the building itself.
>
> There are many formal changes between different urban models. But in fact, only two typical development models and urban planning ideas are related to the topic of outdoor activities: one originated from the Renaissance, the other from the modern functionalist movement. Life in outdoor space is an independent quality and a possible beginning, providing suitable conditions for necessary outdoor activities, for spontaneous and recreational activities, and for social activities.

> Outdoor living is a potentially self-reinforcing process. F-van Klingeren summed up his urban life experience with a formula: 'one plus one equals at least three'. Positive effect process is like: activity occurs because activity occurs. Negative effect process is like: inactivity occurs because inactivity occurs. Public living space is dismembered and streets become empty. This change is an important cause of damage to public facilities and crime in the streets. High-level activities in specific areas depend on two efforts: one is to ensure that more people use public space, the other is to encourage everyone to stay longer. If there is a well-functioning square, people will meet there. If there is no square, there will be no city life. The place around the intersection will become a gathering place, where at least there is something to see. (Jan Gehl, 1971)

"The General Principles of Urban Design and Management of Tianjin Urban Public Open Space System is a relatively characteristic project. Our team took a lot of time and effort to establish the value standard and evaluation system of open space. Firstly, the academic research in the field of open space is sorted out, and three most representative and influential research teams are selected from a large number of research teams. One is Professor Jan Gail of the Royal Danish Academy of Fine Art, one is Professor C.C. Marcus of University of California, Berkeley, and another is Professor C. Francis. Through the analysis and comparison of the results of their research projects for public space, we establishe the research ideas that takes the humanization dimension as the starting point. Five principles of humanization of excellent open space are put forward. The five principles are progressive from basic requirements to high standards."

"The first is accessibility, which is the most basic requirement for open space, guaranteeing the convenient arrival of people. The second principle is security, to make people feel safe in this space. The third principle is comfort, which means the open space should be set up complete service facilities, and provides a good visual landscape. The fourth principle is diversity, to provide diversified activities to meet the needs of different groups of people. Finally, the most difficult principle is the sense of place, through the design of buildings and environmental landscapes to create an open space reflecting the spirit of the place."

"To build the research system of this project, we use the achievements of western academia for reference. However, if we only copy the foreign standards without considering the characteristics of Tianjin, it is difficult to put forward the standards suitable for Tianjin. For example, our activities here are closely linked to the traditional festival, which is exquisite in what season, what to eat and what to do according to local custom. Our project team members have lived in Tianjin for a long time and know Tianjin very well. We studied the universal principle of open space management and control, and made a thorough investigation on the formation and present situation of open space in Tianjin. Different historical periods and important events in Tianjin have impacts on the form and content of open space in Tianjin. The characteristics of Tianjin urban open space are extracted."

"As Chief Planner ZhuGe said in the morning, Tianjin's urban spatial structure has undergone a gradually evolutionary process. From the original development of

Zhigu Walled along the river, the Old Town along the central axis of the cross, to the nine concessions square with the Haihe River, and then to 'three rings and fourteen radiations' in the 1980s, the urban structure was basically finalized. At the beginning of this century, the comprehensive development and renovation of the Haihe River has realized the upgrading and transformation of the Haihe River as the main axis and the most distinctive open space of the city. In the future, combined with the plan named 'A Green Ring Linked Eleven Gardens', a city structure with ecological green space as its skeleton will be formed on the periphery of the city, which will be organically connected with the surrounding large ecological and rural space. The general principles of urban design and management of our public open space system are to further sort out and design the public space under such a large structure, and to formulate guidelines to guide detailed urban design."

"Tianjin has a distinctive waterfront open space system. The waterfront open space represented by the Haihe River gathers the highest level of urban public centers and core service functions, presents the historic and humanistic resources with the most time trajectory, and reflects furthest the 'sense of the times, locality and humanity' of Tianjin. Tianjin has the open space of waterfront beautiful scenery with city parks. In the past, because the terrain was low and the catchment was concentrated, the Jingu (Tianjin and Tanggu) beautiful scenery of "72 gu (bodies of water) with flowers, as beautiful as Jiangnan (the southern China)" was formed. Urban constructions always utilized river and lake wetlands to construct urban parks featuring wetland water to protect water system resources and conserve urban ecological environment. Tianjin has the characteristic streets of pleasant historic blocks. After key renovation project, it has created the main welcome roads, protected and inherited the historic streets of 14 historic blocks, with kindly and pleasant scale and colorful buildings. Tianjin has high-quality urban squares in key areas, with high design tastes and detailed treatment in place, becoming new places for citizens' activities, displaying urban characteristics, gathering urban popularity."

"Starting from the five principles of humanization, the project team investigates the waterfront, street, park, square and other public activity space in the central city of Tianjin, and found that there are still many problems." Director Chen analyzed facing the PowerPoint. "Let's take the Cultural Center for example. For the sake of management, bicycles are not allowed in the Cultural Center. It takes 800 meters to walk to the museum from the intersection of Yuexiu Road and Leyuan Avenue. Once after having dinner in Galaxy Mall, people wanted to go to the library, so we walked 1000 meters from the south gate of the mall to the library. The entrances to these venues are even not open on the same side, causing more inconveniences. In Tianjin, the links between buildings and parks, green spaces are weak, and the interaction between open space and surrounding public buildings is poor. In recent years, a lot of large-scale open space has been built, with the aim to serve the citizens, while pedestrians and cyclists are inconvenient, even the urban traffic has been severed."

"The pilot project of urban design insists on adapting measures to local conditions, highlighting characteristics and strengthening urban design in waterfront areas. Since 2002, urban design work has been carried out at all levels on both sides of the

Haihe River, strengthening the moulding of characteristic features of the Haihe River. Later, systematic urban design of the banks along the main rivers (the Haihe River, the Ziya River, the Xinkai River, the South River and the North Canal) in the central city has been carried out successively. This pilot work continues the theme of river, and puts more emphasis on the General Principles for Urban Design of Public Open Space in Tianjin Central City."

"In the 'General Principles for Urban Design of Public Open Space in Tianjin Central City', we take open space such as waterfronts, parks, squares and streets as research objects, and promotes the practicability of planning and management of open space system from the overall level. Based on the case study of excellent open space in Tianjin, we put forward the layout, target, classification and control requirements of various types of open space, forming the management and control plans, which are mainly based on sections and supplemented by plans."

"In the process of investigation and data analysis, the project team further subdivides the four types of open space, waterfront, street, park and square, and formulates control standards for each type of space. Let's take the waterfront as an example. In the past, Tianjin's urban development has been set around the Haihe River. The traces of Tianjin's life in various periods can be found along the Haihe River. In recent years, various plannings of the Haihe River spring up and are known far and wide, including the waterfront public centers, and waterfront space mainly with natural landscape, so the waterfront area of Tianjin is very suitable to plan and control in sub-thematics. The project team, after investigation and analysis, according to service functions and historical significance, divided them into urban vitality style, historical style, natural type."

"The project team started with the spatial perception of people in different environments. According to the visual characteristics of human beings, from the opposite bank, the basic vertical angle of view to see the whole building is 18 degree, the ultimate vertical angle of view to see the whole building height is 45 degree, which means that the ideal viewing area is between 18 degree and 45 degree. Among them, 27 degree is the best vertical angle of view in the actual experience of human activities. It is more rigid to determine the height of riverside buildings with fixed proportion before. After adjusting it to a control area, more possibilities are created for designers."

"To the historical style waterfront space, it is suggested the ratio of the height of the main building to the distance from the central line of the river to the building be equal to 1:2. Considering many tourists take a cruise ship to enjoy the beauty of the banks every day in the historical waterfront in the urban center, therefore, the viewpoint is defined as the central line of the river, then further control suggestions are put forward. In addition, we strengthen the control of human-scale activity space. To the vitality and the historical waterfront, vertical hard embankment should be the main type, providing more waterfront spaces. Platform should be set on the other side of the landmark building. To the natural waterfront, the soft embankment is the main type. People mainly do their activities on the bank and watch the scenery."

"Tianjin's climate is not generally suitable for outdoor activities in winter and summer.

Inside the hall of the subway station in front of my home, somebody used to kick shuttlecock there in winter and summer. There is really no activity outside. Later, the fence was sealed off the hall to prevent people kicking shuttlecock. Also, there is no tree around some public buildings, making the place too burning in summer and too windy in winter. Spaces for outdoor activity must be planned in combination with climate conditions. Tianjin is hot in summer, cold in winter, short in spring and autumn, windy in winter and spring, so the design goal is not only to provide open space and beautiful visual landscape with human scale, but also to provide opportunities to enjoy positive weather, and to set up complete public service facilities according to the characteristics of different seasons. In addition, all kinds of open spaces have the problems of lack of diversity and sense of place to varying degrees."

Landscape planner Tan Ming

"I think to the planning for Tianjin's open space, we should consider Tianjin people's favorite activities. Indeed , we always don't need to plan. As long as the venue for activities is provided, people will use it spontaneously. We only need to add related facilities according to the type of activities. Tianjin people are very good at playing, who can fully utilize all the resources around. There are fishers and swimmers along the Haihe River in winter and summer, there are also people who walk birds and play chess in the streets, which may be ignored by who have stayed in Tianjin for a long time." As a landscape architect, Tan Ming is very concerned with open space of Tianjin. After the presentation and discussion of "the green ring and eleven parks" project, he entered in this sub-forum from the sub-forum of Theory and Practice of Urban Design.

Director Chen of Urban Design Lab

"I agree with Mr. Tan. The design of public open space is based on human activities. For the new-built public space, it is really possible for the planning management department to guide the public to use the open space, and further improve it according to the actual use situation. However, for areas with historical accumulation, we should make clear their location and put forward a clear direction of control. For example, considering the existing accessibility problem, the design goal from the accessibility principle is to integrate the open space closely with all levels of urban public centers, to coordinate the layout, to give priority to public transport and slow travel, to ensure the accessibility of special populations. Then we derived the control elements and standards from the design objectives. However, to real projects, the control of open space can't be in accordance with a set of standards. To be honest, I have seen a lot of outputs of urban design or guidelines at the level of master plan, which is difficult to be really implemented in urban construction and management. It is difficult for such projects to put forward rigid and flexible control indicators according to different situations."

"Our guidelines for urban design of public spaces pay special attention to the study of urban streets. Allan Jacobs's Great Street puts forward the corresponding design principles and methods on how to build a great street. What kind of street is great?

It should be of clear boundary and friendly scale, safe and comfortable, conducive to neighborhood interaction. It should be old and mysterious, with appropriate street scale. It should be of diverse business, transparent and friendly building bottom interface. It should be of changeable and continuous building interface. It should be adopted walking priority, to meet the needs of a variety of people. It is the symbol of the city and an important public space , gathering a large number of people and rich street life, good street greening and own distinct characteristics which has been inherited.

The texture of streets and blocks can reflect the differences between cities, including the differences on scale, complexity, route selection and spatial attributes. These differences are related to the historical period of urban construction, geographical environment, cultural differences, urban functions and positioning, philosophical attitudes towards design and politics, and technical requirements. Some streets can be regarded as 'game rules makers', which can bring the clear concept and order to a city or an area. They can form a boundary, or a charming center, and they can let you know where you are. Moreover, the texture of streets can bring about an initial order. The texture of streets and blocks is the starting point of some street design.

The texture of urban streets and blocks can bring order and urban structure to a city. Whether through careful design or through the cumulative evolution, the texture of urban streets and blocks can bring order to a city and an area in a city. In a smaller two-dimensional scale, there will also be many most prominent streets to establish order and center for their surrounding areas, depending on the means of comparison, which are usually those longer, wider and more regular streets than surroundings.

It can make it easy or difficult for us to get from one place to another. It can also tell us whether an area is suitable for walking. The more content a square mile contains, the more streets it seems, and more places for people to move around. Although the areas covered by streets are not necessarily better, most of the great streets we encounter are in relatively rich-content areas.

The texture of streets and blocks changes over time. Disasters (wars or fires), changes in social values, obvious changes in economic planning, fashion preferences and design philosophy are all causes of changes in urban scale. The material conditions for shaping great streets include accessibility, the ability to gather human activities, publicity, livability, safety, comfort, and the potential to encourage people to participate and assume social responsibilities. The indispensable conditions include: walking places with comfortable walking scale; climate-related comfort indicators should be an integral part of a great street; clear boundaries; pleasant landscapes; street permeability; coordination; maintenance and management.

Quality doesn't depend on money. In a great street, the key to all problems is appropriateness and fitness, whether it is about materials or maintenance. This is precisely the most advocatable standard in the construction of public space. The quality of icing on the cake includes trees, public space, environmental design elements, accessibility and topography, residential density and land development mode. A more rational urban development mode is imperative: we can no longer rely excessively on large-scale public development, highly centralized property rights and exclusive planning and design patterns, but take the gradually accumulated mechanism of material environment construction and dialogue as legitimate and reasonable. This development attitude is based on acceptance, love and expectation of urban life, encouraging people to live in a healthy public environment.

The street is the most important component of the public sphere, and no other urban space can replace it. It is an important part of public property, community life, and is directly under the jurisdiction of public institutions. It is not only exciting but also challenging to have the privilege of designing streets to meet public needs and create community life. If we find the right road in a street, we will take the right road in the construction of the whole city, so that we can be worthy of the residents in the city, which is the fundamental purpose of all the design. (Allan Jacobs, 1995)

Director He, Department of the Historic City

"Like what we just discussed about the space with historical memory in Tianjin historic blocks, shall we formulate more targeted control standards? What are the project team's considerations in this regard?"

Director Chen

"In each further subdivision of the four types of open space, waterfront, street, park and square, there is always historical style. We have defined the location of historical style space and put forward stricter control standards than other types of open space. However, specific space should not be stipulated one by one on this level. Take the waterfront space as an example, even the historical style waterfront can't be controlled by a set of rigid standards. The concessions in Tianjin were also divided along the Haihe River in that mode. The street texture, architectural style and environmental characteristics of each nation concession are different. As the urban design management at the level of overall planning, the compilation of general rules should guide the making of control guidelines for waterfront, streets, parks and squares in urban design, not to make detailed provisions for each specific situation, but to put forward suggestions on the problems and ideas that should be considered in the compilation of detailed rules, which is the direction of the development of legalization of urban design."

Chapter IX

A More Livable City

51. Urban Design Guidelines for New Communities

Luo Yun, Chief Planner, Urban Design Lab of TUPDI

After a short break, the second half session of the sub-forum entitled of "The Theory and Practice of Urban Design in Tianjin" is to commence. Luo Yun, Chief Planner of the Urban Design Lab of TUPDI, prepares to begin her presentation. Her topic is about Urban Design Guidelines for New Residential Communities in Tianjin. This is the most highlight and sophisticated one in all the urban design pilot projects. In order to fulfill the project with high level, there formed a gargantuan project team with more than 30 members, including urban designers as well as planners , architects, traffic planners, civic engineers, etc. from 9 units, from six different units of TUPDI, such as Urban Design Lab, Vision Company, Binhai Branch, Architectural Design Branch, Transportation Research Center and Civic Planning Branch, and other four companies, such as Tianjin Architectural Design Institute, Bohai Urban Planning and Design Institute, Huahui Architectural Design Co. Ltd. and Huahui Environmental Design Co.Ltd., led by Urban Design Department of the Municipal Urban Planning Bureau. Mr. Dan Solomon is very concerned about the project and present at the sub-forum.

Luo Yun graduated from Tianjin University with a master's degree in urban planning. After graduation, she has been working in TUPDI for more than ten years and participated in many large-scale urban design projects, such as Cultural Center in the central Tianjin and Binhai Cultural Center. The projects are almost urban design about public buildings and public space. As her professional experience accumulates, she gradually realizes that the residential community is the most important field of urban than any others. Because he ultimate goal of urban design is to build a good living environment for residents, therefore, residence and community are the most important design contents deserved more attention, But for many years we have paid less attention in China. With the increase of her age and life experience, she found that many colleagues in her office often suffered and worried about the rocketing housing price and children's schooling. Especially now that she's pregnant herself and will have her first child soon, she feels more and more empathetic.

Many civil servants of the Municipal Urban Planning Bureau attend this sub-forum, including three female directors from the Urban Design Department, the Detailed Planning Department and the Construction Management Department respectively: Madam Liu Ke, Madam Zhao Song and Madam Guo Nan. All of them are mid-ages and have rich career backgrounds and colorful lives. As civil servants and Party members, they are currently carrying out an education activity with the theme of "Never forget the beginning, Remember the mission". The purpose of the activity is following the requirements of The Nineteenth National Congress of the Communist Party of China to find specific measures to satisfy people's yearning for a better life as the Party's mission. To formulate a new design prototype for new residential community is indeed a practical subject closely related to the people's life. During this period, many discussions have been held on the proposal of the new urban design guidelines. It is supposed to be many intense conflicts there because of many radical changes to current regulations. Thanks to the agile coordination of female directors, the project as general goes on promptly. The Municipal Urban Planning Bureau has more than 30 departments, with more than two-thirds of the department directors being female, but less than one-third of the 13 bureau directors are female. In the history of the Bureau, there has never been a female head director. Perhaps in the future, women will assume more and more leadership positions with their more emphasis on communication, care, tolerance, participation and consensus-building.

Director Liu

"Hello! Everyone, thanks for your presence. I'm Liu Ke, director of the Urban Design Department of Tianjin Urban Planning Bureau. Please allow me make a brief opening statement before Luo Yun's presentation. As we know, residence is the main function of the city, and the most basic type of architecture in the city. Apart from every kinds of urban centers at all levels, residence and residential communities determine to a large extent the basic tissue, texture, quality of spatial form and features of the city. Numerous residential buildings comprise the varieties of communities and all of the residential communities forms cities and regions"

"At present, whether in Tianjin or other cities in China, urban design has prevailed. Urban design and urban design guidelines as well have been drawn up for key urban areas, and meticulous management has been carried out according to the guidelines. We can see progresses in these areas in many cities at home. However, as the biggest share of urban construction in China, vast of the residential development projects still follow the traditional residential area planning method with distinct characteristics of planned economy. If we do not discard the outdated traditional residential area planning method thoroughly and make a new one from scratch, our so-called urban design is not the real urban design." planning and design of most

"So, Tianjin, as one of the second batch of urban design pilot cities, put our focuses upon new approach of urban design for new residential communities. Besides a dozen of new community urban design pilot projects, the making of The Urban Design Guidelines for New Residential Communities in Tianjin is the main task as the key field of reform and innovation. Recently, The Ministry of Housing and Urban-Rural Development issued and implemented the newly revised "Standards for Planning and Design of Urban Residential Areas", which made some improvements, but not far enough. Our guidelines of new residential community, on one hand, intends to complement the new standards, on the other hand, it tries to further refine the standards to accommodate to the actual situations in Tianjin, matching various types of housing, communities, in different zones and regions of the city, to avoid a simple one-size-fits-all approach. This guideline also combines with two new standards recently issued by Tianjin Urban Planning Bureau, such as 'Technical Regulations for Detailed Regulatory Planning of Tianjin (Trial Implementation)' and 'Technical Standards for Planning and Management of Construction Projects of Tianjin (Trial Implementation),' which will make the guidelines more practical. Now, I would like to invite Chief Planner. Luo to report first, and please feel free to give your comments afterwards."

Luo Yun, Chief Planner of Urban Design Lab

"Thank you, Director Liu. It is a great honor for me to speak first on behalf of our team. Our project is significant and huge. At first, I would like to report on the background of our research project, especially on our recognition of the important role of houses in city and the perception of the importance of housing diversity. This seems a simple question and an easy work, but it's not. By reviewing the urban planning history, reflecting on the zigzag path of modern urban planning as well as its successes and failures, learning the latest contemporary urban design and architectural theories, and finally combining all of these with the actual situation of our country, we carry out this exploratory work."

"As we all know, in ancient European cities, residential buildings, as background architecture in general, were tightly integrated into blocks, making clear boundaries of streets and squares and setting off brilliance of public buildings of palaces and churches. The unified texture of residential buildings formed not only continuous street walls but also rich urban spaces. Ancient cities in China, such as Chang'an of Tang Dynasty, had a perfect symmetric layout of axis, palaces, markets, city walls and gates, road system, as well as a strict Lifang system of residential areas, which were essential parts of the city. In the past, most of the TV dramas depicted the emperor's life within the palaces. Recently, The Longest Day in Chang'an (Twelve Hours of Chang'an in Chinese), a popular hit on the Internet, was the first one to depict the ordinary people's life on streets markets in ancient Chang'an. It truly reflects the appearance of Chang'an city at that time. In the planning of Beijing in the Ming and Qing Dynasties, the courtyard house is the basic type of residential buildings. The courtyard houses in the shades of green trees, together with the central axis of the Forbidden City, perfectly constitute the marvelous city of 'unparalleled masterpiece' in the history of mankind."

"The main reason for the emergence of modern urban planning is due to the serious problems of housing for working classes. After the Industrial Revolution, the population in Europe rapidly aggregated in cities, resulting in an extreme shortage of housing. The urban problems caused by the overcrowding of dark and narrow dwellings, the lack of basic lighting and ventilation, clean water supply and sewage discharge facilities, and poor sanitary conditions in residential areas with garbage everywhere and sewage crossflow, have led to the emergence of modern urban planning. At the end of the 19th century and the beginning of the 20th century, Ebenezer Howard founded the theory of Garden City. As the cornerstone of modern urban planning, it tried to solve the urban diseases by building garden houses in the integrated areas of urban and rural. But the model of residential autonomy proposed by him can't meet a large number of housing needs."

"The development of modern construction technology at the end of the nineteenth century has made it possible to solve the problem of housing shortage by means of large-scale industrialization. However, the classical architectural design tradition represented by the Academy of Fine Arts in Paris limits the development of new buildings. Le Corbusier, a master of a generation, praised modern industrial designs such as automobiles, ships, planes, and put forward the concept of 'A house is a machine for living in' and the planning idea of 'Radiant City'. Under Le Corbusier 's initiative, modern residential buildings are presented to the world with a new look and layout. The technological progress of modern buildings has greatly improved the construction level of residential buildings and the physical environment such as lighting and ventilation. At the same time, the methods of neighborhood planning characterized by functional zoning and traffic diversion has completely changed the traditional planning and design methods that has been continuing since ancient times. It's peremptorily break from tradition, like Le Corbusier's cry in his book *Towards the New Architecture*: New Architecture or Revolution?"

"Le Corbusier's architectural thought is advanced. He praised the beautiful shape and golden proportion of classical Greece and Roman architecture, but instead of imitating the details of classical architecture with reinforced concrete, he created unprecedented golden ratio and artistic sense with the concise and industrialized design techniques of modern building materials and technologies. That liberates architectural design from the rigid doctrine of classicism, stimulates strong creativity and guides the development of modern architecture. But his idea of dismantling historic blocks in the centre of Paris and planning to build a 'Radiant City' with large green areas and high-rise buildings didn't take into account the historical context of the city. However, under the circumstances of nobody expressing an opinion at that time, Le Corbusier tried to arouse the world's attention through such an audacious move, which has its own historical background and is understandable. He boldly brought the idea of industrialization into urban planning, advocated the transformation of the city with a new planning idea, envisioned the city of high-rise buildings, modern transportation network and large areas of green space, and created a modern living environment full of sunshine for mankind."

"Le Corbusier was concerned very about public housing. In 1928, he led the establishment of an international organization called Congrès International d'Architecture Modern (CIAM). He personally drafted the famous Athens Charter, pointing out the four main functions of work, recreation, transportation and living in modern cities, and the principles of modern urban planning. Many of them seem quite right even today. Earlier in this year, the Tianjin Museum of Art and the Le Corbusier Foundation co-sponsored the Le Corbusier' Exhibition—The Colour Symphony. The exhibition chronologically introduces several stages of Le Corbusier's architectural works: the period of purism, brutalism, lyricism and Bright, accompanied by exquisite drawings and models, including Weissenhof Estate, the Villa Savoy, Unite d'Habitation in Marseille, Chapelle Notre-Dame-du-Haut, as well as Le Corbusier's paintings, books and the original magazines edited by him as the editor-in-chief. Tianjin Urban Planning Society organized collective exhibitions and seminars ad loco to review Le Corbusier's classical ideas. Although it was his voice of a hundred years ago, we are still deeply moved today. At the same time, we frankly defer to the fact that although modern architecture and urban planning movement has played a great role in improving the quality of residential buildings and living standards, it has also led to serious urban problems. There are many reasons for these problems, and Le Corbusier should not take the blame alone."

"It was the Soviet Union who first practiced Le Corbusier's ideas widely. After the October Revolution in 1917, the Soviet Union carried out state-led modernization and industrialization. The great success of 'the First Five-Year Plan' for socialist transformation and industrialization has attracted the admiration not only from communist countries but also from the United States and other Western countries, and it inspired active architects like Le Corbusier. He devoted himself to the construction of the Soviet Union. In 1928, Le Corbusier was invited to participate in a design competition for a new headquarters for the Central Alliance of Consumer Cooperatives in Moscow, and finally his plan won. Later, however, he went away in anger because of the unfair treatment and the failure in competition of Moscow Master Plan. The Soviet architects inherited the modernism creed of 'design for the general public', put forward various types of residential unit buildings, studied the space efficiency, economy and functionality through scientific calculation, interpreted Le Corbusier's concept of 'A house is a machine for living in' in their own way, and built the first modern collective housing in Moscow."

"After the end of the Soviet War of Patriotism, the task of post-war reconstruction was arduous, including the restoration of industrial production and urban life, and a large number of houses were urgently needed. In order to improve the efficiency of construction, they had changed the building type from traditional 4-6 storey walk-up to 12-14 storey high-rise tower and slab. During this period, the emphasis on architectural forms hindered the promotion of new technologies and industrial construction methods. In 1954, the Politburo of Soviet Union criticized Traditionalism, Retro-classicism, Formalism and Decorationism in urban planning and architectural design, emphasizing the importance of functions, social benefits and the use of new technologies in urban construction. After 1955, the Soviet Union entered a period of large-scale construction again, which experienced the industrialization stage of extensive development. As housing construction was invested solely by the state, it was an important means to formulate unified standards. Standard quotas and batches of residential area planning and construction were the most commonly used methods. In fact, as early as the 1920s, the idea of residential area detailed planning had appeared in the Soviet Union. With the promotion of the theory and method of neighborhood units abroad, the theory and the method of residential area planning in the Soviet Union gradually took shape. The standard design of residential buildings played a great role in meeting the needs of housing construction, but at the same time, it caused the monotony of housing and residential areas. I do not know whether this large-scale monotonous housing and residential areas had some attributes to the final collapse of the Soviet Union."

"In the United States, after World War II, large-scale Urban Renewal Campaign and affordable housing construction adopted the so-called Le Corbusier-style modernist urban planning method and built an amount of large-scale affordable housing residential projects, but soon encountered serious setbacks. The famous case is the Pruitt-Igoe Housing Project, designed by architect Yamasaki. This is a slum urban renewal project implemented in the mid-1950s. More than 30 high-rise buildings with 11 floors were planned and constructed to provide affordable housing for 8,000 low-income families. Later, due to the low rental rate, lack of maintenance and serious social security problems, the residential area was demolished by blasting at 3 p.m. on March 16, 1972, whose life expectancy was only 16 years. Architectural historian Charles Jencks described this moment as 'Modern Architecture Died'. At the moment of the explosion, Americans' belief in modern architecture movement began to collapse. Over the next 20 years, most of these large-scale affordable housing were demolished."

"In Europe, the urban renewal and the construction of new towns after World War II have generally adopted Le Corbusier-style planning method of high-rise residential buildings and large-scale construction. They demolished a large number of historical buildings to build ring roads and large-scale residential projects, resulting in drastic impacts on historical urban district, leading to the loss of urban space and the exacerbation of social problems. Since the 1970s, developed countries have stopped building large-scale residential projects. The practice of modern large-scale residential area and neighborhood planning has only a short life span of 50 years in the West, but the impacts which brought to the whole human society are tremendous."

"The development of modern residence and residential area planning and design in China has been influenced by Ebenezer Howard since the beginning of the 20th century. The planning methods of 'Garden City' and 'Neighborhood Unit' are highly recommended by planners and architects of that era, but never widely implemented in practice due to the turmoil of society in modern China. Since the establishment of People's Republic of China, the socialist public ownership and welfare housing distribution system were adopted, simply copying the mode of the Soviet Union. Due to a series of wrong policies, after the Culture Revolution, the whole country faced more serious housing shortage than ever before, and the general living conditions deteriorated badly.

"It is the reform and opening up since 1978 that altered the whole situation. In the past 40 years, through the market-oriented reform of housing and land use system and the quick development of the real estate market, China has basically solved the housing shortage problem and achieved unparalleled results in the world. In this process, with the deepening of reform and the vigorous development of real estate industry, Chinese socialist market economy system has been initially established, but we still use the theory and method of residential area planning with distinctive characteristics of planned economy. In fact, after the reform and opening up, although various genres of contemporary planning theories in the West, including the criticisms of modernism, have poured in, but we have not really realized the harm of modern urban planning, and have repeated the same mistakes again and again." Over the past two decades, with the promotion of current technical standards of residential area planning, detailed regulatory planning, housing construction in China has been growing at an alarming rate. Residential development projects generally adopt simple and uniform types of buildings that can be quickly reproduced, mostly 30-storey, 100-meter-high towers or slabs. dozens of repeated high-rise residential buildings are piled together, surrounded by guarded-gate walls, forming a typical large-scale enclosed residential area, leaving only negative walls to the city space. Monotonous residential buildings and super-enclosed residential estates, ignoring pedestrian and bicycle traffic with extremely wide lackluster streets, have become the mainstream urban form of residential areas in China. This is one of the root causes of urban problems in China today, such as traffic congestion, environmental pollution, property management, etc., leading to the humdrum of urban fabric, the lack of characteristics of cities and the decline of the overall quality of urban built environment ."

"More than a decade ago, we gradually realized the seriousness of this problem began useful attempt. Binhai New Area has been trying to reform the traditional residence area planning method. In 2000, Tianjin Economic and Technological Development Area (TEDA) formulated the urban design scheme by SOM of 'narrow roads, dense network' in its living area and organized its implementation. In 2005, the International Call for Urban Design Schemes of the Central District in Binhai New Area also highlighted the layout mode of 'narrow roads, dense network, small blocks'. After Binhai New Area was incorporated into the national development strategy in 2006, 'narrow roads, dense network' layout was promoted in living areas of several functional districts, such as Central Business District, Airport Economic Zone, Buolong High-tech Zone, Beitang and other areas. In 2007, the governments of China and Singapore cooperated in the construction of China-Singapore Tianjin Eco-city. In its master plan, innovative contents such as the index system of eco-city, the model of 'Eco-cell' planning, so-called eco-valley and the reform of community social management were taken into account. In 2009, the overall urban design of the core area of Binhai New Area, and the urban design of the Southern New Harmony Town named after Robert Owen's New Harmony, also adopted the new urbanism theory and method to create a public transport-oriented layout with 'narrow roads, dense network' as its main characteristics."

We recognize that in order to improve the urban quality of Binhai New Area and highlight the characteristics of its urban space, we must fully consider the relationship between residential buildings and cities, and explore a suitable residential prototype and a new residential community planning and design model. In 2011, under the organization of Tianjin Binhai New Area Urban Planning and Land Resources Bureau, Binhai Branch of TUPDI, together with Mr. Dan Solomon and Tianjin Huahui Environmental Design Co., carried out an urban design study program on 1 square kilometer large residential community named as New Harmony. Located in the northern starting area of the Southern New Harmony Town, this site is used to be the Tianjin Port Bulk Cargo Logistic Center, which will be relocated to the new port area in order to avoiding coal and ore pollution. According to the city master plan of Binhai New Area, it will be transformed into residential community. This urban design is a very useful attempt, and we have followed and witnessed the whole process. At that time, the goal of the work was very clear That it was to change the traditional residential area planning and design methods, to discuss and reflect on the 'conventional' content and norms in depth, and to try to find a new residential community planning prototype, so as to provide a practical model for the promotion of 'narrow roads, dense network' in Binhai New Area." Mr. Dan Solomon did his best and made breakthrough with a marvelous proposal of the project.

"With the deepening of the work, in 2012, the owner of the land, Tianjin Port Bulk Cargo Logistic Center, commissioned Dan Solomon and Tianjin Huahui Environmental Design Company to deepen the urban design and conceptual design of residential buildings in one neighborhood of the New Harmony Community. In the process of the work, extensive investigations were conducted, domestic and foreign cases were analyzed, and two expert seminars were held. Mr. Zhao Guanqian, Mr. Kai Yan, Ms. Zhang Feifei and Prof. Zhou Yanmin, experts in residential design and residential area planning in China, fully affirmed and highly appraised the new community planning and design scheme presided by Mr. Dan Solomon. We have also learned a lot of useful things in this collaborative urban design work. Mr. Dan Solomon also attends our meeting today, and I propose that we all express our heartfelt thanks to him." The auditorium breaks into warm applause, and Mr. Solomon stands up, bows with his both hands and smiled to greet everyone.

Luo Yun

"Through the preliminary work, we are more and more aware that there is a drastic conflict between the new community planning model of the New Harmony demonstrated by Mr. Dan Solomon and the current conventional residential area planning method, as for various norms, relevant technical standards and technical regulations of construction planning management. Therefore, comprehensive reform must be carried out to make the new community planning model feasible. Since the New Harmony did not yet start to develop due to various reasons, in order to clarify the specific content of the reform, in 2015, funded by the Housing Investment Company under the Housing Security Center affiliated to Tianjin Binhai Urban Planning and Land Resources Bureau, the assumed planning and design work, including the regulatory detailed plan, the construction detailed plan and architectural design scheme were entrusted to Binhai Branch of TUPDI and Bohai Urban Planning and Design Institute. The purpose is to test the specific conflict points with the current relevant norms and standards through the actual planning and design work, and to clarify the specific content of residential planning and design norms, related technical standards and regulations that need to be reformed and innovated."

"At the end of 2015, the Central Urban Work Conference was held after 37 years apart. In early 2016, the Central Committee of the Communist Party of China and the State Council issued *Several Opinions on Further Strengthening the Management of Urban Planning and Construction*, outlining 'the 13th Five-Year Plan' or even longer blueprint for Chinese urban development. The document points out precisely the seriousness and reasons of the problems on the planning, design and construction in China, indicating the direction of future development. In particular, it indicates the importance of open communities to urban development and solving urban problems in China: Street network structure should be optimized, which needs, to strengthen the planning and construction of neighborhoods, to clarify the area of new neighborhoods step by step, and to promote the development of open, convenient, appropriate-scale, complete supporting and harmonious neighborhood living blocks. New residential buildings should promote block mode, in principle, enclosed residential areas should no longer be built. The concept of 'narrow roads, dense network' urban road layout needs to be set up."

"So far, the issue of new community planning and design patterns has received the highest level of attention. In my opinion, since the reform and open up, Chinese urbanization has made tremendous achievements, but there are also many problems whose reasons are complex. There are three main reasons if focusing on urban planning. The first is the city master plan guided by resource consumption and expansion; the second is the regulation to meet the needs of real estate development and land finance; and the third is the regulation of residential areas planning and quota indicators, sunshine spacing in the Soviet model. The three levels of planning are the main culprits of creating similarity, large enclosed residential areas, high-rise residential buidings 'corn land' type spreading everywhere in the country. At the same time, it is accompanied by many social contradictions, such as high housing prices, high inventory, unbalanced urban and rural development, urban and rural fragmentation, and the solidification of the vested interests from urbanization."

"The Central Urban Work Conference and the document have sounded the clarion call for urban planning reform and innovation. We should replace the traditional city master plan with the ecological protection oriented spatial planning, improve the regulatory detailed plan with the help of people-oriented urban design, and replace the traditional planned economy-rich residential area planning method with the new pattern of residential community urban design. Among them, the new community urban design pattern is the foundation. To Chinese cities, solving the problems of housing and community is not only related to the vital interests of the common people, the improvement of urban quality, but also to the healthy development of real estate and the sustainable development of Chinese solid economy. It is one of the major subjects of implementing territorial spatial planning, building a well-off society in an all-round way and realizing the Chinese dream of the great renaissance of the Chinese nation."

"Livable is the foundation of a healthy city. Everyday happiness is the source of the city's overall happiness. When a large number of modern residential areas meet people's basic physical needs such as sunshine and ventilation, people begin to recall the traditional life which seems crowded, noisy and lively. Whether it is the alleys in the north, courtyards or rain halls and alleys in the south, it has brought a pleasant life experience to the residents. We believe that in an ideal community, the satisfaction of the basic functions of physical space is only the first step, and the need for emotional quality requires careful consideration by the designer. By meticulous design, we can create charm courtyard life, street life, city life, let residents really enjoy the peace and happiness of life."

"The experiment of New Harmony in Binhai New Area strengthened our confidence. Through rational system design and reform of traditional residential area planning and design methods, we can create affordable well-off housing and higher quality living environment for the majority of middle-income groups. We also realize that the current practice of residential area planning and design in China has lasted for more than 30 years, forming a huge system involving all aspects, and it is very difficult to carry out reform and innovation. However, if we want to implement the requirements of the Central Urban Work Conference and improve our residential buildings and community design, we must carry out systematic reform. It's like Le Corbusier 's cry in his book *Towards the New Architecture*: New Architecture or Revolution?"

"200 years ago, Robert Owen, one of the founders of Utopian socialism, in his own factory in The Town of New Lanark, Scotland, tried to reform the model community and modern personnel management and achieved success. He believed that the environment can change people. In 1824, he purchased 30,000 acres of land in Indiana, the United States, and built the New Harmony, implementing a system of labor for all and distribution on demand. In order to form a good living environment, Owen specially designed the prototype of the community `Parallelogram'. Although the communes collapsed a few years later, Owen's contribution to the exploration of equality for all and the ideal collective housing model was groundbreaking and epoch-making in human history. He was decades ahead of E. Howard in putting forward the theory of Garden City. Some urban planning historians called Robert Owen the father of modern urban planning. Although Owen's own experiment was unsuccessful, his exploratory spirit guided and inspired future generations in pursuit of the ideal living and working environment of mankind. We recognize that the Soviet Union's residential planning with socialist ideals is in line with Owen's ideas. Facts have proved that residential area planning is a failure, but this will not affect our pace of moving forward. At present, Chinese cities are facing a tremendous era of change again. The central government clearly proposes to change the traditional planning methods and promote the street and block system, which requires a thorough reform of the current residential area planning and design theory system and technical norms system in China. We need Robert Owen's spiritual guidance for a new exploration of harmonious communities. The new community research neighborhood of Binhai New Area is named New Harmony, which also shows the attitude and determination to learn from Owen."

"Although for various reasons, the Binhai New Harmony has not been implemented yet, through this project research, we have a very deep understanding of the urban design of new residential communities, and realize that if we want to achieve the new community layout of 'narrow roads, dense network' proposed by the central government, we must reform and innovate. It is not only necessary to adopt new design methods of residential communities in urban design itself, but also to reform the planning and design norms of residential areas and the corresponding technical regulations of urban management. In the past few years, Tianjin has continued to make efforts and made some achievements. For example, in the newly promulgated Technical Regulations for Detailed Regulatory Planning of Tianjin (Trial Implementation) issued by Tianjin Urban Planning Bureau, in order to form 'narrow roads, dense network' layout and humanized streets, the green line of living roads has been cancelled, and green space has been centrally arranged into parks; the red line size of turning radius of roads has been reduced, and the density of residential buildings has been increased. In the Technical Standards for Planning and Management of Construction Projects of Tianjin (Trial Implementation), the distance of building setback is reduced from 8—15 meters to 3—5 meters, and the area of parking garage is not included in FAR, etc."

"In order to collect the experience of these pilot projects and gradual reform into theory and method, we began to design The Urban Design Guidelines for New Residential Communities in Tianjin. The guidelines are formulated by our team members step by step, allowing minor mistakes. In this sense, the presentation I am reporting today is strictly speaking still a stage achievement and will be continuously improved in practice in the future. Based on the traditional and existing different types of residential building and residential communities in Tianjin, we summarize and refine them. Firstly, from the perspectives of quality of life, diversity of housing and community, ecological environment and humanized transportation, we put forward the living vision for 2035 and 2050. Systematic guidelines such as rural-to-urban transect, intensity and density of residential development, types and combinations of residential buildings, diversified residential buildings and residential communities should be formed to guide specific urban design."

"We realize that the layout model proposed in the New Harmony Pilot Project is only one kind of the new community models. It expresses a kind of thought and idea. We can't use one model to plan and design all the communities, otherwise we will return to the old road of modern urban planning. We need the diversity of residential buildings. We need to truly respect people and see them. People are diverse and families are diverse. In order to do a good job in the new community urban design guidelines, we must have diversity, a new understanding of housing, and a examination of the significance of housing as a home."

52. Housing · Home · Mind · People (Housing as Home)

Luo Yun

"This year(2019) is the 70th anniversary of the founding of People's Republic of China and 40th anniversary of the reform and opening up. Our country enters into a new era. The changes of each of our families are the epitome of Chinese cities. For decades, housing has almost become the most concerned topic among ordinary people. Just to think about, what does the housing mean to everyone? There is an old saying: 'don't talk about ghosts or gods but talk about people.' 'Clothing, food, housing and transportation', housing is one of the four elements of life. Ancient said: Housing is also which we can depend on. People live in their house and view it as their home. If they live comfortable in the house, the whole family will be prosperous. In Chinese traditional culture, the family takes the house as the core and achieves a perfect combination with lifestyle, cultural tradition and spiritual pursuit at that time. Housing is not only the carrier of material life, but also the carrier of spiritual aesthetics, even the space of self-psychological healing. From the courtyard to the gardens in Suzhou, the rich forms of residential buildings all express the life pursuit of 'self-cultivation, family unity, country management and world peace' and the spiritual realm of 'harmony between man and nature'.

'Similar in the West, whether Christian culture or Jewish culture, the family also enjoys a noble status. Housing is the basic geographical dividing line, correspondingly, it is also the psychological dividing line of self-consciousness. In the United States, independent detached housing is more regarded as a symbol of the existence of life, a symbol of moral nobility. British and American laws and customs regard everyone's family as their refuge, and the destruction of housing is tantamount to the destruction of individual existence."

"Kenneth Jackson wrote in *Crabgrass Frontier*:

Throughout history, the treatment and arrangement of a particular dwelling place by a nation has shown more significance than other creative works of art. Housing is external expression form of the internal heart of people; no society can fully understand the members without considering their residence. No place can compare with home! The construction of houses is an important achievement of any civilization.

Everyone longs for a house of his own in his heart. Andrew Downing said: We believe that in the secular world, personal housing embodies tremendous strength and noble virtues. In his view, the nuclear family is the best form of society, for a nation, an independent housing has great social value. If setting your home in the countryside, trust, beauty and order will dominate the small world of the family. Commercial land planning can satisfy people's yearning for quiet life, their closeness to nature and their need for convenient urban life by creating an artificial environment, so as to attract residents. (Kenneth Jackson, 1987)"

"From the perspective of the East and the West, the meaning of housing is so similar. Residence is essentially a place for people to project themselves, which is external expression form of the internal heart of people. Some people say that housing is a personal universe, housing is a personal city. Residence has become the most important element of life, expressing people's fundamental position on nature, society and life itself."

"Looking back at the 40 years of reform and opening up in China, as happened in tens of millions of families, along with the housing system reform and real estate development, the housing conditions of urban residents have undergone tremendous changes. Our per capita housing floor area is from only 7 square meters, up to 39 square meters per capita now, ranking at a higher level in the world. However, when the American dream of "a house, a car, decent work" is used to declare the existence of life in the West, the uniform "flat" seems to make us more and more distant from our own heart, social life and nature. We always seem to feel that there is something missing in apartment unit, I think it may be the lack of the "soul" of Chinese residences."

"Over the past decades, China was building about 3 billion square meters of commercial residential buildings annually, and completed and sold more than 1 billion square meters. Some people laugh that the Chinese have built all the houses for the world. Real estate under land finance has become the most successful urban game rule to promote Chinese GDP miracle. But can we build houses to meet people's future needs? What do houses mean to the city?"

"The house is for living, not for frying(speculation). This sentence has a profound meaning. People's demand for housing includes both material and spiritual connotations. Our traditional dwellings not only meet the needs of family life and living, but also meet the needs of social activities of many families, as well as the emotional and spiritual needs of people. After the founding of New China, we copied the welfare housing system of the Soviet Union. The design of housing met the basic needs of life. Standardized design was adopted, which did not give much consideration to people's spiritual life, not alone the inheritance of housing tradition and culture. After the reform and opening up, especially after the reform of housing system, there appeared many types of commercial housing. Besides functional spaces such as bedroom, independent living room, dining room, study and so on to meet higher needs Are imperative in the design of housing. Functions of kitchen, toilet and storage are constantly improved with the progress of technology. The level of housing decoration is also constantly improving, with a variety of styles. But in general, our contemporary housing is still function-oriented, which is the impact of the fast pace of modern life, and little consideration is given to the spiritual needs. Apart from the functions of living, bedroom, dining room, kitchen, bathroom and storage, the future residential buildings need to consider the halls for family worship, study rooms for reading and meditation, space for fitness activities and working, and carefully design to express the individualized and formal cultural meaning of space. Family members have common senses of not only comfort and freedom, but also support and companionship, as well as spiritual and emotional expression. Home is the universe in the heart of man."

"What kind of housing will we live in by 2035 and 2050? This is a crucial question that must be answered If we want to make a correct decision about the real estate market. The current residential floor area of urban residents per capita in China has reached 39 square meters, and the annual sales of newly built commercial housing is 1.7 billion square meters. Considering the stable transition of the real estate market and the growth of the number of urban residents, the per capita housing floor area of urban residents in China should be about 50 square meters in 2035, reaching the level of developed countries. Therefore, in order to meet the needs of future residents for housing, new housing construction should generally be in accordance with relatively high standards, including housing type and form."

Luo Yun

"Apart from the housing type of the residence, the form of the house is also very important. The home should have a form. It is not only about the shape of residential buildings, but also about the deep-seated issues such as people's spiritual needs, family cohesion, community sense of place and urban culture."

"Residential buildings are not isolated buildings, and residential communities are not isolated residential areas. Apart from reflecting the urban form and characteristics, urban residential and residential communities also have deeper social, cultural and spiritual significance. Aldo Rossi's Urban Architecture is the most important work on architectural typology. Rossi's typology divides city and architecture into two levels: physical and image. Physical cities and buildings exist in time and space, so they are historical. They are made up of concrete houses and events, so they are functional. Rosie believes that physical cities are short-lived, changeable and accidental. It relies on image city, which is analogical city. Image city is composed of sense of place, block and type. It is a place of psychological existence and collective memory, therefore it is a form which transcends time and has universality and persistence. Rosie describes this image city as a living body, capable of growth, memory, and even pathological symptoms. In order to maintain the form of a city, the introduction and selection of new building types, especially the selection of a large number of residential types, should be particularly careful in urban renewal and construction. Because the introduction of a completely alienated and exotic housing type will lead to a huge change in the overall urban form and appearance."

"Historically, when Paris was renovated, Georges-Eugene Haussmann chose the prototype of neoclassical three-stage facade house, which made the streets and squares of Paris City orderly, and created a perfect urban space and image. In the planning of Barcelona New District in Spain, Ildefonso Cerda prescribes the form of octagonal enclosure, which makes Barcelona New Area distinctive. These cases show that housing plays a very important and decisive role in the function, quality and characteristics of a city. We assume that if Paris really built a large number of modern high-rise residential buildings in the centre of the city as envisaged by 'Radiant City' of Le Corbusier, the destruction of the city's style and history and culture would be fatal."

Director Guo Nan, Construction Management Department, Tianjin Urban Planning Bureau

"A city should be particularly careful in choosing new types of residential buildings. Mr. Wu Liangyong's exploration of Juer Hutong in Beijing aims to find a new type of residential building with modern function and traditional Chinese style, to coordinate with the overall urban pattern of old Beijing City and to protect the style and features of the ancient capital, which is the reason why he has won the UN Habitat Award. Unfortunately, it has not been widely promoted in Beijing. At the beginning of his book *Global City Blues*, Mr. Dan Solomon fully appreciated Mr. Wu Liangyong's insistence on exploring regional architecture and criticized the master of architectural trends, Mr. Koolhaas. He believes that architecture design must have local characteristics in order to maintain the historical context and characteristics of the city. Historical experience and the lessons of modernism prove once again the importance of residential buildings and community planning in building a good city. Therefore, for nearly half a century, western developed countries have initiated a new type of residential community planning. In the Old Town, more attention should be paid to the integration with the historical environment, and mosaic filling planning and design methods should be adopted. In the new city planning, emphasis should be placed on the shaping of urban space and the continuation of cultural context. In the United States, the new urbanism movement rose and flourished in the 1990s, emphasizing the remodeling of American suburbs in the form of traditional European towns. Mr. Dan Solomon has visited China several times in recent years. He once asked why the Juer Hutong model had not been prevailed in Beijing. Instead, a large number of high-rise apartments with no characteristics was built up, which made us embarrass and stop talking for a while."

"In order to realize the Chinese dream of great rejuvenation of the Chinese nation, we need to find diversified residential community modes through practice that not only inherit Chinese cultural traditions, but also meet the needs of modern people's yearning for quality and spirit need of life. In addition, we have to face another challenge, that is, the current situation in our city has a large number of old communities and houses built before the housing system reform in 1999. Almost all the houses are 6 storey walk-ups. The average household size is small, the layout of rooms does not meet the needs of future, and the appearance is simple. These houses need to be renovated in 15-30 years. To accomplish this task, we need to do more work in advance."

53. Rural-to-Urban Transect: Diversity of Housing and Community

Luo Yun

"As early as the beginning of the 20th century, Patrick Geddes, the famous Scottish biologist, master of humanist planning, founder of Western regional research and regional planning, discovered the regularity of regional ecology and human activities from the interaction between ecology and human beings. From urban to rural areas, people in different positions do different work and have different lives. Anders Duany and Elizabeth Plater-Zyberk, founders of New Urbanism in the United States, in their articles entitled Building Community across the Rural-to-Urban Transect, formulated urban-to-rural transect, describing different densities, heights and forms of housing from urban centre to rural area. The form of residential building reveals the ecological location distribution law of residential building types. With the progress of human urbanization, the types of housing are also evolving, and finally three basic types are formed to meet the specific needs of specific locations in cities and towns. From the rural-to-urban transect, we can see clearly the distribution of three types and seven minor types of residential buildings: In suburban zone, rural zone and natural zone, there are independent low-rise dwellings. In suburban zone and general urban zone, there are low-rise or multi-storey row dwellings. While in urban center zone and urban core zone are multi-storey or high-rise collective apartment. It is a universal law. This law explains the distribution of different types of residential buildings and the diversity nature of residential buildings from one aspect."

"The types and diversity of residential buildings, in a broad sense, refer to different countries, regions, ethnic groups with different characteristics, these distinctive residential buildings are part of the world's cultural diversity. After the international architecture in the 20th century, there has been a consensus internationally to maintain and highlight the architectural characteristics of their respective countries and regions. In a narrow sense, the types and diversity of residential buildings refer to the diversity of different types of residential buildings in a city, such as low-rise detached house, row house, townhouse, multi-storey walk-up, high-rise tower or slab, and super-high-rise tower, which are important components of urban form and spatial characteristics. It is the carrier of urban culture."

"The residential buildings in the developed countries in the world show diversity and distinctive characteristics. After years of evolving, the United States has gradually formed three basic residential types: patio house (single family house), townhouse, condominium and apartment, which correspond to location, architectural form and property rights management. Three types of residential buildings have formed their own relatively mature and perfect construction technology system. According to the site environment, the different needs of residents for function and appearance, there can be endless changes, forming a rich variety of residential types and architectural styles in the United States. At the same time, three basic housing types in the United States correspond to three completely different residential community planning and design patterns, forming a variety of urban blocks and environments with distinct characteristics, different functions and characteristics. Various types, prices and locations of housing offer Americans the possibility of a variety of choices. Different families rent or purchase suitable housing according to their special needs of life and work, which is highly consistent with people's identity and lifestyle. As Wright said, "Architecture is the life of the American people."

"The United States is a country of immigration, and its housing prototype comes from Europe. There are many countries in Europe with different cultural environments that have formed different types and styles of residential buildings. For example, there are dwelling types of cottage, Bungalow, townhouse and detached house, etc. in the United Kingdom. The semi-detached house is the characteristic of England. Government public housing includes Council house, etc. There is typical Tudor style, Queen Anne style of Victoria and Georgia style, as well as thousands of changes in architectural style. Unlike the United States, European cities and towns have a longer history, more prominent features, and relatively more aggregated buildings, with multi-storey residential buildings as the main part of the cities and towns. However, with the entry of post-urbanization stage, urban population growth continues to be declined. According to relevant statistics, in recent years, most of the newly built houses in European countries are detached houses."

"Even in Japan, with its small territory and high population density, residential buildings are rich in diversity. There are three basic types of residential buildings, single-family house, low-rise high-density collective housing, multi-rise and high-rise collective housing, which evolve into various types and rich and diverse residential buildings. In his book *Ten Types of Houses*, Japan's famous architect Kengo Kuma divides Japanese houses into ten types: single apartment type, Qingli Accommodation type, bare concrete type, etc.. It vividly depicts the characteristics, spiritual pursuit and living conditions of these ten types of houses. Kengo Kuma believes that housing has never only the function of living, but also the place of self-projection and the dream of living condition. In this sense, housing is Japanese life, different people need different types of houses. Ten kinds of houses are also the lifestyle of ten kinds of people."

"There are abundant types and diverse forms of housing in Chinese history. After thousands of years of long evolution of Chinese civilization, residential buildings have developed from simple functions such as free of wind, snow, rain, fire, and preventing harmful invasion of animals, poisons and insects to perfect residential functions, which contain rich cultural connotations and metaphors. All parts of the country have formed their own unique residential buildings, which are rich and colorful. They are an important part of the urban characteristics and urban and rural culture of all parts of the country, as well as an important part of Chinese civilization."

"Our country has a vast territory and a large number of nationalities, and has formed various characteristic residential buildings. In fact, the study of ancient residential buildings in China before liberation was carried out by the Society for the Study of Chinese Architecture. After the founding of New China, they summarized and systematically studied the investigation of residential buildings over the years. In 1957, Liu Dunzhen edited and published the book *Chinese Traditional Residence* with the achievements of many years research. He systematically briefed the development and evolution of ancient residential buildings in China and described the main types of Han residential buildings in the Ming and Qing Dynasties. From the architectural layout, Han residential buildings of the Ming and Qing Dynasty are roughly divided into nine types. The Book quotes a large number of ancient documents and field investigation records of existing archaeological materials and architectural objects. It is the first book on systematically sorting out Chinese folk residential buildings, and is considered as the origin of the study of Chinese vernacular folk houses. The Book improves the status of residential buildings and attracts the attention of the architectural community. In the 1960s, there was a climax in the investigation and study of folk houses in China, with rich and colorful results, including the courtyards in Beijing, caves in the Loess Plateau, water dwellings in Jiangsu and Zhejiang provinces, Hakka earthen buildings, coastal dwellings in the south, mountain dwellings in Sichuan Province, etc., as well as dwellings in minority areas such as Yunnan-Guizhou mountain dwellings and Qinghai-Tibet Plateau dwellings, dwellings in dry and hot regions of Xinjiang and grassland dwellings in Inner Mongolia, etc."

"Due to the campaigns of 'breaking the four old, establishing the four new' and the Cultural Revolution, the investigation and research of traditional residential buildings have been hold back. Until the reform and opening up, the study of traditional vernacular buildings revived again. In 1984, Liu Zhiping and Fu Xi'nian, the backbone of the Society for the Study of Chinese Architecture and famous architectural historians, published an article 'An Introduction to the Development of Ancient Chinese Residential Buildings', which made a more systematic study of the residential buildings of different dynasties and nationalities in China. According to the characteristics of different regions and the factors of architectural structure and ethnic customs, the ancient residences in China are divided into seven categories: cave dwellings in the northern Loess plain, Ganlan dwellings in the southern hot wetland zone, towers in the southwest and Tibetan plateau, courtyard dwellings, A-Yi-Wang dwellings of the Uygur nationality in Xinjiang, felt dwellings in the northern grassland, boat dwellings, etc. Each kind of residence has its own characteristics because of the different ethnic customs, which makes the form of the same kind of residence also show a variety of changes."

Luo Yun

"Some experts pointed out that, although there are many kinds of houses in China, compared with the 5000-year history of her civilization, the development and evolution of the ancient housing system in general is relatively slow. Therefore, when modern western residential construction technology entered in 19 century, it could not cope with. In modern China, various types of modern housing introduced from the West appeared in the open port cities and some emerging cities. Due to the slow progress of traditional courtyard dwelling in function, modern residential buildings with bathroom and kitchen have become a new competitive form of housing. The Shi-Ku-Men Residence evolved from the traditional courtyard house in China was originally a courtyard house for one family. Because of its little construction amount, small supply and large demand, it eventually became a multi-family mixed residence. Such a low standard of living, makes it eventually reduced to low-end miscellaneous residential big courtyard. The famous drama '72 tenants' is a true portrayal of the public life in the courtyard in Shanghai at that time of 1940s'."

"After the founding of New China, the welfare housing system of public ownership and planned economy is implemented. In the early period of liberation, there was a serious shortage of housing. Many of the stock of detached houses, row houses, town houses and multi-storey apartments were rearranged according to the way of one family in each room. Owing to overuse and lack of maintenance, the houses were quickly damaged. Tianjin is a typical example. Traditional bungalows in some northern cities, including the courtyards in Beijing, are also used by many families. Due to the disrepair and population growth, they have been built in disorder and gradually reduced to shantytowns. In terms of new housing, starting from the first Five-Year Plan and with the help of the Soviet Union, modern housing construction began, and multi-storey standard unit housing was widely used. Residential district planning is mainly based on the military camp row layout, resulting in the monotony and uniformity of housing type and residential landscape."

"After the reform and opening up, the diversification of housing types in China began to emerge. In the 1980s, the diversification of residential design focused on the diversification of residential shape and appearance. By the end of the 1990s, for the purpose of marketing, real estate development enterprises began to build villas, detached garden houses, multi-storey walk-ups and high-rise high-end apartments. However, during the period of real estate regulation and control after 2003, villas were called off and 90/70 policies, that is, in a development project the units with floor area over 90 square meters cannot exceed 70% of the total units, were issued, and the types of residential buildings tended to be simplified. At the same time, with the increasing volume of land provision, and the strict control indicators such as green space rate, building density and parking space, residential buildings can only develop to air and become 30-storey, 100-meter-high-rise towers or slabs. A dozen or even dozens of single-type, same-looking high-rise buildings are piled together, forming the most typical living environment and urban appearance in China at present. Although the quality of buildings has improved, the overall quality of living environment has declined, and the monotonous housing has also led to an extreme decline in the overall quality of urban space."

"The phenomenon of single housing type domination is firstly influenced by modern architecture movement, and is further strengthened by the inherent common knowledge of large population and small land in China, as well as the psychological deficiency of space caused by long-term housing shortage and the remnants of planned economy. Secondly, the enclosed residential area plus high-rise flat mode is the most beneficial to economic efficiency, and it is the most profitable mode to meet the current market and planning management requirements. A large number of replications of high-rise residential buildings makes everybody happy;

architects can save time in design, construction enterprises are easy to build, developers make money quickly, commercial banks are satisfied with profits, but it has greatly damaged the quality of urban space, and violated the objective law of urban development and housing construction, which laid many hidden dangers for the future. Thirdly, one-sided land policies, residential area planning methods and simplified management of urban planning are also important reasons. For example, the land sector requires that the FAR of housing plot should not be less than 1.0, regardless of location. The national standard *Standards for Urban Residential Area Planning and Design* promulgated in 1993, is a basic legal document for planning and design of residential area. Since its implementation in 1994, although has been revised several times, it has not been thoroughly adjusted to new situation. The *Technical Regulations for Urban Planning and Management* formulated by various cities is based upon the Standard. Generally speaking, the public, including city leaders, planners and designers, are still not aware of the hazards of this single type residential unit buildings due to the restrictions of the stage of economic and social development in which we live."

Jane Jacobs pointed out in his book *The Death and Life of Great American Cities*:

> Cities are the product of human settlements. Thousands of people gather in cities, and their interests, abilities, needs, wealth and even tastes vary greatly. Therefore, from both economic and social perspectives, cities need to be as complex and mutually supportive as possible in order to meet people's living needs. Therefore, diversity is nature to big cities. (Jane Jacobs,1961)

Jacobs pointed out sharply that,

> Modern urban planning theory combines the Garden City movement with the internationalism advocated by Le Corbusier. While advocating zoning, it also depreciates the mixed use of high density, small-scale neighborhoods and open space, thus undermining the diversity of cities. The so-called functional purified areas, such as Central Business District, suburban residential area and cultural intensive area, are actually dysfunctional areas. (Jane Jacobs,1961)

Jacobs abhorred the large-scale urban planning in the United States from 1950 to 1960, mainly referring to public housing construction, urban renewal, highway planning and so on. She pointed out that the lack of flexibility and selectivity in large-scale renovation plans and the exclusion of small and medium-sized businesses would inevitably destroy the diversity of cities. It was a 'natural waste way', which cost a lot of money but contributed little. It did not really reduce slums, but simply moved slums elsewhere to create new ones in a larger scope. The slums make it easier for capital to drain into speculative markets and have a negative impact on the urban economy. Therefore, 'large-scale projects can only make architects blood surge, make politicians and real estate developers blood surge, while the general population is always a victim'. She advocated that 'the use of funds in urban construction must be changed, from the pursuit of drastic changes like floods to the pursuit of continuous, gradual, complex and delicate changes. In the early 1960s, the large-scale old town renewal plan was in full swing in the United States. Jacobs' argument strongly refuted the mainstream theory of planning at that time. It can be said that she sentenced the death penalty for the construction of large-scale industrialized housing and modern residential areas earlier than Charles Jencks, a famous architectural historian."

Jacobs said:

> whether it is the garden house advocated by Howard's 'Garden City' or the tower house of Le Corbusier's 'Radiant City', it is an autocratic and paternalistic plan, trying to solve all the problems with one type and one mode, which is doomed to be unsuccessful. (Jane Jacobs,1961)

Modern architecture and urban planning movement promoted large-scale industrialization and international style, hoping to solve all problems with a universal model, which is contrary to the ecological law, human needs and the objective law of urban development.

Luo Yun

"In the process of making the guidelines, we re-read the classics, *Garden Cities of Tomorrow* by Howard, *Towards the New Architecture*, *Urbanism* and *The Radiant City* by Le Corbusier, *The City in History*, *The Culture of Cities* and *Technics and Civilization* by Lewis Mumford, *The Death and Life of Great American Cities* by Jane Jacobs and *Cities of Tomorrow* by Peter Hall, etc. In fact, many of these works are about the evolution of housing and residential communities in Western cities. Howard and Le Corbusier have made great contributions to the development of modern urban planning, but due to historical limitations, some of their ideas and schemes are one-sided, and future generations must face up to these problems. The critical conclusions of Mumford, Jacobs and Hall on modern residential architecture are strikingly consistent."

"Along with the reform of housing system and land use system, China has created a new era of ever-increasing large-scale housing construction in human history. In 30 years, according to the market-oriented mechanism, through real estate development, tens of billions of square meters of housing have been built, which has solved the long-term housing shortage problem that Chinese people didn't solve in a very long time ago. China created a miracle in mankind history."

"Regretfully, we have vigorously implemented the largest residential and urban construction in human history, but using outdated modern planning and design methods. We did not waken up and really understand the criticism of modernism movement since the 1950s in Western society, nor did we really realize the harm of modern architecture and urban planning movement. Of course, we have paid a huge price, and should learn from the lesson. However, until now, this modern residential area planning and design method is still continuing."

"Diversification of residential types is the imperative requirement of improving living standards and quality of life of residents, as well as the requirement of urban cultural renaissance. With the further development of economy and the continuous progress of society, with the increase of residents' income, the differentiation of residential demand is becoming more and more obvious, which is reflected not only in the differences in the ability to pay, but also in the changes and diversification of life style and functional needs. With the expansion of urban scale, the difference of land value and location conditions has increased. These factors make the types and forms of contemporary urban residential area and residential buildings tend to be diversified. Therefore, reasonable density and concentration, perfect social supporting services, good environment and reasonable distribution, and various types of residential buildings, such as urban apartments, courtyard houses, row houses, including independent houses, are essential conditions for improving our quality of life."

"It's never too late to mend. To change the current situation, improve the environmental quality of the city, adapt to people's yearning for a better life, and meet the diverse needs of residents, we need the diversity of residential buildings. Simply speaking, there are three basic types of residential buildings: low-rise, multi-storey and high-rise, and they should be innovated with the needs of the times. Standards for Urban Residential Area Planning and Design Code contains a kind of low-rise residential land, but there is no clear different types of building."

"Let's talk about low-rise housing first. Maybe somebody have different opinions. Low-rise residential buildings are the most abundant and diverse types of buildings. It not only includes detached housing, semi-detached housing, row housing and other types, but also has a lot of creative space for architects. Especially the low-rise buildings are easy to adopt the courtyard house form, which is a typical pattern of Chinese traditional dwellings, and is conducive to the inheritance and development of Chinese fine traditional residential ideology and culture. Practice has proved that this form is also the most favorite living form of the Chinese people, with a wide market demand."

"There are many questions about whether to build low-rise housing or not. First of all, it's about the land problem. Does China have enough land to build low-rise housing? The answer is yes. First of all, by comparison. China has a large population and relatively scarce land resources. It is an indisputable fact that land conservation should be considered in housing construction. But whether high-rise residential buildings should be built in cities, suburbs and even rural areas is a question we must seriously reflect on. Hong Kong and Singapore adopt high-rise and high-density residential buildings because they are urban countries or equivalent to an area of city in China in size. But even so, they still have a small number of villas. In Japan, where land is more tense than in China, more than 50% of the dwellings are traditional low-rise single-family dwellings, that is, one-family construction. One-family construction is an important part of the diversity of Japanese housing, which also shows that housing is not only a problem of land conservation. The Soviet Union and Eastern European countries built a large number of 9-25 storey high-rise residential buildings in large and medium-sized cities, with a small number of up to 40 stories. For a country as large as the Soviet Union, the construction of high-rise housing must not be due to lack of land and protection of cultivated land, but mainly to industrialization, large-scale production and urban planning layout, the key is the welfare housing system. By the 1980s, the Soviet Union and some socialist countries in Eastern Europe had made great progress in housing industrialization. However, due to the problem of the planned economy itself, problems have arisen in the political, economic and social aspects of the country, leading to the disintegration of the Soviet Union. Franklin Roosevelt said, 'A country is invincible if its people own their own houses and can win a substantial share of their land.' The ultimate goal of social and economic development and urban construction is to improve people's quality of life. Land is only a carrier and means. For saving land, to abolish the diversity of housing is the end of the matter."

"Secondly, we need explain this from the perspective of real estate market smooth transformation and upgrading. At present, Chinese urban residents' residential building floor area per capita is 39 square meters, a set per household in average, In general, China has solved the housing shortage problem. Now it keeps billions of square meters of new residential construction and billions of new commercial housing sales every year. Every year, the per capita residential building area in cities and towns will increase by 2 square meters. In the coming period, implementing the central Supply-Side Structural Reform Policy, the direction of Chinese real estate industry reform and upgrading is to reduce the total amount of development, improve quality, effectively supply, and achieve a smooth transition. Considering the low standard of large stock housing built before 2000, the standard of new commercial housing must be higher than the future planning goal in order to achieve overall balance. If we build it according to the average household floor area of about 100 square meters, our annual sales of commercial housing will reach 17 million units, which can accommodate 45 million people. Even considering the new urban immigrant population, it is far greater than the demand. If the scale is reduced directly, it will affect the real estate industry and the overall economic development. With the improvement of economic level, suburbanization of low-rise housing is an inevitable trend for urban residents to improve their living conditions. However, the old residential housing which has been retained is suitable for the first time home purchase of newly settled families in cities. We can first set the proportion of low-rise residential land in suburban areas, such as no more than 20% of the total per year, while limiting the upper limit of per capita land area, and start trials. At the same time, we should strictly control the encroachment on the ecological red line and increase the punishment."

"Thirdly, we illustrate from the average density of the city. Since modern times, many big cities in China have been facing the problem of population congestion in urban centers. After the founding of New China, the strategies of building satellite cities, dispelling the population of urban centers and old cities, and controlling the growth of urban population were generally adopted in the overall master plan. In fact, with the reform of land system and the development of real estate, the reconstruction of old districts requires a higher ratio of demolition and construction. As a result, the intensity of land development has been doubled instead of reducing population. At the same time, the construction of the new area has also adopted a higher FAR, many of which are high-rise residential buildings. It brought high revenues to the city temporally, but caused serious urban diseases such as traffic congestion and environmental pollution, with a results of too high intensity of land development, too high population density and too few green parks. In the end, city had to spend more money on governance. Therefore, new residential housing and communities need lower density and higher green space rate to achieve the overall balance of the city."

"Fourthly, we illustrate from the aspect of inheriting the cultural tradition of our country's housing. Generally speaking, courtyard house, or patio house, is the most important traditional pattern of residence in China. In the process of historical evolution, various regions have formed their own community patterns corresponding to the different characteristics of residential buildings and layout. The courtyard house and its urban planning model are characterized by safety, quietness, family and culture, which are good living environment, such as courtyard house and Hutong system in Beijing, courtyard house, private garden building, River network, Road double chessboard system in Suzhou, etc. Some areas take environmental factors into account in the overall layout of residential buildings, such as water-adjacent dwellings in Zhejiang, showing a lively and fresh interest. In residential decoration and interior furniture layout, regional characteristics are also prominent. In addition, our country's residential buildings and gardens are closely linked, many of them are equipped with private gardens, which abbreviate the natural scenery landscape between inches, and achieve the organic unity of natural beauty, architectural beauty, painting beauty and literary and artistic beauty. We must inherit and carry forward these fine traditions and not lose them for any reason."

"Fifthly, from the perspective of social harmony. Sir E. Howard thought that we should build an ideal city with both urban and rural advantages. He called it 'Garden City'. He believed that this is a universal key that can solve various social problems in the city. Town-country is essentially a combination of cities and villages, evacuating overcrowded urban population, enabling residents to return to the countryside, and harmoniously combining the advantages of urban life with the beautiful environment of the countryside. The construction of suburbanized low-rise housing is conducive to the integration of urban and rural areas. If there are more middle-income families living in low-rise housing in the suburbs, it will actually help to alleviate the gap between the rich in the city and the poor in country."

"Sixthly, from the perspective of protecting the ecological environment and realizing the harmony between man and nature. Low-rise residential buildings are small in size and flexible in layout. They can be well integrated with the ecological environment and adopt road traffic and municipal infrastructure with low impact. In addition, we can explore the inheritance of the layout of courtyard houses and streets, to continue the traditional Chinese courtyard buildings in 'winter warm, summer cool' ecological construction methods."

"Generally speaking, to alter the bias about low-rise residential buildings, we need to emancipate our minds and consider the diversity of residential buildings in accordance with the scientific laws of urban development. The types of residential buildings described by Professor Pratt-Zybeck in different densities, heights and forms from the city centre to the countryside actually reflect the objective laws of space, including economic activities, differential rent, ecological environment, transportation operation, urban management and social organization. If we want to enrich the diversity of housing, we should first recognize the rationality of low-rise buildings, instead of looking at problems in a simplified way of thinking and formulating rules with a single consideration, so as to form a ever-changing spatial form and a high-quality residential and community environment."

"Obviously, some of our current planning and land policies, such as prohibiting the supply of land with FAR less than 1.0, are one-sided. We have no reason to limit the diversity of residential buildings. Urban population density, land development intensity, urban traffic, environmental landscape, public services and diversity of residential types must be considered as a whole. According to the objective law of urban development, different development intensities and types of buildings are adopted in different locations of the city, including the development of independent residential buildings in suburbs and the development with FAR less than 1.0. In fact, at present, the high-rise and multi-storey residential buildings in urban stock have occupied a dominant position in China. The intensive use of land has been quite high, and the development intensity and population density are beyond a reasonable range. At the same time, there are a large number of inefficient idle land in the periphery of the city. The gradual development of some low-density courtyard dwellings will not have a negative impact on urban land problems. Through rational planning and design, we can create a better living environment with a certain development intensity, rich and diverse types of residential buildings, suitable for local characteristics."

"I want to repeat that today our socialist market economic system has been initially established. The state emphasizes that in order to play a decisive role in the allocation of resources; we must act in accordance with market laws, recognize differences and diversify differences. Of course, housing is not a simple commodity. It is the premise for the state to ensure housing supply for middle-income families. Our colleagues specialize in the study of housing systems, and Binhai New Area has also made reform attempts. The overall idea is to shift from the provision of affordable low-end housing for low-income families in the past to the provision of well-off housing for middle-income families, so as to achieve 'low-end in security, middle-end with supply, high-end by market'. I won't do anything about the details. If you are interested in it, we can discuss it specially."

Luo Yun

"Undoubtly, multi-storey housing is the largest and most extensive form of housing in China, and also the main body of the market for the middle-income families in the future. Before reporting on the urban design guidelines for multi-storey residential buildings, I would like to briefly report on the design requirements for high-rise residential buildings in the guidelines. From Duany's Urban-To-Rural Transect, we can't see 30-storey high-rise residential buildings, even in the city center, but only multi-storey apartment. Considering the current situation, we will have a certain proportion of high-rise residential buildings, such as no more than 20%. Generally speaking, it is to limit the number of high-rise buildings on the whole. Apart from the prescribed location, high-rise residential buildings are not allowed to be built. Secondly, limit the height of ordinary high-rise residential buildings, from the perspective of fire safety, the height should not exceed 80 meters. In the suburban center of the city and near the metro station, high-rise residential buildings not exceeding 54 meters can be built. For a small number of high-end apartment-style residential buildings in the city center, we should limit the scale and improve the level of fire fighting facilities. For the high-end residential houses which are divided and sold, developers are required to maintain the property rights of public parts and assume long-term responsibility for property maintenance."

"It is undeniable that multi-storey residential buildings are the main existing residential forms in China, and the largest number of residential buildings will be built in the future. How to make multi-storey residential buildings have diversity is a long-term topic, but also a more difficult task. In the early stage of reform and opening up, in order to change the monotonous housing layout and building form, the country carried out residential area pilot experiments projects and accumulated some experience. We had been thinking about this issue in the pilot project of New Harmony in Binhai New Area. One of the preconditions for the planning and design of new residential communities is that there must be various types of residential buildings. If we only change the large enclosed area into narrow roads, dense network layout without increasing the diversity of residential building types, the new residential community is doom to fail."

"From the experience at home and abroad, the best way to create diversity is to adopt 'narrow roads, dense network, small blocks' and commission different architects and developers, which plays a more important role than the change of road network itself. In addition, the insertion of some different types of housing in multi-storey community will play a key role. For example, multi-storey residential areas can plan some multi-row houses, by absorbing the layout of narrow face width, multi-courtyard and big depth with patio in traditional Chinese dwellings, which not only has certain density, but also meets the psychological needs of Chinese housing as 'standing upright between heaven and earth'. The guidelines also suggest that although high-rise, multi-storey and low-rise residential buildings are arranged outward from the city centre, there are also some other types of areas in every area. For example, in the high-rise areas in the urban center, some historic blocks are mainly low-rise residential buildings; in the suburbs, in some bus stops, there can be some multi-storey and small high-rise residential buildings. This kind of interpenetration meets the requirements of diversity in different regions of the city, is changeable and avoids monotony."

"Residential building is the most basic type of building, the simplest type of building, but also the most complex type of building. Architects in both ancient and modern architectural history have done classical independent residential design. But there are less public housing designed by masters. The design of Le Corbusier's Marseille apartment is classic. It is not easy to make a good residential building for the public under the constraints of limited area and cost. In order to make the design of multi-storey residential buildings reach a high level, we want to continue the traditional culture and promote the quality and culture of residential buildings."

"There are many meanings in studying Chinese residence, but the most important thing is to make the past serve the present. Liu Dunzhen, Liu Zhiping and Fu Xinian all put forward the same viewpoint in their discussions. As early as in Kunming during the Anti-Japanese War, Lin Huiyin creatively used some methods and styles of folk houses in the design of Yingqiu Yard, the female dormitory of Yunnan University at that time. After liberation, the big roof and other traditional architectural forms were also used in some residential buildings. With the rise of the protection of historic cities, the protection and renovation of traditional residential buildings in old cities has become a very important and meaningful topic. Professor Wu Liangyong of Tsinghua University proceeded continuously the research on Beijing courtyard house. Professor Zhu Zixuan carried out the research on Urban Design in Houhai District of Beijing focusing on protecting the texture of Hutong and courtyards and the research on the protection of Tunxi Old Street in Anhui Province. Professor Shan Deqi carried out the research on Huizhou and other residential buildings. Tongji University carried out the research on the transformation of Tongfang Lane in the ancient city of Pingjiang, Suzhou. All above are very useful attempts and have achieved good results."

"In the 1980s, Mr. Wu Liangyong put forward the theory of 'Organic Renewal'. He advocated that the urban renewal and development of the ancient city of Beijing should be explored on the basis of sustainable development by establishing a new courtyard house system in accordance with the inherent law of urban development and conforming to the texture of the city. This theory has been successfully applied in the pilot project of Juer Hutong reconstruction initiated in 1988. The newly reconstructed Juer Hutong courtyard house is designed according to the 'Semi-courtyard' model, which is the basic courtyard with suitable scale surrounded by the multi-storey unit houses with perfect functions and complete facilities. The building height is less than four stories, which basically maintains the hutong-courtyard system in history, and incorporates the advantages of both unit building and coutyards, which not only rationally arranges the indoor space of each household, guarantees the residents' needs for modern life, but also forms a relatively independent neighborhood structure through the courtyard to provide public space for residents to communicate. The appearance of the building has local traditional symbols and colors. The new courtyard in Juer Hutong and the traditional courtyard form a harmonious group, which retains the spiritual core of Chinese traditional residence that attaches importance to neighborhood friendship. The Juer Hutong Renovation Project in Beijing won the World Habitat Award in 1992. Recently, we paid a special visit to Juer Hutong, and found that the buildings were rather dilapidated because of the unclear boundaries of public spaces such as courtyards. Therefore, there is room for further study on this approach."

Professor Shi Jiafeng, Zhou Enlai School of Government, Nankai University

At this time, someone raised his hand to make a speech. "Hello, everyone. I'm Shi Jiafeng, from Zhou Enlai School of Government, Nankai University. In my opinion, housing is not only a matter of architectural design, but also a matter of urban design and planning, as well as economic, social and political issues. At the same time, housing and community do have a significant impact on the whole society. As housing occupies an absolute proportion in the city, it not only plays a decisive role in the basic composition of the urban form, but also dominates the overall urban social spatial form and the quality of life."

"Over the years, I have been thinking over and over again that although there are many types of housing, from the perspective of sociology and social management, there are ultimately two types: independent housing and collective housing.

Independent housing is a family unit, which can be managed by itself. It has its own land and sky. That is to say, there is no other house above it, 'standing upright between heaven and earth'. Collective housing is a house where multiple families are overlapped up and down. Most of the time in human history, most of the residential form is independent housing. With the emergence of cities and towns and the increasing population, the land inside the walls of cities and towns is becoming tenser and tenser, resulting in the emergence of multi-storey collective housing. Mumford wrote in *Technics and Civilization* that in ancient Rome, there were multi-storey dwellings because of the limited area of the city walls, but at that time there was no corresponding equipment, and the excrement was piled up under the stairwell, which smelled terrible. Until the Industrial Revolution, with the emergence of modern building materials and technology, the collective housing has become more and more perfect, and industrial production has made the collective housing an effective means to solve the housing shortage. Moreover, it can make the city center denser and promote the further agglomeration and development of the city."

"From the experience of all countries in the world, independent housing and collective housing have their own advantages and disadvantages. The property rights of independent residential are clearly demarcated, and the responsibilities of residents and the government are also clear. Generally, owners will maintain their own property to ensure that the housing has a good quality for a long time. Although they can adopt row houses, the building density is relatively low And they occupy still relatively more land. The biggest advantage of collective housing is efficiency and saving land, but there are many problems. Due to the unclear boundary of property rights, apart from a few high-end apartments, there are many difficulties in the property management of most collective housing. Because of the difficulty of maintenance, it eventually becomes a heavy burden for the government. At present, most cities in China have encountered this problem. Moreover, if the design of collective housing is simple, there are also huge problems in the making of community environment. In fact, when independent houses are inhabited by many households, they become collective houses. Many courtyard houses and western-style mansions in our country are difficult to repair because of the deterioration of living environment and housing quality caused by multiple households. Therefore, from a comprehensive point of view, the city's housing should be diverse, and to achieve a balance between the number of independent housing and collective housing and a reasonable distribution of location."

"We used to think that independent housing is not eco-friendly, based on the experience of urban sprawl in the United States. In fact, many historical experiences have proved that in some cases, low-intensity development has a low impact on the ecological environment. From the historical photographs of the Bay Area of the United States, it can be seen that the mountains are bare and unsuitable for growth of trees because of the strong sea winds. But more than a hundred years later, look at the bay area now, the trees are forested and lush. Over the past 100 years, people have not only built cities and developed economy, but also built and improved the ecological environment of the whole Bay Area. We see that the American dream of owning houses, gardens and cars has become a reality. Almost every household has its own garden with a variety of plants. At present, there are 2.44 million dwellings in the Bay Area. We can estimate that 2 million of them are detached houses. 100 million square meters (100 square kilometers) of garden greening has been built according to the calculation of 50 square meters per household, which is maintained by each household without the government spending a penny. This is a huge wealth, playing a major role for the improvement of the ecological environment and the city beautification."

"From this point of view, we can realize that we can't simply regard land as an isolated problem. From the perspective of comprehensive ecological and social management efficiency, it is a good way to construct abundant and diverse low-rise independent houses in the outskirts and suburbs of cities. It not only responds to the pursuit of improving people's living standards, meets the individual needs of residents, but also plays an important role in promoting economic development. Therefore, it cannot be simply prohibited. Moreover, if we can provide effective supply, avoid a large number of vacancies, coupled with careful planning and design, according to the theory of new urbanism, the overall urban area will not be much larger than the multi-storey and high-rise residential construction model."

Luo Yun

"Professor Shi's speech inspired me and made me more aware of the problems of Juer Hutong. In fact, Tianjin also faces the same problem. The Central Urban Area of Tianjin distributes around the Central Ring Road evenly large-scale residential areas constructed by the government before 1999, which can be classified as collective housing. In terms of property management and maintenance, it is difficult for households to reach an agreement, and the implementation of urban renewal ultimately requires the government to pay for. Moreover, due to the unclear boundary, many facilities have just been repaired once, and in a short time they have been damaged again. Practice has proved that the implementation of democracy requires a certain scale, 200 people or less, to have the possibility of consultation. This also shows that our city needs to build a certain number of low-rise independent housing, which can't only reduce government maintenance expenditure, but also cultivate community spirit."

"Finally, I will recapitulate the main contents of the guidelines. The urban design of 'A Ring Linked Eleven Gardens' of Tianjin Central City and the surrounding areas just reported has realized the reconstruction of the urban structure of the central city and the main city in the era of ecological civilization, and formed the overall pattern of residential space which is harmonious with nature and integrated with urban and rural areas. We use urban expressway, outer ring expressway and outer ring boundary to form three circle layers: Central Circle Layer, Ecological Livable Circle Layer and Garden City Circle Layer. According to this structure, the urban design guidelines for new residential communities is organized, which stipulates that different types of residential buildings should be arranged in different circle areas to form diversified communities with different characteristics, and the corresponding control indicators such as FAR, building height, density and green space ratio are formulated. This is the biggest difference between our traditional residential area planning standards. For example, for the green space rate, in the past, new residential areas all require a higher green space rate. However, the quality and level of greening are affected by factors such as underground parking garage, and the effect is not good. Here we seek truth from facts, combined with 'narrow roads, dense network' layout, reasonably reducing the green space rate in small plots, requiring more concentrated small park green space, balancing green space rate in the whole community."

"The central circle in the central city of Tianjin is not only the dynamic central area of Tianjin, but also the gathering of 14 historic blocks, which is the most abundant place of urban life in Tianjin, also the place with the humanistic spirit. There is a complete pedestrian network, pedestrian access to integrated public transport system, high-quality public space, commercial centers and so on. This area, 145 square kilometers, is the most important built-up area and population gathering area at present. It is also the area with the highest building density, development intensity and road network density. Many areas have historically been the pattern of narrow roads network. There are many types of residential communities and residential buildings: Wudadao, Chifeng Avenue, Anshan Road, Italian Style District and other historic and cultural blocks, including three main types of residential buildings, i.e. Gate-yard Style, Courtyard Style and Lane Style. In addition, what remains are the large residential communities built before 1999, which is now old communities that need to be renovated. In the future, the protection of historic districts will be the first step in this region, and population alienation and organic renewal will be carried out in the relatively dilapidated residential areas around Quanyechang Area. A small number of high-standard, high-density high-end residential buildings can be planned and constructed on both sides of the Haihe River in the urban center to serve the business district, including Yujiapu Financial Zone in Binhai New Area."

"Between the Expressway and the Outer Ring Expressway is the Eco-livable Circle. There is a large amount of undeveloped land in this area, which is the main place for new middle-end commercial housing construction in the city. It needs a high level of planning and design. Such communities include the surrounding area of West Water Park, Pillow Park, South Canal Area, Chenglin Park." Those new communities are adjacent to big parks, with a good ecological environment, perfect living facilities, convenient public transportation system, and high-quality schools and hospitals. Here is the choice of more families with children and elderly people, and also the choice of improving the living environment for families in the old community in the city center. They pay more attention to family life, children can run freely in the grass, the elderly can walk in gardens, a family can have a picnic under the tree. Some of them work in the city centre, some work in suburbs, some work in the community, commuting mainly by public transport, leisure travelling mainly by car. The communities are developed at low and medium intensity, and around the Metro station the intensity is increased appropriately."

"Between the Outer Ring Expressway and the Outer Ring Road, there is a Garden City Circle, where people and nature interact, cities and villages blend. People here pursue a peaceful and free life, advocating close to nature and enjoying life. Gardens and courtyards are the center of everyday life of every family, which are places for natural dialogue and inner care. There are small town centers, business streets and schools in the community, and family doctors not far from home. Cars are the main means of travel. The newly built community is a country community with low intensity, low density, of single houses or Chinese courtyards."

"There are large country parks in this area, which reserve a lot of farmland and some villages. On the basis of existing homestead Land, rural housing should greatly improve the design and construction level. The standard area of 0.3 mu homestead per household is not more than 200 square meters, which is enough. The key is to change our current concept and habits of rural housing construction and create a new mode of housing construction. Consideration of the environment should be the focus of rural urban design, including public space such as village center, park and greening, using every inch of land carefully. In the vast rural areas, there is no such awareness. The 0.3 mu homestead per household is basically surrounded by a courtyard wall, 'hiding into a small building as a unified whole'. Historically, courtyards have a tradition of planting trees, which is now rare. Assuming that we increase the requirement for the use of homestead in construction management, requiring each household to green according to a certain standard, and incorporating the greening of each household into the strict municipal management of villages and towns, the change will be historic. Tianjin currently has a rural population of 3.88 million and more than 1.1 million rural dwellings. If each household constructs 50 square meters of gardens and 4-5 trees, it will be a big and real number, which not only improves the ecological environment, the key is to establish awareness of the ecological environment, which is very important, in addition, can be used to develop green horticulture industry."

"'The town is the symbol of human society, and the country is the symbol of God's love for the world. The city and the rural must marry to form a Town-Country Magnet.' Howard's 'Garden City' paints a blueprint of 'Town-Country Magnet'. It not only enjoys the social and cultural achievements of the city, but also enables the residents to live in the beauty of nature, using community-based new culture to cure cerebral hemorrhage in the urban center and paralysis in the remote areas of the city. It coordinates urban and rural elements, forms permeable and natural harmonious urban areas, and balances urban and rural areas in a wider natural environment. At the same time, we should establish community autonomy, provide the freest and richest opportunities equally to individual efforts and collective cooperation, shining with the humanism of independent autonomy. In today's historical coordinates, Howard's classical theory 120 years ago is still a universal key to solving the problem."

Luo Yun

Speaking of this, Luo Yun, who has been engaged in urban planning for more than 10 years in Tianjin, has a complex emotion but firm heart. The purpose of modern urban planning is to reshape the relationship between urban and rural areas. Tianjin in the new era needs a good living environment to realize people's yearning for a better life.

In the process of civilization development, one thing remains unchanged: society should make progress and mankind should pursue a better life. Looking back history, the development of a nation has always been accompanied by a beautiful human settlement, and the ultimate goal of the construction of human settlement environment is social construction. Mr. Wu Liangyong once put forward in Beijing Charter: 'A beautiful human settlements environment and a beautiful human society create together, to build a beautiful and sustainable human settlements environment, which is the common ideal and goal of mankind.' Town-Country integration has laid a healthy pattern of harmonious coexistence between man and nature. The diversified and beautiful community is the basic unit of society. It is a social restructuring from the micro perspective. By caring for people, it can provide possibility and support for the extension and development of each person's will, thus activating a rich and diverse life, safeguarding the social equality and realizing the ideal of a harmonious society ultimately."

"'Man, poetically dwell in the world.' is Heidegger's ideal, but also the ideal of our generation of urban planners."

54. Searching for the Spiritual Home of the Community

Chief Planner. Chu Xia

Luo Yun's presentation receives warm applause. This sub-forum is more like a seminar, with extensive participation and heated discussion. Chief Planner. Chu came to the stage, "The diversity of housing is the key to the construction of new community in the next stage in China, which is the foundation and prerequisite for the realization of the Chinese dream of great national rejuvenation. I am very much in favor of and encouraged by what Luo Yun has just presented. In my opinion, apart from attaching great importance to the spiritual function of the house itself, the community should also bring the residents together and seek the spiritual home of the community."

"I think that the 'new' in new residential communities are embodied in the emphasis on urban space, architectural culture and making of neighborhood places. This is the most important part of the community, as the carrier for the continuation and development of traditional Chinese culture."

"Recent reading of Mumford has inspired me greatly. Unlike other planners, who are keen to interpret Redburn New Town as an example of the theory of neighborhood unit planning, Mumford thought more deeply. He believed that the key to the success of Redburn New Town is not architecture. The key is that the planning provides a civilized core, which can gather people even if it is only embodied in shops, schools and parks. In addition, green belts and streets form a common border, which makes residents feel where they belong to. He said: 'Vigorously to rebuild the stage in the neighborhood community, so that the splendid scenes of social life can be still staged here.' When Mumford wrote Urban Culture, he had just begun to study the urban system, but he had been keenly aware of the inherent cultural significance of the community. Mumford's work on the new residential communities in the United States has prompted him to start a systematic study of urban culture in the United States. He has written two masterpieces, The Cultures of Cities and The City in

History, and has become the greatest master of urban history and theory in the 20th century.

"Mumford's emphasis in his writings are more applied to people, their cultural and spiritual life. Mumford explained that one of the most important motives for the initial formation of settlements was people's spiritual needs. In the initial embryonic period, the formation of settlements often originated not from residence but from sacrificial activities, memorials and memories of the deceased, praying for harvest or enough prey, regular pilgrimage gatherings, worship of the natural forces of the gods, etc. After the Agricultural Revolution, people began to settle down, and villages became shelters and nurturing places for human beings, which already included the functions of shrines, pipelines and granaries which were absorbed by cities in the future. When the ferocious hunting nation ruled the timid farming nation, the ruling power gradually combined with theocracy, enabling the king to possess unprecedented strength, using his peasants to build a series of tall buildings for him, and dividing into a group of strata that praised the country for their spiritual duties, so the city was born. The city first existed as a magnet rather than as a living container. Villages exist on the basis of food and sex, while cities should be able to pursue a higher purpose than survival."

"Contrary to the general view of the dark of the church in the Middle Ages, Mumford believed that the city in the Middle Ages was an era in which human nature was promoted. He often used Venice in the 11th century as the representative of the city in the Middle Ages. From the 5th century, Christianity began to prevail. This new religious culture denies property, prestige and rights, regards poverty as a way of life, eliminates all the material conditions needed for physical survival, and regards labor as a moral responsibility to make it noble. The monastery became the castle of the ideal paradise city. With the slow recovery of the economy, the increase of population and wealth, the role of trade unions had been strengthened, and cities and towns had begun to develop. At that time, the church ruled all over Europe. The first goal people see on the horizon during their journey is the tower and spire of the church. Town centers are centered around cathedrals, with chapels, monasteries, hospitals, nursing homes, schools, and city halls used by trade unions. Later, universities emerged. Rituals such as worship, pilgrimage, costume parade and open-air performances have become regular activities in the city. In the Middle Ages, the housing of the masses was relatively simple, but people were more easily accessible to farmland and natural environment, used to outdoor activities. There were many open spaces in towns for various games. Town scale is pleasant, many small churches become community centers and places for people's spiritual sustenance and communication. Mumford appreciated the city core of the Middle Ages, believing that the urban neighborhoods of the Middle Ages were more humane and cultural than those of Baroque, Renaissance and post-industrial development. Different types of housing form different forms of community. Road network, streets and green environment."

"The design of community center is very important. It is a place for community residents to communicate and a spiritual home. Besides high-quality housing and good community environment, it is also necessary to form good community relations. The community centers play very important role in this aspect. Historically, good community centers have a wide range of samples, but the most spiritually and culturally abundant are medieval towns in Europe. We find that every medieval town will choose a very good geographical location, facing mountains and rivers also keeping a proper distance, which can provide a good pattern for various forces. At that time, the residents' views on urban life were quite consistent, which made people feel as if there was a kind of conscious theory guiding urban planning after seeing many medieval towns continuously."

"But conscious management of urban construction does exist. Descartes mentioned in his Discourse on Method: 'At all times there are officials who are responsible for urging private buildings to contribute to the city's landscape.' Mumford also believed that the medieval towns were unified and diversified, but they were also achieved through efforts, struggle, supervision and control."

Mumford said:

> The city gate was the place where the two worlds of the city and the countryside, the city and the outside of the city met. It later evolved into a checkpoint and control point, and gave rise to a pile of warehouses, inns, hotels and workshops. At the same time, the top of the city wall as an ancient traditional slow walk, can overlook the surrounding fields, enjoy the breeze. The key building at the heart of the town is the cathedral, which exerts great influence on other buildings. The most important means of medieval towns were concealment and unexpected, sudden openness and upward towering. There were meticulous and magnificent sculpture art in the buildings, which contained various religious stories. As the center of the town, the cathedral has a large number of Christians coming in and out, so it needs a front square. According to theological instructions, the altar should face east. The meat market is usually located near churches. (Mumford,2013)

Groups of churches and squares are the equivalent of 'community centers' today, which are activity centers: in grand festivals, they were places for feasts, where religious scripts were presented, and during holidays, great scholars' debates were held. Churches are crowded with people, who usually going to churches before and after a long journey back—praying for and reporting peace. The clergyman here was an important person, whose main duties are undoubtedly psychotherapy and spiritual support. In addition to presiding over daily worship and confession in his parish, he also presided over ritual activities such as baptism, weddings and prayers for the dying. These activities are the information release and exchange platform for urban residents, also the most important spiritual belonging and spiritual sustenance outside their families."

"The location of traditional Chinese towns and villages often follows the realm of harmony between man and nature. In geomantic terms, mountains should be surrounded by water, preventing wind and holding gas. Mountains should form surrounded space from far to near, and flowing water should be planned in order to achieve the goal of seeking good luck and avoiding bad luck, living and working in peace and contentment. It also forms a kind of safe and sheltered relationship between the natural environment and human beings psychologically. The spatial form of the traditional villages and towns closest to the community scale annotates the feelings of homesickness, patriarchal relatives, and the desire to be buried at homeland. 'Harmony of Heaven and Man' and 'ethical concept' are the core ideas of traditional village construction. 'Harmony of Heaven and Man' emphasizes the harmony between man and nature, and 'ethical concept' concerns the harmony between man and man."

"Mr. Liang Shuming, the famous thinker of modern China, has vividly explained the principle of `understanding and reasonable'. 'Reasonable', which is embodied in the symbiotic relationship between villages and towns and the natural environment and the spatial order of the building combination. 'Understanding', refers to the social relationship of people living in harmony in villages and towns based on blood relationship and with patriarchal concept as the core. Residents of villagers and towns without clan relations generally promote in-depth exchanges through coexistence of areas and sharing of public facilities."

"Traditional villages and towns from site selection, overall spatial layout to group combination, single building, etc., can reflect a simple ecological consciousness. The most important buildings should be laid out along a series of ponds or rivers, forming various forms of small squares accordingly. Major roads connect these buildings and squares in series, and zigzag links important land and water terminals or fairs and other destinations."

"The internal layout of villages and towns takes ethical relations as its purpose, lays stress on hierarchy and the superiority and inferiority of the elder and the younger, taking the center as the largest. Therefore, ancestral halls and temples, as the carriers of clan authority, mostly occupy the central position of the village. Architectural group composition often emphasizes a structural order originating from ethical relations. Small temples generally were set by combining functions and terrain, such as Mazu (Sea Godess) Temple in the pier area, while Guandi Temple on a higher ground."

"Ponds and trees play an important role in the public space of villages and towns. In addition to the usage for water storage and irrigation, water pools are usually planted with lotus and euryale feroxs, which have a calm and distant mood, making people feel at ease and peaceful. Xiong Peiyun (the author of My Village, My Country) is the most appropriate description of big trees: without big trees, the village loses its soul, and a big tree has branches and leaves, which supports the public space of the village."

"The elegant rhyme of poems in traditional Chinese architecture is the deep implication of traditional culture. The inscriptions of lintels such as ` Cultivation and Read to Bequeath to the Family', 'Living with good Neighbors', and decorative patterns or brick carvings express auspicious meanings through homophonic sounds such as `Bat' equivalent to `Fortune'(similar in pronunciation in Chinese), which are the most common spiritual and cultural expressions in traditional architecture."

"When it comes to villages, we have to mention squires. They shoulder the responsibility of inheriting culture and educating the people in the countryside. At the same time, they participate in local education and local management, leading the development of a local society. They are the cultural leaders of the countryside, representing local customs and culture. The ancestral temple is the most frequent place for squires to worship and admire their ancestors and wise men; at the same time, it handles family affairs, 'to recognize the origin of ancestors and return to who', including marriage, funeral, longevity and joy; the patriarchal elders who represent the patriarchal power have high prestige and would deal with family internal affairs, establish excellent lists, reward diligence and punish laziness, resolve disputes and deal with contradictions in the ancestral temple. The squires are also the promoter of the family school, opening schools to educate clansmen."

"Gandhi said: 'in terms of material life, my village is the world; in terms of spiritual life, the world is my village.' The villages here, for the urban people, can correspond to the communities where they live. As far as the position of modern people in the world network is concerned, community does not have as great influence and restriction on people as traditional villages and towns, but its role is still profound and far-reaching, because no matter how big the world is, you have to start from home."

"'Home' represents the life heritage of family, ancestors and fathers, as well as the place where 'I' was born and grew up. The sincere respect and gratitude for it is an endless energy system behind a person. No matter how vast the world you enter, it is your origin. You clearly understand—where I come from. At this time, your soul-bound ancestors become a source of inner quiet power, always bless and protect you."

"Dramatic performances in community life are the best way for people to present the meaning of their souls, express their strength and support. Mumford believes that if we remove all the dramatic scenes in urban life, such as rituals, debates, birth, illness and death, and so on, half of the meaningful activities in the city will disappear, and more than half of the meaning and value of urban life will be impaired or even vanished."

"Modern psychology also holds that there is a mirror neuron in everyone's heart—we often mirror the world, obey its needs, and strive to win its love and praise. Whenever we reflect the outside world, there will be a corresponding desire in our hearts to reflect the world back. If this desire is not fulfilled, we will produce a lack of mirror neuron acceptance, which will become a deep pain. The activities and rituals of the community mirror every inhabitant. If it can see the real existence of every individual, it will soothe countless lonely hearts."

"Communication and dialogue will also bring people in the community closer to each other, enhance ties, and enable people to behave moderately, providing various forms of dialogue and rituals is one of the essential functions of community centers, and the expansion of social circles is also one of the key factors of communities. Everyone can participate in the dialogue, and their identity and role need to be identified and seen faithfully."

"Therefore, a good community center with spiritual significance should have some dramatic buildings and spaces, and need specific people to play an important role in psychotherapy and spiritual support. At the same time, functional places in the community should also have spiritual significance: library—place for learning; cafe—place for meeting and working; schools— place for education and communication; station—place for transfer and relocation; stadium—place for exclusive enjoyment and discovery of companions, etc. These places allow people to discover themselves, get mirrored and have the opportunity to communicate and grow. By choosing an easy-to-understand way to organize them, they become the common home of the body and spirit of the community residents."

Chief Planner. Chu

"Finally, I would like to talk about education, medical care, sports and pension, which are of the greatest concern. From the domestic and overseas comparison, this is a distinct difference. The best public service facilities in our city, such as primary and secondary schools, hospitals and so on, are located around the city center. Now some large hospitals have moved to the outskirts of the city because the original land can't meet the development needs, but the overall situation has not changed. Therefore, on the day after school starts, traffic congestion in the city center is unusual, but once during the school vacation, the traffic situation will be greatly improved. This model should be said to be adapted to the current situation that urban employment centers are still in the city center. In western countries, the downtown has declined, so good schools and hospitals went to the suburbs. At present, what we need to do is to vigorously strengthen the construction of high-level public facilities such as education, medical treatment, sports and old-age pension in urban livable circles and rural circles, and encourage market participation, so that our educational, medical, sports and old-age facilities can also be greatly upgraded, reaching the hardware level of foreign countries, to create conditions for software promotion."

55. New Type of Community: Yearning for A Better Life

The urban design of new residential communities is the most salient type in the pilot projects. There includes five urban design projects of new communities, eg. The Pillow Park Community, The West Water Park Community, The Woods Park Community, The West Gate Community and South Liberation Road Community. These new communities, located surrounding eleven gardens along the Outer Ring Road and Green Belt, are the key areas for development of Tianjin's central city in near future. The urban design of each new community has its own characteristics as well as common points. Each project designer's report has its own merits, The new prototype of community and The Urban Design Guidelines for New Residential Communities in Tianjin have become the hottest topic in the final discussion of the sub-forum of urban design theory and practice in Tianjin.

In the afternoon, there were 20 planners, urban designers, landscape architects and architects in total showcasing in the four sub-forums. They presented not only their various pilot projects but also their discrete thoughts upon topics in the discussion. The participants from different fields gave their opinions and supports especially on the reform and innovation of urban design guidelines in Tianjin. Consensus has been reached on many issues.

Old Zhang

While the planners and urban designers are discussing and making comments on presentations of the sub-forums within the planning exhibition hall, the cities outside are operating as usual. After an afternoon rush tour, at sunset, the eldest son drives Old Zhang arriving at The Pillow Park area along Haihe River, where used to be Tianjin Steel Plant, which was relocated to the middle reaches of Haihe River in 2006. After the car pulling off the road, Old Zhang steps fast onto the Chunyi Bridge. The Haihe River here is much broader than the central urban section. As far as he can see, there are trees with luxuriant foliage on both banks. A team of athletes are rowing on the river, like dragonflies sweeping the water. Looking back to the north, the silhouette of the high-rise buildings in the downtown area is full of dynamic under the sunset. It is not too early, so they get into the car and continue to navigate according to the address given by Mr. Zhao, the developer. After driving out of the main boulevard, they enter a small road. Old Zhang fells the same sense in Wudadao a little bit in a moment. The green street trees and beautiful buildings set off one another. In the circular square in front of the entrance of The Pillow Park, they meet with Miss Wang, the real estate agent.

Miss Wang takes the father and son upstairs to the top deck of a small building near the gate of The Pillow Park. Instantly, the full view of the park is displayed before them, which made people relaxed and happy. "The Pillow Park is one of the 11 parks along the Outer Ring Road and green belt in the central city. It covers an area of 1 square kilometer. It is the same size as Wudadao area and two-thirds of the Old Town." "Wow! If this land is used to build houses, how many villas and courtyard houses can be built! How much will it worth!" Old Zhang sighs. "Now people's living standards have improved, more and more attention has been paid to the living environment. Old residential areas such as Tiyuanbei and Huayuan, although living conveniently, are lack of such large parks which makes people close to nature." Old Zhang, who lived in Huayuan for five years, nods his head thoughtfully. "Tianjin Central City has a relatively dense population. In the past, the per capita green space of parks was less than 4 square meters. With the gradual completion of 11 major parks in recent years, the per capita green space of parks in the central city of Tianjin will reach 12 square meters, which meets the national ecological garden city standard. Who get most beneficial of The Pillow Park is the residents of adjacent community. "Is the park free?" Tianjin's parks have been free for many years, but Old Zhang subconsciously mentions it. "That's for sure." Miss Wang directs Old Zhang and his son's line of sight. "The Pillow Park is the only one facing the Haihe River in 11 parks. Because the water surface of the Haihe River is 200 meters wide in this section, The Pillow Park, different from The Water Park, The Meijiang Park and The West Water Park, which has a large area of water, is designed with more green space for activities by people of all ages, and more varieties of plants and flowers are plated, which makes the park seem to be a botanical garden. It is said that the landscape architect is a female French. She combines the artistry of European classical gardens with the artistic conception of Chinese classical gardens and meets the emotional needs of Chinese people with women's unique delicate techniques. "The park is beautiful. What about the apartments?" Miss Wang lead the father and son to the model room of the community. They look back as they walk, hoping to fix the beautiful scenery of The Pillow Park in their minds.

The interior of the model room is totally different from the general real estate sales office. It has no brilliant and magnificent decoration, but friendly and pleasant ambient, like a business space. "Which developer are you from, Miss Wang?" Mr. Zhang, the eldest son, asked. "I'm not from a developer. I'm a real estate agent, a freelancer. Here's my business card." "Tianjin Urban Construction University, Bachelor of Architecture, Master of Real Estate, no wonder she is different from sales ladies of flicker." Old Zhang thought by himself. "Now the real estate development and sales are upgraded and become more professional. In past, the development project is mainly dominated by developers, in the mode of developers taking the land, building luxury sales offices, producing and selling in volume. Now the government takes lead especially in public housing aspects. According to the new regulations, the urban land department builds roads, parks, greening, education and other supporting facilities, and then transfers land. Because the planning is 'narrow roads, dense network', so the land lot sold is not large, The Pillow Park Community now has more than a dozen developers, to ensure the diversity of housing types. Because of market competition, developers are meticulous in quality of products, and promotes Customization, that is, tailored construction . "Customization?" The eldest son's heart moved. He always wanted to design and built his own house according to his own ideas, but he could not yet find any opportunity. Miss Wang introduced the planning and design of the Community. The design attempts to combine the advantages of suburbanized housing with the advantages of vigorous neighborhoods to create a new urbanism community planning model in China." Old Zhang does not understand what new urbanism and what model it was, but he understands what Miss Wang means. "I went to the United States three year ago, and lived in my girl's home. Her house is in the suburbs. The environment is very good. The house is big and beautiful, and there are big courtyards around it. When I first arrived, I really enjoyed it, but after a little time, I felt so depressed that I couldn't speak English and had no friends to chat with. It was really a beautiful but lonely place. I still think it's good to live in China. There's no place better than home where makes us steadfast. But there are so many cars in the city and there are some noises. It would be better if there are houses and environment here like that in the United States. This room is the activity exhibition room of the community center.

"The Pillow Park Community is adjacent to the park and the Outer Green Belt, with good air quality, quiet surroundings. There are famous Thoracic Hospital and Three-A Huanhu Hospital. Pillow Campus of Yaohua Middle School and Pillow Campus of Yueyang Dao Primary School have been built in the community. They are both key schools. Traffic is also very convenient here. There is Metro Line 1, Metro Line 10 under construction, and the future planned municipal express line Z1, as well as driverless buses and taxis for easy transfer, and the community is walkable completely. It's close to South Dagu Road and Outer Ring Road. It's convenient for cars to travel. Considering the needs of many families, we have two parking spaces for most houses." Old Zhang's grandson has two cars in abroad. cars are inexpensive, but it is difficult to find a parking space in the old community, let alone two parking spaces. "Housing in The Pillow Park Community is very diverse, because the planning regulation requires that the house in the same size, shape can't exceed four, and not adjacent. The buildings are mainly six-storey elevator apartments and the medium high-rise flats of 9-storey to 18-storey, there are a small numbers of three-storey row houses and 80-meter high-rise towers. The layout of high-rise residential buildings not only has a good view, but also avoids the depression on streets and squares."

Looking at the beautiful houses with different shapes on the sand table, Old Zhang is already tempted, but is such a good house very expensive? The son saw his father's mind and asks, "Miss Wang, how much is the house per square meter?" "I need to give you a serious explanation about the housing price." Miss Wang paused for a moment: "At present, the state is formulating the Housing Law, which will stipulate that urban middle-income families will have decent housing in 2050. Besides perfect functions and good overall quality, the key to a decent house is to make it affordable, that is, to have a reasonable housing price-income ratio. House price-income ratio refers to the ratio of the total income (including housing provident fund, medical insurance, etc.) of the couple in the year before tax to a suitable housing price, whose reasonable range is about 6. Specifically, if mortgage purchases, in addition to down payment, monthly repayment of bank loans should not affect living standards." Looking at Old Zhang's seemingly confusing expression, Miss Wang further explains, "Do you have grandchildren? We assume that you're buying a house for your grandson. If your grandson and his wife have an annual income of 400,000 yuan, it is feasible to buy a house worth 2.4 million yuan based on the house price-income ratio of 6. Apart from the 30% down payment and 800,000 yuan down payment, mortgage for 20 years, the monthly repayment is about 8,000 yuan, and the remaining living expenses are about 12,000 yuan, which can ensure a better living standard. "How large is the house and how many rooms are there?" "The house here has a larger area of 100 square meters, with three small rooms, which is suitable for young couples and a child or even two children. There are 120 square meters of small four rooms, two children will be no problem, one room for one child, even two children in different genders are not afraid." "100 square meters, 2.4 million, only 24,000 yuan per square meter, is it so cheap? I heard the house here is 30,000 yuan per square meter." The eldest son has some doubts. "Yes, because of the good environment and complete supporting facilities, there are 30,000 commercial housing prices, such as high-end housing with courtyard and good landscape, but also expensive. What I just talked about is the price of well-off housing."

"Well-off housing?" This father and son asked together. "Yes, well-off housing, according to the Housing Law, the government has an obligation to provide decent housing for middle-income families. Apart from perfect functions, good quality and complete supporting, the key is reasonable housing prices. Nowadays, in mega-cities like Tianjin, the house price is too expensive and the house price-income ratio is over 10, which makes it difficult for ordinary families to buy satisfied housing, while the affordable one is too small." This father and son nodded and felt the same. "Housing prices are made up of three parts: land, construction costs and taxes. To ensure the quality of the house, the cost of construction and decoration can't be reduced; in order to ensure the normal operation of the government, the reduction of tax and fee is limited, so the key to reducing house price is to reduce land price, which is composed of land consolidation cost and government net income. In the past, local governments often included the cost of construction of main roads, parks and schools into the cost of land. Increased cost of land led to the increase of government revenue and tax fees, invisibly greatly pushing up land prices, which is an important factor leading to the rise of housing prices. Now, the municipal government has begun to control the cost of land, the main roads, parks, schools, can't be included in the cost of land. At the same time, combined with the implementation of the real estate tax, 'new housing new method' is adopted to reduce the government's net income, that is to say, the government from the previous one-time collection of 70 years of land use right transfer fee to the annual collection. The original net income of the government is 25% of the land price, which can be reduced to at least 5% in a year. If the land price reduces 20% at least, then the housing price will reduce 20%. For example, if the commercial housing in The Pillow Park Area is sold by 70 years' tax in the old method, the price will be 30,000 yuan. Then according to the new method, the price will be about 24,000 yuan, which is now 6,000 yuan cheaper."

"What kind of people can buy this new housing, well-off housing?" Both father and son are eager to know. "Well-off housing is policy housing formulated by the municipal government in order to satisfy people's yearning for a better life, learn from the experience of Singapore Government, and combines the actual situation of Tianjin and the real estate tax reform. The standard of well-off housing is 35 square meters per capita. According to the standard of developed countries in 2035, every urban resident is eligible to buy, but only one set can be bought and sold in a certain well-off housing market in the future, only to the same qualified people, or to be bought back by the government at the market price. The key is to pay real estate tax on an annual basis." "In that case, what kind of housing is appropriate? Well-off housing is cheaper, but it has to pay taxes all the time!" "Let's take the aforementioned well-off housing as an example. The total price is 2.4 million yuan, which is 600,000 yuan cheaper than the commercial housing of 3 million yuan. If the tax rate is 1.5%, it will be 36,000 per year. Let's start with 20 years. It's as long as mortgage. It's 720,000 yuan, which seems more than 600,000 yuan, but the 600,000 yuan is a one-off payment, or as mortgage needed to pay interest. The interest of 20 years is about 300,000 yuan. So anyway, it's appropriate to buy a well-off house." "Won't the 70-year stock house pay property tax?" "It is understood that the National People's Congress is currently legislating that each household may be considered for the stock of housing, within a certain standard per capita can't pay taxes, but this is the process of reform, in the long run, real estate is all needed to pay taxes, which is also an internationally prevalent practice. However, our country now has a strong dependence on land finance in cities, and the reform needs a transition period. Therefore, apart from well-off housing, commercial housing land still charges a one-time land transfer fee of 70 years." After listening to Miss Wang's introduction, the father and son really gained a lot of knowledge. In Old Zhang's mind, the house price of 24,000 yuan is much cheaper than the estimated 30,000 yuan. If his grandson wants two children, he can buy a house of 120 square meters with four small houses. In this case, the total price is 2.88 million yuan and the down payment is about 800,000 yuan. His son and he help to make up for about 10,000 yuan a month, and the income of his grandson couple can afford, so he even doesn't need to use pension to help them. Think of here, Old Zhang suddenly relaxed a lot.

In a moment, he remembers the son of his brother-in-law, who has been working very hard, but with a normal grade. After graduated from a second tier university, he works in a small company with an average income. The income of his family can't afford the 2.4 million well-off housing. "Miss Wang, this well-off house is really good. Our family and my grandchildren can afford it. However, some children's families can earn only about 200,000 yuan." "You are telling the truth, but housing prices are closely related to location. The Pillow Park area is located in the Outer Ring Road of the central city, where the environment is good and the traffic is convenient, so even well-off houses are more expensive than other areas outside the ring. There are cheaper well-off houses in other areas outside the Outer Ring Road. For example, the prices of well-off houses in the core area of Binhai New Area are around 10,000, and families with an annual income of 200,000 are easy to buy such houses. Moreover, education and health resources in Binhai New Area have been greatly improved. Key schools and Three-A hospitals in the city have been opened in Binhai New Area. In order to attract young people, the new district has also introduced the Joint Property Right Policy. For example, a well-off house with a total house price of 1 million square meters and a down payment of 300,000 yuan can be paid only 100,000 yuan if it can't be taken out at once. The other 200,000 yuan can be assumed by the development company, who shares property rights with the buyers, so that the people who settle in the new area can easily buy houses without worries. Miss Wang continues: "in our country, the deposits of the residents have reached 90 trillion, but we are still worried about serious illness, children's education, housing problems and so on and dare not consume. Because there is no other investment channel, so everyone invests in real estate, which results in a bubble, affecting the real economy. At present, we need to improve the medical system, control reasonable housing prices, encourage people to consume, to improve living standards, to develop the real economy."

"You're right!" The eldest son responds, "I have the same feeling. I have a shop in Ancient Culture Street with good returns. I have always wanted to expand the scale or open new shops in tourist areas such as Italian-style District, but I have many worries. One is that in case the elderly and families are seriously ill, the other is that the children are going to get married and buy a house, so I have to save money. House prices have been rising all these years. My heart is full of worries. Today, when I saw this well-off house, I felt much more secure. I decided to open a new shop, set up a studio shop, and make some contribution to the economic development of the city. Old Zhang sees the relaxation and confidence in his eldest son's face that he have not seen for a long time. "Miss Wang, please take us to see the sample room!" Although Old Zhang is over 70 years old, he is still impetuous. "Grandpa, there are no sample rooms now, they are all real houses, and some people already live in them, so we need to make an appointment in advance. You've made a provisional appointment this morning. I haven't made an appointment yet." "That's all right. We came here temporarily. This is to buy a house for our son.

Finally, we have to listen to him. Our son and daughter-in-law have decided to go back to Tianjin for career development. They are currently dealing with various procedures for returning home. When they come back, I'll let them contact you." "OK, no problem!" "Don't feel me annoying, Miss Wang. I have another question. What do you mean by this customized house?" "Oh, come here, let's take a look at the model." Miss Wang pointed a laser pen at the houses on both sides of The Pillow Park on the sand table. "Apart from residential buildings, The Pillow Park should also develop creative design industries to achieve job-housing balance in the region. This block is a mixed commercial and residential block, such as the SOHO block in New York or the design block in Helsinki. In order to reflect the creativity and diverse needs of designers and artists, the houses here are specially designed and constructed according to the needs of the owners. The standard land plot is half an acre, 10 meters wide and 30 meters deep The owners can freely entrust their favorite designers and construction companies, as well as the development companies." The eldest son was interested: "How much investment does this usually require?" "The land is 300 square meters, the FAR is 1.5, the building area is 450 square meters, the land price is more than 5 million, there will be a certain difference in the price of land construction, roughly 3 million or so, totaling 8 million of the two items. The price is not too expensive. It is mainly to attract designers and design companies to settle down. If Mr. Zhang is interested, I can contact the district authorities concerned. They also have preferential tax policies. "Great. I'll go back and discuss it with my wife." The eldest son always wants to build a house by himself. As the eldest son, he not only inherited the family's industry, but also wanted house like the century-old one in the movie, which is shop downstairs and studio and the bedroom upstairs. During festivals, the whole family can come here to gather. The important thing is to build their own house according to their wishes, including layout, shape and interior decoration. "It's getting late. We're leaving. Thank you very much for your explanation."

Farewell to Miss Wang, the father and son returned to the square. They estimate that they would catch up with the rush hour of commuting. Watching the people shopping and sightseeing in the square, they decide to eat here and experience community life. They entera Korean barbecue restaurant facing the park, took a seat on the terrace of second floor, looked at the twilight in the park. While eating delicious barbecue, and thinking about the future life, the father and son are in high spirits.

Part X

Back to the Haihe Waterfront

56. Mr. Bauer · Binhai Juilliard Concert Hall

The seminar is in full swing. Bauer looks at his cell phone. It is already five o'clock. He picks up his bag, says goodbye to everyone and walks out of the conference room. Chief. ZhuGe, Director Song and Chief. Chu follow him out and say goodbye to him. Bauer insists that everyone go back to the seminar as they are all old friends. He says that he will return to Tianjin in a short time. Out of the Planning Exhibition Hall, he sends a message to his wife, Shirley, and then walks to Tianjin Railway Station. Autumn and October are the best seasons in Tianjin, and there are a lot of tourists on the street. Bauer looks at the familiar city and hears the noises of the city. The autumn wind breezes by and he feels refreshed. It is the time to get off work and many people are walking to the Tianjin Station. With the urban pedestrian system transformation and traffic tranquilization measures, the walkway is spacious and clear, and it feels comfortable to walk on.

Mr. Bauer wonders as he walks: Can I start my new career in this city? Can my wife and daughter find their own careers and hobbies in this city? For his wife Shirley, Bauer is very confident–she has been working in Tianjin already. But for his daughter, he can only advise. His daughter has worked in the Juilliard School in Tianjin for six months. Can she stay? As soon as Bauer is seated on the train, he received two replies from Shirley: "I'm almost at Yujiapu" and the location of the restaurant their daughter has reserved. The picture shows the Singapore Ruizhen Restaurant in Yujiapu Financial District. In 20 minutes, the high-speed rail has travelled 50 kilometers. Bauer steps out of the station. The station is bustling, and many passengers are changing trains seamlessly by walking to the subway station at the underground exit. The fast food restaurants and fresh supermarkets in the station are booming with business. When Bauer walks through the crowds to the Yujiapu Global Shopping underground shopping street, he finds that the amount of people walking on the street have increased a lot compared to when he has been here in the past few years. The majority of the people are no longer tourists, but young white-collar workers. They laugh and talk, full of vitality.

At 5:50, Bauer arrives at Ruizhen Restaurant ten minutes ahead of time. The table his daughter has booked is near the window and has a good view of the Haihe River at dusk. Bauer goes to the bathroom to tidy up his clothes. He remembers that a few years ago, every time he went out and went back to the hotel, there would be a layer of floating dust on his shoes. Every day, there would be black stains on his collar and sleeves. But today, the collar and cuffs of his shirt look clean and his shoes are bright. It looks like the air quality of Tianjin and Binhai has greatly improved. Bauer sits down and takes a sip of tea, relaxing both physically and emotionally. After a while, his wife Shirley and his daughter arrive. Shirley is dressed in a dark green evening dress; seems like that she has gone back to the hotel and gotten ready to attend her daughter's school concert in the evening. Their daughter Vivian looks mature and elegant in a long black dress.

Bauer and his daughter gently embrace each other. Having not seen each other for more than half a year, he feels that his daughter is more beautiful than before and his heart rises with pride. There is also a young and handsome Chinese man in a black suit beside her daughter. "This is You Hua, my colleague. This is my father, Bauer." "Nice to meet you, Mr. Bauer." Bauer hears an authentic Oxford accent, and then a gentle and strong hand shakes his. "I'm guessing you're a pianist." "You have a good eye!" It's really rare for the family to meet in Tianjin Binhai New Area thousands of miles away from home. What does that mean? After graduating from the Juilliard School in New York, his daughter has been teaching at the Juilliard School in Tianjin for nearly six months.

Bauer sits opposite to her daughter and sees a trace of weariness on her face. "Vivian, how have you been lately?" "Good. The students here come from different parts of China. They are excellent, talented and diligent. After guidance, they have made rapid progresses. You can see them performing tonight." You Hua continued: "They have to be diligent. There are thousands of children studying music in China. There are so many of them and their parents have put in a lot of time, energy and money, hoping for their children to be successful. For them, the Juilliard School in Tianjin is quite expensive compared with other famous music institutes in China, but they send their children here anyway in order to have such a good platform. "Yes, Juilliard has found a good business, a gold mine." His daughter jokes.

The food is served quickly and is finely cooked and plated. During the dinner, Bauer keeps wanting to know his daughter's future plans. He knows that there is little time for talking today, but there is no better chance. He asks directly, "Vivian, do you want to stay in Tianjin's Juilliard? Do you want to stay in Tianjin?" "Do you really think so? But I don't want to, absolutely not! The school's rule is one rotation per six months. I can't stay any longer. I will go mad if I stay here for too long." "Why?" "Why? I don't have friends here, and there aren't too many social events. It's a cultural desert." "Have you never been to the city?" "I have for several times. The concessions are interesting, but it still feels a little strange. These Western-style buildings in a Chinese city are different from what I imagine. Unlike Beijing and Shanghai, where there are many foreigners and English communication is very convenient, Tianjin gives me a sense of distance." "You Hua, how long have you been in Tianjin's Juilliard?" Shirley asked. "I came at the same time with Vivian. I used to live in Shanghai and have graduated from Shanghai Conservatory of Music. At the time, those of us who studied music, including the teachers, were curious about why Juilliard chose Tianjin Binhai New Area to cooperate with Tianjin Conservatory of Music instead of Beijing's Central Conservatory of Music or Shanghai Conservatory of Music." Everyone looked at each other.

Unaware of the awkward atmosphere, You Hua continued, "I have been to the Juilliard School in New York. It was surrounded by art centers and it was an absolute paradise of music. The Metropolitan Opera House alone has 3,980 seats, and the seven different theaters added up to 18,000 seats, which were always full. The Binhai New Area lacks quite a bit in this aspect." "Shanghai is no good either. Is there a second New York in the world?" Vivian always has a comeback. "Binhai New Area is actually doing pretty well with good infrastructures, and it must be very promising in the future. It's just that I can't wait. Also, the name of Binhai New Area doesn't stand out, or more like it's not a name for a city. It's difficult to say. It's also strange to call it BNA. It seems TEDA or Tanggu would be better." Vivian takes a sip of tea. "There's actually a bar street transformed from an old railway station next to our college. Me and You Hua and the other colleagues at the college often go there. Maybe we can go there after tonight's performance. There are many different bars there and it gets very lively in the evening." Soon, an hour has passed. The food and dishes are good, and everyone has had a good time eating. It's about 15 minutes' walk from here to Juilliard. Vivian gets up and asks for the check. It's time to go. "A good exercise after a full meal." Vivian jokes.

The architecture of the Juilliard School in Tianjin is designed by Diller Scofidio and Rentro, which has designed the renovation and expansion of the Juilliard School in New York. DSR is a partnership of four outstanding designers. Its chief designer and founder is a woman, Elizabeth Diller. In 2018, she was named the most influential architect after Ingels from the Bjarke Ingels Group by the Time magazine. Perhaps it was because she has a unique female perspective, or because she and her partners have artistic creativity in art, architecture, life and other fields, and can see the potential opportunities in challenges. "She made our fantasies possible and turned them into real bricks and stones," wrote Time magazine's commentary. Diller and Scofidio are domestic partners who have been working hand in hand, and the New York High Line Park they designed in 2009 is well known in the United States. Bauer has visited many DSR designs and personally likes their style. He especially admires Diller. Although she was 63 years old when she designed the High Line Park and 68 years old when she designed the Juilliard School in Tianjin, she still had great creativity. Bauer often encourages himself by aspiring to the couple.

When Bauer approaches Juilliard, he finds that the building was of high standard. Although the shape and style are different from the Juilliard's buildings in New York, they project the same affect–echoing tradition in innovation and breeding innovation in tradition. The building is situated on a green area beside the Haihe River, shaped like a lamp, which contrasts with the neat, simple and classical buildings around it. The bottom part of the building is overhead, creating a flowing sense of sequence between the inner and outer space, and the building integrates well into the environment. Today's performance venue is the largest concert hall at the school, with nearly 700 seats. The interior is simple and modern, meaning to position the venue itself as merely a container for music. Juilliard's tradition is that teachers and students perform together. Bauer's daughter and You Hua lead three students to perform two classical pieces, first Schubert's Trout Piano Quintet in A Major, and then Liang Zhu, a classical Chinese piece. This is the first time Bauer hears his daughter playing Chinese music. The beautiful violin flows from her daughter's hands. Bauer feels an emotional sublimation. He holds his wife's hand lightly and Shirley gives a response from which he can feel Shirley's excitement. Although Vivian has inherited more of Mr. Bauer's genes, she has also inherited her mother's delicate emotions.

The two-hour performance ends successfully. The Bauers come to the lobby and wait for their daughter to come out. Several students and parents are also waiting there. When Vivian comes out, they give her small gifts. Bauer's daughter is very happy. She praises the students for their excellent performance tonight and congratulates the students and their parents. "Let's celebrate!" Bauer's daughter, relieved, puts her arms around her parents and they walks to the bar street close by. Sitting on the seats by the Haihe River, the breeze of the Haihe River blows past, gentle and cool, and the whole family is very happy. The Haihe River here is much wider than the downtown area. It has a wider view, and they can see the lights and building outlines in the night of Yujiapu. The environment gives people a good feeling, and it is even better with the family gathered together.

Mr. Bauer orders champagne to celebrate for his daughter. His daughter asks her mother about her work in Tianjin. "Tianjin is a very interesting city. Tianjiners are enthusiastic and loyal. I often get help from friends. A place's unique water and soil raises unique people, so it may be related to this land. My work is closely related to the soil and water environment. The land is saline and alkaline, so it is very difficult to create more green spaces. In the past, there was more water to dilude the saline and alkali, but after the flood problem of Haihe River was eradicated in the 1950s, there was little water in the upper reaches except during the flood season. My Tianjin colleagues themselves say that they are located at the bottom of the nine rivers and can only get floods but not the benefits of being near-water. Historically, Tianjin has retained a large amount of lake water, which can regulate and store rainwater. It is very wise. Today, I visited the newly built Shuixi Park, which retained a large area of water in its design and adopted the style of traditional Chinese gardens. It has some unique characteristics, but the park is not satisfactory either from the ecological point of view or from the human point of view. I understand that there will be eight large parks in the future in the downtown area. I suggested that they should adopt designs that are more ecological and sustainable to make people and nature live in harmony. On this basis, they could pursue individuality and new explorations of traditional gardens.

"Also, I attended a seminar on green ecological barriers between the two cities in the afternoon. This is a major project and it may go on for decades. There is a Chinese proverb: Forefathers plant the trees and later generations enjoy the shade. It is hoped that this project will benefit future generations. I see that they have adopted the traditional platform elevation method in their design, which is both economical and practical. The ancestors of this land must had had many practices adapted to local conditions that deserve careful explorations. To advocate for ecological civilization and paying attention to the construction of green environment in China is just like that advocated by Olmsted, the famous 'American peasant', two hundred years ago. China needs to be led by a master who really understands people and the earth's ecology like him. Relatively speaking, landscape design may be more difficult than architectural design, which requires many people's continuous explorations. Vivian, your grandfather used to work on this ground, and your father wanted to make a difference in this land. I was born in Taiwan, but the mainland is my root. I have a special feeling about this land, as if there was a voice calling me.

It is late, and Mr. and Mrs. Bauer accompany their daughter back to the apartment she has rented not far away. Before she goes upstairs, she turns around and asks, "what do you think of You Hua?" "A good pianist." "Very handsome." "I hope you like him." Their daughter blinks mysteriously, and the Bauers understand. "See you at Christmas then." Bauer's mind wanders to the fact that they have a house in San

Francisco, which he has designed and built in his thirties and was in a good location, but today it seems that the design is rather immature in places at that time. Before leaving for China, Mr. and Mrs. Bauer have already talked about selling their house and staying in China for a long time, as they have found something interesting here. Their daughter likes the bustle of New York and is not willing to settle in California. The next Christmas may be the last Christmas of the family in San Francisco.

Bidding goodbye to their daughter, Mr. and Mrs. Bauer come to stay at the adjacent Intercontinental Hotel. The shape of the hotel is like China's Jinyuanbao (golden boat-shaped ingot). The original meaning is for the Yujiapu Financial District to be prosperous, but the design is also quite good. Five years after its opening, the popularity of the hotel has gradually increased. Standing in front of the window and looking at the quiet river, Bauer and his wife discuss the next steps. Tomorrow they will attend an academic presentation in Beijing and visit Tsinghua University in the afternoon. They decide that after returning to Tianjin, they will go to Wudadao or places by the Haihe River to find a suitable office location. Last week, a friend showed them a historic building at Wudadao, the old American Barracks on Munan Road. This is a U-shaped two-storey building. Although it is not near the street and one has to go through a 50-meter-long alley to get into the inner courtyard, but the sequential arcades on the facades of the building appear very dramatic, and a beautiful glass roof can be added to cover the courtyard to make the center a shared atrium. It will be a very attractive and charming place for socialization, whether it be discussions about projects or chatting among friends.

They are deeply attracted by this idea. They hope to continue negotiating with the Historical Architecture Restoration and Development Company about the rent. If Tianjin, like some European cities, can provide some preferential policies for the renovation and utilization of architectural heritage, such as low or one-yuan rent while the lessee will be responsible for the renovation and transformation, then there would be incredibly rich possibilities in the revitalization and utilization of Tianjin's small villas. Another important thing is to find two or three designers to collaborate with, preferably those with experiences studying and working abroad. Office locations are relatively easy to find, but it is difficult for them to find designers with overseas experiences who are qualified and willing to work in Tianjin for a long time. There are many international talents in Beijing, Shanghai and Shenzhen. Just like New York, all kinds of talents are the foundation of a city's success. They feel that foreign designers are hard to find, but there are still many local designers in Tianjin. Tianjin has a basis for talents. There are dozens of universities and more than 300,000 students in Tianjin. Mr. and Mrs. Bauer believe that as long as they persist, they will be able to select and train the designers they want.

57. Song Yunfei·Binhai Cultural Center

The morning meeting ends after 12 o'clock. Song Yunfei eats a box lunch in the meeting hall. He scans all around and finds no sign of Chief. ZhuGe. When he saw Chief. ZhuGe early this morning, he wanted to talk to him. But Chief. ZhuGe's looked busy and constantly had experts talking with him, so Song Yunfei didn't want to interrupt him and only greeted him. On the stage, Chief. ZhuGe presented the review and summary of the urban design project, and introduced in detail the main contents of the urban design pilot project and their attempts at reform and innovation. Some of the doubts Song Yunfei had when he saw the exhibition in the morning were clear after Chief. ZhuGe's presentation, but Song Yunfei still had some doubts about the strategy of encouraging cars in the periphery of the city.

Whether studying highway and traffic engineering in college in China, or later going abroad to study as a graduate student in traffic planning, all the lessons and books Song Yunfei has taken and read have stated that private cars-based transportation and traffic should limited. The world, mainly Americans, is worried that China and India will begin encouraging private car ownnership, and the Earth would not be able to afford it if both countries reach the per capita carbon emission standards of the United States. Although NEVS, the new energy vehicle brand, is growing rapidly in China, and autonomous driving technology is growing rapidly, Song Yunfei still can't approve of encouraging cars right now. "Every family can have two parking spaces. That's possible, but it's not good to encourage cars. Why is that?" Song Yunfei couldn't find a good answer to the question Chief. ZhuGe asked when he was interacting with the experts. Song Yunfei found that the usually gentle Chief. ZhuGe spoke in a strong tone. He said: We have planned to restrict cars and encourage the development of public transport, but the people "vote on wheels." China has been the world's largest automobile producer and seller for eight consecutive years.

Song Yunfei was somewhat melancholy that he failed to get to Chief. ZhuGe, but also somewhat happy, like a stone in his heart dropped to the ground. In addition to participating in forum and seminar, he has a more important task in the city, which is to close the deal on a planning and design contract with two customers. Before going, he wanted to give Chief. ZhuGe a heads-up. This is also what his wife has been telling him to do, but he has a feeling that it is not appropriate. Song Yunfei has always been a good student since he was young and has never worried his parents. The teachers at school liked him, and he never asked any teachers for sympathy or favors. Later, when he was preparing to study abroad in college, he did not ask for anyone's help at all; even when asking his advisor to write him a letter of recommendation, he hesitated for a long time as the deadline was fast approaching. His classmates said that aside from good grades, good letters of recommendation from famous professors abroad are also very crucial if one wants to get into the Ivys. References and social credits are very important abroad, which he has gradually realized after his going into work, but his personality still makes it difficult for him to act.

Back in China, Song Yunfei knows that the Chinese society is a nepotist society, but he is more willing to believe in his professional abilities. Although the government has been introducing reforming policies, China's nepotist culture is deep-rooted and difficult to change in a short amount of time. Things like drawing upon connections and backdoors are more prevalent in Tianjin, including Binhai New Area. If things that usually need to be done through finding relevant connections are done without doing so, Tianjiners would even feel suspicious or feel like it is a sign of their lack of capability. Song Yunfei's wife told him to drop Chief. ZhuGe a word, and that it is not against regulations to do so, and since Chief. ZhuGe knows more about your abilities and work experience in Binhai, he can help put in a good word for you. But Song Yunfei felt that it was still against the rules. But Song Yunfei still feels like it is against regulations, and he could not get over it. He did not open his mouth when he saw Chief. ZhuGe just now. Now that the meeting is over, he can't find Chief. ZhuGe. It's almost time to meet the customers and he must leave now. He feels like he has made an attempt and it is not his fault that the timing isn't right. He even feels relieved.

It has been three years since he returned to China to start his own business in Binhai New Area. Song Yunfei has witnessed many "entrepreneurship and innovation" companies failing to survive. But there are also a few successful companies. For example, recently, Concino Biology was listed on the Hong Kong Stock Exchange successfully. It was the first vaccine stock to be listed on the Hong Kong Stock Exchange and a company founded by Chinese people. It may be that the nature of the companies are different. New energy, biopharmaceuticals, new materials as well as mobile Internet, new media and other industries are supported by the state and Tianjin government, forming a certain agglomeration effect. However, as an information company that specializes in transportation planning, it is difficult for them to grow. When the company started, it had good momentum. The government provided support and promised them not to collect rent for three years, and also gave them subsidies for renting and moving as well as the employees' living expenses. Most importantly, there were already several consulting and research projects waiting for them, so Song Yunfei was full of confidence at the beginning. In the first year, more than a dozen employees were recruited, all of whom had master's and doctor's degrees in traffic planning. They had strong scientific research abilities, and the projects they completed such as the research on strategies for development of smart transportation in central Binhai area had been well received. But the company's range of services was relatively narrow and its ability to expand the market was not strong, so there was little income growth and they could only break even. Some employees were dissatisfied with the pay, and some left to Chengdu, Wuhan and other places, as all cities were trying to attract talents. The internal management structure of the company was also not well-planned: Employees were only in charge of specific work, and Song Yunfei needed to do all the management work by himself while also having to run projects. His hands were always full.

The company stagnated, but Yujiapu began to prosper. With the merger of TEDA and the central business district, the management committee has moved from TEDA to the southernmost end of the initial area of Yujiapu. The number of companies registered in the initial area of Yujiapu has gradually increased. With the increase of the flow of people, the rent of office buildings also began to rise. Word on the street was that the rent income of office buildings was related to the amount of tax paid by the management company, and so the management committee revised the policy for subsidizing companies. Song Yunfei's consulting company has not yet made stable income, nor has it paid income tax. But they will have to pay rent starting next month, although it is only 1.2 yuan per square meter, but the area rented for the office is 400 square meters large, a total of 500 square meters plus the shared area, so it will be 600 yuan a day, 18,000 a month, and nearly 200,000 a year, coupled with property fees, too large an amount for the company. If they can't land at least one big project in the near future, it will be very difficult for the company to keep going. Therefore, recently, Song Yunfei has been using every means to find projects and striving to sign contracts. It would be the best if they can get the deposit upfront; then he can pay for the rent, the employees' checks and their insurances.

The reason Song Yunfei is going to Tianjin West Railway Station and Xiyingmen this afternoon is to try to sign contracts for these two projects . Song Yunfei has been paying attention to Tianjin West Railway Station for a long time. At the time of construction, he thought that the design had great problems. The layout was the traditional single railway station building in China. The building was an arch of 50 meters high and had great momentum, but felt empty and the architectural shape was loose and chaotic. This situation is very representative in China. Although the rapid development of domestic high-speed rail has created a high-speed rail network in just a few years, which has greatly shortened spatial-temporal distances and promoted regional development, there are two outstanding problems in planning and construction: one is that most of the high-speed railway stations are located far away from the city center; the other is that most of the stations are still built in the old single railway station style.

Tianjin West Railway Station was renovated from an old station and located in the center of the city. The main problems are bad building layout, poor accessibility and inconvenient transferring. Song Yunfei has researched for a long time and imagined the possibility of transforming and optimizing the West Railway Station, and finally came up with an effective scheme. He hoped that through the implementation of the planning and renovation of the West Railway Station, it would become a model for the renovation of the old railway stations in China, but he had not been able to find the right opportunity to realize this idea. He had gotten in touch with the railway department, and the head of the railway department said that they only managed the area within the shade projected by the eaves of the station's building on the ground, and anywhere outside was managed by the local government, that many aspects of Song Yunfei's plan were beyond their control. Song Yunfei went to the station district management committee of the district government. The head of the management committee said that they were a coordinating department without funds and could not initial transformations. Like this, time passed.

Recently, he heard that the municipal government had responded to voices from various sources and planned to comprehensively improve the management of Tianjin's railway stations and airport. It was a rare opportunity and Song Yunfei made every effort to make appointments with the leadership of various departments, including railway, district government, subway, bus and other departments. Song Yunfei presents the concept of his scheme on the spot. He has prepared carefully for the presentation the night before. The designer turned the empty hard-paved square in front of the station into a leisure center to attract tourists and made it a hall with sunshine and plants that improve the underground space environment and guided passengers to transfer. After hearing his presentation, everyone feels inspired. The plan thoroughly considers the fundamental solution to the problems of the West Railway Station. However, all departments have realized that in order to improve the environment of the West Railway Station, comprehensive measures are needed, and the layout of the current building structure needs to be changed comprehensively instead of superficially. After a long discussion, the leader of the Traffic Planning Department concludes: "today we are convening a meeting to look at the plan proposed by Mr. Song to eradicate problems of the two transportation hubs, and to promote the renovation and upgrading work of the two stations. The plan of Mr. Song gives us a fresh point of view. Of course, this plan is more complicated; new investment is needed, and the technical feasibility of underground and square space renovation and the cooperation of various departments also need to be researched. This is the premise for the plan to be carried out. I will report to the my superiors immediately after I go back, and strive to speed up the process. As for the upgrading and renovation of the West Railway Station, it is a mandatory task, and it will need to bear fruit in three months. Now it has been a month since the start of construction. Everything is progressing rapidly as planned, but some work is not in place yet. Everyone has to speed up and complete it according to the schedule so as to avoid being ill-prepared."

After the meeting, the director thanks Mr. Song again and says that he would report back to the superiors immediately. Song Yunfei is touched, but also feels sad. He knows that this project involves all aspects and cannot be settled in a short time, and it cannot help his current situation. At this point, Song Yunfei has nothing to lose: "Director, we have invested a lot of manpower and material resources in the early stage of this project. Now the company has encountered some difficulties in operation. Do you think it is possible to commission us to carry out preliminary research first?" "Mr. Song, thank you very much for your initiative in Tianjin's urban construction. But you know, we have a very standardized project management process now, and scientific research projects are also commissioned in accordance with the plan from the beginning of the year. There is no channel for new projects to be added. You have to wait until next year to apply for new projects to be carried out! I also want to raise this problem to the superiors this time, but based on my experience, the possibility of it getting addressed soon is not very high." "Thank you for your trouble!"

Coming out from the West Station District management committee, Song Yunfei rides a shared bicycle and circled around for a long time to find the Xiyingmen Street Office. The Xiyingmen jurisdiction area is more than 20 square kilometers large and more than 40,000 people live in it, spanning across both sides of the Outer Ring Road. The Daming Road area by the outer ring road has recently demolished company sites that were against regulations and created more space for development. The Daming Road area is three square kilometers large and is in a good location, but the traffic is inconvenient. In the long run, there is planned to be a subway line and a bus station; the coverage rate of public transport stations and the density of the line network are very low. In order to achieve the goal of bus travel taking up 70% of all travels, various forms of public transport must be introduced, but the capacity of ground transportation is limited, and it is difficult to meet the requirements only by conventional bus trips on the ground. Song Yunfei's company has worked out the plan for this area in conjunction with other design companies and proposed to adopt a new type of elevated tram.

Nowadays, many domestic enterprises are working on the research and test line of new elevated trams. He has participated in the evaluation of several projects. Enterprises hope that consulting companies can help recommend their products and technologies. Overall, elevated tramways are much cheaper than the subway, and more flexible, with lower operation costs, but there are no cases of implementation. A area like Xiyingmen would be a small city if it were in the United States, with an independent city government that can undertake its own affairs. The street offices in China are the government's dispatched agencies, which have many functions of social management but are not yet qualified for urban planning and construction. Song Yunfei has explained a lot, including that if there is no capital at present, enterprises can make PPP investments. But the head of the street office said that all of these things were beyond their control. It at least needed to go through the district government. Most importantly, the relevant departments of the municipal government need to put forward some standards. It's good, but when there is no standard, we dare not to run it even if it's built. This is a major issue involving people's livelihood, and we can't take responsibility if there are problems.

Since he came out of the Planning Exhibition Hall at 12:30 p.m., Song Yun has been rushing around all afternoon, barely achieving anything, and he is exhausted. Traffic problems in Chinese cities are very serious. Every year, huge amounts of money are invested in the construction of high-speed railways and subways, and there are many subsidies for the operation of buses. Cars and new energy vehicles are the pillar industries in many cities, but there is no funds for research and improvement in planning and design. These consulting fees are a drop in the bucket compared with the tens of billions of dollars in subway investment, but they can play a huge role. In this way, many urban problems are caused by a lack of understanding and problems with institutional systems, especially in terms of the early stage of research and planning and design, which lacks both funding and time. For example, the state has issued a document encouraging the comprehensive development of high-speed railway stations and their surrounding areas, but more than five years have passed and little progress has been seen.

Song Yunfei feels an invisible pressure on his mind. If everyone is used to driving, it would be difficult to build a bus system to change their travel habits. He has said these words to countless people for countless times. Now he has began to doubt that people have placed too much hope in the vigorous development of public transport; perhaps the way public transport is now is inevitable. October is the best season in Tianjin. It's the best time to ride a bicycle. But Song Yunfei has lost his mood. He calls a cab, goes to the West Station and enters the waiting hall from the underground passage of the back square. His expression is sour, like a downcast scholar, angry and silent.

At 5:30, Song Yunfei boards the Tianjin Intercity Train to Binhai New Area and arrives at Yujiapu High-speed Railway Station after about an hour, transferring through Tianjin East Railway Station halfway as there is no direct train. The train comes out of the East Railway Station and runs smoothly at high speed. With the regular rhythm of the high-speed railway train, Song Yunfei's mood calms down a little. He knows that his experience today is the current situation in China and Tianjin. He anticipates that this entrepreneuring trip to China is not going to end well. The company might be closing down. Maybe the timing is wrong. After getting off the high-speed railway train, Song Yunfei drives quickly to the Binhai Cultural Center. If there was a subway, the Yujiapu High-speed Railway Station would be only one stop away from the Cultural Center, but the Binhai New Area has been planning to build a subway for ten years, and the Z4 line has not yet been completed.

Song Yunfei comes to the corridor of the Cultural Center from the underground garage. It has been finished for more than three years and has been full of vitality as soon as it opened. As an Internet-famous building at home and abroad, Binhai Library has been attracting countless tourists and people need to wait in line in order to visit. The Dutch MVRDV design company has probably earned a lot from the advertisements. When Song Yunfei was in Hong Kong MVA Traffic Firm, he also participated in the traffic organization research of Binhai Cultural Center. What he appreciated the most was the design of the cultural corridor. The walking space connects the five cultural venues. People who go to each venue will first enter the cultural corridor, which gives people more opportunities to see each other, fully utilizing the corridor space. The design of German GMP Design Company's corridor is concise, generous and exquisite. In addition, the parking garage of several venues can be shared, which greatly improves the parking efficiency, and the interior design is also very humanized.

Song Yunfei goes straight to the civic activity center of the Binhai Cultural Center, where his daughter is taking drama lessons from Helen O'Grandy. His daughter is extroverted and likes to perform, which is different from Song Yunfei and his wife. "How's today?" As soon as they meet, the wife asks with concern. "Not so good." Song Yunfei is depressed. Hearing such an answer, his wife feels a little anxious, but realizing that today would not have been easy for Song Yunfei, she resists the desire to keep asking.

Before returning to China to start a business, Song Yunfei first studied for two years in Boston for a Master's degree, and then worked for three years in the Bay Area Transport Commission after graduation. There were many Chinese in the Bay Area, where he met his wife and married and had a daughter. Boston and San Francisco are two of the most culturally distinctive cities in the United States. Song Yunfei liked and was used to the atmospheres there, but in order to develop his career, he found a job in Hong Kong which provided a good salary. His wife studied biochemistry in college and had a good job in the Bay Area, but she quitted her job to take care of the child. But neither she nor their daughter liked Hong Kong, especially because the house was too small. His wife was very supportive of Song Yunfei's return to Binhai New Area to start a business. They settled their home in the Eco-City. They have gone through a lot in three years, which are valuable experiences.

During dinner, the couple are both silent, and their daughter is immersed in her drama world by reciting Dorothy's lines from *The Wizard of Oz* aloud. Song Yunfei's wife takes his hand and says, "Yunfei, I have been thinking. Let's go back to work in the Bay Area. Our daughter is older now, so it will be easier to take care of her. It's okay. We're going to study and get back to work. I believe that we will come back when there is the right opportunity. When my grandfather started his business in Tanggu, during the War of Resistance against Japan, in order to preserve the lifeline of national industry, in a critical and difficult situation, he moved the factory into Sichuan, which not only continued production, but also expanded the market. After liberation, the two factories moved back to Tanggu, recruited workers and started production again. Eventually, they made Jiuda and Yongli well-known brands in Southeast Asia and a symbol of national industry." Hearing this, Song Yunfei's indignant mood transforms into a kind of inexplicable grievance, with tears filling his eyes...

Driving out of the underground garage of Binhai Cultural Center, the Binhai Cultural Center is shining in front of them. His wife's favorite building is the Binhai Science and Technology Museum, which was designed by Bernard Tschumi, a famous American architect. The shape of the building seems to inherit the spirit of Tianjin Alkali Factory. The outer eaves of the perforated bronze metal sheets are very industrial. A light shining at the top of the conical space resembling a copper hotpot shoots directly up into the sky. Song Yunfei feels a force. The car is about to enter the tunnel crossing the In-Port Railway and the Jinbin Light Railway, and his daughter starts singing the song "Over the Rainbow" from The Wizard of Oz: "Somewhere over the rainbow way up high... Somewhere over the rainbow skies are blue... And the dreams that you dare to dream really do come true..." Crossing the tunnel, a bright light appears in front of the car window.

58. Old Zhang · Ancient Cultural Street Waterfront Platform

Old Zhang and his eldest son come home at eight o'clock. After the evening rush, it was very fast to drive on the East Haihe Road and it took less than 30 minutes. The eldest son knew his old man well and opened the window without waiting for his father to speak. The wind along the Haihe River was very pleasant, with a moist atmosphere. Old Zhang narrowed his eyes and took a deep breath. The familiar taste of the Haihe River made him feel steady and satisfied. The daughter-in-law has taken care of Old Zhang's wife for dinner and is watching the TV drama The Thunder. Seeing the two men come in, she greets Old Zhang politely and yells at the eldest son, "What have you been up to this afternoon? Doesn't even reply to my messages... Old Wang has been waiting! Hurry up!" As she talks, she pulls the eldest son by sleeve and goes out. "Dad, mom, we're leaving!" The son's voice disappears in the hallway. Old Zhang says to the backs of his son and daughter-in-law, "take it easy!

Old Zhang does not change his clothes. He sits down on the sofa in the living room next to his wife's wheelchair. He wants to tell what he has seen and heard throughout the day to his wife. But when he sees his wife's focused expression watching the TV, he stops. Since his wife's stroke, although she has made good recovery, it is not the same as before; not only her actions are slowed down limiting her range of activity, but she also has less access to new things. Old Zhang sits for a while. He isn't interested in TV dramas. They are all made up. He peels an orange and hands it to his wife. "I'll go out for a walk." His wife answers. Old Zhang gets off the elevator and walks out of the building. There are people walking dogs and people in a hurry. The night has fallen and everything is quiet. Old Zhang's thoughts wander as he walks, and before he realizes he has returned to the Haihe River again.

It seems that it has become a habit. No matter what happens, Old Zhang would absent-mindedly walk to the Haihe River. It may be that there are people and bustle along the Haihe River. It may be that the Haihe River is open and lighted. It may also be that he can smell the water by the Haihe River. Old Zhang likes the smell of the Haihe River. With the improvement of the ecological environment, especially the water environment, seagulls on the Haihe River have returned. Along the waterfront platform of the Ancient Cultural Street in the daytime, there are many tourists who take photographs and feed the birds, which is a picture of harmony between man and nature. The platform of Ancient Cultural Street is used for activities. In 2004, celebrations were held here during the 600th year of Tianjin's fortification. Later, Mazu worship activities were held many times, and now it has become the main Haihe River cruise ship wharf. As it gets dark, tourists gradually disappear. Old Zhang sits at the far corner of the big footstep, next to the 20-meter-high "Hundred Fu" lamp post on the waterfront platform. There is also a symmetrical lamp post on the other side. Each lamp post is composed of 12 hollowed-out "Fu" characters in different fonts, like a lantern post. During winter, the Haihe River cruise ship is out of service, the river surface is frozen, and many people skate on it, and some people dig holes in the ice to fish. During every Spring Festival, the Ancient Cultural Street is very busy. Like temple fairs, many people on the waterfront platform set off Kongming lanterns to worship ancestors in this way.

Sitting quietly on the platform, Old Zhang stops thinking. Looking at the distant lights, he suddenly feels lonely and tired. Since the death of his parents and uncle, Old Zhang seems to have lost his family, his connection with his grandparents and anything he can reply on. Although he has not experienced war, he has also experienced many things, such as the Great Leap Forward, the Cultural Revolution, the Tangshan Earthquake, Reform and Opening Up, being laid off and finding his own job. In public, he has to act like a man, finding connections and relationships and knowing how to bow to those with money and power; at home, in front of his wife and children, he is the pillar of the family and never dares to complain and show weakness; when the earthquake stroke, he stood up, regardless of his own safety, taking care of the children to protect his family while showing great calmness; then, he had to find a space as soon as possible, and build a simple earthquake shelter with timber and rain cloth; he asked his workmates to help repair the earthquake-damaged house, and poured them tea and gave them cigarettes to show his gratitude. After the Reform and Opening Up, he endured all kinds of obstacles, such as the closure of the factory and being laid off. He had to find a way by himself to support his family. Now, he has to take care of his unhealthy wife, and he has to suppress all kinds of emotions in the fear that his wife would be unhappy. Thinking about it, Old Zhang can't help but shed tears. He has no one to tell, no one to fully confess his feelings. The moonlight is shining on the river, as if watching him and listening to him.

Old Zhang is nearly 70 years old this year. Compared with the people who are 80, 90 years old, he feels young and healthy. For more than ten years, he has been accustomed to such a day-to-day life with no big changes, and he feels numb. Compared with the past, life is very good, with no lack of food and clothes, and supposedly he has nothing to complain about. But there are a lot of emotions in his mind. At home he bears everything, and outside he feels that he is being abandoned by society.

As the head of the family, he has been working hard since he started his family at the age of 23. There has not been one day when he dares to lay down his responsibility and live for himself. He remembers that when he was young, he also had many hobbies. He liked woodworking. He personally made the bedside cabinet at home and the ice cart his son skated with the Haihe River in winter. He also participated in the propaganda team in the factory and performed Tianjin kuaiban (a form of oral rhythmic storytelling performance using wooden instruments to play beats) and crosstalk. He also liked fishing and photography. Thinking of this, his hands want to do something. He recalls that when he went out in the morning, he saw a notice posted at the door of the neighborhood committee saying that the communities in the Gulou street area are planning to upgrade their environments, and designer would come to the communities to guide the residents to renovate the public green space of the communities by themselves. Old Zhang decides to sign up early tomorrow morning.

Also, after seeing the planning exhibition this morning and going to The Pillow Park and the community beside the park in the afternoon, Old Zhang has a vision in his heart. He hopes his children, grandsons and granddaughter can grow up happily and healthily, go to good schools and have good futures. He sincerely wishes those plans could be realized as soon as possible. He hopes that his cousin and nephew would have beautiful houses beside parks. He also dreams that one day, when he is too old to walk by the Haihe River, he could live in a suburban house with a yard, plant some flowers, and have a dog.

He recalls that when he was a child, his father told him that the Zhang family had lived in Tianjin for more than a dozen generations, that is, when Tianjin Wei was just built, his grandfather's grandfather's grandfather... took his family from Anhui to Tianjin, and secured a place on the wharf by doing street performances. Because of his incredible strength, he was well-known far and near, and was nicknamed Zhang Dali. Thinking of this, Old Zhang hears the familiar rhyme he used to say with the naughty children in the alley when he was a child: Don't judge by my thin form, I have travelled everywhere to perform, swallowing iron balls to practice Chi, eating soybeans to take a leak... Old Zhang smiles.

59. Urban Designers of Tianjin

The Tianjin Urban Design Forum and the International Seminar have been successfully concluded. In order to prepare for the conference and plan the Tianjin Urban Design Exhibition, Chief. Chu and his colleagues have made a lot of efforts, and the whole process took more than half a year. Most of the technical staff in the

Institute are women. It happens that three of them are pregnant. One is pregnant with her first child; she has been working hard for several years and is too old to wait any longer. The other two are with their second children; it was not allowed before and they want to grasp the opportunity now, as they can't afford to do it when they get older, or they will hesitate thinking about the pressure of raising children, the need to secure kindergarten spots in the "flash sales" on the Internet, and the need for school-districts housing for elementary schools.

After more than six months of busy work, the urban design pilot project in Tianjin has been completed. To sum up the experience of the pilot project, the book *Back to the Haihe Waterfront: the Tale of Urban Design in Tianjin* was compiled and published, which was first showcased at the international forum; two innovative guidelines in China, Tianjin New Residential Community Urban Design Guidelines and Tianjin Central Urban Area Public Space Design Guidelines, were compiled; the planning, design and arrangement of Tianjin Urban Design Exhibition were carried out; and a presentation of the experience of Tianjin's urban design pilot project was prepared to be presented at the international forum. With these fruitful results, Vice Chief Planner Miao's first son was born. Xiao Fu's stomach has made it inconvenient for her to move around, but she wears an oversized coat and has been busy with work.

Many innovative ideas were put forward in Tianjin's urban design pilot project, which inspired heated discussion and the ending time of the seminar had to be delayed until 6:30. It's seven o'clock when everyone has been seen off. Lights begin to lit up the exhibition hall, and crowds begin to gather on the pedestrian street in the Italian Style District. Director Liu and President Song are ready to leave and bid farewell to everyone in the hall. "The International Symposium on urban design has achieved good results. It not only advertised Tianjin, but also put forward many innovative ideas. Congratulations." Director Liu puts his palms together to the planners and volunteers who has gathered: "Thank you very much for your hard work! On behalf of the leaders of the Bureau, I would like to express my heartfelt thanks to you all. Chief. Chu, you've worked hard. Let's go home early and have a rest! "

After seeing off Director Liu and President Song, Chief. Chu and others gather up the materials in the meeting room and the hall and thank the volunteers. The urban design forum and seminar are over, but the urban design exhibition will last for a month, and volunteers will need to continue to work. Chief. Chu gives the signed copies of *Back to the Haihe Waterfront* to every volunteer personally and says goodbye to them. Then Chief. Chu finds the director of the Urban Planning Exhibition Hall, Director Shi, who has been busy all around, to thank him. After all of this is done, Chief. Chu is relieved at last. Looking at everyone's relaxed face and slightly tired expression, she suggests: "During this period of time, everyone has worked hard. Today's events are very successful, thanks to everyone's efforts and hard work. Well, whoever has something to do at home can rush home. If no one objects, everyone else can go to the Haihe River to grab a bite and relax." Everyone agrees.

Out of the Planning Exhibition Hall and turning left to cross Ping'an Street, they come to the Haihe River. The lights along the Haihe River are dazzling and the golden statues on Bei'an Bridge are shining. Not far north, they come to the Austrian Style District, which is a rare pedestrian section along the Haihe River. In order to solve the traffic problem with the Haihe River and the city center, the two sides of the Haihe River are mostly urban roads, which isolate people and waterfront embankments. But in order to protect the historic buildings such as Yuan Shikai's former residence, and at the same time to give people a chance to get closer to the Haihe River, the traffic roads were moved behind the buildings. Chief. Chu and her colleagues have designed this group of Austrian-style buildings, and at that time they imagined that the bottom floor of the riverside buildings should be cafes and bars. After many difficulties, the group of buildings was finally completed and perfectly displayed by the Haihe River. The buildings are in the traditional style and proportion of the Austrian architecture, but the local shapes and materials are modern, complementing the neighboring former residence of Yuan Shikai.

Many people say that the buildings in the Austrian Style District are beautiful. They also set a benchmark for new buildings in the historic blocks on both sides of the Haihe River. However, due to various reasons, the buildings became office buildings for companies and institutions, and the sides facing the river were all closed, not being able to form a open street of cafes and bars according to the original intention behind the planning and design. There are many tall Chinese parasol trees planted along the river, creating a broad shade. In the evening, neighboring residents come to the riverside to take walks and sit idly. A group of aunts come here to dance in the square every day. But the volume of their dance music played by the tape recorder is like noise to those who want to enjoy the quietness by the riverside. The construction cost of buildings in the Austrian Style District is high, and is difficult for regular companies to bear. Later, several banks took a fancy to this place, as it is believed that the money is at where water flows, but they also thought the surrounding area was too chaotic. So each bank put two huge and fierce lion statues on its doorstep. Since they put these lions on the doorsteps, the aunts have left to dance elsewhere. It seems that the lions are really powerful. In recent years, the municipal government has put a lot of efforts in developing "economy at night" and issued a large number of supportive policies. With the increase of tourists, the Italian Style District and the Haihe River cruise boats are overcrowded, and new development spaces and new forms of business are needed. Finally, fifteen years after their construction, these buildings near the Haihe River in the Austrian Style District have finally formed a nice Western food bar area. Western-style dining tables are lined outside by the Haihe River. Under the riverside space created by the Chinese parasol trees, everyone is enjoying delicious food and watching various kinds of boats glittering on the Haihe River.

Chief. Chu and the group choose a row of seats by the river. After a long day's hard work and having boxed lunch in the meeting hall at noon, everyone is hungry. The female colleagues are ordering giant hamburgers and French fries and gobble them down, not minding their appearances. Someone jokes: "Fu, eat more, don't starve your son in there." "Yeah. My son was so hungry that he kicked me, so I have to eat more today." Looking at the rare happiness on everyone's face for the past few months, Chief. Chu is very happy. Her eyes turns from the river cruise boat to the buildings on the shore. Against the backdrop of the lights, this group of buildings with their rich shapes is very attractive. Looking at the various buildings nearby built from the comprehensive development of the Haihe River and thinking about the hard work and difficulty in the industry, Chief. Chu feels a sudden sense of pride mixed with sadness.

Back in the day, in order to guarantee the quality of the urban design and architectural design for the initial section of the Haihe River, Chief. Chu invited Mr. Bauer to cooperate with the Urban Design Institute. Mr. Bauer was very devoted and paid a lot of painstaking efforts into the project, which won the praise of the customers and leaders. But more efforts needed be made to complete the project. There was a time when problems kept surfacing on the construction site, and Mr. Bauer suddenly returned home for personal reasons. In the face of the angry customers, Chief. Chu suffered a lot of pressure. After a lot of hard work, the building was finally finely completed, and the project had also trained a group of young urban designers who knew both architecture and urban planning. Later, with the passage of time, this group of people also left the Urban Design Institute one after another and took on important positions. Then, young designers came and went, and generations and generations of younger designers came up and needed to be re-cultivated. This is also a strange phenomenon in China's design industry. Designers who have just passed 40 or 50 years of age stop drawing and designing. They all become leaders and officials, and only command others with their mouths. In many foreign design firms, most of the main figures are experienced middle-aged designers who have kept doing design. This is one of the main reasons for the overall gap between China's urban design and architectural design and that of Western countries.

Everyone has eaten, and the young people become energetic again. Xiao Qin takes out a brand-new copy of *Back to the Haihe Waterfront* and comes to Chief. Chu's side. "Chief. Chu, sign your name for me!" "Okay, but I'm not a star, and my signature won't age to be worth a million." Chu says as she writes on the title page of the book handed over by Qin: "This book is dedicated to Tianjin's urban designers! I wish Qin greater progress!" "Chief. Chu, sign one for me, too." Everyone gathers around. "Okay, one by one. Can you all sign on one book and give it to me too?" Xiao Qin hands the book signed by everyone to Chief. Chu. The title page was densely marked with names, just like everyone's face, vivid and different, full of personalities, with handwritings that represent each person. "Thank you all! We held an event last year to celebrate the 15th anniversary of the founding of the Urban Design Institute. We invited officials and colleagues who had worked in the Institute to give speeches. I also told everyone about the history and fine traditions of the Urban Design Institute. Today, I want to talk about it here, by the Haihe River.

At the beginning of the twenty-first century, in order to further Reform and opening up, China decided to join the WTO, that is, to act in accordance with international practices, and to open up to the international market. In the early period of Reform and Opening Up, the architectural design market of our country had been opened to foreign countries, although there were some restrictions, such as requiring cooperation with local design institutes in the design of construction drawings. But the urban planning industry had not been open for various reasons, such as confidentiality. On November 10, 2001, the Fourth Ministerial Conference of the World Trade Organization decided after deliberation to approve China's entry into the WTO, and the urban planning market was about to be open, so there was a panic in the planning circle at that time. In order to cope with this situation, the leadership of Tianjin Planning Institute decided to reform and set up two expert studios in order to improve their competitiveness as a countermeasure preparation for large foreign planning companies entering China to seize the market after China's entry into WTO.

Chief. Chu graduated from the Department of Architecture of Tsinghua University. On the basis of qualifications and abilities, she could have been a director. Before retirement, the old director talked to her about taking over his job, but at that time, the director's job was mainly to make money. Chief. Chu wanted to concentrate more on the actual planning and design, so she declined the old director's offer. The main purpose of setting up expert studios in the institute was to make some achievements in terms of professional work. Chief. Chu decided to take this opportunity. After open competition, she became the chief designer of the expert studio. Only by doing it personally can one truly understand the truth that everything is difficult at the beginning. Chief. Chu worked overtime and painstakingly with three colleagues with whom she had worked together for many years and several newly recruited young designers. When recruiting, she interviewed the candidates personally. No matter how distinctive their personalities were, she took in all young people who were capable and motivated. Chief. Chu hoped to give young people space and opportunities to grow with her mentorship. She believed that if the newly established studio wanted to compete with big foreign design companies, it must first learn and grow. The most straightforward way to learn was to cooperate with good foreign design companies and designers.

Unlike the preferential policies of the non-local branches of the institute, the expert studios received the same proportion of earnings as that of other departments in the institute, and to cooperate with foreign designers and design companies on their own projects meaned that most of the planning and design fees would be paid to the foreign designers and companies. Foreign designers made design plans and sketches, and colleagues in expert studios turned these plans into formal results. However, because most of the fees were paid to the other parties, the income of expert studios in the first few years was always at the bottom among all the departments in the institute, which put Chief. Chu under great pressure. Everyone worked very hard, but their income was not high. Chief. Chu kept encouraging everybody. After several years of perseverance and efforts, things had finally started to improve. The expert studio had made some achievements in urban design and historical protection, and their income had also increased. While doing a good job on the projects, Chief. Chu never forgot to learn. She organized the whole institute to compile and publish the book *Urban Design in China*, and expressed their views on Chinese urban design with their own experiences and design works. Shortly after the publication of the book, officials from municipal governments, administrative committees, planning bureaus and city investment organizations of many cities in China came to them with the book. They said that the content of the book was very real and to the point. From the book, they not only saw Chief. Chu and her studio's abilities, but also their great pursuit, and that they wanted to invite the studio to do urban planning for them. Later, Xiao Qin calculated, and the design fees of the projects that were gained from Urban Design in China added up to more than 20 million yuan. The book even played an advertising role.

In 2008, because of its outstanding performance in urban design, the expert studio was renamed Urban Design Institute. Besides urban design, they also try to use the method of urban design to plan the protection and renewal of historic blocks. In Tianjin, they have researched the historic and cultural blocks of Wudadao for a long time, analyzed the buildings, courtyards and street space of Wudadao using typological methods, organized a planning and design camp for Wudadao in cooperation with the University of California, Berkeley, and published the planning achievements of Wudadao, academic compilation *China, Tianjin, Wudadao: Research on Protection and Renewal Planning of Historic and Cultural Blocks*, and won the first place in National Excellent Urban and Rural Planning and Design. "These are the glorious traditions and successful experiences of the Urban Design Institute. For a long time, we have paid close attention to Tianjin and taken root in Tianjin, which is also an advantage for us to compete with designers from other countries. No matter how great the concept of planning and design is, it should not only stay on paper, but also be able to truly be implemented, and to have a profound understanding of its far-reaching impact on people, cities and the environment.

"At present, the country is undergoing transformation and upgrading, Tianjin is also facing temporary difficulties, and the design industry has also been affected. In this situation, we should persist. We have put a lot of effort into the urban design pilot project. There is no considerable economic profit, but through our efforts, we have achieved good results. Many ideas and methods are worldwide trends, and also stem from real problems. We can feel from the feedback from today's forum and seminar that they have touched many people. This is an inspiration for us and a new beginning."

A bright moon hangs in the sky, and the water of the Haihe River sparkles. Chief. Chu stands up and faces the Haihe River. "The fact that Tianjin's urban planning and construction can reach today's heights is the result of the efforts of generations of planners in Tianjin, and is drawn by countless urban designers stroke by stroke. The old generation of planners has left us a precious tradition, which we should inherit and carry forward for future generations. This path is windy but the future is bright. The development of a city will never be smooth sailing. Tianjin has made great achievements in urban construction, but there are also many problems. We need to face the problems, face the future and be confident.

"When Tianjin formulated the comprehensive development and renovation plan of the Haihe River, it set a goal to build the Haihe River into a world-famous river. Over the past ten years, the water of the Haihe River has become clear and clean, and both sides of the Haihe River have become beautiful and charming. Some people say that the Haihe River is already a world-famous river, more beautiful than the Seine in Paris. This is arrogance. In fact, there is still a big gap between the Haihe River and the Seine. This gap is not only reflected in physical conditions, but also in thoughts and views, with the key being the people. Tianjin's newest version of national territorial space master plan proposes that Tianjin should be built into a world-famous city by 2035. I believe that after 15 years of hard work, Tianjin will surely achieve this goal, from world-famous river to world-famous city. Today, let's make an agreement that we will return to the riverside to celebrate and get together again when that day comes, shall we? "Good!" Everyone all responds with one collective voice. The voice echoes over the silent river, like the whisper of a God.

60.The Eyes of the Dragon

ZhuGe Min

After the forum, ZhuGe Min accompanies Mr. Barnett and Mr. Dan Solomon for dinner, as a gesture of appreciation as well as to bid goodbye. He has chosen the flagship store of the Goubuli steamed buns located in the Italian Style District. He has had hesitations about this: the two experts have been to Tianjin for many times and must have tasted Goubuli steam buns more than three times. As the old saying goes, "good things only happen for three times," is it a good idea for them to have it again? There are other good restaurants in the Italian Style District with better environments and dishes. But the hesitations went as soon as it came, and ZhuGe Min made the decision for them to have dinner at Goubuli. Luckily, the store has just opened, and the two experts have not been here before. ZhuGe Min hopes that Goubuli could improve itself, just like the English brand name it has given itself–Go Believe.

Now things have changed and there is less official-hosted dinner parties, so only ZhuGe Min and Director Song from the Planning Institute are present. It is by chance that the four of them gather here together. As Director Song is ordering, ZhuGe Min asks Mr. Barnett a question: "Mr. Barnett, you were here in Tianjin for the review of the design proposals for North Jiefang Road in 2003, and you proposed two possible directions for the issue of existing, disruptive high-rises in historic blocks. The first is to stop building more high-rises and leave the existing ones be even though they look disruptive, like what Paris have done. The only high-rise building in the old city area of Paris is the 209-meter-high, 59-storey Montparnasse built in 1972, standing there all by itself like an unforgettable lesson. The second is to plan a few new high-rises around the existing ones, forming a group of buildings and a better landscape. Looking back, Tianjin has gone with the second approach. What do you think about the outcome?"

Mr. Barnett has visited Nanjing as the first stop in his trip to China this time. After lecturing at Southeast University, he rode the high-speed rail to Tianjin. He has seen the group of high-rise buildings by Jinwan Plaza as he arrived at Tianjin Railway Station. Compared to when he last visited Tianjin in 2003, the North Jiefang Road and Central Square area have changed greatly, with a few groups of high-rise buildings such as the Jinmen, Tianjin Tower, and Jinwan Plaza. Mr. Barnett does not answer the question directly. His wise eyes turn to ZhuGe Min: "What do you think yourself?" ZhuGe Min pauses as he has not expected a reply like this. "The situation of Tianjin is different from Paris. There was only one high-rise in the city area of Paris while there were a few in Tianjin. The Bank of China on North Jiefang Road, the Ocean Shipping Building at the Central Square and the Telecom Building on the other side of the bank all scattered alongside Haihe River. On top of this, the cost for demolition, moving residents and land re-arrangement was high, so it was out of the circumstances that we had to build more high-rises. Taking into consideration that the river course of this section was only 100 meters long, we controlled the planning of high-rises with the principle that high-rises could only be on one side of the river, avoiding the sense of constriction that came with high-rises on both sides of the river. For example, there were historic blocks along North Jiefang Road and high-rises could not be built there, so the buildings in the South Railway Station business district were higher; the Italian and Austrian Style Districts were multi-level, so the other side was higher; the Ancient Cultural Street was mostly low, so the other side was also higher overall. This way, high-rises are scattered along the Haihe River on both sides and provide some dynamics. Moreover, Tianjin has a strictly controlled high-rise building setback restriction. Except for the few iconic buildings such as the Tianjin Tower, the heights of buildings grow gradually in a stepping style from Haihe River out, and mostly form a continuous building outline along Haihe River. Both of you have ridden the Haihe River Cruise and should have gotten some specific sense."

Mr. Barnett and Mr. Dan Solomon listen carefully, watching as ZhuGe Min sketches on napkins. "After listening to your introduction and based on my understanding of various aspects, you have considered the height control of buildings along the Haihe River, including the height control of buildings in the historic bocks, and the outcome is good." Mr. Barnett comments in a neutral tone. "However, the control of building height is a very important part of urban design, which has a lot more meaning. Sadiq Khan, London's mayor, recently turned down a proposal by the billionaire Joseph Safra to build a305-meter-high tulip-shaped skyscraper, designed by Norman Foster, in the city's financial district. His argument was that the tulip tower would damage the London skyline and would cause significant damage to the Tower of London, which is a world heritage site. Height controls in the city of London are not very strict. A cluster of high-rise buildings has been built, including the Shard designed by Renzo Piano, which is the tallest building in Britain and in Western Europe. At 310 meters, it has 95 floors and is just over a dozen meters shorter than the Eiffel Tower in Paris. The control of building height in Paris has always been very strict, which formed the unified look of Paris today. Can you say that land prices in Paris are not high? In fact, Paris has always had an urge to build high-rise buildings, as Montparnasse shows, but Paris resisted it. Of course, in order to meet the needs of modern office buildings, Paris has planned a new business office district in the west end of the axis away from the main city area. Building a new business district is not easy, and even in an international city like Paris, with the conditions of an old city with strict restrictions on high-rise buildings, the development of LaDefense has hit a crisis. It also shows that there is a lot to height control of buildings in urban design."

Mr. Barnett took a sip of tea. "Mr. He, I was invited to participate in the urban design review of Yujiapu Financial District in the Binhai New Area of Tianjin more than ten years ago. You have talked about the relationships between Yujiapu Financial District and the financial district in the central urban area as well as the financial district in Beijing back then. At that time, it was proposed that the development of North Jiefang Road as a traditional financial district should slow down, and yujiapu as an innovative financial district should accelerate its development. The SOM has created high-level urban design for Yujiapu Financial District. I also visited Yujiapu Financial District and the High-Speed Railway Station under construction in 2012. Now, another seven years have passed. As far as I know, the development of Yujiapu Financial District is facing difficulties. Could you tell me the reason?" Mr. Barnett's question once again leaves ZhuGe Min musing. The control of architectural height or urban form by urban design is, on the surface, the design of urban landscape image, but in fact, it is the external expression of urban spatial structure, form and internal rules, and also an allocation of urban resources. Yujiapao is 50 kilometers away from the central city, 10 times the distance between the old city center of Paris and LaDefense, which has its own disadvantages. However, if the downtown area of Tianjin imposes certain restrictions on high-rise office buildings and other commercial buildings, it will certainly be beneficial to the development of Yujiapu, and more resources can be concentrated. The old city of Tianjin has a long history, with advantages in education, medical care and other aspects, and is more attractive than the new area. Meanwhile, the four districts outside the outer ring road have also developed rapidly in the past ten years, consuming a lot of resources, so that the continuous construction of the central city and the Binhai New Area has become a bigger sprawling situation. Therefore, it is correct and reasonable for the municipal party committee and government to propose to strengthen the planning and control of the area between the two cities and build a green ecological barrier. This is conducive to the optimization of urban spatial structure, the improvement of ecological environment and the development of Binhai New Area. In addition, in the planning of the two-city ecological barrier area, it is also proposed to limit the plot ratio and development intensity and strictly control the height of buildings in the second- and third-level control areas. In the past ten years or more, the regulation of Tianjin has basically given up the control of building height, and the new urban area has also planned and built large areas of high-rise residential areas in consideration of the so-called land price, resulting in a situation of not urban nor rural, which goes against the objective pattern of urban development. Therefore, in the future urban design, the control of urban height and spatial form should be further strengthened. This is also one of the important contents of the overall planning of land space. The master plan of land space should not only plan the shape of ecological red line, basic farmland boundary and urban development boundary, but also determine the development intensity and height. This is a more comprehensive planning and control of space resources. Meanwhile, to follow the distribution rules of urban and rural sections of housing types, different locations must adopt corresponding housing types, so as to better integrate the building height and volume with the environment.

While ZhuGe Min is thinking, Mr. Barnett and Mr. Dan Solomon are whispering to each other. The silence is broken when Mr. Song returns to take his seat after finishing up with ordering. "Sir, I saw on the Internet the year before last a live broadcast recording of the urban design report that you gave at the annual meeting of the China Urban Planning Society. The topic you talked about was: China's urban design–where, when and what, or 3W. How did you think about it?" The question is exactly what Mr. Barnett wants to talk about, as he straightens his body and clears his throat. "I have given this subject serious thought. In recent years, I have been coming to China frequently and paid close attention to urban design in China. I have also done long-term follow-up research on urban design in China. I feel that some problems are more suitable for us to discuss as outsiders who can give alternative perspectives than people who have been working in such an environment for a long time. China's Reform and Opening Up has been going on for 40 years. China's urban design is making continuous progress, and many deep-seated problems need to be solved. Now China has entered a new stage of development, with higher requirements for the ecological environment and quality of living conditions. Your traditional urban planning, such as the 'Residential Planning and Design Standards' and other established norms, is no longer in line with the requirements of the new social environment. I have seen the 2015 'opinions on accelerating the construction of ecological civilization' and the 2016 'opinions on further strengthening the management of urban planning and construction' released by the highest authority of China. Ecological civilization requires that all human activities should live in harmony with nature. The cause is very idealistic, but it is fraught with difficulties. Although the Nordic countries of Denmark, Finland, Iceland, Norway and Sweden are trying to move in this direction, so far no country has achieved this goal. China proposes to accelerate the overall promotion of ecological civilization and build Xiong' an New Area, and puts forward the priority of the protection of natural resources and existing landscape and the development of urbanization. Applying the principle of ecological civilization to urban design means that urban planning and development should be based on the premise of environmental bearing capacity and respect for natural landscape. Ecological civilization is consistent with improving the quality of urban planning and construction." After listening to Mr. Barnett's retelling of two important documents of the CPC central committee and the state council, ZhuGe Min and Director Song felt both admiration for Mr. Barnett and shame. As an 80-year-old foreign scholar, Mr. Barnett studies Chinese central documents in a serious way, which even many of our own planners and planning managers may not have grasped. Mr. Barnett goes on to say, "China's central urban work conference regards urban design as the main means to improve the management level of urban planning and construction and solve urban diseases. It is rare in the world for the central government of a country to explicitly propose the promotion of urban design. The ministry of housing and urban-rural development issued the administrative measures on urban design, laying the foundation for urban design to go on the legal planning track. This time, two more groups of urban design pilot projects were organized, with more than 50 cities nationwide participating, including Tianjin. Through participating in the urban design pilot project, I am confident about the future of urban design in China."

"Of course, in order for urban design to really work, there are several key issues that have to be addressed, so I talked about the three W's in Nanjing. The first W is the question of where to build. This should be a question of the overall urban design scale. As I have discussed in my book *City Design: Modernist, Traditional, Green and Systems Perspectives*, after thousands of years of development, urban design has formed a variety of methods, among which ecological method is the primary method in large-scale urban design. I think it is quite right to focus on a more ecological city in this urban design pilot project in Tianjin. China is now pushing forward the territorial spatial master plan, which will draw ecological red lines and basic farmland boundaries on the basis of the so-called 'double evaluation' of ecological background evaluation and natural landscape environmental carrying capacity evaluation, and then determine urban development boundaries. In this process, the role of urban design should be brought into play. First of all, the determination of ecological red line is not a simple imitation of the natural background; design should be carried out either based on ecological theory or practical experience. We find that a good ecology must have a beautiful landscape, and vice versa. Through the correct intervention of human beings, the ecological environment of some habitats with poor natural background has been greatly improved after years of evolution, making people, cities and natural environment get along more harmoniously (including the optimization of water system and the protection of natural green space, etc.). Therefore, there is also an aspect of design to a good ecological environment system. A famous ancient Chinese example is the Dujiang Dam project carried out by Li Bing, which is a case of outstanding design. Secondly, urban development boundary demarcation also has to go through the process of urban design. I agree with the concept of 'urban artistic skeleton' mentioned by Academician Zhang of Tianjin University in the morning. The city is not just a boundary, and its spatial structure and form have internal rules. Only a good spatial form can ensure the urban planning and construction are on the right track.

"The second question is the question of when to build, which is actually related to the first question. Urban and regional development is dynamic, and urban development boundary is not formed at once, but a dynamic process. Even with the ultimate urban development boundaries, the timing for certain constructions to begin are critical and can have serious consequences if not thought through. For example, there are many ghost cities in China now, which are excessively expanded and developed in many places, with scattered forces and no aggregation effect. These are serious problems that have occurred in China in the past. Therefore, when to build may be more important than the issue of developing boundaries. This time, it is very wise for Tianjin to propose to strengthen planning and control between the twin cities and to build a green ecological barrier. In addition, you have proposed that the planning of 'one ring and eleven parks' in the main urban area is a highly integrated approach of greening system and urban development, which is crucial to optimize the ecological environment of the main urban area, improve the urban structure and relieve the over-dense population in the historic urban area. But there is also a problem of when to build here, which you must carefully measure, combining with the moving of population in the Old City to avoid over-development.

"The third question is how to build. China has made remarkable achievements in the past 40 years of Reform and Opening Up, and a set of corresponding normative mechanisms have been formed in this process, including urban land classification, paid use of land, residential planning and design standards. No matter what the reasons and ways are, I think the model of residential planning and construction of 'large land block and large road network' is Corbusier's model, which is suitable for cars but not for people, and is the root cause of many urban diseases in China today. Therefore, to realize urban transformation and upgrading and promote urban design, it is necessary to change this traditional planning model and method. In my opinion,

Tianjin has grasped the key issues in putting forward the 'humanized city' and 'new urban design guidelines for residential communities' in this urban design pilot project. However, you have proposed the development of car traffic on the basis of emphasizing 'walking-oriented' and 'bus-oriented', which is inconsistent with the current international mainstream ideas. In addition, your proposal in the guidelines for new residential communities to moderately encourage individual housing and low-rise townhouses is also inconsistent with the general international view of China. I have listened to some planners' talks and need some more time to think about it."

"This is exactly what we want to communicate with you." ZhuGe Min carries on speaking, "you said that our traditional high-street, high-road, residential model was making the city adapt to cars. As a matter of fact, we have been promoting the notion of bus-oriented city, and the end results suggested that this model is not suitable for the development of cars. With the rapid development of cars, the traffic congestion problem in our cities is getting very serious. It is because of this kind of large road network and wide road that the traffic is excessively concentrated. At the same time, there is not enough road network to relieve and organize the traffic. Therefore, our traditional residential planning model, which was influenced by modern urban planning and learned from the Soviet Union, only adapted to the large-scale real estate development model at that time. This model could not adapt to the development of cars or human walking, leaving only big walls for the city. After years of tracking and research, we realize that the city must first be people-oriented, and walkability is the core. "

"In urban centers, including urban business centers and community centers, walkability should be the focus of urban design. Measures should be taken to stabilize and tranquilize traffic, provide comfortable pavement width, ensure connectivity, and build walkable cities. At the same time, in these areas, take measures to limit cars, vigorously develop public transport, and make public transport better and more efficiently serve people through 3D methods, that is, appropriate building and population density, diversity and mixed functions, and urban design methods. We recently visited Curitiba, a famous bus city in Brazil, and realized that the urban development model must be closely integrated with the bus corridor, forming a 'hand and glove' relationship, so that the bus can really work and improve the service level.

"We have another amazing finding in Curitiba, which is that even in cities like Curitiba, where public transportation is extremely efficient, public transportation only accounts for 30 percent of the total urban travel, and the majority of the means of travel is by car. Curitiba has the second highest number of cars per capita in Brazil, second only to Brasilia, the capital, because of its economy and high incomes. This also shows a problem: after more than 100 years of development, the automobile technology has gradually matured, and this development has also promoted the progress of human civilization and the development of human nature. Despite the costs, the trend is inexorable, as Peter Hall puts it, 'the people vote with wheels.' If we ignore this trend, we will inevitably be punished. Not to mention that the number of cars owned by per 1,000 people in the United States has reached 800, even in Europe and including Japan, which used to be dominated by public transportation,

, the number of cars owned by per 1,000 people has reached nearly 600. Bus travel only accounts for a little more than 10% of all travel in London, and it has solved the problem of car traffic congestion by introducing a congestion charge in the city center. Tokyo also has a high percentage of car ownership, but most residents still use buses to commute to work, and use cars for shopping, entertainment and travel. So now there is a view that cars should not be restricted, but should be used wisely. At present, due to the limited space in the city center, pedestrians and the development of public transport are the priority, and to restrict cars is the only way. Both London and Singapore have introduced congestion charges to limit the number of cars entering city centers. Curitiba's priority for five bus rapid transit corridors in the city center also limits the number of cars entering the city center. We also recently looked at the Big Dig project in Boston. Now the downtown area is accessible from the Logan international airport and the outer suburbs by the underground expressway, which is very convenient. But once entering the city center and back onto the ground, the ground traffic is mostly one-way and inconvenient, which has also objectively restricted cars from entering into the city center. Therefore, our urban design pilot project in Tianjin also takes the pedestrianization of city center and the development of public transportation as an important aspect, and methods include limiting the number of parking spaces. In the future, the city center will further restrict car traffic. We are firm on this.

"At the same time, we also need to face the further development of private cars. At the national level, the automobile industry is an important pillar industry. At present, the annual production and sales of passenger cars reach 28 million, which takes up a big part of the national GDP and retail sales. The car ownership in China now is about 170 cars per thousand people, compared with developed countries, there is still some room for development. Therefore, from the perspective of economic development and meeting people's yearning for a better life, cities must meet the needs of car development. From the perspective of Tianjin, automobile industry is also the leading industry in Tianjin. From the perspective of Tianjin's urban transformation and development, we should also encourage the development of cars. In order to meet the needs of the development of cars, we have emphasized the use of grid roads in the planning of the area around the 'one ring and eleven parks'. There are 15 subway lines planned in the central city of Tianjin. The first, second, third, fourth, seventh and ninth are radiation lines, the fifth and sixth form ring lines, and the others are filling lines. Although the density of urban rail transit network is higher, but the city's outer ring road surroundings, which is the 'one ring and eleven parks' surrounding area, has a low subway line density. Therefore, in addition to a few areas such as South Jiefang Road and Pillow and Dongli District, most of the surrounding areas of 'one ring and eleven parks' have only one or two railway lines and two or three subway stations. To serve the living communities with a population of nearly 100,000, it needs the cooperation of buses and, more importantly, cars. These areas are close to urban expressways and outer ring expressways, and have a good foundation. The narrow road and dense network layout adapt well to the development of cars."

"What you're saying is true," Mr. Barnett replies gravely, "but I think that even so, there's still a lot of work to be done, not just qualitatively, but quantitatively. In terms of quantitative analysis, traffic planning is relatively mature. Environmental protection, social equity and other issues should also be considered." "I agree," ZhuGe Min says, "the next step is to do quantitative traffic planning and use big data to build traffic models for simulation. In addition to data on land use and road networks, transport planning is an important support for urban design. We have not done well in this aspect in the past and we need to work on it in the future. As for the pollution of cars, new energy vehicles (NEV) are developing rapidly and have been put into the market with a fast growth rate. Smart technologies, such as autonomous cars, are also growing fast, and will make progress in urban traffic congestion and emissions. Of course, the treatment of batteries also involves environmental problems. Overall, environmental problems need to be solved by development and technological progress. Social equity issues need to be considered comprehensively. On the other hand, even if we restrict the development of cars, we cannot solve all the problems of social equity." Mr. Barnett nods. ZhuGe Min continues, "ultimately, transportation is an economic issue. American transportation scientists have done a lot of research in this area. The development of the car is not just as a means of transportation; many policies to rely on economic measures that is in accordance with the patterns of the market. In fact, like cars, public transport consumes social resources. Government subsidies for public transport are all taxpayers' money, and there is a limit. The bus fare in Curitiba, Brazil, is 4.5 reais, or about 9 yuan. Even if you can change buses at will within a day, the fare is not cheap, especially for short-haul passengers. Short-distance bus rides, they argue, are for the affluent. They charge more than they should, but they can help poorer long-distance travelers. In addition, the government gives some shopping malls, newsstands and other government assets to bus management groups to operate, and the rental income is used to subsidize bus operation. In the past, the revenue and expenditure of bus operation were balanced through such policies as rental income of these assets as well as short-distance passengers paying for long-distance passengers. However, the latest calculation shows that the cost of the bus operation can only be balanced by increasing the fare from 4.5 reais to 4.75 reais. The mayor did not approve the price increase due to various considerations, so the government needs to cover the financial gap. The biggest cost is estimated to be the salaries and benefits of bus drivers, which account for one-third of the total cost. And because public transport workers suffer from serious occupational diseases, perhaps autonomous driving technology should be used on public transport first."

"We've talked enough about traffic and cars. The other thing we're concerned about is individual housing. Do you really think China can build individual houses?" Mr. Barnett grows more serious. ZhuGe Min also slows down: "neither foreigners nor we Chinese think Chinese people have the land to build individual houses. Mr. Barnett has also written in his previous article that China and the United States are about the same size (not counting Alaska), and China has five times as many people and half as much site areas as the United States. Although many Chinese people like individual houses, China must consider other forms of housing. In fact, these analyses are somewhat simplistic and do not take into account the reality of China. To discuss the core issue of what kind of houses China will build in the future, we must take into account the reality of China's cities and rural areas, which is not something that can be calculated by simple mathematics. To answer the question of what kind of housing China will build in the future, we must take a systematic and comprehensive consideration. In our plan of the past 40 years, we have built a large number of gated communities, multi-storey and high-rise houses, which cannot be said to be not land-saving. But all of our urban problems today are the result of this so-called intensive land model, while land has been wasted in other ways. We have to change and think about it from a broader perspective. First of all, China's current housing stock has been very large.The per capita housing construction area is 38 square meters in the city, and every family has a unit, but among these houses, more than half are low-standard unit housing built before 2000 and need to be renewed. Therefore, speculating about the standards of 2035 and 2050, new housing, no matter the size or type, should have a higher standard and a larger area. Secondly, we now have a very high urban density, with more than 20,000 people per square kilometer and less than 50 square meters of urban construction land per capita, resulting in a lack of green space and overcrowding. According to the per capita urban construction land of 100 square meters, the area of newly built residential land in the periphery should be larger than that in the urban area, with relatively lower building density and more green and open space. Thirdly, thinking from the aspect of the smooth transition of the real estate market, China has completed and sold 1.7 billion square meters of new commercial residential buildings annually. The average annual building area of urban residents increased by 2 square meters, which is a huge amount. To maintain a smooth transition, this number needs to be reduced. At the same time, we need to take into account the purchasing ability of the majority of middle-income families and limit housing prices. Therefore, if 20% of the newly built commercial houses are used as individual housing and other high-end housing, not only can it meet the demand for housing improvement of some high-income families, but it also ensures the overall stability of the real estate market, and this part of high-end housing land income and tax can be used to subsidize the construction of affordable housing. Fourthly, from the perspective of social management, the vast majority of housing we have constructed is large-scale collective housing, which is difficult to manage. With the growth of housing age, the aging of housing equipment and maintenance structure, it is difficult for residents to form unified opinions and actions. In the end, it is up for the government to fund the maintenance, which becomes a huge burden for the government and actually the whole city. In the future, with the promotion of narrow road and dense network and reduction of the size of the development land, there will be conditions for the owners to coordinate. In addition, the construction of some individual housing can allow owners to manage and maintain their own property, and cultivate a certain sense of independent management, which leads to easy social management and social harmony. Fifthly is the inheritance of Chinese traditional architectural culture. Courtyard house is the basic type of folk houses of all ethnic groups in China, which is perfectly combined with literature, calligraphy and art to form the traditional folk house culture of China. Proper planning of the space carrier of individual housing is conducive to the inheritance and development of Chinese culture. Historically, they were low-rise courtyard houses, and at the level of agricultural productivity at that time, they raised millions of Chinese people. The low-rise individual house is easy to coordinate with the natural environment, forming the ideal state of the house covered in green trees.

Low-rise and low-density residential forms should be adopted in urban peripheries, suburbs, vast rural areas, as well as around ecological zones and scenic spots. Of course, the ecological zone and scenic areas should be cleared of illegal villa construction. The above points, including the diversity of housing, are all reasons why we are considering the construction of some individual houses. Of course, these should be fixed in the form of the housing law, in order to do a good job in providing for the families with housing difficulties and providing affordable housing for the majority of middle-income families, which is the premise."

ZhuGe Min pauses before adding: "I used to hold the simple idea that there were too many people in China and too little land for individual housing. However, in 2003, I was a visiting scholar at the University of California, Berkeley in the United States for one year. According to my professor at the time, some economists believed that the Bay Area was the best urban regional spatial form in the future. Therefore, I conducted some comparative studies between Tianjin and the San Francisco Bay Area, and made some discoveries, which changed my established views. The area of Tianjin City is similar in size to that of the Bay Area, both 12,000 square kilometers. At that time, the total population of Tianjin was 9.2 million, and the Bay Area was 6.7 million. The GDP of the Bay Area was 350 billion US dollars, ranking 24th in the world by country, with a per capita of 50,000 US dollars. Tianjin's GDP was 240 billion yuan, equivalent to 29 billion US dollars, with a per capita of over 3,000 US dollars, which was 1/12 and 1/17 of the Bay Area's numbers respectively. The Bay Area had a better ecological environment, with more green parks. Its ecological land accounted for 45% of the total land, agricultural and other land accounted for 37%, and urban land accounted for 18%. For Tianjin, cultivated land accounted for 45%, forest grass for 3.5%, water area for 26%, construction land for 23%, and traffic for 2.6%. However, in terms of per capita construction land, including urban road traffic, social industry and mining land, the per capita of the Bay Area was 300 square meters, and the per capita of Tianjin was 270 square meters, a difference of 10%. But the vast majority of the American housing form was individual housings with beautiful environments, and our country's housing was multi-storey and high-rise housing. We say that there are too many people and not enough land, so we need to save land. But objectively, we have wasted a lot of land. We plan for 100 square kilometers of construction land per capita, but our recent plan predicts that the population will increase more than the actual growth, and also plans a large number of industrial areas and new cities, so the output inefficiency of industrial land in Tianjin is very obvious. Although we have the same level of construction land per capita as the Bay Area, there is a large gap in the level of residential construction. This got me thinking. The ultimate goal of urban planning and construction is to create a good living environment. If we simply take the so-called saving land as the main goal, we are actually putting the cart before the horse. The reality has proved that this is indeed the case. Over the years, we have been saving land and implementing the strictest land use system, but we have wasted the most land. We cannot simply limit the diversity of housing by prioritizing conserving land. I often use Japan as an example, where land is even more scarce than in our country, but of all the houses in Japan, individual houses (single-family houses) account for more than 50%. Even in densely populated Tokyo, it accounts for more than 45%, which is telling. Of course, Japan has a very compact land per household model. The United States' new urbanism suggests adopting the traditional European urban model with moderately increased development intensity and density, but it does not oppose individual housing either. In addition, some people say that European cities are similar to Chinese ones, and that China cannot learn from America, but it can learn from Europe. However, the situation is that the multi-storey houses in Europe are mostly saturated, and most of the new houses are individual houses. Considering all the factors, and the possible development by 2035 and 2050, it is very necessary for us to try to build a certain proportion of individual houses, including townhouses, in the suburbs of cities."

Mr. Barnett is relieved when ZhuGe Min has finished. "You are talking about the planning of a few individual houses, not all of them, which I think is acceptable and necessary in terms of the diversity of the houses. Of course, taken together, multi-storey high-density housing should be the main housing type in China for some time to come, right?" Seeing that the atmosphere in the room is somewhat serious, Director Song manages to interrupt: "Mr. Barnett, you are quite right, multi-storey high density, including some small high-rise buildings are our main residential building type in the future. Therefore, more than ten years ago, our Planning Institute have cooperated with Mr. Dan Solomon to carry out the planning and design research on new types of communities in Binhai New Area. Mr. Solomon, what's your opinion?" Director Song sees that Mr. Solomon has not spoken, and he wants to give him a chance.

Mr. Solomon, a very wise man, understands the meaning of Director Song. "Thank you, Director Song. You guys have had an intense discussion and I enjoyed listening to it. I remember it was in 2011 when you invited me to participate in the urban design of the harmonious community in Binhai New Area. I was very honored. I had been engaged in residential design and planning, and had been paying close attention to the development of China, but I was not very familiar with the specific urban planning and residential architectural design standards of China. Through two rounds of nearly three years' work on the harmonious community, I now have a deep understanding of the planning and design of Chinese houses and residential communities, and have also discovered the huge problems caused by the currently popular large-scale streets. It completely changes the urban texture of alleyways, streets and other communication spaces composed of courtyard houses in traditional Chinese cities, and forms an inferior urban environment composed of walls and the same houses. The so-called residential planning and design standard is a standard with more problems than achievements. It can be said that it is not a correct planning and design method. It regards people as symbols lacking individuality, and does not consider the shaping of human-scale streets, squares and other urban space, so it has to be fundamentally changed. We adopted the urban design method in the planning of the harmonious community, starting from the shaping of different streets, squares and parks in the city to carry out the community planning. I also learned about sunshine analysis, sunshine spacing, thousand-people indicator, urban road design specifications, urban road intersection design specifications, and so on. We meet the requirements of sunshine as much as possible, concentrate community centers under the condition of meeting the thousand-people indicator, optimize and adjust the standards that are purely based on driving speed such as the radius of road intersection, narrowing the intersection turning radius and forming a kind of space that is convenient for the pedestrians to cross the road, reasonably saving land. Speaking of land waste, in fact, many of your wide roads and large intersections waste a lot of land. We have done a lot of in-depth work, cooperating with the local planning and design institute, creating a conceptual residential construction scheme, and paid a lot of efforts. Unfortunately, due to various factors, the project has not been implemented so far, and the plan has not been tested in practice. I originally thought that the project could be implemented after the central urban work conference in 2016 clearly proposed the construction of a new type of 'narrow road and dense network' community, but at present there is still no sign of starting, which is a pity. Someone told me that people in Tianjin always say that they 'get up early and set up late,' which I feel like is a suitable description for this. Before I came to Tianjin this time, I asked repeatedly whether there was any similar project in Tianjin and whether our research has played a role. Our work is the third key question in Mr. Barnett's 3W, the question of what to build. I think we have an answer."

Dean song chimes in: "yes, the harmonious community is a very important case. Its orientation is exactly 45 degrees, which solves the problem with east-west housings, which is one of the methods recommended by Mr. Barnett." Mr. Barnett says, "I have also seen the design scheme of harmonious community on different occasions. I thought it was a good attempt, which was in line with the characteristics of Tianjin. Too bad it was not implemented." ZhuGe Min picks up the topic: "we are very grateful to Dan for putting so much effort into the planning and design of the harmonious community. It is a pity that the project was not implemented for various reasons, but it did have an impact. Tomorrow we will accompany you to the Binhai New Area, and in addition to visiting the Yujiapu Financial District, the High-Speed Railway Station, Haihe riverside greenbelt, Binhai Cultural Center, the National Maritime Museum and other city-level large public buildings, we also want to take you to visit the Beacon Hill residential community, Tianjian CIC project and Tianjin Port bulk cargo center's initial area as well as the public housing Jianingyuan community in TEDA. These projects all attempt to create narrow road networks and urban Spaces. Binhai new area planning and natural resources bureau colleagues will also accompany. Many old friends also want to see you and hope you can put forward good Suggestions on the construction plan of the new district."

"We are looking forward to it," Mr. Barnett and Mr. Solomon reply together. Mr. Solomon goes on to say, "in order to plan for the harmonious community, I also looked at some good residential community projects in Chinese cities, such as Beijing's Ju'er Hutong and Tianjin's Vanke Crystal City. Of course, most were large communities that were simply reproduced in a large scale. I was asking this question, why did good models such as Ju'er Hutong and Vanke Crystal City fail to be popularized? I have gradually come to understand that this is a structural problem. It involves the whole society. It will be difficult to change it. But the first step must be taken, otherwise the change will never happen." "In fact, we are also very anxious and have made a lot of preparations." ZhuGe Min adds, "in the past two years, combined with the revision of some rules and specifications, we formulated the 'Regulations of Technology of Regulatory Detailed Planning of Tianjin,' 'Technical Regulations of Tianjin Construction Project Planning Management,' which have provided conditions for the narrow road net planning model to be used in new communities, it is just that the market does not know about it yet and we need to promote it more. The year before this we organized the publishing of *The Harmonious Community–Research on the Planning and Design of New Residential Communities in Binhai New Area*, carried out research from theoretical and practical perspectives, and publicize the planning and design of the harmonious community by Mr. Dan Solomon. The Huahui Group has continued to compile the urban design plan of the harmonious community, and held seminars in the China Academy of Urban Planning, with experts from Shanghai, Shenzhen and other locations participating, gradually forming a consensus. Unfortunately, later when the China Academy of Urban Planning presided over the revision of the residential planning and design code, although there were major changes, it still adhered to the objectionable conventions of residential planning, which was unreasonable. It also shows that we need to further promote our ideals. This international seminar on urban design in Tianjin is another attempt to promote the concept and importance of new communities."

"We have all seen the efforts of Chief. ZhuGe and Director Song, which is admirable. I also expressed the same intention in the lecture this morning. I hope this lecture can have some effects." Dan Solomon turns to another topic, "It has been eight years since the planning of the harmonious community of Binhai New Area began in 2011, and I have been thinking about it all the time. I recently had a book come out, titled *Housing and the City–Love vs. Hope*. I brought some copies today for Chief. ZhuGe and Director Song. This book includes more than 50 years of teaching, work, scientific research and practice experiences of mine, studies the housing construction of major countries in the world, and comes to some conclusions. Cities surround us, housing defines our existence. I am full of hope that our inner reality houses, which give us love, our hatred and our hopes, will one day meet our inner needs. There are many cities in the world, such as Rome, Amsterdam, Paris, Berlin, Munich, cologne, Prague, Budapest, etc. We may ask, apart from their distinctive features, what are the differences between these cities? Which city you like, which city you don't like, and why that strange city makes you feel at home? The cities are all made up of houses and these factors are determined by urban texture. Housing is a huge and urgent problem for any city in the world. As technology advances and the accumulation of wealth of change, there have been two different models in the fierce struggle to build housing for the working class and low-income groups .In the book, I analyze the battle between the so-called 'city of hope' and 'city of love' in San Francisco, Paris and Rome over a hundred years, where there has been success as well as failures. The so-called 'city of hope' method is to try to replace the traditional urban housing type and texture with more reasonable housing pattern and housing type. The 'city of love' approach is to love the rich history of the city and respect its complex social fabric. The 'city of hope' projects repeatedly failed, and their goal of social progress and integration is repeatedly frustrated because it wants to put a hot wedge into the underlying social class structure; 'city of love' projects are always successful because they succeed as portals to social absorption and social harmony, as places of urban continuity, resilience and shelter."

Mr. Dan Solomon signs the title page of the books and hands them solemnly to Chief. ZhuGe and Director Song. "As always, the language of my book may not be straightforward, but I hope you will have some understandings after reading it. After listening to your discussion with Mr. Barnett, I am more aware of the seriousness and complexity of the traditional residential planning problem, and that you have already began to approach the root of the problem, which is the start to solving the problem. I'm also thinking, in China's 40 years of Reform and Opening Up, the large number of residential constructions might be in line with what I call the 'city of hope' method, changing the traditional Chinese urban texture. Although it is a great achievement from the perspective of improving living conditions of the Chinese people and solving the problem of housing shortage, but it also caused larger urban problems, which are the root of urban diseases. But these traditions are now the existing texture of most Chinese cities. Are the new communities we are advocating for now also a 'city of hope' method? This requires thinking about how to turn this 'city of hope' approach into a 'city of love' approach, which I think is a big topic and a possible solution for the problem." Dan Solomon's words leave everyone deep in thought. At this time, the dishes come, first a few cold dishes then the steamed buns. Director Song has had experience with this: in the past, whenever eating Goubuli steamed buns, they always had a number of hot dishes first, becoming full and then had no appetite for the steamed buns. Now they are having steamed buns upfront so as to taste the unique delicacy that is the Goubuli steamed buns. After all, Mr. Barnett and Mr. Dan Solomon are in their seventies and eighties and do not eat very much at night. Dinner time soon ends, and ZhuGe Min sees the guests out. Director Song personally accompanies them to the hotel by car.

After bidding goodbye to Mr. Barnett and Dan Solomon, ZhuGe Min walks to the Haihe River. He does not walk to the bustling Italian and Austrian Style Districts, but comes to the Haihe Central Square. The Italian Style District behind and the 'Gold Street' shopping street on Heping Road on the other side are all brightly lit, but there is no direct bridge connecting the two sides of the river, so the Haihe Central Square is quiet and the lights are dim. On the Haihe River, a pedestrian bridge is being built to link the Italian Style District with Heping Road and Gold Street. As a matter of fact, this has been proposed in the "Golden Dragon Dancing" plan of Haihe River in 2002, but it has not been implemented for fear of bad effects on the landscape of Haihe River. The pedestrian city part in the urban design pilot project once again emphasizes the importance of this pedestrian bridge. After international bidding, an excellent design scheme was selected, which not only has perfect functions, but is also likely to become a famous site to attract tourists with its beautiful and light look. Due to the small number of people living on the sides of this part of the Haihe River, the construction is still ongoing at night. ZhuGe Min is a little excited today that the urban design pilot seminar has come to a successful conclusion, and he has just had a thorough discussion with Mr. Barnett and Mr. Dan Solomon. ZhuGe Min is restless; he looks at the city in front of him–Tianjin, the city he has devoted his life to, feeling many thoughts crossing his mind at once.

ZhuGe Min has been working in Tianjin for more than 30 years after graduating from Tsinghua University with a Master's degree. Upon graduation, his mentor, Mr. Wu Liangyong, instructed him: "if you want to contribute to China's urban planning, make a difference, then put down roots in your city." ZhuGe Min did so.He has been engaged in urban planning in Tianjin, in planning compilation and research in Tianjin Planning Institute for the first 13 years, and in planning management in Tianjin

Planning Bureau for the later 20 years. It is the government's trust and training that allow ZhuGe Min as a non-party member to grow from a young college graduate into a professional technical official. ZhuGe Min has estimated: since he started working, he has visited more than 30 countries, nearly 90 different cities, and learned and accumulated a lot of experience. His Master's thesis, 'A Preliminary Exploration of Comparative Urban Planning,' was well received by his teachers, but he had not yet been abroad at that time. Later, he had a deeper understanding of the planning, planning laws and management systems of different foreign countries as he went abroad for surveys and further research. He then went on to pursue a doctoral degree at Tsinghua University, where he spent seven years doing in-depth research on China's strategic spatial planning. From 2003 to 2004, under the organization department of Tianjin Municipal Party Committee, he spent one year as a visiting scholar in the Department of Urban Planning and Environmental Design of the famous University of California, Berkeley. These continuous learning, coupled with the accumulated work experience from the comprehensive development planning of Haihe River, the opening up and development of Binhai New Area and other projects, today ZhuGe Min finally feels that he really has a deep and comprehensive understanding of China's urban planning and spatial planning. He deeply feels the real meaning of what Mr. Wu Liangyong once said, that urban planning is a subject for the seniors. When he really understands urban planning, he is approach the age of retirement of 60. He wants young planners to grow soon, wants to be a mentor for them as much as possible, and is also thinking about his plans after retirement.

A slogan has been in ZhuGe Min 's mind for a while: "Healthily work for 50 years for the motherland." These are the famous words of the old President of Tsinghua University and Minister of the Education Department, Mr. Jiang Nanxiang. It was the same words that used to be broadcasted on the loudspeaker when ZhuGe Min was running on the west sports track at four o 'clock in the afternoon when he was studying at Tsinghua University. He has only been working for 30 years now, and still 20 years short of what President Jiang had wanted, and there is still many things that he wants to do. He has always remembered that Mr. Wu Liangyong often talked with him about himself, who left his post as dean of the Department of Architecture at Tsinghua University at the age of 60 to set up the Institute of Architecture and Urban Planning. Afterwards, he has worked for 30 more years, which are a golden era for his work. Mr. Barnett and Mr. Dan Solomon, both of whom ZhuGe Min has just bid goodbye with, have been working in the discipline of urban planning for more than 50 years and are still working hard. The example of these experts and scholars encourages ZhuGe Min's idea to establish a non-profit research institution after retirement and continue to work hard for Tianjin's urban planning. He recently visited Curitiba in Brazil and Boston in the United States, and talked to local planning agencies. He found that local planning agencies are not government departments and the employees are not civil servants, but are directly led by the mayor and advised on urban development for the long run. Planning permission is the responsibility of the administrative department, but planning does not adapt to the administrative system; it needs high-level talents. The funds for planning are not from government funding, but from the government giving properties such as land, housing and other assets to the planning institutions, to raise funds for planning and pay salaries through income from operation. If the finance is well managed, the pay becomes higher, which just play into and test the courage and talent of the planners. If they cannot even manage some land and housing assets, it means that they would not do good urban planning. This approach should also be very suitable for Chinese cities today in the market economy transition. In short, there will be a variety of possibilities in the future. ZhuGe Min believes that his career will be as brilliant as the future of Tianjin.

The night grows darker, and the lights on both sides of the Haihe River shines brightly. He turns back, his eyes fixed on the green space of the Central Square, now just a few tall trees and a few electrical substations. Haihe Central Square is the geometric center of downtown Tianjin. After the founding of the People's Republic of China, it was once a reviewing stand, a place for parades and parties. In the 1950s, the Soviet planning expert Mochin suggested extending an axis from here to the south, directly to the Water Park, forming the main axis of Tianjin city. Unfortunately, it was not implemented, and there was no more opportunities after that. After the Reform and Opening Up, the reviewing stand was basically abandoned, and the square became a children's playground by the Haihe River. Later, the Hebei District was planning to build the Riverview Place, which was later transferred to the South Railway Station area in Hedong District. In 2002, when planning for the comprehensive development and renovation of Haihe River, proposals for the Central Square were collected, where the Haihe Concert Hall was supposed to be planned and built. Several competitors put forward different bidding schemes, but none of them was implemented. Later, Xiaobailou Concert hall was built, and the matter was no longer mentioned. Another reason may be that this location is too important to know how to deal with. Later, as a site selection for the Planning Exhibition Hall and Haihe Tourist Center, proposals were also collected. Eventually, due to time constraints, the old Planning Exhibition Hall was renovated and became the new one, and there was no possibility for the proposal to be implemented. Today, in the new era of development, it is necessary to find a suitable project positioning for Haihe Central Square and create the design of the highest level in the world. ZhuGe Min envisions that a city-wide discussion can be initiated to find the direction for Tianjin's future urban design through the discussion of the design for Haihe Central Square.

The shape of the Central Square is like an eye. In 2002, when the central district reconstruction of Haihe River was carried out, the plan was called "Golden Dragon Dancing", and the Central Square was like the eye of the dragon. It has been nearly 20 years since the comprehensive development and renovation of Haihe River was launched, and huge changes have taken place in Tianjin over the 20 years, with continuous expansions. Today, we are back by the Haihe waterfront, to re-examine Haihe River from the perspective of the Chinese dream of the great rejuvenation of the Chinese nation in the new century. From the perspective of ecological civilization, Haihe River, especially the middle reaches of Haihe River, should become the core area of ecological environment improvement in Tianjin. From a people-centered perspective, the upper reaches of Haihe River should further improve its pedestrian system, so that people can get closer to Haihe River, for it to become a river of people. From the perspective of Binhai New Area's reform and opening up, the lower reaches of Haihe River will become a river of reform and innovation. As a golden dragon, the dragon head, dragon body, dragon tail of Haihe River echo each other and dance with its whole body. The new planning and design of Haihe Central Square is just like the finishing touch of the eye before the golden dragon starts dancing, which indicates that Tianjin will take off again and rebuild its glory.

重回海河边

天津城市设计的故事

朱雪梅　主编

序
人·天津城

天津是一座因河而生的城市，长 72 千米的海河在城中缓缓流过，形成 600 余年的历史文化积淀。每一个重要历史时期，都在城市发展中铭刻下了印记，都融化在这座城市之中。

在世界城市发展史中，中国与西方是两大主要的脉络，而天津是中西城市规划设计交融的城市。明永乐二年（1404 年），明成祖在天津筑城设卫，采用了典型的正南正北中国传统方城格局。由于海河位于城东，因此虽然是方城，但城市的建设重心偏于海河一隅。400 多年后，第二次鸦片战争爆发，天津被迫开埠，西方列强在城东南海河下游两岸建立了九国租界。租界的规划设计采用了当时西方流行的新古典主义手法。从此，天津城历经一百多年的中西城市融合整合，最终逐渐形成了自己的城市特色以及适合自身的城市规划和城市设计方法。

现代城市设计产生于 20 世纪 50 年代的美国，改革开放后逐步进入中国。天津城市设计工作起步较早，早在 20 世纪 80 年代的震后重建中就开始尝试；20 世纪 90 年代开展了海河两岸、历史街区的城市设计；2002 年启动海河综合开发改造，进行了大规模的城市设计国际征集；2006 年滨海新区广泛推行城市设计；2008 年全市开始了全面推广城市设计。经过近 20 年来城市设计的大规模实践，天津城市面貌发生了巨大变化，拥有大量成功案例，在全国具有较大的影响力。2017 年，天津被列为国家第二批城市设计试点城市，主要目的是总结城市设计编制和实施管理方面的成功经验，完善城市设计体系，针对天津和国内城市当前存在的问题，通过开展具有示范性的、高水平的城市设计试点项目，全面提升城市设计的实践与理论水平，为城市转型升级提供规划保证。

本书全面梳理了天津城市设计的发展历程，通过城市中活生生的人的视角，形象地展示了天津成功的城市设计案例的实践过程与经验，并特别把本次城市设计试点项目是如何满足人们对明天美好生活的向往作为主要内容。本书围绕天津海河两岸公共空间展开，集中探讨人性化的城市设计：通过完善城市中心步行系统与公交系统，提升公共开放空间的品质，以增加天津城市中心的活力，活化城市历史街区；通过大规模增加城市公园、郊野公园和郊区绿化等绿色生活空间，恢复津沽城市生态系统，实现人与自然的和谐共生；通过对天津 2035、2050 年住宅理想类型和新型社区的探索，设计中国人向往的美好生活场景；通过对理想城市的思考，指明城市明天发展的方向，为实现中华民族伟大复兴的中国梦描绘美好愿景。本书既是天津城市设计实践案例的汇编，更是新时代中国城市设计理论重点内容的展示。

书中人物基本上是虚拟的，但故事大部分是真实的。老张和他的家人代表了天津城的广大市民，鲍尔代表了众多参与天津城市规划建设的境外规划设计师，宋云飞代表了在天津和滨海新区创业的年轻人，而诸葛民和褚总是天津几代城市规划设计师的集体画像。

朱雪梅

目 录

第四部分　行万里路·看世界　68

第五部分　金龙起舞——海河　98

第六部分　无城市设计不建设　146

第一部分
两个天津：老城里与租界地

1. 生活在天津的人

老张

清晨6点，红日东升。老张像往常一样出门到海河狮子林桥。他出生在老城里，2002年老城里拆迁，他搬到了华苑的6层楼单元房，虽然楼屋有点高，是五层，但面积从以前的16平方米改善到60多平方米，房子显得宽敞多了，厨卫设施都有，生活配套也算方便，但没了以前的街坊邻居和熟悉的环境，老张心里时常感到空落落的。大儿子继承了老张的手艺，在古文化街开了个卖泥人、风筝的小店。2008年北京奥运会后，海河游船开通，游人络绎不绝，小店的生意特别火爆，逢年过节街上更是挤得水泄不通，一家人忙得抬不起头来。几年下来攒了一些钱，为了就近照顾小店，大儿子一家在老城里买了新房。大儿子心里惦记着老张，为满足老张的心愿，撺掇其他几个兄弟姐妹一起为老张在老城里也买了一套房子，90平方米，两居室，18层。买房是在2012年，那时房价也不便宜，每平方米9000多元。老张卖了华苑的老房，几个孩子分摊20多万元的差额。近两年天津的房价噌噌往上涨，现在老张的房子都快涨到每平方米5万元了。

老张很是得意，不仅房子升值了，关键是又住回了老城里。虽然熟悉的邻居没了，环境也变了，但确实干净多了，马路宽了，建筑的档次也提高了，而且他每天早上可以步行到海河边。海河的变化太大了！可老张依然觉得这还是他熟悉的海河，古文化街、玉皇阁、狮子林桥，还有对岸的望海楼教堂……这些地方见证了他前半生的经历。他每天迎着日出去海河边，不仅是锻炼身体，还仿佛有一种无形的力量吸引着他，那儿是他出生的地方，是他和发小们调皮捣蛋的地方，是他魂牵梦绕的老城里。每一个场景伴随着儿时的感受就像是融入了他的血液，一闻到海河水的味道就会苏醒，每到海河边他都会深深地吸口气。

鲍尔

鲍尔先生每次到天津，都会住在利顺德饭店。虽然隔壁新开的丽思卡尔顿酒店设施更先进，但他还是喜欢利顺德——这座建于1863年的天津第一家豪华酒店，曾经是孙中山、周恩来、袁世凯、梅兰芳等名人下榻之处，张学良和赵四小姐也曾在此翩翩起舞。建造之初它是中国最时髦的酒店，是社会名流和富家子弟的心仪之地，在今天魅力也不减当年。入口处有中国现存最老、尚能使用的推拉式电梯、老楼精致的客房、古旧的木质地板、一流的印度服务生，每处空间都散发着芬芳，那是来自遥远的英格兰的玫瑰气息——兼具绅士的优雅和乡村的舒适。

鲍尔夫妇傍晚时分从北京首都第二机场乘专车抵达酒店，两小时车程。车上他们一直在讨论扎哈设计的机场建筑，用雪莉的话形容就是：巨大、张狂、霸气！入住酒店后稍事休息，他们准备去五大道民园广场，当地设计师朋友要为他们夫妇接风洗尘。由于长时间飞行和时差的关系，二人有些疲劳，但盛情难却，而且他们也一直惦记着五大道，想早一点目睹五大道近年来有些什么变化。鲍尔家族和天津这座城市有着深厚的历史渊源，父亲的老友——奥地利建筑师盖苓，在 20 世纪 30 年代的天津十分活跃，设计建造了数百栋小洋楼。鲍尔的父亲曾跟随盖苓在天津工作 10 余年。鲍尔从小耳濡目染，看到过父亲参与绘制的很多天津业主的建筑设计图。近几年，他供职的美国建筑设计事务所在天津的项目日渐增多，能回到自己父亲年轻时工作过的城市令他十分兴奋。

老张

老张每天早上要步行 1 小时左右，习惯的线路是往北走，那是他小时候习惯的方向。过去老城南面是南市“三不管”，大人怕小孩出事不让去；再往南就是租界地了，虽然那时候已经收回了，但老城里的人一般不敢去。从家里出来，沿北城街东行，经过古文化街北入口牌楼“沽上艺苑”，走向狮子林桥，桥扶手上有 70 多个姿态各异的小铜狮子。下桥走海河东路邻海河的亲水平台。每天都有许多老人在此晨练，还有不少人天天拿个小马扎坐那儿钓鱼。前面是浮雕“潞河督运图”，记载着天津航运码头的繁华历史。再往前是金刚桥和河北区的中山北路，这一带是袁世凯建设的河北新区，当年属于中国地界。大悲院是天津最大的佛教寺庙，信众云集，香火一直很旺，每年腊月初八是祭祀祖先和门神、户神、宅神、灶神、井神等神灵，祈求丰收吉祥的节日，排队喝腊八粥的人挤破脑袋，听说每年都有人出天价在大悲院烧农历新年的第一炷香。

紧邻大悲院的是号称天津之眼的摩天轮，横跨在海河上。据小儿子讲，这是日本著名结构设计师川口卫设计的，永乐桥与摩天轮为一体，摩天轮高 140 米，中间的轴就有 6 米粗，是在法国加工的。过了永乐桥就进入了红桥地界，这一带就是河北大街和南、北运河构成的三条石地区，起初是以铁工业作坊兴起的，是天津民族工业的发祥地之一。老张学过的小学课文“三条石血泪史”就发生在这一带，学校组织参观过三条石博物馆。为了纪念这段历史，海河开发改造时，在海河边新建了一个大的天津近代工业博物馆，造型粗犷，取三条石之意，不知为什么现在变成了天津少年宫，虽然有些不匹配，但总比荒着好。也许是希望这里的孩子不忘血泪史，成为有良知的实业家吧。

老张继续往前走，到了子牙河与南运河转弯处，盼水妈雕像矗立眼前。这个雕像是纪念引滦入津工程为天津人民引来甘甜的清水而建。原来天津人喝的水又苦又涩，滦河水来了天津人喝的水就甜了，吃水不能忘了挖井人。到此，老张早上例行的行程基本结束，他顺着南运河往回走。据说，这里是“天子津渡”之地，也是“天津”名字的由来，即明成祖朱棣“靖难之役”是由此南下攻克南京的。过了南运河，旁边是大胡同小商品批发集散地，过去河北、山西一带的商贩都在这儿批发进货，人来人往，到处是运货的车辆，当年老张店里的一些小纪念品也是从这儿批发的。但现在人去楼空，听说是因为消防不达标，不知未来如何。

今天，老张回家早了一点，准备去规划展览馆看看，想给大孙子选房。老张有三个孩子，除了开店的大儿子，女儿定居国外，小儿子张辉在规划设计院当司机，有两个孙子、一个外孙女。最近，大孙子从美国回来工作、结婚。老张想反正也是闲着，给孙子选婚房就成了他心中的头等大事。小儿子说今天在天津市规划展览馆举办天津城市设计展，里面有许多新理念、新技术，还有新型社区和地铁规划。老张想早点去看展览，让孙子买房时选个好区位，同时作为一个“老天津卫”，市里的什么事他都爱掺和掺和，这些年海河边的变化也让他感到自豪。他对天津城市建设一直很关心，也想看看天津以后要规划成什么样儿。

2. 明成祖筑城设卫

张辉

张辉属马，今年 41 岁，是规划设计院的一名司机，出生在天津，家里从明朝末年起祖祖辈辈都生活在天津，可以说是一个名副其实的“老天津”。虽然张辉年龄不大，但喜欢古旧的、传统的东西，平时喜欢一边开车一边听《话说天津卫》这类广播节目，自己开车走南闯北的时候，也爱打听踅摸老天津的故事。这些天为了准备城市设计国际研讨会和展览，张辉一直帮着忙前忙后的，干了很多司机工作分外的事。张辉挺愿意干，做事能感到自己的价值，而且领导也能看见自己多干了活儿。今天一大早，张辉就开车拉着展板到了规划展览馆，他看见市规划院的褚总、市规划局城市设计处的刘柯处长、规划院城市设计所的陈祺所长和布展人员还在忙碌地做着最后的完善工作。待会儿要去接专家，还有半小时，利用这个间隙，他抽空去看了看展览，开始主要是城市历史沿革部分。

天津位于华北平原海河流域的下游，在海河五大支流汇合处，是九河下梢。东临渤海，北依燕山，西南与河北平原接壤，西北距北京 137 千米。在史前这里是一片浅海。距今约 6000 年，海水开始向渤海方向退落，在地表留下一层相当厚的海相沉积。曾经的黄河可不是今天的黄河，古黄河曾多次改道。3000 年前在天津宁河区附近入海，西汉时在河北黄骅附近入海，北宋时在天津南郊入海。金朝时黄河南移，夺淮入海，天津海岸线才固定下来。古黄河的冲积使旧有海相沉积的表面，出现了大面积的冲积面，逐渐淤积成平坦的土地，同时使海岸线逐步向东推进。天津就是黄河几次迁移改道大泛滥时创造出来的，用千千万万吨泥沙把大海填成陆地；大海退去时反过来将千千万万吨盐碱留在泥土里。盐腌碱烧，造就了天津人一副火热的肠子，素来急公好义，人情味浓重。天津因海河而兴，隋朝南北大运河的兴建、开通使天津成为南北运河上重要的中转站，“晓日三岔口，连樯集万艘”。天津地界日渐兴旺，运输枢纽和军事地位日益突出。1404 年，明成祖在三岔河口筑城设卫。

看着这些老图和文字，张辉想起了自己家的老故事。据家里的老人讲，张家就是这个时候来天津的。那时，天津在人们心里大概就跟今天的纽约、巴黎似的，去过的人回来说天津卫如何如何繁华热闹，没去过的心里总是充满着憧憬。家里祖上是安徽一带的，那会儿种地全靠老天爷，赶上一年旱、一年涝、一年蝗，家里就都填不饱肚子了。祖上一位有闯劲儿的，带着妻儿，就来到了天津卫。自己也没有什么特别的技能，但就是有一膀子力气，看着街头那么多卖艺赚钱的，也开始了卖艺生涯。后来据说在天津卫还小有名气，大家看他身强力蛮，力大没边，便给他起了个艺名：张大力。

老张

老张住在老城厢的龙亭家园。老城厢东面的海河和南北运河交汇后的三岔河口因漕运和贸易发展而自然形成了商贸中心，这里码头云集，还有天后宫、玉皇阁等宗教建筑和宫南、宫北大街，以及城北的估衣街等繁华地带。因此，虽然老城厢是坐北朝南的方城格局，但城里与三岔河口相邻的东北区域发展得比较兴旺，素有“北门富、东门贵、南门贫、西门贱”的说法。老城厢正中有鼓楼，但对老天津人来说更有意义的是天后宫前的广场、戏楼和两根高高的指引海河上航船的灯幡。这里就像一个大舞台，各种新奇的人物、事件轮番上演。1986年，天津市对宫南、宫北大街进行修旧如旧式的整治，改称“古文化街”，以销售天津民俗文化产品为主，生意兴隆，当时在国内影响很大。老张那时刚下岗，为了生计，租下了一家店面。当时，整个街区与老城厢一样，其600年的沧桑，说起来很悠久，却也是相当沉重的历史包袱。中国近现代兵荒马乱，中华人民共和国成立后又是一穷二白，古文化街一带深受影响——房屋年久失修，破败不堪，街道也十分狭窄，很多胡同两人走对面都得侧身而过，救护车、消防车根本进不去，整个地区安全隐患很大。天津市从1994年开始进行了成片危陋平房改造，但由于老城厢及南侧的南市、北侧的三条石地区人口过于密集，居住条件太过拥挤，改造无法实现资金平衡，因此迟迟没有启动改造。到20世纪末，据说老城厢里的居民连一个科长都没有了，有权、有势、有钱、有能耐的人都搬出了老城厢。老张小店的生意一直不温不火，虽然有点积蓄，但没有什么门路，就一直被困在城里。

2002年天津开始启动海河两岸综合开发改造，政府下定决心，筹集资金，实施老城厢、南市和三条石地区的危陋平房改造，改善居民的生活条件，也是对当地居民强烈呼声的回应。老张家过去在一个四合院中有两间房，一家五口人住在里面，大儿子结婚时，在院子里搭了简易棚。赶上下雨天，外边下大雨、屋里下小雨，雨一停就得用搪瓷盆从屋里往外舀水，日子真是艰难。政府的危陋平房改造给老张家在华苑分了两套单元房。接到拆迁通知的那天，老张一家人围在一起坐到半夜才散，老张更是激动得一夜没合眼。他天天盼着赶快搬离老城里的“三级跳坑”，终于等到了这一天。虽说故土难离，但以当时的居住条件哪里还顾得上别的，有房住才是硬道理，因此他从心底里感激政府。

据小儿子张辉讲，1994年天津市已经把老城厢、古文化街、估衣街及五大道等租界划为历史文化风貌区保护起来。当时有领导讲，既然我们花钱把老百姓搬走了，能不能把房子保留下来，先不拆，用围墙圈起来。这个提议在当时被认为是相当不切合实际的。在市中心，1.5平方千米的老城厢，加上南市地区，超过2平方千米，如果成为无人区，不仅存在巨大的安全隐患，而且大量资金投入也无法收回。小儿子说，现在看来，好像可以慢慢腾迁、慢慢改造，专家们称其为“有机更新”。但在当时，老城厢的房屋现状、社会环境、经济实力和人们的认识水平都远没有达到今天的程度，缓慢改造就意味着放弃改造。

因此，老城厢的居民搬走后，房屋拆除已是众望所归的必然选择了。其实，早在1994年，一开发商花大价钱请香港著名的巴马丹拿公司做老城厢的改造规划。在规划前也做了大量的调查分析，明确了规划必须保留老城厢的格局和鼓楼、仓门口基督堂、卞家大院、孔庙等历史建筑。考虑到资金平衡的压力，最后的方案是延续了老城鼓楼十字街格局，中部为2～3层低层建筑，四周围绕着近100米高的高层住宅，四个把角用高150米左右的公共建筑隐喻历史上的角楼。按照习惯的做法，保护历史街区主要就是控制建筑的高度，但老城厢的做法是在权衡了资金压力和历史保护等多方面因素后的折中做法，可以说是一个缩小了的老城厢。开发商为了更好地宣传，命名为“龙城”。但即便起名“龙城”，开发商也无力启动全面改造。最后还是依靠政府筹措资金，按照上面的规划设计实施了拆迁改造。

但在老张眼里，这个缩小版的老城厢与真正的老城厢还真不是一回事儿。过去的老城厢是他们这些底层老百姓居住的地方，虽然破破烂烂，但胡同院里热热闹闹，现在都是围墙围着的高档小区。胡同消失了，走在宽敞的马路上，难得碰上一个熟人。老城的中心建了鼓楼广场，广场上重建了鼓楼。保留了广东会馆，里面经常有京剧演出，看戏的都是外来游客，票价挺贵。老张一直想看看里面还是不是自己小时候的样子，但至今也没进去，感觉自己这个“老城里”倒成了外人。从鼓楼向东、南、西、北四个方向形成了四条步行商业街，有电影院、餐馆、洗浴中心、茶馆等各种商业设施，但不知为什么，人气总是不如古文化街旺。虽然只有一街之隔，最远的入口距古文化街也才1千米，但沿街商业、店面人流稀少，冷冷清清，环境好了人却少了，老张不明白这是怎么回事。他想如果老城里不拆迁现在又会怎么样呢？这不会就是老城厢的命吧。

3. 九国租界的规划建设

鲍尔

小汽车平稳优雅地行驶着，两侧的建筑和街道缓缓退向身后，这一切是鲍尔夫妇熟悉的。一晃100多年过去了。18世纪，西方经过工业革命和资产阶级革命，科学技术迅速发展，开始对外扩张。而曾经是世界中心的中国在清朝末年已经日渐衰落。第二次鸦片战争爆发后，1860 年清政府被迫与英、法、俄签订《北京条约》并批准《天津条约》，天津被开辟为通商口岸。1860 年英国第一个设立了租界，位于海河西岸。紧接着法国在英租界北侧紫竹林一带设立租界。这一时期租界内的建设虽比较缓慢，但租界范围在不断扩大。德租界是 1895 年才设立的。1900 年，借镇压义和团之名，八国联军攻陷大沽口，直取京城。联军攻陷天津之后，老城里一片混乱，到处是被战火焚毁而倒塌的房屋和洗劫后的店铺。八国联军成立了临时政府，也叫都统衙门，开始对城市进行管理。出于军事目的，把拥有 500 多年历史的城墙拆了，将拆城墙的砖料用来铺修道路。东马路、西马路、南马路、北马路，这四条环城马路就是这个时候修的。后来还在这个环路上修了电车线路，这是天津历史上第一条有轨电车。

随着八国联军的占领，原有租界进一步扩张，原来没有租界的俄国等国也强行设立了租界。最后共有九个国家（英、法、德、美、日、意、俄、奥、比）在天津分别画地为界，各有各的政府办事机构、教会、医院、邮局、军队和兵营，这些建筑全是原汁原味各自租界喜欢的样式。有许多外国建筑师在租界内承接设计项目，鲍尔的父亲就是跟随奥地利籍的建筑师盖苓在租界开展设计活动的。因为盖苓比较有名，因此他不仅在奥租界执业，还在英、法等各个租界设计建筑。1945 年后，租界被收回了，但租界的印迹长久地留存了下来。由于当时的历史条件，鲍尔一直没有机会到天津去看一下父亲年轻时工作过的城市，去看一看他设计的房子。直到 1978 年改革开放，鲍尔才在 20 世纪 80 年代初有机会第一次来到天津。时间飞逝， 至今 30 年过去了，这些租界里的老房子再一次换了新的主人，房子里又发生着新的故事。

鲍尔让司机去民园广场，朋友说这是可以带他们体验天津的新去处。到重庆道与河北路交口，一座崭新的建筑呈现在眼前——民园广场。6 年前鲍尔来过这里，当时还是老的体育场。朋友介绍说，天津民园体育场始建于 1925 年，当时英租界工部局邀请苏格兰人埃里克 · 利德尔（中文名：李爱锐，1924 年巴黎奥运会 400 米金牌得主）设计改造民园体育场。一年后落成的民园体育场是当时亚洲最先进的综合体育场。中华人民共和国成立后，全国第一场足球赛是在此举办的，后来成为“王者之师”的中国白队的主场。中超成立后，又做了八年天津泰达足球俱乐部的主场。那时每当周末有比赛时，这里都异常喧闹。2004 年天津开发区建立了第一个专业足球场，承办了中超首届开幕式。2008 年为承办北京奥运会足球比赛，在现在的奥城建成了能容纳 6 万人的水滴综合运动场。由于民园体育场观众席少，而且比赛期间大量观众集散，交通和卫生状况非常糟糕，逐渐就不举办足球比赛了。一个巨大的建筑，曾经人声鼎沸，如今寂静无声，也是很无奈的事情。

2012 年，为增强五大道地区活力，完善旅游配套设施，和平区政府提出了改造更新民园体育场的提议。之前，曾经有香港某开发商看中了民园体育场这块宝地，提出拆除体育场，建设高层住宅的建议，给出的条件是可以无偿为天津建一座现代化的体育场。虽然当时城市资金匮乏，也希望吸引更多的外来投资。但考虑到要保护五大道和历史文脉，政府还是回绝了开发商的提议。这次民园体育场的改造不是一个纯粹的商业开发项目，而是一座对公众免费开放的公共体育场所，并同时完善了五大道的配套，建设了 450 个车位的地下停车场，成为五大道旅游聚客锚地。

鲍尔看到，新建的民园广场是一座仿英式古典主义风格的建筑，基本保持了原民园广场的建筑轮廓，四周的建筑中设置了游客中心、五大道博物馆、体育博物馆、各类餐饮及免税商店等服务功能，中心是草皮为主的广场。今天恰逢假日，这里有音乐家的露天演出，广场上坐满了人。鲍尔夫妇和朋友们从 SMOKI&CO 西餐厅出来。天色渐暗，他们不由自主地加入了在塑胶跑道上逆时针行走的人流。鲍尔匀速走着，他能感受到身边行人的呼吸和热量，还听到人们窃窃私语，闻见阵阵飘来的烤面包香味，灯光忽明忽暗地打在人们的脸上……这是一种奇妙的感觉。鲍尔心想，这些移动的人流形成了宛若宗教朝圣般的景象，真是神奇！看着眼前这座美轮美奂的建筑，鲍尔联想到著名建筑师罗西在他的《城市建筑学》书中的那句话：随着时间的推移，许多建筑的功能发生了变化，但建筑本身的意义无与伦比。显然，民园广场是一个成功的城市建筑。

褚总

褚总是天津城市规划设计院的总规划师，之前与鲍尔夫妇有过多次合作，已经算是老朋友了。鲍尔夫妇到过天津多次，吃过太多次狗不理了，让他们在五大道品尝正宗的西餐，也是天津国际化努力的成果。晚饭后，褚总和同事市规划院城市设计所的总工罗云陪同鲍尔夫妇来到他们位于先农大院小光明里的新办公室——“天津规划院五大道学院”小坐片刻。经过改造的小洋楼很适合中小型设计公司办公，公司内部环境明亮雅致，首层的城市展厅展出了天津五大道、解放北路、社区中心及步行系统的研究成果。可以看出，这是一个致力于做研究型设计的规划设计机构，通过对城市、建筑、环境及社会问题的关注和探讨，强调人们彼此之间的联系、对社会生活的参与。另外，之所以叫“五大道学院”，是因为其还包括了城市设计职业教育的内容，既有理论学习也有动手操作，对城市设计师的培养很有助益。鲍尔又听褚总从这幢小洋楼说起，讲述了天津历史风貌建筑看上去很美，但真正利用活化过程中所体验到的酸甜苦辣。他也看到了褚总眼中闪烁的光芒。看得出来，这些人热爱天津，热爱五大道，热爱设计，热爱生活，他们在用心地体验和融入这座城市。这一切，让鲍尔怦然心动。

与民园西里和庆王府不同，那两处是文艺青年光顾和以精品餐饮住宿为主的地方，而先农大院除一期吸引了星巴克及遇见云台、那家等餐饮店家落户，二期更是吸引了不同的公司总部入驻，整体上业态内容更有厚度，更抗风险、更具持久活力。褚总说，天津市正在推出一系列挖掘和利用历史建筑以发展“洋楼经济”的政策，发挥历史文化资产对城市的多样化价值，使历史街区真正复兴。

鲍尔

回到酒店，鲍尔夫妇睡意全无，二人又在酒店周围转了转。这一片是天津泰安道历史文化街区，曾经是英租界的核心区，中华人民共和国成立后成为天津市委、市政府、市人大、政协及各委局的集中办公地。建于 1890 年的戈登堂，是英租界的工议局，曾是 19 世纪天津体积最大的建筑，在中华人民共和国成立后成为天津市政府所在地。1976 年大地震，这幢大楼受损严重，不久后被拆除，原址建成了新的市政府大楼。2010 年天津市委、市政府搬迁到友谊路新址后，实施了泰安道五大院综合开发建设。市政府大楼被拆除，原址又建成了利兹卡尔顿酒店。鲍尔听介绍说除酒店外，这组新建筑群还包括商场、办公、居住等复合功能，形式上采用围合式布局，形成了完整的城市街道和广场，还呼应传统保留了老的城市肌理关系。的确，布局很有特色，鲍尔夫妇能感受到两侧的红砖建筑在绿树掩映下展现出的高雅格调。但五大院建筑群于 2012 年建成开业，至今已有 7 年了，好像有些人气不足的样子。他们走在街道上，感到一丝冷清。

4. 天津的母亲河

每座世界著名的城市都有一条著名的河流，如巴黎的塞纳河、伦敦的泰晤士河、法兰克福的美因河、墨尔本的亚拉河，举不胜举。海河是天津的母亲河。天津是一座因河而生的城市，长 72 千米的海河在城中缓缓流过，形成 600 余年的历史文化积淀。每一个重要历史时期的重大事件，都在城市发展中铭刻下了印记，都融化在这座城市之中。

鲍尔

一觉醒来，已是早上 7 点。虽然只睡了不足 6 小时，但鲍尔感觉体力已经恢复。拉开窗帘，来到阳台上，映入眼帘的是旭日中的海河，河对岸是南站（六纬路）商务区。鲍尔想起他 20 世纪 80 年代初第一次来天津，就住在这个房间，据酒店服务员介绍，这间房曾住过美国第 31 任总统胡佛，当时看到海河对面的铁路运货南站一片繁忙景象，火车往来频繁。这片地一百年前曾是俄租界，但鲍尔透过利顺德酒店客房的窗子望去，只剩下俄国领事馆这一栋历史建筑，那时该地段的北面是天津卷烟厂。

2002 年天津启动海河两岸综合开发六大节点城市设计方案国际征集，鲍尔第二次来到天津。他还住在利顺德，同一间客房。那时他认识了天津规划局的总规划师诸葛民，诸葛总对专业的执着和知识的广博让他印象深刻，他们一见如故。虽然参加的是中心广场节点的征集，但他注意到了南站（六纬路）商务区的几个征集方案，其中一家法国公司的方案引起了他的兴趣，那个方案中围绕着一片滨河绿地布置了一组精致的玻璃幕墙建筑。后来天津控制玻璃幕墙建筑的建设，为了节能、安全以及避免光污染，方案没有实施。南站地区现在建成的建筑是简欧风格的，看不出和俄罗斯有什么关系。鲍尔心想，如果当时做几个俄罗斯建筑代表性的“洋葱头”该是多么有趣的事，一下子就能辨认出“俄租界”来。

目前从城市整体的效果看，南站（六纬路）商务区的建筑还是严格按照天津市规划设计导则的要求建造的，中规中矩。原来地段北面的天津卷烟厂的厂房质量很好，结构中的牛腿柱非常有特点，可惜没能保留下来。原址上由香港嘉里公司建设的嘉里汇综合体拔地而起，其中包括香格里拉酒店、商业综合体、电影院、滑冰场以及超高层公寓和写字楼。该项目原来选址在海河的中心广场，后来考虑到中心广场的纪念意义和重要性，又担心会对意式风情区形成压迫感，天津市政府做了大量工作，置换到此。

过去天津海河两岸的城市环境差距很大，虽一河之隔，河宽也仅百米，但海河西侧时尚繁荣，海河东侧落后破败。解放北路是传统金融区，历史上被称作“北方华尔街”，位于市中心的和平区。与此处相对的海河东岸有很多工业厂房和仓储用地，房价比西岸低了很多。当年海河规划提出要增强海河两岸的联系，目的就是要平衡两岸的发展。这些年来，海河上新建、改建了许多桥梁，不仅将两岸连成一体，而且诸多形式新颖、品质上乘的桥梁也成为海河上的一道道风景。

鲍尔感觉总体上海河的桥梁设计可圈可点，让他印象颇深。比如他现在所站的位置，视线上有四座桥：最北面的赤峰桥，是一座曲线斜拉桥，这种结构和造型一般用于宽阔的河面，好在赤峰桥位于海河的转弯处，平曲线的处理弱化了与环境的违和感；位于第二序位的保定桥，也是一座斜拉桥，像公路桥，在老城区中显得尺度过大和生硬，在鲍尔眼中这算是一个失败的例子；再往南的大光明桥，与上游的北安桥一样是老桥抬升改造的，造型上也与巴黎的亚历山大三世桥有几分相似；再远处，能够看到法国著名桥梁设计师、建筑师马克 · 曼朗设计的金阜桥，现代流线型，造型轻盈，不愧出自桥梁大师之手，是个好作品。另外，昨天晚上刚到酒店门口，鲍尔就留意到海河上正在新建一座桥，酒店员工介绍说，这是一座步行桥，建成后解放北路与南站（六纬路）商务区的联系会更方便，这座桥还使用了海河上游于 1996 年拆除的金钢桥的一部分零件。听到这里，鲍尔的心弦仿佛被拨动了一下，他感到，中国不仅经济发展得很快，在思想和文化上也成长得很快。

老张

老张吃过早点，换了身衣服，向规划展览馆走去。老张对规划展览馆并不陌生，他去过几次。从家到规划展览馆也就 2 千米，半小时工夫就走到了。记得 2008 年刚开馆时，人头攒动、络绎不绝，老张好奇地随着人流进到规划展览馆。那是他第一次对天津城有了全面直观的了解，把一辈子对天津的感性印象与模型方位联系了起来。

老张喜欢从古文化街穿行，街上熙熙攘攘、摩肩接踵。老张心里有种说不出的愉悦和满足感，人流多，生意就好。不到 9 点，店面还在准备着迎客，已经有许多旅游团抵达。大儿子的店面在宫南大街上，与天后宫距离不远，位置好，生意一直不错。过去古文化街就只有宫南、宫北大街一条商业街。

2002 年再次改造时，拆迁了整个街坊，一整片都成了商业区，从原来的步行街向里延伸出了许多小街、院落，沿海河建了体量气派的高楼，但商业气氛还是老街的好。走到店门口，店员正在整理外摆，见到老爷子，亲切地招呼着。店已是老店，大儿子、儿媳只在节假日客流高峰时才到店里，平时他们事挺多，不知在忙什么？穿过宫南大街南出口的“津门故里”牌坊，是通往海河的水阁大街。这一带的老人都知道，这儿最早是贩卖水的地方，旁边是水阁医院，大儿子就是在这儿出生的。

从水阁大街走到海河边，就上了金汤桥，这座桥能水平旋转开启，是平津战役解放天津时解放军会师的地方。走在桥上，对岸沿海河有一溜漂亮欧式建筑的地方是原奥地利租界，里面的三栋老建筑分别是奥匈领事馆、冯国璋故居和袁世凯故居。海河综合开发改造前，海河东路是条小马路，路边建筑破破烂烂。当时老张就纳闷，袁世凯为了不进租界（估计是嫌丢人），规划建设了中国人自己的河北新区，还为了不经过租界里的老龙头火车站，又建了中国人自己的天津北站，那为嘛他在这奥匈租界中还盖了一栋小楼呢？

老张听说，海河改造时，为了保护奥式风情区这几栋小楼，海河东路绕道建筑后边，海河边变成了石板铺的步行路，沿河种了许多大梧桐，挺拔漂亮，新的建筑也很耐看，屋顶原来是深灰的，后来改成了红色，老张觉得比以前好看了。现在沿河经常有老人带着孩子玩，气氛不错，但往南那半部分有三个银行的营业部，大门两边都摆放着大狮子，走到门口感到后背有点冒凉气，跟环境也不太搭配。在老张看来，这里面向海河的建筑应该开一些小店，支些阳伞，摆放露天座椅，生意肯定好。但他的这个好主意一直没有实现。直到 2018 年天津开始发展“夜市经济”，这三个银行营业部才改成了咖啡馆和西餐厅，海河边也摆起了外摆，渐渐热闹起来。

再往前，位于北安桥头邻奥租界的这块小三角地建成了音乐公园，老张能理解，奥地利盛产音乐家嘛。公园不大，里面有贝多芬、约翰 · 施特劳斯等几位音乐家的雕像，不太起眼，倒是海河上的北安桥，金碧辉煌，引人注目。北安桥始建于 1939 年，原是座木桥，称为“新桥”或“日本桥”，因为通向日租界。1945 年抗战胜利后重建，更名为“胜利桥”。1973 年，又改建成水泥结构桥，称为“北安桥”。北安桥是中华人民共和国成立后新建的板梁桥，河中有两排柱子，上面架上混凝土板做桥面，是为了满足海河通车的要求。海河改造时将桥体进行了提升，当时媒体有报道，老张也跑到现场围观。几十台液压千斤顶在计算机的控制下，同步铰链将桥体抬升，到位后再加固加宽，最后的建成效果出乎老张的预料：大理石桥身，铰金的雕像在蓝天白云和河水倒映下熠熠生辉。有人说北安桥是模仿法国巴黎的亚历山大三世桥，老张不以为然，因为桥头柱上的雕塑是中国传统的青龙、白虎、朱雀、玄武，寓意东南西北四方平安，桥墩雕像为装饰盘龙，桥上的四尊舞姿各异的乐女，造型典雅，栩栩如生，这些怎么可能是抄巴黎的呢？如果像的话，也应该叫中西合璧才对。

过了马路就是规划展览馆和意式风情区。老张对这一带还是有点印象的，小时候他跟小伙伴去过旁边的一宫回力球场看球，“文革”时在中心广场参加过游行，20 世纪 80 年代的时候还带孩子们到中心广场游乐场玩过。但他对租界一直有一种阴森恐怖的感觉，似乎那些洋楼里在发生着可怕的事情，就像曹禺的《雷雨》一样。他在博爱道上行走时，感觉好像有人从小洋楼里监视自己。虽然意租界的路比较宽，但总是觉得不如走在老城厢狭窄的胡同里安全自在。在老张的内心深处有一种印象，住在小洋楼里的人大部分不说天津话，不是天津人。现在的意式风情区成了一个旅游热点，有各式各样的餐厅，但在老张看来都比较花哨，是宰外地人的，他基本不去。他边看边走，抬头就看见了规划展览馆，广场人头攒动，看来今天来看展览的人不少。

5. 天津滨海地区

说到天津，就不得不说渤海湾和滨海地区。滨海新区位于天津东部、渤海之滨，是北京的出海口，战略位置十分重要。历史上，在明万历年间，塘沽已成为沿海军事重镇。到了清末，随着京杭大运河淤积，南北漕运改为海运，从大沽口入海河到达三岔河口，塘沽逐步成为河、海联运的中转站和货物集散地，也是海防要塞，大沽炮台是重要的海防屏障。1860 年，“八国联军”攻打大沽炮台失利，后从北塘炮台偷袭登陆，中国的大门向西方打开。塘沽成为当时军工和民族工业发展的一个重要基地。光绪十一年（1885 年），李鸿章在大沽创建北洋水师大沽船坞。光绪十四年（1888 年），又将开滦矿务局唐（山）胥（各庄）铁路延长至塘沽。1900 年，八国联军攻陷大沽口，占领天津，直杀到北京紫禁城。

1914 年，实业家范旭东和李烛尘等人在塘沽创办了久大精盐厂和中国第一个纯碱厂——永利碱厂，成为中国民族化工业的发源地。抗战爆发后，日本侵略者出于掠夺的目的于 1939 年在海河口开建人工海港。在 20 世纪 30 年代梁思成和张锐在天津特别市规划中就提出发展滨海地区。中华人民共和国成立后，以天津港重新开港为标志，天津滨海地区开始发展。改革开放后，天津作为14个沿海开放城市之一，天津经济技术开发区在塘沽盐场上建设起来，吸引了一大批国外的企业。1994年天津提出10年基本建成滨海新区的目标。经过10多年的艰苦努力，取得了丰硕成果。2006年中央正式将滨海新区纳入国家发展战略。

宋云飞

宋云飞与众多在海外留学、获取硕士学位，从事城市规划设计、建筑设计和道路交通设计的英才一样，为了事业，把自己的发展平台放在了中国。中国是世界上最大的工地，有足够的施展空间和用武之地，与宋云飞经历类似的海归们通常会首选中国香港公司或国外公司在北上广的分支机构。宋云飞凭借自己的专业能力和名校招牌，先是选择去了香港著名的MVA交通顾问公司，公司的市场以香港和内地大城市为主。最初宋云飞参加的项目主要集中在珠三角的深圳、广州和长三角的上海、南京等城市。2006年天津滨海新区被纳入国家发展战略后，MVA公司开始参与天津市和滨海新区的轨道交通等规划，使他有机会更多地接触和了解天津这座与自己身世有关的城市。

在香港工作10多年后，已过不惑之年的宋云飞决定回国创业。下这个决心是经过深思熟虑的。一是考虑到在外资公司发展遇到了瓶颈，二是考虑到滨海有把自己关于中国交通规划的设想付诸实施的空间，三是考虑到这十年妻子一直带着孩子在国外生活不容易，现在一家人需要团聚并安定下来。在选择落脚的城市上，他与妻子反复商量，北上广深等城市，经济发达，人才济济，城市整体发展起步较早，政策环境相对成熟，但竞争异常激烈，生活成本较高，生活节奏也很快，让在北美生活多年的妻子有些担心；而中部大城市，包括自己的家乡郑州，城市规模和影响力不足，对他来讲，在城市布局和交通规划方面难有作为的空间。最后几经权衡，他们下决心选择滨海新区。近两年他在滨海做了几个交通规划项目，有一定的业务基础，同时对滨海新区的整体规划也有一些了解。另外滨海新区的生活环境比较适宜，教育、医疗设施齐全，房子质量不错，房价也不高。而与父辈之间千丝万缕的情感联系，更让他决心在这片土地上扎根。

早在他与妻子相识的大学时代，就听妻子讲述过家族的奋斗史。妻子的爷爷年轻时便追随湖南同乡李烛尘来到天津塘沽，在“实业救国”的理想下，创立天津久大精盐厂、天津永利碱厂，开创民族化工业，并成为世界制碱工业的先导。抗战期间，为了保住民族工业命脉，在局势危机、处境维艰下迁厂入川。在几乎一穷二白的基础上，冲破重重阻力，重头再来，不仅延续了生产，还扩大了销路。中华人民共和国成立后，“久大、永利”两厂再次返回天津。妻子的爷爷一家也随厂迁回天津塘沽，此后岳父也成为碱厂的一名技术员。转眼就过去了半个世纪，碱厂已经停工，整体搬迁到南港工业区。宋云飞还记得第一次跟着岳父去碱厂旧址时，在巨大的“石灰窑”前，老人讲述着父辈传奇又艰辛的创业故事，回忆着自己一生与碱厂的缘分和经历，他纯良、严谨的外表下，饱含着一脉而承的坚毅。

妻子的长兄，虽没有子承父业继续钻研制碱技术，却同样是一个敢为人先的实干者。投身于国家的基础建设是他最大的理想，作为一名工农兵大学生，大学毕业后成了一名土木工程师，一头扎入河北山区从事基建工作，一干就是 8 年。1984 年，天津经济技术开发区成立，他临阵受命，调任开发区建设公司总工程师，遍寻塘沽找石头，在盐碱滩上打出一条路，是滨海新区的首批拓荒者。

与妻子结识，让宋云飞与天津、与滨海新区结下了不解之缘，虽在海外求学、生活多年，但父兄三代的创业信念和精神，冥冥之中牵引着他回归这片土地，也激励着他继续自己的创业故事。

6. 震后重建

褚总

褚总一早就来到规划展览馆，为今天召开的国际论坛和研讨会做准备。她分管的城市设计所就是以城市设计为主业，这次天津城市设计试点由市规划局的城市设计处组织，大部分工作都是所里做的。她很高兴地看到，现在城市设计越来越普及了，许多规划设计单位包括规划院的一些部门都参与了城市设计工作，会议室里常常坐满了人。

褚总可以说是大半个天津人。她出生在浙江，10 岁时随父母来到天津。两年后遇上唐山大地震，天津受灾严重。褚总亲身体验了地震的恐怖，目睹了震后的满目疮痍，她的中学时代是在震后重建中度过的。高考选择专业时，她坚定地选择了建筑系，希望将来能够亲手建设自己的城市。

1976 年 7 月 28 日凌晨 3 点多，褚总在睡梦中被母亲急促的声音叫醒，“地震了，快跑”。褚总被父亲拽着跑到院子里。大地震给唐山带来了毁灭性打击，距唐山 108 千米的天津遭受重创。受灾最严重的四平道、甘肃路、贵阳路一带房倒屋塌，惨不忍睹。后来的数据统计，天津市中心六区有64%的房屋遭到破坏，完全震毁的达14%，有30%以上的学校、医院、文化设施、商业网点倒塌或严重损坏，全市死亡24 296人、重伤21 568人，直接经济损失达39.2亿元。住在楼房里的人和大部分住在平房里的人都不敢回家睡觉，所有的开阔地都搭设了临建棚，学校操场、街心公园全部成了临建区，马路两旁也见缝插针地搭满了小棚子，路中间留出一条窄窄的通道。褚总听同学说塘沽造船厂的职工和家属晚上都住在船上。

从地震惨痛中恢复过来的天津人开始了漫长的灾后重建。1977年，市政府确定和平区贵阳路、河西区大营门、南开区东南角、河北区黄纬路、河东区大直沽后台、红桥区大胡同6片为重建区。1978年改革开放，国务院批准了《天津市震损住宅及配套设施恢复重建三年规划》，在国家和各方面的大力支持下，天津震后重建取得重大进展。完成了引滦入津引水工程，实现了三年煤气化和中心区集中供热，“三环十四射”路网形成基本骨架，“老龙头”火车站实施改造，航空港扩建为国际一级备降机场。城市市政公用设施水平有了较大提高，城市交通状况明显改善。同时，市里修缮了一些有特色的历史建筑，整修或新建了一批沿街建筑和高层住宅，建设了海河两岸的带状公园及街头庭院绿化，城市面貌和环境得到很大改善，天津成为当时国内学习的样板。褚总每次假期回家都能感受到天津日新月异的变化，也总能听到来自其他省市同学对天津的赞美。

在改革开放的契机下，1984年国务院做出了进一步办好特区和开放14个沿海港口城市的决定，天津是沿海开放城市之一。经过多方案比选，天津经济技术开发区最终选址在塘沽东北盐田上，拉开了滨海地区开发建设的序幕。1985年，天津市主要领导亲自主持组织修订天津城市总体规划，编制了《天津城市总体规划方案》、专项附件30项和相应的规划图纸，举办了城市规划展览会广泛征求意见，方案上报国务院。

这个时间正值褚总大学毕业，回到家乡的城市规划局工作。她清晰地记得刚上班没几天，恰逢1986年7月28日唐山大地震10周年纪念日。这一天，她随同事来到南京路黄家花园参加纪念活动，天津抗震纪念碑隆重揭幕。直到这一年，天津的抗震救灾工作才终于圆满完成。1986年8月4日国务院批复同意《天津城市总体规划（1986—2000）》。这是天津历史上第一个国家批准的城市总体规划，结束了中华人民共和国成立以来天津市边编制城市总体规划边实施规划的历史。因为这个规划太重要了，市委、市政府决定再举办一个展览，宣传天津城市总体规划，展览地点定在马场道的天津工业展览馆。

褚总和几位年轻人跟着几位年富力强的老师参与了城市总体规划展览策展、图纸模型制作和布展的全过程。那个时候没有计算机和打印机，大幅面的规划图纸需要人工绘制。他们把图纸裱在大图板上，爬在上面描图、上颜色，每天累得腰酸腿痛。规划展开展后，褚总等几个年轻规划师还去做了讲解员。在这个过程中，她对天津城市总体规划有了初步的了解。

规划展结束后，褚总回到单位，被正式分配到规划局总工办工作。总工办里有刘老总、沈总、陈总，他们都是老专家，中华人民共和国成立前就从事城市规划工作。据说刘老总毕业于著名的天津工商学院（中华人民共和国成立后，工学院并入天津大学，商学院并入南开大学，师范学院扩建为天津师范学院。1970 年原址改为天津外国语大学），在国民党时期的政府里工作过，还是地下党。老专家和蔼可亲，对褚总耐心指导。从他们那里，褚总对天津城市总体规划有了更深的认识。1986 年，国务院批复的《天津城市总体规划（1986—2000）》，实际是在 1952 年以来连续编制的 21 稿城市总体规划方案的基础上形成的。

规划确定了全市范围内城镇和工业发展的布局，提出了“工业东移”的发展战略和“一条扁担挑两头”的城市总体格局，规划符合港口城市由内河港向海口港转移和大工业沿海布置发展的客观规律。城市布局是以海河为轴线，以天津市区为中心，包括了塘沽和海河下游工业区，形成城市主体，与周围的滨海城镇、近郊卫星城、五个县城和建制镇，以及重点乡镇组成多层次城镇网络体系。

规划的目标是，到 20 世纪末，把天津市建设成为拥有先进技术的综合性工业基地、开放型的多功能经济中心和现代化的国际性港口城市。全市总人口将发展到 950 万人左右，其中城市常住人口为 610 万人左右。通过有计划地向沿海和卫星城镇扩散工业和人口，中心市区人口控制在 380 万人。塘沽包括港区和经济技术开发区总人口将发展到 60 万人。整个沿海地区城市人口将达 120 万人。海河将成为贯穿中心市区和滨海地区的天然轴线。二道闸以上的河段以蓄水为主，适当发展游乐设施，沿岸建设大型公共建筑，扩大绿地面积，形成市区的主要风景线。71 平方千米的中心市区要发展科学、文教事业和第三产业，市级商业中心从原有的劝业场、和平路一带适当外延；在丁字沽、中山门规划新建两处市级副中心。以内、中、外三条环线和 14 条放射线组成的主干道系统为骨干，优先发展公共交通和自行车交通系统。继续发展地下铁道和快速有轨交通，加强市区与近郊的联系。海河下游以航运为主，沿岸开辟码头、仓库；北岸发展钢铁、机械加工企业；南岸建设轻纺工业、精细化工企业。滨海地带发展港口和造船、机械、建材、化工、轻纺等工业，并利用沿海口岸优势，积极发展渔业、第三产业，推动海洋工程的开发。以塘沽为核心形成的港口城市，将成为滨海地带的经济、文化、科技中心。北翼的汉沽将建成以海洋化工为主的工业城市；南翼的大港将建成以石油化工为主的工业城市。在其周围发展北塘、葛沽、小站、官港、大港石油区生活基地等小城镇，形成滨海城镇体系，利用其在渤海湾海岸带中心的有利地位，为发展和繁荣渤海湾经济协作圈做出贡献。

规划提出建设“三环十四射”道路系统，解决九国租界分割造成的城市布局混乱、各种设施不成系统等问题。同时，在外环线外围设置了500米宽绿化带，避免城市无序蔓延。这些当时在国内都是领先的做法。这些详细的内容褚总当时都背过，现在还时常拿出来看一看，与今天的实际情况做对比，就更加理解城市规划既要坚持连续性又要随时代发展进步的意义。

当时，乍听起“一条扁担挑两头”褚总感觉很土，缺乏学术气息。刘老总察言观色，对她进行开导。虽然天津的老城厢和各租界本来都有规划，但天津城市发展在总体层面是没有统一规划的，工业区、居住区、商业区混杂，交通拥挤，居住条件差。为减轻中心城区压力，改善居住环境，规划决定将中心城区的工业企业逐步迁移到海河下游的滨海地区，中心城区重点发展居住、生活、商业和文化，滨海地区主要发展工业和港口，这就是“工业东移”战略。“工业东移”战略需要交通等基础设施的支撑，这样海河及周边交通设施就成了“扁担”，“挑”起了中心城区和滨海地区，奠定了城市发展的格局。规划提出“一条扁担挑两头”的城市布局结构，非常形象，易于在广大干部群众中统一思想认识。“工业东移”比“全市工业布局向沿海地区转移”更加简明扼要；“一条扁担挑两头”比“城市空间结构由单核心向双核心发展”更加容易理解。也许就是这些通俗易懂的概念使天津城市总体规划深入人心。于是，在若干年后，人们看到了矗立在那片曾经荒芜的盐碱滩上的天津滨海新区。

7. 两个新天津：中心城区、滨海新区

过去的100年里，有两个天津：老城里和租界地。今天有两个新天津：中心城区和滨海新区。在过去的40年里，中心城区经过震后重建和大规模城市改造，以及历史街区的保护，城市面貌和功能有了很大的提升；港口作

为天津的核心资源，滨海新区作为对外开放和改革的前沿，一直在快速发展，经济总量占天津全市的比重不断提高。虽然是一个城市的两个区域，但城市风貌相差明显；虽然仅仅相距 50 千米，但许多天津人几乎没有去过滨海。随着天津经济的转型，随着双城间交通条件的进一步改善，特别是新区新的景点不断涌现，比如滨海文化中心图书馆、国家海洋博物馆等，在不远的将来两个天津会融为一体。

鲍尔

早上 8 点，鲍尔穿好正装，他今天的日程是参加天津城市设计展开幕式和城市设计论坛，下午参加城市设计实践的研讨会。至于中午的时间，他约了诸葛总，要继续聊早餐时没有聊完的关于海河的话题，日程还有些紧张。妻子雪莉也正准备出门，她今天上午计划去参观水西公园和柳林公园。水西公园是由天津本地设计院设计的一座中式大型公园，柳林公园是由法国著名景观设计师与市规划院景观所合作设计的现代西式园林。雪莉希望了解一些中式园林的设计精髓，将来有机会运用到程林公园等天津项目中。下午她要去海河中游现场参加一个双城间绿色生态屏障景观设计的工作营，晚上还要与鲍尔在滨海新区于家堡会合，到茱莉亚音乐学院听一场女儿薇薇安与学生们演奏的音乐会。薇薇安是小提琴家，在天津茱莉亚音乐学院任教。这次一家人在天津相聚非常巧合。雪莉叫了网约车，手机显示车快到酒店了。因为晚上要穿晚礼服，雪莉带了个小行李箱。鲍尔先生帮忙拿到了楼下。

告别妻子，鲍尔沿解放北路步行走向位于意式风情区的天津市规划展览馆。他查了手机地图，距离 2.5 千米，步行约 24 分钟。时间到了早高峰末，解放北路上机动车虽然不少，但单行路的交通状况井然有序。解放北路最早是用深粗木桩在河滩地上修建的道路，是天津或者说是中国北方第一条柏油路（1918 年，法租界内大法国路）。2008 年天津开展迎奥运环境整治，雪莉与天津规划院的褚总及城市设计所合作做街道空间的整修设计。考虑在兼顾车行的同时，为行人提供最大方便，方案把机动车道由 15 米缩窄至 8 米，用释放出的空间加宽了步行道，同时把机动车道与步行道之间惯常的 15 厘米高差缩小到 5 厘米，柏油路改成了石钉路，既与解放北路历史街道相协调，又限制了机动车车速。过去走在解放北路上，人行道很窄，若遇到建筑台阶，几乎无法通行，而另一侧与机动车道有高差，车速又快，行走时不免提心吊胆，很不安全。

今天再走在解放北路上，虽然还有机动车，但车速慢了，步行道加宽了，特别是与机动车道的高差减小了，没有了过去被逼到墙角的窘迫感，走起来非常舒畅。解放北路的环境改造是令人满意的，施工质量也很好，充分反映了设计师的意图。在一些不规则的道路交叉口，地面铺装成椭圆形图案，设计不仅将车行道和人行道组织得十分巧妙，而且标识功能也很清晰。当时解放北路工程总指挥看到设计图时感叹道："我一辈子干了那么多工程，做这么精细的活还是第一次，简直像绣花一样……我们会把解放北路建成天津工程质量最好的路。"到今天，十多年过去了，车轱辘碾压、风吹雨打，石钉路的色彩质感更加具有历史厚重感，与两侧的历史建筑十分匹配。昨晚鲍尔和夫人散步时，走了一小段解放北路，路灯下，石钉地面泛着粼粼波光，与古典主义建筑交相辉映。

解放北路曾是英、法租界中最重要的一条街道，两侧是古典主义风格的银行建筑，素有中国北方华尔街之称。虽然这里的银行建筑规模都不是太大，但立面比例和建筑细部都很精到，饶有韵味。改革开放后由于银行业务的发展，古老的建筑无法满足新的功能要求，加之交通不便，许多银行都在友谊路等新址建了办公楼。中国人民银行早在20世纪90年代初就在解放北路建设了自己的高层办公楼，与老的街道尺度不太协调，体量比较突兀。虽然这些银行把大量业务转移到了新址，但都把原址保留下来，有的还在首层设立营业厅。

天津把解放北路规划定位为金融街，更多的是出于历史文化传承的目的。曾经有一段时间，市里试图把解放北路改造成高端商业街，但无疾而终，看来还是一直保留金融功能更恰当。鲍尔边走边想，解放北路上还有中国第一家现代邮局——大清邮政津局，发行过中国第一枚邮票——大龙邮票，现今解放北路上还开设了中国邮政博物馆。不远的原法国工部局大楼曾改做美术馆，现在又改成了机关办公大楼，不对外开放。再往前走，原法国俱乐部现在是金融博物馆，博物馆前面是一个崭新的小花园，也是津湾广场的入口，这里的人流渐渐多了起来。

前面就是解放桥，曾经叫万国桥，是可开启的铁桥，保护得很好。由于解放桥有交通功能，不能长时间断交，因此只能在特殊节日开启，每次开启约两小时。据说这座桥是由设计巴黎埃菲尔铁塔的设计公司设计的。鲍尔站在桥上，左手边是津门津塔建筑群，是由SOM设计、北京金融街公司投资开发的，已经成了海河中心广场节点标志。当年鲍尔参加了海河中心广场城市设计方案国际征集，对这个地点很熟悉。这里是海河第一个河湾处，也是天津城的中心，北岸是曾经的意租界。中华人民共和国成立后中心广场是游行集会的地方，后来改成儿童游戏场。海河南岸是日租界及和平路商业街。原来沿海河边有一些仓库，还有一栋 9 层高的出版社办公楼，后来拆除了。当时有几家著名的设计公司入围参加征集，主要内容是设计一座海河音乐厅及周围商业、办公建筑等，最后鲍尔所在的 WRT 公司和天津规划院合作的方案入围。这是一个延续街区格局的方案，比较突出的是在大沽桥头设计了一幢 300 米高的标志性“津塔”，在中心广场设计了一座类似弗兰克 · 盖里做的毕尔巴鄂博物馆的公共建筑，让人印象深刻。另一个方案设计了“津门”等一组大型公共建筑和公园与海河步道。

最后的综合方案把津门、津塔组合在了一起，虽然不一定是最佳方案，但很快实施了。高品质的写字楼、酒店和公寓拔地而起，有条件让一些高收入人群重新回到市中心。后来在和平路与滨江道交口又建成了体量巨大的恒隆购物中心，建筑外观和内部空间品质都很好。但由于项目因故拖延数年，错过了实体商业最好的时机，赶上近年电商迅猛发展，经营受到了一定影响。劝业场和百货大楼所在的和平路步行街更是日渐人稀，其后身的大片区域称为“劝业场地区”，居住人口密集，居住条件恶劣。如何实现劝业场这一地区的城市更新，据说也是这次城市设计试点的重点项目。

鲍尔的视线转到右手边，天津火车站钟塔式建筑的熟悉造型映入眼帘。这个建筑建于 20 世纪 80 年代，采用现代建筑手法，三叉式的建筑布局围合出了多个广场。在 2008 年建设京津城际高铁——中国第一条高铁时，对站房进行了大规模改扩建，设计方案整体保留了建筑面向海河、具有标志性的立面，成为天津的一个重要地标。天津站的海河对面是津湾广场——一组退台式组合的多层和超高层建筑群，多层建筑仿欧式风格，高层建筑则是现代主义融入古典手法和符号。津湾广场位于海河第二个河湾凸起处，成为天津火车站的对景。

津湾广场开业时，正赶上卡梅隆的电影《阿凡达》上映，津湾广场是天津第一个 IMAX 影院所在地，当时形成人山人海、一票难求的景象。但好景不长，不久就迅速衰败下来，恐怕与这里到天津站绕行太远有一定关系。今天，鲍尔有了新发现，在天津东站与津湾广场之间的海河上，他看到了一座造型新奇的步行桥，桥身与车站前的低层波动连廊形成整体，许多人在桥上行走、拍照，津湾广场重现生机勃勃的景象。看来，天津城市步行系统的规划和建设的确给城市带来了实际的效果。

鲍尔继续前行，呈现在眼前的是造型典雅的世纪钟，这是天津为迎接 21 世纪的到来而建造的一个钢结构标识性城雕构筑物，钟摆的摆架为 S 形，取材于太极分割线形状，寓意阴阳交替、互始互终，世纪钟的造型、材质与解放桥相互映衬，也与天津作为近代工业城市的性格很贴合。鲍尔发现世纪钟下的广场道路也改造了，过去围绕世纪钟的机动车道很宽，交通组织混乱，有很多游客想穿过马路在世纪钟前照相，却举步维艰。今天采用了道路稳静化措施，缩窄了机动车道，加宽了步行道，行人通过人行横道到达世纪钟广场方便多了。进入世纪钟广场，鲍尔西行，映入眼帘的是意式风情区入口广场，标识很清楚，他亲身体会到从天津东站出来的人还是很容易被引导到这里的。鲍尔抬起头，天津市规划展览馆快到了，他的心情有一点激动。

宋云飞

今天，宋云飞要去中心城区参加天津城市设计学术研讨会，他要从中心天津生态城的家到于家堡高铁站乘车。从家到于家堡 16 千米的距离，开车 30 分钟，如果乘坐公交加上步行共用时 1 小时 20 分钟。他平时坐公交 9 点到公司上班，在车上听听书。今天妻子让他开车，他一早就出了家门，避免在路上堵车。

7:30 出门，8 点就到了于家堡高铁站停车场，在地下咖啡店买了三明治和拿铁，离检票还有几分钟，宋云飞习惯地抬头仰视于家堡高铁站的穹顶。于家堡高铁站是国内第一个全地下的高铁站，规划有三条地铁线换乘，是滨海新区商务区的交通枢纽。于家堡站地面站房采用“贝壳”建筑设计方案，其灵感来源于鹦鹉螺和向日葵的螺旋线，从圆形双向螺旋网格拉伸出初始平面形态，通过竖直“悬挂”形成初始形体再反转得到贝壳形壳体，形成通透、开敞、明亮、新颖的建筑空间，达到了结构与建筑的完美统一，已成为滨海的地标建筑。

上了 8:15 的高铁，在车上吃完早餐，没一会儿就到了天津东站。宋云飞心里赞叹双城间高铁公交化真是方便，城市区域内的这种出行方式便捷又恰到好处。他用手机地图查了一下，从天津东站到规划展览馆 1.2 千米，步行 12 分钟，时间刚好。他走出车站，来到了站前广场，看到原来空旷的广场经过改造，增加了玻璃连廊，过去散落在广场边缘等候的众多旅客终于有了一个遮荫避雨的地方，在入口处排队上车的乘客也井然有序。天津东站连接津湾广场的网红步行桥在广场的东南方向，有很多人在那儿拍照。此情此景，让宋云飞感到这个城市开始苏醒，充满希望。

宋云飞顺着意式风情街导视系统的指引，来到世纪钟广场。他发现这里也进行了交通重新组织，采取了稳静化措施，有了专门的出租车落客载客区，周围步行环境大为改善。宋云飞感慨，城市设计师和交通规划师们呼吁多年，看来天津真心要在城市中心推行慢行化交通了。他想起不久前《抢街》一书的作者纽约前交通局长萨迪·汗来天津，也是规划展览馆报告厅，她简短介绍了纽约通过三个步骤推行慢行系统的经验，包括：描绘共同愿景、快速行动和用数据说话。萨迪·汗用讲故事的方式描绘了几场战斗，有大有小，比如根据天气状况灵活改变街道，从临时的周末局部步行着手行动，收集反馈后立刻拿出数据说话，等等，以此改变人们根深蒂固的逐利心，改变人们认知上的惰性。《抢街》（*Street Fight*）的书名也挺有意思，乍一看，"Street Fight"，容易让人联想到街头争霸、hip-hop斗舞那种刺激场面，十分抓人眼球，虽然内容并没有预想的火爆但也充满了戏剧性张力，足够吸引专业人员以外的读者。

宋云飞清楚地记得，自己向萨迪·汗提了一个困扰他多年的问题：天津希望推行慢行街道，需要将建筑到建筑之间的整个街道空间进行统筹设计、整体实施，这涉及很多部门，包括市政、交通、规划……因不同的管理部门职责不同，行业壁垒使得街道空间的整合设计比较困难，纽约在这方面是怎么做的？纽约交通局是一个什么角色？在和其他部门沟通时站在什么位置？萨迪汗简短回答了这个问题：任何一项工作都需要合作，在我任纽约交通局长期间，涉及街道管理的有11个部门，每一家都有自己的标准。把11个部门聚合在一起讨论，形成一个共同的标准，非常困难且非常耗时。我们每周都有周会，三年间从未间断。辛苦工作的结果是做出了一个统一的纽约街道设计导则，11家全都认同，这样以后无论谁当市长、谁当交通局长，导则都在那里。布隆伯格市长大力支持这个计划，也非常重视消除各个部门之间的隔阂，比如消防局，传统上消防局不是很支持慢行街道，因为大的消防栓跟自行车转弯道的关系很难处理。在市长的领导下，为了实施慢行街道计划，各位局长坐在一起，通过讨论和博弈最终达成共识。

宋云飞想到了国内的状况和纽约也差不多，想要达成共识，首先需要专业人员自身长期不懈地在专业上进行探索和研究，拿出让人信服的成果和愿景，包括调动市民的感官体验为良好的实施效果背书；同时还要擅于沟通，纽约相关经验中突出的一点就是，专业人员通过非常有技巧的方式激发城市决策者思考如何集结更大的社会资源来推进一种共同的价值、一个愿景的实现，包括最可能支持愿景实现的利益相关方、支持这种愿景实现的基础设施建设，以及如何把它融入项目。他十分钦佩萨迪·汗们有勇有谋、训练有素，擅长在跨领域的嘈杂声音里沟通。

宋云飞边走边想，一抬头看到了意式风情区的入口指引广场。忽然他在人群中发现了一个熟悉的身影——鲍尔先生，他加快脚步赶了上去。

第二部分
人、 住房、公园

8. 城市规划展览馆

天津规划展览馆是一座四层的现代建筑，面向意式风情区广场的主入口，正立面有柱廊，乳白色洞石的墙面上有不规则的海河音调，打破了呆板的感觉。这座建筑原来计划作为意大利商品展示中心，但建成后没能运作起来。2008年天津开展重点规划编制，一批规划成果亟须一个场所展示。原本按规划应该在中心广场靠北的地块上建设规划馆，当时也进行了方案征集，但规划展要求的时间很紧迫，情急之下就选择了这栋现成的房子，略加改造变成了天津市规划展览馆。改造布展只用了半年时间，2009 年 1 月正式开放。展馆面积 1.5 万平方米，布展面积 1 万平方米。展馆空间虽然不大，但恰到好处地展示了天津城市规划的历史和未来的蓝图，既覆盖全面又重点突出，获得各方好评。开放后，人流如织，开馆的前五年每年参观人数都达到百万人。

这次的天津城市设计展布置在规划馆一层的展区，面积不大，但策展内容丰富、紧凑。从天津城市设计发展的历史开始回顾，到近年来城市设计实施情况的经验总结，其中海河综合开发改造规划与评估是一个重点，到本次城市设计试点项目达到高潮。展览以主题分类，包括“海河 · 天津城 - 滨水地区与公共空间”版块、“人性化的城市 - 城市中心区的步行化与老城再复兴”版块、“生态城市 - 城市大型公园与双城绿色屏障”版块、“宜居 · 城市 - 新都市主义与新型社区”版块，这四大主题重点体现城市设计以人为本，从市民的感受出发进行规划设计，以及进一步增加天津城市活力、改善城市公园绿化等生态环境，回应城市人口日益增加后人们对住房、社区及教育、医疗、养老和出行需求的呼声，试图通过城市设计的努力和市民共同携手建设自己的家园，创造美好真实的生活场景，营造培育城市文明的公共空间。

这次展览既是对城市设计成果的展示，又是一次邀请社会各界广泛参与、表达对城市人的深切关怀、共同经历规划价值观转变的进步过程。布展上花了不少心思，既要体现城市设计的理论高度，又要易于观众理解，阳春白雪与百姓生活紧密结合，让普通人产生共鸣。策展以“重回海河边——天津城市设计的故事”为主线，每个主题包括昨天、今天和明天三部分内容，其中最吸引人的是“明天”，“明天你如何出行？”“明天你去哪座公园？”“明天你住什么房子？”和“明天的海河与天津”，每个题目都引人入胜。这是天津城市规划人几十年来第一次尝试以普通人的视角看待我们编制的规划给人的真切感受。

离展览开幕还有 15 分钟，老张已迫不及待地溜进去看了起来。这边鲍尔先生与宋云飞一同进入规划馆门厅。天津市规划院的宋院长迎上来，把他们引导进了贵宾室。里面高朋满座，鲍尔先生看到了许多熟悉的面孔，有天津市规划和自然资源局的刘局、诸葛总，以及天津市规划学会、协会、规划院的老朋友。还有清华大学的周方华院士、东南大学的刘世杰院士、同济大学的李松霖院士、中国建筑设计规划院的邹博学院士和天津大学的张泽平院士。在首排座位上，鲍尔先生见到了预期中的几位国际知名大师。乔纳森·巴奈特，已经 80 多岁，是国际著名的城市设计理论家，他的名著《作为公共政策的城市设计》已经成为世界各国大学的经典教科书。巴奈特先生从 2000 年以来多次参与了天津市区和滨海新区中心商务区、海河历次的城市设计；还有伯纳德·屈米先生，他是世界著名建筑师和建筑理论家，设计了巴黎拉维莱特公园和雅典新卫城博物馆等建筑，也设计了天津滨海新区文化中心的科技和工业馆，在世界上享有盛誉。丹·索罗门先生，他曾是加州大学伯克利分校的教授，积极参与和推广新都市主义，长期从事可负担住房的规划设计，著有《全球城市的忧郁》等著作，他从 2011 年开始参与了滨海新区和谐社区的城市设计工作。见过这些大家，鲍尔先生有些兴奋，对这次研讨会充满了期待，这应是一场城市与设计的思想盛宴。

天津城市设计展的开幕式简洁又有创意，首先刘局致辞："过去我们的规划设计多数是教条式的、家长式的，规划师操纵着画笔，从空中俯视大地，对城市进行着一厢情愿的规划设计。我们这次城市设计试点要转变过去习惯的视角，强调规划师要脚踏实地，以真实的人的体验和感受为出发点来进行规划设计，不仅要有理性分析，更要用身心去感受，要与市民共情。当然，规划师的确不同于一般市民，他经过专业训练，具备一定的规划设计知识和技能，能描绘美好城市的明天。让我们共同期待这美好明天的来临。"随后，刘局邀请几位国际专家和几位参与本次城市设计的规划师上台，扫描二维码后输入自己对明天的愿景与期待，打开了中间象征明天和展示天津的大门。

9. 明天你如何出行?

宋云飞

宋云飞受刘局的邀请作为青年规划师走上了剪彩台，他输入了对明天的期望：清除堵车的烦恼，享受无人驾驶的快乐人生。人是行走的动物，现代人坐的时间太长太久，人需要约会、见面、交往、工作，需要走到郊区田间散步、跑步、骑行、驾驶，人需要多样性的交通方式，如乘坐公交、出租、专车、地铁、高铁、飞机、轮船。宋云飞认为天津在交通方式无缝衔接方面做得比较好，天津站、西站都是历史悠久的老车站，经过改造，成为高铁城际车站，是与地铁、公交、出租车换乘的综合交通枢纽，还包括滨海新区的于家堡高铁站，位于中心商务区内，有三条以上的地铁线交汇。他认为在国内城市中，客观地说天津的交通枢纽规划设计相对还是不错的，当然与东京、伦敦、香港等城市还有很大差距，主要体现在车站的综合开发和复合功能上，这也是宋云飞长期思考的一个问题，自己公司计划投入一定人力在这个研究方向上，目前正处于收集整理资料和基础数据阶段，结合项目做一些前期分析工作。

这一次来市里开会，他也想借机去西站考察调研一下西站地区再改造开发的可行性。在地铁的建设方面，虽然天津在 20 世纪 70 年代就开始利用南京路墙子河改造建设地铁了，是国内除北京外第一个建设地铁的城市，但目前，地铁建设速度较慢，通车里程 215 千米，6 条线，总长国内排名第十位。上海地铁总长已经 705 千米，16 条线；北京 640 千米，22 条线。天津目前的客流量 150 万人次，排名第 13 位，北京、上海、广州已都超过千万人次。是否地铁里程数越大越好呢？宋云飞不以为然。比如香港地铁里程 260 千米，6 条线，但最大客流可达到 560 万人次，平均每千米 2.15 万人次，服务水平很高；天津只有 0.70 万人次，目前地铁的利用率较低；而北京则达到 2.24 万人次，超过香港，高峰时间地铁过于拥挤，乘车体验并不好。过去有人说，天津地铁客流量少是因为不成网，目前基本成环成网，但人流增加没有预计的高。还有人说，天津城市小，适合步行和非机动车出行，宋云飞感觉这可能是真正原因。此外天津地区交通的拥堵程度还可以接受，比北京好很多。交通规划的确要因城施策。

步行化，或者说回归步行化、人性化的城市是当今世界的最新潮流。丹麦建筑师扬 · 盖尔先生从很早就开始关注人性化的城市，他吸收了许多学者的研究成果，在 1971 年写了《交往与空间》一书。他长期从事城市步行化研究和规划设计，许多城市如哥本哈根、墨尔本等在他的指导下取得了重大突破。20 世纪初，伦敦面对全球竞争，为了振兴市中心区，邀请著名建筑师罗杰斯爵士主持制定了城市步行化战略，扬 · 盖尔参与后期具体的技术设计和实施。步行系统的高水平实施对伦敦的振兴发挥了十分重要的作用。前面提到过的纽约前市长布隆伯格和交通局长萨迪 · 汗也是受了扬 · 盖尔影响，在纽约实施了步行化“抢街”计划，获得很大成功，尤其是时代广场步行化改造的前后对比照片，呈现了街道环境改善最真实、最有说服力的直观效果。

今天，天津城市中心的更新改造历经20多年已基本完成，改造后许多大型商业商贸设施集中在市中心，其中部分新建筑质量上乘但总是缺乏人气、经营困难。这其中有电商快速发展冲击了实体店等因素，但城市步行系统不完善、步行设施水平不高、行走不便，以及各商业区、历史街区之间缺乏直接的步行联系也是重要的原因，宋云飞对此深有体会。比如他过去从天津东站去意式风情区，去津湾广场和劝业场，直线距离都不太远，但走路需绕行，步行体验既不安全也不舒适，而这次的体验则完全不同。宋云飞感到天津这次城市设计试点精准抓住了城市步行化这个主题。目前，国内的上海等城市制定了《街道设计导则》，天津可以更加系统全面。

接下来，展板上醒目的标题映入眼帘——“保证每家两个停车位”，宋云飞心里一震。两辆车的车库在美国郊区是标配，但在中国，停车位紧缺一直是个大难题，而且市中心和一些地段地面交通拥堵，北京等城市实施了机动车限购、限号等政策，大力发展以地铁为主的公共交通。无法否认的是汽车是中国重要的产业，对北京、上海、天津和广州这些特大城市，汽车产业都是支柱产业，随着新能源汽车和无人驾驶等智能汽车的发展，环境污染和交通拥堵问题会好转，但现在就提“两个车位”，是否合时宜？对于专门研究交通停车政策的宋云飞来说，他也拿不准。他正准备细看一下展板上的说明，广播通知研讨会即将开始，宋云飞带着疑问向会议厅走去。

10. 明天你住在哪里?

老张

从布展的手法、语汇可以明显看出，这次布展的目的就是希望普通市民都能够理解这次城市设计试点传达的价值观。老张认真地看了很长时间，许多话非常朴实，不像房交会的广告词那么忽悠，也不是那种只有专家才懂的专业词汇。老张看得很清楚，有些话真是说到他心坎里了。比如，人是行走的动物，不能久坐，这点老张十分赞同，老伴生病卧床就是活生生的例子啊，人不能动弹太受罪了，而他每天早晨遛着弯儿浑身就畅快。但有些标题老张觉得有疑问，比如“人性化城市”，他不太理解具体指的是什么，感觉有点虚“城市步行化”，有必要这么讲究吗，不就是走路吗？还有“市中心限制小汽车”，本来商业就不景气，再限制小汽车，开车的人不进来，又有多少店要黄啊！还要把几个景点连起来，在海河上加三座桥，这样会不会太密啊，海河上都快盖盖儿了！老张的自言自语引起了旁边一位志愿者的注意，他解释道："这几座桥都是步行桥，而且设计得非常轻盈，不会影响海河的景观，还会成为网红呢。""网红？"，看着眼前稚嫩的面孔，老张暗自寻思："嘴上没毛，办事不牢。"然后径直走去。

这些年，老张的日子与过去相比真是天壤之别，衣食住行无忧。但有时候，他内心会涌出一股莫名的惆怅，心里空得慌。他常想起当年一家三代七口人住在老城里大杂院两间房的日子，那破破烂烂的房子，好多都是碎砖和着泥砌的，一间屋子半间炕，下雨天家家大呼小叫、鸡飞狗跳，去副食店里买肉没超过两毛钱还都要肥的，但老张却对那种生活有种说不出的怀念。那时街坊邻居常打头碰脸，哪家吃捞面、哪家闹矛盾，大家都心知肚明；王奶奶爱管闲事，那嗓门半条胡同都听得见，常挨她闺女数落；豆豆妈刚生完豆豆的弟弟那会儿，坐在院子中间，一边喂孩子一边东家长李家短地传话；老刘头早晨锅着腰急急忙忙去厕所，小张和几个发小摸准他的时间，故意占着茅坑不起来，把老刘头急得直打转，他们管这叫"憋老头"，就喜欢拿老刘头的痛苦找乐；他每天放学一回家，就能看见奶奶在搭出来的小厨房里忙活着做饭，厨房和房根儿夹角的灶台上有一口铸铁锅，要是刚好赶上蒸窝头，笼屉盖儿一掀热气就升腾起来，热乎乎的蒸汽夹杂着玉米面的味道扑面而来，奶奶挨个扒拉一下窝头，然后一边念叨着"烫"一边掰一小块放到小张的嘴里，那是他最幸福的时刻；在他记忆深处，热气和着玉米面的味道就是家的味道……现在，他很少吃窝头了，偶尔吃一口却再也吃不出小时候的味儿了。

生活条件确实好了，父母都过世了，孩子也大了，他跟老伴单独生活，孙子、孙女、外孙女也都不在身边，只是节假日全家团聚，吃几顿饭，也没有了过去的许多"老例儿"。比如，节前要看望各位亲戚，送节日礼物，感谢一年的关照；从大年初一开始要登门拜年，相互走访，说吉利话；晚辈要给长辈作揖，长辈给晚辈发红包（放红炮）。再如，节前打扫房间，要窗明几净、一尘不染，然后贴福字吊钱，祭拜灶王爷，年三十晚上放鞭炮、烟花和包饺子，大年初一穿新衣、吃饺子；节日中每天吃什么都有讲究，过年一天也闲不下来，累是累，但热热闹闹，心里踏实。

现在，这些"老例儿"都越来越淡了，不光过年，就是日常生活与过去住大杂院、老城里也完全不一样了，亲戚朋友见面的机会越来越少了，说句难听的话，就连吵架的机会都没了，更别提当年的单位、街道组织各种文艺演出，慰问职工、孤老户，学习文件什么的，两口子吵架总有人出来劝和，谁家买个好物件往院里一放，就能引来一帮人指指点点地评论，感觉嘛事都有人管、有人问、有人掺和。现在住单元房，都住好几年了跟对门邻居也没说过几句话。过去平房没那么干净，穿着鞋进屋，邻居亲戚串门是经常的事儿。现在的单元房太私密，邻居串门就像窥探隐私似的，大家都不舒服，所以只能点到为止，见面寒暄两句，不深入交往。

小区里的老人，在小绿地广场上晒太阳聊天，相互交流的主要内容就是国内市里的大事和家长里短的小事，然后就是到菜市场买菜、去医院看病。老张的老伴本来挺活跃，是个急性子，喜欢忙前忙后，几年前突然中风，幸亏抢救及时没落下严重的后遗症，只是右边的手脚不利索，出门不太方便，所以心情一直不好。老张尽力照顾，吃好喝好，但老伴难免心中郁闷，连带着大家心情都不好。孩子们现在都很忙，除了过节能聚在一起，平时也就打个电话问寒问暖，在家的微信群里转发些段子。

前段“五一”小长假，大儿子和儿媳陪老两口出去吃饭，说孙子和孙媳要回国发展，老两口特别高兴。还没高兴够，儿媳就说起现在店里生意不好做，人工费涨得厉害，不好维持。说古文化街上的大店像“泥人张世家”什么的都在做品牌店，他们也想尝试做品牌，开工作室，在新的地方开新店。听儿媳说生意难做，老张不这么看，他认为店的位置最重要，只要店的位置好，其他问题都是经营造成的。听老张这么说，大家就都不说话了，儿子低头看起了手机，儿媳出去打电话。老张知道自己说多了，大家不爱听，就识趣儿地不再讲了，但心里不免有几分郁闷。他管不了孩子们，说了也没用。每天早上遛早，在海河边走走是老张最舒心的事，另外，张罗给孙子买房也是给自己找点高兴的事做。活了一辈子，别自寻烦恼，自己跟自己过不去，老张深知这点道理。

最近让老张心里惦记的还有一件事，前些日子表弟来家，买了许多高档水果，在对老伴的身体状况表示关心后，夸赞了一通老张家的房子，随后表弟就聊起了自己的烦心事。他儿子快结婚了，主要等婚房，原来一直盼着拆迁，拿拆迁款买新房，但最近一段时间拆迁又没信了，儿子有些等不及了。表弟和弟媳商量着把老房子卖了，加上自己的积蓄做首付，给儿子买房结婚，老两口外面租房住。老张听着心里也不是滋味，一时不知说什么好。老张的父亲有三个兄弟，都生在老城里，成人后兄弟俩先后成家，住得不远。老张与表弟跟胡同里的孩子们经常一起玩耍，关系走得很近。中华人民共和国成立后，老张的二伯因为参加过地下党，到机关工作，分到了一大间位于劝业场边上的房子，是一栋二层小楼。虽然几家合住，但有共用的厨房、卫生间，让老张一家羡慕不已。老张在老城厢上学，虽然中营小学是天津最早的现代学校，但学校房子都是平房，表弟的学校都是洋灰盖的楼房，后来表弟还上了有名的汇文中学。

老张住在老城里的四合院，随着人口增加变成了大杂院，人多了但配套条件没什么改变，大冬天还要去上胡同口的公厕，三九天自来水龙头冻住了要用烧开的水烫，一两周才上公共浴池洗一回澡。而表弟家的房子里有厕所，抽水马桶，还可以在家洗澡。老张想自己什么时候能住上这样的小洋楼可就太美了。然而世事多变，从“三反五反”开始，二伯就开始不顺。“文革”时期，二伯二婶被下放到农村，表弟一人在家。虽然他学习一直很好，也参加完了高考，但“文化大革命”突然来临，表弟没上成大学，上山下乡去内蒙古插队锻炼，回城时父母都已故去，单位给“平反”后，表弟回到父亲的单位工作。

表弟虽屡遭打击，但一直心高气傲，可出身又不好，高不成低不就，就一直没结婚。改革开放初期，城市住房十分困难，有一间房子的表弟身价看涨，终于觅得心上人，结婚成家，一年后儿子出生时，表弟已近中年。表弟单位效益一般，单位分房也没轮到他。到住房制度改革后，不再实物分房，他只拿到几万元的货币补偿。随着商品房、二手房价越来越高，表弟一家的住房梦越来越渺茫。现有的房子越来越破旧，眼看着吃喝穿用的水平都提高了，唯有房子成为最大的心病。倒是原来住房条件不如表弟的老张，由于老城厢危陋平房改造住上了新的单元房，加上小店经营收入增加，又买了新的电梯商品房。

老张偶尔去表弟家探望，感觉原来的日租界越来越像当年的老城里，小马路到处停着车，电线挂得哪里都是，房前屋后私搭乱盖，已看不出当年小洋楼的模样。房管站有时也派工人来修修补补，无奈房子已过于老化，管线年久失修，小修小补根本解决不了多大问题，所以拆迁是表弟唯一的希望。但由于位于市中心，人口太密集，人家说拆迁成本太高，还有专家提议把这一片划成历史文化街区保护起来，反正说啥的都有，就是迟迟没动。

前一阵子，国家推动棚户区改造，投入不少资金，一些历史遗留的老大难片区终于拆迁了，但劝业场周边都是二、三层以上的楼房，不能算棚户区，而且现在的大环境也不是哪个开发商想拆就能拆的，只有修地铁等“重大基础设施”才能拆楼房，拆迁费加上早搬迁等奖励，每平方米都快 10 万元了。表弟家虽然只有一间房，但房间面积比较大，有 20 多平方米，加上共用的厨房、卫生间、楼梯等公摊面积，应该有 30 多平方米，按现在的拆迁政策折算下来可以有近 300 万元的拆迁款，表弟就等着这笔钱买房呢。但人算不如天算，这里一直没规划地铁，不能归入拆迁片，如果按老房子卖，又不是学区房，最多只能卖 100 多万元，表弟真是不甘心。老张也多次劝表弟，当断则断，但表弟比较固执，好认个死理。看着小时候与自己一同玩耍、当年春风得意的表弟今天落魄的样子，老张心里不是滋味。人不能太老实窝囊了，这是老张一生经验的总结。

11. 明天你住什么样的住房?

老张

老张一边看着展览，脑海中就会闪现一些事情，天津历史和今天的展览部分，老张都很熟悉。看着过去的老照片，还有与今天对比的新照片，感觉天津真是发生了翻天覆地的变化！看着看着，他来到了展览第四部分：明天你住在哪里？住什么样的房子？标题下面写着: 居住是城市的主要功能之一，随着人类文明的进展，住宅的形式也在演进，功能越来越完善，形式越来越多样，不同的区位有不同的住房。你明天的住房是什么样子的呢？看到这个问题，老张有些疑惑，不都是单元房吗？住别墅的是少数有钱人，而且国家说了，中国人多地少，不许建别墅。但老张到国外女儿家，确实体会过大别墅的好。老张自己这辈子住过大宅院的平房、六层楼不带电梯的单元房、24 层带电梯的高层单元房，女儿家的美国大别墅。每种住房都与老张不同的生活阶段密切联系，成为他人生不可分割的一部分。

老张很庆幸买了现在带电梯的房子，当时房价还不太高。老伴中风后腿脚不便，每天老张要把她搀到楼下走走，幸亏有电梯，要是还住在华苑，五楼没电梯，估计老伴就很难下楼了。老张听说一些老人住在没电梯的楼房里，年纪大了就像蹲监狱。现在听说要加装电梯，但各户反馈的意见不一致，六楼强烈要求装，一楼的又不同意，真是众口难调。装一个电梯要 40 万元，运营修理维护都要花钱，特别是年代久一点的老小区，物业费本来就低得可怜，每户每月才几块钱，还有居民不愿交，物业公司跟走马灯似的换，弄得垃圾都没人收，只能靠政府解决。现在政府正在推动老旧小区整治，刷外墙，更换管线，换防盗门，更换庭院铺装，修补小区路，但有些质量不太好，过些日子又坏了，不能从源头上解决问题。

老张看到展墙上的一张图——天津住宅类型演变图，从老城厢开始的北方四合院，一、二、三进到多进，各式各样；再到租界时期琳琅满目的小洋楼，比四合院造型更丰富；中华人民共和国成立后建设的工人新村、邻里单元小区、苏式围合大院；再后来就是排排坐的兵营式、多层行列式大型居住区；到 20 世纪 90 年代，高层住宅多起来，住宅样子也多了一些；但不知什么原因，近些年盖的房子都成了高层住宅，除了市里、市外、中心、郊区，还有好些县城，建的都是一模一样的 30 层的高层住宅。老张不喜欢这样，怎么四处都住一样的房子呢！

展板上有张漫画，高层建筑连成一片，无边无际，像混凝土森林，每户住宅都像是一个空中的鸽子笼。“鸽子笼”，老张觉得这个词挺形象，的确住在高层之中，有一种莫名的孤独和恐慌感，好像自己跟社会、跟土地失去了联系，总想着万一着火、地震怎么办？虽然一边觉得孤独，可另一边又觉得相互干扰挺麻烦。刚入住那会儿，每家都在装修，拖拖拉拉 2 ~ 3 年都没利索，刚过了头几年噪声伴随的日子，又有人换房了，新买房的又开始装修，难有消停的时候。这样看起来，还是独门独院好，避免互相干扰。

解说员说，根据天津目前存量房规模大、质量不高的实际，天津未来新规划建设的住房要突出多样性，满足居民日益增长的需求。中心城区新规划的社区大部分位于外环线，毗邻大型城市公园，不仅环境好，教育、医疗、体育等设施配套也完善，住房类型以低层的联排住宅、多层洋房、小高层为主，有少量高层住宅，主要是改善型，标准比平均水平略高一些。对于中环线周边从 20 世纪 80 年代开始建设的，到现在有 30 多年房龄的老旧住宅要按照城市更新的方法进行逐步改造，作为外来人口落户和年轻人首次购买的住宅。这样整体上可以实现天津住房水平的升级和多样性的改善。规划的新型社区主要位于外环线的十一个大型公园周边，已建成的有梅江公园及梅江居住区，解放南路公园及新梅江地区。

老张知道，梅江建得比较早，现在没什么新房了，住在那儿生活不如市中心区方便，但环境不错，配套相对成熟，房价比较高。解放南路地区已经建了几年，在外环线边上，全运村周边形成了一组，中间的带状公园和大部分土地还没建设，有些边缘化的空旷感。水西公园，就是过去习惯上叫的侯台公园周边，紧邻快速路，位置好，公园也建成了，还有大型现代化医院，目前政府限制房价，一平方米 3.2 万元左右，出现了一房难求的局面，开发商推出一批房子后立刻就被抢购一空，一套房子拿号报名的就几百人，很难买上。

老张主要看新规划的社区，并一个个记下：海河后五公里的柳林地区在河西，位于津南、东丽交界处；东丽区有程林公园地区和南淀公园地区，剩下的是在北辰的刘园周边和北辰公园周边。这些新社区虽然都在外环线内，但住在老城里的老张还是觉得比较远，最远的是银河风景区，在老外环线外、新外环线内，毗邻京滨高铁北辰站，坐车过去得一个小时，感觉就是在郊外。解说员还说，这些公园和新型生态社区的建设能够增加 10 多平方千米的大型公园绿地，加上各级社区绿地，人均公园绿地可达到 12 平方米，新建住宅 3000 万平方米，近 30 万套，满足 100 万人的居住改善需求，规划期限到 2035 年，也就是未来 15 年是天津中心城区最主要的住宅建设区位。这是什么概念？老张需要再理解消化一下，但他隐隐觉得天津以后邻外环线一圈会有很多好公园和公园边的好房子。

接下来，老张看到一张“住宅类型分布规律图”。解说员介绍，与商业建筑、高层商务楼宇有其内在的分布规律一样，不同类型的住宅也有其内在的分布规律，高层住宅一般位于市、区中心，小高层和多层住宅位于市中心周边的城市地区，低层住宅位于城市郊区，这与城市土地级差地租、人口密度、工作岗位密度分布、交通布局模式等都有关系。老张想起来，自己当年去美国看女儿时就发现，每个城市只有市中心是一堆儿高楼，外边都是多层和联排住宅，郊区都是别墅，从里到外越来越低，从外到里越来越高，市中心一下子就能认出来，特别清楚，可中国好像不是这样，从里到外都一样高，全是一百米，那是不是违反规律了？

解说员还说，人有喜欢亲近自然的本性，随着铁路、电车的普及，城市出现了郊区化，私人小汽车的普及推动了郊区化的蔓延，我们要避免郊区化无序蔓延，同时要满足人们对自然的向往，这里借鉴了新都市主义的理念，城市相对紧凑布局，保证城市与生态环境的平衡。在外环线周边的新型社区规划中，不仅规划了大量的公园，而且保证容积率在 1.2 左右，避免住房过度密集。而在新四区外环线以外的郊区，与美国等国家低密度蔓延不同，我们的现状是规划扩张面积过大，建筑高度太高，开发强度过大，如果任由这样发展下去，天津中心城区将会形成跨过 500 米外环绿化带摊大饼的发展之势，这不是美国的薄煎饼，而是厚厚的发面饼，类似北京的“窒息式蔓延”。

12. 人民的公园

鲍尔

城市设计展开幕式结束后，鲍尔先生才有机会与一直忙碌的诸葛民规划师打了招呼，再次确认了中午两人见面的时间地点。对于天津历史和城市设计的回顾，鲍尔已了然于心。他更关心本次天津城市设计试点的内容。“明天您如何出行？”步行化城市，这是世界性的趋势，鲍尔觉得天津选对了方向，而且今天上午他已经看到了规划成果。“两个车位”鲍尔先生也是赞成的，但前提是地块的容积率要降下来。作为建筑师，他面对汽车的观点与柯布西耶、赖特等大师的观点一致，不反对小汽车，关键是如何进行更智慧的设计。

接下来的标题是“您选择去哪个公园？”，这也是一个有吸引力的题目。鲍尔从妻子雪莉和褚总那儿了解到，天津近几年要建设大量的公园，光中心城区就要建8座，而且在中心城区和滨海新区双城间规划了巨大的生态屏障，规划要使得每个公园都有特点，形成一个绿色项链。他联想到美国城市规划设计和景观设计鼻祖奥姆斯特德在波士顿规划的绿色翡翠项链。公园对人的重要性毋庸置疑，它是城市的必需之物。封建社会皇家都有打猎的林苑，后来逐渐演变成皇家园林。工业革命后，城市快速膨胀，公园成为整治城市病的重要手段。到20世纪末期，全球范围生态意识觉醒，公园成为医治环境污染、改善生态，以及人与自然和谐共生的主要载体。

在1996年第四次纽约三州区域规划《危机中的区域》的五大战略中，也把绿色战略放在首位。现在西方发达国家的生态环境已极大改善，而中国要实现经济的转型升级、生态环境的根本改善以及舒缓市民精神压力等目标，公园建设是一个重点。而设计建造高水平的公园，可能比建筑设计更难。接下来的问题，“您要住什么样的住房？住在哪儿？”鲍尔觉得这句话如同“你是谁？你要到哪儿去？”的哲学问题一样，直接击中了中国城市规划、城市设计的核心问题。他认识到天津城市设计从最初的形体设计、功能设计，到海河综合开发行动规划，到强调绿色生态和智能技术，今天已经进入了一个重视人、重视人的精神需求的新时代。

13. 城市的更新和复兴

老张

来到了“城市中心的复兴”板块，老张看到了一个新名词“城市更新”。他明白“三级跳坑”改造、震后重建，成片危陋平房改造、棚户区改造、地铁拆迁都是指什么，因为亲戚朋友都经历过，给什么政策都心里有数，但“更新”具体是嘛意思？城市更新涉及的老百姓能得到什么好处？解说员说，天津现在没有电梯的多层住宅有7000多万平方米，建于20世纪70年代到90年代，主要分布在中环线周边，量大面广，房龄都在20年以上，设备管线老化，外墙保温按新标准也不达标，这些地区就需要更新。由于大部分建筑是砖混结构，室内分隔墙难于改动，而且单元户型小，没有独立的客厅、餐厅。这些传统住宅里住了近300万城市居民和外来人口，停车更是大难题。但这些多层住宅区位都比较好，环境配套完善，公交服务水平较高。从欧洲等国家的经验看，城市更新改造除了加装电梯、完善管线之外，要对社区平面布局进行减量化和优化，先挑选一部分住宅做公租房，剩下的部分进行合并、减量、增质、改造，简单说就是将三套住房合并改造为两套住房，这是我国当前必须研究的课题，除建筑结构等技术上的问题外，关键是经济平衡如何实现，具体政策如何制定，协商机制如何运作。“这可不容易！”老张脱口而出，他在华苑住过5年，深知这里面的难处，关键是谁走、谁留，众口难调。走的为嘛要走，能得到什么好处，给什么补偿？留下来的房子大了，有电梯了，可能还是学区房，但他不能白得，得花钱，钱数还两边都能接受……这里面复杂着呢。

“天津市中心城区中心商业区劝业场地区城市更新”，题目很长，但是老张一眼抓住了“劝业场”三个字，他此行是给孙子看房的，实际上孙子有出息，不需要自己操心，反倒是表弟的住房困难是老张的心病。虽然年纪大了，但老张骨子里还是有天津人的义气，希望全家人都好，他不希望有些人富得流油，有些人穷得叮当乱响，他想起马三立老先生那句话：“介四干嘛呢！乐和乐和得了。”

劝业场周边地区城市设计的范围是南到南京路，北到海河，东到赤峰道，西到多伦道，围合的面积有1.2平方千米，现在是天津市的中心商业区，有依旧人流不断的和平路和滨江道步行商业街。这里历史上曾是法、英、日租界，当时也是北方最繁华、最时尚的商业消费娱乐区，上到国务总理、军阀官僚，下到富家小姐、公司职员，都能在劝业场、亨德利、盛锡福……或一掷千金或精挑细选买到心爱之物，欧美最新款、最时髦的物件都能同步上市，现代之名享誉全国，离劝业场不远的中国大戏院也走出了许多京剧大家。中华人民共和国成立后，这里依然是享誉三北的时尚中心，当时北京人也觉得天津人会打扮、讲究吃、会生活。到“文革”后期，随着国家经济的衰弱，和平路滨江道风光不再。1976 年唐山大地震，这里受到影响，当时在南京路上、海河边搭满了简易的临时抗震棚。

伴随 1978 年改革开放，天津实施了震后重建，城市规划建设搞得风生水起，可以说是走在了全国前列，许多城市到天津学习取经。市里对海河边的老龙头火车站进行了改造，大厅的穹顶上绘了一幅精卫填海的裸体油画，开风气之先，曾引发热议。在滨江道端头的西开教堂前建了国际商场，展销发达国家的商品，那时滨江道也成为相当热闹的摊贩步行街，销售当时最新潮的从广州贩来的商品。在随后的 30 年间，滨江道、和平路上逐渐新建了许多大型商场，像滨江商场、友谊商厦、劝业场新楼、百货大楼新楼、乐宾百货、伊势丹等。2002 年海河综合开发改造后，海河与和平路之间又兴建了津门津塔和恒隆商厦，以及最近建成开业的天河城，劝业场地区以北的南市“三不管”地区，也得到了全面改造，建成的都是漂亮的高楼大厦，就剩下中间的劝业场周边地区多年没有更新改造。

解说员介绍说，劝业场周边地区位于城市的中心，20 世纪 50 年代，苏联专家穆欣帮助指导天津编制了一版城市总体规划。他试图采用当时流行的古典城市设计手法，通过规划一条从中心广场经劝业场地区通向天津大学、南开大学和水上公园的绿化轴线，以这条轴线作为秩序统领来整合由于各租界区各自为政所造成的城市布局的拼贴和混乱。1986 年国务院批准了《天津城市总体规划》提出了“三环十四射”的城市骨架，天津成为国家第二批历史文化名城；1994 年总规开始修编，这版总规中突出了海河的重要作用，同时确定了哈密道绿轴，也划定了包括海河两岸、中心花园、赤峰道、鞍山道在内的 9 片历史文化保护区和 5 片历史文化风貌区，俗称 14 片，也再次明确了和平路、滨江道作为全市中心商业区和历史文化保护区。

几个规划下来，整个地区只剩下了一片 60 公顷的主要以居住功能为主的用地没有划为特殊地区。多年来，这片也做过多轮规划，多次尝试进行改造，但由于人口密度过高，地块比较小，难以满足日照、停车等技术规定，关键是在经济上无法平衡。20 多年过去了，天津市大规模拆迁式改造已近尾声，在四周雨后春笋般高楼大厦的包围下，留下了这片城市肌理保留相对完整、建筑损坏较严重的非历史保护区。住在这儿的居民虽然靠近市中心，但物价贵、买菜停车都不方便，还有一些住房的主人已不在此居住，房子出租给外来务工人员，或者是旁边血液研究所和医院住院治疗、陪护病人的家属。

市委书记几次微服私访后指出：高楼大厦后面不能有贫民窟；区委、区政府也下决心改变这一现状，到上海、广州等地学习回来准备实施城市更新；规划设计院研究分析国内外经验后，结合现实情况，提出旧城和新区互动的思路及城市设计方案。内容包括两大部分：第一，在劝业场地区进行城市更新，疏解人口，改善居住条件和环境，发展创意商业街区，成为和平路、滨江道线性商业街和大型商业的补充，形成占地面积 1.2 平方千米的辐射三北地区的市级中心商业街区；第二，在外环线周边地区选址规划 2 平方千米左右的“和平劝业新城”，规划布局上考虑恢复一些劝业场地区老肌理的空间尺度和邻里关系，把和平区好的教育、卫生等资源也引过去。“和平劝业新城”不仅作为劝业场周边过于密集人口的疏解地，而且要实现更优质的生活，发展旅游、养老等相关产业，保证居民就业，成为带动区域发展的城市生活发动机。

作为一个新型社区，这里的住房类型非常丰富，形成多样化的社区生活，除了延续部分劝业场地区传统住宅建筑类型，还配合公租房、小康住房和高档住房等多种类型的住房进行规划设计。劝业场周边地区人口 4 万多，规划要疏解出去 2 万人左右。目前大部分居民居住在公租房内，每户面积都很小，平均不足 20 平方米。对于真正的住房困难家庭，即使货币安置，得到的补偿款也难以购买适宜的住房。由于“和平劝业新城”有一系列补贴和鼓励政策，且邻外环线地价便宜，居民比较容易申请到达到国家住房标准的公租房。腾空出来的旧房，区政府可以挑选部分继续作为公租房，另一部分根据规划作为创意、服务业等混合功能使用，既可以租也可以部分销售，销售的部分能够回收适量资金以缓解“和平劝业新城”建设的资金压力。

2 平方千米的“和平劝业新城”预计总投入 300 多亿元，区政府通过平台公司运作，投资和负债应为 100 亿元左右，其中征地费 30 亿元，基础设施建设 30 亿元，40 万平方米公租房建设 30 亿，其他或不可预见费等 10 亿元。“和平劝业新城”可出让的建筑面积 100 万平方米，按楼面地价 8000 元，可收入 80 亿元，能够平衡征地、基础设施及公园建设等费用，而且要分期分部实施。劝业场地区实施小尺度微更新，以居民意愿为主，急需改善居住条件的在和平劝业新城选择公租房，每人不小于 20 平方米，租金水平比过去略有提高，要交一定的物业费。居民搬走后，劝业场地区的老房子结合现实情况进行改建、整修或落地重建，不拓宽道路，不大幅增加建筑高度，保留传统亲切宜人的街道尺度和城市肌理。

这么多内容，都是新概念，与过去习惯拆迁补偿的做法不一样，老张一时还理解不了，图纸上显示的劝业场周边地区依然是现在的模样，没有变成高楼，而旁边“和平劝业新城”的规划图与劝业场地区很像，有中心公园、汇文中学、鞍山道小学，还有哈密道绿轴，好像亲哥俩，既有几分相像又不完全一样。一名志愿者上前向老张解释：“大爷，和平劝业新城之所以规划设计与劝业场地区很像，就是想让搬过来的居民有熟悉的感觉。”老张似懂非懂地点点头，他打心眼里不喜欢人们叫他大爷。“姑娘，那这老房子怎么评估呢？没房本的给补偿吗？多少钱1平方米？”“大爷，这没钱的事儿。”“怎么没钱的事儿呢？我都门儿清，不给现金也得给个存折，或给个卡，把钱打卡里，签拆迁补偿协议就给，谁先搬走，还给十几万奖励呢！”“您说的是过去的货币还迁，现在劝业场地区走的是城市更新政策，根据现有住房和家庭情况发租房券，到和平劝业新城租房，当然也可以选择到其他有公租房的地区去申请。”“和平劝业新城在哪儿呢？你把我从市中心搬到外环线不给补偿吗？只起名“和平劝业新城”就行了？”“大爷，城市更新是自愿，愿意改善您就搬，还愿意住在老地方也没问题。只是想改善住房条件，要等到周围邻居有搬走的，经过房管局改造后，大家抽签，可能等待的时间比较长。”“这能行吗？”“大爷，这不是试点吗？也是想了好多办法，您知道这么多年没拆迁，就是成本太高。要是全靠政府财政，也没有这么多钱，建和平劝业新城政府尽了最大的努力。规划设计做好了您就有获得感了，您看这房型，比单元房都好，我都想去，可惜不符合条件。房子有限，要优先给劝业场周边住房困难的居民。”“那家里要是孩子大了，准备结婚，能分户给两套房吗？”老张想到了表弟的处境。“应该可以，不过我们只是做规划设计的，具体政策和管理办法要向区房管部门了解。这个规划设计刚出来，今天开始展览，然后要征求居民和各方面意见，批准后再实施。”“哦，谢谢你呀姑娘！”老张想，如果能分两套房，表弟和弟妹没问题，不知侄子与未来的媳妇同不同意。为嘛呢？因为婚房是公租房，没有产权。

14. 企业家

开发商

想看的东西差不多了，想了解的问题也有了答案，但这答案让老张心里的疑问更多了。老张心里一阵儿起急，他看到会议厅门口人头攒动，就凑了过去。“大爷，您看展览吧，展览在那边。”门口的工作人员拦住了老张。“我看过了，我要进去。”“大爷，我们是国际研讨会，您不见得听得懂。”“不懂我才要进去学习学习。”看到老张有些着急，旁边一位领导模样的人接话：“大爷您别急，有什么事咱们到旁边慢慢说。”这时一位中年微胖的男人走了过来：“刘局，我来吧！大爷，您有什么事跟我说，您是要买房吧？”“你怎么知道？”“我就是房地产开发公司的，您想买哪儿的房？”老张看了大半天展览，的确就像小儿子提醒他的，展览内容太专业，虽然用了许多简单直白的语言但他还是不太明白，感觉轻飘飘的不实在，也没有售房的具体信息，现在正好碰上了开发商，他索性就问个明白。

“这展览中说外环 11 个公园周边有新型社区，有销售的吗？”“有啊，太有了，柳林公园周边最近就开了好几个盘，您可以去看看，非常不错。我给您一张名片，可以找她——王小姐，她是那一片的房地产经纪人。”老张接过了名片。“大爷，您要是没什么事，我就先开会去了。”“稍等一下，展览说的劝业场周边地区城市更新是嘛意思？和平劝业新城具体在什么地方？开建了吗？”“劝业场城市更新是个新事，我也在学习，具体位置就在外环线边上，从快速路走，如果错开早晚高峰有个 40 分钟就到了，那边的公园已经开建了。”

老张看了看手机，已经 11 点了，得回家给老伴做饭了。他给大儿子发了微信：“下午 2：00 陪我去看房，必须去。”正准备把手机放包里，“哔”的一声，收到大儿子回的微信：得令。

企业家

“窒息式蔓延”是美国新都市主义创始人卡尔·索普对北京的描述，他反对郊区化蔓延，但他对城市郊区高层住宅成片建设更为担忧。天津外环线与外围高速公路环线之间有近 1000 平方千米的土地，60% 是规划的城市建设用地，计 600 平方千米，其中未开发的居住用地 100 平方千米，如果按 1.0 容积率计算，可建住宅 1 亿平方米，按一户 100 平方米计算，会有 100 万套住房，容纳 280 万人。这样的话，天津主城区将会形成一个千万人口、占地 1000 多平方千米的大饼，交通拥堵和环境污染问题会比今天的北京还要严重几倍，滨海新区就没有建设的必要和可能了。因此，必须控制开发强度，降低建筑高度，减少建设用地，增加蓝绿空间，像中心城区和滨海新区双城间生态屏障区规划一样。这个范围涉及四个区，每个区中心为多层住宅，周边为低层联排住宅或合院住宅，绿树掩映之下，与自然完美融合。

参加本次城市设计国际论坛和研讨会的除各位专家、领导之外，大部分是各规划设计单位的规划设计师和大学老师、学生，另外还有少量的开发商，编制城市设计时征求开发企业的意见现在已经成为惯例。除了开发商之外，本次会议特地邀请了几位企业家、大投资商参会。城市设计要能够成功，除了政府的支持和引导外，关键看市场的回应。开发商是具体的操作者，他们了解市场的需要。但开发商一般来说总是短视和唯利是图的，所以需要有战略眼光的企业家。展览上的许多试点城市设计项目都需要巨大的长期投资。参会的几位企业家看着上面的数据，在认真地考虑。不管这些数据分析是否精准，但他们凭直觉判断，思路和方向是对的。

按照这样的规划，外环内新增人口 100 万人，新四区环外 100 万人，滨海新区 200 万人，其余远郊五区近 100 万人，共 500 万人。在未来 15 年中，天津市每年要新增 30 万人才能填满这些房子，了解天津人口状况的人都知道这个难度有多大！而如果只是大量增加外来人口，产业没有发展，其他税收没有大幅增加，城市只能靠房地产销售作为主要税收，完全无法覆盖新增 500 万人口所带来的教育、卫生等各方面的支出。因此，天津在努力吸纳外来人口的同时，还必须内生发展。就像当年的成片危陋平房改造和海河综合开发一样，政府通过制定鼓励政策，拆除危陋平房，释放居民改善住房的需求，形成城市的房地产市场；通过改善海河景观和城市环境，培育旅游项目，带动相关产业，促进城市整体经济发展和振兴。

第三部分
海河十年规划

15. 重新发现海河

诸葛民

诸葛民是天津市规划局的总规划师，清晨他来到了海河边，今天要与参加城市设计国际研讨会的鲍尔先生共进早餐。诸葛民已近花甲，他出生在江苏，博士毕业后来到天津工作，在天津这座城市深耕了 30 年，亲历了天津改革开放以来城市规划发展的全过程，从规划的台前到幕后，许多事件深深地印刻在他的脑海中。他是规划泰斗吴良镛先生的弟子，毕业告别之际，吴先生送他一句话："要想做好一个城市的规划，你就必须扎根于这座城市。"这句话已伴随了他 30 年，他半生的努力和奋斗全都融化在了天津这座城市之中。他在用自己的力量推动天津的城市设计传统传承下去。

鲍尔

早上 7 点，鲍尔来到楼下的餐厅，他看到了熟悉的身影。利用早餐的时间，他约了一位老朋友——诸葛民总规划师，想与他聊聊天津这座城市，聊聊海河。鲍尔最近在写一本介绍天津城市规划的书，这不能没有海河的内容。"诸葛总，我们见过许多次面了，但每次都限于时间，有些事情没有向您深入请教，您也没有时间深入展开介绍。这次我想向您深入请教几个重要的问题，比如天津的母亲河——海河，它贯穿整个城市，对天津影响巨大。我想知道，海河的规划是从什么时候开始的，经历了什么样的演变过程？"

诸葛民

鲍尔先生的问题勾起了诸葛民的回忆。20 世纪 90 年代初，诸葛民从清华大学博士毕业后，来到天津规划院工作。记得参加工作不久，与几位年轻的同事一起被抽调到市规划局总工办，在刘宇骧、沈鹰雏、陈月波等老总的指导下，按照市领导的指示开始编制海河规划。诸葛民曾陪同几位市领导坐船考察海河，金秋十月，他们从三岔河口一直到外环线，用了逾两小时。当时已经过了汛期，水位不太高，就选择了比较矮的船，这样才能通过海河上的十多座桥。

几位市领导都到了知天命的年纪，面容慈祥和蔼，他们边观察海河两岸的情况，边对海河优美的线形、合适的宽度赞赏不已。他也听到、看到，领导们在谈论海河两岸散布了那么多平房、厂房、仓库，具有很大的改造潜力，自己在任上又没有资金去实现时，脸上流露出的遗憾神情。下船前，市领导嘱咐规划局的领导，要做好海河两岸的规划，近期无法开发，要严格管控，为今后的改造留足空间。从各位领导气定神闲、不急不缓的语气中，诸葛民也感受到了一种功成不必在我、宰相肚里能撑船的超然境界。当时天津刚完成震后重建，国家给的支持已经到期，而城市经济转型升级正处于艰难时期，城市运营已非常困难，更不用说投入大规模资金进行海河开发改造了。

不急于开发改造就没有了时间紧迫的压力，这一轮海河规划进行得比较从容和深入。前后用了两年多时间，诸葛民查阅了大量历史资料，反复到现场调研。当时，他只在《天津规划通讯》中查到一篇陈咏扬副总工关于海河中心城区段修建性详细规划的文章。规划是示意性的，附的修详规图也比较简单，内容主要是在海河两岸布置一些文化建筑。诸葛民的毕业论文是《比较城市规划研究初探》，他前后收集了许多国家滨河城市规划和开发建设的经验总结。但毕业前他从未走出过国门，工作后才有机会出国，第一次是经香港到泰国、马来西亚和新加坡考察。所以，当时，对于世界上一些著名的滨水城市像巴黎、伦敦，他都没有亲眼看过，更没有直观的感受，只能看图纸、照片，凭想象。

天津因海河而兴，隋朝南北大运河的兴建开通使天津成为南北运河上重要的中转站。元朝，大运河淤塞，水运改为海运，从大沽口入海河到达三岔河口，天津的战略地位日益突出。1404 年，明成祖筑城设卫，大沽口日后也成为重要的军事要塞。从三岔河口到出海口 72 千米蜿蜒的海河记载了天津 600 年的城市发展演变。与世界上许多临海滨河城市相比，天津既遵循了共同的城市发展规律，也有着自身鲜明的特点。

从交通运输方面看，随着船舶的大型化发展，天津运河、海河的内河航运逐步向河口海港转移，这是世界上许多滨水城市的共同特征。因此，1986 年经国务院批准的天津城市总体规划提出了“工业战略东移，发展滨海地区”的策略，是顺应了城市发展规律的。另外，天津所临的渤海湾和所依的海河又有许多鲜明的特点。渤海 7 万平方千米，是深入大陆的内海，因此天津遭受台风、风暴潮的概率小。但渤海自净能力弱，海水循环一次要 10 多年的时间。天津滨海地区在地质上是退海成陆之地，淤泥质海滩只有 1‰的坡度，退潮时是长数千米的泥滩，海水含沙量高、浑浊，人们常说的“临海不见海”在天津最为典型。天津港是在河口开挖出来的我国最大的人工港，过去的很多年里天津港航道不断淤积，需要常年清淤。

海河是我国七大水系之一。天津位于九河下梢，地势低洼，虽享近水之利，却无法避近水之害。历史上天津时常被洪水淹泡，饱受水患之苦。20 世纪 30 年代的一次大水，劝业场地区被水淹泡的历史照片让人过目难忘。中华人民共和国成立之后也发生了一次大水，河北人民和天津人民奋斗抗洪，才保住了天津城和津浦铁路不被淹。1954 年，毛主席题词：一定要根治海河，打一场人民战争。于是，海河上游大量的水库被修建起来，海河挖深河道，抬高堤防。这样多管齐下，洪水是被堵住了，但上游来水大量减少，由于缺少上游河水冲压，海水在涨潮时随河道上溯，造成河水咸化和两岸土地盐碱化。为避免海水顶托上溯，海河口修建了船闸和防潮闸。

上游来水减少后，饮用水主要靠地下水，但天津地下水含氟量高，所以在天津一些地区（比如塘沽的东沽、西沽）长大的孩子牙齿是黄的，海河流域上游的水治理造成了天津生态环境的巨大改变。1981 年，为解决天津人的饮水问题，国家决定实施引滦入津工程，将唐山滦河水引入天津，从唐山潘家口水库修引滦水渠长 200 多千米，穿山越岭到达天津。至此，天津人终于喝上了甘霖般的饮用水，而海河也成为天津于桥水库水源地之外的唯一的备用水源。为了避免咸化，在海河中游修建了二道闸。1986 年版城市总体规划中明确，海河二道闸“闸上保水，闸下通航”。由此，海河上游受到了严格的环境保护，严控游船等各种活动，防止污染海河水质。

诸葛民在研究海河定位时，看到1986年版城市总体规划中明确指出：“海河是天津中心城市的风景景观轴线。”当时，他还不太理解，海河对天津市如此重要，是天津的母亲河，为什么只被定义为风景景观轴线？随着对海河复合功能的深入研究，他逐渐理解了其中的深层含义。尽管海河根治后就没有发生过特大洪水，但水火无情，不得不防。每年汛期来临，整个城市上上下下如临大敌、严防死守。当时让分管农业水利的市领导和水利部门最头疼的问题是：是把海河等主要河道中的水放掉，等着迎接上游洪水，还是赌上游不发洪水，而把河中宝贵的水留住？他们做梦都想着能避免水放掉了而上游又没来洪水所造成的来年缺水的尴尬局面。

但经常是人算不如天算，按照当年的海河流域防洪规划，海河要承担每秒 2000 流量的泄洪量，也就是说市中心区海河两岸的堤防高度要由现在大沽高程 5 米多，涨到 7 米。据说为此事市领导与天津大学水利系教授有过激烈的争论。如果海河两岸都建成 7 米高的防洪墙，高出地面 2 米多，人的视线都被挡在高墙之下，那还有什么景观可赏？更严重的是，随着常年的淤积，海河河床不断抬升，堤岸也要随着水位不断抬升，此时堤岸已经比城区一般地坪高出了 2 米，堤防要再加高 2 米多，无异于头上悬河，如有意外，对城市的危害不堪设想。

得益于海河规划时间充裕，诸葛民在此期间有半年时间被导师借调回清华大学从事中国特大城市结构与形态比较的课题研究。他现在还清楚记得，当时中国超百万人口的特大城市只有 13 个，唐山刚过百万人口，排在最后。按照导师的要求，他用 tablet 一个点一个点地将中国百万人口特大城市建成区的平面形态和美国主要城市建成区的平面形态输入计算机，用同比例尺叠加，一对比，发现中国的特大城市比美国的城市小很多，他的解释是，美国的城市已经过郊区化进入后工业时代，而我国正处于工业化和城镇化的起步阶段。但这一对比发现让诸葛民陷入深思，我国一贯坚持的“严格控制大城市规模，合理发展中等城市，大力发展小城镇”的基本国策是否正确，这时期也有一部分经济学者提出要充分发挥大城市的聚集效应，不应限制我国大城市发展。然而，对于这些论调我国规划界一律不予理睬，认为基本国策绝对正确，不可怀疑，不可讨论。

从今天的结果看，我国的特大城市，尤以首都北京为代表，严控规模不但毫无成效，反而进一步扩张呈摊大饼发展之势，这还不同于美国低层住宅的郊区化蔓延，而是高楼林立的膨胀式发展，即卡尔·索普所说的“窒息式蔓延”，造成十分严重的交通拥堵、环境污染等大城市病。中国过去的乡村有非常好的人居环境，现在很多人在反对郊区化，然后把郊区都建成城市一样的高楼大厦，让人无语。

诸葛民当年的论文是研究比较城市规划，初衷是希望通过学习借鉴先进经验，能够避免西方发达国家曾经走过的弯路，但我国的实际结果不仅没能避免走西方的弯路、得城市病，而且病得更重。在研究生的政治经济学课上，诸葛民做了简单的数据资料分析后写过一篇论文，建议马上改变中国的计划生育政策，允许一对夫妇生两个孩子，否则中国很快会遇到严重的老龄化问题。想起来也挺可笑，当时的任课老师给打了 85 分，不高不低，不知道他认真看了没有。当时计划生育还属于国策，是不能触碰的禁忌，好在老师没给上纲上线。今天看来，我国确实应该更早地改变计划生育政策，更早地放开二胎政策。

这件事情说明，我国的改革已经进入深水区，就是要对一些原来认为是圣经一样不可撼动的教条进行改革，在城市规划界更是如此。比如，大家一致认为，中国人多地少，我国采取了最严格的土地管理政策，要求高强度开发，但实际上有太多浪费和低效使用的土地，其造成的浪费比土地低强度开发多多了。而对于降低开发强度、改善人居环境，却始终不敢“奢望”。“人多地少”“高强度开发”就像催眠一样蒙蔽了我们的心智，更限制了我们的想象力、创造力和改革的动力、活力。对于这些课题，诸葛民暂时还没有能力深入研究，当时也缺乏工作经验，只是脑海中经常会出现一些念头。因此他在工作之余还是把更多的精力放在了对世界著名滨河城市如巴黎、伦敦的研究上，随着研究的深入，他的思路越来越清晰、坚定。

回到天津后，诸葛民开始整理海河规划的成果。那是一本油印的薄薄的小册子，插图很少，当时还没有普及计算机、打印机，输入文字靠打字机，插图用针管笔绘制，手工排版，规划的题目很简练：海河的定位和两岸规划，内容只有 5000 多字，像个规划纲要，重点是海河的定位，希望人们能够提高对海河对于天津城市重要性的认识。当时的规划范围只到海河中游二道闸，全长 31.8 千米。该规划首先指出：天津城市总体规划确定海河是多功能河道，二道闸建成后在非汛期，闸上保水，闸下通航；汛期将承担排泄上游洪水和市区雨水的功能。然后指出：根据海河现状及总体规划安排，二道闸以上，海河以蓄水为主，汛期兼有泄洪、排雨水、给水、航运旅游、文体活动等功能；非汛期除具有上述功能外，不准排入雨水。

以上这些内容是在几位老总把关下严谨表述的，与总体规划略有不同，旨在增加航运旅游、文体活动功能。非汛期不排雨水的重要原因是保护海河水质，因为北方城市春天头几场雨水会非常脏，污染很严重。虽然在功能上变化不大，但在海河定位性质上与总体规划有重大改变，规划首次明确提出：海河是天津城市中轴线，是贯穿城市政治、经济、历史、文化的中心线，是以经济建设为中心，振兴天津经济、活跃天津的生命线。

这里要注明一下，过去一般认为海河只是风景轴，是次要轴线，因为天津城市重心偏向海河南岸的西南方向发展，所以在人们心中由中心广场指向水上公园的轴线，或由老城厢指向正南的轴线才是天津城市的主轴。多年来，由于海河的切割作用等原因，海河北岸发展相对缓慢。同时，海河边历史上都是码头、仓库货栈和工业用地，这些功能是随着水上货运和工业生产率先发展起来的，而城市生活的发展逐渐背离海河，反向海河发展。

此次海河规划希望将两岸的工业、仓库货栈都调整为商业金融和文化服务设施用地，将海河定为城市的主轴，而与海河垂直的各条街道，包括哈密道绿轴都是城市的次轴，这样就形成两岸以海河为轴，均衡发展的“鱼骨状”结构。1994 年，诸葛民和几位规划师赴澳大利亚考察滨河规划，专程考察了墨尔本市的雅拉河，其宽度、线形都与海河极为相似。当地规划局官员介绍说，为了使雅拉河两岸均衡发展，他们的规划提出墨尔本城市要面向雅拉河发展，诸葛民当时就有种千里遇知音的感觉。

鲍尔

听到这儿，鲍尔先生问："对海河的认识和定位有这么大的变化，各方面能达成共识吗？"之所以问这个问题，是因为鲍尔先后到访中国多次，他的太太是出生在台湾的美籍华人，因此他对中国和天津是比较了解的。诸葛民在美国做过访问学者，他的英文发音虽不纯正，但表达比较清晰、准确，沟通没有障碍，鲍尔与诸葛民用英语交谈时都能互相理解并且心领神会。"您提的问题很关键，吴良镛先生曾对我说，一个好的规划要形成三方面共识，一是学术共识，二是社会共识，三是领导共识。这些话我一直铭记在心，在30年的规划实践中越来越体会到这三个共识的重要性。海河之所以最后能成功实施综合开发改造，就是因为达成了这三个共识，这个共识花了10年时间。"诸葛民特意加重了最后一句话的语气。鲍尔感叹道："这样大的项目，用10年时间做前期决策研究，不算长，应该说是一个合理的周期，我们旧金山湾区从1946年提出建设快捷轨道BART，到1962年获批，花了16年时间。""那是在美国，而中国当时正处于改革开放的高潮期，各地都在加速发展。"诸葛民回应道。

诸葛民

诸葛民继续讲述：对海河定位的提升规划专家们都是赞同的。实事求是地说，当时社会上关注海河的人很少，并且由于规划还不涉及具体实施，各部门并没有太大意见。那些年，天津发展遇到了比较大的困难，震后重建，国家共给了天津20多亿元的资金支持，每年拨几亿元，大家知道几亿元在当时是相当大的数目，这些资金保证了天津三环十四射等道路交通基础设施、住房的建设和环境改善。震后重建结束后，这笔资金就没有了。而天津的现实状况是，工业装备整体落后，效益很差，许多装备技术还是中华人民共和国成立前的，企业经营困难，大批职工下岗。

这时，社会上有许多抱怨的声音。有人说，震后重建中把所有钱都投向了城市建设，没有投资给工业，造成城市发展实力和后劲不足。有人说天津离首都太近，灯下黑，许多大项目被北京抢走了，比如：国家要在华北建个

30 万吨的大乙烯厂，大家都争，最后的结果是北京和天津各建了一个 14 万吨的小乙烯厂，都不成规模。到 1992 年，当时国内也出现了不同的声音，一方说要继续改革开放，一方说要走回头路。1992 年 1 月，小平同志视察深圳等城市，发表重要的南方谈话，指出要继续改革开放，走中国特色社会主义道路，发展市场经济，使中国的改革开放进入一个新阶段、新高潮。

时间过得很快，诸葛民起身，说道："鲍尔先生，时间不早了，我还要到规划展览馆准备一下今天的会议。我们中午接着谈。"

16. 一条扁担挑两头

宋云飞

决定回国发展后，宋云飞比较了天津的中心城区和滨海新区这两个由海河这条扁担挑起来的城区。虽然中心城区配套更加完善，生活更方便，但总给人缺乏活力的感觉。相比之下，他更喜欢年轻充满活力的滨海新区。确定在滨海新区发展后，宋云飞决定把办公场所选在于家堡。这里是自贸区，对创业公司有各种优惠政策，而且对内对外交通都很方便。他参与过于家堡金融区的交通规划，这里大的交通框架已经形成，从于家堡到天津东站、北京南站都有高铁直达。于家堡目前的办公空间和周围环境是一流的，临海河公园，其开阔的视野、宜人的步行街道和完善的地下空间，几乎可以与国内甚至国外许多大城市的商务区媲美。

目前于家堡的生活配套暂时还不完善，他原来想在于家堡或天碱购房，这里有大片公园，还有滨海文化艺术中心、商业中心，房型和环境都不错，但目前的学校比较一般，他听说有关部门在跟一些名校商谈设立分校的事，不知何时才能实现。不远处的开发区生活区生活氛围已经成熟，有国际学校，但习惯了住独栋住宅（HOUSE）的孩子们还是不喜欢。最后他们选择了中新天津生态城湖边的低层联排住宅。这里的环境、学校和住房都有美国近郊的感觉，虽然能看见一些高层住宅，但还不太拥挤。

宋云飞是搞交通规划的，对交通方面的问题比较敏感。中新天津生态城在规划时设计了一套生态城市的指标体系，目前大部分都实现了，唯独在交通上差距较大。原规划技术指标是绿色出行要占整体出行的 80%，但由于轨道建设长期滞后，这一指标远未完成。在国外时，宋云飞一家与美国当地人一样，住在郊区的 HOUSE，家里有两辆车，他一般开丰田混电卡罗拉，家里的另一辆多功能小面包车 VAN 是妻子日常购物、接送和照料孩子时开的。到香港后，宋云飞的出行主要依靠公交和步行。回到内地，他原想全方位真实体验一下绿色出行，就像简·雅各布斯一样，但看来还是太理想化了，目前地铁、公交、步行都不如小汽车来得方便，在新区不开车着实没法生活。他与妻子商量买了一辆特斯拉 S3，既环保又有比较大的内部空间，妻子原来还想延续之前的家庭生活习惯，买一辆 VAN，但国内的停车实在是问题，不管是在自己家的小区还是在公共场合都不方便，后来干脆就放弃了。

17. 世纪危改

诸葛民

诸葛民走出饭店上了汽车，看着窗外海河岸边掠过的景色，他的思绪还在继续追溯。当时全国各地都在加快发展，在深圳特区之后，1992 年国务院批复设立上海浦东新区，天津所在的环渤海地区也在加速发展。1992 年 3 月摩托罗拉天津工厂落户天津 TEDA 经济技术开发区，生产制造手机等产品，加上夏利汽车、康师傅方便面，成为天津当时享誉全国的“一支机、一碗面、一辆车”。虽然对外开放取得重大进展，但无法扭转天津工业整体下滑的态势。为了实施“工业东移”战略，中心城区实施了“双优化”工程，即通过优化产业结构来优化城市用地结构，一些工厂如友谊罐头厂、仁立毛纺厂等工业用地改为房地产开发用地，但数量有限。另外，虽然实施了震后重建，天津城市功能、道路交通和基础设施有了极大的改善，但由于清末民初以来，连年的战乱与纷争，城市建设长期停滞。中华人民共和国成立后，又实施了“先生产后生活”的政策，住房建设等方面欠账很多。随着城市人口的不断增长，住房困难、配套落后的问题越发突出。

诸葛民清晰地记得当年的一个场景：他随市规划局领导参加市政府的一个会议，当时已是初春，仍有些寒意，会议开始时，市长披着绿色的军大衣走进会议室，脸色十分难看。他的目光扫到了市委宣传部的同志身上，立刻来了情绪：“你们宣传部都干什么去了？昨天晚上下大雨，许多居民区都淹泡了，我在现场没看见一个电视台、日报的记者，今天早上的日报新闻只字不提老百姓受淹泡的事儿，讲了许多大道理，有用吗？”

当时，天津大部分管网老化，特别是一些地势低洼的成片平房区，一下雨就被淹，老百姓意见很大，但由于缺乏资金，长期得不到解决。天津学习深圳、广州、上海的经验，也想引入外资解决改造资金短缺问题，但当时住宅市场还未对外开放，所以市里拿出了劝业场、国际商场的几个地块招商引资，地块规划的是商业写字楼等地产项目，但只招来几个外籍华人。市里原本希望在国际商场两侧地块建设钢铁大厦，利用天津冶金工业优势，搞钢铁原材料交易流通的全国或区域市场，但迟迟没有动工。诸葛民曾带队去南方学习经验，回来后也编制了这些地块的控规和技术方案，提出地块的用地性质、容积率、建筑高度等规划指标供招商使用。但城市的整体经济不好，各项工作都受影响。屋漏偏逢连夜雨，1993 年 2 月，天津市委书记、中央政治局委员谭绍文同志因病去世，享年 64 岁，这是天津少有的在任上去世的主要领导。

转眼到了 1994 年，由于小平同志南方谈话后各地加速大干快上，中国经济出现过热现象。1992、1993 年，全国固定资产投资增速分别为 42.5% 和 58.6%，大大超过了往年。投资需求带动消费需求，产生了较为严重的通货膨胀，成为社会稳定的巨大隐患。从 1994 年开始，中央适时提出了“适度从紧”的财政政策，对过热的经济进行宏观调控。这时许多天津人又开始抱怨，说天津总是抓不住机遇，踩不上点，人家已经加快发展了，天津人还没有醒，等醒来想发展了，又赶上宏观调控了。社会上负面情绪比较严重。

实际上，1994 年初，天津新一届市委市政府分析国内外形势，特别是从天津实际、时代要求、历史使命、人民的愿望出发，提出并开始实施“三五八十”的奋斗目标。“三”即提前三年实现国内生产总值翻两番，“五”是用五到七年时间完成市区成片危陋平房改造，“八”是用八年时间将国有大中型企业调整嫁接改造一遍，“十”是十年基本建成滨海新区，应该说这四大奋斗目标准确地抓住了天津城市的关键问题。但当时很多人不理解，包括规划界，规划院的同志们认为“三五八十”这几个字太直白了，太简单，外人看很难理解，还必须再向人解释。

新的市委书记十分重视规划工作，到规划院现场研究临港工业规划布局问题，得知市规划院是正处级单位后，指示将规划院升格为副局级事业单位。听闻此讯，规划院全院上下欢欣鼓舞，既实施了几大战略，又有经济效益，就不再议论了。而规划局的干部职工主要对五到七年完成成片危陋平房改造意见较多，认为城市发展是个长期的过程，五到七年把危陋平房都改造完了，那以后干什么？关键是以当前的水平无法预料未来的发展，要给城市的未来留有空间。的确，刚开始的危陋平房改造存在许多问题。由于没有国家和市里的资金，只能靠地块拆建比达到自平衡，造成还迁住房以能算下账来为标准，设计建造的标准很低。一些还迁房面积比较小，卫生间有的还不到 1 平方米，人进去后甚至连门都关不上。问题不断出现，但也能不断得到解决。

诸葛民对这些事也有些看法，他还记得去香港房屋署了解到的情况，香港在 20 世纪 50 年代遇到住房危机，许多人无房可住，而大部分人只能住在棚户区中，极易发生火灾等安全事故。为此，港英政府实施了棚户区改造和新住房建设。新建的住房都是比较密的七层单元住房，连廊式、无电梯，每户只配有厨房，卫生间是每层楼十几户共用。由于居住标准低，建筑质量较差，从 20 世纪 80 年代起，香港政府开始拆迁这些只有 30 年房龄的多层住宅。为了达到资金平衡，就只能建造更高的超高层住宅。

再看天津实施的危改，新建的大多数是六层砖混住宅，没有电梯，单元户型小，也因为是砖混结构，大部分墙体是承重墙，无法改动。诸葛民总有一种担心，三四十年后如此大量的房子如何升级改造，总不可能学香港吧？香港受用地所限，居住水平在世界上也是比较低的。另外，诸葛民也深切体会到群众的疾苦以及实施危改的必要性和紧迫性。作为一名年轻的规划师，他愿意扎根于这座城市，了解这座城市。他有一个爱好，就是只要有空就在城市各处转悠，他不仅常去海河两岸各处，包括城市商业中心、老租界区，也去过老城厢、南市、谦德庄等区域。当时老城厢、南市的居住条件已经很恶劣了，历史上原有房屋的主要建材是砖加土坯，但到了谦德庄你才能见识到什么是真正的棚户区！那里的房子基本上都不能算作房子。诸葛民记得市主要领导在推动危陋房屋改造时经常说"要带着感情搞危改"，并强调："共产党领导人民闹革命，中华人民共和国成立已经 40 多年了，广大人民群众还住在这样的危陋房屋里，生活在水深火热之中，我们领导干部应该感到脸红！"

在市委、市政府的强力推动下，在各区区委、区政府、各级干部、市区的开发企业和广大居民群众的共同努力下，到 1999 年，市区成片危陋平房改造基本完成。据统计，共拆除危陋平房 1000 万平方米，80 万居民受益，占当时市区总人口的 1/3 左右，这一成绩是十分巨大的。由于 1999 年是 20 世纪的最后一年，所以有人称其为"世纪危改"，世纪危改的完成对天津今后的发展意义重大。诸葛民记得 2002 年准备实施海河两岸综合开发改造前，市委书记讲了一段意味深长的话："如果我们早十年，或者早五年实施海河综合开发改造，一定有人说我们是搞形象工程，因为广大群众还居住在危陋平房中，城市的道路系统还不完善，市政能力不足、设施老化，下雨会经常淹泡。只有到今天，成片危陋平房改造基本完成了，城市基础设施能力提升了，海河综合开发改造才具备必要的条件。"

成片危陋平房改造的确极大地改善了城市居民的居住条件，但是还有老城厢、南市、西于庄等部分成片危陋区改造由于资金无法平衡没有启动实施。改造规划都基本保留了城市的道路格局和街廓肌理，与国内其他大城市相比，天津在危陋房改造的规划理念上走在了全国前列。北京作为千年古都，老城区有大量四合院。除少数作为机关单位保存较好外，大部分居住的四合院由于年久失修，以及居民为解决家庭人口增加的难题而违建私搭乱盖，居住环境是比较恶劣的。但由于涉及古城保护，各方面关注度很高，难以大规模改造。

吴良镛先生通过多年研究，在 20 世纪 90 年代末推动实施了菊儿胡同改造，成功地进行了多层四合院的试点，获得好评，可惜没能大面积推广。上海老城区大部分是两层的石库门和里弄建筑，这些建筑比一般中国城市合院住宅的建筑质量好，但市民每天早上都要到公厕刷马桶。广州老区比较密集，随着城市快速扩张，大量城中村无序发展，生活在其中的城乡居民和外来人口很混乱、居住条件非常拥挤。可以这样说，当时天津在城市危陋平房改造上走在了全国前列，比 2010 年后全国的大规模棚户区改造早了 15 年。

成片危陋平房改造还是带动城市经济发展的重要手段。城市经济学中，房地产是非常重要的组成部分。历史上天津市是一个商贸流通城市，有人讲，靠商贸流通天津养活了 300 万人。中华人民共和国成立初期，天津的轻工产品是比较发达的。后来，随着帝国主义对中国的封锁以及国家将消费城市转变为生产城市的政策，作为对外贸易港口城市，天津既无原材料又缺水，工业发展遇到瓶颈。这种情况下，政府既无资金改造升级老旧企业，又没钱投资大型项目，因而经济发展降速，职工收入减少，服务业也发展不起来。

天津新上任的市领导冥思苦想，终于找到了推动成片危陋平房改造、带动整体经济发展的路子。由于没有财政资金，只能靠市场化的提高拆建比的手段，用新建商品房销售收入覆盖危改投入，政府配合在危陋房的土地税费、配套费减免等方面给予政策支持。随着危改的逐步推开，天津逐渐形成了一定的房地产市场，住房建设也带动了建材和施工等工程企业的发展；百姓住进新房，促进了装修、装饰、家具等商品的销售，房地产这条长长的产业链带动了天津整体经济的复苏。

成片危陋平房改造取得的成功不仅体现在改善了居民的居住条件，改变了城市风貌和功能，拉动了城市经济发展，更重要的是城市风气的转变和思想观念的解放、进步。中华人民共和国成立前，天津是个对外开放的城市，引领着中国北方的时尚潮流，但中华人民共和国成立到“文化大革命”，天津又成为一个闭塞固化的城市。有人说，天津有两种天津人，一种是开放的天津人，一种是落后的天津人。小富即安、不思进取、搞关系、讲义气，这是典型的天津人的特征，冯骥才称之为码头文化。但是，当城市发展遇到危机，遭人嘲弄时，天津人也会爆发出巨大的力量。成片危陋平房改造，推倒的不仅仅是老旧衰败的危陋平房，也打破了旧有的抱残守缺、小富即安的陈旧观念，打破了计划经济、靠单位分房的心理惯性。到危改后期，国家实行深化住房制度改革方案，不再进行实物分房了，此时天津已提前做好了准备。

为了推动危改和房地产发展，当时的市委书记总是反复讲中国和美国老太太买房的故事：美国老太太按揭买房，多年后去世，还清了按揭，自己享用了一辈子的好房子；中国老太太住着很差的房子，一辈子省吃俭用，行将入土时用攒的钱买了自己的房子，但还没来得及好好享受就去世了。市委书记用这个生动的例子告诉大家要对未来有预期，要会使用未来的钱。这不仅是对干部和群众说的，也是说给整个天津市听的。市委书记之前长期从事工业行业，在国有企业做领导，虽说一直都是在做计划经济的事，但他深刻地认识到了市场经济的规律。通过危房改造这项触动居民切身利益和心灵的运动，老天津人的观念又一次开始改变。虽然大家仍对城市怀有抱怨，但干事创业的劲头倍增。

18. 中心商务区 · 中心商业区

诸葛民

持续 6 年的大规模危改可以说是海河综合开发改造的前奏。在这期间，诸葛民做了几件与日后海河综合开发改造规划密切相关的事情。1994 年，他与市规划院的几位年轻人合作，完成了《城市中心商务区 CBD 规划研究》科研课题。课题深入研究总结了中心商务区的基本理论，比较分析了国内外主要城市 CBD 的发展演变、规划建设的规律及各自特色，最后针对天津的实际情况，提出了天津中心商务区规划选址方案，位于城市中心内环线以里的海河两岸，面积 1.8 平方千米，依托解放北路历史上的金融功能和小白楼传统商业商贸功能，同时在海河对岸，利用天津铁路南站等交通工业仓库用地作为新的发展空间。这个选址是从城市的历史脉络中延续和生长出来的，突出了海河作为城市轴线的自然环境特征，新旧结合，得到了各方认可。

当时，国内许多城市都在规划建设 CBD，尤其以上海浦东陆家嘴金融区为代表，绝大部分城市学习巴黎拉德方斯、伦敦道克兰的经验，离开老城中心蛙跳式（FROG-LEAPING）发展，规划建设新的中心商务区。巴黎拉德方斯的建设主要是因为历史城区保护要求禁止建高层建筑，所以跳出老城规划了一个全新的现代化的中心商务区。但诸葛民了解到，新建一个中心商务区需要巨大的投入，历时漫长。巴黎拉德方斯曾经经历了十几年的发展停滞期，伦敦道克兰也曾导致著名开发企业奥林匹亚和约克公司破产。

当时也有人提议在天津中心城区友谊路等地区规划选址新的 CBD，但经过反复研究，诸葛民团队认为不适宜。因为天津是“一条扁担挑两头”的结构，滨海新区要规划一个新的中心商务区，同时从区域的角度分析，北京是我国北方最大的商贸、金融和国际机构中心，天津没有必要也不具备能力在中心城区内再选址规划一个新的 CBD。当时的解放北路还是天津市委、市政府、人大、政协及主要委办局所在地，与中心商务区相邻，同时商务区选址与诸葛民 1991 年主持完成的《海河的定位和两岸规划》中确定的海河分段功能也非常吻合，这再次印证了先前对海河定位的深化方向是符合实际的，更坚定了诸葛民的信心。

19. 城市总体规划修编

同是在 1994 年，为了迎接新世纪的到来，天津市委、市政府决定启动国务院已批准的上版城市总体规划（1986—2000）的修编工作，编制新一轮跨世纪的天津城市总体规划（1994—2010）。当时诸葛民只有 31 岁，硕士毕业后参加工作 5 年，在领导的信任下他被委以重任，担任总体规划修编编制工作组组长。在规划圈子里有这样一个说法：要想成为一名真正的规划师，必须做过城市总体规划，只有做过总规，才能了解城市规划涉及的广度，也才能全面深入地了解一座城市。

诸葛民带领规划编制组的一群年轻人全身心地投入到工作中。大家首先达成一致，跨世纪的城市总体规划应该在理论和方法上具有先进性、创新性。同时学习借鉴联合国环境发展大会提出的可持续发展议程，在城市总体规划的指导思想上提出可持续发展：“既能满足当代人的需求，又不对后代人满足其需求的能力构成危害的发展。”局领导听到这个概念很新颖，觉得提法不错，但也吃不准，就发问道：“现在国内城市都在加快发展，我们提可持续发展合时宜吗？”直到 1997 年党的“十五大”明确将可持续发展作为我国经济社会建设的长期原则，大家才认识到这个概念的先进性。但当时也就仅此而已，包括这一概念在城市总体规划如何真正落实可持续发展的理念也没有成熟的办法，只是加大了对生态环境和环境保护的重视，当时在国内这已经算是走在前列了。

另一个比较领先的做法就是采用了计算机辅助规划编制。此前，城市总体规划图都是规划人员手工绘制，要定制特大图版，正式规划图要把地形图裱在图版上，用水彩渲染上色，晾干后再把图纸从图版上裁下来，保证图纸的平整。当时个人计算机发展较快，运算速度大幅提高，出现了 586 处理器以及较好的数字输入和输出设备，局域网、应用软件也多起来。诸葛民向院里申请购置总体规划计算机辅助系统，领导很支持，花了 17 万元购置了 5 台绘图性能较好的 MICRO 586 计算机，还有平板数字输入仪 TABLET 和大型彩色打印机，联上了局域网 WLAN，购买了 GIS 软件 AUTOMAP、OFFICE 办公软件 POWERPOINT 和动画软件 3D STATION，以及液晶屏和投影仪。那时投影仪的亮度实在太低了，只是把一块连接到计算机的透明液晶屏放到一个普通的薄膜投影仪上，即使用上所谓的增强屏幕，还是看不太清楚。

就是用这样的设备，他们第一次把总体规划的内容做成动画和 PPT 向各级领导包括市人大汇报，所有 30 多张总体规划图纸都是用计算机绘制和打印的，大大提高了表达效果和质量，这不仅在天津城市总体规划的编制历史上是第一次，在国内各城市中也是最早使用计算机辅助总体规划编制的。

成果是巨大的，但成功背后也是巨大的付出，采用新技术首先要学习新的软件、硬件，将大量的数据输入，在工作中还会遇到许多意想不到的问题需要马上解决，加班熬夜、通宵达旦是常有的事。有一次，为了赶图和制作汇报材料，诸葛民连续四天没有回家，想想那时候真是年轻！诸葛民的女儿是在 1994 年出生的，妻子怀孕时他顾不上陪伴照顾，女儿出生后见面的机会也不多。妻子有时带着蹒跚学步的女儿到办公室看望他，女儿坐在原本用来人工画图的大图板上牙牙学语，引来大家阵阵笑声。现在回忆起当时的岁月，诸葛民对妻子和女儿心怀愧疚。当时未能陪伴她们，即使为了事业义无反顾，但作为丈夫和父亲或许可以做得更周到些。

除了在理论和方法上的创新外，诸葛民带领修编小组在规划编制内容上也在考虑如何继承传统并有所创新，特别是把滨河和中央商务区等前期的研究成果纳入了正式的法定总体规划。1986 版天津城市总体规划提出“工业战略东移”和“一条扁担挑两头”的城市总体规划布局，符合世界城市发展的规律，也符合天津城市自身的特点和实际。本次总体规划修编将前述成果全部予以继承。考虑到“一条扁担挑两头”的说法虽然比较形象，但不够正式，用“双心轴向”作为学名。“双心”指中心城区和滨海新区核心区，“轴向”

指双城间由海河、京津塘高速公路、京山铁路等构成的城市发展轴。

本次规划首次提出了中心城市的概念，包括中心城区和滨海新区核心区，北到永定新河，西到防洪线，南到独流减河，东到海岸线，总面积3000多平方千米，是天津城市化的重点地区。规划从可持续发展理念出发，从上版规划确定的中心城区外环线500米绿化带的概念，扩展到3000平方千米的中心城市范围。同时，强化城市安全，由中心城市外围河道堤防和海岸堤防构成天津城市防洪圈。通过疏浚永定新河和独流减河两条河道，加高堤防，提高防洪排涝能力，减少海河下泄上游洪水的压力。这一举措非常重要！在天津城市总体规划获国务院批准后，国家海河流域治理委员会编制了新的海河流域防洪规划，加大了永定新河和独流减河的流量以承接海河转移出来的流量。海河干流行洪流量由每秒2000立方米大幅减少到每秒800立方米，海河两岸的堤防高度大幅下降，这样才为海河两岸综合开发改造提供了基本的条件。所以说，城市总体规划涉及面十分广泛，在许多方面对城市形态、城市设计有决定性影响。

20. 历史文化街区

本次总体规划明确提升了海河的定位。按照1991年海河规划的成果，将海河定义为城市的主轴线，是贯穿城市历史、经济、政治、文化的中心线，除具有保水、通航、泄洪等功能外，还是发展城市经济、展现天津自然和历史文化、发展旅游的重点。按照1994年《城市中心商务区规划研究》的成果，明确天津城市中心商务区位于海河两岸，由解放北路、小白楼和南站（六纬路）商务区构成，面积1.8平方千米，明确滨江道、和平路地区作为城市的中心商业区。同时，考虑到天津作为国家第二批历史文化名城，首次提出并划定了9片历史文化保护区和5片历史文化风貌区，包括：五大道、泰安道、解放北路、解放南路、中心花园、劝业场、赤峰道、一宫花园、鞍山道、古文化街、老城厢、估衣街和海河两岸等。

这些历史文化街区绝大部分位于海河两岸，总体规划都提出了保护要求。这样，通过提升海河定位，明确两岸功能和保护范围，海河两岸规划的战略方向就非常明确了。上版城市总规中海河两岸还有许多工业仓库用地，本轮规划大部分将其改为商业用地。下一步，海河两岸规划的关键是详细规划和高水平的设计。

第四部分
行万里路·看世界

21. 英国伦敦 · 泰晤士河

鲍尔

上午的论坛和研讨结束了，大家还意犹未尽。诸葛民终于抽出身来与鲍尔单独见面。两人边走边聊，来到了位于规划展览馆广场边的意大利餐厅。正值中午用餐时间，许多客人在露台阳伞的庇荫下，吹着微风，品尝着西式佳肴，很是惬意。为了便于交流，两人在屋内较安静的角落里落座。没等点菜，鲍尔就发问了："再次祝贺诸葛总精彩的演讲，天津的这次城市设计试点，包括展览，以市民的视角做规划设计，解决与市民生活息息相关的步行化、公园规划建设和居住问题，非常棒！这也是目前国际上的潮流和发展方向。抱歉，由于时间的关系，我想再问个问题。我想知道海河本身的规划是如何编制的，通过什么方式实现高品质的规划？"

诸葛民

"早餐的时候我向您介绍了海河定位的变化，这是一个前提。后来，我们修编了新一版跨世纪的城市总体规划，花了近 5 年的时间，将海河的定位及相关的内容都落实在法定规划中，为海河规划的编制奠定了扎实的法定规划基础。在这一过程中，我们也在不断进行着与海河相关的规划和城市设计，总体看，这一阶段的城市设计是探索性的。实际上，这段时间我们有机会密集地到国外考察学习，从 1993 到 1996 年四年间，我去了英国、澳大利亚、德国、日本四个国家和我国香港特区，考察参观了近 20 个城市，特别是世界上一些著名的滨河城市，收获很多，增长了见识，开阔了眼界。把过去在书本中学习的东西与实际的城市进行了比对，为高水平的海河规划奠定了理论基础。正好印证了中国传统的价值观：行千里路，读万卷书。"

在总体规划修编繁忙的四年内，诸葛民有机会到国外考察学习，也开始着手海河的规划和城市设计。1994 年中，为做好总规修编，诸葛民与一位年轻处长随局长和总工赴英国学习考察十余天，这是他第一次到西方发达国家。英国，是工业和城市化的发源地，也是现代城市规划的发源地，英方对接机构的规格较高，行程也安排得十分紧张。没有地陪，没有专职翻译，随行的翻译工作由诸葛民担任，他们要自己找考察线路，自己乘坐飞机、火车、地铁、出租车，准时到达约定的政府部门、大学和参观考察目的地，忙得没有片刻喘息的机会。随行的团员中还有一位回民，一闻到某些餐厅的味道就止步不前，更甭说用餐了，所以选择吃饭的地点和方式也成了问题。但那时候大家都过惯了苦日子，出国带着方便面、电热水壶，还有干粮，由于经费限制，大部分时间都住在比较小的旅馆里，吃饭就简单对付了。

团员们考察或者说亲眼看到了著名的泰晤士河，河边的威斯特敏斯特教堂和大笨钟、古城堡和伦敦塔桥；考察了伦敦金融城、圣保罗大教堂，以及正在建设中的道克兰金融区；他们到了泰晤士河南岸的国家大剧院，到了格林尼治，世界的分时线。在伦敦，他们拜访了英国环境部，这是当时负责城市规划事务的部门；拜访了其属下的泰晤士东部走廊规划中心，该机构一直在编制泰晤士东部走廊战略规划。听介绍说，随着船舶运输大型化，英国伦敦的主要港口也由内河港向海口转移，这点与天津很像。当时，为了防风暴潮，英国正在泰晤士河口修建一种新型的防潮闸，建造不惧气候变化的坚强城市，由英国高技派建筑大师诺曼·福斯特设计，其技术、工艺和美学成就，让人大开眼界、叹为观止。顺便说一句，柏林议会穹窿、伦敦金融城的瑞士再保险大厦（腌黄瓜）、苹果新总部等杰出建筑也都出自福斯特爵士之手。

考察团来到伦敦市议会，当时正值铁娘子萨彻尔夫人风光之时，推行新自由主义，去规则化（Deregulation），减少政府对经济和市场的干预。她于 1986 年撤销了包括伦敦市及周围 32 个城市在内的大伦敦议会，相当于一个区域政府。一名官员指着对面一栋建筑向我们介绍，这就是以前大伦敦议会的办公地点，现在是高档酒店。英国城市规划的行政体制和法律在不断变化中。十多年后，诸葛民再次拜访伦敦时，在泰晤士河畔与伦敦塔桥相距不远处，还是由福斯特爵士设计的、好像斜放在地面上的鸭蛋形的玻璃建筑在太阳的余晖下熠熠生辉。这就是恢复的大伦敦政府的新办公楼。恰在这时，远处升起一柱长长的浓烟，办公大楼的人们跑到阳台上观望。不知哪个城市着火了，看来城市大了，操心的问题必然越来越多。

英国是新城、卫星城的发祥地，考察团拜访了英国新城开发公司，该公司位于伦敦内城不起眼的一栋楼中。据接待人员介绍，根据，《新城法（New Towns Act）》，英国新城公司受政府委托开发一座新城，要负责从规划设计、融资、建造，包括建设期的运营管理全过程。待建设全部完成后，将新城再完整地移交给地方政府。在建设期，新城公司实际承担了地方政府的部分职能，这样能够大大提高行政效率。到 20 世纪末，英国的新城建设及后期移交的工作基本结束，中央政府决定解散新城公司。新城公司，这个在 20 世纪城市规划中最有影响的规划模式，已经结束了其历史使命。当时该公司正在结算中。

中国在改革开放后大城市的规划中，一般都会提出严格控制中心城市规模、在外围建设卫星城的构想，但从未成功过。时至今日，还有些领导会提出建设卫星城的想法，看来需要学习一些卫星城的基本知识。

诸葛民一行离开伦敦前的最后一站是拜访著名的伦敦经济学院（LSE: LONDON SCHOOL OF ECONOMICS）。伦敦经济学院位于市中心，半开放式的大学，古老的建筑上落着斑驳的树影，像是诉说着城市和学校的历史。他们来到经济地理系，一位教授接待了他们。这次英国考察主要是一位从该校毕业的天津老乡安排的，他是规划局一位处长的亲戚。教授给我们讲了伦敦城演变的历史，他说伦敦城就像一个洋葱，是一层一层发展起来的。洋葱这个词很形象。

另一个让诸葛民印象深刻的是教授关于城市中心各种功能之间关联性的描述：伦敦市中心历史上就有两个城，西边是皇家的西敏斯特，东边是商人云集的伦敦城，也就是今天 2.6 平方千米大小的伦敦金融城。商人要向皇室纳税进贡，纳多纳少经常会产生纠纷，需要上法院裁决，所以双城之间就出现了法院集中区。法院集中区催生了律师行的繁荣，还衍生出一系列为法官、律师制作服装假发的服务业等。由于法庭是矛盾纠纷最多的地方，所以它也是新闻八卦的信息源，因此，各家报社就在临近的舰队街（Fleet Street）上聚集起来。

诸葛民开始时一边记录一边翻译，有些手忙脚乱，无法全部听清楚教授的每一个词，只能凭感觉理解大意。但听到这儿，他感到茅塞顿开。这段时间他正在做天津中心商务区规划研究，教授讲到城市中各种功能的关联性让他清楚地认识到城市规划必须遵循城市的内在逻辑和发展规律，而不能主观臆断、自以为是。

教授最后讲到，城市转型非常困难，由于伦敦内城有 30 平方千米，是伦敦中心商务区的外延部分，在金融城核心区之外的其他区域受到历史名城保护的严格要求，且管理名目越来越多，没有新的发展空间。所以在 20 世纪 80 年代初英国政府成立道克兰开发公司：LPDC，决定利用道克兰曾经繁华的金丝雀码头区，规划建设新的伦敦中心商务区。美国著名的 SOM 设计公司完成了总体规划设计，加拿大非常有实力的奥林匹亚和约克开发公司开始建设道克兰 200 米的高标志建筑，当时也是欧洲最高的建筑。由于遭遇经济危机，道克兰开发建设陷入停滞，奥林匹亚和约克公司因该项目导致破产。十多年后，当诸葛民再次踏上道克兰码头的时候，经济已经有了很大恢复，市中心舰队街上大的报社印刷厂因为设备需要更新换代，老城区容纳不下，都搬迁到了道克兰。

有人把原因归结为道克兰与老城中心的轨道不通，所以发展不起来。1987 年道克兰轻轨 DLR 已建成通车，但到伦敦市中心还需要在中间换乘地铁，不太方便。因此，英国政府计划将市中心的朱比利线（Jubilee Line）直接延伸到金丝雀码头和狗岛，这项工程耗时十年。到 2010 年诸葛民第二次造访道克兰时，朱比利延伸线已通车。道克兰金融区全面复苏，充满了活力。教授讲完，大家略作讨论，随后院长正式会见了考察团。

会见后，在院长办公室旁边的餐厅，院长宴请了团员，据说这是学院最高的礼遇。点菜时，诸葛民不知道相应的菜名代表什么口味，只知道食材大概是 Beef 或 Fish 。同行的处长曾在英国接受过三个月的培训，略知英餐，正菜点了牛肉，结果证明他的选择是明智的。诸葛民替两位领导主菜都点了鱼，端上桌后才知道是像国内的生煎包一样，将银鳕鱼段包在面包中烤制而成，奶油味很重，三个人都吃不惯。但为表尊重，他们认真吃完，还夸奖厨师的手艺很棒！席间他们了解到院长曾是那位天津老乡的导师，院长说之前他都没有听说过天津这个城市，直到他招了这位学生，然后到中国、天津，才知道北京边上还有这么大一个城市。听院长这么说，大家都有些尴尬。考察团在经济学院走廊看到一张世界地图，上面标的天津依然是用中华人民共和国成立前天津英文名 TIENTSIN。

从学院出来，考察团按计划要去密尔顿·凯恩斯新城。它距离伦敦 70 千米，刚才陪同用餐的一位教授说他家恰巧就住在密尔顿 · 凯恩斯，他愿意陪同考察团一起走。走到不远的纽斯顿车站，是地铁与铁路换乘枢纽（EUSTON），他们坐上了去往凯恩斯的市郊火车。车上人很多，都应该是通勤的人。教授与我们告别，他走到车厢一端站着，从他平静和略显疲惫的脸上，诸葛民看到了英国中产阶级生活的状态。1 小时 20 分钟，到达密尔顿 · 凯恩斯，走出车站，已近黄昏。

这次英国考察历时十余天，行程主要集中在伦敦，并以伦敦为中心向外辐射。因此，作为考察来说时间上是比较充裕的，每个项目都可以看得比较详细，不像后来走马灯似的考察，一次去欧洲四五个国家，每个城市都只能待一两天，多数时间是在路上走马观花。来到英国，就不能不提卫星城，早在 20 世纪 40 年代，阿布克隆比在大伦敦规划中就提出在伦敦外围建设卫星城。诸葛民早在大学时代就学习了解过相关卫星城的内容，那些图纸，从第一代卫星城哈罗新城到第四代密尔顿·凯恩斯新城，可以说如数家珍。前几天，考察团抽空先去了哈罗新城和剑桥大学，乘车出发，一路下着小雨，车窗外满眼绿色。他这才体验到伦敦绿带的56千米宽度，而天津的绿化带规划宽度是500米，不足伦敦绿带的1/100，而且还没全部实施到位，这让诸葛民直观地认识到中国的城市生态环境与发达国家的差距。

路上车很多，许多坐在驾驶座上的都是白发老人，司机在交谈中说，目前独立生活的英国老人越来越多了。哈罗新城到了，新城中心的公共建筑和四周的住宅都是现代主义的方盒子建筑，看上去像北欧或东欧的城市。市中心的住宅有一些是二十多层的塔楼，整个城市包括建筑、广场，似乎没有什么吸引力。按照原规划，这些卫星城要吸纳从伦敦疏解出来的人口，但实际的结果是住在卫星城里的人绝大部分来自其他地方。由于卫星城规模小，缺少就业，实际上是睡城。每天早晚高峰有200万人通勤，给伦敦造成了更大的交通压力。

从哈罗新城继续北上，就到了令人神往的剑桥大学。剑桥城里铺着鹅卵石的小路，两旁都是酒吧、咖啡馆，里面有许多读书的学生。剑桥大学也被称为城市中的大学。剑桥大学有许多学院，经典建筑比比皆是，一派田园风光。剑河曲折蜿蜒穿梭而过，河上架设着20多座设计精巧、造型雅致独特的桥，剑桥之名由此而来。“康桥的灵性全在一条河上；康河，我敢说是世界上最秀丽的一条水。”徐志摩在《我所知道的康桥》中写道，“我的眼是康桥教我睁的，我的求知欲是康桥给我拨动的，我的自我意识是康桥给我胚胎的。”他毫不吝啬地表达对康桥的赞美。还有《再别康桥》中的“轻轻的我走了，正如我轻轻的来；我挥一挥衣袖，不带走一片云彩。”

第二天早上，蓝天白云。诸葛民自认为对密尔顿·凯恩斯的规划耳熟能详。这是英国第四代卫星城，人口规模较大，规划人口40万，以达成职住平衡，与传统邻里中心式布局把社区中心都放在邻里中心内部不同，这里的规划把社区中心、小学都放在社区的边缘上，以增加居民的自由选择，道路网为1千米×1千米的方格网，结合地势起伏，道路线形随之变化，是明确内部交通以小汽车为主的新城。但亲眼看到的景象还是出乎诸葛民的预料。出了新城规整的市中心，马上映入眼帘的是一片田园风光，规划保留了许多河流、水塘湿地，水鸟、鸭子在水中游戏。

最让诸葛民意想不到的是凯恩斯新城的绝大部分住宅都是英国典型的双拼毗邻式别墅住宅和独立住宅，红砖红瓦，充满家的温馨感觉，在城市中心周围有少量多层的公寓住宅。凯恩斯新城没有高层住宅，充满诗情画意般的景象让诸葛民理解了为什么伦敦人每天通勤 70 千米来凯恩斯居住，这里更像霍华德倡导的花园城市，比莱彻沃斯更田园。诸葛民特意买了一套密尔顿 · 凯恩斯的地图，想带回去认真研究。2009 年，诸葛民再次考察英国时，特意又来到密尔顿 · 凯恩斯，发现市中心新建成一个巨大拱形的购物中心，充满了活力，高新技术产业也发展得很好，成为英国一个新兴城市。

在英国十余天的考察马上就要结束了，在到达伦敦回国之前，考察团来到中国驻英大使馆文化参赞处拜访。文化参赞处位于市中心，离考察团住地不远，大家步行前往。路上经过著名的海德公园，远处集聚的人群在听演讲，伦敦依然保持着古老的传统。路上还看到有人骑着摩托，挡风镜下放着一张地图。随行介绍，这些人为了考取出租车执照，需要记忆伦敦的道路。伦敦标志的黑色出租车车厢内宽大敞亮，乘客可相对而坐，司机穿着整洁的黑色制服，彬彬有礼。单从对考取出租车司机的执照上就可以看出伦敦对出租车管理的严格程度。

出租车不仅对一个城市公共交通至关重要，而且对城市的形象有重要影响。一般来说，一个城市的出租车司机有几万人，年复一年、日积月累，他们接待的旅客量非常大，对城市的口碑也影响巨大，国内外城市都会注重对出租车司机群体的管理。国内城市中上海的做法是不错的，成立了几大出租车公司，统筹规范管理。而天津出租车公司有好几十家小公司，一直缺少有效的管理。

中国驻英大使馆文化参赞处的建筑是一栋三层楼，经典的古典建筑。进得门来，第一次不用翻译了，大家都说中文，虽然房间内的装饰还是英式的，但诸葛民仿佛有回家的感觉。午餐时间，参赞知道大家出来十多天了，一定很想念家乡，就请大家吃食堂。食堂在地下室，刚走下楼梯，就闻到一股熟悉的味道，进去发现完全是一个中餐食堂，用的餐具、调料，包括厨师都是国内来的，餐桌上有西红柿炒鸡蛋，锅里还有土豆炖肉，诸葛民感觉这就是在大英帝国里的一个“红色革命根据地”。看来，要转变人的观念和习惯是极其困难的，但城市规划就是要转变人们的观念。

结束了参赞处的拜访，这次考察的商务活动就圆满结束了。晚上诸葛民和年轻处长想看电影，就去向领导请假。由于这次考察活动很成功，领导对大家的表现也很满意，就愉快地批准了，还特别叮嘱：注意安全，早点回来。诸葛民俩人来到了特拉法加广场，这是伦敦文化娱乐的中心，电影院很多。

恰逢周末，电影院排起了长队。诸葛民二人选了由哈里森·福特主演的一部悬疑片《逃亡 FUGITIVE》。电影院的观影厅很大，是容纳近千人的大厅，让诸葛民惊讶的是：观众都穿着礼服，男引导员都穿着精致的制服，袖口、衣领都镶着银色的花边，像是宫廷的侍从。电影的影像和音效让人震撼，不愧是一场视听盛宴。诸葛民扪心自问，国内何时能有这样的电影院，这样的影片和这样的观众。

鲍尔

“我是不是讲太多跑题了？”“没关系，很有意思。”鲍尔正听得津津有味，“这些事情都与海河规划有关系。我听明白了，如果没有城市总体规划对海河定位的提升，没有通过设立城市防洪圈减少海河的泄洪量，就没有今天的海河综合开发改造。如果没有划定天津市中心商务区和十四片历史文化街区，海河边今天就不会有这么多历史建筑，海河的特色就不会这么鲜明。还有，如果没有去英国考察，您对滨河城市的认识就不会这么深刻。您讲的内容太丰富了，我记不全，我可以录音吗能？”在得到诸葛民同意后，鲍尔打开了手机的录音功能。

“您总结得很准确，这也说明，像海河两岸这样重要、这种尺度的城市设计，需要考虑方方面面的问题。上位规划的定位是前提，必须提前谋划做好铺垫；相关专项规划和专业部门的意见需要综合统筹，各方需要达成一致，最后还需要精妙、高超的技巧和手法。以上这些都需要全面的知识和长时间工作的沉淀。”虽然当时诸葛民主持的城市总体规划修编工作繁忙，但他还是挤时间出国学习考察。在去英国后连着几年，诸葛民又去了不少国家和城市考察学习，虽然每次时间都不长。诸葛民研究过的《比较城市规划理论和方法》开始发挥作用。

22. 中国香港·维多利亚海湾

诸葛民

1994 年 8 月，为了做好老城厢改造规划，诸葛民随市规划院工作组在香港待了半个月，配合香港老牌的设计公司巴马丹拿完成了老城厢改造规划，实际上就是一个城市设计。除了参加设计工作之外，巴马丹拿设计公司的董事、香港建筑师协会年轻的副会长林云峰陪同他们参观了香港许多地方。香港的中环是中心商务区，其高架步行系统非常著名，在工作日的午餐时间，林先生带诸葛民一行来到中环，在高架步行平台上，人群熙熙攘攘，都是在写字楼工作的人们中午来此用餐。步行道两侧布满了各式各样的餐厅，餐厅内人声鼎沸，人们边吃边聊。

看来 CBD 的午餐时间是非常重要的社交时间啊！诸葛民想起国内大多数单位自己办食堂，又想到伦敦中国大使馆参赞处的食堂，他哑然失笑。中国要真正建成 CBD，就不可能还是各单位各自办食堂，需要更加开放、高效、国际化，但改起来实在不容易啊。目前在滨海新区开发区的 MSD、于家堡金融区的起步区已经出现了一点苗头，年轻的白领们似乎中午更愿意走出办公楼去用餐、聚会、见朋友。

具有 80 年历史的巴马丹拿设计公司最初办公地点就在中环，中环的快速发展使得房屋租金不断攀升，像巴马丹拿这样的设计机构也负担不起，于是公司就迁到了东面的商务区湾仔，这里汇集了大量的设计公司，以及律师、会计事务所等中介机构。香港的 CBD 让诸葛民对天津 CBD 的规划设计有了进一步的认识。除了参观香港科技大学、香港科学馆等建筑外，林先生也带着大家参观了香港的住宅。诸葛民上一次来香港是在 1991 年，拜访了香港房屋署，参观了大楼内的住房展览，还参观了新界的新市镇，位于轨道站点旁，底层是一个大型商场，商场上面全是高层住宅，当时没机会参观住宅内部。

这次林先生带领参观了位于市区的两个项目。一个项目是高档私有住房，其布局是几栋高层住宅围合一个中心花园，绿化环境做得非常好，住宅室内设计装修标准都比较高，每层 2 ~ 4 户，每隔四层还有空中花园。另一个项目是公共住房，是典型的小户型平面，每层十多户，设内走廊。诸葛民进入走廊，发现因天气炎热及出于通风的需要，每家都关着铁栅门、开着内门，私密性很差，看上去也很简陋，但就是这种住房，香港人也要排很多年队才能申请上。

诸葛民觉得香港的住房不论是高档商品住宅设计，还是公共住房的建筑设计都不是很成功，与西方国家相比标准太低，有很多深层因素在作祟。香港在 20 世纪 50 年代就遇到房荒，有许多铁皮房和棚屋，居住条件十分恶劣，经常发生火灾等事故。为此，港英政府开始了大规模的住房建设，拆除老的铁皮房和棚户，成片建设了 7 层没有电梯的多层条式住宅，设置公共外廊，每户只有厨房，没有独立卫生间，一层几十户共用一个卫生间。这种住房标准很低，在一段时间内作为过渡性住房，到 20 世纪 80 年代就面临着不得不拆除的命运。开发时为了经济上可行、有利可图，拆除后新建的都是密度很大的高层住宅群，走在里面让人心生畏惧。

诸葛民联想到天津当年的成片危陋平房改造时建成的大部分都是 6 层无电梯多层住宅，密度很大，不知道将来怎么办？如果像香港一样拆了建高层也是很恐怖的，有人对天津成片危陋平房改造的担心是有一定道理的。今天来看，天津从 20 世纪 80 年代开始建成的大量成片多层住宅区会成为下一步城市更新的重点和难点。走在香港中环、湾仔等老街区，沿街多是十几层的大板式或 U 形围合的老旧住宅楼，每套房都小得可怜，每家的窗户里伸出晒衣杆，五颜六色的衣服挂在空中，十分凌乱，一个住房单元就像一个鸽子笼。诸葛民认为香港的住房政策和住宅规划设计绝不是中国大部分城市学习的榜样。

23. 澳大利亚墨尔本 · 亚拉河

诸葛民

随着天津国际交往的增加，海河规划也成为一个合作交流项目。天津与澳大利亚的墨尔本是友好城市，1994 年，澳方派出专家协助天津历史建筑保护规划，诸葛民推荐了海河东岸南站（六纬路）商务区作为中澳双方合作的项目。澳方派出了历史建筑保护专家专程来位于南站地区的天津卷烟厂进行调研，编制保护和再利用规划方案。卷烟厂的厂房设计和建造品质都很高，保存得也很好，室内混凝土牛腿结构既增加了厂房的跨度，又体现了工业建筑的力量美学。卷烟厂在天津算是第一个工业遗产保护和改造项目，可惜后来该厂房还是随着南站（六纬路）商务区的建设被拆除了，原址上新建了嘉里中心。诸葛民无奈地想，或许这就是一种两难的选择。

1994 年 8 月，天津市组织代表团回访澳大利亚，主题是学习澳大利亚在滨水城市开发改造的经验。代表团一行六人访问了悉尼、墨尔本、堪培拉、布里斯班等城市及黄金海岸。代表团为了方便考察并节省时间，租了一辆旅行车沿东部海岸一路向北，沿途体验了澳大利亚国土的辽阔。中途休息经常会选择沿途小镇，小镇中间一般会有一条商业街，整洁漂亮，街道两侧是一到二层的商业建筑，有超市、日用品商店、室内装饰、园艺花卉等，他们还会经常看到一家装修非常传统、考究的中餐馆。那个中餐馆室内装修非常中国化，不是新中式，而是一种色彩斑斓的老中式，一对华人夫妇经营着餐馆，既当老板，又兼厨师、服务员，与客人的关系非常融洽，客人看起来也都是小镇上的熟人。餐馆生意很好，还要预约。与一般中餐厅的喧杂不同，这里给人一种繁华阅尽的平静之感。诸葛民感慨中国人无处不在，在白人社区中能有尊严地生存，经营着自己的生意，给当地人提供了一种不同的美食和文化体验。

悉尼是澳大利亚最大的城市，也是一座充满现代感的城市。位于悉尼湾边上的是悉尼歌剧院和悉尼大桥，成为其独特的标志，游人络绎不绝。悉尼湾中心无数的游艇、帆船以及悉尼大桥跨年的烟火表演让这座城市充满了活力。悉尼的中心商务区 CBD 位于海湾南端，在悉尼大桥和悉尼歌剧院之间，1 平方千米范围的中央有一组摩天大楼。为了保证每栋办公大楼都有良好的视野，规划的高层建筑群与海湾垂直，这个做法非常棒，既保证了内部观景，也保护了外部景观，避免滨水地带形成一排高墙。CBD 西侧是著名的达令港（Darling Harbor），过去是悉尼港的码头仓库区。随着港口的外迁，悉尼对这里进行了重点规划改造，建成为文化娱乐中心。包括海洋博物馆等公共建筑和一座中国花园，将保留下来的仓库改造为酒店、商场和餐饮设施，成为悉尼市中心热闹的滨水地带，与旁边安静的金融区形成对照。

诸葛民一行来到悉尼市政府拜访，在顶楼看到了悉尼市中心的规划模型，1 ∶ 500 的比例，软木制作，非常雅致。从中心商务区的规划介绍中他们了解到，为了鼓励公共交通，限制小汽车交通，悉尼严格控制停车位数量，总停车位控制在 5000 个以内，新建项目不允许多配建停车位。诸葛民感到奇怪，这与一般中国城市的做法不同，特别是国内的停车标准，无论什么区位都采用下限控制。悉尼在中心商务区控制停车位的政策非常有效，走在商务区的路上感到非常安全、安静，没有一般城市 CBD 机动交通的喧闹，比如香港中环，即使走在高架步道上，也可以看到脚下机动车道上穿梭如织的车流，听到马达、车轮胎与道路摩擦发出的巨大噪声。

还有一件有趣的事，在悉尼 CBD 南端，诸葛民一行看到了高架单轨电车不时地从楼中穿过。虽说高架单轨也算是公共交通，但根据厢体推测其承载的交通量应该有限。后来听说为了改善城市景观，悉尼的高架单轨已经被拆除了。当年海河综合开发改造时，有几个投资商想在海河边修建类似的高架单轨，考虑到对海河景观的影响，规划最终没有通过。现在国内一些生产厂家为了生产和推广这种所谓新的公共交通形式，会片面夸大高架电车的作用，因此决策时应该慎重。

悉尼市副市长曾筱龙接见了诸葛民一行，曾市长是位华人，并且是一位建筑师，达令港的中国花园就是他推动建造的。他通过个人的努力奋斗，竞选为副市长，成为悉尼建市150年来第一位华人副市长。中午，曾市长请吃饭，饭前大家捧着橄榄球大小的龙虾王照相，体型硕大的海洋生物展示了澳大利亚丰富的自然资源。席间，曾市长不停地用两把勺子给大家夹菜，十分娴熟。他调侃道："我知道你们从来不会打工，我在海外唐人街餐馆打过工，在餐馆洗的盘子多到把一辈子用的盘子都洗完了。"曾市长的普通话讲得非常好，常说些俚语，让人心生敬佩。

在悉尼，他们还参观考察了一些公共建筑，参观了为2000年悉尼奥运会新建的奥林匹克体育中心。体育中心是由澳大利亚著名的建筑师事务所COX设计的，该公司还设计了达令港的海洋馆。COX的设计利用钢的延伸性，采用拉杆技术，使得大跨度的游泳馆的结构变得十分轻盈。跳水池旁的游戏水池采用曲线造型，透过大面积的玻璃，与室外环境融为一体。随后他们又到了悉尼市中心COX事务所拜访了考克斯（COX）先生，他当时应该不到60岁，精力充沛，身边有一位来自北京林学院在澳留学后进入COX事务所工作的非常英俊的得力助手，为考察团做翻译。20年后，诸葛民在滨海新区主持国家海洋博物馆设计方案国际征集时，在两轮竞选中，考克斯先生的方案都获得第一名。

当诸葛民再见考克斯先生时，他已70多岁，看上去从中年人变成了一位慈祥的老人，但创造力依然旺盛。说起当年的拜访，考克斯先生已无印象，当年他身边的那位英俊的中国助手，也不知何处。今天，考克斯先生设计的国家海洋博物馆已在滨海新区中新天津生态城的海湾边竣工。当时，考克斯先生介绍他的设计构思时说过，海洋馆是个停靠在海边的港湾，从建筑自然转变的造型中，可以看到动物从海洋爬上陆地、人与海洋进一步交流等多种象征意义。

不论作为规划师还是建筑师，考察时除了要关注城市中心、公共建筑之外，也一定要看一个国家、一座城市的住房，要去了解住房政策、社区规划和住宅的建筑设计。住宅是量大面广的建筑，也反映了一个国家的建筑的整体水平和人民的居住水准。诸葛民一行在悉尼郊区参观了一栋独立住宅，距市中心30千米，住宅面积200多平方米，价格17万澳元，合80多万元人民币，这个价格当时在国内算高价，折合每平方米4000元人民币，但在澳大利亚的许多新华人都买得起。

随着郊区化的进一步发展，悉尼市中心也出现了空心化现象，但澳大利亚没有像美国那样严重的种族问题，所以市中心没有严重衰败。近年来，许多年轻人喜欢住在繁华的市中心，特别是来自日本和东南亚的有钱人，为迎合他们的需求，市中心建设了不少钢和玻璃为主要建筑材料的高层公寓住宅，出现了贵族化（gentrification）现象。

离开悉尼，下一站是澳大利亚首都堪培拉，堪培拉的规划在世界城市规划史上有着特殊的地位。20 世纪伊始，为解决悉尼和墨尔本都努力争取作为联邦首都的两难局面，1911 年，澳大利亚当局决定另辟第三地，最后选择风景大美的堪培拉作为新首都。1912 年举办了新首都规划的竞赛，36 岁的芝加哥风景建筑师格里芬脱颖而出。他提出的多个六边形构图的设计方案与堪培拉起伏的丘陵地形非常吻合。堪培拉城从 1913 年开始建设，其间由于一战一度停工。到 1927 年，堪培拉正式成为澳大利亚新首都时，已耗时 14 年。

新首都的名字也是经过长时间商讨才最后确定的，最终选择了当地居民的传统名称——堪培拉，意思是“汇合之地”“聚合的地方”。堪培拉距悉尼 230 多千米，考察团乘车到达堪培拉时已是黄昏时分，当地陪同人员赶紧带他们去一家中餐馆吃饭。堪培拉只有十几万人，过了晚上 8 点，所有旅馆、商店、餐厅，包括娱乐场所都关门。这家中餐厅是前中国女篮队员宋晓波开的，生意蛮不错的。车走在回酒店的路上时，已是人迹稀少，周围一片寂静。

第二天早上，考察团赶往国会山，首先参观国会，旁听了一会儿议员的辩论发言，大家一头雾水，就起身来到国会大厦外边。国会大厅位于城市中轴线的中点国会山上，是一个覆土建筑，建筑与山体融为一体。站在柱廊下向北远眺，能看到大片的绿地和湛蓝的格里芬湖，以及壮观的能喷射高达 137 米水柱的“纪念库克船长喷泉”，那景色美轮美奂，让人觉得天堂的景象也不过如此吧。

因为澳大利亚位于赤道以南，大部分建筑是坐南朝北。堪培拉没有工业，总面积有 2400 平方千米，与天津滨海新区相当，但人口只有滨海的 1/7；与深圳市域面积差不多，但人口只是深圳常住人口的 1/25。格里芬的堪培拉规划早在 20 世纪初还是以放射、对称和轴线为特征的巴洛克式古典主义的规划，但他创造性地保留和设计了大片绿化、水面，所以有人称他的规划是生态巴洛克或景观巴洛克。生活在这座天堂般的田园城市中，还有发展的必要吗？

下一站是黄金海岸，是传说中的度假天堂。堪培拉距黄金海岸 1000 千米，为了体验澳大利亚人自驾游的感觉，考察团一行乘旅行车沿著名的沿海高速北上。旅行车的座位原本是比较宽敞的，但由于一位建委领导和建筑设计院的总工二人身材比较魁梧，在车上难以舒展身体，历经逾 10 小时的车程后感觉十分疲惫。好在窗外的美景转移了大家的注意力，每逢中途休息，大家下车徜徉在优美的景色中，沐浴着清新的海风，疲惫消散大半。看到沿海高速上熙熙攘攘的车流，诸葛民预感到，随着生活水平的提高和私家车的普及，国内的自驾游一定会逐步兴起，这也是拉动内需和发展经济的重要方向，城市规划需要及早应对。

早上 5 点从堪培拉出发，到达黄金海岸已是黄昏。预订的度假公寓的厨房中，各种厨具、餐具一应俱全。公寓就紧邻着沙滩，诸葛民站在阳台向外看，在落日余晖下，海水波光粼粼，一派祥和景象。没一会儿，旅途的疲惫和睡意袭来，诸葛民很快进入梦乡。第二天一早，海浪声把大家吵醒，碧海蓝天，几个年轻人在沙滩上狂奔，好不惬意。早餐过后，第一个项目是乘坐旅游直升机。机舱内噪声轰鸣，大家都戴着耳机，通过麦克风交流也要很大声。

黄金海岸有十几个连续排列的优质沙滩，海岸线绵延 42 千米，一眼望不到尽头。但在空中可以看到城市的全貌，近处的沙滩后边就是一排长城般的度假公寓楼群。再往后，水网发达，飞机从低空掠过，眼前是成片的度假别墅区。与一般的别墅区不同，这里的别墅区都是由水路和陆路组成的双路网格局，与我国苏州古城的双棋盘格局相似，苏州建筑是粉墙黛瓦的传统合院民居，而这里全都是二、三层的独立别墅，每户一辆游艇，是滨海地区典型的度假别墅，即所谓的第二居所。

诸葛民联想到 2001 年前后，北京大学经济学家厉以宁教授提议为拉动内需，鼓励国民购买第二居所，小康家庭可以有第二套住房，结果遭到很多“正义”人士的炮轰，说中国还有很多住房困难的家庭，鼓励第二居所是背离了为人民服务的宗旨。真是鸡同鸭讲，让人无言以对。按照这些人的逻辑，大家都很穷，就不能有人先富起来。实际上，随着中国人生活水平的提高和机动能力的增强，拥有第二居所是必然的，是不以某些人的意志为转移的，城市规划需要未雨绸缪，当然也必须同时满足大众的基本住房需求。

黄金海岸是著名的旅游胜地，除了沙滩、度假游艇、别墅、公寓、酒店，还建设了大量的主题公园等旅游设施。诸葛民一行参观了环球主题公园和水世界主题公园，看到 people mountain people sea（网络用语，人山人海字面直译）的人群。当时就在想，天津什么时候能够发展起这样的旅游产业。他们回国后不久，国内的三大城市北京、上海、天津，就开始酝酿引进主题公园了。可能是当时国门打开了，许多人在国外看到了主题公园的惊险刺激的游乐项目，认为中国人也应该体验和享受一下；另一种可能是外国人认为中国有十几亿人口，在中国建设主题公园一本万利。美国迪士尼1955年开业，距今已经40年了，依然人满为患，不知赚了多少钱。

当时北京打算引进环球主题公园，结合首都钢铁公司搬迁和转型，选址在石景山；精明的上海人认准了迪士尼，认为迪士尼内容以动画为主，涉及意识形态内容少，容易操作；天津较晚加入这一阵营，选择了派拉蒙主题公园。虽然派拉蒙影业公司是与环球、迪士尼齐名的电影公司，但在主题公园上与环球和迪士尼相差甚远。派拉蒙主题公园大部分位于美国的二线城市，一般通过收购传统的游乐园并在其基础上增加少量与电影有关的游乐项目，如《古墓丽影》。

由于主题公园要至少占地1平方千米，加上周边的配套设施，投资大，风险也大，所以国家是严格控制的，要经国家发改委、国务院审批，而且还不能用国内资金，等等，申请下来非常困难，所以各地都组成强大阵容参与编制可行性报告等前期工作。天津派拉蒙主题公园最初选址在杨村，以天津电视台拍摄电视连续剧的影视基地为依托，聘请了著名的设计公司和世界银行副行长，拟投资的沙特王子等，阵容十分强大。鉴于意识形态问题，在内容设计上也考虑增加《西游记》等中国元素。

诸葛民也参加了派拉蒙主题公园的前期工作。2002年他作为代表团成员赴美国和日本与派拉蒙公司进行谈判，并考察主题公园，这是他第一次出访美国和日本。但由于时间非常紧张，无暇他顾，只短暂去了美国洛杉矶、辛辛那提和日本东京三个城市。在洛杉矶入住比弗利山庄酒店，第二天到派拉蒙公司参加会谈，并参观了环球等影业公司及拍摄场，体验了环球主题公园和迪士尼主题公园；随后又奔赴位于美国中部俄亥俄州的辛辛那提市，考察了位于那里的派拉蒙主题公园。正值周末，游客很多，据说都是周围城市的居民开车而来。

为了感受亚洲人在亚洲文化背景下对美国主题公园的态度，代表团回国前从美国转道日本，参观了东京环球主题公园和迪士尼主题公园，发现日本人对美国的主题公园喜爱有加，公园里人山人海。主题公园周边建设了大量旅游酒店设施，他们说主题公园的内容是一天玩不完的，需要住下来连续玩几天。公园里每个项目都要排很长的队，但诸葛民一行是VIP，不用排队，加上年轻体能好，他们几乎玩遍了主要的项目，过足了游乐场的瘾。最后，为了了解日本本地主题公园的情况，还到了兵库县的阿童木公园，结果却令人失望。日本的经济和文化很发达，特别是动漫产业，但阿童木主题公园门可罗雀。

时间一年一年过去了，2005 年香港迪士尼开业，2015 年上海迪士尼开业。上海是当年参加竞争的城市中第一个开业的。北京环球主题公园几经周折，最后落户通州，已启动建设多年，目前还没有开业的准确消息。而天津呢？此后便没了下文。后来，派拉蒙主题公园选址变更到航母主题公园旁，靠近海边。尽管 2006 年滨海新区被纳入国家发展战略，迎来了近 10 年的高速发展，派拉蒙主题公园项目也还是没有消息。在中新天津生态城，原滨海旅游区建设了方特欢乐世界、水立方等中资自主开发建设的小型主题公园，吸引了一定的游客。这也许预示着天津和滨海新区下决心搞出拥有完全自主知识产权和品牌的、具有世界先进水平的、中国人自己的主题公园，成为拉动产业发展、促进天津和滨海新区旅游业升级、服务京津冀 1 亿人口的引擎，建设天津渤海湾的黄金海岸。

结束了黄金海岸的行程，考察组驱车 1 小时来到了澳大利亚第三大城市布里斯班，这里是昆士兰州的首府，人口 210 万人，面积约 1200 平方千米，1988 年曾举办过世界博览会。考察团之所以拜访这座城市，因为这里有一条河流——布里斯班河，它穿城曲折而过，就像海河。布里斯班河的河面较宽，有点像伦敦的泰晤士河，河上有许多城市高架路和桥梁通过，尺度比较大。布里斯班市的中心商务区位于河边，诸葛民发现河岸上停靠游船的浮码头设计很独特，他拍下照片，想着日后在海河游船码头的设计上可以借鉴。据规划部门介绍，为了防止中心商务区大楼的阴影长时间对河面产生影响，他们规定高层建筑的阴影投射在河上的面积不能超过总面积的 1/3，这是一个有效的办法。

赴澳大利亚考察学习的最后一站是墨尔本，诸葛民在此前十几天的行程中一直想去墨尔本，为了节约时间，他们从布里斯班直飞墨尔本。诸葛民坐在飞机上，前几天考察的经历像电影般回放：悉尼市中心的滨水区特色十足、充满活力；黄金海岸 42 千米长沙滩，令人流连忘返；布里斯班河水势汹涌，坐在大型游船上品尝海鲜令人惬意……尽管前几个城市给人的印象如此美好，他还是一直盼望着早日到达墨尔本，因为它才是这次滨河主题考察最重要的城市。

墨尔本是澳大利亚第二大城市，是著名的工商业城市，曾经是澳大利亚的首都，也是世界上知名的宜居城市。雅拉河起源于东部山区，全长 240 多千米，在墨尔本市中心穿城而过，最终流入太平洋的霍布斯海湾。雅拉河在市中心的宽度为100米左右，与天津市中心段的海河十分相近；雅拉河也被称为墨尔本的母亲河，这一点与天津海河相同。但雅拉河从东向西流，传统的城市中心位于河的北岸，这些特性与海河恰好相反。同时雅拉河与海河还有许多相似之处。比如：在历史上与海河一样，雅拉河两岸都是码头、仓库和工厂，火车站及货场都位于河边。由于河流的分隔，墨尔本市在雅拉河北岸发展较好，且背离雅拉河，而河的南岸一直没怎么发展。虽然雅拉河只有百米，但对城市的阻隔效应还是很明显的。

墨尔本城市历史悠久，1849年建城，一直是澳大利亚的工商业和文化城市。1901到1927年墨尔本是澳大利亚联邦的首都。由于大洋洲独特的地理位置，20世纪两次世界大战对澳大利亚的影响都不大。1956年墨尔本举办了夏季奥运会，是南半球第一个举办奥运会的城市。墨尔本一直十分重视城市绿化环境和宜居城市建设，1971年，《墨尔本大都市区发展规划》正式出台，以适应二战后经济复苏和城市规模进一步扩张的要求。由于城市规模的进一步扩张和郊区的不断扩大，墨尔本的城市中心因交通拥堵、环境恶化而迅速衰落。随着机动车数量的急剧增长，19世纪由霍都规划的窄路密网式的霍都网（Hoddle Grid）无法承载过大的交通压力，导致机动车与行人之间的冲突越来越严重，墨尔本城市的吸引力急剧下降。

针对上述问题，1982年维多利亚规划部门制定的墨尔本城市设计法令中做出了一项日后对墨尔本产生重要影响的决策：墨尔本需要继续保持并不断发展适宜步行的、安全的、具有商业与休闲娱乐吸引力的城市环境。法令从城市形象、建筑与土地开发、交通与步行三个方面提出了具体的设计指导原则。据此法令，墨尔本于1985年制定了《人性化的街道——墨尔本市中心活力区步行策略》，即便以现在的视角看，这份30余年前的城市设计法令仍极有远见地奠定了墨尔本人性化城市的基础：一是实现了将街道设计的重点从机动车转向人，并激发步行对提升城市文化、商业活力方面的重要价值，街道因此成为城市设计的关键对象和重要内容；二是促进了城市建设的重心从最初空间开发转移到创造更舒适、更宜人的城市环境，塑造高品质的城市环境，从而吸引和提升步行能力，这是已取得广泛共识的城市可持续发展的正确方向；三是首次从人性化的视角研究制定了城市街道设计策略。

诸葛民一行考察墨尔本时已是法令颁布的10年后，墨尔本在打造人性化城市方面成绩显著，他也是近几年才了解到。在他们到访的那一年，墨尔本市首次同国际著名的人性化城市设计大师——丹麦的扬·盖尔合作制定了《墨尔本人性化空间规划》。继街道之后，墨尔本对城市公园、城市广场、公共绿地、公共建筑等展开了又一轮人性化改造。此后，墨尔本再次与扬·盖尔合作，于 2004 年编制了新一轮人性化空间规划，从此形成惯例，每 10 年编制一次。2015 年墨尔本编制了第三轮人性化空间规划，将研究范围扩大到南岸区和滨水新城，突破了传统的城市中心区 2 个多平方千米的霍都网。

鲍尔

“鲍尔先生，我们国内城市，包括天津，近两年才开始关注人性化城市和步行系统，比澳大利亚晚了 30 年，是不是说明我们的城市设计比发达国家的整体水平晚了 30 年？”鲍尔还没有完全从诸葛民前面的长篇讲述中缓过神来，“这是一个有意思的问题。不过我不这么认为，中国在 30 年前已经很开放了，与世界各国的交流非常多。按照比较研究理论，城市间的影响有历时性和共时性，历时性有先后影响之分，但共时性是相互影响。现在中国已经站在世界的舞台上，我认为不会存在所谓 30 年的绝对差距，可能要从侧重城市扩张转向提高城市品质，这个转变的确非常困难。如果能够实现这个转型，包括观念上和机制上，我认为中国的城市设计就可以与发达国家齐头并进。诸葛总，在比较城市规划研究方面，您是专家，比我有发言权。”

对于鲍尔的看法诸葛民是非常认同的，这也是他最近一直在思考的问题，要实现中华民族的伟大复兴，在 2050 年达到中等发达国家水平，中国的城市规划和设计就必须先行一步，找到正确的方向，这点至关重要。虽然心里这么想，但诸葛民嘴上却谦虚地说：“鲍尔先生，您客气了，我们与发达国家在城市设计上的差距还是非常巨大的，需要更多的实践去学习提高。”

诸葛民

在考察墨尔本时，诸葛民一心想从雅拉河规划中学到能为海河规划直接所用的东西。当时墨尔本的同行除介绍雅拉河的规划外，也介绍了与扬·盖尔的合作与人性化城市相关的内容，但他当时并不了解扬·盖尔，竟没留下任何记忆。倒是墨尔本同行提出的要改变城市背向河流发展的趋势、城市要面向雅拉河发展的战略构想，引起了诸葛民的共鸣，印象极其深刻，他当时脱口而出：我们海河规划也是一样！

墨尔本也面临着与天津相同的另一个问题，如何跨越雅拉河的阻碍，促进南岸区发展。诸葛民一行到雅拉河南岸考察时，发现墨尔本把唯一的赌场——皇冠赌场布置在南岸，还有水族馆等公共建筑。他们去的时候，皇冠赌场基本建设完成，但尚未开业，所以诸葛民一行没能亲眼看到皇冠赌场设在雅拉河畔的 8 根巨大喷火柱每晚的喷火表演，他猜测可能是学习了拉斯维加斯赌场，以吸引大量游人观看、提高亚拉河畔的人气。

后来在海河综合开发改造规划中，规划团队中的连院长，因他曾在墨尔本培训过三个月，体验过喷火柱的刺激和活力，多次建议在海河岸边中心广场上也建造这样的喷火装置，后来由于多方原因没有实现，最终安装了一排喷火柱形状的景观灯。如果有一天真能装设喷火装置，海河的活力和吸引力一定会进一步增强。

除了喷火表演外，雅拉河上还有许多活动，如游船、赛艇等。雅拉河两岸还有许多公园，如奥林匹克公园、植物园等，与雅拉河一道形成一个完整的绿地系统。雅拉河堤岸两侧都建设了亲水平台和步道，人们可以在水边散步。在雅拉河上共有 20 多座桥梁，造型各异，其中一座造型独特的步行桥斜跨在河上，从河南岸上桥，桥的方向直指墨尔本中心商务区——一组现代化的高层建筑，墨尔本全城，包括雅拉河两岸，建筑给人的整体感觉是现代的、充满活力的。

同时，墨尔本在历史建筑和历史街区的保护上也不遗余力。考察团漫步在墨尔本中心区，虽然每个地块都是现代化商场，但规划却完整地遵循了历史上的城市路网和肌理。诸葛民和同伴们一连走过几个大百货商场，当进入下一个建筑时，发现里面是一个巨大的、圆锥形的共享大厅，大厅下面是一栋红砖建造的、有一座高塔的历史建筑。原来为了保护这栋历史建筑，新建筑专门设计了一个共享大厅。

诸葛民后来了解到，这栋现代化建筑是日本著名建筑师黑川纪章设计的墨尔本中心（Melbourn Central），建于 1991 年，中间的历史建筑是具有百年历史的库伯子弹工厂塔楼（Coop’s Shot Tower），始建于 1880 年，塔楼高 50 米，是专门用来做铝子弹的。塔楼内有 300 多级台阶，在 20 世纪 40 年代是墨尔本最高的建筑，库伯工厂曾是南半球最大的子弹工厂。二战后，铅制子弹被淘汰，子弹工厂最终于 1961 年关闭，1973 年被澳大利亚政府列为“具有历史价值建筑”得到保护。当时，墨尔本城市环线（City Loop）正在建设，1981 年建设中央火车站（Melbourn Central Station）时，刻意保留下了这栋建筑。火车线和车站的建成，使周边土地升值，墨尔本市政府决定将车站边建成商业中心。

从 1986 年开始历时 5 年，耗资 12 亿澳元的商业建筑墨尔本中心拔地而起，黑川纪章设计了一个世界上最大的玻璃圆锥形穹顶，将库伯子弹工厂塔楼完整地包含在其中。墨尔本中心被日本公司大丸百货占据了 1/3，内部共 6 层，铺一开张就吸引了大量人流，一时风头无两，但 2002 年这家日本公司却因经营不善等各种原因撤出了，现在里面是澳大利亚本土的百货公司。

还有一件事让诸葛民印象很深，多年后还一直记得。当时他看到在市中心雅拉河畔有两栋现代主义风格、20 多层的住宅塔楼，墨尔本同行介绍说，为了保护雅拉河的景观，规划建议拆除这两栋住宅。诸葛民有些惊讶，他在雅拉河边多角度观察，发现这两栋建筑并不难看，风景还很好，只是体量有些突兀。诸葛民怀疑这两幢建筑能否拆除，所以后来他只要听说有人去了澳大利亚就会询问这两栋建筑是否被拆掉了，但许多人并不了解这件事情。

直到 2007 年，一位澳大利亚 LAB 事务所的著名建筑师在滨海新区做项目时，介绍了自己设计的墨尔本联邦文化广场，诸葛民忽然意识到这个联邦广场就是拆除了两栋住宅楼后在原址上建造起来的。天津的海河边在中华人民共和国成立后由于缺乏规划，无序建设了一些高层建筑，对海河的景观影响较大，但对当时的天津来说，全市的高楼加一块儿也不多，为了景观而拆高楼是想也不敢想的事情。不是没想过，但觉得不可能，这会顶着多大的压力！但在墨尔本学习后，开始有了一点勇气。2002 年天津开始综合开发改造时，的确拆除了一栋位于中心广场对面、海河南岸的 9 层板式大楼——天津出版大楼，这是当时天津拆掉的最高建筑。诸葛民体会到，有时候不破就不立，纠正错误或许也要付出代价。

鲍尔

“不破不立，准确地说是什么意思？”鲍尔对诸葛民表述的这个成语吃不太准。“噢，这是中国的一个成语。从字面上讲就是不拆除旧的建筑，就无法建造新的建筑，但还有更丰富的含义，包括哲学上的，是说如果你不破除头脑中旧的观念，就很难接受新的思想。当然，这里有一个度，不太容易把握。有时候能够采取折中的策略，如墨尔本中心的库伯子弹塔楼就很好地与外面的新建筑和谐共存。有时候‘破’与‘立’是对立的。就墨尔本联邦广场来说，如果不拆除位于雅拉河边的两栋高层建筑，后来的联邦广场就不会有这么好的效果。”“看来，您在澳大利亚的考察收获是很大的，尤其是对墨尔本雅拉河的考察，不仅学到了具体的规划设计，也学到了规划设计的实施、历史街区和建筑保护等多方面内容。还有什么事情给您留下了深刻的印象吗？”“有啊，就是坚定了我对天津建设成为国际化城市的信心。”

虽然 1994 年时，天津城市整体上比较破旧，追赶伦敦、悉尼、墨尔本等国际城市看似遥遥无期，但天津有海河发展的自然条件，有悠久的历史文化，而且正是因为有大片的破旧平房、仓库、工厂，天津才有更大的发展空间。澳大利亚是一个年轻的国家，1770 年成为英国殖民地，航海家库克船长发现东海岸，将其命名为“新南威尔士”，最早是流放囚犯的地方。1788 年第一批共 736 名囚犯登陆，1790 年第一批自由民移居，国家只有 200 多年的历史。1901 年澳大利亚成立联邦，通过第一部宪法，至今只有 100 多年。悉尼、墨尔本城市的快速发展让我们相信，拥有海河和 600 年历史的天津一定能够跻身为世界知名的国际大都市。

24. 德国法兰克福 · 美因河

诸葛民

每座著名的城市都有一条著名的河流，这句话千真万确。随后几年，诸葛民有机会考察了其他国家的城市，许多城市都有河流穿过。1995 年诸葛民随天津规划局代表团赴德国考察，第一站是柏林。1989 年柏林墙倒塌，东西德国合并，结束了东西柏林 40 多年的分裂和对峙。诸葛民一行考察的时候，正是柏林大规模建设的时候，为了把位于伯恩的联邦政府尽快搬到柏林来，正在进行联邦议会、总统府的改造和建设，同时也在推动柏林规划建设新的公共建筑。

诸葛民一行到柏林的波兹坦广场参观，现场是塔吊林立、热火朝天的工地。在一栋临时办公楼里，戴姆勒奔驰公司的项目负责人介绍了波兹坦广场（POTSDAMPLATZ）规划和奔驰公司的项目及悉尼广场项目的情况。波兹坦广场历史上曾是柏林的商业中心，二战前是欧洲最大的广场，是德国乃至欧洲的文化和商务中心，二战中毁于战火。战后这里成为东西柏林的分界线，建立起了柏林墙，繁华的城市中心沦为人迹罕至的隔离区。许多人为逃到西柏林在这里被岗哨的机枪扫射而丧生。东西德国统一后，柏林墙被拆除，柏林成为新的首都。

为了缝合饱受创伤的东西柏林，德国政府决定在原来的分界线处做点事情，建设波兹坦广场成为最佳选项。作为德国二战后最大的项目，其建筑面积为 120 万平方米，土地拍卖给了奔驰、克莱斯勒、日本索尼等跨国公司。在此之前，就整个地区规划进行了国际方案征集，邀请了包括诺曼 · 福斯特爵士等知名的建筑师和设计事务所，包括当时还不太出名的丹尼尔 · 里伯斯金（DANIEL LIBERSKIND）参加。参赛方案风格迥异，基于对柏林文脉的考虑，在竞赛组织者托马斯 · 西等施加影响和多方运作下，一些超高层方案被严格筛选掉了，德国慕尼黑建筑师 Heiz Hilmer 和 Christoper Sathermin 的方案最终胜出，气得身为评委之一的雷姆 · 库哈斯甩手而去。

获奖规划恢复了相邻波兹坦广场的莱比锡广场著名的八角形，采取了整齐划一的传统街区形式，没有高层建筑，整体高度不超过 35 米，实施时又降到 28 米，地块的大小是 50 米 ×50 米。在总体方案确定后，进行了开发商地块的城市设计方案招标，奔驰地块中标方案的建筑师是伦佐·皮亚诺（Renzo Piano），索尼地块的是赫尔穆特 · 扬（Helmut Jahn）事务所。当时只能看到项目的模型和效果图，索尼中心椭圆形的平面和奔驰中心整齐划一的造型，给诸葛民留下了很深的印象，至今记忆犹新，但只看模型是无法真正体会其中奥妙的。

直到 5 年后，诸葛民再访柏林时，亲临现场、亲眼所见，才真正体会到波兹坦广场两组建筑的高超水准和精细构思。索尼中心整体由赫尔穆特 · 扬这一家事务所设计，最突出的特点就是由数栋建筑围合而成，张拉膜结构覆盖一个巨大的、11 层楼高的半室外空间，里面包括商店、剧场、办公、IMAX 剧院等多种功能，露天餐厅拾级而上，成为十分吸引人的公共空间，据说这个设计与原规划出入较大。

而伦佐 · 皮亚诺的奔驰中心设计则严格遵循了总体规划，一条室内步行商业街将各地块串联起来，每个地块都具有复合功能：办公、住宅、餐厅、剧院、音乐厅及赌场等。奔驰中心在伦佐 · 皮亚诺的统领下，由多个著名的建筑师或事务所参与设计，包括理查德 · 罗杰斯、矶崎新和阿尔多 · 罗西等，是一种统一中有变化的理想建筑群模式。在总体建筑高度、街道尺度和功能定位的导则约束下，每位建筑师在各自的设计中，使用金属、玻璃和褐色条陶板作为立面材料，各自确定设计的主题和细部，每座建筑都没有以自我为中心，而是尽可能去参与公共空间的营造，对公共空间有所贡献。

波兹坦广场是反映城市设计实施全过程的优秀案例，由总体城市设计到详细城市设计，再到最终的建筑设计，在不同的设计阶段中，规划师、工程师、建筑师之间既有分工又有合作，统合规划分区、分期建设，相互尊重、充分沟通。不仅有内部沟通，在与民众沟通上也做了大量工作，获得了社会广泛认可。在 5 年建设期内，就有数百人前来参观，建成后，波兹坦广场不负众望地成为一个颇具文化娱乐特色的城市中心，同时也是一个以文化娱乐为特色、多种功能高度复合的、转型中的城市商务区（CBD）。

诸葛民后来在滨海新区于家堡金融区和滨海文化中心的规划设计中，一直有强烈的愿望，希望将波兹坦广场的规划设计特色和规划设计组织方式应用其中，虽然很努力地去推动，但他认为只学到了一些皮毛。比如，于家堡起步区由 SOM 负责总体城市设计方案及导则，由 9 位国内新锐建筑师分别负责 9 个地块的建筑设计，形体组合和建筑设计上相对是成功的，但功能不够复合，包括宣传和大众参与的程度也不够，加上区位和整体环境等各种影响，目前人气还不足。

与之相比，滨海文化中心好一些，在城市设计阶段做了大量多方案比选工作，也邀请了曲米、扎哈等大师从城市设计阶段就参与进来，包括 MVRDV 等新锐的事务所，拟增加文化中心的活力。为了真正学习波兹坦的经验，诸葛民也邀请了皮亚诺、罗杰斯和墨菲 · 扬等参与过德国波兹坦广场项目的大师，他们具有工业感的建筑风格与滨海新区的城市特征十分吻合，只是后来由于各种原因，只有扬参与了文化交流大厦的设计，他付出了许多心血，但遗憾的是该项目没有上马，如果能够建设的话，扬设计的极具工业感的圆形大楼一定会为滨海文化中心建筑群增添光彩。

波兹坦建筑群不仅在空间肌理上延续了城市文脉，缝合了东西柏林，而且在建筑风格、城市功能上发挥了聚合创新的作用。诸葛民一行当时还考察了原来位于东柏林的亚历山大广场建筑群，经过 40 年的社会主义建设，其建筑都烙上了鲜明的苏联和东欧建筑风格的印记，广场中心是 1959 年建成的东柏林电视塔，369 米高，比巴黎埃菲尔铁塔还高 45 米，比当时西柏林的电视塔高出 220 米。很多时候，建筑也发挥着政治作用。在电视塔 200 米高处是一个 7 层高的球体，里面是旋转餐厅和观光厅，可俯瞰柏林全貌。当然，从西柏林也能够看到电视塔高耸入云，想不看都不行，这也是一种意识形态的宣传。

东西德国统一后，民众要拆除电视塔的呼声越发强烈，但最终决定还是将电视塔保留下来，成为记忆东西柏林分裂的历史标志，这可能比保留下来的柏林墙让一些人更不舒服。东西柏林除建筑风格明显不同外，东西柏林人的差异更加明显，不论是在衣着还是在语言和行为举止上，真正的融合任重道远。

德国作为两次世界大战的元凶和战败国，在进行历史性反思的同时，又十分重视对城市历史和古建筑的保护。二战中大量古建筑被炸毁，包括位于西柏林繁华商业大街布赖特沙伊德广场上的威廉皇帝纪念教堂，建于 19 世纪 90 年代，炸毁后被人称为“断头教堂”。为了警告后人不要战争，位于市中心的教堂一直没有修复，十分醒目。1957 年举办了教堂重建设计竞赛，获奖方案拟全部拆除教堂残迹，建设一座全新的教堂，这引起了巨大争议。最后的妥协方案保留了 68 米高的“断头”——主钟楼残骸，周围依旧按照竞赛获奖方案建造了 4 栋新建筑。建筑是灰色的，只有在夜晚由超过了 3 万块玻璃窗组成的格状墙壁透出的蓝光使教堂显得无比美丽。

类似的项目还有德国议会大厦，它始建于 1884 年，采用古典主义风格，并融合了哥特、文艺复兴和巴洛克等多种建筑风格，是德国统一的象征，最初为德意志帝国的议会，1918 年成为共和国的议会。1933 年 2 月 27 日大厦失火，部分建筑被毁。“国务纵火案”曾成为纳粹统治者迫害政界反对派人士的借口。1945 年 4 月 30 日苏联红军把红旗插在国会大厦的屋顶，宣布了二战的胜利。1961 至 1971 年间，大厦重建时对建筑进行了简化，去掉了 1945 年被炸毁的穹顶。东西德国统一后，国会大厦成为联邦议院驻地。

诸葛民一行访问柏林期间，大厦正在重建，由英国著名建筑师福斯特爵士设计，他保留了大厦的外观，新建了一个有供人游览等多种功能的玻璃穹顶，与被炸毁的穹顶尺寸相同，1998 年联邦议院正式迁入这座现代化的议会大楼。2001 年诸葛民再次访问柏林时，又去参观了新的议会大厦，大厦的外观看起来非常诱人，集古典与现代于一体。诸葛民十分想进入玻璃穹顶内部看一下，但由于要提前一段时间预约，此行未能达成愿望。数年后，天津准备选址在中心广场建城市规划馆，为了与后边的意大利历史街区呼应，曾设计了一座新旧交融的建筑，有点类似于德国议会大厦，方案也设计了一个玻璃穹顶，后来由于时间仓促等原因没有建设。现在看来中心广场就应该留白，没建真是万幸！

除了对历史建筑珍视外，德国对二战以后新建的现代主义建筑也十分尊重。与波兹坦广场西侧紧邻的就是建于 20 世纪 60 年代的西柏林文化广场（Kultur Forum），由著名德国建筑师汉斯 · 夏隆主持规划，并设计了著名的德国爱乐音乐厅等一组建筑。皮亚诺在主持波兹坦广场奔驰中心时，就与其中的柏林图书馆有良好的呼应。文化广场南侧是由密斯 · 凡 · 德 · 罗设计的德国新美术馆（Neue Nationalgallerie）。建筑界都知道，密斯是现代主义四位建筑大师之一，该作品也是他的代表作，被列为 20 世纪最有影响的建筑之一。

诸葛民一行怀着崇拜的心情参观了新美术馆，其简洁到无以复加的经典造型以及充满力度的钢与玻璃的细部构造令人叹为观止。这座建筑是在密斯逝世后的 1968 年才落成的，是密斯毕生探索钢与玻璃纯净建筑艺术风格的绝唱，被人们称为钢与玻璃的帕提农神庙，从中可以体味德国的工业技术水平及严谨、精益求精的民族精神。在勃兰登堡门西北，他们发现了一个用印有建筑立面的橙色帆布围合起来的建筑。据当地建筑师介绍，这是为了考量建筑的体量是否与周围环境和谐，而在现场搭的一个体量模型，德国人做事情的一丝不苟由此可见一斑。

柏林不愧是世界最著名的城市之一，市中心有大片绿色的植物园，整个城市的建筑低缓，密度适中，与周围自然环境融为一体，但斯普雷河两岸确实算不上十分出众。要说德国著名的滨河城市，还得是法兰克福，法兰克福的全称就是美因河畔的法兰克福。法兰克福是德国第五大城市，德国和欧洲乃至全球很多重要的工商业、金融、会展、文化和交通中心都在这里。

法兰克福市中心和古老的内城位于美因河北岸，美茵因河是莱茵河支流，100 多米宽，从市中心顺直流过。河上有众多的桥梁，把内城与河南岸连接在一起。在内城和南岸有众多博物馆，许多是由著名建筑师设计的杰出建筑，内城还有德国大文豪歌德故居等众多文化建筑。法兰克福的中心商务区在内城外，是一组现代化高层建筑，包括由福斯特设计的欧洲最高的、世界上第一座高层生态建筑——德国法兰克福商业银行大厦，当时正在建造中。法兰克福汇聚了欧洲 80% 的新建高层建筑，并妥善处理了与老城保护的关系，也与美因河相互映照，避免产生对河道的压抑感。

诸葛民与代表团中的几位年轻人住在了法兰克福老城边的公寓里，房间是利用坡屋顶下的空间形成的夹层，一个钢制旋转楼梯和天窗让室内空间充满变化和灵动。这是诸葛民第一次住进欧洲城市的现代住宅，室内设施齐备，门窗的质量和封闭性都非常好，暖气可调节温度，让人真切体验到了德国住宅的品质。联想到中国的住宅建筑质量真是相形见绌，我们一直太强调经济，不重视品质，品质才是百年大计之先。诸葛民还抽空专程考察了市中心区内一些新建的住宅区，都是小地块，围合式的布局，建筑一般为三层，每个院子内部和建筑造型各异，显示了丰富性和多样性。诸葛民随机走进一个三层住宅围合的内院，大约是 30 米 ×30 米见方，院子内部完全是一个供孩子们玩耍的沙坑，也许是时间不对，他没有看到一个孩子。

代表团还考察了慕尼黑，这是德国第三大城市，是巴伐利亚州的首府。慕尼黑历史悠久，有 850 年建城史，有深厚的德国传统文化，也曾是纳粹的据点，是德国重要的工商业城市，其汽车、机械、化工产业在德国经济中扮演重要角色，金融产业居法兰克福之后排第二位。代表团体验了慕尼黑的传统美食——啤酒以及民族歌舞，参观了宝马博物馆和“四缸”造型的宝马总部，各种优质的汽车和商品展示让大家流连忘返。

在宝马总部不远处就是著名的慕尼黑奥林匹克中心体育场，它独特的张拉网和上面透明的亚克力材料覆盖了 7.5 万平方米的面积，是世界上最大的屋顶，相当于 10 个足球场大。由于透光性好，各种植物生长旺盛，让人仿佛置身于晶莹剔透的绿色艺术世界当中。规划体现了技术的创造力，也少见地展现出德国人的想象力和浪漫。1972 年第 20 届奥运会在此举办，这次奥运会上也发生了震惊世界的巴勒斯坦恐怖分子袭击奥运村事件。

在德国的考察中，代表团体验到了德国交通的发达。法兰克福机场是世界十大机场之一，全国铁路网密集。诸葛民一行乘坐高速铁路从柏林到慕尼黑，新开通的高铁车厢内装修得富丽堂皇，座椅是皮制的，时速 300 多千米，这也成为他们后来提议建设北京和天津之间城际高铁的初始体验。在高速公路建设方面，德国更是令人钦佩，许多公路是战争期间用钢筋混凝土修筑的，质量有保障。在高速公路上汽车不限速，在诸葛民的印象中好像这是唯一一个高速公路不限速的国家。

德国的城市轨道交通也十分发达，有 S-bahn 城市快轨，大部分在地面运行；有 U-bahn 地铁、Stadtbahn 轻轨、Tram 有轨电车和 Region 区域火车。不论哪种形式的城市轨道，车厢内都干净整洁，座位都非常舒适，只是不时遇到骑自行车的人带车上地铁，偶尔遇到特别大的狗也随人被带入车厢。总的感觉是德国地铁中旅客人数比伦敦少，相对比较安静舒适。总体看，与一年多前英国和澳大利亚之行相比，德国考察的内容范围更广，收获更大，美中不足的是在滨河城市的建设上收获不多。

促成这次德国之行的是一位柏林的前规划官员。从 20 世纪 80 年代起，中科院生态所就与德方开展合作，以天津为例进行城市生态研究，作为联合国教科文组织人与生物圈规划 MAB 的组成部分，天津环保局和规划局都参与了相关工作，这位官员也参与了这项工作。他退休后，想进一步加强与天津的合作和开发。考察团在德国参观了一些城市更新开发项目，了解了德国相关的法律 LAND ORDER，它的字面意思是土地收购，即政府组织或信托公司投资一个地区的道路交通和市政基础设施、改善绿化环境后，相关土地受益者根据土地价格增长的幅度缴纳一定的费用。如果不缴，政府有权按照实施前的土地和房屋价格进行收购。

经综合考虑，天津当时选定了意式风情区作为研究对象，探寻可行的路径。一位年轻的高个子女士在规划局工作了一段时间，她的德国名字发音为“舒瓦茨”，大家为了便于记忆，都诙谐地叫她“湿袜子”。这个项目由于各种因素最终没有成功。但后来随着海河综合开发改造规划的实施，意式风情区通过另一种途径实现了更新改造，就是诸葛民和鲍尔现在坐着聊天的这个区域。整个国外考察过程中，诸葛民和伙伴们学习借鉴了许多经验，在实践中结合自身实际，不断探索、创新。

鲍尔

听到这儿，鲍尔先生总结道：“虽然在德国没有学到太多滨河城市的规划经验，但更重要的是学习了像波兹坦广场这种大型项目的规划设计组织模式，以及像德国土地收购 Land Order 这种城市运营机制。”“您理解得很准确！实际上滨河城市只是一种城市类型，世界上所有城市在很多方面都是共通的，要做好海河综合改造，需要方方面面的知识和能力。事实上，历史文化街区保护和历史建筑的整治就是我们后来实施的海河“十大工程”之一。我们现在的历史风貌建筑使用和产权状况与德国完全不同。政府投资进行了海河环境提升，但无法向两侧受益的土地和房屋产权方收增值税费。

为什么呢？一是还没有这样的法律；二是与德国土地私有制不同，我们的土地均为国家所有，大部分房屋的产权也是政府的，自己给自己交增值税费没有意义。中华人民共和国成立后，部分历史风貌建筑用作机关办公，多数还是用于居住。一栋小洋楼中的每个房间分给一户，一栋楼可以住十几户，他们最迫切的需求是改善居住条件。因此，政府按照规划，对这些建筑进行保护，花钱请住在小洋楼中的居民搬迁出去，改善居住条件，由于没有专门的法律法规，就按照一般的拆迁补偿政策与居民签订协议，与住在危陋平房中的居民享受一样的待遇。

设想到的是，居民搬走后，平台公司刚开始修缮这些小洋楼，他们就又找回来了，说你们与我们签的是拆迁协议，为什么楼不拆了？后来针对这种新情况，2005 年天津人大批准颁布了《天津历史风貌建筑保护条例》，把有保护价值的小洋楼或类似建筑确定为历史风貌建筑，依法保护、腾迁、维护、更新利用，对居住在历史风貌建筑中的居民采取腾迁而非拆迁的方式。当然，除了租用公产房的以外，还有许多产权情况，包括有些房屋是教会产。当年住成大杂院时没有人申请权利，可当政府把居民腾迁走了教会却来了，拿出房产证，说这房子的产权是我们教会的。遇到了各种各样想都想不到的问题，一言难尽，经过艰苦卓绝的工作，终于使包括意式风情区、五大道等历史街区部分恢复了活力，为海河两岸增添了色彩，也成为天津城市的一大特色。

天津共有 14 片历史文化街区，分为中国传统和九国租界两大种类型。多数保护得不错，如五大道、意式风情区、解放北路等，也有少数不够理想，如鞍山道，也是我们这次城市设计试点项目的重点。另外，要进一步通过城市设计与历史街区保护规划相结合的方式，增加五大道、解放北路、中心花园等地区的活力。我们把这些历史街区的步行系统品质提升作为增加城市活力、复兴城市中心的重要手段，还请鲍尔先生多提宝贵意见。

为进一步增强海河两岸步道的连通性，我们在海河上新增加了三座步行桥，把意式风情区与鞍山西道原日租界连接起来，把解放北路泰安道五大院与河东南站（六纬路）商务区连接起来，把津湾广场与天津火车站交通枢纽连接起来，形成完整的步行系统。三座桥一座已建成投入使用，就是联系东站与津湾广场的步行桥。“这个做法很好，我认为您要坚定地做下去。您讲了意式风情区等历史文化街区腾迁和修旧如故的艰辛历程，我能从一些老照片与今天现实场景的对比中体会到。海河综合开发改造一方面改造了海河堤岸环境，更重要的是对历史街区的保护、恢复和复兴。”

25. 日本东京 · 东京湾

诸葛民

1996 年诸葛民还在规划设计院工作时，单位组织去日本进行专业考察，包括东京、大阪、名古屋等，拜访了国土局等规划部门和同行。参观了秋叶原、银座等老的城市中心，也参观了新宿、临港、大阪 OBP 等新的规划建设。1990 年日本经济泡沫破裂，经济进入了长期滞涨期，他们在参观中看到了当年雄心勃勃的大手笔规划都陷入停滞，包括已经通过立法的迁都计划。由于土地价格和房价大跌，过去计划依靠出让政府机关土地建新首都的方案已经难以操作。

诸葛民当时很好奇日本的土地价格和房价为什么能涨到那么高，再看今天中国的许多城市房价飞涨，他对房产泡沫就更产生了担忧。在大阪，他们考察了震后重建工作，虽然严重的阪神地震刚过去几个月，但城市基本恢复了常态。诸葛民感慨，如果这种强度的地震发生在国内，死亡人数绝不仅仅是 6000 人。日本城市规划对城市防灾减灾的重视，以及国民教育和日常演习培训在应对灾害时发挥了重要作用，值得中国借鉴。

在日本，他们还特意考察了日本公共住房的设计、建造和分配。他们访问了一家大的住房公司的设计部，看到住宅的施工图，竟与机械加工图一样精准，施工的误差以毫米计！他们在日本同行带领下，进入了一栋新建的 50 层高的公共住房，内部设计是西式与和式结合，开放厨房与餐厅组合，严谨而精妙。日本的相关规范并不要求餐厅有独立的采光通风。他们也参观了大阪的一个居住项目，以高层和板式小高层为主，均为外廊式，整栋建筑的管线布局和施工安装规整而精密，其精细程度就像前面看到的精准的施工图。垃圾收集在地下室，地下一层空间用来收集和压缩垃圾，地下二层有专门的垃圾车将压缩后的垃圾运走，干净、整洁、环保。

他们还参观了一处公团住宅的售楼处和样板间。据了解，日本的公共住房政策与我国香港的相近，30% 公共住宅由政府提供，采用申请抽签制，申请者要排队等待一定时间。除了公营住房和公团住房外，在城市外围总会看到大片的独立式低层住宅。虽然面积不大，但都是一家一户一院。据了解这种叫作“一户建”的住宅在东京也占到了总住宅套数的 45%，这让诸葛民一行非常惊讶。要讲人多地少，日本的人口密度 338 人 / 平方千米，是我国人口密度 138 人 / 平方千米的 2.5 倍，日本特别是东京等特大城市的人口密度比我们国家大多了，但他们仍然有大量的低层独户住宅，全国住房中甚至有一半都是独户住宅。为什么我们国家就能简单粗暴地说住宅容积率不能低于 1 呢?

“你们觉得我们的菜怎么样？”餐厅的意大利籍老板送上了一份熔岩蛋糕，打断了诸葛民的激动情绪。这家意大利餐厅已经在意式风情区开了 3 年多了，老板与诸葛民很熟悉。规划展览馆搞学术活动后，总有规划师喜欢在此小聚。“您刚才顺着时间脉络讲到了 1995 年，那距 2002 年开始的海河综合开发改造还有漫长的 7 年时间呢，这期间又发生了什么故事呢？”鲍尔先生追问到。“这几年我没有再出国，这期间做的一些事情看似与海河规划没有直接关系，但回过头看都是在为 2002 年的海河规划做储备。”

26. 夜色中的巴黎

诸葛民停顿下来，吃了口蛋糕，他喜欢甜的味道。“我给你讲个小故事。世界上最著名的滨水城市是哪里？对！毫无疑问，是法国塞纳河上的巴黎。多年以来，我搜集了巴黎塞纳河的大量资料，魂牵梦绕，但始终没有机会来到塞纳河畔。也许是天意，那次德国之行在登陆德国之前意外地来到巴黎，看到了塞纳河，只是由于夜深人静、灯火阑珊，没能清晰目睹塞纳河的芳容。”

当时欧洲的申根协定刚开始实施，考察团的签证国家是德国，计划在德国入境，只是要在法国戴高乐机场转机，从一个航站楼转移到另一个航站楼。在海关他们体验到了法国边境官员的官僚作风。边境官说，“出了这个航站楼就进入法国领土了，必须有法国签证，而你们没有法国签证，不许通行”。不管有无申根协定，他说必须要有落地国的签证。几经交涉无果。团员们看到另一个航站楼就在目及范围，可就是过不去。时间一小时一小时地过去，他们眼巴巴地看着航班飞走了。最后无奈考察团每人交了 200 法郎，办了临时签证，改签下一个航班。

出了机场，已是黄昏，到达酒店时天已经全黑了。考察团订的是第二天一早的飞机，已不可能有时间在白天看巴黎了。经大家一撺掇，领导发话了：既然已经到了巴黎，就连夜考察一下吧。同伴们立刻从酒店叫了两辆出租车，夜游巴黎。两位出租车司机是东方面孔，却听不懂中文，原来是越南人。接近深夜，喧闹的巴黎已入睡，香榭丽舍大街上只有孤单的路灯，塞纳河畔几乎漆黑一片，跟大家设想的塞纳河两岸亮丽的夜景灯光完全不同。到 2001 年诸葛民再次游览塞纳河时才发现，建筑上的灯光是游船的探照灯打上去的，船上的游客能清晰地看到两岸明亮的建筑，而游船驶过后，建筑就消失在黑暗之中了。

他们正想让司机开车去蒙马特高地，想从高处一览巴黎的全貌，突然后面响起警笛，一辆警车跟在后面。停车后，警察检查了两位出租车司机的证件，不知说了些什么。警察折腾完了，夜色已深，团员们也睡意渐浓，考虑到明天还要早起赶飞机，就只好作罢。这就是诸葛民的第一次法国之行，欣赏了短短几小时夜幕下的巴黎。诸葛民没想到，他看清塞纳河的真颜，已是 6 年之后了！

第五部分
金龙起舞——海河

27. 海河规划的前奏

1998 年规划局换了新的一把手，搞建设出身，因过去受够了规划局这帮“小鬼儿”的气，曾立志说如果他到规划局当局长，一定好好治治这帮“龟孙子”。新局长上任伊始，就提出规划局不能只顾着审件（审批项目），还要给市委、市政府出谋划策，因此，布置给规划院研究和策划各类规划设计新项目。诸葛民负责组织海河两岸中心商务区小白楼、南站、解放北路的城市设计。南站方案以 1997 年澳大利亚专家方案为主，为了方便推敲方案、做好设计，他们制作了一个包括天津内环线在内的大的工作模型。领导觉得效果很直观，也很新颖，询问是什么规划，回答说叫城市设计，就沉吟了一下若有所思，然后布置诸葛民在规划局办个讲座，给大家介绍一下城市设计。由于没有现成的讲座材料，诸葛民用两小时的时间介绍了城市设计的 10 本经典著作：培根的《城市设计》，库伦的《市镇景观》，林奇的《城市意象》，芦原义信的《外部空间设计》，特雷西特的《寻找丢失的空间》等。诸葛民在上面讲着，自己能感觉到讲得比较枯燥，看下面听众也没什么反应。这是他第一次在规划局做学术讲座，也是他一生中最不成功的一次讲座。现在想想，当时如果介绍一个具体的城市设计项目，会比讲城市设计理论更让大家易于接受。

1999 年 8 月，国务院正式批准了修编后的城市总体规划（1999—2010 年），这一版总体规划就是诸葛民 1994 年开始主持修编的。修编工作于 1996 年基本就完成了，但正式批下来已是 1999 年，这就是我国总体规划的状况，从修编到正式批复整整用了 5 年时间，还算是常态。为了贯彻落实城市总体规划和就建设项目现编控规的状况，市规划局决定开展中心城区控制性详细规划全覆盖工作。开始大家有畏难情绪，诸葛民与规划院同仁一道制订工作计划，划定控规单元，制定技术标准，经过各部门各单位的共同努力，用了两年时间，圆满完成了中心城区 180 多个单元、334 平方千米的控规，实现了全覆盖，获得市政府批复，是天津市历史上第一个中心城区全覆盖的法定规划，位于全国前列。即使到了 20 年后的今天，我们走遍全国各地，还是常常发现多数大城市都没有一张完整的中心城区道路定线图，而天津市中心城区早在 20 年前就完成了这项基础工作，后续就是进行动态维护、不断调整完善。当时有人怀疑规划的水平，但实践证明，这版控规是成功的，虽然或多或少存在一些问题，但确实对天津的城市建设发挥了重要的引导和管控作用，使城市规划逐步走上规范化、法制化的轨道。

20 世纪 90 年代末，在天津市推动城市设计和控规全覆盖的过程中，海河两岸规划在逐渐成熟、成型。1996 年国家建设部与澳大利亚工业出口和旅游部开展滨河城市合作项目，选择了天津市，由澳大利亚政府资助。澳大利亚专家与天津市开展了海河东岸中心商务区南站地区的概念性规划，1997 年完成了规划成果，1998 年结合中心商务区解放北路、小白楼和南站地区城市设计对海河两岸规划持续深化，这段时间海河两岸规划的总体结构基本稳定。恰在此时，天津画家邓家驹先生 1996 年在联合国展示了根据海河历史漕运绘制的《海河漕运图》，引起很大反响，联合国教科文组织致函天津市政府推动海河规划。由此，1999 年天津规划院组成代表团赴法国巴黎专程考察塞纳河，与著名的机场设计公司和保罗 · 安德鲁接洽海河两岸规划事宜，后由于各种原因没能合作。到了 2001 年，在市引智办和法国大使馆推动下，法国康赛普特（Concepetal）设计公司一行来天津考察，开始编制海河两岸概念规划。

28. 海河综合开发改造规划——金龙起舞

2000 年，全世界在担心计算机操作系统“千年虫”爆发的恐慌中欢欣鼓舞地迎来了新世纪，天津也进入了一个新的历史时期。经过上一阶段的艰苦努力，“三五八十”奋斗的阶段目标基本实现，特别是市区成片危陋平房改造基本完成，城市面貌和功能得到极大提升，天津跨入了新世纪。未来如何发展，成为摆在全市上下面前的首要问题。在市规划局的布置下，在与国外开展合作的同时，市规划院立足自身，开始编制新一轮海河规划。诸葛民负责这个项目，在院里组成工作组。这个规划是开放式的，归纳起来要回答三个关键性问题：为什么搞海河开发建设？如果搞开发建设，要包括哪些内容，达到什么标准？规划如何才能实施？应该说海河规划编制已经有了近 10 年的积累：完成了总规定位、控规全覆盖、重点地区城市设计、国内外参观考察学习、翔实的现状资料整理等，为下一步编制出一个好的规划奠定了坚实的基础。但光有这些还是远远不够的，如果仅仅按习惯套路照本宣科，是不可能做出高水平规划的。项目组采用了开放式的规划方式，经常坐在一起信马由缰地头脑风暴，相互激发、群策群力。同时，组织相关部门和单位编制了《世界名河》《海河史话》《海河两岸历史》《建筑汇编》等资料集。

通过不断的研讨，大家认识到要回答上面三个问题的关键是要先解决数个影响海河规划的关键性问题。海河规划是一个牵一发而动全身的项目，所以首先给这个规划起了个响亮的名字：天津市海河两岸综合开发改造。俗话说：名不正，言不顺。与一般的成片危陋平房改造不同，该名称突出了综合开发，虽然只有短短的四个字，但包括了许多内容，大家对这个名字都表示认可，但觉得还是有点长，读起来不上口。工作组里的连副院长和齐所长两位才子一商量，起了个响亮又形象的名字：金龙起舞。朗朗上口，大家都说好，也感觉稍稍有点用力过猛。但作为这种鼓动性的规划，学名应该叫“社会动员（Social Mobility）”式的规划，需要一个响亮的名字。“金龙起舞”的寓意也很明确：海河蜿蜒流过天津城，绵延 72 千米入海，宛若一条沉睡的巨龙俯卧在津沽大地上，海河综合开发改造的实施就是要带动天津城市经济、社会、文化事业的整体腾飞，像一条充满活力的金色巨龙舞动起来。有人提问：金龙的龙头在哪里？传统上龙头肯定是中心城区三岔河口的海河源头。也有人说：随着新时代对外开放，中国进入海洋时代，龙头是否应该倒过来，位于海河入海口的天津港。还有人说：历史上位于海河边的天津最早的火车站就叫“老龙头”。但外界似乎也没有太多人回应这个问题。本次规划将海河 72 千米分为上、中、下三段，上游段 19 千米的从三岔河口到中心城区的外环线，是本次海河综合开发改造的重点。

关于海河的定位和功能，由于有先前研究讨论达成共识的基础，包括以此形成的 1999 年经国务院批准的天津城市总体规划的支持，没有太多的争议。一开始，对海河的定位用了许多描述性的语言，比如：海河是天津的母亲之河、创世之河、希望之河，是城市布局的蓝色轴线，是天津的希望和生命线，是贯穿天津城市政治、经济、历史、文化的中心线及绿化、旅游和风景景观轴线，是服务型的经济带和景观带，是体现天津作为我国北方经济中心和国际港口大都市水平的核心区域，等等。从这些表述中能够看出规划师们对海河的深切感情和无限期待，所以用词到无以复加，生怕遗漏了什么。后来，在海河规划多轮汇报、征求意见的过程中，特别是最后市委研究室搞文字的同志帮忙，选择了其中“海河是服务型的经济带、景观带”的表述为基础，考虑各方意见和因素，又加上了“文化带”，这样才形成了一个既高度概括又清晰明确的海河定位。

鲍尔

“经过 20 年实践，我越来越认识到文字对规划设计之重要，而且中文与英语不同，有自身的力量和魅力，鲍尔先生您怎么看？”“我与您有同感，在我们西方国家，许多重大规划也都有简短响亮的口号，这是规划非常重要的内容。”

诸葛民

海河的功能相比海河的定位，是个硬碰硬的科学问题，来不得半点含糊。海河的泄洪和保水两大功能与海河综合开发改造存在着对立的矛盾。好在经过多年的努力和沟通，配合多种技术手段，各方面基本达成一致，并落实到 1999 年经国务院批准的天津城市总体规划中，海河规划才有了上位法定规划的支撑，海河综合开发对功能的确定才有了依据，海河综合开发改造才有了成功的基础，这是诸葛民常津津乐道的经验之谈。诸葛民 1994 年开始主持总规修编，到获批的时候，他已不具体负责这项工作了，但他对总规文本进行了逐字逐句的认真修改，特别是一些与海河相关的文字，不能遗漏，也不能有错。因为总体规划要经过住建部组织的部级联席会审查，规划中的图纸和相关文本都经过国家相关部委的严格审查，各部委无意见后国务院才最后批准。要在天津建城区的城市防洪圈内降低海河泄洪标准，是个很大的事情。如果在天津总体规划中明确了，就等同于水利部同意了。因此，国务院批准《天津城市总体规划（1999—2010 年）》后，国家海河流域管理委员会组织编制并报批了《海河流域防洪规划》，增加了永定新河和独流减河的洪水泄洪量，抬高了相应堤防标高，以此为基础，降低了海河干流的洪水径流量，从每秒 1200 立方米到每秒 800 立方米，下降了 1/3。同时也明确了海河两岸堤防的高度，比原规划高度平均下降了 1 米多。这些上位规划的调整落地为海河规划提供了保障，诸葛民团队成员也吃下了定心丸。

1986 版天津城市总体规划文本中对海河的定位是：“海河是多功能的河道，二道闸建成后，非汛期闸上保水、闸下通航，汛期所承泄上游洪水和市区雨水的功能。”从文字间可以看出，表述得十分严密。天津人习惯说闸上保水、闸下通航，但国家明确加上了“非汛期”的限定。新版总体规划在 1986 版的基础上，经过与市里相关部门的沟通和与国家相关部委的博弈，去掉了“海河是天津社会政治经济文化历史中心线”等描述性的文字，最终文本表述为“坚持统筹兼顾，综合治理，充分发挥海河防洪、排涝、供水、旅游、航运等多种功能。”文字比以前简洁了，功能上新增了旅游。能增加这个实在不容易，难度很大，意义更大。“统筹兼顾，综合治理”八字方针则从更高层面概括了海河的功能、作用和下一步治理改造的方向。如果单纯片面地按照水利专家的说法“水火无情，天津城都淹了，还谈什么景观”，那海河两边只能修更高的堤防；如果单纯说环保、保护水质，有环保专家说了“天津市水都被污染了，没水吃了，生存都成问题，还搞什么旅游”。然而城市规划和城市设计的高明之处就是统筹兼顾、综合施策。以此为基础，海河综合开发改造规划明确了：海河市区段功能以旅游、景观为主，成为一条生活性的河道，兼有排沥、供水、防洪、航运等功能。同时，要求通过相应的工程技术措施和规范管理，从根本上解决供水与旅游的矛盾和问题。由于海河还是引滦入津的备用水库，也是引黄济津、南水北调水源的输水河道，结合南水北调工程，从河北省修筑干渠至外环，市区内采用管道输送至水厂。通过这样的调整，海河不再作为南水北调的输水河道，相对减少了对海河旅游功能的限制。此外，在海河上行驶要求采用电动游船，不能向海河排污，避免污染。从 10 多年的实际效果看，虽然游船越来越多，并且随着游客的增长游船活动频率不断增加，但海河水质得到了保证。

海河的定位和功能确定后，接下来的主要工作是具体的规划设计。两岸的堤岸如何改造成亲水平台？断面是什么形式的，标高多少？如何沿海河两岸形成连续不间断的步道？规划要设法解决许多具体的问题。发展旅游，海河上必须通游船，那游船高度应该是多少？海河上游许多不同时期建造的桥梁，桥下净空不一样，怎么协调？有一个关键性问题是海河两岸的道路交通。海河两岸实施综合开发改造，发展服务业，首先必须解决交通问题，否则其他事情无从谈起。人都到不了海河边，或者到达很不方便，如何发展旅游？但当时的情况是，海河上的桥梁少，宽度窄，仅有几个过河通道，在上下班高峰时间非常拥堵。另外，沿河路在桥头都是平交，车辆与红绿灯设置造成交通更加混乱、拥堵。规划为了促进城市面向海河发展，加强海河两岸的联系，要增加桥梁数量，如果仍然是平交，交通拥堵口会进一步增加。如果建成高架桥，又会对海河景观造成破坏。诸葛民想起几年前为缓解中山路的交通压力，拆除了老金刚桥，建成了双层的新金刚桥，各方反应强烈，集中在两个方面，一是很多老天津人在感情上接受不了，二是有人认为桥体过高破坏了海河景观。经过认真分析和多方案比选，工作小组初期提出的方案是，考虑结合海河通航标准的提高，抬升桥梁标高，让沿河道路下穿桥头路，这是一个大胆又巧妙的方案。据了解，巴黎为解决老城区的交通问题，在塞纳河一侧占用部分河道建成了城市快速路，但有些技术细节不知是如何处理的。海河市内河道原本就不宽，不可能占用河道，但交通问题又必须解决。

另一个关键问题是海河两岸综合开发改造的规划目标。在 1999 年国务院批准的新版城市总体规划中，已经提出了“利用 10 ～ 15 年时间，将海河两岸建成贯穿天津城市政治、经济、历史、文化的中心线及绿化、旅游和风景景观轴线，成为体现天津作为我国北方经济中心和国际港口大都市水平的核心区域”的奋斗目标。在实施过程中，领导要求时间改为 3 ～ 5 年。显然时间太短，实际上是无法完成的，但作为一种社会动员，也无可厚非，事总要启动，然后一件一件地做。

比较吃不准的是，他们提出了打造世界名河的想法，有过多种表述，开始是：“用 3 ～ 5 年的时间，将海河建成独具特色的、国际一流的服务型经济带、景观带，弘扬海河文化，创世界名河。”后又修改为：“用 3 ～ 5 年时间，将海河建成独具特色、国际一流的服务型经济带、景观带和文化带，用较长时间的努力，创建世界名河。”许多人反对，嗤之以鼻。的确，从当时海河两岸的现状，根本看不出世界名河的影子。

但诸葛民内心十分笃定。经过 10 年的规划研究及国内外考察的认识和积累，他坚信海河有巨大的潜能、天津有巨大的潜能，这一信念从来未动摇过。在 20 世纪 90 年代初，许多人不看好天津。社会上有许多抱怨的声音的时候，诸葛民写了一篇文章《天津的战略地位和自信心》，在《天津日报》上发表。他认为从世界城市发展的客观规律以及我国城镇化、工业化和京津冀协同发展的未来看，天津一定会重振北方经济中心的雄风。对于把海河打造成为世界名河，他有十分的信心。但在他的心中始终有一个未竟的愿望，那就是亲赴世界上最有魅力的滨河城市——巴黎，他要到塞纳河两岸亲身体会一下，给自己一个最终的确认。

29. 世界名河——塞纳河

时间恰好到了 2001 年 5 月，按照天津市委市政府指示，市规划局组织赴欧洲考察，同时与正在编制海河概念规划的法国康赛普特设计公司全面讨论规划方案。代表团由局长带队，一行 6 人用 15 天时间走了欧洲 8 个国家。说是 8 个国家，其中有梵蒂冈、卢森堡这样的小国，重点考察的还是法国巴黎。康赛普特设计事务所就在拉德方斯自己设计的楼中。讨论过海河概念规划方案后，事务所的总设计师带领他们参观了拉德方斯——这座全新规划的、现代化的中央商务区。可能是法国人对巴黎的保护太严格，巴黎人的建设热情被压抑了太久，因此当规划建设一个全新的、不受高度和尺度约束的中心商务区时，明显感觉尺度过大、用力过猛。

拉德方斯商务区核心区占地 1.6 平方千米，开发量近 300 万平方米，是一个多功能的商务区，包括办公、金融、商业等。作为巴黎城市主轴线西北部的交通枢纽，规划采取了完全立体化的组织形式：所有建筑全部架在大平台上。一行人站在占地 40 公顷的大平台上，正赶上万里晴空，阳光直射下，人在硬质铺装的广场上无处躲藏，只好来到大拱门下。大拱门与凯旋门相呼应，长、宽、高均为 110 米，占地 5.5 公顷。大台阶很高，一行人好不容易爬上大台阶，勉强躲在张拉膜下，逃避烈日的煎熬。大拱门的顶层，就是建筑水平连楼部分，有一个展览，讲述了拉德方斯及大拱门规划建设的过程，他们看得兴致盎然。但在广场上被太阳晒得无处躲藏的不好感觉一整天都挥之不去，他亲身体验的拉德方斯在他心中成为一个太大的负面典型，在以后的规划管理和建筑方案、城市设计方案审查中经常会被他提起，提醒设计师们用相同的比例横向对比一下，以找到合适的尺度。

塞纳河水奔流不息，站在塞纳河边，眺望着西岱岛上的巴黎圣母院，诸葛民激动的心情渐渐平静。2000 多年前，在不到半平方千米的西岱岛上生活着数百个居民，公元 4 世纪，罗马人的一个部落强占岛上高卢人的村庄，建立了“巴黎吉”人的首府，巴黎从此得名。从公元 6 世纪起，巴黎成为法兰西王国的首都。1400 多年来，历代王朝均以巴黎为国都。1789 年 7 月 14 日，巴黎人民捣毁巴士底狱，由此爆发资产阶级大革命，发表人权宣言，废除君主制。自此，巴黎进入快速发展期，尤其是拿破仑三世时，委任巴黎行政长官豪斯曼实施了著名的大规模巴黎重建，使巴黎形成了今天的面貌。

尽管豪斯曼拆除了大量老旧建筑、拓宽了街道，引起很大争议，然而，今天巴黎严整的城市格局确是豪斯曼奠定的，他规划了重要的街道、广场和大量公共建筑，确定了道路行道树、路灯，确定了两侧的建筑高度和形制。此外，他极富创造力地改造了巴黎的上下水系统，建设了大量的开敞空间和公园绿地。可以说，没有豪斯曼改造就没有今天的巴黎。

天津考察团从多视角考察塞纳河两岸。首先查看了滨水岸边亲水步道。巴黎的塞纳河畔（Banks of the Seine In Paris）1991 年被列入世界遗产目录。世界遗产委员会的评价是：从卢浮宫到埃菲尔铁塔，或是从协和广场到凡尔赛宫，巴黎的历史变迁被看作源于塞纳河。豪斯曼的宽阔的广场和林荫道影响了 19 世纪末 20 世纪初全世界的城市设计。巴黎圣母院、圣心教堂、卢浮宫等古建筑代表了不同时期不同艺术风格的建筑，名胜古迹沿塞纳河畔汇集。巴黎的美，与塞纳河分不开。有了这条缓缓流过的河，巴黎才显得温情脉脉、风情万种。塞纳河是巴黎的母亲河，是巴黎的魂，是一幅流动的画卷。

对于时间有限的游客，一般都选择乘游船观赏塞纳河风光，能大致浏览塞纳河两岸的重要建筑和景观。而想要真正了解巴黎、了解塞纳河，必须步行在河两岸，走走停停，或进入著名景点参观，或在岸边徜徉，看着画家给游客画像；也可以在 200 多个巴黎著名的“绿书籍”旧书摊上寻找宝藏和纪念品；还必须下到亲水平台步道，或在岸边席地而坐，看着来往的游船，甲板上的游客与岸边的人相互打招呼。

塞纳河全长 700 多千米，流域面积 7.8 万平方千米，长度是海河干流的 10 多倍，流域面积是海河流域面积的 1/4。塞纳河发源于勃艮第朗格勒高原，最高山 470 米，从源头出发 40 多千米高度就下降了 250 米。巴黎海拔 24 米，位于塞纳河中段，距出海口 365 千米。塞纳河因而总体流势平缓，适合航运，又因其水质无特殊要求、涨落有规律而更加适航。塞纳河流经的巴黎高地是法国最富饶的农业地区。塞纳河从盆地东南流向西北，到盆地中部平坦地区，流速减缓，形成曲河，穿越巴黎市中心，最后在芒特拉若利下方穿过诺曼底奔向位于英吉利海峡的河口湾。

在巴黎城区段，塞纳河已经过整修，两岸码头之间的河道已经变窄。市中心沿河十多千米都是石砌码头和堤岸。由于平时水流与洪水位相差有 6 米左右，所以河边码头和亲水堤岸距城市路有 5 米多的高差。亲水堤岸总体感觉非常自然简朴，桥下都是贯通的，上层堤岸或通过台阶或通过大的坡道相连。塞纳河两岸的亲水堤岸步道和陆上堤岸步道是巴黎最富有诗意的地方。

巴黎人将塞纳河以北称为右岸，右岸有香榭丽舍大街、凯旋门、卢浮宫和众多高级百货商店、精品店及饭店；而河以南被称为左岸，这里有许多大学、学院等文化教育机构，由于文艺知识界集聚在左岸，于是各种书店、出版社、小剧场、美术馆、博物馆等逐渐建立起来。咖啡馆、啤酒馆也应运而生，成为左岸知识分子重要的聚会场所。当你随便走进一家咖啡馆，一不留神就会坐在海明威坐过的椅子上，遇到萨特写作过的灯下、毕加索发过呆的窗口。巴黎人说笑话，“左岸用脑，右岸用钱”。在左岸散步，从你身边匆匆而过的不是教授就是大学生；而在右岸，走路时注意不要踩到别人的脚——那十有八九是一双穿着高级皮鞋的银行家的脚。

塞纳河上有 30 多座桥梁，建于不同的历史时期，千姿百态，有许多的故事。不止一座桥的栏杆上挂满了情人锁，密密匝匝，大大小小，各式各样。情侣们来到巴黎，在一把锁上写上两人的名字，然后锁在桥上，相拥亲吻，将钥匙扔到塞纳河里，代表着永久的爱情。最早开始挂锁的桥是艺术桥，也被称为爱桥，1802 年，拿破仑下令，该桥的风格要仿英式设计，这是巴黎最早的金属建筑物。2014 年，为了减轻桥上的负荷，政府拆除了 45 吨重的爱情锁。塞纳河上的另一座桥阿尔玛桥则见证了皇室的罗曼蒂克。

为了考察塞纳河桥与沿河道路的关系，考察团请陪同的法国建筑师带领参观一处立体交叉的现场，这位建筑师马上认为他们是要去观看戴安娜王妃的车祸现场。他开车带他们来到位于右岸阿尔玛桥头下的隧道，他放慢车速指着路中间的柱子说："在那儿。"这条很短的隧道是双向的，每边两车道。1997 年 8 月 31 日，为了躲避"狗仔队"的围追堵截，戴安娜王妃与男友多迪——著名哈罗德百货公司老板的儿子，在此发生车祸而身亡。汽车以高速撞到了中间的柱子上，痕迹仍然清晰可见。从此以后，小报记者们来此收集各种材料，拼凑出一个个王室阴谋论，而粉丝们不断在桥头悼念、献花。

连接西岱岛与左右岸的新桥，是巴黎现存最古老的桥，名字却是"新桥"。始建于 16 世纪末，建设了近 30 年，长 232 米，宽 22 米，于 1607 年建成，成为巴黎第一座没有建房的桥，因此叫"新桥"。塞纳河上最壮观、最金碧辉煌的是亚历山大三世桥，建于 1900 年，为了庆祝俄国与法国的结盟，是沙皇尼古拉二世出资捐赠给法国的厚礼，以上代沙皇的称号命名。俄法两国在百年前还是世仇，拿破仑曾挥师攻打莫斯科，进行大屠城，百年后，两国一笑泯恩仇。欧洲大陆一直上演着这样的戏剧。发生在更早的英法百年战争，持续了 116 年，是人类历史上最长的战争。

亚历山大三世桥是跨度为 107 米的单拱桥，将两岸的香榭丽舍大街与荣军院广场连接起来，广场上有拿破仑墓。桥两端各有两座 17 米高的桥头堡，堡顶是女神勒住飞马的金色雕塑。这些雕像，即使在阴暗的天气里，也都金光闪闪，熠熠生辉。

来巴黎的游客最终还要坐一下塞纳河上的游船，旅游旺季的游船很密集，20 分钟一班，主要线路集中在塞纳河最精彩的部分，来回约 1 小时，船票十几法郎，约合人民币 100 元。据说有夜游的船，带晚餐，要着正装，约 1000 元人民币，当时团员们觉得价格有些贵，便放弃了。游船一般是两层的，顶层是开敞的，站在上面方便观赏两岸的风光。团员们惊讶地发现，由于船身较长、驾驶舱较高，经过桥下时驾驶舱居然是可以升降的。桥的净空是大家非常关心的问题，经多方询问得知，塞纳河桥下净空 9 米才能保证双层游船通行，而天津海河上桥的净空都比较低，只有 3 米左右，这是必须研究的课题。

考察团登上了埃菲尔铁塔，俯瞰整个巴黎，充满艺术气息又整齐划一的巴黎城尽收眼底、令人赞叹，规划的严格管控和设计的精美组合让巴黎成为世界时尚之都。考察团来到了卢浮宫广场，看到由贝聿铭设计的玻璃金字塔，近距离感受到大师的精湛造诣，也理解了为什么密特朗总统力排众议，坚持让贝聿铭做卢浮宫改扩建设计。在各个展厅内，他们匆忙浏览一幅幅出自大师之手的名画杰作，挤到一群人头攒动的前面，终于看到了卢浮宫镇馆之宝——《蒙娜丽莎的微笑》。原来这幅名画，只有八开纸大小，完全不是想象中的大幅尺寸，可能越小越显示价值吧。诸葛民体会到，巴黎之所以成为世界名城、塞纳河之所以成为世界名河，除河流本身和两岸优美的建筑及城市环境外，巴黎所拥有的文化艺术内涵发挥着至关重要的作用。

除了参观塞纳河和巴黎的历史建筑，考察团特意参观了老城区的新建筑——著名的蓬皮杜文化中心，它在建设之初也曾引起了轩然大波。蓬皮杜文化中心独具个性和开创性的钢结构与各种管线暴露在外的造型，在各种书刊上很容易看到，亲身走近时，仍然让人感到十分熟悉和亲切。虽然它的造型与周边的历史建筑相比就是另类，但它的体量和尺度与周围建筑环境还是十分协调的，像是在向世人宣告，巴黎的历史是悠久的，但当今的巴黎是时尚、新潮、先锋的。蓬皮杜文化中心给国内做历史名城和街区保护做了一个独特的示范。

考察团特意参观了位于巴黎市区东北部的拉·维莱特公园，占地 35 公顷，过去是屠宰厂及批发市场，现在是巴黎市区最大的公园之一。当诸葛民置身其中，看到运河北岸矗立的未来科技馆和散布在公园中的红色构筑物，以及斜插在绿地中巨大的自行车轮胎、把手等时，他体会到了设计大师屈米心中 21 世纪城市公园的样子。业界称拉·维莱特公园开创了城市设计的新纪元。1982 年，密特朗总统提议举办公园方案竞赛，标题是“21 世纪的城市公园”，邀请了库哈斯、扎哈等国际上知名的建筑大师参赛。最后美国建筑师伯纳德·屈米夺冠，他以一种别致的、非同寻常的方式表达了公园作为城市空间延续的理念，通过点——红色构筑物（folly）、线——不规则的道路、面——绿色景观区域组合的网格体系及解构主义手法，营造出了与周围城市、公共建筑以及社区互动的场所。

考察团来到了巴黎区东南塞纳河左岸的法国国家图书馆，其造型是位于一个木制大平台的 4 本打开的书，它们围合成一个下沉的内庭院。这个项目也是密特朗总统推动的。1989 年法国举办国家图书馆竞赛，来自世界各地的 244 位建筑师投标，最终令人吃惊的是 36 岁的法国建筑师多米尼克 · 佩罗赢得了设计权。该项目 1995 年竣工，获得了 1996 年欧盟密斯·凡·德·罗奖等一系列奖项。法国国家图书馆是世界上屈指可数的大型图书馆之一，佩罗的设计颠覆了传统图书馆的设计理念和方法，成为塞纳河畔一个新的文化地标。

在离开巴黎的傍晚时分，考察团一行又来到香榭丽舍大街，树影斑驳，游人如织，两侧商店的商品广告灯光通明，宽阔的步行道和梧桐树与中间喧嚣的机动车道恰到好处地分隔开。眺望着凯旋门和远处的拉德芳斯新区的天际线，诸葛民心中感慨万千。几天来，对巴黎和塞纳河的密集参观考察，近距离看到了巴黎许多很棒的经验和做法，也切身体会到海河与塞纳河、天津与巴黎的相似点和巨大差距。正是这些相似点和巨大差距，让诸葛民进一步看到了海河和天津城的发展潜力和巨大的空间，更加坚定了将海河建设成为世界名河的决心。

离开法国，考察团接着访问了德国、荷兰、意大利等国，考察了法兰克福、阿姆斯特丹、威尼斯、佛罗伦赛和罗马等六大城市和若干城镇。因为有了巴黎塞纳河之行，在法兰克福，诸葛民第二次来到美因河畔时，感觉和第一次有所不同，城市中心段的河畔不仅可以做生态，还可以更城市化。阿姆斯特丹的运河、威尼斯的水城让他印象深刻，城与水可以和谐共处，经典的圣马可广场让人流连忘返。

佛罗伦萨是 500 年前文艺复兴的发祥地，是欧洲文化艺术中心，一条阿尔诺河贯穿市区。市中心有 7 座桥梁，其中建于 1345 年的廊桥最为著名，桥分两层，下层为店铺，上层是一条长廊，连接两岸的是旧宫和比蒂宫。廊桥也叫作旧桥（Ponte Vechio）， Vechio 是古老的意思，廊桥是佛罗伦萨很古老的圆弧拱桥。历史上，桥上都有商业功能，佛罗伦萨的廊桥与我国一些地方的廊桥，如湖南凤凰古城的虹桥非常类似。看着佛罗伦萨廊桥，诸葛民开始琢磨在天津的哪个地方适合建这样一座廊桥。佛罗伦萨只有 100 平方千米，40 多万人口。他们来到位于高处的米开朗琪罗广场，广场上矗立着大卫雕像。回身俯瞰佛罗伦萨老城，老城保存得十分完整，在落日斜阳映衬下，老城建筑红色的屋顶、米色的墙面散发出温暖的感觉，穿城而过的阿尔诺河的河水也染上了金色霞光。河上的桥梁留下了金色的剪影，如诗如画！

最后一站来到有 2500 年历史的世界文化名城罗马。罗马作为意大利的首都和最大的城市，其市区也只有 200 平方千米，人口近 300 万人。罗马城建于台伯河下游的七座小山丘上，有“七丘之城”的说法，是古罗马帝国的发源地。罗马因历史悠久，有“永恒之城”“万城之城”之称。市内的梵蒂冈是天主教教皇和教廷的驻地，有700多座教堂与修道院、7所天主教大学，与老城内众多的文物古迹相比，不足百米宽的台伯河稍显逊色。1980年，罗马的历史城区被列为世界文化遗产。行走在历史古迹中间，不论是万神庙，还是大斗兽场，耳边回响起历史的声音，心中体验到历史的厚重。

30. 国外考察技术报告

回到天津后，诸葛民开始执笔写调研报告。几易其稿，最后定的题目是“观念创新，实现天津城市规划建设跨越式发展”。提笔之初，诸葛民就决定不写学术文章，而要写一篇通俗易懂的技术调研报告。文章反复斟酌，写了1万多字，分为三部分。

在文章的开头，诸葛民写道：在考察中，我们经常换位思考，假如让我们来规划管理欧洲今天的城市，或反过来让欧洲人来规划建设我们的城市，结果会怎样？尽管这种可能不存在，但是通过换位思考，我们认识到城市规划建设的一些基本规律、原则是共通的。我们需要用历史的眼光看欧洲城市发展的经验教训，同时我们也不得不用今天与西方同样的标准来衡量我们的城市规划建设工作。

借鉴欧洲的经验，我们提出要树立“景观环境经济”“城市特色”“城市规划建设即城市文化建设”“城市经营”和“分区规划建筑管理”等五种观念。在本次考察中，我们印象最深刻的，就是所有先进的城市无不把景观环境建设放在非常重要的位置，而且持之以恒。这是21世纪人类文明建设特别是城市建设的大趋势。首先，只有营造“天更蓝、地更绿、水更清”的清新环境，才能使市民生活得心旷神怡，才能广泛吸引人才，让人们安居乐业，从而大大促进经济发展。其次，环境景观本身就能直接创造巨大的经济效益。

我们看到，在巴黎、罗马、威尼斯、佛罗伦萨等全球著名的城市，世界各地的观光游客络绎不绝。据称巴黎年旅游收入高达40亿美元。埃菲尔铁塔、塞纳河、卢浮宫和凡尔赛宫聚集着世界各地的游人。美丽的城市景色和著名的景点就像开足马力的印钞机一样为巴黎带来源源不断的财富。巴黎人为保护城市景观环境所做的努力得到了巨大的回报，为子孙后代留下了一个聚宝盆。此外，景观环境是提高市民文化品位的重要媒介。

丘吉尔说：“人们塑造了环境，环境反过来塑造了人们。”除去各种文化设施的建设外，优美整洁的城市环境同样能够约束和规范人的行为、陶冶人的情操、提升居民文化品位。单纯追求眼前的经济效益、为部门的利益而不惜破坏景观环境，长远上将给我们造成更大的损害。如果巴黎、罗马、威尼斯等城市不注意城市历史和景观环境的保护，就不会有今天的辉煌。天津城市规划建设要全面提升水平，就必须以创造优美的景观环境为最终目标。要形成牢固的观念和制度，使天津的城市形象向国际性城市转变。

强调个性和特色一直是人类社会追求的目标。20 世纪以来，伴随着工业化的发展，出现了城市和文化趋同性现象。特别是进入 21 世纪，随着知识经济和全球经济一体化的不断发展，这一问题更加突出。欧洲城市的特色美又一次让我们体会到了个性与特色对城市景观环境的重要性。比如巴黎的旧城和塞纳河、罗马的古迹、威尼斯水城、佛罗伦萨的桥、音乐之都维也纳、阿姆斯特丹的运河，它们独特巧妙、因地制宜的城市布局和特色鲜明、多样统一的建筑风格，让我们留恋、赞叹。

一个国家、民族乃至一座城市强调个性与特色是其在全球舞台上立足的资本，是在当今全球城市竞争中的一个重要法宝。让自己的特色成为金字招牌，成为城市强大的凝聚力和号召力。城市特色由多种因素组成，自然的、人工的和历史的等，其中，一座城市的历史文化遗产是城市特色最重要的组成部分，是非常宝贵的财富。欧洲各个城市都把文物古迹的保护放到了非常重要的位置。城市的文物古迹是城市文化的重要组成部分，反映了城市的文化品位。

对城市历史遗产的发掘、整理和开发利用不仅可以带来旅游收益，同时也是教育市民的重要手段。天津作为我国近代史上最重要的城市，留有大量的历史遗产，像五大道、意租界、解放北路等人文景观，以及海河这一独特的自然景观。我们应大力挖掘、弘扬天津城市历史文化传统，使其成为发展旅游业、创造城市特色和提升城市文化品位的重要手段。

考察欧洲的城市，回顾世界文明发展的历史，我们看到城市规划和建设一直是世界文明的重要组成部分。城市是石头书写的历史，建筑像凝固的音符，让我们去读、去欣赏、去聆听。而这一点我们有时会因为眼前的问题、困扰而淡忘。20 世纪，城市的飞速扩张带来众多的城市问题，而现代工业技术的发展有使许多幻想成为可能。错综复杂的社会问题、五光十色的现代技术手段，使我们对“建筑既是技术，又是艺术”这一千古命题产生了怀疑。因此，才出现了“功能主义”“住宅是居住的机器”“国际式建筑风格”等潮流和现象。

进入 21 世纪，人类的认识得以回归，欧洲又一次兴起城市文化复兴运动。历史再一次证明，城市规划建设不单单是简单的物质劳动，更是人类的精神追求和创造。今天，一座城市如果没有文化探索而单纯追求经济发展，就不会有真正长远的经济发展。同样地，一座城市的规划建设如果单纯以经济效益为最高目标，就一定不会是一座好的城市。城市规划和建筑对城市文化的影响是极其深远的，而城市道路交通等大型基础设施对城市长远发展的影响同样是巨大的，这种影响经常成为城市历史、文化的一部分。

我们可以从欧洲城市布局、开敞空间与大型基础设施之间不断变化的关系看出这种影响。因此，对于城市道路、桥梁和基础设施的规划建设也必须从城市文化影响的角度进行深入的思考。城市规划不仅仅是对形态和功能的摆布，城市建设也不仅仅是盖一座房子、修一条路。城市规划和建设是一种城市文化的探索。要全面提高城市规划建设的水平，就必须把城市规划建设作为城市文化建设的重要组成部分，把城市建设提升到文化建设的高度来认识和对待。

一个特大城市，要保持城市经济长期繁荣和较高的建设水平，就必须树立城市经营的观念，把城市发展做通盘的考虑。交通拥挤、环境污染、住房短缺和犯罪率高是西方大城市曾经遭遇的"城市病"，也是几十年来我们在城市规划建设中一直试图避免的。今天，我们再访欧洲，发现这些城市问题逐步得到了解决。

交通拥挤曾经是欧洲各大城市最为严重的痼疾，同时带来空气污染。今天，随着社会经济发展和科技进步，欧洲各大城市都建立了城市综合交通系统。通过规划布局调整、大力发展公共交通、建立现代化交通管理系统和技术进步（如建立汽车排放标准）等手段，较好地解决了交通拥挤这一难题。另外，随着欧洲统一的一体化进程，欧洲各国取消了边界，纵横欧洲大陆的高速公路网使得越境交通非常便捷，大大促进了各国之间的人员交流和经济交往，缩短了欧洲的时空距离。交通便捷已成为欧洲经济发展的一大优势。

所以说，面对城市问题，不能采用"头疼医头，脚疼医脚"单打一的方法，而必须以社会经济发展、科技创新为基础，用综合的手段加以解决。对城市的经营，不能仅仅局限在建成区，还应该包括城市周围的土地和环境资源，必须建立土地资源经营和环境资源经营的观念。在欧洲，我们所到之处，不论是在城市还是郊区，满眼翠绿，碧水蓝天。要使城市规划建设达到一个高水准，单单的城市黄土不见天是不够的，必须考虑城市周围的区域环境。奥地利的小镇凡尔登，不仅有漂亮的建筑，更表现出对土地和自然环境的尊重，让人感受到人间天堂的境地，这才是制胜的根本。

分区是城市发展的必然产物。随着城市的发展，不同时期建设的城市形成了各种不同的历史分区，各类城市用地功能相对聚集而产生了不同的城市功能分区，每个分区都有各自显著的特点。在欧洲，城市规划建筑管理基本上是根据分区不同特点而制定相应的管理规定，不像我们目前简单的“一刀切”。罗马对古城进行严格的保护控制，新建筑必须尊重周围的历史环境，所以基本保持了古城历史的原貌。

巴黎总体上采用严格保护老城区、另辟拉德方斯新区的规划布局，新区与老区采用完全不同的规划和建筑管理手法。老城区制定了严格的规划建筑管理规定，以保持其传统风貌格局。规划严格限高，不允许新建筑高度超高，街道建筑退线一致，檐口高度一致，建筑体形和风格一致。通过百余年的严格管理，使巴黎成为整体感很强、非常优美的世界著名城市。法兰西民族是个浪漫的民族，但到了巴黎之后，我们透过城市的外观，也看到了这个民族内在的秩序美。巴黎老城区唯一一栋超高层建筑——巴黎邮政局，已成为破坏巴黎古城风貌典型的反面教材。

分区管理的结果体现出城市街区的完整统一和城市景观的多样变化，其内涵反映出规划管理法规的严肃性、连续性和不可动摇性。随着时代的发展和变迁，建筑的功能、性质和现代化程度可以随之改变，建筑形式也可以在严格遵循历史文脉的前提下不断发展创新，但严格的建筑高度、整齐划一的建筑退线是永远不可逾越的，即使像法国巴黎“最前卫”的蓬皮杜文化中心，什么都变了，但其与周围建筑的退线和高度却没有变。

“百花齐放，百家争鸣”是我国文艺创作的一贯方针。在建筑创作上，我们也长期引用了这样一个方针。总体上对鼓励建筑创作的繁荣是有利的。但是，建筑又具有自身的特殊性，它必须置身于城市一个特定的场所，必须考虑与周围环境的协调，否则只能产生混乱。我们经常看到一组建筑，既有中式的，又有欧式的，还有现代的，毫无关系地堆积在一起。欧洲的经验证明，分区规划建筑管理是行之有效的规划管理方法，其本质反映了城市的秩序和多样统一的美学原则，它强调历史文脉对城市和建筑的重要性，对创造富有特色的城市整体形象是非常重要的。天津曾是八国租界，每个分区都有各自的特点。因此，提倡分区规划建筑管理对我市来讲尤为重要。

在总结欧洲经验和树立新的观念的基础上，我们提出天津的跨越式发展是经济的跨越式发展，是城市空间的跨越式发展，同时也是城市文化的跨越式发展。城市规划建设应融入世界城市化的潮流中去，去提升自己的文化品位。由此提出了构筑天津历史文化遗产的“白金项链”、建设中心城区高架快速的“金色通道”、建设中心城区的“水上新城”、建设滨海新区“拦海大坝”、形成40千米“黄金海岸”、实施规划建筑分区管理、营造“万国之城”、以“公建带开发”、形成城市建设“公私合营”新机制等具体的规划建议。其中构筑天津历史文化遗产的“白金项链”是本考察报告提出的主要的建议。

在我们赞赏欧洲各国名城的历史文化遗产时，总是不断地联想到天津。天津历史可上溯几千年，设城建卫600余年，是一个集近代中西建筑文化于一体的城市，有大量的历史文化遗存，大有文章可做。学习欧洲城市的经验，我们建议将天津老城区著名的景点通过规划的道路和河流连接起来，构成一条反映天津历史文化遗产的“白金项链”。它既是市区发展旅游业的主线，同时也是展示天津历史遗产、文化品位和城市形象的窗口。

具体将形成两条线路。第一条线路是水上线，突出天津北方水都的自然特色。以海河为主轴，延伸到北运河、南运河、子牙河和新开河，通过津河、卫津河、月牙河形成一个整体，将沿线景观串接起来，组成“蓝色珍珠项链”。第二条线路以天津市核心区为主体，以张自忠路和海河东路、北马路、城厢中路、南门内大街、南京路、成都道、马场道、曲阜道组成环路，全长约15千米，将老城厢、南市、和平路、滨江道、五大道、小白楼、解放路、意租界和天津站等著名景点联合起来，形成一条“白金项链”，并向中山路、大直沽等地区延伸。沿路开辟旅游公交专用线和旅游配套设施。

整个线路围绕的地区约8平方千米，目前已改造整理了五大道、和平路、滨江道和意租界等地区。近期工作的重点主要是海河两岸的整治，沿岸以突出老城厢和各个租界的建筑特色为主题构思，形成天津独特的滨河景观。同时，集中力量拓宽张自忠路和大沽北路部分路段，解决该地区交通不畅的问题。通过对天津历史遗产的挖掘、保护和再利用，充分展示天津的历史文化遗产和文化品位，提升老城区的综合功能，使天津成为国际知名的旅游城市，打造我们自己的“印钞机”。

同时，对于如何打造这样的“印钞机”，考察报告也提出了具体的实施思路。城市发展建设和城市经营必须走“以发展求出路，以综合求提高”的道路。实践证明，我市“以路带危改”是非常成功的经验。政府投资道路和基础设施，一方面拉动了成片危陋平房改造，另一方面也改善了城市功能，吸引了投资，促进了经济发展，是一举多得的胜招。从国外城市发展的历史中，我们能够得到相似的印证。巴黎原本是中世纪从塞纳河中心岛——斯德岛上发展起来的城市，城市内的道路蜿蜒曲折。工业革命使得巴黎以老城为中心快速、无序地向外膨胀，导致城市拥挤、混乱不堪。18世纪中期，时任巴黎行政长官的豪斯曼进行了大规模的城市改造，将蜿蜒曲折的小路拓宽为大马路，建设了环城道路、放射状的城市轴线和广场，形成城市良好的骨架。豪斯曼拓路的真正目的是为了方便军队进城镇压平民起义，但客观上起到了促进城市发展的作用，为巴黎成为当今世界之都打下了坚实的基础。

在欧洲，城市建设的资金主要来自政府和民间两大部分。具体的操作是政府通过规划引导、投资城市大型基础设施和公共建筑，以及营造良好的投资环境，吸引非政府的民间投资。因此，政府的城市政策和公共投资可以起到促进地区发展的重要作用，而民间投资则是城市建设的主体，城市开发建设必须“公私合营”，才能取得最好的效果。法国国家图书馆的建设极大地促进了周围地区的建设热潮，就是一个很好的例子。

如何解决交通拥堵问题是考察报告的另一个重点内容。市区内交通拥挤是摆在我们面前的一个越来越急迫的问题。近几年，结合危陋平房改造，我市中心城区拓宽改造了20余条道路，大大改善了市区的道路交通状况。然而，随着城市社会经济活动的增长和机动车保有量的不断增加，以及开发强度的加大，交通拥挤问题仍然十分突出，而且有进一步恶化的趋势。从国外的经验看，解决城市交通问题有3种主要模式：一是限制私人小汽车的发展，大力发展公共交通，如中国香港地区；二是鼓励私人小汽车的发展，如美国洛杉矶；三是大力发展公共交通，同时合理发展私人小汽车，如欧洲大部分城市。

过去几十年来我市与国内其他城市一样，一直采用第一种模式，即采取严格限制私人小汽车发展、大力发展公共交通的城市交通发展战略。近年来，我市通过多种改革措施，大力发展公共交通，取得了很大的进步。目前，地铁一号线工程已经开工，必将进一步促进公共交通作用的发挥，自行车过多的问题也会随之逐步得到解决。小汽车交通已经成为我们面临的主要问题。从当前的实际情况看，限制私人小汽车的发展已明显地不合时宜。我们应从实际出发，调整思路，在大力发展公交的同时，适应私人小汽车合理发展的需求。

从欧洲的城市看，在依据不同的城市分区采取不同的交通政策的前提下，主要通过采用现代化的交通管理和提高机动车车速两个手段来解决小汽车发展问题。城市老区由于道路狭窄、停车紧张，从客观上限制了小汽车的进入，通过交通组织管理和现代化的信号系统能基本解决老城区的问题。在城市外围的新区主要是采用建设快速路的方法，车速的提高既提高了城市的效率，又减少了拥挤和汽车尾气污染。天津必须建立快速路系统。

经过研究分析，我们认为，就中环线的位置、现状情况和重要性来看其维持现有功能为好，而目前准备实施的东南和西南两个半环现规划为准快速路，与城市现有道路不封闭，虽然建成后可以分担中环线的部分压力，但作用不明显。我们建议将东南和西北两个半环连接，建成高架、全封闭的环城快速路系统。高架环路初步规划40千米长，设计时速80～100千米，双向六车道，估算造价40亿元，全线两年即可建成通车。高架快速环路的建设，在保持原三环十四射道路网的基础上，叠加了一层空中的快速路系统，将极大地补充和完善现有的道路系统规划，改善道路交通状况，形成具有天津特色、国内领先的道路网系统。

规划高架快速环路连接北辰、河北、河东、东丽、河西、津南、南开、西青、红桥等九个区，各区之间的时空距离因此大大缩短，从高速公路进出市区不超过半小时。高架环路的建设不仅解决了交通问题、大大提高了城市的速度，同时改变了中心市区落后的单中心放射的规划结构，以高架快速环路为骨架形成环状组团式结构，将极大地推动我市中环线与外环线之间新区的发展。高架环路同时将天津中心城区一些重要的公共设施和新型居住区联系起来，如丽苑居住区、华苑居住区、梅江居住区、华苑产业园区、水上公园、堆山公园、天津体育中心、卫南洼风景区、侯台风景区等，形成城市新的景观带，并极大地促进沿线土地的开发和房地产的升值，增加城市收益，成为一条促进城市发展的“金色通道”。

同时，将城市总体规划中确定的从中环线到外环线的放射状快速路采用现有道路封闭或利用地铁预留线建设高架路的方式一直延伸到内环线，形成从城市中心区到高架快速环路的9条放射状通道，与高架快速环路共同组成轮辐状的快速路系统，使市中心车流能够便捷地上下高架快速环路，解决中心区与外围区的联系问题。建设中心城区高架快速路系统将综合解决城市交通、小汽车进入家庭和新区发展问题，将成为天津中心城区综合解决交通问题和新区发展的制胜一招。

在考察报告的最后，诸葛民用精心规划、科学决策、放眼未来12个字来总结。城市规划建设是一项繁杂的工作，影响面很广，无论巨细，都是重要的政府行为。今天我们所做的任何事，一方面要尽量满足市民的现实要求，贯彻党的宗旨；另一方面要为未来的发展做铺垫，把握方向，使天津不断向国际性城市迈进。要进一步强化城市规划工作，提高城市设计水平。城市规划建设实现跨越式发展的关键是进一步提高城市规划设计水平。

城市规划作为城市发展建设的龙头，必须在规划理念、思路、方法上不断创新和提高。考察国外城市的各项建设，研究我市的城市建设和市容环境建设工作，无不与规划设计有直接关系。尽管我们一直把城市规划放在极其重要的位置，尽管规划工作不断探索创新，取得了一定成绩，但是离实现跨越式发展、建设国际性城市的要求依然有较大的差距，还需要进一步完善提高。要深化调整城市总体规划，进一步研究和把握城市总体发展的方向和思路。

从城市规划学科本身的发展看，进入20世纪90年代以来全球城市理论、可持续发展理论、生态城市理论，以及数字城市和信息技术的发展使城市总体规划的理论和方法上升了一个台阶。与此同时，国内改革开放以来城市规划建设的经验提高了城市规划的水平，也对我国传统的城市总体规划理论和方法整体滞后提出了改革的要求。目前，我国正处于社会经济的转型期，许多城市规划问题是深层次的，需要结构的调整和完善。城市总体规划是城市建设的总纲，但不是一成不变的教条。根据不断出现的新情况、新问题，根据认识的不断提高、深入，规划必须做相应的深化调整。我们上面提到的一些思路，正是对城市总体规划深化调整的一些思考。我们要在借鉴先进规划经验的基础上，对我市城市总体规划深化进行研究，进一步明确城市规划建设跨越式发展的目标。

首先，大大强化城市设计。要实现城市建设跨越式发展的目标，加强城市设计工作是行之有效的措施。我市的城市设计刚刚起步，目前正在编制14条放射线以及10余片城市重点地区的城市设计。城市设计以创造宜人的城市景观环境为主要目标，城市设计质量的好坏，关系到城市形象和建设水平的优劣。要大大加强组织工作，集中一批力量强化城市设计的编制、实施和控制管理。在目前的基础上，编制城市海河两岸、城市中心商务区、中心商业区、历史风貌区等重点地区的城市设计，做到全市重点地区城市设计的全覆盖，为分区规划建筑管理提供依据。城市设计要坚持以人为本的原则，全方位处理好城市特色景观、城市空间和功能的关系，特别从细部上满足人的活动需求。要强化城市设计对城市轮廓、街景、建筑群组、公共活动系统和城市环境小品的控制手段的研究，提出城市公共空间整理改造的实施办法。

其次，抓好总体规划的实施。规划管理的核心工作就是积极实施规划，而不是单纯被动的管理。要主动地促进和推动规划的实施，促进城市社会经济发展。要创造性地把握法律、政策、措施和管理几个层面的手段，服务城市规划建设。因此，首先要抓好队伍建设，规划管理人员必须有强烈的事业心、高超的业务能力，才能应付城市不断提高的建设要求。再者要大力研究规划实施的主体问题，研究政府公共投资与非政府组织投资的互动关系，寻找互动的途径。政府投资起到重要的引导和带动作用，民间投资是实施规划最重要的力量。最后要充分发挥各行政区的作用，在总体规划的指导下，使每个行政区都有自己新的城市中心每个中心，都建设成为城市的建设热点和亮点，真正形成中心城区理想的多中心布局结构，使城市整体实现跨越式发展。

第三，完善管理体制，促进城市规划建设综合提高。建立科学高效、协调有序的规划建设管理体制，强化市政府对城市规划建设的统一领导，加强各部门之间的综合、协调和协作。在操作上，借鉴国外经验，一是要成立城市规划委员会及其下属的城市规划艺术委员会，包括计划、规划、财政、交通等部门，发挥其综合议事的职能，提高决策效率和水平，增强实施的可操作性。二是集中力量共同制定城市建设策略，指导各部门工作。三是要进一步协调和明确规划建设管理各部门的职能，建立固定的协作制度。

第四，加强城市规划研究，为城市建设的计划和决策提供依据。城市中大量而繁杂的社会经济活动，需要城市空间的合理组织，城市规划建设的决策和计划实施必须建立在两个条件上：一是认识城市运作的各种行为的现状和发展，二是认识城市建设自身的规律。城市科学是一个理论与实践相结合，并涉及多个领域的学科群。各项研究通过政策制定、规划编制和公众参与等方式，促进城市建设决策与实施科学化。我市社会经济持续快速发展，不断向城市建设提出新的要求，但目前城市研究还没有得到足够的关注，缺乏组织。要综合政研、计划、规划、财政、交通等相关部门的业务与研究力量，组织开展政策、经济、社科、法律、科技、城市行为等多学科综合研究，摸清我市状况，认识发展规律和方向，提出应对措施，为决策提供参考，为编制规划和制定政策提供依据。

在考察报告的最后，诸葛民写道："随着我市城市经济能力、社会文明、规划管理水平和建设质量的不断提高，我们具备了面向世界水平，直接借鉴国际先进经验的条件。通过考察学习，我们提高了认识，找到了一些思路，更加坚定了把天津建设成为世界性城市的信心。只要我们真抓实干，不懈努力，一定会把天津城建设得更加美好，实现新世纪的新跨越。"

鲍尔

"诸葛总，您这个报告是 2002 年写的，到现在 17 年过去了，您还记得如此清晰，说明这个报告凝聚了您的许多经验，是有感而发。从今天的视角看，您觉得这个考察报告怎么样？当时这个考察报告发挥作用了吗？另外，这一段太长了，您能把原文发给我吗？ ""好，我发给你。这个考察报告不仅是对欧洲八国的考察报告，也包含了我多年来对西方发达国家研究的成果，以及多年来对天津规划的思考。为了引起重视，的确在许多建议上有些激进。

那一年，我 38 岁，比较年轻，血气方刚，有些问题考虑得还不够全面。从今天的视角来看，许多观点有待商榷。不过，这个报告当时的确引起了足够的重视。”

诸葛民

考察报告交上去时间不久，一天局长在走廊上碰到诸葛民，说：“报告领导看过了。”看到局长表情神秘，本想追问领导看法的诸葛民欲言又止，忍住没有发问。欧洲考察归来后，海河综合开发改造规划工作的节奏越来越快。市领导 6 次听取汇报，从规划思想、设计思路、市场化运作和启动策略等方面做指示和提具体要求。到 2002 年 7 月，规划方案基本成熟，市领导在规划局召开规模较大的会议，听取汇报，审查工作方案。参会的有市属国有开发建设企业的主要领导。可能是对海河综合开发的重要性还不明了，也可能是对海河开发的市场缺乏信心，有的单位一把手没来，派了副手参会。市领导很不高兴，直言“把他给我叫来”。会议最后，市领导提出规划要广泛征求意见。

31. 七彩的海河 · 统筹化规划

公众参与，听取各方意见，海河两岸综合开发改造规划可能是天津规划史上汇报次数最多的一次。规划小组分别向市人大、市政协领导汇报，向部分人大代表和政协委员汇报，主动上门征求相关区政府意见，多次向文化界、历史界、企业界、金融界进行专题汇报，听取各方意见。还邀请天津市著名的院士们进行评议和论证，据统计各种汇报、征求意见会不下数十场，面对面汇报达 2000 人。相貌英俊、口才好的齐所长成为汇报海河规划的专业户，不停地汇报，广泛听取各方意见，规划越来越完善，文字也越来越流畅、优美。规划将海河描绘为一条绚丽多姿的七彩之河：

海河是红色的，她是我们城市沸腾的血脉，哺育着天津从出生到成长，从幼年到成熟；

海河是橙色的，在她身边聚集着取之不尽、用之不竭的宝贵资源；

海河是黄色的，好像一部原版的史书，记载着天津从一个漕运码头发展成为我国现代化大都市的沧桑巨变；

海河是绿色的，为我们这个巨型的城市岛屿注入了清新的自然气息；

海河是青色的，她融入了大海的气魄，在天津形成了独树一帜的地域文化；

海河是蓝色的，孕育着我们建设未来生态城市的希望；

海河是紫色的，她孕育出众多独领风骚的艺术大师，展现着全天津人自豪的艺术魅力。

不久的将来，当海河两岸矗立起一座令世界瞩目的城市之时，海河也将欣然地微笑，这座水晶般的城市更会折射出海河更加耀眼的七彩光芒！

从历史文化、产业经济、景观环境、生态建设、道路交通、旅游休闲等方面，我们确定了海河两岸综合开发改造的六个主题目标：

通过海河开发改造，展现悠久历史文化——如歌岁月，魅力之河；

通过海河开发改造，发展滨河服务产业——经济腾飞，动力之河；

通过海河开发改造，突出亲水城市形象——城河一体，标志之河；

通过海河开发改造，建设生态城市依托——绿色浸润，自然之河；

通过海河开发改造，改善道路交通系统——整合两岸，凝聚之河；

通过海河开发改造，开发旅游休闲资源——人民共享，活力之河。

根据海河两岸的历史发展沿革和建设情况，规划将外环线以内的海河中心城区段划分为四个段落：自北洋桥至南马路为传统历史文化区（CHD）；南马路至赤峰桥为都市消费娱乐区（CRD）；赤峰道至奉化道为中央金融商务区（CBD）；对于奉化桥至外环线起什么名字大家费了不少脑筋，但始终不得其所，这时美国易道公司（EDAW）开始参与海河两岸规划，他们提出把这段定义为智慧城（STD，Smart City），大家异口同声地赞成。

传统文化商贸区，这一地区是天津的发源地，包括三岔河口、老城厢、大胡同、大悲院、望海楼以及古文化街等众多传统地区，集中展现了天津发源和初期发展的历史风貌。规划强化各地段之间的关系，使之成为一个相互联系的整体。同时对这一地区的整体空间形态和景观环境特色加以整顿，形成与其历史文化氛围相称的特征，使该地区成为延续天津城市历史脉络的核心区域。依托传统的文化资源，大力开展旅游和商贸活动，赋予这一地区活跃的生命力。

都市消费娱乐区，包括和平路和滨江道共同构成的中心商业区、中心广场以及铁路交通枢纽——天津站，在这一地带集中体现了现代化城市的消费娱乐功能，是我市的黄金地带。规划依托现有设施的基础，增强文化、娱乐等休闲活动设施建设，增强活力，形成现代化城市独有的中央休闲区。由于这一段也是海河水线变化最丰富的一段，规划与之相应开辟了大型的绿地与广场，创造出与城市中心相称的空间形象，使这一地区成为我市的形象象征。

中央金融商务区，这一段海河沿岸有解放北路金融区、南站和小白楼商务办公区，是城市 CBD 的核心部分。现代化经济中心城市的功能在这一带集中展现。规划在这一带则侧重于商务、办公、信息、金融和展览设施的建设，创造吸引国际化企业的良好商务环境，成为现代化经济中心的突出标志。在这一带还要发掘很多近代城市发展的重要历史遗迹，形成一个小型特色博物馆群落，展现天津近代发展过程中的丰富成果。

智慧城，从奉化道开始，海河沿岸还有大量的可改造地段，也是中心城区未来建设生态城市的宝贵资源。规划以可持续发展和生态建设为主题，在这一带构造新型的城市形态，重点建设以先进的网络技术与智能化技术为核心的、以松散的城市形态创造具有高产出和高附加值的新产业区。结合柳林风景区建设集娱乐设施和风景园林于一体的综合性游憩设施，使整个地段展现出国际化大都市所具有的高环境质量和先进居住设施水平。

在广泛地征求意见的过程中，项目组成员也受到了一次又一次教育，逐渐在许多方面与社会各界达成共识。天津知名文学评论家滕云说：“经济是海河的血脉，景观是海河的容貌，文化是海河的灵魂。”在天津市文联主席冯骥才眼里，海河又是天津的“魂”。天津人一直非常重视海河，多年来，在方方面面的关心和共同努力下，海河两岸的城市建设取得了很大成绩。但是，与世界上著名的河流相比，海河还潜存着进一步发展的良好条件和机遇。通过反复研究海河两岸开发改造规划所面临的形势和问题，大家深深地领会到这一重大工程实际上是天津宏伟发展战略体系中的一个重要步骤。改革开放以来，天津市先后实施的一系列重要工程已经为海河开发改造打下了坚实基础。

首先，城市中心地区的防洪圈和海挡的实施和建设，作为城市的“生命线工程”，从区域范围内协调了海河的防洪、泄洪功能，使海河有条件成为一条生活景观型河道。第二，危改工程解决了市民最紧迫的居住环境的改善问题，为海河开发改造这一城市环境景观建设工程打下了扎实的群众基础。第三，工业战略东移使得海河两岸大量的工业企业有了新的发展空间，腾出了开发改造的土地和空间资源。第四，对城市地下管网特别是排水管网的改造改善了排水体系，减轻了海河的负担，也使海河开发改造的投资成本大大降低。第五，津河、卫津河等二级河道的改造整治改善了城市景观，解决了水系连通的问题，为海河的开发改造积累了经验，做好了舆论准备。第六，经过多年经济快速发展，天津逐步把产业发展的重点转到了第三产业，而海河正是城市第三产业发展的主线，产业发展的时机也已经成熟，景观建设与经济发展也就有机地联系在了一起。以上工程从各个方面为海河的开发改造做好了准备，使海河两岸的开发改造"水到渠成"。如果没有这些工作做基础，海河两岸的开发改造就会面临众多难以解决的矛盾和问题，难以从构想变为现实。

按照"高起点规划、高水平设计、高标准建设、高效能管理"的要求，海河综合开发改造规划的高起点首先就是要突出海河的独特性，突出天津城市的特色，创建世界名河。通过与社会各界不断沟通，在树立更高标准、创建世界名河方面社会各界基本认可。

经过各方群策群力，确定了本次海河综合性开发改造规划的宗旨："统筹化规划、市场化开发、法规化管理"。通过借鉴天津市危改工程和国内外河流建设的先进经验，从城市经营出发，统筹规划海河两岸的资源，充分发掘海河两岸的土地资源、文化资源、景观资源、品牌资源和经营资源，大力发展旅游、交通、商贸等服务性第三产业，让海河成为天津人民共建共享的河流，成为独具特色、国际一流的服务型经济带和景观带，成为展示天津现代化国际港口大都市的标志性区域。

规划突破了规划专业领域的限制，更加突出、强化了规划的全局性、战略性和综合性，力求通过这一规划使城市整体功能得到优化和提高。海河综合开发改造的六个主题目标很好地反映了统筹化规划。

第一个目标：通过海河开发改造，展现悠久历史文化。海河就是天津城市发展的历史之河，是中国近代史之河，它记载了天津城市的历史，孕育了天津的城市文化。由于对历史风貌的保护力度不够，致使海河的历史风貌没有完整地体现出来。规划以海河自然风景轴线为中心，将天津发祥地三岔河口、老城厢与各个时期的近代风貌建筑保护区、城市风景副轴线等相关景观元素串联和组织起来，以此反映传统的城市空间风貌和街市生活面貌，展示城市的持续发展和文化特色。保护和恢复海河沿岸的文物和风貌建筑，保护和恢复沿线的历史桥梁、构筑物和历史遗存，保护传统历史风貌街区。建设海河历史博物馆、三条石中国近代工业博物馆、城市历史展览馆等纪念设施，建设海河音乐厅等文化设施，建设展现历史风貌的雕塑等艺术作品长廊，使海河成为天津历史活的博物馆和市民文化活动的中心场所，弘扬海河文化，展现出天津的历史文化神韵，提升城市文化品位。

第二个目标：通过海河开发改造，发展滨河服务产业。在海河沿线周围有和平路、滨江道商业区，有中心广场、东站、古文化街、解放北路金融街、小白楼商务区、十一经路商贸区等城市的公共中心，但商贸经济活动与海河仍有一路之隔，城市仍然是背向海河。海河沿线用地构成中，商业、文化设施占地较少，居住、工业仓储用地所占的比重较大。同时，土地的使用效率较低，建筑质量较差。规划创建以商业、贸易、服务、文化、娱乐、金融等公共设施为主的，充满活力且繁荣的滨河经济开发带，带动第三产业发展，增强经济活力，增加就业岗位。加快沿线城市中心商务区、中心商业区的建设，完善海河的经济和创业功能，进一步优化调整海河两岸的用地结构，建设标志性的经济中心区，吸引国内外经济活动的开展，提高城市经济地位。

第三个目标：通过海河开发改造，突出亲水城市形象。经过多年的城市建设，海河两岸已形成了较好的环境景观。海河水质得到改善，堤岸得到整治，修建了沿海河的带状公园。基本保留了和平路上劝业场、百货大楼面向海河的景观展示线，恢复了望海楼，新建了天津站及附属建筑，改造了赤峰桥东岸和大光明桥西岸，构成了海河两岸景观的基本骨架。同时，海河两岸存在的问题也是相当突出的。如景观点较少，缺乏高潮，较平淡。在市中心段，一些滨河高层建筑如凯悦饭店、国际航运大厦等退线不够，板式建筑沿河布置，且建筑体量较大，使海河过于封闭。河岸由于防洪墙的设置使得海河岸线呆板，缺乏闪光点。细部处理没有特点，河岸处理较粗糙，桥梁设计没有特点（如桥的造型、灯饰等），两岸还存在一些卫生死角急需改造。海河纵深景观处理不好，缺少中景的灰色过渡空间。河道水质较差，水面上的娱乐开发项目较原始、简易。沿河交通阻塞严重，城市噪声过大，也是影响城市形象的一大因素。海河东岸赤峰桥到刘庄桥段大部分为工厂、仓库建筑，厂房破旧，景观很差，夜间景观更显沉闷、死寂。刘庄桥以下景观质量较差，从刘庄桥到沙柳南路段为工业区，两岸绿化较少。沙柳南路以下除少数树木外，基本没有绿化，沿河一些村庄为土坯平房，景观更加暗淡。规划确定海河的整体景观风格、景观分区和重要的景观节点，完善城市整体空间形态控制。在环境设计层次上，按照“绿起来、亮起来、美起来、活起来”的思路，进行海河沿线的绿化、灯光、环境设施和景观建设，力争用较短时间，使海河景观焕然一新，使海河两岸地区新建的每一栋建筑、每一座桥梁、每一项设施都成为一个新的滨河景观。

第四个目标：通过海河开发改造，建设生态城市依托。天津市地处海河流域下游，素有“九河下梢”之称，北有蓟运河、潮白河、北运河、永定河，西有大清河，南有子牙河、南运河诸水系，各水系汇集天津入海。早期海河是条潮汐河道，由于上游来水量充沛，足以冲污压咸，保证城市用水。进入20世纪50年代以来，由于上游来水量逐渐减少，加之城市日益发展，用水量逐年增加，用水量明显不足，特别是冬春季节表现更为严重。1958年，市政府通过了《关于海河改造工程的报告决议》，决定兴建以“清浊分流、咸淡分家”，保证城市用水为目的的“海河改造工程”。工程主要是使海河干流由潮汐河变为蓄水河，海水不再上溯，实现“咸淡分家”。兴建了海河防潮闸，同时进行排水设施改造。海河水系历史上曾发生过360多次水灾，其中有45次殃及天津，仅20世纪上半叶就有1917年和1939年两次特大洪水淹没天津市区。中华人民共和国成立后，1954年和1963年又出现两次特大洪水，全市人民进行了空前规模的防洪斗争，保证了市区的安全。海河在历史上是北运河、大清河、永定河、子牙河、南运河的入海尾闾，20世纪六七十年代根治后，部分河系另辟入海通道。从可持续发展的角度看，要成为世界名河，海河必须建设成为生态河流。天津中心城区有一级河道与二级河道和水面组成的城市水系，海河是其中的主要河流。此次规划通过科学调度手段，使城市水系循环起来，逐步恢复河道的生态功能。规划通过河道这一重要载体建立人与自然相互协调的关系，通过建设形式多样的滨河公园，创造丰富多样的生态空间，为生物物种的丰富创造有利条件，为多种多样的植物提供良好的生存环境。滨河绿地是城市的绿色通风走廊，在滨河绿地建设中，可以充分考虑物种间的生态特征，合理选配植物种类，形成结构合理、功能齐全、种群稳定的复层群落结构，既能保护物种多样性又能形成优美的景观。同时通过改善水边环境和创造独具特色的景观来增强城市给人的愉悦感；通过海河河道生物活动的物质循环和能量流动产生生态效益；通过植物景观所构成的美化城市环境和为人们提供游憩空间，提高居民生活质量，创造减灾条件所产生的城市安全效益，改善城市投资环境和促进旅游的发展所派生的多项经济效益，达到对海河综合开发、建立生态健全环境的目标，创建生态之河。为了实现这一目标，首先，规划从流域范围综合解决海河防洪问题，保障城市的生态安全。其次，以海河为主干，沟通支流与二级河道，建立网络状的河流系统。第三，严禁两岸排污，治理海河水体，保证海河水质。第四，依托河道系统，在两岸建设网络状的绿化系统，使海河绿起来，为城市增添自然气息。

第五个目标：通过海河开发改造，改善道路交通系统。天津因漕运而生，应海河海运而兴。海河曾经是城市的交通中枢，两岸是城市经济活动最繁华的区域。近代，随着海河市区段航运功能的消失，海河已由城市的交通轴线演变为分割城市的障碍。我市正在编制中心城区综合交通规划，2000 年完成了我市有史以来最大的一次交通出行调查，获取数据近 20 万条。其中，对海河的道路交通问题从全市的范围进行了研究。海河的交通影响到天津市中心城区 75% 的区域。海河穿越核心区，包括大悲院、大胡同、古文化街、和平路、滨江道、解放北路、小白楼等各种中心以及铁路交通枢纽——天津站。海河沿线交通问题相当突出，尤其是沿河道路不畅、跨河桥梁过少、桥与顺河路的平面交叉等问题突出，造成人去不了、车没地儿停、去了留不住等问题，海河因此活不起来。规划以中心城区综合交通系统规划为基础，提出海河两岸地区交通整体解决方案。海河两岸地区规划有 5 条地铁线路通过，共设置地铁车站 18 座（车站距河岸 200 ~ 800 米），每处地铁车站都将成为向海河沿岸输送人流的枢纽。在海河两岸地区规划了多条与海河平行及垂直的道路，在海河沿岸地区形成由道路、地铁、公交、水上巴士、停车及集散广场、滨河散步路组成的比较完善的道路交通网络，保证海河沿岸地区交通便捷畅通。规划调整海河两岸的滨河道路，修建沿河机动车专用路，增强海河沿岸地区的可达性；明确主要过河通道，新建、改建桥梁，加强两岸的交通联系；开通海河水上巴士，开辟水上游览观光线路，建设客运码头，并与其他交通方式相衔接，减轻地面交通的压力；兼顾交通与生活岸线对海河两岸的需求，创造具有吸引力的滨河步行系统和开敞空间。

第六个目标：通过海河开发改造，开发旅游休闲资源。海河是旅游观光、亲水和市民回归自然的景观轴线，是天津旅游资源的“王牌”。海河具有十分丰富的历史文化资源，具有丰富多样的自然景观和人文景观。目前海河的旅游开发还处于初级阶段，除单一的海河一日游外，没有像样的旅游项目，尚有很大的发展空间。旅游作为新兴的朝阳产业，对城市的发展将起到十分重要的作用，海河旅游的大发展对于促进天津旅游业整体的发展、促进城市经济发展都将起到十分重要的作用。规划通过海河两岸道路、桥梁、游船、码头和市政基础设施的建设，改善海河的旅游环境；通过海河两岸岸线、绿化、雕塑小品的建设，通过传统历史风貌街区的保护和恢复，通过沿线商业、文化设施的建设和经济文化活动的增加，丰富海河旅游的内容，创造新的旅游资源。系统建设滨河自然风光旅游产业链，形成区域性独特旅游资源。

上述六大目标构成了海河综合开发的总体目标。按照这六大目标，我们对海河两岸堤岸景观、道路交通、城市用地布局与空间形态进行系统的规划设计和优化调整，保证海河两岸地区的经济中心功能和环境景观水平都得到很大提升。海河两岸的开发改造在沿岸有组织地形成了一系列经济活动中心，如运河文化商贸区、古文化街商贸区、和平路中心商业区、中心商务区，以及智慧城地区的先锋南高新技术创业区等，为城市的经济发展提供了核心空间，从而使景观环境的改造与经济功能的发展有机地联系在一起。同时，海河沿线的开发可以通过这些核心区域进一步带动城市纵深地区的发展。海河沿岸的 8 个行政区都可以以相关的区域为龙头，促进区内的城市建设发展。海河开发改造也就成为“牵一发而动全身”的龙头工程。

32. 十大工程 · 市场化建设

诸葛民

经过长时间的探讨，确定海河的改造采取综合开发、政府引导、市场运作的方式，首先通过统筹性规划明确海河两岸综合开发改造的目标。为实现打造世界名河的总体目标和以上六个主题目标，以综合性、系统性为出发点，规划确定了海河两岸综合开发改造必须实施的水体治理、堤岸改造、道路交通、桥梁隧道、通航、绿化广场、环境景观、灯光夜景、公共建筑、整修置换等十大基础工程。通过政府的部分先期投入和银行的滚动资金，先期实施十大基础工程，使周围环境和基础设施配套得以改善，从而提升和显现海河两岸土地价值，带动两岸的招商引资和经济发展。通过开发改造，充分挖掘海河两岸的土地资源、文化资源、景观资源、品牌资源和经营资源，大力发展金融、旅游、交通、商贸、娱乐、中介等服务业，促进经济进一步发展和社会全面进步，力争在开发体制、启动方式、融资机制、管理模式、规划设计等方面开拓思想，进行创新。同时，通过严格的规划管理、高水平的规划设计以及高标准的建设，使海河的开发改造达到世界领先水平，使天津发生翻天覆地的变化。

海河两岸开发改造是城市实现跨越式发展的标志性工程，必须本着“出大思路、动大手笔”的原则规划好其中的建设内容。特别是作为先期启动的基础设施，必须具有高水准，才能保证海河整体建设的高质量。实际上，先期实施的十大基础工程是由许多内容构成的，是规划局、规划院牵头，与相关委办局和专业规划设计配合完成的，为了容易记忆，归纳为十大工程，实际上每个工程内又包括许多子项目。

十大工程之首：水体治理工程。其中包括海河水资源和水系规划、水环境保护工程、排水工程以及清淤工程。海河是天津中心城区的骨干河道，要充分发挥其调蓄作用。同时也要发挥二级河道和水面的调蓄作用。目前海河上游的各支流中，只经过屈家店闸每年在春季灌溉和汛后两次向海河补充引滦水，每次大约 2000 万立方米。丰水年经海河干流泄洪。海河每年绝大多数时间为水库式河道，基本上无流量。规划海河二道闸以上为引滦入津水源的备用水库，为引黄济津水源的输水河道；南水北调实现后，海河将作为引江水的备用水库。天津市二级河道陆续改造后，与一级河道贯通。为保证二级河道景观需要，让水活起来，每年需增加对河道的蓄水，同时为保证海河水位常年保持在较高蓄水位（1.5 米），满足海河景观亲水需要，规划每年增加海河的蓄水。

目前海河水质以有机污染和水质咸化为主，汛期排入的雨污水是海河的主要污染源。海河对外部污染的缓冲能力小，每次较大的降雨后会使海河水质在短期内受到较重的污染，其中氨氮和高锰酸盐指数最为严重。海河入境水量少，平时河水滞留河床，各种污染物通过吸附、沉降等作用逐渐积累于沉积物中。随着时间的推移沉积物中各种污染物的浓度逐渐增加，并通过生物、化学和物理等因素作用，使污染物逐渐向水体中扩散，这是非汛期的主要污染源。要实现生态河流恢复和保持水体功能的目标，市区各泵站除汛期雨水外，非汛期严禁向海河排污，严禁沿河各工厂向海河排放废水。水环境保护目标水质标准执行国家地面水环境质量标准（GB3838—1988），二道闸以上非汛期为三类，汛期为四类；二道闸以下为五类。

海河两岸在市中心段的老城区，各种市政设施和管线大部分年久失修，无法满足发展要求。海河两岸地区排水体制大多为雨污水合流制，部分地区为排水空白区。合流制排水，旱季时雨污水入各地区所属的现状污水处理厂和排污河，汛期时入海河，造成海河污染。结合天津市城市总体规划，为保护海河水质，规划加强排水设施建设，将雨污水合流制改造成分流制。雨水就近由现状及规划的雨水泵站提升排入海河，污水、初期雨水，按照污水排水系统规划，分别排入系统内现状及规划污水泵站，之后排入污水处理厂。要切断海河两侧所有的污水管道，从根本上保证中心城区水系的水质。

由于对海河干流本身缺乏治理，加上两岸地面大幅度下沉，河道普遍淤积，海口段淤积尤为严重，致使河道行洪能力骤减，因此，要经常性地清淤，保持河道畅通。

十大工程之二：堤岸改造工程。现状海河护岸、河堤大多为近代修建，是随着城市发展逐步建设起来的，有的已经有近百年的历史，需要治理改造。金汤桥以上为坡式护砌；金汤桥以下为垂直高桩承台及板桩（后有拉锚）。中华人民共和国成立后，陆续对部分河道进行了旧有码头加固。1996 年对营口道至南京路段的海河右岸堤防进行加固改造，新建了刘庄桥以下段的护岸。目前，仍有部分河堤、护岸由于年久失修需要改造。本次规划综合沿线道路建设，亲水岸线、旅游、通航码头的建设，对旧有海河大堤进行加固、翻建，对沿河道路管线进行迁改。通过对海河护岸局部进行改造，采用堤防后移加高等措施，试图从根本上解决海河市区段防洪标准问题，解决防洪与城市景观的矛盾。

海河现状堤顶的高程为 4.63 ~ 5.43 米（大沽高程），高出地面 2 米左右，海河最高蓄水位 2.5 米（大沽高程），正常通航水位 1.5 米。为使人们更容易看到水面，增强亲水性，同时减少噪声，防洪河堤的断面设计成退台式，通行机动车的一侧紧邻河堤的机动车道设计标高 3.0 米（大沽高程），机动车道外侧为 20 ~ 30 米宽的绿化带，逐渐高出机动车道 1.5 ~ 2.5 米后再放坡与 7 米宽的辅道接顺；以人行为主的一侧紧邻河堤的散步道设计标高 3.0 米（大沽高程），为 15 ~ 20 米宽，与防洪设计堤顶高统一的绿化、广场平台及滨河散步道用台阶接顺，平台临 12 米宽辅道一侧设置半地下的吧廊与停车场，同时作为防洪的紧急通道。

十大工程之三：道路交通工程。海河沿岸没有贯通的平行道路。海河西岸有海河西路、张自忠路、台儿庄路，从金钢桥到第二棉纺厂贯通。海河东岸有海河东路（从金钢桥到六纬路），赤峰桥到光华桥之间只有间断道路。光华桥以下海河两侧基本没有沿岸道路。规划调整海河两岸的滨河道路，修建沿河快速路，增强海河沿岸地区的可达性。适当调整海河两岸滨河道路的线型及竖向标高，把两岸目前都能通行机动车及自行车调整为一侧做机动车专用路，另一侧为绿地、散步路及自行车道，其中在海河的核心区段尽可能地把滨河地段让出来作为开敞的亲水空间。在海河东岸，北洋桥至新开河，沿河为加高的防护绿带，其次是滨河道路；新开河至北安桥，规划为下沉式机动车专用路；北安桥至大光明桥，为地下（或路堑式）机动车专用路，彻底解决东站前广场地面交通紧张状况；大光明桥至海河大桥（东南半环），为下沉式机动车专用路；海河大桥至外环线，为 40 米宽机动车专用路，路中设置 10 米宽的绿化带，道路两侧为 30 ~ 50 米宽的绿化带，强化景观效果。在海河西岸，金钢桥至南马路，规划为下沉式道路；南马路至大沽北路，调整道路线型，将张自忠路合并到兴安路的线位，规划道路红线宽 30 米。在道路与河岸之间预留出 100 米宽的绿化带用地，该路向南与大沽北路连通至曲阜道。大沽北路道路红线由 30 米加宽至 40 米，疏解南北向的穿越交通，减轻解放北路地区交通压力；赤峰桥至光华桥（中环线），沿河为亲水平台、散步与绿化平台、半地下的自行车停车场及吧廊、辅道等一体化设计。光华桥（中环线）至外环线，为游览性道路，并与环外的道路接顺。下沉式机动车专用路段的海河防洪堤岸后退设置，与跨海河桥相交处规划桥头立交，折返交通利用南站等地下交通枢纽及部分桥头立交桥实现。在沿河道路抵达中心区的滨河广场，结合停车设施及地铁车站设置交通集散枢纽，提供换乘服务。规划分别在东北角三角地、兴安广场、东站、南站、天钢附近设置多处大型交通集散枢纽。其中南站地区规划设置 5000 个泊位的大型地下停车场，可疏解大部分通过海河东岸快速抵达中心商务区的机动车流。在建设的同时，重视道路交通管理。建立行人优先步行区，开辟公交专用路（专用道）系统，包括建国道、赤峰道、元纬路、东马路、新华路、建设路、鞍山道等专用路或专用车道，合理组织单向交通。

十大工程之四：桥梁隧道工程。现状海河上桥梁过少，现有桥梁 12 座，平均间距 1.6 千米，绝大部分为机非混行桥，而且海河交通过度集中在有限的几座桥梁上。跨海河的可通行机动车的桥梁全天交通量为 35.53 万辆，平均每座通行 2.963 万辆。其中，大光明桥高达 7.05 万辆。大光明桥、光华桥、狮子林桥、北安桥四座桥梁就集中了 57% 的过河交通量。同时桥头与沿河路交叉处绝大部分为平面交叉口，使跨海河的交通在海河两岸经常严重受阻。

为保证两岸之间交通的密切联系，分流大光明桥等交通量特别集中的桥梁的流量，规划增加主要过河通道，新建、改建桥梁，增加桥头立交，减少与沿河交通的干扰，确保穿越交通快速通过。规划桥梁和隧道共 28 座，平均间距减少到 0.68 千米，新增加桥梁 16 座，其中包括根据需要增加的非机动车和人行桥 6 座。

桥梁是城市的标志。天津由于历史上多国租界曾分别在其租界地建桥，桥梁类型齐全，有“万国桥梁博览会”之称。今后海河上新建的每座桥梁都设计建造成精品桥梁。桥梁的设计应结合城市周围环境，进行创造性的发挥，形成海河之上一道独特的风景线。

十大工程之五：通航工程。海河目前已有游览线，但由于金汤桥、解放桥等桥梁梁底中心标高过低，仅为 4.47 米，无法满足海河通航标准，在高水位时，只能断航。同时，由于航道标准低，游船船身矮，观光感较差。规划完善海河游览路线，更新海河游船，建设观光游览码头和配套设施。开通水上巴士，西可至杨柳青，北可达西沽公园、桃花堤，在二道闸设换乘码头，东可抵塘沽出海口。在大悲院、古文化街、和平路、中心广场、东站、南站、大直沽、柳林风景区等处设客运码头，与其他交通方式相衔接，减轻地面交通的压力。

通航标准，金钢桥至光华桥段规划为Ⅵ级航道，桥下净空 5.0 米；光华桥至外环线段为Ⅳ或Ⅴ级航道，桥下净空 5.5 米。按以上通航标准，金钢桥至光华桥段的大部分桥梁需要改造。其中金汤桥、解放桥提升 2.00 米后作为人行观光桥使用，提升后的桥下净空达到 5.0 米（正常通航水位为大沽标高 1.5 米）。光华桥上游新建和改建桥梁净空要求不低于 5.0 米（梁底标高 6.5 米），光华桥下游至外环线新建桥梁净空要求不低于 5.5 米（梁底标高 7.0 米）。

十大工程之六：绿化广场工程。海河两岸的绿地是开敞空间，海河带状公园是其中的主要部分。在 20 世纪 80 年代天津震后重建中，通过艰苦的努力，拆了大量的破旧房屋和临建，建设了海河带状公园，使天津城市中心滨河面貌发生了巨大变化。规划在原海河带状公园的基础上，在几个主要节点上规划了比较集中的绿地，形成点线面相结合的滨河绿化系统，规划新增绿化面积 210 万平方米。在子牙河与南运河交口处规划河口公园，在中心广场对岸规划大片带状绿地，在大沽北路至吉林路间规划集中绿地，在南站地区和天津站的站前规划比较集中的绿地和广场，增加人们交往和活动的外部空间。规划在海河中心城区段，结合周围环境和绿化，重点开辟各具特色的主题广场和绿地公园，通过广场的建设，改善周围环境，提升周边土地价值。

规划三岔河口思源广场、金汤桥胜利广场、中心广场、和平广场、天津站对岸津门广场等，规划广场总面积 20 万平方米。同时，结合沿河散步道、游览路系统的建设及亲水设施的建设，形成完善的滨河开敞空间系统。

海河两岸绿化设计由河堤、地面和垂直绿化三部分组成。这三部分绿化分别展现在三个不同标高层面上，形成层层向上的绿化景观。沿墙种植攀缘和藤本树种，上垂下攀， 郁郁葱葱，提升绿视率。为了避免遮挡沿河建筑的观赏视线，海河两岸规划道路以内的带状绿化范围内不宜种高大乔木和树冠呈尖塔状的树木，以免与沿河建筑的轮廓相重复而产生单调乏味感，设计中须采用横向的绿化界面，以衬托竖向建筑。各路段进行不同特色的设计，以达到三季有花、四季常绿的绿化景观效果，使海河绿带与沿河建筑相互辉映，成为天津市游览观赏的一大亮丽景观。海河两岸的绿化设计宜采用自然式布局，局部进行地形改造，加强其景观效果。在景域空间的总体组织上，以水、绿为景观主体，以植物造景为特色造园手法，以休闲游览、闲暇休息为主要活动项目，合理安排服务设施，配置绿化及导游路线，形成以“静”为主的特色绿地，为忙碌的城市居民提供安静、娱乐的休闲场所。

十大工程之七：环境景观工程。海河两岸的广告、雕塑、小品，包括路灯、电话亭、报刊亭、公交站、出租车站、公厕等都要有地域特色、历史文化特色，要与沿河的亲水设施、沿河空间、标志性建筑、景观建设等总体上相协调。其中，重点在沿岸结合历史文化遗迹，设立一系列环境艺术作品，如反映近代史实的浮雕、表现解放天津成功的群雕等。再补充由现代艺术家创造的反映新世纪生活的艺术作品，使海河沿岸成为艺术作品的长廊。海河沿岸还要结合现代化城市的需要，系统建设环境标志和指引设施，为旅游者和市民提供方便。海河是天津的象征，海河沿线的景观环境最直接地反映出天津的城市形象和品质。海河两岸的建设，从护岸、桥梁、绿化、小品到两侧的建筑，是天津城市人文景观的重要组成部分，它的内容反映了城市的历史、文化，要充分体现出天津城市的特色美。

十大工程之八：灯光夜景工程。制定海河灯光照明专项规划。海河灯光照明规划结合海河两岸重要节点和标志性建筑，确定重点照明区、照明段和层次，确定照明的总体色彩、风格和形式，做到既热烈繁华又井然有序。多层次的灯光夜景是海河夜景的特色。根据海河两岸规划，共分为建筑物、广场、道路和桥梁、园林绿地四类夜景灯光。总体上，桥梁是海河夜景中主要的景观，要把桥梁打亮，突出桥梁的造型。滨河绿化和小品以泛光照明为主，强调整体和私密性。广场照明是地面照明的精华，要展现华丽、跳跃的色彩。建筑物照明突出建筑的整体轮廓，突出打亮屋顶部分。海河宽度100米左右，两岸建筑灯光的倒影可以反映在河面上，河面基本不照亮，游船照明是海河灯光的重要组成部分。

十大工程之九：公共建筑工程。规划在海河中心市区段，重点建设几处标志性公共建筑，通过公共设施的建设，完善城市功能，丰富海河景观，提升周边土地价值，带动海河沿线主要功能节点的开发建设。规划十大标志性公共建筑：三岔河口天津城市历史博物馆，工业博物馆等，规划总建筑面积约3万平方米。结合周围环境和绿化，形成海河沿线各具特色的地标性建筑，形成海河景观中最亮丽的音符。

十大工程之十：整修置换工程。对海河沿线建筑和环境的整治是海河综合开发改造的重要方面，整治工程要按照海河统一的景观规划实施，保证建设的高水平、高标准。整修包括对风貌建筑的维护性整修，恢复原有的历史风貌，特别是依托原租界建筑区形成不同风情的异国建筑风貌区，整体上再现天津“万国建筑博览会”的独特景观。对于危陋房屋进行拆迁改造，协调整体建筑形象。对于沿河重要区域的建筑，还要调整置换使用功能。大量增加经营性的文化、服务设施，如咖啡屋、酒吧等，增加沿岸地区的活力。

诸葛民

“鲍尔先生，我向您介绍得这么详细，是想向您说明海河的规划设计是比较扎实的，是各相关部门和专业规划设计单位共同工作的结果。有了这样详细明确的项目目录，才使得我们对工程的规模有了清楚的认知，也才能够比较准确地测算出投资规模，包括未来的土地出让收益。”

海河两岸的整体规划建设工程宏伟、投资巨大。经过测算，十大工程总投资约为 195 亿元（含拆迁补偿费），其中：公益性设施投资 60 亿元，道路桥梁投资 80 亿元，市政工程投资 46 亿元，不可预见费用 10 亿元。另外，资金利息成本 30 亿元。这个数字对于设计院年轻的规划师来说就是天文数字，第一次算出一个项目需要这么大的投资。海河两岸综合开发改造规划所确定的控制范围，总用地面积为 42 平方千米。初步测算两岸完全实现规划更是需要投入 1800 亿元的建设资金。

按市领导提出的“按照市场机制，多渠道筹集资金，进一步加快基础设施建设”的思路，借鉴危改工程的宝贵经验，走“以基础设施建设带动开发”的路子，通过对两岸的道路、管网等基础设施和文化设施的建设，改善投资环境，提升土地价值，进一步吸引市场力量共同建设海河。确定了启动资金不完全依靠财政，以市场化手段筹集的操作策略。市领导曾多次指出，要改变用钱的观念，“原来是存足了钱再干事，发展到现在是先干事，钱以后一点一点地还”。海河两岸的开发建设具有非常好的前景，完全有条件吸引各方面的资金参与到其中来。通过一段时间以来与开发银行和商业银行的洽谈，银行方面普遍表现出很大的兴趣和热情，虽然还没有找到具体可行的路径，但坚定了他们的信心，确定了市场化开发的思路。

市场化开发的总体思路是：海河两岸的开发改造工程分为两个阶段，第一阶段用 3 ~ 5 年时间（2003—2007 年）通过银行贷款进行基础工程和文化设施建设，改善投资环境，使海河改变面貌。这一阶段主要为投入建设期。其后的 15 年（2007—2022 年）通过市场投入完成两岸的全面开发改造，通过多渠道的收益逐步还清银行贷款。这一阶段主要为回收期。

具体运作方式和步骤由以下几个方面构成：首先建立组织机构，成立领导小组，由市主要领导任组长，各相关委办局主要负责同志为领导小组成员。在领导小组下成立海河开发改造工程办公室。下设一个海河开发投资公司，按市场化运作，作为引资载体，负责融资和规划范围内建设的招商。海河开发投资公司采取以土地开发获得建设资金的模式，获得的建设资金交给各个建设部门实施海河两岸的基础设施建设。

具体操作的最初设想是：海河投资公司将预开发土地预征，按政府指定收购价格与目前的土地所有者签订土地收购合同，给付定金，取得土地的使用权和土地使用证。市规划国土局利用规划管理手段、土地整合手段，编制海河两岸综合开发改造统筹规划。依据政府批准的海河两岸综合开发改造规划，海河投资公司委托著名公司编制海河综合开发改造的可行性研究报告，以基础设施建设项目向银行申请 15 年期低息贷款，提供海河改造建设的资金。贷款以海河投资公司取得的土地作为抵押标的物，建议市财政提供担保。另外作为补充，还可以通过发行债券的方式筹集资金，通过海河冠名权、经营权获取部分资金，民间投资机构也可投资入股。将获取的资金用于道路、桥梁、市政工程等基础设施以及绿化、广场等公益性设施的建设，集中在 3 ~ 5 年内完成海河综合开发改造工程，改善海河环境，促进沿岸经济发展和招商引资，提升土地价值。然后，对抵押土地重新进行评估，将重新评估的土地以转让、拍卖等形式重新获得土地出让金、补偿资金和土地增值收益。在 15 ~ 20 年的期限内，海河投资公司将获得的土地增值收益和经营性收入，政府将相应的土地出让金、大配套收益以及每年一定的财政投入，用于还贷（包括利息）、支付民间投资机构的投资和收益。另外，海河两岸各项市政公益设施的管理采取招标的方式，市场化运作管理。

启动资金筹措主要通过项目包装，以“海河两岸基础设施及土地开发项目”作为项目名称向开发银行贷款。海河投资公司以规划范围内的基础设施评估后作为自有资金的主要部分，各个建设部门对海河两岸地区的投入也作为自有资金的一部分。两者构成约 80 亿元的自有资金数额，以 2 ：3 的比例向开发银行贷款约 120 亿元。

环内 42 平方千米控制用地范围内，可开发用地为 29 宗，可出让土地面积为 15 平方千米，通过对其整理、市场运作，可开发土地的收益总额为 256 亿元，其中：土地出让金总额 37 亿元；市政配套费总额 28 亿元；土地净增值总额 166 亿元；建议市财政每年投入 1 亿元， 20 年共 20 亿元；经营项目收益额 6 亿元。

按照静态投入产出分析，海河上游段综合开发改造纯收益为 10.4 亿元。按各种风险对收益折扣系数 90% 计算：项目总收益水平为 9.36 亿元。对国民经济增长贡献方面，按每年投资 39 亿元计算，可拉动经济增长 1.5%，就业岗位可增加约 30 万人。

33. 持之以恒 · 法规化管理

俗话说：三分建，七分管。一个好的城市，必须在城市管理上做到位，这方面年纪长的市领导比年轻的规划师有更多的经验。因此，当他们对海河两岸综合开发改造规划最终形成统筹化规划和市场化开发两大部分的结构欢欣鼓舞时，市领导明确要求增加法规化管理章节，要特别制定针对海河两岸的开发建设规划的管理规定。在海河已经做的数轮规划的基础上，借鉴国外好的规划管理经验，通过编制详细蓝图，对控制性详细规划进行深化，编制建设管理和景观控制性详细规划等，提出海河两岸开发建设的“纲”。规划管理的内容不仅包括对两岸土地使用、建筑高度、建筑形式的控制，还包括绿化、水体、桥梁、景观、岸线、灯光等方面的规划设计要求，主要控制内容包括建筑风格、建筑高度、通航标准、桥梁建设标准和海河水位标准等。

海河两岸建筑风格大致可以形成一个序列，从三岔河口的具有中国传统特色的建筑，到一宫地区、解放北路地区具有西洋近代风格的建筑，以及下游新区现代风格的建筑等。新建筑与保护建筑之间要有过渡，体现发展的延续性。

海河比较窄，两岸要限制建筑物的高度。依照现状及空间景观效果分析，制定海河两岸建筑高度控制规划，确定标志建筑、景观轮廓线、建筑退线及坡度控制线。

通航是海河综合开发改造的主要成果，必须保证航道净空。金刚桥至光华桥段规划为Ⅵ级航道，桥下净空 5.0 米；光华桥至外环线段为Ⅳ或Ⅴ级航道，桥下净空 5.5 米。

桥梁建设必须满足通航等标准，以后光华桥上游新建和改建桥梁净空要求不低于 5.0 米（梁底标高 6.5 米），光华桥下游至外环线新建桥梁净空要求不低于 5.5 米（梁底标高 7.0 米），确保通航要求。

海河水位控制对通航、旅游、景观等方面十分重要。市区二级河道陆续改造后，与一级河道贯通，形成天津市水系风景。为保证二级河道景观及游船通行需要，让水活起来，每年需增加对河道的蓄水，同时为保证海河水位常年保持在较高蓄水位（1.5 米），满足海河景观亲水要求。

以上规划控制内容是从海河两岸综合开发改造规划中提炼出来的认为最重要的内容。虽然没有形成正式的文件或法律规章，但在海河综合开发改造的实施过程中真正成为了海河开发建设管理的“基本法”。今天我们认识到，类似海河这样重要的城市区域，应该制定专门的规划管理法规。

鲍尔

“诸葛总，我十分赞同您的这个观点。在这方面我们的法律法规比较完善。当然，规矩多了，也会限制发展或降低行政审批的效率，这其实是把双刃剑。听您系统而详细地介绍了海河综合开发改造规划，我觉得深受触动，这的确是一个既有高度又可操作的规划，那么这个规划属于哪种类型的规划呢？”“我当年就在思考这个问题，海河两岸综合开发改造规划不属于任何一种现行的规划类型，从它发挥的作用来看，是一个包含开发策划、城市设计、交通规划、市政规划、旅游规划的发展规划，我们也可以这样理解，这是一个‘高维城市设计’，是为城市发展建设提供的一项顶层城市设计。这十几年来我一直在琢磨，这样一个宏伟的、引领性的城市设计为何只出现了这一个？天津开创了一种新的城市设计类型，以空间为载体，统筹经济、社会、文化等全要素的城市运营设计，但没有将其发展为天津模式，实属可惜。不过，是金子总会发光，今天海河两岸的实际效果证明了这个城市设计是举足轻重的、高水平的城市设计。在这样一个强调协同发展的新时代，在部制改革的背景下，也许海河的创新之举会成为城市设计变革的方向，我想这也是鲍尔先生对海河规划如此有兴趣的原因。也是我不厌其烦地讲述这一段十几年前的城市设计经历的原因。”

34. 海河近期建设规划

经过长时间规划和审慎准备，2002 年 10 月 29 日，天津市委常委会听取了市规划局海河两岸综合开发改造规划的汇报，做出了“综合开发改造海河”

的战略决策：用3～5年时间，将海河建成独具特色、国际一流的服务型经济带和景观带；弘扬海河文化，创建世界名河，使海河成为展示天津现代化港口大都市的标志。决议同时指出，这是天津建设现代化国际港口大都市的迫切需要，是实现天津跨越式发展的重大举措。通过服务业发展，带动海河两岸综合开发，集聚金融保险业，促进商贸、交通、旅游、娱乐、餐饮和中介服务业的发展，有力地拉动市场，吸引更多的人流、物流、资金流、信息流，其作用是全局性的、战略性的、历史性的，是天津继“三五八十”四大奋斗目标基本实现后的又一历史性工程，是天津进一步加快发展、向更高目标阔步前进的必然选择。

为了准备这次重要的汇报，规划人员废寝忘食。原本以为总体规划通过后可以放松一下，没想到市委的决议是要求马上启动建设。诸葛民及规划院的同事们马不停蹄地加快进度编制海河两岸综合开发改造近期项目的规划设计。本着“解放思想，开拓创新，走市场化的路子，充分发挥市场机制的作用”和“基础设施与开发地块同步建设”的原则，采取市场运作、统一规划、分区分步实施的方针，经与各有关委办局和六区政府研究，反复论证，并征求各有关部门的意见，确定海河综合开发改造一期启动“四二一六”工程。其中“四”代表海河东路、海河西路、台儿庄南路、五马路－三条石横街等四条城市道路，“二”代表永乐桥和大沽桥两座跨海河大桥，“一”代表一条轻轨线，“六”代表运河经济文化商贸区、大悲院商贸区、南开区天后宫旅游商贸区、海河广场及和平路地区、中心商务区南站地区、海河五环广场等海河沿线六处重点地区。按照市委市政府“与时俱进，力争上游，抢抓机遇，跨越发展”的要求，市规划和国土资源局组织规划院、建筑设计院、测绘院等单位进行了大量深入细致的工作，对近期启动的道路交通、桥梁、轨道交通和重要节点的规划建设方案等进行了深化编制和经济测算。

为了充分宣传和加快这一阶段海河服务型经济带的综合开发改造，除先后多次向在津的两院院士、市人大部分代表、市政协部分委员，以及市计委、市建委、市商委、财政局、文化局等部门和单位汇报，详细阐述海河两岸综合开发改造规划的重大意义和美好愿景外，我们还借助多家新闻媒体单位开展了宣传报道，天津日报、今晚报刊发了200多篇新闻报道，天津电视台和天津电台开展了60多次新闻专题和专访。此外还利用现代化信息技术，通过互联网进行信息发布和广泛宣传，社会反响十分强烈。天津泰达公司、滨海快速交通发展公司、天津市房地产开发集团、天津市房产总公司等纷纷提出了投资意向。国内外一些投资公司、知名企业等不断来津洽谈合作事宜。中国银行、国家开发银行、深圳发展银行、中国人民银行等各大银行纷纷表示支持该项目。

市委常委会决议后，一些先期批准的项目的拆迁等前期工作已经开始提速，启动工程于2003年开始全面实施。2002年12月17日，天津市委八届三次全会召开前夕，市领导视察市内六区启动现场，再次专题听取了市规划局海河两岸综合开发改造规划和近期建设工作的汇报。原则同意通过实施“四路、二桥、一条轻轨”等基础设施建设，带动海河沿线六处节点的开发建设，进而促进海河两岸综合开发的全面建设，形成良性循环。

35. 海河六大节点城市设计国际征集

要将海河打造成世界名河，前提是提高海河两岸的规划设计水平。如何提高水平，就是要开放设计市场，吸引世界一流的规划设计师参与海河规划，高起步搞好海河开发规划设计。在海河综合开发建设启动前期，我们着眼于国际一流的策划、规划和设计，请国际知名专家绘制海河开发蓝图，注重运用世界一流智力资源和国际通用的规划设计理念，高起点高标准启动海河综合开发建设工程。在海河两岸总体城市设计过程中，市规划院与美国泛亚易道公司合作成立了天津海河规划工作营，应用先进的规划思想与设计理念，不断推进规划方案深化完善。针对一期启动工程中的六处节点和一座桥，我们报请市委市政府同意，拟举办高水平的概念规划设计方案国际征集活动。2002年12月15日发布报名消息。2002年12月18日，天津市政府专门召开新闻发布会，公布了海河六大节点城市设计国际征集的有关内容，各方面反响强烈，国内外规划设计公司踊跃报名。到截止日期，共有56家国内外知名规划设计机构报名，其中境外公司42家，来自美国、法国、德国、加拿大、日本、韩国以及我国香港等十几个国家和地区。这么多的境外规划设计公司参与天津的规划设计这在历史上是第一次。

2002年12月24日，天津市委召开八届三次全会，提出了“三步走”战略和五大战略举措，勾画了天津发展的宏伟蓝图。第一步，2003年，人均国内生产总值达到3000美元，实现全面建设小康社会的主要经济指标；第二步，提前3～4年，实现国内生产总值和城市居民人均可支配收入、农民人均纯收入分别比2000年翻一番，使经济总量和群众收入水平再上一个大台阶；第三步，到 2010 年，人均国内生产总值达到 6000 美元，把天津建设成为现代化国际港口大都市和我国北方重要的经济中心，建立起比较完善的社会主义市场经济体制，成为全国率先基本实现现代化的地区之一。五大战略举措是大力发展海河经济、大力发展海洋经济、大力发展优势产业、大力发展区县经济、大力发展中小企业和个体私营经济。把大力发展海河经济作为五大战略举措之首。

2002 年 12 月 31 日报名截止日期过后，市规划局马上进行了资格预审，2003 年 1 月 2 日出预审结果。然后各区政府在 1 月 7 日确定了各节点的参赛规划设计单位。最后从 56 家报名单位中筛选出了 31 家参赛。原本应该有 35 家设计单位，但在各区确定设计单位时，几个区同时选定了某个设计单位。经过协商，最终有四个单位同时参与了两个节点的征集。市规划局 1 月 10 日下发了规划设计任务书，规划设计单位查看现场，正式开始了征集工作。为了赶时间，确定征集时间从 2003 年 1 月 10 日到 2003 年 3 月 5 日，约 2 个月的时间，然后组织专家评审。由于时间紧，而且海河六大节点和一座桥梁的规划设计要求各具特色，规划设计工作非常具有挑战性。

第一节点是运河经济文化商贸区，位于红桥区子牙河、南运河与海河起点三条河流的交汇处——三岔河口西岸，规划用地面积 64 公顷。这里是海河的起点，也是天津城市的发祥地。规划依托传统的文化资源，通过中国近代工业博物馆和天津城市历史博物馆等项目的建设，结合传统文化商贸区建设，强化自然特征，充分挖掘运河文化，采用现代和传统相结合的方式，创造海河第一景观。

第二节点是大悲院商贸区，位于河北区，地处子牙河、南运河、海河三河交汇的三岔河口东岸。该节点以商业、娱乐、餐饮、休闲设施等为主，以改造城市环境和功能体系为目标，形成该地区独特的区域优势和特色，为天津的旅游和商贸活动注入新的内容，建设“庙、市”合一的传统商贸区。

第三节点是天后宫商贸区，位于南开区北部，与海河源头三岔河口仅 500 米之遥，是天津市最早的漕运码头之一，规划用地面积约 14 公顷。规划要求利用该地区在全国的影响力和知名度，结合天津市公共设施建设，保留玉皇阁、天后宫等文物古迹，重建海河楼，恢复水阁，使天后宫地区成为充分反映天津风貌特色、景观优美、建筑风格独特和经济繁荣的传统文化商贸区，成为海河沿岸的一个亮丽的景观节点。

第四节点是海河广场商贸娱乐区，海河广场地区是天津市的几何中心，规划用地面积 72 公顷。规划要求充分利用横跨海河、水面辽阔的特点，发挥中心商业区黄金地带的区位优势，利用现有劝业场、百货大楼、和平路金街以及众多现代城市商业基础，结合天津市商业中心区的整体建设，新建海河北岸广场、海河音乐厅等设施，再创繁华盛景，建设成为标志性景观区和都市消费娱乐区，充分展现天津国际港口大都市的风采。

第五节点是南站中心商务区，该地区是天津市中心商务区的重要组成部分，规划用地面积 30 公顷。规划通过合理组织滨水空间，强化自然特征和亲水特点，塑造优美宜人、景观独特、环境优美、交通便捷、设施先进完善的国际一流水平的中心商务区，以商务办公、展览、会议、商业、娱乐、餐饮、休闲设施、高级公寓等建筑为主，建设海河艺术中心等重点项目，为天津建设北方重要经济中心及国际港口大都市创造物质条件。

第六节点是海河五环城，位于天津市河西区挂甲寺地区，规划用地面积 42 公顷。规划对该地区自然环境资源、土地资源、文化资源进行合理利用和保护，以体育运动、商业、娱乐、餐饮、休闲设施、高级公寓等设施为主。通过水上大世界、五环体育城的建设，结合危陋平房改造，完善该地区的服务功能。建设成为环境优美、交通便捷，社会、经济、环境协调发展，充分体现新天津特色的城市滨水旅游度假胜地。

规划设计的一座桥是永乐桥，位于河北区、红桥区交界处的子牙河上，连接运河商贸区和大悲院商贸区。要求桥宽为双向六车道，一跨过河，桥下净空 4.5 米。规划设计要求桥本身除满足交通功能外，还要具有商业和观光休闲功能，通过桥体独特的建筑造型，为海河增姿添彩。

2003 年 1 月 2 日市委市政府主要领导再次视察了海河两岸市内六区的启动现场，再次来到规划局专题听取市规划局海河两岸综合开发改造规划和近期建设工作的汇报。按照市领导“尽快把二期实施方案拿出来，千方百计扩大开工面”的指示要求，市规划局及早动手，提前策划完成了二期启动项目，其中包括 13 条道路、6 个广场、2 座桥梁建设、4 座桥梁整修改造、10 项文化建筑和 14 个地块开发的策划。从领导几天就一个活动的频率看，市委市政府对这次海河综合开发改造是真下了决心。

2003 年 1 月 24 日市第十四届人民代表大会第一次会议上，新当选的市长在政府报告中提出：大力发展海河经济，对海河两岸进行综合开发，完善基础设施，优化整体环境，带动服务业和相关产业发展，促进老城区改造，形成独具特色的服务型经济带、文化带和景观带，逐步使海河成为世界名河，带动天津成为世界名城。天津市人大和政协的会议都在建于 20 世纪 50 年代、建筑风格与人民大会堂相似的天津大礼堂举行，在本次人大会议议程结束后，市委市政府马上召开了专题听取市规划局海河两岸综合开发改造规划和近期建设工作汇报的市委扩大会议，包括人大新选举出的市主要领导。200 多人的大会议室内坐满了人，大家都为海河综合开发改造的美好愿景所欢欣鼓舞。

2003 年 2 月 1 日是中国的传统节日——春节，这一年是羊年，春节来得早，预示着五谷丰登。中国人迎来了农历新年的长假，与家人团聚，欢度春节。但想必这时候参加海河六大节点城市设计方案征集的国外设计公司正在加班加点。到 3 月 5 日，参赛单位共提交了六个节点和一座桥梁的 35 个规划设计方案。各区政府组织进行评审。经专家评审和向社会公示，这些设计方案各具特色，竞相争艳，达到较高水准，得到各级领导和社会各界的认可。

鲍尔

鲍尔一直认真、费力地聆听着诸葛民的叙述，说实在的，听一位母语非英语的人长篇大论，特别是包含许多投资数据和相关分析，还是非常需要集中注意力的。听诸葛民讲到这儿，他的表情变得轻松起来，因为鲍尔参与了海河中心广场的方案征集，并参与了天津其他项目的规划设计，正是从那时起，他开始关注天津，对天津后来发生的事情比较了解。多年来，他对天津在城市规划设计方面取得的成绩很是称道，同时对一些做法有看法，如鲠在喉，他决定说出来:“诸葛总，六大节点的城市设计方案征集这种做法非常好，让许多国际著名的设计公司能够认识了解天津，参与到天津的城市设计中来，但评审中好多非常有创意的方案没有入围，而且最后的综合方案有的不十分理想。您怎么看这件事？”

诸葛民

对于鲍尔的问题，诸葛民好像早有准备，看来这也是他早已经思考过的问题。“鲍尔先生，您提出的问题非常中肯。的确，许多有创意的方案最后没能入围，这说明评审团还比较保守。一般来说，方案征集与方案竞赛最大的不同也在于此。综合方案没有能够达到最理想的效果，我想这也可能与天津当时所处的发展阶段有关系，有其必然性。”受天津市政府委托，市规划和国土资源局负责组织、管理、监督本次海河两岸综合开发设计方案征集工作。各节点所在地的区政府和投资商为承办单位，负责各节点的具体方案征集工作。这时诸葛民已经从规划院调职到市规划局。为了保证规划设计的水平，市规划局督促各区政府明确方案征集的补偿费和奖金数额。经济条件好的区非常爽快地答应了。而经济比较困难的区拿出钱来搞方案征集，还是舍不得。经过不断做工作，最后各区都明确了资金数额，落实了资金来源渠道。六个节点和一座桥梁方案征集的补偿费和奖金共计 1000 万元人民币，这在天津历史上还是第一次。“说是这些资金调动了参加征集单位的积极性，实际上不尽然。1000 万元人民币的资金，共计 35 个方案，平均每个设计方案不到 29 万元人民币。许多设计单位投入巨大，总体模型和主要单体模型，包括动画，加上国际差旅费，应该刚够成本。参加征集的各单位的主要目的是为了承揽后面的实际工程设计。这方面鲍尔先生您有亲身体会，对吧？”

“为了保证规划设计方案征集活动的公正与科学，我们要求各个节点的评审邀请 7 ~ 11 名包括规划、建筑等方面的国际国内著名专家及市政府有关部门代表、区政府代表、投资商代表等共同组成规划设计方案评审委员会，对每一个征集项目进行评审，投票确定优选方案，向市委市政府汇报后，宣布规划设计方案征集结果并进行公示。这就是关键，因为专家本身就比较注重实施的可行性，将专家推荐的方案再向市委市政府汇报后，有创意的方案大多没能入围。永乐桥是个例外，永乐桥就是现在的天津之眼，这也是一位市领导力排众议的结果。他询问规划局的意见，大家都不太赞成这个方案，但看到领导的倾向又不好直说。诸葛民你怎么看？领导对着我发问，我说桥主要还是桥的功能，暗示不赞成摩天轮的方案，让领导一顿抢白，说你们规划局太保守。

“竞技赛场上人们记住的往往只有冠军。规划征集没有入围的得以存档，流传下去的往往是最终实施的综合方案。这两种情况看上去相近，但有本质的不同。冠军是通过战胜其他对手产生的，而规划最终实施的综合方案往往是从很多方案的融合与演变而来。因此我们看待一个地区的规划建设，往往只看一个最终实施的综合方案与建设效果是不足的，更应该在时间维度中清晰地把握其方案发展的脉络与变化原因。六个节点后续的深化方案，结合工程投资、开发建设难度等因素，在技术上不断深化落位，最终形成实施方案。从结果看，大悲院商贸区、天后宫商贸区、海河广场商贸娱乐区和海河五环城（后改名为海河水世界）按综合方案实施度比较高，达成预期目标。运河经济文化商贸区和南站中心商务区后期变化比较大，目前也还没有完全实施完成。总体看，城市设计方案国际征集活动是成功的。

“为了做好征集工作，我们事先组织天津市规划院、建筑设计院、华汇设计公司、博风设计公司等设计单位开展了海河六大节点概念性规划设计工作。概念性规划设计方案广泛征求社会各界的意见和建议，对道路交通、规划建设规模等进行了深入论证和经济测算，为国际征集做好了前期准备。设计任务书中要求在借鉴国内外成功经验的同时要突出天津地方特点，充分体现天津历史文化的底蕴，吸纳世界文化的精华，展示都市文化的风貌，集功能性、艺术性、观赏性于一体，体现世界水平。这些要求是合理和恰当的。但后来我们发现一些要求过于细致，限制了设计师的想象力。比如，在永乐桥确定设计条件时要求采用塔楼式拱桥和斜拉桥两种形式，规划设计要求在塔楼基座四周立面布置大型雕塑，反映天津近代民族工业发展史、天津漕运发展史、天津危改业绩等重大纪事。现在看这些过细的要求是多余的。”

36. 改革创新 + 持之以恒 + 细节决定成败

鲍尔

鲍尔：“诸葛总，很感激您向我坦诚地讲述了海河规划动人的故事，特别是您的许多真实的想法，听罢让我对海河、对天津有了更深的认识。我还有最后一个问题，海河这样一个高水平的宏大规划，最后能够完全实施，您认为什么是最关键的？”

诸葛民

“我认为，一个城市规划要取得成功，人是最重要的因素，这里的人是一个广义的概念，包括方方面面的人。比如，城市的领导、规划管理者、规划设计师、工程师、城市居民、外来的参与者，等等。要规划建设一座世界名城，首先，要有信心，持之以恒，不断求索，探索规律，适时决策；其次，要广泛学习和借鉴，利用各方面的资源，明确具有自身特色的规划设计，培养自己的规划设计人才； 第三，要靠改革创新，破解规划设计和建设中技术和资金的困难，找到跨越鸿沟的路径；最后，细节决定成败。

“海河规划是天津几代人持之以恒、不屈不挠、接续努力的结果，是方方面面、上上下下齐心协力奋斗的结果。天津人视海河为母亲河，十分珍惜。在历来的规划中，都围绕海河进行，逐步调整海河的功能定位，为海河的转型逐步奠定基础。在海河不具备大规模开发改造条件时，能保持定力，将海河两岸保护控制好，持续进行分析研究，通过成片危陋平房改造等为海河全面开发建设积蓄力量。在具备条件时，举全力把规划设计做好，树立高标准，广泛学习借鉴国内外先进经验，努力达成更高水平。充分讨论，科学决策。在目标明确后，勇往直前，攻坚克难。通过改革创新，破解技术、资金等各方面看似无法解决的难题，完成不可能完成的任务。在综合开发改造的实施过程中，注重细节，精益求精，一丝不苟，确保规划设计建设达到一流水平。

“天津因海河而兴。从世界城市的发展历史看，滨河城市虽然在地域、人文等方面不尽相同，但城市结构形态发展演变的客观规律是相同的。滨河城市中的大部分是由于河流的交通运输功能而发展起来的，最初河流沿线均是码头、货栈、仓储和工厂。随着工业化的发展和船舶运输向大型化发展，内河航运功能逐步衰退，内河港口向河口转移，形成海港。河流沿线的码头、货栈、仓储和工厂用地逐步调整为第三产业用地，成为展现城市形象和特色的中心区。当然，这种演变必须以城市经济发展为基础。总体看，1986 年版天津城市总体规划确定的“工业战略东移”和“一条扁担挑两头”的总体布局是符合滨河城市发展客观规律的，随后的几版城市总体规划都坚持了这样的布局结构。随着“三五八十”阶段性目标的完成，天津经济发展上了一个层次，城市产业结构调整大势所趋，第三产业的比重不断增加，为海河两岸综合开发改造提供了必备的条件。

“对海河两岸的规划我们持之以恒地做了 10 年的时间。我们深入研究了海河及天津城市的历史演变过程，深入分析了现状情况和存在的问题，学习借鉴国外的先进经验，明确了海河的功能定位、用地布局、道路交通等关键问题，处理好了海河旅游开发与防洪和保水的关系。在开展海河两岸的总体城市设计规划时，市规划院与美国易道公司合作主持工作营，邀请相关专业的公司参与。起步段六大节点城市设计组织国际征集。在永乐桥、大沽桥方案设计上，请国际著名的日本“川口卫”事务所和世界知名的桥梁设计师邓文忠院士主持设计。同时，充分发挥本地规划设计单位的力量。总体看，坚持搞标准，开放规划设计市场是方向。近百个规划设计单位参与了海河两岸综合开发改造的规划设计。天津市城市规划设计研究院、天津市建筑设计院、天津市市政设计院、法国康赛普特设计事务所、美国易道公司等国内外规划设计机构进行了深入的前期研究、概念规划和总体城市设计，为后续的规划设计工作奠定了坚实的基础。参与开发建设的大部分是天津国有开发公司，是天津自己的公司，从建成的效果看说明天津自己的公司也能干好。

“如果说以上这些是海河综合开发改造成功的必要条件，那么改革创新是海河综合开发改造成功的充分条件。如果墨守成规，不改革创新，就不可能有今天的海河。首先在规划设计上，如果延续过去片面单一的做法，机械地将防洪和保水作为不可触碰的红线，不从流域和区域的角度解决防洪问题，对立环保与旅游的关系，则无法实现将海河建成独具特色、国际一流的服务型经济带和景观带，弘扬海河文化，创建世界名河的发展目标。”

规划确定后，实施规划必须有资金支持，如果沿用传统的依靠财政投资的办法，是不会有这种类型和这么大规模的资金的，必须改革创新，依靠市场化的手段。虽然海河综合开发改造规划中提出了市场化的路子，大的方向是可行的，但最初所提的具体路径不可行，包括预征地、发放土地白证等，都是违规的，无法操作。贷款归还主要依靠土地升值收入等收益，是有保障的。但贷款需要抵押，用土地抵押的话，土地还需要资金先期征用，而征用土地的资金也没有。最后还是国家开发银行搞金融创新的专家和天津搞土地市场的专家一同研究，创新性地提出“以土地出让政府预期收入质押的方法进行贷款”的新路径。按照此路径，天津市向国家开发银行贷款 500 亿元，为 15 ~ 20 年长期贷款，除海河综合开发改造的 196 亿元外，还用于城市地铁建设、快速路建设和公园建设。当时分别给这四大工程起了好听的名字，海河综合开发改造是“金龙起舞”，城市地铁建设是‘时空通达’，快速路建设是“环城银链”，公园建设是“翠屏叠绿”。这种新的政府融资模式为天津的发展奠定了坚实的基础，也成为其他城市竞相效仿的对象。

规划有了，资金有了，如何达到海河综合开发改造的高水平，关键是细节，细节决定成败。从 2003 年春节过后，我们就进入海河综合开发改造现场指挥部。指挥部原来是河北区房管局的办公楼，位于袁世凯故居旁，是个四层楼。每天我们能够看到海河边热火朝天的劳动场面和工程进展。一天中午吃过饭，诸葛民像往常一样看向窗外，忽然发现正在建设的海河亲水平台堤岸与正在实施的海河下沉路不在一个标高上。他赶紧赶到河对面的施工现场，从现场了解到水利设计院设计亲水平台用的高程是黄海高程，而市政设计院道路设计用的是大沽高程。两者相差近 1 米。幸亏发现及时，没有造成大的损失。

泛亚易道公司进行了上游起步段堤岸环境的综合规划和设计，在现场派驻了一位年轻的法国设计师，市规划院也派了一位同样年轻的规划师谭鸣进驻现场指挥部，配合施工，做好深化设计工作，他们在现场工作时非常认真负责。当时，老城厢等地区正在拆迁，谭鸣和法国设计师找到许多现成的砖石，把它们用到了大悲院滨河步道上。具有历史沧桑感的灰砖与整个景观设计的历史定位和基调非常吻合。从此，从天津大学城市规划专业毕业的谭鸣喜欢上了景观设计。后来市规划院成立景观所，谭鸣当了所长。以上两个例子充分说明天津海河两岸综合开发改造在细节上的成功。

鲍尔

从12：30到13：30，整整一小时，诸葛民向鲍尔讲述了海河规划的主要内容，回答了鲍尔关心的问题，这些内容是大家关心的，诸葛民一直也没有机会与大家分享。这是第一次，诸葛民畅所欲言地讲述了海河规划背后的故事，心情轻松起来，他从鲍尔手里拿过账单，“鲍尔先生，我们学习现在的年轻人，AA制吧。”“好啊！听您的。”鲍尔附和着。意大利籍老板走过来，看到餐桌上的食物还剩下一部分，就问：“二位对我们的菜不满意吗？”“菜很好，只是我们光顾着谈话了，下次我们专门来品尝。”与老板告别，两人起身向规划展览馆走去，只有不到百米的距离。到门口，诸葛民对鲍尔说：“听说您下学期到天津大学任教，到时候欢迎您给我们多做些讲座。”“很高兴受到邀请，谢谢您。诸葛总，我除了在天津大学教书外，也想参与一些科研课题和天津的规划设计项目。我妻子雪莉也计划到天津来，她是个优秀的景观设计师，我们想在天津开始新的事业。”“太好了！天津就是缺好的设计师！”

第六部分
无城市设计不建设

37.2008 年迎奥运的城市变化

天津城市设计国际论坛以两位世界大师的演讲作为开始，乔纳森 · 巴奈特先生介绍了城市设计最新理论和方法，以及他对中国城市设计的建议。丹·索罗门先生介绍了世界最新的居住社区规划的趋势。两人的报告十分精彩，掌声不断，这在学术演讲中不多见。进入茶歇时间，在会场外休息厅，大家喝着咖啡、茶水，相互寒暄，自然而然地形成了两组人群。一群年轻的规划师、城市设计师围绕着巴奈特教授，继续请教重要的城市设计问题：智慧的城市建设该在什么时候建、在什么地方建以及怎么建。另一群人，主要是建筑师，还有一些开发公司的管理和技术人员，围绕着丹 · 索罗门先生询问关于新型社区的相关问题。

时间很快过去，时针指向 11 点，主持人招呼大家入座："各位嘉宾，刚才我们请巴奈特教授和丹 · 索罗门先生分别为我们做了主题演讲，十分精彩，下面进入本次论坛的天津时段，我们先请天津市规划局的诸葛民总规划师介绍天津城市设计试点情况，然后请清华大学建筑学院周方华院士、同济大学李松霖院士东南大学建筑学院刘世杰院士、天津大学张泽平院士和中国建筑设计研究院建筑设计大师邹博学院士上台点评。首先有请诸葛总！"

诸葛民

诸葛民走上讲台，心情有些激动，甚至还有些紧张，他脑海中浮现出 30 年前他给市规划局的干部们讲城市设计的一幕。当时天津规划局的职工对城市设计还比较陌生。今天，30 年过去了，城市设计在天津已经十分普及，并且城市设计项目在天津的实施度很高，使得这个城市在空间品质和城市风貌上有了很大提升，诸葛民也从 30 岁的年轻人到了花甲之年。今天，台下的听众也与 30 年前完全不同，有国际上的城市设计大家，有国内城市设计顶尖人物，在他们面前，诸葛民要给 30 年来的天津城市设计盖棺定论。更有挑战性的是，结合本次天津城市设计试点工作的重点，诸葛民将首次在国内提出一些有关中国城市设计的全新观点，这些观点与国内几十年来普遍视为经典的观点相左，可能会引起激烈的反应。不过，城市设计试点的意义不就是在于改革创新及探索吗？想到这儿，诸葛民心情平复了许多。

"城市设计与城市一样古老。我们从中外古代城市规划史中可以清晰地看到这一点，比如古埃及卡洪城、古罗马，包括中国的明清代北京城。这些城市是一次性规划建设的人工城市，而大部分城市不是一次性建好的，而是不断生长变化的自然城市。天津设城筑卫 600 多年来，城市发展经历了四个主要阶段，每个阶段都能看到城市设计发挥的重要作用，从而对城市空间形态产生深刻的印记和深远的影响。明永乐二年设卫时，老城厢采用了中国古代典型的县城的布局，长方形城廓，正南北布局，这一格局延续了 450 年。由于海河在城东北面，所以方城内的发展也偏向东面和北面。第二次鸦片战争后，天津被开辟为通商口岸，帝国主义在老城下游沿海河设立租界。租界大规模建设始于 19 世纪末。租界的规划各自为政，它们的城市设计方案均采用了当时西方国家流行的方法，我们现在还能体会到豪斯曼巴黎改建规划及后来的霍华德田园城市思想的一些影响。租界的道路以方格网为主，有少量的圆形广场、花园和放射路。这些租界有一个共同特点就是路网布局与海河的走向平行或垂直，这样就形成了一个位于三岔河口正南北轴线的天津，以及一个在海河两岸垂直或平行于海河走向的天津。包括袁世凯建设的河北新区，虽然道路以《千字文》中的“天、地、元、黄”等命名，但其方格路网也是与海河垂直或平行的。

“到 20 世纪初，经过三个阶段扩张形成的租界总面积达到 15 平方千米，相当于 1860 年天津建成区的 3.5 倍。到 1949 年时，建成区面积约 70 平方千米。如何整合老城厢为主的中式城区和租界为主的西式城区是天津城市空间形态设计上面临的主要问题。1938 年梁思成和张锐编制的《天津特别市物资建设方案》中，采用天津老城南北中轴线统领全市的整体结构，通过增加一些圆形广场和放射路，将租界区融合于城市之中。中华人民共和国成立至今整整 70 年，总体来看，这 70 年是天津飞速发展的阶段，大体可分为五个时期。

“最初的 30 年是战后恢复时期，天津从消费型城市向生产型城市转型，在城市外围规划建设了 10 个工业区和配套的居住区，同时各种政治运动也使城市备受煎熬。这一时期编制了 21 轮城市总体规划方案，基本没有脱离 20 多年前梁思成、张锐的构思。

“20 世纪 50 年代苏联规划专家穆欣帮助天津编制了城市总体规划，他以中心广场为城市中心，规划了一条与海河垂直、指向西南水上公园的轴线。他尝试用这条轴线，将轴线以西的中式的天津与轴线以东的西式的天津整合起来，以达成均衡。这与贝聿铭在旧金山规划的市场街作用类似。可惜这条轴线没有做实，以至于后来再没有机会实施。

“1976 年唐山大地震天津受损严重，使原本已经难以为继的城市雪上加霜。中央除对天津震后重建给予资金支持外，建设部还派来专家组协助天津编制震后重建规划。20 世纪 80 年代，是天津城市定型的重要时期，借改革开放的东风，天津震后重建取得重大进展。在城市引水、给水、燃气及道路交通方面极大提升了城市功能。建设了天津东站、食品街、古文化街、国际商场等一批公共建筑，开展了城市环境综合整治，对老旧建筑进行整修美化，破墙透绿，城市面貌和城市管理水平得到极大改善，成为全国各地学习仿效的榜样。同时，天津城市规划不断完善，逐步稳定，在前 21 版规划方案的基础上，制定了新一版城市总体规划（1986—2000 年）。1986 年经国务院批复，这版规划明确提出了“工业东移战略”和“一条扁担挑两头”的城市总体布局。中心城区则明确梳理出“三环十四射”的道路骨架，内环线和中环线以现有路网为主，联结成环线，接顺了中式和各租界的路网，但也进一步丢失了方向感，这也是为什么许多人说到天津容易迷路。外环线几乎是全新规划的，适当考虑了与海河走向的关系，形成了一个以海河为轴的鸭梨形。但实际上在中环线与外环线之间绝大部分新规划的地区路网都尽可能规划成正南北向或正东西向。客观地说，这一阶段的规划受功能主义规划思想影响较大，从城市设计的视角考虑得较少，几乎不涉及城市轴线、公共空间、城市轮廓线等内容。

“进入 20 世纪 90 年代，天津实施了中心城区大规模危陋平房改造，这是对过去半个多世纪欠账的偿还。同时启动了滨海新区的发展。这一时期天津的城市设计开始起步，在城市结构与形态方面进行了深入思考和探讨。首先是在 20 世纪 90 年代初对海河的功能和定位进行研究，明确了海河是城市的主轴线。这实际上是对梁思成和张锐版、穆欣版的继承和深化，是对 1986 年城市总体布局的明确，提出城市由背向海河改为面向海河发展，将海河南岸与北岸不均衡发展调整为均衡发展。到 1994 年新一轮城市总体规划修编，明确天津城市中心商务区位于海河两岸的解放北路、小白楼和河东南站地区；中心商业区位于海河南岸的和平路和滨江道地区，同时集中在海河两岸划定了 14 片历史文化街区进行整体保护。虽然这一时期还没有编制中心城区总体城市设计，但城市设计的整体艺术骨架已经确定。对于中环线与外环线之间过渡地带，规划初步提出建设 10 个职住平衡的各具特色的功能组团，如奥林匹克体育中心组团、会展区组团等。

“进入 21 世纪，我们编制了海河两岸综合开发改造规划，并开始实施。

海河两岸综合开发改造规划实际上就是城市设计，是一个实施性的城市设计。在第一阶段的总体城市设计中，不仅包括规划目标、策略和规划实施的‘十大工程’，还包括综合性规划、市场化开发和法定化管理等内容。海河综合开发改造的总体城市设计确定后，举行了六个节点详细城市设计方案的国际征集，最终形成可实施的综合方案，还同步编制了起步段的景观城市设计方案。随着海河综合开发改造的深入，我们又组织了第二轮重要节点的城市设计方案征集，包括英法租界区、德租界区、意租界区、西沽地区等。

“为了迎接 2008 年北京奥林匹克运动会，天津作为协办城市启动建设了奥林匹克体育中心。由于海河两岸周边地区当时还不具备这样的空间，所以也从没想到过将奥林匹克体育中心安排在海河轴线上。为了加强京津冀区域协调发展，天津会同北京市和原铁道部启动了我国第一条高速铁路——京津城际铁路的建设，城际站的终点站就在海河边的天津东站。城际车站靠近市中心，这是一个睿智的选择，结合车站改建，天津东站成为有三条地铁线换乘的综合交通枢纽。有了海河的经验，在 2006 年滨海新区被纳入国家发展战略后，我们启动了九大功能区的城市设计方案征集。2007 年，我们又组织了滨海中心商务区海河两岸城市设计的国际研讨活动。”

诸葛民

“2008 年，北京举办第 24 届夏季奥运会、天津是协办城市，同年夏季达沃斯论坛要在天津举办，为了以亮丽的城市面貌迎接奥运会和达沃斯论坛两大盛事，天津实施了城市环境综合整治，重点部位是‘一带三区五线’。海河是“一带”，奥体中心、天津站和小白楼地区为“三区”，“五线”主要是迎宾线，包括复康路和卫国道、卫津路、南京路、解放路及友谊路以及东南半环快速路。‘一带三区五线’串联城市交通枢纽、历史风貌区、商业中心区、科技文化区、奥体中心区等重要功能区，构成了横跨东西、纵贯南北的两大轴线，形成天津中心城区的景观骨架。这次迎奥运市容环境综合整治工程实际上是对城市的美化和更新，天津采用市容环境综合整治的手段，在时间短、任务重的情况下，寻求行之有效的途径，选择具有可操作性的方法，保证了工程的全面实施和最终效果。

“市容环境综合整治作为一项系统工程，城市设计统筹了建筑立面整治、绿化景观、街道绿地、环境设施、道路交通、灯光广告等街道空间全要素，妥善处理局部与整体、形式与功能、环境整修与业态提升、市容与繁荣的关系，从细微处入手，精雕细刻进行设计，保证了城市面貌在短期内有重大变化，改善了市民生产生活环境，增强了城市吸引力，初步展现了天津大气洋气、清新亮丽、中西合璧、古今交融的城市特色风貌。

“迎奥运环境整治是在一个集中的时间段内，全市上下按照统一要求，选择重点路段与地区，在突出特色的同时，对路面、绿化、建筑、灯光、店招牌匾、街道家具进行统一提升。编制了路内公共设施设置导则、店招牌匾设计导则、街道家具选型建议等导则，保证了整体效果的协调和水平的一致。迎奥运环境整治投入的人力、物力、财力是巨大的，效果也是明显的。比如拆除了城市建筑屋顶凌乱的广告，消除了公交车上不良的广告，全市出租车统一了外观，清除了大量的垃圾和卫生死角。很多项目在景观绿化节点上精益求精，个别的为了达到理想效果甚至多次返工。解放北路和南京路道路景观环境整治以步行化优先为原则，实施后整体达到较高水平。我们不否认，综合整治也引来一些非议。比如某些路段对底商的牌匾做了统一尺寸和样式的规范要求，店面缺乏特色；街道灯杆采用了统一的样式，缺乏个性；原来快速路上精心设计的以天津工业为特征的路灯杆也差点儿被换成统一的样式；海河边一些墨绿色的栏杆，都被刷成了白色，一些建筑的灰色屋顶也被统一刷成了红色。

“天津迎奥运环境整治有些工作似有矫枉过正之嫌。整治前的城市环境的确太乱了，各种做法八仙过海、鱼龙混杂。通过统一的整治，保证了总体的水平，也可能埋没了一些有创新的优秀设计。这需要未来逐步放开设计市场，提供创新和张扬个性的环境。城市环境的改善一定是一个稳定上升的过程，实际上，天津环境整治的做法既有必然性，也有国际上的先例。19 世纪末，美国为了改变脏乱差的城市环境，结合芝加哥世界博览会召开的契机，开展著名的城市美化运动。当时为了在污浊的空气中显得城市更整洁，所有的建筑、栏杆、灯杆都被刷成了白色。天津在 2003 年启动海河综合开发改造，到 2008 年的 6 年时间里，海河两岸基础设施和景观建设基本完成，一些主要的公共建筑也建成了，但当时还没有形成整体的面貌，城市其他地区的建设品质也良莠不齐。迎奥运环境综合整治是对天津城市面貌和整体环境的一次系统梳理和提升。

“也就是在 2008 年，天津市成立了重点规划指挥部，集中力量编制了 119 项规划，其中近半数为城市设计，从中心城区总体城市设计、滨海新区核心区总体城市设计、各区总体城市设计，到重点地区、功能区城市设计，直到重要建筑群的详细城市设计。如中心城区文化中心城市设计及建筑设计，举办了两轮国际征集；于家堡金融区起步区编制了城市设计导则来指导建筑设计；中心城市 14 片历史文化街区都编制了城市设计，使得保护规划在城市设计和导则的指引下展开，促进了历史文化街区的整体保护和有序更新。一批高品质项目，如中心城区文化中心建筑群、于家堡金融区起步区、津湾广场、民园广场、五大院、滨海文化中心等建设完成，加上铁路西站、于家堡高铁站、邮轮母港、天津机场第二航站楼等，城市功能与城市环境整体迈上了一个台阶，显示出了城市设计的力量和作用。”

38. 历史街区的城市设计

诸葛民

“一个好的城市一定是美的城市，城市特色是城市美的重要特征。天津是国家第二批历史文化名城，同时具有山河湖海等自然特征，在我国近代史上拥有重要地位。天津历史城区既有中国传统建筑文化遗存，又有异国风貌的租界环境和建筑，城市空间形态多元，路网街巷尺度宜人，城市格局和风貌特色突出。为保护好天津的历史文脉和城市特色，彰显城市魅力，我们运用城市设计方法，编制了历史街区的城市设计，并转化为法定的保护规划（等同于控制性详细规划）。到 2012 年，包括海河两岸历史街区在内的 14 片历史文化街区都编制了城市设计和保护规划。

“五大道历史文化街区是我市保存最为完整的历史街区。早在 20 世纪 90 年代初，天津规划局组织编制了五大道地区建设管理保护规划，明确了要保护的文物建筑和历史建筑，确定了可改造的地块，规定新建建筑要与周围环境协调，建筑檐口高度不得超过 12 米。这些规定简洁清晰，保证了五大道地区的整体空间形态没有因新建设而破坏。实际上，中华人民共和国成立后五大道虽然建筑形态未变，但住在里面的主要居民已发生更迭。特别是 1976 年地震后，五大道违章搭建，人口拥挤，环境恶化。1999 年，市、区政府开始对环境进行综合整治，拆除私搭乱盖的违法建筑。2003 年，结合海河综合开发中的腾迁整理工程，市委、市政府提出打造‘近代中国看天津’品牌，制定风貌建筑保护利用条例，成立风貌公司，对由多户居民聚居的历史风貌建筑进行腾迁置换。按照修旧如故的原则进行整修。同时，推动相关规划的实施。

“2007 年，五大道地区开始编制城市设计。当我们对整个区域内的建筑进行普查时惊奇地发现，五大道地区从 1901 年启动大规模建设，到 1946 年基本建设完成，至今前后历时 100 多年。但历史文化街区范围内的 2159 幢房子中，只有 1037 幢是中华人民共和国成立前建设的，占总数的 47%，其余的 53% 都是中华人民共和国成立后新建或插建的。我们在五大道里是体会不到这么多新建筑的，一直觉得五大道里的老建筑应该占绝大多数。这主要是因为这些新建筑的规模、体量和高度延续了五大道地区的旧有模式，整体环境保持得很好。改革开放前，五大道的建设规模都比较小。20 世纪 50 年代规划建设了苏联式的友好里、团圆里等住宅小区。这些小区的建筑只有三层，采用小的围合式院落布局，保留了五大道原有路网，其尺度和规模与整体环境也是和谐的。

“改革开放后，由于天津外来投资少，五大道改造压力小。更重要的是天津很早就意识到要整体保护这片区域，特别是 20 世纪 90 年代规划局制定了严格的建设管理保护规划，要求建筑檐口高度不得超过 12 米，保证了五大道的整体和谐。同时，我们通过对五大道建筑空间环境的深入分析，发现一些新建筑虽然满足了建筑高度控制要求，但建筑布局和形体设计上比较随意，破坏了五大道的内部空间和肌理。为了保证新建筑不破坏历史环境、与整体环境相协调，必须进一步完善原有的规划管理规定，这首先需要一个好的城市设计。

“为了做好五大道的城市设计，我们采用了城市形态学和建筑类型学的分析方法。通过对所有街坊进行分析，我们发现五大道的建筑可以归纳为门院式、院落式和里弄式三种主要类型，并且找到它们在规划布局上呈现出的分布规律。这些研究让我们对五大道有了更加科学理性的认识，为高水平的城市设计编制提供了保证。

“由于五大道过去是英租界的高级住宅区，房屋主人为了保证居住的私密性和安全性，多数建筑都有比较高的围墙，它们也在表达着主人的身份和喜好，因此围墙的样式和品质是十分讲究的。城市公共空间除了睦南花园、民园体育场外，最突出的是街道和里弄空间。它们可以归纳为五种类型。另外，以对五大道的每一栋建筑，以及每一个地块、街坊的细致分析为基础，我们编制了五大道地区的详细城市设计，建立了五大道建筑形体和街道空间的三维模型。通过这次城市设计，各方达成一致，将原来规划要拓宽的道路红线都缩回到现状道路的宽度。同时提出了精细的城市设计导则，对每一个地段提出了新建筑介入历史环境的详细管控要求，在管理上更加有的放矢。其作用也是显著的：第一，通过时间和空间对应在城市管理中；第二，重新发现街道、街区的定位；第三，少量化的原型研究使得管理更加简明。

“在城市设计编制的同时，风貌公司在五大道地区实施了庆王府以及先农大院一期置换和修旧如故工程，在实施过程中听取了城市设计方面的意见建议。庆王府原为宫廷太监小德张所建的寓所，后被清室第四代庆亲王载振购得并举家居住于此，中华人民共和国成立后一直是天津市外办办公楼。外办迁走后，庆王府作为展示馆，山益里改造为小型特色精品酒店。先农大院为原先农地产公司开发的院落式联排职工住宅。置换整修后一期作为以餐饮为主的街廓。先农二期为适应现代办公功能，做了进一步的尝试，部分风貌建筑在不改变外观的情况下，对内部进行优化。同时修建了部分新的办公建筑，在体量上与原有肌理保持协调。这些整修和更新项目规模都不大，但无论投资运营方还是规划师、建筑师、规划管理部门，都花费了大量心血。2012年，毗邻庆王府和先农二期，和平区投资公司实施了民园体育场更新改造。项目事先经过反复论证，新建的建筑没有过多地扩大建筑规模，而是延续了老建筑的类型，设计了三条塑胶跑道，保留了运动的功能，增加了许多餐饮、展示、游客服务的功能，沿河北路打开形成开放的柱廊，实施后的效果很好。”

诸葛民

诸葛民停顿了一下，目光由投影的屏幕转回到观众。“每次讲到这儿，我都会想到意大利著名建筑设计家阿尔多·罗西在《城市建筑学》中著名的论述，他指出：建筑不是孤立存在的，建筑是城市的建筑，建筑的形式非常重要，它是人们的集体记忆。虽然随着时间的推移，建筑的功能可能发生了变化，但建筑的形式是不变的。罗西用符号学、现象学等文化哲学思想对建筑的本质进行了分析，他超越了现代主义建筑单纯强调功能、遵循‘形式追随功能’的做法，提出了建筑作为独立的艺术形式的客观实在。天津就有许多这样的例子。除五大道历史街区外，过去以居住为主的原意大利租界，现在叫意式风情区，大量的花园住宅建筑今天作为餐厅、展示馆、办公等功能使用，成为天津著名的旅游目的地。意式风情区的城市设计也达到了较高水平，为形成一个功能复合的历史街区提供了规划支撑。

“五大道地区城市设计基础扎实，有理论和实践支撑，达到较高的专业水平，曾获得全国优秀城乡规划设计一等奖。规划院城市设计所长期坚持开展深入研究，几年前与美国加州大学伯克利分校合作，以五大道周围的建控地带为题目进行学生课程设计，为城市更新探讨新的可能性。他们还把多年研究和实践的丰硕成果汇编成一本书《中国 · 天津 · 五大道——历史文化街区保护与更新规划研究》，这种专题研究与对历史街区实体的保护同等重要。虽然五大道城市设计水平比较高，成就有目共睹，但依然有一些问题没有找到适合的答案。五大道历史上是达官显贵聚居的高档居住社区，院墙都比较高，很封闭。中华人民共和国成立后，许多较大体量的小洋楼用于机关办公，居住的人群也发生了变化，但仍然是相对安静的居住和办公地段。虽然城市设计考虑旅游业的发展增加了餐饮、酒店等复合功能，但在定位上还是以复合型的居住社区为发展目标，旅游与居住社区相互之间的矛盾并没有妥善解决。另外，五大道地区居住人口约 4 万人，人口密度比较高，还有一些非历史风貌建筑的建筑质量和居住品质不高。如何形成比较良性的市场机制实现地区更新，目前暂时还不得要领。

“近年来，随着五大道配套功能完善和知名度不断提高，来五大道旅游的游客不断增加，但游客们只能在路上走一走，看一看洋楼的外观，无法深度体验。为此，城市设计所继续主动开展了相关规划研究。他们借鉴美国波士顿‘自由之路’的做法，在五大道内规划了一条完整的体验线路，把五大道内主要的旅游设施、名人故居、博物馆以及独特的建筑和空间串联起来，推动五大道特色旅游发展，同时探讨五大道地区新的成长结构。

“本次天津作为全国城市设计试点城市，我们再次把五大道地区的城市设计列为试点项目。结合当前全域旅游发展，五大道申报了国家 5A 级景区，以及天津市政府推动的‘洋楼经济’和‘夜间经济’活动。本次城市设计将五大道地区定位为：天津市历史文化街区的保护典范，全域旅游和洋楼经济的示范区，世界文化遗产的候选区。结合天津历史城区步行化和城市中心文化复兴，将五大道体验之旅纳入全市的步行体系。考虑到游客数量的不断增加，进一步完善游客服务设施，增加公共交通出行的比例。结合近期地铁规划，除目前在使用的小白楼地铁站之外，新增两个规划地铁站，同时预留了恢复 M1 线新华路地铁站的可行性。最后，受和平劝业场周边地区城市设计的启发，我们建议结合城市外围和平劝业新城建设，疏解五大道过密居住人口，改善部分居民的居住条件。从巴黎老城区和伦敦威斯敏斯特区发展演变的经验看，五大道地区将来会有类似效果，成为城市复合中心的组成部分。”

39. 天津市文化中心及周边

诸葛民

“文化中心建筑群在城市总体空间结构上通常具有决定性的影响，不仅是歌剧院、音乐厅、图书馆、美术馆、博物馆等这些场馆建筑，更重要的是由这些场馆建筑围合成的广场、公园等公共开放空间，都会成为一座城市的活动中心和人们对城市意象的焦点。

“多年来，天津没有形成集中的文化中心建筑群。原来的规划中没有预留出现成可供建设的场地，而且当年城市财政预算有限，每建一栋公共建筑都会有许多区来争取。为了平衡各区的诉求，以及为了带动落后区的发展，天津仅有的几座新建文化建筑分散在各区。天津市图书馆和周邓纪念馆建在南开区，科技馆在河西区。平津战役纪念馆放在红桥区，选址与历史事件发生地关系不大。而且它的规划设计就像在一个单位大院内建了个文化建筑，与城市公共空间也没什么关系。

“2000 年天津拟建一座大型的博物馆，以纪念建城设卫 600 周年，选址在河西区友谊路与儿童乐园之间多年保留下来的一块空地上。路对面是 20 世纪 50 年代建的大礼堂和 20 世纪 80 年代建的展览中心。博物馆建筑设计方案采用了国际征集的方式，最终日本著名结构建筑设计师川口卫和高松伸的方案获得优胜。川口卫在结构设计方面的造诣很深，曾配合丹下健三完成了著名的东京代代木体育馆的结构设计，也正是他后来设计了天津三岔河口标志性的摩天轮与永乐桥。天津博物馆的造型像一只展翅飞翔的天鹅，预示着天津新的腾飞。后来据说天津博物馆是世界上最完美的、跨度最大但高度最低的大跨度壳体结构。但作为博物馆建筑，特别是天津博物馆，在功能和造型不十分契合。因此，在后来天津文化中心的整体设计时，改作为自然博物馆。当年配合博物馆建了一个占地 17 万平方米的巨大的银河广场。银河广场的名称源自‘天津’是银河的别名，在屈原《离骚》中有‘朝发轫于天津兮’，同时该名称又具有时代气息，与整个广场日月星辰的设计理念相得益彰。但从实际建成效果看，对天津城市空间形态改善作用有限。

“2002 年海河综合开发时，也初步拟定了十大公共建筑。在红桥区规划建设三条石近代工业博物馆，在河东区规划建设了海河图书馆，在河西区建设了儿童剧院。原来在中心广场规划了海河音乐厅，这个名字很好听，但一直找不到合适方案。后来将小白楼的原小白楼电影院改造成了纯欧式的天津音乐厅。总体看，由于海河市区段比较窄，以及河两岸用地的局限，也没有形成集中的文化建筑群。2008 年，天津规划北京奥运会后的新一轮发展，提出建设新的文化中心，选址在现有一定基础的友谊路东侧，包括现在的天津科技馆、博物馆、青少年活动中心、中华剧院等，总用地 90 公顷。在建成区能有这样规模和规整的用地是十分难得的，这主要是占了儿童乐园 30 公顷的用地，让儿时在乐园有许多美好记忆的人们唏嘘不已。

“天津文化中心举办了两轮高水平的规划总体布局和建筑设计方案国际征集，被誉为达到了‘明星级的总体水平’。综合方案总体布局糅合了‘山、水、塔’的中国园林布局和‘大轴线、林荫道’的西方园林手法，从大剧院的公共平台向西望去，以高度 60 米的迎宾塔为中景标志物，以高度 350 米的天塔为远景标志物，形成近、中、远的多层次景观效果，并将基地内的东西向轴线进一步延伸与城市的开放空间系统连接。城市设计确定了建筑组群中每个建筑的位置、高度、界面、主次关系、新旧关系以及空间处理要求，形成和谐有序的空间体验。整个建筑群组突出了建筑形态的整体协调。在从设计到建造的全过程中，发挥城市设计的统筹与指引作用，正视争论、尊重个性、主动协调、寻求共识，最终基本实现了城市设计最初的构想，是城市设计一张蓝图干到底的样板。天津文化中心建筑群及其围合的城市公园、广场等公共空间成为城市南部地区最重要的公共空间，对这一地区的空间形态起到提纲挈领的作用。

“天津文化中心已经投入使用 7 年了，总体看效果良好，基本上实现了‘人民的殿堂’的目标。同时也存在一些问题：一是在功能布局上不够复合，虽然同步规划建设了银河购物中心，但商业类型不够丰富，缺少酒店、办公等多种业态；二是在总体布局上，缺乏对人的细致考虑，比如从图书馆到银河购物中心步行有 800 米距离，这可能与管理有关，也是气派的布局本身所无法避免的。有时鱼与熊掌不可兼得。虽然有以上的遗憾，但文化中心对天津城市空间结构和形态所起的作用还是巨大的。

“前面谈到，天津 1986 年版的城市总体规划对中心城区的城市设计内容考虑得不多，缺乏真实的城市空间结构的描述，重点是在明确‘三环十四射’的道路骨架。到了 1999 年版城市总体规划，则开始明确了海河作为城市空间结构的主轴线，以及中心商务区、中心商业区、14 片历史文化街区的位置和形态，基本确定了历史城区的空间结构和总体形态，但对城市中环线以外地区的空间结构和形态也还没有明确。在 2008 年各区的总体城市设计中对这部分考虑也不多。2008 年天津空间发展战略提出中心城区‘一主两副’的城市中心结构，‘一主’是将小白楼传统城市中心与文化中心周边地区连接起来，形成天津中心城区的一个中心，虽然实际上两区之间相距较远，无法真正形成一个中心。根据空间发展战略以及文化中心城市设计，市规划局邀请 SOM 编制了文化中心周边地区城市设计。

“文化中心周边地区占地面积约 2.41 平方千米，规划总的开发量 650 万平方米，规划建成中心城区主要的中心商务区。SOM 主持，市规划院、上海城建院等单位多专业配合，形成了一个高水平的城市设计，最后形成了详细城市设计导则和市政府批复的控制性详细规划。2012 年文化中心建成后，市城投把目光转向了文化中心周边地区，开始实施文化中心周边地区城市设计，启动了老八大里的拆迁。老八大里是 20 世纪五六十年代建成的、以 3 ~ 5 层单元楼为主的小区。由于区位好，拆迁难度比较大，只拆迁了两个街坊就陷入了停滞。今天再看，文化中心是天津绝无仅有的、最集中的一组文化建筑，对友谊路及周边的空间结构起到了整合梳理的作用。然而，从目前的经济形势看，文化中心周边地区新规划的商务中心很难按规划实现。一是因为目前天津对写字楼市场无需求。二是因为拆迁难度极大，几乎没有可能性。这一地区的多层住宅楼建筑密度非常高，二手房市场的价格也很高，拆迁需要海量资金。如果没有可以平衡海量拆迁资金的房地产市场，这个很好的城市设计方案恐怕要束之高阁相当长时间了。”

40. 于家堡金融区

诸葛民

“有历史学家说，城市是人类历史上最伟大的发明，是人类文明集中的诞生地。在 21 世纪信息化高度发达的今天，城市的聚集功能依然非常重要，特别是高度密集的城市中心。陆家嘴金融区、罗湖和福田中心区，对上海浦东新区和深圳特区的快速发展起到了至关重要的作用。2006 年被纳入国家发展战略伊始，滨海新区就开始研究如何选址和规划建设新区的核心——中心商务区。这是一个急迫需要确定的课题，而困难在于滨海新区并不是一张白纸，实际上是一个经过 100 多年发展的老区。经过深入的前期研究和多方案比选，最终确定在海河下游沿岸规划建设新区的中心。这片区域由码头、仓库、油库、工厂、村庄、荒地和一部分质量不高的多层住宅组成。当时有市领导到现场考察中心商务区的选址，看到当时如此破败的景象，直问没有走错地方吧。于家堡要规划建设成为滨海新区世界高水平的中心商务区，在真正建成前会一直有怀疑和议论，就像十几年前我们规划把海河建设成世界名河所受到的非议一样，也是情理之中的事情。规划需要远见卓识，更需要深入工作。滨海新区中心商务区规划明确了在区域中的功能定位，明确了与北京金融中心、天津老城区城市商务金融中心的关系。通过对国内外有关城市中心商务区的研究，确定了新区中心商务区的规划范围和建设规模。我们发现，于家堡金融区半岛与伦敦泰晤士河畔的道克兰金融区形态上很相似，这冥冥之中揭示了滨河城市发展的共同规律。

“为提升新区中心商务区海河两岸和于家堡金融区规划设计水平，2007年我们聘请国内顶级专家吴良镛、齐康、彭一刚、邹德慈四位院士，以及国际城市设计名家、美国宾州大学教授乔纳森 · 巴奈特（Jonathan Barnett）等专家为顾问，为规划出谋划策。邀请美国 SOM 公司、易道公司（EDAW）、清华大学和英国沃特曼公司（Waterman Inc.）开展了两次工作营，召开了四次重大课题的咨询论证会。历时一年之久的‘滨海新区中心商务区海河两岸城市设计’国际工作营，是城市设计作为设计和决策过程在天津滨海新区的实践与应用。通过工作营，确定了高铁车站位置、海河防洪和基地高度、起步区选址等重大议题，并会同国际建协进行了于家堡城市设计方案国际竞赛。于家堡地区的规划设计，汲取纽约曼哈顿、芝加哥壮丽一英里、上海浦东陆家嘴等地区的成功经验，通过众多规划设计单位的共同参与和群策群力，包括国际建协组织的城市设计竞赛，通过多方案比选，最终采用了窄街廓、密路网和立体化的规划布局，将京津城际铁路车站延伸到金融区地下，与地铁形成交通枢纽。规划以人的感受为主，形成了完善的地下和地面人行步道系统，建设了中央大道隧道和地下车行路以及市政共同沟。规划沿海河布置绿带，形成完善的滨河景观和城市天际线。于家堡的规划设计充分体现了功能、人文、生态和技术相结合，达到了很高的空间品质，具有时代性，为充满活力的金融创新中心的发展打下坚实的空间基础，营造了美好的场所，希望它成为带动新区发展的‘滨海芯’。

“从 2006 年确定选址开始，十几年过去了，于家堡发生了翻天覆地的变化。百年前的破旧小渔村变成了现代化的金融创新运营示范区，这种变化确实可以用沧海桑田这四个字来形容。2008 年底城市设计方案确定后，开始拆迁。在短时间内完成了于家堡村、郭庄子、老湾道、二府街、港务局第三作业区码头，包括南岸东西沽的拆迁。2009 年以京津城际高铁延伸线于家堡高铁站动工为标志，于家堡起步区启动建设。2015 年以京津城际高铁延伸线于家堡高铁站开通为标志，于家堡起步区 0.8 平方千米初步建成。海河中央大道隧道开通，14 个地块基本完工，部分楼宇已经投入运营，地下商业街环球购已经开街，国际金融会议中心洲际酒店已投入使用。高铁站公园、中央大道景观主轴线、起步区滨河公园、起步区北公园已建成。一个环境优美、尺度宜人、高效生态的金融区初具规模，矗立在海河岸边。”

诸葛民

“总结于家堡金融区规划设计的成功经验，最重要的就是系统的城市设计编制和严格按城市设计导则实施的管理。于家堡金融区的规划设计一直在‘窄街密网’的城市布局结构和空间形态的总体控制下不断深化，主要突出了四方面的特色。一是突出滨水、人文、生态特点，注重塑造与金融区定位相符的城市布局和城市形态。二是从功能完善、集约土地利用、优化土地结构着手，大力发展公共交通，建立高铁、地铁和常规公交组成的公共交通体系，倡导绿色出行。三是地下空间系统开发，集地铁、地下车行、地下人行、商业、共同市政管廊等于一体。四是分别从城市布局、交通运输、建筑、环境、能源等方面展开研究，探索低碳城市发展技术路线。同时，于家堡金融区注重历史人文景观的延续性，保留历史文物和风貌建筑，如塘沽南站、《塘沽协定》遗址、太古轮船公司遗址、三菱油库遗址等，展现于家堡的历史遗存。优美的自然环境、优越的区位条件、较为完整的用地条件和深厚的历史积淀，为于家堡金融区的城市设计奠定了坚实的基础。

“按照城市设计，中心商务区管委会组织编制了于家堡金融区控规和起步区城市设计导则，作为规划管理的法定依据。于家堡金融区起步区城市设计导则非常严格，除一般导则对地块街墙贴线率、出入口方位、地下空间利用等控制外，还对建筑塔楼和裙房形体做出严格控制，即国外所谓的‘信封控制（Envelop Control）’或‘包络控制’，即建筑设计方案必须放入‘信封’内。

“于家堡起步区 9 个写字楼地块的建筑设计齐集九位国内知名的实力派中青年建筑师——崔恺、周恺、胡越、齐欣、张颀、崔彤、王辉、张雷和姚仁喜，为中国优秀建筑师提供了创作的舞台。在城市设计导则的统一控制引导下，整个建筑群体从高度、体量、材料、地下空间、交通组织、色彩等多方面进行了一体化设计，最终实现了国内首个整体性地下空间实施案例，是一项大胆的创新和试验。目前 9 栋建筑已完工投入使用。开始一些建筑师不理解，不考虑导则要求，做出了许多有独特想法、个性很强的建筑，对建筑群整体效果和城市环境考虑不足。在设计过程中，于家堡城市设计和城市设计导则编制负责人、SOM 事务所合伙人菲尔 · 恩奎斯特（Philip Enquist）以及建筑设计团队共同协作，经过说服、沟通，还有市领导亲自参与与设计师的沟通，最后达成一致。SOM 事务所主创建筑师设计国际金融会议中心方案时，最早的方案也有比较怪的，最后才形成现在的方案，既保证了建筑单体方案建筑设计的高水平，又保证了城市街道、广场的整体形象以及绿地、公园等公共空间的品质，建筑真正成为城市的建筑。

“于家堡起步区 9 个写字楼地块加上后来的国际金融会议中心和其他两个地块建筑，形成了起步区 9+3 建筑群，总占地 20 万平方米，地上总建筑面积 180 万平方米，共 12 栋 120 ~ 300 米高的塔楼和一个水平体量的地标建筑。各位建筑师按照城市设计导则，精心设计，既体现了各自的建筑风格，也保证了建筑群整体的协调统一，最后的效果非常有说服力。于家堡起步区建筑群的建筑风格统一又多样，与金融中心的气质十分匹配，甚至可以与柏林波茨坦广场建筑群相媲美。作为金融建筑，建筑风格‘经典不复古，出新不出奇’。整体建筑体量适度、高低有致、疏密得当。建筑材料材质相对规整，建筑色彩冷暖适度，稳妥且富有新意，形成了自身鲜明的建筑特色。于家堡金融区起步区的建筑设计代表了当前国内建筑群设计的最高水平和方向。

“在规划实施的过程中，于家堡的规划管理者和建设者们不断遇到新的问题。先进的规划理念、设计手法与国内相关规范的冲突时有发生；土地的混合使用与规划管理审批确权矛盾重重；超前的规划设想与相对滞后的城市管理手段对话困难。为此，我们开展了滨海新区城市设计规范化和法定化改革，制定了改革方案和相应的管理规定。通过不断的改革创新及探索，以上问题都得到了较好解决。”

诸葛民

“从城市设计和建筑设计的角度讲，于家堡达到了很高的水准，但从市场经营的角度讲，目前面临诸多困难，其原因是多方面的。我想在城市规划建设方面有三个主要原因：

一是滨海新区中心商务区建设太分散，相互竞争。开发区现代产业服务区建设了100多万平方米的办公建筑，包括530米高的周大福中心已经封顶。在于家堡起步区对岸，规划建设了响螺湾商务区，40栋楼宇，总建筑面积500万平方米，其中1/3是写字楼，1/3是公寓。开发区MSD、于家堡金融区、响螺湾商务区三者相距不超过3千米，虽然定位略有不同，但建设的商务建筑都是写字楼，几百万平方米的写字楼一块儿投入市场，对于以第二产业为主的滨海新区无疑是场灾难。当时由于三个区分属不同的政府或管委会，而且也认识不到市场的容量有限。响螺湾土地整理成本比较低，当时定位为外省市驻津商务区，为了招商引资，给予了很优惠的政策，但招来的客商良莠不齐，建筑的质量整体不高。曾经有一位新区领导汇报说，响螺湾可以做试验，如果搞不好，我们还有于家堡。让市里主要领导大怒，这么大的开发怎么可能搞试验？如果当时响螺湾规划为以居住为主的生活配套区，商务区今天的日子会好过很多。

“二是中心商务区内交通联系不便，缺乏相互间的支撑。滨海新区中心商务区由开发区MSD、于家堡金融区、响螺湾商务区和天碱文化商业区组成。由于海河和铁路的分割，使得几个区之间缺乏有效的交通联系。最初各方面专家提出拆除进港二线铁路，但最后的结果是不但没拆除，还建设了复线。尽管建设了穿越进港二线铁路的中央大道地道、春光路地道、北海路地道，以及在建的洞庭路地道，但通道数量还是比较少。由于通航标准问题，海河在建桥梁迟迟未能合拢，过河交通一直不便。滨海新区中心商务区内各区之间由于缺乏便捷的联系，导致相互缺少支持和互动，没有体现聚集和规模效益。虽然近期通航标准获批，但五六年时间过去了，在建桥梁一直没有合拢，未发挥应有的作用，对新区中心商务区发展的影响还是很大的。

“三是生活配套设施缺乏，地铁建设缓慢。当代的中心商务区功能已经更加复合，生活配套设施是必备的条件。但于家堡金融区、响螺湾商务区都是以办公建筑作为起步项目，缺乏住宅项目和配套的教育、医疗等设施，无法形成人气的聚集效应。相比之下，开发区MSD在这方面就做得好一些。开发区经过20年的建设，具备了完善的居住社区和完备的配套，如心血管医院、泰达医院、南开泰达学院等。另外，新区公共交通还不够发达。早在2009年新区就制定了轨道网规划，并上报纳入天津市轨道网。直到2015年国家批准天津地铁近期建设规划，其中包括滨海新区B1、Z4、Z2三条线。B1、Z4已经启动建设，但多少年时间过去了，进展缓慢，直到最近B1才通车，Z4还在建设中。没有快速的轨道交通，就无法让原塘沽区、中心生态城的旅客便利到达于家堡，无法发挥于家堡中心的作用，无法给于家堡提供人流支撑。

“虽然存在这些问题，但滨海新区中心商务区正逐步形成行政、商务、商业、文化、娱乐、居住、交通七大功能。滨海新区政府位于中央大道旁，在对面的天碱地区、滨海文化中心和万达商业广场已经运营。在响螺湾，滨海新区 20 多个委办局入驻，新区出入境服务大厅运营，响螺湾极地海洋馆、气膜体育场日均人流量 1 万人次，酒店式公寓已租售 3000 余套。在于家堡，宝龙国际中心、环球购商业街、CGV 影院已经运营多年，加上新区行政审批局，工作和办事人流量日均可达 1 万人次，区内办公企业商务人流量 5 万人次。于家堡高铁枢纽站使京津乘客往返便利顺畅，日均人流量接近 1 万人次。

“越来越多的政策在于家堡金融区落地。2014 年 12 月，天津自贸区挂牌成立，于家堡金融区是三个片区之一。天津市政府响应国家‘大众创业、万众创新’的号召，在于家堡建立双创特区。‘金融创新运营示范区 + 自贸试验区 + 双创特区’的政策和功能优势不断显现。同时，天津简政放权，实施一颗印章管审批等改革，使投资服务贸易便捷化，使于家堡注册企业呈指数型增长。我查了一下相关资料，2017 年于家堡新增市场主体 8452 家，合计注册资本金 2383 亿元，市场主体总数超过 3.38 万家，其中注册资本超过 5000 万元的 2433 家。聚集金融及类金融机构近 2000 家，涵盖了几乎所有金融细分业态，管理资产规模超过 2.6 万亿元。在近三年新增的市场主体中，30% 来自北京。同时，更多的资源在于家堡汇集。世界著名的美国茱莉亚音乐学院滨海于家堡校区建筑已经投入使用。开发区管委会也已经入驻于家堡。轨道 B1 线的通车对于家堡的影响是巨大的，目前已经看到效果。如果滨海轨道 Z4 线近期能够通车，将生态城、北塘、开发区与于家堡高铁站连通，作用将更加明显。轨道 B1 线连通滨海高铁站，与已经启建的京津间第二条高铁连通，未来可以从于家堡起步区、高铁站直达天津滨海国际机场，与北京通州新的行政中心和国贸中心商务区方便联系，进一步提升滨海新区对外辐射水平。

“尽管有这些利好的消息，但于家堡面临的困难依然很大。设想中的办公商业总量受目前世界性经济环境的恶化影响巨大；起步区办公、商业设施、公共服务设施先行建设，生活设施配套严重滞后。即使市场主体数量和办公人数不断增加，于家堡车站和地下商业街中人流涌动，但在偌大的商务区中，这些办公人口很快被淹没在楼宇中，尤其在周末更看不到多少人影，因此受到质疑在所难免，而且这种状况恐怕还会延续一段时间。

“城市的产生和发展是一个历史性的过程，而城市设计也是在每天都进行的一连串的决策制定过程中产生的。美国城市设计师乔纳森 · 巴奈特先生提出城市设计作为“公共政策”的现代城市设计概念，即‘真实的城市设计是一连串的行政决策过程和城市形态塑造过程，是一个既有创意又有发展弹性的连续变化过程，应当使设计具有更大的自由度和弹性，而不是建立完美的终极环境或是提供一个理想蓝图’。如果说传统城市设计更多地注重‘目标取向’的话，那么现代城市设计则应是‘目标取向’和‘过程取向’的综合，城市设计师不只是将包装精美的产品交出去就万事大吉了，而是要制定一个可以随着需求和时间的改变而弹性修订的规则。

“与巴黎的拉德方斯、伦敦的道克兰一样，于家堡金融区的发展现在也遇到了困难和挫折，这似乎就是命运。不管是经济上还是其他方面的问题，从城市规划和城市设计上我们也在思考，查找原因。我们的城市规划，包括城市设计，是终极蓝图式的，或者像建筑施工图，希望毕其功于一役。实际上，城市发展、商务区建设，需要依靠市场，是一个逐步发展形成的过程。著名建筑大师库哈斯因为设计了中央电视台新楼‘大裤衩’为中国人民所熟悉和诟病，但作为哈佛大学建筑系主任，他长期对纽约曼哈顿进行研究，很有造诣，出版了一系列书籍。在《疯癫的纽约》一书中，库哈斯真实生动地记载了曼哈顿的发展过程。从纽约最早最简单的方格网规划，到 1919 年出台区划法，纽约的发展一直依靠市场的力量，充满生机和活力，成为美国的中心。实际上，曼哈顿的每一块土地上的建筑都不止建了一次，都是根据市场的变化发展不断改造重建，才形成了今天的曼哈顿的样子。即便是世界贸易中心地块，1966 年在拆除原址的情况下启动建设，1975 年建成，不到 30 年，由于 2001 年的‘9 · 11’事件，被全部摧毁，又再次进行了重建，当然这是一个极端的案例。虽然曼哈顿的建筑不断变化，但曼哈顿基本的路网骨架一直没有根本的改变。纽约的经验证明，地块规划要根据市场变化不断调整，但基本的规划结构、道路骨架和基础设施是不能随意改变的。

“于家堡金融区窄路密网的方针规划就具有这样的弹性。但在规划上也有一个大的问题，就是于家堡最高的建筑都规划布局在高铁站周围，这从交通可达性和地价水平的角度来讲具有合理性，但从市场发展的过程来讲，起步的建筑应该是低成本的，而且目前的市场也不支持建设超高层的综合性建筑，所以高铁站周边如果按照终极的规划设计，那么在相当长一段时间内就会是空地，这对商务区的发展，甚至对滨海新区的发展都是不利的。因此，于家堡高铁站周边应该采用过渡的建设方案，近期应建设有市场回应的、成本合理的建筑群，与周边的南站滨河酒吧娱乐区一道形成在京津冀甚至全国都有影响的打卡目的地。

“罗马不是一天建成的。城市发展需要时间。上海陆家嘴金融区发展到今天已经有 30 年了。巴黎的拉德方斯和伦敦的道克兰有过痛苦的经历，而现在都起死回生，充满生机。随着现代技术的进步，一次性规划也能够适应未来发展的变化，贵在坚持。持续 10 年的于家堡金融区城市设计和实施工作，更是城市规划管理部门和城市设计人员长期坚持的结果。我们对滨海新区于家堡金融区的未来充满信心。

“目前，根据于家堡近 10 年的规划实施情况和经济、社会环境的变化，公众持续对金融区居住配套环境的关注等，于家堡的决策者、规划者、管理者、建设者们正在进行新一轮的规划优化调整和提升。中心商务区城市设计的总体布局结构和路网不能变化，于家堡高铁站周边过渡方案可以举办国际征集，发现有创意、有吸引力的方案。要开始高水准的教育医疗等配套设施建设，住宅遇到与新居住区标准矛盾的问题要通过改革创新来破解，但高档住宅的投放量不能大，高档贵在精和稀缺。这些规划提升将为于家堡金融区下一步发展做 好准备。”

41. 滨海新区文化中心

诸葛民

“一个城市的文化中心就像是一个城市的客厅，通常位于城市的核心，优雅大方，邀请城市的居民和外来游客来此聚会做客，体现出一个城市的文化底蕴和特色。国内外城市的文化中心都是一个城市的主要空间场所，产生许多好的艺术作品，展示出城市的格调、品位，陶冶着居民的情操，延续城市的精神。天津滨海新区文化中心的规划设计，学习借鉴国内外先进经验，与中心城区文化中心错位发展，通过创新型的规划设计，努力孕育出一个功能复合、尺度宜人、充满活力、富有多样性、具有国际水准的 21 世纪的文化中心。

“从 2010 年 12 月开始，在吸取天津市文化中心规划建设成功经验的基础上，经过充分准备，我们会同新区文化广播电视局、中心商务区管委会组织开展了滨海新区文化中心建筑群概念设计征集工作，邀请了英国扎哈 ·

哈迪德建筑事务所与天津市城市规划设计研究院建筑分院、华南理工大学建筑设计研究院何镜堂院士工作室、伯纳德 · 屈米建筑事务所与美国 KDG 建筑设计有限公司、荷兰 MVRDV 建筑事务所与北京市建筑设计研究院这四组国内外一流设计单位联合体参与。根据各位设计大师的特长和意愿，每位大师负责一个主要场馆的设计，同时对建筑群的总图提出方案。扎哈 · 哈迪德流线型的滨海大剧院及建筑群总图的方案很时尚；伯纳德 · 屈米现代工业博物馆的方案体现了新区的工业历史和科技感；何镜堂院士覆土美术馆的设计使文化中心建筑群与公园融为一体；MVRDV 航空航天博物馆未来建筑的设计令人印象深刻。2011 年 2 月我们邀请了李道增、刑同和设计大师和马国馨院士等国内外专家及相关单位对方案进行了方案征集的评审和研讨工作。本次概念设计征集活动为后续文化中心规划设计深化奠定了坚实的基础。

“四位大师的方案，不论是建筑设计，还是总图设计，都各具特色。如何把四位大师的方案综合起来，难倒了规划院的年轻设计师。最后以何镜堂院士的建筑方案和总图方案为主，勉强把其他三位大师的方案纳入。虽然图面上还看得过去，但实际上就是拼盘，就是屈米大师总体方案的四种几何形状的拼盘方案，似乎昭示了大师的预见性。后来，情况发生了变化，新区区委、区政府决定先行建设急需的文化场馆，市规划院城市设计所开始编制文化中心及起步区的城市设计方案。未曾想，这一编就编制了 12 版方案。经过不断的探索尝试，到 2012 年底，天津市文化中心建成，为了与市文化中心有所不同，同时吸收它的优点，避免相同的问题出现，最终在规划院的城市设计所和建筑分院等规划师和建筑师的努力下，形成了文化长廊的方案。文化长廊的方案改变了习惯的围绕中心广场和中轴线布局的做法，把文化场馆集中在一侧，另外一侧形成大面积的文化公园。同时，在城市设计中试图把文化事业的发展与文化产业的方针结合起来，形成功能复合、充满活力和创造力的文化中心。最终这个新颖的方案得到各方的认可。

“滨海新区文化中心的规划布局，经过两年的不断深入完善，逐步稳定。同时，滨海新区文化中心的建设被列入新区‘十大民生’工程，准备正式启动建设。为此，2013 年 7 月开始，按照最新的文化长廊的城市设计和规划设计条件，开展了滨海新区文化中心（一期）建筑设计方案第二次征集工作。我们邀请了曾经参与第一次征集的伯纳德 · 屈米建筑事务所和荷兰 MVRDV 建筑事务所，又邀请了美国赫尔穆特 · 扬、德国 GMP、加拿大谭秉荣等境外知名设计大师和公司分别设计 5 个文化场馆和文化长廊，并形成一个团队。

如果说第一次征集是自由发挥，这次征集则是命题作文，要严格按照文化长廊的规划来设计建筑。因此，在设计师的选择上我们也做了一些考虑。

“原本对于文化长廊的主创设计师我们有两个选择，一是参加过首次征集的扎哈·哈迪德建筑事务所，二是设计了天津文化中心的德国 GMP。在滨海文化中心首次征集中，扎哈·哈迪德建筑事务所付出了巨大的努力，展示了他们的实力，完成了高水平的总图方案和滨海大剧院概念设计方案。为了做好大剧院的声学设计，他们邀请了专业的公司参与。从完成的图纸看，扎哈·哈迪德建筑事务所是花了巨大的人力物力。我们只给了 10 万美金保底费，远远不够。扎哈·哈迪德本人虽然没有能够参加第一次征集评审会，但她后来专程来到新区，来到于家堡考察，其敬业精神令人钦佩。但这次是一次命题作文，而且几位建筑大师和著名设计事务所合作，我们担心扎哈·哈迪德建筑事务所的风格比较难与其他设计师的协调，另外，我们也认识到滨海新区目前还不具备建设扎哈·哈迪德式建筑的能力，主要是资金，包括管理、施工、配套方面的人力资源，所以最后只能忍痛割爱。其实扎哈早在 10 年前就为开发区设计了一座文化建筑，但由于各种原因也没有实施。这次又因为各种原因，我们最终没有选择扎哈，看来扎哈注定与天津无缘。事实证明，选择德国 GMP 做主创设计是正确的。德国 GMP 事务所工作作风和设计风格如出一辙，认真严谨。他们在北京有上百人的分公司，可以随时到现场。在设计过程中，与天津本地设计院一道，很好地发挥了统筹协调作用，保证文化长廊与其他五个文化场馆的设计无缝衔接，而且文化长廊设计精致，展现了工业化的美，与滨海新区的整体风格也很契合。

“征集工作历时 10 周，期间设计单位三次到现场进行‘工作营’式的协同设计，并多次召开视频会议，深入研究各种问题，协同各场馆设计。既要求各场馆的方案各具特色，又必须做到整体协调统一。2013 年 9 月，我们邀请马国馨院士等国内专业权威人士与参与项目的国际设计大师共同研讨评审，为滨海新区文化中心（一期）建筑设计方案进行评审和提出建议。以文化长廊构筑文化综合体的方案得到以马国馨院士为组长的专家组的充分肯定。本次征集活动并非一般性的征集，而是一次协同式的系统设计，除以上著名的设计大师和公司外，还有许多公司单位在各专项设计中担任顾问，包括日本日建设计株式会社、香港 MVA 交通咨询公司、天津华汇景观设计有限公司、第一太平洋戴维斯策划公司，以及天津市规划设计院、天津市建筑设计院、天津市渤海规划设计院、天津市市政设计院等，他们在设计过程中提供了各专项顾问服务和技术支持，确保了规划设计方案高水准且可操作性强。

“如果比较的话，滨海文化中心项目相对于于家堡起步区项目的设计，更像是波兹坦广场的设计组织模式。滨海文化中心 470 米的长廊把五个不同的场馆统一了起来，统一中有变化。文化长廊是本设计的亮点，重点是从人的使用和感受出发，以人为本。在征集包括后期设计的过程中，德国 GMP 的设计师多次提出把长廊做成开放型的，不封闭。考虑到北方气候的特点，我们坚持做成封闭的市内空间，从现在的效果看，还是封闭的长廊更舒适。

“滨海文化中心一期建成后，滨海图书馆成为网红建筑，吸引了大量的目光和人群，文化长廊内图书馆前总是有人在排队入场。比较遗憾的是，由于各种原因，文化交流大厦最终没同步建设，缺少了酒店、公寓、办公、会议等复合功能。如果借文化中心的影响， 尽快采用市场化手段建设文化交流大厦，完善文化中心功能，将会是一个不错的选择。另外，文化中心规划是要发挥缝合滨海新区核心区的作用，要下决心打通洞庭路与于家堡、开发区的联系，打通解放路与海河外滩的联系，尽快建成滨海新区地铁B1线和Z4线，建设滨海中心区步行系统，将于家堡、天减地区、文化中心和开发区 MSD 连成整体。要加快天碱居住社区的建设，要完善配套，建设高水准学校、医院等设施。要超前谋划文化中心二期、三期市场化建设的项目策划和可行性研究。

“滨海文化中心一期建成两年后，滨海文化中心旁边的文化公园终于建成了，虽然举办了设计竞赛，最终由一家国际著名的景观公司设计，但设计水平实在不敢恭维。手法太过花哨，与文化中心主体建筑争当‘主角’，在功能上也不便于使用。这个位于市中心 20 公顷的大公园，让人感觉不到静谧幽深，过于浅显直白，缺乏让人停留的诸多宜人空间，也无法吸引足够的人气，似乎背离了公园建设的初衷。日后若有机会应加以改造、提升。”

42. 天津的城市设计与管理体系

诸葛民

“以上我回顾了 2008 年以来天津城市设计发展的历程，介绍了几个重点项目的城市设计，并做了简单的评价。能够取得这样的成绩，原因是多方面的，但关键是我们建立了完备的城市设计编制和实施管理体系。2008 年以来，天津市成立重点规划指挥部，集中力量编制了大量的、类型多样的城市设计。一批高品质项目的实施促使城市功能与城市环境整体迈上了一个台阶，显示出了城市设计的力量和作用。在这一过程中，天津初步形成了较为完善的城市设计法规体系和管理机制。

“早在 2008 年，《天津城市规划条例》第三十七条明确指出：市人民政府确定的重点地区、重点项目，由市城乡规划主管部门按照城乡规划和相关规定组织编制城市设计，制定城市设计导则；前款规定以外其他地区，由区、县城乡规划主管部门组织编制城市设计，制定城市设计导则。同年，作为滨海新区综合配套改革试验总体方案中的一项改革内容，滨海新区开展了“探索城市设计规范化、法定化编制和审批模式，做好重点区域和项目的城市设计”的改革试验。改革试验持续了三年多时间，主要完成了三方面的工作。一是选定了试点地区，二是 2010 年市规划局印发了《天津市滨海新区重点地区城市设计导则管理暂行办法》，三是编制试点地区城市设计及城市设计导则，并依据导则开展建设项目审批和行政许可，取得了明显成效，改变了过去依据城市规划技术管理规划、一刀切的建设项目管理模式。比如，道路交叉口红线与转角（转弯半径）细化，地块绿地的精细化处理，对建筑退线的调整等。

“在总结实际工作经验的基础上，市规划院制定了城市设计及城市设计导则编制规程，将城市设计与控制性详细规划有效衔接。重点地区控规依据城市设计编制，保证城市设计落地，在此基础上构架了‘一控规两导则’规划实施管理体系和机制。控规作为规划管理的法定依据，为了避免经常调整，保证其严肃性，还要具有一定的弹性和兼容性。控规主要对城市主要道路市政基础设施、结构性的用地布局和整体开发强度进行控制。土地细分导则是控规的实施方案，这部分借鉴美国的土地细分方法（Subdivision），确定用地的二级支路网，而具体地块的用地性质和开发强度在不改变控规原则的前提下可进行调整。城市设计导则通过空间环境和建筑或建筑群体的控制，强化对城市风貌、区域建筑特点、公共空间和环境景观等城市空间环境的控制和引导。为了确保城市建筑的高品质，我们从横向、纵向两个维度加强管控。在横向层面，突出城市设计的统领作用，统筹协调建筑设计、景观设计、交通组织、地下空间、绿色生态等专项设计；在纵向层面，强调城市设计的切实落实，从项目筹划、建筑方案设计、施工图设计、建筑施工选材直至最后竣工验收，从头至尾，全过程跟进，保证项目按城市设计导则和建筑设计、景观设计的方案实施。

“具体讲，在城市规划工作的每一个阶段，我们都非常重视城市文化特色的传承与发展。我们形成了以总体城市设计为纲，以重点地区和重要节点城市城市设计为先导，以‘一控规两导则’为实施依据的覆盖各层次、各类型的城市设计编制和管理体系，为全市高质量城市规划建设提供了强有力的支撑和保障。城市设计因其在城市环境和空间形态设计上的整体性和表达的直观性，在城市空间形象、风貌特色塑造和宜人环境创造上发挥着重要作用。我们认识到城市设计法定化是发挥城市设计对城市建设进行控制引导的关键。我们注重城市设计在理念、管理和实施过程中的衔接。在前期规划阶段，城市设计结合市场需求和建筑布局进行概念设计；在法定控规阶段，重点地区城市设计将地下空间、生态、交通、市政工程、景观、建筑方案设计等内容落实成控规和细分导则、城市设计导则，最后在建设实施阶段，将城市设计导则作为土地出让合同和规划设计条件的附件，通过规划管理确保实施效果。

“同时，除重点地区城市导则外，编制专项控制导则，形成管理的技术规范。如《天津市规划设计导则》从建筑布局、建筑高度、建筑风格、建筑色彩等八个方面，对居住、商业、办公等各类建筑设计提出控制要求，规范和引导建筑设计。制定了主要河流边、公园边、历史街区边的‘三边管理规定’，严格控制居住区开发强度；快速路内一般容积率不超过 2.5，快速路外容积率不超过 2.0；对建筑形体、高宽比提出控制要求；对建筑风格也提出控制要求，要与周边建筑风格协调；为了避免幕墙的光污染，提出了窗墙比的要求。另外我们将城市设计成果转化为三维数字模型，中心城区实现了建筑体块模型全覆盖，重点地区实现精细化模型全覆盖，作为单体建筑项目审查的附加手段。通过不断的实践探索，我们认识到，良好的城市风貌特色和高品质的城市环境不是单单依靠规划的理念就能达成的，而必须建立在各层面统筹管控，以及精细化的具体设计和实现工作之上。”

诸葛民

“天津的城市设计起步早，取得了许多成绩，但也存在许多问题。”讲到这儿，诸葛民提高了嗓音。“在这次城市设计试点中，我们进行全面反思，作为我们本次城市设计试点研究探讨的重点。通过分析，我们发现，有些问题是天津特有的，有些是有共性的，因此解决这些问题具有一定的示范作用。我们把共性的问题归纳为五个方面。”

“第一个方面是市场的问题。我们国内的城市设计师、规划管理者大部分受的教育是功能主义城市规划的教育，在体制机制上计划经济的烙印还很深，所以我们做的城市设计大部分是自上而下、家长式的。虽然我们嘴上一直说市场导向，发挥市场在资源配置上的决定性作用，但在真正关注市场、尊重市场、回应市场上还十分欠缺。对市场的重视反映在许多方面，特别是对人的需求、提高生活质量和房地产市场的考虑和重视不够。比如许多按照城市设计建成的项目，整体环境尚可，建筑造型也算漂亮，但市场接受度差，活力不足，如五大院、津湾广场、于家堡起步区等。当然，我们可以解释说有些可能是时间问题，比如法国的拉德方斯和伦敦的道克兰都经历过萧条后重新复苏，但实际上市场的决定力量是我们必须重视和学习的。

“第二个方面是对人考虑不够。虽然我们嘴上一直说以人为本，但实际上眼中只有物，眼中的人也只是没有个性的数字。与人尺度接近的细节决定城市设计的成败，要支持城市中心商业的复兴，就必须进一步完善步行系统，这是世界各国的成功经验。城市的步行化建设在天津大部分地区特别是中心城区，还存在较大差距，整体上不宜于人的步行。步行道被占用、景观缺乏设计的情况比比皆是。还有就是公园建得太少。从 20 世纪 50 年代建了水上公园后，后来的几十年就没有建大型公园。近十几年，建设了梅江公园，近几年建设了水西公园、柳林公园，但跟城市发展建设速度和总量比起来，公园并没有显著增加，无法满足人们的多种需求。我们在外环线周边规划了 11 个公园，但只建成了 4 个，还有 7 个未建。另一个是居住问题，我们发展的最终目的是提高人民的生活水平，城市设计的目的除了提高城市中心地区、公共空间的品质外，就是要面对量大面广的居住和社区问题。但我们的城市设计在这方面考虑得不多，一直沿用计划经济的居住区规划方法，建设了大量大规模封闭的小区和没有个性的住宅，不能完全适应未来人民对美好生活的向往。

“第三个方面是对生态文明的思想和景观设计的方法不熟悉，缺乏对城市大的空间艺术骨架和大生态系统的管控。现代城市规划的产生主要是解决工业化后城市建设扩张带来的城市问题。1791 年，在朗方的华盛顿规划中，城市与自然浑然一体。1824 年，英国空想社会主义者罗伯特 · 欧文，在美国印第安纳规划建设了自给自足的新和谐村。虽然结果并未成功，但他探索了有组织的规划，避免工业化对农村地区的肆意侵害，因此有人称欧文为现代城市规划之父。19 世纪 30 年代，随着火车的运营，美国随之出现了郊区化，人们向往乡村的自由和美丽。1860 年，被称为‘美国农民’的奥姆斯特德用自然的方式设计了纽约中央公园以及波士顿‘翡翠项链’绿地系统，使人工化的城市中融入了自然的气息。1898 年，埃比尼泽·霍华德提出城市与乡村磁铁的田园城市，为世人展现了通向未来的和平之路，开启了现代城市规划的新篇章。1915 年，盖迪斯出版《进化中的城市》，认为城市规划应该是城市和乡村结合在一起的区域规划。因此，现代城市规划诞生的目的就是使城市与自然乡村融合。反观我们的城市规划和城市设计，却主要以城市规模的扩张为目的，过度片面考虑设计实体的城市和建筑，而较少考虑生态环境和大尺度的公园绿地系统。实际上，中国传统的城市规划一直强调天人合一、道法自然，即使严格按照《周礼·考工记》形制规划的北京城，依然保留了自然水系和北海、中海、南海等三海系统，自然灵活的三海与严谨的中轴线交相辉映。

“第四个方面是道路和基础设施对规划的影响没受到足够的重视。城市的聚集促进了科学技术的爆炸式进步，道路交通和市政基础设施反过来又促进了城市文明的进步。城市道路交通和市政基础设施是密不可分的整体。而我们的城市规划师、城市设计师习惯认为道路交通和市政工程是城市文明的‘配角’，在思想上没有给予应有的重视，同时道路交通规划师和市政工程师也认为其是‘配角’，认为只要符合规范标准即可，不考虑它们对规划整体上的重要影响。基础设施也有美的问题、文化的问题。很多人习惯用一些观念口号来处理一个具体的交通问题，以为高喊‘大力发展公交，限制私人小汽车’就万事大吉、天下太平了，无视城市交通拥挤的现状，无视人们出行的实际需求。而实际上，计算机辅助规划设计，包括大数据的应用，在道路交通领域应用最为成熟，反倒在城市设计上应用不多，我们还没有真正树立起城市道路交通和市政基础设施是城市文化的观念。

“第五个方面是对待居住建筑过于简单化。有些小区虽说是在做城市设计，实际上一直沿用计划经济和现代主义的居住区规划设计手法，大规模重复性设计。这些正是简·雅各布斯在 1961 年发表的《美国大城市的死与生》中深恶痛绝和严厉批评的。我们口头上盛赞雅各布斯的理论，却大张旗鼓地干着雅各布斯反对的事情，以独立高层或多层单元住宅的封闭大院式开发，扼杀了居住建筑的丰富性和多样性，其后果是扼杀了所有居住的人的个性。

“巧合的是，上述五个方面的问题实际上对应着柯布西耶在《雅典宪章》中提出的城市的工作、游憩、交通和居住这四大功能。我们现在的所作所为只能说明，我们依然受现代功能主义城市规划思想方法的控制。如果思想方法上不与现代主义决裂，我们大部分所谓城市设计就是‘一只披着羊皮的狼’。因此，必须改变居住区规划模式，结合现代住房制度改革，真正用城市设计的理论、方法来研究城市的历史文化，用综合统筹的城市设计措施，创造高品质的城市形式、功能和环境。本次天津城市设计试点就是结合具体项目的实践，探索在市场、居住、生态、交通等四个方面形成具体的解决思路和方案。”

诸葛民

“以上是我对天津城市设计工作的回顾和总结。过去我们说‘无规划不建设’，现在我们说‘无城市设计不建设’。城市设计不是灵丹妙药，但没有城市设计，城市规划建设的水平难以提高。天津的城市设计覆盖率已经比较高，在城市设计的指导下城市建设的水平的确有了较大幅度的提升。我们通过实例，努力客观评价城市设计实施后的效果，在肯定成绩的同时，我们也发现了以上五个方面的主要问题，作为本次试点工作要解决的重点内容。由于时间的关系，我现在简单介绍一下试点工作得出的主要观点，请在座的嘉宾点评指导。待下午分论坛时我们的设计师再向大家做详细介绍，请大家讨论。

“另外，这里我没有介绍天津中心城区总体城市设计和滨海新区核心区总体城市设计的情况。总体城市设计意味着对城市空间形态的总体把握，很重要，也很难做。特别是在当前国家推动国土空间规划的情形下，如何更好地发挥大尺度城市设计的作用，是我们城市设计专业面对的一个关键问题。”

第七部分
一座更生态的城市

43. 国土空间总体规划——空间艺术骨架

下午共有四个分论坛，一是城市设计理论和实践，二是生态化的城市设计，三是历史名城保护与城市设计，四是城市设计教育。巴奈特先生和诸葛总都十分重视城市设计教育，他们不约而同地来到城市设计教育分论坛的会议室。论坛由天津大学教授、博导、中国科学院院士张泽平先生主持，许多著名高校城市规划和建筑学院的教授与会。大家把研讨的重点放在了国土空间总体规划、控制性详细规划与城市设计关系的理论内容上。随着国家强调“多规合一”和建立国土空间规划体系，在大学的城市规划学院里弥漫着一种不安的气氛，不管是教授还是学生，都关着心自己专业的未来发展。

张泽平院士

张院士已是花甲之年，但在中国科学院院士中算是年轻的，特别是在空间规划领域。他在美国学习工作了近 20 年，主要从事国际城市规划研究，2006 年被天津大学盛情邀约回国任教。回国后又从事了 10 多年以欧盟为主的空间规划研究，成为这方面的专家。由于他是建筑和城市规划专业出身，因此更关注城市和空间环境。“大家知道，城市总体规划是城市发展的蓝图，除了指导城市社会经济发展和城市建设外，其主要的作用是指导城市美的塑造，为人们塑造优美的宜居环境。从城市规划诞生那天起，建设美的城市是城市规划的最高追求和境界。虽然伴随着现代城市的快速发展，城市越来越复杂，规划要考虑的问题越来越多，但不能因此忽视了城市总体规划根本的目的和内容。我们在规划工作中，特别是在城市总体规划中，必须强调优美人居环境的创造。”

“30 年前，针对当时国内对天津城市环境整治的议论，吴良镛先生发表了《城市美的创造》一文，指出：美好的城市应具备舒适、清晰、可达性、多样性、选择性、灵活性、卫生等要素，人在其中生活，要有私密感、邻里感、乡土感、繁荣感。城市美包括城市自然环境之美，城市历史文化、环境之美，现代建筑之美，园林绿化之美，城市中建筑、雕塑、壁画、工艺之美等诸的方面。城市美的艺术规律包括整体之美、特色之美、发展变化之美、空间尺度韵律之美等方面。

“美是人居环境建设的最终目标，也是手段。美是人类最高的精神体验，人居环境是美学的物质和空间体现。人们建设了城市，城市环境改变着人们。

要创造美好的人居环境，除经济发展、生活水平提高、社会进步和环境改善外，还需要人居环境美的设计和创造。有人说我们的规划是传统工艺美术的规划，过于注重物质形体空间的创造。实际上，国际经验和最新的理论研究表明，城市和城市区域物质环境的创造越来越重要，这也是从我国近几十年城市规划建设的深刻经验教训中得出的结论。城市总体规划应该而且必须把城市美的塑造作为规划的重点内容之一。

“现在国家建立了国土空间规划体系，目的之一是解决规划类型和层次众多、政出多门不统一的问题；另外一个目的主要是提高我国城市化建设的水平，解决城市病。许多从事城市设计的同仁感到困惑，不知道国土空间规划与城市规划设计是什么关系，感觉所谓三区三线、图斑、双评价等内容将取代城市规划。实际上，空间规划（spatial planning）在 20 世纪 80 年代在欧洲兴起，由区域规划演变而来，作为欧盟各种政策的空间导向。新区域主义理论倡导者、加州大学伯克莱分校的斯蒂芬 · 威勒（Stephen M. Wheeler）教授在这方面有许多研究。我在加州大学伯克利分校做访问学者时，与他聊过，他认为，近几十年来区域规划过于注重经济地理和经济发展（economic geography and economic development），以忽略区域科学的其他内涵作为代价，损失很大。21 世纪城市区域物质形体快速演进，区域在可居住性、可持续性和社会公平方面面临更大的挑战。未来的区域规划因此需要更加整体的方法和观点（holistic perspectives），新区域主义（New Regionalism），除考虑经济发展外，应该包括城市设计（urban design）、物质形体规划（physical planning）、场所创造（place-making）、社会公平（equity）等主要内容，并作为研究的重点。不仅有定量分析，还要有定性分析，要建立在更加注重直接的区域观察和区域经验的基础上。之所以这样，根本的原因是要重新评价区域发展的重要目标，找到经济发展目标、社会发展目标和优美人居环境的平衡点。

“美国经过百余年城市和区域规划发展的历史，空间规划的机制和方法已经比较成熟，它较好地解决了国家意志和地方发展的关系问题，各个州和地方相对独立。在空间规划目标的选择上继续采取了相对单纯的方法，如‘生活的质量’，基本延续了‘美国梦’的概念。生活的质量由社会、环境和经济所决定，强调社会公平、环境保护和经济发展三者并重，这也与新的发展观相一致。纽约区域规划协会（RPA）1996 年第三次区域规划《纽约：危机的区域》（*Third Regional Plan, A Region at Risk*）中的规划目标‘三原色’，非常有代表性，在这里经济被有意地放在最下边。生活质量实际上与我们提的宜居环境是一致的，包括社会公平的内涵。美国其他州的规划尽管题目不同，基本都采用了相同的总体规划目标。

“与美国聪明增长规划的背景不同，欧盟空间规划目前的主要目的是欧洲的联合，实现领土和社会的融合，形成合力，共同发展。在欧盟1999年《欧洲空间发展展望（ESDP）》中空间规划的目标三角中：总体目标是追求平衡和可持续的空间发展，三原色强调社会、经济和环境三者的统一。在《欧洲空间规划宪章》等文件中，空间规划的基本目的是寻求平衡的区域社会经济发展，改善居民的生活质量，管理自然资源和保护环境，合理利用土地。从中可以看出，尽管生活质量同样是规划追求的目标，但相比较而言，追求平衡的区域社会经济发展更重要，是第一位的，这是欧盟存在的基础。在欧盟其他层次的空间规划中也反映出同样的趋向。因此，我们说，尽管各国的情况不同，但国土空间规划的最终目的也是在更大的空间尺度上创造更加平等、优美的人居环境。

“在国土空间总体规划中，如何创造优美的人居环境呢？最终都离不开传统的城市规划，特别是城市设计。当然，这里首先需要大尺度的城市设计，或者可以叫作区域城市设计。除了经济发展和社会进步外，国土空间总体规划的重点是区域空间艺术骨架的塑造。建立大尺度区域的空间艺术骨架，实际上包括两大系统和广大农村地区的整体设计，一是由区域城镇和道路交通市政基础设施组成的人工系统，二是由区域自然生态斑块和河流等生态廊道组成的生态系统，以及广大的农村地区。一个区域好的艺术骨架就是要有良好的生态环境、富裕的农村地区作为本底，以及优美的城镇和便捷的交通、市政基础设施作为保证。实现人工系统与自然系统和谐共生的理想状态。要做到这一点，就需要多方面的知识，包括对自然生态系统的认识、对道路交通市政基础设施系统的认识，以及对区域城镇网络体系的认识，还有相关的各种公共政策。因此，要求我们的规划师、城市设计师有更广泛的知识体系，这对我们的教育体系也提出了新的要求。

“在国土空间总体规划的尺度上创造优美的人居环境，要讲究地区特色。如何避免千城一面是关键的课题。我们把大尺度的城市设计作为提升国土空间和城市美的主要抓手。在国土空间总体规划编制过程中，应该开展总体城市设计研究，探讨全市域的总体空间形态和城市特色。同时，同步开展分区总体城市设计和重点地区城市设计等一系列规划工作，形成了自上而下、自下而上、相互交融、相互镜映的局面。规划工作在深度和广度上同步展开，共同探讨城市美的课题。

“如果我们仍然按照传统的城市总体规划、控制性详细规划方法编制和实施规划，管理城市区域发展，可以预见最后的结果也一定是一样的，会形成千篇一律、毫无特色的城市和混乱的郊区。我们需要一种改变，改变传统的规划编制和管理模式，改变我们已经习以为常的建筑和空间的理念，开展城市区域整体城市设计。建立城市区域的艺术骨架，必须从城市总体规划和城市设计的手段和理念上改革创新。”

美国城市设计大师巴纳特

巴纳特先生非常赞同张院士的观点。“我在 2015 年出版了一本书《城市设计——现代主义、传统、绿色和系统的观点》，书中提到，迄今为止，我们看待每一个城市设计问题至少可采取四种方法，即现代主义城市设计、传统城市设计、绿色城市设计和系统城市设计等四种理论，设计师可以自由地选择最适合城市和目前问题的城市设计方法或各种方法的综合。但现代城市变得如此广阔，既有的四种方法是不够的，应该有第五种方法，要涵盖变化如此之迅速而复杂的社会经济条件。”他的新方法包括：从自然景观开始、强化互通性交通系统、理解新经济体下的设计含义、推广紧凑型商业中心和适宜步行的邻里社区、满足现代建筑类型和停车需求的设计、通过公共程序制定大型公共设计决策等内容。

“我认为，现在并不需要一个通用的城市设计新概念，而是要将城市设计整合到社会经济变革过程中，同时能够与自然保持可持续关系。城市空间品质的塑造通过公共政策对人们的生活产生了越来越深远的影响。美国不同于欧洲，没有统一的空间规划体系及其机制。但美国的一些州编制了精明增长规划，实际就是空间规划，重点是抓住大的空间结构，就是空间艺术骨架，这里首先是自然生态系统的保护，以及城镇空间良好形态的塑造。”

诸葛民

诸葛民也非常认同张院士的观点，他补充道“吴良镛先生的《广义建筑学》和《人居环境科学导论》均强调城市规划、建筑设计、景观设计的‘三位一体’，这是空间规划的核心。我们研究学习发达国家的空间规划，如纽约三州的区域规划，也都反映出大尺度的规划与城市、建筑模式的一致性和协调性。我国要完善空间规划体系，建设优美的人居环境，不论是作为制度体系，还是规划设计的内容和方法，空间规划、城市设计和建筑设计一定是三位一体、密不可分的。即使是全国范围的国土空间规划，或跨省市的大尺度空间规划，也必须明确绝大多数人的住房模式、交通出行模式、城市的空间形态，包括街道、广场和公园，特别是城市的风貌特色、美好的天际轮廓以及乡村的自然风光。只有这样，我们中华民族伟大复兴的中国梦才能越来越有形、越来越清晰、越来越触手可及。”

“天津是京津冀世界级城市群的重要支点，是全国先进制造业研发基地、北方国际航运核心区、金融创新运营示范区、改革开放先行区，也是京津冀重要生态功能区。在正在编制的天津市国土空间总体规划中，以优化生态格局作为确定城市空间艺术骨架的重点，提出市域空间结构为‘三区两带中屏障、双城双廊多节点’。‘三区两带中屏障’是描绘天津的生态重点区域：‘三区’是指北部的蓟州生态区、中部的七里海大黄堡生态区和南部的北大港团泊生态区；‘两带’是指东部沿海生态屏障带和西部防护林带；‘中屏障’就是天津中心城区与滨海新区核心区之间的绿色生态屏障。为优化天津中心主城空间布局，对接国土空间发展战略，我们本次城市设计试点项目‘天津市外环城市公园及周边城市设计深化’将外环城市公园地区作为天津中心主城转型升级发展的战略空间，衔接双城绿色屏障，连接外环绿带、城市公园及郊野公园，形成绿色骨架，营造城园融合的人居环境。”

44. 双城生态屏障——融入京津冀

规划师郭源

在“城市设计教育”分会场讨论正酣的同时，在“生态化的城市设计”分论坛会场，来自天津规划院的规划师郭源正在介绍双城屏障区的总体规划。郭源是 90 后，看上去意气风发、神采飞扬。她说：“党和国家高度重视生态文明建设和区域协调发展，党的‘十八大’明确提出大力推进生态文明建设，努力建设美丽中国，实现中华民族永续发展。2015 年中央出台《京津冀协同发展规划纲要》，明确了京津冀的空间结构和发展方向，提出在生态、产业、交通等方面率先取得突破。同时明确提出要构建京津冀区域生态体系，实现京津冀区域生态协同发展。为切实贯彻实施京津冀生态协同发展国家战略，天津市委市政府开展了一系列卓有成效的工作，积极落实纲要中提出的京津保生态湿地过渡带，谋划构建环首都东南部生态屏障带。”

“天津市第十一次党代会从实施京津冀生态协同发展国家战略的高度出

发，做出加强滨海新区与中心城区中间地带约 736 平方千米土地面积的规划管控、建设绿色生态屏障区的战略决策。去年 5 月 28 日市人大常委会审议通过了《关于加强滨海新区与中心城区中间地带规划管控建设绿色生态屏障的决定》，将绿色生态屏障区的规划建设纳入法制轨道。市政府组织编制了《天津市双城中间绿色生态屏障地区规划（2018—2035 年）》。规划提出，到 2035 年，实现绿色生态屏障区蓝绿空间面积占比不低于 70%，形成‘生态屏障，津沽绿谷’。通过绿色生态屏障区，北连天津七里海和大黄堡生态湿地保护区与盘山和于桥水库生态保护区，并进而与北京通州生态公园和湿地公园相接；南接北大港和团泊生态湿地保护区，并进而与雄安新区生态公园和湿地公园相连，积极融入京津冀区域生态环境体系，成为环首都东南部生态屏障带的重要组成部分。

“双城间生态屏障涉及东丽区、津南区、西青区、滨海新区与宁河区等五个区。到 2017 年，区内现状总人口 115 万人，其中户籍人口 55 万人。现状建设用地为 198 平方千米，占总用地面积的 28%；已平整尚未建设用地为 77 平方千米，占总用地面积的 10%；其余为蓝绿空间用地，包括水域、农用地等，包含城市绿地，总面积为 462 平方千米，占总用地面积的 62%。

“这个地区的发展经过一个曲折的过程。1986 年版天津城市总体规划提出‘工业东移’空间发展战略和‘一条扁担挑两头’空间布局。伴随着城市发展和历版城市总体规划调整，双城间地区也在不断发展变化。1986—1996 年，地区处于自生发展时期。按照总体规划，在中心城区外环线外规划 500 米绿化带，避免城市蔓延发展。同时规划了军粮城、咸水沽等卫星城镇，但没有什么发展。这个时期，双城间土地以农业用地为主，分布较多村庄。最大的空间变化是实施天钢工业东移和建设大无缝选址在海河中游，荣钢等项目也开始建设。这些项目对海河中游的生态环境影响是巨大的。

“1996—2006 年，是天津发展由慢到快的逐步转变期。1992 年伴随小平同志南方谈话后，全国进入新一轮发展热潮。1994 年总规启动修编，提出可持续发展这一理念。结合津滨轻轨、津滨高速建设和京津塘主轴线，全市提出‘双心轴向’布局，继续实施工业东移战略。考虑到东丽、津南等区的发展建设和津塘公路沿线连绵发展趋势，结合全市生态圈的连通，在双城间新规划多条生态廊道，连通南北生态地区。

“2006 年以后，结合滨海新区开发开放，沿京津塘高新技术走廊规划空港高新区、开发区西区、滨海高新区等，空客 A320 总装线、大火箭等重大项目落户带动了地区产业升级。2008 年城市空间发展战略提出‘双城双港，相向拓展'的布局思路，双城中间地带被作为城市拓展区进行规划建设，双城之间城镇发展提速。这一时期，双城之间建设了海河教育园，形成了创新和研发中心，规划了国家会展中心，预留了体育中心等大项目用地。实施了以华明镇为代表的示范小城镇建设，城镇化进程提速。同时，城市功能开始外溢，随着东丽湖、天嘉湖'两湖'建设，小站、八里台等示范镇建设，城市进入相向发展阶段。由于通过土地平衡建设资金，造成地区内城市建设用地增加，开发强度过大，建筑高度过高。

“目前，区内拥有滨海高新区、空港经济区等国家和市级重点园区，地区工业总产值约 2800 亿元。以海河为界，北部形成了以大飞机、大火箭等为代表的国家重大项目集聚区。海河南部，中小企业和民营经济的簇群特征明显。城镇空间主要集中于津南主城区、开发区西区，以及军粮城、葛沽、小站示范镇等。区内拥有以高等教育和职业教育为主的市级功能区——海河教育园，以及航空航天、新能源新材料、生物医药等高新技术产业，科技创新基础较好。区内历史文化资源丰富，拥有国家历史文化名镇葛沽、天津市历史文化名镇小站、津东第一镇咸水沽等历史文化资源。

“虽然经济发展和城镇建设取得一定的成绩，但从区域生态角度分析问题严重。该地区是海河流域尾闾，拥有一级河道 4 条，二级河道 20 条，湖库 5 座。水域面积约 189 平方千米。目前，河道水质较差，除海河为 Ⅳ 类水体外，其余均为 Ⅴ 类和劣 Ⅴ 类水体。区内生态空间以农用地为主，占地 166 平方千米，林草用地 45 平方千米，建设用地内绿地 56 平方千米。总体看，地区内蓝绿空间受挤压，碎片化。资源性、水质性严重缺水，水质较差，生态用水匮乏，水系沟通不畅，缺乏循环。地面沉降严重，造成生态退化，土壤盐渍化污染严重，其中轻度盐渍化地区占 60%，中度及重度盐渍化地区占 40%。生态本底水平低，林地只占 5%。农作物种植量大、作物单一，种植时间短，生物多样性差。区内仍有大量钢铁冶金企业，造成大气等污染。许多制造企业占据大量宝贵的生产空间，地均产出低。城乡规划建设水平参差不齐。许多城镇建设强度低，建筑高度过高，缺乏特色。大型道路交通和市政基础设施对生态空间随意切割。此外，区域内仍有军粮城热电厂等传统高影响设施，村庄建设水平低，缺乏田园特色风貌。

“通过现状分析，我们进一步认识到，加强双城中间规划管控、构筑绿色生态屏障，是修复我市生态环境、避免城市盲目扩张、拓展绿色空间的战略决策。同时也是我市落实生态文明要求，肩负起天津在京津冀协同发展中的职责使命，主动融入京津冀区域生态体系，构建完整的区域生态空间布局，做好首都‘生态护城河’的重要举措，也是优化天津城镇布局和生态环境体系的重要举措。为了更好地编制规划，我们组织了专家工作营，开展了生态、水系统等七个专题研究。

“绿色生态屏障区规划形成‘一轴两廊两带五湖多线’的总体生态空间格局，呈现‘水丰、绿茂、成林、成片’的景观特色。其中，‘一轴’是海河生态走廊，‘两廊’是东西两条古海岸与湿地走廊，‘两带’是永定新河生态带、独流减河生态带，‘五湖’是指东丽湖、黄港水库、官港水库、天嘉湖、大孙庄湿地，‘多线’是指规划区域内密布河道与交通防护线型绿带形成的网状廊道。规划重点在八个方面下功夫。一是重塑津沽水生态环境。践行海绵城市理念，建立多源共济的水源保障系统，保障生态和农业用水量；改善地表水质量，从现状劣Ⅴ类主要指标达到Ⅳ类水标准；加强水系连通，实现南北水系贯通成网。二是重构津沽绿色森林屏障，加强植树造林，大幅度提高森林覆盖率。采取传统‘台田’法式，在规划生态廊道内集中造林绿化。加强林相设计，因地制宜、适宜生境，选取林、柳、榆、槐、椿等主干乔木。三是重现津沽鱼米之乡风貌。保持屏障区现有的基本农田规模不变。积极发展农业经济，推广小站稻种植，鼓励种植经济作物。四是重塑湖泊湿地涵养功能。充分发挥湿地生态净化功能，消减有机污染物，新增金钟河湿地、古海岸生态湿地廊道和卫南洼湿地等人工湿地，总用地面积约29平方千米。五是再现生物多样性。积极营造适宜动植物栖息活动的生境，构建滨水栖息地，保留并新建生态通道，为生物提供活动空间。六是重归诗意田园。保障生态廊道形成，拆除已纳入城镇化建设的旧有村庄，复垦复绿。对尚未纳入城镇化建设的村庄，实施乡村振兴战略，留住乡愁。七是建设高质量发展绿谷。取缔散乱污企业和低效工业园区，实施生态修复工程，以绿色业态发展引领区域创新转型发展，实施低效制造业转移。八是建设低冲击影响的基础设施。塑造亲近自然、绿色出行的环境，构建绿道系统。围绕绿色生态景观规划建设五横两纵的绿道主骨架。同时，为保障蓝绿空间规模，通过缩减规模、调整断面等措施优化原规划干道系统。建设小型生态基础设施，实现低碳排放。

“为加快环首都东南部生态屏障带规划实施步伐，计划用3年的时间实施绿色生态屏障区植树造林、水环境治理、大气和土壤环境污染防治、生态修复、人居环境整治与乡村振兴、高标准农田建设、道路交通和基础设施建设等10项工程，重点开展海河生态芯滨河绿廊、古海岸生态绿廊、津南卫南洼湿地片区、滨海官港生态森林、津南绿芯、金钟河湿地片区和西青生态绿廊等重点地区建设。上述工程实施后，绿色生态屏障区蓝绿空间面积占比将达到 65%，成为环首都东南部生态屏障带的重要组成部分，对于构筑环首都生态屏障带、改善京津冀区域生态环境具有十分重要的作用。”

景观规划师屠总

接下来，景观规划师屠总发言，他进一步介绍了双城生态屏障区中海河两岸的生态景观设计。“总体看，海河干流在大的海河流域中是一个相对独立的封闭区域，自金刚桥以下 72 千米，大概接纳北至永定新河、南至独流减河这一区域的降水。一直以来，该区域形成了以海河干流为主干，各支流为支干，各小型河道、沟渠为毛细，各湖泊、水库、鱼池、坑塘为蓄洪区的水网系统，日月恒久，生生不息。改革开放以来，城市化建设突飞猛进，随着一个个大型建设单元落位海河干流所在区域，各坑塘沟渠相继被填平，原有致密的水系结构遭到部分破坏。由于缺水，各种生态问题相继而至，区域的生态发展到了一个关口。”

“《天津市双城间绿色生态屏障区规划》明确了大的生态和城市建设的布局结构，制定了‘一轴两廊两带五湖多线’的总体生态空间格局。结合生态屏障区规划的编制，我们编制了海河一轴两侧绿心的景观设计方案。海河位于双城间绿色生态屏障区范围内长度约 23 千米，即海河中游区域。该区域以海河为主脉，形成具有宽窄变化的生态景观主轴；结合海河各支流，确定景观带支脉，整体形成鱼骨状空间结构。

“海河中游区域未来应该是怎样的片区呢？我们这里借鉴巴黎的案例。巴黎市区东西两边各有一个很大的森林公园：西部的是布洛涅森林公园，位于塞纳河畔讷伊和布洛涅－比扬古之间，南北最长处 3.5 千米，东西最宽处 2.6 千米，面积 8.46 平方千米，相当于整个巴黎城区面积的 1/12；东部的为文森森林，东西最长处 5 千米，南北最宽处 3 千米，面积 9.95 平方千米。两森林公园林内道路纵横交错，树木郁郁葱葱，具有涵养水源、保持水土、净化空气、降低城市热岛效应、维持生物多样性等多种生态功能，发挥良好的生态调节作用，被视为巴黎的两个肺。公园内包含动物园、游乐场、赛马场、足球场、体育馆、自行车赛场、临时展览馆、花卉公园等功能区域，涵盖休闲、旅游、体育、艺术等方面活动。除具备城市基本功能，公园同时承担具有国际影响力的活动，建有著名建筑师弗兰克·盖里设计的路易威登基金会艺术中心、玛摩丹美术馆、罗兰·加洛斯网球中心、隆尚赛马场等。”

“由《费加罗报》发起的‘费加罗越野赛’迄今已经在布洛涅森林跑了

40 多年。40 年来，参赛人数累计达 120 万人，是世界上规模最大的群众性越野障碍赛。”讲到这儿，屠总抬起头看着大家，脸上露出自豪和兴奋的神情。他是从法国留学的海归，喜欢极限运动，亲身体验过“费加罗越野赛”，回国后也经常参加全国各地的越野赛，保持了长跑运动员的修长体型。在法国他学习的专业是景观规划，他的导师大力倡导与自然和谐的景观设计。

“海河中游是双城生态屏障功能的核心区域，从管控区西侧边界到葛沽约有 12.5 千米，这 12.5 千米段海河两岸北至先锋东路，南至天津大道，超过 4 千米宽度的区域现状全是未建设区域，给未来发展预留了无限可能。对比美国中央公园 4.1 千米的长度，海河 12.5 千米具有至少 3 个中央公园的体量，未来海河这 12.5 千米将依托生态屏障区规划，成为确保城市生态安全的生态屏障，成为具有国际品质的都市绿心。以生态格局为基础，以生态服务功能为指导，加强生态基础设施建设，进行多方位生态修复，形成清晰的景观结构，即在由鱼骨状空间结构划分的景观单元中，依据城市发展和功能需求，确定不同功能的生态园区。

“作为核心的海河景观带，未来将形成沿海河单侧 200 ~ 700 米宽、具有宽窄变化的景观带，成为双城之间城市绿心。北岸占地 3.91 平方千米，南岸占地 1.97 平方千米，河道面积 2.12 平方千米，形成总计超过 8 平方千米的绿心，具体包含：水域，林地，田地，游览、展出活动的花园，休闲、运动的活力公园，足球、网球、篮球运动场，赛马、骑马的马场等活动区域，通过游览路加强南北岸联系，成为生态与活力统一的综合景观带。结合北岸 7.7 千米堤顶路、南岸 7.3 千米堤顶路，建设堤顶游线，满足骑行步行活动需求；结合河道建设水上游线，通过码头、驿站实现水上、路上游线结合，并将游线与周边活动场地和节点结合，涵盖整个海河中游区段，并向上下游和两岸其他区域延伸。12 条景观生态廊道是海河中游区域沿海河支脉向南北延伸的生态廊道，通过将支脉河道景观化提升，构建区域生态与游憩网络，达成绿带向海河南北的渗透，增加区域生态承载力。该区域依靠海河、海河支流和主要道路，划分出 24 个相对独立的生态建设单元，以水、路为边界，形成面积大小不一的生态岛屿。依据现规划，位于海河中游南岸西端的国家会展中心，就是一个重要独立的生态建设单元，将对其他建设单元形成强力带动。在海河南岸，生态单元未来将成为承担不同国际级别或城市级别的活动的区域；在海河北岸，生态单元将成为未来天津大事件预留地。生态单元近期作为生态涵养区域，进行大面积覆绿，成为城市苗圃。

“未来海河中游南北岸具有无限可能。海河南岸可以依托国家会展中心，成为举办中小型国际级别活动的区域，成为天津对外交流的核心区域。总面积 180 公顷的大型生态单元群，可以类比上海世博文化园 187.7 公顷的占地，成为未来可能举办世博会的场地。总面积 33 公顷的中型生态单元群，可以类比上海旗忠森林体育城网球中心。总面积 11 公顷的小型岛屿，可以类比巴黎罗兰 · 加洛斯球场，成为未来举办网球赛事的场地。

“海河南岸也可以依托双城之间的区位，成为未来天津双城的城市活动中心，生态建设单元可以通过不同邻近岛屿的整合，灵活应对未来的可能性。例如类比巴黎动物园 14.5 公顷的占地面积，15 公顷的生态单元可以成为未来的天津动物园。类比现天津动物园 63 公顷的占地面积，67 公顷的生态单元群可以成为未来天津动物园新址。类比厦门园博园单个岛屿 20 ~ 80 公顷的占地面积，67 公顷的生态单元群也可以成为未来进行花园展、艺术展的区域。类比上海辰山植物园 217 公顷的占地面积，118 公顷的生态单元群可以成为未来的天津植物园。

“海河北岸则成为未来天津大事件的预留地，成为未来天津可能举办奥运会、园博会等重大国际赛事活动的区域。例如类比北京奥林匹克森林公园的 680 公顷占地面积，海河北岸生态单元总面积约 770 公顷，可以成为天津的奥林匹克森林公园。类比伦敦奥运会场地占地 275 公顷，海河北岸西区的 250 公顷占地可以成为未来天津举办奥运会的场地。类比国际花园展 66 公顷的占地，海河北岸东区的 145 公顷占地可以成为未来天津举办园博会的场地。”

“在近期，我们采用逐步修复的生态策略。海河北岸是生态涵养区域，进行大面积植树造林，成为双城之间生态屏障区的核心生态区域。由于天津整体地势低洼，且处于盐碱地带的地理因素，直接种植的乔木经历数年生长后，根系将扎入盐碱含量较高的浅层地表水层，难以存活。因此我们借鉴天津地区传统台田做法，将低地的土堆到高地上，低地更低变成鱼塘水域，高地更高变成旱地台田，台田上种植树木，保证乔木存活率。在现阶段，我们采用高密度种植低胸径乔木的办法，将生态涵养区建设成为储备林地，随着林地乔木长大，再逐步进行间苗，将区域建设成为屏障区的苗圃基地，保障其他区域苗木供给，以园养园，营造生产型的低碳高效景观，实现公园与城市协同共生。这样我们参照在海教园地区的实践经验，1 平方米造价约 174 元，海河南北两岸近 15 平方千米的生态建设单元总投资不超过 26 亿元。

“双城生态屏障区坚持世界眼光、国际标准、高点定位和天津特色，融绿色发展和创新驱动于一体，规划设计以生态理论和以人为本为理念，形成具有天津特色的高水平的‘双城生态屏障、津沽绿色之洲’。”

45. 设计结合自然——一环十一园 + 人民的公园

景观规划师谭鸣

在“城市设计理论和实践”分论坛会场，天津规划院的景观规划所所长谭鸣正在介绍一环十一园及周边地区城市设计。这个内容恰巧也是关于生态规划的，好像是景观规划师的一个交流盛会。谭鸣是 80 后，从海河综合开发后一直从事景观设计工作，常年在现场跑，风吹日晒，皮肤黝黑。近些年做了大量的实际项目后，他开始思考中国景观设计的未来方向，就如同中国建筑师在不停地问自己，什么是中国的当代建筑。

“一环十一园及周边地区城市设计项目是本次天津城市设计试点中的关键项目，我们很荣幸能够承担这项工作。所谓‘一环十一园’，是指天津市中心城区外环线绿化带及绿化带内沿线的 11 个城市公园。这个规划其实我们已经做了 5 年的时间，从最初的外环绿带和步道的建设，后来到 11 个公园及周边地区的城市设计，包括对建设公园和基础设施配套的测算等。但这次列入城市设计试点，其目的和意义超越了简单的绿化和步道的建设，也超越了公园周边地区的城市设计，而是要结合天津市国土空间总体规划的编制，从天津主城区城市空间重塑的角度来展开。为了做好这个规划，我们进一步查阅了历史资料，重新阅读了一系列大师的经典著作，同时做了大量的实地调查，采访了当地的群众，还专程拜访了参与外环绿化带规划建设的老规划师。

“天津位于九河下梢，虽然由于盐碱地的地理因素，树木长势普遍不好，但有许多坑塘洼淀，所以有‘七十二沽’的美誉，是城水相融的津沽大地。过去由于发展缓慢，许多人说天津不像是直辖市，更像是大农村。近 20 年，在‘大农村’向大都市的蜕变过程中，津沽自然的美景渐渐褪色，城市越来越大，承载快乐童年的绿色乐园愈来愈远。在实地调研中我们采访到 A 哥。现在就从 A 哥家的故事开始说起吧。”

A 哥

“A 哥是一个土生土长的天津人，出生在天津近郊，就是今天快速环路边上的一个村庄。由于 20 世纪 80 年代末修建了外环线，A 哥所在的村子就成了城中村，那时候城中村在城市中随处可见，这也可能就是那时天津被戏称为‘大农村’的原因之一。虽说是郊区，但往里走不远，就是天津的城市商业中心，而他家后院不远处就是一片拥有水塘和绿树、蝉鸣蛙叫的菜地，再远处就是连片的河流湿地、农田林地。谈及那时家门口的绿色乐园，A 哥唏嘘不已，满满幸福、美好的回忆。后来由于拆迁，他家就住楼房了，虽然居住条件改善了，但也没了出家门进菜园的乐趣了。

“A 哥家的变迁，其实是天津大规模城市扩张、城中村改造以及城市更新后，城市居住生活环境和自然环境变迁的一个缩影，那片淹没在城市‘混凝土’森林的菜园、坑塘水面，代表着大规模城市建设后，城市与生态空间渐行渐远、相互分离的状态。今天，中心城区的建设早已漫过快速环路，并即将填满快速环路与外环线之间的区域，再往外就与环城四区连为一体了，摊大饼式蔓延的趋势越来越明显。外环内原有的坑塘湿地、河流滩涂等城市生态空间被进一步压缩，生态系统的完整性、稳定性、多样性遭到了很大破坏。同时，在新一轮城市更新后，城区建设密度和开发强度日趋提高，已接近每平方千米 3 万人的水平，而同时期新建的城市公园数量却很少，城市人均公共绿地不到 4 平方米，远低于国家标准，与国外发达国家的差距更大。另外新建绿地中道路防护绿地占了很大的比重，防护绿地并不能算作适合人使用的公园绿地，这进一步加剧了人多绿少的矛盾。

“人多绿少、人口过度密集、社区老化，这些问题已成为迫切需要解决的现实问题。在大规模城市建设后，生态空间的缩减，已成为制约中心城区高质量发展的重要因素。中心城区的空间结构有待调适，城市的人居环境亟待提升。在天津提出后 30 年将坚持绿色生态发展战略背景下，通过整合提升中心城区现存的生态空间，连通环内外的生态体系，重塑城市生态系统，恢复往日津沽美景——九河下梢的城市特色，将成为当下的城市空间发展的重要战略举措。

“进入新世纪，海河的综合开发战略带动了津城的经济腾飞；进入新时期，下个 30 年，到 2050 年，天津主城区以一环十一园为骨架的自然生态体系的建立，将推动城市的转型和高质量发展。参与过 20 世纪 1986 版天津市城市总体规划以及新世纪总体规划修编的老专家 B 总，在一环十一园项目之初，提出了这样一个观点，‘一环十一园’应是‘波士顿的翡翠项链’，而非‘伦敦绿环’。城市与自然生态应是一个互为支撑的整体，而非分割与对立。”

老专家 B 总

B 总在回顾 1986 版总规编制时说：“当初规划外环线及外环 500 米绿带，是学习借鉴了伦敦绿环规划的经验。最初的目的在于划定中心城区明确而清晰的增长边界，建设中心城区的绿色屏障。这在当时具有前瞻性和现实意义，在国内城市中天津是第一个。外环 500 米绿带以及同时沿主要河道规划的绿楔成为中心城区城市结构的重要组成部分，也建立了基本的生态空间框架。多年来，历版天津市城市总体规划一直延续和完善了外环绿化带及绿楔规划，逐步形成了今天以外环绿带及绿楔共同构成的城市公园体系——‘一环十一园’。30 年过去了，尽管外环线绿化带的建设不理想，大部分区段只建成了 200 米宽的林带。500 米范围内还有许多村庄、工业厂房，以及许多违法建设，11 个公园也只建成了 4 个，但规划的基本功能还是实现和发挥作用了。”

“在今天看来，在规划方面也存有遗憾。外环绿带过于强调防护职能，且自成体系，作为‘增长边界’的角色始终独立于中心城区的生态与空间系统，未能发挥‘增长边界’之外的更大职能。‘一环’虽学习‘伦敦绿环’，但只有 500 米宽，仅是大伦敦绿环宽度的 1/100，空间尺度不足，难以孕育完整的生态系统。对于天津绿色生态战略，建设生态宜居城市，‘一环’应有更高的定位，发挥更大的作用。

“随着外环线交通职能的调整以及双城之间绿色屏障的确立，中心城区与环城四区一体化成为必然。高速环线以内约 1400 多平方千米的空间需要重塑结构体系。在这个更大的空间尺度上，容纳了更多的自然生态要素，既有外环线内的 11 个大型城市公园，也包含了外环线外的 5 个郊野公园以及大面积的农业生态用地，还有多条河流绿化走廊，为构建一个更大的城市生态系统提供了机会，也为中心主城区建立一个生态系统引导的健康空间结构体系提供了可能。

“从这个意义上讲，‘一环十一园’不再是‘伦敦绿环’，而是‘波士顿翡翠项链’，是一个生态结构更稳定、更多样，与城市空间系统更加契合，且为人居环境提供有力支撑的‘翡翠项链’。在这样一个绿色生态系统的支撑下，高速环路以内的主城区的空间结构将进一步整合重塑。”

谭鸣

“为了编制好城市设计，我们参考了许多发达国家成熟的经验，重温了几部经典著作，包括麦克哈格的《设计结合自然》、雅鲁的《危机中的区域——纽约三州地区第四次区域规划》。我们认识到，重建人与自然、城市与自然和谐共生的关系，以及公园和运动场地对提高居民生活品质的重要性。在开展天津国土空间总体规划的背景下，城市空间结构的优化首先是生态系统结构的重塑和优化，将生态空间与城市建设空间放在一个同等重要的位置，甚至生态空间居于更重要的位置。一环十一园及周边地区城市设计实际上是一个基于城市主城区的生态系统重建，是建立生态导向的中心城区以及环城四区空间整合与结构优化的途径，是重塑城－绿关系和城乡关系的规划实践。

“麦克哈格在《设计结合自然》中写道：

不要问我你家花园的事情，也不要问我你那区区花草或你那棵将要死去的树木，关于这些问题你尽可以马虎对待。我们是要告诉你关于生存的问题，我们是来告诉你世界存在之道的，我们是来告诉你如何在自然面前明智地行动的。我们并不关心岩石本身或植物本身，因为它们能好好照顾自己，我们关心的是人，人是自然也是自己最危险的敌人。

“我们既要承认人是自然中不可分割的一部分，也必须尊重人的独特性，从而赋予人以特殊的生存价值、责任及义务。而这正是生态学的思想，或是人类生态规划……这些自然要素现在与人类一起，成为宇宙中的同居者，参加到无穷无尽的探索进化的过程中去，生动地表达了时光流逝的经过，它们是人类生存的必要的伙伴，现在又和我们共同创造世界的未来。我们不应把人类从世界中分离开来看，而要把人和世界结合起来观察和判断问题。愿人们以此为真理。

“让我们放弃那简单化的割裂地看问题的态度和方法，而给予应有的统一。愿人们放弃已经形成的自我毁灭的习惯，而将人与自然潜在的和谐表现出来。世界是丰富的，为了满足人类的希望仅仅需要我们理解、尊重自然。人是唯一具有理解能力和表达能力的有意识的生物。他必须成为生物界的管理员。要做到这一点，设计必须结合自然。”

“人与自然的关系：人，不仅是一个独一无二的物种，而且有极具天赋的自觉意识。这样的人，知晓他的过去，和一切事物和生命和睦相处，持续不断地通过理解而尊重它们，谋求自己的创造作用……我们应当相信，自然是进化的，自然界的各种因素之间是相互作用的，是具有规律的；人类利用自然的价值和可能性是有一定限制的，甚至对某些方面要禁止。最为合理的途径也许就是对不同的环境进行调查，弄清不同的环境作为人们一般使用及特殊使用，其可能承受和不能承受的程度。

“大自然像一张大网，包罗万象，它内部的各种成分是相互作用的，也是有规律的、相互制约的。它组成一个价值体系，这个体系本身具有供人类可利用的多种可能和种种限制。我们要将一些社会的因素、美学的因素、生物的因素等有层次地量化起来综合考量……从生态价值体系中，从‘我与它（I–it）’到‘我与你（I–you）’关系的转变，这是一大进步，但是‘我们’似乎是一个更合适描述人与生态关系的词汇。经济价值体系必须扩大成为一个包括所有生物物理的进化过程和人类渴望意愿的相关体系。

“人应成为生物界的酶，即生物界的管理人员，提高人–环境之间创造性的适应能力，实现人的设计与自然相结合……自然想象是相互作用的，是动态的发展过程，是各种自然规律的反映，而这些自然现象为人类提供了使用的机遇和限制。人类适应自然不仅带来了利益，也要花费代价，但自然演进过程不总是具有价值属性的，也没有一个综合的计算体系来反映全部费用和利益。自然演进过程是整体的，而人类的干预是局部的和不断增加的。轻视自然演进过程不太可能会有长远的收益增长，但可以肯定和证明轻视自然的结果一定是负担大量的费用。

“城市是一个形式，它是由地址和生物演进而来的，它是自然演进和人工改造以适应自然的综合产物，还需要把城市历史发展看作反映在城市规划和构成城市的个体及群体建筑中的系列文化适应，有些适应是成功的并得以保存，其他一些则不然……寻找城市特性的基础——从自然的特性和人造城市的特性中，选择有表现力和有价值的、对新发展起限制作用和提供机会的诸要素，它揭示出城市形式的基础。

“我们在一环十一园及周边地区的城市设计中，着重重建中心城区及主城区生态系统的完整性、多样性。依据现存生态要素，进行自然保护地、保护要素的选择，构建高速生态环 + 郊野农林生态以及外环绿带环 +11 个城市公园的双环体系，促进内疏外联、绿城融合。一是一环十一园及周边地区与中心城区内部水系连通，发挥海绵城市的调蓄功能；二是一环十一园及周边地区与城市绿道、绿廊的联系。通过河道、绿廊与外围郊野公园、外环绿带与高速生态环相连通，强化与双城之间绿色屏障的联系，形成内外相连、绿色相渗的生态格局。未来将建成 300 千米长的 3 条环城绿道，外环 500 米绿带、5 个郊野公园、11 个城市公园，通过 12 条绿化廊道实现与外围郊野公园连接。”

“一环十一园地区将成为建设双城间绿色屏障的重要补充，是城市级生态功能区建设以及城市人居环境质量提升的关键区域之一。外环绿带和城市公园是连接外环郊野公园，以及优化中心城区和主城区布局的绿色艺术骨架，是天津市中心城区的‘绿色翡翠项链’，是天津之链。

“目前，一环十一园地区建设水平比较低，作为一环的外环绿化带绿化水平不高，缺乏人的活动空间。11 个公园只建成了 4 个，其中柳林公园刚刚建成投入使用。还有 7 个公园需要建设，它们目前还是荒地或待拆迁整理的土地。从长远看，外环绿化带也应该是城市公园。要加快外环绿化带城市公园和其他 7 个公园的建设，通过实施投资绿色的战略，带动周边新型社区的建设。这些新型社区提供的住房和公园绿色开放空间是为了满足人民对美好生活的向往。加文（Alexander Garvin）在《美国城市规划设计的对与错》一书中指出，城市公园对人和城市的作用非常重要，要把公园建设成为人民的公园。”

46. 城市—乡村磁铁——郊区化 + 城乡共生

谭鸣

“天津主城区由市内 6 区和环城 4 区去掉双城生态屏障部分组成，面积有 1400 平方千米。现状人口 900 万人，占全市总人口的近 60%。以中心城区快速路和一环十一园为界，可划分为中心圈层、生态宜居圈层、田园城市圈层三个圈层，与自然生态系统相呼应。中心圈层是快速环以内的城市建成空间，是城市活力中心区，以城市更新为主要特征，是高强度、高密度的空间形态；生态宜居圈层是快速环路与外环线之间的区域，目前还有很大的发展空间，这里突出居住空间与大型城市公园结合的特征，采用中低强度开发和 TOD/SOD 开发模式，局部高强度；田园城市圈层，是外环线至高速环线之间的区域，以存量、减量发展为主要特征，居住空间与城郊生态空间结合，采用低强度、低密度开发，是郊区化发展的主要地区，也是解决天津中心城区目前存在的各种问题的出路。

“现代城市规划之父霍华德在《明日的田园城市》中写道：苍穹笼罩、微风吹拂、阳光送暖、雨露滋润下的我们的美丽土地，体现着上苍对人类的爱。使人民返回土地的解决办法，肯定是一把万能钥匙，因为它能打开入口。即使是入口微开，也能看到光明的前景。人口集中的一切原因都归纳为‘引力’，必须建立‘新引力’来克服‘旧引力’。可以将每个城市当作一块磁铁，每一个人当作一枚磁针，只有找到一种方法能够使引力大于现有城市的磁铁，才能有效、自然、健康地重新分布人口。

“城市—乡村磁铁就是这个新引力。要打破城乡的二元对立，除了‘城市生活、乡村生活’，还有第三选择‘城市—乡村生活’。它可以把一切最生动活泼的城市生活的优点和美丽、愉快的乡村环境和谐地组织在一起。这种生活的现实性将是一种‘磁铁’。

“城市是人类社会的标志，乡村是上帝爱世人的标志，城市和乡村必须‘成婚’，构成一个城市一块乡村磁铁。它可享有与城市相等的，甚至更多的社交机会；可使居民身处大自然的美景之中；可使高工资与低租金、低税收相结合；可保证所有人享有丰富的就业机会和光辉前途；可吸引投资、创造财富；可确保卫生条件；可到处见到美丽的住宅和花园；可扩大自由的范围，使愉快的人民享有一切通力协作的最佳成果。

“我国的城市规划自诩师从霍华德的田园城市思想，但在过去几十年的实践中，我们只注重城市，忽视农村，割裂城乡关系。对郊区化存有许多偏见，这可能就是我国目前城市问题突出的原因之一。在国家转型发展的新时期，我们必须转变旧的城市规划观念和方法，注重城市—乡村磁铁。国土空间总体规划作为全域的规划，更是要从城乡一体来进行统筹规划。

“郊区化是促进城绿融合、城乡共生，提升城乡人居环境品质的主要途径。过去我们的思想僵化，简单地认为郊区化就是无序蔓延、占用大量土地、不生态环保。实际上，除了美国郊区化存在蔓延的问题外，郊区化是人们向往自然环境的必然选择。美国著名城市历史学家肯尼斯·杰克逊在其著作《马唐草边疆》中写道：

郊区作为安家之所，作为城市以外的居住区和商务区，其历史几乎与人类文明一样悠久，郊区是城市的一个重要组成部分。然而郊区化却不等同于郊区，它起源于美国和英国，可以追溯到1815年前后，指的是一种进程。外围地区经历了系统化的增长，其速度远远超过中心城市。当风力、人力和水力仍然是文明主要的推动力时，郊区无论社会地位还是经济实力都远在城市之下。这时的‘郊区’代表了习俗之低劣、视野之狭隘和躯体之肮脏。当铁路、地铁、小汽车高度发展的时候，郊区成为人们实现向往田园生活本性的美丽家园。对于美国大城市日益分层和分散的社区地理现象，不能只看到工业革命所释放的交通技术和强有力的机械力量，还应看到新的文化价值观的发展。

“郊区化是城市发展的客观规律。过去我们过度人工控制的郊区化，采用了与城市中心无异的规划设计模式，郊区也高楼大厦遍布，造成‘城不城、乡不乡'的局面。从空间规划的角度看，郊区化不单是郊区化本身的问题，而是涉及整个天津主城区城市空间结构优化、解决城市病的大战略。

“天津主城区可以看作由三个圈层组成，每个圈层存在各自的问题，但都无法靠自身解决。快速环路以内的中心圈层，面积 145 平方千米，现状人口 340 万人，人口密度 2.34 万人 / 平方千米，人均公共绿地 3 平方米，包括五大道等 14 片历史文化街区和城市中心商务区、中心商业区、文化中心、体育中心等主要功能区。经过成片危陋平房改造后，城市整体建筑质量得到极大改善。除了城市各种公共建筑外，其他大部分是 20 世纪 80 年代以来建设的以 6 层为主的板式多层和 100 米以下的塔式高层居住区。其中在 20 世纪 90 年代末住房制度货币化改革前建设的老旧居住区问题比较严重，这些房屋总量约 7000 万平方米，居住着 300 万人，占该区域人口的 80% 多，占中心城区总人口的一半。这些老旧小区一是房龄在 30 年以上，房屋老化，建筑和管道陈旧破损；二是大部分多层住宅没有电梯，随着老年人口比例上升，行动愈发不便；三是小区停车位严重短缺。为了解决老旧小区存在的老化问题，从 2012 年起，天津市分两批进行了大规模的老旧小区整治，改善了居住条件和社区环境。日前，国务院也下发通知，要求各地区开展老旧社区更新改造工作，包括更换管道、增加外保温层、增设电梯等。这些举措会极大地改善老旧小区居民的生活质量，是一项民心工程。

“但是，这种改造无法改变房屋的结构和户型。由于建设年代早，这些住房的面积比较小，房型设计以满足基本的功能需求为主，一般没有起居室，只有一个小的过厅，厨房、卫生间都比较局促，不适应生活水平提高的要求。这种房型的小区，人均住宅建筑面积小，人口密度相对比较高。未来，按照城市总体规划，2035 至 2050 年，人均住宅建筑面积需要增加，居住条件需要改善。我们初步测算，这个圈层要疏解约 90 万人口，使人均住宅建筑面积从 24 平方米增加到 33 平方米。如果我们学习欧洲的经验，对这种多层住宅进行结构性改造，那么主要是通过两套并一套或三套改两套的方式，改善户型结构，增加建筑面积，从而提高居住条件。这是未来 15 ~ 30 年天津中心城区面临的最大挑战，也是城市实现转型升级的关键之战。要实现这一目标，必须解决两个关键性问题：一是这种多层砖混住房改造的技术可行性；二是人口疏解的动力机制，这个机制必须依靠市场，政府从政策上给予支持。这次城市设计试点中有老旧社区更新项目，体院北项目组做了很多深入的研究，他们会做专题介绍。大家有兴趣可以做深入了解。

“对于快速环与外环线之间的生态宜居圈层，面积 255 平方千米，现状人口260万人，人口密度每平方千米1万多人。这个地区在规划新建的7个大型城市公园旁，还有比较多的开发用地，可建设住宅4500万平方米左右，按照人均45平方米的高标准，可容纳100万人。这些人口可能部分是原居住在中心圈层老旧社区中的居民，为改善居住条件而购买新住房；另一部分可能是外来人口，希望有比较好的居住条件。这个圈层规划的关键是按照2035、2050年的标准，实现高水平的规划建设。不仅要建设满足未来需求的新型住宅，而且在社区教育、医疗、养老等方面配套完善，品质良好。新建商品房的户均面积为130平方米，以四居室为主。社区建设与公园融为一体；以大型公园为核心，组织社区开放空间系统，形成步行街区；开放空间网络同时作为海绵城市的载体。围绕公园绿心，建设多种类型、开放强度适宜、整体高度有序、多元混合、绿色生态宜居的新型社区。靠近公园的内圈层为低层高密度、联排、多层住宅。地铁站点周边为适当高强度、中高层住宅。外圈层为中强度、小高层、高层住宅。强化公园周边城市空间控制要求，结合公园的出入口，合理布置公共服务设施，形成生活圈。

"对于外环线以外900平方千米的田园城市圈层，现状人口300万人，主要集中在四个郊区政府所在地的城区和靠近外环线的大型居住区。未来规划人口增长到410万人，新增110万人。这个区域生态功能突出，蓝绿空间占50%，有5个郊野公园以及大面积的农业生态用地，还有多条河流绿化走廊，为构建一个更大的城市生态系统提供了机会，也为主城区建立一个生态系统引导的、健康的空间结构体系提供了可能。规划明确按照自然演进过程保护开放空间系统，安排城镇用地，进行与自然融合的居住空间整体设计，形成城镇与乡村的平滑过渡，并强化城乡与自然融合的开发强度和建筑密度控制。按照保护—游憩—城市化地区适宜度分析进行总量与分布控制，形成与自然生态融合的低层低密度组团式居住空间分布。田园城市圈层就是我们的郊区，要让人们广泛认识郊区，体验生活环境与自然环境的融合之美，使得大量既想改善居住条件又想接近大自然的中心圈层的居民自然而然地选择郊区，实现中心圈层的人口疏解。按照集约节约利用土地、不浪费土地和塑造高品质人居环境的原则，借鉴新都市主义的规划设计方法，建设适合中国特点的郊区花园住宅。设计以小镇中心为核，外围是具有中国传统特色的乡间院落式建筑，建筑掩映在绿树之中，人与自然和土地和谐共生。

"麦克哈格在《设计结合自然》中写道：

从乡村到城市的旅行，代表了从依赖于土地的生活向社区生活的进化，

而经过一定时间后，反过来又由城市转向乡村，这里面反映出最早的人与土地的关系。人们深深地感到需要乡村，这种需要最有力的证据是人们纷纷涌向郊区，这是历史上最大的移民活动。城市的确有许多有价值的东西，如广泛的人际交往、强大的社会组织机构、竞争、各种刺激、多样化和各种机遇，但是人们古老的记忆使他们坚持要回到土地上去，到大自然中去。这种交替都是必要的和有益的。

“我们中的许多人来自农村，特别是我们的父辈，更多的人生长在农村。虽然我们长期生活在城市中，但我们对农村都有深厚的感情。我们对美好生活的向往，不仅体现在物质空间方面，更重要的体现在对土地和自然的向往。在论及满足人民日益增长的对美好生活的向往的时候，郊区是不可或缺的。另外，通过郊区化的探索，可以为远郊更广大的农村地区村庄规划建设和人居水平的提升提供技术和适宜产品的支持，真正实现城乡共生。

“郊区化需要一定的支撑条件，交通是基础。历史经验证明，交通的发展是实现郊区化的前提，城市轨道网规划合理向外延伸，形成沿轨道线葡萄串式的布局。除了轨道和公共交通外，郊区主要的交通模式是小汽车，具有门到门的灵活性和便捷性。郊区布局要与城市交通规划紧密结合。天津主城区不同的圈层需要采用各自不同的规划战略、出行方式构成和道路交通布局模式。”

中新天津生态城管委会马科长

生态城的马科长专程来参加研讨会和论坛。参会之前她认为自己在生态城工作了 10 年，对生态城市已经足够了解了。她最关心的是如何使城市具有活力，就报名参加了“城市设计理论和实践”分论坛，没想到听的第一个发言还是与生态城市有关的议题，她忍不住举手发言。

“大家知道，中新天津生态城是中国、新加坡两国政府战略性合作项目。生态城位于滨海新区永定新河以北，毗邻北塘和天津经济技术开发区，交通便利，距滨海新区核心区 16 千米，距天津中心城区 45 千米，距北京 150 千米，总面积约 31 平方千米，规划居住人口 35 万人。生态城是为滨海新区功能区配套服务的重要生活城区。

“从用地上看，生态城用地是盐田和废弃的土地，包括一个污水库。从位置和规模上看，生态城位于滨海新区核心区的边缘，区位应该算郊区。按照传统的规划理论，生态城是一个大的卫星城，在滨海新区总体规划中确定的是组团式网络化布局中的一个组团，相当于一个小城市。最初的选址在汉沽与唐山交界处，后逐步南移，靠近天津经济技术开发区、天津港和滨海新区核心区，有产业支撑。

“生态城的总体规划由中国城市规划设计院、天津规划院与新加坡方共同编制。为了编制好规划，我们首先确定了中新天津生态城的指标体系。依据选址区域的资源、环境、人居现状，突出以人为本的理念，涵盖了生态环境健康、社会和谐进步、经济蓬勃高效等三个方面 22 条控制性指标和区域协调融合的 4 条引导性指标，将用于指导生态城总体规划和开发建设，为能复制、能实行、能推广提供技术支撑和建设路径。

“生态城总体规划依据不规则的用地形状和生态本底，保留了蓟运河故河道和污水库，将污水库改造为清净湖，规划贯穿全城的生态谷，串联生态居住组团。居住组团由 400 米 ×400 米的居住单元组成，中间保留十字形步行系统。生态城学习新加坡的经验，规划建设集中的社区中心。规划了国家动漫园、高新产业院和生态环保产业园，努力实现经济发展和产业融合。在道路交通和市政工程方面也广泛采用了新的绿色理念和技术。

“中新天津生态城 2007 年 11 月确定选址，2008 年启动建设。10 年过去了，生态城的建设取得了有目共睹的巨大成绩，媒体有很多宣传，这里我就不占大家时间赘述了。我想重点讲一下对前一阶段规划实施进行的总结。我们有许多成功的经验，也有许多困惑和问题。

“第一个问题是空间形态规划，也就是城市设计。客观地讲，生态城的总体规划虽然说法比较新，技术标准高，但实际的规划还是比较传统，是传统居住区规划及卫星城规划的手法。虽然后来也完成了城市设计和控规，但对城市空间形态的考虑不多。比如，生态城城市街道的绿化很好，但没有形成生活性的街道。原规划的步道系统没有能够连通，最后形成 400 米 ×400 米见方的大型封闭居住小区。住宅建筑高度普遍比较高，形态单一，这也可能是住宅用地的容积率过高造成的。实际上这里的低层和多层洋房更受欢迎，大家选择住在生态城主要是希望有一个好的宜居环境。我想这主要与生态城的定位有关，它是个独立的中等城市，还是卫星城，还是一个郊区的田园城市？

“第二个大的问题是区域协调和城乡共生。与新加坡的完全城市化不同，滨海新区也有广大的农村地区，包括在生态城周围。当年为了加快发展，减少与周围的矛盾，生态城把内部 30 平方千米封闭起来，当时有它自身的合理性。生态城东部就是原滨海旅游区，南部是北塘，北边还有中心渔港，今天都划归生态城统一管理。生态城西部和北部是大片的农村。实际上，城市不是孤立存在的，区域协调、城乡共生是一个优势，但我们没有利用起来，包括在大的生态环境上，还有产业上。生态城现在实际是个卧城，尽管我们不想承认，努力实现职住平衡。但目前居住的 8 万人口中，绝大部分是在开发区和滨海核心区就业。随着第三产业发展，生态环保和全域旅游是个大的产业。目前滨海旅游区的航母主题公园和国家海洋博物馆是旅游的热点。除了这些点位外，我们需要形成完整链条，让游客可以住在生态城一段时间，包括农业地区的民宿等。我们现在认识到，位于城市边缘郊区的小城镇、卫星城，应该注重与周边区域在自然环境、社会文化、经济及政策上的协调，包括建筑形态的协调，实现区域协调与融合。

“最后，交通是我们经常在思考的另外一个主要问题。中新生态城依托天津市区和滨海新区，对外交通比较有优势。经城际高速铁路与北京、天津两大城市及其他省市联系方便。有数条高速公路从四面接驳生态城，包括通向中新生态城南部的京津高速（京津塘二线）、通向生态城北部的津汉快速路以及通向中新生态城东部的中央大道和海滨大道等。但生态城与滨海核心区和中心城区的联系不方便。在生态城选址之初，就提出要规划建设轨道交通，初步考虑将津滨轻轨延长到生态城。但延长后津滨轻轨长度已经超出合理范围，运营时间过长，不很适用。2009 年滨海新区规划了轨道交通网，沿中央大道规划市域 Z4 线，在生态城范围设 3 站，可直达于家堡高铁车站，生态城内部沿生态谷设置电车喂给线。这个方案最终也征得新加坡方面的同意，被纳入了天津市轨道网规划。但直到 2015 年，国家发改委才批准天津轨道近期建设规划，开工建设以来，由于各种原因建设缓慢，现在还没有通车，所以现在生态城的交通拥挤问题比较突出。经各方努力，Z4 线通车在即了。我们了解到，天津国际机场和空港经济区、高新区和开发区西区的 Z2 线也在启动建设。但我在想，轨道通车后能彻底解决生态城的交通拥挤问题吗？我认为肯定会有改善，但恐怕不能根本解决。原因是什么？一方面是轨道建设缓慢，造成大家已经形成开车出行的习惯。另外，也有职住不平衡的直接原因。根本上说，还是我们没有正视小汽车发展的客观事实，规划过于理想化。因此，我认为，既要大力发展公共交通，同时必须正视小汽车的发展。随着新能源汽车技术的成熟，小汽车污染排放的问题已经解决了。关键是道路交通政策和规划设计怎样智慧应对和精明组织。”

第八部分
一座更人性化的城市

47. 超越机动性——公交都市

滨海新区渤海规划院交通规划所钟所长

生态城马科长的即席发言引起了钟所长的共鸣。按照会议议程，他本来是接下来发言的，发言前一直在心中默默地做着准备。钟所长是交通规划专业出身，十几年来参与了滨海新区许多重大规划，积累了丰富经验。他喜欢读书且量多面广，实际上他读过的城市规划方面的书比交通规划的还要多。前段时间，他做了一个重要决定，向设计院领导请假，停薪留职，自费到加州大学洛杉矶分校做一学年的访问学者。他用了 10 个月时间，考察了美国许多城市和区域交通的实际状况，又读了很多书，收获满满。同时，经过认真思考，针对中国当前的道路交通规划提出了自己的思路。会前，他曾与滨海新区云飞交通咨询中心的宋云飞博士交流过，两人在一些关键问题上观点还不一致，所以钟所长对自己的发言有些忐忑。刚听到马科长的发言，他感觉有了一些底气。他想用更形象的语言来介绍自己的观点。

“目前国内城市交通拥堵排行榜非常流行，拥堵指数成为评价一个城市交通功能或者交通模式好坏的标准，有些城市还将拥堵排行结果视为城市治理水平高低的参考。在同一时期，国外许多城市在人性化城市建设方面大刀阔斧，如纽约的“抢街”行动。去年在这个报告厅，纽约交通局前局长、《抢街》一书的作者珍妮特·萨迪汉女士做了生动的讲座。这次我到美国游学，特地到纽约时代广场体验一下，感觉很不一样。不知大家如何看待这两种现象？

“交通对城市发展至关重要。从以美国为代表的西方发达国家当代城市交通发展的历史看，都经过了从公共交通到小汽车再到步行化的历程。我们可以清楚地看到这一发展规律。然而，我们好像总是不愿意承认小汽车的作用，或者一味地用美国当下的交通规划理念来解决我们当前的问题。美国等西方发达国家的交通已经过了小汽车高速发展的年代，进入平稳期，而我国的小汽车交通还处于高速发展期。我国的交通规划多少年来就像是皇帝的新衣，一直高喊大力发展公共交通、限制小汽车，却堂而皇之地无视小汽车的存在。实际的结果是，我国的小汽车产销量持续保持世界第一。这种不一致导致我们的城市特别是大城市的城市规划、交通规划与实际的交通发展情况相背离，交通拥堵问题在一些城市非常突出。

“现在看来，综合公共交通战略、小汽车战略和步行化战略，三者并重并同时发力，才是解决我国城乡交通问题的方向和出路，甚至可以说是唯一出路。城市的聚集为交通技术的进步提供了沃土，而现代交通技术的突破促进了城市的大发展，促使城市形成特定的空间形态。城市空间形态在交通系统影响下经历了由‘步行城市’到‘轨道城市’直至‘汽车城市’的转变过程，有学者将其划分为传统步行城市、公交城市和汽车城市三个阶段。在不同的阶段，实际上都存在着各自相应的问题。发展到今天，大部分城市已经形成三种形式交融的状态，因此需要综合的城市交通发展战略。

“首先说公共交通战略。我们一直提倡以公共交通为主导的现代交通体系，但在公共交通的规划设计上还不够深入细致。我们简要回顾一下公共交通发展的历史。现代公共交通出现在 19 世纪，其起源至少可追溯至 1826 年，在法国西北部的南特（Nantes）市郊一个浴场出现了提供接驳市中心的四轮马车服务。1829 年，英国人乔治 · 施里比尔（George Shillibeer）的公车（Omnibus）出现于伦敦街头。1831 年，世界上最早的蒸汽公车开始运营。不久，德国奔驰公司制造出以汽油发动机为动力的公共汽车，代替了蒸汽机公共汽车。不到 10 年的时间，公交服务在法国、英国及美国东海岸各大城市，如巴黎、里昂、伦敦、纽约得到普及。公共汽车使市民体验到彼此间前所未有的接近，也缩短了城市和邻近村镇间的距离。

“1879 年，德国工程师维尔纳 · 冯 · 西门子在柏林的工业博览会上首先尝试使用电力带动轨道车辆。此后，俄国的圣彼得堡、加拿大的多伦多都进行过开通有轨电车的商业尝试。匈牙利的布达佩斯在 1887 年创立了首个电动电车系统。1888 年，美国弗吉尼亚州的里士满也开通了有轨电车。电车在 20 世纪初的欧洲、美洲、大洋洲和亚洲的一些城市风行一时。天津是中国第一个有公交运营的城市。1904 年比利时世昌洋行获准投资经营天津第一条也是中国第一条有轨电车路线。经过 20 多年的发展，到 1927 年底，天津有轨电车共有 6 条线路，全长大约 22 千米，运行区域覆盖了天津华界、奥租界、意租界、日租界、法租界以及部分俄租界。

“1863 年，世界上第一条地下铁路在伦敦建成并通车，它的成功运行为人口密集的大都市如何发展公共交通提供了宝贵的经验。19 世纪 60 年代至 20 世纪 30 年代，欧美的城市轨道交通发展速度很快，有 13 个城市相继建成了地铁。当时，旧式的有轨电车仍是主要的公共交通设施。不过，相比于地铁，其运行速度慢、噪声大、正点率低等缺点已经显露出来。城市轨道交通的发展历程也是曲折的。20 世纪 30 至 50 年代，由于二战的爆发，城市轨道交通发展停滞不前，而汽车凭借自身便捷灵活的特点得到快速发展。

这一时期，世界上只有 5 个城市发展了城市地铁。而有轨电车则渐渐被淘汰，主要是由于路面交通车辆增加，与电车的矛盾加剧。洛杉矶是个典型例子，它原本是一个以铁路和电车为骨架的城市，后来随着小汽车的发展拆除了电车，成为世界上无出其右的小汽车城市区域。天津的有轨电车线路于 1964 至 1973 年由于各种原因也全部拆除。

“20 世纪 50 至 70 年代，由于汽车制造业的高速发展，小汽车进入家庭，得到普及。城市交通量暴增，使得交通日渐拥挤，严重时会导致交通瘫痪。再加上空气污染、噪声大等问题，使得人们重新认识到公共交通和轨道交通的重要性。巴黎、伦敦等欧洲城市，包括美国纽约、日本东京及大阪等建成了最完善的轨道网。从此，城市进入轨道交通时代，严格说是进入了以轨道交通为主的公共交通时代。1970 年，天津决定利用墙子河改造工程同步建设地铁，称为‘7047 工程’。期间，由于国家实行停缓建政策以及唐山大地震的影响，再加上资金限制，‘7047 工程’被迫停建。直到 1984 年，经过第三次扩建之后，天津地铁正式通车，时称‘天津地铁一号线’，天津成为国内继北京之后第二个有地铁的城市。

“虽然表面上我们也参与了公共交通发展演变的全过程，但我国对公共交通的认识还比较局限，公共交通整体上还比较滞后。加州大学伯克利分校的罗伯特 · 塞文路（Robert Cervero）教授在 1998 年出版的著名的《公交都市》一书中说过，用 transit 这个术语表示城市或区域中所有运送旅客的公共交通服务，从多起点和多目的地的、没有规定路线的小客车和小巴、中巴，到点对点的、有固定轨道的现代重轨列车，从车辆的类型、运送旅客的能力，到运营的环境，公共交通的各种类型构成了一个连续统一体（continuum）。普通的公交形式大家都很清楚，主要包括以下类型：小公交（paratransit，也称为类公交，包括 vans、jitneys、shuttles、microbuses、minibuses 等出租汽车类公交）、公共汽车（bus transit）、电车和轻轨（trams and light rail transit）、重轨和地铁（heavy rail and Metros）、通勤和郊区铁路（commuter and suburban railways）等。从长远发展看，小型商务或公用飞机、高速火车、高速轮船、高速长途公路客运等都会成为城市区域的公共交通形式。目前在发达国家，城市都形成了比较完善的公共交通体系，以上不同类型的公共交通根据各自的特性被规划使用，并相互配合，与城市布局有机结合起来，发挥着越来越重要的作用。

“我国几十年来口头上一直坚持以公交为主的城市规划模式，但在实际工作中一直是以道路为主的规划布局，至于路上行驶什么车似乎不是规划要管的事情。‘公交为主’更多像是一个口号，也从没有进行过以公交为主的试验性的城市规划和建设。在小汽车拥有量快速增长、城市交通不断恶化的情形下，几乎所有城市在城市总体规划修编或新的国土空间总体规划中都继续将发展公共交通作为缓解交通拥挤的唯一出路。国家也出台了支持公交发展的政策。当前，我国城际高速铁路发展迅速，开始发挥城际联系的重要作用。同时，许多城市在加快地铁建设，建设公交都市。普遍采用公交专用道、新能源公共汽车，发布‘掌上出行’等公交运营 APP 等措施，提高公交运行的准时率和服务水平，取得了一系列成绩。但是，在公共交通发展战略、城市公共交通与土地利用关系等关键问题上仍然缺乏深入考虑；在公共交通的服务和换乘等细部处理上仍然粗放，没有做到以人为本。如果没有进一步有效发挥公共交通作用的办法，没有统筹交通发展战略的新范式，发展公共交通注定是‘事倍功半’。

“总结国内外城市的经验，可以看出，公共交通战略成功的关键因素之一是处理好公共交通供给总量与需求的关系，以及各种公共交通的比例。目前，在国内的交通规划中，公共交通的类型不够丰富，对各种公共交通的使用思路不清晰，虽然提出了轨道快线和城区线的区别，但一些城区线路盲目延伸拉长，超出了合理运营距离。同时，对不同种公共交通工具之间的规划配合研究不够，缺乏细节处理，没有能够形成丰富多样的公交网络。

“关键因素之二是处理好公共交通线路与城市空间布局的关系，建设以公共交通为导向的公交捷运大都市。公共交通与城市形态的关系就如同手与手套的关系，一定要匹配。塞文路教授在《公交都市》一书中，将公交捷运大都市划分为四类：适应公交的城市（adaptive cities）、适应城市的公交（adaptive transit）、强核城市（strong-core cities）和杂交城市（hybrids）。

“适应公交的城市就是以公交为导向的大都市，通过对轨道交通系统的投资来引导城市发展，同时实现更大的社会目标，如保护开敞空间、提供有轨道交通服务社区的低收入住房等。所有适应公交的城市都以紧凑的、混合使用的郊区社区和以公交节点为中心的新城为特征。适应城市的公交是指以蔓延、低密度为发展模式的城市，通过寻求合适的公交服务和新技术，更好地服务整个城市区域。强核城市提供综合的公交服务，这种服务以混合交通电车和轻轨系统为中心，成功地把公交和城市发展整合在一个范围更加限定的中心城市内。布置在道路上的电车与人行和自行车共存。这些城市的首位度重要性和健康的公交爱好者是将城市中心区复兴和电车复兴成功融合的基础。杂交城市在沿主要公交走廊的集中开发和使公交有效服务它们蔓延的郊区及远郊区之间获得了一种可行的平衡。慕尼黑把重轨干线服务与轻轨和常规公交服务结合起来，两种公交由一个区域公交权威协调，强化了中心城市，同时也服务于郊区的增长轴。渥太华和库里提巴都引入了有灵活专用车道的公交，同时在主要公交站周围增加区域商业的比重。灵活的以公交为基础的服务和沿公交走廊混合使用的开发的结合，导致了不寻常的高人均公交出行量。

“从上面以公交为主的四种城市类型看，不论是土地利用追随公共交通，还是交通适应土地使用，交通和土地的利用关系虽然不是问题的全部，但仍然是问题的关键之一。要在空间规划中处理好公共交通和土地利用的关系，不仅体现在大尺度的规划结构上，而且体现在具体的城市设计上。以捷运为导向的开发（Transit Oriented Development ，--TOD）是现在城市规划师和交通规划师常挂在嘴边的话，但我们的做法过于简单和概念化。实际上，TOD 包括以捷运为导向的居住开发以及以捷运为导向的商业和就业中心开发两种主要类型，而且城市的不同区位内具体的开发强度和布局形式也不一样，不是简单地以公共交通站点为中心，周围以高密度的开发围绕就可以的。

“今天发达国家的公共交通体系都比较完善，但公共交通也不是包医百病的灵丹妙药。公共交通虽然至关重要，但单一的公共交通战略并不能完全解决城市交通问题。小汽车交通还在不断增长。英国伦敦一直强调公共交通战略，但发展到今天，公共交通出行只占总出行的 10% 左右，80% 还是小汽车交通。我国许多特大城市，大力发展轨道交通，地铁建设速度举世无双，但交通问题仍然突出。另外轨道交通的出行条件也比较恶劣，高峰时人挤人。因此，不能以为有了公共交通就可以一本万利、高枕无忧。在当年以马车为主的公共交通时代，大量的马粪曾经要了纽约的命。今天公共交通的补贴问题可不比马粪问题简单。如果都要政府补贴，太多的公共交通对于政府财政来说也是巨大的负担。因此，现在在西方学术界已经形成共识，公共交通发展战略主要是对现有公共交通线路的优化、与城市布局的配合，而不是一味地加大线路和线网规模。

“细节决定成败。各种形式公共交通的站点和各种规模换乘枢纽规划设计的成败直接决定着城市公共交通系统的成败。换乘枢纽通常位于城市中心或次中心，是商务商业活动与交通枢纽的综合体，是现代大城市中心所应该具备的基本功能。随着城市郊区化的发展，在近远郊区出现了新的城市换乘枢纽。这些枢纽是城市区域快速轨道交通的站点，位于城市区域发展轴上，是快速轨道交通与私人小汽车和地方公交的换乘点，也成为地方生活的一个中心。与城市中心换乘枢纽建筑不同，新的城市换乘枢纽建筑较为简单、方便、人性化，运行成本低，是新型城市区域节点，成为社区和邻里的中心。

“要充分发挥公共交通的作用，发达国家有比较成熟的“3D”经验，即密度（Density）、多样性（Diversity）和城市设计（Design）。增加密度是提高公交出行的最好和最有效的规划方法，但也要适度合理，不是密度越大越好。混合土地使用的多样性也能够鼓励公交出行，还可以促进资源的有效利用，比如停车场的共用。除了合理的密度和土地多样化使用外，城市设计是更有效的手段。

“目前，普遍认为城市设计也是改善城市交通功能的有效方法。世界上有许多紧凑、混合土地使用的城市，如曼谷、雅加达、圣保罗等，虽然具有活跃的街道生活，但它们都不是可持续发展的范例，普遍缺乏高质量的城市设计。北美的新都市主义倡导紧凑布局、混合土地利用，利用传统的方格路网、公共场所、路边停车、开敞空间、步行范围内的商业设施等设计手法，努力将以私人小汽车为主的郊区改变为对步行者友善和公交支持的城市和社区。

“发展公共交通战略，必须要形成系统的规划策略和可操作的方法，与小汽车战略、步行化战略相配合。因此，我们这次城市设计试点把交通规划设计作为一项重要的研究内容。”

48. 用轮子投票——适应小汽车发展

钟所长

“下面我想重点说一说小汽车战略，实际上小汽车交通与城市设计的关系更加紧密。我们同样先简要回顾一下小汽车的发展历程。1879 年，德国工程师卡尔 · 本茨首次试验成功了一台二冲程试验性发动机。1883 年戴姆勒和迈巴赫发明了汽油内燃机。1885 年，他在曼海姆制成了第一辆发动机三轮汽车。1885 年末，戴姆勒将马车改装，发明了第一辆四轮汽车。19 世纪末，这种车传到美国后，也只有纽约、费城等少数大城市中的富人才有资格享用。1896 年，亨利 · 福特试制出第一辆汽车。1903 年，福特建立福特汽车公司，立志要让美国人都能够买得起汽车。1908 年，福特及其伙伴将奥尔兹、利兰以及其他人的设计和制造思想结合成为一种新型汽车——T 型车，这是一种不加装饰、结实耐用、容易驾驶和维修、可行于乡间道路、满足大众市场需要的低价位车，车身由原来的敞开式改为封闭式，其舒适性、安全性都有很大提高。大批量流水生产的成功，不仅使 T 型车成为有史以来最普遍的车种，而且使家庭拥有轿车的理想变为现实。福特发明的流水线生产方式的成功，大幅度降低了汽车成本，扩大了汽车生产规模，创造了一个庞大的汽车产业。

“20 世纪大地景观最大的改变是私人小汽车的普及、交通量的增长和高速公路网的建设。小汽车的发展和普及是人类文明历史性的进步，它不仅促进了经济发展、技术发明和科技的进步，而且极大地增强了人的机动性和能动性，可以说，小汽车是人性化的交通工具。

“汽车及其相关产业的发展构成了全球经济非常重要的、不可或缺的一个方面。我们可以从石化工业的产业链中看到，石油的产品如汽油、柴油以及汽车零件等需要汽车的消耗。其他许多产业，如钢铁、橡胶等同样如此。据统计，美国的汽车消耗了其石油产量的 50%、钢铁产量的 50%、橡胶产量的 30%、皮革产量的 20%。在美国的总就业人口中，30% 的就业与汽车有关。美国汽车制造和销售公司一年花费 110 亿美元的广告费推销汽车，单是通用汽车公司（GM）就用了 20 亿美元。私人小汽车的发展极大地推动了第三产业的发展，产生了许多新的经济形式。汽车及相关工业的发展促进了美国经济的发展和国力的增强。汽车产业带动一系列和汽车有关行业的进步，带动国家经济的增长，解决就业问题，促进消费和创新。

“小汽车的发展是文明的进步。回顾历史，人类文明的发展伴随着交通方式和手段的进步。工业革命以来出现的轮船、火车、汽车、飞机及城市轨道交通等运输工具，极大地促进了社会的进步和经济的发展。与其他科技的发展一道，现代交通运输方式成为当代人类文明的重要组成部分，其中私人小汽车的普及是一道亮丽的风景线。对于个体来说，私人小汽车是最便捷、最灵活自由、最舒适的交通方式，实现了自由出行的理想。人们称美国是轮子上的社会，这不仅体现在字面上，而且有更深层次的含义。‘花园洋房和小汽车’所构成的‘美国梦’，代表了所谓民主、自由、平等、个性，是美国快速发展和日益强大的助推器之一。我国全面建设小康社会的过程中很重要的一个方面是机动性和出行能力的提高。航空、铁路、公路、水运、管道运输以及信息的传输，对经济发展和人民生活水平的提高都具有重要的作用，而私人小汽车也是其中必不可少的一个关键指标，也反映了经济的发展和繁荣。

“没有免费的午餐，任何事情都是要付出代价的。汽车在带来便利的同时也给人类的生活带来种种问题。这些问题集中表现在三个方面：能源、交通、污染。第一，汽车产业高度消耗自然资源。除了使用钢铁外，现代的汽车还需要使用能耗很高的铝材和难以回收的塑料。另外，汽车大量消耗石油，全世界一半以上的石油用于运输，而其中 1/3 的燃油被用于驱动汽车和卡车的内燃机。第二，汽车导致交通拥挤，交通事故频发。汽车引起的交通事故是当今世界上导致人类死伤数最多的原因之一，每年约有数百万人受到车祸的伤害。同时大量车辆的停放也日益压缩人们的生活空间。第三，汽车的生产和使用污染环境。汽车排放的尾气对城市大气污染非常严重，由尾气引发的光化学烟雾是世界上许多大城市共同面临的难题。另外，汽车造成噪声污染。

“汽车对城市的影响也是巨大的，以美国最为突出。为解决城市交通拥堵问题，一些高速公路深入城市中心，对城市肌理和景观环境造成严重破坏。同时，交通的发展助长了郊区化，促进了旧城人口和商业中心外迁，最终导致城市中心的衰败和城市的蔓延。

“20 世纪 70 年代，石油危机使人们开始对小汽车进行反思，包括在经济、社会、城市蔓延方面的问题，尤其是生态的问题。美国人开始了对小汽车的持久战争。反对派对私人小汽车深恶痛绝，认为其是造成城市无序蔓延、生活质量下降、环境污染、资源浪费、社会问题的罪魁祸首。《沥青国度》一书的作者简 · 凯伊指出：这是一条通向环境毁灭之路。

“今天美国共有 2 亿辆汽车，‘美国人驾驶的汽车正在操纵着美国整个国家’。从 20 世纪 20 年代开始的以小汽车为主的郊区化、1956 年开始实施的跨州高速公路法案（Interstate Highway Act），等等，进一步促进了私人小汽车的加速发展。以 1991 年完成历经 35 年的跨州高速公路计划为标志，美国已经进入所谓的‘后汽车时代’。正是在这时，曾经是报社记者的简 · 凯伊带着三个孩子从波士顿郊区来到市中心居住生活，她下决心卖掉了自己的汽车，开始写关于小汽车的书。她共花了 6 年时间，于 1997 年出版了《沥青国度》，发表了反对小汽车的宣言。

“站在美国历史的地平线上，福特 T 型车和‘规划的城市’的出现，就像一个剪影，背对着 20 世纪最初的愿景。无论今天我们在后汽车时代有怎样的反思，汽车文化都在过去的近一个世纪里永久地侵蚀了美国的公共交通系统、他们的公民意识以及美学修养……城市发展已然在 20 世纪 20 年代走到了一个巨大的分水岭。将明日城市与我们祖先塑造的美丽家园区别开的，是以人为本，或是以汽车为本的价值观的鸿沟。

“在公路上，美国人似乎又希望去开拓找寻消失的新土地。它再一次刺激着美国人的想象力和进取心——这些深入骨髓的美国性格。罗斯福及其新政把美国从经济与社会灾难的深渊里拉了出来，并对国家公园体系、公路体系的建设等产生里程碑式的影响。然而，在这个过程中，公路已经遍布美国的城市、乡村，住房政策促进了美国的进一步蔓延，公共道路管理局的权力更加根深蒂固，它们像一台永动机，不停地拉着美国走在继续沥青化的道路上。

“汽车是拯救我们的工具，也是毁灭我们的根源。尽管我们的汽车行业犯下了愚蠢的错误，但它依然根深蒂固地存在着，3/4个世纪以来，我们的生活和景观都是以满足汽车的要求为前提的。身在此山中，我们自然无法看到个人乃至全球的困境，或者觉察出我们应当逃离它。我们必须开始援救行动了。结束汽车统治时代的课题有很多，不会仅仅靠一个改变汽车依赖的宣言、一个计划或一个项目就可以实现的。

“放弃汽车，塑造一个符合人性尺度的活动空间和鼓励步行的城市结构，这项工作就像环境保护运动一样，需要在微小的尺度上，在充分尊重地方性、传统性和尊重个人的前提下进行工作。这项工作及其目标可能并不像登月计划那样激动人心，也不是历史上‘大跃进’那样的心态，但它一定拥有可持续的前景。

“因为同样是女性，也叫简，而且她的著作同样重要，所以人们把简·凯伊与写出举世名作《美国大城市的死与生》的简·雅各布斯相提并论。在简·凯伊出版《沥青国度》10年后，一位叫珍妮特·萨迪汉的女士在纽约实施了从小汽车手中抢街的行动。汽车文化风向标开始轻微转向。三位女性对城市规划都作出了巨大贡献。”

“我们从简·凯伊的书中，能够看到美国因小汽车而产生的深层次问题的广度和深度。在今日美国，大量的高速公路、坡道和大片停车场占用了大量的城市土地，形成的景观十分单调、空旷。在城市郊区，以汽车为基础的超大建筑破坏了城市的宜人尺度。美国是一个交通拥挤的国家，人们对局部拥挤习以为常，整个地区的交通同时堵塞，即所谓的‘网格锁定’（grid lock）也已经是普遍的现象。虽然道路和停车场面积已经占了美国城市建成区面积的40%左右，但还是经常找不到停车位，交通状况仍然没有根本好转。

美国一年里堵车造成通勤者多消耗80亿小时。拥有美国2/3人口的沿海地区，如旧的波士顿－华盛顿和新的洛杉矶－圣迭戈是主要的交通拥挤通道。

“美国是一个被汽车锁定生活的国家。依靠汽车的生活方式和以私人小汽车为基本出行工具所构成的‘出行链’，把美国人的生活紧紧地与小汽车锁在一起。随着城市向外蔓延，郊区的人口增加了3倍，密集、可步行和公交服务的城市数量萎缩。城市的蔓延造成距离和空间的分离。这儿居住，那儿购物，别处上班，没有汽车便无法工作和生存，而且使得每次出行都是独立的。出行不便，里程和交通量成倍增长。在美国，91%的家庭拥有汽车。平均每个家庭每天6次往返出行。据调查统计，每辆车每年平均16 093 ~ 19 312千米的里程数中超过60%与生活有关，生活锁定于汽车。

“小汽车的发展加剧了环境污染问题，汽车尾气排放成为城市空气最严重的污染源。内城交通拥挤，污染严重，人们捂着鼻子跑出城，搬到郊区，进一步增加了交通量，也就增加了尾气的排放。美国人均汽油消耗量是欧洲的5倍。同时，汽车在其他方面的污染折磨着美国的一些贫困社区和邻里，如废弃的轮胎、泄露的电池、报废的汽车等，大量没有处理的汽车废弃物污染着美国的环境。

“在简·凯伊看来，小汽车的罪恶罄竹难书。也许上文对私人小汽车问题的阐述还不全面。实际上，客观地看，许多问题并不是私人小汽车本身的问题，而是人类社会发展中长期面临的社会政治经济问题。经济发展就需要消耗资源，生产会产生污染。解决问题的最终出路是技术进步。今天新能源汽车产业发展已经如火如荼，所以尽管有这么多的问题，私人小汽车的发展仍是时代潮流。车流滚滚，势不可挡。在这个潮流背后的巨大动力就是私人小汽车满足了人们的需求和推动了社会经济的快速发展。这就是人们为什么选择用轮子投票的最终原因。

“实践证明，在大城市地区，更加完善的公共交通系统可以发挥更大的作用，包括城市布局的优化，可以减少小汽车通勤出行的比例。但在周末超市购物、郊外休闲远足等出行需求方面，私人小汽车是无法被公共交通替代的。因此，在大城市地区，私人小汽车和公共交通系统，包括步行慢行系统可以组成互为补充的城市交通体系。今天，在日本东京，城市公共交通十分发达，但私人小汽车的拥有量已经达到了让人吃惊的水平，平均每两个人拥有一辆小汽车。人们通勤坐公交，生活休闲用自家车。因此，小汽车不是洪水猛兽，不是道德问题，只是需要我们合理地使用。

“一座好的城市必须有好的交通体系，好的交通体系需要好的交通规划指引。早在1977年，英国交通专家J·M·汤姆森在《城市布局与交通规划》中提出公共交通与小汽车交通、轨道与道路的斗争是交通规划中两个相互竞争的主要方面。好的交通规划需要结合自身城市结构、交通系统以及经济发展水平，协调好公交与小汽车的关系。汤姆森根据世界五大洲、30个不同发展阶段的城市交通问题处理经验，提出解决城市交通问题的五种应对战略及其适宜布局形式，并强调每种战略均不完满，需要结合地区实情、所处发展阶段灵活运用。

“第一种布局形式：充分发展小汽车战略。方格网形态路网结构，没有轨道，也没有真正的市中心，均值化的郊区中心沿干道网布置，其设计思想是小汽车可以在全城各处畅通无阻，适合人口较少的小城镇，英国、美国的新镇广泛采用。采用这种布局形式的小镇很少有交通拥堵、停车难以及环境问题。它们真正的交通问题是没有小汽车，不能开车的人出行活动受限，男人开车上班后，老人、小孩只能待在郊区的家里。另外一个问题是，采用这种战略的城市规模不能发展得很大，否则交通问题会很严重而难以弥补，比如洛杉矶。

“第二种布局形式：限制市中心战略。市中心规模较小，有放射形的道路网为市中心服务。城市的大部分工作岗位分布在郊区和边缘地带，交通主要靠小汽车，有通过能力很强的圈层式环路为其服务。此外，还有一个规模不太大的公共交通系统，几条简单的放射形轨道线供进出市中心的通勤人员使用。这种布局结构既不稳定又浪费资源，过多的圈层式快速环路则是不稳定和浪费资源的根源所在。城市修建快速环路的初衷是分离穿越市中心的过境交通，缓解市中心交通拥堵。在环路通车初期，确实有一定的交通改善作用。但从城市系统的角度来看，这种改善效果只是短暂的，环路对沿线地区交通的改善，提高了土地价值，刺激了沿线地产开发。随着时间推移，环路周边与市中心地区空地得以填充开发，曾经的郊区被并入市中心，曾经的外围环路转变为城市内部路，市中心没限制住，反而像摊大饼一样无序扩张。市中心规模的扩大会带来交通量的剧增，密集的中心区也会要求加剧环路的进出口，出入口的增多会让进出快速环路变得困难，拥堵增加也成为必然。对城市环路功能的不恰当认知和设计初衷，一定程度加剧了城市交通的拥堵，巴黎、伦敦、哥本哈根都经历过，但它们及时采取了补救措施，伦敦停止新建环路，哥本哈根则采取强力的管制措施。北京、上海、武汉、郑州等城市也正在经历拥堵，问题是我们没有及时总结经验教训，反而期望修建更多的环路缓解拥堵，陷入越修环路越堵的恶性循环。值得注意的是，环路修建热潮正在朝二、三线城市蔓延。

“第三种布局形式：保持市中心强大战略。有一个强大的市中心，道路与轨道组成放射形交通网，次中心设置在轨道网上，轨道网容量很大，到市中心上班的人大部分乘坐公共交通。在交通线路上，公共交通与私人小汽车在竞争中达到平衡。这种布局形态在欧美顶级大城市得到广泛采用，例如巴黎、纽约、悉尼、多伦多、雅典等。

“第四种布局形式：少花钱战略。这种布局形式的特点是：没有高快速路及轨道网，城市密度很大，城市次中心集中在放射性公共交通走廊上，并与市中心保持适当的距离。战略达成的核心是立足于管理好现有道路，由公共汽车承担大部分运输功能，另外需要将土地利用规划好、管理好、配合好。波哥大、库里蒂巴等城市采用这种布局形态。

“第五种布局形式：限制小汽车交通战略。这种战略的城市结构有一个强大的市中心，以及很好地为市中心服务的公共交通系统。轨道网呈环放结构，次中心设置在轨道线交叉点，环线连接各次中心。道路也呈环放结构，通往市中心的放射型道路通行能力逐渐降低，靠近市中心时，减少车道条数，或者改为收费道路，把小汽车疏导到其他环形干道上去，最理想的是将车辆疏导到仅仅环绕市中心的环路。城市外边还有条过境环线，截留城区边缘过境交通。这种战略适合前小汽车时代就已建立起来的城市结构，另外还需要配合限制小汽车计划使用，包括实行停车收费，设置小汽车禁行街道，公交、步行、骑行优先等。伦敦、香港、新加坡等城市多采取这种交通发展战略。

“上述五种城市布局形态各异，但蕴含一些基本原理。一是城市活动中心必须具备良好的通达性，无论是位于道路相交处，还是公交走廊枢纽点。二是市中心无论大小，得配套一定水平的公共交通设施。三是强大的市中心需要大容量的公共交通来支撑，连接市中心和次中心的轨道网是维持强大市中心的基础。四是圈层式快速环路容易导致摊大饼式无序蔓延与扩张，对于构建良好的城市多中心体系负面影响较大，需谨慎对待圈层式快速环路的修建。

“今天，在我国城市规划、土地规划和环保领域，以及空间规划领域，有一个奇怪的现象，就是普遍反对发展私人小汽车，也反对以私人小汽车为主的交通模式，但事实上我们正走在这条路上。单纯以公交为主的、紧凑的城市和区域空间规划布局结构和形态无法满足私人小汽车迅速发展对交通的需求。造成驾车人长时间的驾驶和堵车，以及乘客在公共交通站长时间等车的双重困扰。要想避免这种情况的发生，必须转变观念，实事求是，采取公共交通、小汽车交通和步行交通并重的战略，在城市中心区、城市边缘地区和郊区等不同地区采取适合各个地区特点的规划布局和交通策略。

“20 世纪末开始，新都市主义活跃，它强调公交、步行、自行车和小汽车等多种出行方式共存，是较成功的规划设计手法。需要注意的是，许多近现代建筑设计和城市规划的国际大师们并不反对私人小汽车交通，如赖特、柯布西耶等，他们试图寻找新的空间规划模式以期找到解决问题的答案。实践证明，社会政治经济模式和生活方式与城市空间模式紧密相关。今天，我国在经济发展上选择了私人小汽车这样一条道路，也就决定了我们的空间发展模式。在小汽车时代开始来临的时刻，如果我们不能面对现实，而是像看待皇帝的新衣一样无视私人小汽车的存在，试图单纯地以发展公交来解决城市交通问题，我们就会犯历史的错误。因此，我们需要客观理性地看待私人小汽车的发展，规划设计出既适合公交、步行，又满足私人小汽车发展的城市和区域空间模式。”

钟所长

“国外交通领域的专家已基本形成共识，交通问题最终是一个经济问题。交通方式的抉择取决于人们愿意付出多大代价。交通供应不同于水、电、煤气等其他服务设施，它有两个不同的特点：城市交通不仅为城市服务，它也是城市的组成部分之一；交通为各种活动服务，但它与建筑物和建筑物内的活动是互相依存的，不可分割。

“城市交通设施把组成城市生活的各种活动连接起来，这些活动必须依靠交通设施。城市的结构、城市的大小及其扩展、城市生活的方式及特点都是由城市交通系统的性质和服务质量来决定的。因而，选择什么样的交通方式和交通系统等，能够影响城市的功能和布局。交通与城市布局之间的相互影响又使确定交通功能这一任务大大复杂化。

“一个城市的目标是多种多样的，有些目标是互不相容的。人们要有宽敞的住宅、小花园、公园，同时要把工业、商业有效地集中起来，要求缩短上班的路程，但这些目标是互不相容的。人们希望享受小汽车的舒适方便，也想享受大城市市中心的许多乐趣，这两个目标也是互不相容的。因为后者就需要把大量的人运到一个小的地区内，所以必须有一个折中的办法，在方便、快速、廉价的交通与城市生活的质量之间做出选择。孤立地把不同城市中生活或交通的某一点来做比较而不把其他相关的各点来做比较，那是没有现实意义的。例如，仅仅拿车速来做比较而不把人们在这种车速下所需要行驶的距离长短做比较；仅仅比较居住密度，而不去比较人们到工作地点或去当地商店所花的时间与金钱。又如，仅仅把去超级市场是否方便来做比较，而不把去市中心和郊区是否方便以及诸如此类的各点来做比较。

“人们习惯用一套标准去评价事物的好坏，评价结果可能直接影响后续的行动，对于城市交通也是如此。单纯的交通功能认知导致单一的评价指标选取和片面的行动指南导向。评价一个城市交通模式好坏应该采取更加完善的标准。交通与城市活动、城市建筑、城市本身的互动特点，决定了一个城市交通系统的优劣，不能只看道路是否拥堵、车速是否够快，还应该植入社会活动的因子去进行系统评判。比如，交通在多大程度上能满足人们参加各种活动的需要？是否由于交通困难或费用太高而妨碍了人们去喜欢的学校、商店或工作单位，或妨碍了其他的社会活动？在经济条件许可下，在多大程度上能满足人们居住在喜欢的地区？人们的社会活动与家庭生活是相互关联的，两者之间必须权衡利弊，进行选择。例如，在现在的交通条件下，有些人为了上班等方便，宁可住得差些，也要住在离各种社会活动场所近些的地方；也有些人为了要住得宽敞，环境好些，宁可住到较远或交通不大方便的地方去。对于城市中的绝大部分居民来说，与社会活动有关的交通，特别是上班、上学、购物、去市中心以及去办事的交通，如何方便、费用低、舒适、时间短（不是指车速快），这是交通规划的核心，是政府补贴的公共交通要做到的。同时，对于以小汽车为主的个人交通出行，政府可以提供道路交通条件，许多方面要靠收费等市场的手段来解决。

“交通系统对人们居住、工作、购物和娱乐的环境影响如何？这些需要新的交通模型来分析判断。影响这些内容的因子不像基于交通工程角度，强调机动车行程、速度等指标那样容易收集、量化，但利于客观地评判城市交通系统是否合适，应该是以后的一种改进方向。国外的很多城市交通价值观经过小汽车导向、公共交通导向、移动性导向后，正在向场所性导向演进。加州大学伯克利分校的塞沃瑞教授在去年出版了一本新书，名叫《超越机动性》，目前还没有中文版。我们知道，建筑师、城市设计师经常讲场所，现在国外交通专家也在讲场所性。场所性更多地体现交通与城市活动、建筑的相互依靠性。不同性质的城市，或者同一城市不同地区，评价标准的侧重点应该是不一样的。我们不能简单用大力发展公共交通、限制私人小汽车来解决所有问题。该发展公共交通的地方要大力发展，该限制小汽车的地方要严格限制，该鼓励小汽车的地方应该大张旗鼓地鼓励。”

市工业和信息化局规划处孙处长

听到这儿，市工业和信息化局规划处孙处长带头鼓起掌来。由于他一直鼓掌，周围的人也开始附和起来。掌声落下，大家把目光聚集到他身上。

“大家好，我是市工信局规划处的孙承刚，副处长。今天很荣幸能参加这么专业的城市设计论坛。我非常赞成钟所长的观点，要客观公正地看待小汽车的发展。

“在中国城市规划领域，以西方发达国家私人小汽车发展带来严重城市问题为借鉴，几十年来一直呼吁限制私人小汽车的过度发展，坚定地鼓励以公共交通为主导的城市发展模式。在 1997 年亚洲金融危机之后，在如何选择拉动内需的经济增长点上，曾爆发过商品住房和私人小汽车的比争。表面上商品住房占据了上风，实际上是汽车工业卧薪尝胆地发展。在 21 世纪的今天，汽车工业已经成为推动我国国民经济快速发展的非常重要的主导产业之一，我国成为世界汽车产销量第一大国，年产销 2800 万辆，而且新能源汽车产销量也是世界第一，拉动了众多相关产业和服务业的快速增长。这段时间，也正是环境保护、可持续发展、循环经济等观点的讨论最为热烈的时期。

“事实表明，尽管小汽车在环境保护和资源消耗等方面存在着严重的问题，这些问题越来越被大家所认识，但政府和公众还是选择了用轮子投票。私人小汽车拥有量的快速增长反映了人们生活水平提高之后对机动性的强烈需求。随着私人小汽车大量在城市出现，传统的以公交为主的城市布局模式被滚滚车轮碾碎，交通拥挤成为最严重的城市问题之一。在车流滚滚的情形下，城市规划却仍然坚持无视小汽车发展。几乎所有城市的总体规划修编和综合交通规划依然遵循着以公交为主的城市布局和交通规划模式。这种做法在理念上无可厚非，但在实际操作中，必须面对私人小汽车不断发展、城市交通问题不断恶化的现实。现在国内许多特大城市包括北京、上海、天津等采取了限购、限行等措施，小汽车产销量也出现下滑的趋势，都对小汽车作为负责的规划工作者，包括我们主管汽车产业的部门，必须客观地面对这样的局面，找到解决问题现实可行的办法。

“发展小汽车必须解决交通拥堵问题。钟所长，我想问一个问题。长期以来，城市中心普遍被交通问题所困扰，您认为引起交通问题的根本原因是什么，合理的应对方式是什么？”

钟所长

“要采取公共交通、小汽车和步行并重的交通战略，不同的区域采取不同的策略。我一直在思考如何解决中心城区和滨海核心区的交通拥挤问题，是要进一步拓宽道路吗？再建高架路立交桥吗？不管从历史名城的保护，还是从经济的角度看，基本不可行了。目前发达国家的趋势是许多城市把市中心的高架路、立交桥都拆除了，我们不可能重蹈覆辙。但交通量还在不断增长，怎么办？最近读了《抢街》和交通规划的几本经典著作，我感到豁然开朗。著名交通规划专家当斯有句名言：‘通过不断拓宽马路来解决交通拥堵问题，就如同给一个减肥的人放松裤腰带！’要解决市中心的交通拥挤问题，只能在进一步发展轨道等公共交通、实施步行化和人性化城市战略的基础上，通过合理收费限制小汽车交通进入市中心，这是唯一而且正确的出路。

“城市交通问题主要是经济问题，是因为收费不当，对使用道路、停车、公共交通、造成的环境污染缺乏调控。长期以来的收费不当影响着各种活动地点的分布，影响着提供什么样的新的交通设施、是建设道路还是轨道。二战后，西方大多数城市把钱花在修建道路上，不断地试图缓和拥塞，而轨道却一直停滞衰退。公共交通公司在非常艰难的境况中坚持下来，与小汽车竞争。随着越来越多的人不乘坐公共交通工具，而改乘私人小汽车，政府受到的压力越来越大，一方面，小汽车多了要修建道路，需要花钱；另一方面，公共交通乘客少了，收入少了，要防止公共交通破产，也要花钱。经济压力之大，以致许多城市陷入绝望的地步：一方面道路经常拥塞，另一方面公共交通濒于崩溃。

“城市交通问题的根源可以说是由于长期以来，在城市交通的管理和发展上未能按经济规律办事，导致公共交通在与小汽车交通的竞争中全面落败。而人口和收入的增加，加剧了问题的严重度。这些收费不当，在过去的 40 多年里，有技术原因，也有尚未达成共识的原因，但现在情况在慢慢发生改变。比如伦敦、新加坡等城市中心区已经利用成熟技术在收取道路通行费、环境污染费等，利用经济和技术手段调节城市交通，提高公交竞争力，实践效果良好。需要注意的是，经济手段是城市用地成熟、文明高度发达后，为实现可持续发展所采取的应对措施。对于快速发展的城市地区，需要在规划阶段协调城市交通模式与城市空间结构、用地布局之间的关系，用一个短的‘行程’来解决人们参加活动的问题，引导形成健康、可持续的城市活动模式。“

孙处长

“过去 10 多年是滨海新区城市建设和机动化快速发展期，规划在引导城市建设方面发挥了重要作用，它在交通层面是如何考虑的？”

钟所长

“从世界大城市交通发展经验来看，有三种不同的交通发展模式、五种不同的城市布局形式。滨海新区在借鉴先进经验的基础之上，也结合自身特征进行了相应的创新探索。空间布局特征方面，与一般城市区不一样，滨海新区为典型的超大型城市尺度、多极核生长状态，是一个城市区域的概念。各片区之间的空间距离普遍在 30 千米左右，对机动性需求较高，超出普通公交适宜的服务范围。传统的城市交通组织形式难以适应这种超大尺度空间布局。

“片区内的空间布局也呈现多样化特性，具有明显的差异性，既有集中发展的城市区，又有多组的生态城、空港经济区、渤龙湖高新区、开发区西区等片区。不同类型的区内空间布局各异，对于交通组织模式、交通网络布局等的要求具有较大的差异性。”

“由于其特殊的空间布局形态与发展特征，滨海新区的交通出行模式不应仅仅局限于某一种，而应是在不同层面体现不同的交通出行模式。滨海新区在总体发展思路方面，提倡以公共交通为主，充分考虑小汽车的发展，鼓励步行的出行模式，但分片区、分层次又具有差异性，从而充分体现滨海新区的布局特点。片区之间，采取‘双快’连通战略，主要依托快速大运力轨道和高快速路承担对外交通，公交与小汽车模式并重。片内出行，整体鼓励以公共交通、自行车、步行等为主导的绿色交通出行方式，但各片区又有差异性。比如在于家堡、天碱等片区，落实‘小街区、窄马路、密路网’‘完整街道’等理念，注重围绕公交枢纽建造一体化的绿色接驳体系，小汽车出行比例会很低。而在产业园区，其机动化出行的比例会明显高于公共交通出行的比例。对于广大城市外围地区和郊区，不光是滨海新区，还有主城区，地铁和公交线路密度大幅降低，还是要以小汽车为主要的出行方式。新能源车 NEVs 可能成为郊区家庭的标配。”钟所长向周边扫视了一下，没有看到宋云飞。他多少有些担心在论坛上公开谈论反对宋博士的观点，不知他什么感受。

“当然，现在世界上最流行的还是步行化城市战略。简·凯伊在《沥青国度》中说：我们不用一直沿着高速公路来寻找解决小汽车问题的出口，我们可以从城市的复兴和步行尺度的修复中找到新的方法。首先是我们必须明确‘场所’的价值和交通所扮演的角色——让我们更方便地抵达。我们的目的是运送人，而不是运送汽车本身，这种理解能让我们站得更高、看得更远。在许多发达国家的城市里，特别是城市中心区，用得最多的交通方式是步行和骑自行车。在不同的用地布局下，这些简单而价廉的交通方式是最健康的，而且步行化对城市中心的复兴发挥着巨大的作用，这一点很重要，应仔细考虑。当然，这是以市中心强大的公共交通，以及对私人小汽车的限制，准确地说是拥堵收费做保障的。”

49. 步行化城市——老城区的再复兴

规划院历史名城所何所长

在“历史名城保护与城市设计”分论坛，天津规划院历史名城所的何所长刚好开始汇报老城区步行化和城市复兴，好像与隔壁会场心有灵犀似的。何所长从同济大学毕业后一直从事城市设计工作，对历史街区的保护更新尤为关注。天津在历史上是一个比较温和的城市，中国传统的老城厢、河北新区，以及舶来的各国租界区，建筑和街道的尺度亲切宜人，使天津老城区具有良好的建筑和城市空间传统。近几年，面对老城区活力不足的问题，何所长一直在思考能做些什么。去年，他也参加了萨迪汉《抢街》的讲座，并做了主持。会场气氛热烈，他学习到很多。只是由于经验不足，在主持中出现了一个小纰漏。在介绍完萨迪汉女士丰富的简历后，他忘了邀请萨迪汉女士上台，让见过世面的萨迪汉有点小尴尬。想到这儿，何所长自嘲地笑了一下。

“近年来，天津历史城区保护更新取得了有目共睹的成绩，但由于受到外围大型商业区的分流，特别是电商冲击的影响，老城商业区出现萧条的迹象，新建的津湾广场、五大院等商业设施也日渐萧条。如何实现老城区的再次复兴？我们提出步行化城市的策略。有人可能怀疑，步行化就能使老城区复兴吗？为了回答这个问题，我首先从三方面讲起，第一是天津老城区商业现状概况，第二是天津老城区与交通发展关系的历史，第三是对国内外步行化城市成功经验的借鉴。

“天津老城区是历史城区，是由14片历史文化街区构成的天津历史城区的核心部分，集中在海河上游两岸，面积10.6平方千米。本段海河长6.7千米，这个区域南北、东西各5.3千米宽。近年来，市商务局组织编制了《海河国际商业中心发展规划（2010—2035年）》。该规划统计，这个范围内年零售商品销售额350亿元，规模以上商业70家，总建筑面积236万平方米。其中购物中心9家，占全市的1/4；规模百货26家，占全市的1/3。互联网数据进一步显示，有各类商家9001家，其中超市、商店3525家，餐饮商户5332家，各类酒店144家。滨江道是各类APP点击量最高的区域，其次是东马路、大悦城、小白楼。大众点评最高的是滨江道和南京路交口，可以说这个区域仍然是天津商业最集中的地区。目前区域内有M1、2、3、9四条地铁线，11个车站，29个出入口；步行10分钟可覆盖65%的区域；公交车站97个，也是公交相对最发达的地区。但随着电商的发展，传统商业业态受到巨大冲击，劝业场等传统商业销售额不断下降，和平路北端的百盛百货、万达中心、苏宁电器等商场相继关闭，当然地铁建设长时间封闭道路交通也是原因之一。与此同时，随着旅游和体验经济的发展，区域中的五大道、意式风情区等历史街区出现了很好的势头，人流量持续增长。但这种现象还不足以扭转整体下滑的态势。如何实现天津老城区的复兴呢？我们的观点是实施老城区的步行化。

“回顾历史，可以发现，天津城市中心的兴盛与交通的发展密切相关。18世纪前，天津的城市中心在三岔河口周围，依靠河运的便利，非常繁荣兴旺。到第二次鸦片战争后，各国开辟租界，天津城市中心南移。到20世纪30年代，天津是闻名华北的商业中心。1904年，比利时世昌洋行获准在天津投资运营公共交通公司，兴建了天津第一条有轨电车路线。之后，陆续开辟红牌、黄牌、蓝牌三种电车线路，线路起点均为北大关。其中，红牌经北马路、东北角，沿河马路，过金汤桥，经建国路至天津火车站。黄牌经北马路、东北角、东马路、东南角、中原公司、四面钟、劝业场，至天津海关。蓝牌前段与黄牌共线运行，至劝业场后拐向滨江道，过万国桥至天津火车站。1921年，天津有轨电车增设绿牌，从当时天津法租界老西开沿滨江道，过万国桥至天津火车站。直至1921年，天津共有5条电车路线修成通车。1927年底，有轨电车增设花牌，由东北角至天津海关。至此，天津有轨电车共 6 条线路，运行区域覆盖了天津华界、奥租界、意租界、日租界、法租界以及部分俄租界，大部分与劝业场、老龙斗火车站连接，电车对商业中心起到极大的促进作用。

“这里我讲一个插曲。天津第一条有轨电车在刚开始运营时生意惨淡，原因是当时的中国人对电不甚了解，担心乘坐电车会触电而亡，因此，没人敢去尝试这个新鲜事物。之后，电车电灯公司采取了免费试乘、降低票价等手段吸引乘客，并在电车的头等车厢特加设绒垫座位，还配备地毯、痰盂、电扇等公共设施。一些天津市民试乘后，发现有轨电车不但安全、便利，而且比乘马车和轿子出行更加舒适。此后，随着客流量增多，电车的生意日渐火爆。公司便取消了车厢等级，票价也开始上涨。到 1912 年，该公司就收回了之前的全部投资成本。1916 到 1928 年间，天津电车电灯股份有限公司从公交线路上赚取了 2500 万银圆的利润，并且每周都要把高额利润汇往比利时首都布鲁塞尔。到 19 世纪 30 年代，‘坐电车逛劝业场’成为当时天津流行的都市生活时尚。有轨电车也促进了天津城市中心的发展和兴旺。

“由于各种原因，有轨电车线路在 1964 至 1973 年间全部被拆除。同时‘文化大革命’对天津城市商业的冲击也是巨大的。改革开放后，天津的商业中心开始逐步更新。地铁一号线的通车和国际商场的建设使滨江道商业兴旺，中心从劝业场南移。到 2002 年实施海河综合开发，拓宽了一些道路，改善了城市中心区的道路交通条件。新建了恒隆等大型商业设施。地铁三号线的通车进一步确定了这一区位的优势。

“多年来，天津一直在历史街区保护上下功夫，除保护好文物和风貌建筑外，也较好地保持了历史街区的空间尺度和完整性。随着近期环境综合治理和旅游工作的推进，历史街区重现活力。人们越发地认识和热爱宜人的建筑和城市空间环境，成为天津的一大特色。历史文化街区步行系统城市设计是本次城市设计试点的重点项目，我们要探索步行化与市中心活力复兴的结合。通过步行网络，串联文化遗产保护传承和合理利用，从而更好地延续历史文脉，展现天津的城市风貌，促进历史文化街区的活化和市中心的复兴。

“为了做好这项工作，我们进行了认真的学习研究。国外许多著名的城市近年来都采取了步行化策略，取得了巨大的成功。在规划理论方面，步行化城市的理论方法已经相对成熟。丹麦著名建筑师杨 · 盖尔多年来从事人性化城市研究和实践，出版了数部专著，参与了伦敦、纽约、墨尔本、莫斯科等城市的步行化城市改造，具有丰富的经验。他在《人性化的城市》一书中写道：

创造适合步行的城市。一个婴儿能够迈出第一步的那一天，是他人生中一个重要的日子。生活始于足下！但行走不仅仅是行走，步行的目的有很多，既有快速、有目标的前进，也有漫步闲逛。无论什么目的，人走在城市中，就会与周围的各种社会活动融为一体，从而产生一种广场效应。步行不仅是一种交通方式，而且是很多其他活动的潜在起点与发生场合。

建设适合停留的城市。一个城市中有很多步行者，并不一定说明城市质量高——有很多步行可能是被迫的。但是，如果一个城市中有很多人自在地停留着，通常说明城市有良好的品质。人们选择停留的地方，通常会是空间边界地带，这种现象叫作'边界效应'，站在边界的人感觉那里安静、谨慎。还有一个停留现象叫作'钢琴效应'，是说人们喜欢找个地方依靠。背靠墙壁让人觉得低调、安全。威廉·怀特曾在《小城市空间的社会生活》中称威尼斯'整个城市都可坐'。

鼓励人们在城市空间中自我表现、嬉戏和锻炼，有助于创造健康和有活力的城市。城市中没有好地方和好的人性化尺度，就会缺乏至关重要的质量。城市规划者应该极为仔细地处理宏观和局部气候，对于微观气候，需要在城市的人性化维度上创造必需的环境条件。城市空间的魅力不仅是城市生活本身，还来自各种感官印象的巧妙汇集……城市生活的最重要主题就是人本身。'人是人的最大乐趣'，城市的最大吸引物是人。人性化城市的核心目标是：建造充满活力的、安全的、可持续的、健康的城市。

充满活力的城市需要不同的且复杂的城市生活。城市生活是一个潜在的自我加强的过程，'有人的地方就会有人来'，人们自发地感受到活动和他人存在的鼓舞、驱动和吸引。城市活力包含两个维度的问题——数量和品质。'数量'可以从密度上来反映，而更重要的是'品质'方面。充满活力的城市真正所需要的是将有吸引力的城市空间与使用它的人群结合起来。提高一个既定的环境的城市活力，从量上是吸引更多人来，从质上是吸引更多人长时间停留，实现减速交通。二者比较，更简单、更有效率的解决方法就是提高品质，从而使人愿意来度过时光。

影响城市活力的另一个重要因素就是城市边界的处理，临街面该如何做，即把要做逗留区域的边界，做成'柔性边界'。任何单一的主题都不如活跃的、开放且充满活力的边界，对城市空间的生活产生更大的吸引作用。充满活力的城市，是精细规划的产物，时间因素和吸引人的柔性边界是这项工作的关键词。

安全性包含交通安全和社会安全两方面。如果给汽车提供更多的空间与更好的条件，结果就是，步行变得更加困难且更加缺乏吸引力。在设计街道类型和交通方案时，最重要的是从人性化的维度考虑。安全性在设计中体现为：步行在各种交通中必须具有优先权。同时，安全的体验和安全的感知对城市生活是至关重要的。城市空间中其他人的存在意味着更加安全。街道两边的建筑同样对安全性起到重要作用，夜晚居住区中窗户透出来的灯光给附近的人们送来舒服祥和的信号。同时，建筑底部如果是柔性边界，也会让人感觉更安全。此外，清晰的城市结构也意味着安全，我们更容易找到周围的路。

可持续包含物质环境、社会环境两个角度。良好的城市空间、城市景观对公共交通起到非常重要的作用，车站无论白天还是夜晚都要保证舒适与安全。社会环境的可持续是指社会不同人群进入公共的城市空间并且到达各处的同等机会。它的前提是具有可达的、可进入的、吸引人的公共空间。同时，关注贫穷差距问题，需要让被贫穷限制的边缘人群能够得到机会。新的城市规划必须起步于设计最便捷的、最具吸引力的步行和骑行的联系，然后再阐述其他交通的需求。

“本次城市设计以整体历史文化街区为研究对象，通过对天津历史城区现状特别是步行系统现状的认真分析，以目标和问题为导向，完善历史文化街区步行系统，通过高标准步行网络系统的建设，提升城市公共空间的品质和适步性，让人回归市中心，让城市生活回归市中心，让城市中心充满活力。

“我们采取了四个策略。第一个策略叫‘静街区’，就是通过收取拥挤费、控制停车位规模和提高停车收费等措施，限制小汽车交通进入市中心；通过交通稳静化措施，营造宜步街区。该地区现有社会停车场 9 处，3000 个车位。另外商场配建停车位约 5000 个。每天缺口 7000 个停车位，采取稀缺性供给。第二个策略叫‘畅路线’，就是明确将沿海河两岸步道作为步行系统的骨架，完善鱼骨状步行主路，增加过河步行桥，加强两岸步行系统的联系，保障历史城区内步行畅通，连接成网。我们希望像塞纳河两岸步道作为世界文化遗产和品牌一样，把海河两岸步道向世界文化遗产的方向努力营造。第三个策略叫‘用遗产’，进一步利用文化遗产、盘活存量资产，形成新的业态。第四个策略叫‘增品质’，精心提升公共空间的品质，建设世界一流宜步行城区，实现以步行带动城市中心的全面复兴。

“在历史文化街区步行系统城市设计引导下，选取五大道、解放北路、中心花园、鞍山道地区开展历史文化街区建设整治城市设计，按照总体步行系统设计，将历史街区步行系统连接成网络。五大道历史文化街区依据建筑类型来保护建筑空间关系，深化细化保护规划。充分利用历史文化资源，规划一条旅游线路——互联网 + 五大道体验之旅。建立三级步行体系，衔接区域步行网络。同时，规划以民园广场为中心，向南扩展，形成和平宾馆、第一工人疗养院的置换更新；向西扩展，对河南路进行改造，将第六十一中学地块建设成为一个与五大道地区总体形态协调的、具有中国元素的、高品质的新时代建筑，探索租界建筑风格的当代化。努力将五大道地区建设成为天津历史文化街区保护典范、洋楼经济和全域旅游示范区以及世界文化遗产的备选区。

“解放北路历史文化街区将解放北路与天津站地区作为整体统筹规划，完善步行系统，激发津湾广场商业活力。通过改造天津站广场和世纪钟节点、改善出租车停靠场地和广场候车环境等一系列措施，实现天津站的人性化。通过海河步行桥建设、津湾广场商业店面改造，提升街道空间品质，激活津湾广场。这个项目目前已经实施，效果十分明显。天津站作为天津的窗口，体现出与之匹配的水准。津湾广场访客人数由过去的每日 3 万人，激增到每日 10 万人。

“中心花园历史文化街区在遵循保护原则的基础上，通过渐进的微更新模式、梳理引导车行公交系统、优化慢行交通空间、塑造多层次景观绿化体系、建设多元混合城市空间等手段，对中心花园及周边地区进行有序的功能提升和业态更新。中心花园已经恢复了当年的原貌。赤峰道等街道作为世界银行贷款项目，已经实施了慢行系统改造，有显著效果。

“鞍山道历史文化街区保持鞍山道特有历史风貌，进一步提炼历史文化资源要素，引入多元混合业态，建设以辛亥革命之路为特点的文旅街区。结合历史建筑设置主要开发节点，与社区公园、街坊绿地形成三级空间景观体系。特别突出了鞍山道改造的断面和交通稳静化设计。

“我想解释一下交通稳静化。《文明的街道——交通稳静化指南》一书出版于 1992 年，哈德克劳博士把德文术语翻译成‘交通稳静化’。最早提出这一思考的是荷兰城市规划师和交通工程师，他们注意到人们的生活质量不仅与住房条件有关，还受到周边街道的影响。20 世纪 60 年代末，荷兰规划师提出了一种被称为 woonerf（生活化道路）的设计思想，一反习惯上将车行道与人行道分开的做法，转而将二者合并成一个路面。在这样的路面上，行人可以使用车行道，而机动车辆必须以‘马漫步的速度’行驶。这种设计看上去非常像居民的院子，又被翻译成‘居家庭院式设计’。1976 年，该设计获得荷兰法律认可，同时迅速在德国、英国等欧洲国家流行开来。交通稳静化不是对付道路堵塞的利器，而是一种带有局限性的工具，只有与其他交通工程方法、交通限制策略和城市规划相结合，才能取得效果。交通稳静化作为交通发展策略的重要组成部分，主要工作是限制道路空间为机动车行驶和停放服务，优先考虑公共运输、自行车和步行，只有私人汽车受到限制时才能有效实现交通稳静化。

“交通稳静化不仅仅涉及安全问题，虽然其极其重要，但用孤立和惩罚性的方法处理交通问题是有局限性的。交通稳静化必须灵活，它是一种审视街道设计的新方法，不是静止的，而是在不断演变和发展的。新建地区大多进行交通稳静化设计，道路比没有设计得更畅通。但是在老城区，交通稳静化被寄希望于改善城市环境，却与回流到市中心的富裕阶层的停车产生冲突。

“在这些城市设计中，我们在非常注重宜人公共空间和伟大街道的塑造的同时，也强调历史文化街区的整体保护和有机更新。杨·盖尔在《交往与空间》中写道：‘更合理的城市发展模式势在必行：不能再过度依赖大规模的公共开发、高度集中的产权和专擅的规划设计模式，而是将逐步积累的物质环境建设和对话的机制看作正当合理，从而改变那种大规模拆除现有建筑环境的开发方式。’这种发展态度建立在对城市生活的接受、热爱和期望的基础之上，鼓励人们生活在健康的公共环境中。城市中心的步行系统是必备条件。”

50. 公共开放空间城市设计导则

规划院城市设计所陈所长

在“历史名城保护与城市设计”分论坛，规划院城市设计所陈所长代表项目组做汇报。陈所长与何所长是高中校友，大学毕业后一直从事城市设计工作。他不仅对历史文化街区的保护更新特别关注，而且还十分关注城市公共空间的塑造。他曾参加了海河综合开发改造的节点城市设计，也参加了 2008 年迎奥运环境综合整治中解放北路的景观设计，了解城市公共空间细节之重要。近几年，他一直在做城市公共空间的城市设计，从天津五条河流开放空间到全市开放空间的城市设计。这次城市设计试点，他主持了开放空间的城市设计导则编制。去年，他从在世界银行工作的同学那里了解到，萨迪汉正在上海、广州等地做《抢街》的巡回演讲，便马上联系各方面，与同学共同努力，促成萨迪汉中国之行增加了天津站，反响热烈。

“回顾人类历史，不管是在西方，还是在东方，城市一直有着良好的建筑和城市公共空间的关系，公共空间包括河流、街道、广场、公园等。20世纪的现代主义运动和美国的郊区化蔓延式发展，打破了这个悠久的传统。勒·柯布西耶技术至上的高层花园住宅让城市成为孤独塔楼的停车场、汽车的天堂；而美国郊区的低密度住宅和‘棒棒糖式’的道路系统让城市空间丢失了。特里希特在1986年出版的《寻找丢失的空间》一书中所描述的城市景象，成为20世纪城市发展的真实写照。同时，书中提出了城市设计的三种方法——图底方法、联系方法和场所方法，这三种方法都强调对城市公共空间的设计。

“21世纪，在新的奇异建筑不断涌现的情形下，美国出现新都市主义思潮，试图从蔓延的郊区中寻找欧洲传统城镇美好的城市记忆。中国开始寻求新的建筑方向，探索注重城市空间、人的感受的新的建筑设计和空间设计模式。随着城市设计的逐步普及，这些理念被逐步地带入各个城市的规划设计中，逐步形成了小街廓、密路网，注重城市街道、广场空间、公园等公共空间的空间设计模式，开始强调人的交往和人性化设计。

“杨·盖尔在《交往与空间》中提及，城市设计和城市规划涉及若干不同的尺度层面。大尺度是整体城市布局的规划，是从航拍角度看到的城市。中等尺度也叫开发尺度，是各个城市各个部分的设计，建筑与城市空间的组织原则，是从低空的直升机视角看城市规划。而小尺度也就是人性化景观的规划，是人们在行走、停留时直观看到的人性化景观质量，这是视平层面的城市。这三个层面有不同的法则和质量准则，理想情况下三个尺度应该融合为一个有机整体，天际线、建筑布局、空间比例等要与视平层面的空间序列、建筑细部和城市设施结合在一起设计规划。设计人性化的城市只有一条成功之道，那就是以城市生活和城市空间为出发点。要始于视平层面，止于鸟瞰视角，同时关注三个尺度。

“视平层面的城市质量在于小尺度。在视平层面享受良好的城市质量，是人的一项基本人权。人与城市之间最切身的接触存在于小的尺度内，存在于以每小时5千米的速度步行时所见的城市景观之中。规划中考虑人性化，就是考虑最普通的人类活动，必须为人们的行走、站立、坐下、观看、倾听和交谈提供良好条件。

“城市公共空间或住宅区中的见面机会和日常活动，为居民间的相互交流创造了条件，使人能置身于众生之中，耳闻目睹人间万象，体验到他人在各种场合下的表现。室外空间的交往是一种接触形式，是一种深化交往的可能方式，是一种保持已建立起来的接触的机会。通过接触，还可以得到有关社会环境的信息，受到启发，产生多种感受。其中，活动是引人入胜的因素，只要有人存在，就会吸引人，人会寻找与他人靠近的位置，新的活动便在进行中的事件附近萌发了。对活动与游戏的习惯、活动与座位选择等方面要给予高度的关注。人们对街道上形形色色的人的活动有更大的兴趣，而建筑内外的生活比空间和建筑本身更根本、更有意义。

“不同城市模式之间在形式上的变化很多。但实际上只有两种典型的发展模式和城市规划思想与户外活动这一论题有关：一种源于文艺复兴，另一种源于现代功能主义运动。建筑的室外空间的生活是一种独立的质量和一种可能的开端，为必要性的户外活动提供适宜的条件，为自发的、娱乐性的活动提供合适的条件，为社会性活动提供合适的条件。

“室外空间生活是一种潜在的自我强化的过程。范 · 克林格里用一个公式总结了他的城市生活经验：‘一加一至少等于三’。正效应过程：有活动发生是由于有活动发生。负效应过程：没有活动发生是由于没有活动发生。公共生活空间被肢解，街头成了空寂之地。这种变化是导致在街道上破坏公共设施和犯罪的重要原因。在特定地区高水平的活动依赖于两个方面的努力：一是保证有更多的人使用公共空间；二是鼓励每一个人逗留更长的时间。如果有功能完善的广场，人们就会在那里聚集；如果没有广场，也就没有了城市生活，交叉路口周围便成了聚会的场所，在那里至少还有一点东西可看。

“天津城市公共开放空间系统的城市设计及管理通则是比较有特点的项目，我们团队在开放空间的价值标准和评价体系的建立上花了不少心血和时间。先是对开放空间领域的学术研究进行梳理，从众多的研究团队中选取三个最具代表性和影响力的研究团队，一个是丹麦皇家艺术学院扬 · 盖尔教授，一个是加州大学伯克利分校的 C · C · 马库斯，还有一个是 C · 弗朗西斯教授。我们对他们研究的公共空间项目（Project for Public Space）的成果进行分析比对，确立了以人性化维度作为出发点的研究思路，归纳出优秀开放空间的五个人性化原则，是按从基本要求到高标准要求的顺序逐步递进的。首先是可达性，这是对开放空间的最基础要求，保证人可以方便地到达；第二个原则是安全感，要让人在这个空间里感觉到安全；第三个原则是舒适性，开放空间内应设置完备的服务设施，并且提供良好的视觉景观；第四个原则是多样性，要提供多样化的活动，满足不同人群的使用需求；最后也是最具难度的原则是场所感，通过建筑、环境景观的设计来营造体现场所精神的开放空间。

“这个项目的研究体系借用了西方学术界的成果，但如果只是照搬国外的标准而没有考虑天津市的特点，是很难提出适合天津市的管控标准的。比如，咱这儿的活动都是跟节日紧密挂钩的，以前在什么时令干什么事、吃什么东西、搞什么活动，是有讲究的。我们项目组成员都在天津生活了很长时间，对天津很了解。我们对开放空间管控的普适性原则进行研究，也对天津市开放空间的形成和现状进行深入调研，不同历史时期和重要事件对天津开放空间的形式和内容都会产生影响，从而提炼出天津城市开放空间的特质。

“正如上午诸葛总所述，天津的城市空间结构经过了一个逐渐演变的过程。从最初沿河发展的直沽寨、十字中轴线的老城厢，到垂直海河发展的九国租界，再到 20 世纪 80 年代的‘三环十四射’，城市结构基本定型。在 21 世纪初，海河的综合开发改造，实现了海河成为城市主轴线和最具特色的开放空间的提升转变。未来，结合‘一环十一园’规划，城市外围形成以生态绿色空间为骨架的城市结构，与周围大的生态和乡村空间进行有机的衔接。我们本次公共开放空间系统城市设计及管理通则就是在这样一个大结构下，对公共空间进行进一步的梳理和设计，并制定指导详细城市设计的导则。

“天津有特色鲜明的滨水开放空间系统。以海河为代表的滨水开放空间，汇聚了最高级别的城市公共中心与核心服务职能，呈现最具时间轨迹的历史人文特色资源，最大限度地体现了天津城市的‘时代感、地方性、人情味’。天津有沽上美景的城市公园开放空间，过去地势低洼，汇水集中，历史上形成‘七十二沽花共水，一般风味小江南’的津沽美景。城市建设多利用河湖湿地建设以湿地水体为特色的城市公园，保护水系资源，涵养城市生态环境。天津有尺度宜人的历史街区、特色街巷，经过重点整治，塑造了主要迎宾道路，保护并传承了 14 片历史文化街区内的历史街道，建筑多姿多彩。天津有高品质的重点地区，设计品位较高、细节处理到位，建成了许多城市广场，作为市民活动的新场所，展示城市特色，聚集城市人气。

“项目组从五项人性化原则出发，针对天津中心城区滨水空间、街道空间、公园空间、广场空间等市民活动空间进行考察，发现还存在许多问题。就拿文化中心说吧，为了管理，文化中心里面不让骑车，从越秀路和乐园道交口去博物馆，要走 800 米。有一次在银河广场吃饭，大伙说去图书馆看看，从南门出来走了 1000 米才到。这几个馆的入口还不在同一侧，别提多不方便了。天津很多城市建设与公园绿地的联系太弱，开放空间和周边的公共建筑互动性差，这些年还建了不少大尺度的开放空间，本来目的是服务于市民的，结果人行、骑行都不方便，反而割裂了城市交通。

“本次试点坚持因地制宜，突出特色，强化滨水地区城市设计。自 2002 年起，开展海河两岸地区各个层次的城市设计工作，强化了海河特色风貌塑造。之后，相继开展了中心城区主要河流（海河、子牙河、新开河、南运河、北运河）两岸系统性城市设计。本次试点工作延续以河为脉的主题，编制子牙河两岸地区城市设计和中心城区开放空间系统城市设计及管理通则。

“《天津市中心城区公共开放空间城市设计及管理通则》以水系、公园、广场、街道等开放空间为研究对象，从总体层面提升开放空间系统规划与管理的可实施性。在对天津现有优秀开放空间案例进行研究的基础上，提出各类开放空间的布局、目标、分类及管控要求，形成以断面图为主、以局部平面图为辅的管控图则。

“项目组在调研、资料分析等过程中把滨水、街道、公园、广场这四类开放空间进一步细分，为每一个类型的空间制定控制标准。拿滨水空间举例。过去天津城市发展一直是围绕海河展开的，海河边遗留了各个时期天津人生活的痕迹。近年来海河边的各项规划也是风生水起、远近闻名，包括汇集城市公共中心的滨水空间，还有以自然景观为主的滨水空间。因此，天津的滨水地区是非常适合分主题规划、控制的。项目组深入调研调查分析后，将其按照服务职能和历史意义分为都市活力型、历史风貌型、自然生态型三类。

“项目组从人在不同环境中的空间感受入手，根据人的视觉特征，从河对岸能看清建筑全貌的基本垂直视角是 18°，能看清整个建筑高度的极限垂直视角为 45°，介于 18° ~ 45° 之间是理想的观景区域。其中，在人的实际活动体验中，27° 是最佳垂直视角。之前用固定的比例来确定滨河建筑的高度较为僵化，调整为控制区间后也给设计单位提供更多创造的可能性。

“历史风貌型滨水空间在都市活力型滨水空间的控制基础上，建议主体建筑高度与河流中心线的距离之比等于 1 ∶ 2，这主要是考虑到天津历史风貌型滨水空间在城市中心，每日都有很多游客、部分市民乘坐海河游轮观赏海河边的美景，因此将视点定为河流中心线，进一步提出控制建议。另外，对人尺度的活动空间加强控制和指引。对于都市活力型和历史风貌型滨水空间，应以硬质垂直堤岸为主，提供更多亲水空间，在地标建筑所在位置的对岸应设置观景平台。对于自然生态型滨水空间，则以软质护坡堤岸为主，人们主要在岸上活动、观景。

"天津的气候，冬天和夏天就不太适合到室外活动。我们家门口的地铁站，以前一到冬天夏天就总有人在那儿踢毽子，外面实在没有活动的地儿啊，后来就拿围栏给封上不让人去了。还有就是有的公共建筑外围连棵树都没有，夏天的时候晒死，冬天时又被风吹死。规划室外的活动空间，必须结合气候条件。天津夏季炎热、冬季寒冷、春秋季短、冬春季多风，所以设计目标不仅是提供具有宜人尺度的开放空间和精美的视觉景观，还要提供享受积极天气的机会，根据不同季节特点，设置完备的公共服务设施。另外，各类开放空间都存在不同程度的多样性缺乏、场所感缺失的问题。"

景观规划师谭鸣

"我觉得做天津开放空间的规划，就应该考虑天津人喜爱的活动，往往不需要刻意规划，只要提供活动的场地，人们就会自发地利用起来，再根据活动的类型补充相关的设施就好。天津人很会玩，周边有什么资源都能充分利用。海河边上不管冬天还是夏天都有钓鱼的、游泳的，街道上有遛鸟的、下棋的，对于这方面在天津待久的人可能反而熟视无睹。"谭所长十分关心城市公共开放室内设计，他汇报完"一环十一园"及周边地区城市设计后，从城市设计理论与实践分论坛赶到了历史名城保护与城市设计分论坛。

陈所长

"我同意谭所长的观点。公共空间的设计关键从以人的活动出发。对于新建的公共空间确实可以这样，由规划管理部门来引导市民利用开放空间，并根据实际的使用情况来进一步完善。但是对于已经有历史积淀的地区，就应该明确其所在位置，由规划部门主导，以控制为主，提出明确的管控方向。比如，考虑到现存的可达性问题，从可达性原则出发得出的设计目标就是开放空间与各级城市公共中心紧密结合、统筹布局，公交与慢行优先可达，保障特殊人群的可达性。之后，再根据设计目标提出管控的要素和标准。但对于要建设的项目，开放空间的管控就不能按照一套标准来。说实话，我看过不少这种总体规划层级的城市设计或者导则的成果，确实很难真正落实到城市建设和管理上，根据不同情况提出硬性和弹性的控制指标是这类项目的难点。

“我们这次的公共空间城市设计导则特别注意了对城市街道的研究。艾伦·雅各布斯的《伟大的街道》一书，对如何塑造伟大的街道提出了相应的设计原则和方法。什么样的街道是伟大的？边界清晰、尺度亲切、安全舒适、利于邻里交往；古老而又有神秘感，具有合适的街道尺度；拥有多样的商业，以及通透而又友好的建筑底层界面；具有富有变化而又连续的建筑界面；步行优先，适应多种人群的需要；是城市的标志及重要的公共空间，汇集大量的人流并具有丰富的街道生活；具有良好的街道绿化；具有自身的鲜明特色，并一直传承下去。

“街道与街区的肌理能够反映出城市之间的差别，包括尺度、复杂度、路线的选择以及空间属性的差别。而这些差异与城市修建的历史时期、地理环境、文化差别、城市的功能与定位、对设计与政治的哲学态度、技术要求相关。一些街道可以看作‘游戏规则制定者’，它们能将明晰的概念与秩序赋予一座城市或一个区域。它们可以形成一个边界，或者也可以形成一个有魅力的中心，它们能让你知道自己身在何处。另外，街道的肌理可以带来一种初始的秩序。街道与街区的肌理是个别街道设计的出发点。

“城市街道与街区的肌理实际是一座城市的空间秩序与结构。不论是经过精心设计，还是通过日积月累的演进，城市街道与街区的肌理都能给一座城市、城市中的一个区域带来秩序。在一个较小的二维尺度中，也会有许多杰出的街道为其周边区域建立起秩序与中心，这种秩序的建立通常有赖于对比的手法，比如那些比周边街道更长、更宽且更规矩的街道。

“它能告诉我们从一个地方到达另一个地方容易或困难，也能够告诉我们某个区域是否适于步行。每单位面积的区域中包含的内容越多，其中的街道似乎也就越多，供人活动的场所也越多。虽然街道所覆盖的区域不一定更优秀，但是我们所遇到的伟大街道多数是出现在内容相对丰富的区域中。

“街道与街区的肌理会随着时间的流逝发生变化。灾难（战争或火灾）、社会价值的改变、明显的经济规划的变化、时尚喜好与设计哲学，都是引起城市尺度发生变化的原因。塑造伟大街道的物质条件包括：易达性、汇集人流活动的能力、公共性、宜居性、安全性、舒适性、具有鼓励民众参与并承担社会职责的潜力。不可或缺的条件包括：散步的场所具有舒适的步行尺度；

与气候相关的舒适指标应该是伟大街道的一个组成部分；明确的边界；悦目的景观；街道的通透性；协调性；维护与管理。

“品质不取决于金钱。在伟大的街道中，一切问题的关键就是恰当和得体，不管是材料问题还是维护问题。而这也恰恰是公共空间领域建设中最值得提倡的标准。锦上添花的品质包括：树木、公共空间、环境设计要素、可达性与地形、居住密度与土地开发方式。更合理的城市发展模式势在必行：不能再过度依赖大规模的公共开发、高度集中的产权和专擅的规划设计模式，而将逐步积累的物质环境建设和对话的机制看作正当合理。这种发展态度建立在对城市生活的接受、热爱和期望的基础之上，鼓励人们生活在健康的公共环境中。

“街道是组成公共领域的主要因素，其他任何城市空间都望尘莫及。它们是公共财产的重要组成部分，社区生活，并受公共机构的直接管辖。有幸进行街道设计，通过这一过程满足公共需求并创造社区生活，不仅令人兴奋，也充满挑战。如果我们在一条街道上找到了正确的道路，我们将在整个城市的建设中走上正确的道路，这样，我们才能对得起城市里的居民，而这才是所有设计的根本宗旨。”

何所长

“对于我们刚刚讨论到的天津历史街区中具有历史记忆的空间，是不是应该制定更有针对性的控制标准，项目组在这方面有什么考虑呢？”

陈所长

“滨水、街道、公园、广场这四类开放空间的进一步细分中都有历史型空间的类型，我们明确了历史型空间的位置，并提出了较其他类型开放空间更为严格的控制标准，但是这个层面上不应对具体的空间一一做出规定。还是拿滨水空间来说，就算是历史风貌型滨水空间也不能用一套硬性标准来控制。天津当年的租界区是沿着海河划分的，每个国家的街道肌理、建筑风格、环境特点都不尽相同，作为总体规划层面的城市设计管理通则编制，应该指导城市设计细则中的滨水空间、街道、公园、广场等控制导则编制，并非针对每一种具体情况做出细致的规定。应对细则编制中应当考虑的问题和思路提出建议，这是城市设计法制化发展的方向。”

第九部分
一座更宜居的城市

51. 宜居性——新型社区城市设计导则

规划院城市设计所总规划师罗云

“城市设计的理论和实践”分论坛会场在“一环十一园”城市设计的讨论后，开始进入论坛的下半场，由规划院城市设计所罗总代表项目组汇报本次城市设计试点的重头戏：新型居住社区城市设计导则。这是一个庞大的项目组，包括了来自规划院城市设计所、愿景公司、滨海分院、建筑分院、交通研究中心、市政规划所、天津建筑设计院、华汇建筑设计公司、华汇环境设计公司 9 家单位的 30 多位规划师、城市设计师、建筑师、交通规划师、市政工程师等，由市规自局城市设计处牵头组织。丹 · 索罗门先生非常关心这个项目，也来到了会场。

罗总毕业于天津大学。她的硕士论文是《城市公共空间的规划设计研究》，毕业后在天津规划院工作了 10 多年，参加了包括滨海文化中心在内的许多大型城市设计项目。随着经验的增加，她越来越感到居住社区是城市中非常重要的内容。经济发展、社会进步的最终目的是满足人民对美好生活的向往，城市规划建设的最终目的是为居民提供良好的人居环境，其中住宅和社区是最重要和最直接的内容，但多年来我们一直缺乏清晰的思路和成熟的手法。随着年龄和生活阅历的增加，她发现周围同事会经常为住房、孩子上学而苦恼和焦虑。特别是现在她自己怀孕了，就更加感同身受。这个论坛分会场在座的有许多规划局的公务员，包括城市设计处、详细规划处、建管处的三位女处长：刘柯、赵颂、郭楠。作为公务员，她们会定期进行学习，目前正在开展“不忘初心、牢记使命”主题教育活动。党的十九大提出，要满足人民群众对美好生活的向往，新型居住社区是一个与人民群众生活密切相关的实际课题。这段时间大家多次对新型社区城市设计导则进行研讨，期待在论坛上发布后看到与会各方的积极反应。市规划有 30 多个处，女处长超过了 2/3，但在 13 名局领导中，女性却不到 1/3。天津规划局历史上还从来没有过女性一把手局长。或许在未来，女性会以她们更加注重沟通、关怀、包容、参与以及建立共识的特质越来越多地出现在重要的领导岗位上。

刘处长

“我是规划局城市设计处的处长刘柯，我来做一个开场白。居住是城市的主要功能，居住社区在城市用地中占据最大的比例；住宅是城市最基本的建筑类型，也是城市中最多的建筑。除了各级综合性城市中心，住宅和居住社区在很大程度上决定着城市基本的肌理、空间形态的品质和风貌特色。住宅组成了居住社区，居住社区构成了城市。

“目前，不论是天津还是国内其他城市，都开始推行城市设计，对于城市重点地区编制城市设计及城市设计导则，并依据导则进行精细化管理。但是，城市设计中的大部分居住用地的规划设计还是沿用具有计划经济特点的居住区规划方法。如果不改变传统的居住区规划方法，我们的所谓城市设计就不是真正的城市设计。

“我们本次城市设计试点，把新型居住社区作为重点，不仅有水西公园周边地区等数个新型社区城市设计，而且特别将新型居住社区城市设计导则的编制作为改革创新的重点。新型居住社区导则首先从生活品质、住宅与社区的多样性、生态环境、交通出行等多角度，提出面向 2035、2050 年的居住愿景。在天津市传统和既有不同类型居住社区、住宅建筑类型的基础上，进行总结和提炼，形成城乡断面、住宅开发强度与密度、住宅类型与组合方式、多样化居住社区等系统性的导则。最近国家发布实施了新修订的《城市居住区规划设计标准》，我们这次导则的编制是贯彻落实新标准，并同时结合天津的实际进一步细化，落实到各类住宅、社区，落实到城市不同的区域和分区，避免简单的一刀切。该导则与规划局最近出台的《天津市控制性详细规划技术规程（试行）》和《天津市建设工程规划管理技术标准（试行）》等新标准相结合，使得新型社区的城市设计导则更加实用。下面有请罗总首先做汇报，也欢迎各位专家发表意见。”

罗云

“感谢刘处。很荣幸能够代表项目首先发言。我先汇报一下我们研究的背景，特别是对住宅的作用以及住宅多样性的认知。这个问题看似容易，但其实并不简单。我们通过回顾历史，反思现代城市规划走过的道路，学习当代最新的城市设计和建筑理论，结合我国的实际状况，进行这一探索性的工作。

“大家都知道，在欧洲古代城市中，是住宅建筑围合成了完整的街道和广场，居住建筑统一而富有变化的肌理不仅形成了连续丰富的城市空间，而且烘托出广场中教堂等公共建筑的辉煌。我国古代的城市，如唐长安城，其完善的里坊制度和整齐的居住建筑是城市重要的组成部分。过去的电视剧大多描写的是宫廷，最近在网上热播的《长安十二时辰》是第一个描写长安街市的电视剧，比较真实地反映了长安城的面貌和当时人们的生活。在明清北京城的规划中，四合院是基本的居住建筑类型，绿树掩映中的四合院与紫禁城中轴线交相辉映，共同完美地构成了人类城市历史上‘无与伦比的杰作’。

“住宅在古代的城市中占有重要地位，而现代城市规划产生的主要原因正是因为住宅和居住社区的问题。欧洲工业革命后，人口快速向城市聚集，导致住房极度短缺。阴暗狭小的住宅人满为患，缺乏基本的采光、通风条件，缺乏干净的供水和污水排放等设施，垃圾遍地、污水横流，居住区恶劣的卫生条件所造成的城市问题引发了现代城市规划的产生。19世纪末20世纪初，霍华德创立了‘田园城市理论’，作为现代城市规划的奠基石，他试图通过规划建设城乡一体的花园住宅，解决工业革命带来的城市病。但他提出的居民自治模式无法满足大量的住房需求。

“19世纪末现代建筑技术的发展，为采用大工业化的手段解决城市住房短缺问题、改善居住环境提供了可能。但以巴黎美术学院为代表的古典建筑设计传统限制了新建筑的发展。柯布西耶以超越常人的眼光，赞颂汽车、轮船等现代工业设计，提出‘住宅是居住的机器’的理念和光明城市的规划设想。在他的倡议下，现代住宅以崭新的面貌展现在世人面前，现代建筑在技术上的进步极大地提升了居住建筑的建造水平和采光、通风等物理环境。同时，以功能分区、人车分流为特征的居住邻里等规划方法，彻底改变了自古以来一直延续的古典主义建筑和居住社区的规划设计传统。就像柯布西耶在《走向新建筑》一书中发出的呐喊：不搞（现代）建筑，就革命。

“柯布西耶的建筑思想是先进的，他赞美古典建筑的优美造型和黄金比例，但没有用钢筋混凝土去模仿古典建筑的细部，而是用现代建筑材料和技术的简洁、工业化的设计手法，创造出前所未有的、具有同样黄金比和艺术感的现代建筑，使建筑从古典主义的教条中解放出来，激发了强大的创造力，指引了现代主义建筑的发展。但他关于在巴黎市中心拆除历史街区、规划建设大片绿地及高层建筑的‘光辉城市’设想，的确欠缺对城市历史文脉的考虑。但在当时万马齐喑的情况下，他试图通过这样的举动引起世人的关注，有其时代背景，也是可以理解的。他将工业化思想大胆地带入城市规划，主张用全新的规划思想改造城市，设想在城市里建造高层建筑、现代交通网和大片绿地，为人类创造充满阳光的现代生活环境。

“柯布西耶关心城市规划和大众的住房问题，1928年他牵头成立了国际现代建筑协会的国际组织（Congrès International d'Architecture

Modern，CIAM），亲自提笔起草了著名的《雅典宪章》，指明了现代城市以人民为中心的居住、工作、游戏、交通四大功能，以及现代城市规划的原则。我们今天看来，许多还是非常正确的。今年初，天津美术馆和柯布西耶基金会合作主办了大展——柯布西耶·色彩交响乐。展览按时序介绍了他的建筑作品的几大分期：纯粹主义时期、粗野主义时期、抒情与光明时期，配以图纸和模型，包括威森霍夫双宅、萨伏伊别墅、马赛公寓、朗香教堂等，还有他的绘画和主编的杂志书籍原稿。天津市规划学会组织了集体观展和研讨会，重温其经典思想，虽然是百年前的声音，但今日我们仍然被深深地打动。我们已经清醒地认识到，虽然现代主义建筑在提高住宅建筑质量和居住水平方面发挥了巨大的作用，但也导致了严重的城市问题。造成这些问题的原因是多方面的，不应该单单由柯布西耶承担。

“最早实践柯布西耶思想的是苏联。1917 年十月革命后，苏联实施了国家主导的现代化和工业化。‘一五’社会主义改造和工业化计划取得了巨大的成功，引起了美国等西方国家的羡慕，包括柯布西耶这样的建筑大师，也投身于苏维埃的规划建设。1928 年，柯布西耶受邀参加了为莫斯科的消费者合作社中央联盟设计一座新总部的设计竞赛，最终他的方案胜出。但后来，由于参加莫斯科总体规划竞赛未如所愿而离去。苏联建筑师继承了现代主义‘为大众设计住宅’的传统，提出多种居住单元类型组合，通过科学计算研究其空间使用效率、经济性和功能性等特性，用自己的方式诠释‘住宅是居住的机器’这一现代建筑理念，并在莫斯科建成了公共集合公寓。

“苏联卫国战争结束后，战后重建任务繁重，包括工业生产和城市生活的恢复，急需建设大量住宅。为提高建设和配套效率，新建单元住宅从传统的 4 ~ 6 层的多层为主转为 12 ~ 14 层的塔式和板式高层。这一时期，由于过于强调建筑形式，阻碍了新技术和工业化施工方法的推广。大量的住宅需求，必须按工业化方法施工，对建筑设计和居住区规划提出了新的要求。1954 年，苏联批判了城市规划和建筑设计中的传统主义、复古主义、形式主义和装饰主义，强调在城市建设中重视功能、社会效益和使用新技术。1955 年以后，苏联再次进入大规模建设时期，经历了 1955—1960 年建筑工业化阶段、20 世纪 60 年代粗放型发展阶段和 20 世纪 70 年代以后的集约化发展阶段。由于住宅建设都由国家投资，因此制定统一的标准是重要的手段，标准定额和成片的居住小区规划建设是常用的方法。实际上，早在 20 世纪二三十年代，苏联的某些居住区详细规划中就已经产生了居住小区的思想，随着国外邻里单元理论的推广，苏联居住小区理论逐步成型。住宅建筑设计注重功能和新技术使用，推行标准设计和建筑构件定型化，对满足住宅建设速度的需求起了很大的作用，但同时也造成了建筑单体和居住区面貌的刻板和单调。如果归因，不知这种大规模单调的住宅与苏联的最后解体有无关系。

“在美国，二战后大规模的城市更新运动和保障性住房建设，也采用了所谓柯布西耶式的现代主义城市规划方法，建造了大量大规模保障房居住区，但不久就遭遇严重挫折。著名的案例是由建筑师雅马萨奇（Yamasaki）设计的普罗蒂－艾戈低收入者公寓居住区（Pruitt－Igoe Housing Project）。这是一个在 20 世纪 50 年代中期实施的贫民窟改造项目，规划建设了 30 多栋 11 层的高层建筑，为 8000 户低收入家庭提供可负担住宅。后期由于出租率低、缺乏维护、社会治安问题严重，这片住宅区在 1972 年 3 月 16 日下午 3 点被爆炸拆除。寿命只有短短的 16 年。建筑历史学家查尔斯 · 詹克斯（Charles Jencks）将这一时刻描述为‘现代建筑死亡（modern architecture died）’。在炸毁的那一刻，美国人对现代主义建筑的信念开始瓦解。后来的 20 年里，这些大规模保障房多数被拆掉了。

“在欧洲，二战后进行的城市更新和新城、居住区建设也普遍采取了柯布西耶式的高层居住建筑和大规模建设的规划手法，拆除大量历史建筑而建设环路，建设大规模的居住区，导致对历史城区的冲击和城市空间的丢失以及社会问题的产生。20 世纪 90 年代末，巴黎发生的城市暴乱与这种居住模式不无关系。从 20 世纪 70 年代开始，发达国家已经不再进行集中的大规模居住社区建设。现代主义城市规划和大规模居住社区规划的实践在西方只有短短 50 年的寿命，但它带给整个人类社会的影响和变革是巨大的。

“我国近现代住宅和居住社区规划设计的发展，从 20 世纪初开始，受到霍华德田园城市和柯布西耶现代主义城市规划理论的影响，田园城市、居住邻里等规划方法是我国近现代规划师、建筑师所推崇的模式。中华人民共和国成立后，我国实行社会主义公有制和福利住房制度，照抄苏联计划经济时代的住房建设和居住区规划理论体系及方法。改革开放后，虽然西方的各种当代规划理论流派，包括对现代主义批判的观点涌入，但我们没有真正认识到现代主义城市规划的危害，一次次重蹈覆辙。

“改革开放 40 年来，通过住房和土地制度改革，发展房地产市场，我国基本解决了城镇人口的住房短缺问题，取得卓越的成绩。在这一过程中，住宅建筑设计和社区规划一直延续了苏联住宅建筑定额指标、标准图和居住区规划方法，延续了柯布西耶的‘集合住宅’‘阳光城市’等功能主义的现代城市规划理论，没有能够取得突破。伴随着改革的深入，房地产业蓬勃发展，我国社会主义市场经济体制初步建立，但我们依旧一直使用计划经济特色鲜明的居住区规划理论方法。虽然局部内容有所改进，但核心理念并没有改变。在过去的20年间，中国的住房建设以惊人的速度成长，在市场的推动和现行居住区规划、控制性详细规划、城市规划技术标准的控制下，住宅建筑普遍采用了简单划一、可以快速复制的建筑类型，多为30层、100米高的塔式或板式高层，不断重复的数栋甚至几十栋高层住宅堆积在一起，围墙一围，形成了目前典型的大型封闭居住区，留给城市的只有围墙。重复单调的住宅建筑、漠视行人与自行车交通的超大封闭小区、极宽的缺少活力的街道，成为中国住区建设甚至城市建设的主流都市形态，是交通拥挤、环境污染、物业和社会管理等城市问题形成的根源之一，造成全国各地城市形态的千篇一律、特色缺失和整个城市环境品质的下降。

“10多年前，在规划界大家逐步认识到这个问题的严重性，开始了有益的尝试。滨海新区一直努力尝试住宅和居住社区规划设计的改革创新。2000年，开发区生活区就制定了窄路密网的城市设计方案，并组织实施。2005年，滨海新区城市设计国际征集的综合方案也突出了‘窄马路、密路网、小街廓’的布局模式。2006年，滨海新区被纳入国家发展战略后，在几个功能区的生活区以及中心商务区、北塘等地区推广窄路密网布局。2007年，中国和新加坡两国政府合作建设中新天津生态城，在生态城的规划中考虑了生态城市指标体系、生态细胞规划模式和社区社会管理改革等创新内容。2009年编制的滨海新区核心区总体城市设计和南部新城规划，采用新都市主义理论和方法，营造以公交为导向的、以窄路密网为主要特色的布局形式。

“我们认识到，要提升滨海新区的城市品质，突出城市空间特色，就必须充分考虑住宅建筑与城市的关系，探索一种适宜的住宅类型和新型居住社区规划设计模式。2011年，在天津市滨海新区规划和国土资源局的组织下，天津规划院滨海分院与丹 · 索罗门先生及天津华汇环境设计公司一起合作开展了位于南部新城北起步区的1平方千米和谐社区的规划研究。这是一次非常有益的尝试，我们跟踪和见证了这一过程。当时的工作目标非常明确，就是要改变传统的居住区规划设计方法，对‘约定俗成”的内容和规范进行深入讨论与反思，尝试构建一种新型居住社区规划原型，为滨海新区推广窄路密网提供一种切实可行的模式。”

“随着工作的深入，在 2012 到 2013 年，又由这片土地的业主天津港散货物流中心委托丹 · 索罗门事务所和天津华汇环境设计公司就和谐社区内的一个居委会规模的邻里进行了深化规划和住宅建筑概念设计，形成了比较详细的规划设计成果和模型。工作过程中也进行了广泛的调研，对国内外案例进行分析，召开了两次专家研讨会。我国住宅设计和居住区规划方面的专家赵冠谦、开彦、张菲菲、周燕珉等对丹 · 索罗门主笔的和谐社区规划设计方案给予充分肯定和高度评价。我们这些城市设计师也在这次与丹 · 索罗门先生的合作设计中学习到了许多有用的东西。今天丹 · 索罗门先生也参加了我们的会议，我提议大家对他表示感谢。”会场上响起了热烈的掌声，丹 · 索罗门先生起身，双手合十微笑着向大家致意。

罗云

“通过前期工作，我们越来越清楚地认识到，丹 · 索罗门先生示范的和谐新城新型社区规划模式与现行的居住区规划设计规范、相关的技术标准和城市规划管理技术规定存在着巨大的冲突，要使新的社区规划模式具有可行性，必须进行相应的全面改革。为了明确改革具体的内容，2015 年，和谐社区作为滨海新区新型社区规划设计试点项目，由天津市滨海新区规划和国土资源局住房保障中心下属的住房投资公司出资，委托天津规划院和渤海规划院进行了控制性详细规划、修建性详细规划和建筑设计工作，目的是通过实际规划的具体落位来检验与现行相关政策、标准的冲突点，明确居住区规划设计规范、相关技术标准和规定需要改革创新的具体内容。

“恰逢此时，平地起春雷。2015 年底，中央城市工作会议时隔 37 年举行。随后不久的 2016 年初，中共中央、国务院印发了《关于进一步加强城市规划建设管理工作的若干意见》，勾画了‘十三五’乃至更长时间中国城市发展的蓝图，点明了目前我国居住区规划设计建设中存在问题的严重性和原因所在，指明了未来发展的方向。文件特别提出开放社区对我国城市发展和解决城市问题的重要性。要优化街区路网结构，加强街区的规划和建设，分梯级明确新建街区面积，推动发展开放便捷、尺度适宜、配套完善、邻里和谐的生活街区。新建住宅要推广街区制，原则上不再建设封闭住宅小区。树立‘窄马路、密路网’的城市道路布局理念。

“至此，关于新型社区规划设计模式的问题获得了最高层面的重视。在我看来，改革开放以来，我国城市化取得了巨大成绩，但也存在许多问题。

原因是复杂的，聚焦在城市规划方面主要有三方面原因。第一是以资源消耗和扩张为导向的城市总体规划；第二是作为满足房地产开发和土地财政工具的控规；第三是苏联模式的居住小区和千人指标、日照间距等规范。这三个层面的规划是造就‘千城一面’的元凶。全国随处可见的大型封闭居住小区、高层住宅‘玉米地’式的蔓延，也是三种长期不变的城市规划管理工具综合效应的外化形式。同时，还伴随着住宅价格高企、高库存，城乡发展不平衡、城乡割裂，城市化既得利益阶层的固化等诸多社会矛盾。

“中央城市工作会议和中共中央的文件吹响了城市规划改革的号角。要用以生态保护为导向的国土空间规划取代传统扩张性城市总体规划，以以人为本的城市设计完善控制性详细规划，以新型社区设计代替传统的计划经济色彩浓厚的居住区规划。其中新型社区设计是基础。对于中国城市而言，解决好住宅和社区规划问题，不仅关系到老百姓的切身利益，关系到城市品质的提升，更关系到房地产的健康发展和我国社会经济的可持续发展，是落实国土空间规划、全面建成小康社会、实现中华民族伟大复兴中国梦的重大课题之一。

“宜居的住宅是健康城市的基础，日常生活中每天感受到的点滴幸福就是城市整体幸福的源泉。当大量现代化居住小区满足了人们日照通风等基本生理需要的时候，人们却开始怀念起看似拥挤嘈杂却又热闹亲切的传统生活，无论是北方的胡同、四合院还是南方的雨厅、小巷，都曾给居住者带来愉悦的生活体验。我们相信在一个理想的社区中，物质空间基本功能的满足只是第一步，情感的需求更需要设计者的精心考量，通过减小街廓尺度、建立精明交通、住宅围合布局、户型精心设计、改革和创新社会管理等措施，营造富有魅力的庭院生活、街道生活、城市生活，让居民在此真正享受到生活的安心与快乐。

“滨海新区和谐社区试验坚定了我们的信心。通过合理的制度设计，通过改革传统的居住区规划设计方法，我们可以创造出广大中等收入群体能够负担得起的小康型住宅和更高品质的人居环境。我们也意识到，由于我国居住区规划设计现行做法延续了60多年，形成了一个涉及方方面面的庞大体系，要进行改革创新难度很大。但是，若要落实中央城市工作会议的要求，改善我们的居住建筑和社区规划，就必须进行系统性改革。

“空想社会主义创始人之一欧文在200年前，在苏格兰新拉纳克（The Town of New Lanark）自己的工厂尝试模范社区和现代人事管理改革，取得了成功。他相信环境可以改变人。1824 年，他在美国印地安纳州购买了一大片土地，建设了‘新和谐村（New Harmony）’，实行人人劳动、按需分配制度。为了形成良好的居住环境，欧文专门设计了社区的原型‘方形院落（Parallelogram）’。虽然几年之后‘公社’瓦解了，但欧文对人人平等和理想集体居住模式探索的贡献在人类历史上是开创性的，具有划时代的意义。他比霍华德提出田园城市理论早了几十年，有城市规划历史学家称欧文为现代城市规划之父。虽然欧文自身的试验没有成功，但他为追求人类理想居住工作环境的探索精神指引和激励着后人。我们认识到，苏联具有社会主义理想的居住区规划与欧文的思想一脉相承。事实证明，居住区规划是失败的，但这并不会阻止我们继续前行的脚步。当前，我国的城市再次面临巨大的变革时代，中央明确提出改变传统规划方法，推广街区制，这需要对我国现行居住区规划设计理论体系和技术规范体系进行彻底变革。我们需要欧文的精神指引，进行一场对和谐社区的新探索。滨海新区新型社区研究的街坊命名为和谐社区，也是表明向欧文学习的态度和决心。

“虽然由于各种原因，滨海和谐社区目前还没有实施，但通过这个项目研究，我们对新型居住社区的城市设计有了非常深入的认知，认识到如果要实现中央提出的窄路密网的新型社区布局，就必须改革创新。不仅需要在城市设计中采用新型居住社区的设计方法，还要改革居住区规划设计规范和相应的城市管理技术规定。在过去的几年中，天津一直在努力并取得了一些成绩。比如，规划局最近出台的《天津市控制性详细规划技术规程（试行）》中，为了形成窄路密网布局和人性化的街道，采取了一系列改革措施，如取消了生活性道路的绿线，将绿地集中布置为公园，减小了道路转弯半径的红线尺寸，提高了居住建筑密度等。在《天津市建设工程规划管理技术标准（试行）》中，减少了建筑退线距离，停车库面积不计入容积率等。

“为了把这些试点项目的经验和渐进的改革汇集成系统的理论和方法，我们开始编制《新型居住社区城市设计导则》。导则的编制以改革创新为指导思想，允许小的失误。从这个意义上讲，我今天汇报的导则严格说是一个阶段性成果，今后要在实践中不断完善。在天津市传统和既有不同类型居住社区、住宅建筑类型的基础上，进行总结和提炼。首先从生活品质、社区的多样性、生态环境和人性化交通出行等多角度，提出面向 2035、2050 年的居住愿景，形成城乡断面、住宅开发强度与密度、住宅类型与组合方式、多样化住宅与居住社区等系统性导则，以指导具体的城市设计。

“我们认识到，和谐社区试点中提出的布局模式只是新型社区中的一种模式，它更多地表达的是一种思想和理念，我们不能用一种模式来规划设计所有的社区，否则我们就又回到了现代城市规划的老路上去了。我们需要住宅建筑的多样性，需要真正尊重人、看到人。人是多样的，家庭是多样的。若想做好新型社区城市设计导则，就要使其具有多样性，要有对住宅的新认识，要重新审视住宅作为家的意义。”

52. 住宅——家和心灵的港湾

罗云

“今年是中华人民共和国成立 70 周年、改革开放 40 余年。几十年来，房子几乎成为老百姓嘴里、心里最挂念的事儿。我们每个家庭的变迁，就是中国城市的缩影。大家想一想，这房子对每个人究竟意味着什么？衣、食、住、行的住，是人生四大要素之一。古人云：宅，所托也。人以宅为家，居若安，则家代昌吉。中国传统文化中，住宅以家庭为核心，与当时的生活方式、文化传统、精神追求达成了完美的结合。住宅，不仅是物质生活的载体，还是精神审美的载体，更是自我心理疗愈的空间。从四合院到苏州园林，丰富的民居形式，无不表达了‘修身、齐家、治国、平天下’的人生追求，以及‘天人合一’精神境界。”

“同样地，在西方，无论是基督教还是犹太教文化，家庭也享有尊贵的地位。住宅是基本的地理分界，相应地，也是自我意识的心理分界。在美国，独立住宅更被认为是生命存在的象征。英美法律和习俗将每个人的家庭视为自己的避难所，破坏住房就相当于破坏个体的存在。

“肯尼斯·杰克逊在《马唐草边疆》中写道：纵观历史，一个民族对特定住所的处理和安排，与其他富有创意的艺术品相比，彰显了更多的意义。住房是人们内心的外在表现形式；任何一个社会，如果不考虑其成员的住所，就无法充分地理解他们。哪里都不能与家相比！房屋的建造是任何一种文明的重要成果。

“每个人在心底都渴望拥有一套属于自己的房子。安德鲁·唐宁写道：我们相信，在世俗人间，个人住房体现的是无比的力量和高尚的美德。他认为，核心家庭是最好的社会形式，对一个民族而言，一座独立的住房有巨大的社会价值。把家安在乡间，这样在家庭这个小世界中，信任、美和秩序便会占据主导。商业化的土地规划可以通过人为创造环境的方式满足人们对宁静生活的向往、对自然的亲近和对城市便捷生活的需要，这样才能吸引居民。

“东、西方视角下，住宅的意义竟如此相似。住宅本质上是一种人们自我投射的场所，这是人内心的外在表现形式。有人说，住宅是个人的宇宙，住宅是个人的城市。住宅成为生命最重要的生活元素，表达的是人对自然、社会以及生命本身的根本立场。

“回看我国改革开放 40 余年，像发生在千家万户中的故事一样，伴随着住房制度改革和房地产发展，城市居民的住房条件发生了巨变。我们从人均住房建筑面积仅 7 平方米，提升到现在的人均 39 平方米，位居世界较高水平。然而，当西方用“一栋房子，一辆汽车，一份体面的工作（a house，a car，a decent work）”的美国梦来宣示生命存在的时候，千篇一律的‘单元房’似乎让我们与自己的内心、与社会生活、与自然的关系越来越远。我们似乎总感觉缺了点什么，我想可能就是缺了中国人住宅的‘魂’吧。

“几十年来，我国每年商品住宅开工建设量约 30 亿平方米，竣工量和销售量约十几亿平方米。有人笑称，中国人把全世界的房子都盖完了。在土地财政下的房地产成为推动中国 GDP 发展的最成功的城市游戏规则。但我们建的房子能满足人们未来的需求吗？房子对城市究竟意味着什么？

“房子是用来住的，不是用来‘炒’的。这句话有深刻的含义。人对住宅的需求包括物质和精神双重内涵。我国传统的民居，不仅满足了当时家庭生活起居的需要，也满足了许多家庭的社会活动需要，还有人的情感和精神需要。中华人民共和国成立后，我们学习苏联的福利住房制度，住房的设计满足了基本的生活需求，采用了标准化设计，对人的精神生活考虑不多，也缺乏对住宅历史文化传承的考虑。改革开放后，特别是住房制度改革后，商品房出现了很多类型，房型设计上也出现了客厅、独立餐厅、书房等满足更高需求的功能空间，厨房、卫生间、储藏室等功能空间随着技术的进步不断完善。住房装修水平也不断提高，风格多种多样。但总体看，我国当代的住宅还是功能主导的，这是现代生活快节奏带来的影响，对精神需求鲜有考虑。未来的住宅，除了目前有的起居室、卧室、餐厅、厨房、卫生间、储藏室等功能区外，需要考虑家族祭祀的厅堂、名堂、读书冥想的书房、健身活动和工作的空间等，并精心设计，表达空间的个性化和形式上的文化含义。家庭成员既有安适自由，又有支持陪伴，还有精神和情感表达。家是人心中的宇宙。

“到 2035、2050 年我们居住什么样的住房？这是一个必须回答的、非常关键的问题。我国城镇居民现状人均居住建筑面积已经达到 39 平方米，2018 年新建商品房销售量 17 亿平方米。综合考虑房地产市场平稳过渡、城镇居民数量增长等因素，2035 年我国城镇居民人均住房建筑面积应该在 50 平方米左右，达到发达国家水平。因此，为了适应未来居民对住房的需求，新的住宅建设总体上应该按照比较高的标准，包括房型和形态。”

罗云

“除了住宅的房型外，住宅的形态也非常重要，家是要有一个形态的。它不只是住宅的造型问题，而且涉及人的精神需求、家庭的凝聚力、社区场所感的共同营造和城市文化等深层次的问题。

“住宅建筑不是孤立的建筑，居住社区也不是孤立的居住区。除了体现城市形态和特色外，城市住宅和居住社区还有更深层次的城市社会、文化和精神意义。阿尔多 · 罗西的《城市建筑学》是有关建筑类型学最重要的著作。罗西的类型学将城市及建筑分成实体和意象两个层面。实体的城市与建筑在时空中真实存在，因此是历史的，它由具体的房屋、事件构成，所以又是功能性的。罗西认为实体的城市是短暂的、变化的、偶然的。它依赖于意象城市，即类似性城市（Analogical City）。意象城市由场所感、街区、类型构成，是一种心理存在和集体记忆的所在地，因而是形式的，它超越时间，具有普遍性和持久性。罗西将这种意象城市描述成一个活体，能生长，有记忆，甚至会出现病理症候。为维持一座城市的形态，在城市更新和构造时，对新建筑类型的引进和选择，尤其是对大量性住宅类型的选择要格外慎重。因为引进一种完全异化和异域的住宅类型会导致整体城市形态、面貌的巨大改变。

“历史上，巴黎改造时，豪斯曼选择了新古典主义的三段式住宅原型，使得巴黎城市的街道、广场建筑整齐有序，塑造了完美的城市空间环境和形象。在西班牙巴塞罗那新区的规划中，伊尔德方索 · 塞尔达（Ildefonso Cerda）规定了八角形围合住宅的形式，使巴塞罗那新区特色鲜明。这些案例都说明，住宅对一个城市的功能、品质和特色起着非常重要的、决定性的作用。我们假设，如果巴黎真的按照柯布西耶的‘光明城市’设想在市中心建设了大量现代风格的高层住宅，那么对巴黎城市风貌和历史文化的破坏会是致命的。”

规划局建设管理处郭楠处长

“一个城市要特别慎重选择新的居住建筑类型。吴良镛先生对北京菊儿胡同探索的重要目的就是试图寻找既具有现代居住建筑功能，又延续中国传统的居住建筑类型，与北京城市整体格局相协调，保护古都风貌，这也正是其能够获得联合国人居奖的原因所在。可惜没有能够在北京广泛推广。丹 · 索罗门先生在《全球化城市的忧虑》（*Global City Blues*）一书的开篇，对吴良镛先生坚持对地域建筑的探索给予充分的赞赏，对建筑潮流大师库哈斯进行了批评。他认为建筑要具有地域性，要保持城市的历史文脉和特色，建筑设计必须具有地方特点。历史的经验和现代主义的教训再次证明，住宅建筑及社区规划对规划建设良好城市的重要性。因此，近半个世纪以来，西方发达国家开启了新型居住社区规划。在老城中，更加注重与历史环境的融合，采用镶嵌式的填充规划设计方法；在新城规划中，强调城市空间的塑造和文脉的延续。在美国，20 世纪 90 年代新都市主义运动兴起并蓬勃发展，强调以欧洲城镇的模式重塑美国的郊区。丹 · 索罗门先生近几年数次到中国，他曾问为什么菊儿胡同模式没有在北京实现，而是建设了大量毫无特点的高层住宅，让我们一时语塞。

“要实现中华民族伟大复兴的中国梦，要通过实践寻找到既传承中国文化传统，又满足现代人物质和精神需求的多样性住宅模式和多样性社区模式。另外，我们还必须面对一个挑战，就是我们的城市中已经有大量 1999 年住房制度改革前建设的老旧社区和住宅，户均面积小，房型不满足未来发展的需求，外观简陋，这些住房需要彻底的更新改造，要完成这个任务比设计新的住宅模式更难。”

53. 城乡断面——住宅和社区的多样性

罗云

“早在 20 世纪初，苏格兰生物学家、人文主义规划大师、西方区域研究和区域规划的创始人帕特里克 · 盖迪斯（Patrick Geddes）就从生态和人的互动中发现了区域生态和人的活动的规律性，即从城市到乡村，处在不同位置上的人从事不同的工作，过着不同的生活。美国新都市主义发起人安德鲁斯 · 杜安伊(Andres Duany)和伊丽莎白 · 普拉特 - 兹伊贝克（Elizabeth Plater-Zyberk）在题为《建设从乡村到城市的社区》的文章中，制定了城市乡村横断面，描述了从市中心到乡村不同密度、高度和形态类型的居住建筑形式，揭示了住宅建筑类型的生态区位分布规律。随着人类城镇化的进步，住宅类型也在不断演进，最终形成三种比较稳定的基本类型，适应城镇特定位置的特定需求。从城市乡村横断面和平面模型中可以看到三大类、七小类住宅的分布规律：城市外围和郊区以及乡村是独立低层住宅，城市外围和城市中心之间的中部是低层或多层联排住宅，城市中心及核心是多层或高层的单元集合住宅，这是一个普遍性的规律。这个规律说明了不同类型住宅的分布规律，也从一个方面说明了住宅建筑的多样性本质。

“住宅建筑的类型和多样性，广义上是指不同国家、地区、民族各具特色的住宅建筑类型，这些特色鲜明的住宅建筑是世界文化多样性的组成部分。经过对 20 世纪国际主义风格建筑的反思，国际上对保持和突出各自国家和地区的建筑特色已形成共识。狭义的居住建筑类型和多样性，是指一座城市内部不同的布局、高度和造型等住宅种类的多样性，如低层独立式住宅、联排式住宅、多层单元式住宅、高层单元式住宅、超高层住宅等，是城市形态和空间特色的重要组成部分，是城市文化的载体。

“世界上发达国家的住宅建筑都呈现出多样性，有鲜明的特色。经过多年的实践，美国逐步形成了与区位、建筑形制、产权管理等相对应的独立花园住宅（Patio House，Single Family House）、联排住宅（Townhouse）、出售公寓（Condominium）和出租公寓（Apartment）三种基本住宅类型。三种类型的住宅形成了各自相对成熟完善的建造技术体系。根据场地环境，以及住户对功能和外观的不同需求等可以有无穷无尽的变化，形成了美国丰富的住宅类型和建筑风格的多样性。同时，美国三种基本的住宅类型对应着三种完全不同的居住社区规划设计模式，形成具有明显特色、有不同功能和特征的、多种多样的城市街区和环境。多种类型、价位和不同区位的住宅为美国人提供了多样选择的可能性。不同家庭根据各自生活、工作等特殊需求，租住或购买适合自己的住房，与人的身份和生活方式高度契合。正如赖特所说：建筑就是美国人民的生活。

“美国是移民国家，它的住宅原型来源于欧洲。欧洲国家众多，不同的文化环境形成了各具特色的住宅类型和风格各异的住宅建筑。比如，英国住宅类型有度假别墅 / 农舍式（Cottage）、平房别墅式（Bungalow）、联排式（Townhouse）、独立式（Detached House）等，其中双拼独立式（Semi-detached House）是英国的特色。政府公共住房有市议会住房（Council

House）等。建筑风格上有典型的都铎风格、维多利亚的安妮女王风格和乔治亚风格等，以及万千的变化。与美国不同，欧洲的城镇历史更加悠久，特色更加突出，而且建筑相对更加聚集，城镇中以多层住宅为主。但随着进入后城市化阶段，城市人口增长持续低迷。据有关统计数据，近年来，欧洲各国新建的住宅中绝大部分是独立住宅。

“即使在国土面积狭小、人口高度密集的日本，住宅建筑也具有丰富的多样性。基本类型有‘一户建’独户住宅、低层高密度集合住宅和多、高层集合住宅三种类型，在这三种基本类型基础上演变出各种类型和丰富多样的住宅。日本著名建筑师隈研吾在其著作《十宅论》中，将日本住宅分为单身公寓派、清里食宿公寓派等10种类型，生动描绘出这10类住宅的特点、居者的精神追求和生活状态。隈研吾认为住宅从来都不仅仅有居住功能，而是人们自我投射的场所和对生存状态的梦想。从这个意义上讲，住宅是日本人的生活，不同的人需要不同类型的住宅。10种住宅，也是10类人的生活方式。

“我国历史上有丰富的住宅类型和多样性的住宅形式。中华文明经过数千年漫长的演进，居住建筑从具有不怕风、雪、雨、火和防止野兽、毒虫等危害侵扰的简单功能，发展到完善的居住功能，且包含丰富的文化内涵。全国各地形成了各具特色的民居建筑，争奇斗艳，丰富多彩，是各地城市特色和城乡文化的重要组成部分，也是中华文明的重要内容。

“我国地域广阔，民族众多，形成了各具特色的民居建筑。实际上，中华人民共和国成立前对我国古代住宅建筑开展研究的是中国营造学社的同仁们。中华人民共和国成立后，他们将多年来的民居建筑调查进行总结和系统的研究。1957年，刘敦桢用多年研究的成果编辑出版《中国住宅概说》一书，简要而系统地叙述了我国古代住宅建筑的发展演进和明清时期汉族住宅建筑的主要类型。从建筑布局上，将明清汉族民居建筑大致划分为9种。该书引证了大量的古代文献和对既有考古资料和建筑实物的实地考察记录，是第一部对中国民间住宅建筑进行系统梳理的著作，被认为是中国民居研究的滥觞。该书提高了民居建筑的地位，引起了建筑界的重视。我国在20世纪60年代对民居的调查研究出现了一个高潮，成果丰富多彩，包括北京的四合院、黄土高原的窑洞、江浙的水乡民居、客家的围楼、南方的沿海民居、四川的山地民居等，还有少数民族地区的云贵山区民居、青藏高原民居、新疆旱热地带民居和内蒙古草原民居等。

“由于‘破四旧、立四新’和‘文化大革命’等运动，传统居住建筑的调查研究受阻。直到改革开放后，对传统居住建筑的研究再次复苏。1984 年，原营造学社骨干、著名建筑历史学家刘致平和傅熹年发表文章《中国古代住宅建筑发展概论》，对中国各个朝代、各民族的居住建筑进行了更为系统的研究。按照不同地区特征和建筑构造、民族风俗等因素，将我国古代住宅划分为 7 类，即北方黄土高原的穴居、南方炎热湿润地带的干阑、西南及藏族高原的碉房、庭院式宅第、新疆维吾尔族的阿依旺住宅、北方草原的毡房、舟居及其他。每类住宅因民族风俗不同而有各民族自己的特点，使得同一类宅第的形制也呈现出多样性变化。”

罗云

“有专家指出，虽然我国住宅种类丰富繁多，但与5000年的文明史相比，我国古代住宅形制总体看发展演变比较缓慢。因此，当近现代西方住宅建造技术进入时，无法应对。近代在我国的开埠城市和一些新兴城市，出现了从西方引入的各种现代住宅类型。由于传统合院式住宅在功能上进步缓慢，带有卫生间和厨房的外来现代住宅成为极具竞争力的新住宅形式。而从我国传统合院住宅演变来的石库门式住宅，原本是给一家人居住的天井住宅，由于建设量小，供应量少，但需求量太大，最后都成为多户家庭混居。这样太低的生活标准，使石库门式住宅最终沦为低档的杂院住宅。电影《七十二家房客》是当时大杂院中大众生活状态的真实写照。

“中华人民共和国成立后，实施公有制和计划经济的福利住房制度。建国初期，住房严重短缺，许多存量的独立住宅、联排住宅、多层住宅都被按照每个房间一个家庭的方式重新安排使用，由于过度使用、缺乏维护，房屋迅速破损。天津是个典型的例子。北方一些城市的传统平房区，包括北京的四合院，也是多家使用，由于年久失修和人口增长，私搭乱盖，渐渐沦为棚户区。在新建住房方面，从第一个五年计划开始，在苏联帮助下开始现代住宅建设，普遍采用多层的标准单元住宅。居住小区规划以军营行列式布局为主，造成房型和居住环境景观的单调和千篇一律。

“改革开放后，我国住宅类型的多样化开始涌现。20 世纪 80 年代试点小区时期，住宅的多样化创作集中于住宅形态和外观造型的多样化。到 20 世纪 90 年代末，出于市场营销的目的，房地产开发企业开始建设别墅、花园洋房、多层和高层的高档住宅。但是，到了 2003 年以后房地产调控期，叫停别墅，出台“90/70”政策，住宅建筑类型又趋向单一化。同时，用地的容积率不断升高，加上绿地率、建筑密度、停车位等严苛的控制指标，住宅建筑只能向高处发展，都成为 30 层、100 米高的塔式或板式高层。几栋、十几栋甚至几十栋类型单一、外观相同的高层住宅堆积在一起，形成了目前最具代表性的居住环境和城市面貌。虽然建筑和小区绿化环境的质量提高了，但居住的整体品质下降了，单调乏味的住宅也导致城市整体空间品质极度下降。

规划局详细规划处赵颂处长

我以为“我国住宅类型单一现象的出现，首先，是受现代主义建筑思潮的影响。人多地少的固有认知、长期住房短缺形成的匮乏心理和计划经济思想的残余进一步强化了这种影响。其次，封闭小区加高层住宅模式最有利于获取经济效益，是满足当前市场和现行管理要求的最赚钱的模式。高层住宅楼的大量复制，建筑师设计省事，施工企业作业高效，开发商赚钱快，商业银行很满意，却极大地损害了城市空间的品质，既违反了人性，也违反了城市发展和住宅建设的客观规律，给未来埋下了许多隐患。第三，片面单一的某些土地政策、居住区规划方法和城市规划简单化的管理也是重要原因之一。比如，土地部门要求住宅容积率不得低于 1，不论区位地一刀切。我国居住区的规划设计理论和方法师从苏联的计划经济和福利住房制度，具体代表即 1993 年颁布、1994 年开始实施的国标《城市居住区规划设计规范》，以及各个城市制定的《城市规划管理技术规定》。《城市居住区规划设计规范》是居住区规划设计的基本指导文件。自 1994 年实施以来，虽经几次修订，但没有做大的调整。总体看，还是受我们所处的经济社会发展阶段的限制，公众、领导，包括规划设计人员，还没有意识到这种单一住宅建筑类型的危害。

罗云

“简 · 雅各布斯(Jane Jacobs)在《美国大城市的死与生》一书中指出:城市是人类聚居的产物，成千上万的人聚集在城市里，而这些人的兴趣、能力、需求、财富甚至口味又千差万别。因此，无论从经济角度，还是从社会角度来看，城市都需要尽可能错综复杂并且各功能相互支持的多样性，来满足人们的生活需求，因此，‘多样性是大城市的天性’（ Diversity is nature to big cities ）。雅各布斯犀利地指出，现代城市规划理论将田园城市运动与柯布西耶倡导的国际主义学说杂糅在一起，在推崇区划（ Zoning ）的同时，贬低了高密度、小尺度街坊和开放空间的混合使用，从而破坏了城市的多样性。

而所谓功能纯化的地区如中心商业区、市郊住宅区和文化密集区，实际都是功能不良的地区。雅各布斯对20世纪五六十年代美国城市中的大规模计划（主要指公共住房建设、城市更新、高速路计划等）深恶痛绝。她指出，大规模改造计划缺少弹性和选择性，排斥中小商业，必然会对城市的多样性产生破坏，是一种'天生浪费的方式'，其耗费巨资却贡献不大，并未真正减少贫民窟，而仅仅是将贫民窟移到别处，在更大的范围里造就新的贫民窟，使资金更多更容易地流失到投机市场中，给城市经济带来不良影响。因此，大规模计划只能使建筑师们热情高涨，使政客、地产商们热血沸腾，而广大普通居民则总是成为牺牲品。她主张必须改变城市建设中资金的使用方式，从追求洪水般的剧烈变化到追求连续的、逐渐的、复杂的和精致的变化。20世纪60年代初正是美国大规模旧城更新计划甚嚣尘上的时期，雅各布斯的论点是对当时规划界主流理论思想的强有力批驳。可以说比著名的建筑历史学家查尔斯 · 詹克斯（Charles Jencks）更早地宣判了大规模工业化住宅和现代主义居住区建设方式的死刑。

"雅各布斯说：不论是霍华德的田园城市所倡导的花园住宅，还是柯布西耶的光明城市的塔楼住宅，都是专制的、家长式的规划，都试图用一种类型、一种模式解决所有的问题，它注定是不会成功的。现代主义城市规划和现代主义建筑推广大规模工业化和国际式风格，希望用一种通行的模式解决所有问题，是违背生态法则、人性需求和城市发展的客观规律的。"

罗云

"在导则编制的过程中，我们重读经典著作，霍华德的《明日的田园城市》柯布西耶的《走向新建筑》《明日的城市》《光辉城市》芒福德的《城市历史》《城市文化》《技术与文明》雅各布斯《美国大城市的死与生》彼得 · 霍尔《明日的城市》等，实际上这些著作的许多篇幅和内容是在讲述西方城市住宅和居住社区发展的演变。霍华德、柯布西耶对现代城市规划的发展贡献巨大，但由于历史的局限，他们的一些设想、方案具有一定的片面性，而后人必须正视这些问题。芒福德、雅各布斯、霍尔等对现代主义住宅建筑的批判结论惊人得一致。

"我们大张旗鼓地实施了人类历史上最大规模的住宅和城市建设，用的却是过时的现代主义规划设计方法。我们没有真正理解西方社会20世纪50年代对现代主义运动的批判，没有真正认识到现代主义建筑运动和城市规划的危害。当然，我们付出了巨大的代价，收获也将是巨大的。伴随着住房制度和土地使用制度的改革，中国开创了人类历史上大规模住宅建设的新时代，在 30 年的时间里按照市场化机制，通过房地产开发，建设了数百亿平方米的住宅，解决了中国人过去长期无法解决的住房短缺问题，创造了人间奇迹。但是时至今日，这种现代主义的居住区规划设计方法的确不能再继续下去了。

“住宅类型的多样化是提高居住水平和生活质量的要求，也是城市文化发展的要求。随着经济的进一步发展和社会的不断进步，随着居民收入的提高，居住需求的分化越来越明显，不仅体现在支付能力上的差异，也表现在生活方式、功能需求等方面的变化和多样化。随着城市规模的扩大，土地的价值和区位条件差异加大，这些因素都使得当代城市住区和住宅建筑类型和形态趋于多样化。因此，合理的密度和聚集程度、完善的社会配套服务、良好的环境和合理分布、多种多样的居住建筑类型，如城市公寓、合院住宅、联排住宅、独立住宅等，是提高我们生活质量的必备条件。

“亡羊补牢，犹未迟也。要改变目前的状况，提高城市的环境品质，适应人们对美好生活的向往，满足居民多样生活需求，就需要住宅建筑的多样性。简单讲，低层、多层、高层三种基本住宅类型都要存在，而且要随着时代的需要创新。《城市居住区规划设计规范》中有低层住宅的一类居住用地，却没有明确的建筑类型。

“我们先谈低层住宅，可能大家有不同的意见，可以讨论。低层类住宅是建筑类型最丰富、最多样的。它不仅包括独立住宅、半独立住宅、联排住宅等类型，且给建筑师很多的创作空间。特别是低层建筑便于采用合院形式这一中国传统民居的典型形式，有利于将中国优良的人居传统思想文化传承发扬。实践证明，这种形式也是中国人最喜爱的居住形式，有广泛的市场需求。

“对于是否要建设低层住宅，社会上存在的疑问比较多。首要是土地问题，中国是否有足够的土地建设低层住宅？答案是肯定的。

“首先，我们可以用比较的方法。中国人口众多，土地资源紧张，住宅建设要考虑节约土地是不争的事实。但不论城市还是郊区，甚至农村，是否都要盖高层住宅，是我们必须认真反思的问题。新加坡采取高层高密度住宅，因为它是城市国家，土地狭小。但即便如此，也依然有少量别墅住宅类型。在土地资源比我国还紧张的日本，有超过 50% 的住宅是传统的低层独户住宅——一户建。一户建是日本住宅多样性的重要组成部分，这也说明住宅不单单是节约土地的问题。苏联及东欧国家在大中城市中大量建造 9 ~ 25 层住宅，少量的高达 40 层。对于苏联这样地大物博的国家，建设高层住宅一定不是因为缺少土地和保护耕地，主要是因为工业化、大规模生产和城市规划布局，关键是福利住房制度。到 20 世纪 80 年代，苏联及东欧一些社会主义国家，在住宅工业化方面取得很大的进展。但由于计划经济本身的问题，导致国家经济社会等大的方面产生了问题，这里面不能说没有住房的问题。富兰克林 · 罗斯福说：‘如果一个国家的人民拥有自己的住房，并能够在自己的土地上赢得实实在在的份额，那么这个国家是不可战胜的。’社会经济发展和城市建设的最终目的是提高人民的生活质量和水平，土地只是载体和手段，为了节约土地而取消住宅的多样性，是本末倒置。

“其次，我们从房地产转型升级的角度来说明这个问题。目前，我国城镇人均住宅建筑面积 39 平方米，户均一套，总体看已经解决了住房有无的问题。现在每年保持住宅几十亿平方米的新开工量、十几亿平方米的新建商品房销售量。每年城镇人均住宅建筑面积要增加 2 平方米。未来一段时期，贯彻中央供给侧结构性改革方针政策，我国房地产业改革升级的方向是：减少开发总量，提高质量，有效供给，实现平稳过渡。考虑到 2000 年前建设的大量存量住房标准比较低，因此新建商品房标准必须比未来规划目标更高，才能总体上平衡。如果再按照户均 100 平方米左右来建设，我们每年的商品房销售量会达到 1700 万套，能够容纳 4500 万人，即使考虑新增城镇人口，也远大于需求。如果直接减少规模，会影响到房地产行业和整个经济的发展。随着经济水平的提高，郊区化低层住宅是城镇居民改善居住条件的必然趋势。而保留下来通过城市更新的老旧小区住房则适宜刚在城市落户家庭的初次置业。我们可以先设定在城市郊区每年供应低层住宅用地的比例，比如不高于 20%，同时限制户均用地面积上限，开始进行尝试。同时，要严格控制侵占生态红线，加大处罚力度。

“第三，我们从城市平均密度和绿化水平来说明。我国从近现代以来，许多大城市都面临城市中心人口拥挤的问题。中华人民共和国成立后，在城市总体规划中，普遍采取了建设卫星城、疏解城市中心和老城人口、控制城市人口增长的策略。实际上，随着土地制度改革和房地产行业的发展，旧区改造要求比较高的拆建比，结果非但没有疏解人口，反而成倍地提高了土地开发强度。同时，新区的建设也采用了较高的容积率，许多是高层住宅。近期可以为城市带来土地出让的高收益，但过高的土地开发强度、过高的人口密度、过少的绿地公园是造成交通拥挤、环境污染等城市病的主要原因。到头来再花钱治理，得不偿失。新建的住宅需要比较低的密度、比较高的绿地率来求得城市整体的平衡。

“第四，我们从传承我国住宅的文化传统方面来说明。总体看，合院住宅，或称庭院、天井住宅是我国传统上主要的住宅形制。在历史的演进过程中，各地区都形成了与各具特色的民居和城市规划相对应的社区规划设计模式。合院住宅及其城市规划模式具有安全、安静、突出家庭性、文化性等特征，是良好的人居环境，如北京四合院和胡同体系，苏州合院住宅、私家园林建筑和河网、道路双棋盘系统等。一些地区在住宅整体布局上考虑环境因素，如浙江一带的邻水民居，呈现出生动而清新的风趣。在住宅装修和室内家具等布置上，地区特点也很突出。另外，我国的住宅与园林联系密切，许多宅邸配建私家园林，把自然风景山水浓缩在方寸之间，达到了自然美、建筑美、绘画美和文学艺术美的有机统一。我们必须继承和弘扬这些优秀传统，不能因为任何理由借口丢失传统。

“第五，从社会和谐的角度来说明。霍华德认为应该建设一种兼有城市和乡村优点的理想城市，他称之为‘田园城市’。他认为此举是一把万能钥匙，可以解决城市的各种社会问题。田园城市实质上是城和乡的结合体，疏散过分拥挤的城市人口，使居民返回乡村，把城市生活的优点同乡村的美好环境和谐地结合起来。郊区化低层住宅的建设有利于城乡融合。如果有较多的中等收入家庭入住郊区的低层住宅，实际上有利于缓解贫富差距。

“第六，从保护生态环境、实现人与自然和谐的角度。低层住宅体量小，布局比较自由灵活，可以与生态环境很好地结合，采取低影响冲击的道路交通和市政基础设施建筑掩映在绿色之中。另外，可以探讨传承合院住宅和街巷的布局，延续中国传统合院建筑冬暖夏凉的生态建造方法。

“归纳起来讲，对待低层住宅建筑，需要我们解放思想，按照城乡发展的科学规律来考虑住宅的多样性。普拉特－兹伊贝克教授所描述的从市中心到乡村不同密度、高度和形态类型的住宅建筑类型，实际反映了空间的客观规律，包括经济活动、级差地租、生态环境、交通运行、城市管理和社会组织等。我们要丰富住宅的多样性，就首先要承认低层建筑的合理性，而不是以一种简单化的思想方法看问题，用单一方面的考虑来制定规则，这样才能形成千变万化的空间形态和高质量的住宅及社区环境。

“很显然，我们目前的一些规划和土地政策，如禁止别墅和容积率低于1.0 的用地供应和规划审批，有以偏概全之嫌。我们没有任何一种理由，来限制居住建筑类型的多样性。必须统筹考虑城市人口密度、土地开发强度与城市交通、环境景观、公共服务配套及住宅类型多样性等问题。按照城市发展的客观规律，在城市不同的区位采用不同的开发强度和建筑类型，包括在郊区允许独立住宅和容积率低于 1.0 的开发。实际上，目前我国城镇存量住房中高层和多层住宅已经占据主导地位，土地的集约利用已经相当高了，开发强度和人口密度都超出了合理的范围。同时，在城市的外围地区有大量的低效闲置土地。逐步发展一些低密度合院式住宅，不会对城镇土地问题带来负面影响。通过合理的规划设计，我们可以创造出既有一定的开发强度，又具有丰富多样居住建筑类型的、适合地域特点的、更好的人居环境。

“我想再重复一遍，今天我国社会主义市场经济体制已经初步建立，国家强调要发挥市场在资源配置中的决定性作用，就要按照市场规律办事，要承认差异，差异产生多样性。当然，住房不是简单的商品。国家要做好中等收入家庭的住房保障，这是前提。我们院有同事专门做外国住房制度和体系的研究，滨海新区也做过改革尝试，总体思路是从过去的保障低端住房和收入困难家庭转向对广大中等收入家庭的保障，做到‘低端有保障，中端有供给，高端有市场’。具体内容我就不班门弄斧了。大家如果感兴趣，接下来可以专门讨论。”

罗云

“多层住宅是城市中量大面广的住宅形式，也是未来中端有市场的主体。在汇报多层住宅建筑城市设计导则前，我先简要地汇报一下导则中关于高层住宅建筑的设计要求。首先，从杜安伊的住宅城乡断面图中，我们看不到 30 层以上的高层住宅，即使在市中心，也只是以多层住宅为主。考虑到我国的现状，我们还会有一定比例的高层住宅，比如不大于 20%。概括起来讲，就是总体上限制高层建筑的数量，除规定的区位外，不得建设高层住宅。其次，限制普通高层住宅建筑的高度，从消防安全的角度，高度不超过 80 米。在城市外围中心，轨道站点附近，可以建设不超过 54 米的高层住宅。对于市中心少量的高档公寓式住宅建筑，要限制规模，提高消防等设施的等级。对于分割销售的高档住宅，要求开发商必须保持公共部位的产权，并承担长期的物业维护责任。

“不可否认，多层住宅是我国现存最多的居住形式，也是未来一段时期建设量最大的住宅形式。如何使多层住宅建筑具有多样性，是个长期的课题，而且是个比较困难的课题。改革开放初期，为改变行列式单调的住宅布局和建筑形式，国家开展了住宅试点小区试验，积累了一定的经验。我们在滨海新区和谐社区试点中也一直在思考这个问题。新型居住社区规划设计的前提之一是必须有多种多样的住宅建筑类型。如果只是把大院式封闭小区改成窄路密网式布局，而不增强住宅建筑类型的多样性，新型居住社区是不会成功的。

“从国内外的经验看，采用窄路密网，划小地块，委托不同的建筑师设计和不同的开发商开发，是创造多样性的最好方法，比地块和路网本身形态的变化更起作用。另外，在以多层住宅为主的社区中插入一些不同的住宅类型，会起到画龙点睛的作用。比如，多层住宅区可以规划联排住宅为主，吸收我国传统民居中窄面宽、多重院落、天井的大进深布局，既有一定的密度，又满足中国人住房‘顶天立地’的心理需求。导则也提出，虽然高层、多层和低层住宅从市中心向外依次布置，但在各自的区域，也有一些其他区域的类型。比如，在城市中心区以高层为主的区域，一些历史街区就是低层住宅为主；在外围郊区，在一些公交站点附近可以有部分多层和小高层住宅。这种穿插满足城市不同区域多样性的要求，富有变化，避免单调乏味。

“住宅建筑是最基本的建筑类型，是最简单的建筑类型，但也是最复杂的建筑类型。无论是古代还是近现代建筑历史上的建筑大师都做过经典的独立住宅设计。但做大众集合住宅的不多，柯布西耶的马赛公寓设计是经典。在有限的面积、造价等约束条件下，做出面向大众的、好的住宅建筑不是一件容易的事情。要使多层住宅建筑设计达到高水平，我想最终要延续传统文化，促进居住建筑品质和文化性的提升。

“研究中国民居的意义有很多，但主要的还是要古为今用。刘敦桢、刘致平、傅熹年在他们的论述中都提出了同样的观点。早在抗战时期的昆明，林徽因为当时的云南大学女生宿舍映秋院所做的设计中，创造性地运用了一些民居的手法和风格。中华人民共和国成立后大屋顶等民族形式也用在了一些居住建筑上。随着历史名城保护的兴起，对老城内传统民居建筑的保护及更新改造成为一个非常重要和有意义的课题。清华大学吴良镛教授组织对北京四合院的研究，朱自煊教授开展的以保护胡同和四合院肌理为主的北京后海地区城市设计研究、安徽屯溪老街保护研究，单德启教授对徽州等地民居的研究，同济大学关于苏州平江古城桐芳巷改造的研究等，都是非常有益的尝试，取得很好的效果。

“20 世纪 80 年代，吴良镛先生提出‘有机更新’理论，主张按照城市内在的发展规律，顺应城市之肌理，通过建立‘新四合院’体系，在可持续发展的基础上，探求北京古城的城市更新和发展。这一理论在1988年启动的菊儿胡同改造试点项目上得到成功实践。重新修建的菊儿胡同新四合院住宅按照‘类四合院’模式进行设计，即由功能完善、设施齐备的多层单元式住宅围合成尺度适宜的基本院落，建筑高度为四层以下，基本维持了历史上的胡同—院落体系，同时兼收了单元楼和四合院的优点，既合理安排了每一户的室内空间，保障居民对现代生活的需要，又通过院落形成相对独立的邻里结构，提供居民交往的公共空间。建筑外观具有本地传统的符号和色彩。菊儿胡同新四合院与传统四合院形成协调的群体，保留了中国传统住宅重视邻里情谊的精神内核。北京的菊儿胡同改造项目于1992年获‘世界人居奖’。但是我们发现也有一些问题。最近我们专程去考察了菊儿胡同，发现建筑比较破败，其原因是院落等公共空间界限不清。因此，这种做法还需要进一步研究推敲。”

南开大学周恩来政府管理学院施家丰教授

这时，台下有人举手示意要发言。“大家好，我是施家丰，来自南开大学周恩来政府管理学院。我认为，住宅不单是建筑设计问题，也不单纯是城市设计问题、规划问题，更是经济问题、社会问题和政治问题。与此同时，住房的房型设计和社区设计确实对整个社会有重大影响。从整体上看，住宅在城市中占绝对比例，对城市形态的基本构成起着决定性作用，它主导着街区形态，进而主导着城市整体空间形态和生活的品质。”

“多年来我一直在思考，尽管住宅有许多类型，但从社会学和社会管理的角度看，最终分为两大类型：独立住宅与集合住宅。独立住宅是以家庭为单位，可以自己管理的住宅，有自己的土地和天空，自己的上方没有其他住宅，即‘顶天立地’。集合住宅是多个家庭上下叠在一起的住宅。人类历史上的大部分时间里，大部分住宅形式是独立住宅。随着城镇出现，人口的不断增加，城镇围墙内的土地利用越来越紧张，由此出现了多层的集合住宅。芒福德在《技术与文明》中写到，古罗马时由于城墙内面积有限，所以就有多层住宅，但当时不具备相应的设备，粪便都堆积在楼梯间下面，恶臭难闻。直到工业革命后，随着现代建筑材料和技术的出现，集合住宅越来越完善，工业化大生产使得集合住宅成为解决住房短缺的有效手段，而且可以使城市中心更加密集，促进城市进一步聚集发展。

“从世界各国的经验看，独立住宅与集合住宅有各自的优势，也有各自的问题。独立住宅产权界限清楚，居民与政府的责任也很清楚，一般情况下业主会自行维护自己的物业，保证住房长期保持良好的品质，但建筑密度比较低，虽然可以采取比邻式和联排式，但占用土地相对比较多。集合住宅的最大优势是节约土地，但产生的问题也非常多。由于产权界限不清，所以除少数高档公寓外，大部分集合住宅在物业管理上都存在很多困难。因维修难度大，最后都成为政府的沉重负担。我国大部分城市目前已经遇到这个问题。此外集合住宅如果设计简易，在社区环境营造上也存在巨大的问题。实际上，独立住宅被多户居住时，就变成了集合住宅。我国许多的四合院、小洋楼，由于多户居住，造成居住环境和房屋质量恶化却难以修缮。因此，从综合的整体角度看，城市的住宅应该有多样性，而且要取得独立住宅与集合住宅数量的平衡和区位的合理分布。

“我们过去一直认为独立住宅是不生态的，实际上许多历史经验证明，某些情况下低强度的开发是对生态环境的低影响冲击。从美国湾区开发之初的历史照片看，由于海风很大，树木不宜生长，所以山上是光秃秃的一片。但 100 多年之后再看现在的湾区，树木成林，郁郁葱葱。100 多年来，人们不仅建成了城市，发展了经济，也建设和改善了整个湾区的生态环境。我们看到，拥有洋房、花园和汽车的美国梦已经变成现实，几乎户户都有特色各异的花园，植物种类繁多。湾区目前有 244 万套住宅，我们权且估算其中 200 万套是独立住宅，每户绿化按 50 平方米计算，则共建设了 1 亿平方米（100 平方千米）的花园绿化，由每家自己养护，不用政府掏一分钱，这是一笔巨大的财富，对生态环境的改善和城市美化发挥了主要的作用。

“从这点延伸开来，我们可以认识到，不能简单地将占地大小看成一个孤立的问题，而要看综合的效果。从综合的生态环境、社会、城市社会管理效能等方面看，在城市外围和郊区建设丰富多样的低层独立式住宅是一条好路子，既回应了人们生活水平提高的追求，满足居民的个性需求，又对促进经济发展具有重要作用。因此，不能简单地禁止。此外，如果可以提供有效供给、避免大量空置，再加上精心规划设计，按照新都市主义的理论，城市区域的总城市用地规模不会比以多层和高层住宅为主的建设模式多很多。”

罗云

“施教授的发言让我很受启发，也更清楚了菊儿胡同的问题所在。其实天津也面临同样的问题。天津中心城区在中环线周围均匀分布着 1999 年前由政府建设的大型居住区，都可以归为集合住宅，现在是老旧社区和大量需要改造的多层无电梯住宅。在物业管理和维护上，住户很难达成一致，如果实施城市更新，最后都需要政府买单。另外由于界限不清，好多设施刚修过一次，过不长时间又坏了。这也说明我们的城市需要建设一定数量的低层独立住宅，不仅可以减少政府维护的支出，还可以培养社区精神。

“最后，我把导则的主要内容再概括一下。刚才谭所长汇报的天津中心城区一环十一园及周边地区城市设计，实现了生态文明时代下中心城区和主城区城市结构的重塑，形成了与自然和谐、城乡融合的居住空间整体格局。我们以城市快速路、外环快速路、环外环为边界，形成中心圈层、生态宜居圈层、田园城市圈层三个圈层。按照这个结构编制新型居住社区城市设计导则，规定在不同圈层区域，布置不同的住宅类型，形成各具特色的多样化社区，并制定了各区域相应的容积率、建筑高度、密度、绿地率等控制指标。这是我们与传统居住区规划等标准一刀切的最大不同。比如，对于绿地率，过去新的住宅小区都要求较高的绿地率。但是绿化的质量和水平受地下停车库等因素的影响，效果不佳。这里我们实事求是，结合窄路密网布局，合理降低小地块内的绿地率，要求更多地集中小公园绿地，绿地率在整个社区实现平衡。

“天津主城区中心都市圈层，既是天津活力中心区，同时也聚集了 14 片历史文化街区，这里是天津城市生活最丰富的地方，也是最具人文精神的场所。这里吸引着天津最活跃的人群，他们大多在市中心工作，追逐潮流，热衷社交，有丰富的业余生活，追求生活品位。这里具有完备的步行网络，步行可达一体化的公共交通系统、高品质的公共空间、商业街区等。这个区域面积为 145 平方千米，是目前最主要的建成区和人口聚集区，也是建筑密度、开发强度、道路网密度高的地区，许多地区在历史上就是窄路密网的格局。现状的居住社区和住宅建筑类型有许多种：五大道、赤峰道、鞍山道、意式风情区等历史文化街区和其中的居住建筑，包括门院式、院落式和里弄式三种主要类型。剩下的是在 1999 年前建设的大型居住社区，现在是老旧社区和大量需要改造的多层无电梯住宅。未来，这个区域内首先要做的是对历史街区的保护，对劝业场周边比较破败的住宅区实施人口疏解和有机更新。在城市中心的海河两岸可以规划建设少量的高标准、高强度、高密度的高层高档住宅，为商务区配套服务，包括滨海新区的于家堡金融区。

“这个区域内最大的任务是老旧小区的更新，这是一个长期的工作。在中环线周围，均匀分布着 20 世纪 80 年代以来由政府建设的大型居住区。当时主要作为工业区的配套，经过多年的发展，这里社会结构较为稳定，邻里关系和谐，中老年家庭占比较高。人们更习惯传统的街道生活，下楼可以买早点，离家不远就有菜市场、便利店，午后天气好的时候，下楼找邻居们下棋、聊天。步行距离内有学校和医院，出行主要依靠公共交通。比较典型的有体院北社区。但问题也比较突出。除近期实施环境综合整治、加装电梯外，未来需要按照 2035 年的标准实施城市更新。这是中心都市圈层最大的挑战，必须及早面对，做好准备。我们这次城市设计试点中有体院北城市更新的项目，做了专题研究。

“在快速路和外环快速路之间是生态宜居圈层，结合 11 个公园周边地区，形成生态宜居社区——城中城。这里紧邻公园，有良好的生态环境、完善的生活配套、便捷的公共交通系统，周边有优质的学校和医院，是更多有小孩、有老人家庭的选择，也是市中心老旧社区中的家庭提升改善居住环境的选择。他们更注重家庭生活，孩子们可以在草地中自由地奔跑，老人可以在林荫下漫步，一家人可以在树下野餐度假，共享天伦。他们有的在市中心工作，有的在社区内工作，通勤以公共交通为主，休闲出行以小汽车为主。社区中低强度开发，轨道站点周边强度适当提高。这一区域还有大量的土地是城市新建中端商品房的主要场所，需要高水平的规划设计。这类社区包括水西公园周边地区、西营门南运河地区、程林公园周边地区，以及滨海新区和谐社区等。

“在外环快速路和环外环之间是田园城市圈层，是人和自然对话、城市和乡村交融的地方。这里的人们追求宁静自由的生活，崇尚亲近自然，享受人生。花园和庭院是每个家庭日常生活的中心，是对话自然、关照内心的场所。社区里有小镇中心、商业主街和学校，离家不远有家庭医生。出行以小汽车为主。新建社区是低强度、低密度、独栋住宅或中国合院的郊野社区。

“这个区域有大的郊野公园，保留大量的农田和部分村落。农村住房要在现有宅基地的基础上，大幅提高住宅建筑的设计和建造水平。每户三分宅基地面积标准是不大于 200 平方米，面积足够，关键是要改变我们目前农村住宅建设的观念和习惯，创造新的住宅建筑建设模式。对于环境的考虑要作为乡村城市设计的重点，包括村落中心等公共空间、公园绿化，精心用好每一寸土地。在广大的农村地区，目前还没有这样的意识。每户三分宅基地基本上是院墙一围，‘躲进小楼成一统’。历史上四合院有植树的传统，现在已经少见了。假设我们在建设管理上增加对宅基地使用的要求，要求每户按一定标准进行绿化，将每户的绿化纳入村庄、城镇严格的市政管理，变化会是历史性的。天津目前有 388 万农村人口，110 多万户农村住宅，如果每户营造 50 平方米的花园、4 ~ 5 棵树，就是一个很大的、很真实的数字。不仅改善了生态环境，而且树立了生态环境的意识，这是很重要的，另外可以发展绿化园艺产业。所以，要坚决避免把现有村庄拆除，建成单元住房的做法。

“城市是人类社会的标志，乡村是上帝爱世人的标志，城市和乡村必须‘成婚’，构成一个城市—乡村磁铁。霍华德的田园城市为我们描绘了一幅‘城市—乡村磁铁’的蓝图，它既可享有城市社会文化成果，又可使居民身处大自然的美景之中，用以社区为基础的新文化来医治城市中心区的脑溢血和城市边远地区的瘫痪病。它统筹城乡要素，形成可渗透的和自然和谐相处的城市区域，使城市与乡村在更大范围的自然环境中取得平衡。同时建立社区自治，把最自由和最丰富的机会同等地提供给个人努力和集体合作，闪耀着以人为本的光芒！站在今天的历史坐标，120 年前霍华德的经典理论依然是一把解决问题的万能钥匙。”

罗云

讲到这里，作为在天津从事城市规划工作 10 多年的罗云，内心复杂但坚定：“重塑城乡关系是现代城市规划的宗旨，新时代的天津，需要美好的人居环境，去实现人民对美好生活的向往。”

“文明发展的进程中有一点是始终不变的：社会要进步，人类要追求更美好的生活。回顾历史，一个民族的发展始终与美好的人居环境相伴随，人居环境建筑的最终目标是社会进步。吴良镛先生曾在《北京宪章》中提出：‘美好的人居环境与美好的人类社会共同创造，建设一个美好的、可持续发展的人居环境，是人类共同的理想和目标。’城乡融合奠定了人与自然和谐相处的健康格局，而多样化的美好社区是社会的基本单元，是从微观出发进行社会重组，通过对人的关怀，为每个人的意志得以伸展和发展提供可能和支持，从而激活丰富多样的生活，维护社会公平，最终实现和谐社会的理想。

“‘人，诗意地栖居在大地上’是海德格尔的理想，更是我们一代代城市规划建设者的理想。”

54. 社区中心——寻找社区的精神家园

褚夏总规划师

罗总的演讲迎来了热烈的掌声。这个分论坛更像是一个研讨会，大家广泛参与，讨论热烈。褚总走上台：“住房的多样性是我国下一阶段新型社区建设的关键，是实现中华人民民族伟大复兴中国梦的基础和前提条件。刚才罗总讲的内容我非常赞同并深受鼓舞。我认为，除了住宅本身要高度重视其精神功能外，社区也要把居民凝聚在一起，要寻找社区的精神家园。”

“我想，新型居住社区的规划设计的‘新’字就体现在对城市空间、住宅建筑文化和社区邻里场所规划设计的重视，对街区、街坊、邻里文化的塑造。这是城市特色的重要组成部分，更是中华优秀传统文化延续和发扬光大的重要载体。

“最近读芒福德的理论书籍让我深受启发。与其他规划师热衷于将雷德朋新城作为邻里单位规划理论的实例进行解读不同，芒福德进行了更为深入的思考。他认为，雷德朋新城成功的要诀不是建筑，关键在于规划提供了一个文明核心，即使这核心仅体现为商铺、学校、公园，但它可以聚集人群。此外绿带、街道构成了共同的边界，使居民有归属感。他说：大力在邻里社区重搭戏台，让社会生活的精彩场面还能在这里上演。在写作《城市文化》一书的时候，芒福德对城市系统的研究刚开始，但他已经敏锐地发现社区内在的文化意义。20 世纪 20 年代关于美国新型居住社区的工作促使芒福德开始系统研究美国的城市文化，写出了《城市文化》和《城市历史》两部巨著，成为 20 世纪最伟大的城市历史和理论大师。

“芒福德对历史描述的重点和笔墨更多地用在人的身上，讲述人的文化和精神生活。芒福德解释说，聚落最初形成的主要动力之一是人的精神需求。在最初雏形期，聚落的形成往往不是源于居住而是源于祭祀活动，如对死者的祭奠和回忆、祈求丰收或是有足够的猎物、定期朝拜集会、对天神自然力等的崇拜等。农业革命后，人类开始定居下来，村庄成为了庇护、养育人类的场所，其中已经包含了日后为城市所吸纳的圣祠、管道、粮仓等。当凶猛强悍的狩猎民族对坚忍的农耕民族实施统治的时候，统治权逐渐与神权相结合，使国王拥有了空前的力量，役使他的农民为他建造一系列的高大建筑物，并且分化出一批司精神之职的赞颂国家的阶层，于是城市就此诞生。城市最初是作为一个磁体而不是一个居住的容器存在。村落的存在基础是食物和性，而城市则应该是能够追求一种比生存更高的目的。

“与一般人对中世纪教会统治黑暗的看法不同，芒福德认为中世纪是一个人性得到弘扬的时代，他常用 11 世纪的威尼斯作为中世纪城市的代表。从 5 世纪开始，基督教开始盛行，这种宗教文化否定财产、威望和权力，把清贫当作一种生活方式，消减肉体生存所需的全部物质条件，把劳作当成一种道德责任，使之高尚化。修道院成为理想的天堂城市的城堡。随着经济的缓慢恢复、人口和财富增加，行业公会的作用增强，城镇开始发展。这时候，教会统治遍布整个欧洲。旅途中的人们从地平线上看到的第一个目标就是教堂的塔尖。城镇都是以大教堂为中心，周边有小教堂、修道院、医院、养老院、济贫院、学校，以及行业工会使用的市政厅等，后来出现了大学。礼拜、朝圣、盛装游行、露天表演等仪式成为城市中定期进行的活动。中世纪大众的住房比较简陋，但人们更便于接近农田和自然环境，习惯户外活动，城镇有很多空地供各种游戏活动。城镇尺度宜人，众多的小教堂成为社区中心，成为人们精神寄托和交往的场所。芒福德赞赏中世纪的城市核心，认为中世纪的城镇邻里比巴洛克、文艺复兴时期君权主义和工业革命后大工业发展时期的城市都更有人情味，更具有文化气息。不同类型的住宅形成不同形态的社区、道路网、街道、绿化环境。

“社区中心的设计非常重要，它是社区居民交流的场所，是精神的家园。除了具备高品质的住房和优美的社区环境，还必须形成良好的社会关系，社区中心能够非常重要之作用。历史上好的社区中心的样本琳琅满目，但在精神文化上最为丰盛的是欧洲中世纪城镇。我们发现，每一个中世纪的城镇都会选择一个非常好的地理位置，山水相望，进退有据，可以为各种力量提供很好的格局。那时候，居民对城镇生活的看法相当一致，这种一致性使人在连续看了许许多多中世纪城镇后，感觉好像事实上存在着一种指导城镇规划的自觉理论。

“但对城市建设有意识的管理的确是存在的。笛卡尔在《方法论》中提到:任何时候都有一些官员，他们负责督促私人盖的建筑物必须有助于城市的景观。芒福德也认为: 中世纪的城镇，统一而又多样化，但它也是经过努力、斗争、监督和控制而取得的。

“芒福德说，城门是城市与乡村、城里与城外两个世界相遇的地点，后来演变为检查和控制点，并衍生出一堆仓库、客栈和酒店、作坊。同时，城墙顶上作为古老传统的慢行步道，可以俯视四周田野，享受清风。城镇核心的关键建筑是主教堂，它会对其他建筑物施加很大影响。中世纪城镇建设最重要的手法，就是隐蔽和出其不意，突然开阔或向上耸立，建筑上有细致和富丽堂皇的雕刻艺术，蕴含着形形色色的宗教故事。大教堂作为城镇的中心，有大量教徒进进出出，所以需要一个前广场。按照神学上的指示，圣坛应向东。肉菜市场一般设在教堂附近。成组出现的教堂和广场相当于现在的‘社区中心’，是个活动中心：在盛大节日里，它是举行欢宴的场所，宗教剧目也在此上演，节假日，大学者们在这里进行辩论。教堂的人川流不息，通常在出远门前和长途返回后都会去趟教堂——祈福平安和禀报平安。这里的牧师是一个重要人物，心灵治疗和精神支持毫无疑问是他的主要职责。除了主持所在教区的日常礼拜、接受忏悔之外，他还要主持为孩子洗礼、婚礼、为垂危者祷告等仪式性活动。这些活动是城镇居民的信息发布和交流平台，也是他们在家庭之外最重要的心灵归属和精神寄托。

“中国传统城镇和村落选址往往遵循天人合一的境界，人与自然和谐共处。讲究山环水抱，藏风聚气——山峦要由远及近形成环绕的空间，规划区域内要有流动的水，以达到趋吉避凶、安居乐业的目的。这也在心理上形成了一种自然环境给人以安全感和庇护的关系。与社区规模最为接近的传统村镇的空间形态诠释了乡情、宗亲、叶落归根等情感。‘天人合一’和‘伦理观念’是传统村镇建设的核心思想。‘天人合一’讲究人与自然的和谐，‘伦理观念’关系到人与人的和谐。

“梁漱溟先生曾生动阐释了乡土中国的‘通情达理’原则——理，表现在村镇与自然环境的共生关系上和建筑组合的空间秩序上；情，是指基于血缘关系基础上的、以宗法观念为核心的村镇中人们和谐生存的社会关系。而无宗族关系的村镇居民，一般通过共处地域和共用公共设施来促进深入交流。

“传统村镇从选址、空间整体布局到群体组合、单体建筑等，都能体现出一种朴素的生态意识。最重要的建筑会沿着一系列水塘或河流布置，相应地形成各种形态的小广场。主要道路串联这些建筑和广场，曲曲折折地连通重要的水陆码头或集贸市场等目的地。

“乡镇的内部布局以伦理关系为宗旨，注重等级制度和长幼尊卑，居中为大。因此，祠堂、宗庙作为宗族权威的载体，大多占据村落的中心位置。建筑的群体组合往往强调一种源于伦理关系的结构秩序。小型寺庙一般结合功能和地形设置，比如妈祖庙设置在码头一带，而关帝庙会设置在地势较高的特殊位置。

“水塘和大树在乡镇的公共空间中扮演重要角色。水塘除具有蓄水、灌溉的功效之外，通常还栽种莲芡，有静心和悠远的意境，让人感到安心与平和；关于大树，熊培云（《一个村庄里的中国》的作者）的描述最为贴切：没有大树，村庄就失去了灵魂，一棵大树拥有枝繁叶茂的万千气象，它支撑起了村庄的公共空间。

“中国传统建筑上的诗风雅韵是传统文化深藏的意蕴，像‘耕读传家’‘地接芳邻’这样的门楣题字，以及装饰图案或砖雕花饰通过谐音如‘蝠’同‘福’的方法表达吉祥的寓意，是传统建筑中最常见的精神文化表达。

“说到乡镇，就不得不提到乡绅。他们在乡间承担着传承文化、教化民众的责任，同时参与地方教育和地方管理，引领着一方社会的发展。他们是乡村的文化领袖，代表着一方的风气和文化。祠堂是乡绅最频繁的活动场所，在这里供奉祖先、瞻仰德能；同时，处理家族大事和‘正本清源、认祖归宗’，兼有办理婚、丧、寿、喜等大事，代表宗族权力的族长德高望重，会在祠堂中处理家庭内部事务、树优立榜、赏勤罚懒、化解纠纷、处理矛盾。乡绅也是家塾和门馆的推动者，他们开馆设学，教化族人。

“甘地说过：就物质生活而言，我的村庄就是世界；就精神生活而言，世界就是我的村庄。这里的村庄，对城市人而言，可以对应为他所居住的社区。以现代人在世界网络中的位置来讲，社区并不像传统村镇对人有那么大的影响和约束，但它的作用依然是深刻和深远的。因为无论世界多大，你都要从‘家’出发。

“‘家’代表家族、祖先、父辈的生命传承，以及‘我’诞生和成长的地方，对它由衷的尊重和感激是一个人背后源源不断的能量系统，不管你走进多么广大的世界，它都是你的原点，你清楚地了解——我来自那里。这时，你灵魂牵系的祖先成为一种内在宁静力量的源头，时刻祝福和护佑着你。

“社区生活中的戏剧化表演是人们呈现心灵的意义、表达力量与支撑的最佳形式。芒福德认为，如果把城市生活中的戏剧场面都去掉，像仪式、辩论、生老病死等，城市中有意义的活动的半数都会消失，城市生活的意义和价值半数以上都会减损，甚至化为乌有。

“现代心理学也认为，每个人的内心，都有一个镜像神经元——我们时常镜像映照这个世界，顺从它的需求，努力赢得它的爱和嘉许。每当我们映照外部世界的时候，心中就会涌起相应的渴望，想要世界映照回来。如果这种渴望没有得到满足，我们就会产生‘镜像神经元接受匮乏’，这种匮乏会变成深深的痛楚。社区的活动和仪式镜映到每位居民，如它能看到每个个体真实的存在，它便将安抚到无数孤寂的心灵。

“交流和对话也会使社区的人们相互亲近，增进联结，使人们具备温和的举止。提供各种形式的对话和仪式是社区中心的本质功能之一，社交圈子的扩大也是社区的关键因素之一。所有人都能参与对话，其身份角色需要一一如实地予以认定和看到。

“因此，一个良好的拥有精神意义的社区中心，应具备某种戏剧化的建筑和空间，并且需要有特定的人行使心灵治疗和精神支持的重要角色；同时，社区中的功能场所应同时具有精神的意义：图书馆——学习的场所，咖啡厅——会面和工作的场所，学校——教育和沟通的场所，车站——转移和再出发的场所，运动场——独享和发现同伴的场所，等等。这些场所使人发现自我、得到镜映，并获得交流和成长的机会，它们通过选择一种容易理解的方式组织起来，便成为社区居民身体和精神的共同家园。”

褚总

“最后，我想讲一讲教育、医疗、体育、养老等内容，这是大家最关心的。从国内外产生的对比看，这是一个明显不同的地方。我们城市中最好的中小学校、医院等公共服务设施都位于市中心，现在一些大的医院由于原来的用地满足不了发展需求，搬迁到了城市外围。但总体的形势没有改变，所以在学校开学后的日子，市中心交通异常拥堵，但学校一放假，交通状况就会有很大改观。这种模式适应了当前城市就业中心仍然在市中心的状况。西方国家市中心衰退，好的学校和医院去了郊区。目前我们要做的是大力加强城市宜居圈层和田园圈层高水平教育、医疗以及体育、养老等公共设施的建设，还要鼓励市场广泛参与，使得我们的教育、医疗、体育、养老设施也有一个大的提升，达到国外的硬件水平，为软件的提升创造条件。”

55. 对美好生活的向往

下午的四个分论坛共有 20 位规划师、城市设计师、景观设计师、建筑设计师汇报，与会各单位、各行各业的嘉宾踊跃参加讨论，内容丰富多样，又精彩纷呈。城市设计的理论和实践分论坛汇报了“一环十一园”及周边地区城市设计、城市一乡村磁铁、超越机动性－公交都市和小汽车、新型居住社区城市设计导则，以及水西公园周边地区、西营门地区、程林公园周边地区和柳林公园周边地区的城市设计方案。生态化的城市设计分论坛研讨了双城绿色生态屏障、海河中游景观设计，讨论了绿色出行和低影响基础设施等技术在城市设计中的应用。历史名城保护与城市设计分论坛研讨了步行化城市、老城区的复兴、公共开发空间城市设计导则、劝业场地区城市更新等。城市设计教育分论坛研讨了国土空间规划与城市设计、城市设计、教育与就业、城市设计作为二级学科的建立等课题。延续时间最长的还是城市设计的理论和实践分论坛，主持各个新型居住社区建设的规划师、建筑师们都参加了新型居住社区城市设计导则的编制，所以对许多问题已经达成共识。

新型居住社区的设计是本次城市设计试点中最多的类型，之所以会这样，是因为主要是结合“一环十一园”周边地区的深化设计，包括柳林公园周边地区、成林公园周边地区、水西公园周边地区、西营门地区，以及解放南路地区五个新型社区的城市设计。这些新型社区都位于“一环十一园”周边，是天津中心城区未来发展的重点地区。每个新型社区的城市设计既有共同点，又有各自的特点。各位主持设计师汇报的也是各有千秋，成为城市设计理论与实践分论坛最后讨论的热点。除去新规划的社区外，本次试点对体院北等老旧社区的更新改造也进行了初步的研究。

规划院城市更新所筹备组耿总

“大家好。我今天汇报的题目是《体院北等老社区更新升级》。这类社区非常有代表性，可以说与新型社区同样重要，是我市未来 15 ~ 30 年必须面对的问题和挑战，需要未雨绸缪。

“目前天津城市的居住社区形态反映出天津城市扩展和居住社区发展演变的过程。现在天津中心城区、滨海新区核心区主流的住区平面布局形态有三种。一是历史街区保留下来的居住社区，是窄路密网、相对开放的街区，如五大道地区、劝业场地区等，包括新建的北塘地区。二是中华人民共和国成立后政府投资规划建设的居住区，包括最初的工人新村和后来为工业区配套建设的大型居住区，如天拖南、真理道、体院北居住区等，多数位于中环线附近。这些居住区采用了典型的居住小区规划方法，是以多层条式住宅为主的行列式布局，有少量高层住宅点缀。三是改革开放后规划建设的大型居住区，如华苑居住区等，还有以房地产开发为主的大片居住区如梅江南地区，以及正在建设的解放南路地区等。第三类居住区主要位于外环线边，采用居住区—居住小区—组团规划模式。这些布局形态中的第二种——多层居住区，多年来基本形成了满足日常生活的居住环境，道路系统、服务配套与整个城市衔接得比较好。因为建成时间比较久，需要及时维修和改造升级。

“据初步统计，中心城区内 1999 年前建设的老旧小区总建筑面积 7000 多万平方米，居住人口约 300 万人，是中心城区现状常住总人口的一半。从 2012 年起，天津市委、市政府从民生出发，实施了 2012—2014 年和 2015—2017 年两轮老旧社区整理改造，取得了明显成效。最近国务院决定，在全国开展老旧社区整治，涉及全国 17 万个社区，要更换老旧管线，增装电梯等。这项工作已经步入正轨。应该说老旧社区引起了国家和社会各界的重视，有了良好的开端。从老旧楼区整治具体工作看，主要是住建部门的工作，规划部门积极配合。从城市设计方面看就是城市更新，我们主要的任务应该是研究未来 15 ~ 30 年内，也就是到 2035 年和 2050 年，这些老旧社区的规划和城市设计问题。

“我们选择了体院北地区作为研究对象。体院北居住区建于 20 世纪 80 年代，现在属天塔街，目前居住 8 万人，绝大部分是 20 世纪 80 年代后到 2000 年份建设的老旧楼房。我们这项工作已经开展了三年时间，多次查看现场，与街道和社区管理人员座谈。按照市区部署，体院北这几年一直在不间断进行老旧社区改造，供暖、排水等设施得到改善，总体环境还不错。大部分准物业管理，平均每户每月交 8 元物业费。由于建设时间早，除个别外，大部分房型设计比较紧凑，没有起居室和独立的餐厅，厨房、卫生间面积比较小。绝大部分多层住宅没有电梯。停车问题也比较突出。日前，国务院要求开展老旧社区整治改造，增加电梯等设施。我们预计，经过整治改造，近期居住条件和社区环境会进一步改善，能够满足居民一段时间的需求。但规划要考虑得长远一些。如果再过 15 年、30 年，到 2035 或 2050 年，应该怎么办呢？据我们团队的建筑师讲，这些住宅的设计寿命为 50 年，那时这些房子将达到设计年限。目前能做的就是环境整治，包括一些市政管线的更换和加装简易电梯。但住房结构、户型和室内的更新将来会是比较大的问题。目前人口密度已经是 3.6 万人 / 平方千米，除去环湖医院等公共设施，人口密度太高，不可能再走先全部拆除然后提高容积率来平衡改造资金的老路。

“要从根本上改善这些老旧小区，必须解决两个关键问题，一是这种多层砖混住房改造的技术可行性，二是依靠市场和政策支持的人口疏解动力机制。首先是建筑技术方面的问题，如何对砖混结构为主的房屋结构和户型进行舒适性改造。由于建设年代早，这些住房采用标准图设计，套型面积比较小，房型设计以满足基本的功能需求为主，一般没有起居室，只有一个很小的过厅，厨房、卫生间空间都很局促，无法满足现代生活水平提高的需求。欧洲对这种多层住宅进行结构性改造的经验是，通过两套合并成一套、三套改两套等方式，改善户型结构，增加建筑面积，提高居住舒适度。还要对结构进行拆改和加固，满足新的设计规范和抗震烈度要求，并对管线重新布置，外檐重新设计建造，还可通过局部装配式方法，满足新的保温节能要求，形成新的面貌。第二个主要问题是资金和机制。体院北地区人口密度相对是比较高的。未来，按照城市总体规划，到 2035 年甚至 2050 年，要疏解约人口才能使住房和居住环境得到改善。要实现这个目标，必须依靠市场的手段，根据居民意愿来逐步实施。随着时代发展，居民改善居住条件的意愿一定会越来越强烈。我们需要通过制度设计，使房地产开发公司能够实现盈利，有参与的积极性。这是我们这次城市设计总的思路，由于时间关系，我就不展开了，展览有全部内容，大家如果感兴趣，我们可以会后进一步交流。”

老张

规划展览馆中的论坛和研讨会上的规划师、城市设计师在激烈讨论，外面的城市在按部就班地运行。经过一下午的奔波，在夕阳西下的时候，大儿子载着老张到了柳林地区，这里是曾经的天津钢厂。停好车，老张健步走上了春意桥。这里的海河比市里要宽阔许多，极目望去，海河两岸绿树成荫，河面上一队运动员正在划赛艇训练，像蜻蜓掠水；回头北望，夕阳下市中心的高层建筑轮廓充满了灵动气息。时间不早了，两人回到车上，他们按赵先生给的地址继续导航，驶出大马路后进入了一条小路，老张一瞬间觉得这里有五大道的感觉，绿树掩映着一栋栋漂亮的建筑。来到了柳林公园入口前的圆形广场，他们与房地产经纪人王小姐会合。

王小姐带父子俩来到了柳林公园大门旁一幢小楼的屋顶，瞬间，柳林公园完整地展现在他们面前，让人心旷神怡。“柳林公园是规划的中心城区沿外环绿带的 11 个公园之一，占地 1 平方千米，与五大道面积相近，是老城厢占地面积的 2/3。”“哇！这块地要是用来盖房子，那得盖多少小洋楼、四合院啊！得值多少钱啊！”老张感叹道。“现在人们生活水平提高了，对住房环境越来越关注，过去的体院北、华苑等老居住区虽然生活便利，但缺少这样大的公园让人接近自然。”曾在华苑住过五年的老张若有所思地点点头，“天津中心城区人口比较密，过去人均公园绿地面积不足 4 平方米，随着近年 11 个大公园逐步建成，我市中心城区人均公园绿地将达到 12 平方米，达到国家生态花园城市标准。柳林公园建成了，最受益的还是我们柳林社区的居民，出门就进大公园，这叫近水楼台先得月。”“公园是免费的吧？”天津市的公园已经免费好多年了，但老张还是下意识地找补这么一句。“那一定。”王小姐手指引着父子俩的视线，“公园远处就是海河，柳林公园是 11 个公园中唯一临海河的，由于海河这段水面有 200 米，比较宽阔，所以柳林公园与水上、梅江、水西公园这种有大片水面的公园不同，景观设计上就布置了更多的绿地供各年龄段的人活动，有更多植物和花卉品种，像个植物园。柳林公园的设计是由著名的法国景观设计师查莉主笔的，她把欧洲古典园林的艺术性与中国古典园林的意境有机结合起来，以女性特有的细腻手法去满足当代中国人的情感需求。”“公园真漂亮，再看看房子是啥样的？”王小姐引领着父子俩向屋内沙盘模型走去，父子俩一边走一边不约而同地回头眺望，希望把柳林公园的美景定格在脑海中。

室内与一般的售楼处不同，没有金碧辉煌、气宇轩昂的气势，而是亲切宜人，像一个商务空间。“请问王小姐是哪家开发公司的？”大儿子张先生发问。“我不属于开发公司，我是房地产经纪人，自由职业者，这是我的名片。”“天津城市大学建筑学学士、房地产硕士，怪不得感觉与过去的售楼小姐的忽悠劲儿不一样。”老张暗自寻思。“现在的房地产开发销售模式已经升级了，过去以开发商为主，开发商拿地后，盖豪华售楼处，批量生产销售。现在是市场和政府共同主导，城市土地管理部门按照规定先建设道路、公园绿化和教育等配套，然后再出让土地，咱们现在的房间就是社区中心的活动展示室。由于规划是窄路密网，所以出让地块都不大，柳林社区现在就有十几家开发商，保证了住房类型的多样性。由于市场竞争，各家开发商对产品、质量都是精雕细琢，而且还可以定制。”“定制？”大儿子张先生心中一动，他一直想按照自己的想法设计和建造一个属于自己的房子，但一直找不到机会。对照着沙盘，王小姐向父子俩深入浅出地介绍了柳林社区规划情况及设计理念。“柳林地区规划以 2050 年跻身世界发达国家的居住社区为目标，就是实现中国梦的住房。设计尝试把临近郊区住房的优点与活力街区的优点相结合，创造出中国新城市主义规划模式。”老张虽听不懂什么新城市主义、什么模式，但他理解王小姐的意思。“我大前年去美国的闺女家住过，她家就在郊区，环境特别好，房子又大又漂亮，还有前后大院子。刚去时真享受，但时间长了就觉得闷得慌，不会说英语，也没朋友聊天，真是好山好水好寂寞。还是觉得国内好，‘金窝银窝不如自己的狗窝’，住着踏实。但城里车真多，有些闹，咱这儿要有美国那样的房子和环境就好了。”

“柳林社区靠近大公园和外环绿化带，空气好，环境安静，而且周边配套好，有著名的胸科医院、‘三甲’环湖综合性医院，社区新建了耀华中学柳林校区和岳阳道小学柳林校区，都是重点名校，可以说是学区房。这里交通也很方便，有地铁一号线、正在建设的十号线和将来规划的市域快线 Z1 线，还有无人驾驶公交、出租车方便换乘，社区内可以完全步行。这里靠近大沽南路和外环线两条城市快速路，小汽车出行方便，考虑到许多家庭的需求，我们为大部分住房配了两个车位。”老张的大孙子家在国外就两辆车。现在车不贵，但在老小区一个车位都不好找，更何况两个车位了。“柳林社区的住房非常多样，规划要求相同体型、造型的住房不能超过 4 栋，且不相邻，建筑以 6 层电梯洋房和 9 层以上、18 层以下的小高层为主，还有少量 3 层联排住宅和 80 米的高层。高层住宅的布置既有很好的视线，又避免造成街道和广场庭院的压抑感。”

看着沙盘上形态各异、造型漂亮的住宅，老张已然心动，可这么好的房子价格是不是很贵啊？儿子看出了老爸的心思，问道：“王小姐，这房子 1 平方米多少钱？”“房价的问题我需要给你们认真解释一下。”王小姐停顿了一下，“目前国家正在制定《住房法》，将规定到 2050 年城镇中等收入家庭都有体面的住宅。所谓体面的住宅，除了功能完善、整体品质好以外，关键是市民能够负担得起，也就是有合理的房价和收入比。房价收入比是指夫妻均有工作，一套适宜住房价格与两人的税前年总收入（包括住房公积金、医疗保险等）的比率，比较合理的范围是 6 左右。具体地说，如果按揭买房，除首付外，每月还银行贷款不会影响生活水平。”看着老张似懂非懂的表情，王小姐进一步解释：“您有孙子吧？哦，您就是给孙子买房。假如您孙子夫妇年收入 40 万元，按照房价收入比 6 计算，买一套价值 240 万元的房子比较可行。除去首付 30%，即 80 万元外，按揭 20 年，每月的还款大约 8000 元，除去各项税费外，余下的生活费用约 10 000 元，可以保证比较好的生活水平。”“房子多大面积？几室几厅？”“这里的房子 100 平方米建筑面积的比较多，小三室，适合年轻夫妇带一个孩子，两个孩子也勉强够用。还有 120 平方米的小四室，两个孩子就没问题了，一人一个房间，即使是异性子女也不怕。”“100 平方米，240 万元，每平方米只有 2.4 万元，这么便宜吗？我听说这里的房价是 3 万元。”大儿子有些怀疑。“对，这里由于环境好、配套全，商品房价格 3 万元起，像联排住房和景观好的高档房还要贵，我刚才讲的是小康房的房价。”

“小康房？”父子俩异口同声。“对，小康房。按照《住房法》，政府有义务为中等收入家庭提供体面的住房，除功能完善、质量好、配套全外，关键是房价合理。现在像天津这样的特大城市房价太贵，房价收入比在 10 以上，造成一般家庭很难买到称心如意的住房，而买得起的房子面积又太小。”父子俩连连点头，感同身受。“房价是由三部分构成的，土地、造价和税费。要保证房子的品质，土建和装修费用不能减；为了保证政府的正常运转，税费减少的幅度也有限，因此降低房价关键是土地降价，土地价又由土地整理成本和政府净收益构成。过去，地方政府经常把主干道、公园、学校建设的成本纳入土地成本，土地成本增加导致政府收益和税费增加，无形中大大推高了地价，这是导致房价上升的重要因素。现在，市政府已经开始控制土地成本，将主干道、公园、学校等由财政投资，不得纳入土地成本。同时，结合房地产税的实施，采用“新房新办法”，降低政府净收益，将政府从前一次性收取 70 年的土地使用权出让金改为分年度收取。原来的政府净收益是土地价格的 25%，改为分年度后，至少可减至 5%，土地价格降低 20%，住房价格也会降低 20%。比如现在柳林地区的商品房，按 70 年税收出让的老办法住房，房价是 3 万元。那么按照新办法，每平方米住宅价格就是 2.4 万元左右，便宜 6000 元。”

“什么样的人可以买这种小康住房呢？”父子俩都很急切。“小康住房是市政府为满足人们对美好生活的向往，学习借鉴新加坡政府组屋的经验，结合天津的实际情况和房地产税改革而制定的政策性住房。小康住房的标准是人均 35 平方米呢，比商品房面积标准低一些，但也接近，2035 年发达国家的标准，每位城市居民都有资格购买，但只能买一套，而且将来也只能在特定的小康住房市场买卖，只能卖给同样资格的人，或由政府按市场价回购。关键是要按年缴纳房地产税。”“那这样的话，买哪类住房合适呢？小康住房虽然价格便宜，但要一直缴税！”“我们就拿上面那套小康住房为例，总价 240 万元，比商品房价 300 万元便宜 60 万元。如果按 1.5% 的税率纳税，每年 3.6 万元。我们先算纳税时间为 20 年，与按揭一样长，总共 72 万元。

好像超过了60万元的差价，但这60万元是一次性交，或者按揭，要还利息，20年的利息大概是30万元。因此不管怎么算，买小康房还是合适的。”“那70年的存量房就不交房产税了吗？”“据了解，现在全国人大正在立法，对于存量房可能会考虑每户一套，人均一定标准以内可以不缴税。但这都是改革的进程，从长远看，房地产都是要缴税的，这也是国际通行的做法。只是我们国家现在各城市对土地财政依赖性强，改革需要一段过渡，所以除小康房外，商品房土地还是一次性收取70年的土地出让金。”听了王小姐一番介绍，父子俩还真涨了不少知识。老张心里合计，2.4万元的房价比以前估计的3万元多便宜了不少，孙子想要2个孩子，可以买120平方米的小四居房子，这样的话总价288万元，首付80多万元，我和儿子帮着凑一下，月供1万多元，即使加上每年的房地产税以孙子孙媳的收入也能够负担，我就不用拿退休工资补贴了。想到这儿，老张一下子轻松了许多。

一转念，他又想起了堂弟的儿子。孩子一直很努力，只是成绩一般，考上二本，毕业后在一个小公司工作，收入一般，他家的收入可承受不了240万元的小康房。“王小姐，这小康房确实不错，我们家、我孙子家都能承受。不过，有些孩子家庭年收入达不到40万元，也就20万元左右，那还是承受不起啊。”“您说的是实情，不过房价与区位关系密切，柳林地区在中心城区外环线以里，环境好，交通方便，所以即使是小康房也比环外其他地区要贵。环外其他地区的小康房有便宜一些的，比如滨海新区核心区外围的小康房每平方米价格在1万元左右，家庭年收入20万元购买这样的住房轻轻松松，而且现在新区的教育和卫生资源有了大幅改善，市内的重点学校和三甲医院都已在新区开设了或即将开设分校和分院以促进发展，并且新区为了吸引年轻人，还推出了共有产权政策。比如一套100平方米的小康房，总房价100万元，首付30万元，如果一下子拿不出30万元，可以只付10万元，另20元万由开发公司承担，与购房者共有产权，这样就让在新区落户的人能轻松购房，没有后顾之忧。”王小姐继续说道，“咱们国家居民存款已达90万亿元，但大家还是担心大病、子女教育、购房等问题，不敢消费，又没有其他的投资渠道，所以人人都投资房产，就出现了泡沫，实体经济受影响。目前需要完善医疗体系，控制不合理的房价，鼓励大家消费，用于提高生活水平，用于发展实体经济。”

“您说得太对了！”大儿子回应道，“我深有同感。我在古文化街有个店，收益还行，一直想扩大规模，或者在意式风情区等旅游区开新店，可又有很多顾虑，一是担心家人生大病，二是孩子要结婚、买房，我得攒钱，这些年房价一直涨，我心里发毛。今天看了这小康房，心里就踏实多了，我决定再开个新店，搞个工作室，也为市里的经济发展做点贡献。”老张从大儿子的脸上看到了许久不曾见到的轻松和自信。“王小姐，带我们去看下样板间吧！”老张虽然70多岁，但一辈子的急性子改不了。“大爷，现在已经没有样板间了，都是实体房，已经有人居住了，看房需要提前预约。你们今天上午临时定的，还没来得及预约呢。”“没关系，我们是临时起意，这是为儿子买房，最后还得听他们的。儿子儿媳已经决定回天津发展，目前正在办理回国的各种手续，等他们回来后，我让他们跟您联系。”“好，没问题！”“王小姐，您别嫌烦，我还有一个问题，您这定制房是什么意思？”“哦，您来这边，咱们一起看一下模型。”王小姐用激光笔指着沙盘上柳林公园两侧的房子。“柳林公园地区除了住宅之外还要发展设计创意产业，在区域内实现职住平衡。这一街区就是商住混合的设计街区，像美国纽约的SOHO区或芬兰赫尔辛基的设计街区。为了体现设计师、艺术家的创意和多样需求，这儿的房子根据业主的需求特别设计建造，标准用地单元是半亩，也就是300平方米，10米面宽，30米进深，都是联排式，3—4层，业主可以自由委托喜欢的设计师和建筑公司，也可以委托开发公司建造。”大儿子很有兴趣：“这一般需要多少投资？”“土地300平方米，容积率1.5，建筑面积是450平方米，地价500多万元，土建造价会有一定差异，大体在300万元左右，两项总计800万元。合每平方米1.8万元，单价不贵，主要是想吸引设计师、设计公司落户。如果张先生有意向，我可以与街道有关部门联系，他们还有税收优惠政策。”“太好了，我回去与爱人商量一下。”大儿子一直想亲手建造一座房子，作为长子，他不仅继承了家里的产业，更希望有一处像电影里那样的百年老宅，楼下是店，楼上是工作室和居室。逢年过节，全家人可以来这里聚会，重要的是能按照自己的心愿建造自家的房子，包括布局、造型和室内装修。“时间不早了，我们告辞了，十分感谢你的讲解！”

告别了王小姐，父子俩回到广场上，他们估计现在往市里走会赶上下班交通高峰，看着广场上逛街观光的人群，决定就在这儿吃饭，体验一下社区生活。他们进了一家临着公园的朝鲜烤肉馆，到二层的露台落座，眺望公园的暮色，一边吃着香喷喷的烤肉，一边畅想着今后的生活，父子俩兴致盎然。

第十部分
重回海河边

56. 鲍尔·滨海茱莉亚音乐厅

研讨会进行得很热烈，鲍尔看了一下手机，已经 5 点了，他拎起包，与大家打个招呼，走出了会议室。诸葛总、宋院长、褚总跟了出来，与鲍尔先生告别。鲍尔执意要大家回去继续开会，不要送他，都是老朋友了，他说自己过一段时间就会回到天津。走出规划展览馆，他给妻子雪莉发了信息，然后向天津站走去。金秋十月，是天津最好的季节，路上游人如织。鲍尔看着这座熟悉的城市，耳边响着城市的嘈杂。秋风袭来，他感到神清气爽。接近下班时间，许多人步行走向天津站。有了城市步行化改造和交通稳静化措施，步行道宽敞又顺畅，走路感觉很舒服。

鲍尔先生边走边思索：我能在这个城市开展新的事业吗？妻子、女儿能在这个城市找到她们自己的事业和爱好吗？对妻子雪莉，鲍尔非常有把握，她已经在天津开展工作了，而对于女儿，他只能建议。女儿在天津茱莉亚音乐学院工作半年了，能否留下来呢？鲍尔刚在火车上落座，就收到了雪莉的两条回复：我快到于家堡了，以及女儿定的餐厅位置。图中显示的是于家堡金融区的新加坡瑞珍酒楼。20 分钟时间，高铁已经把 50 千米的距离抛在身后。鲍尔走出出站口，车站里熙熙攘攘，许多旅客在地下出口直接跟地铁站无缝换乘，站内的快餐店、生鲜超市生意兴隆。鲍尔穿过人流走向于家堡环球购地下商业街时，发现商业街中的人流量比他前几年来时增加了不少，大部分人也已经不是观光的游客，而是年轻的白领，他们说说笑笑，活力四射。

下午 5 点 50 分，鲍尔到了瑞珍酒楼，提前了 10 分钟。女儿定的位置临窗，视线很好，可以看到黄昏的海河。鲍尔去洗手间想整理一下自己的衣着，他记得几年前，每次临时出门再回到酒店，皮鞋上一定会有一层浮土，一天下来领口和衣袖都有发黑的污渍，但今天，衬衣衣领和袖口还挺干净，皮鞋也很亮，看起来天津市里和滨海的空气质量改善了不少啊。鲍尔落座，喝了一口茶，身体和心情都放松了下来。一会儿，妻子雪莉和女儿到了。雪莉身着深绿色晚礼服，看来已经回过酒店了，为晚上出席女儿学校演奏会做好了准备。女儿薇薇安一袭黑色长款演出礼服，看上去既成熟又优雅。

鲍尔与女儿轻轻拥抱，分别了半年多，他感觉女儿更漂亮了，心中升起骄傲的感觉。女儿身边还有一位同样穿着黑色演出服的华人男子，年轻、英俊、沉稳。“这位是尤华，我的同事。这位是我的父亲，鲍尔。”“幸会，鲍尔先生。”鲍尔听到了纯正的牛津腔，接着握到了一只纤长的手，温柔而有力。

"我猜你是钢琴家。""您真是好眼力!"一家人能够在千里之外异国他乡的天津滨海新区会面，还真是难得，这意味着什么呢？女儿从纽约茱莉亚音乐学院毕业后，就来到了天津茱莉亚音乐学院任教，已经快 6 个月了。

鲍尔坐在女儿的对面，感觉女儿的脸上有一丝疲惫。"薇薇安，最近过得好吗？""不错，这里的学生来自中国不同的地区，都十分优秀，很有天赋，又很勤奋，经过指导后进步很快。你们今晚就能看到他们的表演。"尤华接道："他们必须要勤奋，在中国学音乐的孩子太多了，父母都付出了很多时间、精力、金钱，望子成龙。天津茱莉亚音乐学院对他们来说，与国内其他著名的音乐学院相比，学费不菲，但他们为了有这样一个平台，还是将孩子送到这里。""对，茱莉亚找到了一个好生意，一座金矿。"女儿调侃着。

菜上得很快，做工也很精细。席间，鲍尔一直想知道女儿未来的打算，他知道今天时间不充裕，但也没有更好的机会了。他直接发问："薇薇安，你想在天津茱莉亚留任吗？你想留在天津吗？""你真这样想吗？可我不想，绝对！学校规定，6 个月轮换，我不能再待下去了，时间久了，我会发疯的。""为什么？""为什么？我在这里没有朋友，也没有太多的社交活动，对我来说简直就是文化沙漠。""你没有去过城里吗？""我去过几次，那些租界区还挺有意思，不过感觉还是有些奇怪，这些西方风格建筑与东方人在一起，与我的想象不同。不像北京、上海，有很多外国人，英语交流很方便，很国际化，天津反而让我有距离感、违和感。""尤华，你在天津茱莉亚多长时间了？"雪莉问道。"我与薇薇安同时来的，我过去一直生活在上海，毕业于上海音乐学院。我们这些学音乐的，包括学校的老师，当时都好奇，为什么茱莉亚选择了天津滨海新区，与天津音乐学院合作，而没有选北京的中央音乐学院，或上海音乐学院。"大家面面相觑。

尤华没意识到气氛有些尴尬，继续说："我去过纽约茱莉亚音乐学院，周围是艺术中心，是音乐的殿堂，光大都会歌剧院就有 3980 个座位，7 个剧院的座位加起来有 18 000 个，演出场场爆满，这个滨海新区现在还差挺多的。""上海也不行啊，全世界有第二个纽约吗？"薇薇安喜欢打抱不平。"其实滨海新区还是挺不错的，基础做得很好，未来一定很有希望。只是我等不了了。还有就是滨海新区的名字太没有特点了，或者说就不是一个城区的名字，读起来很拗口，很麻烦。读成 BNA 也很奇怪，不如叫泰达或者塘沽好了。"薇薇安喝了一口茶，"其实我们学院旁边有一个老火车站改造成的酒吧街，我和尤华还有学院的同事们经常去，也许今晚演出后我们可以去坐一下，那儿有许多不同的酒吧，晚上很热闹。”很快，一个小时过去了，食物菜品不错，大家吃得很开心。这里到茱莉亚大约步行15分钟，薇薇安起身结账，该出发了。“吃多了，正好在演出前活动一下。” 薇薇安调侃道。

天津茱莉亚音乐学院的建筑由设计纽约茱莉亚音乐学院改扩建的 DSR（Diller Scofidio Rentro）事务所领衔设计。DSR 由 4 位出色的设计师合伙组成，其首席设计师，同时也是创始人是位女性，伊丽莎白 · 迪勒。2018 年她被《时代周刊》评为继 BIG 事务所的英格尔斯之后又一最具影响力的建筑师，也许是因为她拥有女性独特的视角，又或许是她与合伙人在艺术、建筑、生活等领域都有像艺术家一般的创造力，能够在挑战中看到潜在的机遇。《时代周刊》的评语写道：“她使我们的幻想成为可能，将幻象建成真实存在的砖石。”迪勒与斯科菲迪奥（Scofidio）是伴侣，一直携手前行，2009 年他们设计的纽约高线公园在美国家喻户晓。鲍尔参观过 DSR 设计的许多作品，十分喜欢他们的风格。他尤其佩服迪勒，虽然她设计高线公园时已经 63 岁，设计天津茱莉亚音乐学院时已经 68 岁，却有着旺盛的创造力，鲍尔经常用迪勒夫妇来激励自己。

当鲍尔走近茱莉亚时，他发现这个建筑达到了很高的水准。虽然造型与纽约茱莉亚不同，但表达出同样的气质，在创新中呼应着传统，在传统中孕育着创新。建筑坐落在海河边的一片绿地上，像一盏灯的造型，与周边规整、简洁、经典的建筑形成对比。建筑底层局部是架空的，使得内外空间产生了流动的序列感，建筑与环境融为一体。今天的演出地点是学校最大的音乐厅，有近 700 个座位，室内简洁现代，寓意为音乐的容器。茱莉亚的传统是老师与学生共同演奏。女儿和尤华带领 3 名学生，演出了两首经典曲目，先是舒伯特的《A 大调钢琴五重奏（鳟鱼五重奏）》，第二首是中国经典曲目《梁祝》。鲍尔是第一次听女儿演奏中国的音乐曲目，凄美的琴声从女儿的手中流淌出来，使他感到一种情感的升华。他轻轻握了一下妻子的手，雪莉给了一个回应，他能从中体会到雪莉的激动。薇薇安虽然外形更像鲍尔，但内心还是继承了母亲细腻的情感。

两个小时的演出圆满结束了，鲍尔夫妇来到前厅，等女儿卸妆出来。有几位学生和家长也等在那儿。薇薇安出来时，他们送上了小礼物。女儿很高兴，夸奖学生今晚表现优异，向学生和家长表示祝贺。“让我们去庆祝一下！”女儿如释重负，搂着父母的肩膀，向旁边的酒吧街走去。坐在海河边的座位上，海河的风吹过来，柔和又清凉，一家人都很高兴。这里的海河比市中心的宽了很多，视线更开阔，能够看到于家堡楼宇中的灯光和建筑轮廓线，环境给人的感觉很好，而一家人聚在一起的感觉更好。

鲍尔点了香槟为女儿庆祝，女儿关心地询问母亲，在天津的工作怎么样？“天津是个很有意思的城市，天津人热情、仗义，我常常得到朋友们的帮助，一方水土养一方人，这可能与这片土地有关。而我从事的工作恰好和土壤与水环境有密切关系。这里退海成陆，土地的盐碱水平很高，绿化非常困难，过去水比较多，可以压盐碱，但 20 世纪 50 年代根治海河后，上游除了洪水期很少有来水。天津的同行自己讲，他们位于九河下梢，只能承受洪水之害，而无法享受近水之利。历史上，天津保留了大量湖泊水面，可以调蓄雨水，是非常智慧的。我今天参观了新建的水西公园，设计上保留了原有的大片水面，采用了中国古典园林的手法，有一些特点，但这个公园不论从生态角度还是从人的角度看，都还不能令人满意，有很多遗憾。我了解到未来中心城区还要建 7 个大型公园。我向他们提了建议，应多采用生态和可持续的设计，让人与自然和谐共生，在此基础上追求个性和传统园林的新风格。”

“另外，我下午还去参加了一个关于双城间绿色生态屏障的研讨会，这是个大工程，可能会持续几十年。中国有句俗话：前人栽树，后人乘凉。希望这个项目能惠及子孙后代。我看到他们在设计上采用了当地传统的台地法，既经济又实用，这片土地的先人一定还有许多因地制宜的做法值得认真挖掘。在中国倡导生态文明，重视绿化环境建设，真的很像 200 年前著名的‘美国农民’奥姆斯特德（Olmsted）所倡导的一样，中国需要像他一样真正了解人类和地球生态的大师引领。相对来讲，做好景观设计可能比做好建筑设计更难，需要许多人持续的探索。薇薇安，你的爷爷曾在这片土地上工作过，你父亲也想在这片土地上有所作为，我出生在台湾，但大陆是我的根，我对这片土地有特别的感觉，好像有一个声音在召唤我。”

夜深了，鲍尔夫妇送女儿到不远处租住的公寓楼下，女儿上楼前回身问道：“你们觉得尤华怎么样？”“是个优秀的钢琴家。”“人很英俊。”“希望你们能喜欢他。”女儿神秘地眨眨眼，鲍尔夫妇心下了然。“那我们圣诞节再见。”鲍尔又想到，他们在旧金山有一处房子，那是他 30 多岁时自己设计建造的，位置很好，但今天看来当时的设计有许多稚嫩之处。在动身来中国前，鲍尔夫妇已经商量好把房子卖掉，以后要久居中国，他们在这里找到了感兴趣的事情。女儿喜欢纽约的繁华和丰富，不愿意去加州，下一个圣诞节可能是全家在旧金山家中的最后一个圣诞节了。

告别女儿，鲍尔夫妇来到相邻的洲际酒店下榻。酒店的外形像中国的金元宝，原来的寓意是于家堡金融区财源滚滚，但设计还是相当不俗的，开业5年了，酒店人气逐渐旺起来。站在房间落地窗前，望着静静的海河，鲍尔和妻子商量着下一步的安排。明天去北京做一个学术报告，下午去清华大学拜访。他们决定回天津后，去五大道或海河边找一个适合的办公地点。上周一位朋友带他们看了五大道的一处历史风貌建筑，睦南道43号，已经由风貌整理公司腾迁完毕。这是一个U形的两层建筑，虽然不临街，要穿过一条50米深的胡同才能进到内院，但院子内侧在建筑立面上的序列拱廊颇具戏剧性，将来改造时在院子上方加上漂亮的玻璃顶，中央设计成一个共享中庭，无论项目交流或朋友聊天，都会是一个极有魅力和风情的交往场所。

这个想法深深吸引着他们，他们希望继续与风貌公司协商租金的事宜，如果天津能像欧洲一些城市一样，对建筑遗产的修缮利用提供一些优惠政策，比如低价或一元钱承租，然后由承租人负责修缮、改造，那天津小洋楼的活化利用该会有多么丰富的可能啊。另一个关键是要找两到三个合作的设计师，最好是有国外学习和工作背景的设计师。办公地点相对好找，但他们很难找到有一定能力又愿意在天津长期做事的有国外工作经历的设计师。北京、上海、深圳的国际人才很多，如同纽约，各式各样的人才才是一个城市成功的基础。他们感觉，国外设计师不好找，但天津本地设计师还是不少的。天津是有人才基础的，有几十所大学，30多万在校生，鲍尔夫妇相信只要自己努力坚持，一定能从中筛选和培养出他们想要的设计师。

57. 宋云飞 · 滨海文化中心

上午的会议到12点多才结束，宋云飞在会场吃了盒饭，他扫视四周，没有发现诸葛总的身影。今天一早见到诸葛总时，他就想聊几句，但看到诸葛总忙碌的样子，不断有专家在跟他交流，就没好意思打扰，只是礼貌性地跟他打了个招呼。诸葛总在台上汇报了城市设计的回顾总结，详细介绍了这次城市设计试点的主要内容和改革创新的尝试。上午看展览时，经诸葛总解释，原本的一些疑惑就清楚多了，但要归纳提出在城市外围地区鼓励发展小汽车的策略，宋云飞还是有些疑惑。

不管是在国内大学学习道路交通专业，还是后来出国读交通规划方向的研究生，上过的课、读过的书都异口同声地要限制私人小汽车的发展。全世界，主要是美国人，担心中国和印度如果都鼓励发展私人小汽车，达到美国的人均碳排放标准，地球会承受不了。虽然目前新能源汽车 NEVs 在中国增长迅速，无人驾驶技术也快速成长，但现在就提出鼓励发展小汽车，宋云飞还是不能赞成。“每家可以有两个车位，这是可能的，但发展小汽车不行，这是为什么？”对诸葛总在与专家互动时的反问，宋云飞也找不到好的回答。宋云飞发现，原来温文尔雅的诸葛总讲起话来语气逼人，他说：“我们规划一直单方面地限制小汽车，鼓励发展公共交通，但老百姓‘用轮子投票’，中国已经连续 8 年蝉联世界第一大汽车产销国了。”

没能找到诸葛总，宋云飞有些惆怅，也有些庆幸，像心里的一块石头落地。他这次到市里，除参加论坛和研讨会外，有一项更主要的任务是找两个甲方谈规划设计合同的事儿。本来想去之前请诸葛总帮忙打个招呼，这也是妻子不断叮嘱他的，但他总觉得这样不太合适。宋云飞从小到大学习好，没让父母操过心，学校里的老师都喜欢他，他也从未向哪位老师求过情。后来上大学期间联系出国留学，他也完全不靠别人，包括请导师写推荐信，他也会犹豫很久，眼见快到截止日期才去求情。听同学说，要上国外常春藤学校，除了成绩外，有国外名校的老师写推荐信是很加分的。在国外，熟人的信用价值是很重要的，这也是他在工作以后慢慢认识到的，但他的性格还是让他难以行动。

回到国内创业，宋云飞知道中国更是很重人情，但他也更愿意相信自己的专业能力。虽然政府一直在推出政策进行改革，但中国的“关系文化”根深蒂固，有些局面很难在短时间内改变。天津在这方面更突出，也包括滨海新区。妻子说，你找诸葛总说一下，这不是违反纪律，只是因为诸葛总比较了解你的能力和在滨海工作的经历，别人说他们也不信，诸葛总帮你说一句，甲方容易相信。但宋云飞感觉这还是有违规之嫌，感到面子上过不去，刚才见到诸葛总时他也没张开嘴。现在散会了，他又找不到诸葛总了，与甲方约的时间快到了，他必须马上出发。他总觉得自己努力了，时机不对那怪不了他，他甚至有如释重负的感觉。

回国在滨海新区创业已经 3 年了，宋云飞目睹了许多双创企业最后没能活下来。但的确也有少数成功的企业，比如最近，康希诺生物股份有限公司成功在香港证交所上市，是首只疫苗股，也是国人创办的企业。可能是由于企业性质不同，新能源、生物制药、新材料等产业以及移动互联网、新媒体等产业受到国家和天津市较多的支持，形成一定的集聚效应。但宋云飞他们作为交通规划资讯类企业，成长比较困难。公司初创时势头还不错，有政府支持，承诺 3 年不收租金，还给了置办费，员工租房有补贴等，关键是有几个咨询研究的项目垫底，开始时信心满满。第一年招收了十几个员工，都是交通规划专业硕士、博士，科研能力比较强，他们完成的滨海核心区智能交通发展战略等课题受到好评。但公司的业务面比较窄，拓展市场能力也不强，收入增长缓慢，只能维持。有员工对收入不满，逐渐有人离开去了成都、武汉等地，各地都在吸引人才。公司内部的管理结构也不太合理，员工都只干具体工作，只有宋云飞一个人经营、管理，还要做项目，手忙脚乱。

公司发展停滞不前，而于家堡却开始蒸蒸日上。开发区与中心商务区合并，管委会也从开发区搬到了于家堡起步区的最南端，在于家堡注册的公司逐渐多了起来。随着人流的增加，写字楼的租金也开始上涨。听说写字楼租金收入多少与管理公司上缴税收的多少挂钩，管委会也修改了补贴企业的政策。宋云飞的咨询公司还没有稳定的项目收入，也没交过企业所得税。可下个月就要开始交租金了，虽然每平方米只有 1.2 元，办公室租用的建筑面积有 400 平方米，加上公摊一共 500 平方米，一天 600 元，一个月 1.8 万元，一年要 20 多万元，再加上物业费，公司已难以承受。如果近期不能至少谈成一个大项目，公司就很难维持下去了。所以，最近一段时间，宋云飞一直在千方百计找项目，争取能签下合同，当然能打一部分定金就更好了，下月可以给员工发工资，上“五险”，交租金。

宋云飞今天下午去天津西站和西营门，就是争取签下这两个项目的合同。他关注天津西站很长时间了。在建设之时，他就认为当时的设计有很大的问题，布局模式是国内传统的单一铁路站房模式，建筑是一个 50 米高的拱形，很有气势。但旅客感觉空旷，流线松散、混乱，这种情况在国内很有代表性。虽然国内高铁发展突飞猛进，几年之内就形成了高铁网，极大地缩短了时空距离，促进了区域发展，但在规划建设上，存在两个突出问题：一是绝大多数高铁站选址在远离城市中心的地方，二是绝大部分站房仍然是陈旧的铁路站房模式。

天津西站是由老的车站改造的，位于中心城区，主要问题是布局不合理，可达性差，换乘不方便。宋云飞做了长时间研究，设想西站改造优化的可能性，终于找到了一个行之有效的方案。他希望西站通过规划改造的实施，成为一个国内老铁路车站改造的样板，但苦于一直没有找到合适的机会实现这个想法。他曾找过铁路部门，铁路部门领导说我们铁路只管站房屋檐在地上投影以内的部分，外面是当地政府管，你的方案涉及的很多方面不是我们能定的。他又去找了在区政府的站区管委会，管委会领导说我们是个协调部门，没有资金，做不了具体改造。时间就这样过去了。

最近听说市政府响应各方面的反映，要对天津铁路车站和飞机场进行综合提升治理，好不容易盼来机会，宋云飞千方百计约好了各部门的领导，包括铁路、区政府、地铁、公交等各部门。在现场，宋云飞汇报了方案构想，他前晚做了充分的准备，设计师把站前空旷的硬铺广场变成吸引旅客的休闲中心，成为指引旅客换乘和改善地下空间环境的阳光植物大厅。听完他的汇报，大家都认为很受启发，方案深入考虑了解决西站老大难问题的治本之法。各部门都认识到要改善西站的站区环境，需要综合施策，要对现状建筑结构布局有较大改变，不能表面化处理。经过较长时间的讨论，交通规划部门的领导总结道："今天我们召集大家开会，一是研究一下宋总提出来的根治西站交通枢纽问题的方案，二是推动一下西站整治提升工作。宋总的方案给我们大家耳目一新的感觉。当然，这个方案比较复杂。需要新增投资，还要研究地下和广场空间改造的技术可行性以及各部门协作的问题，这是方案成立的前提。我回去后马上向领导汇报，争取加快推进。至于西站提升整治，这是个硬任务，3 个月要见成效。现在已经开工一个月了，各项工作都在按计划快速推进，但有的工作还没到位，大家都得抓紧，按预定的时间表完成，避免被动。"

会议结束后，主任向宋总再次表示感谢，并表示回去马上向领导汇报。宋云飞有些感动，同时也感到无奈。他知道，这项工程涉及方方面面，回去研究不是短时间内能定下来的，远水解不了近渴。事到临头，宋云飞也顾不上面子了："主任，我们对这个项目前期投入了很多人力物力，现在公司经营遇到了困难，您看有没有可能先委托我们开展前期研究？""宋总，非常感谢你为天津的城市建设主动开展工作。但是你知道，我们现在项目管理非常规范，包括科研课题也都是要按照年初的计划，当年新增加项目没有渠道，得等到明年才能申请新的计划！我这次也向领导一起反映一下这个问题，不过根据我的经验，可能性不大。""您费心了！"

从西站站区管委会出来，宋云飞骑上共享单车绕了一大圈，找到了西营门街道办事处。西营门管辖地区有 20 多平方千米，共 4 万多人，横跨外环线两侧。外环线内大明道地区近期拆除了"散乱污"企业，有较多的开发空间。大明道这个区域有 3 平方千米，区位还不错，但交通不方便，规划远期建一条地铁线，一个车站，车站覆盖率和线网密度都非常低。要达到公交出行占70% 的目标，必须增加多种形式的公共交通，可地面交通容量又有限，仅靠地面常规公交难以满足要求。宋云飞的公司联合其他设计公司编制了该地区的规划，提出采用高架新型电车形式。

现在国内许多企业在做高架新型电车研究和试验线，他参加过几个项目的评审，企业希望咨询公司帮助推荐使用他们的产品和技术。总体看，高架电车道比地铁便宜很多，而且比较灵活，运营成本低，但目前没有实施的案例。像西营门这样的地区，在美国就是一个小城市，市政府是独立的，可以承揽自己的事务。而国内的街道办是政府的派出机构，社会管理职能很多，要做城市规划建设还不具备条件。宋云飞做了很多解释，包括投资方面，如果目前没有资金，企业可以做 PPP 投资。但街道办领导说："这些我们都说了不算，至少要区政府决定，关键是市政府有关部门要提出标准。这东西虽然好，但没标准，建成了我们也不敢用。这可是涉及民生的大问题，出了问题负不起责任。"

从中午 12：30 从规划展览馆出来，宋云飞奔波了一个下午，几乎一无所获，他身心疲惫。中国城市的交通问题十分严重，每年投入巨额资金在高铁、地铁的建设上，对公交运营的补贴也不少，小汽车、新能源汽车是许多城市的支柱产业，但就是没有资金做一些研究和提升规划设计。这些咨询费与地铁动辄几百亿的投资相比，九牛一毛，但能发挥巨大作用。这样看来，很多城市问题是思想认识和体制机制造成的，特别是在前期研究和规划设计上，无论是在经费上还是在时间上都严重不足 。比如，国家曾经发文鼓励高铁车站及周边的综合开发，但 5 年多过去了，也没看到什么进展，主要是缺乏配套的政策和前期规划研究。

宋云飞心里感到无形的压力，就像如果大家都习惯开车了，再建公交去改变出行习惯就难了。这些话他已经对无数人说过无数次了。现在他开始有些怀疑，大力发展公共交通被寄予了过多的美好愿望，也许公共交通的发展现状就是一种必然。十月是天津最好的季节，最适合骑单车，但宋云飞没了心情，他叫了一辆快车。来到西站，从后广场的地下通道进入候车厅。他的脸色十分难看，就像一个落魄的书生，悲愤而沉默。

5 点 30 分，宋云飞登上了开往滨海新区的天津城际，到达于家堡高铁站约 1 个小时，中间必经天津站换乘，没有直达车。火车开出来，高速平稳运行。随着高铁有节奏的声音，宋云飞的心情平复了一些。他知道，他今天的经历就是天津和国内普遍的现状。他预感到这次回国创业凶多吉少，公司可能要关门了，或许时机不对吧。下了高铁，宋云飞驾车很快就到了滨海文化中心。本来，有地铁的话，从于家堡高铁站到文化中心只有一站距离，但好事多磨，滨海新区规划建设地铁已经 10 年了，Z4 线还没建成通车。

宋云飞从地下车库来到文化中心的长廊，这里建成 3 年多了，一开业就活力十足。作为国内乃至国际上的网红建筑，滨海图书馆一直有人慕名前来，需要排队进入，看来荷兰 MVRDV 设计公司赚足了广告费。宋云飞当年在香港 MVA 交通事务所时，也参与了滨海文化中心前期的交通组织研究工作。他最欣赏的还是文化长廊设计，步行空间把 5 个文化场馆连接起来，去每个场馆的人都会先进入文化长廊，使得人们有了更多见面的机会，长廊空间也得到了充分利用，德国 GMP 设计公司的长廊设计既简洁大方又精致严谨。另外，几个场馆的停车库可以实时共享，大大提高了停车效率，内部设计也十分人性化。

宋云飞径直来到滨海文化中心的市民活动中心，女儿在这里上海伦 · 奥格雷迪（Helen O’Grandy）的戏剧课。女儿性格外向，喜欢表演，这点与宋云飞夫妇不同。“今天怎么样？”一见面，妻子就关心地问。“不太好啊。”宋云飞情绪低落。听到这样的回答，妻子有些着急，但想到宋云飞这一天的奔波肯定不容易，就忍住了。

回国创业前，宋云飞先是在波士顿读了 2 年硕士，毕业后到湾区交通委员会工作了 3 年。湾区华人多，在那儿他认识了妻子并结婚育女。波士顿和旧金山是美国两个最有文化特征的城市，宋云飞很喜欢也很适应那里的氛围，但为了事业发展，他应聘到了香港，待遇还不错。妻子是学生物化学的，在湾区工作原本不错，但为了照顾孩子，辞职在家。可她和孩子都不喜欢香港，尤其是房子太小了。妻子是非常支持宋云飞回内地到滨海新区创业的，他们把家安在了生态城。3 年时间内经历了很多，是非常宝贵的经验。

吃饭时，夫妇俩都沉默着，女儿在大声朗诵着《绿野仙踪》中多乐西的台词，沉浸在她的话剧世界里。妻子拉住了宋云飞的手：“云飞，我想好了，我们回湾区去工作吧，女儿现在大了，也好带了。没关系，我们重新学习，重新就业。我相信等有适当的机会，我们还会回来。当年我外公他们在塘沽创业，抗战期间，为了保住民族工业的命脉，在局势危急、处境艰难的情况下，把工厂迁进了四川，不仅延续了生产，还扩大了销路。中华人民共和国成立后，两厂又搬回到塘沽，招兵买马，重新生产，终于把‘久大’‘永利’做成了声名远扬的知名品牌，成为民族工业的象征。”听到这儿，宋云飞心里生起了一股莫名的委屈，泪水涌满眼眶……

车开出了滨海文化中心地下车库，眼前的滨海文化中心光彩照人。妻子最喜欢滨海科技馆，是由美国著名建筑师伯纳德·曲米设计的。建筑的造型像是传承了天津碱厂的精神，古铜色的穿孔金属板外檐十分具有工业感，形似铜火锅的锥形空间的顶部有一束灯光直射苍穹。宋云飞感到一股力量。汽车马上就要进入穿越进港铁路和津滨轻轨的地道，这时女儿唱起了《绿野仙踪》中的插曲《飞跃彩虹》：“彩虹深处，天空湛蓝，有一些梦想，你敢憧憬就真会实现……”穿过地道，车窗前面出现一片光明。

58. 老张 · 海河

老张和大儿子 8 点回到家。晚高峰过去了，车在海河东路上开得非常快，只用了不到 30 分钟。大儿子了解老爷子的心情，没等父亲开口就打开了车窗，海河边的风让人十分痛快，夹杂着一股湿润的气息。老张眯着眼睛，深深地吸了一口气，海河那熟悉的味道让他感到踏实、满足。儿媳妇已照顾老伴儿吃过晚饭，在看电视剧《破冰行动》。看到爷儿俩进来，客气地与老张见过面，冲着大儿子大声喊：“你一下午干什么去了？微信也不回，人家老王还一直等着呢，赶紧赶紧！”边说边拉起儿子的袖子就往外走。“爸，妈，我们走了啊！”儿子的声音消失在楼道里。老张对着儿子儿媳的背影说了句：“慢点！”

老张也没有换衣服，在客厅的沙发上坐下来，挨着老伴儿的轮椅，他想把一天的见闻向她念叨念叨。不过他看到她看电视剧的专注表情，就打住了。自从老伴儿中风后，虽然恢复得不错，但还是与以前不一样了，不仅行动慢了，活动范围有限，而且接触新事物的机会也少了。老张坐了一会儿，对电视剧他不感兴趣，那都是瞎编的。他剥了一个橘子，递到老伴儿手里：“我出去转转。”老伴儿答应了一声。老张下了电梯，走出楼门。小区里有人遛狗，有人行色匆匆，夜幕笼罩，一片安静。老张边走边想，不知不觉又来到了海河边。

好像已经成了习惯，老张现在不管遇到什么事儿，都会下意识地往海河边走。可能是海河边有人、热闹，也可能是海河边开阔、有灯光，还可能是海河边能闻到湿润的水气，老张就喜欢海河水的味道。随着生态环境特别是水环境的改善，海河上的海鸥重新回来了，白天在古文化街亲水平台沿岸，有许多市民游人投食、拍照，一幅人与自然和谐的画面。古文化街平台是用来做活动的，2004 年天津筑城设卫 600 周年时在这儿举行了庆典活动，后来又多次举办了妈祖祭拜活动，现在成为主要的海河游船码头。天黑了，游客逐渐减少，老张坐在远处大台阶一角，旁边就是亲水平台上的 20 多米高的“百福”灯柱，那边对称的地还有一根，每个灯柱由 12 个镂空的不同字体的“福”字组成，像是一种灯笼柱。冬季来临，海河河面结冰，游船停运，许多人在上面滑冰，还有人在冰上凿洞钓鱼。每到春节，古文化街都很热闹，像庙会一样，亲水平台上许多人燃放孔明灯，用这种方式祭拜祖先。

静静地坐在平台上，老张的思绪停止了。望着远处的灯光，他忽然觉得很孤独，感觉心很累。自父母和大伯先后去世后，老张好像失去了家，失去了与祖辈的联系，失去了依靠。虽然没有经历过战争，但他也经历了许多，“大跃进”、“文化大革命”、唐山大地震、改革开放、企业员工下岗、自主就业。在外面，他要像条汉子，托关系，找门路，在权贵面前也要会鞠躬、作揖；在家里，在老伴和孩子面前，他是家里的顶梁柱，从来不敢诉苦、示弱；地震来袭时，他挺身而出，顾不上自己的安危，左拉右扯着孩子们，保护家人的安全，还要表现出镇定自若；然后，他还要想办法尽快找到一块空地，用木材、雨布搭起简易抗震棚；他要找工友帮忙，修复震损的房子，干完活给人家沏茶递烟表示感谢。改革开放后，他承受了工厂关闭、自己下岗的各种憋屈，要自谋出路养活一家大小；现在，他要照顾身体不好的老伴，各种情绪都得压抑下去，生怕老伴不高兴。想着想着，老张泪流满面。他没有地方去诉说，没有可以痛痛快快表达自己情绪的对象。月光洒在河面上，泛起粼粼水波，好像在看着他，听他说话。

老张今年快 70 岁了，与 80、90 岁的老人比起来，他感觉自己还年轻，身体没大毛病，这十几年他习惯了这样日复一日的生活，没有什么大的变化，感觉自己都麻木了。与过去相比生活已经很好了，不缺吃不缺穿，按说他不应该有什么抱怨，但他心里就是会生出很多不满，在家里他忍辱负重，在外边他感觉自己正在被社会抛弃。

自打 23 岁成家立业以来，作为一家之主，他一直绷着劲儿，没有一天敢放下责任为自己活。他想起自己年轻的时候也是有许多爱好的，他喜欢做木工活，家里的床头柜是他亲手打的，儿子冬天在海河上滑的冰车是他给做的；他还在厂里参加过宣传队，说过天津快板和相声；他还喜欢钓鱼和摄影……想到这儿，他手心有点痒痒。他想起早晨出门时，看到居委会门口贴了张告示，说鼓楼街道下辖的社区要进行环境升级改造，设计师进入社区指导，居民自己动手，改造小区的公共绿地。老张决定明天一早就去报名。

还有，今天上午看了规划展，下午去了柳林公园和公园旁的社区，老张心里有了一丝憧憬。他希望自己的儿子、孙子、孙女都能快快乐乐、健康成长，上个好学校，有个好前途。他衷心祝愿那些规划能早日实现，希望表弟和侄子一家都有公园边的漂亮房子。他也梦想着，有一天，等他老了没法儿在海河边遛弯时，能住在郊区的一套带院子的房子里，种点花草，养一条狗。

他想起小时候父亲跟他讲，张家上溯十几代人都在天津生活，也就是刚有天津卫那会儿，他爷爷的爷爷的爷爷……带着一家人从安徽闯荡到天津，靠耍把式卖艺在码头上站住了脚跟，因身大力蛮，远近闻名，艺名就叫张大力。想到这儿，老张耳边仿佛响起小时候他和胡同里的小皮孩们常说的那段顺口溜：别看小生胳膊细，跑过江湖卖过艺，吞过铁球练过气，吃过黄豆放过屁……老张笑了。

59. 天津的城市设计者

天津城市设计论坛和国际研讨会圆满结束了，为了准备这次会议和策划天津城市设计展，褚总和同事们付出了很多，整整花了半年多的时间。所里的技术骨干多数是女将，恰好有 3 位都怀孕了。

经过忙忙碌碌的 6 个多月，天津城市设计试点工作全部完成。为总结试点经验，诸总及团队编写了《重回海河边——天津城市设计的故事》，在国际论坛上首发；编制了在国内创新的两个导则：《天津新型居住社区城市设计导则》和《天津市中心城区开放空间设计通则》；完成了天津城市设计展的策划、设计和布展；准备了在国际论坛上介绍天津城市设计试点经验的成果汇报。伴随着这些丰硕的成果，苗副总的第一个儿子呱呱落地，小付因肚子已经行动不便，她穿了一件宽大的外套，一直忙前忙后。

天津城市设计试点提出了许多创新的观念，引起大家热烈的讨论，不得不延时到 6 点半才结束。送走了各位嘉宾，已经 7 点了。展览馆华灯初上，意式风情区步行街上的人流开始多起来，刘局长、诸葛总和宋院长准备离开，在大厅向大家告别。“这次城市设计国际研讨会效果不错，既宣传了天津又提出了许多创新的思想，祝贺。”刘局向围拢上来的规划师和志愿者双手合十，“十分感谢大家的辛苦付出！我代表局领导向大家表示衷心的感谢！褚总，您辛苦了，让大家早点回家休息吧！”

送走了刘局长、诸葛总和宋院长，褚总几人将会场上的材料收拾起来，然后向志愿者表示感谢。城市设计论坛和研讨会结束了，城市设计展区还要持续一个月，还需要志愿者继续服务。褚总把签好字的新书《重回海河边》亲自送给每一位志愿者，与志愿者告别，又找到一直在忙碌的城市规划展览馆史馆长表示感谢。这一切都做完了，她终于松了口气。看着大家放松下来的脸色和略显疲惫的神情，她提议："这一段时间大家辛苦了，今天的活动很成功，多亏了大家的努力和辛勤付出。这样吧，谁家里有事儿，可以赶紧回家，如果大家不反对，其他伙伴们，我们就到海河边吃点儿东西，放松一下。"大家表示赞同。

出了规划馆，左转穿过平安街，他们来到海河边。海河边的灯光绚丽夺目，北安桥上的金色雕像熠熠生辉。再向北走不远，来到奥式风情区，这里是海河边少有的步行段。为了解决海河和市中心的交通，海河两岸一般都是城市道路，隔离了人与亲水堤岸的联系，但这一段为了保护袁世凯故居等历史建筑，同时更是为了让人有机会亲近海河，交通性道路移到了建筑后边。褚总和同事们设计了这组奥地利风格的建筑，当时就设想滨河的建筑底层应该是咖啡馆和酒吧。这组建筑经过许多困难，终于建成了，完美地呈现在海河边。建筑是奥式建筑传统模式、比例，但局部造型和材料是现代的，与邻近的袁世凯故居相得益彰。

很多人都说奥式风情区的建筑漂亮，它也为海河两岸历史街区的新建筑设立了标杆。但由于各种原因，这组建筑成为公司和事业单位的办公楼，开始时临海河的一侧都是封闭的，没能按规划设计的意图形成开放的咖啡馆和酒吧街。河畔种了很多高大挺拔的梧桐树，绿树成荫。到了傍晚，周边的居民来到河边散步、闲坐。一群大妈每天都来此跳广场舞，但录音机播放《最炫民族风》的音量对想在海河边静一静的人来说就像噪声。奥式风情区建筑的造价较高，一般公司难以承担。后来几家银行看中这个地方，因其水流生财，但又嫌周边太乱，所以每家银行都在门口摆上两个硕大凶煞的大狮子雕像。自打摆上这些狮子，大妈们的跳舞阵地就转移了，看来这狮子的威力还真大。近年来，市政府大力发展夜间经济，出台了大量扶植政策，随着游客的增多，意式风情区和海河游船人满为患，需要新的发展空间和新的业态。终于，奥式风情区濒海河的这些建筑在建成 15 年后终于形成了品位极佳的西餐吧。沿海河是外摆的西式餐桌，在梧桐营造的临河空间下，人们品尝着美食，看着海河上流光溢彩的各式游船，十分惬意。

褚总一行选择了一排靠河的座位，大家辛苦了一整天，中午在会场吃盒饭，现在早已饥肠辘辘了。几位女同事都点了巨型汉堡、薯条，狼吞虎咽，忘记了吃相。有人打趣："小付，多吃点，别把肚子里的儿子饿着。""好，儿子饿得都踢我了，今天得多吃点。"看着大家几个月来少有的欢笑，褚总十分欣慰。她的目光由河上的游船转回到岸上的建筑上，在灯光的映衬下，这组造型丰富的建筑十分迷人。浏览着这附近海河综合开发建起来的各种建筑，联想到在行业里打拼的辛苦和不易，褚总感到既自豪又心酸。

当年，为了高质量完成海河起步段的城市设计和建筑设计，褚总代表城市设计所邀请了鲍尔先生合作。鲍尔先生非常用心地投入，付出很多心血，方案赢得甲方和领导等各方好评。但要把这个项目实施建成，还要付出更多的努力。有一段时间，工地上问题不断，而鲍尔先生有事突然回国。面对恼怒的甲方，褚总承受了很大压力，经过千辛万苦，这座建筑终于高质量地完成了，也锻炼了一批年轻的既懂建筑、又懂城市规划的城市设计师。后来，随着时间的推移，这一批骨干也都陆续离开了城市设计所，奔赴重要岗位。然后，不断有年轻设计师来了又走，走了又来，每有一批更年轻的设计师加入，都需重新培养。这也是中国设计行业的一个怪圈，通常设计师刚过 50 岁，就不再动手画图搞设计了，都去当领导，指挥别人；而国外的事务所中，绝大多数都是富有经验、一直坚持做设计的中年设计师。这就是我国城市设计、建筑设计与西方国家有总体差距的主要原因之一。

大家都吃过饭了，年轻人又来了精神。罗云拿出了一本崭新的《重回海河边》，来到褚总身边："褚总，帮我签个名吧！""好，我又不是什么明星大腕儿，签名也不会升值哈。"褚总边说边在罗云递过书的扉页上写道："谨以此书献给天津的城市设计者！祝罗云总取得更大进步！""褚总，也给我签一个吧。"大家都围了上来。"好，一个一个来，你们大家也都在一本书上签上名，送给我好吗？"罗云把大家签好名的书递到了褚总手里，扉页上密密麻麻签满了名字，就像每一个人的脸，生动而不同，充满个性，字如其人。"谢谢大家！我们去年搞了个活动，庆祝城市设计所成立 15 周年，邀请了对我们有帮助的领导和曾经在所里工作过的同事寄语，我也向大家讲述了城市设计所成长的历史和优良传统。今天，我想在这儿，在海河边，再讲讲。"

21 世纪初，为了进一步改革开放，我国决定加入世界贸易组织，就是要按国际惯例办事儿，关键是要开放市场。改革开放初期，我国的建筑设计市场就向国外开放了，虽然也有一些限制，要求国外与当地设计院合作设计施工图。但城市规划行业以涉及保密等各种理由，一直没有开放。2001 年 11 月 10 日，世界贸易组织第四届部长级会议审议通过中国加入 WTO。面对马上就要开放的规划市场，当时在规划界一片惊恐的声音。为了应对这种局面，天津市规划院的领导决定改革，成立两个专家工作室，意在提高自身竞争力，作为应对中国加入 WTO 后国外大规划公司进入中国抢占市场的反制准备。

褚总毕业于清华大学建筑系，论资历、论能力，她早可以当所长了。老所长退休前跟她谈过，要她接班，但当时的生产所以挣产值为主，褚总希望把更多的精力放在规划设计技术本身，所以她回绝了老所长的好意。这次院里成立专家工作室，主要目的是在专业上有所建树，褚总决定当仁不让。经过公开竞争，她成为专家工作室的首席设计师。事非经过不知难，只有亲力亲为，才能真正理解万事开头难的道理。褚总带着多年一起工作的 3 位同事和新招收的几个年轻设计师通宵达旦、全力以赴。招人时她亲自面试，无论个性多么鲜明，只要是有能力、有追求的年轻人，她都收下。褚总希望用自己的包容给年轻人以成长的空间和机会。她认为刚成立的工作室要想与国外的大设计公司同台竞技，首先必须要学习和成长。如何学习？最直接的就是与国外好的设计公司、设计师合作。

不像院里的外地分院有优惠政策，专家工作室与院里其他所是一样的分成比例，自己的项目与国外设计师和设计公司合作，就意味着要把规划设计收费的大部分给合作方。国外设计师做出了设计方案、草图，专家工作室的同仁们就把这些方案变成正式的成果。但由于大部分收费给了合作方，所以专家工作室最初几年收入总是在全院垫底，让褚总承担了很大的精神压力。大家干得很辛苦，但收入不高，褚总不停地鼓励大家。经过几年的坚持和努力，终于收到成效，专家工作室在城市设计和历史保护方面小有建树，收入也提高了。在做好实际项目的同时，褚总不忘学习和总结，她组织全所编写《城市设计在中国》一书，用自己的亲身经历和设计作品表达了他们对中国城市设计的立场和观点。书出版后不久，国内许多城市的市领导、管委会的领导、规划局的领导、城投的领导带着这本书找来，说：“这本书写得太真实了，切中要害，从书中不仅看出了你们的能力和水平，也看出了你们的追求，我们就邀请你们做城市设计。”最近陈祺所长统计了一下，因《城市设计在中国》找上门来的项目，设计费加起来达到了 2000 多万元，书竟然起到了广告的作用。

2008 年，由于在城市设计方面的突出表现，专家工作室更名为城市设计所。在城市设计之外，他们还尝试用城市设计的方法做历史街区的保护与更新规划。在天津，他们长期对五大道历史文化街区进行研究，用类型学的方法对五大道的建筑、院落、街巷空间进行分析，还与加州大学伯克利分校合作举办五大道规划设计工作营，编写了五大道的规划成果及学术专著《中国 · 天津 · 五大道——历史文化街区保护与更新规划研究》，并获得全国优秀城乡规划设计一等奖。“这些是我们城市设计所的光荣传统和成功经验。长期以来，我们关注天津，扎根天津，这也是我们与外方竞争的一个优势。无论多么了不起的规划设计理念，都不能仅仅停留在纸面上，而是要能真正实施落地，还要能深刻理解它对人、城市和环境所产生的深远影响。”

“当前，国家正在转型升级，天津面临着一些暂时的困难，我们设计行业也受到了影响。越是在这种形势下，越是要坚持。这次城市设计试点，大家付出了很多，虽然没有可观的经济回报，但经过大家的努力，达到了很好的效果，许多观点和方法是世界性的趋势，也是我们国家当前面对的最现实的问题。我们从今天的论坛和研讨会的反馈中可以感受到，它们触动了很多人。这是对我们的激励，也是新的开始。”

一轮明月挂在天空，海河水泛起粼粼波光。褚总站起身来，面向海河方向：“天津的城市规划建设能取得今天的成绩，是天津一代又一代规划人努力的结果，是无数城市设计师一笔一笔画出来的。老一代规划师给我们留下了宝贵的传统，我们要一代一代传承下去，发扬光大。道路是曲折的，前途是光明的，城市发展也绝不会一帆风顺。天津的城市建设取得了成绩，也存在很多问题。我们要直面问题，面向未来，充满信心。”

“当初天津编制《海河综合开发改造规划》时，定下目标，要把海河打造成世界名河。十几年过去了，海河的水变得清澈洁净，海河两岸变得美丽动人。有人说，海河已经是世界名河了，比巴黎的塞纳河还要美。这叫自高自大、自以为是，实际上海河与塞纳河相比还有很大差距。这差距不仅体现在硬件上，也在软件上，关键在人。天津市新一轮国土空间总体规划提出，到 2035 年要把天津打造成世界名城。我相信，再经过我们 15 年的奋斗，天津一定会实现这一目标，从世界名河到世界名城。今天，让我们约定，到那时，我们再重回海河畔来相聚、庆祝，好不好？”“好！”大家异口同声地回应道，激昂的声音在寂静的海河上空回荡，像是一个神灵的低语。

60. 画龙点睛

论坛结束后，诸葛民陪同巴纳特先生和丹·索罗门先生一起到酒店共进晚餐，既是感谢也是饯行，聊表心意，他选择了意式风情区的狗不理包子店。他曾犹豫过，两位专家到天津多次，狗不理吃了也不下 3 回，俗话说“好事不过三”，还吃吗？意式风情区这里有其他好餐厅，环境更好、菜品也更精致。不过，这犹豫也就是一闪念，诸葛民还是下决心选定了狗不理。好在这店新开张不久，两位外国专家还没到过这家店，诸葛民打心里希望狗不理再有些改进，就像它给自己起的英文名——Go Believe。

现在风气改变，请客吃饭、陪客少了，陪同的只有诸葛民和规划院的宋院长。4 个人相聚在一起也是机缘，趁着宋院长点菜，诸葛民首先向巴纳特先生提了个问题。“巴纳特先生，2003 年您到天津参加解放北路城市设计评审，针对当时在历史街区中已经存在的几栋高层建筑比较突兀的状况，您曾经给出了两个方向：一是不再建新的高层，即使那几栋突兀的高层不好看，也让它留在那里，就像巴黎的做法一样。巴黎老城区唯一的高层是建于 1972 年的蒙巴纳斯大厦（Montparnasse，59 层，209 米高），孤零零地矗立在那儿，像是一个不可忽视教训；第二个方法，就是围绕现有高层，再规划新建几栋高层，形成一组建筑群，构成比较好的景观轮廓。今天看，天津是采取了第二种方法，您对今天的效果如何看？”

巴纳特先生这次来中国先到的南京，在东南大学讲学后，坐高铁来到天津。他从天津站下车，看到了津湾广场这组高层建筑，与他 2003 年来时相比，解放北路与中心广场地区都发生了巨大变化，形成了津门、津塔、津湾广场等几组高层建筑群。巴纳特先生没有直接回答，他用睿智的眼光看着诸葛民回问道：“你们自己怎么看？”有点出乎意料，诸葛民怔了一下：“天津与巴黎的情况还是不同的，巴黎的城区只有一栋高层，而天津当时已经有了数栋，沿海河分布着解放北路上的中国银行、中心广场的航运大厦和对岸的电信大厦等。加上房屋拆迁和土地整理成本比较高，所以建高层也是形势所迫。考虑到海河这段的河道只有 100 米宽，所以我们对高层建筑布局进行了控制。原则上只能在海河一侧有高层，避免出现两岸同时是高层建筑形成的压迫感。比如，解放北路沿线是历史街区，不能建高层建筑，对岸的南站（六纬路）商务区就高一些；天津站是低的，对岸的建筑群就高一点；意奥风情区是多层区，对岸就是高的；古文化街地区总体以低层为主，对岸整体也高一点。

这样，高层建筑在海河两岸穿插布局，也带来一点灵动。另外，天津严格控制高层建筑退线，除津塔等少数标志性建筑外，建筑高度从海河向外梯次升高，而且尽量沿海河形成连续的建筑界面。您二位都坐过海河游船，应该已经有具体的感受。”

巴纳特先生和丹·索罗门先生在认真听，还不时看着诸葛民在餐巾纸上画示意草图。“听了你介绍以及我通过各方面的了解，你们在海河沿线建筑高度控制上，包括历史街区内建筑高度控制上考虑周全，效果是不错的。”巴纳特先生语气中肯，“不过，建筑高度的控制作为城市设计非常关键的内容，包含更多含义。最近，伦敦市长萨迪克·汗拒绝了亿万富翁约瑟夫·萨佛拉（Joseph Safra）提出的在伦敦金融区建造一座 305 米高郁金香形状的摩天楼方案，方案的设计者是诺曼·福斯特。否定的理由是郁金香塔将损害伦敦的天际线，还会对作为世界遗产的伦敦塔造成重大伤害。伦敦城地区对建筑高度的控制并不十分严格。已经建成了一组高层建筑群，包括由佐伦·皮亚诺（Renzo Piano）设计的英国第一高、也是西欧最高的碎片（shard）大厦，高 310 米，共 95 层，比巴黎的埃菲尔铁塔只矮十几米。巴黎对建筑高度的控制一向十分严格，才形成了今天统一的风貌。你能说巴黎的土地价格不高吗？实际上，巴黎一直以来也有建高层的冲动，蒙巴纳斯大厦就是例证，但巴黎坚持住了。当然，为了满足现代办公建筑的需求，巴黎在远离主城区的轴线西端规划了德方斯新的商务办公区。新建一个商务区并不容易，即使像巴黎这样的国际大城市，有老城区严格限制高层建筑的支持条件，德方斯在发展过程中也遇到了危机。同时，这也说明城市设计对建筑高度的控制有许多内涵。”

巴纳特先生喝了一口茶，“诸葛总，我 10 多年前受邀到天津滨海新区参加了于家堡金融区城市设计评审，记得当年你们论证了于家堡金融区与中心城区金融区以及与北京金融区的关系。当时提出解放北路作为传统金融区，不再大发展，于家堡作为创新金融区加快发展。由 SOM 主设计的于家堡金融区城市设计的水平是较高的。我也曾在 2012 年来过，现场考察了正在建设中的于家堡金融区与高铁站。现在，又一个 7 年过去了，据我所知，于家堡金融区发展遇到了困难，您能告诉我是什么原因吗？”巴纳特先生的发问再次让诸葛民陷入沉思。城市设计对建筑高度或城市形态的控制，表面看是对城市景观形象的设计，实际上是城市空间结构、形态、内在规律的外化表现，也是对城市资源的一种分配。于家堡距中心城区 50 千米，比德方斯距巴黎老城中心远 10 倍，有自身的不利条件。但如果天津中心城区对高层写字楼等商务楼宇进行一定限制的话，对于家堡的发展一定是有利的，可以集中更多资源。天津老城区历史悠久，有教育、医疗等各方面的优势，相对新区更吸引人，同时外环线外的环城四区在近十多年也高速发展，消耗了大量资源，以至于形成了中心城区与滨海新区连绵建设成了更大的“摊大饼”态势。因此，市委市政府提出加强双城中间地带的规划管控，建设绿色生态屏障的战略决定是正确合理的。这有助于城市空间结构的优化、生态环境的改善，也有助于滨海新区的发展。而且在双城生态屏障区规划中，也提出了二、三级管控区中要限制容积率与开发强度，严格控制建筑高度的要求。在过去的十多年，天津的控规基本放弃了对建筑高度的控制，城市新区也从所谓土地算价考虑，规划建设了大片的高层住宅区，搞得“城不城、郊不郊”，违背了城市发展的客观规律。所以，在未来的城市设计中，要进一步加强对城市高度和空间形态的控制。这应该也是国土空间总体规划的重要内容之一。国土空间总体规划不仅要规划生态红线、基本农田边界和城市开发边界等平面的形态，还要确定开发强度、高度等三维控制线，这才是对空间资源更全面的规划把控。同时，要遵循住宅类型的城乡断面分布规律，不同的区位必须要采用相应的住宅类型，使建筑高度和体量与环境更好地融合。

诸葛民陷入思考的时候，巴纳特先生和丹·索罗门先生在低语交流。这时，宋院长安排完菜品后回来落座，他向巴纳特先生提出了一个问题，打破了室内的安静。“大师，我前年在网上看到您在中国城市规划学会年会上做关于城市设计报告的视频直播，您讲的题目是“中国城市设计——建于何处，建于何时，建设什么”，就是“3W”（where，when，what），您是怎么考虑的？”这个问题正是巴纳特先生想讲的，他正了正身形，清了清嗓子：“我讲这个题目经过认真的思考。这些年我经常到中国来，对中国的城市设计非常关注，也做了长期的跟踪研究，感觉有些问题由我们这些外人讲出来，比长期在这种环境下工作的人们更适合，显得更客观。中国的改革开放已经 40 年了，中国的城市设计在不断进步，也积累了许多深层次问题需要解决。现在中国进入了新的发展阶段，对生态环境和人居品质要求更高了。你们传统的城市规划，比如《居住区规划设计标准》等强有力的规范，已经不适应新社会环境要求了。我看到了中国最高层公布的 2015 年的《关于加快推进生态文明建设的意见》和 2016 年的《关于进一步加强城市规划建设管理工作的若干意见》。生态文明要求所有人类活动应与自然和谐相处，这项事业非常具有理想色彩，但困难重重。尽管丹麦、芬兰、冰岛、挪威和瑞典这些北欧国家正努力朝这个方向发展，但迄今为止没有任何国家达到这个目标。中国提出全面加快推进生态文明，建设雄安新区，提出了自然资源和现有景观的保护优先与城市化发展，把生态文明的原则应用到城市设计中就意味着城市规划发展应以环境承载能力和尊重自然景观为前提。建设生态文明与提高城市规划建设品质是一致的。”听巴纳特先生如数家珍般地复述中共中央和国务院的两个重要文件，诸葛民和宋院长既心生敬佩又感到汗颜。巴纳特先生作为一名 80 多岁高龄的外国学者，还认真学习中国的中央文件，而我们的许多规划师、规划管理者可能都没有认真学习领会。巴纳特先生继续讲到：“中国的中央城市工作会议把城市设计作为提高城市规划建设管理水平、解决城市病的主要手段。作为一个国家的中央政府，明确提出推广城市设计，在世界上也是不多见的。住房和城乡建设部出台了《城市设计管理办法》，为城市设计走上法定规划的轨道奠定了基础。这次又组织了两批城市设计试点，全国有 50 多个城市参与，包括天津。通过参与城市设计试点活动，我对中国城市设计的未来充满信心。”

“当然，要想让城市设计真正发挥作用，还必须解决几个关键问题，所以我在南京讲了 3 个“W”。第一个 W 是在哪儿建。这应该是总体城市设计尺度的问题。正如我在《城市设计：现代主义、传统、绿色和系统的观点》一书中提出的，经过数千年的发展，城市设计形成了多种方法，其中生态的方法是大尺度城市设计中的首要方法。我看天津这次城市设计试点中把更生态的城市作为一个重点，是非常正确的。现在中国正在推动国土空间总体规划，在开展生态本底评价和自然景观环境承载力评价的所谓‘双评价’的基础上，划定生态红线、基本农田界线，然后确定城市开发边界。在这个过程中，要发挥城市设计的作用。首先，生态红线的确定不是简单的模仿自然本底，无论是从生态理论还是实践经验中，都要进行设计。我们发现，好的生态，其景观一定是优美的，反之亦然。一些自然本底并不好的生态环境，通过人类的正确干预，经过多年的演变，得到很大改善，使人、城市与自然环境更和谐地相处（包括对水系的优化、自然绿地的保护等）。所以，一个好的生态环境系统也有一个设计问题，中国古代著名的例子就是李冰所做的都江堰工程，是一个杰出设计的代表。其次，城市开发边界划定也要经过城市设计的过程。我同意上午天津大学张院士讲的‘城市艺术骨架’的概念，城市不只有一个边界，其空间结构与形态有内在的规律，只有好的空间形态才能保证城市规划建设走上正确的轨道。

“第二个 W 是何时建，实际上这个问题与第一个问题是相关联的。城市和区域发展是动态的，城市开发边界也不是一下子形成的，是个动态发展的过程。即使是终极式的城市开发边界，在什么时间启动什么地方的建设也是非常关键的，如果考虑不周，就会产生许多严重的后果。比如现在中国有许多“鬼城”，过度的扩张发展、多点开发，力量分散，无法形成聚集效应。这些都是中国过去曾经发生的严重问题。所以，何时建可能比开发边界的问题更重要。这次天津提出了加强双城间规划管控，建设绿色生态屏障的方案就十分明智。另外，你们提出的主城区“一环十一园”的规划是一个绿化系统与城市开发高度结合的做法，对于优化主城区生态环境和改善城市结构、疏解历史城区过密的人口非常关键。但这里也有一个何时建的问题，必须认真把握好，与旧城人口疏解配合，避免同时遍地开花。

“第三个 W 是如何建。中国改革开放 40 年取得了举世瞩目的成就，在这个过程中也形成了一套相应的规范机制，包括城市用地分类、土地有偿使用、居住区规划设计规范等。不管什么原因和途径，我认为居住区规划建设‘大地块、大路网’这种模式就是柯布西耶的模式，是适应小汽车不适应人的模式，是造成今天中国许多城市病的根源。因此，要实现城市转型升级，推进城市设计，就必须改变这种传统的规划模式和方法。我看天津在这次城市设计试点中提出了‘人性化城市’和‘新型居住社区城市设计导则’，抓住了关键问题，只是你们在强调‘步行化’和‘公交导向’的基础上提出了发展小汽车交通，与目前国际的主流思想不一致。另外，你们在新型居住社区导则中提出适度鼓励独立住宅和低层联排住宅，这与国际上对中国的一般看法也不一致。我听了一些规划师的发言，还需要一点时间考虑。”

“这正是我们想与您沟通的问题。”诸葛民接过了巴纳特先生的话题，“您说我们传统的大街廓、大马路居住区模式是城市适应小汽车的模式，实际上我们当年一直讲城市以公交为主，而且最终的结果也说明这种模式不适合小汽车。随着小汽车的快速发展，我们的城市交通拥堵问题很严重，就是因为这种大路网大马路造成了交通的过度集中，同时又缺乏足够的路网来疏解和组织交通。所以说，我们传统的师从苏联、受现代城市规划影响的居住区规划模式只是适应了当时大规模房地产开发的模式，这种模式既不适应小汽车的发展，也不适应人的步行，留给城市的只有大围墙。经过多年的跟踪和研究，我们认识到，城市首先必须以人为本，步行化是核心。”

“在城市中心区，包括城市商业中心、社区中心，都要把步行化作为城市设计的重点，采取交通稳静化措施，有舒适的人行道宽度，保证连通性，建设宜步行的城市。同时，在这些区域，采取限制小汽车的措施，大力发展公共交通，并通过“3D”方法，即适宜的建筑和人口密度、多样性混合功能和城市设计的方法，使公交更好、更高效地为人服务。我们最近到巴西著名的公交城市库里提巴考察，真正认识到城市开发模式必须与公交走廊紧密地结合，形成“手和手套”一样的紧密关系，这样公交才能真正发挥作用，才能提高服务水平。

“在库里提巴我们还有另一个惊人的发现，即使像库里提巴这样把公交做到极致的城市，其公交出行在城市整体出行中也只占到30%，其他主要还是小汽车出行。由于库里提巴经济好，居民收入高，其小汽车人均拥有量位居巴西第二，仅次于首都巴西利亚。这也说明了一个问题，小汽车经过100多年的发展，技术逐渐成熟，这一发展也促进了人类文明进步和人性的发展。尽管有许多代价，但其发展趋势不可阻挡，就像彼得·霍尔说的‘人民用轮子投票’。我们如果无视这一规律，必然要受到惩罚。不要说目前美国的千人小汽车拥有量已达到800辆，就是过去以公交为主的欧洲，还有日本，目前千人小汽车拥有量也已经达到近600辆。伦敦的公交出行只占10%多一点，通过采取市中心收取拥堵费的方法，解决了市中心小汽车交通拥堵问题。东京小汽车拥有量也很高，但大部分居民上班通勤还是用公交，购物、娱乐、出游用小汽车。所以现在有观点说，不能限制拥有小汽车，而是要合理使用小汽车。目前来看，城市中心区由于空间有限，以行人为先和发展公交为主，只能限制小汽车。伦敦和新加坡都已经对进入市中心的小汽车收取拥堵费，以此限制进入市中心小汽车的数量。而库里提巴由于市中心以5条快速公交走廊为优先，实际客观上也限制了小汽车进入市中心的数量。最近我们也考察了波士顿大开挖（BIG DIG）项目实施的情况，现在从波士顿洛根（Logan）国际机场和城市外围郊区到达市中心都可以走地下快速路，非常便捷；但进入市中心，出了地面，地面上都是单向交通组织，绕来绕去，非常不方便，这也客观上限制了小汽车进入市中心。所以，我们这次天津城市设计试点也把市中心步行化、发展公共交通作为一项重要内容，包括限定停车位的数量，未来市中心一定会进一步限制小汽车交通。这一点我们是坚定不移的。

“同时，我们也要正视私人小汽车的进一步发展。从国家层面看，汽车产业是重要支柱产业，现在年均乘用车产销量达2800万辆，在GDP和零售商品销售额中占很大比重。我国现千人拥有轿车170辆左右，与发达国家相比还有一定发展空间。所以从经济发展和满足人民对美好生活的需要讲，城市必须满足小汽车发展的需求。从天津层面看，汽车产业也是天津的主导产业。从天津城市转型发展看，也应该鼓励发展小汽车。我们在“一环十一园”周围地区的规划中强调采用方格网道路，就是为了适应小汽车发展的需要。天津中心城区共规划有15条地铁线，1、2、3、4、7、9为放射线，5、6形成环线，其他为填充线。虽然市中心轨道交通网密度较高，但城市外环线周边，也就是“一环十一”园周边地区地铁线网密度并不高。因此，除了解放南路、柳林地区等少数地方外，“一环十一园”周边的大部分地区只有一到两条轨道线、两到三个地铁车站，要服务近10万人口的居住社区，还需要公交的配合，而更重要的是依靠小汽车交通。这些地区靠近城市快速路和外环快速路，有比较好的基础，采用窄路密网布局正适应小汽车的发展。”

“你讲的这些都是事实。”巴纳特先生语气沉重地回应道，“但我以为，即使如此，还需要做很多工作才能最后下定论，不光有定性分析，还必须有定量分析。在定量分析方面，交通规划是比较成熟的。另外，还要考虑环境保护，社会公平等方面的问题。”“我赞同。”诸葛民回应道，“下一步我们马上要做定量的交通规划，用大数据建立交通模型进行模拟分析。除了土地利用和道路网数据外，交通规划是城市设计的重要支撑，过去我们的这方面偏弱，今后要加强。关于小汽车的污染问题，现在新能源汽车（NEV）发展迅速，并且已经投入市场，增速很快。另外，无人驾驶汽车等智能技术也在快速发展，这些技术都会在解决城市交通拥堵和尾气排放等方面予以助力。当然，电池的处理也涉及环境问题。总体看，环境问题要靠发展和技术进步来解决。社会公平问题更需要综合的考虑，反过来想，即使我们限制小汽车发展，也解决不了社会公平的所有问题。”巴纳特先生点了点头。诸葛民继续说道：“交通问题最终还是经济问题，美国交通学家在这方面做了大量的研究工作，小汽车不只是交通工具，许多政策要靠采用符合市场规律的经济手段。实际上与小汽车一样，公共交通消耗的也是社会资源。政府给公交的补贴是所有纳税人的钱，这种补贴也是有限度的。巴西库里提巴的公交票4.5元巴币，相当于9元人民币，即使一天内可以随意换乘，票价也不便宜，特别是对短途乘车的旅客来说。当地人认为，乘坐短途公交的是富裕人口。它们的票价超过了实际价格，但它们可以为相对贫困的长途旅客提供帮助。另外，政府将一些商场、报刊亭等政府资产交给公交管理集团运营，出租收入用于补贴公交运营。过去通过这些资产出租收入，短途旅客补贴长途旅客等政策，公交运营收支是平衡的。但最近一次测算表明，公交票价要从目前的4.5元巴币涨到4.75元才能收支平衡，市长出于各方面考虑不批准涨价，这块不足还是要政府补贴的。据测算，成本中最大的还是公交司机的工资和福利，占到总成本的1/3。而且公交工人职业病很严重，也许无人驾驶技术首先应该用于公共交通。”

“对于交通和小汽车的问题我们讨论的够多了，另一个我们比较关心的问题是独立住宅。你们真的认为中国可以建设独立住宅吗？”巴纳特先生的神情越发严肃起来。诸葛民也放慢了语速：“不论是外国人，还是我们中国人自己，都认为中国人多地少，不能建独立住宅。巴纳特先生在上篇文章中也写道，中美两国国土面积相当（美国不计算阿拉斯加），中国人口是美国的5倍，可用地面积是美国的一半，虽然许多中国人喜欢独立住宅，但中国必须要考虑其他形式的居住形式。实际上，这些分析都有些简单化，没有考虑中国的现实情况。要探讨中国未来建什么样的住宅这个核心问题，必须考虑中国城市、乡村的现实情况，它们不是一张白纸，用简单的算术就可以算出来。要回答这个问题，必须进行系统综合的考虑。”

“我们过去40年的规划中所建的大量封闭小区和多层、高层住宅，不能说不节约土地。但我们今天所有的城市问题就是这种所谓‘集约土地’模式所造成的，而同时在其他方面浪费土地的现象严重。我们必须改变，必须更统筹考虑。首先，我国目前住宅存量已经很大，城镇人均住房建筑面积39平方米，户均一套，但在这些住宅中，有一半以上是2000年前建设的标准比较低的单元住房，需要更新。因此，考虑到2035年、2050年的标准，新建的住房，不论是面积还是类型都要标准高一些，面积大一些。第二，我们现在城市人口密度非常高，每平方千米2万以上，人均城市建设用地不到50平方米，造成绿地少和过于拥挤。按人均城市建设用地100平方米考虑，外围新建住宅用地面积应比市区大一些，建筑密度相对低一些，拥有更多的绿化和开放空间。第三，从房地产平稳过渡考虑，目前我国年竣工和销售新建商品住宅17亿平方米。城镇居民每人年均增加2平方米建筑面积，数量巨大。为保持平稳过渡，要减少数量。同时，还要考虑广大中等收入家庭的购房能力，限制房价。因此，如有20%新建商品房做低层住宅等高档住房，既可满足部分高收入家庭住房改善需求，又保证房地产市场总体平稳，而且这部分高档住房的土地收入和税收可用来补贴保障性住房建设。由于低层住宅占用了相对较多的土地，我们可以设定较高的税费，以保证社会公平和共同富裕。第四，从社会管理的角度，目前我们建设的绝大部分是大规模的集合住宅，难于管理。随着房龄增长，住房设备、维护结构老化，居民难以统一意见行动，最后只能由政府出资维修，成为政府的巨大负担，实际也是整个城市的负担。今后，随着推广窄路密网，缩小开发地块规模，有条件使业主协调一致。另外，政府允许建设一些低层联排和合院住宅，业主可以管理和维护自己的物业，培育一定的独立自主管理意识，易于社会管理和社会和谐。第五，利于中国传统建筑文化的传承。合院住宅是中国各民族民居的基本类型，与文学、艺术完美结合，形成了中国传统民居文化。适当规划独立式住宅的空间载体，有利于中国文化的传承与发扬光大。历史上，我国都是低层合院住宅，而且在当时的农业生产水平情况下，最多时养育了 4 亿中国人。低层住宅易于与自然环境相协调，形成住宅掩映在绿树中的理想状态。在城市边缘、郊区、广大农村地区，以及生态区、风景名胜区周边都应该采用低层低密度的住宅形态。当然，要坚决清除生态区和风景区内部的违规别墅建设。以上这些要点，包括住宅多样性，都是我们考虑规划建设部分低层住宅的理由。当然，这些都应该以《住宅法》的法律形式固定下来，以做好住房困难家庭住房保障和为广大中等收入家庭提供可供给的小康住房为前提。”

诸葛民停顿了一下，接着说道：“我过去也简单地认为中国人多地少，不宜建低层住宅。但 2003 年我在美国加州大学做了一年访问学者，听老师讲，有经济学家认为湾区是未来最好的城市区域空间形态，因此做了一些天津与旧金山湾区的对比研究，有了一些发现，改变了自己的一些既定看法。天津市域与湾区面积相当，都是 12 000 平方千米，当时天津市总人口 920 万，湾区 670 万。在经济上相差还是比较大的，湾区 GDP 3500 亿美元，全球按国家排第 24 位，人均 GDP 5 万美元，天津 GDP 2400 亿人民币，相当于 290 亿美元，人均 GDP 3000 多美元，分别是湾区的 1/12 和 1/17。湾区生态环境更好，有更多的绿化公园。其生态用地占总用地的 45%，农业及其他用地占 37%，城市占 18%。天津耕地占 45%，林草占 3.5%，水域占 26%，建设用地占 23%，交通占 2.6%。但在人均建设用地方面，包括城镇道路交通、社会工矿等用地，湾区人均 300 平方米，天津人均 270 平方米，相差 10%。但从美国住宅形态上，绝大部分是环境优美的独立住宅，而我国的住房都是多层和高层集合住宅。我们说人多地少，要节约土地，客观上我们却浪费了大量土地。我们规划人均建设用地 100 平方千米，但我们近期做的规划中预测的人口超过实际增长，又规划了大量工业区和新城，天津的工业用地产出低效是非常明显的。虽然我们与美国湾区人均建设用地水平相当，但我们居住建设水平差距很大。这引起了我的深思。城市规划建设的最终目标是创造宜居的人居环境，如果我们单纯地以所谓的节约用地作为主要目标，实际上是本末倒置，现实也证明了这一点。我们多年来坚持节约用地，实施所谓最严格的土地使用制度，但我们浪费的土地却很多。我们不能简单地以节约用地来限制住宅的多样性。我现在经常用日本作例子，日本土地比我国还紧张，但在日本的所有住宅中，独立住宅（即“一户建”）占了 50% 以上。包括在人口稠密的东京，也占到 45% 以上，这很能说明问题。当然，日本一户建采用了户均用地十分紧凑的模式。美国新城市主义建议采用欧洲传统城镇模式，适当增加开发强度和密度，但它也没有反对独立住宅。另外，有人说欧洲城市与中国比较像，中国学不了美国，却可以学学欧洲。但实际情况是，欧洲的多层住宅基本饱和，现在新建住宅中绝大部分都是独立住宅。综合各种因素，考虑到 2035 年、2050 年的发展，我们现在开始尝试在城市郊区规划建设一定比例的低层住宅，包括联排和合院住宅，是非常必要的。”

听诸葛民一口气讲到这儿，巴纳特先生紧张的心情得到缓解，舒了一口气：“你是讲规划少部分低层住宅，不是全部，我想这是可以接受的，而且从住宅的多样性看，这是必需的。当然，综合看，多层高密度集合住宅应该是中国未来一段时间内主要的住宅类型，对吗？”看到屋内的气氛有些严肃，宋院长好不容易插进话来：“巴纳特先生，您说得很正确，多层高密度，包括部分小高层是我们未来的主要居住建筑类型。所以，10 多年前，我们规划院就与丹·索罗门先生合作，开展了滨海新区新型社区的规划设计研究，丹·索罗门先生，您的看法呢？”宋院长看到丹·索罗门先生一直没讲话，他想给递个台阶。

索罗门先生是个十分聪明的人，他理解宋院长的意思：“谢谢宋院长，刚才你们讨论得非常热烈，我听得也很有滋味。我记得是 2011 年，你们邀请我参与滨海新区和谐社区的城市设计，我非常荣幸。我一直从事住宅设计与规划工作，持续关注中国的发展，但对中国具体的城市规划、住宅建筑设计规范不十分了解。通过“和谐社区”两轮近 3 年的工作，我对中国住宅和居住社区的规划设计有了深入了解，也发现了现在流行的大街廓大马路所带来的巨大问题。它彻底改变了中国传统城市由合院住宅构成的胡同、街道等交往空间的城市肌理，形成了由围墙和相同的住宅构成的品质低劣的城市环境。所谓的居住区规划设计规范是一个问题多于成绩的规范，可以说不是一种正确的规划设计方法，它把人看成缺乏个性的符号，也不考虑以人为尺度的街道、广场等城市空间的塑造，实际上也不考虑交通问题，所以必须从根本上改变。我们在和谐新区的规划中采用城市设计方法，从城市不同的街道、广场、公园空间的塑造开始进行社区的规划。我也学会了日照分析、日照间距、千人指标以及城市道路设计规范、城市道路交叉口设计规范等。我们尽可能满足日照要求，配套在满足了千人指标的情况下集中布置社区中心，对道路交叉口半径等单纯以汽车行驶速度为出发点的规范进行优化调整，缩小道路交叉口转弯半径，形成亲切的空间，便于行人过马路，合理节约用地。讲到土地浪费，实际上你们许多道路交通的大马路、大路口就浪费了大量土地。我们做了非常深入的工作，与当地规划设计院合作，包括制定概念性住宅建筑方案，付出了大量心血，只可惜由于各种因素，迄今项目没有实施，没能得到实践检验。原来以为2016年中央城市工作会议明确提出建设‘窄路密网’新型社区以后，这项目能够实施，但目前看还是没有启动的迹象，甚是遗憾。有人告诉我，你们天津人总说自己“起个大早，赶了晚集”，这件事我看就是个例子。这次到天津来之前，我一再问，天津有没有类似的项目实施，我们的研究是否发挥了作用。我们的工作就是巴纳特先生所讲的3W中第3个关键问题，建什么的问题。我想我们给出了一个答案。”

宋院长插话进来：“对，和谐社区是一个非常重要的案例，包括它的朝向正好是45°角，解决了东西向住宅问题，这正是巴纳特先生推荐的一个方法。”巴纳特先生接过话：“我也在不同的场合看到了和谐社区的设计方案，我以为它是一种很不错的尝试，符合天津地方的特点，只可惜没有实施。”诸葛民接过话题：“丹在和谐社区的规划设计中，倾注了大量心血，我们非常感谢。的确有些遗憾，由于各种原因项目没能实施，但它确实发挥了一定的影响作用。我们明天陪你们去滨海新区考察，除了看于家堡金融区、高铁站、海河滨河绿带、滨海文化中心、国家海洋博物馆这些大型公共建筑外，也想带你们参观一下开发区的贝肯山住宅社区、天碱的中投项目和天津港散货中心起步区，以及保障房佳宁苑。这些项目都尝试了窄路网的道路布局和城市空间的创造。滨海新区规划和自然资源局同行也会陪同。有许多老朋友也希望见到你们，希望你们对新区的建设规划提出好建议。”

巴纳特和丹·索罗门先生异口同声地说：“我们很期待。”丹·索罗门先生继续讲道：“当年为了做好和谐社区规划，我也考察了中国一些城市好的居住社区项目，像北京菊儿胡同、天津万科水晶城等，当然更多地看到的是大量简单复制的大型社区。我当时在问这个问题，为什么菊儿胡同、万科水晶城这样好的模式没能推广。到今天，我慢慢明白了，这是一个结构性问题，牵一发动全身，涉及全社会方方面面，改变起来很难，但终究要迈出第一步，否则难以改变。”“其实我们心里也是非常焦急的，也做了许多准备工作。”诸葛民接着说，“近两年，我们结合一些规章规范的修订，制定了《天津市控制性详细规划技术规程》《天津市建设项目规划管理技术规定》，已经为新型社区窄路密网规划提供了一定的条件，只是市场上还不了解，需要加大宣传。我们前年组织编写了《和谐社区——滨海新区新型住宅社区规划设计研究》，从理论和实践方面进行研究，宣传丹·索罗门先生主持的和谐社区规划设计，华汇公司也继续编制了和谐社区的城市设计，并在中国城市规划研究院召开研讨会，上海、深圳等专家参与，逐步形成共识。只可惜，后来中规院主持修订《居住区规划设计规范》，虽然有较大改动，但仍然坚持了居住区规划传统劣根，让人费解。这也说明我们还需要进一步宣传。这次天津城市设计国际研讨会也想再一次宣传新型社区的理念及其重要性。”

“诸葛总和宋院长你们的努力我们都看在眼里，令人钦佩，我上午在讲座上也表达了这样的意思，希望这次讲座能够起一定作用。”丹·索罗门的话锋一转，“从 2011 年开始制定滨海新区和谐新城规划，到今天已经 8 年过去了，我也一直在思考。最近我写了一本书，书名是《住宅与城市，爱对抗希望》，今天我带来了，送给诸葛总和宋院长。这本书，结合我 50 多年的教学、工作、科研和实践，研究了世界主要国家的住房建设，得出了一些结论。城市围绕着我们，住房限定了我们的存在。我们的内心面对现实的住房，它赋予了我们的爱、我们的恨和我们的期许，希望某一天我们的住房能够满足内心的需要，我对此充满着希望。世界上许多的城市，如罗马、阿姆斯特丹、巴黎、柏林、慕尼黑、科隆、布拉格、布达佩斯等。也许我们会问，这些城市除了各自鲜明的特色外，到底有什么不同？你喜欢哪座城市？不喜欢哪座城市？那座陌生的城市为什么让你有宾至如归的感觉？这些都是由住宅构成的，由城市肌理决定的。住房对全世界任何一座城市来说，都是一个巨大而急迫的问题。随着技术进步和财富的积累变化，为工人阶级、为低收入阶层创造住房一直有两种不同的模式在激烈斗争，我在书中重点分析了旧金山、巴黎和罗马百年来在所谓‘希望的城市’与‘爱的城市’方案之间的战斗，各有胜负。所谓‘希望的城市’方法就是试图用更加合理的住房模式、住房类型替代传统的城市住房类型和肌理；‘爱的城市’方法，就是热爱城市丰富的历史，尊重它复杂的社会结构肌理。而‘希望的城市’项目不断重复失败，它希望实现社会进步和融合的目的不断遭遇挫折，因为它想把一个热楔子放进潜在的社会阶级构成中；而‘爱的城市’项目总是成功的，因为这些项目成功地成为社会吸收和社会和谐的入口，作为赋予城市连续性、弹性和庇护的场所。”

丹·索罗门先生在书的扉页上签上名，庄重地递到了诸葛民和宋院长手中：“我的书一如既往，语言可能不够直白，但希望你们读后有所体会。刚才听了你们与巴纳特先生的讨论，我更认识到传统居住区规划问题的严重性和复杂性，也意识到你们开始触碰到了问题的深层次根源，这就是解决问题的开始。我也在想，中国改革开放40年，建设了大量住宅，改变了中国传统城市的肌理，虽然从改善中国人的居住条件和解决住房短缺问题的角度看，成绩是巨大的，但也存在更加巨大的城市问题，是城市病的根源之一。不过，这些传统目前已经成为中国大部分城市现有的肌理。我们现在推广新型社区，是否也是一种'希望的城市'的方法呢？这需要思考，如何把这种'希望的城市'方法变成'爱的城市'的方法，我觉得是一个大课题，也是解决问题的办法。"丹·索罗门的话让大家都陷入沉思。这时，凉菜上来了，之后就上了包子。宋院长已经有了经验，过去吃狗不理包子前，先吃了数道热菜，已经饱了，再吃包子就没了胃口。现在直接吃包子，就可品味狗不理包子特有的精妙美味了。况且巴纳特、丹·索罗门先生都是耄耋老人，晚上吃不了太多。晚饭时间很快结束了，诸葛总送客人到餐厅外面，宋院长不放心，亲自上车陪同去宾馆。

告别了巴纳特和丹·索罗门两位大师，诸葛民信步来到了海河边，他没有向人声鼎沸的意奥风情区走，而是不由自主地来到了海河中心广场。身后的意式风情区和对岸的和平路金街商业街都灯火通明，只是由于两岸没有直通桥梁联系，所以海河中心广场很清静，灯光暗淡。在海河河面上，正在建设一座步行桥，目的是将意式风情区和和平路金街联系起来。其实这在2002年，海河"金龙起舞"规划中就已经确定了，但担心对海河景观有不好的影响，一直没有实施。这次城市设计试点中的步行化城市再次强调了这座步行桥的重要性。经过国际招标，有关部门选出了一个出色的设计方案，该方案不仅功能完善，而且建成后飘逸轻盈的造型八成能成为"网红打卡地"。由于这段海河两岸居住人口不多，所以虽然到了晚间，但仍在施工，桥面上闪烁着点点电焊花。今天诸葛民有点兴奋，城市设计试点研讨会圆满结束，刚才又与巴纳特和丹·索罗门大师进行了深入透彻的讨论沟通。他似乎意犹未尽，看着眼前的城市——天津，这座他投入了毕生心血的城市，心潮起伏。

诸葛民从清华大学硕士毕业后到天津工作已经30多年了。毕业时，导师吴良镛先生教诲他："如果你想为中国的城市规划做出贡献，有所作为，就要扎根你的城市。"诸葛民做到了，他一直在天津从事城市规划，前13年在天津市规划院从事规划编制和研究工作，后20年在天津市规划局从事规划管理工作。是组织的信任和培养，让诸葛民作为党外人士，从一位年轻的大学毕业生成长为一名专业技术干部。诸葛民初步统计过，自参加工作以来，他出访过30多个国家，近90个不同的城市，学习积累了很多经验。他的硕士论文是《比较城市规划初探》，得到老师们好评，但他那时还没出过国。

后来，随着不断出国考察和持续研究，他对国外不同国家的规划编制、规划法律、管理体系都有了较深入的了解。后来他又在清华大学读在职博士，花了7年时间，对中国战略空间规划进行了深入的研究。2003—2004年，在天津市委组织部组织下，他在美国著名的加州大学伯克利分校城市规划与环境设计系做了一年访问学者。这些持续的学习，加上海河综合开发规划、滨海新区开发开放等实际工作经验的不断积累，诸葛民感到直到今天才真正对中国的城市规划、空间规划有了深刻而全面的认识。他深深地体会到吴良镛先生曾经说过的“城市规划是老年人学问”的真正含义。当诸葛民真正理解城市规划时，自己也马上就到60岁退休年龄了。他希望年轻规划师们尽快成长，自己做好“传帮带”，同时也在考虑自己退休后的打算。

诸葛民脑海中一直回想着一句口号：“为祖国健康工作50年。”这是当年清华大学老校长、教育部部长蒋南翔先生的一句名言，也是当年他在清华读书，每天下午4点在西操场锻炼跑步时，大喇叭里一直播报的声音。现在刚工作了30年，离蒋校长的要求还差20年，还有许多事情想做。诸葛民也一直记着，吴良镛先生经常与他聊起，吴先生自己60岁从清华大学建筑系系主任岗位上卸任，开始组建建筑和城市规划研究所，以后又一直工作了30多年，这30多年是他工作成果最丰硕的30年。包括刚告别的巴纳特和丹·索罗门先生都已经从事城市规划工作50多年了，还在不懈努力。这些专家学者的榜样，鼓励诸葛民坚定信心，退休后创办一个公益性的研究机构，为天津的城市规划继续努力。他最近去巴西库里提巴和美国波士顿考察，与当地规划机构交流，发现当地规划编制机构都不是政府组成部门，不是公务员，但却直接由市长领导，长期为城市发展出谋划策。规划许可由行政部门负责，但规划编制不适应行政这种体系，它需要高层次的人才。规划编制的经费也不是来自财政，而是政府给规划机构一定的土地、房屋等资产，通过资金经营收入筹集规划编制的经费和发放工资。资金经营得好，就多发工资，这也正好发挥和考验了规划师的胆量和才识。如果一点土地和房屋资产都经营不好，也就说明城市规划不会做好。这种做法应该也很适合我们中国今天处于市场经济转型期的城市。总之，未来会有多种可能性，诸葛民相信自己的事业与天津市一样，一定能再创辉煌。

夜色更加深沉，海河两岸灯火通明。诸葛民转身往回走，他的目光停留在中心广场现状的绿地上，现在只是一些高大的树木和少量变电站等设施。海河中心广场是天津中心城区的几何中心。新中国成立之后，这里曾经是检阅台，是游行和聚会的地方。20世纪50年代，苏联规划专家穆欣曾建议就从这里向南延伸出一条轴线，直通水上公园，形成天津城区的主轴线，可惜没有实现，后来就再也没有机会。改革开放后，检阅台基本废弃不用，广场成为海河边的儿童游乐场，喧闹一时。后来河北区招商嘉里中心，又被转移到河东南站。2002 年，海河综合开发改造规划时，做了中心广场的方案征集，中心广场曾设想规划建设海河音乐厅，几个参赛单位都提出了不同的竞标方案，但都没能实施。再后来因小白楼音乐厅，此事便不再提起，另外的原因可能是因为这个地方太重要了，不知如何处置。后来，又曾经作为规划展览馆和海河游客中心的选址，也做了方案征集，最终由于时间紧迫，选择了意式风情区的现状建筑改造成规划展览馆，这个项目也没有再实施的可能了，因此这块用地就成为临时绿地，一晃又过去 10 年。今天，进入发展的新时代，要为海河中心广场找到合适的项目定位，做当今世界最高水准的设计。诸葛民设想，可以发起一个全市范围的海河中心广场设计的大讨论，寻找到天津未来城市设计的方向。

中心广场这个位置的形状就像是一只眼睛，2002 年，实施中心城区海河改造时，规划叫“金龙起舞”，中心广场就像龙的眼睛。海河综合开发改造启动实施已经快 20 年了，天津城市 20 年来发生了巨大的变化，不断扩张。今天我们又回到海河边，就是要站在新世纪中华民族伟大复兴中国梦的高度，重新审视海河。从生态文明的高度看，海河，特别是海河中游段，要成为天津生态环境改善的核心地区；从以人民为中心角度看，海河上游应该进一步完善步行系统，让人们更加亲近海河，使之成为人民之河；从滨海新区改革开放的角度看，海河下游段要成为改革创新之河。海河作为一条金龙，龙头、龙身、龙尾首尾呼应，全身舞动。而海河中心广场的新规划设计就像金龙舞动前的画龙点睛一般，预示着天津将再次腾飞，再造辉煌。